encyclopedia of instrumentation and control

other reference works of interest

American Society of Mechanical Engineers ASME HANDBOOKS:
Engineering Tables; Metals Engineering—Processes; Metals Engineering—Design; Metals Properties

Baumeister and Marks STANDARD HANDBOOK FOR MECHANICAL ENGINEERS

Beeman INDUSTRIAL POWER SYSTEMS HANDBOOK

Brady MATERIALS HANDBOOK

Burington and May HANDBOOK OF PROBABILITY AND STATISTICS WITH TABLES

Callender TIME-SAVER STANDARDS

Carrier Air Conditioning Company HANDBOOK OF AIR CONDITIONING SYSTEM DESIGN

Carroll INDUSTRIAL INSTRUMENT SERVICING HANDBOOK

Considine PROCESS INSTRUMENTS AND CONTROLS HANDBOOK

Considine and Ross HANDBOOK OF APPLIED INSTRUMENTATION

Crocker and King PIPING HANDBOOK

Dudley GEAR HANDBOOK

Emerick HANDBOOK OF MECHANICAL SPECIFICATIONS FOR BUILDINGS AND PLANTS

Emerick HEATING HANDBOOK

Emerick TROUBLESHOOTERS' HANDBOOK FOR MECHANICAL SYSTEMS

Factory Mutual Engineering Division HANDBOOK OF INDUSTRIAL LOSS PREVENTION

Fink and Carroll STANDARD HANDBOOK FOR ELECTRICAL ENGINEERS

Flügge HANDBOOK OF ENGINEERING MECHANICS

Gartmann DE LAVAL ENGINEERING HANDBOOK

Harris HANDBOOK OF NOISE CONTROL

Harris and Crede SHOCK AND VIBRATION HANDBOOK

Heyel THE FOREMAN'S HANDBOOK

Kallen HANDBOOK OF INSTRUMENTATION AND CONTROLS

King and Brater HANDBOOK OF HYDRAULICS

Klerer and Korn DIGITAL COMPUTER USER'S HANDBOOK

Koelle HANDBOOK OF ASTRONAUTICAL ENGINEERING

Korn and Korn MATHEMATICAL HANDBOOK FOR SCIENTISTS AND ENGINEERS

LeGrand THE NEW AMERICAN MACHINISTS' HANDBOOK

Machol SYSTEM ENGINEERING HANDBOOK

Magill, Holden, and Ackley AIR POLLUTION HANDBOOK

Manas NATIONAL PLUMBING CODE HANDBOOK

Mantell ENGINEERING MATERIALS HANDBOOK

Maynard INDUSTRIAL ENGINEERING HANDBOOK

Merritt BUILDING CONSTRUCTION HANDBOOK

Morrow MAINTENANCE ENGINEERING HANDBOOK

Perry CHEMICAL ENGINEERS' HANDBOOK

Perry ENGINEERING MANUAL

Rossnagel HANDBOOK OF RIGGING

Rothbart MECHANICAL DESIGN AND SYSTEMS HANDBOOK

Shand GLASS ENGINEERING HANDBOOK

Society of Manufacturing Engineers:
Die Design Handbook; Handbook of Fixture Design; Manufacturing Planning and Estimating Handbook; Tool Engineers Handbook

Staniar PLANT ENGINEERING HANDBOOK

Streeter HANDBOOK OF FLUID DYNAMICS

Touloukian RETRIEVAL GUIDE TO THERMOPHYSICAL PROPERTIES RESEARCH LITERATURE

Truxal CONTROL ENGINEERS' HANDBOOK

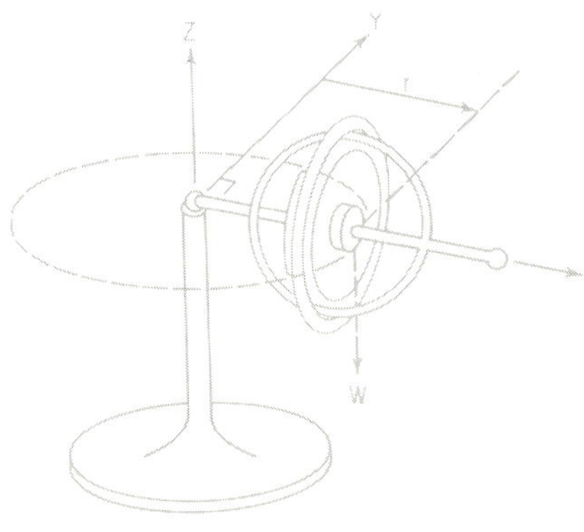
Z
Y
r
W

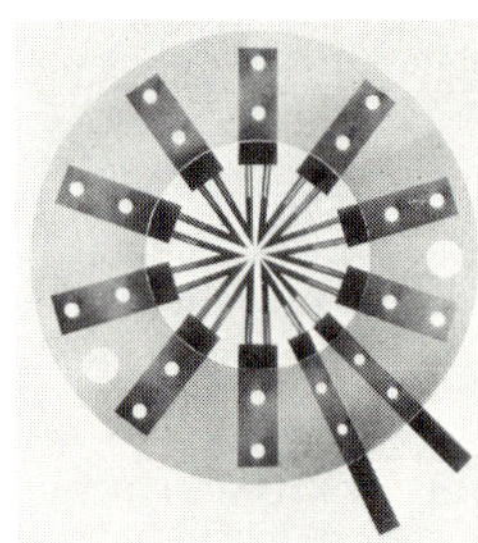

editor-in-chief

Douglas M. Considine

Consulting Engineer, Los Angeles, California

encyclopedia of instrumentation and control

new york st. louis san francisco **McGRAW-HILL BOOK COMPANY**
düsseldorf johannesburg kuala lumpur london mexico montreal
new delhi panama rio de janeiro singapore sydney toronto

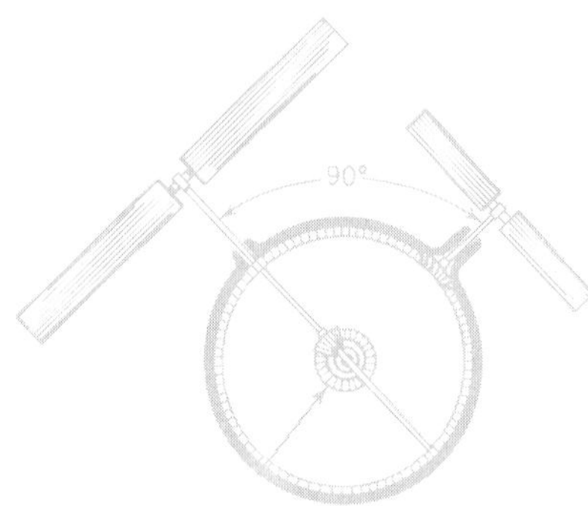

encyclopedia of instrumentation and control

 Library of Congress Catalog Card Number 77-116661

07-012424-8

1234567890VHVH754321

This book was set in Alphatype Baskerline by University Graphics, Inc., and printed on permanent paper and bound by Von Hoffmann Press, Inc. The designer was Naomi Auerbach. The editors were Harold B. Crawford, Daniel N. Fischel, and Lila M. Gardner. Stephen J. Boldish supervised production.

contents

contributors

L. T. Akeley, B.S. *Project Engineer, Research and Development Department, Process Instruments Division, Beckman Instruments, Inc., Fullerton, California. Member: American Society of Mechanical Engineers.* (Liquid level, measurement of)

C. Foster Aldridge, B.S. *Chief Engineer, Taylor Instrument Consumer Products Division, Sybron Corporation, Arden, North Carolina. Member: Association for the Advancement of Medical Instrumentation; Engineering Society of Western North Carolina.* (Barograph; Barometer; *and* Barostat)

Carl W. Alsen, B.S.E.E. *Engineer, General Radio Company, West Concord, Massachusetts. Member: Institute of Electrical and Electronics Engineers; Society for Technical Communications.* (Stroboscope; *and* Wave analyzer)

Anthony F. Amante, B.A., M.S. *Industrial Products Division, Texas Instruments Incorporated, Houston, Texas. Member: European Association of Exploration Geophysicists; Society of Exploration Geophysicists.* (Geophysical instruments)

George M. Aries, B.S.E.E. *Staff Engineer, Inertial Components Department, Aerospace and Defense Group, Honeywell Inc., Minneapolis, Minnesota.* (Flight-control systems)

R. W. Armstrong *Staff Engineer, Aerospace and Defense Group, Honeywell Inc., Minneapolis, Minnesota. Member: Society of Automotive Engineers.* (Altimeter)

Alex J. Arthur, B.S. *Staff Programmer, IBM Corporation, Boca Raton, Florida. Member: Association for Computing Machinery; British Computer Society.* (Computer programming terms)

Vernon H. Aske, B.S.E.E. *Senior Staff Engineer, Aerospace and Defense Group, Honeywell Inc., Minneapolis, Minnesota. Member: Institute of Electrical and Electronics Engineers.* (Accelerometer)

Allan J. Atrubin, B.S. *Advisory Engineer, IBM Corporation, Rochester, Minnesota. Member: Association for Computing Machinery; Institute of Electrical and Electronics Engineers.* (Character recognition systems)

Lyle K. Barnes, B.S. *Manager, Analytical Instrumentation, The Hays Corporation, a Milton Roy Company, Michigan City, Indiana.* (Combustion/combustibles analyzer; Oxygen analyzer; *and* Thermal-conductivity analyzer)

A. M. Bartz, B.S., M.S. (Physics) *Project Leader, Chemical Physics Research Laboratory, The Dow Chemical Company, Midland, Michigan. Member: Instrument Society of America; Research Society of America.* (Infrared process analyzers)

Gordon B. Baty, B.S., M.S., Ph.D. *President, ICON Corporation, Cambridge, Massachusetts. Member: The Institute for Management Science; Numerical Control Society; Society of Manufacturing Engineers.* (Numerical control)

Paul F. Bechberger, B.E.E., M.Sc. *Consultant and formerly Manager, Technical Services, Navigation and Control Division, The Bendix Corporation, Teterboro, New Jersey. Member: Institute of Aeronautical Sciences.* [Air-data systems (aircraft); Air-traffic control; Aircraft, instrumentation for; Flight-control systems; Navigation systems; *and* Space and missile-systems instrumentation]

J. Malvern Benjamin, Jr., B.S., M.S.E.E. *President, Bionic Instruments, Inc., Bala Cynwyd, Pennsylvania. Member: Institute for Cybercultural Research; Institute of Electrical and Electronics Engineers; Society for Social Responsibility in Science.* (Medical instrumentation topics)

William O. Bennett, B.S. *Vice President in Charge of Research and Engineering, Bulova Watch Company, Inc., Flushing, New York. Member: American Association for the Advancement of Science; American Institute of Watchmaking; American Geographical Society.* (Chronograph; Clock, atomic; Crystal oscillator; *and* Time and time measurement)

Robert C. Berger, B.E.E. *Manager, Systems Engineering, Speed Variator Department, General Electric Company, Erie, Pennsylvania. Member: Institute of Electrical and Electronics Engineers. Registered Professional Engineer (Ohio).* (Electric motor drives, control of)

Henry Berring, M.S. *Deceased. Formerly, Manager of Patents and Documentation, Weston Instruments, Inc., Newark, New Jersey.* [Power (electrical), measurement of; *and* Power-factor meter]

John H. Bickford, B.S. *Director, Research and Development, Veeder-Root Company, Hartford, Connecticut. Registered Professional Engineer (Connecticut).* (Counter)

W. S. Bloor, B.S.E.E. *Director, Engineering Coordination and Services, Leeds & Northrup Company, North Wales, Pennsylvania. Member: Institute of Electrical and Electronics Engineers; Instrument Society of America.* (Electric controller; *and* Radiation pyrometer)

John E. Bobbin, B.S.E.E. *Product Manager, Branson Instruments Company, Stamford, Connecticut. Member: American Society of Mechanical Engineers; American Society for Nondestructive Testing; American Welding Society; International Institute of Welding.* (Thickness gaging, ultrasonic)

Henry L. Boppel *General Manager, Automation and Measurement Division, The Bendix Corporation, Dayton, Ohio. Member: Aerospace Industries Association of America; American Society of Mechanical Engineers; American Society for Quality Control; Metric Association; Numerical Control Society; Society of Automotive Engineers; Society of Manufacturing Engineers.* (Metrology; *and* Thickness gaging, X-ray)

John S. Boynton, B.S. *Senior Products Engineer, The Foxboro Company, Foxboro, Massachusetts. Registered Professional Engineer (Massachusetts).* (Flowmeter, electromagnetic; *and* Flowmeter, turbine-type)

Arthur J. Burke, B.S.M.E. *Vice President and Director of Engineering, Howe Richardson Scale Co., Clifton, New Jersey. Member: American Society of Mechanical Engineers; Scale Manufacturers Association.* (Batching systems, automatic-weight; Scale, beam; Scale, counting; Scale, industrial; Scale, pendulum; Scale, spring; Weight, measurement of)

Howard Cannon *Texas Instruments Incorporated, Houston, Texas.* (Gravimeter; Magnetometer, metastable-helium; *and* Magnetometer, rubidium-vapor)

John J. Carpenter, B.S.E.E. *Vice President and General Manager, Electronics Division, Bulova Watch Company, Inc., Woodside, New York. Member: American Management Association; Institute of Electrical and Electronics Engineers.* (Chronograph; Clock, atomic; Crystal oscillator; *and* Time and time measurement)

Hall Cary *B & K Instruments Inc., Cleveland, Ohio. Member: Institute of Electrical and Electronics Engineers; Institute of Environmental Sciences; Society of Automotive Engineers.* (Acoustic instrumentation; *and* Audiometer)

Anthony Y. Chiang, B.S., M.Sc. *System Engineer, Bristol Division, American Chain & Cable Company, Inc., Waterbury, Connecticut. Member: Instrument Society of America.* (Adaptive process control; Automatic control; Cascade control; Direct digital control; Feedforward control; Loop; Process optimization; *and* Supervisory control)

Genung L. Clapper *Fellow, IBM Corporation, Research Triangle Park, North Carolina. Member, American Association for the Advancement of Science; Institute of Electrical and Electronics Engineers.* (Voice recognition systems)

Richard N. Clarke, B.S.M.E. *Project Engineer, Servonic Division, Gulton Industries Inc., Costa Mesa, California.* [Linear variable-differential transformer (LVD); *and* Linear-displacement transducer, potentiometric]

Douglas M. Considine, B.S.Ch.E. *Consulting Engineer, Los Angeles, California. Formerly, Manager of Advanced Control Engineering Staff, Hughes Aircraft Company, Culver City, California. Fellow: American Association for the Advancement of Science; Instrument Society of America. Member: American Institute of Chemical Engineers; Numerical Control Society.* (Control system; Process variable; *and* Variable)

Leon J. Corcoran *Product Line Manager, Hydro Products Division, Dillingham Corporation, San Diego, California. Member: Institute of Environmental Sciences.* (Current meter, ocean; *and* Tide and wave recorder)

James R. Cosby, B.S., M.S. (Physics) *Manager Engineering, Environmental Science Division. The Bendix Corporation, Baltimore, Maryland.* (Anemometer; Hygrometer; Hygrothermograph; Psychrometer; *and* Weather balloon)

Lee E. Cuckler *Staff Executive, Robertshaw Controls Company, Anaheim, California. Member: Instrument Society of America; National Management Association.* (Moisture, measurement of)

J. B. Damrel, Jr., B.A. *Manager, Scientific Instruments Branch, Industrial Products Division, Texas Instruments Incorporated, Houston, Texas. Member: American Cryogenics Society; Instrument Society of America; Society of Exploration Geophysicists.* (Gravity, measurement of)

Paul H. Dannenberg, B.S.E.E. *Radar Applications Engineer, Aerospace and Defense Group, Honeywell Inc., Minneapolis, Minnesota.* (Altimeter)

Frank G. Daveler, B.S.M.E. *General Manager, AMETEK/Instruments and Controls, Feasterville, Pennsylvania. Member: Instrument Society of America.* (Force transducer, proving-ring; *and* Linear displacement transducer, variable-reluctance)

P. D. Dieterich *President, Ellison Instrument Division, Dieterich Standard Corporation, New Buffalo, Michigan. Member: Instrument Society of America.* (Manometer)

Henry A. Dinter, Jr., B.S., M.A. (Physics) *Staff Engineer, Inertial Components Department, Aerospace and Defense Group, Honeywell Inc., Minneapolis, Minnesota. Member: Institute of Electrical and Electronics Engineers.* (Flight-control systems)

Frank Dostal *Chief Engineer, American Time Products, Electronics Division, Bulova Watch Company, Inc., Woodside, New York. Member: Acoustical Society of America; American Institute of Physics; and Optical Society of America.* (Chronograph; Clock, atomic; Crystal oscillator; *and* Time and time measurement)

Harry A. Downer, B.S. *Vice President, The Bin-Dicator Company, Port Huron, Michigan.* (Level instrumentation, bulk solids)

L. S. Dysart, B.S.E.E. *Control Equipment Division, Fulton Sylphon Division, Robertshaw Controls Company, Knoxville, Tennessee. Member: Instrument Society of America.* (Self-actuated regulators)

E. E. Escher, B.S.Ch.E. *Senior Application Engineer, Instrument Products Division, E. I. DuPont de Nemours & Co., Inc., Monrovia, California. Member: Air Pollution Control Association; American Chemical Society; American Society for Testing and Materials; Instrument Society of America. Registered Professional Engineer (California).* (Mass spectrometer)

C. E. Fahlgren, B.S., M.S.Ch.E. *Senior Process Systems Engineer, The Dow Chemical Company, Midland, Michigan. Member: American Institute of Chemical Engineers.* (Analog computer, process)

Charles E. Fees, B.S.M.E. *Product Manager, Fischer & Porter Company, Warminster, Pennsylvania. Member: Instrument Society of America.* (Flowmeter, variable-area; *and* Rotameter)

Urban Floor, B.S.M.E. *Vice President, Engineering, Woodward Governor Company, Rockford, Illinois. Member: American Society of Mechanical Engineers.* (Governor)

Paul W. Flynn *Senior Applications Engineer, Applied Research Laboratories, Sunland, California.* (Optical emission spectrochemical analysis)

Lloyd M. Geertz, B.S., M.S. *Manager, Large Optical Systems, Fecker Systems Division, Owens-Illinois, Pittsburgh, Pennsylvania. Member: American Society of Mechanical Engineers; Optical Society of America. Registered Professional Engineer (Pennsylvania).* (Astronomical instrumentation; Coronagraph; Schmidt camera; Solar telescope; Spectroheliograph; *and* Telescopes)

Philip W. Harland, B.S.M.E. *Director of Engineering, AMETEK/U.S. Gauge, Sellersville, Pennsylvania. Member: Instrument Society of America. Registered Professional Engineer (Pennsylvania).* (Bourdon tubes; *and* Pressure gage, elastic-element)

Thomas J. Harrison, B.S., M.S.E.E., Ph.D. *Senior Engineer, IBM Corporation, Boca Raton, Florida. Member: Institute of Electrical and Electronics Engineers; Instrument Society of America; Technical Association of the Pulp and Paper Industry.* (Data acquisition terms)

Gordon W. Hickenlooper *The Ohmart Corporation, Cincinnati, Ohio.* (Density meter, radiation-type; Level gage, radiation-type; Thickness gaging, nuclear-radiation; *and* Weighing, radiation methods for)

Jerome L. Holcomb, B.S., M.S. *Manager, Chemical Spectroscopy, Varian Associates, Palo Alto, California. Member: American Chemical Society.* [Nuclear magnetic resonance chemical analyzer (NMR)]

L. F. Holder, B.S., B.A. *Mass Flowmeter Product Manager, Black Sivalls & Bryson, Inc., Bethany, Oklahoma.* (Flowmeter, mass)

John R. Huebner, B.S.E.E. *Director of Engineering, Singer, GPE Controls, Inc., Morton Grove, Illinois. Member: Instrument Society of America.* (Hydraulic controller; *and* Hydraulic relay)

Gordon Ferrie Hull, Jr., Ph.D. *Hull Associates, Concord, Massachusetts. Fellow: American Association for the Advancement of Science; American Physical Society.* (Radio telescope)

Russell D. Irwin *Application Engineer, Industrial Division, Honeywell Inc., Fort Washington, Pennsylvania. Member: Instrument Society of America.* (Differential-pressure transmitter)

Howard R. Jaquith, B.M.E., M.B.A. *Manager, Mechanical Division, Engineering Department, Taylor Instrument Process Control Division, Sybron Corporation, Rochester, New York. Member: Instrument Society of America.* (Pneumatic controller; *and* Transmission, pneumatic)

Richard F. Jarrell, B.S. *Jarrell-Ash Division, Fisher Scientific Company, Waltham, Massachusetts. Member: American Society for Testing and Materials; Optical Society of America; Society of Applied Spectroscopy.* (Flame photometer-spectrophotometer; Spectro terminology; *and* Spectroscope)

Anthony M. Johnson, B.S., M.P.I. *Apollo Lasers, Inc., Los Angeles, California. Director, Laser Industry Association. Member: American Institute of Astronautics and Aeronautics; American Institute of Chemical Engineers; Instrument Society of America.* (Laser, instrumentation of)

William N. Johnson, B.E.E. *Staff Engineer, Aerospace and Defense Group, Honeywell Inc., Minneapolis, Minnesota. Member: Institute of Electrical and Electronics Engineers.* (Flight-control systems)

Harold C. Jones, M.S.E.E. *Electrical Engineering Department, University of Maryland. Member: Institute of Electrical and Electronics Engineers; Institute of Environmental Sciences. Registered Professional Engineer (Maryland).* (Vibrometer)

Thomas J. Kehoe, B.S. *Manager, Application Engineering, Process Instruments Division, Beckman Instruments, Inc., Fullerton, California. Past President, Intrument Society of America. Member: American Chemical Society; American Institute of Chemical Engineers; American Institute of Chemists.* (Oxidation-reduction potential, measurement of; *and* pH, measurement of)

Richard E. Kehrli, B.A. *Tektronix, Inc., Beaverton, Oregon.* (Cathode ray tubes; *and* Oscilloscope)

Donald B. Kendall, B.S.E.E. *Chief Scale engineer, Toledo Scale Company, Division of Reliance Electric Company, Toledo, Ohio. Member: American Railway Engineering Association; Institute of Electrical and Electronics Engineers.* (Dynamometer)

Richard H. Kennedy, B.S.M.E. *Manager, Systems Engineering, Bristol Division, American Chain & Cable Company, Inc., Waterbury, Connecticut. Member: Institute of Electrical and Electronics Engineers; Instrument Society of America.* (Control-system terms)

William Kes, B.S.M.E. *Vice President, Engineering and Research, Veeder-Root Company, Hartford, Connecticut. Member: American Ordnance Association; American Society of Mechanical Engineers; Institute of Electrical and Electronics Engineers; Instrument Society of America; Scientific Apparatus Makers Association.* (Counter)

Joost H. Kiewiet de Jong, Ph.D. *Associate Professor of Astronomy, University of Pittsburgh, Pittsburgh, Pa. Member: American Astronomical Society.* (Astronomical instrumentation; Coronagraph; Schmidt camera; Solar telescope; Spectroheliograph; *and* Telescopes)

Elias Kimmel, Ph.D. *Director of Research and Development, Tempil° Division, Big Three Industries, Inc., New York, New York. Member: American Chemical Society; Instrument Society of America.* (Pyrometric cone; *and* Temperature-sensitive materials)

George H. Krell, B.S.Ch.E., L.L.B. *Project Engineer, Operations Headquarters, AMETEK/Process Design Group, Paoli, Pennsylvania. Member: Instrument Society of America.* (Deadweight tester)

Bruce W. Landeck, B.S., M.S. *Staff Planner, IBM Corporation, Poughkeepsie, New York. Member: Association for Computing Machinery.* (Computer programming terms)

Robert J. Lasher, B.A. *Fisher Controls Company, Marshalltown, Iowa.* (Actuators and control valve terms)

George Robert Leavitt, B.S.M.E. *General Section Head, Mechanical Product Engineering, Taylor Instrument Process Control Division, Sybron Corporation, Rochester, New York. Member: American Society of Mechanical Engineers; Rochester Engineering Society.* (Thermometer, filled-system)

Malcolm D. Lefcourt, B.S.M.E., M.Sc. (Thermodynamics) *Engineering Section Manager, Transducer Division, United Control, a subsidiary of Sundstrand Corporation, Redmond, Washington. Member: Instrument Society of America; Marine Technology Society.* [Air-data systems (aircraft); *and* Pressure transducer, vibrating-wire]

Gabor B. Levy, Ph.D. *Vice President, Photovolt Corporation, New York, New York. Member: American Association for the Advancement of Science; American Chemical Society; Electron Microscope Society; Instrument Society of America.* (Color, measurement of; Colorimeter; Densitometer; Gloss meter; *and* Photometers)

Edmond B. Lynch, M.E. *Manager, Industrial Meters and Control, B I F, Unit of General Signal Corporation, Providence, Rhode Island.* [Dall tube; Flow metering devices, primary; Flow nozzle; Flume; Gentile tube (Foster); Orifice plate; Piezometer ring; *and* Venturi tube]

Robert A. McCarron, B.S.E.E. *Leeds & Northrup Co., North Wales, Pennsylvania. Member: Institute of Electrical and Electronics Engineers; Instrument Society of America. Registered Professional Engineer (Pennsylvania).* (Integrator)

Elmer W. Madsen, B.S.E.E. *Product Engineer, The Superior Electric Company, Bristol, Connecticut. Member: Institute of Electrical and Electronics Engineers; Numerical Control Society; Society of Manufacturing Engineers.* (Stepping motor)

Ernest C. Magison, B.S.E.E. *Supervisor, Sensors Development, Industrial Division, Honeywell Inc., Fort Washington, Pennsylvania. Member: Instrument Society of America; National Academy of Sciences Advisory Committee to the United States Coast Guard. Registered Professional Engineer (Pennsylvania).* (Pyrometer; Thermocouple; *and* Thermopile)

Theodore F. Masse, B.S.M.E. *Director of Product Engineering, B I F, Unit of General Signal Corporation, Providence, Rhode Island. Member: Society of Automotive Engineers.* (Metering pump)

David L. Mellen, M.S. *Section Head, Advanced Flight Systems, Aerospace and Defense Group, Honeywell Inc., Minneapolis, Minnesota. Member: American Institute of Astronautics and Aeronautics; Society of Aeronautical Engineers.* (Flight-control systems)

R. Michel, B.S.M.E. *Photronic, Inc., Jenkintown, Pennsylvania.* (Turbidimeter)

John W. Miller, B.E.E. *Principal Staff Engineer, Aerospace and Defense Group, Honeywell Inc., Minneapolis, Minnesota. American Institute of Astronautics and Aeronautics; Institute of Electrical and Electronics Engineers; Society of Automotive Engineers.* (Navigation systems; *and* Space and missle-systems instrumentation)

Gary D. Moon, B.S.E.E. *Project Engineer, Inertial Components Department, Aerospace and Defense Group, Honeywell Inc., Minneapolis, Minnesota.* (Flight-control systems)

David G. Moulton, B.S., Ph.D. *Associate Professor of Physiology, University of Pennsylvania, Monell Chemical Senses Center, Philadelphia, Pennsylvania. Member: American Association for the Advancement of Science; American Institute of Biological Sciences; American Physiological Society; American Zoological Society; Institute of Biology; Institute of Food Technologists; New York Academy of Sciences.* (Taste and odor instrumentation)

Otto Muller-Girard, B.S., M.S. *Consulting Engineer, Rochester, New York. Member: American Society of Mechanical Engineers; Instrument Society of America; National Society of Professional Engineers; New York State Society of Professional Engineers. Registered Professional Engineer (New York).* (Measurement-system and transducer terms)

Domenic A. Narducci, Jr., B.S.E.E. *Product Manager, Bristol Division, American Chain & Cable Company, Inc., Waterbury, Connecticut. Member: Institute of Electrical and Electronics Engineers. Registered Professional Engineer (Connecticut).* (Potentiometric instrument)

Robert J. O'Kane, B.A., M. S. *Vice President, Satec Systems, Inc., Grove City, Pennsylvania. Member: American Society for Testing and Materials. Registered Professional Engineer (Pennsylvania).* [Testing (materials), destructive]

W. D. Perkins, S.B. Chem., A.M. Physics, Ph.D. *Product Specialist, Infrared Products Department, The Perkin-Elmer Corporation, Norwalk, Connecticut. Member: American Chemical Society; American Physical Society; Optical Society of America.* (Infrared spectrophotometer)

R. A. Pontefract, A.M.E., B.B.A. *Chief Engineer, Jerguson Gage and Valve Company, Burlington, Massachusetts.* (Gage glass)

William R. Polye, M.S. *Senior Staff Scientist, Navigation and Control Division, The Bendix Corporation, Teterboro, New Jersey. Member: American Association for the Advancement of Science; American Institute of Astronautics and Aeronautics; American Physical Society.* (Flight-control systems; Navigation systems; *and* Space and missile-systems instrumentation)

David N. Rager, B.S.E.E., M.S. *Algorex Data Corporation, Hicksville, New York. Member: Association for Computer Machinery; Institute of Electrical and Electronics Engineers; Institute of Printed Circuits; Numerical Control Society; Society for Information Display.* (Drafting system, automatic)

Alfred G. Ratz, Ph.D., M.A.Sc., B.A.Sc. *Director of Engineering, MB Electronics, a Textron Company, New Haven, Connecticut. Member: Institute of Electrical and Electronics Engineers; Instrument Society of America; Professional Engineers of Province of Ontario.* (Shock, measurement of; *and* Vibration, measurement of)

Jack C. Riley, B.S.E.E., M.S. (Physics) *Metrologist and Consulting Engineer, Portland, Oregon. Member: Institute of Electrical and Electronics Engineers; Instrument Society of America; National Society of Professional Engineers; Precision Measurement Association. Registered Professional Engineer (Oregon).* (Bridge circuits, electric)

F. E. Romberg, B.A., M.A. *Chief Geophysicist, Science Services Division, Texas Instruments Incorporated, Dallas, Texas. Member: European Association of Exploration Geophysicists; Geological Society of America; Society of Exploration Geophysicists.* (Magnetometer, airborne; Richter scale, seismic; *and* Seismology)

H. D. Ruhl, B.S.E.E., M.S. (Physics) *Chemical Physics Research Laboratory, The Dow Chemical Company, Midland, Michigan. Member: Instrument Society of America.* (Infrared process analyzers)

Erwin J. Saxl, Ph.D., M.D. *President and General Manager, Tensitron, Inc., Harvard, Massachusetts. Fellow: American Association for the Advancement of Science; Instrument Society of America. Member: American Institute of Physics; American Physical Society; Engineer's Club of New York City; Society for Experimental Stress Analysis; Society of Photographic Scientists and Engineers; Wire Association. Registered Professional Engineer (Massachusetts).* (Tension, measurement and control of)

Peter Schindler, Ph.D. *Varian Associates, Palo Alto, California.* (Electron paramagnetic resonance chemical analyzer)

Harvey A. Senior *Unimation Inc., Danbury, Connecticut. Member: Society for Technical Communications.* (Manipulator; *and* Robot)

Paul D. Senstad, B.E.E., M.S.E.E. *Research Section Chief, Aerospace and Defense Group, Honeywell Inc., Minneapolis, Minnesota. Member: Institute of Electrical and Electronics Engineers; National Society of Professional Engineers.* (Accelerometers)

Edgar B. Sharpe *Technical Manager, Electronic Associates, Inc., West Long Branch, New Jersey.* (Hybrid computer)

Israel Sheinberg, B.S. (Physics) *Vice President, Recognition Equipment Incorporated, Dallas, Texas. Member: Optical Society of America.* (Optical character recognition)

W. G. Shequen, B.E. *Manager, Field Engineering, Applied Research Laboratories,*

Sunland, California. Member: Instrument Society of America; Society for Applied Spectroscopy. Registered Professional Engineer (California). (X-ray analyzer, electron microprobe; *and* X-ray fluorescence analysis)

James K. Sparrell, B.S.M.E., S.M.M.E. *Sparrell Engineering Research Corporation, Marblehead, Massachusetts. Member: American Society of Mechanical Engineers.* (Calorimeter; *and* Thermal conductivity instruments)

Joseph P. Spetzer, Jr., B.E.E. *Chief Engineer, Synchros, Clifton Division of Litton Industries, Clifton Heights, Pennsylvania.* (Linear transformer; Resolver; Synchro; *and* Transolver)

Robert C. Staats, B.S.E.E. *Staff Engineer, Aerospace and Defense Group, Honeywell Inc., Minneapolis, Minnesota. Member: Institute of Electrical and Electronics Engineers.* (Flight-control systems)

Tracy S. Storer, B.S.E.E., M.S.E.E. *Engineering Manager, Cupertino Division, Hewlett-Packard Company, Cupertino, California. Member: Institute of Electrical and Electronics Engineers.* (Gamma-ray spectroscopy)

Charles L. Strickland, B.S. *Hydro Products Division, Dillingham Corporation, San Diego, California. Member: American Society for Oceanography; Marine Technology Society; Society for Photographic Instrumentation Engineers.* (Depth measurement, oceanographic)

Norman J. P. Taylor *Graduate Royal Naval Engineering College. Manager, Feeder and Weigher Products, B I F, Unit of General Signal Corporation, Providence, Rhode Island. Member: British Institute of Management; Institute of Materials Handling; and Institute of Mechanical Engineers.* (Feeder, weigh)

Robert D. Thompson, Ph.D. *Coordinator Industrial Development, Taylor Instrument Process Control Division, Sybron Corporation, Rochester, New York. Member: American Chemical Society; American Society for Quality Control; American Society for Testing and Materials; Instrument Society of America; Society of Glass Technology.* (Temperature scales; Thermometer; *and* Thermometer, liquid-in-glass)

W. F. Ulrich, B.S., Ph.D. *Manager, Applications Research, Scientific Instruments Division, Beckman Instruments, Inc., Fullerton, California. Member: American Chemical Society; Canadian Spectroscopy Society; Society for Applied Spectroscopy.* (Ultraviolet spectrophotometry)

E. Van Haaften, B.A. *Director, Timer Laboratory, Bulova Watch Company, Inc., Flushing, New York. Member: American Ordnance Association.* (Chronograph; Clock, atomic; Crystal oscillator; *and* Time and time measurement)

R. W. Van Tilburg, B.S.Ch.E. *Corning Glass Works, Raleigh, North Carolina. Member: Instrument Society of America.* (Fluidics)

Richard Villalobos, B.S. Chem. *Principal Application Engineer, Process Instruments Division, Beckman Instruments, Inc., Fullerton, California. Member: American Association for the Advancement of Science; American Chemical Society; Instrument Society of America.* (Chromatography)

James R. Walker, B.S., M.S.E.E. *Chief Electrical Engineer, De Vlieg Machine Company, Royal Oak, Michigan. Member: American Mathematical Society; American Physical Society; Institute of Electrical and Electronics Engineers. Registered Professional Engineer (Michigan).* (Adaptive control; Lead screw; Position measurement and control systems; *and* Rack and pinion)

W. T. Walker, B.S. *Application Engineer, Sensor Division, Leeds & Northrup Company, North Wales, Pennsylvania.* (Optical pyrometer; *and* Resistance thermometry)

E. Gordon Watson *Gaertner Scientific Corporation, Chicago, Illinois.* (Autocollimator; *and* Polariscope)

Seymour Weiner, B.S.E.E. *Applied Physics Department Head, Barnes Engineering Company, Stamford, Connecticut.* (Infrared detector, thermistor)

Walter W. Welz, B.S.E.E. *Advisory Engineer, IBM Corporation, Boca Raton, Florida. Member: Institute of Electrical and Electronics Engineers.* (Computer terms)

Leon J. Weymouth, B.S., B.A.E. *Staff Consultant, BLH Electronics, Inc., a subsidiary of Baldwin-Lima-Hamilton Corporation, Waltham, Massachusetts. Member: Instrument Society of America; Society of Experimental Stress Analysis.* (Load cell, strain gage; *and* Strain gage)

Robert L. Wilson, B.S.M.E. *Head, Evaluation and Materials Engineering, Industrial Division, Honeywell Inc., Fort Washington, Pennsylvania. Member: Instrument Society of America.* (Ambient conditions; Instrument performance terms; *and* Reference junction compensation)

L. E. Wise, B.S.M.E. *Staff Programmer, IBM Corporation, Boca Raton, Florida. Member: Association for Computing Machinery.* (Computer programming terms)

T. E. Wolk, B.S., M.S.E.E. *Engineering Supervisor, Test Instruments Division, Honeywell Inc., Denver, Colorado. Member: Institute of Electrical and Electronics Engineers.* (Oscillograph)

Wallace Dean Wood, B.A. *Director, Standards Laboratory, Taylor Instrument Process Control Division, Sybron Corporation, Rochester, New York. Member: Instrument Society of America; National Conference of Standards Laboratories; National Society of Professional Engineers; Rochester Conference of Standards Laboratories. Registered Professional Engineer (New York).* (Error, measurement; Hysteresis; *and* Parallax error)

A. W. Young *Rank Scherr-Tumico, Des Plaines, Illinois.* (Goniometer; *and* Theodolite)

O. John Zamparo, B.S.M.E., M.A.E. *Manager, Electromechanical Engineering, MB Electronics, a Textron Company, New Haven, Connecticut. Member: Connecticut Society of Professional Engineers; Institute of Environmental Sciences.* (Vibration measurements)

preface

This encyclopedia represents a major step toward bringing together the many facets of scientific information which, when grouped together, may be described as *instrumentation and control technology.*

Although the nearly 700 topical entries are arranged alphabetically, the information content was delineated first in terms of major functions and systems. There are five principal information categories: (1) *measurands*—with ten subclassifications, including the chemical composition, electrical, force, geometric, physical property, quantity, radiation, rate, thermal, and time variables; (2) *measurement systems*—with subclassifications concerned with the basic building blocks used (mechanical, electrical, electronic, optical, and so on) and with performance characteristics, errors, and compensation; (3) *instrument data processing*—with the subelements of acquisition, computing, programming, communication, and display; (4) *control systems*—with the important subcategories of control actions, configurations, and modes; control media (electric, electronic, fluidic, hydraulic, pneumatic, and so on); final controlling elements, such as valves and motor drives; and switching elements; and finally (5) *applications*—with in-depth attention given to uses in aerospace, astronomy, design and drafting, geophysics, machine control, medicine and biochemistry, meteorology and hydrology, navigation, oceanography, process control, and testing and inspecting.

Because instrumentation and control technology is one of the most interdisciplinarian of the applied sciences, reaching down as it does in octopus fashion to practically all the sciences for principles to incorporate into instruments and fanning out in all directions to tens of thousands of scientific and industrial applications, the user of this book is urged to study and to refer frequently to the **Classified Index**—as well as to the comprehensive alphabetical index at the end of the volume. To assist the user further, cross references from one entry to the next are used extensively.

On the Nature of Instrumentation and Control Technology

The growth rate of the instrument, computer, data processing, and automatic control industries over the past two decades, particularly during the last 5 years, is startling on any scale of comparison. The rate continues to exceed the average of the technically oriented industries. In the expansion of instrumentation and control, we witness at work a large, closed-loop feedback system—if we may use a term right out of the technology.

New instruments, for example, by making measurements possible that previously had to be postulated—by greatly speeding up the measurement and data-interpretation process—by providing an order-of-magnitude betterment in precision—have accelerated research, thus constantly tempting the scientist to reach for new knowledge. Each time the researcher finds himself on the threshold between the known and the unknown, he finds that he must measure entirely new phenomena or measure phenomena with which he may be familiar, but under new conditions and environments. The researcher feeds these new needs back to the instrument designer—better instruments are built—and so the cycle repeats time and time again.

The impact of computers and data-processing equipment on nearly all of man's endeavors requires little detailing here. Experts agree that only a start has been made. As the costs of these machines spiral downward through the application of new techniques, components, and materials, literally thousands of additional applications for this equipment will become economically justifiable. Within a decade, one will be hard pressed to name an industry, a commercial business, or a branch of government that is not vitally dependent upon computers and data processors for its day-to-day operations, let alone for management and direction.

Automatic controllers, the predominant ingredient of automation, are beginning to take on—in large measure—the really tough assignments that have been left until last. By comparison with a group of cutting and sewing machines in an upper loft for making garments, the mammoth, glamorous chemical plant or petroleum refinery has been a pushover to automate. But, these long-neglected pockets of technical impoverishment now are being improved one by one—greatly aided by instruments and control devices that originally may have been developed for a space probe. Ironically, the assembly of electronic equipment, including instruments and controls, per se is among these vestigial situations where gross improvements in production automation remain to be made.

It seems that the science of instrumentation and control has everything in its favor. Research would be helpless without accurate measurements and quick and reliable data interpretation. In fighting the profit squeeze, where can business and industry better turn than to instruments, computers, and controllers to help in lowering costs? This causes feedback too because automated equipment requires more exacting maintenance which, incidentally, has given rise to an entirely new branch of instrumentation—machine and process diagnostics.

Man appears to have a never-ending thirst for faster and faster transportation and for more and quicker communications. It could be stated with some plausibility that a supersonic aircraft is a flying instrumentation and control system with some space allowed for passengers and cargo. Nearly half the cost of a military aircraft goes for instruments and controls. Traffic control, of both aircraft and land vehicles, certainly must be on the eve of tremendous technological improvement and expansion. Most of what is needed to make the flow of traffic faster and safer will come out of instrumentation and control technology. And, what is a communications system but a long series of black boxes, most of which perform some measurement, computing, data-handling, or control function—and connected by some form of electromagnetic link—falling well within the sphere of instrumentation and control technology.

One only can conclude that instrumentation and control indeed is a ubiquitous technology. Thus far, a brief picture has been given of the way instrumentation and control, as a science, penetrates the universe around it—manufacturing, processing, business, medicine, biology, meterology, space exploration, transportation, oceanology, research, communications, geophysics—to mention but some of its end uses. From the other direction, what do we see when we probe deeply into the inside of an instrument, a computer, a data processor, or an automatic controller? Are these devices the result of a pure instrumentation science at work? Obviously not. Instrumentation and control maintains its identity as an applied science because of the final functions that it performs—to measure, to display information, to compute data, to interpret data, and to control.

The distinctive identity of instrumentation and control cannot stem from what is inside the equipment because the hardware represents a fabulous agglomeration of scientific inputs from nearly all the scientific disciplines. A pneumatic chromatographic controller is a good example. The design of such an apparatus must call upon mathematics, information theory, chemistry, electronics, hydraulics, pneumatics, chemical engineering, mechanical engineering, system engineering, and precision machining—because the device is a system comprised of an instrument, a computer, a process, and a controller all condensed into a package.

Thus, we find instrumentation and control in the middle of a large matrix—a matrix in which row upon row of applications fan out and, radiating in the other direction, column after column of the sciences upon which instrumentation depends.

In this encyclopedia, an attempt has been made to bring these many facets together in an organized fashion and in sufficient depth to provide initial orientation on each topic.

Classified Index

The structure of the classified index is shown in the table on page 762. *Each* entry of this encyclopedia is listed *once* and *only once* in the classified index. Cross references are used extensively throughout the book, and these references, coupled with the detailed alphabetical index at the end of the volume, provide a webbing to hold all topics together even though all topics are entered into the book alphabetically. Admittedly, some of the decisions involved in placing topics under the major categories of the table were arbitrary. This situation could have been avoided only by listing some entries under several categories. Such multiple listings would have resulted in a much longer and less convenient topical classification.

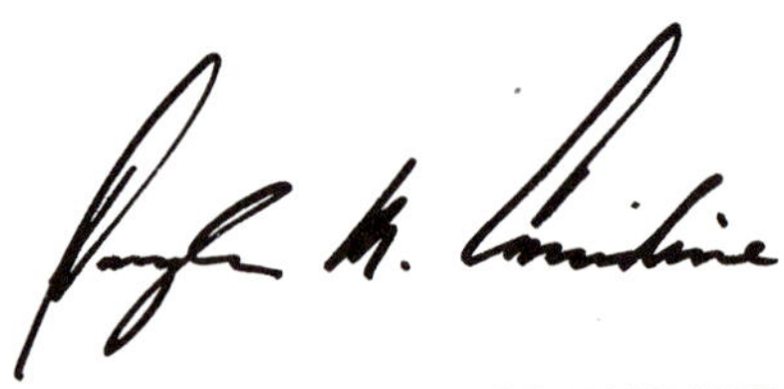

DOUGLAS M. CONSIDINE
Editor-in-Chief

acknowledgments

The preparation of an encyclopedia, particularly a treatise that is concerned with such a diverse body of scientific information as instrumentation and control technology, is a rather massive task. While care and thoughtfulness are prerequisites to the generation of a useful, long-term reference, a relatively short time span for preparation and production also is an important directive. Thus, the broad scope, the complexities and interrelationships of the science, and the emphasis on timing dictate that an effort of this kind must be one of extensive cooperation among hundreds of experienced authors and advisers. The pleasant interchanges enjoyed by the editor with these hundreds of interested individuals, both in the United States and abroad, prove once again how people will cooperate, how they will take time from otherwise very busy schedules to further a cause that is dedicated to the advancement of their professional pursuits. Thus—this general acknowledgment of assistance and of appreciation for that help rendered by numerous individuals who in some way sponsored and encouraged this work.

The important roles of the over 124 contributors who signed the majority of entries in the volume are rather obvious. There is a list of contributors and their affiliations included in the book. But, the roles of the contributors should not be glossed over quickly simply because they are self-evident. The tasks did not involve simply writing down a few facts and opinions, but rather in a large majority of cases also involved searching out new information and particularly organizing information in the highly condensed, yet comprehensive format required by an encyclopedia as contrasted with other reference-book configurations. The contributors were very helpful in making decisions as regards those topics which would become self-contained alphabetical entries—and those associated and interrelated subjects that would give necessary visibility through cross referencing and indexing. By way of example, a few instances shall be cited.

The topic of *measuring instrument characteristics* was coordinated by Otto Muller-Girard, Consulting Engineer in Rochester, N.Y.; of *various temperature scales* by Dr. Robert D. Thompson of the Taylor Instrument Process Control Division, Sybron Corporation in Rochester, N.Y.; of *photometric instrumentation* by Dr. Gabor B. Levy of Photovolt Corporation in New York City; of *acoustic instrumentation* by Hall Cary of B & K Instruments, Inc., in Cleveland, Ohio; of *scales and weight measurement* by Arthur J. Burke of the Howe Richardson Scale Company in Clifton, N.J.; of *geophysical instruments* by Howard Cannon of Texas Instruments Incorporated in Houston, Tex.; of many aspects of *flow measurement* by John S. Boynton of The Foxboro Company in Foxboro, Mass.; of *primary flowmetering devices* by Edmond B. Lynch of BIF (a unit of General Signal Corporation) in Providence, R.I.; of *electronic laboratory instruments* by Carl W. Alsen of General Radio Company in West Concord, Mass.—and by Richard E. Kehrli of Tektronix, Inc., Beaverton, Ore.; of *humidity and meteorological instrumentation* by James R. Cosby of The Bendix Corporation Environmental Science Division in Baltimore, Md.; of several aspects of *chemical analytical instrumentation* by Thomas J. Kehoe and his associates of the Process Instruments Division of Beckman Instruments, Inc., in Fullerton, Calif.; of *X-ray analytical equipment* by W. G. Shequin and his associates of Applied Research Laboratories in Sunland, Calif.; of *electrical measuring instruments* by Henry Berring (deceased), formerly of Weston Instruments, Inc., in Newark, N.J.; of *automatic-control terminology* by Richard H. Kennedy of the Bristol Division, American Chain & Cable Company, Inc., in Waterbury, Conn.; of *time measurement* by John J. Carpenter of the Electronics Division of Bulova Watch Company, Inc., in Woodside, N.Y.; of *medical instrumentation* by J. Malvern Benjamin, Jr., of Bionic Instruments, Inc., in Bala Cynwyd, Pa.; of *astronomical instrumentation* by Dr. Joost H. Kiewiet de Jonge, Associate Professor of Astronomy of the University of Pittsburgh, and Lloyd M. Geertz of the Fecker Systems Division of Owens-Illinois in Pittsburgh, Pa.; of *aircraft instrumentation and navigation* by Paul F. Bechberger of The Bendix Corporation Navigation and Control Division in Teterboro, N.J.; of *accelerometers* by John W. Miller of Honeywell Inc. in Minneapolis, Minn.; of *instrument performance characteristics* by Robert L. Wilson of Honeywell Inc. in Fort Washington, Pa.; of *in-*

strument-system errors by Wallace Dean Wood of the Taylor Instrument Process Control Division of Sybron Corporation in Rochester, N.Y.; of *electric controllers* by W. S. Bloor of Leeds & Northrup Company in North Wales, Pa.; of *computing, programming, and data-acquisition terms* by Dr. Thomas J. Harrison, Alex J. Arthur, Walter W. Welz, and L. E. Wise of IBM Corporation in Boca Raton, Fla.—and their associate, Bruce W. Landeck of IBM Corporation in Poughkeepsie, New York; of *major automatic control* systems by Anthony Y. Chiang of the Bristol Division of American Chain & Cable Company, Inc., in Waterbury, Conn.; of *final controlling elements* by Robert J. Lasher of Fisher Controls Company in Marshalltown, Iowa; of *hydraulic controllers* by John R. Huebner of Singer, GPE Controls, Inc., in Morton Grove, Ill.; and of *oceanographic instrumentation* by C. L. Strickland of the Hydro Products Division of Dillingham Corporation in San Diego, Calif.

Special mention is made of the cooperation of John Markus, editor of the "Electronics and Nucleonics Dictionary" (McGraw-Hill); Byron K. Ledgerwood, editor of *Control Engineering Magazine;* and numerous officers and committee chairmen of the Instrument Society of America, of the Scientific Apparatus Makers Association, and of the American Petroleum Institute. In organizing and editing this encyclopedia, there were numerous secretarial, typing, and administrative tasks for which the cooperation and very helpful assistance of Mrs. Sharon Habbick, Mrs. Tash Nashino, and Miss Gerlinde Traut are gratefully acknowledged.

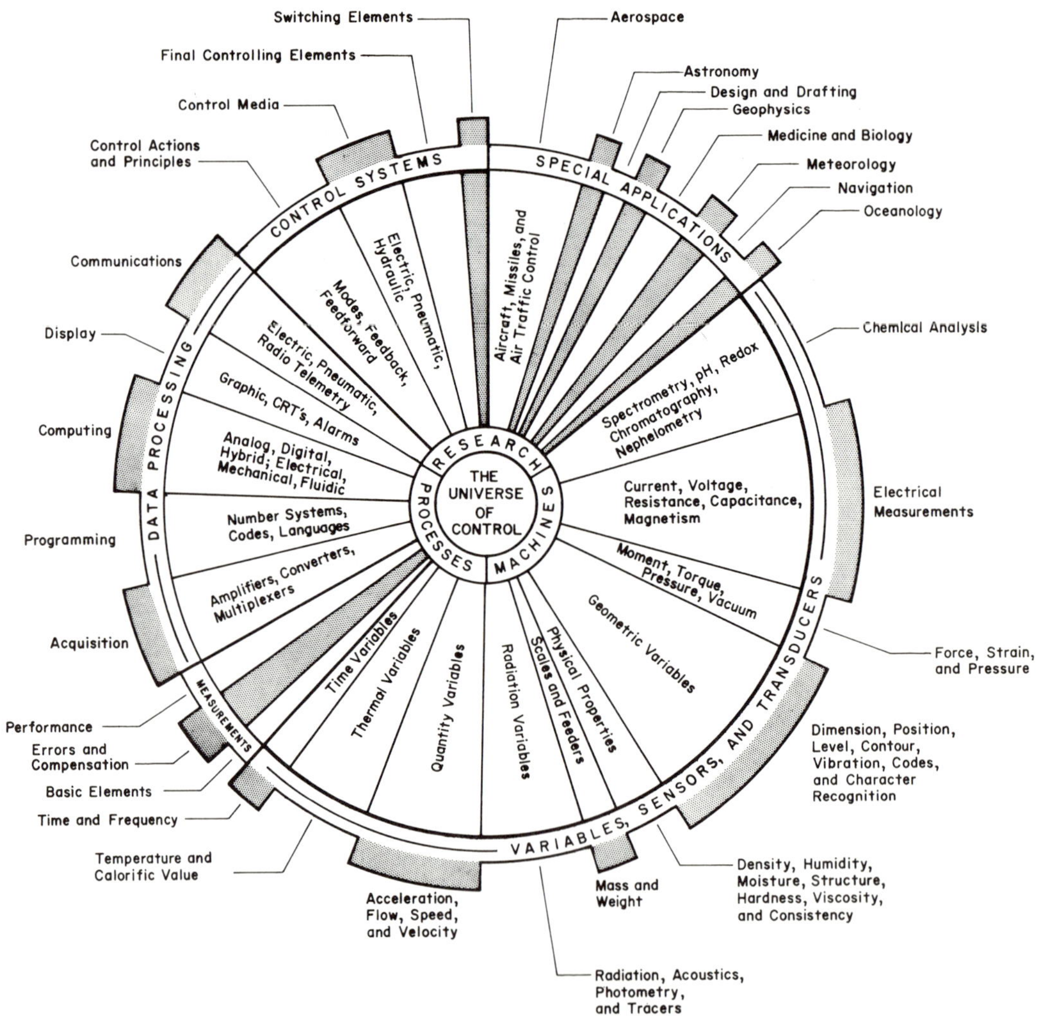

THE UNIVERSE OF CONTROL
RESEARCH
PROCESSES
MACHINES
CONTROL SYSTEMS
SPECIAL APPLICATIONS
DATA PROCESSING
MEASUREMENTS
VARIABLES, SENSORS, AND TRANSDUCERS
Switching Elements
Final Controlling Elements
Control Media
Control Actions and Principles
Electric, Pneumatic, Hydraulic
Modes, Feedback, Feedforward
Aerospace
Astronomy
Design and Drafting
Geophysics
Medicine and Biology
Meteorology
Navigation
Oceanology
Aircraft, Missiles, and Air Traffic Control
Communications
Display
Computing
Programming
Acquisition
Electric, Pneumatic, Radio Telemetry
Graphic, CRT's, Alarms
Analog, Digital, Hybrid; Electrical, Mechanical, Fluidic
Number Systems, Codes, Languages
Amplifiers, Converters, Multiplexers
Chemical Analysis
Spectrometry, pH, Redox Chromatography, Nephelometry
Electrical Measurements
Current, Voltage, Resistance, Capacitance, Magnetism
Force, Strain, and Pressure
Moment, Torque, Pressure, Vacuum
Dimension, Position, Level, Contour, Vibration, Codes, and Character Recognition
Geometric Variables
Density, Humidity, Moisture, Structure, Hardness, Viscosity, and Consistency
Physical Properties
Mass and Weight
Scales and Feeders
Radiation, Acoustics, Photometry, and Tracers
Radiation Variables
Acceleration, Flow, Speed, and Velocity
Quantity Variables
Temperature and Calorific Value
Thermal Variables
Time and Frequency
Time Variables
Basic Elements
Errors and Compensation
Performance

graphic representation of contents

The total content of this book is illustrated graphically by the accompanying pie chart or "contents wheel." Within plus-or-minus a percentage point or two, the number of pages devoted to each major topic relative to the total number of pages of the book is represented by the size of the sector. An attempt to break down the coverage of instrument and control applications for the major functions of research, processes, and machines has not been made because such information appears throughout the entire book. The contents wheel dramatically illustrates how the scope of instrumentation and control has extended over the past decade or two, particularly in connection with data processing which embraces computing, communications, and displays. Not many years ago, less than ten percent of the book would have been required for adequate coverage of these aspects of instrumentation. The wheel also accentuates the growing importance of chemical analysis, both in research and process control. The impact of automation and automatic machine control is emphasized by the extensive coverage of the geometric variables, such as dimension, position, and contour measurements, and of the growing emphasis on code reading and character recognition.

Although the design of an encyclopedia does not depend wholly upon the frequency of appearance or of use of the many topical segments of a total field, there is indeed a rough relationship between topics indicated and their application in the universe of control.

A

Absolute Control System (See Drafting system, automatic; Numerical control.)

Acceleration, Standard (See Standards, measurement.)

Accelerator A device that accelerates charged particles to high velocities so that they have high kinetic energy. The device can be used for electrons, protons, deuterons, and ions. The term *particle accelerator* also is used. A number of different device configurations are in use, including the betatron, cyclotron, linear accelerator, synchrocyclotron, synchrotron, and Van de Graaff generator.

Accelerometer Acceleration is the rate at which the velocity of a body changes. It cannot be measured directly but must be computed by measuring the force exerted by restraints that are placed on a mass to hold its position fixed in an accelerating body. Thus, *restraint meter* would be an apt term to use in describing an accelerometer. The relationship between restraint and acceleration is defined by Newton's second law, namely, force = mass × acceleration.

The simplest mechanization of an accelerometer is a spring-restrained mass as shown in Fig. A-1. The important elements are a constant seismic mass, a calibrated restraint, a support bearing, and a pickoff. Some displacement of the mass must occur for the pickoff to work. Thus, a transient error is introduced during the time the mass is acquiring displacement. A block diagram and the mathematical relationships that apply to the classical spring-mass device are given in Fig. A-2.

Attempts to reduce this transient error have led to the development of devices that increase the

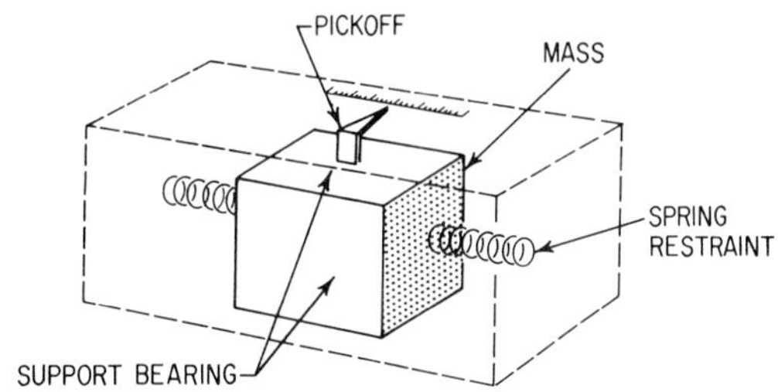

Fig. A-1. Simple accelerometer using a spring-restrained mass.

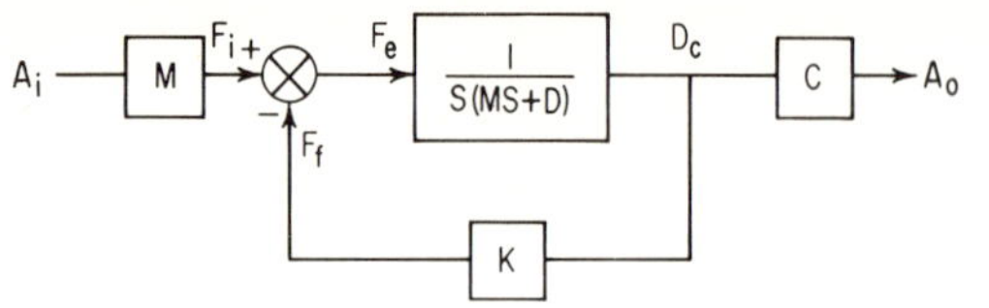

where K = spring constant
C = readout device characteristic, e.g., volts/inch for a potentiometric pickoff excited by a fixed voltage

Transfer function is

$$\frac{A_O(s)}{A_i(s)} = \frac{MC/K}{(M/K)S^2 + (D/K)S + 1}$$

or

$$\frac{A_O(s)}{A_i(s)} = \frac{C/\omega_n^2}{(S^2/\omega_n^2) + 2\zeta(S/\omega_n) + 1}$$

where ω_n = the undamped natural frequency
ζ = the damping ratio D/D_C

Fig. A-2. Block diagram and transfer function of the classical spring-mass-type accelerometer. *(United Control.)*

stiffness of the system. Low displacements result from the increased stiffness and demand sensitive pickoffs—often requiring that the pickoff work over a very small range. The useful output then becomes the restraint upon the mass. A configuration of this type is shown in Fig. A-3. The

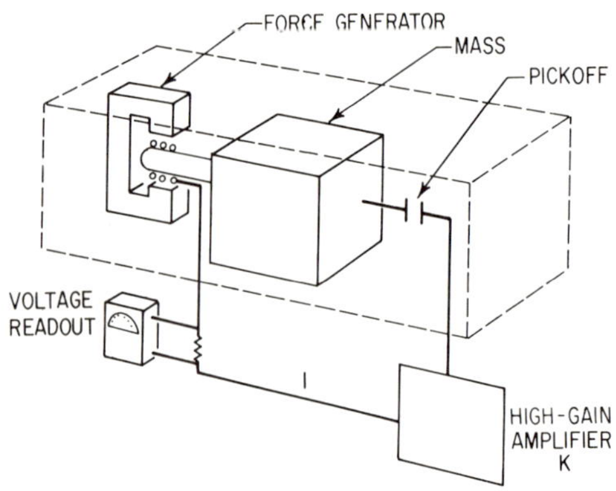

Fig. A-3. Basic servoed accelerometer.

block diagram and mathematical relationships for a servoed system are given in Fig. A-4.

The readout is proportional to the restraining force. The position pickoff output no longer directly represents the restraining force. The pickoff simply detects small changes in mass position and commands a proportional change in a current which is sent through the forcer. This current is used as a direct measure of restraining force. The accuracy of this current measurement is dependent upon maintaining the magnetic field around the forcer coil constant. Often a precision resistor is used to convert the current into a voltage signal. Where forcer current is the quantity that is read out, long-term stability of the electronic amplifier is not required. Instability of the forcer, however, cannot be tolerated. This will be evident from the block diagram. Mass displacement with respect to the case is held extremely small by the very high gain of the amplifier. The velocity and acceleration of the mass relative to the case, therefore, also are very small, approaching zero. Any forces of an extraneous nature acting on the mass

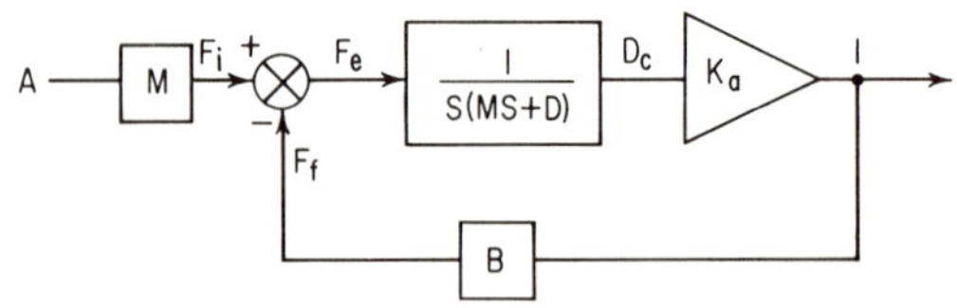

Transfer function is

$$\frac{I(s)}{A(s)} = \frac{M/B}{(M/K_aB)S^2 + (D/K_aB)S + 1}$$

or

$$\frac{I(s)}{A(s)} = \frac{M/B}{(S^2/\omega_n^2) + 2\zeta(S/\omega_n) + 1}$$

Fig. A-4. Block diagram and transfer function of a servoed accelerometer. *(United Control.)*

will cause an erroneous reading of acceleration. Elimination of such forces requires rather sophisticated designs.

Classification of Accelerometers: Like many other kinds of instruments and transducers, accelerometers can be classified in numerous ways. One useful categorization is:

1. *Dynamic-acceleration (vibration) instruments:* These include crystal types of accelerometers that do not respond to static acceleration. Such devices are used primarily in the control of vibration-test systems and in monitoring localized vibration response on an item under test. Also see **Vibration, measurement of.**
2. *Nonservoed static and dynamic instruments:* These include strain-gage devices and spring-mass units with either potentiometric or inductive pickoffs. These devices are used in instrumentation and control systems and are accurate from 1 to 5 percent, dependent upon the temperature environment.
3. *Servoed static and dynamic instruments:* This class includes active devices that essentially measure the force required to accelerate the proof mass with the case rather than

measure the deflection of the mass at the end of a spring. Generally, such servoed devices are regarded as being accurate within 0.1 percent, although some are good for only 2 percent from − 65 to + 200°F. These types also are used in instrumentation and control systems.

4. *Gyroscopic instruments:* Higher accuracy is attainable with these devices because inertia can be increased by spinning the proof mass and making use of gyroscopic-precession principles. The devices are used for navigation and guidance applications where the acceleration signal is integrated twice to obtain distance traveled. Accuracies obtainable are 0.001 to 0.01 percent.

Vibrating-string accelerometers and velocity meters would be placed in a category of special types.

The following specific design descriptions are in order of increasing complexity and sophistication.

Strain-gage Accelerometer: This device is based upon measuring the strain in a material whose stress is used to provide the rebalance force of an accelerometer mass. The accuracy is dependent upon the linearity of stress and strain in a material. If the tension in a small wire is increased, its resistance increases; or if the tension is decreased, the resistance decreases. See **Strain gage.** In a strain-gage accelerometer, the moving mass causes the resistance to increase in two resistive elements and to decrease in two resistive elements. The elements are placed into a bridge arrangement, as shown in Fig. A-5, to provide an output that is proportional to acceleration. The accelerometer shown in Fig. A-6 uses a single-crystal silicon beam to obtain linear stress vs. strain. Two piezoresistive strain elements are diffused into both sides of the beam to provide the basic sensing elements. Interconnection of the

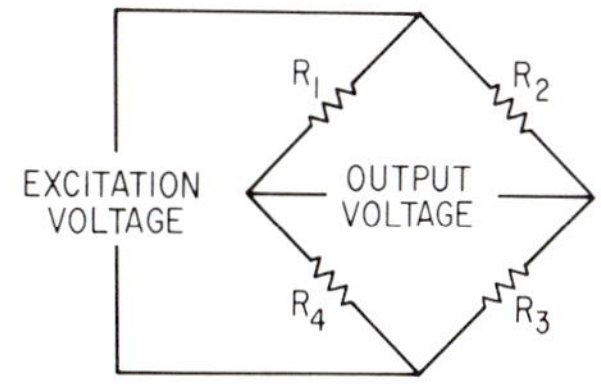

Fig. A-5. Bridge arrangement of strain-gage accelerometer.

four elements into an electric bridge circuit permits measurement of beam stress as a function of acceleration input. The planar-diffusion approach also is used in this device. This process allows the placement of local strain-sensing elements at appropriate locations on the silicon beam while electrically isolating the four active elements. As a result, the strain-gage elements are physically an integral part of the beam, yet electrically isolated. Tensile strain applied to any strain element will, because of piezoresisitive properties, increase its resistance. Conversely, a compressive strain will decrease the resistance. Locating two elements on one side of the beam and two elements on the opposite side provides a bridge wherein the resistance of two elements increases while the resistance of the other two elements decreases when acceleration is applied. The single-crystal silicon beam is nearly perfectly elastic and therefore provides a spring whose de-

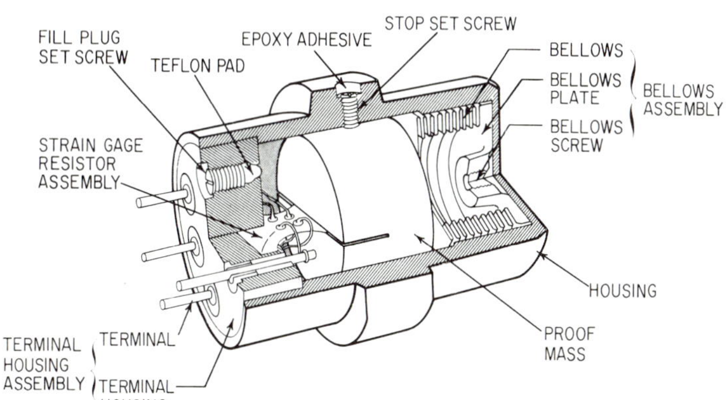

Fig. A-6. Strain-gage accelerometer that uses a single-crystal silicon beam to obtain linear stress vs. strain. *(Honeywell.)*

flection is linear with applied acceleration. The pendulous mass is attached to one end of the beam. Surrounding fluid provides viscous damping. Devices of this type are used for instrument applications as well as for control and stabilization.

Force-balance Hinged-pendulum Accelerometer: In contrast to the open-loop strain-gage accelerometer, the force-balance unit additionally includes a torquer or forcer which performs the function of restraining the mass near its null position. This accelerometer is used with external electronics to provide the appropriate gain or amplification. The block diagram of Fig. A-7 indicates the essential functions of the device. In response to an acceleration *a,* the mass will be displaced, which causes a signal from the signal generator. This signal is amplified and provides power to the forcer which restrains the mass close to its null position. The feedback current flowing through the readout resistor provides an output voltage proportional to the input acceleration. The feedback-type accelerometer provides versatility inasmuch as the scale factor and frequency response can be readily changed. The frequency response can be changed by changing the amplifier gain; the scale factor can be changed by changing the magnitude of the readout resistor. The force-balance principle also minimizes errors introduced by the signal generator and suspensions means, since the mass moves over a very small range and nonlinear restraints of the signal generator or suspension means produce correspondingly small errors.

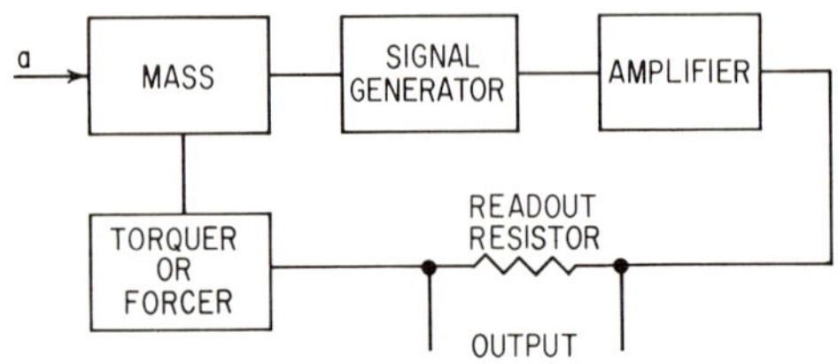

Fig. A-7. Block diagram of force-balance hinged-pendulum accelerometer.

A force-balance accelerometer without the external amplifier is shown in Fig. A-8. The small pendulous mass is supported on a pair of flexural pivots. These pivots deflect freely in one axis but are highly resistant to displacement in the other axis, thereby reducing cross-coupling errors. Torque rebalance of the pendulum is accomplished by means of two permanent-magnet torquing elements operating in a push-pull manner to provide extremely linear torque vs. current. A moving-coil-type pickoff element is used. The device is filled with fluid to provide viscous damping of the pendulous element and to provide high vibration and shock capability. The expansion bellows provide positive pressure in the fluid over a wide temperature range. For

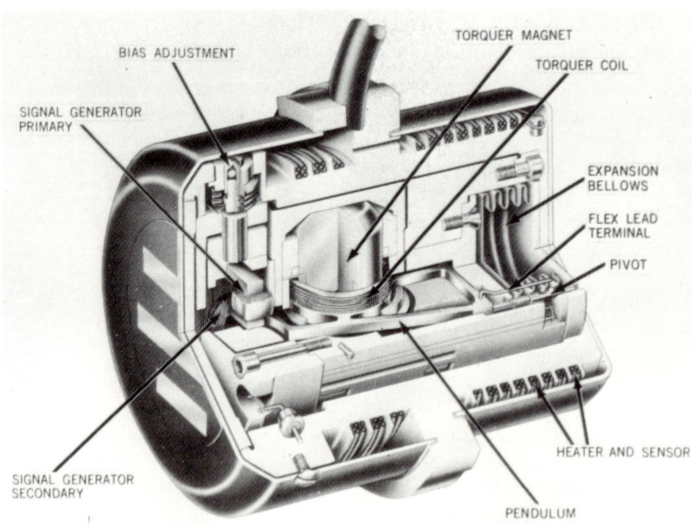

Fig. A-8. Force-balance accelerometer without external amplifier. *(Honeywell.)*

optimized performance, a self-contained heater and sensing element are available. This type of accelerometer is used for inertial navigation and is capable of sensing accelerations of less than 1 micro-*g*.

Pendulous Accelerometer: Stiff cross suspension is difficult to achieve when combined with the requirement for frictionless restraint on the sensitive axis. The effort to reduce friction has led to pendulous accelerometers which use rotational bearings. Further reduction of friction has been achieved by floating the pendulum to reduce bearing loads. An advantage of the pendulous accelerometer is its cross-axis suspension, usually achieved by floating a pendulous cylinder in a viscous heavy liquid. The cylinder is pivoted on its axis and is pendulous about it. Because the cylinder is floated, cross-axis accelerations do not displace it. The flotation liquid also provides some damping.

Operation of the device is diagramed in Fig. A-9. Acceleration along the sensitive axis causes

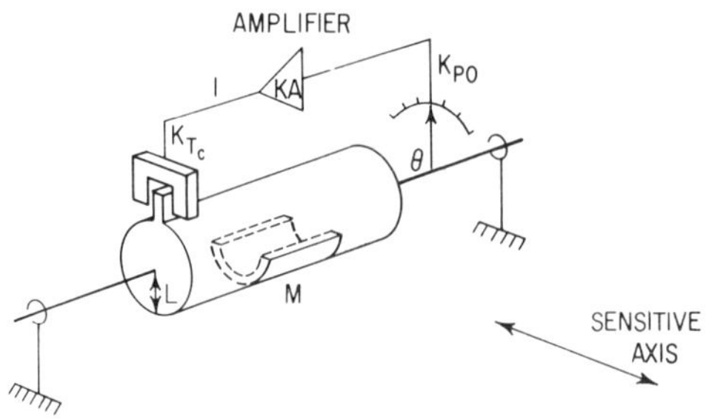

Fig. A-9. Pendulous accelerometer.

deflection of the pendulum. This deflection is sensed by the pickoff, amplified, and applied to a torquer in a direction to reduce the deflection. The mathematics of response is similar to the linear case

$$A = \frac{IK_{T_c}}{P} \tag{1}$$

where P = pendulosity = ML

A = acceleration

K_{T_c} = torque-generator transfer function

The pendulous accelerometer permits some cross-axis sensitivity when the pendulum is deflected. The cross-axis sensitivity is proportional to the sine of the pickoff angle θ. The higher the loop gain on the sensitive axis, the smaller will be θ. As θ is kept near zero, so is the cross-axis sensitivity. The cross-coupling error is proportional to the product of the magnitude of the cross acceleration and sensitive-axis acceleration. Therefore, this is a g^2 type of error.

Vibrating-string Accelerometer: The strings of the device shown in Fig. A-10 are caused to vibrate. Their frequencies of vibration are measured. Acceleration along the sensitive axis causes an increase in tension and consequently in the frequency on one string, while producing a decrease in tension and lowering of the frequency on the other string. The readout is nonlinear and the device is sensitive to ambient temperature changes. No position servo is used.

Magnetic-field-supported Accelerometers: Servocontrolled electrostatic or magnetic fields are capable of providing precise frictionless support and restraint for an accelerometer seismic mass. Pulse-rebalance techniques permit a digital readout where desired. Field supports are inherently nonlinear but can be linearized by push-pull schemes. The dynamic range of field-supported devices is about 10^6, but this range can be conveniently shifted to nearly any portion of the acceleration spectrum by switching the electronics. This also makes possible multiple-range devices. The electrically suspended accelerometer is an example of this class of accelerometer.

The electrically suspended type incorporates a seismic mass which is a conductor that is held floating, either in a vacuum or in gas under pressure, by an electrostatic-field attraction. The position of the seismic mass is sensed, and changes are used to command acceleration of the mass back toward its null position. The accelerating force is generated by changing the voltage gradient between the mass and electrodes. The force is proportional to the voltage gradient squared and is always a force of attraction. Therefore, on each axis, fields must be established in balanced opposition to one another. As one is increased, the other is decreased. The net force on the seismic mass thus is proportional to voltage change, and the readout of restraint voltage vs. acceleration is linear. The mathematical relationships are given in Fig. A-11.

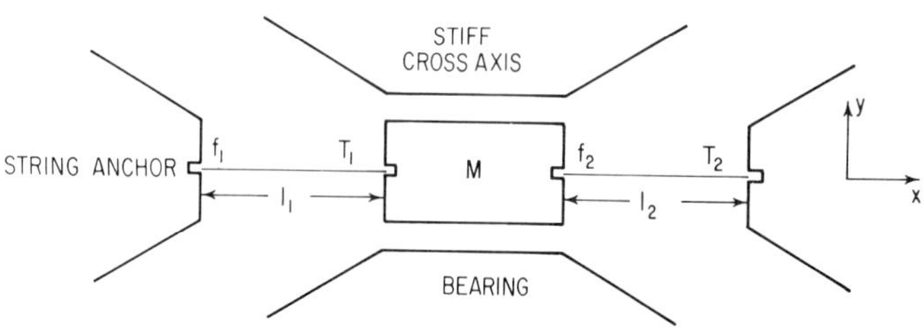

Fig. A-10. Vibrating-string accelerometer.

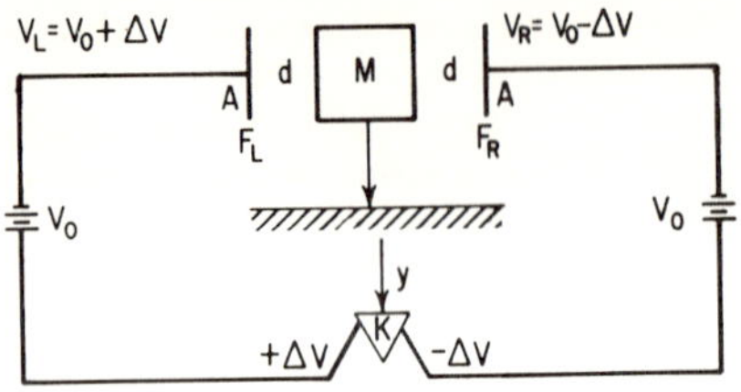

At the left force: At the right force: Net force:

$$F_L = \frac{kk_0}{2} A \left(\frac{V_L^2}{d}\right) \qquad F_R = \frac{kk_0}{2} A \left(\frac{V_R^2}{d}\right) \qquad F_N = F_L - F_R$$

but

$$V_L = V_0 + \Delta V \quad \text{and} \quad V_R = V_0 - \Delta V$$

thus

$$F_N = \frac{kk_0}{2} A[(V_0^2 + 2V_0\Delta V + \Delta V^2) - (V_0^2 - 2V_0\Delta V) + \Delta V^2)] = 2kk_0 A \frac{V_0 \Delta V}{d^2}$$

but

$$\Delta V = Ky \quad \text{and} \quad F_N = \frac{2kk_0 A V_0 K}{d^2} y = Ma$$

F_N now is proportional to y, as a linear spring would be. In the process a bias voltage V_0 had to be added. Linearity is destroyed if $\Delta V > V_0$. Note, too, that stiffness is proportional to V_0. The g spectrum of response is readily tailored or changed by switching the magnitude of V_0, K, and M.

Fig. A-11. Block diagram and mathematical relationships of electrostatic accelerometer.

High-g Accelerometers: The accelerometers described thus far all have thresholds of 10^{-6} to $10^{-4}g$ — with upper g limits of 10 to 20g. Further development work on inertial-grade accelerometers has a design goal of thresholds and linearities of 10^{-6} to 10^{-7} g. At the other extreme are accelerometers which will survive and measure many hundreds of g's (see Fig. A-12). Most

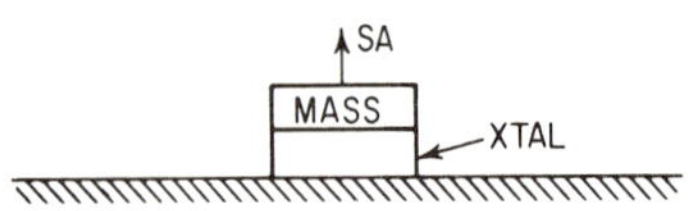

Fig. A-12. Piezoelectric accelerometer.

important of these are piezoelectric and piezoresistive accelerometers. In the first type, the seismic mass is cemented to a piezoelectric crystal, such as barium titanate, which in turn is cemented to the case. Acceleration stress causes the crystal to generate a proportional voltage. The device is limited because it cannot measure steady acceleration. The piezoresistive accelerometer was previously described.

Pendulous-gyro Integrating Accelerometer: In the devices described thus far, the induced force or acceleration has been compared with a measurable force produced by a spring or by current flowing in a magnetic field. Another measurable force or torque is available, namely, that developed by a processing gyroscope. This produces a very broad range of measurable restraints. The overall device bucks the torque of an accelerating pendulum against the reaction torque of a precessing gyro. A pendulous-gyro accelerometer essentially is a single-axis gyro with an unbalanced gimbal mounted on a single-axis platform controlled by the gyro. Reference to an elementary gyro (Fig. A-13) will assist in the explanation. Imagine a toy gyro placed on a pedestal with the wheel spinning. As soon as the gyro is released, a torque Wr exists about the Y axis because of the gyro weight. The Y-axis torque is initially resisted by only the gyro Y-axis moment of inertia. Thus, the gyro starts to fall and an angular velocity starts to build up around

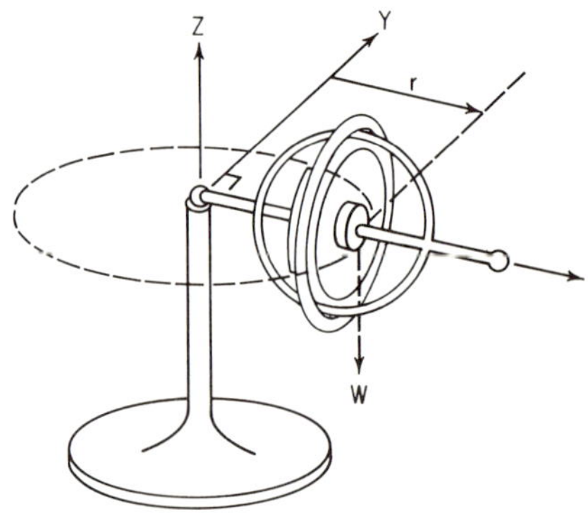

Fig. A-13. Elementary pendulous-gyro accelerometer.

the Y-axis. But this angular velocity ω_y causes a gyroscopic reaction torque $H\omega_y$ around the Z-axis. There is nothing to oppose this torque except the moment of inertia of the gyro about the Z axis. Thus, the gyro starts to turn around the Z axis. The angular rate ω_z then builds up until the gyroscopic reaction torque $H\omega_z$ exactly opposes the unbalance torque Wr, at which time the turning rate about the Y axis is decreased to zero. This is a pendulous gyro with the pendulosity, $P = Wr/g = mr$. An equality can be written between the gyroscopic torque and the pendulous torque:

$$H\omega = mgr = Pg \tag{2}$$

The turning (or precession) rate of the gyro about the Z axis is

$$\omega = \frac{P}{H} g \tag{3}$$

In this example, the torque was produced by the gravitational-attraction force, but a torque would also have been produced if the toy gyro of Fig. A-13 were imagined turned so that the Z axis was horizontal and the gyro was given an acceleration along the positive Z axis. See the pendulous two-axis gyro of Fig. A-14.

The spin motor is mounted on the inner gimbal with an off-balance mass or pendulum along the spin axis. The inner gimbal is free to turn about an axis perpendicular to the spin axis and is mounted in an outer gimbal. The outer gimbal is free to turn about an axis perpendicular to the spin axis and the inner-gimbal axis. If the base is accelerated along the input axis, the inertial reaction of the pendulum mass creates a torque about the inner-gimbal axis. Then, the gyro will precess around the outer-gimbal axis, that is, the input axis. The gyroscopic torque then balances the pendulous torque.

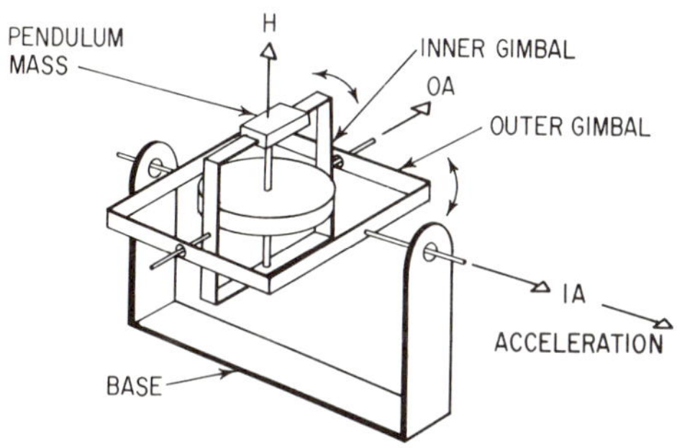

Fig. A-14. Pendulous two-axis gyro.

If friction and gimbal inertias are assumed equal to zero, then the gimbals remain nearly orthogonal and the gyro precession rate will be such that

$$\begin{aligned} H\omega &= PA \\ \omega &= \frac{P}{H} A \end{aligned} \tag{4}$$

The preceding equation shows that the output of this hypothetical pendulous two-axis gyro would be a tumbling rate proportional to the acceleration of the base. If both sides of the equation are integrated with respect to time, then

$$\int \omega \, dt = \frac{P}{H} \int A \, dt \tag{5}$$

$$\theta = \frac{P}{H} V \tag{6}$$

and the angle through which the outer gimbal turns is directly proportional to the base or vehicular velocity. If this angle is taken to be the output of the device, then it is an integrating accelerometer or velocity meter.

The practical problems of this device are gimbal friction and the inertias, particularly around the outer-gimbal axis. These two factors cause the inner gimbal to turn about its axis, and therefore, the desired orthogonal condition deteriorates. A solution to these problems is to mount a single-axis pendulous gyro on a single-axis platform or turntable. This arrangement is shown in Fig. A-15. The turntable takes the place of the outer gimbal of the two-axis gyro. The output of the pendulous gyro is used to energize a servomotor which drives the turntable. The servomotor overcomes the friction of the turntable and drives it at a rate that causes the gyroscopic torque to equal the pendulous torque except during periods of transient accelerations. Since the gyro is an integrating gyro, the device will integrate the input accelerations during transient conditions until the turntable has time to catch up. The output of the accelerometer can be either analog or digital. The analog type has a helical multiturn potentiometer mounted on the base with a wiper mounted on the turntable. The potentiometer voltage output is directly proportional to the vehicle velocity. One digital type has a glass digital code wheel which rotates with the turntable. A photoelectric circuit produces output pulses which are added algebraically to determine the total vehicle velocity. A block diagram of the pendulous-gyro accelerometer is given in Fig. A-16.

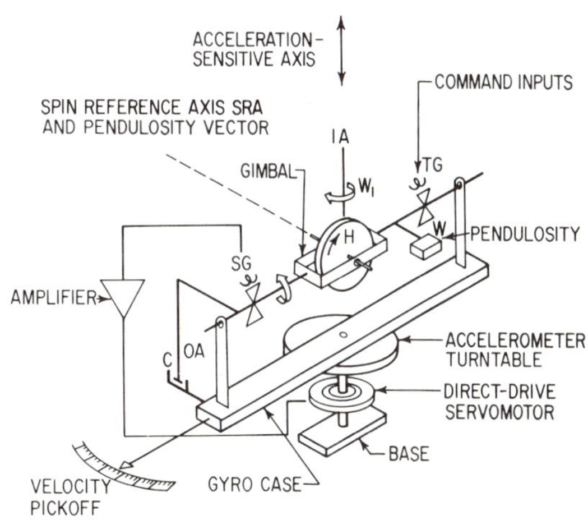

Fig. A-15. Pendulous-gyro integrating accelerometer.

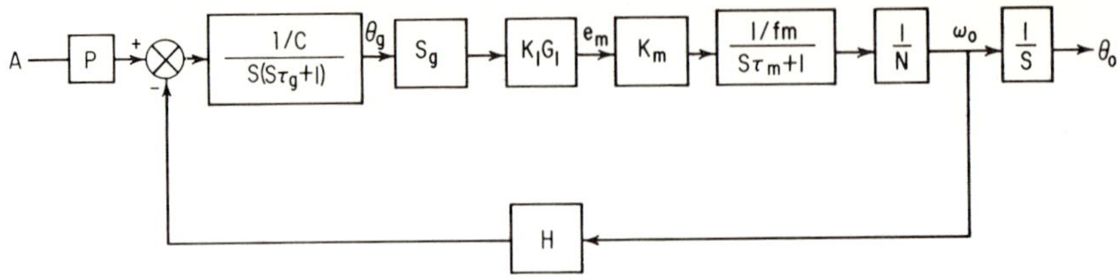

A = input acceleration
P = gyro gimbal pendulosity
$\frac{1/C}{S(S\tau_g+1)}$ = gyro gimbal transfer function
θ_g = gimbal deflection
S_g = gyro signal generator sensitivity
K_1G_1 = transfer function EG8002A servoamplifier
e_m = servomotor control phase voltage
K_m = servomotor torque-to-voltage circuit
$\frac{1/fm}{S\tau_m+1}$ = servomoter transfer function
$\frac{1}{N}$ = gear train reduction
ω_0 = accelerometer turntable turning rate
H = gyro wheel angular momentum
θ_0 = accelerometer turntable angular position

Fig. A-16. Block diagram of pendulous-gyro integrating accelerometer.

Advantages of the pendulous-gyro accelerometer as a precision accelerometer include: (1) the spinning rotor provides a tiny dither which reduces the null of the pendulous gimbal; (2) readout is in units of velocity; (3) misalignment errors are cyclical and bounded; (4) precision of scale factor is associated with constancy of wheel speed, a relatively easy factor to hold precisely; and (5) turntable striction can be nullified by imposing a known component of gravity always along the sensitive axis.

Viscous-drag Integrating Accelerometer: A force proportional to velocity gradient is developed when a body is moved through a newtonian fluid. Thus, if a floated gimbal is surrounded by a turning motor, the gimbal will experience a torque which is proportional to the relative speed between the rotor and gimbal (see Fig. A-17). This accelerometer consists of a floated pendulous gimbal with pickoff and torquer and a motor which drives a rotor (which partially encloses the floated gimbal). The gimbal pickoff signal is amplified and drives the motor-rotor combination to establish the relative velocity between the rotor and the pendulous gimbal. The shear torque rebalances the acceleration torque and is proportional to the speed of the rotor. Therefore, rotor position is proportional to velocity attained. This type of accelerometer is especially subject to temperature-dependent errors. The viscosity of the flotation fluid is sensitive to temperature changes which alter the integration scale factor.

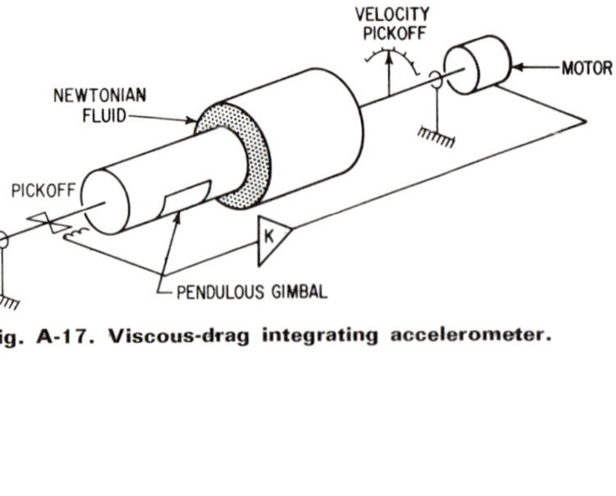

Fig. A-17. Viscous-drag integrating accelerometer.

Wave-train Integration: Several pickoff/restraint methods produce a wave-train output with a frequency that is proportional to acceleration. The total count of waves or pulses then is proportional to velocity. Each pulse has a well-defined velocity weight. Examples of frequency or pulse-rate devices include (1) those which use pulse rebalance; and (2) those which use vibrating strings. In the first, the rebalance pulses are of controlled shape and the rate at which they are fed to the rebalance torquer or forcer is proportional to the acceleration as sensed by the accelerometer pickoff. The vibrating-string unit has already been described.

In one type of pulse-rebalanced velocity meter,[1] the device senses vehicle thrust acceleration, integrates this acceleration to obtain vehicle velocity along the thrust axis, compares this velocity with a preset velocity, and generates cutoff signals when the measured velocity is equal to the preset velocity. The device uses a pulse-on-de-

[1] Honeywell velocity cutoff switch.

mand pulse-rebalance loop in conjunction with a hinged-pendulum accelerometer to provide a pulse rate proportional to input acceleration. This pulse rate is integrated (counted) in a binary register (accumulator). A digital comparator determines coincidence between the register velocity and the preset velocity and furnishes an output signal.

Double Integrating Accelerometer: If a motor is mounted inside a pendulous gimbal with the spin axis along *OA*, the device is called a double integrating accelerometer or distance meter. In this unit, the pendulous torque is balanced by accelerating the wheel so that

$$I\alpha = PA \tag{7}$$

where I is the wheel inertia around the spin axis and α is the angular acceleration of the wheel. Integrating once,

$$I\omega = Pv \tag{8}$$

so that the wheel angular velocity is proportional to vehicular velocity. Integrating again,

$$I\theta = PS \tag{9}$$

and the wheel angle is proportional to the distance traveled.

Also see **Aircraft, instrumentation of;** and **Navigation systems.**

REFERENCES

Bouche, R. R.: Understanding Accelerometers, *Electron. Eng.*, April, 1967, pp. 90–94.

Considine, D. M., and S. D. Ross (eds.): "Handbook of Applied Instrumentation," sec. 5-2, McGraw-Hill Book Company, New York, 1964.

Gruenberg, Elliot L. (ed): "Handbook of Telemetry and Remote Control," McGraw-Hill Book Company, New York, 1967.

—Vernon H. Aske, John W. Miller, and Paul D. Senstad, *Honeywell Inc.*

Access Time, Computer Access time is the time required to obtain data from a storage device. (See **Storage, computer.**)

On an immediate-access storage device, such as a ferrite-core storage, access time is the time that elapses from the start of the storage cycle to the availability of the addressed data at the output register of the memory. This corresponds to the read cycle of the storage unit. (Also see **Memory-cycle time.**)

On a disk storage unit (see **Disk storage**), access time is the time required for the read-write head to be moved to the addressed track plus the required rotation time of the disk. On the average, the required rotation time is one-half the disk rotation time, since this is the average time to arrive at a particular record on the track.

—W. W. Welz, *IBM Corporation*

Accuracy This term is by far the most widely quoted parameter used to describe the relative performance capability of a measuring device. It is probably also the most frequently misunderstood and misapplied parameter and may even be totally invalid when taken by itself as a relative figure of merit.

General Definition: For the general case, American National Standard C85.1-1963, Terminology for Automatic Control defines accuracy as "Conformity of an indicated value to an accepted standard, or true, value."

This is interpreted by the user as a quantitative statement of the ability of a device to measure, indicate, or otherwise respond to a true value. To determine the accuracy of a specific device, then, requires that the true value be established in terms of a more accurate reference standard. In practice, this reference is sometimes chosen arbitrarily, but in the truest sense, the measurement, or calibration standard (usually called a "working" standard), that determines the true value can be further referenced through secondary standards and primary standards back to the National Bureau of Standards to the four "basic units" of length, mass, time, and temperature.

Accuracy should be expressed as an "error," i.e., either as a deviation from the true value expressed in the unit of measurement (error $= \pm 1\,^\circ$F) or as the ratio of the deviation from the true value to the true value (error $= \pm 1$ percent).

A Practical Definition: In the more practical case, particularly with regard to measurement in the general field of instrumentation (as compared with precision laboratory measurements), normal practice partially modifies the general definition. For example, the ability to reference a particular measurement back to an absolute true value is not usually critical, and concern seldom goes beyond the ability to reference a particular measurement to a stable reference value. That is, as long as the bias or deviation between the standard reference value being used and the absolute true value is relatively constant (within reason), this is usually quite satisfactory.

In the practical case, accuracy is generally and necessarily expressed as the "limit of error" of a measuring device or system under certain operating conditions that may or may not be specified. The reason here is that the user is primarily interested in the maximum error that is probable over the entire operating range of the device under a defined set of operating conditions.

Limit of Error: A practical expression of accuracy (limit of error) can be in output units or percent of reading as previously expressed, but for most measuring devices, it is much more commonly expressed as a percent of the upper range value (percent of full scale) or as a percent of output span.

"Percent of reading" is seldom used, since for most measuring devices, the magnitude of the error in output units is usually relatively constant over the operating range of the device. Examples of devices where the magnitude of the error is directly proportional to the magnitude of the input signal would be something approaching a primary measuring standard (such as a manometer) or an electrical device whose operating principle is digital rather than analog (such as a digital voltmeter).

In some measuring devices, there is no quantitative difference between the output span and the upper range value (for example, a millivoltmeter with an output span of 0 to 100 mV). Consider two other devices, however, both with a scale range of 50 to 100 mV. The less sophisticated device may not include an integral biasing device and, therefore, would require an input signal in excess of 50 mV to overcome the mechanical bias of the readout mechanism. This type of device would necessarily have its limit of error specified as a percentage of its upper range value. A more sophisticated device might utilize a biasing source of relatively high accuracy, such that the readout mechanism would not "see" a positive-direction signal until the input exceeded 50 mV. Here the limit of error could be expressed as a percentage of the output span (50 mV), and therefore, the limit of error would be reduced to one-half that of the former case.

Complementary Parameters: For any specific example, the accuracy (or limit of error) can be defined as the total influence of errors due to conformity, hysteresis, and repeatability.[1] The conformity error (nonconformity) is defined as "the maximum deviation between an average curve (the average between two or more full-scale, up-scale and down-scale traverses) and a specified functional (true value) curve."[1] Hysteresis (maximum) is "the maximum difference for the same input between the up-scale and down-scale indications of the measured signal during full range traverses."[1] Repeatability is "the closeness of agreement among a number of consecutive measurements of the output for the same value of the measured signal under the same operating conditions, approaching from the same direction, for full range traverses."[1] If the independent conformity ("the value of non-conformity determined after any translation and/or rotation of the actual curve is made to minimize the maximum deviation"[1]) is known, along with the maximum hysteresis and repeatability errors, the best accuracy that can be obtained with optimum calibration under a fixed set of operation conditions can be predetermined.

Sources of Error: The user of a specific piece of measuring equipment too often operates on the assumption that the total limit of error in determining a true value is influenced only by the stated accuracy of the measuring device. The total error, however, is a combined effect of the relative accuracy of the measuring device under the prevailing operating conditions, the accuracy of the reference standard and the methods used to calibrate the reference device, the methods used in making the specific measurement, and human errors attributable to the observer (observer bias, ability to eliminate parallax). Such sources of error can be either "random" (nonpredictable but tending to follow a probability distribution function) or "systematic" (usually predictable in direction, or qualitatively if not quantitatively). Since these sources of error can seldom be determined accurately or eliminated, it is important to recognize that they probably exist in all cases, and precautions should be taken to reduce their influence.

Operating Conditions: A simple statement of accuracy is totally inadequate as a relative figure of merit, since such a figure usually represents optimum performance and the actual performance during a specific measurement is substantially influenced by the prevailing environment.

For this reason, it has become conventional in the instrumentation industry for a manufacturer to specify significant static performance characteristics, such as accuracy, dead band, and hysteresis, under a set of tightly toleranced reference operating conditions (as might be found in a calibration laboratory) and then the effect of several operating influences, such as ambient temperature, humidity, overload, stray, shock, vibration, time, and many others depending upon the type of measuring device. In the extreme case, the to-

[1] RC20-11-1964, Measurement and Control Terminology, Scientific Apparatus Makers Association.

tal effect of the influences due to varying operating conditions can deteriorate the performance of a device many times over what might be implied by a simple statement of accuracy under reference operating conditions.

These additional sources of error also tend to be both random and systematic, and it is therefore impossible for the user to predict them accurately and compensate for them. To the degree that compensation is not possible, it is essential that the user at least understand and anticipate them.

System Accuracy: Particularly in the instrumentation industry, the user is ultimately concerned with total system accuracy. For example, this might mean his ability to measure and possibly control the temperature in an oven where the system may consist of sensors, transducers, transmitters, signal-conditioning devices (computing elements), receivers (indicators, recorders, or other types of readout devices), and possibly control stations, error generators, controllers, final control elements (valves, silicon controlled rectifiers), and even digital computers. Not only must the potential inaccuracies of all these devices be taken into account, but the methods employed—particularly at the sensor end—must be fully considered.

Predictions of system accuracy are determined as much by art and experience as by science. This involves a detailed knowledge of the error sources, the probable operating influences that will prevail, the degree of compatibility of the devices, the planned ability of the system to correct for certain sources of error (for example, the predictable nonlinearity of a thermocouple can be compensated for by an offsetting nonlinearity of a measuring circuit or readout device in the receiver), and many other factors, such as frequency of maintenance.

Even if the predominant limits of error can be predicted, total system inaccuracies still do not lend themselves to calculation by arithmetic sum or probability distribution functions. It might be considered reasonable, if somewhat conservative, to predict the probable system inaccuracies by taking the square root of the sum of the squares (RSS) of the predominant influences. Even this technique can be somewhat liberal, however, if a relatively few sources of error overshadow all others, or where the interaction of certain influences causes a greater effect than would be predicted by the sum of the individual influences taken separately.

In process-control applications, the operator is frequently far more concerned with the ability of the control loop to repeat or maintain a value of the process variable under similar operating conditions (see **Repeatability**) than he is with accuracy. In this case, the word accuracy may be used incorrectly.

On the other hand, other types of feedback control systems may require the output to achieve an intended value to a high degree of accuracy. This is particularly true of many position-output systems, such as positioning gun turrets.

In either case, the analyst must recognize the beneficial influence of the loop gain of the system in reducing the effects of errors (or other types of disturbances) in the forward loop of the system. Analysis will show that disturbances in the input (input signal and transducer) which are outside the control loop, in the sensor or feedback transducer which are in the feedback portion of the loop, or in the error generator which provides the signal to the controller will have a 1:1 effect on the accuracy of the output. Disturbances in the controller or final control element will approximately be reduced by the product of the gains of the forward- and feedback-loop elements.

"Dynamic" accuracy generally relates to the ability of the measuring device or system to follow a transient in the measured variable. See also **Error, measurement.**

—Robert L. Wilson, *Honeywell Inc.*

Accuracy, Digital-data-acquisition System In digital-data-acquisition systems and, in particular, those designed to measure data at relatively high rates, accuracy must be interpreted in terms of statistical distributions. This is due to the nonrepeatability of readings because of noise or other short-term disturbances (see **Repeatability**). Although there is not presently a standard written in these terms, the following discussion represents the common usage in the digital-process-control and data-acquisition field.

If a large number of readings are taken using a stable input signal under reference environmental and other operating conditions, the resultant digital values vary because of noise and other disturbances. In a properly adjusted system, the distribution of these readings approaches a normal probability distribution, as shown in Fig. A-18. For the purposes of specification, it is assumed to be a normal distribution of random events. The difference between the "true value" and the mean value of the distribution is the mean error and is the number quoted as the "mean accuracy" of the system.

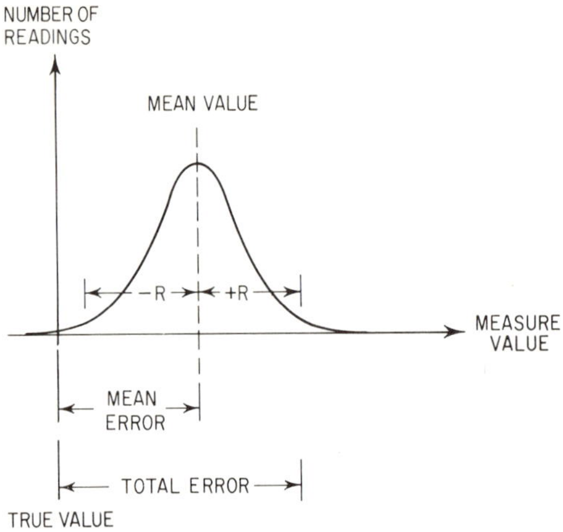

Fig. A-18. Statistical-accuracy definition.

The variations in the readings around the mean value are measured in terms of the repeatability parameters $+R$ and $-R$. If the distribution is symmetrical, $+R = -R$. The "total error," normally quoted as either "total accuracy" or simply as "accuracy," is the difference between the "true value" and the limit used to define the repeatability. Typically, this limit is set as the 3σ confidence level, where σ is the standard deviation of the distribution. Since the distribution is assumed to be normal, σ is unity for a normalized distribution and the σ limit means that 99.7 percent of the readings will fall within these bounds.

As in all cases concerning specifications, the user should verify the meaning of all terms. In the case of accuracy, the distinction between mean accuracy and total accuracy is often not stated and a single accuracy parameter is stated without explanation. Similarly, in a few cases the statistical confidence level is not indicated. This is significant, since a 2σ level corresponds to only 95.4 percent compared with the 99.97 percent for 3σ.

—Thomas J. Harrison, *IBM Corporation*

Acidity, Measurement of (See pH, measurement of.)

Acoustic Instrumentation Acoustics is the study of compressional oscillations about an equilibrium pressure within a continuous medium. The field can be divided into *objective acoustics,* concerned with the physics of sound generation and propagation, and *subjective acoustics,* concerned with human perception of sound.

Since acoustical disturbance propagates in waves, any acoustical variable must be defined as a function of space as well as time. The most important type of acoustical motion is the plane wave, wherein acoustical pressures and particle displacements have common phases and amplitudes at all points on any plane perpendicular to the direction of wave propagation. Acoustical plane waves obey the usual laws of refraction and reflection. When a plane sound wave strikes a plane interface between two media with densities ρ_1 and ρ_2 and propagation velocities C_1 and C_2, the incident, reflected, and transmitted rays, all in the same plane, make angles θ_i, θ_r, θ_t with the normal to the interface, then, according to Snell's law:

$$\theta_i = \theta_r \quad \left(\begin{matrix}\text{angle of}\\ \text{incidence}\end{matrix} = \begin{matrix}\text{angle of}\\ \text{reflection}\end{matrix}\right)$$

$$C_2 \sin \theta_i = C_1 \sin \theta_t$$

With such reflection at normal incidence, the reflected and incident waves interfere to reinforce or cancel each other and the resulting points of minimum and maximum amplitude form a standing wave which is stationary in space.

Just like light waves, sound waves can be diffracted and scattered, and like electromagnetic waves they can be guided by tubes or ducts and focused by acoustical reflectors.

Sound-pressure level is the rms time-averaged value of the pressure fluctuation expressed in decibels relative to a reference pressure of 0.0002 microbar. This reference pressure is approximately the threshold of human hearing for a 1,000-Hz tone.

Sound level is a frequency-weighted sound-pressure level obtained by the use of special frequency-weighting networks, A, B, and C specified in American National Standard S1.4–1961, Sound Level Meters for Measurement of Noise and Other Sounds. The weighting network used in measuring sound level must always be stated. The networks follow the equal-loudness contours developed by Fletcher and Munson and compensate for the frequency response characteristics of the human ear for pure tones at different levels.

Sound-level measurement is done with the aid of electronic instruments. The basic instrument is the sound-level meter (see Fig. A-19). The instrument consists of four functional components: (1) a calibrated measurement microphone, (2) a frequency-weighting network, (3) a calibrated amplifier, and (4) an rms rectifier and meter which indicates sound level directly in decibels.

More complete measuring systems may include

frequency analyzers and a graphic level recorder. In using these instruments, one should know (1) the desired measurand, (2) the anticipated amplitude range, (3) its anticipated frequency range, (4) the anticipated measurement environment, and (5) the required measurement accuracy.

In the case of sound, the measurand is pressure. The sound-pressure level or amplitude in decibels of a sound is 20 times the logarithm to the

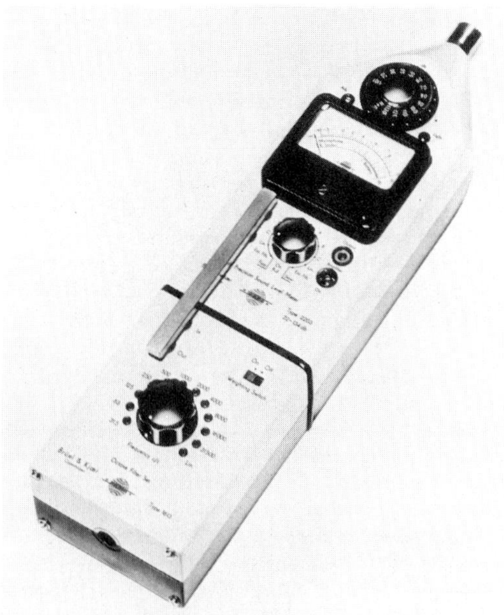

Fig. A-19. Precision sound-level meter with octave filter set. *(B & K Instruments.)*

base 10 of the ratio of the pressure of this sound to a reference pressure. The reference pressure in general use for measurements concerned with hearing (and with sound in air and liquids) is 2×10^{-4} microbar.

Although a number of different artificial ears are commercially available, technology has not developed to the point that all the variables of human perception of sound can be exactly duplicated.

Of the three main parts of the ear, the outer ear serves to match the impedance of the eardrums to the air. The matching is remarkably good at 800 Hz and remains fairly good even at higher frequencies. Only at frequencies below some 400 Hz is the matching rather poor. The vibrations of the eardrums are transferred mechanically by the middle ear to the inner ear. The lymph filling the inner ear provides further impedance "matching." Simultaneously, the vibration amplitudes of the eardrums are transferred to the much smaller vibration amplitudes but higher pressures of the inner ear. The perception of the sound by the nerves finally takes place along the basilar membrane of cochlea in the inner ear.

The complete hearing process seems to consist of a number of separate processes which, in themselves, are fairly complicated, so that no simple relationship exists between physical measurement of a sound-pressure level and the human perception of sound. The loudness of a certain pure tone, for example, may be judged to sound different from another pure tone, and different again from a combination of tones or complex sound, although the sound-pressure level in each case may be the same. The equal-loudness contours show how the loudness levels of pure tones with constant sound-pressure level (SPL) vary with frequency.

Further investigations by Fletcher produced a subjective loudness scale. The relationship between the loudness level in decibels (1,000 Hz) and the subjectively judged loudness is shown by the Fletcher-Munson curve. It reveals that the relationship between sound level (phon) and loudness (sone) is approximately linear over the commonly experienced range of sound levels (20 to 120 phons). A twofold change in loudness equals approximately a 10-db change in sound-intensity level.

Through combined use of the equal-loudness contours and the subjective loudness scale, it is possible to construct a set of loudness curves from pure tones.

Other variables involved in the perception of sound, in addition to loudness and pitch, include timbre of a complex sound, masking effect, and blocking. Also, research has revealed the existence of certain "critical" bands of frequencies which have a definite relationship to the vibration maxima on the basilar membrane of the ear.

In the absence of a frequency analyzer that divides the spectrum into these critical bands, a set of correction curves has been derived that relate the one-third-octave-band frequency analyzer to the critical bands. To simplify the calculation process, Stephens has published certain diagrams which make it possible to take all the variables including masking effect and frequency response of the ear into account directly from one-third-octave data (see Fig. A-20).

Acoustic principles are used in a number of laboratory and industrial instruments. See **Flow, measurement of; Liquid level, measurement of**; and

Fig. A-20. Octave-, one-third-octave, and narrow-band spectrum recorder. *(B & K Instruments.)*

Nondestructive inspection. Also see **Microphone.**

REFERENCES

Specifications and Tests for Piezoelectric Pressure and Sound-pressure Transducers, *ISA Publ.* S37.10, 1969.

Albers, V. M.: Underwater Acoustics Instrumentation, *ISA Publ.* UAL, 1969.

Acoustic Telemetry and Control System Tested on Offshore Rig, *Instr. Technol.*, vol. 16, 1969.

Considine, D. M., and S. D. Ross (eds.): "Handbook of Applied Instrumentation," sec. 3-5, McGraw-Hill Book Company, New York, 1964.

Kinsler, L. E., and A. R. Frey: "Fundamentals of Acoustics," 2d ed., John Wiley & Sons, Inc., New York, 1962.

Broch, J. T.: "Acoustic Noise Measurements," B & K Instruments, Inc., Cleveland, Ohio, February, 1969.

Beranek, Leo L.: "Noise Reduction," McGraw-Hill Book Company, New York, 1960.

Stephens, R. W. B., and A. E. Bate: "Acoustics and Vibrational Physics," Edward Arnold Ltd., London, 1966.

—Hall Cary, *B & K Instruments, Inc.*

Actuator, Electric An electric actuator is an electrically powered device operated by some form of electric control system and used to position a control element. Electric actuators are used most frequently to position control valves of various types, but they also are used to position numerous other devices, such as dampers, burners, rheostats, or adjustable autotransformers for the regulation of motor speeds or electric-heating loads. The electric actuator and the control valve frequently are supplied by the manufacturer as an integral assembly. Also see **Electric controller**

Electric actuators may be classified as (1) electric-solenoid-operated types (see **Solenoid valve**) and (2) electric-motor-operated types. Solenoids provide only two-position operation while the electric-motor-operated actuator may provide two-position, floating, or proportional operation, dependent upon their construction and the control system with which they are used.

Electric-motor-operated actuators consist basically of a suitable driving motor, a reduction gearing, and an output shaft or lever arm incorporated in a common housing. Driving motors are of various sizes, dependent upon the power-output requirements of the actuator, and they may be ac or dc, dependent upon the available power supply. The driving motors of most actuators are of the reversible type. Induction motors generally are less costly and produce greater torques than synchronous motors of comparable size. Consequently, induction motors are used most frequently.

There are four general types of electric actuators: (1) two-position, (2) multiposition, (3) reversing floating, and (4) proportioning. Selection depends on the manner in which the final control element, operated by the actuator, is required to function in the control system.

Two-position electric actuators may be of either a unidirectional or a reversing type. In either case, the actuators travel toward and stop only at either of the extremes of travel. This action is accomplished by holding and limit contacts which are a part of the motor circuit.

Multiposition electric actuators are of the reversing type and are similar to the two-position type except that multiposition actuators are provided with one or more additional sets of contacts which may be adjusted to operate at any position of the actuator travel. Multiposition actuators must be operated by multicontact controllers.

In floating control, the valve moves at a constant rate toward either the open or the closed position. Electric actuators for valves for floating control use a reversible motor. This type of control generally requires a relatively slow speed actuator, and occasionally an interrupter may be interposed in the common lead to the motor to obtain even slower operating speeds.

Proportioning-control electric actuators use a reversing driving motor and are provided with limit switches and a motor slidewire or control rheostat. The position of the slider on the motor slidewire is related to the position of the actuator —and moves from one end of the slidewire to the other while the actuator moves from one extreme of its travel to the other. Electronic balancing relays usually provide a means of adjusting the proportional band of the control system. Proportional control with automatic reset action also may be accomplished.

Electric actuators generally are available for operation on 24, 115, or 230 V, 50 or 60 Hz, single-phase ac. Some actuators are available for operation on 6 or 440 V, 50 or 60 Hz; 6, 24, or 115 V dc; and for 230 or 440 V, 50 or 60 Hz, two- or three-phase ac. Operating power requirements vary from 10 to 100 W. Operating speeds can be obtained in a range of 1 to 240 s/in. of travel.

Electric actuators require only electric power for operation. Long-distance transmission of the controlling signal can easily be accomplished, and the control valve need not be located near a source of pneumatic pressure. Most electric actuators, in the event of power failure, will remain in the last controlled position.

Adjustable limit switches are provided on most electric-motor actuators to deenergize the driving-motor winding at the limits of travel of the actuator to prevent damage to the actuator or actuated device. Some electric actuators have adjustable auxiliary switches which are used to operate signal lights or alarms, or to actuate other control devices. Manual operating devices are provided in event of power failure or control-system failure. Explosion-proof construction is available.

—Robert J. Lasher, *Fisher Controls Company*

Actuator, Electrohydraulic An electrohydraulic actuator is an electrically powered device used mainly for operation of control valves of various types, but it may also be used for positioning dampers, louvers, and rheostats. Unlike pure electric actuators which position the valve directly by an electric motor and gearbox, electrohydraulic actuators utilize an intermediate, self-contained hydraulic transmission system. In its simplest form, the actuator has an electric-motor-driven pump which provides a continuous source of high-pressure hydraulic fluid for positioning the actuator main operating piston. The piston is attached directly to the valve-plug stem. The flow of oil to either side of the piston is controlled by a sutiable pilot system—servovalve, force motor, flapper nozzle, or jet pipe, with feedback from the stem. Some electrohydraulic actuators are single-acting, with the hydraulic pressure acting on one side of the piston only and a main spring used to furnish power for positioning the piston and valve plug in the opposite direction. Most models, however, are double-acting, with the hydraulic pressure acting on both sides of the piston to move it in one direction or the other, dependent upon the pressure differential across the piston. The operation of an electrohydraulic actuator is shown schematically in Fig. A-21.

Electrohydraulic actuators usually are self-contained, requiring only electric power to the motor and an electric control signal from an automatic controller. These devices make long-distance transmission of a control signal practical and are well suited for isolated locations where pneumatic supply pressure is not available. Self-dampening is another feature. The hydraulic piston is positioned by a noncompressible fluid that dampens vibrations and prevents "jumping" of the valve plug. Electrohydraulic actuators tend to be complex and cost more initially than pneumatic actuators. Maintenance costs also are higher. Double-acting actuators do not "fail safe" upon loss of power—the piston and valve plug will go to either the open or the closed position, dependent upon the force acting on the valve plug. Models that lock in the last controlled position upon power failure are available.

—Robert J. Lasher, *Fisher Controls Company*

Actuator, Electropneumatic An electropneumatic actuator is a combination of an electropneumatic transducer and a standard air-operated diaphragm or piston actuator. The transducer receives an electric control signal, generally in milliamperes, and converts that signal into a pneumatic output pressure for operation. The output pressure is directly proportional to the input current, which is the output signal (milliampere signal) of the automatic controller. The distinct advantage of electropneumatic actuators is the combining of fast electrical transmission of control signals with the various operating advantages of pneumatic actuators.

Electropneumatic transducers generally incorporate a high-capacity pneumatic relay to permit

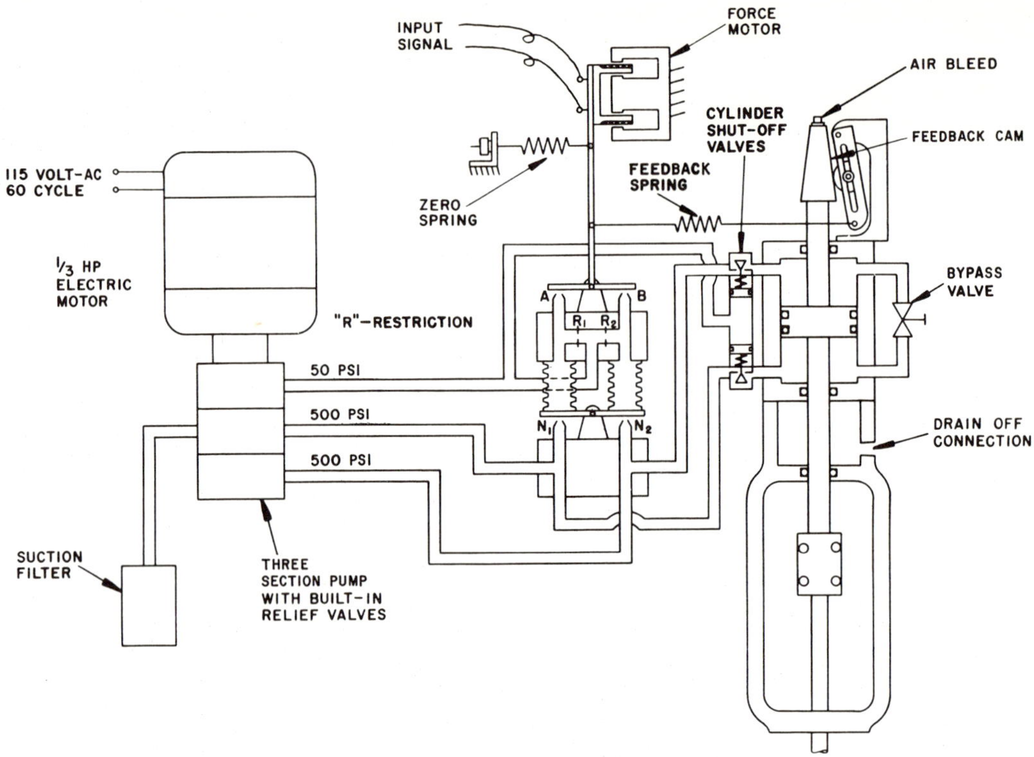

Fig. A-21. Schematic illustration of operating principle of an electrohydraulic actuator.

the operation of a diaphragm control valve directly. However, most transducers are sufficiently sensitive and accurate to provide stable operation when the transducer output pressure is transmitted to small terminal volumes, such as the control bellows in valve positioners. It is not uncommon to see both an electropneumatic transducer and a valve positioner attached to the same control valve.

Two methods of converting the electrical milliampere signal to a pneumatic output pressure are widely used, namely, (1) the use of a force-coil motor and (2) the use of a torque motor. Each of these methods uses a nozzle-flapper signal amplifier. The flapper is attached to the coil in the force-coil motor system, and to the armature of the torque motor in the other system. Major advantages of a torque-motor system over a force-motor coil are a much more rugged construction, development of greater force, and less susceptibility to vibration—allowing the transducer to be valve-mounted. (see Fig. A-22). Also see **Electric controller.**

On applications where a valve positioner is required on the actuator, an electropneumatic positioner is available. The latter consists of an electropneumatic transducer and a standard pneumatic valve positioner and mounts directly on the control-valve actuator.

—Robert J. Lasher, *Fisher Controls Company*

Actuator, Hydraulic Hydraulic actuators basically are cylinder (piston) devices that use oil or fire-resistant liquid as the operating medium. The units are capable of developing very high thrusts (up to about 50,000 lb) because of the high allowable working pressure of the hydraulic fluid. Hydraulic actuators require a source of high-pressure oil at the valve location, and the distribution system for this oil becomes a problem for remotely located control valves. A solution has been the use of a local pump, powered by an electric motor, to form what is commonly known as an electrohydraulic actuator. See **Actuator, electrohydraulic.**

Actuator, Pneumatic A pneumatic actuator essentially is a pressure-to-stroke transducer—the

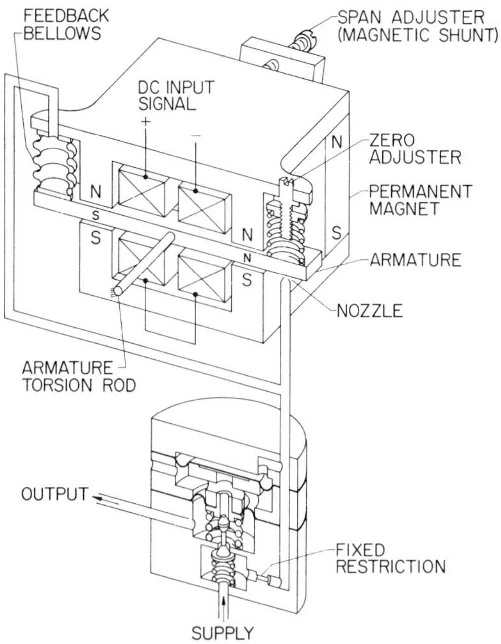

Fig. A-22. Schematic illustration of operating principle of an electropneumatic transducer.

device translates an air-pressure signal into valve-stem motion. The signal received by the actuator is transmitted by the controller directly to the actuator or indirectly through an auxiliary device, such as a valve positioner or pneumatic amplifier. Standard air-signal ranges are 3 to 15 psi and 6 to 30 psi. This pressure, acting on the area of the actuator, creates the force necessary to position the control valve. The two major types of pneumatic actuators are (1) diaphragm actuators and (2) the cylinder (piston) type.

Diaphragm Actuators: Basically, this actuator consists of a flexible diaphragm in a pressure-tight housing. The actuator stem connects the diaphragm to the valve stem of the control-valve body. The diaphragm acts as a pressure seal between the upper and lower case (see Fig. A-23). Diaphragms are made from Neoprene, Buna N, or other suitable elastomers and generally are molded instead of constructed from a flat sheet of material. A properly molded diaphragm provides a nearly constant effective area and gives linearity between loading pressure and travel, makes the actuator easy to assemble and to service, and provides the maximum effective area for a given case size.

Although the change in effective area is greatly minimized by the proper molded shape, the effective area does not remain constant over the rate of stem travel. Thus, it is necessary to know the variation in effective diaphragm area in order to determine the force available from the actuator throughout the stroke. Most manufacturers publish nominal values—which are neither the maximum nor the minimum areas. To simplify spring selection, a usable effective area can be substituted in the calculation—and the maximum and minimum values can be ignored. This is true only when a molded diaphragm is used.

The supporting yoke must have sufficient strength and rigidity to ensure that proper alignment is retained between the actuator and the valve body. Improper stem alignment can cause bending, high dead band, and excessive wear in the packing box and guide areas. Cast iron still is considered the best material for use in the yoke. With this material, careless or rough handling is not likely to cause misalignment of parts. Any damage to the yoke will be obvious since cast iron breaks clean rather than bending under severe loads. Consequently, there is no danger that damage will show up only in the form of poor operation.

In the spring-return diaphragm actuator, the spring opposes the force developed by the loading air pressure acting on the effective area of the dia-

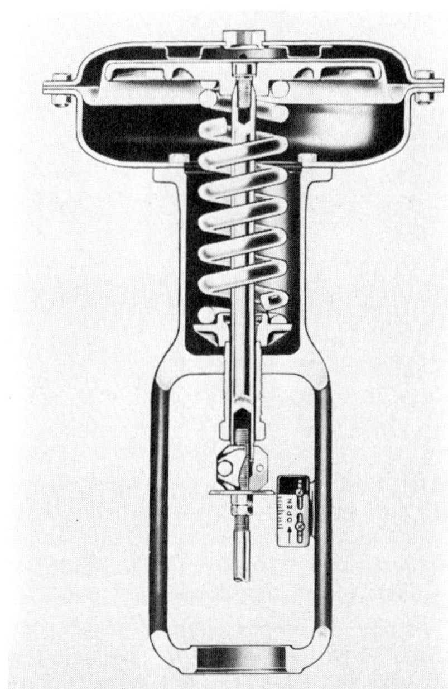

Fig. A-23. Direct-acting diaphragm actuator. *(Fisher Controls.)*

phragm. This provides a given actuator stem position for a specific value of air-pressure signal received from the controller. Diaphragm actuators can be direct-acting (Fig. A-23) or reverse-acting (Fig. A-24). A direct-acting unit strokes down with increasing air pressure; a reverse-acting actuator strokes up with increasing pressure. The spring moves the actuator stem in the opposite direction as the air pressure is decreased. Spring-return actuators will "fail safe" in either the extreme up or the extreme down position upon loss of air pressure, dependent upon the actuator type, that is, direct or reverse. The valve plug, in turn, will open or close fully, dependent upon this action, that is, push down to close or push down to open.

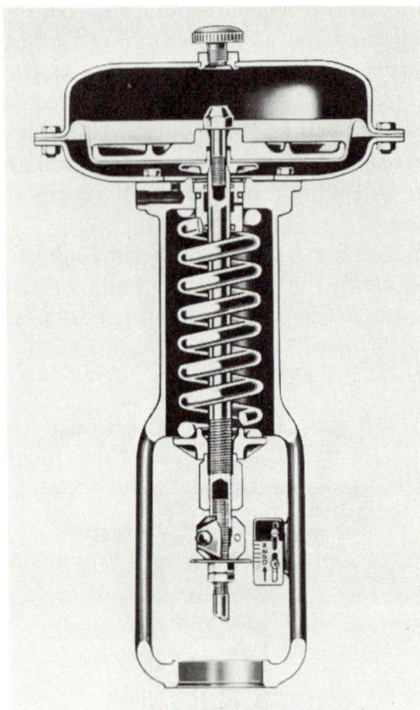

Fig. A-24. Reverse-acting diaphragm actuator. *(Fisher Controls.)*

In the springless diaphragm actuator, stem position is also proportional to the air-pressure signal, but the opposite side of the diaphragm is pressure-loaded with an independently regulated air supply, or by operation of a double-acting valve positioner. Springless actuators can be either direct- or reverse-acting. They develop more force than spring-return-type actuators but have the disadvantage of not "failing" in any one position. Auxiliary equipment is required to move a springless actuator to the desired position upon loss of the operating-medium pressure.

Diaphragm actuators are available in various sizes to deliver thrusts up to 12,000 lb. They also can be provided with top- or side-mounted handwheels to allow manual operation.

Cylinder (Piston) Actuators: This device consists basically of a piston operating within a pressure-tight cylinder (see Fig. A-25). Operation is similar to that of a springless diaphragm actuator except that air-pressure differential is impressed over a piston instead of a flexible diaphragm. The use of pneumatically operated piston actuators has increased—they develop more thrust, up to 25,000 lb, using high-pressure air supply to 150 psi. The devices are fast-stroking and generally are double-acting to provide maximum thrust in both directions. For throttling applications, a valve positioner becomes an integral part of the actuator, mounted on top of the cylinder. See **Valve positioner, pneumatic.** The positioner provides the necessary loading pressure on either side of the piston to position the valve plug in accordance with the dictates of the automatic controller. For on-off applications, the positioner is omitted; the loading pressure is piped to one side of the piston while the other side is exhausted to travel the valve in the required direction.

The basic piston actuator has no spring to absorb piston power. However, the lack of a spring prohibits desired positioning of the valve upon supply-pressure failure. To compensate for this, most manufacturers offer spring-return models and accessories, such as trip valves and volume tanks which function upon loss of supply pressure to move the piston (and valve plug) to a required position. Integral, side-mounted handwheel constructions also are available for manual operation. Variations of piston actuators without a yoke operate butterfly valves, louvers, and Saunders valves. Spring-return-type piston actuators also are available.

In quality models, the cylinder and piston are aluminum, with the cylinder honed to a mirror-like finish. All dynamic seals are low-friction synthetic-rubber *O* rings. The mirror finish on the cylinder wall, the close tolerances, and the low-friction seals all contribute to minimizing the breakaway friction. If this friction is high, cycling may occur. The actuator yoke is cast iron to provide a rugged, rigid support for the cylinder and a connecting member for mounting the actuator on a valve body.

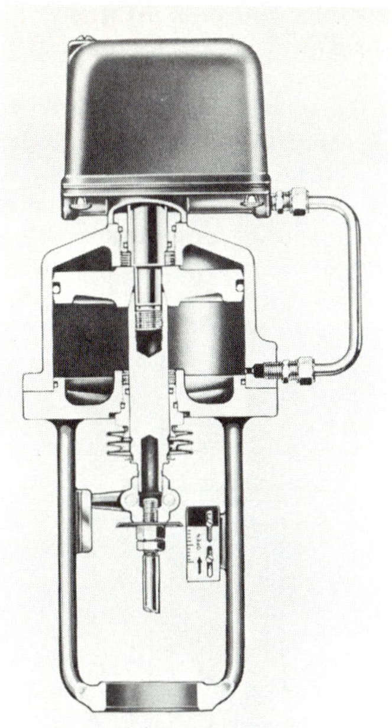

Fig. A-25. Pneumatically operated piston actuator. *(Fisher Controls.)*

REFERENCE
Lasher, Robert: "Control Valve Handbook," Fisher Governor Company, Marshalltown, Iowa, May, 1965.

—Robert J. Lasher, *Fisher Controls Company*

Adaptive Control The concept of an *adaptive control* is based upon the measurement and evaluation of auxiliary control variables for the purpose of modifying principal control functions so that selected control-system performance criteria or indexes can be more fully realized.

Certain literature references to adaptive control systems are concerned with "fuel-optimal" or "time-optimal" control systems used in aircraft-pursuit problems. In the former system, fuel economy is optimized by modifying aircraft-engine fuel-air mixture and airframe attitude for varying conditions of airspeed, air pressure, aircraft loading, and altitude. An autopilot or flight-control system providing fuel-optimal control necessarily requires the addition of at least two minor control loops for accomplishing auxiliary control functions.

Such systems utilize empirically determined and often nonlinear functions to relate the auxiliary-system control variables.

More recent control systems employing the optimal system concept include a small-scale digital computer to update performance-index-related functions according to varying system parameters and thus remove the necessity for arbitrarily fixed constraints. Although such systems in a sense adapt their operating parameters according to system environmental parameters, the general criterion for system performance is based upon the achievement of an optimum condition of system function.

Another class of plant control that utilizes the concept of adaptivity provides automatic compensation for variations in system dynamics. With this concept of plant control it is necessary first to measure suitable dynamic variables that are sensitive to system stability and then, with an auxiliary actuating means, a modification of the plant dynamics can be effected to accomplish system stabilization.

Often a combination of optimum system performance and stability results in a complex interrelationship of these two requirements. Summarily the necessity arises in an adaptive control system for the automatic measurement or continuous updating of selected control variables in order to assess the system dynamical stability for each point of a control trajectory or path of operation. To provide a continuous redesign of the control system for accomplishment of selected stability and performance criteria, it is necessary that some form of digital computer be employed to minimize system updating time during the system trajectory.

Adaptive control systems must contain the two essential elements:

1. An *identification* means for the purpose of measuring the dynamic characteristics of the plant process to be controlled.

2. An *actuation* means that provides an appropriate actuation signal for the plant-process input by means of an auxiliary signal generator or by modification of the principal control-system error. System identification can be expensively sophisticated to the point where complete process dynamics, including the impulse response and system transfer function, are obtained with a suitable test system.

Practical considerations generally limit the complexity of such measuring means to some indirect variable that is available within the plant control loops. This must involve the judicious choice of a variable that does not introduce control components that further complicate the required stability evaluation.

The second characteristic member function of an adaptive control, the actuation signal, is used for process actuation either directly or in tandem with the principal process path.

An example of an ideal direct-path control is shown in Fig. A-26, where the process computer generates an actuation signal from observation of the input signal, system response, and plant dynamics. Applications of the basic two-element adaptive system requirement necessitate a continuity of plant operation by conventional control for the condition of adaptive-control-section failure.

The utilization of a "decision computer" element in the adaptive control system of Fig. A-26 leads quite naturally to a comparison of this control with the human controller. Further, the extension of the adaptive controller to include adaptation from a learning routine would represent the ultimate realization of one important aspect of the human controller.

Shown in Figs. A-27 and A-28 is an adaptive principle applied to a two-axis milling machine where a cutter-velocity signal ω and torque signal τ are used in a simple multiplier to obtain a signal proportional to the power delivered to the workpiece by the cutter.

From a previously determined relationship between cutter power and workpiece feed rate (axis velocity), a model signal that is useful in determining maximum cutter deflection is obtained. This signal is used to modify position-loop command (axis velocity) for optimum conditions of stock removal.

An example of the complexity related to automatic control of a plant necessitating an adaptive loop concerns the control of neutron flux level in a nuclear reactor. Therein a nonlinear relationship exists between the control variables as a result of their dependency upon output-power level and varying conditions of damping. Reference to the typical system of Fig. A-29 reveals that the control rod for a nuclear reactor is positioned by an actuator depending upon the error e defined by $e = n_D - n$.

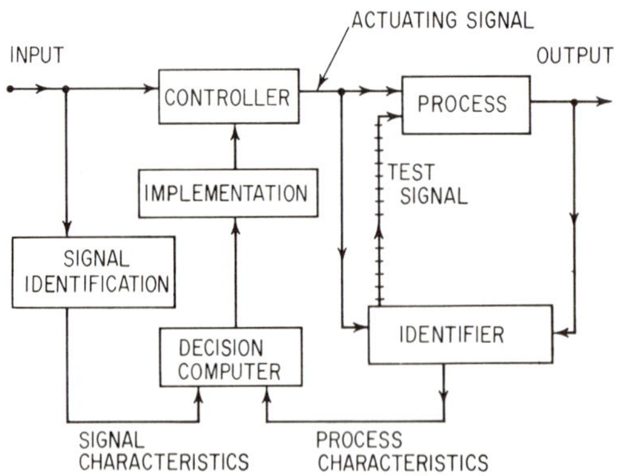

Fig. A-26. Adaptive control system utilizing a "decision computer."

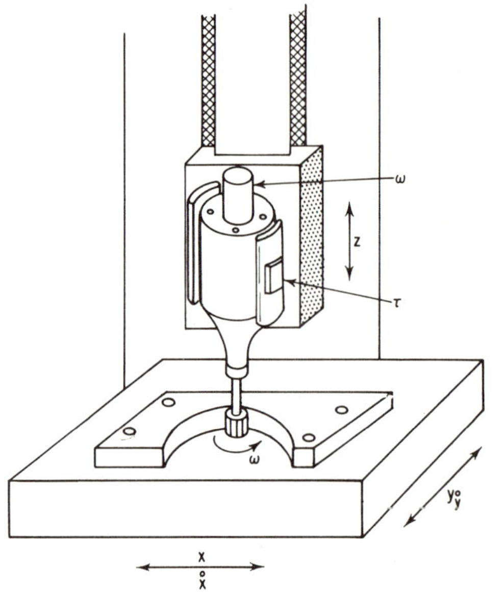

Fig. A-27. Two-axis milling machine showing variables involved in adaptive control system.

The transfer functions G_A, G_C, G_R are defined as follows:

$$G_A = K_A$$
$$G_C = K_C$$
$$G_R = \frac{K_R\,(S + a)}{S\,(S + b)}$$

The relation between p and n is nonlinear in that K_R is a function of neutron level and the relation between X and $\rho(K_C)$ is a function of rod position as well as the positions of other rods and the reactor history.

Variations in K_C and K_R result in a variation of loop gain which also changes system dynamic behavior. When K_R is large at high neutron flux levels, the system is lightly damped, while the reverse is true at low flux levels.

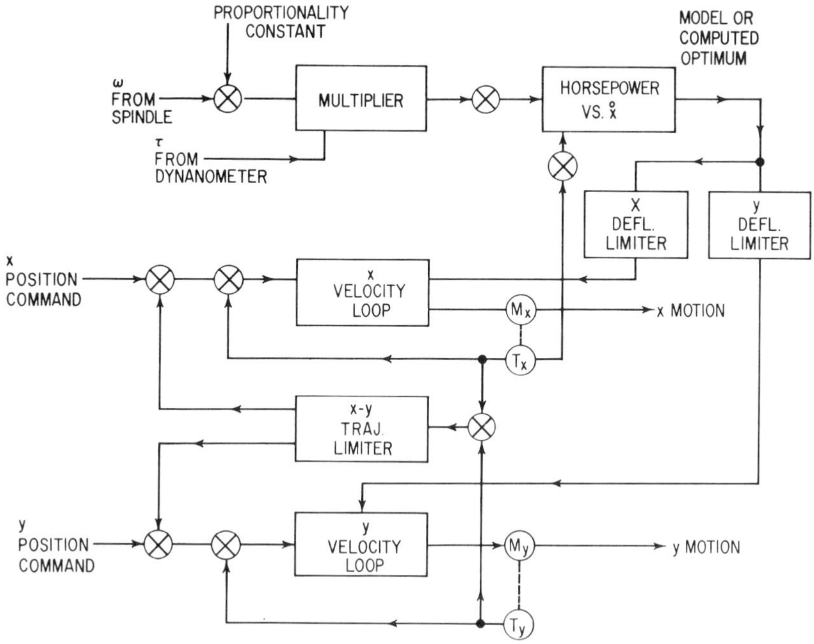

Fig. A-28. Adaptive control loop for a milling machine.

The system shown in Fig. A-30 utilizes a computer to evaluate the forward loop gain.

$$K_m = K_A \, K_C \, K_R$$

This gain is compared with a chosen gain factor K_D which is known to give good dynamic performance, and the difference is used to vary the gain

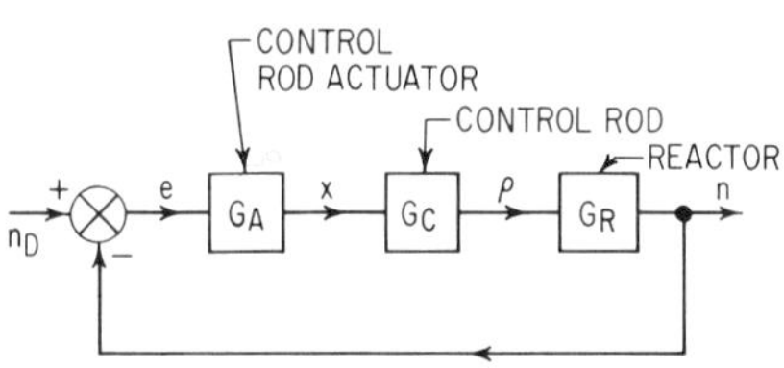

Fig. A-29. Control system for a nuclear reactor. $e = n_D - n$. $n_D =$ **desired neutron level.** $n =$ **actual neutron level.**

of the variable-gain amplifier K so that the total forward loop gain is K_D.

Particularly it has been the desire to introduce levels of learned response in control systems for the purpose of modifying plant parameters and/or dynamics to achieve both optimality and stability for a continuously changing trajectory with its attendant variations in plant parameters. Programmed updating of learned information must be dependent upon the dynamics of the original plant as well as the dynamics of the adaptive sampling loops associated with this controller.

Traditionally the use of feedback in a control system has been to control the sensitivity of the system to included parameter variation. An ideal adaptive system, however, will provide zero sensitivity when its elements operate ideally and instantaneously without disturbance.

The introduction of a learning mechanism within the adaptive portion of a system can result

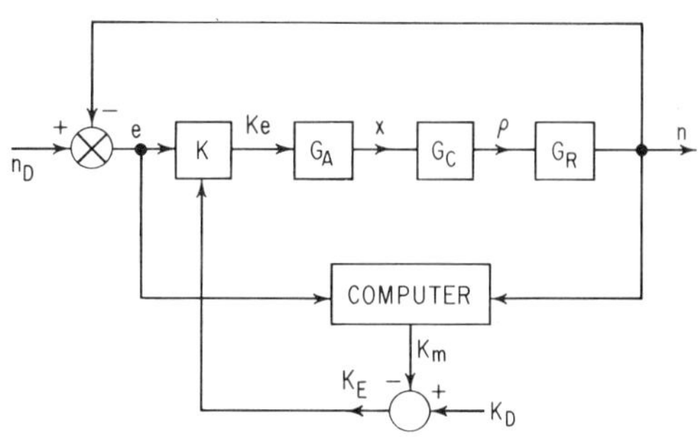

Fig. A-30. Adaptive system for a nuclear reactor.

in a significant approach to the characteristics of a human controller.

A form of an adaptive control called "self-organizing control" provides for readaptation of the controller many times within a single closed-loop response period of the plant by modification of possibly several signal paths. Such a system combines statistical decision theory to determine instantaneous plant performance, prediction theory to determine the performance trend, and rapid trial generation to determine a correction to improve this performance trend.

—James R. Walker, *De Vlieg Machine Co.*

Adaptive Process Control In the field of process control, the same problems exist as are found in the aircraft and milling-machine applications (as described under **Adaptive control**) except that instead of environmental changes, adaptability is required because of the nature of the process. If enough information is known about the process being handled, it is possible to compute the necessary adjustments and thus provide the adaptability for the system. A good example is the heat-balance problem of a semibatch reactor as shown in Fig. A-31.

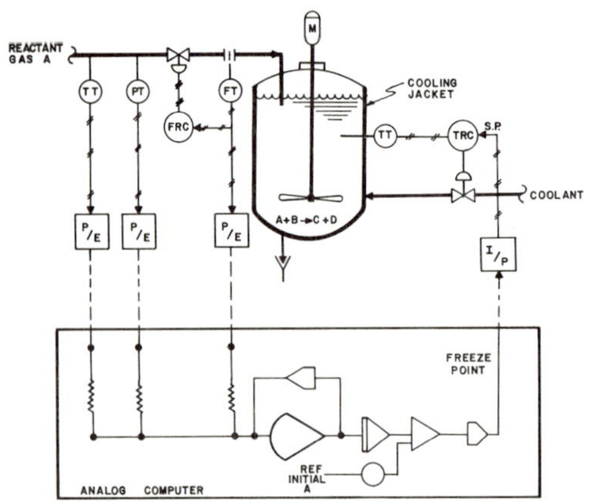

Fig. A-31. Adaptive control as applied to the heat-balance problem of a semibatch reactor.

The reactant gas A is added to liquid B to make liquid product C along with liquid by-product D. It is desired to operate the reactor near the batch freeze point for maximum production of C. The set point of the temperature control loop has to be adjusted to follow the freeze point, which keeps increasing with the changing chemical composition during the batch. The problem was solved by utilizing an analog computer which actually computes the incoming flow and thus determines the freeze point of the batch system to be used as a set point of the temperature controller.

The scheme presented in Fig. A-32 is an entirely different approach to the problem. The main advantage of such a system is its nondependence on external data. Instead, the adaptive control signal which controls the primary loop gain is obtained by analyzing the error signal (i.e., separating into high and low components, multiply by ratio factor, combining, and then passing through a filtering and integrating network). Since the system is designed for somewhat universal purposes, the performance characteristics and the reliability cannot be expected to be competitive to the programmed controller, which is specially "tailored" for a particular process.

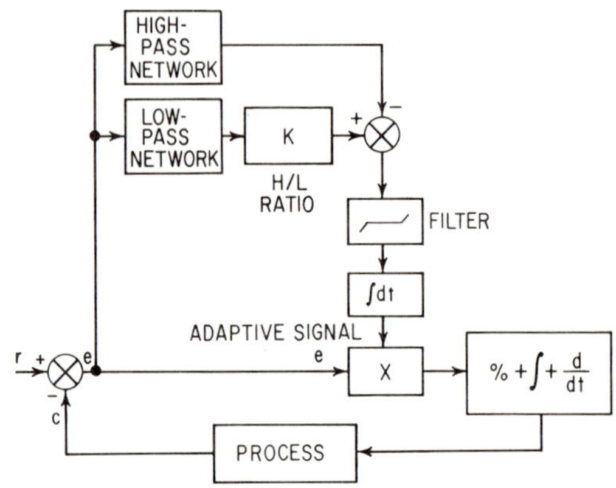

Fig. A-32. Adaptive control system based upon analysis of error signal.

—Anthony Y. Chiang, *Bristol Division, American Chain & Cable Company, Inc.*

Adder An adder is a circuit which provides the sum of two or more input numbers as an output. An example of a binary adder is shown in Fig. A-33, where A and B are the input bits and C and NOT C are the carry and no carry bits from the previous position (see **Bit**).

An adder may be serial or parallel. In a serial adder, only one adder position is required and the bits to be added are sequentially gated (see **Gate**) to the input. The carry or no carry from the previous position is remembered and pro-

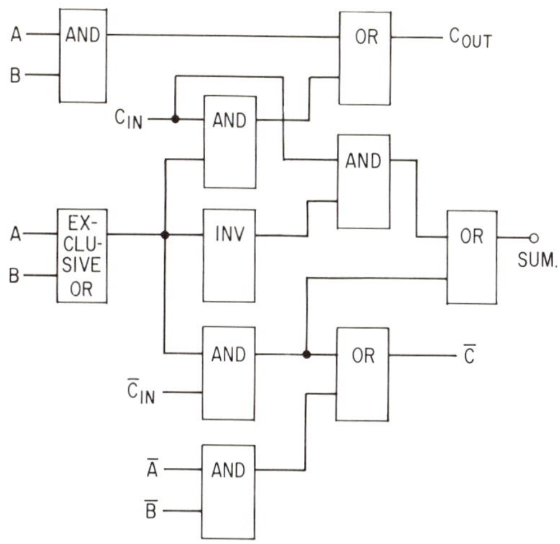

Fig. A-33. Binary adder.

vided as an input along with the bits from the next position. In a parallel adder, all the bits are added simultaneously with the carry or no carry from the lower-order position propagated to the higher position. A significant factor in the speed of a parallel adder is the propagation delay due to the carry.

Usually the adder is used to perform the subtraction as well as the addition of two numbers. This is generally accomplished by complementing one of the numbers and then adding the two factors. See Fig. A-34 for an example of a two's-complement binary-subtraction operation. The

(*a*)	0 1 1 0	+6	(True)
	(+) 1 0 1 0	−6	(Complement)
	1 0 0 0 0	0	(True)
(*b*)	0 1 0 1	+5	(True)
	(+) 1 0 1 0	−6	(Complement)
	1 1 1 1	−1	(Complement)
(*c*)	1 1 1 1 (Complement) = − 0 0 0 1 (True)		

Fig. A-34. Two's-complement subtraction.

two's complement of a binary number is obtained by replacing all 1s with 0s, all 0s with 1s, and adding 1 to the units position. Figure A-34*a* shows 6 substracted from 6, and the result is all 0s; the carry implies that the answer is in true form. Figure A-34*b* shows 6 subtracted from 5, and the result is all 1s with a no carry. The no carry signifies the result is in complement form and that the result must be recomplemented as shown in Figure A-34*c*.

—W. W. Welz, *IBM Corporation*

Address Register An address register is a collection of flip-flops (see **Flip-flop**) or similar logical storage elements which is used for the temporary storage of an address. A computer typically contains a number of address registers, each with a specific purpose. An adjective modifier such as in "storage address register" is commonly used to indicate the specific use of the register. (See also **Register, computer.**)

The storage address register contains the address of the data currently being accessed in storage (see **Storage, computer**). This register and its associated decoding logic provide the signals to the storage-selection circuits while storage is being cycled.

The instruction address register contains the address of the instruction which is to be read from storage. When the instruction is requested, the contents of this register is typically transferred to the storage address register.

The data address register contains the address of the data to be obtained from storage or the address at which the data are to be stored. When data are being read from or written to storage, the contents of this register is transferred to the storage address register. In machines which provide the function of moving data from one storage location to another storage location with a single instruction, two data address registers may be provided. One register contains the address of the present location of the data and the other contains the address of the location to which the data are to be transferred.

—W. W. Welz, *IBM Corporation*

Adjustable-speed Drives (See **Electric-motor drives, control of.**).

Admittance Measurement (See **Bridge circuits, electric.**)

Air-data Systems (Aircraft) An aircraft can extract much useful information from the atmosphere through which it is passing. Essentially, all air-data computations are based upon four sensed parameters: (1) static pressure, (2) total pressure, (3) total temperature, and (4) angle of attack. Air-data systems may range in complexity from an unheated pitot-static tube, altimeter, and indicated-airspeed instrument aboard a light airplane—to numerous and complex instuments and computers aboard an advanced commercial or military aircraft. With ever-increasing emphasis on safety, the trend is toward more instrumentation even for light airplanes.

U.S. Standard Atmosphere: Pressure and temperature of the atmosphere are functions of altitude above sea level, latitude, season, and time of day. Because static pressure is easily and accurately measured and because the actual altitude for a given pressure varies only slowly with respect to time and distance, the United States government has established standard atmospheric altitude vs. static pressure and temperature values (see Table A-1). Thus, all assigned altitudes, altitude reporting, and altitude control—essential to traffic separation and the systematic use of the airspace—are based upon these values.

TABLE A-1. *U.S. Standard Atmosphere* Altitude versus Static Pressure and Temperature*

Altitude, Geopotential, feet above sea level	Temperature, °F absolute	Pressure, in. Hg
−1,000	522.24	31.019
0	518.67	29.921
+1,000	515.10	28.856
3,000	507.97	26.817
5,000	500.84	24.896
10,000	483.01	20.577
15,000	465.18	16.886
20,000	447.35	13.750
30,000	411.69	8.885
40,000	389.97	5.538
50,000	389.97	3.4247
60,000	389.97	2.1178
80,000	397.86	0.8155
100,000	408.83	0.3219
200,000	454.91	0.00407
300,000	340.06	0.02948×10^{-3}

*U.S. Standard Atmosphere, 1962.

Pitot-static Tube: Openings on the smooth side of a pitot-static tube, as shown in Fig. A-35, provide a source of the undisturbed atmospheric pressure—termed *static pressure.* The open end of the pitot-static tube is headed into the airstream and provides a source of total pressure resulting from the impact of a body traveling through the atmosphere. Most pitot-static tubes are electrically heated to melt off any ice that might form and partly or completely seal off the openings and give erroneous and possibly very dangerous readings of airspeed, altitude, and rate of climb.

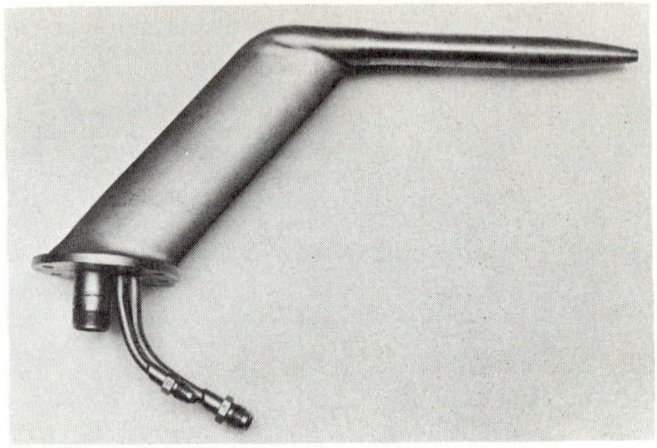

Fig. A-35. Electrically heated pitot-static tube. *(Rosemount Engineering.)*

Altimeter: Most aircraft altimeters operate like the aneroid barometer, except that altimeters are calibrated in height instead of pressure. Static pressure surrounding one or more evacuated-diaphragm capsules operates a pointer through a system of linkages and gears. Corrections must be made for changes of barometric pressure. One inch of mercury-pressure change corresponds roughly to 1,000 ft of altitude at sea level. Sensitive altimeters will show changes in pressure altitude of 10 ft or less. The face of a sensitive altimeter is shown in Fig. A-36. Since atmospheric pressure reduces in half for approximately each 18,000-ft increase in altitude, nonlinear diaphragm-linkage systems are used to provide linear altitude vs. pointer deflection over the complete altitude range. Pressure-altitude information may be supplied by air-data computers to

Fig. A-36. Sensitive altimeter. *(Kollsman.)*

various forms of indicators and displays. Also see **Altimeter.**

Airspeed Indicator: The difference between total pressure and static pressure is a measure of indicated airspeed. Airspeed indicators of the type shown in Fig. A-37 usually are calibrated in knots (nautical miles per hour) at standard pres-

Fig. A-37. Airspeed indicator. *(Bendix.)*

sure (29.92 in. Hg) and temperature (59°F). The indicated airspeed at which a given aircraft with a given load stalls is a constant over wide ranges of pressure and temperature. A knowledge of this is of the utmost importance during takeoffs and landings, and is also important during flight. (See Table A-2.)

True-airspeed Indicator: True airspeed is primarily useful in navigation. This factor is a

TABLE A-2. *Calibrated Airspeed vs. Differential Pressure* (Standard Temperature and Pressure Conditions)*

Indicated Airspeed, knots	Differential Pressure, in. Hg
0	0.000000
50	0.1200
100	0.4821
150	1.0924
200	1.9616
300	4.5407
400	8.3973
500	13.7967
600	21.1088
700	30.8159
1,000	73.6613

*Battelle Memorial Institute, 1953.

function not only of difference in total and static pressure, but also of temperature. A true-air-speed indicator, as shown in Fig. A-37, is in effect a computer that operates on a rather complex relationship of two pressure measurements and the air-temperature measurement to give the correct indication. This relationship is expressed by

$$V = 38.97M \sqrt{\frac{T_i}{1 + 0.2KM^2}} \quad (1)$$

where V = true airspeed, knots
T_i = indicated air temperature, °K
K = recovery factor of temperature probe (usually 1.00 or slightly less than 1.00)
M = Mach number

Machmeter: The ratio of true airspeed to the velocity of sound is called *Mach number.* The safe top speed of fast aircraft usually is expressed in terms of Mach number and is indicated on a Machmeter (see Fig. A-38). This instrument usually is a mechanical computer which operates on the relationship of static and pitot pressures to

Fig. A-38. Combined true-airspeed and Mach-number computer-indicator. *(Bendix.)*

give the required indication. The relationship for Mach number is

$$\frac{dP}{P_s} = (1 + 0.2M^2)^{3.5} - 1 \quad \text{for } M < 1 \quad (2)$$

$$\frac{dP}{P_s} = \frac{166.9215M^7}{(7M^2 - 1)^{2.5}} - 1 \quad \text{for } M > 1 \quad (3)$$

where M = Mach number
dP = differential (pitot less static) pressure
P_s = static pressure

Rate-of-climb Indicator: The instrument shown in Fig. A-39 comprises an enclosed volume of air connected to atmospheric (static) pressure through a constriction. As the altitude changes, the enclosed pressure lags the outside pressure. The pressure difference is measured in terms of rate of change of altitude, or rate of climb. Because any change of temperature causes a proportional change in pressure of an enclosed volume of air, the instrument enclosure must be a good thermal insulator to prevent all but very gradual temperature changes of the enclosed air. Otherwise, corrections must be introduced for both static pressure and temperature to provide the correct rate-of-climb information.

Angle-of-attack Transducer:[1] The angle of attack is the acute angle between the velocity vector of the aircraft through the surrounding air and some reference, such as the fore-aft axis of the aircraft or the chord of an airfoil. The lift developed by an airplane wing increases as a function of angle of attack and airplane velocity. Velocity information is readily available on all aircraft, but angle-of-attack information is not. A knowledge of both factors enables pilots to fly airplanes more economically (by flying at the optimum angle of attack for a given velocity) and more safely by avoiding angles of attack leading to airflow separation and hence to stall. The outputs of an angle-of-attack transducer may supply information to go-around systems, speed-command systems, and auto-throttle systems to optimize airplane performance during these maneuvers. In addition, angle-of-attack transducers provide essential information to stall-warning systems, which in some cases, will actually push the stick forward if the sensor indicates angles of attack approaching stall. One transducer of this type, shown in Fig. A-40, utilizes the pressure difference over a cylinder to align the cylinder with the airstream. Two rows of static-pressure ports, 45° on either side of the stagnation line at null, detect pressure differences due to nonalignment. The pressure difference is converted to an electrical error signal in a sensitive differential-pressure transducer. The error signal is amplified and then drives a motor which servos the cylindrical probe into the wind. The amount of probe rotation is a measure of the angle of attack.

[1] Information on angle-of-attack transducers was prepared by Malcolm D. Lefcort, Transducer Division, United Control, a subsidiary of Sundstrand Corporation.

Fig. A-39. Rate-of-climb indicator. *(Bendix.)*

Fig. A-40. Angle-of-attack transducer. *(United Control.)*

A schematic of the system is shown in Fig. A-41. The output of the device is a potentiometer voltage. Synchros can be used in place of the potentiometer if required, or a combination of potentiometers and synchros may be used. A failure monitor detects failures by monitoring the output of the bridge of which the variable-reluctance differential-pressure transducer forms one-half. Since all error signals should decrease with time because of the servoing action of the device, a failure is indicated if the bridge output exeeds a nominal value for longer than a prescribed time. The normal 6 V dc output of the failure-monitor circuit then is reduced to 0 V dc. A self-test capability also is provided. Energizing the self-test relay removes the bridge error signal from the servoamplifier and injects an artificial signal corresponding to a predetermined angle of attack. The probe then servos to the false null, and the output potentiometer reads the self-test angle.

Ice Detector: The principle of operation of

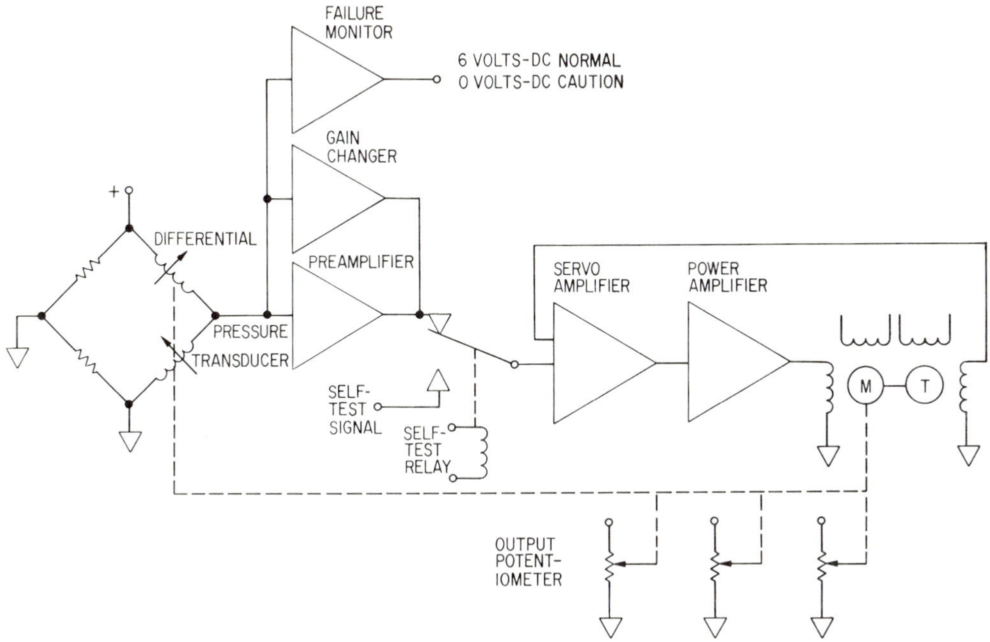

Fig. A-41. Schematic of angle-of-attack transducer. *(United Controls.)*

one form of ice detector[1] is shown in Fig. A-42. This device utilizes the absorptive properties of ice when subjected to beta-particle radiation to indicate the presence of ice. An airfoil-shaped probe containing a radioactive source, a Geiger-Mueller tube, and a heater is positioned in the airstream. An amplifier-detector, which is packaged with the probe or placed in a remote location dependent upon specific application, receives ice information from the probe and causes the probe heater to turn on when ice accretions of greater than 0.015 in. occur. The rate at which the heater turns on and off can be used as a measure of the severity of icing conditions. Beta particles, emitted by a strontium-90 source, are directed in a collimated beam along the leading edge of the sensing portion of the airfoil toward the Geiger-Mueller tube. As ice accumulates, the beta-particle pulse rate falls. When the rate falls below 20 pulses per second (equivalent to 0.015 in. of ice), the amplifier-detector energizes a relay which turns the heater on. The heater is turned off when the pulse rate exceeds 40 pulses per second. There is a visual indication when ice is present.

[1] Information on ice detector prepared by Malcolm D. Lefcort, Transducer Division, United Control, a subsidiary of Sundstrand Corporation.

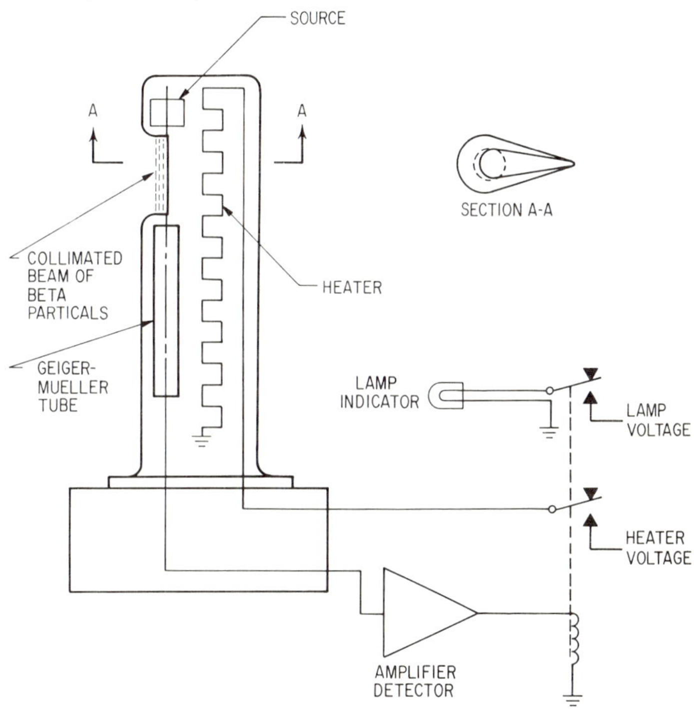

Fig. A-42. Schematic of radiation-type ice detector. *(United Control.)*

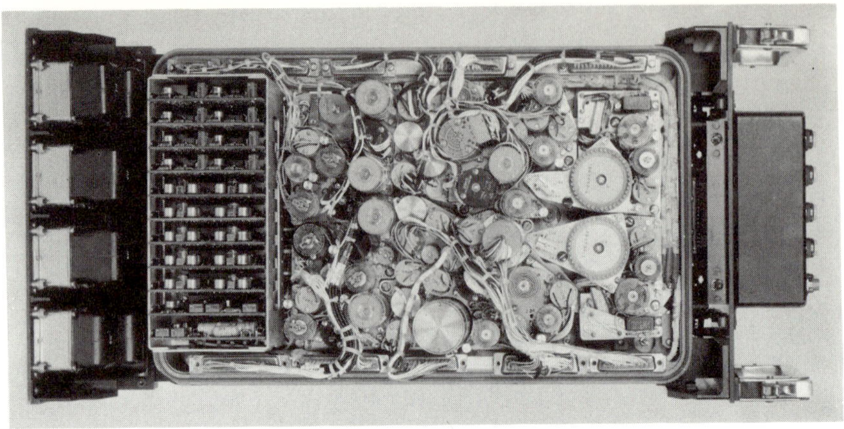

Fig. A-43. Central air-data computer. *(Bendix.)*

Air-data Computers: The computer shown in Fig. A-43 centralizes the computation of air data from a number of inputs, including static and total pressures, air-stagnation temperature, and angle of attack. Among the numerous possible outputs are pressure altitude, indicated airspeed, true airspeed, true air temperature, Mach number, air density, and rate of climb. This information is transmitted to indicators and displays and is used by various aircraft subsystems, such as the flight-control system, fire-control system, and navigation system. Computation may be mechanical, electromechanical, digital, or a combination of these methods. Transmission may be by synchro, potentiometer, or digital encoder.

—Paul F. Bechberger, *Navigation and Control Division, The Bendix Corporation*

Air-pollution Instrumentation The basic needs for instruments and controls as these relate to air-pollution problems are two: (1) the devices required by pollution-control researchers and engineers—to determine degrees and sources of pollution and to enforce pollution regulations; and (2) the systems required by the organizations that generate the pollutant materials, principally manufacturers of various types, but also including various modes of transportation that depend upon combustion processes for their motive power. An identifiable body of science specifically slanted toward pollution instrumentation has been slow in forming. Consequently, the instruments and controls used largely are adaptations of apparatus that has been developed for other reasons. Pollution engineers have become quite expert in adapting conventional instruments to their specific needs. The reader is referred to other entries in this encyclopedia, including: **Analytical instrumentation (chemical); Anemometer; Barometer; Chromatography; Combustion/combustibles analyzer; Dew point, measurement of; Flow, measurement of; Humidity, measurement of; Medical instrumentation; Meteorological instrumentation; Temperature, measurement of;** and **Turbidimeter.** Also, consult the alphabetical index.

Sources of atmospheric contaminants provide clues to the kinds of instrumentation required. Three factors are critical in source testing and analysis for air-pollution control: (1) measuring gas properties, including suspended particulate materials, *in situ*—in the atmosphere or in ducts, chimneys, and so on; (2) withdrawing representative samples; and (3) accurate analyses (see Table A-3).

Considerably more automation is employed in the measurement and control of air pollution by the manufacturers (sources of pollutants) than by pollution enforcers. The latter depend heavily upon laboratory techniques rather than continuous analysis. The mobile laboratory, often well instrumented, has added much to the pollution engineer's efficiency.

Estimation of pollutant-emission rates is extremely important in studying an air-pollution problem. These rates are determined in two

TABLE A-3. *Major Air Pollutants and Sources*

Pollutant	Source
Beryllium dust	Ore preparation; metalworking
Carbon monoxide	Fuel-burning operations (vehicles; power plants; incinerators; dump burning)
Carbonaceous smoke, soot, and fly ash	Fuel-burning operations; waste recovery; field burning; smudge pots
Chlorinated hydrocarbons	Crop dusting and spraying; chemical plants
Fluorides	Chemical and metallurgical plants
Hydrocarbons	Fuel-burning operations; paint spraying; solvent cleaning; printing; chemical and metallurgical plants
Metal fumes, such as arsenic, lead, and zinc	Metallurgical plants; crop dusting and spraying
Nitrogen oxides	Fuel-burning operations; chemical plants
Organic phosphates	Crop dusting and spraying
Organic vapors	Fuel-burning operations; chemical plants; paint spraying; waste-recovery operations
Particulates (mineral and organic)	Milling, crushing, screening, grinding, and demolition operations
Radioactive fallout, such as carbon-14, cesium-137, and strontium-90	Nuclear-device testing
Sulfur oxides	Chemical and metallurgical plants
Uranium dust	Ore preparation

basic ways: (1) by making a materials balance across an entire plant and by reference to shelf information which indicates the weight of contaminants that results from fuel burning, for example; and (2) by direct measurement. Because of the volume involved, generally it is impractical to "meter" total pollutants. Hence, accurate sampling techniques must be developed. Accurate estimating of duct velocities, for example, becomes very important. In some cases, traverse sampling can be used where ports are located to provide a representative cross section of total flow. This introduces the interesting concept of *isokinetic rate,* which can be defined as a sample rate that precisely matches the average velocity of the main stream.

The measurement and analysis of particulates has progressed immensely because of the emphasis on air-pollution control. The general properties of particulates of interest to the pollution engineer are listed by Lapple (Ref. 1) and summarized in Table A-4. A short summary of particulate-analysis techniques is given in Table A-5.

TABLE A-4. *Factors in Physical Characterization of Particulates*

- I. Particle size
  - *A.* Significance (discriminating property used)
  - *B.* Size distribution or uniformity
    - 1. Bases
      - *a.* Count or number
      - *b.* Surface
      - *c.* Mass or volume
      - *d.* Conversion between bases (uniform shape assumed)
    - 2. Representation of distribution data
      - *a.* Graphical
        - (1) Frequency
        - (2) Cumulative
      - *b.* Analytical
  - *C.* Representation of average sizes
    - 1. Median size
    - 2. Mean size
    - 3. Conversion between averages
  - *D.* Special problems
    - 1. Mixture of phases
    - 2. Flocs and agglomerates
- II. Particle shape
- III. Particle density and structure
- IV. Pore structure and distribution

TABLE A-5. *Particulate-analysis Techniques*

Size-discriminating Property		Type of Measurement Technique
Type	Character	
Geometric	Physical barrier	Sieving Ultrafiltration
Mechanical or dynamic (in fluids)	Inertia	Impaction on surface Pressure pulse (sonic)
	Terminal settling velocity	Elutriation Sedimentation Decantation Photosedimentation
	Diffusion	Particle displacement Particle deposition
Optical	Imaging	Light microscopy Ultramicroscopy Electron microscopy
	Transmission	Spectrum
	Scattering	Single-particle count Macroscopic
	Diffraction	Light X-ray Laser
Electrical	Resistance	Current alteration
	Capacitance	Potential pulse
	Charge	Tribo-charging Induction charging Corona charging
Magnetic	Magnetic particles	Particle migration; pulse
Thermal	Particle deposition	Particle migration
Physical-chemical	Condensation	Growth of nuclei

REFERENCES

1. Lapple, C. E.: Particle-size Analysis and Analyzers, *Chem. Eng.,* May 20, 1968, pp. 149–156.
2. Rossano, A. T., and H. B. H. Cooper: Air Pollution Control Sampling and Analysis, *Chem. Eng.,* Oct. 14, 1968, pp. 142–146.
3. Methods for Determination of Velocity, Volume, Dust and Mist Content of Gases, *Western Precipitation Corp. Bull. WP-50,* 6th ed. Los Angeles, 1958.
4. Leonard, J. S.: Stack Emission Sampling, *TAPPI,* vol. 49, no. 10, October, 1966.
5. Blosser, R. O., and H. B. H. Cooper: Analytical Equipment and Monitoring Devices for Gases and Particulates, *National Council for Air and Stream Improvement Tech. Bull.* 35, March, 1968.
6. Nishikawa, K.: Portable Gas Dilution Apparatus for the Dynamic Calibration of Atmospheric Analysers, State of California Air Pollution Methods Conference, Los Angeles, Jan. 31, 1963.
7. Adams, D. F.: Air Pollution Instrumentation Monograph, *ISA Pub.* APIM, 1966.
8. Chapman, R. L.: Environmental Pollution Instrumentation, *ISA Pub.* EPI, 1969.

Air-traffic Control Position reporting by the pilot to the air-traffic controller over a voice-radio link is the basic source of air-traffic-control position data. The pilot, of course, can use information from any of the navigation aids (VOR, DME, TACAN, inertial navigator, and so on) —or from visual sighting of landmarks. Traffic, however, is largely routed along paths defined by the VORTAC system (radials from VOR stations, intersections of radials or DME, and range check points). See **Navigation control systems.**

Ground controllers have two major independent methods of checking on the aircraft location: (1) *primary radar,* which operates on reflection by the aircraft of the pulse signals which the radar transmits; and (2) *secondary radar,* which operates on replies from pulsed radio beacons (on the aircraft) verified by the secondary-radar pulses.

The generally used primary radars are of two types: (1) relatively short range, known as *airport-surveillance radars,* or ASRs; and (2) longer-range systems, known as *airport-route-surveillance radars* or ARSRs. Both types of systems employ pulse transmissions and rotating antennas (see Fig. A-44). Both systems have moving-target indicators which compare the rf phase of successive returns in order to determine whether the target is moving or stationary. The stationary targets are canceled from the display, using either a delay-line analog technique to make the comparison or a digital-storage system. In some designs of either type of system, the pulse-repetition frequency may be "jittered" or staggered in order to avoid "blind speeds" where the aircraft has moved by precisely one or more integral wavelengths between successive pulses. Typical parameters of an airport-surveillance radar system are given in Table A-6; and of an airport-route-surveillance radar system in Table A-7.

Secondary radars usually are co-located with the primary radar and use a line-array antenna fastened on top of the primary-radar antenna structure, as shown in Fig. A-44. The frequency of operation is L-band (1,030 MHz). Because one-way transmission elicits a reply from the aircraft transponders, secondary radars pose a special problem. The difference in antenna gain from main lobe to side lobe is not as effective in suppressing false side-lobe replies as the primary-radar system, which uses the antenna gains on both transmission and reception to enhance the difference between main-lobe and side-lobe signals. To alleviate this problem, the secondary-radar system transmits a signal from an omnidirectional (or moderately directional) antenna which has slightly more gain than any of the side-lobe signals.

Fig. A-44. Ground-radar antenna used for air-traffic control. *(Bendix.)*

The beacon in the aircraft receives the omnidirectional transmission and the signal from the high-gain directional antenna in a standard time sequence. If the pulse from the omnidirectional antenna is not smaller than the other pulses from the high-gain antenna, the beacon assumes that the reception is from a side lobe and does not reply. The reply frequency from the transponder is 1,090 MHz; so there is no problem with ground echo from the transmitted signals. No moving-target indicator is used in a secondary radar. The secondary-radar system uses multiple pulses (in addition to the side-lobe suppression pulse) in order to elicit different types of response from the beacons.

Some airports are equipped for precision-approach radars. These are primary radars and employ two very narrow beams scanning a relatively narrow sector aligned with the approach

TABLE A-6. *Characteristics of Airport-surveillance Radar System (ASR-4)*

Peak power	425 kW
Average power	425 W
Transmitter tube	Magnetron
Pulse-repetition frequency	1,200
Pulse width	1 μs
Noise figure	8 db
Antenna gain	34 db
Beam width (azimuth)	1.5°
Antenna speed	12 rpm
Duty cycle	0.001
Preselector filter	Yes
Video mapping	No
Consoles	No
Cancellation ratio	35 db
Range (1 m² target)	32 miles (40 km) ±
Bandwidth (normal)	2.1–2.7 Hz
Bandwidth, MTI	4.5 Hz
Bandwidth, if	2.5 Hz
Minimum discernable signal (MDS), normal	103 db
MDS, moving-target indicator (MTI)	101 db
Duplexer	Solid-state
Radar performance monitor	Yes
Slow time constant (STC)	Yes
Subclutter visibility (SCV)	30 db
Identification friend or foe (IFF) video	Yes
Dual channel	Yes
Rotary joint	Dual
Polarization	Vertical/circular
Staggered pulse repetition frequency (PRF)	Yes
Cancels second-time-around targets	No
Power requirement	25 kW

course to a particular runway. One beam, broad in vertical dimension and narrow in horizontal dimension, scans at a relatively high rate (several sweeps per second) in the horizontal dimension. The controller, watching the cathode-ray-tube display, is able to tell the pilot whether the pilot is right or left of the true approach course. The other beam is narrow in the vertical dimension and broad in the horizontal dimension and scans vertically. An expert controller, watching the cathode-ray-tube display from this beam, is able to advise the pilot whether the aircraft is above or below the correct approach path. More extensive use of these systems has been limited because of the high costs of maintenance and operational personnel required to implement the systems.

Low-approach System (ILS): Near the touchdown end of an airport runway, a transmitter is set up to radiate a composite signal with two different audio modulations. One (150 Hz) decreases in depth of modulation with increasing elevation angle; the other (90 Hz) increases with increasing elevation angle. If an aircraft flies the glide slope established by equal depths of modulation between the two audio signals, the aircraft will make a proper letdown toward the runway (see Fig. A-45). A transmitter located beyond the stop end of the runway radiates a composite signal similarly modulated, but with the relative depths of modulation varying in azimuth —with the surface of equal 90- and 150-Hz modulation located along the runway centerline extended. This information is displayed as localizer and glide-slope data, typically along with other data on a flight-data indicator.

Automatic Altitude Reporting: This system, as diagramed in Fig. A-46, eliminates many of the manual steps required in air-traffic control systems and consists of an air-data computer that corrects static-pressure errors and provides synchro-driven altitude information to the pilot's altimeter and altitude in digital form to the aircraft transponder in high-performance aircraft,

TABLE A-7. *Characteristics of Airport-route-surveillance Radar System (AN/FPS-100A)*

Range	190 miles on 1 m² target (75% hits)
Output power	2 MW peak; 4.3 kW avg
Pulse length	6 μs, PRF 360 cycles, duty 0.00216
Power amplifier	Klystron L3035
MTI cancellation ratio	35 db
Noise figure overall	4.5 db; noise-figure parameter only, 3 db
Channels	Two, 1 active, 1 standby. Each channel includes transmitter and receiver
Antenna gain	35 db
Azimuth beam width	1.3°
Vertical beam width	4.5°, cosecant-squared pattern to 18°

Shaped reflector and single-horn feed. Polarizer permits horizontal and circular polarization. IFF antenna located on top of radar system. Rotary joint used for radar and IFF

and a direct readout to the pilot and digital altitude information to the aircraft transponder in lower-performance aircraft. The digital information then is transmitted to the ground interrogator and presented on the radarscope in alphanumeric form.

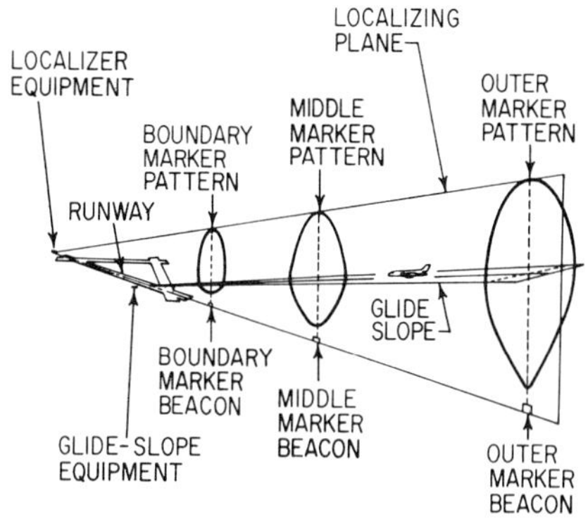

Fig. A-45. Instrument-landing system.

An interrogation pulse group is transmitted from the interrogator-transmitter unit through a directional-interrogator antenna assembly. The pulse group triggers an airborne transponder, causing a multiple-pulse reply group to be transmitted. The transponder transmission is received by the ground interrogator-receiver, processed through a computer, and then displayed in alphanumeric form on the controller's radar. The length of the round-trip transit time determines the range of the replying aircraft; the mean direction of the main beam of the interrogator antenna during the reply determines the azimuth; and the encoded signal from the transponder provides the aircraft's altitude in 100-ft increments. Also, it is possible to identify the specific aircraft. The system lessens the need for voice communication and the workload of the air-traffic controller.

—Paul F. Bechberger, *Navigation and Control Division, The Bendix Corporation*

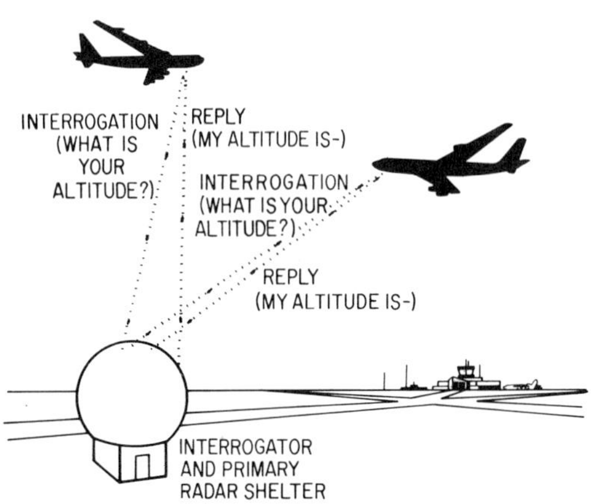

Fig. A-46. Automatic altitude-reporting system.

Aircraft, Instrumentation of There are few, if any, applications where instruments and controls are so concentrated as aboard a modern aircraft or spacecraft. Because each device must operate in any attitude and over a wide range of pressures, temperatures, and accelerations, careful design is required to avoid errors due to these causes. Most aircraft instrumentation involves some form of remote indication, usually electrical and often synchro. Gyroscopic instruments are widely used. Some of the application categories of aircraft instrumentation include: (1) flight and attitude instruments and controls; (2) navigational aids and systems; (3) engine instruments and controls; (4) cabin controls; and closely associated topics, such as air-traffic control and simulators for training. These topics are covered elsewhere, and reference to the **Classified Index** as well as to the alphabetical index at the end of this volume is suggested.

Aircraft Instrument Displays: The problem of displaying information is common to all the foregoing application categories. The tremendous amount of information needed to fly properly or even monitor an aircraft presents extensive and difficult problems in display and great attention to human-factors engineering. The information must be presented as needed in an easily assimilated fashion, under wide variations of illumination, and often against a background of distraction and in an environment of anxiety. Most indicators are hermetically sealed, with the edge of the coverglass metallized and soldered to the metal case. True hermetic-seal electrical connectors are used, and the cases are filled with a dry, inert atmosphere, usually partly helium, to permit detection of leaks with a sensitive mass spectrometer.

A pointer moving against a scale is a widely used and simple presentation. The pointer may be driven by a meter movement, as shown in Fig. A-47*a*, or by a synchro (Fig. A-47*b*), or it may be servoed (Fig. A-47*c*). Readability is about 1° on a 3-in. dial. Greater accuracy of reading can be obtained by use of an additional geared-up pointer (Fig. A-47*d*).

Another popular form of indication involves moving tapes against a fixed scale, sometimes with tape-moved command markers. Instruments of this type for monitoring multiengine performance are shown in Fig. A-48. Moving tapes may contain graduations, readable against fixed reference lines—again sometimes with settable command markers. Scale lengths of several feet are possible with tapes having graduations. A digital range and range-rate indicator for a space vehicle is shown in Fig. A-49. Digital encodings are on the back of the tapes, directly behind the graduations.

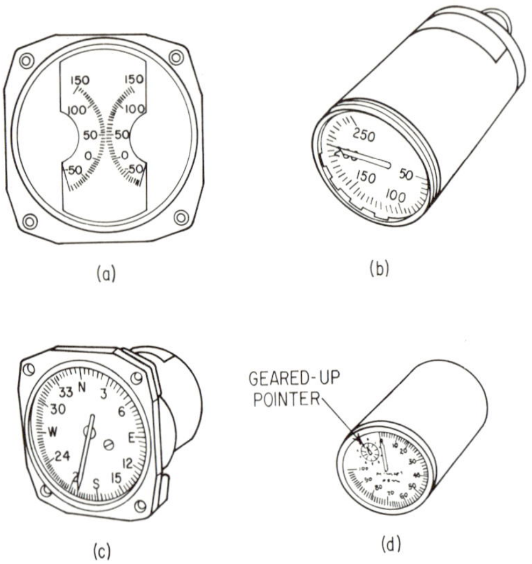

Fig. A-47. Dial-type moving-pointer instruments: (a) driven by meter movement; (b) synchro; (c) servoed; and (d) with separate geared-up pointer. *(Bendix.)*

Flight attitude usually is shown by positioning a sphere (colored to represent the sky and earth meeting at the horizon) against the fixed image of an aircraft (see Fig. A-50). Cathode-ray tubes often are used for display purposes, including weather radar.

Headup displays present information in the viewer's line of sight, usually focused at infinity. Such displays are used in gunsights, landing systems, and wherever a one-to-one correspondence of information with the outside world is required. These displays consist of one or more cathode-ray tubes, lenses, and a combining mirror through which the pilot looks at the outside, yet sees the cathode-ray-tube images and possibly fixed or moving reticles reflected to him in his line of sight. High accuracies are required of such sys-

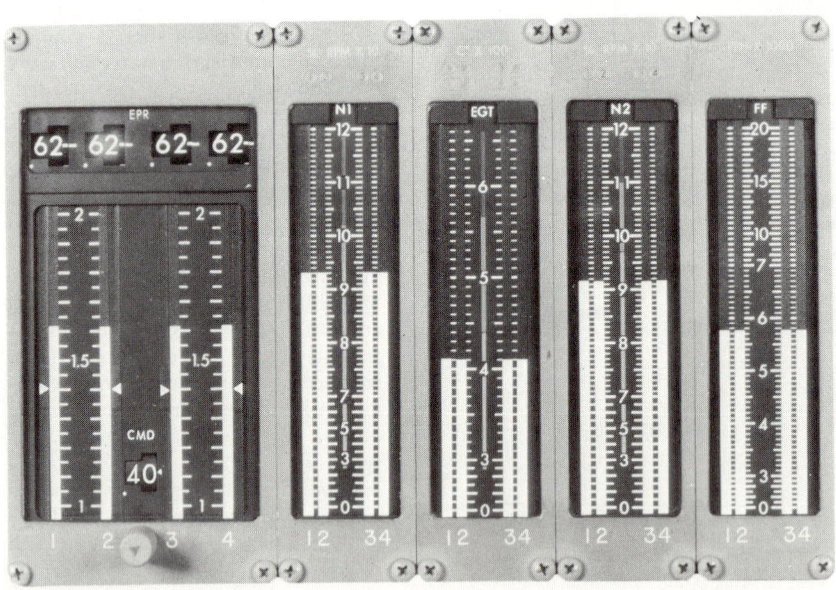

Fig. A-48. Bank of moving-tape indicators for monitoring multiengine performance. *(Bendix.)*

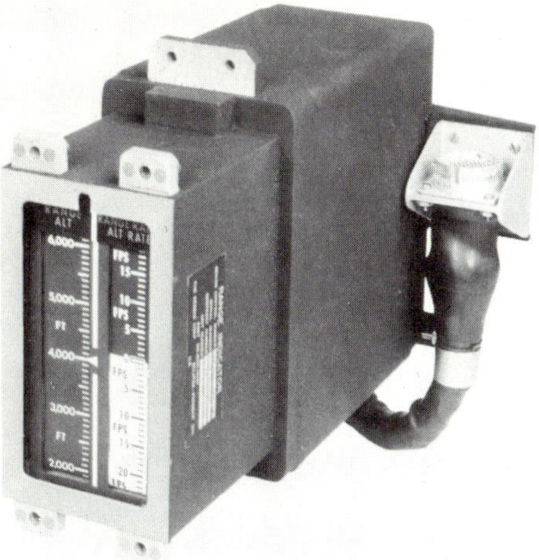

Fig. A-49. Digital range and range-rate indicator for space vehicle. *(Bendix.)*

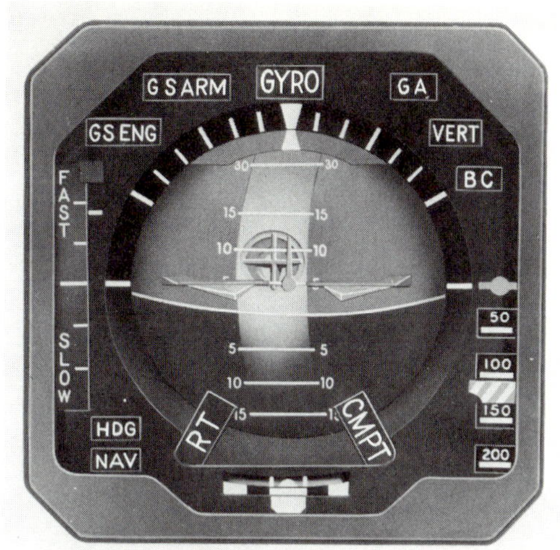

Fig. A-50. Attitude direction indicator. *(Bendix.)*

tems to obtain an effective one-to-one overlay. Nonoverlay information often is presented on the periphery. A headup projector is shown in Fig A-51.

Numerals often are shown on electromechanical counters—either mechanically driven or attached to miniature motors. Components used for alphanumeric displays include rear-projection readouts, segmented filament bulbs, and gas-glow character tubes. Meter-operated flags and back-illuminated warning or operating-mode annunciators often are used.

Examples of how a lot of information can be packaged into a small display space are shown in the two small panel instruments illustrated in Figs. A-52 and A-53.

Fig. A-51. Headup projector. *(Bendix.)*

Fig. A-52. Horizontal situation indicator. *(Bendix.)*

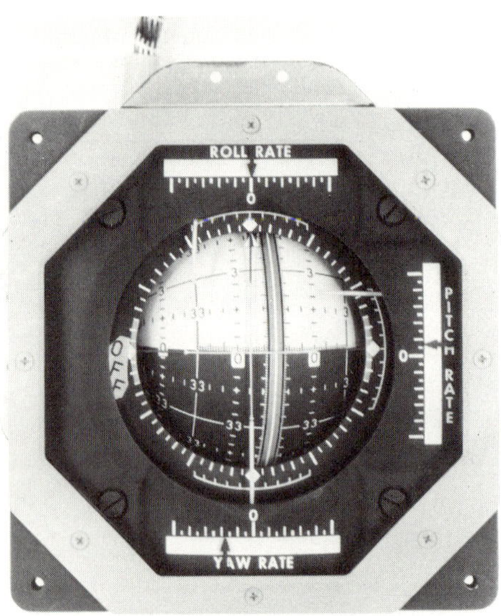

Fig. A-53. Attitude direction indicator for spacecraft. *(Lear Siegler.)*

REFERENCES

Burns, R. C.: Experience with Integrated Circuits in Control Systems, *Control Eng.*, vol. 12, no. 3, p. 76, March, 1966.

Snodgrass, R. P.: On-the-Windshield Displays Improve Flight Control, *Control Eng.*, vol. 12, no. 1, January, 1966.

Otten, I. D. D.: Strap-down Guidance Design, *Control Eng.*, vol. 12, no. 5, p. 93, May, 1966; vol. 12, no. 10, p. 61, October, 1966.

Johnson, R. M.: Artificial Feel for Servo-boosted Manual Controls, *Control Eng.*, vol. 12, no. 3, p. 67, March, 1965.

Aronson, R. L.: Simulators at Langley, *Control Eng.*, vol. 12, no. 10, p. 105, October, 1965.

Lindahl, J. H.: Fluid Amplifiers—Vortex Senses Aircraft Pitch Rate, *Control Eng.*, vol. 11, no. 9, p. 99, September, 1964.

Inertial Navigation, *Control Eng.*, vol. 11, no. 8, p. 26, August, 1964.

Proc. 15th Natl. ISA Aerospace Instrumentation Symposium, ISA Pub. AS691, Las Vegas, Nev., 1969.

Povejsil, Donald J., Robert S. Raven, and Peter Waterman: "Airborne Radar," D. Van Nostrand Company, Inc., Princeton, N.J., 1961.

Perry, M. A. (ed): "Flight Test Instrumentation," Pergamon Press, New York, 1965.

—Paul F. Bechberger, *Navigation and Control Division, The Bendix Corporation*

Aircraft-engine Controls The extent and sophistication of engine instrumentation vary widely with the type of aircraft and intended use. In small single-engine private airplanes, most instruments are simple and a mechanically connected throttle suffices. It is highly desirable to keep fuel and oil under pressure out of the cockpit. Thus, most engine parameters are remotely indicated in the cockpit from transmitters mounted near or on the engine. Direct connection is inhibited by the fact that most lubricating oils congeal sufficiently to prevent accurate pressure readings at the end of a long connecting tube.

Engine-mounted transmitters are used on both reciprocating and jet engines. The transmitters may be exposed to temperature as high as 450° F and to vibrations of 20 *g* up to 2,000 Hz. High-temperature magnet wire and leads are used. Essential indications and typical ranges and environmental peculiarities for reciprocating engines are: (1) manifold pressure: 10 to 100 in. Hg abs; (2) fuel pressure: 0 to 50 psi; (3) oil pressure: up to 600 psi occurring on cold start; (4) engine speed: 0 to 3,600 rpm; (5) hydraulic pressure: 0 to 5,000 psi; and (6) cylinder-head temperature: from −50 to 300°C. Electric-resistance elements normally are used for the latter temperature measurements. Also, the torque pressure, which is proportional to the torque exerted on the propeller, must be measured.

Instrumentation for a typical jet engine will

include: (1) low compressor-rotor speed: 0 to 110 percent (3,600 rpm = 100 percent); (2) high compressor-rotor speed: 0 to 110 percent (7,800 rpm = 150 percent); (3) fuel flow: 750 to 16,000 lb/hr; (4) exhaust-gas temperature: + 100 to 800° C (usually thermocouples); (5) engine pressure ratio, that is, the ratio of turbine-discharge total pressure to engine inlet-air pressure; (6) engine inlet-air pressure; (7) engine inlet-air temperature; (8) fuel-pump inlet temperature: from −54 to + 64° C; (9) fuel-pump inlet pressure: 5 to 50 psig; (10) fuel-deicing air-shutoff-valve position indication; (11) fuel-filter pressure-differential warning; (12) engine oil pressure: 0 to 100 psi, vented to accessory gearbox or overboard; (13) engine-oil inlet temperature: +25 to 165°C; and (14) engine radial vibration.

Low-pressure sensors (up to 100 psi) use bellows or diaphragms. High-pressure sensors (above 100 psi) use bourdon tubes. If the maximum pressure exceeds that which the sealed case can withstand (should there be an internal leak), a blowout or overpressure relief must be provided, preferably by means of a vent connection through which leakage can be disposed of safely. For pressures above 1,000 psi, pressure changes inside a sealed case can be neglected, and thus one pressure element usually is satisfactory. Below 1,000 psi, two pressure elements—differentially coupled—with one element connected to a pressure reference normally are used. Small synchros driven by pressure sensors can operate various types of synchro indicators—with resolutions of ±1° for nonservoed systems and ±¼° for servoed types. Some of the display techniques used in connection with aircraft-engine measurements are described under **Aircraft, instrumentation of.**

Temperatures up to 300° C normally are measured by resistance bulbs, the resistance values of which are given in Table A-8. Chromel-alumel thermocouples are used for the higher temperatures.

Rotor-speed indication commonly uses a system consisting of a permanent-magnet three-phase generator with output voltages and frequency proportional to the speed of the driving engine. Connected to this generator is a three-phase motor which will start and synchronize at all except the very lowest speeds. This motor, in the indicator, usually drives a magnet whose field pulls an alloy rotor against a hairspring. The deflection is linearly accurate. Two-pointer presentation (fast and slow) is common. The alloy rotor is made of a material having a low change of conductivity with temperature. Such compensation as may be required is accomplished by shunting the magnet with a magnetic alloy

TABLE A-8. *Temperature-resistance Characteristics for Aircraft-engine Cylinder-head Bulbs*

Temperature, °C	Resistance, Ω	Tolerance, ±Ω	Temperature, °C	Resistance, Ω	Tolerance, ±Ω
−50	37.95	0.6	130	89.20	0.6
−40	40.15	0.6	140	92.80	0.6
−30	42.45	0.6	150	96.50	0.7
−20	44.85	0.6	160	100.30	0.7
−10	47.35	0.6	170	104.20	0.7
0	50.00	0.6	180	108.20	0.7
+10	52.60	0.6	190	112.30	0.7
20	55.25	0.6	200	116.50	0.8
30	57.95	0.6	210	120.80	0.8
40	60.70	0.6	220	125.20	0.8
50	63.55	0.6	230	129.70	0.8
60	66.50	0.6	240	134.30	0.8
70	69.55	0.6	250	139.00	0.9
80	72.65	0.6	260	143.80	0.9
90	75.80	0.6	270	148.75	0.9
100	79.00	0.6	280	153.85	0.9
110	82.30	0.6	290	159.10	0.9
120	85.70	0.6	300	164.50	1.0

having its Curie points in the operating-temperature range.

A typical complex automatic engine control for supersonic augmented turbofan jet engines is shown schematically in Fig. A-54. Figure A-55 illustrates the interrelationship of the engine, the main fuel control, and the afterburner-exhaust nozzle control (A/B-ENC). The simple per-

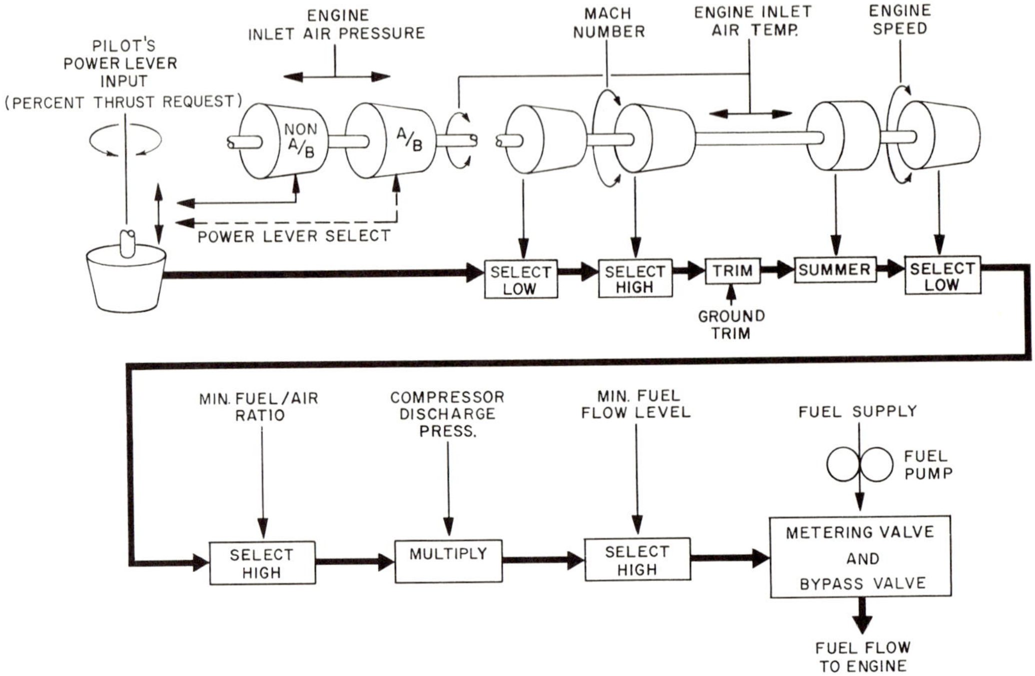

Fig. A-54. Automatic control system for turbofan engine using a three-dimensional cam-type hydromechanical computer. *(Bendix.)*

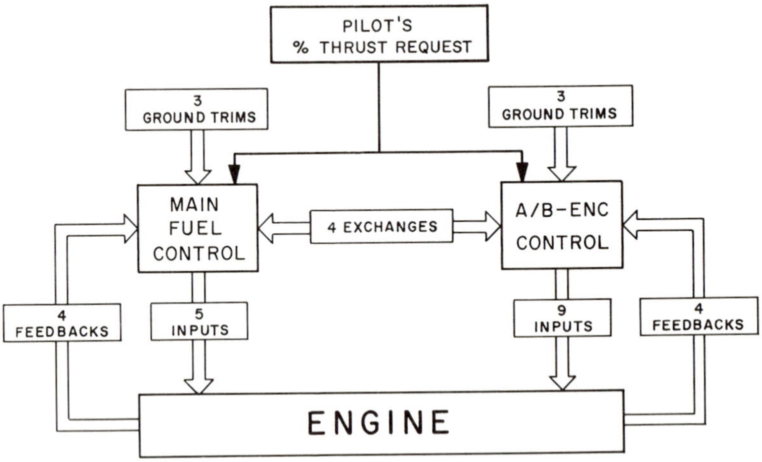

Fig. A-55. Interrelationship of engine, main fuel control, and afterburner-exhaust nozzle control. *(Bendix.)*

cent thrust request from the pilot also is shown. In all cases, the percent thrust is automatically related to the maximum allowable for the given flight condition, such as air temperature, pressure, and aircraft speed. The system further has to protect the engine and the inlet duct from excessive transients that may be caused by rapid power-lever manipulation (also see Fig. A-55).

The main fuel control contains a hydromechanical computer which selects the desired engine speed by means of a series of three-dimensional cams (memory bank) which are positioned by various engine parameters. Some of this information is used to program the exhaust nozzle to maintain automatically the proper turbine pressure ratio. Percent augmentation is automatically related to the maximum allowable for the given flight condition by use of a series of three-dimensional cams in a separate hydromechanical computer. The computer also contains a number of hydromechanical logic circuits to sequence afterburner ignition automatically under various power-lever manipulations and to protect the engine in the event the afterburner flame blows out.

Aircraft-engine control must be integrated with other aircraft-instrumentation systems that are described elsewhere in this encyclopedia. Particularly, see **Air-data systems (aircraft)**, and **Flight-control systems.**

—Paul F. Bechberger, *Navigation and Control Division, The Bendix Corporation*

Aircraft Simulator (See **Simulator, aerospace-flight.**)

Alarms, Control-system An alarm is a protective device for maintaining critical points in a system under constant surveillance. In basic principles, alarm devices differ little from other forms of indicating and controlling instruments. A very high emphasis, of course, is placed on reliable performance. Such devices also must remain operable over long periods of inactivity because, aside from periodic, scheduled tests, the equipment does not function except in the case of an emergency. Most alarm installations are of the audiovisual type, with illuminated nameplates or bull's-eye lights with identifying nameplates. Upon actuation of an alarm point, the alarm light flashes and the horn or audio device sounds. A means for silencing the alarm and switching the light to a steady "on" state is provided. Another pushbutton usually is provided for testing other components of the system. The silencing and test device can be made common to a number of annunciator cabinets.

ALGOL (ALGebraically Oriented Language.) This is a compiler language (see **Compiler**) designed for ease of computation of algebraic expressions. The language is described in Ref. 1, which is a syntax description and the guide for publishing algorithms in the language. However, most implementations differ somewhat from the published standard, and, indeed, as very little provision for input/output is made in the basic language, there are considerable differences in how these functions have been implemented. The user must therefore study the manuals for a particular implementation in order to use it. Many books and manuals are available which are easier for the beginner to understand and follow than the report, for example, Ref. 2.

The following example illustrates some features of ALGOL. The COMMENT statements have no effect on the execution of the program.

```
   begin
 comment   This is a sample ALGOL program which il-
           lustrates some features of the language. Under-
           lined words are 'keywords' in the language:
 integer   A, B, C, EROR;
    real   D;
           EROR  := -1;
           A     :=  1.0;
           B     :=  2.0;
           C     :=  3.0;
 comment   The above statements declared the variable
           type for A, B, C, EROR and D and assigned
           values to A, B, C and EROR. Next D is cal-
           culated and some logic done testing its value;
       D   := A/B × C;
    if D   = 4 then go to over;
    if D   = 3 then go to right
           else go to Exit1;
   Over:   EROR  := 1.0;
           go to Exit15;
  Right:   EROR  := 0;
  Exit1:   end
```

REFERENCES

1. Naur, P. (ed.): Report on the Algorithmic Language Algol 60, *Commun. ACM.* vol. 3, pp. 299–314, 1960; *Numer. Math.*, vol. 2, pp. 106–136, 1960.
2. McCracken, D. D.: "A Guide to ALGOL Programming," John Wiley & Sons, Inc., New York, 1962.

—A. J. Arthur, *IBM Corporation*

Algorithm A detailed logical procedure representing a solution of a particular problem. This term is most commonly used to indicate an analysis procedure such as that used for the evaluation of a square root or for sorting a data file. Programming algorithms are published regularly by, for instance, the Association for Computing Machinery (ACM). For publication purposes, the ACM requires that the algorithms be written in ALGOL (see **ALGOL**).

—A. J. Arthur, *IBM Corporation*

Alkalinity, Measurement of (See **pH, measurement of.**)

Alphameric (See **Alphanumeric code.**)

Alphanumeric Code The set of symbols consisting of the alphabetic characters A through Z and the digits 0 through 9. The definition is sometimes extended to include some special characters such as \$, %, and &. It is quite common for a programming system to restrict the user-defined symbols to be only those using alphanumeric characters and for the system to take special action on the occurrence of a nonalphanumeric character. For example, cancellation of the job currently in progress may result when the \$ character is encountered.

—A. J. Arthur, *IBM Corporation*

Altimeter An instrument generally used in air navigation to indicate altitude above sea level or above ground level. Altimeters are of two major types: (1) the *barometric altimeter,* which is a pressure-measuring instrument designed to indicate altitude in feet from an adjustable datum to a standard scale; and (2) the *radar altimeter,* which is based on the precise measurement of the time required for electromagnetic energy to travel from an aircraft to the ground terrain and return. Automatic altitude reporting is covered under **Air-traffic control.**

Barometric altitude is used for aircraft separation rather than radar or radio altitude because the datum can be set to a smooth reference of mean sea level and because the measuring function from that datum is smooth and accurate as a relative measure of several aircraft occupying the same general airspace. In radar-type altitude, the datum is the ground, which is very rough except over water or such surfaces as aircraft landing strips. Radar altitude, therefore, is used for precision low-visibility landings and for terrain avoidance.

Barometric Altimeter: This instrument has developed into several forms from the original simple evacuated (aneroid) pressure capsule that is mechanically linked by a nonlinear-function mechanism to a pointer on the instrument dial. Relatively large capsules were used to drive geared displays having several pointers, drums, or counters to enhance the resolution or readability of the instrument. A barosetting mechanism geared to a counter made it possible to set the datum of the instrument to mean standard sea level or to compensate for local barometric conditions relative to meteorological conditions and airport elevation. More advanced instruments use the evacuated capsule basically as an electrical-signal generator rather than as a prime mover. Servomechanisms electrically follow or rebalance the function of the capsule with less force offset. The result is improved accuracy and a much greater capacity to drive output sources such as flight control, navigation, and automatic altitude-reporting systems. The use of solid-state components and integrated circuitry has practically eliminated moving parts and provides computer-compatible outputs. In all barometric altimeters, the measuring function is provided by a capsule (bellows) from which all air (down to 10 microns or less) is evacuated, after which the unit is sealed. The reference bellows is contained within another pressure-sensing element which is connected to the static-pressure ports external to the aircraft. Also see **Aneroid barometer.**

Radar Altimeter: The two most common designs are (1) the frequency-modulated continuous-wave (FM-CW) device; and (2) the pulsed-radar altimeter. The FM-CW altimeter measures altitude by comparing the transmitter frequency —which is shifted as a function of time in a controlled manner—with the frequency of the returned signal. The echo signal is heterodyned with a portion of the transmitter signal, thus producing a beat frequency. The amount that the transmitter changes frequency while the wave travels to the terrain and back is proportional to the rate of change of transmitter frequency multiplied by the time of travel. Therefore, the resulting beat frequency is a measure of altitude.

The pulsed-radar altimeter transmits discrete pulses of rf energy and computes altitude by measuring the time of the echo's return with respect to the time of transmission. The time difference is multiplied by a constant proportional to the velocity of propagation (in feet per second) which produces a direct measurement of absolute altitude. Radar altimeters are used primarily as altitude sensors in automatic landing systems, terrain mapping, terrain-following systems, low-level missile control, and other applications where precise fast-responding altitude information is required. The accuracy of a radar altimeter is not affected by barometric changes or aircraft forward velocities. The typical radar altimeter system consists of a receiver/transmitter unit for signal processing, an indicator for visual pilot display, and two antennas for simultaneous transmission and reception of rf energy. The altitude-measuring range is limited only by the transmitted power and system sensitivity available to produce an adequate return signal-to-noise ratio. For special applications, radar altimeters have been built for measuring to several hundred thousand feet, although the most common versions have a maximum measurement range of 5,000 ft or less.

Characterisitics of a typical state-of-the-art radar altimeter system are summarized: altitude range, 0 to 5,000 ft; altitude accuracy (analog), 2 ft + 2 percent of altitude, (digital), 2 ft + 0.2 percent of altitude; analog altitude signal (linear), 0 to −40 V dc; tracking rate, 2,000 ft/s; maneuverability (roll and pitch), ±35°; transmission frequency, 4.3 GHz; input-power consumption, 28 V dc, 35 W; mean time between failures, 5,000 hr.

—R. W. Armstrong and Paul H. Dannenberg, *Honeywell Inc.*

Ambient Conditions, Compensation for Ambient-condition compensation is the technique of offsetting the effects of changes in the environment on the input-output relationship of an instrument or other device. The environmental influences are caused by such conditions as temperature; pressure; supply voltage, frequency, and waveform; shock and vibration; and position. Compensation may operate on a principle that corrects the effects of environmental changes, which prevents the environment from changing, or which makes the instrument (or other equipment) insensitive to changes that do occur. The method may be either automatic or manual. It may be electrical, electronic, thermal, or mechanical. One or more of these means may be applied to correct or prevent the effects of these conditions on an instrument regardless of the instrument's principle of operation. In general, however, electronic, electrical, or thermal compensation is used in electrical and electronic instruments, and mechanical compensation in mechanical instruments.

—Robert L. Wilson, *Honeywell Inc.*

Ammeter The basic unit of electric current is the *absolute ampere,* derived from the centimeter, gram, and second. See **Standards, measurement**. The unit is defined in terms of the dimensions of a conductor and the strength of the magnetic field produced by the current flowing in it. Absolute measurements are made at the National Bureau of Standards with an instrument known as the Rayleigh current balance, which employs two horizontal fixed coils with a movable coil placed halfway between them and suspended from one arm of a gravity balance.

Direct-current Measurement: Three fundamental methods are used: (1) the coulometer (see **Coulometer**); (2) the potentiometer; and (3) the ammeter, usually the permanent-magnet moving-coil type.

Alternating-current Measurement: Commercially available ac instruments (ammeters) are of the following main types: (1) permanent-magnet moving-coil type when used in conjunction with a rectifier; (2) iron-vane type; and (3) electrodynamometer type—all described under **Electrical quantities, measurement of.** Although simple, low-cost, and less accurate ammeters of the hot-wire type were once made—in which expansion of a wire in the instrument through which the current is passed causes a pointer to move—these devices now are obsolete because of their inadequate performance.

Panel and Switchboard Meters: DC ammeters are obtainable with ranges from 0 to 1—to 0 to 500 A; milliammeters from 0 to 1—to 0 to 500 mA; and microammeters from 0 to 5—to 0 to 500 μA. The ranges for ac ammeters, milliammeters, and microammeters are roughly similar. The meters are obtainable in both circular- and rectangular-front formats. Scale lengths from 1½ to 7½ in. are available in stock models. Panel meters wherein the scale (PMS = projected moving scale) is optically projected; or wherein the pointer (PMP = projected moving pointer) is optically projected are obtainable. The accuracy of stock-panel ammeters ranges from ±2 to ±3 percent of full scale. Ammeters also are available in portable form for testing and laboratory use where the accuracy generally will range from ±1 to ±2 percent of full scale, although special, higher-cost instruments are available with greater accuracy. Ammeter ranges also are obtainable in multimeter form where voltage and resistance as well as current may be measured. Most multimeters are portable. Recording ammeters also are available. Shunts and instrument transformers frequently are used in connection with ammeters to stretch their range. (See **Meter, electrical.**)

High-frequency Alternating Currents: The well-designed dynamometer-type instrument is satisfactory for audio-frequency measurements, while thermocouple ammeters can be used for all frequencies from power frequencies up to and including radio frequencies. Through careful design, the errors of a thermocouple ammeter due to frequency can be kept smaller than 1 percent for up to as high as 50 MHz.

Portable and Laboratory Instruments: For applications requiring that an instrument be carried to the place of measurement, there are portable instruments. These generally are of higher accuracy than panel and switchboard instruments so that the accuracy of the latter can be checked on location. While very few electrodynamic instruments are found in panel-instrument form (for reasons of cost), they are quite frequently encountered in portable form where their greater accuracy and flexibility of use justify the higher cost.

Laboratory instruments usually are of the permanent-magnet moving-coil type or the electrodynamic type, both in the interest of highest achievable accuracy. Iron-vane-type instruments are not in the same accuracy class because of inherent limitations of their operating principle.

—Henry Berring, *Weston Instruments, Inc.*

Ampere, Standard (See **Standards, measurement.**)

Amperometric Oxygen Analyzer (See **Oxygen analyzer.**)

Amperometry An instrumental method for the analysis of electroreducible or oxidizable ions, molecules, or dissolved gases in solution, including most metal ions and organic materials containing oxidizable or reducible groups. Range of concentration is from 0.01 to 1,000 ppm. The method is useful for determination of free available or total available chlorine, especially for water-chlorination control where the range of concentration is from 0 to 50 ppm Cl_2. The method also responds to free iodine. Identification of unknowns is not determined except by selection of fixed potential between polarized microelectrode and reference electrode to cause oxidation or reduction of the particular component sought. Concentration of substances is proportional to diffusion-limited current that flows in the electrode circuit—as in polarography.

Amplifier An amplifier is a device in which an input signal controls a source of power and is therefore capable of providing an output which is a function of the input signal. In the case of a linear amplifying device, the output is a reproduction of the input signal. Nonlinear amplifiers produce an output controlled by the input but not necessarily proportional to the input signal.

Amplifiers may be mechanical, electromechanical, electronic, or hydraulic (including pneumatic). In instrumentation systems, most amplifiers are electronic, although pneumatic and hydraulic amplifiers are becoming more common with the increased interest in the use of "fluidics" in the instrumentation industry. Electronic amplifiers may be constructed of vacuum tubes, solid-state devices, or magnetic devices. Most commercially available instrumentation amplifiers use either transistors or magnetic devices, although vacuum tubes are still used in some special applications.

Within the general category of amplifiers, there are a number of subclassifications which relate to the particular configuration of components which is used in constructing the amplifier. These different configurations generally result in unique input/output characteristics or signal-handling capabilities. Examples of these subclassifications include bridge amplifiers, carrier amplifiers, chopper-stabilized amplifiers, comparators, differential amplifiers, floating or isolated amplifiers, sample-and-hold amplifiers, and single-ended amplifiers. Each of these amplifier configurations is discussed in more detail elsewhere.

—Thomas J. Harrison, *IBM Corporation*

Amplifier, Bridge A bridge-amplifier configuration utilizing four subamplifiers is shown in Fig. A-56. With the usual assumption that the open-loop gains G_1, G_2, G_3, and G_4 of the individual amplifiers are very large, the expression for the output voltage is

$$V_0 = \frac{R_1 + R_2}{R_1} V_1 - \frac{R_2}{R_2'} \frac{R_1' + R_2'}{R_1} V_2$$

Voltage V_2 is the applied common-mode voltage V_{cm} and V_1 is the sum of the differential voltage $V_{sig} = V_1 - V_2$ and the common-mode voltage (see **Differential-mode voltage; Common-mode voltage**). With these substitutions, the output voltage is

$$V_0 = \frac{R_1 + R_2}{R_1} V_{sig} + \left(1 - \frac{R_2 R_1'}{R_2' R_1}\right) V_{cm}$$

Thus, the closed-loop gain of the amplifier is $(R_1 + R_2)/R_1$ and the common-mode rejection ratio is $|G/[1 - (R_2R_1')/(R_1R_2')]|$. [See **Common-mode rejection ratio (CMRR)**.] Theoretically, infinite

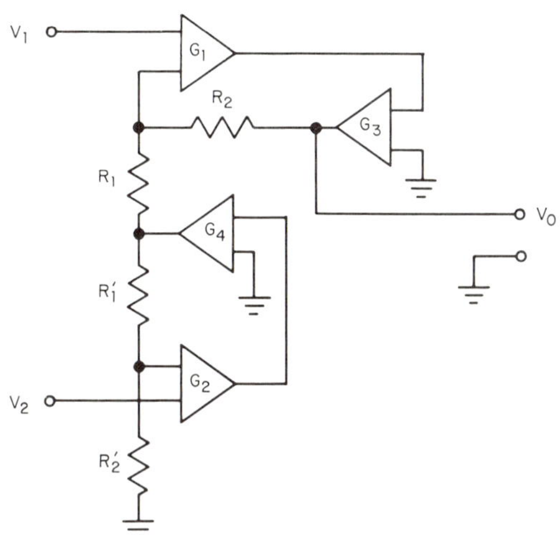

Fig. A-56. Dynamic bridge amplifier.

common-mode rejection can be obtained if $R_1'/R_1 = R_2'/R_2$, which is the condition for a balanced resistive bridge.

Because of the limitations in matching the resistors and errors which are not revealed by this simplified analysis, the actual common-mode performance is finite, but values in excess of 120 db can be readily achieved. A unique feature of this amplifier is that the common-mode rejection ratio (CMRR) is directly proportional to the gain. In most differential amplifiers, the CMRR is essentially independent of the gain.

Although not extensively used in digital-data-acquisition systems, the bridge amplifier is commercially available for instrumentation purposes. It is generally a direct-coupled amplifer (see **Amplifier, dc**) which offers moderately wide bandwidths (up to 50 kHz) at gains from near unity to 1,000.

—Thomas J. Harrison, *IBM Corporation*

Amplifier, Carrier A carrier amplifier is generally an amplifier with a bandwidth extending from zero to some finite frequency (i.e., it is a dc amplifier) in which the signal is first modulated and then demodulated during the amplification process. (See also **Amplifier, chopper.**) In most cases, the modulation consists of "chopping" the signal using either electromechanical or electronic switches. The chopping action is equivalent to square-wave modulation of the signal.

There are two primary reasons for using the carrier technique. The first of these is to minimize the effects of zero-offset drift, a critical performance parameter in any dc amplifier. A conceptual illustration of a carrier amplifier designed primarily for this purpose is shown in Fig. A-57. The input signal is modulated or chopped to provide an ac signal. This signal is amplified by an ac amplifier whose output is then demodulated to provide a dc output signal. Since only the ac component is amplified, zero-offset drift in the amplification section of the amplifier does not affect the value of the output signal. Offsets in the modulator, however, can produce the equivalent of an offset in the output signal if they increase or decrease the magnitude of both the positive and negative peaks of the modulated signal. If the input signal is high-level (greater than about 1 V), these offsets are not a serious problem in most cases. In low-level amplifiers, however, they can be a source of significant error. Since the output demodulator is generally operating at

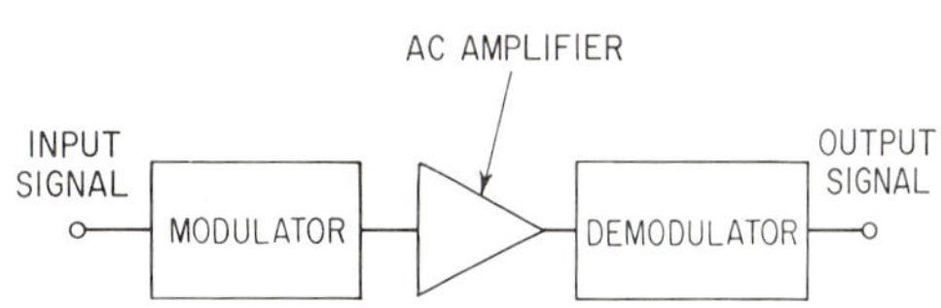

Fig. A-57. Carrier amplifier.

a high level, demodulator offset does not constitute a severe limitation on the overall amplifier performance. Carrier amplifiers designed primarily for the reduction of zero-offset drift are not used as extensively at present as they have been in the past. This is primarily due to improved components and techniques for implementing low-drift direct-coupled amplifiers. (See **Amplifier, dc.**)

The second primary reason for utilizing a carrier technique is to provide isolation between the input and the output of the amplifier. In this case, the amplifier is usually called a "floating amplifier." (See **Amplifier, floating.**) An example of such an amplifier which incorporates an overall feedback path is shown in Fig. A-58. The same basic carrier concept is employed here in that the input signal is modulated and demodulated by chopper circuit. In the case shown, an ac amplifier is not utilized, however, since the primary purpose is one of isolation rather than drift reduction. The four-terminal isolation property of the transformer allows the input signal to be referenced to a ground point which is independent of the output-signal reference point. Floating-carrier amplifiers are used in instrumentation and digital-data-acquisition subsystems to provide amplification of signals in the presence of high common-mode voltages. (See **Common-mode voltage.**) Since the common-mode voltage is primarily limited by the breakdown voltage of the coupling transformer, amplifiers of this type can operate with up to several hundred volts of common mode, compared with the 10- to 20-V limit imposed on most direct-coupled amplifiers by the breakdown limitations of most semiconductor devices. The common-mode voltage capability of the floating-carrier amplifier is probably its most outstanding characteristic.

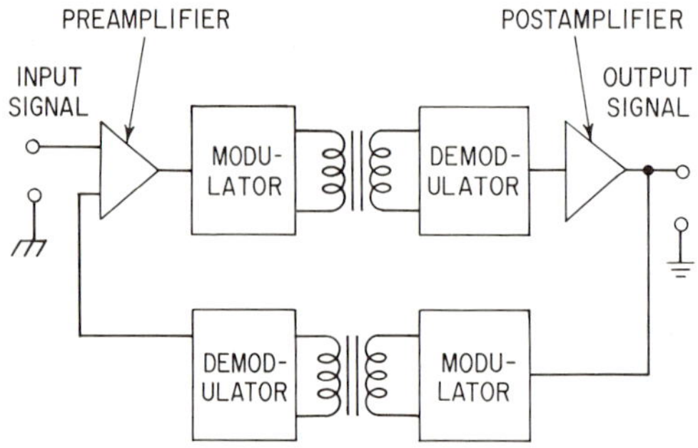

Fig. A-58. Floating-carrier amplifier.

—Thomas J. Harrison, *IBM Corporation*

Amplifier, Chopper In common usage, this term describes one of two types of amplifiers. The first (described separately in the entry **Amplifier, carrier**) refers to an amplifier in which the input signal is modulated or chopped, amplified by an ac amplifier, demodulated, and filtered to obtain a dc output signal.

In the second meaning of the term, it refers to an amplifier in which an error signal is chopped to provide stabilization of gain and offset. This type of amplifier is more properly described by the term "chopper-stabilized amplifier." A schematic representation of a chopper-stabilized amplifier having an overall gain A is shown in Fig. A-59. The output of the amplifier is attenuated by a factor of $1/A$ and compared with the input signal. If an error exists because of a change in the gain of the dc amplifier, zero-offset drift, or noise, the error is modulated (chopped), amplified by an ac amplifier, demodulated, and summed with the input to the dc amplifier in such a way as to compensate for the disturbance causing the error.

The chopper-stabilized amplifier provides the favorable wide-bandwidth characteristics of the direct-coupled amplifier together with excellent zero offset and gain characteristics. It is not a widely used technique in instrumentation and digital-data-acquisition amplifiers at the present time, however, because comparable performance can now be obtained with conventional direct-coupled amplifiers without the relatively complex modulator-demodulator circuitry. In addition, the chopper-stabilized amplifier often exhibits a

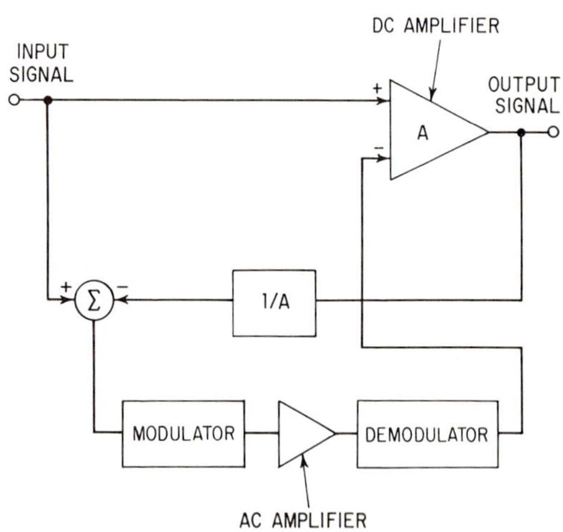

Fig. A-59. Chopper-stabilized amplifier.

long saturation recovery time, which is detrimental in time-shared systems.

—Thomas J. Harrison, *IBM Corporation*

Amplifier, Comparator A comparator amplifier is a nonlinear amplifier used to sense either the polarity or the magnitude of an input signal. The output of the amplifier is one of two states, generally used to represent the binary states 1 and 0. If the input signal exceeds a predetermined level (often zero), the output of the amplifier is in one state; if the input signal is less than the predetermined level, it remains in the other state. When the predetermined level is other than zero, the comparator amplifier is sometimes called a threshold detector (see **Detector, threshold**).

The transfer characteristics of an ideal and a practical comparator are shown in Fig. A-60. The heavy trace indicates the ideal characteristic; the amplifier output is, in this case, ±6 V if the input signal is greater than zero, and it is zero if the input is less than zero. In a practical case, a linear region of operation exists between these two output levels corresponding to a small "uncertainty" in the operation of the comparator. Thus, the input level must exceed a threshold level V_t to ensure that the output reaches the +6-V level. This threshold voltage can be made arbitrarily small by increasing the gain of the amplifier, within the limitations imposed by amplifier-stability considerations. Typical comparator circuits have gains exceeding 10,000.

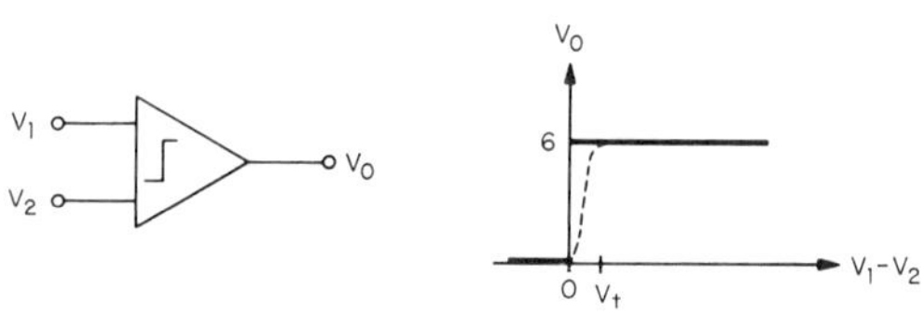

Fig. A-60. Comparator transfer characteristics.

Physically, the comparator is merely a high-gain differential or single-ended amplifier. In most applications, it is designed for fast recovery from saturation to allow it to follow the state of a rapidly varying input signal. In precision applications, it must also be designed to have low zero offset since zero offset is equivalent to a shift in the threshold level. Monolithic integrated amplifiers specially designed for use as comparators are commercially available.

—Thomas J. Harrison, *IBM Corporation*

Amplifier, Differential In general, a differential amplifier is any amplifier which amplifies a signal between two input terminals, neither of which is connected to the signal reference point (ground) (see **Differential-mode voltage**). This amplification is essentially independent of the potential which exists between each of the input terminals and the signal reference point. In practice, the potential between the input terminals and the signal reference point, known as the common-mode voltage, may cause an error whose magnitude depends on the characteristics of the amplifier (see **Common-mode voltage; Error, common-mode**).

In terms of this definition, a number of amplifier configurations commonly used in instrumentation and digital-data-acquisition systems are differential amplifiers. This includes configurations such as the bridge amplifier, the carrier amplifier, the floating amplifier, and in some cases, the chopper-stabilized amplifier. Each of these amplifier types is discussed in a separate entry.

The term differential amplifier usually refers to an amplifier whose input circuit is designed to amplify differential voltages, and this is the sense of the definition given above. It is also possible, however, to have an amplifier which has a differential output with or without a differential input. This is less common, however, and when referring to such an amplifier a term such as "differential in—differential out" is sometimes used. If this type of terminology is not used, it can usually be assumed that the term differential amplifier refers to an amplifier with a differential input circuit and a single-ended output circuit.

—Thomas J. Harrison, *IBM Corporation*

Amplifier, DC The term dc amplifier has two related meanings when applied to amplifiers used for instrumentation and digital data acquisition. It can mean either "direct-coupled amplifier" (implying that the connection between each stage and the succeeding stage in the amplifier is made via a low-resistance connection) or an amplifier whose frequency response extends to zero frequency (dc).

Although all direct-coupled amplifiers have a frequency response extending to dc, not all amplifiers with this type of frequency response are direct-coupled. For example, the carrier amplifier (see **Amplifier, carrier**) has a response extending to dc but it is not direct-coupled.

An example of a direct-coupled differential amplifier is shown in Fig. A-61. Since the most important amplifier characteristics are usually determined by the first stage of the amplifier, it is the only stage which is shown explicitly. The remaining stages (usually two or three) are lumped together and shown as a single symbol. The characteristic of these stages is that they provide high gain, typically greater than 1,000, and the output-drive capability of the amplifier. In most cases, the conversion from differential input to single-ended output is accomplished in these stages as well.

The input stage of this direct-coupled differential amplifier consists of a pair of transistors whose bases are biased by current sources. The emitter current of the transistor pair is also determined by a current source. The gain of the stage is determined by the feedback resistor R_{f1} and the emitter resistor R_e. The differential-output signal generated at the collectors of the first-stage transistors is directly coupled to the bases of the transistors in the second stage.

Typically, the first-stage gain is on the order of 100, so that the first stage largely determines the drift and noise characteristics of the overall amplifier. Because of the first-stage configuration, the drift characteristics are determined by the difference in the drifts of the two input transistors (see **Drift**; **Noise**). In most cases, the major contribution to zero-offset drift and zero-offset temperature coefficient is the base-emitter voltage drop. By use of matched transistors in this differential configuration, the overall amplifier drift can be reduced to only a few microvolts per degree Celsius. By using additional temperature compensation in the first stage, a drift of 1 μV/° C or less can be obtained. The noise characteristics are primarily related to the intrinsic noise of the input transistors. Although this can be controlled somewhat by selection of the biasing point, the usual procedure is to select low-noise transistors, since biasing is often dictated by other considerations.

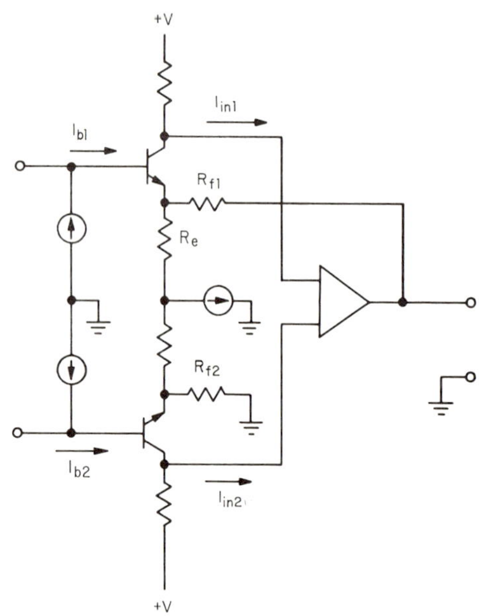

Fig. A-61. Direct-coupled differential-amplifier configuration.

The common-mode characteristics of the amplifier are also affected by the design of the first stage (see **Error, common-mode**). In general, every attempt should be made to maintain the configuration symmetrical with respect to ground. This is the primary reason for including the dummy feedback resistor R_{f2}. In addition, the input base-bias sources should be balanced. High-frequency common-mode characteristics are dependent on the balance between the RC constants of the base-to-collector capacitance and the collector resistance. In critical designs, an adjustment of one of these parameters may be necessary. Common-mode feedback between the first-stage emitters and a common-mode point in a subsequent stage can also be used to enhance the common-mode performance of the amplifier. This feedback is often injected at the ground side of the emitter-bias source.

The direct-coupled differential amplifier is widely used in instrumentation and digital-data-acquisition systems. Because of the direct-coupled design, it is capable of wide bandwidth, typically 0 Hz to frequencies in excess of 50 kHz at relatively high gains. In addition, it is relatively inexpensive when compared with some other forms of differential amplifiers. Its primary disadvantage in some applications is that the common-mode voltage capability is relatively low. Because of the breakdown characteristics of the input transistors, the common-mode voltage is generally limited to 10 to 20 V.

—Thomas J. Harrison, *IBM Corporation*

Amplifier, Floating A floating or isolated amplifier is an amplifier in which the input and output signals need not be referred to the same signal reference point (ground). Although differential-input amplifiers generally satisfy this criterion, the term "floating" generally excludes any amplifier in which the input reference point and the output reference point are common through either a signal conductor or a power supply. Thus, the term specifically applies to amplifiers which include a four-terminal coupling device such as a transformer or light-coupled signal-transmission element.

The common floating amplifier used in instrumentation and digital-data-acquisition subsystems is the carrier amplifier (see **Amplifier, carrier**). In this configuration, the input signal is converted to an ac signal by means of a modulator. The modulated signal is coupled to the output amplifier, demodulated, and filtered to produce an output signal. Feedback from the output to the input may be provided by a similar modulator-demodulator combination as shown in Fig. A-62.

Because of the four-terminal isolation property of the transformer, the amplifier input and output sections are completely isolated in the sense that the signal reference point (ground) for the input can be completely independent of the output-signal reference point. This implies that the amplifier acts as a differential amplifier even though the preamplifier and postamplifier may be single-ended. The maximum difference in the ground

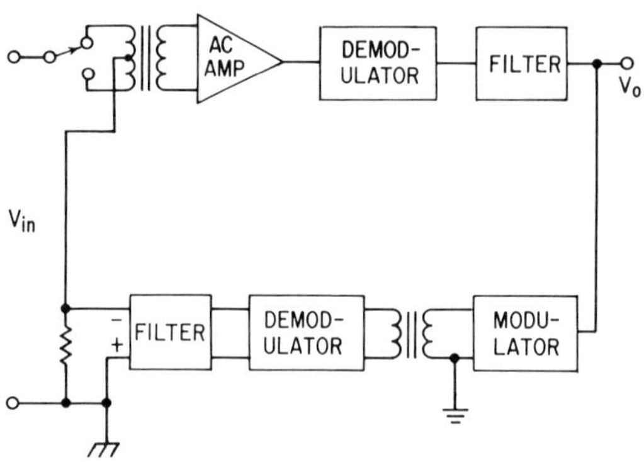

Fig. A-62. Floating-carrier amplifier.

potential between the input and output circuits is generally a function of the breakdown-voltage limitation of the transformer. Since this is usually a relatively high voltage (hundreds of volts), this type of amplifier is well suited to applications requiring the amplification of signals in the presence of high common-mode voltages (see **Common-mode voltage**). A number of commercially available instrumentation and data-acquisition amplifiers use this technique.

Because of the unbalanced nature of the preamplifier and unbalances in the coupling transformer, maintaining a high common-mode rejection ratio is a significant design problem [see **Common-mode rejection ratio** (CMRR)]. Specially shielded transformers having extremely low interwinding capacitances are typically used for this application. In addition, the guard-shielding technique is usually required if optimum performance is to be attained (see **Guard shield**).

The isolated carrier amplifier is relatively expensive compared with other differential-amplifier techniques. Its primary advantage is its input/output isolation. Its major disadvantage is a relatively narrow bandwidth (usually less than 10 kHz) imposed by carrier-frequency limitations and by the characteristics of the transformer. In addition, the ac common-mode rejection ratio is often seriously degraded if the guard shield is not properly connected to the source of common-mode voltage. Thus, its use is restricted to three-wire systems if optimum performance is to be realized.

—Thomas J. Harrison, *IBM Corporation*

Amplifier, Sample-and-Hold The sample-and-hold, or track-and-hold, amplifier is an amplifier in which the output is proportional to the input until a "hold" signal is given. At that time, the output remains essentially constant even though the input signal may be changing. This action is illustrated in Fig. A-63, which shows the input and output waveforms for such an amplifier.

Conceptually, a sample-and-hold amplifier consists of two independent amplifiers connected by a switch as shown in Fig. A-64. When an input is applied to the first amplifier with the sampling

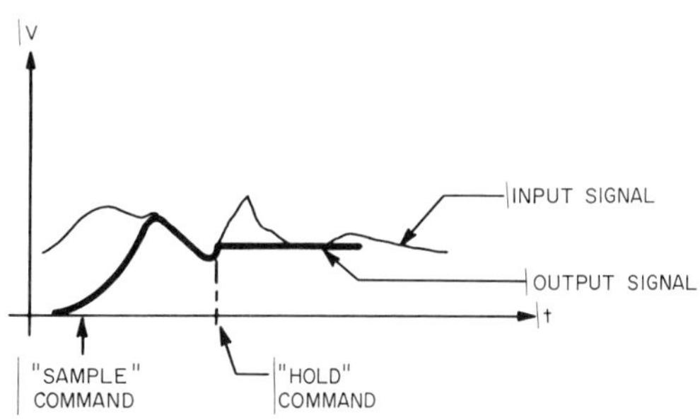

Fig. A-63. Sample-and-hold amplifier action.

switch S_1 closed, the amplifier charges the holding capacitor C. When the "hold" command is given, switch S_1 is opened, leaving the capacitor C charged to the instantaneous value of the input signal. Since the second amplifier is designed to have a high input impedance, capacitor C is not discharged and the output of the second amplifier remains essentially constant for some period of time. Depending on the application, the "hold" command may be generated by a digital control unit or computer under control of a stored or hardware program, or it may be generated by an external circuit coupled to the process or experiment being monitored.

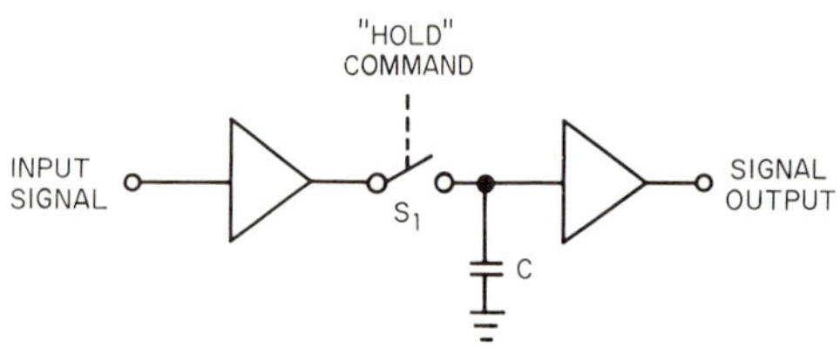

Fig. A-64. Sample-and-hold amplifier circuit.

Sample-and-hold amplifiers are generally used in specialized digital-data-acquisition applications. Their use is dictated by one, or both, of the following requirements: Either the value of a single signal must be determined at a precise moment of time and/or the value of two signals must be compared at a precise moment of time. Neither of these requirements can be fulfilled with a time-shared amplifier and analog-to-digital converter system because of the finite settling and conversion time required by these systems (see **Analog input**). If a sample-and-hold amplifier is used, the instantaneous value of one or more input signals can be retained for the time interval necessary to convert them to digital values.

—Thomas J. Harrison, *IBM Corporation*

Amplifier, Single-ended A single-ended amplifier is an amplifier designed to amplify signals applied between a single input terminal and the ground or signal reference point (compare **Amplifier, differential**). In most cases, the output signal is generated between a single output terminal and the same signal reference point. Although differential amplifiers are often preferred for use in instrumentation systems, single-ended amplifiers are used extensively in some applications.

The primary disadvantages of the single-ended amplifier relate to installation problems caused by the common connection between the input and output circuits. Any potential existing between the amplifier signal reference point (ground) and the reference point for the input signal appears in series with the input signal. As such, it is amplified with the signal and is indistinguishable from the true signal voltage. Since a potential between two signal reference points is a common-mode voltage, this means that the common-mode rejection of the single-ended amplifier is negligible. [See **Common-mode voltage; Common-mode rejection ratio (CMRR).**]

If single-ended amplifiers are used in instrumentation or digital-data-acquisition systems, detailed attention must be paid to the grounding circuits. The primary requirement is that the signal reference point for the input signal must be at precisely the same potential as the signal reference point for the amplifier input. This requires a low-impedance connection between these reference points at all frequencies of interest. This is difficult to guarantee if the signal source and the amplifier are physically separated by an appreciable distance. For this reason, single-ended amplifiers are not commonly used in systems where the signal source is far removed from the amplifier input. They are, however, used in physically small subsystems where the ground structure is well defined and controlled.

—Thomas J. Harrison, *IBM Corporation*

Amplier, Track-and-Hold (See **Amplifier, sample-and-hold.**)

Amplitude Modulation Modulation in which the amplitude of the carrier-frequency current is varied above and below its normal value in accordance with the instrumentation, audio, picture, or other intelligence signal to be transmitted.

Analog Computer A computer that solves problems by setting up equivalent electric circuits and making measurement as the variables are changed in accordance with the corresponding physical phenomena. An analog computer gives approximate solutions, whereas a digital computer gives exact solutions. However, exactness as used here is not to be confused with accuracy or correctness. With any computer, the output can be no better than the input—a condition which has given rise to the expression, "garbage in—garbage out!"

Mechanical and other nonelectrical analogs may be used as the basis of an analog computer. An analog is the representation of numerical quantities by physical variables—electrical or nonelectrical. Very simple forms of analog computers include the familiar engineers' slide rule and the odometer of an automobile. Analog computers of various forms are used in many scientific and industrial instruments.

REFERENCES

Fifer, S.: "Analogue Computation," 4 vols., McGraw-Hill Book Company, New York, 1961.

Korn, G., and T. Korn: "Electronic Analog Computers," 2d ed., McGraw-Hill Book Company, New York, 1956.

Karplus, Walter J.: "Analog Simulation," McGraw-Hill Book Company, New York, 1958.

Jackson, A. S.: "Analog Computation," McGraw-Hill Book Company, New York, 1960.

Smith, G. W., and Wood: "Principles of Analog Computation," McGraw-Hill Book Company, New York, 1959.

Rogers A. E., and T. W. Connolly: "Analog Computation in Engineering Design," McGraw-Hill Book Company, New York, 1960.

Eterman, I. E.: "Analogue Computers," Pergamon Press, New York, 1960.

Simpkin, L. J., and L. F. Wooden: Use of Analog Computers in Investigations of Power Plant Automatic Control Systems, *Proc. 9th Natl. Power Instrumentation Symposium, ISA Publ.* PID-66, 1966.

Scrivens, Donald B.: Special Purpose Analog Computers as Applied to the Pulp and Paper Industry, *Proc. 7th Natl. Pulp and Paper Instrumentation Symposium, ISA Publ.* PP-66, 1966.

Shackcloth, B., and R. L. Butchart: Synthesis of Model Reference Adaptive Systems, Symposium on Theory of Self-adaptive Control Systems, *ISA Publ.* SACS-66, 1966.

Analog Computer, Process Analog computers, both general- and special-purpose, offer convenient solutions to process-control problems. General-purpose analog computers have flexibility to serve in the analysis and evaluation phase of varied control and measurement jobs. Once the analog configuration is selected and proved out, a permanent, low-cost, special-purpose computer can be built or purchased to carry out continual on-line calculations. Analog computers are particularly useful for material balance, ratio control, and heat-balance problems, as pointed out by the following examples.

Finding Batch Freeze Point and End Point: Sometimes it is easier and more practical to compute a variable than to measure it directly. In this example, the analog computer programs the set point of a temperature controller to follow the batch freeze point. The computer calculates the freeze point from the flow rate of reactant added to the batch.

Earlier, an analog simulation revealed that an exothermic semibatch reactor required adaptive temperature control for proper operation. In this process (see Fig. A-65), gas A is added to liquid B; the mixture reacts to make liquid product C and liquid by-product D. The kinetic coefficient k defines the reaction rate: $A + B^k \rightarrow C + D$. The simulation study dictated reaction at relative low temperatures and operation near the batch freeze point for maximum production of C. During the batch, the freeze point progressively increases with changing chemical composition. So the temperature set point must follow this freeze point for optimum yield. An indication of batch end point also was desirable. In addition, normal sampling for a check on the freeze point put the operator close to hazardous materials. Thus, operator safety was a consideration in determining sample periods.

The subsequent proposal to install a small analog computer was supported by the following: development of an automatic temperature-control method and reduction or elimination of sampling. Three possible control approaches were considered: (1) feed back a signal from a sensor for freeze point or for some process variable that is a function of freeze point; (2) feed forward a signal representing a running total of reactant A after correlation with the freeze point; or (3) feed forward the A flow signal to a model having the kinetic coefficients and initial composition conditions; this material-balance model would produce a continuously updated composition that would be a function of freeze point.

The feedback method proved difficult; direct measure of freeze point had been tried and remained unreliable. Correlations between output from a radiation device for density and freeze point also provided unsatisfactory results, ruling out this particular indirect measuring technique. These considerations left the material-balance method.

A good assumption in this case is that reactant A immediately and completely reacts; the instantaneous total of reactant A added to the batch corresponds to the batch's composition and its associated freeze point. Since A is a gas, compensation for pressure P and temperature T is necessary to get a mass-flow value F_A from orifice pressure drop ΔP. $F_A = C_0\sqrt{\Delta P(P/T)}$.

Integration of F_A gives a running indication of A added during reaction time. This integral, when summed with the A in the original charge, gives an instantaneous total of reactant A in the batch. This total, representing instantaneous

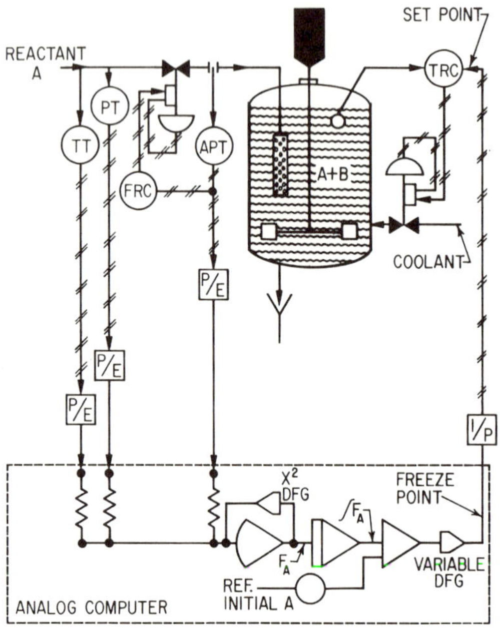

Fig. A-65. Analog computer calculates the mass-flow rate of gas *A*, and then integrates and scales this value to represent the batch freeze point. After conversion to a pneumatic signal, the freeze point serves as a set point to the temperature controller.

batch composition, is converted by a variable diode function generator to correspond to batch freeze point.

The freeze point in turn, after conversion to a pneumatic signal, serves as the set point of the reactor temperature controller. Freeze point also indicates degree of reaction, making it useful for finding the batch end point. Thus freeze-point computation serves two functions: temperature-controller set point and end-point indication.

One flaw in the system turned out to be wildly fluctuating feedstock composition, invalidating the integrated total of A. Analytical data on feed composition from the laboratory arrived too late to be useful. To solve this problem, the plant operator checks the freeze point manually and uses this information to update computer output. After one manual check, the computer is remarkably accurate, following the batch to completion within about 2° C of the freeze point without further correction. The feedforward systems gives increased production from less raw material for each batch, resulting in less undesirable by-product. Also, operation is safer because chemical sampling is minimized.

Controlling Multiple Flow Ratios: Ratio computation and control offers numerous opportunities for special-purpose analog computers. The system described here is successfully maintaining a key process ratio important to product quality and operating cost.

Simple stated, the particular problem in this example is to add up four flow rates, divide by a fifth, then close a control loop on one of the four to maintain the ratio set point. The calculation is simple and well within the capabilities of an operator. But an analog computer dedicated to this job offers reliable around-the-clock operation and relieves the operator of a time-consuming, repetitious duty.

The normal procedure for attacking the problem was followed: analysis of the problem, implementation of temporary evaluation hardware, evaluation of performance, and implementation of permanent hardware.

The analysis phase showed the need for linearizing the flow signals from the ΔP transducers, by taking the square root of these signals. After this operation, the computer adds flows F_1, F_2, F_3, and F_4, and takes the logarithm of the sum (see Fig. A-66). Another amplifier on the fifth flow signal F_5^2 feeds a diode function generator to find half its logarithm (square root). The last amplifier subtracts this last logarithmic signal from the logarithm of the summed flows and takes the antilog of this difference, resulting in a computer output that corresponds to

$$\frac{F_1 + F_2 + F_3 + F_4}{F_5}$$

Output currents of the electronic ΔP transmitters feed appropriate resistors to ground, producing voltages on which the computing elements operate. Computer output is converted back to current, calling for a power-boost module capable of delivering 50 mA for remote set of a controller on one of the flows to maintain the desired ratio. Evaluation of this concept for ratio computation

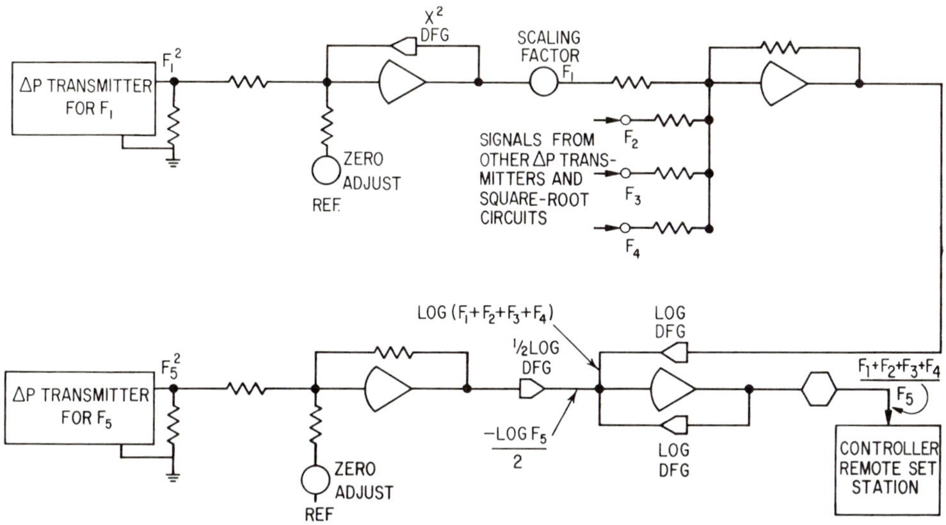

Fig. A-66. Analog computer linearizes all flow signals, sums four of the flow rates, and divides by a fifth flow rate.

and control proceeded to a successful conclusion without any notable upsets.

A static-test system aids checkout convenience. This static-test mode replaces all the process inputs with known constant voltages through relays. Then the operator can simply check output voltage of the analog-computer components. An operator panel that consists of a digital voltmeter and a selector-switch array further eases the static test. Thus the plant operator merely pushes buttons to check out the analog-computing system.

Computing Batch Heat Balance: In the study of chemical reactions, the analog computer can relieve the research chemist of time-consuming and inherently difficult measurements. In this example, the researcher desired accurate data for reaction conversion vs. time. Traditionally, the researcher had studied reaction rates by loading bottles or small reactors, reacting for a given interval, killing the reaction, and venting the system. Next, he treated the product formed by filtering, washing, and drying.

Two main problems affect the accuracy of this method for finding conversion-vs.-time data. Measurement of the time of reaction is difficult because the researcher cannot always kill the reaction completely. Also, the researcher has a difficult time maintaining catalyst uniformity from batch to batch. Sampling presents additional problems. It is difficult to get uniform samples from a pressurized, multiphase system during the run. Having obtained the sample, the researcher must inhibit its reaction quickly. Chemists tried several sampling devices but abandoned them because of inadequate results. In this example, a high reaction exotherm suggested an on-line computation of conversion based upon system heat balance.

During initial studies under manual operation, an isothermal reaction, controlled by a flow of cooling water, gave encouraging results. In these studies, flow and temperature differences were recorded; hand calculations gave rates and conversions.

Mechanization of these ideas amounted to changing the reactor into a well-instrumented calorimeter (see Fig. A-67). Controller TC1 fixes the temperature of the coolant water to the reactor. TC3 controls the bath-water temperature that surrounds the reactor. A highly sensitive thermistor serves as the sensor for the reactor mass. This thermistor feeds TC2, controlling reactor temperature by varying coolant flow. The set-point temperature of the water bath is slightly higher than that of the reactor, adding a constant-bias heat to the cooling water. The coolant water primarily soaks up the heat generated by the reaction's exotherm. This heat Q may be represented by the formula

$$Q = \int Fc_p \, \Delta T \, dt$$

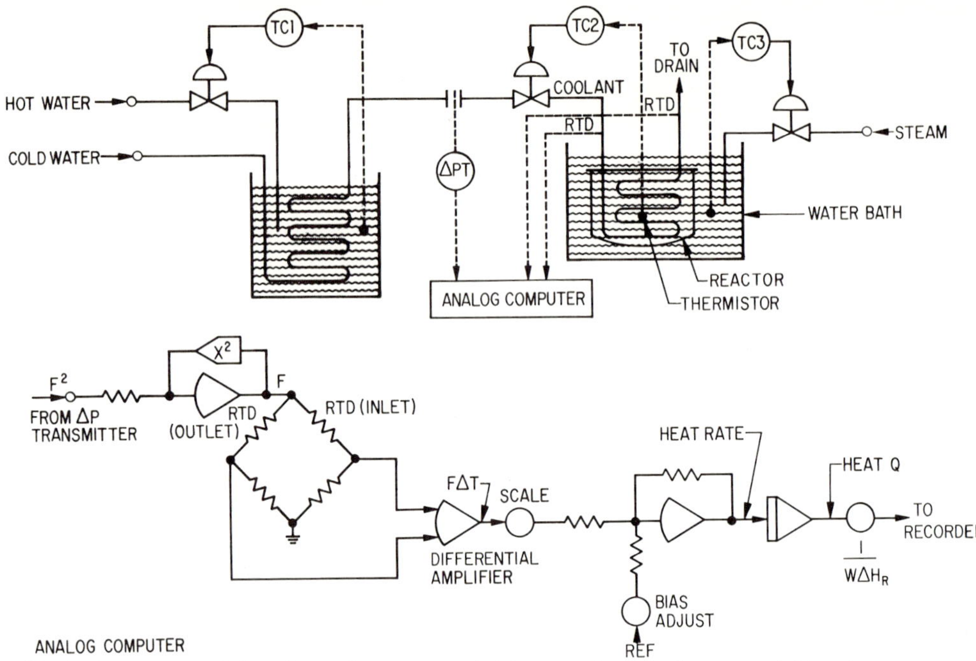

Fig. A-67. Analog computer finds and sums instantaneous batch-conversion rates in a research laboratory.

where F = coolant flow rate
c_p = coolant specific heat
ΔT = difference in coolant temperature at inlet and outlet of reactor

The analog computer accepts signals for coolant flow and inlet and outlet coolant temperatures. An electronic ΔP transmitter senses differential pressure across an orifice to give coolant flow. A pair of resistance thermometers sense the coolant's inlet and outlet temperature differences. The computer circuit first takes the square root of the ΔP signal. The linearized flow signal serves to power a resistance bridge that contains the resistance thermometers for inlet and outlet reactor coolant temperature differences. Bridge output represents the product of flow and the difference between the resistance-thermometer outputs. This product then is scaled to give instantaneous heat rate, which in turn is fed to an integrator that accumulates total heat. This result feeds a potentiometer having a setting that corresponds to the reciprocal of the product of weight charged and heat of reaction (enthalpy). Potentiometer output then represents cumulative fractional conversion. The reactants must be charged quickly with good mixing for accurate results.

Note: Adapted from paper given at 23d Annual ISA Conference, New York. With approval of Instrument Society of America.

—C. E. Fahlgren, *The Dow Chemical Co.*

Analog Electrical Instrument (See **Electrical quantities, measurement of.**)

Analog Input Analog input is a generic term used to describe the equipment which is attached to a digital computer for the collection of analog data. Taken collectively, this equipment forms a subsystem which performs selection of the analog signal to be sampled, modification or conditioning of the analog signal, analog-to-digital conversion to provide a digital representation, and the digital controls necessary for subsystem control and communication with the digital computer. The pieces of equipment generally included in the analog input are signal conditioners such as attenuators and filters, analog signal multiplexers, amplifiers, analog-to-digital converters (ADCs), and control logic.

There are two basic analog-input subsystem configurations which are used for data-acquisition or process-control computers. Within each basic configuration, however, there are many possible variations based on the use of different types of multiplexers, amplifiers, and analog-to-digital converters. The selection of a particular configuration is based primarily on cost/performance considerations.

Figure A-68 illustrates the basic configuration for a high-level analog-input subsystem. High-level signals, typically in the range of 0 to ± 10 V, are connected to the subsystem input terminals. (See **Signal conditioning; Multiplexer, analog; Amplifier; Analog-to-digital converter.**) Signal conditioning of the signals is done on a per-channel basis and may include filtering, limiting to protect the analog-input subsystem from overvoltage, attenuation, or a combination of these. The signal to be converted is selected by the multiplexer under the control of the subsystem control logic. A buffer amplifier provides isolation between the analog-to-digital converter and the input-signal source. This buffer amplifier, sometimes included as an integral part of the analog-to-digital converter, may be a sample-and-hold amplifier to reduce aperture-time errors. The amplifier output signal is converted to a digital representation by the analog-to-digital converter. The control logic transmits the digital value to the digital computer as well as providing the timing and sequencing of the subsystem operation.

Subsystems of this configuration are typically single-ended, although some differential systems are available. The differential subsystems are generally more expensive, but they have the advantage of easier installation since noise and ground loops are less of a problem. (See **Amplifier, single-ended; Grounding.**) This can be of particular importance in high-speed high-resolution applications or in cases where input signals are derived from signal sources located long distances from the subsystem.

The analog multiplexer (see **Multiplexer, analog**) generally uses solid-state switches and may be capable of sampling signals at rates in excess of 250,000 samples per second. Bipolar transistors and field-effect transistors are both used in these multiplexers. The use of bipolar-transistor switches restricts signal levels to high level because of voltage-offset errors. Even in these cases, however, the Bright-switch configuration is often employed (see **Switch, analog**).

The analog-to-digital converter may be any of several major types (see **Analog-to-digital converter**). The successive-approximation and parallel-serial types are most common because their

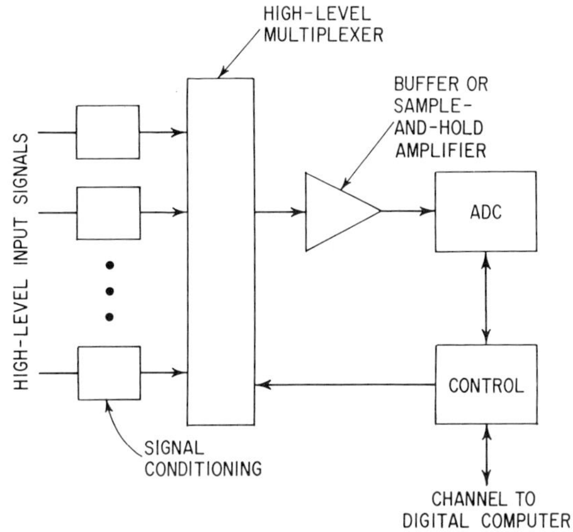

Fig. A-68. High-level analog-input subsystem.

greater speed matches the solid-state multiplexer characteristics. Ramp and integrating-ramp analog-to-digital converters are used in lower-speed applications, however.

The control logic provides the necessary timing and channel-communication functions. The operation of the system is generally initiated by a control word received from the digital computer, which specifies one or more input addresses. The control unit selects the appropriate multiplexer switches and, after a predetermined delay time, provides a "start convert" signal to the analog-to-digital converter. The analog-to-digital converter operates asynchronously with respect to the computer and signals the control logic at the end of the conversion. (See **Asynchronous.**) The control unit may reformat the analog-to-digital converter output word (e.g., add parity bits or convert from a parallel to a serial form) before transmitting the result back to the digital computer.

Depending on the characteristics of the multiplexer, amplifier, and analog-to-digital converter, the subsystem sampling speed may be as high as 100,000 samples per second at resolutions up to 14 bits (see **Resolution**). Higher speeds are possible at reduced resolution. A total measurement error (see **Error, measurement; Accuracy**) of less than 0.05 percent of full scale is typical under specific operating conditions.

Subsystems of this configuration are widely used in high-speed data-acquisition applications such as rocket-test-stand monitoring, hybrid computation, wind-tunnel studies, and many types of research activity. These applications are characterized by wide channel-bandwidth requirements, high-level signals, and/or a relatively small number of signals. The configuration is not widely used in process-control applications because it is not suited for low-level signal handling. A high-speed low-level system can be obtained by adding a high-gain amplifier for each input signal. Although this is done in applications involving only a few data points, it is very expensive in cases involving many low-level signals, a typical situation in process control.

The second basic configuration, shown in Fig. A-69, is widely used in process-control applications. It is similar to the previous configuration except for the use of a time-shared low-level amplifier. (See **Amplifier.**) The input signals are conditioned as before and are multiplexed by a low-level multiplexer. The multiplexer output is amplified by a time-shared amplifier before being converted into a digital representation by the analog-to-digital converter. Total system measurement accuracy for a 50-mV signal range is typically 0.1 percent, and a resolution of 10 to 14 bits is common (see **Error, measurement; Accuracy; Resolution**).

Because of the low-level signals, typically less than 100 mV and sometimes as low as 10 mV full scale, this configuration generally provides a differential-signal input (see **Differential-mode voltage**). The signal-conditioning and multiplexer circuits are differential, and depending on the type of multiplexer, the amplifier may or may not be differential. When the multiplexer provides common-mode isolation, as in the case of a transformer-coupled multiplexer, the amplifier may be single-ended [see **Common-mode rejection ratio (CMRR)**]. In almost all cases, the output of the amplifier and the analog-to-digital converter input are single-ended.

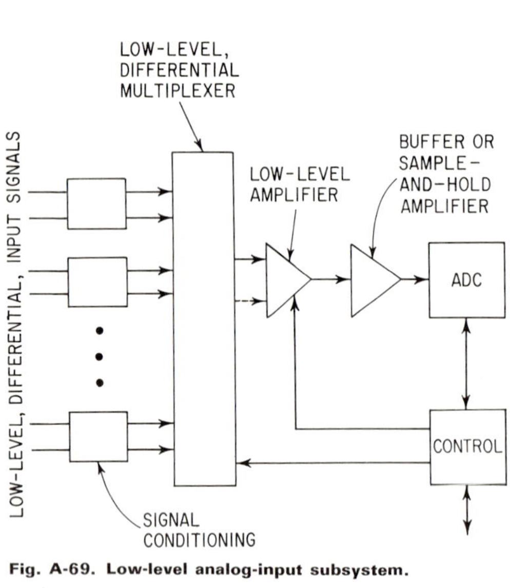

Fig. A-69. Low-level analog-input subsystem.

The common differential-multiplexer configurations for low-level applications are the double-pole single-throw differential, flying-capacitor, and transformer-coupled (see **Multiplexer, analog**). The switching devices are typically dry-reed relays, mercury-wetted contact relays, or field-effect transistors (see **Relay; Switch, analog**). Bipolar transistors are occasionally used, but the additional circuitry and adjustments required to minimize errors due to voltage offset make them less attractive. The electromechanical devices are limited in speed to less than about 300 samples per second, but field-effect-transistor switches can be used at speeds in excess of 100,000 samples per second. Usually, however, amplifier limitations restrict the subsystem speed to less than 10,000 samples per second. Depending on the multiplexer configuration and capabilities of the amplifier, common-mode voltages (see **Common-mode voltage**) up to several hundred volts can be tolerated by the subsystem. Multiplexers utilizing solid-state switches are typically limited to less than 30 V of common-mode voltage.

The amplifier used in this configuration may be differential or single-ended, depending on the characteristics of the multiplexer (see **Amplifier**). The amplifier may have a fixed gain, or the gain may be selected by the control logic either automatically or under control of the computer program. In the case of a differential amplifier, the common-mode performance of the amplifier is a critical parameter in determining the common-mode rejection of the subsystem. In addition, the low-level application requires that amplifier noise be minimized.

This configuration is widely used for analog subsystems in process-control and large data-acquisition applications. These applications are characterized by relatively low channel-bandwidth requirements and a large number of low-level signals. The use of the time-shared amplifier makes this system more economical than an amplifier-per-channel system, and the low bandwidth application characteristic makes the limited system-sampling speed acceptable in a large number of applications.

—Thomas J. Harrison, *IBM Corporation*

Analog Output Analog output is a generic term applied to those features or subsystems in a data-acquisition or process-control computer system which provide the capability of providing a continuous voltage or current output which can be controlled by the digital computer. Applications involving digital computers often require this capability. For example, closed-loop control utilizing a digital process-control computer requires that current signals in the range of 4 to 20, 1 to 5, or 10 to 50 mA be generated for the control of process actuators such as valve positioners. Similarly, strip-chart recorders and oscilloscopes require analog-input signals for the generation of visual displays.

The primary component in the analog-output subsystem of a data-acquisition or process-control computer system is the digital-to-analog converter (DAC) (see **Digital-to-anlog converter**). In essence, the digital-to-analog converter is an electronically controlled attenuator network and a constant reference source. Digital-input signals to the digital-to-analog converter activate analog switches, either electromechanical or solid-state, which determine the attenuation factor of a passive network. A constant-voltage or -current reference source is used as the input to the attenuator network so that the output of the network is proportional to the settings of the attenuator switches and, therefore, to the digital-input signals.

The digital-to-analog converter is connected to the digital computer through control logic which provides addressing for each digital-to-analog converter and the necessary control and timing signals. A typical analog-output system for a data-acquisition or process-control computer system is shown in Fig. A-70. Digital-control words received from the digital computer specify which digital-to-analog converter is to be adjusted. Similarly, a digital data word specifies the desired output value. The control logic decodes the address information and routes the digital data to the input of the addressed digital-to-analog converter. Setting the input switches of the digital-to-analog converter results in the desired analog-output value. The type of subsystem illustrated in Fig. A-70 is most often used in applications which require that the analog-output value be altered relatively often, typically at a rate of more than 1,000 samples per second. Depending on the characteristics of the digital-to-analog converter and digital-computer channel, this subsystem configuration is capable of adjusting a num-

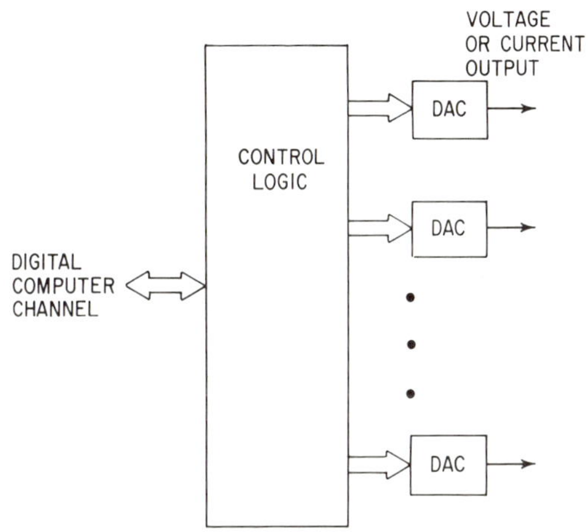

Fig.A-70. Analog-output subsystem.

ber of digital-to-analog converters at rates in excess of 100,000 samples per second. The resolution of the digital-to-analog converters used in this type of application generally ranges from 8 to 13 bits, and the total error of the output voltage may be less than 0.01 percent under specific operating conditions (see **Resolution; Accuracy**).

In many applications, the high-speed high-accuracy performance obtainable with the subsystem of Fig. A-70 is not required. In addition, if many analog outputs are required, the system cost can be quite high because of the cost of the digital-to-analog converters. An alternate subsystem which is more economical and entirely adequate for many applications is shown in Fig. A-71. The system is similar to that discussed above, and the control logic performs the same basic function. The major difference is that the digital-to-analog converter is time-shared through the use of an analog multiplexer (see **Multiplexer, analog**). The output of the digital-

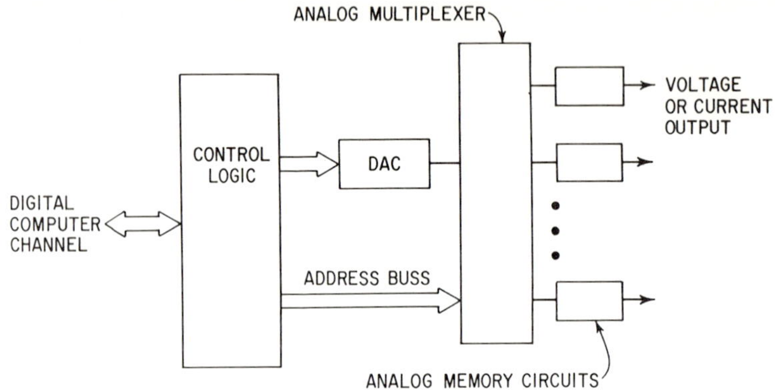

Fig. A-71. Time-shared digital-to-analog converter analog-output subsystem.

to-analog converter is set according to the digital data received from the digital computer, and the resulting analog-output signal is switched by the multiplexer to the output specified by address data received from the digital computer.

The multiplexer output, however, remains constant only as long as the multiplexer switches are activated, and many applications require continuous analog signals. For this reason, an analog memory circuit is included in the subsystem to "remember" the analog value and to provide a continuous analog-output signal (see **Memory, analog**). The analog memory circuit may consist of a sample-and-hold amplifier (see **Amplifier, sample-and-hold**) or, more commonly, a simple circuit consisting of a capacitor on the input of a field-effect-transistor amplifier. The output from the digital-to-analog converter deposits charge on the capacitor during the period that the multiplexer switches are closed. Because of the high input impedance of the field-effect-transistor amplifier, the charge and, therefore, the voltage across the capacitor remain essentially constant for a long period of time; a "droop" of less than 0.1 percent/hr under a wide range of environmental conditions can be obtained. Eventually, however, the charge on the capacitor must be regenerated or "updated" by the digital-to-analog converter.

As in the case of the subsystem of Fig. A-70, the time-shared digital-to-analog converter subsystem can provide either voltage or current output. The applications of this type of subsystem typically are exemplified by the requirement for many output signals, moderate resolution (10 to 12 bits), and relatively slow changes in output value (less than 2,000 Hz). Examples of these applications include most industrial process-control applications such as valve actuators and strip-chart recorders.

—Thomas J. Harrison, *IBM Corporation*

Analog-to-Digital Converter An analog-to-digital converter, often abbreviated ADC or A/D converter, is an apparatus which provides a digital representation of an analog quantity such as position, voltage, or current. Since most real quantities are analog, at least macrocosmically, the concept of analog-to-digital conversion is inherent in most measurement systems. For example, in measuring the length of a line with a ruler, a digital representation is obtained by comparing the length of the line with the calibrated scale on the ruler. In this case, the human observer is the analog-to-digital converter, the analog quantity is the length of the line, and the digital representation is the length expressed in measurement units such as inches or feet.

The two common analog-to-digital converter types in data-acquisition subsystems are electromechanical and electronic analog-to-digital converters. The electromechanical converters, often called shaft- or position-to-digital encoders (see **Encoder**), generally consist of a mask attached to the movable mechanical element and a means of reading the information on the mask. Electrical, magnetic, and optical reading means are commonly used. In electrical sensing, the mask consists of a conducting pattern on an insulated substrate. Fixed conducting brushes in contact with the pattern are used to read the code represented by the mask pattern. Similar arrangements are used in optical and magnetic sensing, except that optical sensing involves the use of a light source and photodetectors, and magnetic sensing uses inductive-pickup coils.

Because of the finite width of the sensing means, there can sometimes be an ambiguity in the digital output when, for example, the conducting brushes are in simultaneous contact with two adjacent portions of the mask pattern. This is avoided by the use of special codes and arrangements of the sensing detectors. Gray code is often used, since only one bit in the digital output changes at a time as the position of the mask is monotonically varied. An example of the use of a special arrangement of detectors is the V-scan technique in which two detectors are used for each track on the mask. By suitably decoding the outputs from the pairs of detectors, an unambiguous digital representation can be obtained.

Electronic analog-to-digital converters utilize two basic techniques: In one class of analog-to-digital converter the input quantity (usually voltage or current) is converted into another form such as a pulse duration or a frequency. The intermediate quantity is then measured to provide a digital representation of the input signal. In the other class of analog-to-digital converter, the input is directly compared with a known reference signal which can be varied under control of the analog-to-digital converter logic.

Within these two classes, many different techniques are available for analog-to-digital conversion. The five major techniques most often used in data-acquisition and process-control computers are the ramp, the integrating-ramp, voltage-to-frequency, successive-approximation, and the parallel-serial methods. The first three of these methods involve the conversion of the input signal into an intermediate quantity which is then measured; the latter two are direct-comparison methods. (See subsequent entries under **Analog-to-digital converter.**)

In the ramp analog-to-digital converter, the input voltage and a linear ramp voltage $V = Kt$, where k is the slope, are compared using a comparator amplifier (see **Amplifier, comparator**). Since the slope of the ramp is known, the time duration between a zero ramp voltage and equal input and ramp voltages is a measure of the magnitude of the input signal. This time duration is measured by counting constant-frequency pulses with a counter circuit. (See **Analog-to-digital converter, ramp.**)

In the integrating-ramp analog-to-digital converter, the input signal is integrated for a fixed period of time. A constant reference voltage of opposite polarity is then integrated until the output of the integrator reaches zero. Since the duration of this second integration period is proportional to the magnitude of the input signal, a digital representation is obtained by measuring the length of this period using a counter and a constant-frequency pulse source (see **Analog-to-digital converter, integrating-ramp**).

In the voltage-to-frequency technique, the input signal controls the frequency of a variable-frequency oscillator. The average frequency of the oscillator over a fixed time interval can be measured using digital methods to obtain a digital representation of the input signal (see **Analog-to-digital converter, voltage-to-frequency**).

The successive-approximation analog-to-digital converter is an example of a direct-comparison conversion technique. The analog-to-digital converter consists of a digital-to-analog converter, which is used as a digitally controlled voltage reference source (see **Digital-to-analog converter**), a comparator, and control logic. The conversion consists of a sequence of comparisons between the input-signal and reference-voltage values derived from the digital-to-analog converter. For example, in a binary successive-approximation analog-to-digital converter, the input signal is first compared with a reference signal equal to 0.5 of the full-scale value. If the comparator output indicates that the reference signal is greater than the input signal, the digital-to-analog converter output is reduced to 0.25 of the full-scale value for the next comparison; if the reference is greater than the input signal, the digital-to-analog converter output is increased to 0.75 of the full-scale output. By continuing in this manner, the value of the digital-to-analog converter output is made to converge to the value of the input signal. The number of comparison steps required for a binary successive-approximation analog-to-digital converter is equal to the number of bits in the digital representation (see **Analog-to-digital converter, successive-approximation**).

The parallel-serial technique is a modification of the successive-approximation method which is used for high-speed analog-to-digital converters. In this method, three or four bits of the digital-output representation are determined simultaneously by using multiple comparators. The digital-to-analog converter output is then adjusted to the value represented by these bits and is subtracted from the input signal. The resulting difference signal is then used to determine the next three or four bits in the output word. The number of steps necessary to complete the conversion depends on the number of bits in the digital-output representation and the number of bits determined in each step. For example, if three bits are determined at each comparison, four steps are required to complete a 12-bit conversion (see **Analog-to-digital converter, parallel-serial**).

Although all these methods of analog-to-digital conversion are used in data-acquisition and process-control computer systems, there are cost/performance trade-offs between the methods. The ramp-type converters and the voltage-to-frequency converters are relatively slow because of the necessity for serial counting. In general, they are used at speeds of less than several thousand samples per second and resolutions of less than 12 bits. The successive-approximation technique can be used up to about 100,000 samples per second at a resolution of 14 bits and at speeds in excess of 250,000 samples per second at resolutions of 8 bits or less. The speed limitations in this type of converter are primarily due to the recovery characteristics of the comparator and the settling time of the digital-to-analog converter. The parallel-serial analog-to-digital converter is primarily used for high-speed applications which require conversion rates in excess of 100,000 samples per second at resolutions of 8 to 14 bits.

—Thomas J. Harrison, *IBM Corporation*

Analog-to-Digital Converter, Integrating-ramp The integrating-ramp analog-to-digital converter is a quantizing method (see **Quantize**) which is based on the concept of converting the unknown input signal into an equivalent-pulse-duration signal which is measured by counting pulses generated by a precise clock pulse generator. Although several variations of the integrating-ramp converter are known, the most popular is the dual-slope or dual-ramp integrating analog-to-digital converter. In this converter, the input signal is integrated for a precise period of time; the resulting integral signal is then integrated for a variable length of time using a reference signal of opposite polarity. Since the integral of the unknown input signal is proportional to the input signal, the duration of the second integration period is also proportional to the input signal. The length of this second integration period is measured by counting constant-frequency pulses to provide a digital representation of the input signal.

The basic hardware for a dual-integrating-ramp analog-to-digital converter is shown in Fig. A-72. The comparator at the output of the integrator changes state when the polarity of the in-

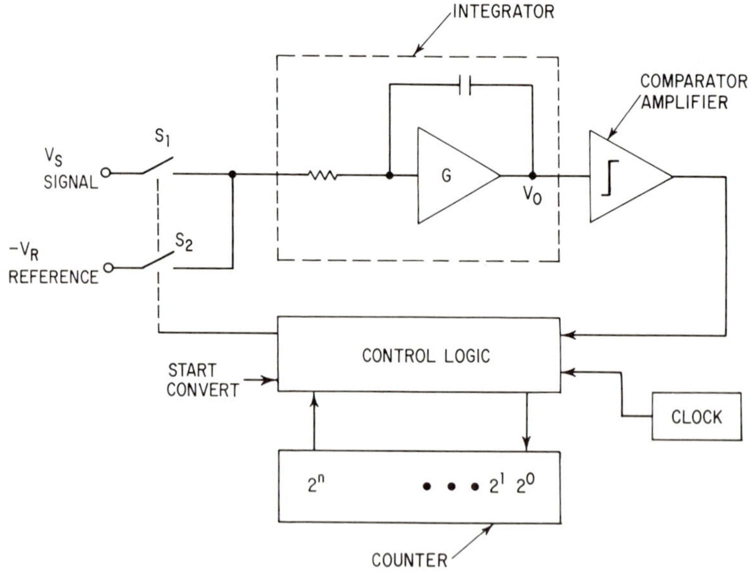

Fig. A-72. Dual-integrating-ramp analog-to-digital converter.

tegrator output changes. The switches at the input of the integrator control which signal is to be integrated. The control logic sequences the operation of the switches and provides the necessary gating of clock signals into the counter. The counter also serves as an output register for the analog-to-digital converter.

In operation, the "start convert" signal clears the counter and closes switch S_1. Since the input signal V_s is assumed essentially constant, the output of the integrator increases as a linear ramp signal starting from some initial integrator offset voltage $-V_i$ as shown in Fig. A-73. This integration is continued for a fixed time interval $(0, t_1)$ which is usually determined as the time required to fill the counter once. At the end of the interval, as indicated by an overflow pulse from the counter, switch S_1 is opened and S_2 is closed to apply a reference signal $-V_r$ to the integrator input. This causes the output of the integrator to decrease as shown in Fig. A-73. During this second integration interval clock pulses are counted by the counter. When the integrator output reaches its initial level $-V_i$, the comparator changes state and pulses are inhibited from entering the counter. Since the integrator output at the time t_1 was proportional to the average value of the input signal during the time interval $(0, t_1)$, the length of the second integration interval is also proportional to V_s and, therefore, the count in the counter at time t_2 is a digital representation of the input signal.

The dual-ramp integrating analog-to-digital converter is typically limited to conversion speeds of about 1,000 or 2,000 samples per second at a resolution of 10 or 12 bits because of the limitation on counter counting speed. Other techniques, such as the use of a two-step integration during the second integration period, can be used to increase the conversion rate to about 30,000 samples per second at 14 bits resolution without significantly increasing the logic requirements (see Reference).

The primary advantages of the integrating-ramp analog-to-digital converter are its averaging characteristics and the cancellation of errors which normally limit the performance of a ramp-type analog-to-digital converter. Since the input signal is integrated over the first integration interval $(0, t_1)$ and integration is equivalent to averaging, the digital representation obtained from the analog-to-digital converter represents the average value of the input signal over this time interval. As a result, spurious noise pulses or other disturbances do not appreciably affect the result of the conversion. It can be shown that the integration is equivalent to a low-pass filter with a 6-db rolloff and points of infinite attenuation at harmonics of $f = 1/t_1$.

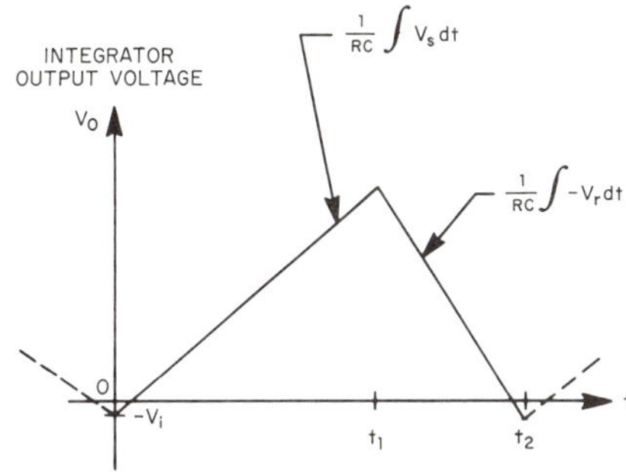

Fig. A-73. Integrator-output voltage.

A detailed analysis of the integrating-ramp analog-to-digital converter also shows that some other errors cancel. For example, in the final expressions describing the operation, the integrator time constant RC and the value of the clock frequency f_c do not appear. This means that if RC and f_c remain constant during each conversion, their exact values do not affect the conversion result, even if they change from one conversion interval to the next. Thus, long-term drift in the time constant due, for example, to temperature changes or aging does not affect conversion accuracy. Similarly, long-term changes in the clock frequency have no effect. It can also be shown that if the initial offset voltage $-V_i$ is the same at the beginning and end of the conversion, its value does not affect the result. This can be assured by running the converter continuously or by providing special circuitry which establishes the level before each conversion.

REFERENCE

Aasnaes, H. B., and T. J. Harrison: Triple Play Speeds A-D Conversion, *Electronics*, Apr. 29, 1968, pp. 69–72.

—Thomas J. Harrison, *IBM Corporation*

Analog-to-Digital Converter, Parallel The parallel analog-to-digital converter is the simplest type of electronic analog-to-digital converter. For an n-bit converter, a comparator is provided for each quantization level (see **Quantize**). By suitably decoding the output of these multiple comparators, a digital representation of the input signal is obtained (see **Amplifier, comparator**).

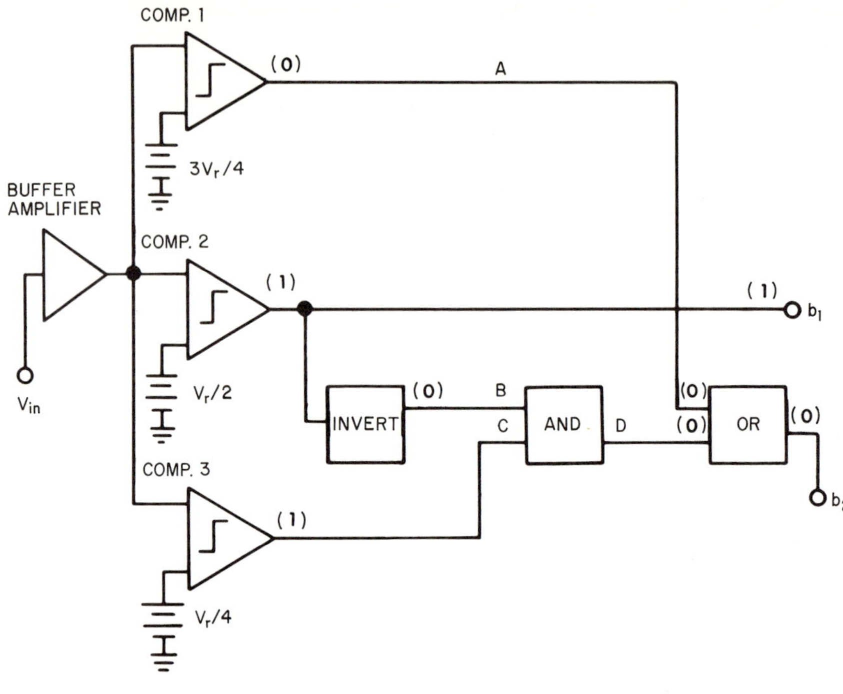

INPUT SIGNAL	A	B	C	D	b_1	b_2
$3V_r/4 < V_{in}$	1	0	1	0	1	1
$V_r/2 < V_{in} < 3V_r/4$	0	0	1	0	1	0
$V_r/4 < V_{in} < V_r/2$	0	1	1	1	0	1
$V_i < V_r/4$	0	1	0	0	0	0

Fig. A-74. Parallel analog-to-digital converter.

An example of a 2-bit parallel analog-to-digital converter is shown in Fig. A-74. Comparators 1, 2, and 3 are biased with reference voltages of 0.75 V_r, 0.5 V_r, and 0.25 V_r, where V_r, the full-scale reference voltage, is equal to the full-scale input range of the analog-to-digital converter. When the input signal is applied, all the comparators which are biased at a level less than the level of the input signal provide a binary 1 output. Thus, for example, in the analog-to-digital converter of Fig. A-74 an input signal of 0.6 V_r results in a 1 output from comparators 2 and 3; the output of comparator 1 will be a 0. By suitable decoding, the binary digital representation of 1 0 is obtained. The binary state at the input and output of each logic block is shown in the figure for this example.

Although the example of Fig. A-74 provides only 2-bit resolution, the resolution can be increased by adding additional comparators and decoding logic (see **Resolution**). Because each additional bit of resolution results in doubling the required hardware, the parallel analog-to-digital converter is generally used for converters of less than 5 or 6 bits resolution. In addition to increased hardware, increasing the resolution places severe requirements on the threshold stability of the comparators if excessive errors and output ambiguity are to be avoided.

The primary advantages of the parallel analog-to-digital converter are its simplicity for low-resolution requirements and its high speed. The speed is primarily determined by the settling time of the comparators and logic delays. Speeds in excess of 50,000,000 samples per second can be obtained with relatively simple hardware.

—Thomas J. Harrison, *IBM Corporation*

Analog-to-Digital Converter, Parallel-Serial The parallel-serial analog-to-digital converter is one of the conversion techniques used to perform relatively high speed quantization of analog signals (see **Quantize**). Although not as fast as the parallel analog-to-digital converter (See **Analog-to-digital converter, parallel**), the attainable resolution in practical cases is much better: 12- to 14-bit resolution at speeds in excess of 100,000 samples per second is readily achieved (see **Resolution**).

The parallel-serial analog-to-digital converter is similar to the successive-approximation analog-to-digital converter in that the conversion is accomplished as a series of distinct steps. The difference, however, is that more than one bit is determined at each step. An example of the hardware of a parallel-serial analog-to-digital converter is shown in Fig. A-75. In this example, the analog-to-digital converter consists of four 3-bit digital-to-analog converters (DACs) and four amplifiers, each with a gain of 8, corresponding to the 3-bit resolution of each section of the analog-to-digital converter ($2^3 = 8$). The general operation is that a 3-bit conversion is made by means of

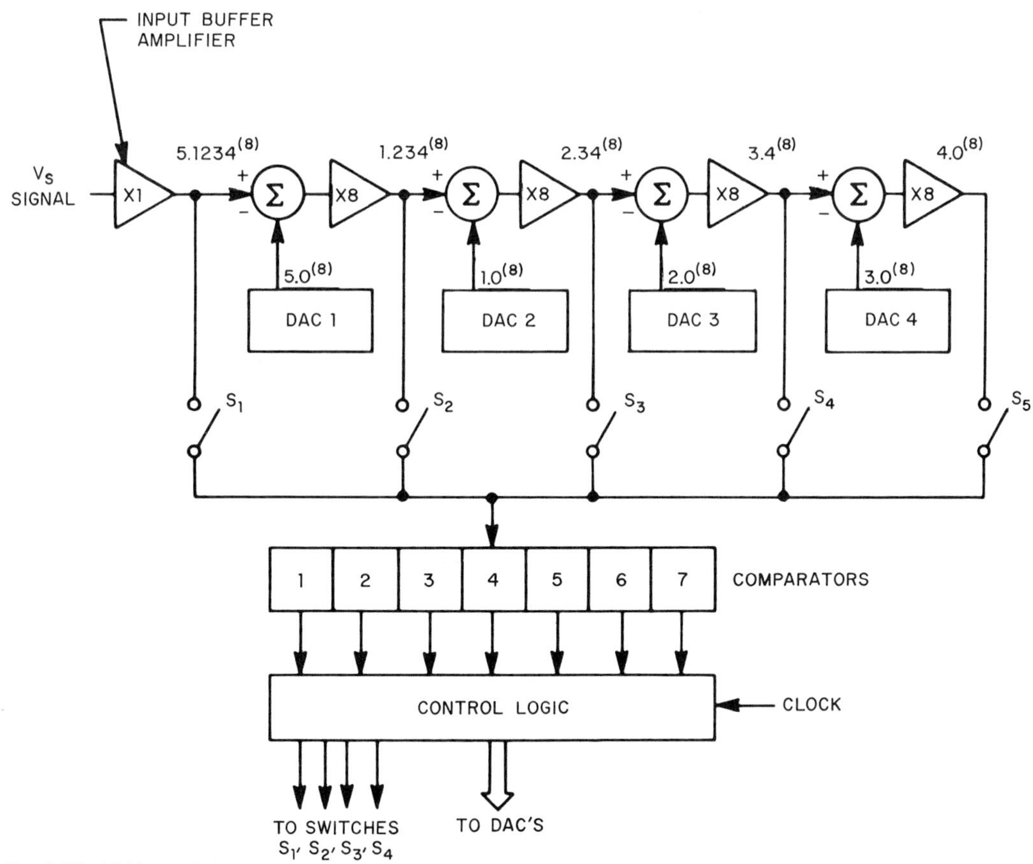

Fig. A-75. 15-bit parallel-serial analog-to-digital converter.

the seven parallel comparators (see **Amplifier, comparator**) in a manner similar to that used in the parallel analog-to-digital converter. The voltage corresponding to this 3-bit conversion is then subtracted from the input signal by means of DAC 1, and the difference signal is amplified by a factor of 8. The next stage of the analog-to-digital converter then determines the next three bits in a similar manner. Since three bits are determined essentially simultaneously, five steps result in a 15-bit quantization of the input signal. This compares with the 15-step procedure that is required in the successive-approximation analog-to-digital converter.

As a specific example, suppose that the input voltage is 5.163086 V and that the full-scale range of the analog-to-digital converter is 8.0 V. Since the operation of the analog-to-digital converter is more easily explained in terms of octal numbers, this input voltage is expressed as 5.1234^8, where the superscript 8 indicates that the number is an octal code. The first step in the conversion process is to close switch S_1 and apply the input signal to the parallel comparators. The output of the comparators indicates that the input exceeds 5^8 but is less than 6^8. As a result, DAC 1 is set to 5^8 and a voltage of 0.1234^8 is applied to the first gain of 8 amplifier. The output of the amplifier, 1.234^8 V, is applied to the parallel comparators by opening S_1 and closing S_2. The result of this parallel conversion is the determination that the amplified difference voltage is greater than 1^8 and less than 2^8 V. As a result, DAC 2 is set to 1.0^8 V to provide a difference signal of 0.234^8 V. Subsequent steps, involving the closing of switches S_3, S_4, and S_5, result in the octal result of 5.1234^8 or the 15-bit binary result 101001010011100.

In the analog-to-digital converter illustrated in Fig. A-75, each stage of the converter must settle completely before the next 3-bit conversion is made. In actual devices, additional circuitry is added to provide error correction. Thus, if a particular stage has not completely settled and this results in an incorrect 3-bit conversion, the conversion in the subsequent stage provides a signal which corrects the previous 3-bit representation.

The parallel-serial analog-to-digital converter provides relatively high speed conversion but uses more hardware than is required in a successive-approximation analog-to-digital converter of equal resolution. Although not as fast as the parallel analog-to-digital converter, it requires much less hardware for resolutions of more than 6 bits and avoids the problems associated with precise comparator adjustments in a high-resolution parallel analog-to-digital converter.

REFERENCE

Gorbatenko, G. G.: High Performance Parallel-Serial Analog to Digital Converter with Error Correction, *IEEE Intern. Conv. Record,* vol. 14, pt. 3, pp. 39–43, 1966.

—Thomas J. Harrison, *IBM Corporation*

Analog-to-Digital Converter, Ramp The ramp-type analog-to-digital converter quantizes an analog-input signal by converting the signal into a time-duration pulse which is measured by means of a counter and a constant-frequency pulse generator. The basic configuration of the ramp analog-to-digital converter is shown in Fig. A-76.

For an n-bit analog-to-digital converter, the "start convert" signal resets the n-bit counter and initiates the operation of a function generator which produces a linear ramp output signal $V_r = kt$, where t is time and k is a constant. This reference ramp signal is continuously compared with the value of the unknown input signal V_s by the comparator. The output of the comparator represents a binary 0 if $V_r > V_s$ and a binary 1 if $V_r < V_s$. During the period of time that the comparator output is 1, pulses from a clock pulse generator are counted in the n-bit counter. When the input signal V_s and the ramp signal V_r are equal, the output of the comparator changes to a binary 0, and through the action of the AND gate, clock pulses are inhibited from entering the counter. The length of time that the comparator output remains in the 1 state is proportional to the magnitude of the input signal; similarly, the count in the counter at the time the comparator changes state is proportional to the length of time that the comparator output is 1; therefore, the count in the counter is a digital representation of the input signal.

The ramp-type analog-to-digital converter is one of the simplest forms of electronic analog-to-digital converters and is useful for conversions of less than 10 to 12 bits at speeds of less than several thousand conversions per second. At lower resolution, higher speeds can be obtained (see **Resolution**). The speed limitation is primarily a matter of logic speed, since the speed of the converter is equal to the clock frequency divided by the number of quantizing intervals. For example, in a 12-bit converter a total of 4,096 pulses must be counted for a full-scale input signal; at a 10-MHz

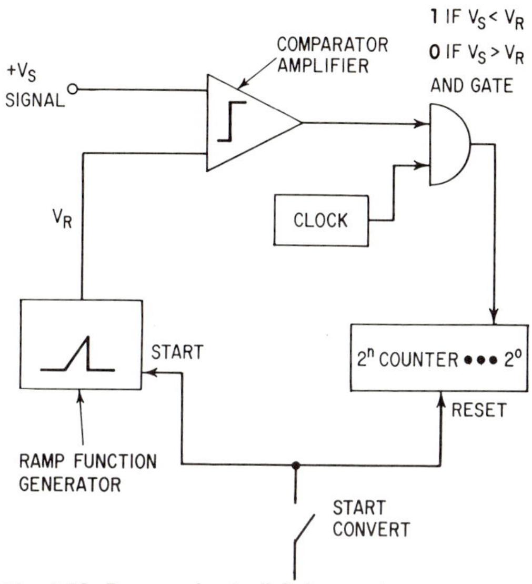

Fig. A-76. Ramp analog-to-digital converter.

clock frequency, this requires 1,024/10 MHz, or about 0.4 ms, with a resultant converter speed of 2,500 samples per second.

A primary design consideration is the speed and overload characteristics of the comparator, although this is not as severe as in the case of some other types of analog-to-digital converter such as the successive-approximation analog-to-digital converter. A major problem is the design of a suitable ramp generator which provides the required linearity and stability (see **Linearity**).

In addition to its simplicity, an advantage of the ramp analog-to-digital converter is its excellent differential linearity (see **Linearity, differential**). This is one of the primary reasons that this type of analog-to-digital converter is extensively used in applications requiring accurate quantization for the generation of histograms, such as in the nuclear-experimentation field.

—Thomas J. Harrison, *IBM Corporation*

Analog-to-Digital Converter, Successive-approximation The successive-approximation analog-to-digital converter is an example of an analog-to-digital converter in which quantization (see **Quantize**) is accomplished by a direct comparison between the unknown input signal and a reference signal. A block diagram of this type of analog-to-digital converter is shown in Fig. A-77. The digital-to-analog converter (DAC), an electronically controlled precision attenuator (see **Digital-to-analog converter**), provides an output voltage V_r which is a precise fraction of the reference source voltage. The attenuation factor of the digital-to-analog converter is determined by the setting of analog switches which are controlled by the digital outputs of the output register. The output register also stores the digital-output information of the analog-to-digital converter. The conditional-reset gates are simply two input AND logic gates, and the shift register is used as a step counter in this application.

For the purposes of illustration, the successive-approximation analog-to-digital converter of Fig. A-77 is assumed to be a binary n-bit converter. In operation, the "start convert" pulse clears all registers and sets a 1 into the first position of the shift register. This, in turn, sets the first stage of the output register to 1 and turns on the first switch in the digital-to-analog converter. Since the digital-to-analog converter is binary, the output of the digital-to-analog converter is equal to one-half the full-scale digital-to-analog converter output voltage (usually equal to the value of the reference voltage). The digital-to-analog converter output voltage is directly compared with the signal voltage V_s by the comparator (see **Amplifier, comparator**), whose output is a binary 1 if $V_r > V_s$ and a binary 0 if $V_r < V_s$. Depending on the output of the comparator, the first bit in the output register either is reset to 0 or remains as a 1. When the next clock pulse occurs, the 1 in the shift register moves to the second bit and the comparison procedure is repeated, except that the output of the digital-to-analog converter is either 0.25 or 0.75 of the full-scale digital-to-analog converter output, depending on whether the first bit was reset to 0 or remained as a 1, respectively.

As a specific example, suppose that the input voltage is 0.626 V, that the converter has 4-bit resolution, and that the reference source is 1 V (see **Resolution**). The "start convert" pulse sets a 1 into the output register and causes the output of the digital-to-analog converter to be 0.5 V. Since this is less than the input signal, the output of the comparator is a 0 and the first bit in the output register is not reset but remains as a 1. On the next clock pulse, the second bit of the output register is set to 1, and since the first bit is still a 1, the output of the digital-to-analog converter becomes $0.5 + 0.25 = 0.75$ V. Since this is greater than the signal voltage, the output of the comparator is 1 and the second bit of the output register

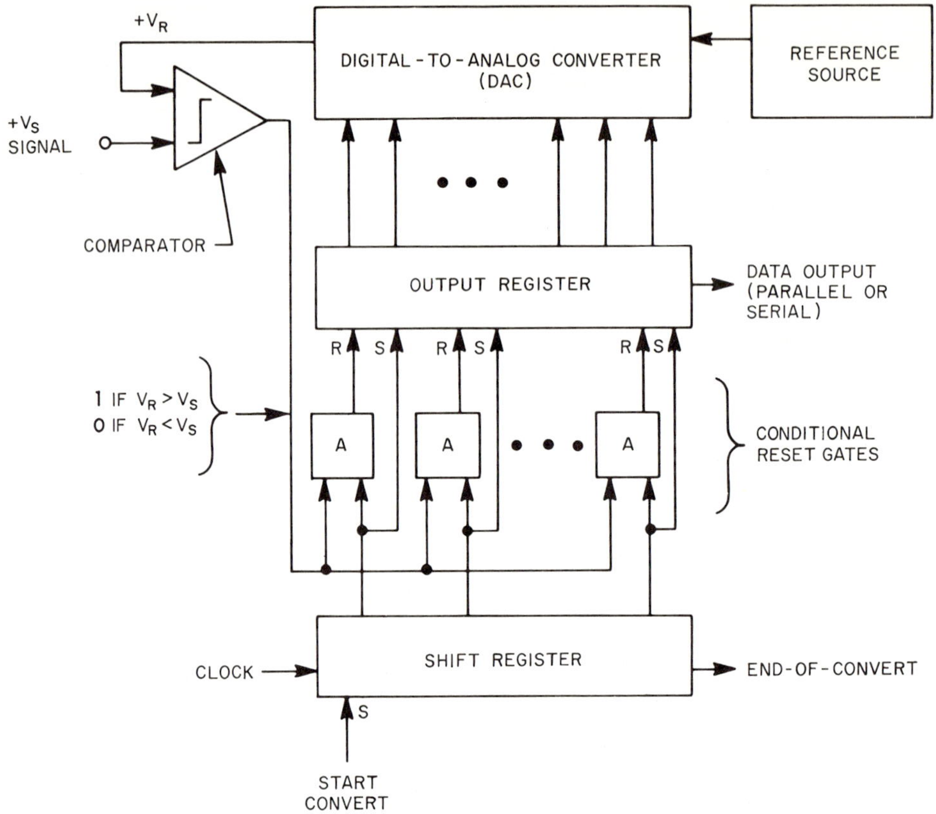

Fig. A-77. Successive-approximation analog-to-digital converter.

is reset to 0 through the conditional-reset gate. On the next clock pulse, the third bit of the register is set to 1 and the output of the digital-to-analog converter becomes $0.5 + 0.125 = 0.625$ V since only the first and third bits are 1. The output of the comparator is a 0 and the third bit remains a 1. The next clock pulse sets a 1 into the fourth and last bit position of the output register and shift register and causes a digital-to-analog converter output of $0.5 + 0.125 + 0.0625 = 0.6875$ V. Since this is larger than the input signal, the comparator output is 1 and the fourth bit is reset to 0. The next clock pulse shifts a 1 out of the shift register to indicate that the conversion is completed. The binary word stored in the output register is 1 0 1 0, which is a digital representation of the input voltage.

The successive-approximation analog-to-digital converter can be built using codes other than binary. All that is required is a digital-to-analog converter which produces an attenuation proportional to the weighting value of the code digit. For example, a BCD (see **Binary-coded decimal**) converter can be built by using a BCD digital-to analog converter. The digital-to-analog converter provides attenuation ratios of 0.8, 0.4, 0.2, 0.1, 0.08, 0.04, etc. Suitable coding of the output register is required to store the information in coded form.

The successive-approximation analog-to-digital converter is suitable for use at conversion speeds up to about 100,000 samples per second at resolutions up to 14 bits, not including sign. At lower resolution, speeds in excess of 250,000 samples per second can be obtained. Primary design considerations include the overload and recovery characteristics of the comparator, the stability and regulation of the reference source, and the design of the digital-to-analog converter, including the analog-switch characteristics and the speed and response of the ladder network.

—Thomas J. Harrison, *IBM Corporation*

Analog-to-Digital Converter, Voltage-to-Frequency In the voltage-to-frequency type of analog-to digital converter, the unknown input signal controls the frequency of an oscillator or multivibrator circuit. In this way, the magnitude of the analog-input signal (generally a voltage or a current) is converted into a frequency which can be measured by digital circuitry. For example, the analog-input signal could be used to control the bias of a varactor diode. If this diode is the frequency-determining component of a resonant circuit in an oscillator, the frequency of oscillation is dependent on the value of the input signal.

An example of a voltage-to-frequency analog-to-digital converter based on the use of an integrator is shown in Fig. A-78. At the beginning of the conversion, the counter is set to zero and the integrator is reset. The input signal V_s is then integrated until the integrator output reaches a value V_r. This point is detected by a change in state of the comparator output (see **Amplifier, comparator**). When the comparator changes state, the integrator is reset and the process is repeated.

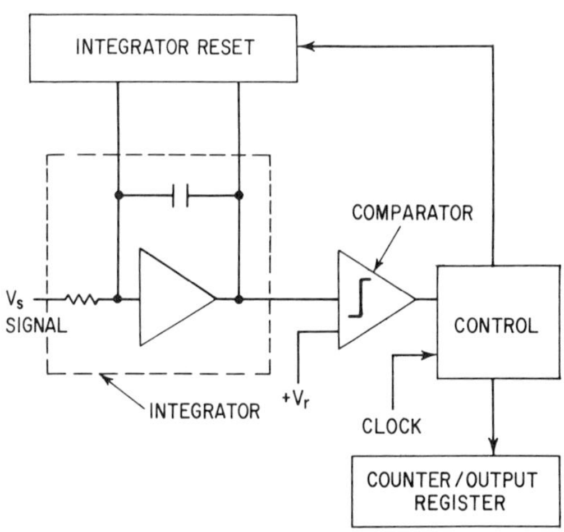

Fig. A-78. Voltage-to-frequency analog-to-digital converter.

Since the rate of integration depends on the value of the input signal, the frequency of the sawtooth waveform generated by the integrator is a function of the input signal. By counting the number of integration cycles which occur in a fixed period of time, a digital representation of the input signal is obtained. The counting interval is determined by counting clock pulses in a separate counter.

The voltage-to-frequency analog-to-digital converter has the advantage of providing a digital representation of the *average* value of the input signal over the conversion interval. Therefore, input-noise transients have little effect on its performance. Because of the serial counting, however, this type of analog-to-digital converter is relatively slow, usually less than 1,000 samples per second at a resolution of 10 bits or less (see **Resolution**). Laboratory instruments which utilize this principle are available at higher resolutions, but these are not generally used in data-acquisition systems because of their low speed.

—Thomas J. Harrison, *IBM Corporation*

Analytical Instrumentation (Chemical) Scores of different types of industrial and laboratory research instruments are available for determining the chemical composition of materials. The principal uses of these instruments are given in Table A-9.

Measurement Methods: Chemical-composition variables are measured by observing the basic *interactions between matter and energy.* All physical and chemical analysis techniques, whether instrumental or strictly manual, fundamentally are based upon these relationships. This follows directly from the fact that all known matter is made up of a complex but systematic arrangement of particles having mass and electric charge. To generalize, these particles consist of (1) neutrons having mass but not charge, (2) protons having essentially the same mass as neutrons with a unit positive charge, and (3) electrons having negligible mass with a unit negative charge. The number of protons, which fixes the number of electrons, determines the chemical properties and the physical properties, except mass, of the resulting atom. Chemical combinations of atoms into molecules involve only the electrons and their energy states. Chemical reactions involving both structure and composition usually occur by loss, gain, or sharing of electrons among the atoms. Every configuration of atoms in a molecule, crystal, solid, liquid, or gas can be represented by a definite system of electron energy states. Moreover, the particular physical state of the molecules, as represented by their mutual arrangement, also is reflected upon these electron energy states. These energy states, which are characteristic of the composition of any particular substance, can be readily inferred by

TABLE A-9. *Uses of Chemical-composition Measurements*

RAW MATERIALS

1. Composition analysis to check purchase specifications
2. Detection of contamination by trace impurities
3. Analysis check on materials priced on an active-ingredient basis
4. Continuous analysis of materials delivered by pipeline; water analysis

PROCESS CONTROL

1. Speedup and improved control by automation of, or replacement of control laboratory tests on "grab" samples
2. Improved control by replacement or augmenting of inferential measurements, such as temperature or pressure, with more significant composition data
3. Use of continuous processes permitted that could not be controlled except by continuous-analysis instrumentation

PROCESS TROUBLESHOOTING

1. Temporary use of analysis instruments for process studies aimed at overcoming occasional upsets
2. Determination of leaks and contamination; bad atmospheric and environmental conditions

YIELD IMPROVEMENT

1. Continuous analysis of process streams to measure effects of variables influencing yields
2. Analysis of overflow or purge streams, recirculated material, sumps, and similar equipment to determine product losses and to detect buildup of undesirable by-products that affect yield

INVENTORY MEASUREMENTS

1. Analytical monitoring of material flowing between process steps and plant areas to establish consumption and in-process inventory on the basis of active or essential ingredients
2. Periodic materials and energy balances

PRODUCT QUALITY

1. Determination of product-composition specifications and adjustment of process to meet those specifications desired
2. Assessment of structurally dependent attributes, such as color, melting or boiling point, and refractive index

SAFETY

1. Detection of presence of hazardous materials
2. Survey of operating areas for escape of toxic materials from leaks or spills, especially materials not readily detectable by the human senses
3. Detection of flammable or explosive mixtures in atmospheres and process lines

WASTE DISPOSAL

1. Monitoring of plant stacks for accidental discharge of toxic or nuisance gases, vapors, or smokes
2. Analysis of waste streams for toxic and other objectionable materials
3. Control of waste-treatment or product-recovery facilities

RESEARCH AND DEVELOPMENT

1. Continuous analysis to speed up research and optimize results
2. Provision of structural and compositional information not otherwise obtainable
3. Production of results in a more directly usable form

---

observing the consequences of interaction between the substance and an external source of energy. This external energy source may be (1) electromagnetic radiation, (2) chemical affinity or reactivity, (3) electric or magnetic fields, and (4) thermal or mechanical energy.

Interactions with Electromagnetic Radiation: These techniques involve the measurement of the quantity and quality of electromagnetic radiation emitted, reflected, transmitted, or diffracted by the sample. Electromagnetic radiation varies in energy with radiation frequency—that of the highest frequency or shortest wavelength having the highest energy and penetration into matter. Radiation of the shortest wavelengths (gamma rays) interacts with atomic nuclei; x-rays interact with the inner-shell electrons; visible and ultraviolet radiation interacts with valence electrons and strong interatomic bonds; infrared radiation and microwaves interact with the weaker interatomic bonds and with molecular vibrations and rotation. Most of these interactions are structurally related and completely unique. These methods may be used to detect and measure the elemental or molecular composition of gas, liquid, and solid substances within the limitations of available equipment. Instrumental techniques based upon interaction with electromagnetic radiation are given in Table A-10.

Interactions with Other Chemicals: These techniques involve the measurement of the results of reaction with other chemicals—in terms of amount of sample or reactant consumed, product formed, or thermal energy liberated, or by determination of equilibrium attained. The selectivity inherent in the chemical affinity of one element or compound for another, together with their known stoichiometric and thermodynamic behavior, permits positive identification and analysis under many circumstances. In a somewhat opposite sense, the apparent dissociation of substances at equilibrium in a chemical solution gives rise to electrically measurable valence potentials, called oxidation-reduction potentials, whose magnitude is indicative of the concentration and composition of the substance. Instrumental techniques based upon interaction with other chemicals are given in Table A-11.

TABLE A-10. *Analytical-instrumentation Techniques Based upon Interaction with Electromagnetic Radiation*

EMITTED RADIATION
1. Thermally excited
   a. Emission spectroscopy
   b. Flame photometry
2. Electromagnetically excited
   a. Fluorescence and phosphorescence
   b. Raman spectrophotometry
   c. Induced radioactivity
   d. X-ray fluorescence

TRANSMISSION AND REFLECTION MEASUREMENTS
1. X-ray analysis
   a. Absorption
   b. Critical-edge absorption
   c. Diffraction
2. Ultraviolet spectrophotometry
3. Transmission colorimetry
4. Colorimetry
5. Light scattering (nephelometry; turbidimetry)
6. Optical rotation (polarimetry)
7. Refractive index
8. Infrared spectrophotometry
9. Microwave spectroscopy (molecular rotational resonance)
10. Nuclear quadrupole moment

Reaction to Electric and Magnetic Fields: These techniques involve the measurement of the current, voltage, or flux changes produced in energized electric and magnetic circuits containing the sample. The production of net electric charges on atoms or molecules by bombardment with ionizing particles or radiation, or by electrolysis or dissociation in solution, or the induction of dipoles by strong fields establishes measurable relationships between these ionized or polarized substances and electric and magnetic energy. Ionized gases and vapors can be accelerated by applying electric fields, focused or deflected in magnetic fields, and collected and measured as an electric current in mass spectroscopy. Ions in solution can be transported, and deposited if desired, under the influence of various applied potentials for coulometric or polarographic analysis and for electrical-conductivity measurements. Inherent and induced magnetic properties give rise to specialized techniques, such as oxygen analysis based on its paramagnetic properties and nuclear magnetic resonance, which is exceedingly precise and selective for the determination of the compounds of many elements. Instrumental techniques based upon reaction to electric and magnetic fields are given in Table A-12.

TABLE A-11. *Analytical-instrumentation Techniques Based upon Interaction with Other Chemicals*

CONSUMPTION OF SAMPLE OR REACTANT
1. Orsat analysis
2. Automatic titration

MEASUREMENT OF REACTION PRODUCTS
1. Impregnated paper tape
2. Continuous chemical reactions

THERMAL-ENERGY LIBERATION
1. Combustion
2. Quantitative exothermic reaction of unknown with reactant

EQUILIBRIUM-SOLUTION POTENTIALS
1. Oxidation-reduction potential
2. pH (hydrogen-ion concentration)
3. Metal-ion equilibria

Interaction with Thermal or Mechanical Energy: These techniques involve the measurement of the results of applying thermal or mechanical energy to a system in terms of energy transmission, work done, or changes in physical state. The thermodynamic relationship involving the physical state and thermal-energy content of any substance permits analysis and identification of mixtures of solids, liquids, and gases to be based upon the determination of freezing or boil-

TABLE A-12. *Analytical-instrumentation Techniques Based upon Reaction to Electric and Magnetic Fields*

MASS SPECTROSCOPY

ELECTROCHEMICAL
1. Controlled-potential electrolysis
2. Polarography
3. Coulometry
4. Amperometry
5. "Dead-stop" methods

ELECTRICAL PROPERTIES
1. Electrical conductivity
2. Dielectric constant and loss factor
3. Oscillometry
4. Gaseous conduction

MAGNETIC PROPERTIES
1. Paramagnetism (oxygen meters)
2. Nuclear magnetic resonance
3. Electron paramagnetic resonance

REFERENCES

Proc. 15th Natl. Analysis Instrumentation Symposium, ISA Pub. AID69, New Orleans, La., 1969.

Proc. 14th Natl. Analysis Instrumentation Symposium, ISA Pub. AID68, Philadelphia, Pa., 1968.

Proc. 13th Natl. Analysis Instrumentation Symposium, ISA Pub. AID67, Los Angeles, Calif., 1967.

Note: Adapted from C. M. Albright, Jr., Chemical Composition Variables, in D. M. Considine (ed.), "Handbook of Applied Instrumentation," sec. 8, McGraw-Hill Book Company, New York, 1964.

ing points and upon the quantitative measurement of physically separated fractions, as in vapor-phase chromatography. Useful information often can be derived from thermal conductivity and viscosity, measurements involving the transmission of thermal and mechanical energy, respectively. Instrumental techniques based upon interaction with thermal or mechanical energy are given in Table A-13.

Numerous analytical instruments are described elsewhere in this volume. Refer to Table A-9 as well as to the alphabetical index.

TABLE A-13. *Analytical-instrumentation Techniques Based upon Interaction with Thermal or Mechanical Energy*

EFFECTS OF THERMAL ENERGY
1. Thermal conductivity
2. Melting point
3. Boiling point
4. Ice point (crystallization)
5. Dew point
6. Vapor pressure
7. Fractionation
8. Chromatography
9. Thermal expansion

EFFECTS OF MECHANICAL ENERGY OR FORCES
1. Viscosity
2. Sound velocity
3. Density

AND Circuit A logical decision element which provides an output if and only if all the input functions are satisfied. With reference to the three-variable AND element of Fig. A-79, the function F is a binary 1 if, and only if, A *and* B *and* C are all 1s. When any of the input functions is in the 0 state, the output function is in the 0 state. In terms of Boolean algebra, this function is generally represented as $F = A \cdot B \cdot C$, or $F = ABC$ (see **Boolean algebra**).

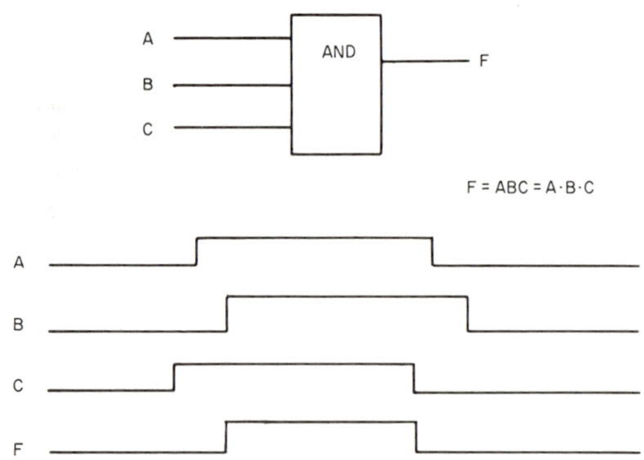

Fig. A-79. AND circuit.

Diode and transistor-circuit schematics for the two-variable AND function are illustrated in Fig. A-80. For the diode AND circuit, the output F is positive only when both inputs A and B are positive. If one or both inputs are negative, one or both diodes are forward-biased and the output is also negative. The transistor AND circuit performs in a similar manner in that if an input is negative, the associated transistor is conducting and the output is negative.

The maximum number of input functions for a given circuit configuration, generally referred to as "fan in," is determined by the leakage current of the active element. The number of circuits which can be driven by the output, called "fan

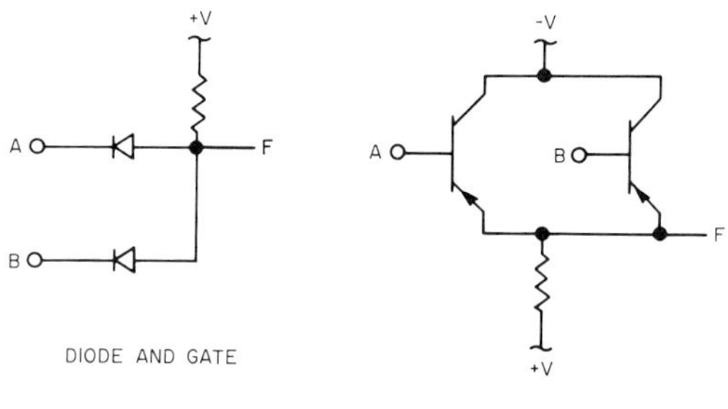

Fig. A-80. Diode and transistor AND circuits.

out," is a function of current which can be supplied by the AND circuit.

—W. W. Welz, *IBM Corporation*

Anderson Bridge (See **Bridge circuits, electric.**)

Anemometer An instrument for measuring the velocity of a gas stream, such as air in a ventilating duct or wind tunnel, or to indicate wind speed as in meteorology. Anemometers often are combined with a wind vane which indicates wind direction.

Propeller-type Anemometer: This instrument, as shown in Fig. A-81, incorporates wind velocity and direction sensors that are mounted in a common housing. A wind vane at one end of the housing keeps the assembly pointed into the wind. A propeller-type rotor at the other end turns a dc generator whose voltage output is directly proportional to the speed of rotation. The

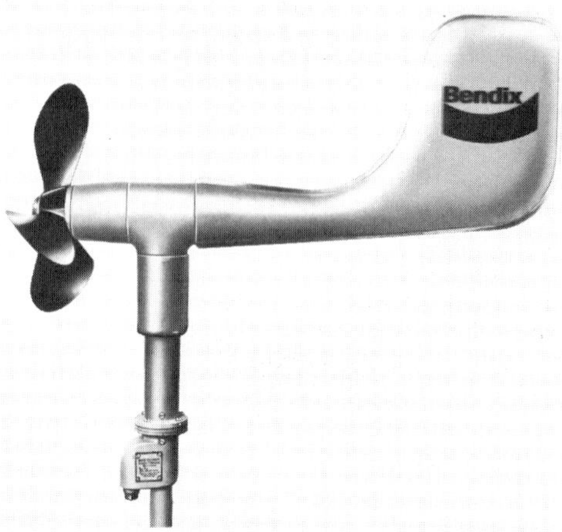

Fig. A-81. Propeller-type anemometer. *(Bendix.)*

assembly is attached to the rotor of an angular-position synchro generator and mounted on a stand or post so that it can rotate freely around its center of gravity. The wind-speed voltage and the angular-position signal are transmitted over wires to an indicator or recorder. Wind speed is displayed on a voltmeter dial scaled in miles per hour or knots. The synchro signal is applied to a dial scaled in azimuthal degrees. The indications are continuous.

Cup-type Anemometer: A number of cups are mounted about a vertical axis so that the cups rotate freely because of wind pressure (see Fig. A-82). The rotating cups may drive an impulse

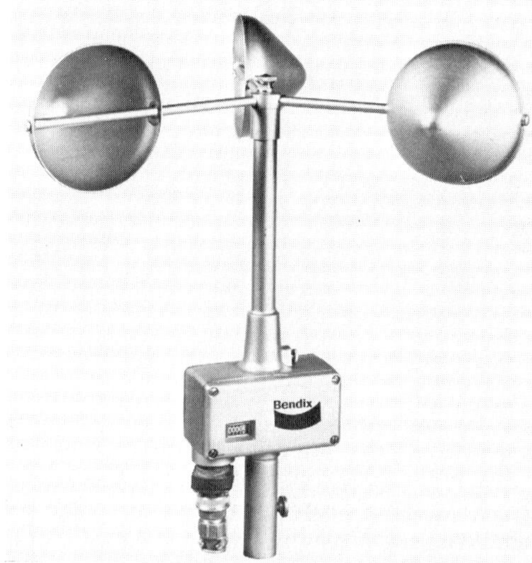

Fig. A-82. Cup-type anemometer. *(Bendix.)*

generator where the instrument is calibrated to provide one impulse for a known wind-speed increment (for example, 0.1 mile per hour or knot). In a totalizing anemometer, each impulse steps a counter one count. Counter readings are taken at given intervals, converted to counts per hour, and multiplied by the given increment of wind for which the counter is calibrated to arrive at the average wind speed for that period of time. This instrument can be designed to provide recording of wind speed by use of an electric circuit that is activated at each rotation of the cup assembly. The electrical output is used to actuate a pen to scribe a mark on a chart mounted on a revolving drum. Counting the marks made during a specified interval of time gives an indication of the

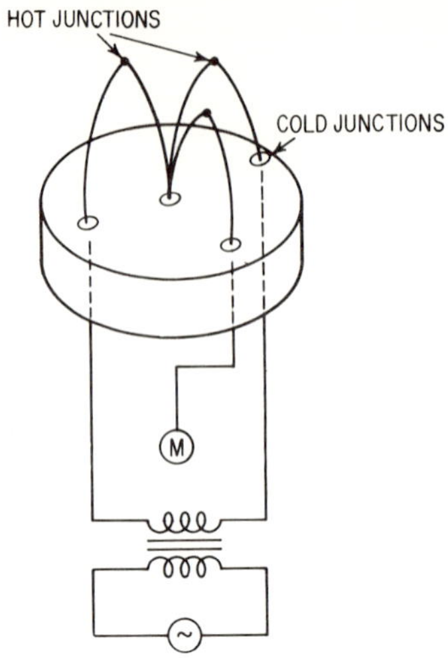

Fig. A-83. Hot-wire anemometer utilizing thermopile and millivoltmeter.

wind speed. Impulses having a discrete time period independent of impulse rate can be summed electrically and applied to a direct-reading voltmeter scaled in miles per hour. Wind-direction sensors used with cup-type anemometers usually are separate assemblies but sometimes are mounted on the same mast. With some signal conversion, the wind data from propeller and cup-type anemometers can be applied directly to computerized equipment.

Vane-type Velocity Meter: This device consists of a series of vanes radially mounted about a shaft. Impact of moving air or gas against the vanes moves the mechanism against a hairspring. Extension of the spring is approximately proportional to velocity. A device of this type is used in open air or with special jet tubes in ducts or restricted locations.

Hot-wire Anemometer: One type of hot-wire anemometer is shown in Fig. A-83. This device uses a thermopile heated by alternating current. Cooling by flow is proportional to flow rate and is measured as the dc output of the thermopile by a millivoltmeter.

—James R. Cosby, *Environmental Science Division, The Bendix Corporation*

Aneroid Barometer The basic element of an aneroid barometer consists of a capsule fabricated from thin corrugated metal which is hermetically sealed after the removal of practically all air. A small, precise amount of air is retained in the capsule to compensate for temperature changes which affect the elastic properties of the capsule and cause expansion and contraction of the connecting mechanism. As the barometric pressure increases, the capsule is compressed. Conversely, decreasing pressure causes capsule relaxation. Delicate lever systems transmit the relative motion of the two faces of the capsule to a chain or thread wrapped around a small pulley or to a sector and pinion that provides a rotational motion which moves a pointer across the dial.

Angle-of-attack Meter [See **Air-data systems (aircraft).**]

Angular-momentum Flowmeter (See **Flowmeter, mass.**)

Angular-position Transducer (See **Position-measurement and control systems; Resolver;** and **Synchro.**)

Annunciator A signaling device, usually electric, but sometimes pneumatic, that provides an audio and/or visual alarm of some off-limit condition within a process or piece of equipment. Annunciators can be field-mounted, but normally the devices are located on a central control panel or console. Also see **Alarms, control-system.**

Anticipatory Control [See **Derivative (rate) control.**]

Approach Systems (See **Flight-control systems.**)

Area, Standard (See **Standards, measurement.**)

Arithmetic and Logical Unit (ALU) The ALU (Fig. A-84) is that part of a central processing unit [see **Central processing unit (CPU)**] which performs the arithmetic operations such as add, subtract, multiply, and shift and the logical operations such as AND, OR, and EXCLUSIVE OR. The ALU generally consists of an adder (see **Adder**), registers for the storage of operands and results, and control logic for implementing the desired operations.

—W. W. Welz, *IBM Corporation*

Asdic (See **Sonar.**)

Assembler A program which processes (or "assembles") a program written in assembler language into direct machine code.

Assembler Language: A computer language characterized by a one-to-one relationship be-

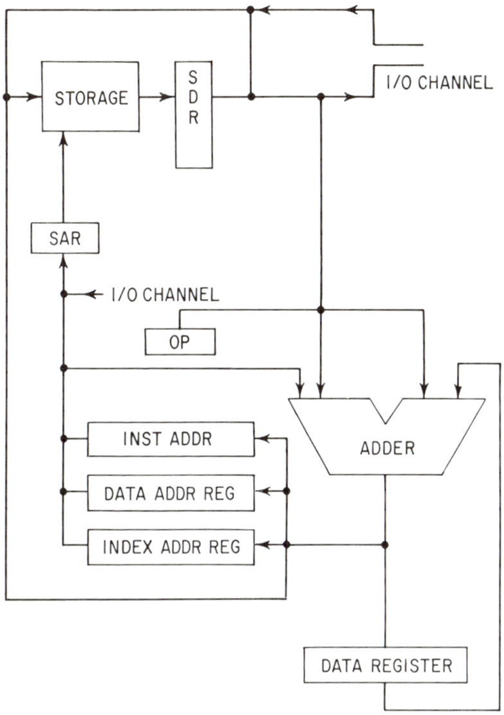

Fig. A-84. Arithmetic and logical unit.

and the actual machine instructions performed. The programmer thus has direct control over the efficiency and speed of the program. The language normally allows the use of mnemonic names instead of numerical values for the operation codes of the instructions and similarly allows the user to assign symbolic names to the locations of the instructions and data. For the first feature the assembler simply contains a table of the permissible mnemonic names and their numerical equivalents, and for the second, it commonly builds such a table on a first pass through the program statements. The table is then used to replace the symbolic names by their numerical values on a second pass through the program. Certain dummy operation codes (or "pseudocodes") are normally required by the assembler to pass control information to it. For example, an origin statement is normally required as the first statement in the program. It gives the numerical value of the desired location of the first instruction or piece of data so that the assembler can, by counting the instructions and data, assign numerical values to their symbolic names.

The format of the program statements is normally rigidly specified and only one statement, per input record to the assembler is allowed. A typical statement form is: symbolic name, operation code (or pseudocode), modifiers and/or index registers, symbolic name of data.

The mnemonic names used for the operation codes are normally uniquely defined for a particular computer type with little standardization between manufacturers even for the most common operations. The programmer must thus learn a new language for each new machine that he works with.

An example of a program written in an assembler language follows. The explanatory comments following the REM (remark) mnemonic and those to the right of the other program statements are ignored by the assembler program and therefore do not affect program execution.

```
       ABS
       ORG   100
       REM   THIS IS AN EXAMPLE OF A PRO-
       REM   GRAM IN AN ASSEMBLER LAN-
       REM   GUAGE. "ABS", "ORG", "CALL",
       REM   "REM", "DC", "DEC" AND "END"
       REM   ARE PSEUDOCODES. THE OTHER
       REM   MNEMONICS IN THE SAME COL-
       REM   UMN ARE OPERATION CODES.
       REM   THE NAMES SUCH AS A, B,
       REM   START, AND OVER ARE SYM-
       REM   BOLIC ADDRESSES ASSIGNED BY
       REM   THE PROGRAMMER

START  SLT   32       THESE INSTRUCTIONS
       LD    A        COMPUTE A/B*C AND
       D     B        PUT THE VALUE IN D.
       M     C
       SLT   1
       STO   D

       CMP   D4B1     THESE INSTRUCTIONS
       BSC   Z+       COMPARE D WITH 4 AND
       MDX   OVER     3.2 AND BRANCH OUT IF
       CMP   D32B1    IT DOES NOT LIE BE-
       BSC   Z-       TWEEN THEM
       MDX   UNDER

EXIT   CALL  EXIT     THE ASSEMBLER SETS
       REM            UP A BRANCH TO THE
       REM            SYSTEM EXIT SUBROU-
       REM            TINE (NAMED "EXIT")

OVER   LD    D1       THESE INSTRUCTIONS
       STO   EROR     SET THE VALUE OF
       MDX   EXIT     EROR TO 1 BEFORE
       REM            EXITING.

UNDER  SRA   16       THESE INSTRUCTIONS
       STO   EROR     SET THE VALUE OF
       MDX   EXIT     EROR TO ZERO BEFORE
       REM            EXITING.

       ORG   200      THIS SET OF PSEUDO-
A      DC    1        CODES IS USED BY THE
B      DC    2        ASSEMBLER TO SET UP
C      DC    3        THE DATA AND DATA
```

```
D       DC    0      WORKSPACE FOR THE
EROR    DEC   −1.0   ABOVE PROGRAM.
D1      DEC   1.0
D4B1    DEC   4.0B1
D32B1   DEC   3.2B1
        REM          LAST STATEMENT IN
        END          PROGRAM.
```

—A. J. Arthur, *IBM Corporation*

Astrographic System (See **Astronomical instrumentation.**)

Astronomical Instrumention Unlike physics, which is fundamentally an experimental science, astronomy is largely dependent on observation. The primary instrument of observation in optical astronomy is the telescope. In its basic form, the telescope may be considered a visual device variously modified to accept specific accessories whose operation calls for specific detectors. The resulting optical systems then meet the rather specialized demands of astronomical observation. While these demands vary greatly, astronomical observation is of necessity restricted to (1) measurement of position and (2) measurement of incoming radiant energy. Observations in the first category are *astrometric* in nature, while those in the second category are *photometric* in nature.

Astrographic Systems: Measurement of position in astronomy is either one- or two-dimensional. Thus, the combination *telescope–doppler spectrograph* is used to establish photographically line-of-sight velocities of stars through the precise measurement in one coordinate of the position of stellar spectral lines (relative to comparison lines); while a type of telescope, termed the *astrograph,* yields very precise photographic relative positions of stars on the celestial sphere in two coordinates. To be effective, astrographs or astrometric accessories, such as a doppler spectrograph, must be stringently free of coma, although some astigmatism may be tolerated so long as focal images remain uniformly illuminated.

These instruments must in addition be very stable. The classical solutions, therefore, call for large aperture ratios to satisfy even moderate field requirements and incorporate all-refractive optical elements. Only flat or very nearly flat fields are acceptable, and this in turn limits the wavelength bandpass of astrographs. Thus, some observatories are equipped with twin astrographs on a common mount, one design for photographic (blue bandpass), the other for photovisual (yellow-green bandpass) use. For a given *f* ratio, the change in focal length with color in classical achromats becomes increasingly severe with increasing aperture. Hence astrographs always have moderate apertures. Recourse to large triplet astrographic objectives has not been successful (they are unstable), though small triplet camera lenses in doppler spectrographs are in use.

Visual refractors in combination with the filar micrometer are used to measure the separation and position angles of binary stars too little separated to be resolved photographically. These instruments are also ideal for the visual study of planetary and lunar surfaces and when equipped with a Herschel wedge and suitable filters may be trained on the sun. Refractors are generally superior to reflectors in the visual observation of extended surfaces. This is because reflectors suffer from the contrast losses induced by (1) central obscuration and (2) pitting of aluminum reflective coatings. On the other hand, in applications where even minimal contrast loss is critical, refractor objectives must be wholly free of air bubbles and scratches. This is achieved in coronagraphs. See **Coronagraph.**

The largest visual refractor in existence has an aperture of 40 in. The largest astrograph has an objective measuring 30 in. in diameter. While somewhat larger refractors could be designed and built with improved transmission and color correction over existing instruments, the gain in light-gathering power is nearly offset by the greater thickness of glass involved. Considerations of tube length and lens support further argue against such a course. These limitations do not apply to the reflector, which, though also subject to certain size limitations, can be made in much larger apertures. As a result, demands for increased light-gathering power have led to the concept of astrometric reflectors. One such instrument may be described as a classical Cassegrain system with zero-power secondary mirror. A moderately large aperture ratio is required to suppress coma at minimum acceptable field, and the flat secondary ensures a measure of optical stability against inevitable tube flexure. The existing instrument has an aperture of 60 in.

Attempts to increase the speed of doppler spectrographs have also led to the design and construction of partly or wholly reflective systems. At a given linear dispersion the speed of a spectrograph is primarily a function of camera aperture. The larger the aperture the faster the system; hence reflecting gratings are used to overcome the light losses inherent in thick prisms.

The use of classical astrographic systems includes the precise determination of stellar parallax, the proper motion of stars, radial velocity, relative and absolute positions of components of resolved binaries, and the position of planets and satellites relative to the stellar background. Astrographic reflectors are employed primarily in the observation of objects too faint for study by classical astrographs. Small, specialized catadioptric wide-field instruments are used in the photographic tracking of artificial satellites at moderate precision levels.

Photometric Systems: Instruments for photometric studies include refractors and reflectors. With suitable filters photoelectric magnitudes can be obtained with both types of telescopes since color aberration is not critical to the measurement of radiant energy at the collecting surface of the photoelectric photometer or radiometer. The latter is always located on the optical axis of the telescope where the imagery is adequate for purposes of tracking. Reflectors become prominent in the photoelectric photometry of very faint stars, since large apertures are then called for. This has led to the development of so-called "light collectors," or zero-field compound reflectors, whose on-axis imagery is not fully corrected.

The photographic determination of stellar magnitudes is based on both refracting and reflecting telescopes. Thus photographic refractors in conjunction with blue-sensitive emulsions have been extensively used in the wholesale measurement of the photographic magnitudes of stars. Again, the study of very faint stars is based on the use of reflectors. While photographic photometry is then often done at prime focus, the fields available are limited. As a result, wide-field Schmidt-type catadioptric telescopes have been used for photographic photometry. Unfortunately these and related instruments suffer from vignetting, and their typically high aperture ratios lead to rather small and hard star images which are not considered ideal for precision photographic photometry. Schmidt cameras are ideal for nonmetric sky mapping and morphological studies of star systems and nebulae. They can serve effectively as scouting instruments for very large small-field reflectors. See **Schmidt camera.**

Spectrographs are used in conjunction with reflectors, since reflecting optics are free of color aberration. Low-dispersion spectrographs used in the classification of stellar spectra are often operated at prime focus, and this includes the high-speed slitless or nebular spectrographs. Moderate-dispersion spectrographs are generally fed at focus of compound, most often Cassegrain-type reflectors, while high-dispersion spectrographs are always stationary instruments operated at the so-called Coudé focus of the compound reflector. Slitless spectra of stars obtained for purposes of spectral classification are most efficiently obtained by mounting a full-aperture shallow-angle objective prism in front of the telescope objective. Every star image in the field of the telescope is thus transformed into a spectrum of the star. Types of telescopes best suited for this purpose include the Schmidt catadioptric telescope and small astrographs. Slitless objective-prism spectra obtained for the purpose of measuring doppler shifts involve so-called Fehrenbach objective prisms in combination with small-aperture astrographs.

The use of photometric systems includes the determination of stellar magnitudes, color indices; the study of variable stars, eclipsing binaries; the classification of stellar spectra, the physical interpretation of spectra through spectrophotometry; the measurement of albedo and infrared planetary temperatures, and the study of polarized sources.

—Joost H. Kiewiet de Jonge, *University of Pittsburgh,* and Lloyd M. Geertz, *Owens-Illinois*

Asynchronous A property of a device whose operation is not a direct function of the clock cycles in the system. Instead, the operation is started as a result of the completion of a previous operation. Compare with **Synchronous.**

Regarding a computer channel [see **Channel (CPU)**], an asynchronous channel is not dependent upon the computer clock pulses to transmit information to and from the input or output device. The information is transmitted under the control of interlocked control signals. That is, when a device has data to send to the channel, it activates a service request signal. In response to this signal, the channel activates a "service out" signal which, in turn, activates a "service in" signal in the device and deactivates the request signal. The data are transferred to the channel in coincidence with "service in," and the channel acknowledges receipt of the information by deactivating "service out."

Another example of an asynchronous operation occurs in the operation of an analog-to-digital subsystem. In this case, the system issues a command to the subsystem to read an analog point and then proceeds to the next sequential operation. The analog subsystem proceeds with the analog-to-digital conversion, and when the conversion is complete, the subsystem interrupts the system to signal the completion.

—W. W. Welz, *IBM Corporation*

Atomospheric Moisture, Measurement of (See **Hygrometer; Hygrothemograph;** and **Psychrometer.**)

Atomic-absorption Spectrophotometer (See **Flame photometer-spectrophotometer.**)

Atomic Clock (See **Clock, atomic.**)

Attitude Indicator (See **Flight-control systems.**)

Audiometer A pure-tone audiometer is a device for generating tones of relative purity, that is, tones with a very low harmonic content. These pure tones cover the range from 125 to 8,000 Hz—the important range of human hearing. Most sounds of human speech, especially the consonant elements, fall between 500 and 2,000 Hz. The audiometer can produce these frequencies over a very wide and critically calibrated intensity range. The instrument is calibrated in decibels relative to average normal threshold of hearing. With this instrument, it is possible to determine with great precision the existence, type, and severity of a given hearing problem. Tests conducted include the air-conduction audiogram and the bone-conduction examination. In audiometers, such as that shown in Fig. A-85, calibrated white-noise masking is incorporated for temporarily deafening one ear while the other is being tested. A reading of 0 db on the audiometer dial equals the level at which the existence of that sound is barely perceptible to the average young adult. The true sound pressure relative to 0.0002 dyne/cm^2 is frequency-dependent. Most pure-tone audiometers have intensity steps of 5 db, thus making threshold determination relatively easy. Also see **Acoustic instrumentation,** and **Ear, instrumental measurements of.**

—Hall Cary, *B & K Instruments Inc.*

Autocollimator An optical instrument for directly measuring small angles of tilt of a reflective surface. The instrument effectively combines the functions of a collimator and a viewing telescope into a single system. Diffused light is directed at a partially silvered mirror which diverts part of the beam forward past a target cross hair to the reflective surface to be observed. The reflected beam is passed back into the autocollimator and focused by the objective so that the shadow of the cross hair appears in a plane which coincides with the image of the cross hair. A measuring device, such as a filar micrometer, is used to determine the displacement of the shadow of the cross hair with respect to the cross hair per se. This displacement then is related to the angle of tilt of the reflector. Accurate readings can be taken direct to 0.5 of arc—within a 10-min range, up to a distance of 30 ft.

—E. Gordon Watson, *Gaertner Scientific Corporation*

Automatic Control Generally, automatic control may be defined as that branch of science and technology which is concerned with the development of devices, plants, and systems that operate without direct human intervention and that assume

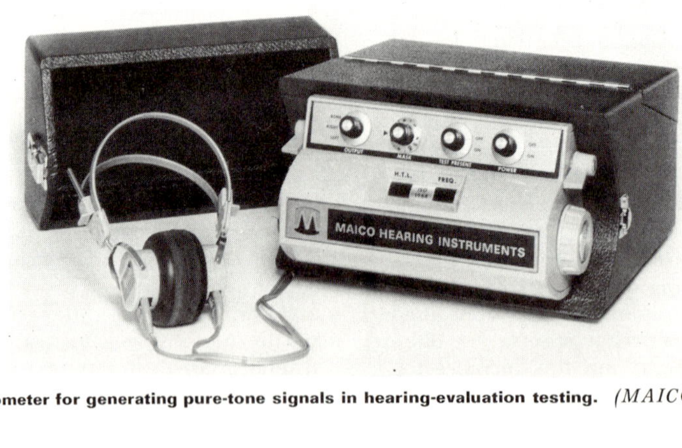

Fig. A-85. Audiometer for generating pure-tone signals in hearing-evaluation testing. *(MAICO.)*

the responsibility for performing certain types of human mental work.

Problems involved in automatic control generally pertain to the regulation of one or more variables of a control system at certain constants or within certain desired limits despite the fluctuations caused by disturbances or parameter changes. In the field of trajectory control and servomechanisms, the variables to be controlled may be position, velocity, acceleration, or jerk (a derivative of acceleration). In process control, the variables may be temperature, pressure, flow, level, chemical concentration, and many others. In power and other electrical and communications systems, important variables will include voltage, current, power, and energy. Also see **Control system.**

Early Windmill Control: Automatic-control history goes back at least as early as the middle of the eighteenth century, when Andrew Meikle invented his first automatic direction controller for windmills. The problem involved was to keep the huge windmill always facing into the wind regardless of which direction the wind blew. This was accomplished, as shown by Fig. A-86, by building another relatively smaller auxiliary windmill perpendicular to the main windmill—so that whenever the wind changed its directions, the auxiliary windmill would drive the main windmill, by means of gearing, until the windmill was facing into the wind.

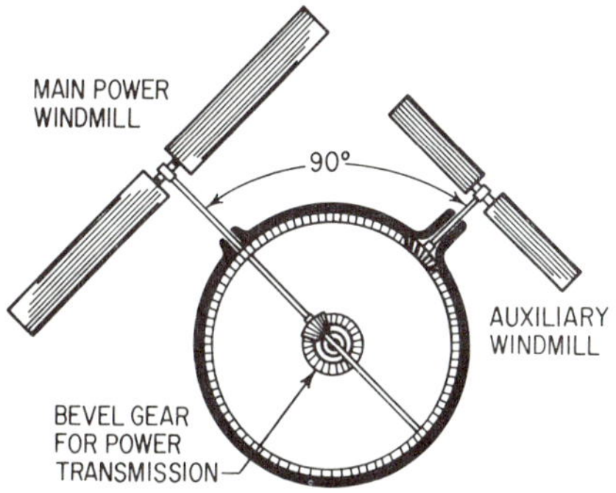

Fig. A-86. Windmill-direction control (1750).

Steam-engine Controls: In 1788, James Watt invented the first speed controller for his steam engine, marking another milestone in automatic-control history. The two flyballs, which are geared to the engine-output shaft, as shown in Fig. A-87, rise or fall, dependent upon whether the speed of the engine increases or decreases. The motion of the flyball, in turn, controls the position of the steam governing valve. It should be noted that this arrangement is a closed-loop system. The output-shaft speed is fed back through the gear train and spindle, as shown in Fig. A-88. Also see **Governor.**

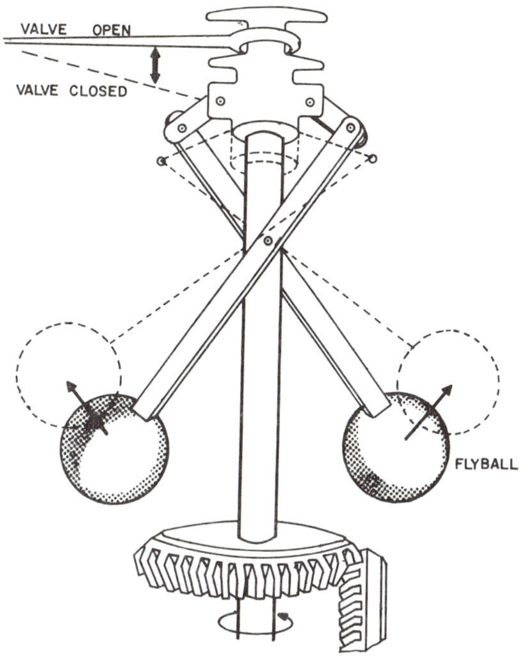

Fig. A-87. Steam-engine-speed control (1788).

Further History: The two inventions just described, together with others such as the steering servoengine on ships and a simple voltage regulator, covered man's achievements in the field of automatic control up to about the first quarter of the twentieth century. Other related activities during this earlier period included the developments of mathematical theories which have had an immense influence on the control theories developed later. These include Fourier's transformation (1779), Cauchy's theory of functions of a complex variable (1789–1857), and Routh's stability criterion (1877).

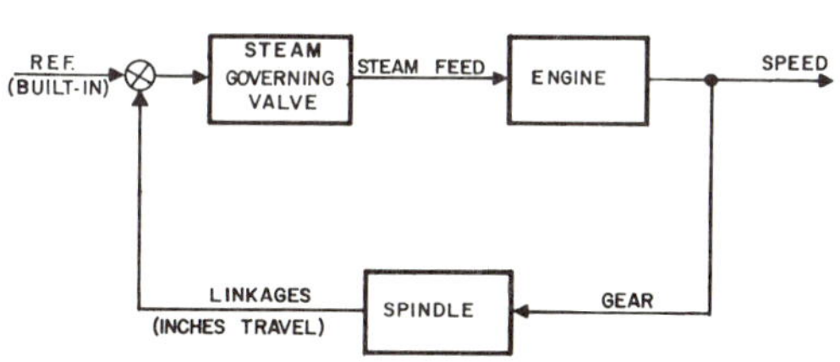

Fig. A-88. Block diagram of steam-engine-speed control.

In the second quarter of this century, science and technology developed rather rapidly in almost all fields because of necessity. For instance, more accurate temperature regulators were needed for more and more complicated chemical plants; more sophisticated automatic machine tools were required in order to control product quality by eliminating human factors, and thus increase the product's quantity as well; more accurate and faster response servosystems were needed for high-speed aircraft guns; and above all, more powerful electronic amplifiers were required for the enormously expanding radio-communicating industry.

At first, no connections between these various developments were recognized. However, the developments became more rapid in the late 1940s and 1950s once it was realized that all these control systems, different in nature as they are, depend upon common control principles—namely, the feedback-control and compensation theories developed during the second quarter of this century for radio-amplifier design. It was these theories that blazed the fields of automatic control for the decades that followed. Today, the design of control systems in a modern chemical plant is based on the same principles used in aircraft flight control or guided missiles.

Feedback-control theory is employed to achieve the goal of automatic control by means of "comparing and correcting" the errors (Fig. A-89) between the current variable reading and the actual desired target point. This process continues until it reaches a point where error theoretically should be zero and practically is considered tolerable (within the band of tolerances, e.g., $\pm$ 0.5 percent).

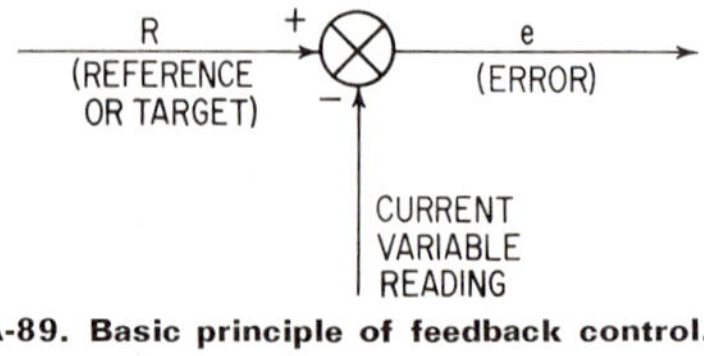

Fig. A-89. Basic principle of feedback control.

However, the same goal can be achieved by the technique developed and applied in various areas in recent years. The means is to predict, based on the up-to-date information collected from process streams, the control signal before it happens. This is generally known as *feedforward control.* Not like feedback control, feedforward control requires a reasonable amount of knowledge of the process being handled, and often a certain mathematical model is needed in order to have a reasonably precise calculated prediction.

In most applications, the best results can be expected when a combination of both feedback and feedforward is applied, by taking advantages of both. A typical example in process industry is pH control.

While feedback is taking care of local flow-control loops, feedforward, as shown in Fig. A-90, is applied to predict and overcome the nonlinearity of pH by actually calculating $[H^+]$ or $[OH^-]$ ions required and thus to determine the amount of reagents needed to neutralize the process stream. In addition, another feedback loop consisting of a pH transmitter and a controller at the process-output end is included to take care of a possible drift resulting from the accumulation of errors in calculations.

The recent trends in the development of automatic control have been in the direction of optimizing control and adaptive control.

The former involves the problem of determining a best control policy which fulfills certain criteria by optimizing the controlled parameters. A very simple example to illustrate this effect is the time-optimal control. This very special case, as shown in Fig. A-91, involves the determination of cutoff time T_2 so that the total time required to bring the process to a step-input change will be minimized. This scheme often is referred to as *bang-bang* control.

Adaptive control, on the other hand, is concerned with the situations of a control system that are under the influences of environmental changes. This concept was initiated in the aircraft industry, which was involved in the design of hypersonic and supersonic aircraft flight-control systems. Adaptive control then spread to applications in other fields. An adaptive-control system should be capable of handling all situations in spite of the fluctuations of variables caused by the surroundings (e.g., the fluctuations of aerodynamic parameters as the aircraft flies from the heavy atmosphere at low altitudes to the rare atmosphere at very high altitudes). Generally, this will require a nonlinear controller with adjustable gain; in other cases of greater complexity, this may require modeling control (including mathematical models as well as real models). See **Adaptive control.**

REFERENCES

Newton, George C., Jr., Leonard A. Gould, and James F. Kaiser: "Analytical Design of Linear Feedback Controls," John Wiley & Sons, Inc., New York, 1957.

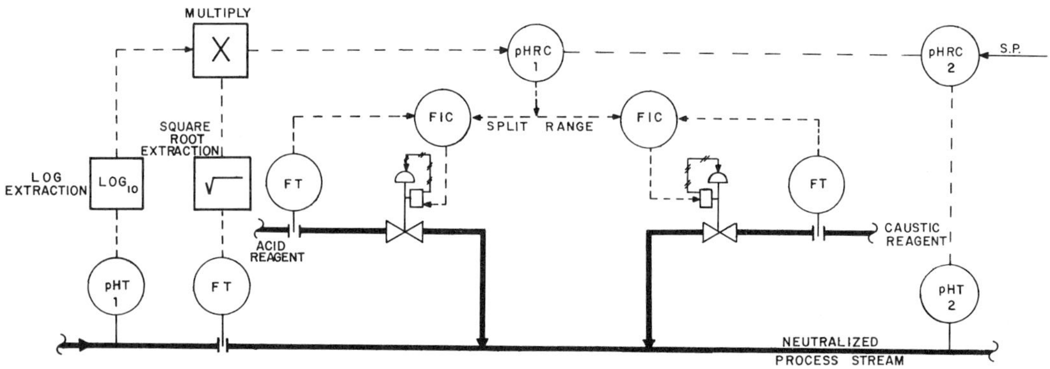

Fig. A-90. Application of feedforward and feedback control to a pH-control system.

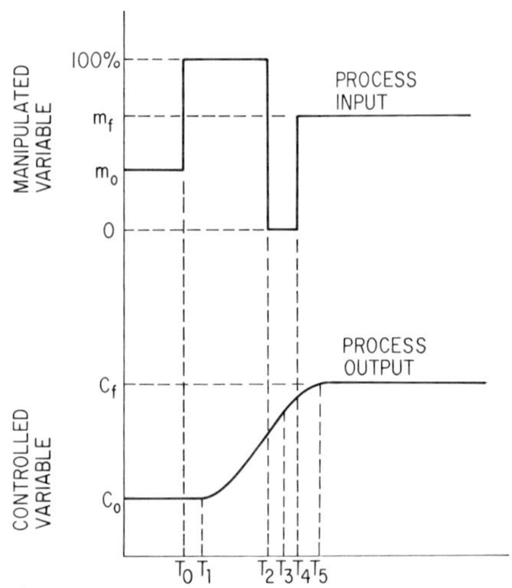

Fig. A-91. Time-optimal control.

Grown, G. S., and D. P. Campbell: Control System, *Sci. Am.,* September, 1952.

Tou, J. T.: "Modern Control Theory," McGraw-Hill Book Company, New York, 1964.

Li, Y. T., and H. P. Whitaker: Performance Characterization for Adaptive Control System, *Intern. Fed. Autom. Control* (IFAC), Symposium on Optimizing and Adaptive Control, Rome, Italy, 1962.

Braun, L., Jr.: On Adaptive Control System, doctoral dissertation. Brooklyn Polytechnic Institute, New York, 1959.

Shinskey, F. G.: "Process-control Systems," McGraw-Hill Book Company, New York, 1967.

Toro, V. D., and S. R. Parker: "Principles of Control Systems Engineering," McGraw-Hill Book Company, New York, 1960.

Chestnut, H., and R. Mayer: "Servo-mechanisms and Regulating System Design," vols. I, II, John Wiley & Sons, Inc., New York, 1963.

Courant, R.: "Calculus of Variations and Supplementary Notes and Exercises," New York University, 1956–1957.

Pontryagin, L. S., V. G. Boltyanskii, R. V. Gamkrelidze, and E. F. Mishchenko: "The Mathematical Theory of Optimal Processes," Interscience Publishers, Inc., New York, 1962. (Translated from Russian.)

Coughanowr, Donald R.: New Algorithms for Control, *Chem. Eng.,* June, 1969.

Aris, R.: "The Optimal Design of Chemical Reactors," Academic Press, Inc., New York, 1961.

Bellman, R.: "Dynamic Programming," Princeton University Press, Princeton, N.J., 1957.

Fahlgren, C. E.: Process Analog Computer, *Instr. Technol.,* November, 1968.

Ryan, Frank M.: Supervisory Control Systems, *Control Eng.,* January, 1963.

Holmes, R. B.: Application of Supervisory Control Systems, 19th Annual Symposium on Supervisory and Direct Digital Control Techniques, ISA, New Jersey and New York sections, April, 1967.

Wherry, T. C., and J. R. Parsons: Guidelines for Profitable Computer Control Systems, Texas A. & M. Symposium on Instrumentation for the Process Industries, January, 1967.

Shinskey, F. G.: Feedforward Control Applied, *ISA J.,* November, 1963.

Considine, Douglas M. (ed.): "Process Instruments and Controls Handbook," McGraw-Hill Book Company, New York, 1957.

Considine, Douglas M., and S. D. Ross (eds.): "Handbook of Applied Instrumentation," McGraw-Hill Book Company, New York, 1964.

—Anthony Y. Chiang, *Bristol Division, American Chain & Cable Company, Inc.*

Automatic Drafting Machine (See **Drafting system, automatic.**)

Automatic-Manual Transfer (See **Display-command operator/machine interface.**)

Automation The term *automation* first appeared in the literature during the early 1930s and, at that time, simply meant "the automatic transfer of parts and materials from one production operation to the next." In essence, this meant an extension of mechanization of individual machines to include the mechanized interconnection of machines. The meaning of the term has broadened over the years and now generally is interpreted as a technique for minimizing or eliminating the need for manual intervention in the operation of a system.

Autopilot (See **Flight-control systems.**)

Average-position Control (See **Proportional-input control.**)

B

Backlash Backlash, or mechanical hysteresis, is defined as that lost motion or free play which is inherent in mechanical elements, such as gears, linkages, or other mechanical-transmission devices that are not rigidly connected. Backlash causes an effect similar to that of a hysteresis loop in that the output magnitude will assume a different value for a given value of input, dependent upon whether the input is increasing or decreasing.

Ball Valve (See **Control valve.**)

Ballistic Galvanometer (See **Galvanometer.**)

Ballistocardiography (See **Cardiography.**)

Bang-Bang Control A control system in which corrective action is always applied to the full extent of servomotion to give two-position or on-off control.

Barograph A recording barometer which makes a continuous record of the ambient pressure. As shown in Fig. B-1, the instrument consists of (1) an aneroid-type sensing element, the output of which is transmitted by a practically friction-free linkage to (2) a pen which, in turn, creates a record on (3) a rotating chart. The chart drum may be driven by a spring-wound or electric-clock mechanism. Usually 7-day charts are used, since relatively long-term pressure characteristics and trends are needed to make day-to-day weather forecasts. Mercury barographs have been made but are not as well accepted as the aneroid instrument, because the latter is more portable and can withstand rougher treatment. The microbarograph is a refinement of the standard barograph and is designed to record small and rapid pressure variations on a chart having a magnified scale.

—C. Foster Aldridge, *Taylor Instrument Consumer Products Division, Sybron Corporation*

Barometer An instrument used mainly to measure atmospheric pressure (commonly defined as the weight of a vertical column of air extending upward from the measuring instrument to the top of the atmosphere). Atmospheric pressure at sea level is approximately 14.7 psi; this value often is used as a unit of pressure—called 1 atmosphere. The mercury barometer, when properly corrected for several variables, generally is considered the most fundamentally satisfactory type of barometer.

Fig. B-1. Barograph with aneroid pressure-sensing element at right and chart drum at left. *(Taylor.)*

Essentially, a mercury barometer is a clean glass tube approximately 3 ft long with uniform bore (internal diameter), closed at one end and filled with clean mercury. The open end is temporarily closed and submerged in clean mercury in an open vessel or cistern. When the temporary seal is removed and the tube is held vertically, the mercury recedes—creating a vacuum in the upper end of the tube. At sea level, the mercury-column height will be approximately 30 in. (29.921 in.), or 760 mm above the surface of the mercury in the cistern. The weight of the mercury in the tube is exactly balanced by the weight of the air (or atmospheric pressure) on the open cistern. Thus, as the atmospheric pressure varies, the height of the mercury column will change. Because of the temperature coefficient of expansion of mercury, barometer readings must be temperature-adjusted.

The three main types of mercury barometers are:

1. *Fortin type*—wherein the barometer scale position is fixed and a portion of the cistern is flexible so that the mercury level in the cistern can be adjusted to the exact zero position on the scale each time a reading is taken.

2. *Newman type*—in which the cistern is a rigid container and the scale is vertically adjustable to accomplish the zero-set requirement.

3. *Kew type*—which has no zero adjustment because the nonlinear layout of the scale satisfactorily compensates for the varying level of the mercury in the cistern brought about by pressure changes.

In addition to temperature, barometer readings should be corrected for variations in gravimetric force due to the latitude and altitude of the instrument location. Also see **Aneroid barometer.**

—C. Foster Aldridge, *Taylor Instrument Consumer Products Division, Sybron Corporation*

Barometer Aneroid (See **Aneroid barometer.**)

Barometric-pressure Error The change in reading of an instrument due to change in barometric pressure. This occurs in highly sensitive thin-bulb mercury-in-glass thermometers. Also see **Error, measurement; Thermometer, filled-system;** and **Thermometer, liquid-in-glass.**

REFERENCE

Swindells, J. F.: Calibration of Liquid-in-Glass Thermometers, *Natl. Bur. Std. (U.S.) Monograph* 90, p. 21, 1965.

Barostat A device used to calibrate pressure-measuring instruments and, in particular, to calibrate surveying altimeters which are sufficiently sensitive to indicate a change of 1 ft elevation. To obtain the most accurate results, surveying altimeters should be calibrated at frequent points throughout their range immediately before and after use. To do this in a conventional pressure chamber, using mercurial or aneroid pressure standards, would be time-consuming and would require unusual precautions. The barostat operates on the pneumatic force-balance principle in which a series of precisely machined weights, usually calibrated to produce constant-pressure steps equivalent to 100-ft changes in altitude, is used. The consistent repeatability and degree of precision achieved by a barostat exceed that which in normal practice can be obtained from mercury or aneroid barometers.

—C. Foster Aldridge, *Taylor Instrument Consumer Products Division, Sybron Corporation*

BASIC BASIC (Beginner's All-purpose Symbolic Instruction Code) is a computer language developed at Dartmouth College, Hanover, N. H., by Kemeny and Kurtz (Ref. 1), for time-sharing use by students and other nonprofessional programmers. The language has been adopted by some of the major computer manufacturers and is offered through time-sharing services. It has been made widely available and has become much more popular than any other similar language. It is characterized by a very simple statement form, normally requiring that the first word in the statement be one of a small number of key words and having a restricted naming convention for variables which allows only a two-character name, the first alphabetic and the second numeric. The input/output statements are simple but quite sufficient for the needs of most casual users. An example of a program written in BASIC follows:

```
05 LET E = -1
10 READ A,B,C
15 DATA 1,2,3
20 LET D= A/B*C
30 IF D=4 GO TO 60
40 IF D=3/2 GO TO 80
45 PRINT A,B,C,D,E
50 STOP
60 LET E = +1
70 GO TO 45
80 LET E = 0
90 GO TO 45
100 END
```

REFERENCE

1. Kemeny, J. G., and T. Kurtz: "Basic Programming," John Wiley & Sons, Inc., New York, 1967.

—A. J. Arthur, *IBM Corporation*

Batching Systems, Automatic-weight Increased productivity, close control of ingredient proportions, savings in labor, and elimination of material giveaway are all benefits deriving from automatic-weight batching systems. Information on ingredient usage and on formulas delivered can be recorded and tabulated for production-control and accounting use. The system control panel can program the routing of incoming ingredients to designated bins, the feeding of ingredients to the batching scale, the weighing, volumetric addition of material, discharge of batch to mixer, and routing of mixed batch to storage or bagging line or delivery truck.

To specify an automatic-weight batching system, the objectives to be achieved by the system must first be clearly defined. Then, a list should be made of the number and the flow characteristics of each ingredient; the number of formulas to be used, together with the number and weights of ingredients in each; the frequency of formula usage; the maximum-minimum formula-weight range; the frequency of formula changing; the optimum batch size; the number of batches required required per hour; the required accuracy; the type of weight-information readout; and the other control functions such as material-level indication in bins, controls for loading supply bins, start-stop controls for mixers, take-away conveyors, liquid injection by meter or pump, batch-stop counters, timers for mixing cycles, indicating lights to monitor equipment operation, and motor-starter centers. Good economy and electrical uniformity result from including controls of associated equipment in the system control panel, because the system designer is so familiar with the total process that he can readily extend the design. For highest system accuracy and reliability, bin and feeder designs must be matched to the flow characteristics of the ingredients handled to assure a uniform feed stream into the scale hopper. The scale hopper can use either a lever system or an electronic load cell for weight sensing.

Weight-programming devices include an individual weight selector where formula changes are frequent and nonrepetitive; a formula capsule where formulas are repetitive and information must be confidential, with each capsule set for a different formula; and a punched card where formulas are repetitive and where nonweighing functions must also be controlled. Automatic-weight batching systems can offer such refinements as dual feed rates for highest accuracy, monitoring of accuracy, and monitoring with interlocking to prevent any unprogrammed events in the batching cycle. The principal elements of a batching system are shown in Fig. B-2.

Indication and recording in these systems give a visual, printed, or graphic record of performance; this information can be used to check batch quality, for inventory control, or for accounting purposes. These readout devices can be located at the scale or in the control panel or office. Where numerical records are needed, a digital printer can be supplied. To tie in with bookkeeping records, an automatic typewriter can be used to give weight printout and space for adding other information. To check on ingredient usage, a multiple-ingredient totalizer can be used which has as many counters as ingredients. Each counter records the weight of ingredient drawn from its bin. Interfacing with data-processing equipment and computers is also possible.

Circuit Logic in Automatic Batching Systems: Circuit logic is the basic plan behind the system control which causes it to operate as planned. In

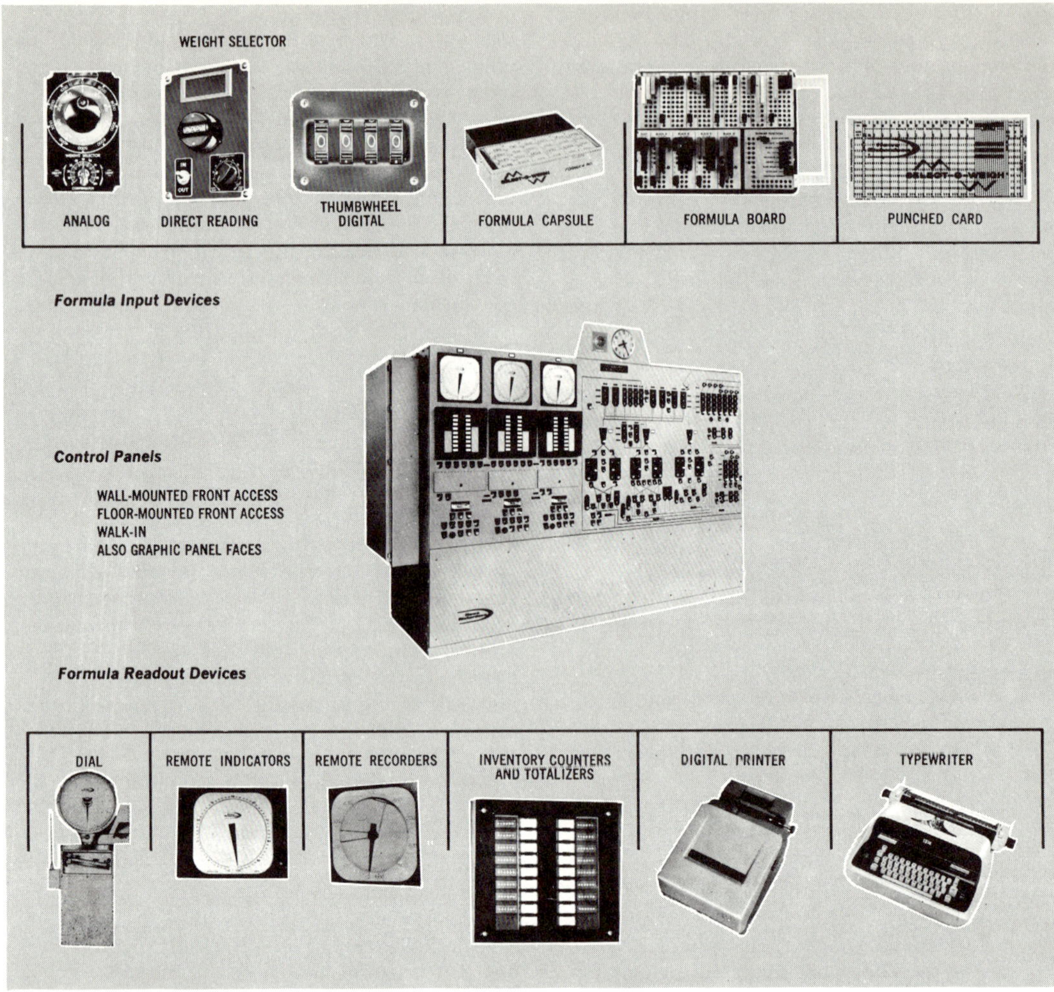

Fig. B-2. Principal elements of an automatic-weight batching system of the type widely used in batching rubber, plastics, glass, baked goods, candies, ceramics, concrete, asphalt, and other formulated products. *(Howe Richardson.)*

the case of an automatic batching system this logic must encompass the automating of ingredient feeding and discharging in a programmed sequence and also, usually, provide a record of events and amounts batched. In a typical example:

Using a pendulum dial for weight sensing and a lever system for weighing, a low-torque transducer (potentiometer) is attached to the dial spindle. This unit emits a position signal that is fed into a control relay, where it is compared with the preset weight signal coming from the potentiometer of the weight-programming device (weight selector, formula capsule, or formula card). As an ingredient is fed into the scale hopper, a point is reached where the dial potentiometer signal and the preset weight signal are of the same value. At this point the control relay will cause the feeder to stop. To allow for the inertia of the mechanical components and the material in motion, an adjustable amount of "compensation is introduced into the feeding cycle which will cut off the feeding slightly ahead of time. The amount of compensation will vary with the flow characteristics and feed rate of each ingredient. Where highest accuracy at high output is required, a two-step feed rate can be used—90 percent full (or fast) flow, 10 percent dribble (or slow) flow.

Where remote analog indicators and recorders

are required, they can be driven by a servomechanism which in turn is controlled by a transducer driven by the dial. For on-site or remote printing, calculator printout, typewriter printout, inventory totalizing, and for interfacing with data-processing equipment and computers, digital signals are needed. To get the dial's analog signal to a digital form, an analog-to-digital converter is mounted on the dial which will give a discrete type of electrical signal. Further, electrical signals can be obtained from the weight programmer to identify ingredients being fed, the order in which they are added, and the feeder (bin) numbers.

A monitoring subsystem should be part of any automatic batching system. For monitoring individual weightings, another control relay is used which compares weight delivered (from the dial signal) with the preset weight (from the programmer signal) to determine if it falls within the preset tolerance. If not, the relay can be made to stop the cycle and actuate an alarm device. For monitoring the occurrence and sequence of events in a weighing cycle, many different combinations of relays can be used. Monitoring subsystems must be adequate enough to signal and identify unprogrammed events but must not restrict system flexibility. The system designer should strive to use plug-in-type units and modular design for simplicity of construction and servicing.

Also see **Scale, industrial;** and **Weight, measurement of.**

REFERENCES

Considine, D. M.: "Industrial Weighing," Reinhold Publishing Corporation, New York, 1948.

Considine, D. M.: Process Weighing, *Chem. Eng.,* Aug. 17, Sept. 14, 1964.

Kendall, D. B.: Weight and Weight Rate of Flow, in D. M. Considine and S. D. Ross (eds.), "Handbook of Applied Instrumentation," McGraw-Hill Book Company, New York, 1964.

—Arthur J. Burke, *Howe Richardson Scale Co.*

Baud A baud is the unit of telegraph-signaling speed, derived from the duration of the shortest signaling pulse. A telegraphic speed of one baud is one pulse per second. The term *unit pulse* is used sometimes for the same meaning as the baud. A related term, the *dot cycle,* refers to an on-off or mark-space cycle in which both mark and space intervals have the same length as the unit pulse.

Baudot Code A teleprinter code that uses combinations of five and six marking and spacing intervals of equal duration. The five-unit code gives 32 possible characters; and the six-unit version gives 64 possibilities. The code is used in radio and wire teleprinter operations.

Bayard-Alpert Gage (See **Vacuum gaging.**)

Beam Scale (See **Scale, beam.**)

Belt Feeder (See **Feeder, volumetric; Feeder, weigh;** and **Weighing, radiation methods for.**)

Belt Scale (See **Scale, industrial.**)

Betatron (See **Accelerator.**)

Bias, Observer (See **Error, measurement.**)

Binary-coded Decimal (BCD) A coding system for characters using 6 binary digits to represent each character. (See **Data-conversion codes.**)

Binary Number System The number system created by using the number 2 as the base (radix instead of 10 as in the decimal system (see also **Number Systems**). The only digits are thus 0 and 1; the successive binary integers are 0, 1, 10, 11, 100, 101, 110, 111, 1000, etc.; and the binary fractions are 0.1 (1/2), 0.01 (1/4), 0.001 (1/8), etc. Although requiring many more positions to express a given number than the decimal system, the binary system has the advantage that there are many more physical devices or systems which may be used to represent binary numbers than there are for any other base. For example, any system with both an "on" and an "off" state, a positive and negative state, or a high and a low state can be used to represent binary numbers. The implementing engineer thus has a wide choice of mediums and systems with which to work. Successful binary digital computers have been built using mechanical, relay, electronic, and fluidic logical elements and using mechanical, relay, electronic, acoustic, magnetic, photographic, and other storage elements. Most so-called "decimal" computers actually are internally coded in binary (see **Binary-coded decimal**).

—A. J. Arthur, *IBM Corporation*

Binary State The property of a Boolean (see **Boolean algebra**) variable is that it can assume either of two mutually exclusive alternatives which are called binary states. It is common engineering practice to assign the symbols 1 and 0 to the two possible values of the Boolean variable.

In general practice, 1 represents the YES, ON, or TRUE alternative and 0 represents the other alternative, NO, OFF, or FALSE. In the logical circuits in a digital computer, the two states are represented by discrete voltage or current levels.

—W. W. Welz, *IBM Corporation*

Bit A *bi*nary dig*it*, that is, a 1 or 0. (See **Number Systems; Binary state.**)

Blackbody (See **Radiation pyrometer.**)

Blind, Sensory Aids for (See **Vision, sensory aids for.**)

Blind Controller (See **Self-actuated regulators.**)

Block Diagram, Computer A graphical representation of the hardware (see **Hardware computer**) in a computer system which indicates the paths along which the data or control information flows between the various parts of the system is referred to as a block diagram. Figure B-3 depicts an example of a block diagram of a central processing unit.

In a block diagram, each block generally represents a functional subdivision of the system being considered. The function represented by a single block may range from a very simple one, such as a single AND block, to a very complex function, such as a total computer. The example in the figure illustrates blocks of intermediate complexity.

—W. W. Welz, *IBM Corporation*

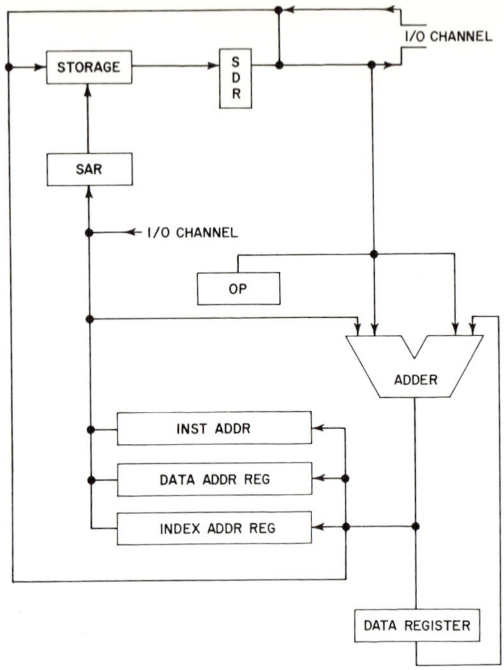

Fig. B-3. Central-processing-unit block diagram.

Blood, Measurements of (See **Cardiovascular instrumentation.**)

Boiling-point Determination An instrumental technique for (1) identifying various liquids, mostly organic, including azeotropes, and (2) checking the purity of liquids and the concentration of solutions. Essentially a laboratory tool, the method can be batch or continuous. To determine identity, the constant temperature of boiling under partial reflux, at known absolute pressure, identifies pure materials. However, additional physical measurements may be needed to identify unknowns, since different substances can have the same boiling point. The amount of impurity in a known substance can be related to variation in boiling point of the sample as compared with the known boiling point of pure material. Similarly, the concentration of solutions in evaporators can be determined by comparing the boiling-point temperature of the sample with that of a known concentration. The average batch analysis requires from 15 to 45 min with a sample of from 10 to 51 ml. Standard temperature and pressure measuring equipment can be used.

Bolometer A device used to measure microwave and infrared energy. The device contains a resistance element that changes in resistance when heated by the radiant energy. Used in infrared search and homing equipment in missile programs. Also used in waveguides to measure microwave power and standing-wave ratio. Two types of bolometer are in use, (1) the barretter and (2) the thermistor. Also see **Infrared detector, thermistor;** and **Resistance thermometry.**

Bonded Strain Gage (See **Strain gage.**)

Boolean Algebra Boolean algebra, originated by George Boole (1815–1864), is a mathematical method of manipulating logical relations in symbolic form. Boolean variables are restricted to two possible values or states. Possible pairs of values for the Boolean algebra variable are YES and NO, ON and OFF, TRUE and FALSE. It is common practice (especially in engineering) to

Logical Operation	Symbol	Definition
AND	$\bullet$	$A \bullet A = A,\ A \bullet 0 = 0,\ A \bullet 1 = 1,\ A \bullet \overline{A} = 0$
OR	$+$	$A + A = A, A + 0 = A, A + 1 = 1, A + \overline{A} = 1$
NOT	$-$	$A\overline{A} = 0,\ A + \overline{A} = 1$
EXCLUSIVE OR	$\oplus$	$A \oplus B = \overline{A}B + A\overline{B} = \overline{A \odot B}$
COINCIDENCE	$\odot$	$A \odot B = \overline{A}\,\overline{B} + AB = \overline{A \oplus B}$
NAND (OR SHEFFER STROKE)	$/$	$A/B = \overline{A} + \overline{B} = \overline{AB}$
NOR (OR PEIRCE)	$\swarrow$	$A \swarrow B = \overline{A}\,\overline{B} = \overline{A + B}$

Fig. B-4. Boolean algebra symbols.

use the symbols 1 and 0 as the Boolean variables. Since the digital computer typically employs signals having only two possible values or states, Boolean algebra enables the computer designer to combine mathematically these variables and manipulate them in order to obtain the minimum design which realizes a desired logical function.

Figure B-4 is a table of definitions and symbols for some of the logical operations which are defined in Boolean algebra.

—W. W. Welz, *IBM Corporation*

Booster, Valve (See **Valve positioner, pneumatic.**)

Boron Thermocouple A device for the detection of slow neutrons. This detector is made by coating a thermocouple junction with boron. The heating of the junction by the energy given off from the neutron-boron reaction is a measure of the slow neutron flux [neutrons/(cm^2)(s)]. In use, alternate junctions of a series of thermocouples are boron-coated. An assembly of chromel-alumel couples joined in series and having 10 coated junctions gives an output of about 5 mV in a neutron flux of 2×10^{12} neutrons/(cm^2)(s). Fluxes of this magnitude are found only in nuclear reactors.

Bourdon Tubes Patented by Eugene Bourdon in 1852, the bourdon tube still finds wide application in instruments, notably for pressure and force measurement. Although made in various forms, the principal configurations are the C shape, helical, and spiral. In the C-shaped bourdon, commonly used in dial-type pressure gages, pressure is applied at the fixed end, causing movement of the free end as the result of deformation of the cross section of the tube. Also see **Pressure gage, elastic-element.**

Brain, Measurements of (See **Central nervous system, intrumental analysis of.**)

Bridge Circuits, Electric In 1833, S. H. Christie discovered the Wheatstone-bridge circuit. At that time, the concept of resistance had not yet been thought of, but in 1843, Sir Charles Wheatstone applied Christie's circuit to the measurement of resistance. The circuit and its many modifications have been the major tools for the precise measurement of electrical quantities ever since. Bridge circuits are used to compare impedance elements by comparing the voltages or currents associated with them. This works both ways. The voltages and currents can be compared by using known impedance ratios to make potentiometers and current comparators. Bridge circuits can be thought of as arrangements of dividers for comparing the equivalent circuits of impedance elements. Thus, bridges also can be used for the comparison of divider ratios.

The term *bridge* originally referred only to the detector connection between the divider tap points. By present practice, the bridge circuit includes the dividers, the generators, and the null detectors required for supplying power and for finding the bridge balance. The characteristics of several bridges are summarized in Table B-1.

Impedance: Impedance is the ratio of voltage to current at a single frequency. Thus, in a series circuit with the same current in all elements, the voltage will be proportional to the impedance. In a parallel circuit with the same voltage across all elements, the current will be inversely proportional to the impedance. Because of these relationships, impedance elements can be connected in bridge circuits so that their impedances can be compared by comparing their voltage or current ratios.

Two-terminal Impedance: Impedance elements usually are thought of as having only two terminals. These terminals, however, also have stray impedances associated with them. In use there may be various amounts of lead and contact impedances connected in series between the impedance element and its connection point in the circuit. There are leakage impedances from

TABLE B-1. *Circuits and Balance Equations of Electric Bridges*

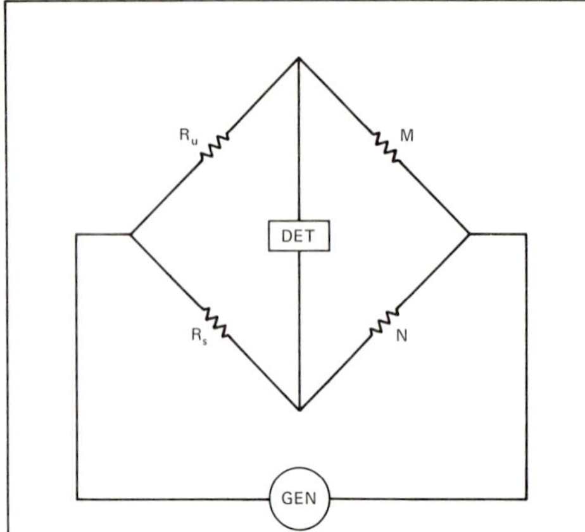

Purpose: Two-terminal resistance

Balance equation:

$$R_u = R_s\left(\frac{M}{N}\right)$$

Wheatstone Bridge

Remarks: Both ac and dc resistance measurement.

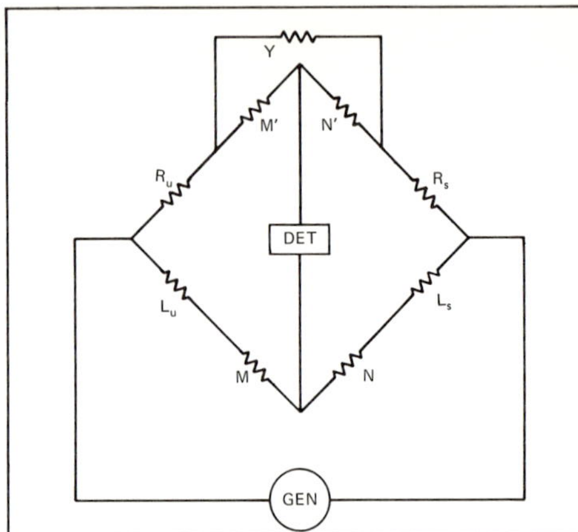

Purpose: Four-terminal resistance

Balance equation:

$$R_u = R_s\frac{M}{N} + \underbrace{\frac{N'Y}{M'+N'+Y}\left[\left(\frac{M}{N} - \frac{M'}{N'}\right) + \frac{L_u}{N}\left(\frac{M'}{N'} - \frac{L_s}{L_u}\right)\right] + L_u\frac{R_u}{N}\left(\frac{R_s}{R_u} - \frac{L_s}{L_u}\right)}_{\text{Error term}}$$

Kelvin Bridge

Remarks: Measures four-terminal resistors. Used for low-value resistors or precision measurements. Lead resistances of the unknown, R_u, and standard, R_s, resistors are included in M′, N′, Y and the generator leads. The remaining lead resistances are L_u and L_s. When $R_u:R_s = M:N = M':N' = L_u:L_s$, then $R_u = R_s(M/N)$

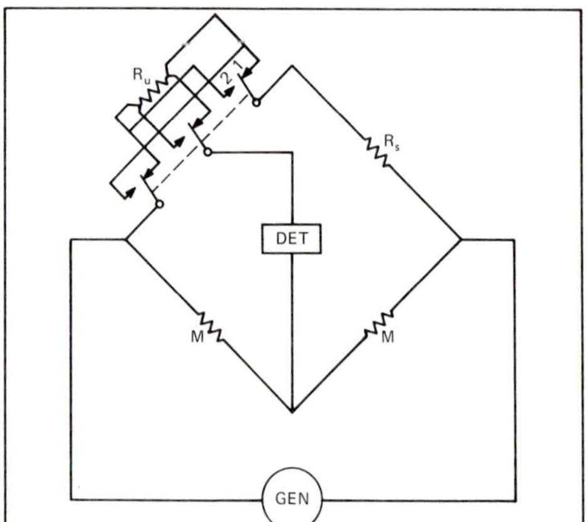

Purpose: Resistance thermometers

Balance equation:

$$R_u = \frac{R_{s(1)} + R_{s(2)}}{2}$$

Mueller Bridge

Remarks: Measures four-terminal resistors by averaging two readings with lead resistance effects reversed. Note that $R_u = R_s(AVE)$.

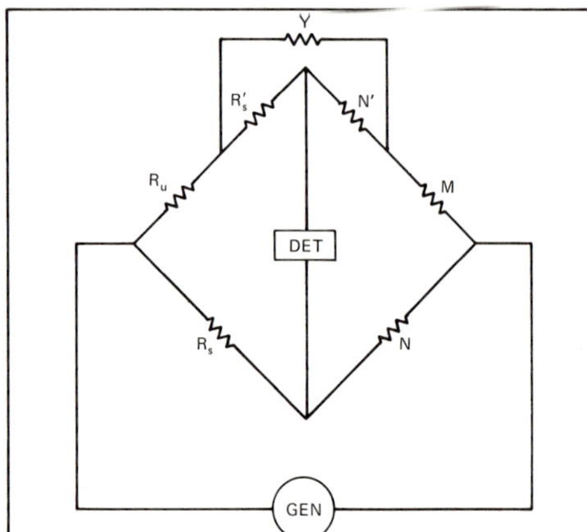

Purpose: Resistance thermometers

Balance equation:

$$R_u = R_s\frac{M}{N}$$

Smith III Bridge

Remarks: Measures four-terminal resistors.

TABLE B-1. *Circuits and Balance Equations of Electric Bridges (Continued)*

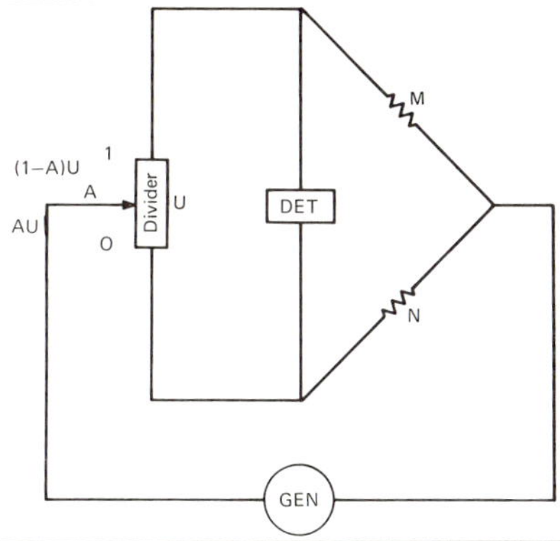

Purpose: Fault locating

Balance equation:

$$A = \frac{N}{M+N}$$

Murray Bridge

Remarks: Measures the distance from the near end of a line to a line fault. The distance is found in terms of the line length L and the ratio M/N.

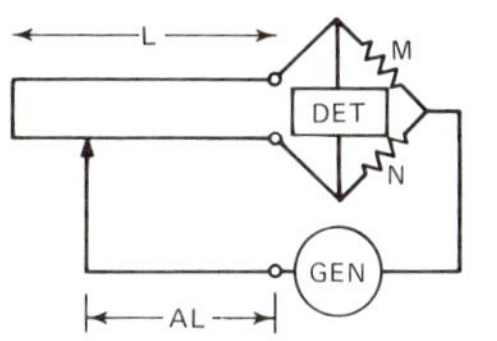

$$AL = \frac{2\,LN}{M+N}$$

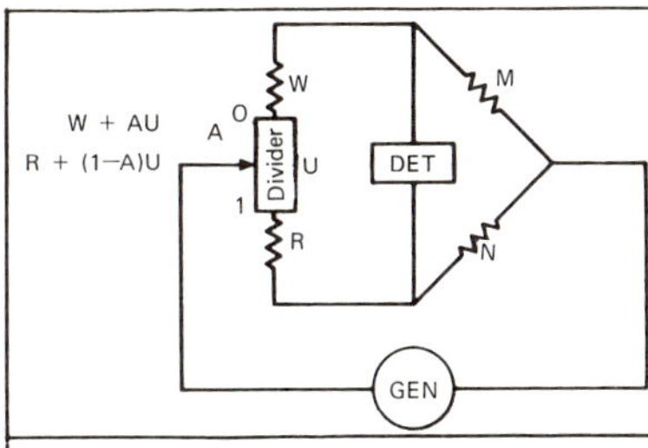

Purpose: Fault locating – resistance

Balance equation:	For $\frac{M}{N} = \frac{W}{U}$
$A = \frac{R}{U}\frac{M}{M+N} + \frac{MU-WN}{(M+N)U}$	$A = \frac{R}{U}\left(\frac{M}{M+N}\right)$
	$R = AU\frac{(M+N)}{M}$

Varley Bridge

Remarks: Measures the distance from the far end of a line to a line fault. The distance is found in terms of the line length L, line resistance U, the ratio M/N and the resistance R.

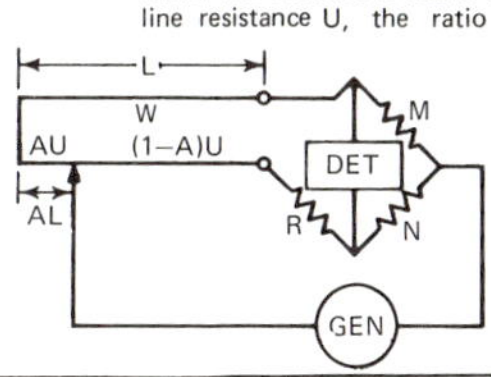

$$AL = \frac{RLM}{(M+N)U}$$

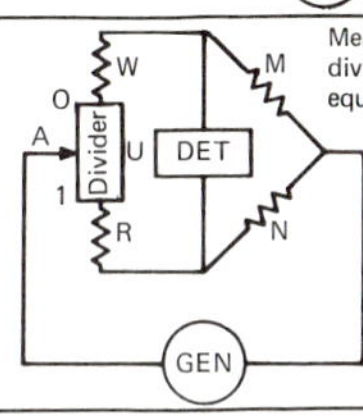

Measures the resistance R in terms of the divider setting A and the ratio M/N, after the equality M:N = W:U is established.

$$Z_u = R_u + \frac{1}{j\omega C_u}$$

$$D_u = R_u\,\omega C_u$$

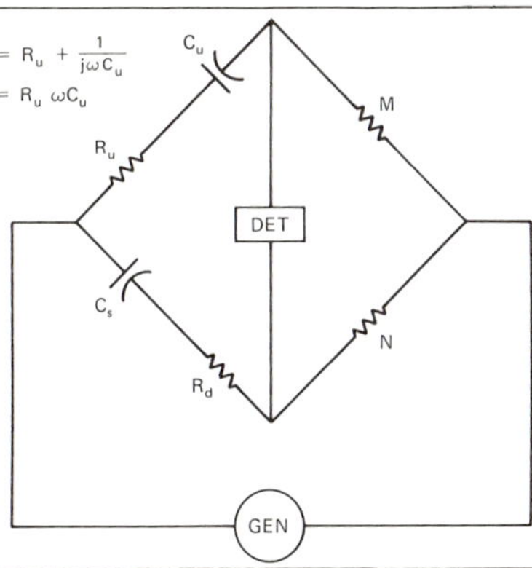

Purpose: Series capacitance

Balance equation:

$$C_u = C_s\frac{N}{M} \qquad R_u = R_d\frac{N}{M}$$

$$D_u = D_s = R_d\,\omega C_s$$

General-purpose–equivalent-circuit Bridge

Remarks: Measures unknown impedance in terms of equivalent circuit. Equivalent circuit is a capacitor and resistor in series. Used for capacitors which have a low dissipation factor.

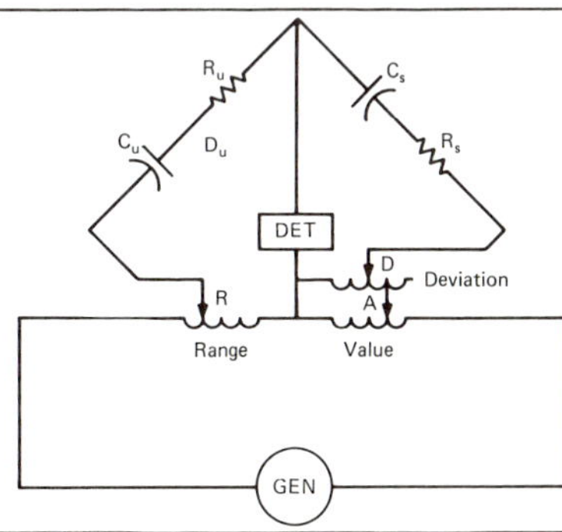

Purpose: Series capacitance

Balance equation:

$$C_u = C_s\frac{A}{R}\,(1 + D)$$

$$D_u = \omega C_s R_s$$

General-purpose–equivalent-circuit Bridge

Remarks: Measures unknown impedance in terms of equivalent circuit. Equivalent circuit is a capacitor and resistor in series.

TABLE B-1. *Circuits and Balance Equations of Electric Bridges (Continued)*

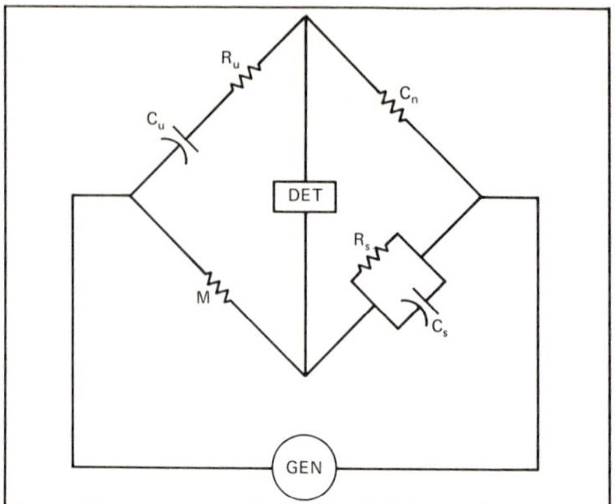

Purpose: Series capacitance

Balance equation:

$$C_u = R_s \frac{C_n}{M} \qquad R_u = C_s \frac{M}{C_n}$$

$$D = \omega R_s C_s$$

Schering Bridge

Remarks: Measures unknown impedance in terms of equivalent circuit. Equivalent circuit is a capacitor and resistor in series.

Used for precision capacitance measurement and for high voltage insulation leakage measurement.

$$Z_u = \frac{R_u}{1 + j\omega R_u C_u}$$

$$D_u = \frac{1}{\omega R_u C_u}$$

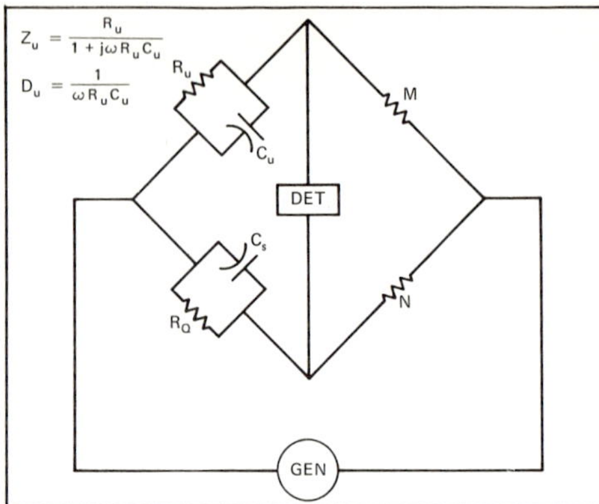

Purpose: Parallel capacitance

Balance equation:

$$C_u = C_s \frac{N}{M} \qquad R_u = R_Q \frac{N}{M}$$

$$D_u = \frac{1}{Q_u} = \frac{1}{R_Q \omega C_s}$$

General-purpose–equivalent-circuit Bridge

Remarks: Measures unknown impedance in terms of equivalent circuit. Equivalent circuit is a capacitor and resistor in parallel. Used for capacitors which have a high dissipation factor.

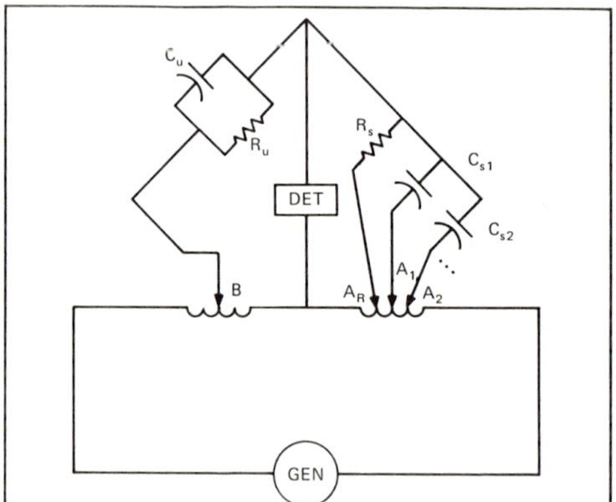

Purpose: Parallel capacitance

Balance equation:

$$C_u = \frac{1}{B}(A_1 C_{s1} + A_2 C_{s2} + \cdots)$$

$$\frac{1}{R_u} = \frac{1 A_R}{R_s B}$$

General-purpose–equivalent-circuit Bridge

Remarks: Measures unknown impedance in terms of equivalent circuit. Equivalent circuit is a capacitor and resistor in parallel.

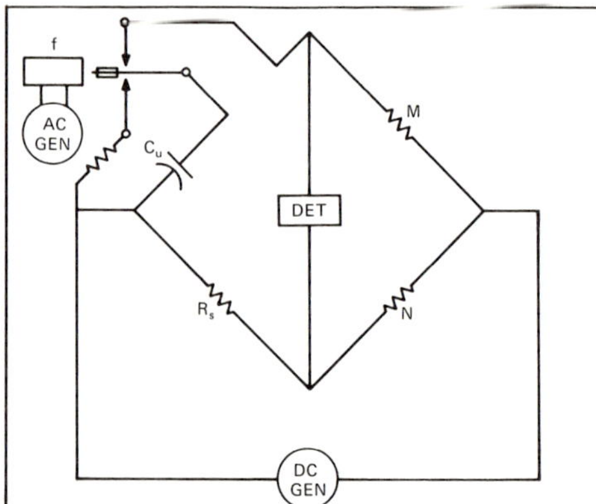

Purpose: Capacitance

Balance equation:

$$C_u = \frac{N}{f R_s M}$$

Maxwell-Commutator Bridge

Remarks: Measures an unknown capacitance. Used for precision measurement of capacitors with low dissipation factor.

TABLE B-1. *Circuits and Balance Equations of Electric Bridges (Continued)*

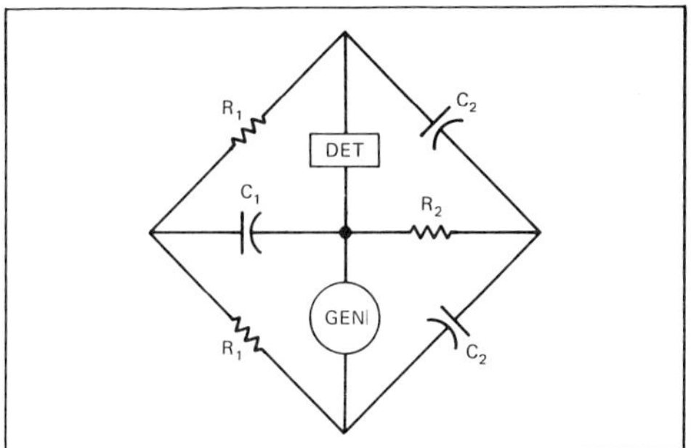

Purpose: Frequency, capacitance

Balance Equation:

$$\omega^2 C_1 C_2 = \frac{2}{R_2^2} \qquad \text{for } R_2 = 2R_1 \text{ and } C_2 = 2C_1$$

$$\omega^2 C_1^2 = \frac{1}{2R_1 R_2} \qquad f = \frac{1}{2\pi R_1 C_2} = \frac{1}{2\pi R_2 C_1}$$

$$C_2 R_2 = 4C_1 R_1$$

Parallel T Bridge

Remarks: Measures an unknown frequency or low dissipation factor capacitor. Used at higher frequencies.

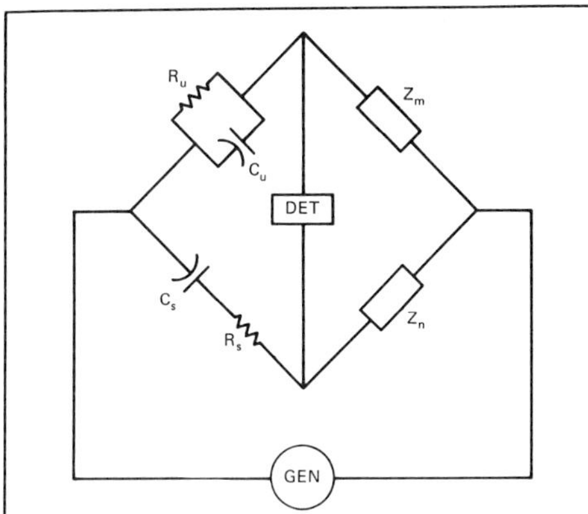

Purpose: Frequency or series or parallel capacitance

Balance equation:

$$\frac{C_u}{C_s} + \frac{R_s}{R_u} = \frac{Z_N}{Z_M} \qquad C_u = C_s \frac{Z_N}{Z_M}\left(\frac{1}{1 + \omega^2 C_s^2 R_s^2}\right) \qquad C_s = C_u \frac{Z_M}{Z_N}\left(1 + \frac{1}{\omega^2 C_u^2 R_u^2}\right)$$

$$\omega^2 = \frac{1}{R_u C_u R_s C_s} \qquad R_u = R_s \frac{Z_M}{Z_N}\left(1 + \frac{1}{\omega^2 C_s^2 R_s^2}\right)$$

$$R_s = R_u \frac{Z_N}{Z_M}\left(\frac{1}{1 + \omega^2 C_u^2 R_u^2}\right) \qquad f = \frac{1}{2\pi\sqrt{R_u C_u R_s C_s}}$$

Wien Bridge

Remarks: Measures unknown frequency in terms of resistors and capacitors. Often used as frequency determining element of oscillators. Was used for capacitance measurement before development of the Schering bridge.

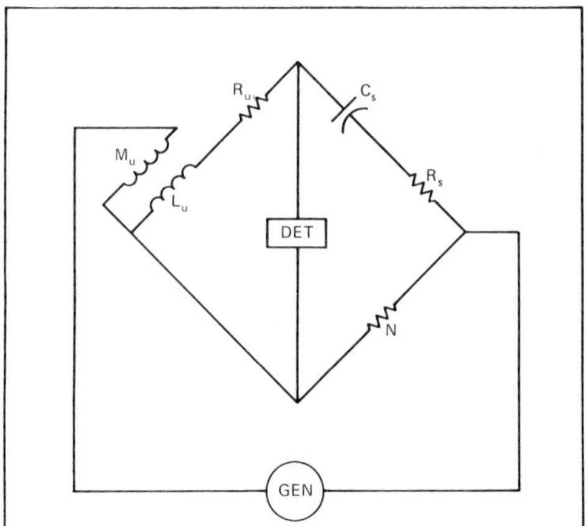

Purpose: Capacitance, mutual inductance

Balance Equation:

$$M_u = C_s R_u N \qquad C_s = \frac{M_u}{R_u N}$$

$$L_u = C_s R_u (R_s + N) \qquad R_s = \frac{N(L_u - M_u)}{M_u}$$

Carey-Foster, Heydweiler Bridges

Remarks: The Carey-Foster bridge measures unknown impedance in terms of equivalent circuit. Equivalent circuit is a capacitor and resistor in series.

Heydweiler bridge measures mutual inductance.

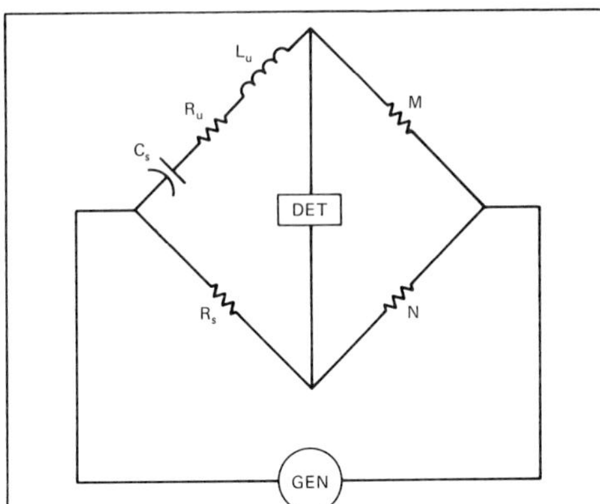

Purpose: Series inductance

Balance equation:

$$L_u = \frac{1}{\omega^2 C_s}$$

$$R_u = R_s \frac{M}{N}$$

Resonance Bridge

Remarks: Measures unknown impedance in terms of equivalent circuit. Equivalent circuit is an inductor and a resistor in series.

TABLE B-1. *Circuits and Balance Equations of Electric Bridges (Continued)*

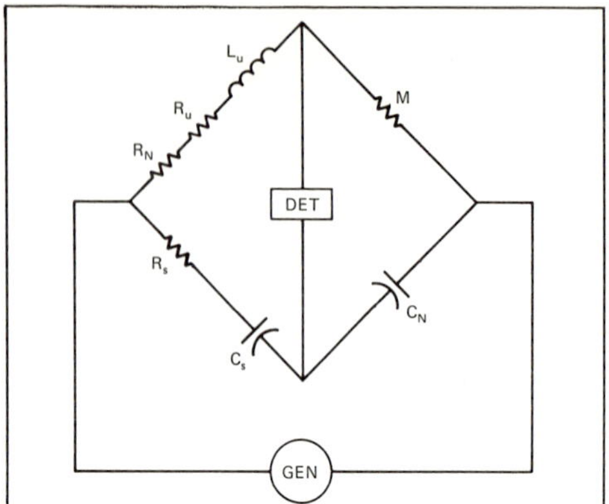

Purpose: Series inductance

Balance equation:

$$L_u = R_s C_N M$$

$$R_u = \frac{C_N M}{C_s} - R_N$$

Owen Bridge

Remarks: Measures unknown impedance in terms of equivalent circuit. Equivalent circuit is an inductor and a resistor in series.

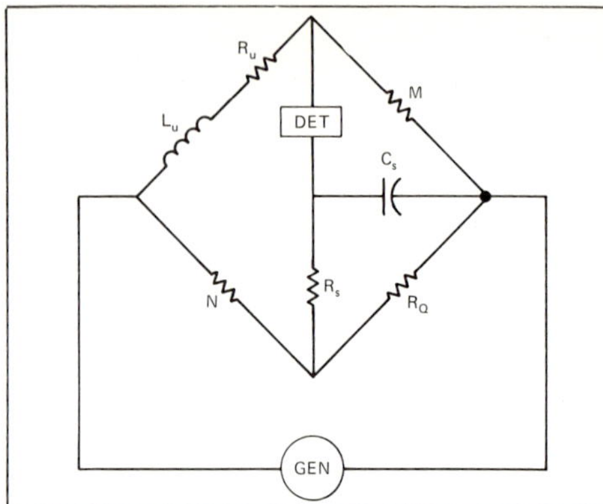

Purpose: Series inductance

Balance equation:

$$L_u = C_s M\left[R_s\left(1 + \frac{N}{R_Q}\right) + N\right] \qquad R_u = \frac{MN}{R_Q}$$

$$Q_u = \omega C_s\left[\frac{R_Q R_s}{N} + R_Q + R_s\right]$$

Anderson, Stroud and Oates Bridges

Remarks: Measures unknown impedance in terms of equivalent circuit. Equivalent circuit is an inductor and a resistor in series. Used for precision inductance measurement.

With the generator and detector interchanged, this is the Stroud and Oats bridge.

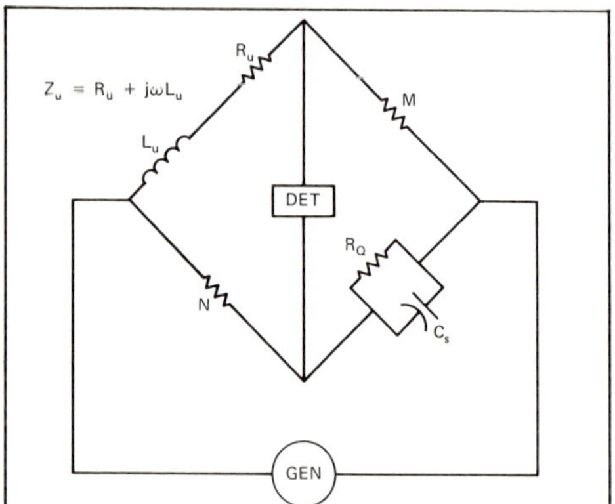

Purpose: Series inductance

Balance equation:

$$L_u = MNC_s \qquad R_u = \frac{MN}{R_Q}$$

$$Q_u = R_Q \omega C_s$$

Maxwell Bridge

Remarks: Measures unknown impedance in terms of equivalent circuit. Equivalent circuit is an inductor and a resistor in series. Used for inductors with a low quality factor.

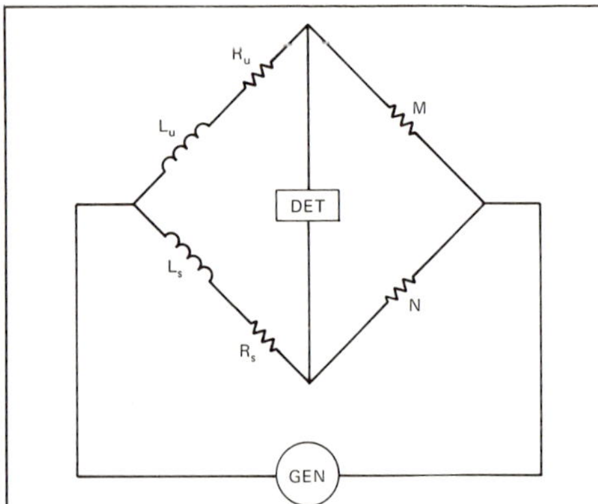

Purpose: Series inductance

Balance equation:

$$L_u = L_s \frac{M}{N}$$

$$R_u = R_s \frac{M}{N} \qquad Q_u = \frac{\omega L_s}{R_s}$$

Inductance Comparison Bridge

Remarks: Measures unknown impedance in terms of equivalent circuit. Equivalent circuit is an inductor and a resistor in parallel. Used for high frequency measurements.

TABLE B-1. *Circuits and Balance Equations of Electric Bridges* (*Continued*)

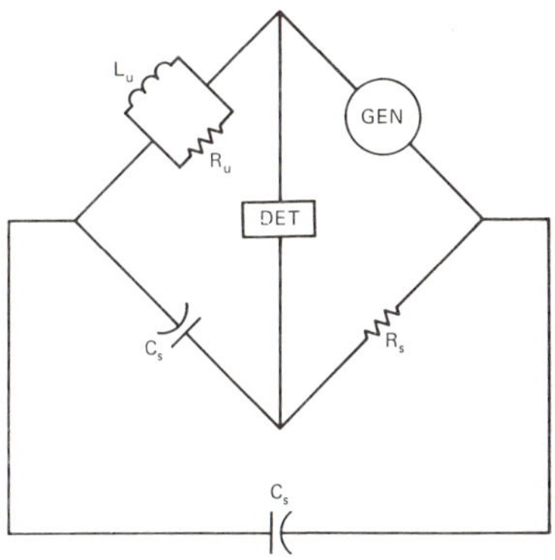

Purpose: Parallel inductance

Balance equation:

$$L_u = \frac{1}{2\omega^2 C_s} \qquad D = \omega R_s C_s$$

$$R_u = \frac{1}{\omega^2 C_s^2 R_s}$$

Bridged T Bridge

Remarks: Measures unknown impedance in terms of an equivalent circuit. Equivalent circuit is an inductor and resistor in parallel.

Used at higher frequencies. Circuit is very frequency sensitive.

$$Z_u = \frac{j\omega L_u R_u}{R_u + j\omega L_u}$$

$$Q_u = \frac{R_u}{\omega L_u}$$

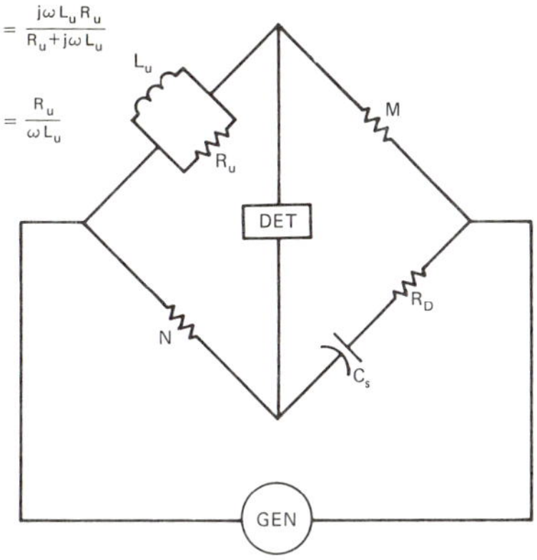

Purpose: Parallel inductance

Balance equation:

$$L_u = C_s MN \qquad R_u = \frac{MN}{R_D}$$

$$Q_u = \frac{1}{R_D \omega C_s}$$

Hay Bridge

Remarks: Measures unknown impedance in terms of equivalent circuit. Equivalent circuit is an inductor and a resistor in parallel. Used for inductors with a high quality factor.

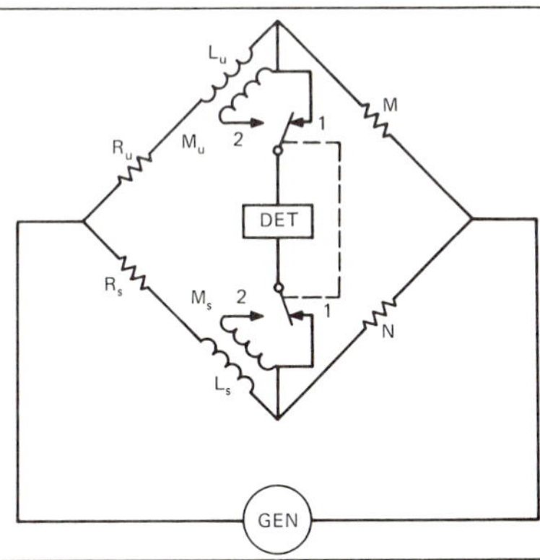

Purpose: Mutual inductance

Balance equation:

$$M_u = M_s \frac{M}{N} \qquad L_u = L_s \frac{M}{N}$$

$$R_u = R_s \frac{M}{N}$$

Campbell Bridgei

Remarks: Measures unknown impedance in terms of equivalent circuit. Equivalent circuit is a mutual inductor with a resistor and inductor in series on the primary side.

With the switches in position 1 balance for R_u and L_u. With the switches in position 2 balance for M_u.

TABLE B-1. *Circuits and Balance Equations of Electric Bridges* (*Continued*)

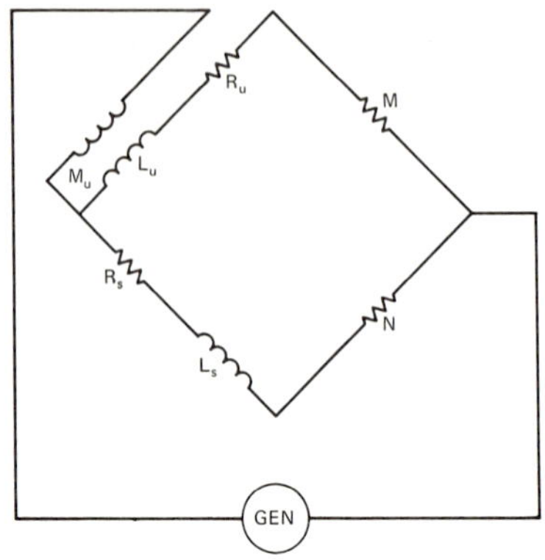	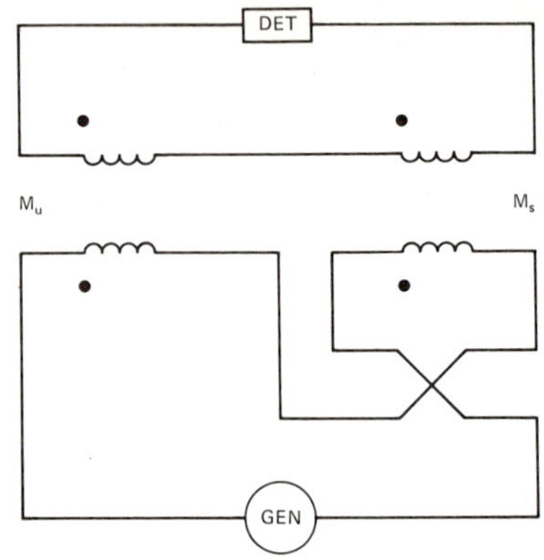
Purpose: Mutual inductance	Purpose: Mutual inductance
Balance equation: $M_u = \frac{ML_s - NL_u}{M + N}$ $R_u = R_s \frac{M}{N}$	Balance equation: $M_u = M_s$
Heaviside Bridge	Series-opposition: Bridge
Remarks: Measures mutual inductance.	Remarks: Compares two equal mutual inductances.

terminals to surrounding conductors. These stray impedances often make it desirable to use three- and four-terminal impedance elements. Special bridge configurations are available for measuring the resulting three- and four-terminal impedance values.

Three-terminal Impedance: In Fig. B-5, note the third terminal, which is connected to the leakage associated with the usual two terminals. This third terminal intercepts the leakage path between terminals. Undesirable leakage effects often can be avoided by using the third terminal

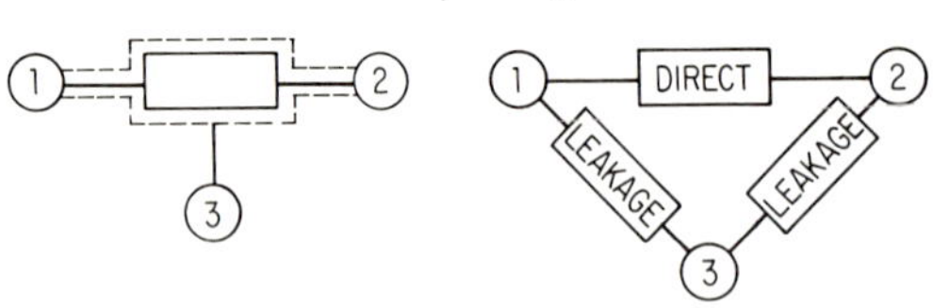

Fig. B-5. Three-terminal impedance.

in circuit applications. Special bridge circuits are available to measure the direct impedance even in the presence of low-impedance leakage paths.

Four-terminal Impedance: Note in Fig. B-6 that a four-terminal impedance has two terminals connected to each end. The three-lead junction which connects the end of the impedance element to its two leads provides a unique terminal point for the direct impedance. Special bridge circuits are available for measuring the direct impedance of a four-terminal element even in the presence of high-impedance leads.

Five-terminal Impedance: The configuration of Fig. B-7 combines the advantages of the three- and four-terminal connections. Special bridges permit the measurement of the direct impedance when both lead and leakage impedances are excessive.

Equivalent Circuits: Bridges measure the impedance connected to their terminals. Bridges often present the result in terms of an equivalent circuit. The components of simple equivalent circuits can be either in series or parallel. The

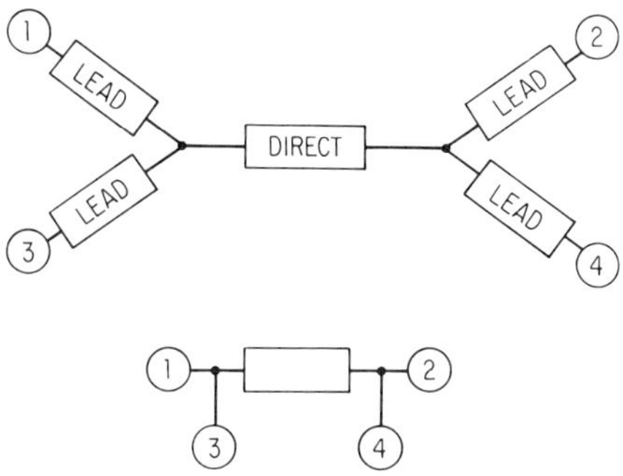

Fig. B-6. Four-terminal impedance.

relationships of the four most common equivalent circuits are given in Table B-2.

Many ac bridges measure the inductance or capacitance of the equivalent circuit and the dissipation factor D or quality factor Q of the impedance Z. $Z = R + jX$; $Q = \arctan X/R$; $D = 1/Q$. Circuit applications of bridge measurements often need values other than those measured. Frequently used values in terms of measured values are given in Table B-3.

Voltage Divider: A bridge circuit can be developed from a voltage divider. Two equal ratio voltage dividers can be bridged by a detector to compare impedances. The resulting circuit can be used with a constant-voltage source to maintain a fixed voltage across unknown impedances of various values.

The voltage drops across impedance elements through which equal currents pass is proportional to the impedance values. The voltage-divider circuit of Fig. B-8 uses this relation to produce a known ratio between its input and output voltages in terms of two known impedance values:

$$\frac{E_{\text{out}}}{E_{\text{in}}} = \frac{Z_1}{Z_1 + Z_2} \tag{1}$$

Two such dividers can be used to compare voltages or impedances as shown in Fig. B-9. By using equal generator voltages and adjusting impedance values to give equal output voltages, an impedance ratio is established.

$$\frac{E_{G1} - E_1}{E_1} = \frac{Z_2}{Z_1} = \frac{Z_4}{Z_3} = \frac{E_{G2} - E_3}{E_3} \tag{2}$$

Bridge Circuit: A voltage divider uses known impedances to establish known voltage ratios. An impedance bridge can use known voltage ratios to compare impedances. The generator voltages of Fig. B-9 can be made equal by using the same generator for both circuits, and the equality of the output voltages can be checked with a null detector. The resulting bridge circuit is shown in Fig. B-10. If the voltage-divider ratio is established for the circuit of Fig. B-10 by impedances Z_1 and Z_2, an unknown impedance Z_3 is found in terms of a variable Z_4. The voltage across Z_3 will always be the same if the generator voltage and the ratio Z_1/Z_2 are left the same. By using this constant-voltage ratio, a wide range of impedance values can be measured with the same applied voltage.

Transformer Bridges: The voltage-divider ratio for bridge circuits also can be supplied by a transformer as shown in Fig. B-11.

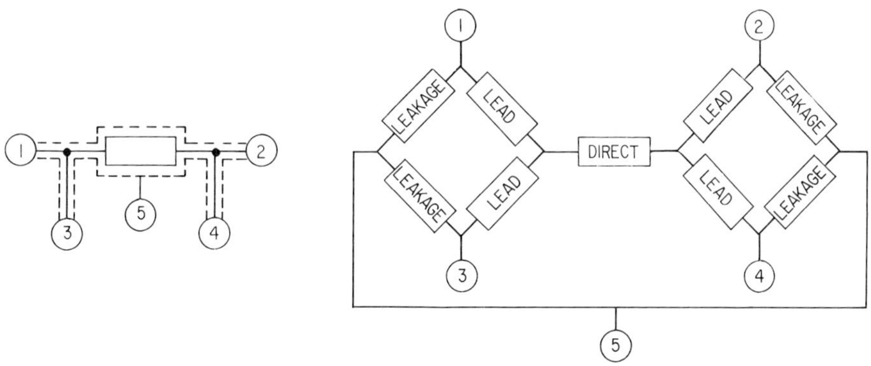

Fig. B-7. Five-terminal impedance.

TABLE B-2. *Equivalent Circuits for an Impedance*

Circuit	Impedance Z	Admittance Y	Q	D
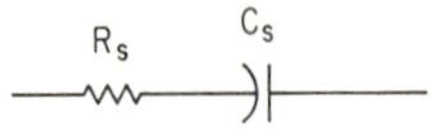	$R_s - \dfrac{1}{j\omega C_s}$	$\dfrac{R_s + j(1/\omega C_s)}{R_s^2 + 1/(\omega^2 C_s^2)}$	$\dfrac{1}{\omega R_s C_s}$	$\omega R_s C_s$
	$R_s + j\omega L_s$	$\dfrac{R_s - j\omega L_s}{R_s^2 + \omega^2 L_s^2}$	$\dfrac{\omega L_s}{R_s}$	$\dfrac{R_s}{\omega L_s}$
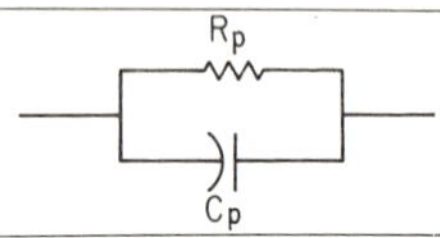	$\dfrac{R_p - j\omega C_p R_p^2}{1 + \omega^2 C_p^2 R_p^2}$	$\dfrac{1}{R_p} + j\omega C_p$	$\omega R_p C_p$	$\dfrac{1}{\omega R_p C_p}$
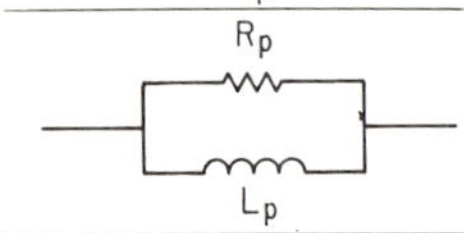	$\dfrac{R_p + j(R_p^2/\omega L_p)}{1 + R_p^2(\omega^2 L_p^2)}$	$\dfrac{1}{R_p} - j\dfrac{1}{\omega L_p}$	$\dfrac{R_p}{\omega L_p}$	$\dfrac{\omega L_p}{R_p}$

R_s = equivalent series resistance, ohms
C_s = equivalent series capacitance, farads
L_s = equivalent series inductance, henrys
R_p = equivalent parallel resistance, ohms
$\omega = 2\pi f$ = frequency, radians/seconds
C_p = equivalent parallel capacitance, farads
L_p = equivalent parallel inductance, henrys
Q = quality factor in proportional parts
D = dissipation factor in proportional parts
f = frequency, Hertz

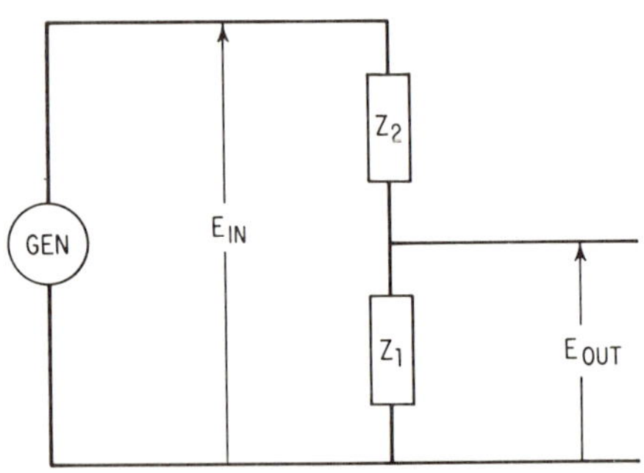

$$E_{OUT} = \frac{Z_1}{Z_1 + Z_2} E_{IN}$$

Fig. B-8. Voltage divider.

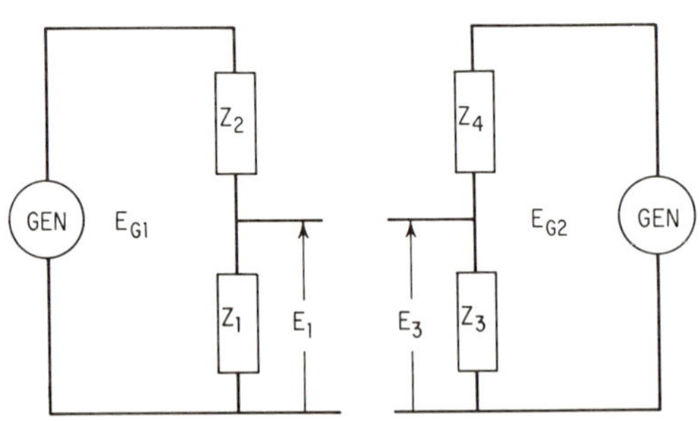

Fig. B-9. Two-divider comparison.

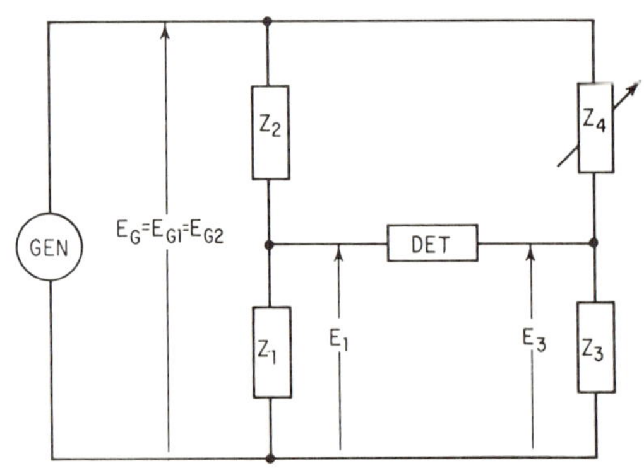

$$\frac{Z_1}{Z_2} = \frac{Z_3}{Z_4} = \frac{E_1}{E_G - E_1}$$

Fig. B-10. Impedance bridge.

Current Divider: A bridge circuit can be developed from a current divider. The resulting circuit can be used with a constant-current source to maintain a fixed current through unknown impedances of various values. A current-divider circuit is shown in Fig. B-12. The same voltage drop occurs across each branch

$$I_1 Z_1 = I_2 Z_2$$

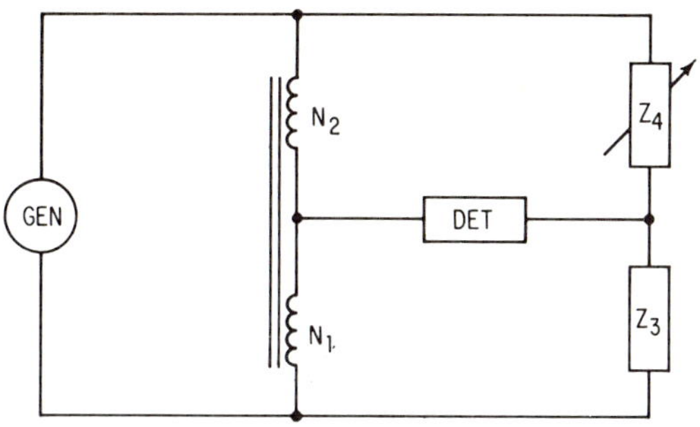

$$Z_3 = \frac{N_1}{N_2} Z_4$$

Fig. B-11. Transformer impedance bridge.

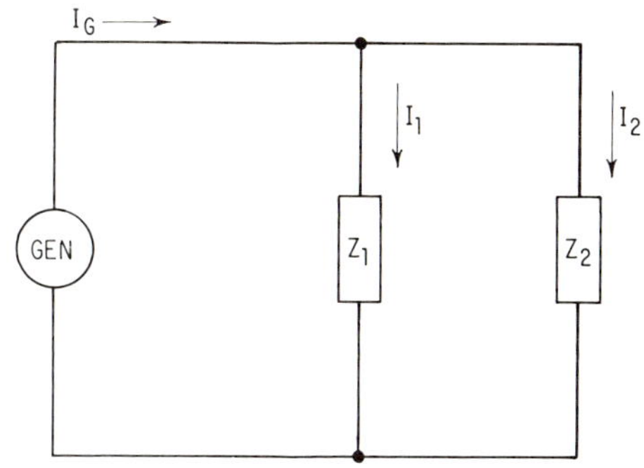

Fig. B-12. Current division.

TABLE B-3. *Equivalent Circuit Conversion Equations for Bridges*

If the bridge measures C_s and D	If the bridge measures C_s and Q
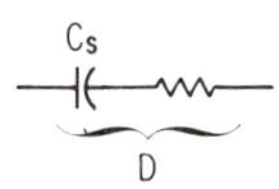 $R_s = \frac{D}{\omega C_s}$ $R_p = \frac{1}{D\omega C_s}$ $C_p = \frac{C_s}{1 + D^2}$ $Q = \frac{1}{D}$	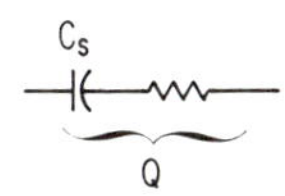 $R_s = \frac{1}{Q\omega C_s}$ $R_p = \frac{Q}{\omega C_s}$ $C_p = \frac{C_s Q^2}{1 + Q^2}$ $D = \frac{1}{Q}$
If the bridge measures C_s and R_s	**If the bridge measures C_p and D**
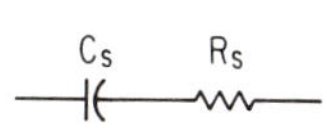 $R_p = \left(\frac{1}{\omega^2 R_s^2 C_s^2} + 1\right) R_s$ $C_p = \frac{C_s}{1 + \omega^2 R_s^2 C_s^2}$ $D = \omega R_s C_s$ $Q = \frac{1}{\omega R_s C_s}$	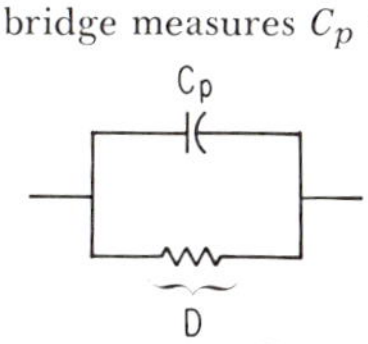 $R_s = \frac{D}{\omega C_p(1 + D^2)}$ $C_s = (1 + D^2)C_p$ $R_p = \frac{1}{D\omega C_p}$ $Q = \frac{1}{D}$

TABLE B-3. *Equivalent Circuit Conversion Equations for Bridges (Continued)*

If the bridge measures C_p and Q

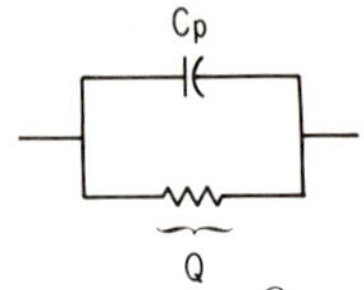

$$R_s = \frac{Q}{\omega C_p(1 + Q^2)}$$

$$C_s = \frac{1 + Q^2}{Q^2} C_p$$

$$R_p = \frac{Q}{\omega C_p}$$

$$D = \frac{1}{Q}$$

If the bridge measures C_p and R_p

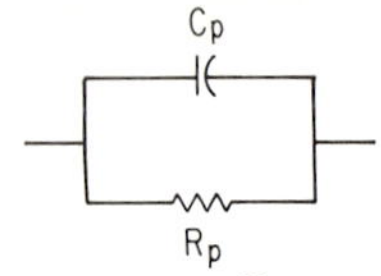

$$R_s = \frac{R_p}{1 + \omega^2 R_p^2 C_p^2}$$

$$C_s = \left(\frac{1}{\omega^2 R_p^2 C_p^2} + 1\right) C_p$$

$$D = \frac{1}{\omega R_p C_p}$$

$$Q = \omega R_p C_p$$

If the bridge measures L_s and D

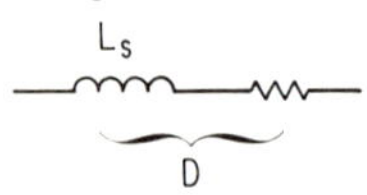

$$R_s = D\omega L_s$$

$$R_p = (1 + D^2) \frac{\omega L_s}{D}$$

$$L_p = (1 + D^2) L_s$$

$$Q = \frac{1}{D}$$

If the bridge measures L_s and Q

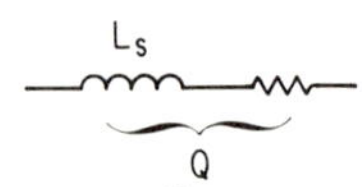

$$R_s = \frac{\omega L_s}{Q}$$

$$R_p = (1 + Q^2) \frac{\omega L_s}{Q}$$

$$L_p = \frac{1 + Q^2}{Q^2} L_s$$

$$D = \frac{1}{Q}$$

If the bridge measures L_s and R_s

$$R_p = \left(\frac{\omega^2 L_s^2}{R_s^2} + 1\right) R_s$$

$$L_p = \left(\frac{R_s^2}{\omega^2 L_s^2} + 1\right) L_s$$

$$D = \frac{R_s}{\omega L_s}$$

$$Q = \frac{\omega L_s}{R_s}$$

If the bridge measures L_p and D

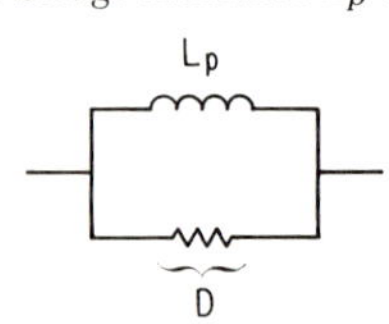

$$R_s = \frac{D\omega}{1 + D^2} L_p$$

$$L_s = \frac{1}{1 + D^2} L_p$$

$$R_p = \frac{\omega}{D} L_p$$

$$Q = \frac{1}{D}$$

TABLE B-3. *Equivalent Circuit Conversion Equations for Bridges (Continued)*

If the bridge measures L_p and Q	If the bridge measures L_p and R_p
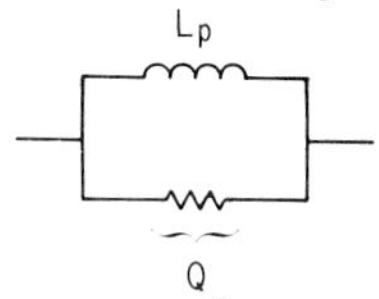	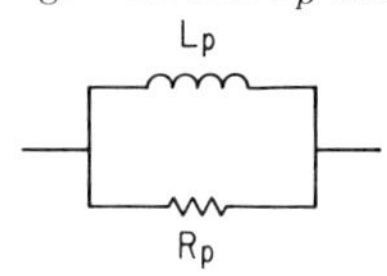
$R_s = \frac{Q\omega}{1+Q^2} L_p$	$L_s = \frac{1}{1+\omega^2 L_p{}^2/R_p{}^2} L_p$
$L_s = \frac{Q^2}{1+Q^2} L_p$	$R_s = \frac{1}{1+R_p{}^2/(\omega^2 L_p{}^2)} R_p$
$R_p = Q\omega L_p$	$D = \frac{\omega L_p}{R_p}$
$D = \frac{1}{Q}$	$Q = \frac{R_p}{\omega L_p}$

The branch currents add to give the generator current

$$I_1 + I_2 = I_G \tag{4}$$

I_2 is chosen as the divider output current and I_1 is eliminated from the foregoing equations:

$$\frac{I_2}{I_G} = \frac{Z_1}{Z_1 + Z_2} \tag{5}$$

The same pair of impedances that make a voltage divider can be used for a current divider, as shown in Fig. B-13. The input and output terminals are interchanged and the output load is a short rather than open circuit, but the division ratio is the same.

Bridge Circuit: A current divider uses known impedances to establish known current ratios. An impedance bridge can use known current ratios to compare impedances. A current divider can be used to establish a current ratio that can be used by another divider to establish an impedance ratio as shown in Fig. B-14.

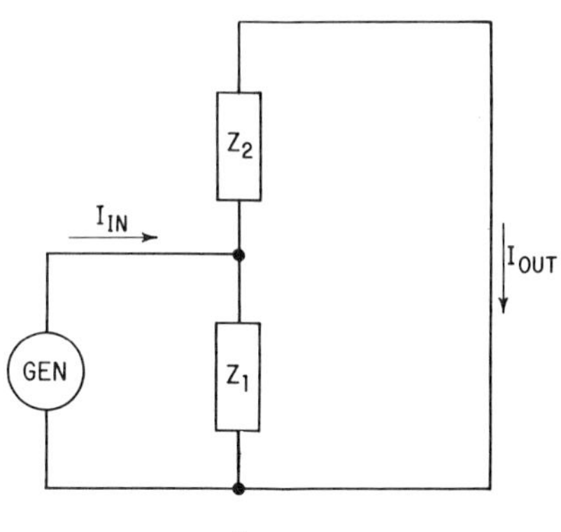

$$I_{OUT} = \frac{Z_1}{Z_1 + Z_2} I_{IN}$$

Fig. B-13. Current divider.

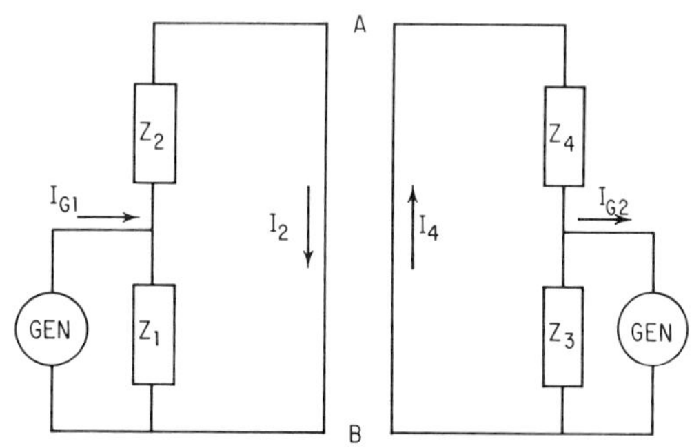

$$\frac{Z_1}{Z_1 + Z_2} = \frac{I_2}{I_{G1}}$$

$$\frac{Z_3}{Z_3 + Z_4} = \frac{I_4}{I_{G2}}$$

If $I_2 = I_4$ AND $I_{G1} = I_{G2}$

THEN $\frac{Z_2}{Z_1} = \frac{Z_4}{Z_3}$

Fig. B-14. Bridge requirements.

Currents I_2 and I_4 can be forced to be alike by connecting the circuits at A and B and removing the short circuits between A and B. If I_2 and I_4 were equal before the connection, no current would flow in the short circuits after the connec-

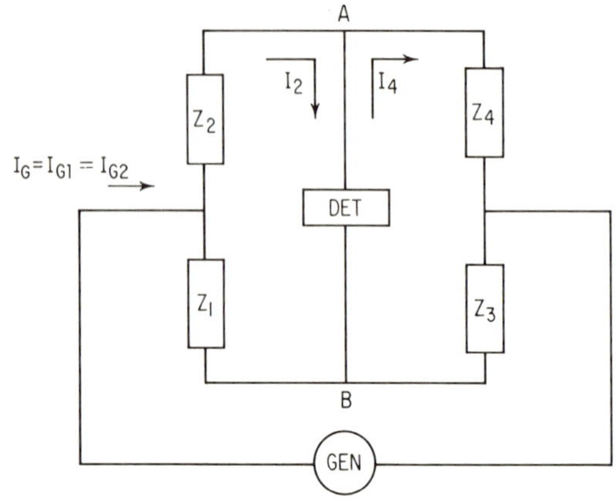

$$\frac{Z_1}{Z_2} = \frac{Z_3}{Z_4} = \frac{I_2}{I_G - I_2}$$

Fig. B-15. Bridge circuit.

tion. The currents I_2 and I_4 must be "short-circuit" currents for their respective dividers. This condition can be indicated by a voltage null between points A and B after the short circuits are removed. Currents I_{G1} and I_{G2} can be forced to be the same by connecting the generators in series. The resulting circuit is the bridge circuit shown in Fig. B-15. All the currents in the im-

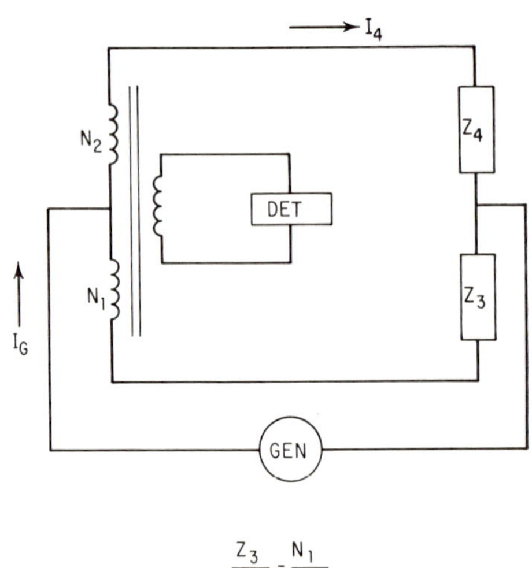

$$\frac{Z_3}{Z_4} = \frac{N_1}{N_2}$$

Fig. B-16. Transformer bridge.

pedance branches will be the same as those indicated in Fig. B-14 when the bridge is balanced.

The current ratio also can be established by a transformer, as shown in Fig. B-16. The current-divider transformer circuit also can be used for direct current by using a separate ac generator and a detector tuned to its second harmonic. This circuit is shown in Fig. B-17. If the transformer hysteresis curve is symmetrical for positive and negative excursions, no second-harmonic signal is generated. The curve will be symmetrical if the dc ampere-turns add to zero.

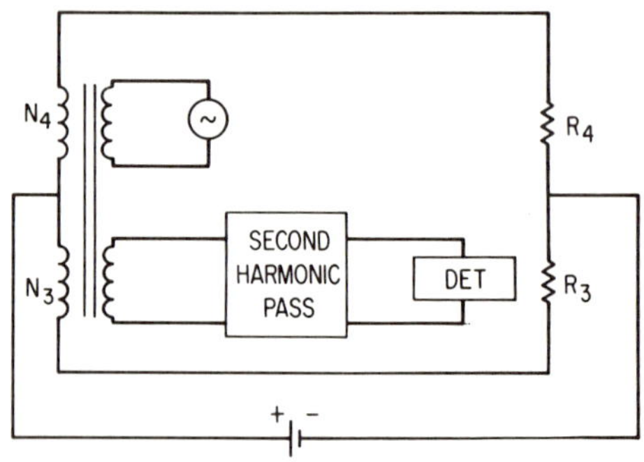

$$\frac{R_3}{R_4} = \frac{N_3}{N_4}$$

Fig. B-17. Transformer dc resistance bridge.

Bridge Circuits: There are several basic bridge principles, and many variations of these have been made for special applications. See Table B-1.

Two-terminal Bridges: Many bridges, both dc and ac, are made in the Wheatstone configuration. Ac bridges require a double balance because both the magnitude ratios and the phase-angle differences of the impedances must be compared.

The Wheatstone-type impedance bridge (Fig. B-18) has four impedance elements, a generator, and a detector. Two of the bridge-arm impedances are fixed and determine a ratio which is matched by the ratio of the unknown impedance to a variable impedance. Any of the three permanently connected bridge arms may be the variable element, and the other two may be the ratio-determining pair. The impedance magnitudes are compared by the usual Wheatstone-ratio equation, but the impedance angles are compared in terms of their differences.

The Wheatstone impedance bridge circuit can be used in three modes: (1) constant-voltage;

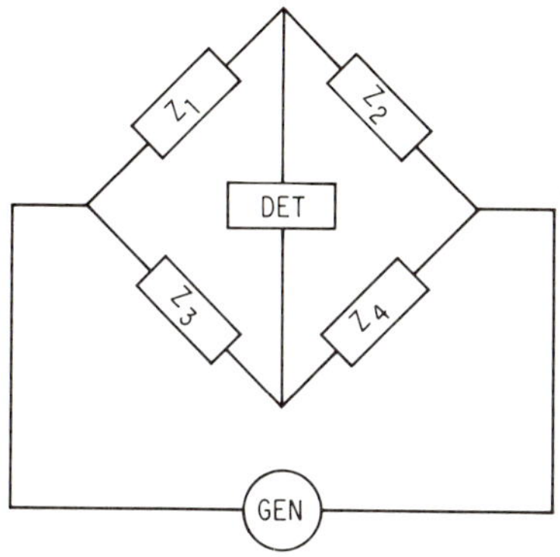

$$\frac{Z_1}{Z_2} = \frac{Z_3}{Z_4} \quad \frac{|Z_1|}{|Z_2|} = \frac{|Z_3|}{|Z_4|} \quad \angle Z_1 - \angle Z_2 = \angle Z_3 - \angle Z_4$$

Fig. B-18. Wheatstone bridge.

(2) constant-current; and (3) impedance-admittance, as shown in Fig. B-19. When the bridge is balanced, the constant-voltage mode maintains the same voltage ratio between the bridge input voltage and the voltage across the unknown impedance for any impedance measured. This fixed voltage ratio is predetermined by the fixed bridge arms.

The constant-current mode maintains a fixed current ratio between the input current and the current across the unknown. The constant-voltage and constant-current modes can be interchanged by interchanging the generator and detector. The four impedance arms are connected the same way in both configurations. The impedance-admittance mode is used to balance an admittance value by an impedance value. Thus, an unknown conductance can be measured in terms of a calibrated variable resistance.

Bridge Power: The power supplied to a bridge circuit divides in the same ratio as the voltage and inversely as the current ratio—as shown in Fig. B-20. Interchanging the generator and detector can cause great changes in the voltage, current, and power in the various bridge arms. Bridge circuits often can be arranged either to supply most of the power to the impedance to protect

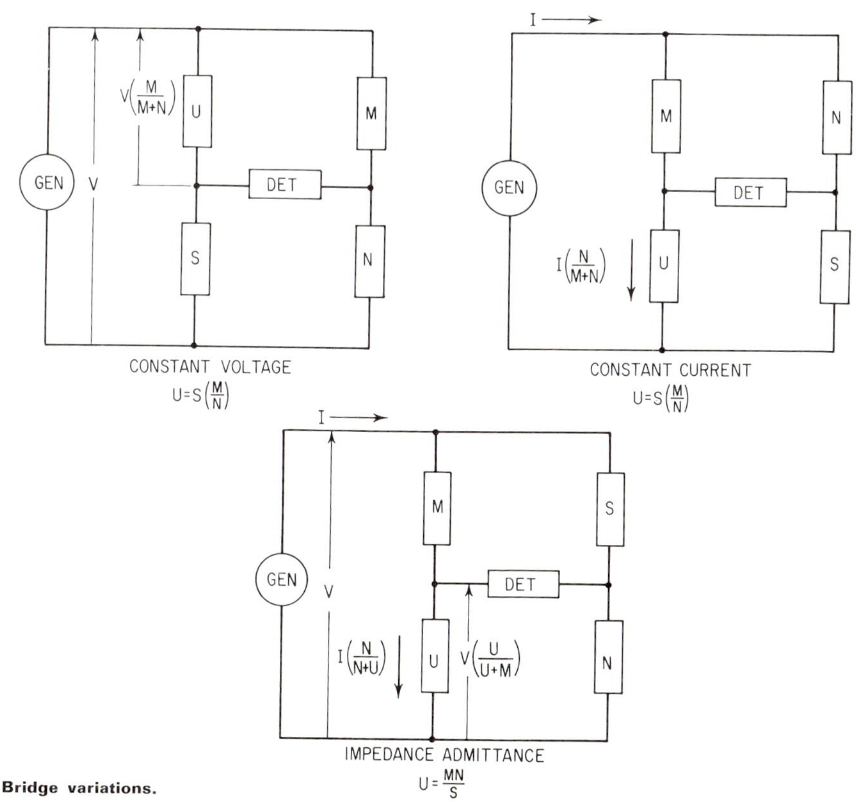

Fig. B-19. Bridge variations.

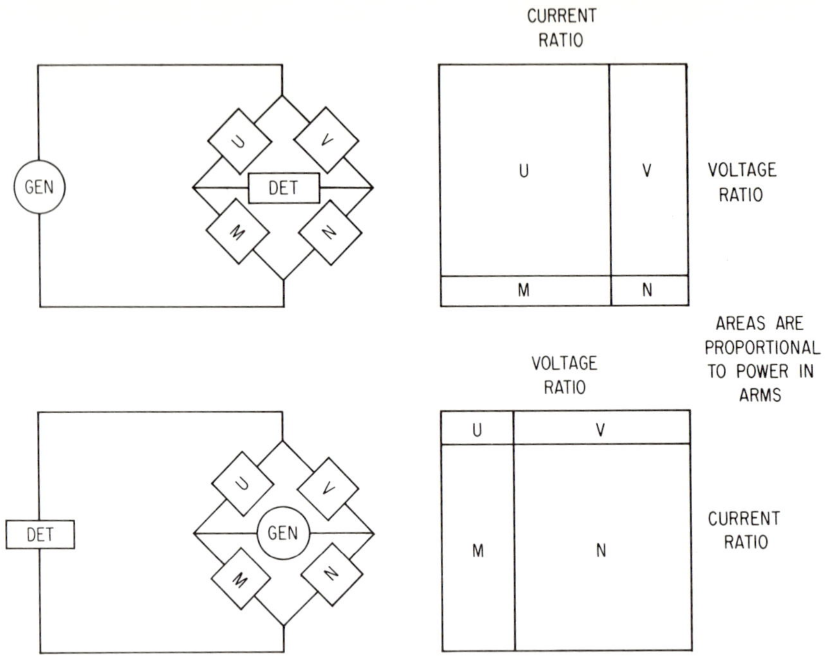

Fig. B-20. Bridge power.

the bridge or to supply most of the power to the bridge to protect the unknown impedance.

Three-terminal Bridges: These bridges are used for measuring three-terminal impedances. The direct impedances shown in Fig. B-21 are connected in a conventional Wheatstone arrangement. The third terminals are connected together at a common point. This common point is adjusted by means of four additional impedance elements so that the common point has the same voltage at balance as the voltage at the detector terminals. The common point also is balanced with the generator and detector interchanged. With this circuit, almost any amount of terminal leakage can be ignored even though the leakage may have a very much lower impedance than the direct impedances being compared.

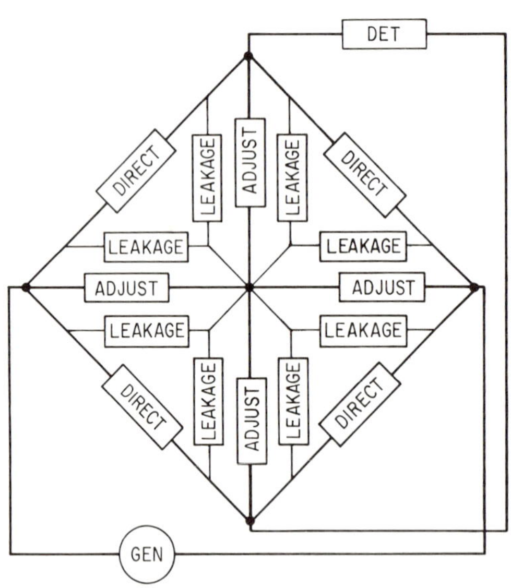

Fig. B-21. Three-terminal impedance bridge.

Usually, the leakage is a sufficiently high impedance relative to bridge impedances so that the leakage can be handled in much simpler ways. By connecting the third terminal to the detector terminal, as shown in Fig. B-22, leakages are removed from the direct-impedance bridge arm. One is shunted across the detector and will cause a slight reduction in detector sensitivity. The other is shunted across the *S* bridge arm and will change its impedance. Usually, this change is smaller than other measurement uncertainties and can be ignored. When a transformer is used for the bridge ratio, as shown in Fig. B-23, the leakages are across the detector and across the very low output impedance of a transformer winding.

Many modern impedance bridges use a floating generator and connect one detector terminal to the bridge chassis so that the latter can act as a guard for the bridge circuits. A chassis connection can be used for an external shield connection. Thus, by connecting the third terminal of a three-terminal impedance to the chassis, most leakage effects are eliminated. Since the third terminal of a three-terminal impedance often is

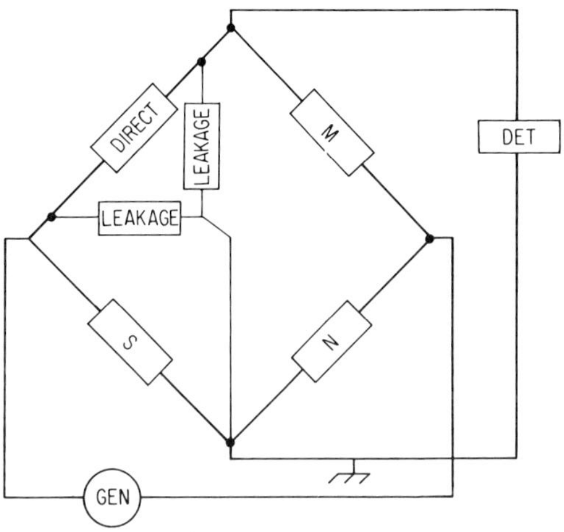

Fig. B-22. Three-terminal measurement.

its chassis, the use of shielded leads, with the shield connected to the bridge and impedance chassis, avoids external capacitance and resistance leakages.

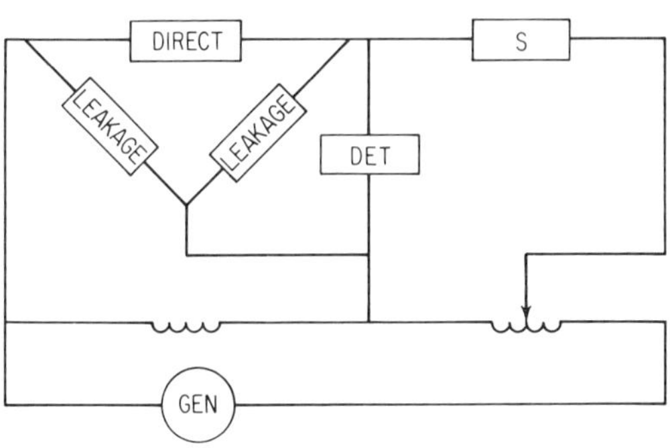

Fig. B-23. Transformer bridge three-terminal measurement.

Four-terminal Bridges: These bridges are used for measuring four-terminal impedances. The measurement of four-terminal impedances with the circuit of Fig. B-24 takes advantage of two circuit characteristics. The first is a property of the ratio of sums, and the second is the delta-wye

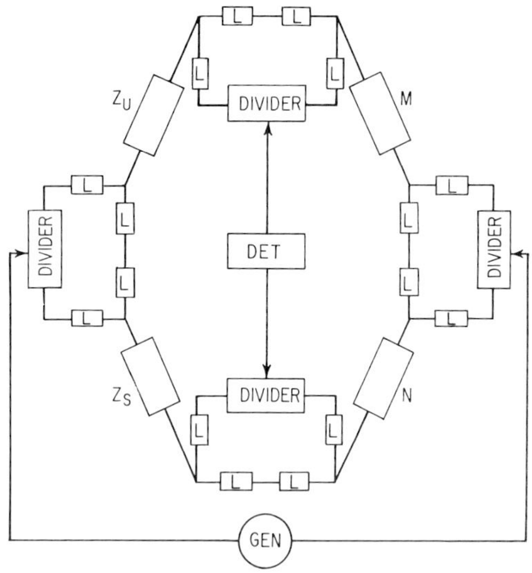

Fig. B-24. Four-terminal impedance bridge.

transformation. If all the lead impedances can be adjusted, as shown in Fig. B-25, so that the lead impedances are proportional to all the impedances being compared, the lead impedances will have no effect upon the comparison.

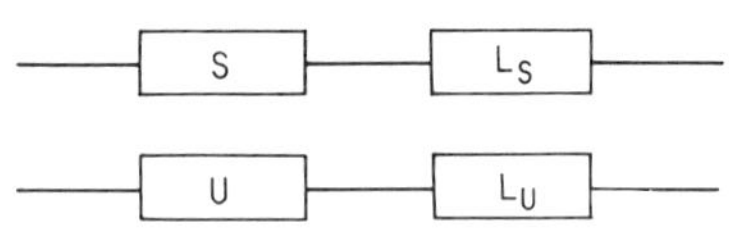

IF $L_S = KS$ AND $L_U = KU$,
THEN $\frac{S+L_S}{U+L_U} = \frac{S(1+K)}{U(1+K)} = \frac{S}{U}$

Fig. B-25. Ratio relations.

The delta-wye transformation shows that a shunted high-impedance divider, as shown in Fig. B-26, operates as a low-impedance divider of the same ratio with a high resistance in series with its tap. The complete four-terminal bridge uses such a divider in each corner—each balanced in the ratio of its adjacent impedances. These dividers place small, properly balanced impedances in series with each of the bridge impedances being considered. The resulting balance gives the desired measurement. With this circuit, almost any amount of lead resistance can be ignored even though the lead resistance may have a much higher impedance than the impedances being compared. Usually lead-impedance problems are not serious enough to require the full

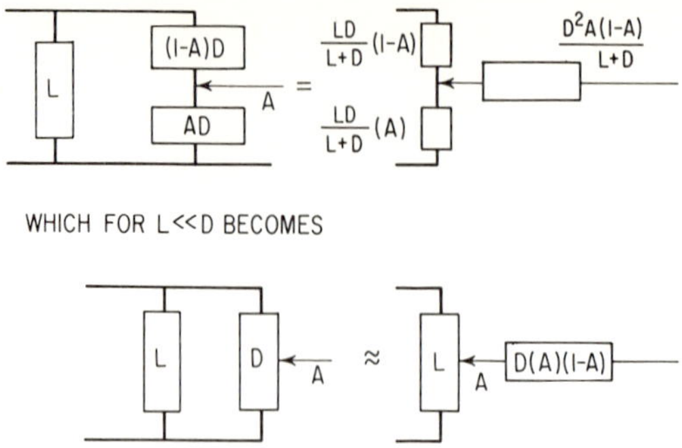

Fig. B-26. Shunted divider.

four-terminal bridge. By using high-impedance bridge arms M and N, two low-value impedances Z_u and Z_s can be compared by the Kelvin-bridge circuit shown in Fig. B-27.

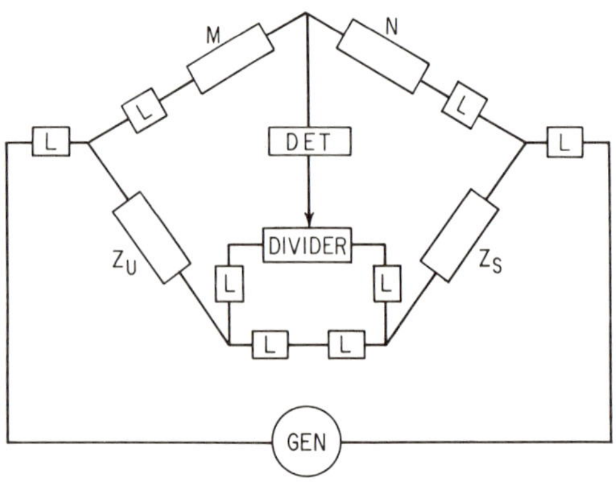

Fig. B-27. Kelvin-bridge circuit.

Transformer Bridges: A transformer can be used as a divider, an isolator, or an impedance-level-changing device. All these functions can be used in bridge circuits. Transformer functions can be analyzed in terms of the equivalent circuit of the transformer shown in Fig. B-28. An open-circuit impedance Z_{oc} can be measured at one pair of terminals when the other pair of terminals is open-circuited. A short-circuit impedance Z_{sc} can be measured at the second pair of terminals when the first pair is shorted. These two impedances, plus the open-circuit voltage ratio A' (which usually is not exactly the turns ratio A), determine most of the operating characteristics of a transformer.

A transformer bridge can be made by using transformers to supply the bridge voltages in a desired ratio or to compare bridge currents in a desired ratio, or both. In the circuit of Fig. B-29, the detector transformer ratio A_D is the ratio of the currents through the unknown and standard impedances at balance. Since at balance there is zero voltage across all windings of the detector transformer, all the voltage from the generator transformer will be across the impedances being compared by the detector. The resulting balance

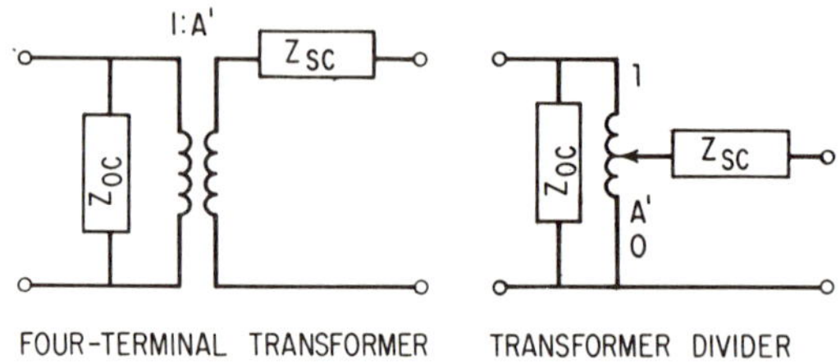

Fig. B-28. Transformer equivalent circuits.

equation gives the unknown impedance as the product of the standard impedance and the two transformer ratios.

$$\frac{I_{\text{unk}}}{I_{\text{std}}} = \frac{1}{A_D}$$

$$I_{\text{unk}} = \frac{EA_G}{Z_{\text{unk}}} \qquad I_{\text{std}} = \frac{E}{Z_{\text{std}}} \tag{6}$$

$$Z_{\text{unk}} = A_G A_D Z_{\text{std}}$$

where unk = unknown
std = standard

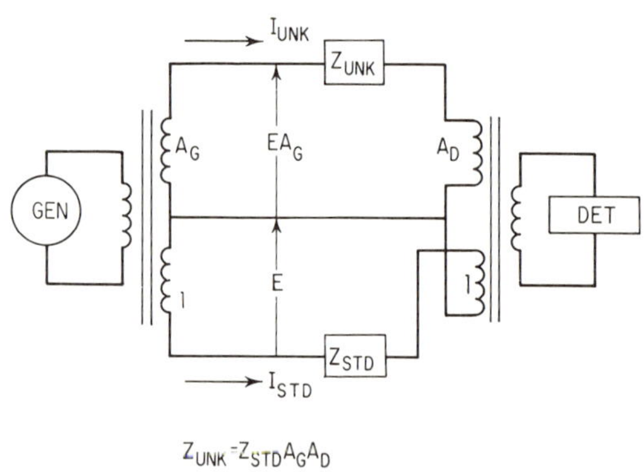

Fig. B-29. Transformer bridge.

Modern transformer core materials and winding techniques have resulted in transformer dividers of extreme accuracy. As a result, transformer bridges are being used widely, especially for capacitance measurements.

Generators: Bridge-generator choice usually is a compromise between high power for maximum measurement sensitivity and low power for maximum accuracy and bridge protection. For protecting the bridge and the unknown impedance from accidental overloading, a power-limiting resistor can be used between a constant-voltage source and the bridge, as shown in Fig. B-30. For example, a 6-V battery output can be limited to 1 W by applying a 1-Ω, resistor in series. With such a resistor, the bridge input power will be a predictable maximum. Maximum bridge power will result when the input impedance is resistive and equal to the protective resistance. For high sensitivity, the voltage should be chosen so that the power-limiting

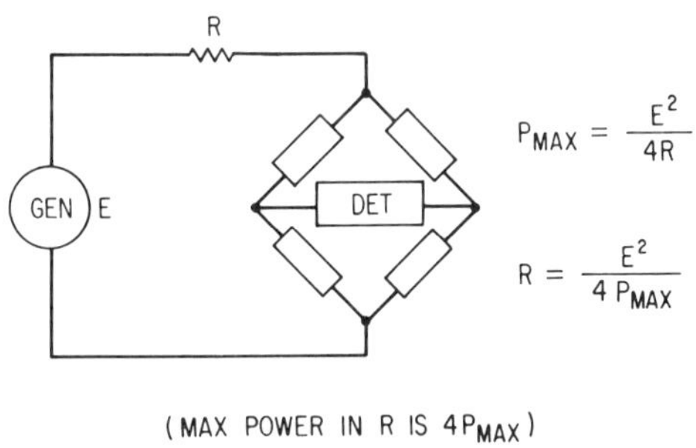

Fig. B-30. Power limiting of bridge.

resistor will be similar in value to the input impedance of the bridge circuit.

Detectors: Bridge-detector choice usually is a compromise between maximum sensitivity and measurement speed. Sensitive detectors respond to thermal noise, and long-term averaging is required for higher sensitivity. The bridge output signal can be thought of as a Thévenin equivalent generator as shown in Fig. B-31. The output impedance at balance can be calculated or

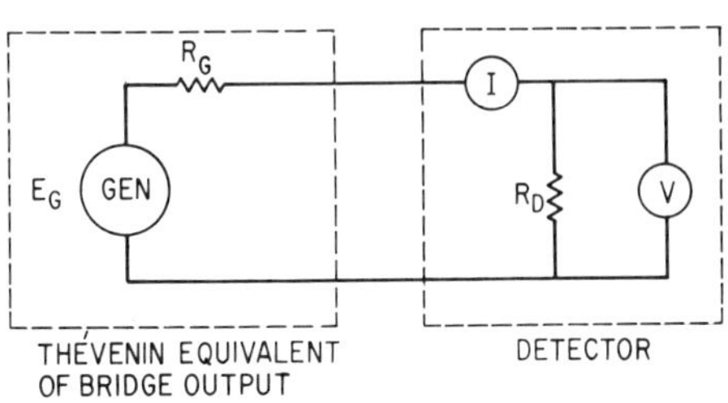

Fig. B-31. Detector circuit.

measured. The voltage can be found in terms of proportional parts of unbalance. Detectors can be calibrated in terms of the Thévenin equivalent generators, which will provide barely detectable signals to give the response curve of Fig. B-32. The source curve represents those equivalent generators which will give the same detector indication. A point on the source curve can represent the resistance and open-circuit voltage of a Thévenin equivalent generator or the resistance and short-circuit current of a Norton equivalent circuit.

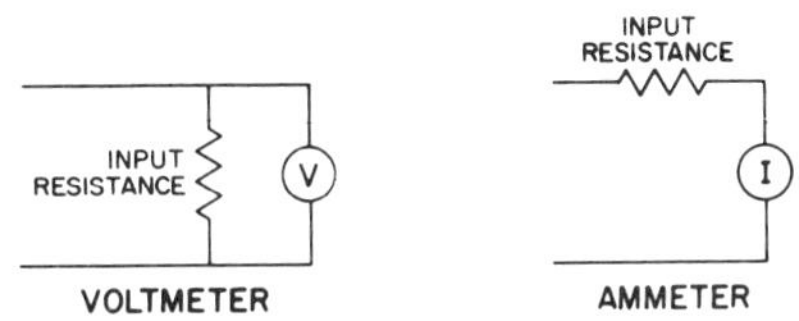

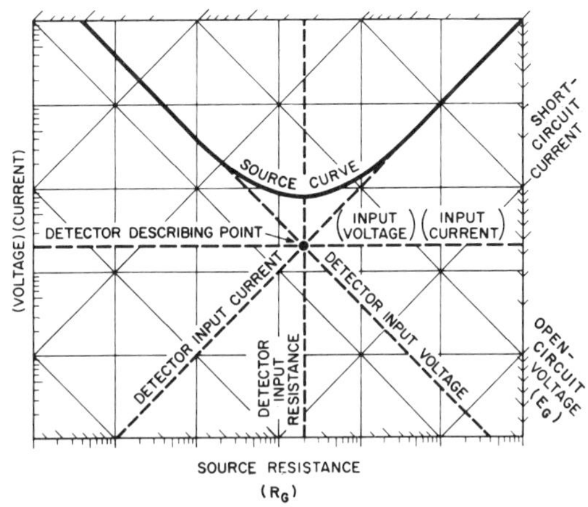

Fig. B-32. Detector response curve.

Isolation: The generator and detector seldom have a common terminal. Each must be well isolated from the other. Even a small leakage signal from the generator to the detector will cause a measurement error. Many detectors also respond to stray fields from power lines and other sources. Measurements often require a shielded environment.

REFERENCES

Brookes, A. M. P.: "Basic Electric Circuits," Pergamon Press, New York, 1963.

Fitzgerald, A. E., and D. E. Higgenbotham: "Electrical and Electronic Engineering Fundamentals," McGraw-Hill Book Company, New York, 1964.

Suffern, M. G.: "Basic Electrical and Electronic Principles," 3d ed., McGraw-Hill Book Company, New York, 1962.

Gray, A., and G. A. Wallace: "Principles and Practice of Electrical Engineering," 8th ed., McGraw-Hill Book Company, New York, 1962.

Lynch, W., and J. Truxal: "Principles of Electronic Instrumentation," McGraw-Hill Book Company, New York, 1961.

Knowlton: "Standard Handbook for Electrical Engineers," 9th ed., McGraw-Hill Book Company, New York, 1957.

Smith, A. W., and M. L. Wiedenbeck: "Electrical Measurements," 5th ed., McGraw-Hill Book Company, New York, 1959.

Thwaits, R. R.: "Fundamentals of Electricity," McGraw-Hill Book Company, New York, 1964.

Harris, F. K.: "Electrical Measurements," John Wiley & Sons, Inc., New York, 1952.

Studer, J. J.: "Electronic Circuits and Instrumentation Systems," John Wiley & Sons, Inc., New York, 1963.

—Jack C. Riley, *Technical Consultant to Electro Scientific Industries, Inc.*

Brightness Meter (See **Photometers.**)

Bubble Chamber [See **Radiation (nuclear), detection of.**]

Buffer Solution (See **pH, measurement of.**)

Buffer (Storage) A buffer is a device which provides temporary storage (see **Storage, computer**) for information which is being transferred between two storage or data-handling devices. For instance, if information is being transferred from an input device to storage in the central processing unit [see **Central processing unit (CPU)**], the data are sampled from the input-data bus into a buffer. While the next byte (see **Byte**) of information is read by the input device, the contents of the buffer is transferred to storage. Buffers are also used in conjunction with input-output equipment (see **Input/output devices**) in that a central processor usually transfers an entire line of print information to a printer buffer and the actual printing occurs from the buffer after the central processing unit has disengaged itself from the device.

The term buffer is also used in reference to intermediate storage areas in conjunction with input or output areas of a program (see **Program computer**). In this respect, a program which is to perform an output operation moves the data to an output buffer area to be acted upon by the device or channel program. On an input operation, the device or channel program reads the data into an input area or buffer. The information is then moved to another area, where the data are acted upon.

The term buffer amplifier is often used to refer to an amplifier which provides impedance transformation between two analog circuits or devices. For example, a buffer amplifier is often provided at the input of an analog-to-digital converter to provide a low source impedance for the analog-to-digital converter and a high load impedance for the signal source.

—W. W. Welz, *IBM Corporation*

Bulk Feeder (See **Feeder, vibratory; Feeder, volumetric;** and **Feeder, weigh.**)

Bulk Materials, Level Measurement of (See **Level instrumentation, bulk solids.**)

Business Machine (See **Computer; Data communications.**)

Butterfly Valve A butterfly valve consists of a body, a disk supported on a shaft, and a suitable packing box to allow the shaft to protrude for operation by manual and/or automatic actuators. The flow path in a butterfly valve is straight through the body with only the disk to obstruct the flow—resulting in relatively high capacity. These valves are self-cleaning, thus permitting their use for the control of stocks, slurries, and sludges. The valves exhibit a nearly equal percentage-flow characteristic in throttling applications.

Byte A collection of bits transmitted and processed as one unit within a data-processing or communications system. The term is most commonly used today to describe an 8-bit unit which is a convenient size to represent an alphanumeric character. The major difference between "byte" and "word" in the computer field is in the processing of instructions. A byte-organized computer contains its instructions in a number of bytes, the number used varying with the type of instruction, and the data fields being operated on are of controllable length. In a word-organized computer, all instructions are contained in one, or at most two, words and almost all data operations take place on one or two words of data; that is, both instructions and data paths through the machine are of the same fixed number of bits, this being large enough to describe and specify a complete instruction. Both machine organizations have advantages and disadvantages. The byte-organized machine is somewhat more flexible, but it is also more complex. Modern medium- and large-scale computers commonly exhibit a mixture of the two approaches, with only the smaller machines typically being specialized in one direction or the other.

—A. J. Arthur, *IBM Corporation*

C

Calibration Determination, by measurement or comparison with a standard, of the correct value of each scale reading on a meter or other measuring instrument; or determination of the settings of a control device that correspond to particular values of voltage, current, frequency, or other output. See **Standards, measurement.**

Calibration, Electrical Instrument (See **Electrical quantities, measurement of.**)

Calibration Curve A graphical representation of the calibration record; or a curve relating standard values of input or measurand to actual values of output throughout the operating range of a sensor or transducer and generally under stated operating conditions. A calibration curve properly prepared should give information on (1) the absolute static errors to which a given device is subject, (2) the extent of the device's linearity or conformity, and (3) the hysteresis and repeatability of the device. When the latter values are very small, it may not be practical to extract this information from the calibration curve. Typically, the output is plotted as the ordinate and the input or measurand as the abscissa. Additionally, input and output may be expressed as percent of full-scale input and output, respectively. Alternately, the calibration data may be presented as a curve which is plotted such that the difference between percent actual output and percent ideal output is the ordinate—and percent full-scale input or measurand is the abscissa. This type of deviation presentation facilitates the rapid visual assessment of the accuracy of the described device.

—Otto Muller-Girard, *Consulting Engineer, Rochester, N.Y.*

Calomel Electrode (See **pH, measurement of;** and **Oxidation-reduction potential, measurement of.**)

Calorimeter An instrument designed to measure the thermal energy released during, or required to effect, a change in thermodynamic state. Calorimeters either may establish static or equilibrium states or may explore rates of change of thermodynamic properties during nonequilibrium processes. Generally, calorimeters are quantitative instruments, although certain calorimeters are useful in qualitatively detecting material transitions. Calorimeters find use in determining the heating values of fuels, the caloric content of foods, heats of reaction, the heat of transition or phase change, heat of degradation, the specific heats of materials, chemical-reaction rates, and in various stoichiometric determinations.

Basic Components: Common to most calorimeters are (1) a sample container, (2) an environment enclosure, and (3) a means for determining an internal-energy change of the container and enclosed specimen. In the simplest form, the internal energy of the sample and container is determined through a precise measurement of temperature change. Since the heat capacity of the container is previously known, the change in temperature precisely establishes the change in internal energy during the process. Various environmental enclosures may be used. The most common forms are the constant-temperature jacket or the adiabatic jacket. The former, by being held at a known temperature, permits a calculation and correction for thermal interchange between the container and jacket. The latter is controlled so as to negate thermal interchange.

Because the application of calorimeters is so diverse, there are several types of calorimeters. Five of the most common types are described here. This summary is by no means all-inclusive, as there are numerous other special-purpose calorimeter designs.

Bomb Calorimeter: A pressure vessel forms the sample container of a bomb calorimeter and is capable of operating at high internal pressures. The instrument is useful for determining heats of combustion of materials, usually in a pure-oxygen environment. The specimen is ignited within the container, and the temperature rise of the container, usually suspended in a water bath, produces a measurement of the exothermic heat released. The sample container or bomb normally is enclosed in either an isothermal or adiabatic jacket.

Drop Calorimeter: This is an instrument which determines the heat capacity of a specimen between two temperatures, one being the temperature of the calorimeter receiver and the other some external furnace temperature. The specimen is heated to the external furnace temperature and then dropped into the sample receiver, producing a temperature rise. The system is allowed to equilibrate, and the heat transferred to the receiver during this process provides knowledge of the heat capacity of the specimen. This information determines the specific heat of the material. Heats of vaporization or fusion may be determined in a similar manner.

Variations in the types of drop calorimeter include the type of sample receiver used. A large heat capacity for the receiver is required to produce a small temperature rise, thus simplifying the environmental guarding. Commonly, the receiver is either a large copper block or a water bath. These materials are used because their heat capacities are well known and quite reproducible. The receiver usually is surrounded by an isothermal or an adiabatic jacket.

The ice calorimeter is the most sophisticated of the drop calorimeters. In this instrument, the sample receiver consists of a finned tubular section enclosed in a sealed glass container. Distilled, deaerated water surrounding the finned receiver is frozen by lowering dry ice into the receiver, thus forming an ice mantle over the fins. When a heated specimen is lowered into the receiver, the volumetric contraction caused by the melting of a portion of the ice mantle displaces a volume of mercury from the glass container. The displaced mercury is weighed externally and is proportional to the heat transferred to the receiver from the specimen. Heat transfer from the environment to the receiver is minimized by surrounding the receiver with an ice-water bath. The accuracy of an ice calorimeter may be as great as a few tenths of a percent. The reported accuracy of the instrument shown in Fig. C-1 is better than 1 percent.

Adiabatic Calorimeter: The sample container of this device consists of a small electrically heated enclosure into which the specimen is placed. The container is electrically heated while the responding temperature rise is measured. Heat transfer to the surroundings is negated by (1) surrounding the sample container with an adiabatic jacket whose temperature is controlled to match that of the container, (2) operating in a vacuum, (3) using highly polished surfaces, and (4) minimizing mechanical contact between the container and the environment. The specific heat of the specimen may be determined from the electrical-energy input and from the resulting temperature rise, that is, after subtracting the predetermined heat capacity of the sample container.

Flow Calorimeter: This instrument is used to determine the specific heat of a gas. Usually a stream of gas having a constant flow rate is passed through a guarded heater and the temperature rise is precisely determined. The flow rate, temperature rise, and electrical-energy input permit a determination of the specific heat of the gas. A modified version of the flow calorimeter is used to determine the heat of vaporization of liquids.

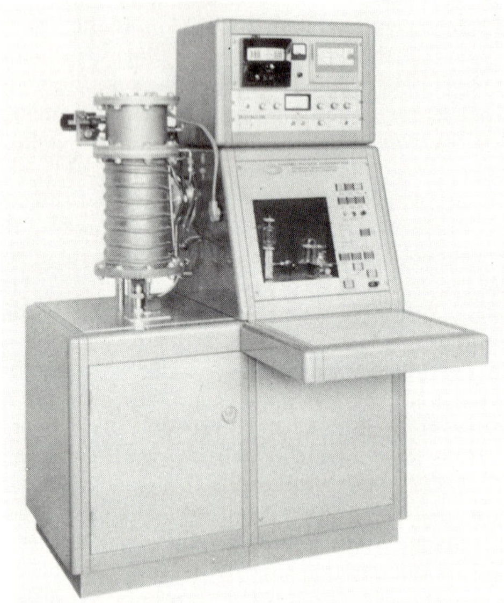

Fig. C-1. Ice calorimeter.

Differential Calorimeter: In this instrument, two specimens are employed—one being a standard of known properties and the other the test specimen. The specimens are heated electrically so that identical temperatures are maintained. Environmental heat transfer within the two specimens is assumed to be identical. A comparison of electrical input relates the heat capacity of the unknown to that of the standard.

REFERENCE

Wilhoit, Randolph C.: Recent Developments in Calorimetry, *J. Chem. Educ.*, vol. 44, no. 7, July, 1967, through no. 10, October, 1967.

—James K. Sparrell, *Sparrell Engineering Research Corporation*

Camera, Schmidt (See **Schmidt camera.**)

Campbell Bridge (See **Bridge circuits, electric.**)

Candela (See **Standards, measurement.**)

Capacitance Measurement (See **Bridge circuits, electric.**)

Capacitance Transducer A device whose capacitance is caused to change when exposed to the variable being measured. A simple configuration comprises two parallel plates that are separated by a small distance—with the space between the plates occupied by a dielectric medium.

Carbon-button Microphone (See **Microphone.**)

Card Reader Punch The punched card is a storage medium used for data input or output for digital computers. The information is stored in punched holes in the card, which is divided into rows and columns, with each column having as many hole positions as there are rows. An IBM card, for example, has 80 columns and 12 rows and therefore provides 960 hole positions (see **Hollerith code**).

The card reader is a device for reading the information from the card and transmitting the data to the computer. The card reader consists of a read station which senses the holes in the card electrically or photoelectrically as it is being transported from the input stacker to the output stacker. It also includes a buffer (see **Buffer storage**) which holds the information for transmission to the computer while the next card is being positioned for the sensing operation. The information from the card may be transmitted to the computer exactly as it is punched in the card, that is, *Card Image,* or it may be translated from card code into machine code in the reader and then transferred to the computer. If translation of the Card Image code is required, it may also be done within the computer with hardware or software.

The card punch is an output device (see **Input/output devices**) which receives information from the computer and punches the data into cards. The information from the computer is held in a buffer until the card is under the punch station, at which time the information is gated (see **Gate**) to the punch magnets. When the punching is complete, the information to be punched in the next card position is transferred to the buffer.

The reading or punching of a card can be accomplished column by column, row by row, or the entire card at one time, depending upon the design of the unit. Since the mechanical controls for reading and punching are similar, the two

Fig. C-2. IBM 1442 Card Reader Punch.

functions are frequently combined, as in the IBM 1442 Card Reader Punch shown in Fig. C-2.

—W. W. Welz, *IBM Corporation*

Cardiography Cardiography is concerned principally with the functional activity of the heart, the means for measuring that activity, and the ways of using those measurements diagnostically. The instrumentation pertaining to other aspects of the cardiovascular system is discussed separately. See **Cardiovascular instrumentation.** See **Medical instrumentation** for a summary of instrument applications in other areas of medicine.

Electrocardiography: The heartbeat normally is controlled by a rhythmically occurring electrical discharge that emanates from a spot in the right auricle known as the sinoatrial (SA) node (Ref. 1). This wave of electrical activity is spread through the heart muscle by a network of fibers known as Purkinje's network. As the wave spreads, the muscle of the heart rhythmically contracts and then gradually relaxes. The contraction, or squeezing, of the muscle forces blood from the heart into the blood vessels, producing the pumping action.

As the wave of electrical activity spreads across the surface of the heart, the wave produces an electrical field which spreads through the conducting medium of body tissue to the surfaces of the body. For analytical purposes, this field can be considered to be produced by an electric-dipole moment which rotates and varies in amplitude in the course of the cardiac cycle. A typical pattern of this electrical activity, as recorded with electrodes placed on the left and right arms, is shown in Fig. C-3. The characteristic features of this electrocardiogram are labeled as conventionally described by physicians. The key wave is due to electrical activation of the auricles of the heart; the *QRS* complex occurs at the time that the ventricles (or main pumps) of the heart contract; the *T* wave is caused by the repolarization or resetting of the electrical system following contraction. The contracting phase of the heart cycle is called *systole,* and the relaxed phase is known as *diastole.* Changes from the normal electrocardiographic pattern can be used to diagnose a heart attack and the progress and recovery from it. The pattern also can be used to diagnose and follow irregular heart rates (arrhythmias) and conditions of faulty conduction.

At the surface of the body, the voltage amplitude of a typical *QRS* complex will vary between 100 μV and 2 mV. Faithful reproduction of an electrocardiogram can be achieved with a frequency response flat between 0.25 and 100 Hz. There are some pathological situations which produce "splintering" of the *QRS* complex that can be easily detected only with a frequency response to approximately 1,000 Hz. This is not a common requirement, and very few electrocardiograph (ECG) recorders have a response above 100 Hz.

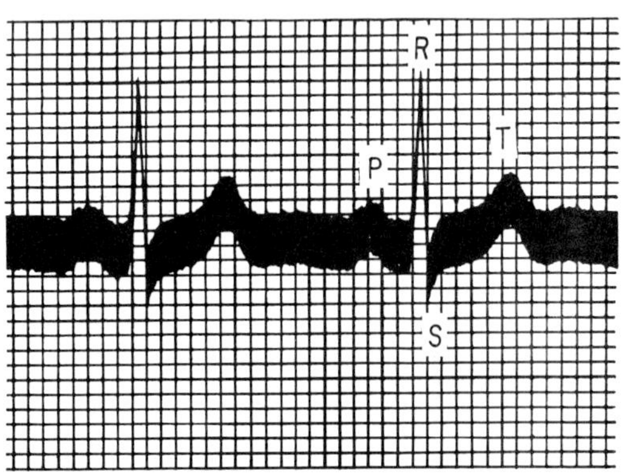

Fig. C-3. ECG of a normal subject.

To produce a faithful electrocardiogram, the voltage-amplifier input impedance should be above 0.1 MΩ and preferably above 1 MΩ.

In order to avoid motion artifacts in the record, care must be taken in the choice and application of the electrodes. Chemical action will cause a gradual change in the impedance of the skin-metal interface which slowly will reduce the amplitude of the ECG and more seriously will produce an abrupt change in the dc level, as this resistance changes with relative motion between the skin and the electrode. If silver is used as the electrode, chloride ions from the sodium chloride in body perspiration gradually will deposit on the silver, reducing the effective surface area and thus increasing electrode resistance. By coating the silver with a silver chloride surface and by further introducing a paste salt bridge between this surface and the skin, the process can be reduced by several orders of magnitude. The skin has a high resistance compared with the subdermal tissues. Thus, for good results, the skin usually is rubbed briskly before application of the electrode paste and the electrodes. Skin resistance typically can vary from 5 kΩ to 0.2 MΩ under varying conditions of perspiration and surface condition—hence the need for a high-impedance input.

Since the body usually is in a 60-Hz electric field and the room in which an electrocardiogram is made is not shielded, the 60-Hz pickup between two points on the body frequently will be one or two orders of magnitude larger than the desired ECG signal. To reduce this effect, ECG amplifiers always employ several differential-input stages and a third electrode attached to an arbitrary point on the body. This latter electrode is used as a "ground," or common-mode reference point. A good ECG amplifier will have a common-mode rejection ratio of approximately 3,000 to 1.

Impedance Cardiography (Ref. 2): An electrical measure of the mechanical activity of the heart may be made by passing 4 to 5 mA of 100-kHz current vertically through the heart, using electrodes at the neck and wrist. The change in this current as a function of the varying amount of blood in the heart through the cardiac cycle is measured with a pair of electrodes placed along the left and right midaxillary lines at the level of the heart, that is, along the sides of the trunk about 4 in. below the armpits of an adult.

Since blood has appreciably lower resistivity than the body tissues, the impedance between the electrodes of approximately 25 Ω will be seen to decrease during diastole by about 0.1 Ω as blood flows into the heart and to increase sharply during systole as it is forced back out of the current path. Anomalies in the pumping cycle will show up if the blood moves out of the electrical field in an abnormal fashion.

Ballistocardiography (Ref. 3): One important measure of heart activity is the force with which the heart ejects blood. If one stands quite still on a spring scale, one will note a small pulsation in the dial reading with each heartbeat. This ballistic recoil of the body as the blood is propelled from the heart can be measured more carefully if the subject lies on a specially constructed bed. The platform or bed on which the patient rests is coupled to outside supports through a flat or coil-spring suspension having a very low resonant frequency so that the response in the frequency range of interest will be far enough from resonance to be fairly flat in the frequency range of interest. The motion of the platform can be measured photoelectrically, with a linear variable-differential transformer, or with a coil and magnet combination.

Historically, the photoelectric measurement came first. A mirror is placed on the platform, and the deflection of a light beam reflected from this mirror is measured with a photocell and a slit of variable width. This method, while quite precise, is subject to vibration sensitivity and is difficult to keep in adjustment. A much simpler approach is to unbalance the current flowing through the coils of a linear variable-differential transformer (LVDT). See **Linear variable-differential transformer (LVDT).**

The LVDT is a quite convenient motion transducer and also is used frequently in other biomedical instrumentation because it gives high levels of output. The frequency response of the LVDT is high enough for many purposes, and it imposes virtually no load on the system being measured.

For analysis of cardiac output and mechanical heart function, the first- and second-derivative (velocity and acceleration) curves are generated as well as the curve of instantaneous position. These curves can be formed electronically by using high-pass filters, but such filters also increase the effect of noise in the system. To reduce this problem, some ballistocardiographs are made with a velocity pickup made by passing a magnet through a solenoid coil. The voltage appearing across such a coil will be proportional to the velocity with which the magnet is being moved through the coil. It then requires only one differentiation to get the acceleration curve and one integration to get the motion curve, thus markedly improving the signal-to-noise ratio for the system.

The ballistocardiogram is quite difficult to quantize because of the inherent difficulty involved in calibration and the variability in coupling between the body and the platform. Nevertheless, qualitatively, it gives some predictive indication of cardiac malfunction.

Phonocardiography: A very old technique, useful for inferring the condition of the valves and some of the other internal heart structures, is to listen to the sound that the blood makes as it passes through the heart. This is most often done with a stethoscope; many characteristic sounds have long since been identified with characteristic conditions. Electronic stethoscopes have been available for many years, but except for the occasional hard-of-hearing doctor or for teaching and demonstration purposes, they have not gained wide acceptance because the old-fashioned stethoscope is quite simple, inexpensive, and reliable; and because the physician learns to identify heart sounds through a conventional stethoscope while in medical school.

The stethoscope is a low-pass filter the exact acoustical characteristics of which vary significantly from one manufacturer to another; thus once a physician has learned heart sounds with a particular type of stethoscope, he will find it difficult to switch to an electronic device, even though it has greater fidelity.

An electronic system for heart-sound amplification should have a passband between 10 and 1,000 Hz. The low-frequency end is too low for a conventional tape recorder or a simple audio amplifier; it is also too low for many microphones. Hence either unusually high-quality conventional equipment is used, or occasionally special equipment which multiplies the lower frequencies by a factor of 3 to 10 to make them more readily audible.

The high-frequency end causes trouble when visual recordings are made. One can thus get only an extremely crude direct-writing or ECG-type recording—usually oscillographic or oscilloscopic film recording is used for this purpose. Such recording is done primarily for research purposes; for clinical use, the ear is still relied upon more than the eye.

Vector Cardiography: An interesting variant on the conventional electrocardiographic presentation attempts to show the locus of the basic dipole moment which is producing the electrocardiogram. For this purpose, instead of the usual pair of electrocardiographic electrodes, a minimum of two pairs of electrodes are placed on the patient's thorax so that the vector sum of the instantaneous voltages appearing across each pair will show at any instant the direction and magnitude of the voltage emanating from the heart. The two output signals thus derived are passed through appropriate weighting networks and then displayed on an oscilloscope, presenting the locus of the tip of the potential vector as it rotates with time through the cardiac cycle as a closed loop.

The difficulty with this approach is that the normal vector cardiogram will vary with body build, electrode placement, and patient position, as well as with the condition of the heart, so that this technique is still considered by most to be a research tool.

Procedures Used in Further Processing and Analysis of the Electrocardiogram (Ref. 4): Because of the importance of the electrocardiogram as a diagnostic tool and the difficulty of interpreting it, three basic means have been developed to increase the efficiency of interpretation:

Experimental telephone links have been established to allow the taking of an electrocardiogram by a general practitioner or a small hospital and the transmission of this electrocardiogram over a telephone line to the office of a heart specialist, where a facsimile ECG is recorded for his immediate analysis and telephone diagnosis. The usual device consists of a frequency-modulated audio tone generator connected to the transmitting ECG machine, which a telephone microphone transmits to the receiving telephone, where it is demodulated and fed into a conventional ECG recorder.

Several manufacturers have developed analog-computing equipment incorporated in an ECG machine to diagnose abnormalities in heart rate and a few somewhat more subtle deviations of pattern from the norm.

Several digital-computer programs have been written which are appreciably more flexible and capable of doing a more thorough job. Because of the expense of the digital computer and of the program, it is necessary to think of using this approach with remote transmission of many ECGs by telephone lines from a network of hospitals. This is probably the ultimate solution to rapid reading of the ECG.

The electrocardiogram taken on a patient in a doctor's office may not accurately reflect the patient's situation in daily life; so several systems have been developed for monitoring ECG on a long-term basis. In one approach the patient wears a small, battery-operated electrocardiograph which telemeters by FM radio over a distance of up to ¼ mile, where the signal is picked up, demodulated, and recorded on a conventional ECG. This allows examination of the patient's heart during exercise and other usual physical and psychologically stressful situations in the patient's life. A variant on this approach is to record the ECG on magnetic tape instead of on paper for later analysis at high speed or for spot checking.

REFERENCES

1. Glasser, O., et al.: "Medical Physics," vol. 1, Year Book Medical Publishers, Inc., Chicago, 1944.
2. Geddes, L. A., and L. E. Baker: "Principles of Applied Medical Instrumentation," chap. 10, pp. 150–205, John Wiley & Sons, Inc., New York, 1968.
3. Fidler, A., A. N. Bhargava, and O. Parent: Clinical Value of Ballistocardiography, *Am. Heart J.*, vol. 55, pp. 881–904, December, 1958.
4. Computer Analysis of Electrocardiograms, *Dig. 7th Intern Conf. on Medical and Biological Engineering*, Session 4, pp. 99–107, August, 1967.

—J. Malvern Benjamin, Jr., *Bionic Instruments, Inc.*

Cardiovascular Instrumentation Some applications for instrumental techniques involving the heart, blood, and circulatory system include cardiography; diagnosis of pericardial effusion and measurement of heart size; determination of blood pressure and blood flow; and measurements and controls required for various therapeutic procedures, such as heart-lung temporary bypass machines and heart-assist devices, pacemakers, and defibrillators. This review is confined to the highlights of the foregoing areas, but it should be emphasized that there are numerous other applications for instrumental techniques in the research, diagnosis, and therapy of cardiovascular problems. Cardiography is discussed separately. See **Cardiography.** Other areas of medical instrumentation are presented elsewhere in this encyclopedia. See **Medical instrumentation** and classified and alphabetical indexes.

Pericardial Effusion and Heart Size: The heart is surrounded by a thin sac called the pericardium. Between this sac and the heart muscle is a thin layer of fluid. Sometimes an injury produces a large secretion of additional fluid into this space—between the pericardium and the heart itself. This condition (pericardial effusion) can be diagnosed by means of an ultrasonic transducer placed against the patient's chest. Some of the sound waves generated are reflected back to the transducer by structures inside the patient's body. The fluid around the heart has a density different from that of surrounding tissue, thus affecting the intensity of the returned echoes. The output is displayed on an oscilloscope which indicates the border of the pericardium, the border of the heart, and the thickness of the space between the two. The same technique is used to measure the size of the heart, since blood—like pericardial fluid—has a density different from that of tissue. A sound reflection will occur at the boundary between the dense blood and the less dense heart muscle and between the heart muscle and the surrounding tissue within the chest cavity. Although less precise than x-ray measurements, the technique does not require a dosage of x-radiation to the body.

Cardiac Output: The product of the blood pressure times the blood volume rate of flow is the cardiac output. There is no fully satisfactory direct way of measuring this parameter in human beings. The long-accepted technique is to infer cardiac flow by measurement of respiratory gases, using a technique developed by Fick (Ref. 1). Other indirect measures are based upon the ballistocardiograph, pulse-wave-height analysis, and indirect measures of flow rate, coupled with indirect measures of pressure, as later described.

Blood Pressure: As the heart contracts, the blood pressure rises until it reaches a peak at the end of systole, producing peak systolic blood pressure (Ref. 2). As the heart relaxes, the blood pressure falls until it reaches its lowest point during late diastole—a point that is the diastolic blood pressure. High-frequency components of the pressure curve are reduced as one moves away from the heart—because of the elasticity of the blood vessels. Many attempts have been made to produce an accurate hydrodynamic model of the blood circulatory system. Although not completely satisfactory, some of these models can serve various predictive purposes (Ref. 3).

Direct Blood-pressure Measurement: The simplest measurement procedure is to puncture an artery or vein with a needle, one end of which leads to a small chamber containing a diaphragm (tambour) and some form of transducer. Sufficient energy is available to allow magnification of the motion with a lever arm which, in turn, is connected to a pen for recording. This simple, low-cost instrument (kymograph) is useful for teaching and demonstration purposes. However, the procedure is tedious to set up and it has poor high-frequency response. These disadvantages are overcome by the use of a strain-gage pressure transducer.

For arterial-pressure recording, a full-scale sensitivity of approximately 300 mm Hg or approximately 5 psig is appropriate. For venous-pressure recording, the sensitivity should be about ten times greater. For arterial-pressure recording, frequency response should be from dc to 100 Hz; while for venous recording, no high-frequency components are present and a 1- or 2-Hz response is adequate.

Cardiac Catheterization: To measure the blood pressure within the chambers of the heart or at any other place within the body which is not directly accessible from the surface without surgery, a catheterization technique is used. A long, thin tube (catheter) is inserted into an artery or vein and, usually under fluoroscopic guidance, is moved into the location of the desired measurement. The catheter is filled with saline solution and connected to a pressure strain gage. The long tube, however, acts as a low-pass filter, thus damping out the higher-frequency components of the blood pressure. Thus, useful information, particularly with regard to the functioning of valves in the heart, is lost. To increase the frequency response, the catheter is made as large in diameter as possible, the pressure transducer is made to operate with as little motion as possible, and great care is taken to eliminate entrapped air from the saline solution. With good technique, it is possible to attain a frequency response of 100 Hz. The chief instrumentation problem is temperature instability of the pressure transducers, since the sensitivity is pressed practically to the limit of what such a transducer can be expected to do. The same technique also can be used for the measurement of pressure at various points in the urinary tract and in the cerebrospinal-fluid system.

Indirect Blood-pressure Measurement: For routine, less critical determinations, indirect blood-pressure measurements avoid the need to puncture an artery (Ref. 2). The sphygmomanometric technique (Ref. 4) in which an air-pressure cuff is wrapped around the arm is the best known. By pumping up the pressure cuff, the physician balances the air pressure within the cuff against the blood pressure in the brachial artery which is squeezed. With a stethoscope, the physician listens to the sound of the blood as it passes into the forearm below the restriction imposed by the cuff. A characteristic sound indicates that diastolic pressure has been reached; while total absence of a sound pulse occurs when systolic pressure has been reached. This system is automated by placing a microphone above the artery and allowing the cuff to receive a pulse of air each time the microphone detects a heartbeat. When the pressure completely closes the vessel, no further pulses of air are injected. A slow air leak is introduced into the system to allow the system to follow changes in pressure. The system is not useful for continuous monitoring of systolic blood pressure because the circulation should not be restricted for more than a few seconds out of a minute. If the whole system is adapted to a finger instead of an arm, however, it can be used safely on a continuous basis. The automatic location of the diastolic pressure point is much more difficult, since it requires careful electrical filtering to discriminate the appropriate sound. This has been accomplished using an arm cuff on a discontinuous basis. Maximum monitoring speeds are approximately one reading per minute.

Pulse-pressure Wave: Many efforts have been made to relate the pulse-pressure wave to absolute blood pressure (Ref. 2). Although such correlations are still unreliable because of other physiological variables, the technique is still of qualitative use. Light can be transmitted through the earlobe, the web between the bases of the fingers, and other thin body parts—and also can be reflected from the surface of the skin. The amount of red light reflected, of course, is proportional to the amount of blood in the path. Thus, a photoelectric method can be used for measuring pulse waveshape. One instrument clamps to an earlobe (Ref. 5). In another form, a grain-of-wheat lamp and small photocell are placed next to each other and embedded in a small, clear epoxy block (approximately ½ by ¼ by ¼ in.). The device can be attached to the surface of the skin with adhesive tape or elastic to pick up the pulse at any convenient location. The instrument usually is referred to as a pulse *plethysmograph*.

Blood Flow in Individual Vessels: The volume rate of flow to and from the major organs of the body is an extremely important parameter. No continuous, indirect, accurate measurement of volume flow rate in individual blood vessels in the intact animal or man is available. The most direct approach for research and training purposes is to sever the vessel and interpose a flowmeter. The "waterwheel" type of flowmeter can be used, but it must be designed carefully so that the blades will not break the red-blood cells. Furthermore, the response time of the meter is not sufficiently fast to demonstrate pulsatile flow with any precision. Also, cutting a vessel sometimes imposes sufficient trauma on the system to change significantly what one is attempting to measure.

A series of dilution techniques has been developed for discontinuous flow measurement. In the simplest of these, a measured amount of blue dye is injected into an artery and the density of the dye is read photometrically at a suitable distance downstream to allow for mixing. Knowing the injection rate, the absolute amount of dye injected, and the photometric-density curve, the blood volume flow rate can be calculated. Only a limited number of measurements can be made before the dye level will reach an objectionable limit. The technique usually requires withdrawal of blood from the measurement site and passage through a measuring chamber (cuvette), which is inconvenient and causes loss of blood.

Radioactive materials sometimes are used, but these have the same disadvantage of maximum tolerance level, with the added inconvenience of handling radioactive materials.

When it is possible surgically to "cut down" on the individual vessel being measured, a pulse of heat can be introduced upstream instead of a physical material. This device, known as a *thermostromuhr* (Ref. 6), consists of a small, C-shaped channel into which a section of intact blood vessel is placed. Upstream, a small, fast-time-constant heater injects pulses of heat. Downstream, a thermocouple or thermistor measures the temperature. The device requires calibration against a more accurate device to be useful. Because of its simplicity, the thermostromuhr finds use for teaching purposes and relatively crude research.

When it is possible to clamp something around the vessel itself, the electromagnetic flowmeter finds the widest acceptance for continuous measurement of blood flow. See **Flowmeter, electromagnetic.** Early electromagnetic flowmeters used sine-wave magnetic fields, but a higher output can be achieved with a higher rate of change of magnetic field. Thus, square-wave and trapezoidal-wave flowmeters were developed. Polarization of the electrode must be avoided. Care must be taken to keep the electrode input impedance high and the surface material clean. Since probes are not adjustable, a series of C-shaped channel probes must be used with a single meter to cover a variety of blood-vessel diameters.

The hot-wire anemometer principle also has been used for blood-flow measurement within a vessel. A heated thermistor is placed at the end of a catheter, which is inserted into the blood vessel. See **Anemometer**. Another cooler thermistor, placed upstream, measures the temperature of the blood to compensate for changes. The technique is simple but has limitations. The method is quite nonlinear; it basically measures rate of flow rather than volume rate of flow; and unless one knows where the thermistor tip lies within the vessel, it is not possible to infer what flow streamline one is measuring. In a blood vessel, the flow rate varies approximately parabolically from the side to the center and back to the other side. Thus, large errors can occur, depending upon the details of probe placement.

Ultrasonic techniques also are used for blood-flow measurement (Ref. 7). Two ultrasonic transducers are placed diametrically opposite each other across a vessel wall. A 5-MHz acoustic signal produced by one is received by the other. Motion of the blood through the sound field will produce a doppler shift linearly related to the flow rate. In another version of this principle (Ref. 8), the two transducers, each D-shaped, form the complete circular end of a catheter which is inserted into a blood vessel. One of the transducers is the transmitter, the other the receiver. In another use, the probe is placed on the surface of the skin and the sound is directed through the skin into a superficial blood vessel (Ref. 9). In this embodiment, the device is used to locate constrictions in blood vessels near the surface and to make relative flow measurements.

Peripheral Vascular Bed: A technique known as plethysmography is used to measure the volume flow rate of blood into body segments, such as arms, legs, toes, and fingers. The venous return from the body segment is stopped momentarily, and as arterial blood continues to flow into the segment, the rate at which the part swells is a measure of the volume rate of flow of arterial blood. The problem then becomes one of measuring the small swelling that takes place.

In one method, the part is placed in water and

the water displaced is measured. There are a number of obvious drawbacks to this method. In another version, the part is placed in air instead of water. Despite many problems, air plethysmographs can be made sufficiently stable for calibration.

If cylindrical symmetry can be assumed, it is sufficient to measure the increase in diameter at any convenient cross section and infer the volume swelling of the part from this. A simple, ingenious strain-gage device can be used. This consists of a thin rubber tube filled with mercury, at the ends of which are inserted platinum wires to form electrodes. Electronically, this mercury strain gage is treated like any other strain gage. The technique is in common use but has the disadvantage of taking a significant amount of time for attachment and for balancing the strain-gage bridge.

A technique known as impedance plethysmography (Ref. 10) measures the amount of blood in a volume by passing an electric current through it. Usually the current is on the order of 50 μA to 5 mA at a frequency between 20 kHz and 1 MHz. Sometimes the same pair of electrodes is used for both introducing the current and measuring the current change. In more critical experiments, separate pairs of electrodes are used for introducing the current and for measuring voltage change. Although simple to apply, the method is not easily calibrated, and it is difficult to quantify results on a theoretical basis.

Heart-Lung Temporary Bypass and Heart-assist Machines: The heart-lung machine is used to bypass the patient's heart and lungs for the time necessary to perform an operation on the heart. The machine must be carefully designed and instrumented. The pump that substitutes for the heart must be designed not to damage the blood corpuscles, heat the blood, or cause coagulation. The blood aerator, which substitutes for the lungs, must have a large surface area and very thin, semipermeable walls to permit gas exchange without physically damaging the blood. The machine must not alter blood temperature or blood pH significantly and must not use more blood for priming than absolutely necessary. The machine must be easily disassembled for cleaning and total sterilization. Controls for the machine monitor and control flow rate, temperature, oxygen, and carbon dioxide.

Ventricular Assist Devices: To reduce the load on a heart recovering from a heart attack (coronary occlusion), several devices have been developed that provide additional pumping action during the resting (diastolic) phase of the heart cycle (Ref. 13). Some of these devices require a major operation to connect them to the heart or blood vessels, while some require only a minor operation to insert a balloon into the descending aorta (Ref. 11), or the external application to the legs and abdomen of an inflatable suit. To time the inflation of the device properly, the patient's electrocardiograph triggers an adjustable delay period, at the end of which a solenoid valve is opened for another adjustable time period.

Pacemakers and Defibrillators: To correct faulty electrical action of the heart, a small electronic battery-operated pacemaker is placed by operative procedure inside the body and connected by electrodes to the surface of the heart. The device is set to pace the heart at a fixed rate and contains a battery which usually will last for over a year before another operation is required to replace it. Some pacemakers are rechargeable —with batteries that can be inductively charged through the skin.

A severe electric shock, physiological shock, or massive heart attack may cause the natural pacemaker within the heart to lose control. When this occurs, small muscle bundles within the heart will contract at random, thus stopping the heart from contracting as a whole, or "beating" properly. This condition, called fibrillation, can be corrected by sending a heavy pulse of electric current through the heart.

The instrument, known as a defibrillator—which generates this pulse, consists of two electrodes, a timer, and a voltage or current controller. If fibrillation occurs while a patient is on the operating table or within a hospital alerted for such an emergency, the chest can be opened and the electrodes placed directly on the heart. Sometimes several pulses of 1 to 5 A at 60 Hz are required to start the heart and to keep it going, but the technique almost always is successful. The heart must be restarted within 5 to 7 min. or irreversible brain damage will occur.

When the heart cannot be reached directly, electrodes are placed on the chest. A current of 4 to 7 A at 60 Hz is used. Too little current can cause greater fibrillation. Chest electrodes are less reliable than direct heart connection, because for most reliable results the heart also should be massaged.

REFERENCES
1. Glasser, O., et al.: "Medical Physics," vol. 1, p. 576, Year Book Medical Publishers, Inc., Chicago, 1944.
2. NASA: "The Measurement of Blood Pressure in the Human Body," NASA SP-5006, Washington, D.C., 1964.
3. Hemodynamic Models, *Dig. 7th Intern. Conf. on Medical and Biological Engineering*, Session 1, pp. 50–62, August, 1967.
4. Burch, G. E., and N. P. DePasquale: "Primer of Clinical Measurement of Blood Pressure," The C. V. Mosby Company, St. Louis, 1962.
5. Wood, E. H., et al.: Measurement of Blood Content and Arterial Pressure in the Human Ear, *Proc. Staff Meeting, Mayo Clinic*, vol. 25, pp. 398–405, July 5, 1960.
6. Gregg, D. E.: "Thermostromuhr: Methods in Medical Research," vol. 1, Year Book Medical Publishers, Chicago, 1948.
7. Franklin, D. E., N. W. Watson, K. E. Pierson, and R. L. Van Citters: Technique for Radio Telemetry of Blood-flow Velocity from Unrestrained Animals, *Am. J. Med. Electron.*, vol. V, pp. 24–28, 1966.
8. Stone, H. L., H. F. Stegall, V. S. Bishop, and C. Laenger: Continuous Measurement of Blood Flow Velocity with an Intravascular Doppler Flowmeter, *Dig. 7th Intern. Conf. on Medical and Biological Engineering*, p. 215, August, 1967.
9. Harding, D. C., D. W. Baker, and R. F. Rushmer: Isothermal Cutaneous Blood Flowmeter, *Proc. 17th Ann. Conf. on Engineering in Medicine and Biology*, p. 77, 1964.
10. Allison, Dr. Robert D.: The Role of Impedance Plethysmography in Bio-medical Research and Diagnostic Medicine, *IEEE Trans. Bio-Med. Eng.*, vol. 3, no. 11, pp. 500–503, November, 1968.
11. Welkowitz, W., D. A. Molony, J. L. Potter, E. M. Grabowy, D. Jaron, W. Z. Yahr, and A. Kantrowitz: A Portable Electronic-Pneumatic Ventricular Assist Driving System, *IEEE Trans. Bio-Med. Eng.*, vol. BME-16, p. 58, 1969.
12. Pacemakers and Defibrillators, *Dig. 7th Intern. Conf. on Medical and Biological Engineering*, Session 2, pp. 63–82, August, 1967.
13. Cooper, T., and P. J. Dempsey: Assisted Circulation, *Modern Concepts of Cardiovascular Disease*, vol. 37, nos. 5 and 6, pp. 95–106, May and June, 1968.

—J. Malvern Benjamin, Jr., *Bionic Instruments, Inc.*

Carey-Foster Bridge (See Bridge circuits, electric.)

Cascade Control Control action where the output of one controller is the set point for another controller. This is a control-configuration technique used to increase the stability of a complex circuit. Cascade control accomplishes two important tasks: (1) it reduces the effect of load changes near their source; and (2) it improves the control circuit by reducing the effect of time lags. The block diagram of a cascade-control system in shown in Fig. C-4.

The outer loop containing a primary controller and a primary measuring device sets the control point of the inner loop, which contains a secondary controller and a secondary measuring device. Usually both inner and outer loops have measuring devices, but only the primary controller has an independent control point and only the secondary controller has an output to the final element.

It is believed that the best way to illustrate the operations and applications of a cascade-controlled system is by diagrammatical examples.

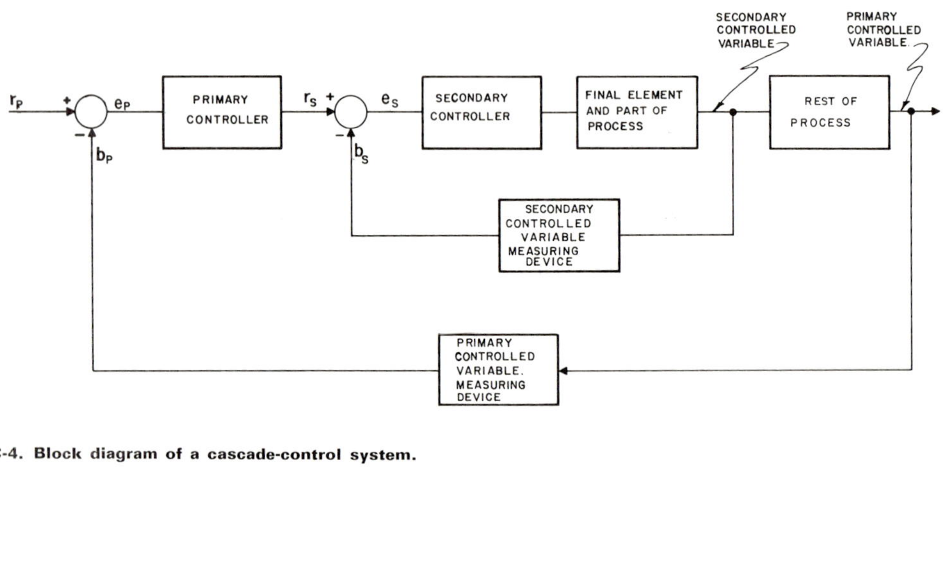

Fig. C-4. Block diagram of a cascade-control system.

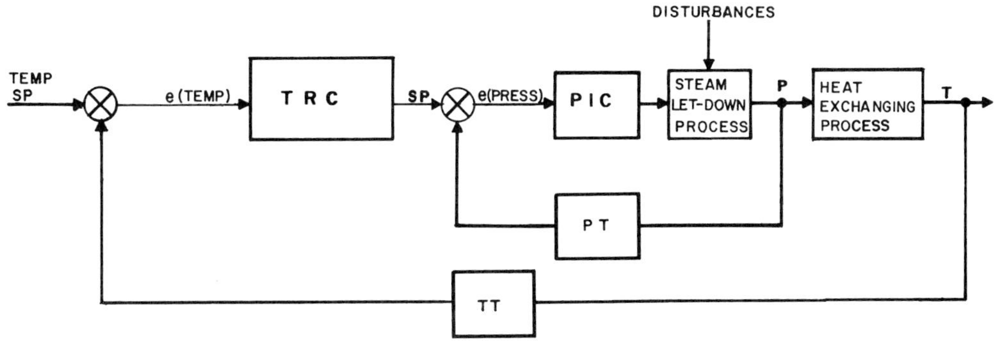

Fig. C-5. Block diagram of cascade-control system as applied to a fractionating column.

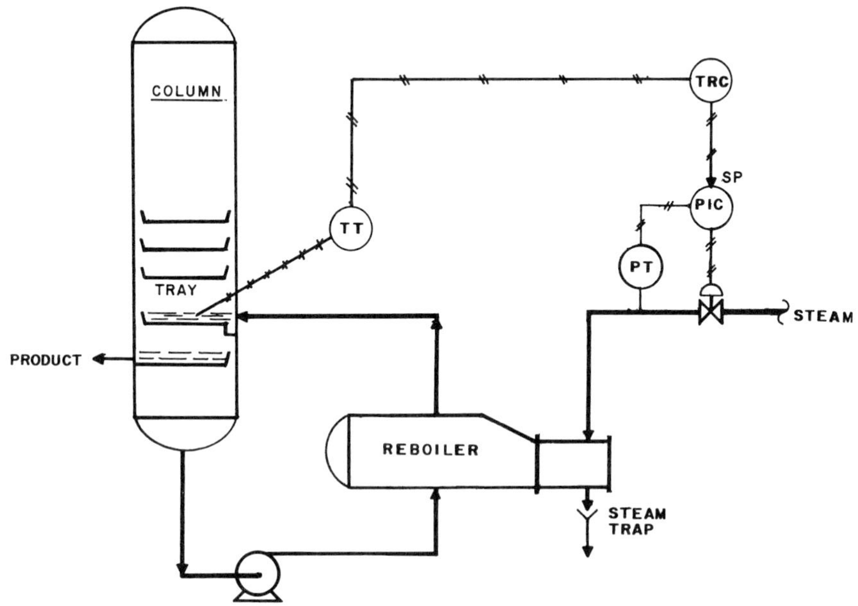

Fig. C-6. Cascade control of fractionating column.

Figures C-5 and C-6 show a typical application of fractionator control. The temperature of a certain tray is used to control the amount of steam entering the reboiler. The block diagram as shown in Fig. C-5 could be helpful in appreciating the basic concept of a cascade-control system—namely, to absorb most process disturbances in the inner loop before they affect the primary controlled variable. This can be easily recognized in the fractionator example. The pressure controller *PIC* in Figs. C-5 and C-6 is employed to "guarantee" a constant-pressure downstream of the control valve regardless of the fluctuations of upstream steam pressure.

The same principle is applied to the control systems schematically shown in Figs. C-7 through C-9.

REFERENCES

Principles of Automatic Process Control, *ISA Publ.* PAPC, 1969.

Tucker, G. K., and D. M. Wills: "A Simplified Technique of Control System Engineering," Honeywell Inc., Fort Washington, Pa., 1958.

Considine, D. M. (ed.): "Process Instruments and Controls Handbook," McGraw-Hill Book Company, New York, 1957.

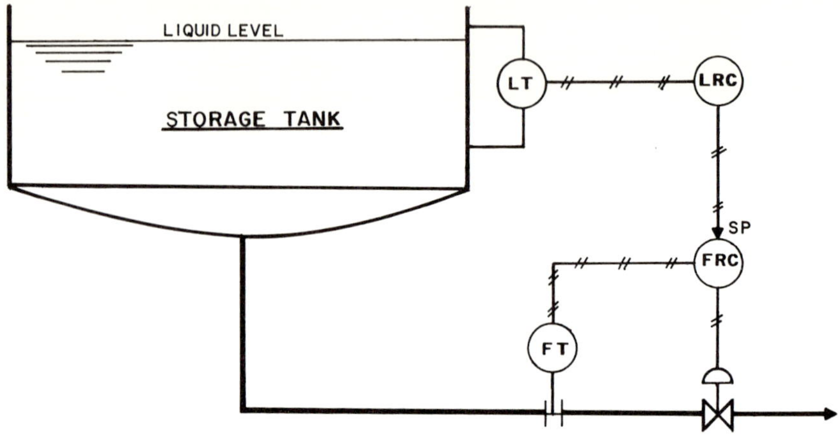

Fig. C-7. Cascaded liquid-level control system.

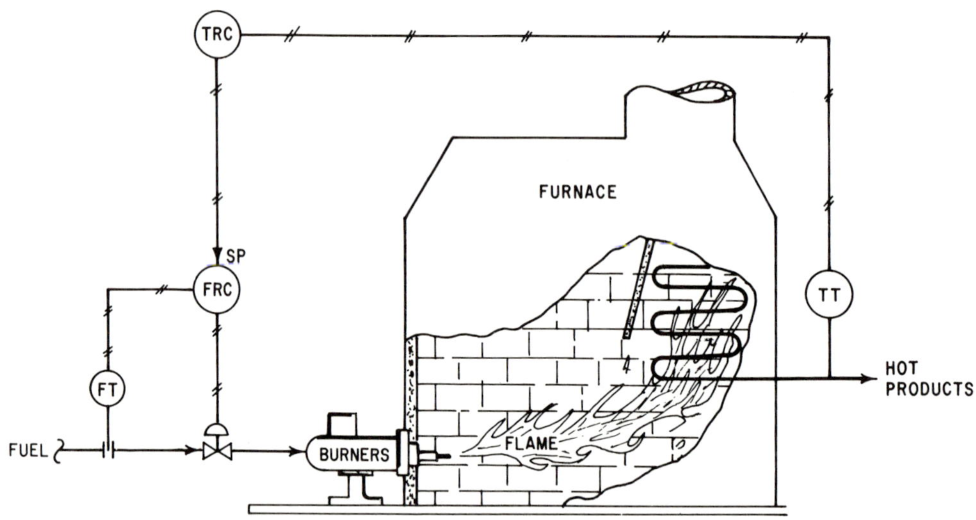

Fig. C-8. Cascaded furnace control system.

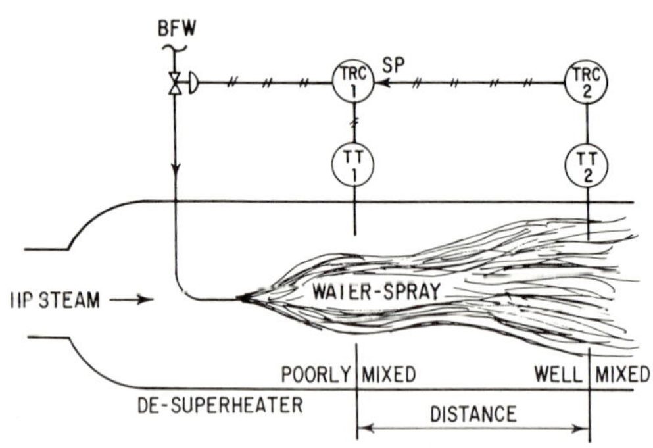

Fig. C-9. Cascaded desuperheater control system.

"Manual on Process Instrumentation and Control," 2d ed., American Petroleum Institute, Division of Refining, API RP 550, New York, March, 1965.

—Anthony Y. Chiang, *Bristol Division, American Chain & Cable Company*

Cassegrain Telescope (See Telescope, reflecting.)

Cathetometer A form of optical comparator used for the accurate measurement of vertical distances. Cathetometers are well suited for inspection and layout work. The vertical cathetometer consists of a telescope (or, for close work, a microscope) that is mounted horizontally on a guide bar whose length is parallel to the displacement to be measured. The height of the telescope (hence of the object) is read on a precision scale attached to the guide bar. A typical precision cathetometer, which utilizes an accurately calibrated guide bar, will have a measuring range of 100 cm and can be read to 0.01 mm.

Cathode-ray Oscilloscope (See **Oscilloscope.**)

Cathode-ray Tubes Cathode-ray tubes (CRTs) convert electron-beam energy into light energy. Electrons are emitted from a thermionic cathode and controlled by a triode section. The electrons then are formed into a beam and accelerated into the focus section of the tube. The deflection section of the tube deflects the beam (typically on vertical and horizontal axes) by internal electrostatic deflection plates or by external electromagnetic deflection coils. The acceleration or drift area controls the electron beam until the energy reaches the CRT screen. The electrons strike a light-emitting phosphor on the inside face of the CRT screen, which fluoresces and emits visible light. CRT phosphors possess the characteristic of phosphorescence, emitting light energy after the electron beam is removed. This effect of image persistence allows a repetitive pattern to appear as a stationary display. In addition to presenting x- and y-axis information on the deflection plates, CRTs use the cathode or grid of the CRT gun to present z-axis intensity information. The major industrial use of CRTs is in oscilloscopes. They are also widely used in television receivers, in radar displays, and as readout devices in computer terminals. Storage CRTs also are available to permit the retention over long periods of time of signals which may occur only once (transients) or which have low repetition rates. Storage tubes eliminate much of the need for photographing such transients.

—Richard E. Kehrli, *Tektronix, Inc.*

Cathode-ray-tube Recorder (See **Oscillograph.**)

Cavity-type Wavemeter [See **Frequency (electrical), measurement of.**]

Celsius Temperature Scale This is a temperature scale named in honor of Anders Celsius, who was the first to report (1742) on the use of thermometers in which the fundamental interval, ice point to steam point, was 100 degrees. Celsius designated the ice point as 100° and the steam point as 0°. Subsequently Christin (1743) and Linnaeus (1745) independently interchanged the designations. Up to 1948 the scale was known as the centigrade scale. By international agreement it was renamed in 1948. It is used world-wide in scientific work and for general purposes in non-English-speaking countries. Temperatures are denoted as degrees Celsius and by the symbol °C.

—Robert D. Thompson, *Taylor Instrument Process Control Division, Sybron Corporation*

Centrigrade Temperature Scale (See **Celsius temperature scale.**)

Central Nervous System, Instrumental Analysis of The central nervous system consists of the brain, the spinal cord, and all the nerves that interconnect these two structures with the muscles, vital organs, and sensory organs of the body. Also see **Medical instrumentation.**

The electrical activity of the brain is manifested at the surface of the scalp by small potential changes on the order of 5 to 200 μV in a frequency band from 1 to 50 Hz (Ref. 1). A sample of these "brain waves," or electroencephalogram, is shown in Fig. C-10. On casual observation, the usual electroencephalogram (EEG) would appear to be simply random noise, but spectral analysis and some direct encephalograms show pronounced components at several frequencies. The lowest, at 1 to 3 Hz, has been called the delta rhythm. The theta wave appears next—from 4 to 7 Hz. The most pronounced rhythm appears between 8 and 13 Hz and is called the alpha rhythm. A third pronounced rhythm appears between 13 and 30 Hz and is called the beta rhythm. As shown by Fig. C-10, the alpha rhythm appears most pronounced during light sleep, while the delta rhythm appears during deep sleep. Thus, one of the values of the electroencephalogram is to determine stages of sleep for sleep research work and other investigations.

The EEG also is useful for the diagnosis of epilepsy and the location of brain tumors and other lesions. However, there are many gross abnormalities of central-nervous-system functioning and of thinking which in no way alter the pattern of the EEG. Furthermore, because of its very noisy nature, it is extremely difficult to interpret for diagnostic purposes, except in the situations outlined above. It is also useful for monitoring the depth of anesthesia during operations.

A conventional electroencephalogram is taken by placing a series of 6 to 10 electrodes symmetrically on each side of the head from the front to the back, with a reference electrode on the mastoid bone or some other similar spot. Sometimes the electrodes consist of small pins which puncture the skin to make better contact, but more usually the scalp electrode is a small sponge soaked with saline solution and held down with tape. It is not usually necessary to shave the hair.

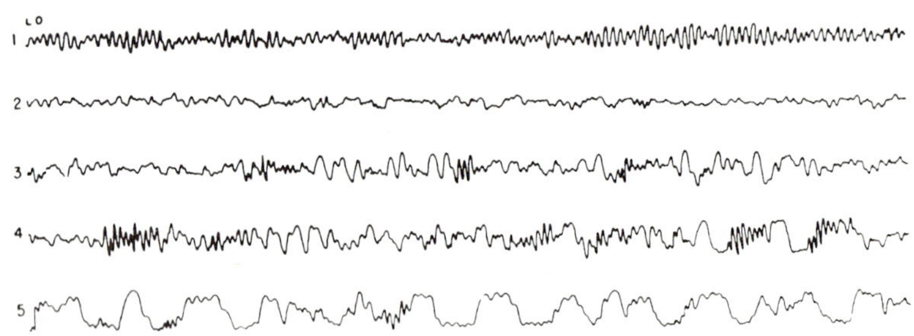

Fig. C-10. Electroencephalogram tracings of normal subject during sleep. Numbers indicate increasing depth of sleep.

The usual EEG machine uses a multichannel direct-writing recorder with 6 to 10 channels and speeds vary from 2 to 50 mm/s, the width per channel typically being about 25 mm. Because the EEG amplitude is an order of magnitude smaller than the ECG amplitude, the criteria for the design of an EEG amplifier (see Electrocardiogram under **Cardiography**) apply with greater force to this situation. Common-mode rejection should be at least 10,000:1, and the recording must be taken in a shielded room. On the other hand, low-frequency response need not be as good for the EEG as for the ECG; so polarization of the electrodes is not so critical.

When making an EEG recording, it is necessary that the shielded room be relatively free of external stimuli, such as changes in ambient light, sound, or vibration, as sudden changes will produce startle-response artifacts that, while easily recognized, are best avoided.

Evoked Potentials (Ref 2): A large sensory stimulus will have a large effect on the EEG, but a small stimulus, such as a small sound or a small light change, will not produce any obvious effect. However, a technique known as the evoked-cortical-potential technique makes it possible to see the effect on the EEG of a sensory stimulus. At the same time as the stimulus is initiated (a flash of light or a burst of sound), an EEG recording is made and stored in a computer memory, usually on a digital basis. Each time a fresh stimulus is applied, the storage is updated, so that after 50 or 100 trials the digital memory contains an average EEG synchronized with the stimulus. A typical memory may contain 2,500 bits: 50 elements in one direction could represent fifty 10-ms periods, starting with the onset of the stimulus, while the 50 elements in the orthogonal direction would represent the average amplitude of the EEG for each time period. This technique is useful for research work on sight, sound, touch, and other sensory stimuli and, under some circumstances, in the location of disorders causing blindness or deafness.

Nerve-pulse Amplifiers: Work on the brain and on individual fibers frequently requires the use of microelectrodes to stimulate and detect the responses from very tiny spots. Microelectrodes, consisting of very fine wires surrounded by thin glass insulation, are made down to an active diameter of approximately 1 micron. The impedance of such an electrode can be 10 to 100 MΩ, and the frequency of the signal can have components up to 10 kHz; at the same time, the signal-to-noise ratio can be very low. Thus the requirements of very high common-mode impedance, very high input impedance, and virtually no input capacitance dictate a special class of amplifier in which the input capacitance is neutralized by feedback.

Apparatus exists that allows the simultaneous use of an electrode for producing a stimulus and and recording the response. The input and output circuits to the electrode must be carefully isolated to make this possible.

REFERENCES

1. Allemandi, R. J.: Biomedical Measurements, *Meas. Data News,* vol. 3, no. 1, pp. 56–57, January–February, 1969.
2. Larson, S. J., A. Sances, Jr., and P. C. Christenson: Evoked Somatosensory Potentials in Man, *Arch. Neurol.,* vol. 15, pp. 88–93, 1966.

—J. Malvern Benjamin, Jr.,
Bionic Instruments, Inc.

Central Processing Unit (CPU) The central processing unit, or main frame, is that part of a computing system exclusive of input or output devices (see **Input/output devices**) and, sometimes, main storage. With reference to Fig. C-11, it includes the arithmetic and logical unit [see **Arithmetic and logical unit (ALU)**], channels, storage and associated registers (see **Register, computer**), and controls. Information is transmitted to and from the input or output devices via the channel [see **Channel (CPU)**]. Within the central processing unit, data are transmitted between storage (see **Storage, computer**) and the channel and between each of these and the arithmetic and logical unit. In some computing systems, if the channel and storage are separate boxes of equipment, the central processing unit includes only the arithmetic and logical unit and the instruction interpretation and execution controls.

The functions of a CPU are described with reference to Fig. C-12, which is a simplified example of a central-processing-unit data flow. The operations performed in the CPU can generally be divided into three phases: instruction fetch, indexing (if appropriate to the instruction format), and execution.

To initiate an operation in this example CPU, the contents of the instruction address register (see **Address register**) is transmitted to the storage address register (SAR), and the addressed instruction is fetched from storage to the storage data register. The operation code portion of the instruction is set into the operation code register (OP), and the data address part is set into the data address register. If indexing is defined in the instruction format, the indexing address is set into the index address register (see **Index register**).

During the indexing phase of the operation, the index address register is transmitted to the SAR and the contents of the storage location is set into the storage data register. The contents of the storage data register and contents of the data address register are summed in the adder, and the result is placed in the data address register.

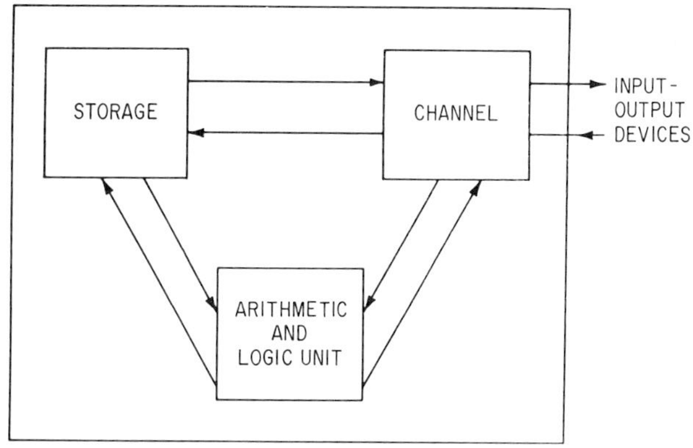

Fig. C-11. Central processing unit of a computer.

On an arithmetic operation, the data address register is transmitted to the SAR, and the data in the referenced location are gated through the adder along with the contents of the data register. The result replaces the contents of the data register. If the arithmetic operation causes a carry out of the high-order position of the word being operated on, the overflow indication is saved in

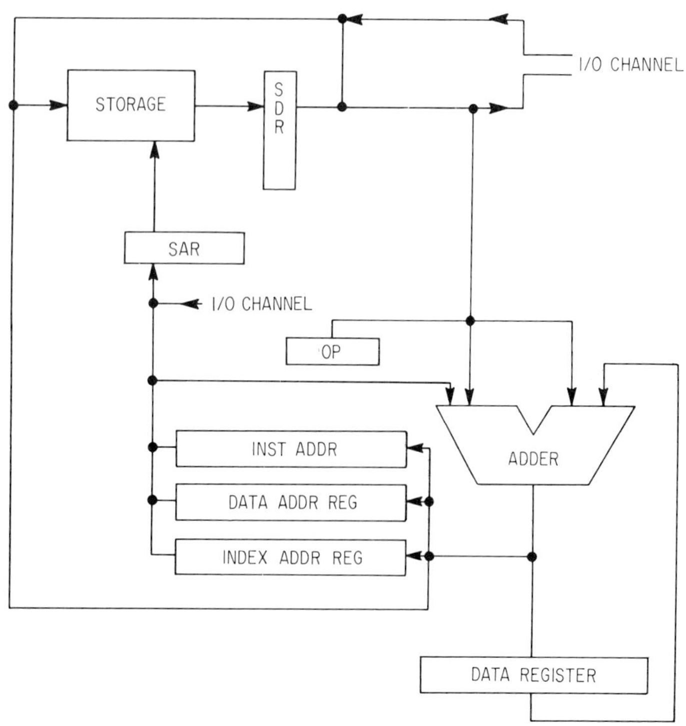

Fig. C-12. Central-processing-unit data flow.

a condition code trigger. This condition may be tested on a subsequent operation, such as a transfer on condition. At the completion of the specified operation, the next instruction is requested from storage and previously described phases are repeated.

—W. W. Welz, *IBM Corporation*

Centrifugal Governor (See Governor.)

Cerenkov Detector [See Radiation (nuclear), detection of.]

Cesium-135 Beam (See Clock, atomic.)

Channel (CPU) A channel is that portion of the central processing unit which executes the instructions relating to the input or output devices (see **Input/output devices**). It also provides the interfaces and associated controls for the transfer of data between storage (see **Storage, computer**) and the input-output devices attached to the central processing unit [see **Central processing unit (CPU)**]. It generally maintains the storage address for the device in operation and includes the buffer registers [see **Buffer (storage)**] required for synchronizing with storage. For high-speed devices, a serial or selector channel may be used for attachment, while for slower devices a multiplex channel is generally used.

The multiplex channel is used to attach lower-speed devices to the central processing unit. It has the capability of concurrently servicing several devices. The storage address for each operating device is controlled by the channel and is maintained in main storage or in the logic of the channel. If the address is maintained in the logic, it increases the maximum data rate of the channel and reduces the number of storage references required to service the device. An adaptation of the multiplex channel is the storage-access channel which requires that the device maintain the storage address. With this type of channel, the device transmits both the data and the storage address to the central processor when it requires servicing.

A serial channel is one in which several devices may be physically attached to the channel interface but only one of these devices is logically connected at any time. As a result, the total channel data-rate capability is available for the selected device. For a given device data-rate requirement, the serial channel is less expensive than either dedicated channels per device or a multiplex channel. In the latter case, the saving results from the fact that the maximum required data rate is determined by the data rate of a single attached device. In the multiplex channel, the maximum rate is the sum of the data-rate requirements of several devices plus the data rate required for device addressing.

—W. W. Welz, *IBM Corporation*

Channel, Communication (See Data communications.)

Character-recognition Systems Because of the remarkable growth in the number of computer installations and in the range of applications to which computers are applied, a problem has evolved which concerns converting huge amounts of original input data on which the computer *will* operate into a form on which the computer *can* operate.

The nature of the original data can take many forms—the spoken word, transducer outputs—but by far the largest source of data is printed paper. Paper, as a storage medium, is inexpensive to buy, transport, and store. It can be easily printed or written on without much cost, and there is a tangible, humanly readable record. The problem was that the record was readable only by people. To go from the printed record into the computer required a human intermediary such as a keypunch operator.

With the rapid expansion of computer applications, it became economically feasible to develop and manufacture a new kind of input device—character-recognition systems—systems for reading information printed, typed, or hand-printed on paper directly into the computer without human assistance. Such systems are applicable where there are voluminous amounts of printed input data. For example, post offices are faced with the problem of sorting mail—outgoing to the appropriate city or town; incoming to the appropriate postman for delivery. Banks must read checks and enter the data thereon into their accounting system. Insurance companies, warehouses, department stores, and gasoline companies are among those who record their transactions on paper and then process it by computer. The United States Social Security System maintains some 140 million insurance accounts and updates these quarterly from employers' reports. These reports are partially contained on 6½ million pieces of paper, variously printed, which required the employment of over 400 keypunch operators for data conversion. The character-recognition systems described here are one solution to the computer-input problem.

Basic Systems: Character-recognition systems are often put into two classifications depending on the kind of ink with which the printing is done. Systems based on magnetizable ink and magnetic-field-detecting reading heads are referred to as MICR (magnetic-ink character recognition). Those based on ordinary inks which are read by optical principles are called OCR (optical character recognition). The similarities between MICR and OCR systems far outweigh the differences, and they are discussed together here. Both systems are composed generally of a document transport, a scanner, a character-recognition logic section, and an ouput medium (see Fig. C-13). The transport moves the paper (on which the information is printed) into the field of view of the scanner. The scanner locates the data to be scanned and converts the printed representation of the letters or symbols to a form which can be internally processed by the recognition device.

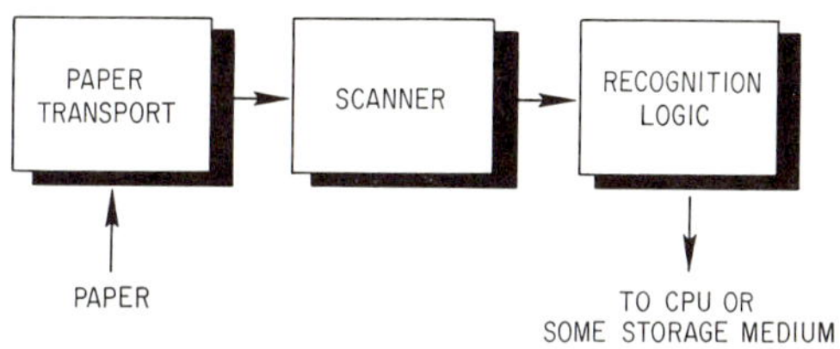

Fig. C-13. Block diagram of a character-recognition system.

The purpose of this process is to make measurements on the internal representation of the character which will enable the scanned character to be recognized, that is, to be distinguished from all other possible characters. Identification is then made, and the result is sent to the central processing unit (CPU) of a computer or is recorded on some storage medium such as magnetic tape. The result may also be used to control subsequent scanning of characters or the disposition of the paper on which the material is printed.

The nature of the characters to be read (the input) is all-important in determining the kind of reading technique that will be used and the degree of reading accuracy that will be achieved.

Magnetic and Nonmagnetic Ink: Characters can be printed in inks containing particles which can be magnetized. Single-gap reading heads, such as those found on magnetic-tape recorders, or multigap heads with as many as 30 separate gaps are used to scan the magnetized characters. This technique has the advantage that overprinting in ordinary nonmagnetic inks does not interfere with machine reading. Magnetic inks are used on bank checks; the first commercially available character-reading machines were designed for this application. The cost of preparing documents for successful reading by this method is high, and subsequent work for nonbanking applications has been with nonmagnetic inks. In these cases, the printed character is distinguished from the background paper by the different reflectivities which each presents to a light source that illuminates the character (or a part of it) during scanning.

The Input Set: The difficulty of distinguishing between characters is partially determined by the number of different classes which must be distinguished. The simplest case is that where only the 10 numerals are involved. It becomes much more difficult when 26 each of uppercase and lowercase letters and a large number of punctuation marks and other special symbols are added to the 10 numerals. Most commercially available machines handle some mixture of classes between the two extremes. In many cases, even when a machine can handle a large number of classes, there may be format or context information which restricts the number of possibilities for a particular character which must be read. Substantial improvement in performance can often be made by utilizing this information.

Mark-reading devices—machines which look only for the presence or absence of a mark as the fundamental unit of data—have been excluded from this discussion.

Standard and Nonstandard Type Fonts: The shape of the printed characters substantially affects their distinguishability. Most certainly the letter O and zero are usually indistinguishable, but even pairs such as 5 and S or 2 and 7 can be very difficult to separate in cases of reasonable amounts of perturbation caused by print-quality deterioration. To aid in making the distinguishability less sensitive to such perturbation (and perhaps to make the distinguishing features conform to the peculiarities of a particular reading technique), considerable work has been done and is being done to design standard fonts. To improve the distinguishability of look-alike pairs, one or both characters in each pair must be made less conventional. A compromise must be made between distinguishability and appearance.

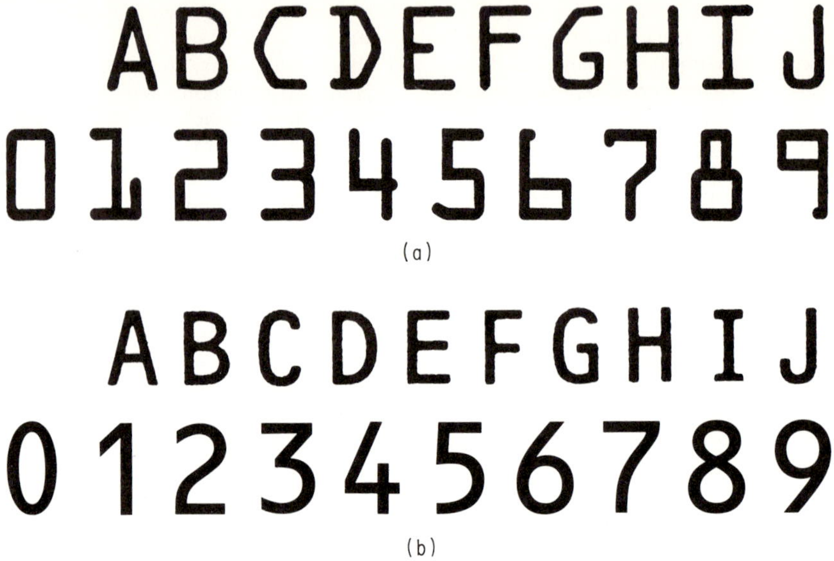

Fig. C-14. Examples of standard optical character-recognition (OCR) fonts: (a) International and American National Standards Institute A font; (b) International Standards Organization B font.

Standards have been set at the corporate, industrial, national, and international levels. Also, standardization activity is being carried on with respect to shapes for hand printing. Some examples of standard fonts are shown in Figs. C-14 and C-15.

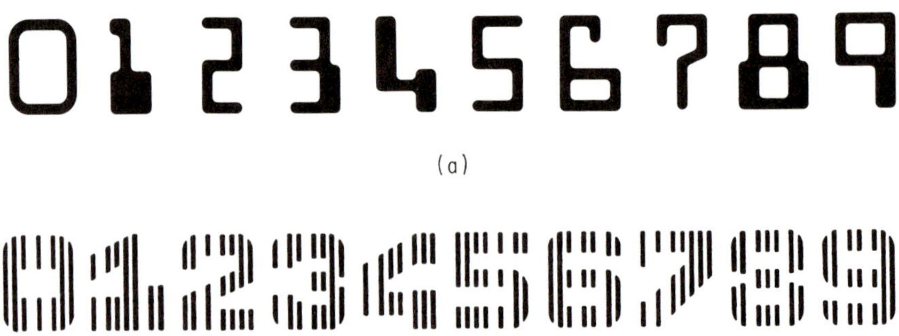

Fig. C-15. Fonts designed for one-dimensional distinguishability: (a) standard font for United States banking industry (E13B); and (b) bar-coded font (CMCT).

Print Quality: The quality of the printed character substantially determines its readability. This quality is affected by the kind of device used to do the printing (for example, typewriters, business machines, credit-card imprinters) and its state of adjustment, by the age and kind of ink ribbon used, by the quality of paper used, and by the amount of handling the document has had which can cause smearing or deposition of dirt. For reading hand printing, the care with which the letters are formed and the condition of the writing instrument are important.

The performance of reading machines is usually specified by the rate at which they fail to recognize characters (rejects) and by the rate at which they misrecognize characters (substitutions). It is not unusual, in a given machine, for these rates to vary by one or several factors of 10 over a realistic range of input quality. In systems designed around reading machines, provision is usually made to handle the exceptional characters by employing context, redundancy, or manual correction.

Paper Transport: The paper transport separates the individual sheets of paper from a stack, moves them into and out of the scanner region, and then stacks them in one or more compartments called stackers. Machines vary in their capabilities to handle intermixed sizes and thicknesses of paper. Some transports handle cash-register rolls or fan-folded stacks.

Paper Sizes: Typical paper sizes handled vary from about 2 in. up to legal-size sheets. Transports are classified roughly by the paper size they will accommodate. Document scanners handle small papers up to check size or somewhat larger, with one or, at most, a few lines of print; page scanners handle larger sheets. Separating documents from a stack requires work to be done not only against the weight of the sheet but also against the frictional and electrostatic forces which tend to make one document adhere to the next. Typical methods used involve vacuums and counter-rotating belts.

Motion Past the Scanner: In some scanners the document must be stationary during the scanning process. The transport must stop the document at least once and sometimes more than once in the region of the scanner. This is usually the case with page readers where a substantial amount of scanning must take place. Where fewer characters are to be read (especially from a single line), the scanner can be designed to read while the document is moving by. The motion is utilized by the scanner and usually must be carefully controlled or linked to the scanner. Some page scanners wrap the sheet around a rotating drum and read the lines from the rotating document.

After scanning, the document may be stacked into a single pocket or into one of two or more pockets as controlled by information read from the document. A reject pocket is often provided for documents which are known to have been improperly read. Document sorters have 10 or more pockets for stacking.

Format: Scanners vary in their capabilities to handle documents with information printed in different formats. Single-line document readers normally require the information to be printed in a fixed location on the document corresponding to that area which passes under the scanning head. Sometimes a little flexibility is provided by using several scanning heads or a movable mirror.

Page readers are sometimes built to locate information by looking at specific locations on the document as measured from printed fiducial marks. The location (relative to the fiducial marks) of the data to be read is often programmable. Other designs allow the scanner to search the entire document for printed data. Sometimes identifiable symbols or heavy lines are used to denote beginnings or endings of data fields and so direct the search.

Scanning and Recognition: One-dimensional: Although characters are two-dimensional distributions of ink on paper, some special fonts have been designed in which sufficient information is available to make distinctions when one of these dimensions is ignored (Fig. C-15). A long, narrow vertical slit is passed horizontally across the characters. The signal obtained is the amount of material sensed through the slit—a single function of time corresponding in duration to the time required for the character to pass by the slit. This method is often used to recognize the magnetic-ink characters on bank checks, with the reading head actually giving the derivative of the signal described above. Correlation techniques are used to distinguish between characters. Where the shapes are formed by bars (Fig. C-15), the decoding is done by the sequence of wide and narrow intervals between corresponding positions on adjacent bars.

These one-dimensional methods are limited in usefulness to small numbers of accurately printed characters. Other machines scan so as to obtain a two-dimensional representation of the character for recognition. More data are available, but the data are much more difficult to handle than in the one-dimensional case.

Two-dimensional: Locating characters on the paper and then processing them for recognition usually depends upon the movement of the character or of its machine representation relative to some part of the recognition apparatus. In one-dimensional systems, the paper motion usually supplies this movement. In two-dimensional systems, the two degrees of freedom imply that at least one of the motions must be artificially supplied. This considerably complicates the recognition system.

Optical masking has been suggested as a particularly simple two-dimensional recognition system (see Fig. C-16). Here, an image of the character is projected on a set of masks and the total amount of light passing through the masks is collected by a photodetector. The mask designs are related to the expected character shapes so that the mask giving the largest signal indicates the actual character identity. Typically, the second dimension of motion is supplied by a moving mirror. In general, this recognition method requires good print quality for acceptable error rates.

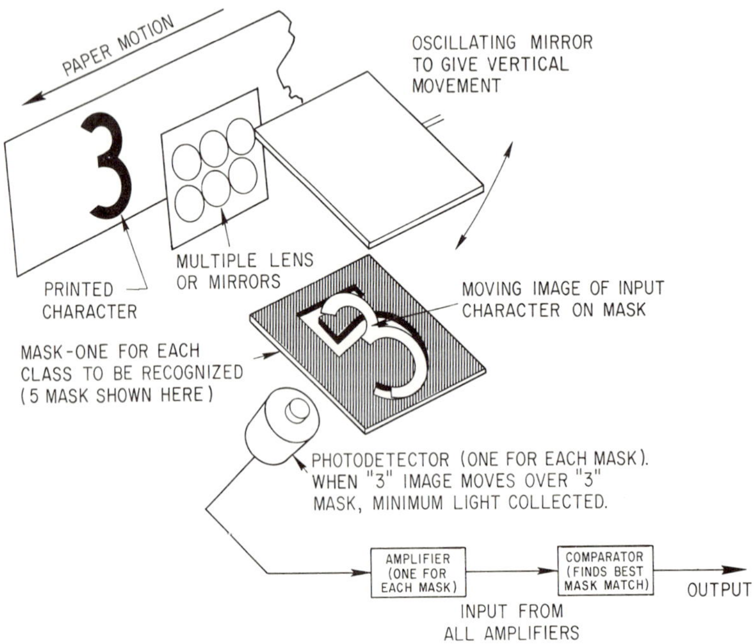

Fig. C-16. A masking system.

Most available machines are based on having, within the machine, some electrical analog of the density distribution of the character on the paper. This often is a quantized (black-white) representation held in an array of binary storage elements. Sometimes these representations are obtained by considering a number of adjacent points for each quantizing decision. In some machines, a continuous electrical signal of the density at each point is used in the recognition process.

The scanner may be of the total-array type, which uses a two-dimensional array of photodetectors for capturing the character. It may also be of the line type, where the character is swept past a vertical line of photodetectors. (At least one magnetic-ink character-recognition machine uses a vertical row of small sensing heads in this manner to read the United States banking industry standard font—E13B.) Spot scanners are also used—either of the mechanical Nipkow-disk design (the illuminated character is dissected spot by spot by a hole in a moving disk; see Ref. 8) or of the flying-spot cathode-ray-tube type (see Fig. C-17). The latter technique allows considerable format flexibility.

One technique used for recognition is to perform an operation on the internally stored character which is analogous to the optical masking method described earlier. In this case, nonlinearities can be more easily introduced to emphasize small but significant differences.

Another popular recognition technique involves making measurements of the character features, for example, the presence or absence of a long left vertical stroke. The output of such a measurement can be either binary (yes or no) or graduated to indicate the likelihood that the character has such a stroke. The outputs of the individual feature tests are combined in a decision logic to identify the character.

Special Types of Scanning: In some cases, where a small set of characters is involved (for example, numerals only), recognition is possible by sweeping a scan across the character at a few locations. These sweeps may be horizontal, vertical, or both. This saves having to store

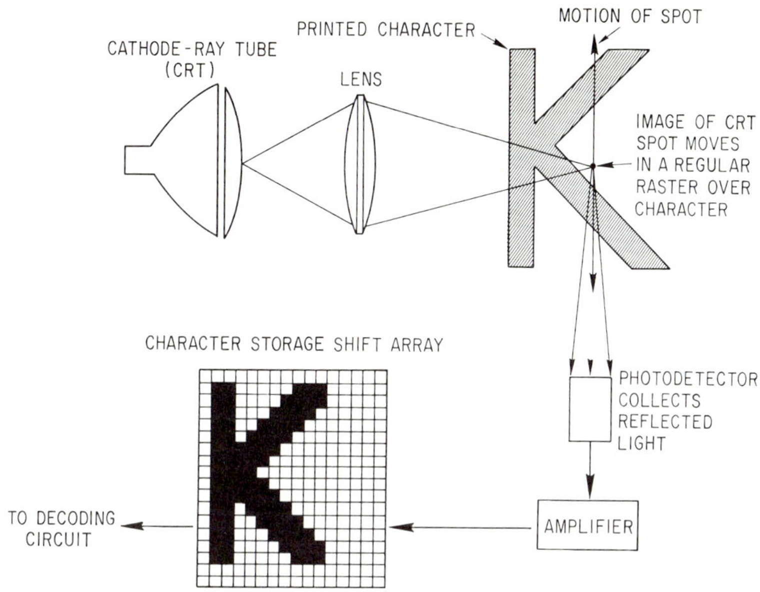

Fig. C-17. A spot-scanning system.

the entire character, and recognition is determined by the relative locations of a few line crossings.

Another technique that is used is called curve following (see Fig. C-18). A cathode-ray-tube spot is made, by appropriate feedback techniques, to trace around the periphery of the character. The sequence of location and direction information which is extracted is used to identify the character. This technique is particularly useful for characters which vary widely in size, such as hand printing.

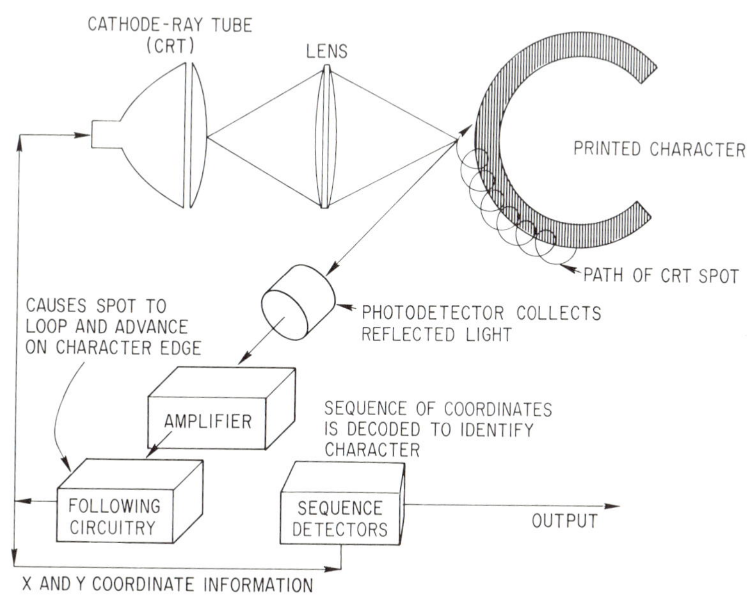

Fig. C-18. A curve-following system.

REFERENCES

1. Andrews, D. R., A. J. Atrubin, and K. C. Hu: The IBM 1975 Optical Page Reader, Part III: Recognition Logic Development, *IBM J. Res. Develop.*, vol. 12, no. 5, pp. 364–371, September, 1968.
2. Bartz, M. R.: The IBM 1975 Optical Page Reader, Part II: Video Thresholding System, *IBM J. Res. Develop.* vol. 12, no. 5, pp. 354–363, September, 1968.
3. Drattell, A.: Scanning the World of OCR, *Business Automation,* vol. 15, no. 1, pp. 32–54, January, 1968.
4. Feidelman, L. A., and J. L. Katz: Scanning the Optical Scanners, *Data Processing Mag.*, vol. 9, no. 10, October, 1967.
5. Hennis, R. B.: The IBM 1975 Optical Page Reader, Part I: System Design, *IBM J. Res. Develop.*, vol. 12, no. 5, pp. 346–353, September, 1968.
6. Holt, A. W.: Comparative Religion in Character Recognition Machines, *IEEE Computer Group News,* vol. 2, no. 6, pp. 3–11, November, 1968.
7. Kanal, L. N. (ed.): "Pattern Recognition," Thompson Book Company, Washington, D.C., 1968.
8. Stevens, M. E.: Automatic Character Recognition: A State of the Art Report, *Natl. Bur. Std. (U.S.) Tech. Note* 112, May, 1961.
9. Wilson, Robert A.: "Optical Page Reading Devices," Reinhold Publishing Corporation, New York, 1966.

—Allan J. Atrubin, *IBM Corporation*

Characteristic, Valve (See **Control valve.**)

Chemical-analysis Instrumentation [See **Analytical instrumentation (chemical).**]

Chemical Dosimeter (See **Dosimeter, chemical.**)

Chemiluminescence, Measurement of (See **Fluorometry.**)

Chlorine Analyzer (See **Amperometry.**)

Chromatography A procedure based on physical-adsorption principles for separating various components from a mixture of chemical substances. In its broad interpretation, chromatography is a combination of separation, identification, and quantitative measurement.

The procedure is broadly applied to mixtures of organic and inorganic materials and is particularly useful with mixtures of compounds whose chemical characteristics (composition and molecular structure) and whose physical properties (boiling point, density, etc.) are so nearly identical as to make other separation and analytical techniques difficult or impractical. Some forms of chromatography also offer the distinct advantage of requiring only very small samples—in terms of milligrams. Consequently, chromatography is used widely in the laboratory for organic chemical, biological, and medical studies—and in industry, notably in chemical, petroleum, and petrochemical plants, for quality and process control.

The recovery of pure compounds from mixtures in the laboratory often is as important as or more important than making quantitative determinations; hence, chromatography is widely used to separate and collect pure compounds for study by other methods.

For industrial applications, a chromatograph is a complex transducer which not only puts out a signal that identifies the types and amounts of given substances but also must first separate out the target substances from a stream that may contain numerous other substances, some of which may be closely related physically or chemically. The transducer output can be used for off-line quality control, or on-line so that the chromatograph becomes part of the total control loop.

In 1906, a Russian botanist (Mikhail Tswett) used the chromatographic principle to separate plant pigments. He filled a vertical glass tube with an adsorbent, and as a sample of the pigments was washed through the tube with a solvent, a series of colored adsorption bands was produced—in essence, a graphic presentation of colors—thus the term *chromatography.* The term no longer is meaningful in this way, however, because chromatography now is used with colorless materials and often with the operator no longer visually witnessing the separation process.

Basic Principle: Chromatography depends upon selective retardation and separation of substances by a bed of porous sorptive media as they are transported through the bed by a moving fluid. The degree of retardation, and hence the rate of migration of each substance, is determined by its relative affinity for the sorbent.

The sorbent bed often is referred to as the *stationary phase,* the moving fluid as the *moving phase.* The moving phase may be a liquid or a gas. The stationary phase may be a solid adsorbent which sorbs sample components on its surface, or it may be a liquid dispersed on a porous, inert solid support in which the sample components are soluble and are partitioned in equilibrium with the moving phase, hence the *solvent*[1] or *partition liquid.*

[1] The term *solvent* is used in both gas chromatography in reference to the stationary-liquid phase, and liquid chromatography in reference to the moving-liquid phase.

Chromatography may be classified according to a number of schemes:

1. *The form of technique:* (*a*) frontal analysis, (*b*) displacement, (*c*) development, and (*d*) elution.

2. *Some distinctive feature of the apparatus or method,* as shown schematically in Fig. C-19. Most commonly used for referring to a specific technique, e.g., paper, thin-layer, and gel-permeation chromatography.

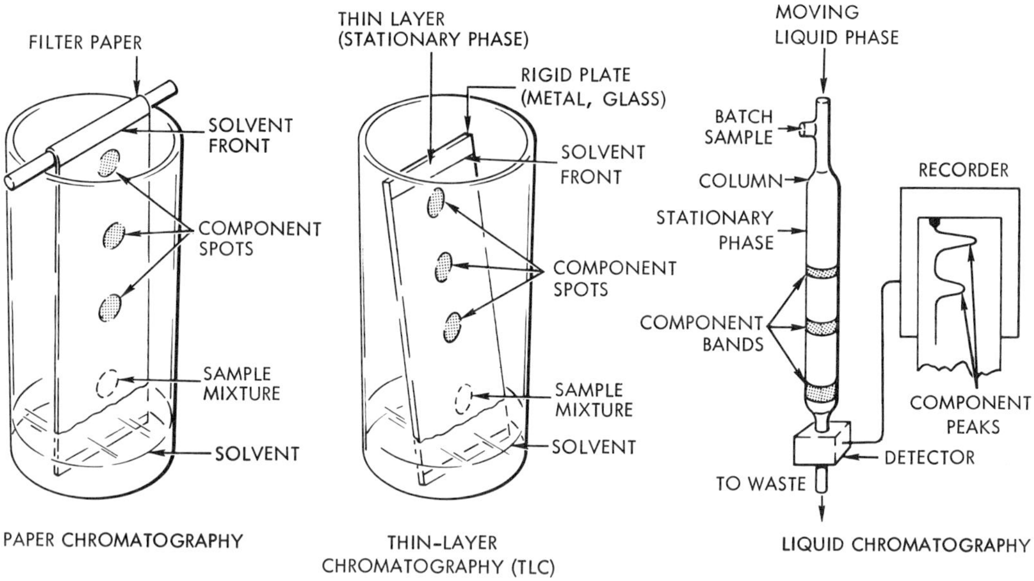

Fig. C-19. Basic chromatographic techniques.

3. *The nature of the moving and stationary phases,* as shown diagrammatically in Fig. C-20.

Frontal analysis and displacement chromatography are not discussed here because they are of limited value and use.

Development Chromatography: The sample is introduced onto a dry column or sorbent bed and washed through the bed with a solvent that is less strongly sorbed than the sample components. The solvent-washing process is continued until the solvent reaches a point just short of the exit or opposite end of the bed. The migration rate of each sample component is dependent upon its *partition coefficient* or distribution between sorbent and solvent. The most useful forms of development chromatography are paper and thin-layer chromatography. These have gained wide acceptance because of the ease and convenience of the method, and the modest cost and availability of materials. The use of columns for development chromatography has declined because of difficulties in visualizing and documenting the developed column and in recovering the pure compounds from the sorbent bed.

Elution Chromatography: In this method, the moving phase is passed through the sorbent bed until all sample components have been washed or eluted from the bed. In contrast to development chromatography, elution chromatography uses columns almost exclusively. In gas-elution systems and in most type of liquid-elution chromatography, the column can be reused many times and without interruption of flow of the moving phase. Components move through the column at rates dependent upon the partition coefficient and are separated into bands which elute at characteristic times. A detector at the end of the column can generate an analog signal that is proportional to the concentration of the sample component in the moving phase. A time record of the detector signal is called a chromatogram and, as shown in Fig. C-21*a*, contains the characteristic gaussian peaks.

Gas and liquid chromatography are the most popular forms of development chromatography. The theory, technique, and apparatus for both methods have been highly developed. Completely automated versions (see Process-gas Chromatography, later) have been developed and widely applied to the continuous, on-line analysis

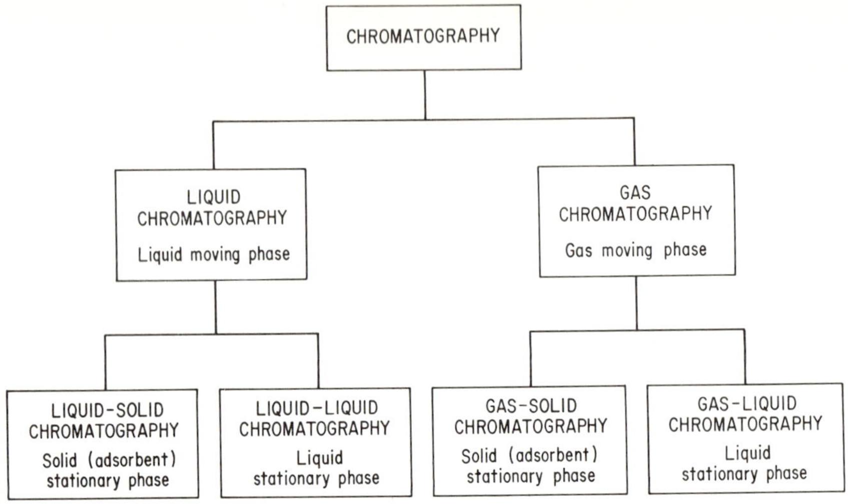

Fig. C-20. Chromatography techniques as classified according to nature of moving and stationary phases.

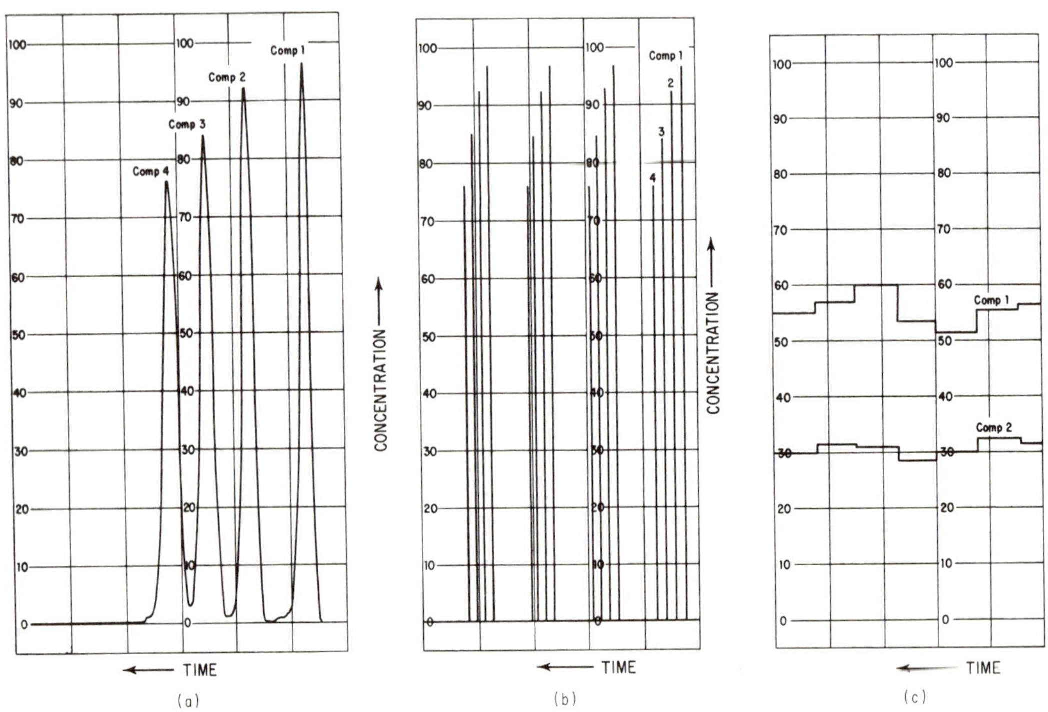

Fig. C-21. Typical chromatograph readouts: (a) chromatogram; (b) bar graph; and (c) trend.

of chemical and petroleum-refining process streams.

Paper and Thin-layer Chromatography: Both methods are similar in apparatus and technique. The sorbent bed is in the form of a thin sheet of paper or a thin layer of a finely divided sorbent material deposited on a supporting metal, glass, or plastic plate. The sample is spotted near one end of the bed, which is then brought in contact with a source of solvent. As the solvent moves through the bed by capillary action, the sample components are washed through the bed at different rates and are separated into spots of pure compound, which can be recovered after the solvent is allowed to evaporate. Spots can be detected visually if the substances are colored, or with ultraviolet light if they are fluorescent. Components can be reacted to give colored or fluorescent derivatives by spraying with reagents. Conventional quantitative determinations can be made after recovery of the substance of interest with a solvent, or roughly estimated by comparison with spots produced with known quantities and concentrations. Recording densitometers are used for quantitative measurement *in situ.* The paper is drawn automatically past a slit in front of a photocell, and the transmitted or reflected light is recorded. Radioactive compounds are detected by contact exposure with x-ray film, which after development, can be measured with a densitometer. Quantitative measurements *in situ* can be made with apparatus similar to a densitometer in which a radiation detector is used in place of the photocell.

Gas Chromatography: Almost any organic or inorganic compound that can be vaporized can be separated and analyzed with a gas chromatograph. As shown in Fig. C-22, the minimum requirements for a system include (1) a column which contains the *substrate* or stationary phase, (2) a supply of inert *carrier gas* (moving phase) which is continually passed through the column, (3) a means of *admitting* or *injecting* the sample into the carrier-gas stream, (4) a *detector* which senses the sample components as they elute, and (5) a *recorder.* The carrier gas may be any gas that does not react with the sample or adversely affect the detector. Helium, hydrogen, and argon are most commonly used.

Gas Chromatography in the Laboratory: A typical application is illustrated and described in Fig. C-23. Gas chromatography is so flexible that it has many end uses and apparatus configurations in the laboratory, some of which include:

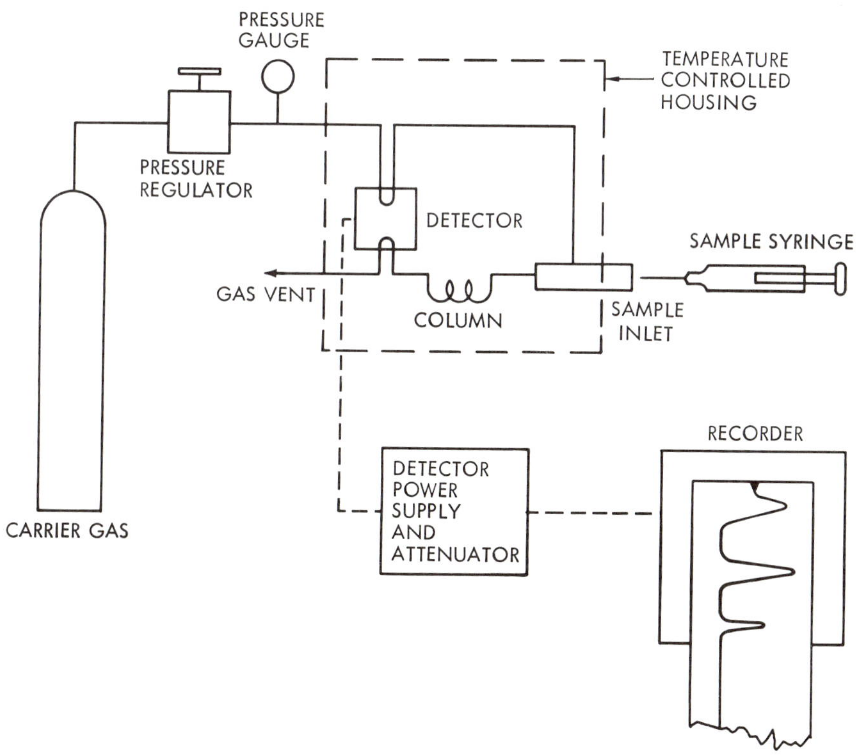

Fig. C-22. Basic elements of a gas chromatograph.

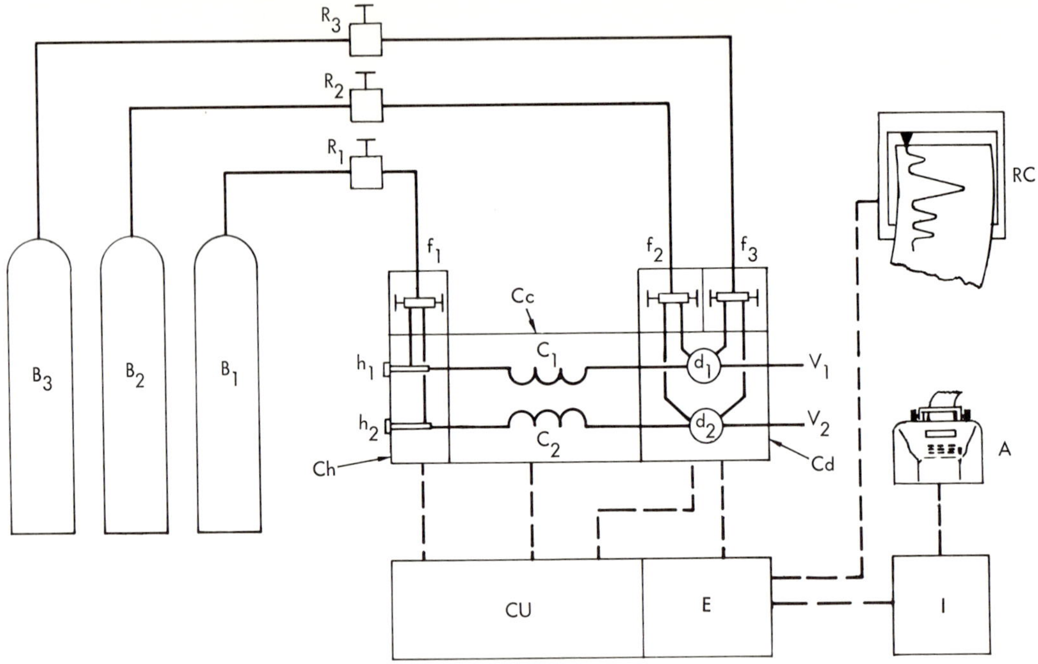

Fig. C-23. Basic elements of a gas chromatograph for research and laboratory purposes. Dual column/ dual detector gas chromatograph shown is suitable for anlaysis of complex organic mixtures. Columns C_1 and C_2, with different characteristics, are enclosed in temperature-controlled column compartment *Cc,* which may be isothermal and/or temperature-programmed during analysis for rapid elution of high-boiling compounds. Individual flash vaporization heaters h_1 and h_2 for injection and rapid vaporization of liquid samples are enclosed in temperature-controlled inlet compartment *Ch.* Carrier gas, supplied from high-pressure cylinder B_1, is maintained constant by pressure regulator R_1 and flow controller f_1. Sample components are detected by hydrogen-flame ionization detectors d_1 and d_2 in constant-temperature detector compartment *Cd.* Hydrogen fuel B_2 and combustion air B_3 are supplied to detectors through pressure regulators R_2 and R_3 and flow controllers f_2 and f_3. Combustion products are vented through v_1 and v_2. Control unit *CU* gives individual temperature control of sample inlets and detectors, and programs temperature of column compartment at rate and over range required. Electrometer amplifier *E* permits recording chromatogram on potentiometric recorder *RC.* Electronic integrator *I* detects peak maxima, integrates and digitizes peak areas, and prints out area and elution time of individual peaks on adding machine *A.*

1. *Programming of column temperature* to permit analysis of wide boiling-range fractions. A low starting temperature (sometimes subambient) affords separation of low-boiling components, whereas a later, higher column temperature reduces the elution time of high-boiling components and, in the overall, reduces analysis time. This technique is used widely for complex samples, such as essential oils and flavor and aroma concentrates and, in the petroleum industry, for simulated distillation analysis of crude oil and petroleum fractions.

2. *Multiple columns and column switching* to effect separations where a single column would be impractical and also to reduce analysis time.

3. *Fraction collecting or preparative chromatography* where separated compounds are collected for analysis by other methods.

4. *Pyrolysis chromatography* wherein the thermal degradation of high-molecular-weight materials, which yield characteristic fragment patterns, can be studied.

5. *Reaction chromatography* where compounds may be converted to other related compounds in a reaction zone ahead of the chromatograph column.

6. *Subtraction chromatography* wherein a reaction zone in the system preferentially reacts with and removes compounds of a class of compounds, such as *n*-paraffins on a molecular sieve.

7. *Physicochemical measurements,* such as determination of activity coefficients, partition

coefficients, K values, second virial coefficients, heat and entropy of solutions, adsorption, vaporization, and other physical and thermodynamic properties.

8. *Hyperpressure (supercritical) chromatography* in which a fluid above the critical point is used as the moving phase and which enhances the volatility of sample components by as much as 10^4. This permits the analysis of compounds of extremely high molecular weight (mol. wt. = 1,000) and of high boiling point (1000° C) at column temperatures below 300° C.

A large number of detectors have been used, many of which are highly specific and of limited application. The most widely used are the thermal-conductivity, hydrogen-flame-ionization, argon-ionization, and electron-capture detectors.

Chromatograph Data Reduction: For a given substrate under given conditions, each compound has a characteristic retention time which can be used for tentative identification. However, two or more compounds may have the same time on a particular column. In such cases, the compound may be rerun on a different column with other characteristics to reduce ambiguity. Extensive compilations of individual compound-retention times on different substrates are available for reference. Positive identification can be made only by collecting the compound as it elutes and analyzing it by other means.

Quantitative analysis is based on the proportionality of detector response to the amount of component in the elution band. The most widely used measure of detector response is the area under the chromatogram peak. However, the peak height (amplitude of detector signal at peak maxima) also may be used. The area under the peak may be determined with a planimeter or from dimensions of the height and width of the peak. A ball-and-disk integrator, attached to the recorder, can be used to provide a permanent analog record of the area on the recorder chart.

Electronic integrators may sense the detector signal directly and provide digital printout of peak area and time of peak maxima. Special circuits detect peak maxima, the beginning and end of the peak, and minima between overlapping or incompletely resolved peaks, and compensate for zero drift and offset. This eliminates the need for operator attention other than for sample injection. Some systems may include an intermediate stage for converting the detector signal to a pulse-frequency-modulated signal and then recording this on magnetic tape for subsequent high-speed playback into an electronic integrator. This arrangement allows one integrator system to service several chromatographs.

An on-line computer may continuously monitor the output of several chromatographs and reduce all the collected data. The operator simply uses a console to identify the chromatograph and sample type, inject the sample, and signal the start of analysis. The output of each detector is sampled at 0.1- to 0.2-s intervals via high-speed multiplexer and analog-to-digital converter. The computer performs the integration of each peak by summation of individual amplitude measurements during elution of the peak. The computer identifies the component by comparing the time of peak maxima with a table of known values (stored in the computer memory), selects the applicable relative response factor, and calculates and prints out the composition of the sample. The program may include subroutines for resolving overlapping peaks and correcting for tailing, baseline drift, and other factors. A system may use a large computer for handling from 20 to 40 chromatographs, or small, low-cost computers tied in with fewer instruments.

Process-gas Chromatography: A system for continuous, repetitive, and fully automatic on-line analysis of process streams similar in all essential elements of the basic technique to the laboratory chromatograph, but different in design and appearance. Factors affecting design include (1) the need to comply with the National Electrical Code for operation in hazardous atmospheres, (2) the need to automate the procedure, and (3) the need for ready adaptability to closed-loop process control and communication with computers. Demand for maximum reliability and minimum maintenance has emphasized simplicity of hardware and methodology. Emphasis is placed on analyzing for a few rather than a number of components and on minimizing analysis time. These design targets have resulted in extensive use of multicolumn techniques for rapid separation of selected components, with large portions of the sample being discarded. As shown in Fig. C-24, the major components of a process-gas chromatograph are (1) the *analyzer,* (2) the *programmer,* and (3) one or more *recorders.* In some cases, (4) a *sample conditioning* system and (5) a *stream selector* are required.

Analyzer: This usually is located close to the process sample point in a housing that meets NEC Class I, Group D, Division 1. The controlled-temperature compartment will contain the

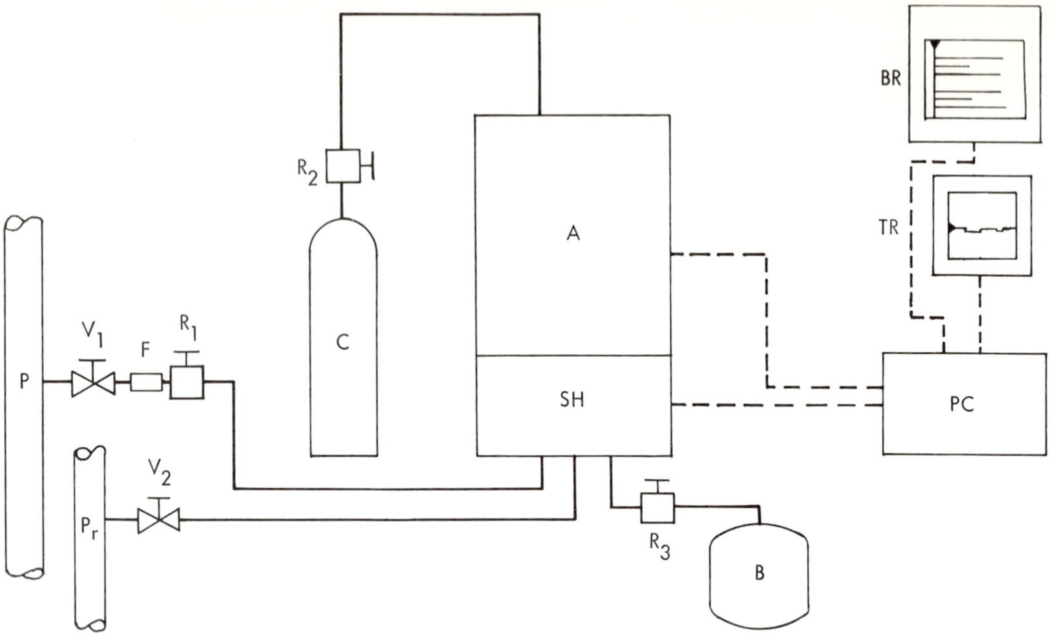

Fig. C-24. Basic elements of a process-gas chromatograph. Vapor sample is continuously withdrawn at high rate from process line *P*, circulated through sample conditioner *SH*, and returned to lower pressure point P_r, through shutoff valves V_1 and V_2. Particulate matter is removed by filter *F*, and pressure reduced to constant low level by regulator R_1. Sample conditioner contains flow control and other conditioning components and valve for switching to synthetic calibration blend *B* through pressure regulator R_3. Sample slipstream is circulated to sample valve in analyzer *A*, which also contains columns, detectors, and temperature-control system. Carrier gas *C* is controlled by regulator R_2. Programmer *PC* contains detector power supply, component attenuators, timer for controlling analyzer functions, and means for operating calibration blend valve in sample conditioner on demand. Programmer converts signal to form suitable for recording as bar graph on recorder *BR* or trend on trend recorder *TR*.

column, sample- and column-switching valves, and detector.

Programmer: Commonly, this is located near or in the process-control room. The programmer sequences all analysis functions in the analyzer by means of a mechanical or electronic digital device. Peak-height measurement is most common, but electronic operational-amplifier integrators are also used, especially for summation of two or more components. The programmer may include a "peak picker" to convert the peak value of the component peak to a continuous analog signal for trend recording, for closed-loop control with conventional recorder-controllers, or for priority-interrupt transfer to a digital computer. The programmer also includes manual controls (sample valve and column-switching valve, attenuator, and zero adjustment) and a means for obtaining conventional chromatogram presentation for manual operation of the analyzer during start-up and maintenance periods.

Recorder: The simplest form of record is a bar graph, as shown in Fig. C-21*b*. The record consists of a series of bars, one for each component, of height proportional to the component concentration. An attenuator for each component is adjusted to give a full-scale reading equivalent to a convenient concentration value. Each component may have a different full-scale range. A large number of components in one or many streams may be recorded on the same instrument. Different streams may be identified by the height of a flat-top bar that precedes the series of bars for that stream.

Trend recording uses a conventional voltage, current, or pneumatic recorder or recorder-controller to display a continuous analog signal from the programmer. The reading is held constant at the last value recorded until a new value is recorded, causing a stepwise change in the record as the signal is updated (see Fig. C-21*c*). A separate pen is required for each component trend so recorded.

Valves: Electrically or pneumatically operated

valves are used for sample injection and column switching. Rotary, spool-and-*O*-ring, diaphragm, and sliding-plate valves with from 4 to 12 ports are used. The sample valve meters a fixed volume of liquid or gas into the column with a repeatability of ±0.25 percent. The sample size may vary from approximately 0.1 microliter to 50 ml (with external sample loop).

Detectors: Thermal conductivity (for general use) and hydrogen-flame ionization (for trace organic analysis) are the most widely used detectors. Helium ionization (radioactive and photoionization) and thermal conductivity with amplification are used for trace inorganic and inert-gas analysis. Halogen detectors are used for applications that require sensitivity and specificity for halogen compounds (see Table C-1).

Computer-controlled Chromatograph System: Large-process chromatograph systems may use a dedicated low-cost computer to control all analyzer functions and perform all data reduction, thereby eliminating the individual programmers. The computer monitors each individual chromatograph detector, performs integration of peak areas, corrects for zero offset and drift, identifies components from elution times, applies response factors, and computes the composition of the sample by normalization of peak areas or by comparison with calibration standards for select components. Partial or complete stream analysis may be printed out on a teletypewriter, continuously or on demand, in accordance with program format. The computer may simultaneously provide analog trend output to conventional recorders for closed-loop control. The computer may also communicate analysis results to a larger super-

TABLE C-1. *Characteristics of Process-gas Chromatography Detectors*

Detector	Compounds Detected	Best Full-scale Sensitivity*	Advantages	Disadvantages
Thermal conductivity (low cost), two-element	All	0–1%	Low cost, rugged, reliable. Low maintenance	Limited sensitivity
Thermal conductivity, four-element	All	0–0.1%	Moderate cost. Rugged, reliable. High sensitivity	Requires excellent stability in balance of system
Thermal conductivity with amplifier	All	0–100 ppm	Rugged, reliable. Superior sensitivity	Requires more sophisticated system, special components. Higher cost
Gas-density balance	All	0–1%	Similar to thermal conductivity, but wider choice of carrier gas	Lower sensitivity. Greater internal volume
Hydrogen-flame ionization	All organic except formic acid and formaldehyde	0–1 ppm	Ultrahigh sensitivity and rugged. Wide choice of carrier gas	Cannot detect inorganic compounds and inert gases. Requires more complex system
Helium-ionization detector (radioactive and photoionization)	All except neon in radioactive form	0.3 ppm	Ultrahigh sensitivity. Universal	Requires ultrapure helium carrier and extreme sophistication. High cost

*Referred to concentration in process sample with nominal 1 cm^3 sample size; component peak approximately 0.25 min wide at base; He or H_2 carrier. Under special conditions, sensitivity can be increased up to fifty times by use of large samples. See Process Gas Chromatography Detection Systems, *Chem. Eng. Progr.,* vol. 64, no. 4, p. 55, 1968.

visory computer (e.g., for material-balance calculations or for direct digital control). High- and low-level alarms for select components may also be set by the computer. The computer program may also include subroutines for detecting and alarming in case of malfunction or failure in chromatograph analyzers.

Liquid Chromatography: This method is particularly useful for separation and analysis of high-molecular-weight compounds which are beyond the range of gas chromatography. It is generally classified according to type of stationary phase: (1) liquid-solid (adsorption), (2) liquid-liquid (partition), (3) ion-exchange, and (4) gel-filtration (gel-permeation). Each type imposes specific requirements on apparatus design. The liquid is moved through the system by gravity or constant-flow pumps. Gradient devices make stepwise or continuous changes in the composition of moving phase during analysis (gradient elution). The differential refractometer probably is the most widely used detector; other detectors are based on visible, infrared, or ultraviolet photometry. Eluting compounds may be reacted with reagents to form colored substances for ease of detection. Other detectors used include a heat-of-adsorption detector and a modified hydrogen-flame-ionization detector.

REFERENCES

Purnell, J. H.: "Gas Chromatography," John Wiley & Sons, Inc., New York, 1962.

Heftmann, E.: "Chromatography," 2d ed., Reinhold Publishing Corporation, New York, 1967.

Ettre, L. S., and A. Zlatkis: "The Practice of Gas Chromatography," Interscience Publications, a division of John Wiley & Sons, Inc., New York, 1967.

Kobayashi, R., P. S. Chappelear, and H. A. Deans: Physico-chemical Measurements in Gas Chromatography, *Ind. Eng. Chem.*, vol. 59, no. 10, p. 63, 1967.

—Richard Villalobos, *Beckman Instruments, Inc.*

Chronograph A wristwatch or pocket watch that displays time of day and incorporates the functions of the stopwatch or sports timer. The chronograph is equipped with a special internal mechanism, controlled by an external hand-operated plunger system, that permits the measurement of elapsed time—either continuously or interruptedly for periods of up to several minutes or hours. Elapsed time is displayed by separate hands on the dial, or by separate hands on subsidiary dials on the main dial. Normally, the elapsed-time reading is in fifths of a second, but if a "fast-beat" (36,000 beat-an-hour) balance-wheel movement is used, time can be shown in tenths of a second. For certain sports and industrial applications, a dial is scaled into hundredths of a minute instead of seconds. Certain chronographs, like some stopwatches, can display more than one elapsed-time reading; others also incorporate tachometer or other special-scale dials that report computed velocities or distances.

—William O. Bennett, John J. Carpenter, Frank Dostal, and E. Van Haaften, *Bulova Watch Company, Inc.*

Chronometer A special type of mechanical clock used as a reference by a ship's navigator. Also known as a marine chronometer. The clock, hung on gimbals in a box to minimize the balance wheel's position error, is powered by a fusee-equipped mainspring that drives a temperature-compensated balance wheel and hairspring which work a specialized escape-wheel system called a detent escapement. A chronometer displays Greenwich Mean Time (GMT) and is characterized by a very constant daily rate (usually within 0.1 s/day), though the timepiece may gain or lose several seconds per day.

—William O. Bennett, John J. Carpenter, Frank Dostal, and E. Van Haaften, *Bulova Watch Company, Inc.*

Circular Interpolation (See **Drafting system, automatic.**)

Climatology (See **Meteorological instrumentation.**)

Clinical Thermometer (See **Thermometer, liquid-in-glass.**)

Clock, Atomic A clock that incorporates an atomic frequency standard, such as the (1) ammonia maser, (2) hydrogen maser, (3) rubidium-gas cell, and (4) cesium-133 beam. All atomic-clock systems involve the use of atoms or molecules in certain energy states that are exposed to microwave energy to permit measurement of the quantum changes in the energy states of atoms. Even though highly accurate, such systems do not maintain strictly uniform time under all conditions because their *running rates* (as predicted by the theory of relativity) are affected by their position relative to the sun and moon, latitude, height, and the gravitational field in which they are placed or through which they move.

Cesium-133 Atomic System: This and other

atomic clocks have replaced quartz-crystal clocks as primary standards in major timekeeping observatories. The cesium-133 system is the basis of the National Bureau of Standards (NBS) frequency standard. See **Time and time measurement.** The cesium primary standard maintains a resonance frequency accurate to 5 parts in 10^{12} and also is used as the primary reference in maintaining the atomic time scale designated NBS-A. This time scale is based on two quartz-crystal oscillators and three commercial cesium-beam devices that are calibrated, or frequency-referenced, to the primary cesium atomic standard. See **Crystal oscillator.**

Earlier ammonia-maser clocks were less accurate than the cesium-133 system, which was used to define the *atomic second* as 9,192,630,770 oscillations of the atom of cesium-133—the international definition of *atomic time* provisionally introduced in 1964 and finally accepted as the international standard in 1967. This value for the atomic second very closely approximates the ephemeris time (ET) second and also fairly closely approximates UT_2. See **Time and time measurement.** For example, the National Bureau of Standard's *coordinated universal time* scale, designated UTC (NBS), is referenced to the NBS frequency standard and closely approximates UT_2, because UTC incorporates frequency offsets and shifts in epoch announced by the Bureau International de l'Heure at the Paris Observatory—as required by the International Astronomical Union.

Rubidium-gas-cell Systems: These now serve as secondary standards, like quartz-crystal oscillators, because they must be referenced to more accurate systems. Rubidium systems have a characteristic resonance at 6,835 MHz and are useful because, unlike other atomic-frequency standards, they require little power and are relatively compact. Portable rubidium atomic clocks weighing as little as 44 lb and occupying about 1 ft^3 of space were introduced by the U.S. Army Electronics Command in 1965. These clocks operate on standard 110-V current, the 24-V output of military vehicles, or both. The clocks are used to set precise radio-broadcasting frequencies, to synchronize radar nets and to assist in the accurate tracking of missiles and satellites. Exceptional accuracy is required by tracking stations following a satellite moving at 25,000 ft/s in earth orbit because of time error of a tenth of a second is the equivalent of a position error of about ½ mile.

Other atomic-frequency standards developments include (1) gas-cell frequency standards using alkali metals, such as sodium, cesium, and rubidium; (2) thallium-beam devices; and (3) the hydrogen maser, first developed at Harvard University in 1960.

REFERENCES

Markowitz, William: The Atomic Time Scale, *IEEE Trans. Instr.*, vols. I–II, nos. 3, 4, December, 1962.

Mason, W. P.: "Quartz Crystals for Electric Circuits," D. Van Nostrand Company, Inc., Princeton, N.J., 1946.

"Explanatory Supplement to the Astronomical Ephemeris and the Nautical Almanac," H. M. Stationery Office, London, 1961.

"NBS Standard Frequency and Time Services," National Bureau of Standards, Washington, D.C., 1967.

Beehler, R. E.: A Historical Review of Atomic Frequency Standards, *Proc. IEEE*, vol. 55, no. 6, June, 1967.

Frequency and Time Standards, *Hewlett Packard Co. Appn. Note* 52, Palo Alto, Calif., 1965.

—William O. Bennett, John J. Carpenter, Frank Dostal, and E. Van Haaften, *Bulova Watch Company, Inc.*

Closed-loop Control A feedback loop or closed loop is a signal path which includes a forward path, a feedback path, and a summing point, and which forms a closed circuit.[1] When the automatic controller is combined with a process, as shown in Fig. C-25, closed-loop control results. A change in set point or controlled variable produces a change in the manipulated variable, which in turn produces a change in the controlled variable. This recirculation is the chief characteristic of closed-loop control.

Because of the recirculation or feedback, the control-action computation, which determines the corrective action required to keep the measured variable at the set point, is critical to prevent cycling and instability. If the feedback is positive, that is, in phase with the deviation, the error is magnified and the process cannot be controlled. Only with negative feedback can the deviation be reduced. A reversal of phase must take place at the summing point so that the effect of e, the control input, is opposite to that of c, the controlled variable. All feedback controllers produce this phase shift of 180° to give a negative feedback.

[1] Extracted from American National Standard C85.1–1963, American Standard Terminology for Automatic Control, with permission of the publisher, The American Society of Mechanical Engineers, New York.

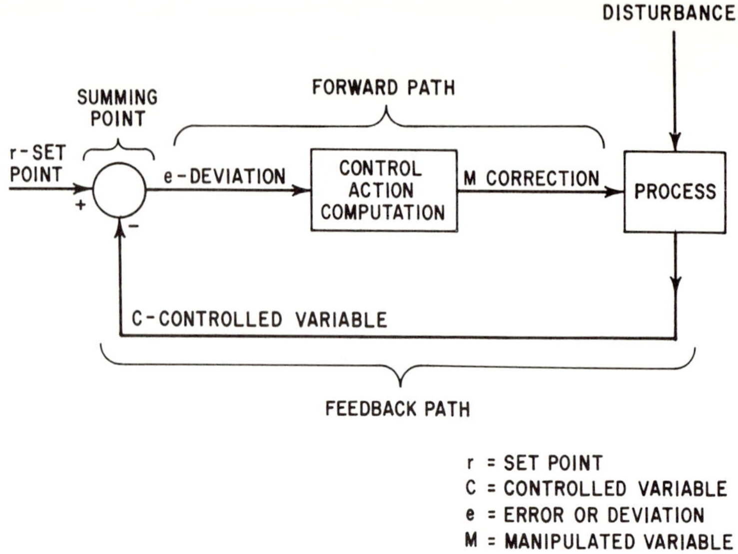

Fig. C-25. Block diagram of a closed-loop or feedback control system.

For a process with an overdamped second-order lag response—after an increasing step change in input, the output will not change immediately, but after a short period of time it will exhibit an S-shaped response which levels off at the final steady-state value without overshoot. Most processes exhibit this response, with examples being two resistance-capacitance networks series-connected, a thermocouple in a well, or two pressure tanks separated by a resistance loop.

For a process with a critically damped second-order lag response—after an increasing step change in input, the output responds similarly to an overdamped second-order process, but the rate of increase to the final steady-state value is the fastest obtainable without overshoot.

—Richard H. Kennedy, *Bristol Division, American Chain & Cable Company, Inc.*

Cloud Chamber [See **Radiation (nuclear), detection of.**]

COBOL (Common Business Oriented Language.) This computer language was developed in the late 1950s and early 1960s to assist in the processing of large quantities of character-oriented data such as name and address files. It is heavily used in business and government, for example, for payroll and inventory programs. Its chief characteristics are a good clear method of file and record description and powerful data-manipulation facilities. An example of a program written in COBOL follows:

```
IDENTIFICATION DIVISION
  PROGRAM-ID.  SMPLPRGM.
    REMARKS.  THIS IS A SAMPLE PRINT PRO-
              GRAM.
    REMARKS.  IT ILLUSTRATES ONLY SOME FEA-
              TURES OF COBOL.
    REMARKS.  THERE ARE 4 DIVISIONS OF A
              COBOL PROGRAM IDENTIFICA-
              TION, ENVIRONMENT, DATA AND
              PROCEDURE.
    REMARKS.  IN THE ENVIRONMENT DIVISION
              DETAILS ARE GIVEN OF THE CPU'S
              AND I/O DEVICES NEEDED.
    REMARKS.  IN THE DATA DIVISION DETAILS
              ARE GIVEN TO THE FILES AND OF
              THE RECORD LAYOUTS FOR BOTH
              INPUT AND OUTPUT FILES.
    REMARKS.  THE KEYWORD PICTURE IS USED
              TO DESCRIBE THE FORMAT OF A
              FIELD WITHIN A RECORD.
    REMARKS.  FOR EXAMPLE, IN THE # FILE
              "READ-IN", THE RECORD "SOURCE"
              HAS 3 FIELDS—"AUTHOR", "INI-
              TIAL", "TITLE", OF 20, 3 AND 57
              CHARACTERS RESPECTIVELY.
    REMARKS.  THE PROCEDURE DIVISION IS THE
              ACTUAL PROGRAM.

ENVIRONMENT DIVISION.
  CONFIGURATION SECTION.
    SOURCE-COMPUTER. IBM-360 F50.
    OBJECT-COMPUTER. IBM-360 F50.
  INPUT-OUTPUT SECTION.
    FILE CONTROL
    SELECT READ-IN       ASSIGN TO UNIT-RECORD
                         2540 UNIT.
```

```
      SELECT PRINTOUT  ASSIGN TO UNIT-RECORD
                       1403 UNIT.
DATA DIVISION.
  FILE SECTION.
FD READ-IN
  RECORDING MODE IS F, DATA RECORD IS
  SOURCE.
  01 SOURCE
     02 AUTHOR PICTURE X(20).
     02 INITIAL PICTURE X(3).
     02 TITLE   PICTURE X(57).
FD PRINTOUT
  RECORDING MODE IS F, DATA RECORD IS
  OBJECT.
  01 OBJECT
     02 FILLER  PICTURE A(20).
     02 AUTHOR PICTURE X(20).
     02 FILLER  PICTURE A(3).
     02 INITIAL PICTURE X(3).
     02 FILLER  PICTURE A(9).
     02 TITLE   PICTURE X(57).
     02 FILLER  PICTURE A(20).
PROCEDURE DIVISION.
  OPEN INPUT READ-IN, OUTPUT PRINTOUT.
  NOTE THIS PROGRAM READS CARDS AND LISTS
  THEM IN A NEAT FORMAT.
  NOTE AFTER THE LAST CARD IT SHUTS DOWN
  THE READER AND PRINTER THEN STOPS.
  LOOP.  READ READ-IN RECORD AT END GO TO
         COMPLETE.
         MOVE CORRESPONDING SOURCE TO
         OBJECT.
         WRITE PRINTOUT RECORD, GO TO LOOP.
  COMPLETE. CLOSE READ-IN, PRINTOUT, STOP
            RUN.
```

REFERENCE

McCracken, D. D.: "A Guide to COBOL Programming," John Wiley & Sons, Inc., New York, 1964.

—A. J. Arthur, *IBM Corporation*

Code Readers and Systems A code[1] may be defined as a "system of symbols and rules for expressing information, such as the Morse code of telegraphy, the EIA color code for specifying electrical values and terminal connection of electronic components, and the binary and other machine languages used in digital computers." In essence, the elements or characters of a code and the rules governing their combination comprise a language. Admittedly, the resultant language can be very simple or quite complex, dependent upon the amount of information the code must convey and whether or not the designer of the code intends that the code be easy to interpret and apply, as would be the case of most industrial and commercial uses; or difficult to decipher, as in the case of military and diplomatic coding systems.

The topic of coding and symbolism comprises a subscience in its own right and, of course, cuts across many scientific and engineering disciplines, including communications, information processing, instrumentation, and control. A brief account and a few examples of code systems are included in this encyclopedia to illustrate the important role that instrumentation plays in the practical implementation of a code system and, conversely, the importance of a code system to the overall operation of some control and automation systems.

Need for Codes: Man has become accustomed to transferring information with a code system that satisfies the requirements of not one, but two of his natural sensors or detectors, namely, (1) a code system or language that can be communicated to his ears, that is, a code that can be sensed *acoustically;* and (2) a code system or language that can be communicated to his eyes, that is, a code that can be sensed *visually.* To a somewhat lesser extent, but still important, is a code format that can be sensed by man's *touch*—as one finds in the braille system. Thus, English, German, French, Russian, and other languages have been refined over centuries of use to maximize the capabilities of man's sensors. Further, man's languages have to be comprised of symbols that are easy to generate—symbols that can be detected quickly by his eye and interpreted by his brain, but also symbols that are easy to create with his hands; and likewise symbols that can be distinguished clearly one from the other by his ears, yet symbols that can be generated easily by his voice box into sounds. Man's languages are full of subtleties, particularly in the audio form, where loudness, pitch, tone, and other acoustic characteristics give additional information content to the raw symbols.

The fundamental elements of language, of course, are letters and numbers. The letter-code system is the alphabet; the number-code system most commonly used is the arabic decimal-numbering system. With these two basic coding systems, man can communicate and store all his knowledge and transact all his business. But, with the advent of the computer and electronic data processing, the limitations of man's basic languages for machine use became apparent. In essence, the sensors—eye and ear, the interpreter

[1]Adapted from Markus, John: "Electronics and Nucleonics Dictionary," 3d ed., McGraw-Hill Book Company, New York, 1966.

—the brain, and the code generators—the hand and voice box, are not easy to duplicate instrumentally. Further, machine data-processing systems require response speeds that far surpass the parameters of the human mechanisms. Thus, there has been much activity in the development of code systems that can be generated, sensed, and interpreted by machines. Various aspects of this topic are covered by other entries in this encyclopedia, notably, **Character-recognition systems; Magnetic-ink character recognition (MICR); Optical character recognition (OCR); Pattern recognition;** and **Voice-recognition systems.**

Because data to be conveyed via a coding system frequently originate from documents, packages, parts, pieces, vehicles, and other discrete items that are moving, the ability of code sensors to respond quickly and, in some instances, to lock on and follow the code element for a brief instant is very important. Coding systems provide a great vent for scientific imagination and ingenuity. There are scores of coding schemes—some proved and practical, others quite clever but costly and impractical to implement.

Even for manual use, where machine processing may not be a factor, a coding system can introduce brevity to the problem of information transfer. Part number 1583745 states with seven digits that the piece is a gear with a long list of dimensions, tolerances, and an additional long list of metallurgical and mechanical specifications. Information condensation is particularly important in total-automation systems. Consider, for example, the concept of where a discrete piece or component of a larger product may contain all the machine instructions that may be required to shape, manipulate, and otherwise treat that particular piece during the course of manufacture. If the discrete piece or part is relatively small, as so often is the case, the codified instructions which it incorporates must be contained within a small physical space. Further, the coded information must be capable of being read very quickly (often while the piece is moving) and reliably because of the high speed of production machinery.

Basic Code Formats: The ingredient of a code system that must be generated, sensed, and interpreted must comprise some specific physical characteristic or combination of physical characteristics. The ingredient may be (1) *shape*—as typified by an alphabetical letter or arabic numeral, but including rarer forms, such as Chinese characters, Egyptian hieroglyphics, and Babylonian cuneiforms; (2) *mass*—if a code element can be made in three dimensions; or (3) *area*—if in two dimensions; (4) a *spectral quality,* notably color, although other portions of the electromagnetic spectrum are employed in some coding systems; (5) a *magnetic property;* or combinations of these ingredients. Because a code system must contain numerous elements so that sufficient data can be transmitted, either the foregoing property ingredients have to be discretely *alterable* to accomplish the required variety; or the ingredients have to be *arranged in time and/or space* so that their pattern of presentation to the sensor carries the required information. The Morse code, for example, contains only two basic elements—the dot and dash, but with varying combinations of these elements, all letters of the alphabet and numbers 0 through 10 can be communicated. A dot-dash = A; a dash-dot-dot-dot = B; a dash-dash-dot-dot-dot = 7; and so on.

Code elements can be *passive* or *active.* Characters that are read optically, for example, must be illuminated by light to be "seen" by the sensor. In this case, the code element can be considered inactive or passive. On the other hand, a code element that contains a radioisotope and emits gamma radiation that is picked up by the sensor can be considered an active element. Thus, the transfer medium (between code element and sensor) can be provided by the code element per se—in the case of active code elements, or provided externally wherein the code element exhibits a specific, measurable, recognizable reaction (such as reflection or absorbance) when "illuminated" by an external medium or source of energy. The principal transfer media used in code systems are light, magnetic or electromagnetic fields, sound and ultrasonic radiation, and gamma radiation. Thus, it is found that code readers often are very similar to instruments that are used for entirely different purposes.

Cold-cathode Vacuum Gage (See **Vacuum gaging.**)

Collimator A device for confining the elements of an optical, electron, or other beam within an assigned solid angle. Also see **Autocollimator.**

Color, Measurement of Everybody has definite judgments of color. It is also obvious that color is associated with light, since in the absence of light everything looks black. Therefore, the objective measurement of color must involve some sort of photometer, i.e., light-measuring device. As it turns out, the measurement of color is a very complex problem, involving physics, physiology, and even psychology (Ref. 1).

Light is an electromagnetic radiation, and different wavelengths have dramatically different properties. Those of the shortest wavelength (a billionth of a meter and shorter) are gamma and x-rays. Those of the longest wavelength (up to 1,000 m) are radio waves used in broadcasting. In between is a narrow range of 390 to 720 nm (billionth of a meter) which is light visible to the human eye. The well-known rainbow colors are associated with this narrow spectral domain, with the color gradually changing: violet-blue-green-yellow-red, as the wavelength increases. It would be natural to imagine that the eye is a sort of biological spectrophotometer (see **Photometer**) which responds to these various wavelengths of light (the spectrum) in an orderly linear way. Nothing could be further from the truth. This becomes obvious when the various colors, as sensed, are plotted in a two-dimensional diagram, the so-called "chromaticity diagram" (Fig. C-26).

The eye seems to operate as if it contained three distinct color receptors that give rise to three

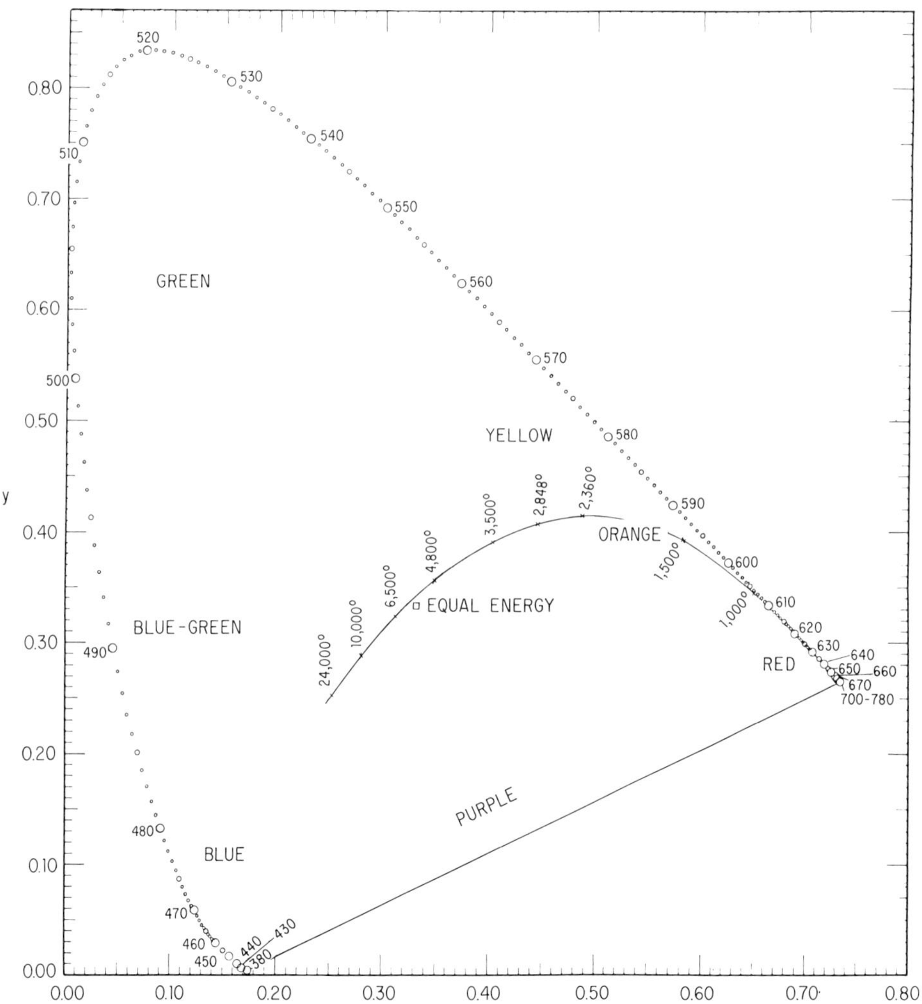

Fig. C-26. Chromaticity diagram showing the three parameters of color in two-dimensional form ($x + y + a = 1$). Consequently, an x, y representation is complete because any value of $x + y$ subtracted from $1 = z$.

stimuli. It is the interplay or balance of these that creates color sensation. At equal intensity, we sense no color, i.e., white to gray depending on the intensity. Any other intensity ratio evokes color in a well-defined numerical way.

The three tristimulus weighting functions are shown in Fig. C-27. x, y, z indicate *chromaticity*, or balance of these parameters.

The horseshoe-shaped area of Fig. C-26 contains all colors. The point in the middle, where x and y (and z) have a value of one-third, is white. The upper circumference corresponds to pure spectrum colors, and the corresponding wavelengths are indicated in nanometers. The bulk of the colors enclosed in the area are not those of the rainbow, which has neither "pale" colors such as pink nor the purples and browns contained near the base. The significance of the area outside the line is nil. It seems peculiar indeed that the colors are arrayed in such an odd manner, but this oddity is the result of human physiology, the organization of the eye and the nervous system.

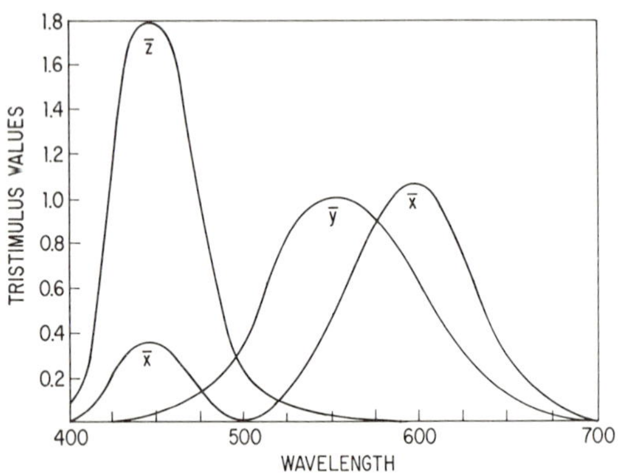

Fig. C-27. Weighting functions required to reduce spectrophotometric data to tristimulus values.

There are, of course, *total intensities* to be considered *(X, Y, Z)*. They can be calculated from a single term, once the relative intensities, i.e., chromaticity, are known. Consequently, the three data x, y, and Y are sufficient, and they are usually given. The capital Y term is commonly chosen because the y function, the luminosity function, corresponds to the relative spectral sensitivity of the human eye and serves as the basis of such photometric units as footcandle and lux.

The tristimulus system is a complete and practical description of color. For instance, if light of a given chromaticity is mixed with light of equal intensity and another chromaticity value, the resultant color can be predicted precisely. A straight line is drawn between the two points on the diagram, and the new color will be that corresponding to the midpoint. Thus, the appropriate red and green, or blue and yellow mixtures (complementary colors) will give white, as well as a complex mixture of all "rainbow" colors. Yellow and red gives orange, white and red yields pink, etc. Furthermore, if light is analyzed for spectral distribution (intensity vs. wavelength) in a spectrophotometer, the resulting curve can be weighted according to the three curves shown in Fig. C-27. This results in three integrated values, and these now completely describe and predict the color which will be seen. An amazing corollary is that two quite dissimilar spectral distributions can result in identical tristimulus values. If this is the case, the colors are identical and, in fact, are indistinguishable by eye even though they are dissimilar in a physical sense and appear so spectrophotometrically. Thus, color-measuring devices described below must work along the three parameters of human color perception. In some ways, this is much simpler than physical spectrophotometry, because instead of a complex curve, only the three values X, Y, and Z are required. There is real beauty in this simplicity.

As a matter of fact, light sources can be completely specified by simple numbers. For instance, an automotive turn-signal lamp is defined as having a minimum intensity of 20 and maximum of 500 candlepowers, and a color of $y = 0.39 - 0.44$ and $z = < 0.01$ (Ref. 2). The specification of "white" light sources is even simpler; a single figure of color temperature is used. This definition is based on the radiation of a theoretical blackbody, and for each temperature there is a corresponding chromaticity, as shown in Fig. C-26. Incandescent bulbs are always reddish (tungsten filaments melt below 3600°), but fluorescent bulbs and the sky are bluer than pure white. Thus, light sources can be defined free of ambiguity by two or three numbers relating to chromaticity and intensity.

Unfortunately, most daily requirements for color specifications do not relate to colored signal lights or light bulbs but rather to the "color" of objects, or of paints and dyes to color them. Much of the color research has been directed toward this problem. The tristimulus system has great advantages in this field, too. It is rational and maintains additivity, but it has an inherent drawback. It is not at all linear with respect to detectable *color differences*. Thus, in the blue and red areas (toward the base of the horseshoe in Fig. C-26) the detectable differences are small, while in the green areas, where the eye has its peak sensitivity, they are large. This is an irreconcilable fact; one always has either uniform spread of colors or additivity. It is often practical and desirable to have even steps, and a great variety of systems have been developed to accomplish this (Refs. 1, 3, and 4).

Figure C-28 is a representation of the tristimulus system codified by international agreement (Commission Internationale de l'Éclairage) and called the CIE system. The direction of the arrow represent the *hue* and its length the *saturation*. The vertical direction represents lightness, black to white. As lightness is increased or decreased, the horseshoe-shaped area changes and shrinks, eventually to a point, since total blackness allows no color and maximum lightness gives a white sensation. In everyday language, lightness and saturation are often lumped. "Paler" or "whiter" is lighter and less saturated; "more brilliant" or "cleaner" is lighter and more saturated. Conversely, "duskier" or "dirty" is darker and less saturated, and a "deeper" color is darker and more saturated. A very popular grading of color of surfaces is done with the Munsell system, which arrays colors in roughly even steps with perception, and the company (Ref. 5) offers colored chips that serve as physical standards for comparison. In this system, the color solid (totality of all colors) is a sphere radially sliced up by hues (blue, blue-green, green, etc.). From a central axis representing neutral, the *chroma* (saturation) fans outward, while the *value* (lightness) is dispersed vertically. Several systems are used to space the colors in convenient ways. They can be interconverted by more or less complex mathematical manipulations (Refs. 1, 3, and 4), and computer programs exist for some conversions (Ref. 6).

However, the crucial problem is that fate has decreed that physical objects shall have no inherent, actual color. When light strikes a surface, a part is reflected at the interface (this is gloss—see **Gloss meter**). Another portion may penetrate the surface and be scattered and back-reflected in all directions. As this diffusely reflected light goes through the material, it may be attenuated spectrally; i.e., light of some wavelengths is absorbed more strongly than that of others, and this gives rise to "color." Obviously, then, the spectral characteristic of the illuminat-

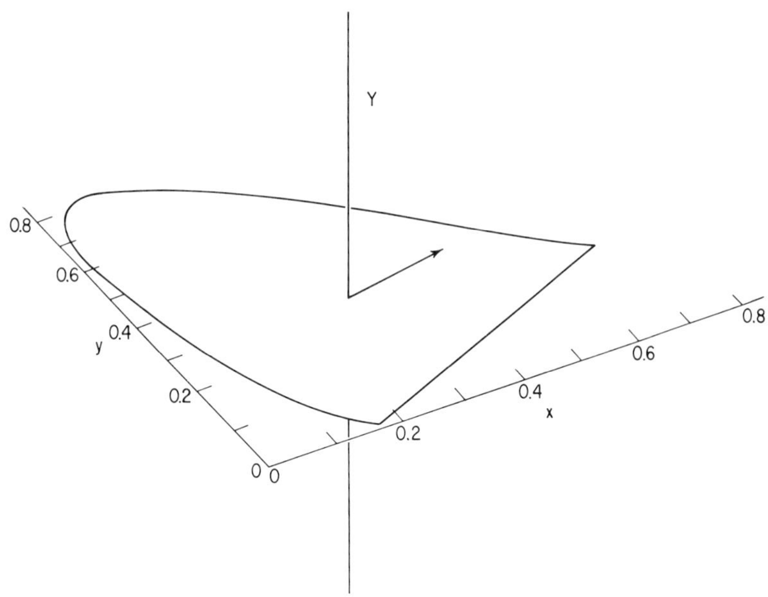

Fig. C-28. Three-dimensional representation of a combination of chromaticity and lightness.

ing light is decisive to the color appearance of the object. The old joke: "Sam, put on the green light, the man wants a pair of green pants" is literally true. An excellent illustration is offered by two printing-ink samples prepared by the General Electric Company. Their spectral-reflection curves are shown in Fig. C-29. When the two samples are viewed in daylight, they appear to be a perfectly matched olive color. The reason is that the two curves yield exactly the same tristimulus values. However, when they are illuminated with redder, incandescent light, sample *A* appears distinctly brown because its ability to reflect the red component is much stronger, and this now comes to the fore. Colors such as these are called *metameric* colors, and they plague manufacturer and consumer alike. Every woman known that when she is matching a scarf to a dress, it is not enough to match them in the store under one kind of light, but they must also match in daylight. Woe to the motorist whose car was carefully repaired with a beautiful matching paint job over a dent. The mercury-vapor boulevard light shows up the damaged areas like a medical x-ray. And so ad infinitum.

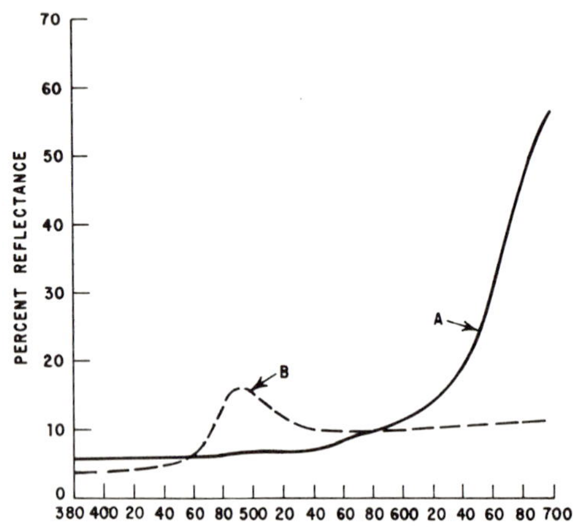

Fig. C-29. Spectral reflectance of two metameric printing inks. *(General Electric.)*

The human vision has many other vagaries. The cones, which are the organelles responsible for color vision, are concentrated in the central area of the retina in the back of the eye. Furthermore, over the most sensitive spot, a yellow protective pigment is spread, sort of built-in "sunglasses." Consequently, it makes an appreciable difference in color vision whether the image is spread out or concentrated, whether an object is viewed from far or from near.

A variety of psychophysical effects enter color vision. There is a tremendous "surround effect." Artists and interior decorators play on this quite consciously. A given yellow will appear "colder," greener, when surrounded by red, gray will take on a reddish cast when surrounded by green, etc. Moreover, the weave, texture, blemishes, and similar surface effects will modify the color impression. In general, we see what we suppose an object should look like rather than what it really is.

If we consider that the normal human eye can distinguish about 10 million surface colors under optimum conditions, it is a valid question whether instrumental evaluation of color is justified. The answer is affirmative. First of all, essentially all difficulties enumerated above apply to visual color matching as well. The color of the illuminating light, the surround, and the size of the sample need to be defined. More importantly, all visual work is *comparative;* it requires a sample to match to, because human color *memory* is singularly poor. It has been shown that even with highly saturated colors, only about 24 can be identified with assurance from memory (Ref. 7). Because physical samples notoriously fade and change with time, an instrumental measurement is invaluable to classify a color by reliable figures as a permanent record. Instrumental color matching to an existing sample is also valuable because it is objective and because it may give a quantitative or semiquantitative clue to corrective measures. It is understandable then that color-measuring instruments clear up to the $30,000 class have become commonplace. Instrumental color measurement requires certain ground rules. The nature of the illumination must be pegged. CIE provides three light sources *A, B,* and *C* with equivalent color temperatures of 2854, 4870, and 6740° K, respectively. These typify incandescent light, noon sunlight, and average overcast daylight, respectively. The last, illuminant *C,* is usually employed, but the others can be used in research problems and to assess metamerism. Fluorescence is usually not evaluated, but special fluorescence meters (see **Photometer**) are used if this is at issue. The sample in most instruments is illuminated by a light beam at, or near, 90° to the surface. The diffusely reflected light is picked up at 45° according to many standard techniques (Ref. 8). However, many instruments are built to pick up all the light (usually excluding the specular reflection, gloss). The geometry makes little difference in many samples which diffuse light uniformly in all directions, but this is not always the case. It can be argued which of these two systems is preferable.

For standardization, diffuse reflectance of color is expressed by comparison with freshly prepared magnesium oxide (or a working standard calibrated in these terms). The accepted value for it is 1.00 or 100 percent. This brings about the peculiarity that excellent white pigments such as titanium white pigments can exceed 100 percent, but this causes no practical difficulty. For color matching, a standard color sample or chip is used, and the results are expressed as *color difference* rather than tristimulus values.

The simplest color-measuring devices are based on a three-filter system originally developed at the National Bureau of Standards (Ref. 9). The spectral response of the instrument with the three filters approximates the three curves of Fig. C-27. A commercial instrument of this type is shown in Fig. C-30. The sensing head contains a stabilized light bulb and optics to direct a parallel beam of light onto a sample. The filters are then interposed one after another. The diffusely reflected light is picked up at 45° by an annular photocell. The output of the cell is amplified and read off the meter or a digital accessory. This simple instrument is suitable for routine color classification of many diverse products, e.g., coffee and instant coffee (Ref. 10), anodized structural panels, and painted or printed surfaces.

More precise work is possible using high-class, and expensive, spectrophotometers equipped for reflectance work. The illumination in these instruments sweeps through the entire spectrum, and the instrument records the spectral-reflection curve as in Fig. C-29. Such a curve can form the basis of color evaluation if the result is integrated according to the three functions (Fig. C-27). In modern instruments, this is done automatically and the three figures x, y, z are simply read off in digital form.

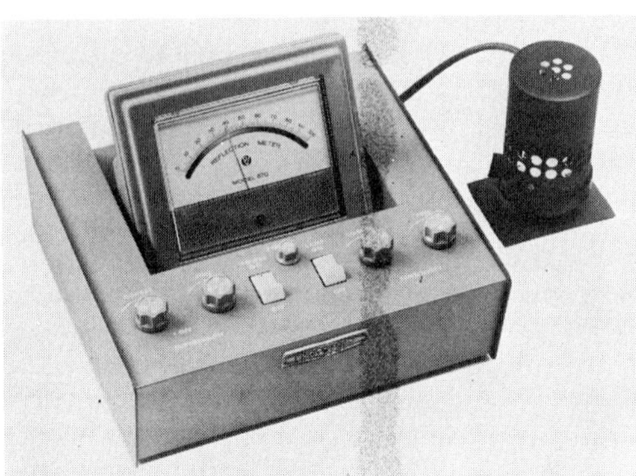

Fig. C-30. Three-filter reflection meter for color measurement. *(Photovolt.)*

A color-difference meter is shown in Fig. C-31. In this instrument, three data are obtained in digital form, L, a, and b. L is a measure of overall lightness, i.e., the Y parameter; a indicates redder if positive and greener if negative; b indicates bluer if positive and yellower if negative.

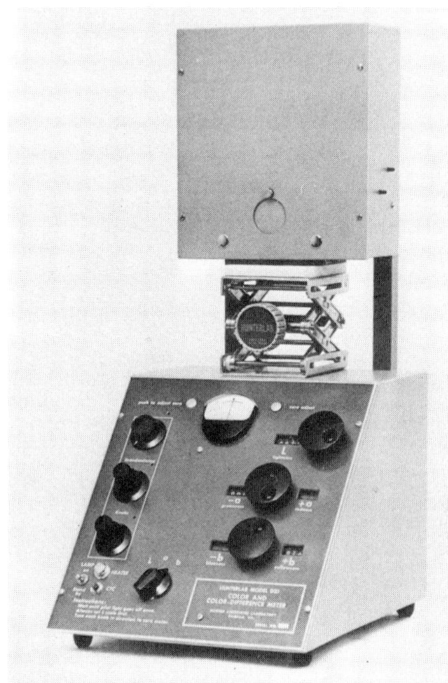

Fig. C-31. Color-difference meter. *(Hunter Associates.)*

There is always a desire to boil down color differences into a single term, and considerable effort has been expended in this direction. Using the instrument described above, the overall color difference between the standard and the sample can be expressed as

$$\Delta E = \sqrt{(\Delta L)^2 + (\Delta a)^2 + (\Delta b)^2}.$$

There are a number of commercial instruments of this type and a variety of systems, and they yield different numerical results (Ref. 3). All can be used with discrimination. In simpler problems, color differences have been expressed by a single figure with success. In the case of near-white samples, the "yellowness" is expressed by using the three readings obtained with a filter instrument, such as shown in Fig. C-30:

$$\text{Yellowness} = \frac{\text{amber} - \text{blue}}{\text{green}}$$

Yellowness, or lack of yellowness, of paper is designated as paper "brightness." This is a simple diffuse reflectance reading with a specified blue filter (Ref. 11). There are a variety of special color scales which include only certain circumscribed areas of the total color solid that is of particular pertinence to special industrial problems. The color-temperature scale of light sources has already been mentioned. Petroleum, oil, and rosin scales are of this type. In industrial practice, visual comparators are used to some extent for grading, and the Lovibond instrument equipped with color filters is mentioned in this connection.

Returning to the most difficult applications, the formulation of colorants for an exact match in the final product is the toughest. This used to be an art. When a mixture of pigments diffusely scatters light giving rise to color, the simple logarithmic law of spectral absorbance does not hold. A good approximation has been worked out to compute the resulting color, but the mathematical manipulations become rather too complex for daily use. Some manufacturers such as the American Cyanamid Co. and Imperial Chemical Industries have developed computer programs as a customer service, and recently a specialized colorant-mixture computer has become available. The purpose is batch control, where a paint formula is adjusted to conform to specifications. With the aid of computers, this can be done in a rational way with much less trial and error.

When it comes to materials which are produced in a continuous stream by a machine, complete control and automation are possible. There are several commercial color-measuring devices which are intended to monitor the product continuously. These can record the three color coordinates, and they can sound an alarm when the color difference exceeds a predetermined limit. Corrective measures can then be taken by the operator. Complete automation is certainly possible, but the installation must be tailored to the process at hand. A system of this type would consist of a color-measuring instrument which monitors the product as it comes off the line. The output of this instrument is fed to a computer, which initiates changes in flow of several colorants to maintain the standard color of the product. Such a closed-loop operation represents the ultimate in color control. More commonly, control is maintained by manually taking samples and presenting them to standard instruments. In most exacting applications, a high-class color-difference meter or specially equipped spectrophotometer is used. For the more common color-classification problems, inexpensive reflection meters equipped with three filters are widely employed.

It is understood that the subtle psychological aspects of color sensation elude instrumentation. However, many of these are virtually impossible to define because of individual reactions of the observer. Even the CIE system is functional only in the sense that it considers the "average" observer from a physiological point of view. Ignoring these evanescent factors, there is a basic color sensation as defined by the standard observer at a standard illumination. It is then reassuring to have an objective scientific color system which is free of ambiguity. Color-measuring devices are built on this basis, and they can furnish reliable results when used with discretion.

REFERENCES

1. Judd, D. B., and G. Wyszecki: "Color in Business, Science, and Industry," John Wiley & Sons, New York, 1952.
2. SAE Standards J578a (April, 1965) and J914 (February, 1965).
3. Billmeyer, F. W., Jr., and M. Saltzman: "Principles of Color Technology," Interscience Publishers, a division of John Wiley & Sons, Inc., New York, 1966.
4. Wyszecki, G.: "Farbsysteme," Musterschmidt Verlag, Göttingen, 1960.
5. Munsell Color Co., Inc., Baltimore, Md.
6. Dlydesdale, F. M., and C. H. Podlesny, Jr.: A Computer Program for the Interconversion of Color Data, *Color Eng.*, vol. 6, no. 3, p. 55, 1968.
7. Hellmig, E.: Versuche über das Farberinnerungsvermögen, *Farbe,* vol. 7, p. 65, 1958.
8. Harris, J. C.: Detergency Evaluation and Testing, Federal Test Method Standard 141/6121; ASTM D985, E97.
9. Hunter, R. S.: Photoelectric Tristimulus Colorimetry with Three Filters, *Natl. Bur. Std. (U.S.) Circ.* C429, 1942.
10. Lockhart, E. E.: Roasted Coffee Color Measurement and Classification, *The Coffee Brewing Institute Inc., Pub. 53, Food Technol.,* vol 14, p. 597, 1960.

11. Brightness of Paper and Paperboard, TAPPI Standard T452–48. 1948.
12. Wyszecki, G., and W. S. Stiles: "Color Science," John Wiley & Sons, Inc., New York, 1967.

—Gabor B. Levy, *Photovolt Corporation*

Color-change Temperature Indicator [See Temperature-sensitive materials (indicators).]

Colorimeter Historically, colorimetry is a method of chemical analysis in which the intensity of color of a sample solution is compared with that of a standard solution. To determine concentration in the sample with reasonable precision, the two solutions were matched in a visual field by changing either the light path (Duboscq) or the size of an aperture (Pulfrich). The concentration (or a figure proportional to the concentration) was read from a scale after the fields were matched. Visual methods largely have been replaced by photoelectric colorimeters except for screening or rough approximations. Simple color comparators still are available for quick estimation of pH and a few other chemical tests.

Photoelectric Colorimeter: This instrument is similar to a spectrophotometer. These two terms overlap. Typically, a colorimeter operates only in the visible part of the spectrum (400 to 800 m), the instrument uses nondispersive optics for wavelength isolation, that is, by means of color filters, and is less costly than a spectrophotometer. In contrast, spectrophotometers reach into the ultraviolet and infrared regions, employ dispersive optics (prisms and diffraction gratings), and often include extra features, such as automatic wavelength scanning and chart recorders.

With all instruments of this general type, the basic scientific principle is the same. Compounds which have specific absorption in certain wavelength domains (or which can be treated or converted with reagents to absorb light) can be determined quantitatively by measuring the attenuation of a beam passing through the solution. The basic law of absorptiometry, according to Lambert's and Beer's laws, is

$$\log \frac{I}{I_o} = edc$$

where I_o = intensity of original beam
I = intensity of attenuated beam after passage through the solution
e = a constant characteristic of the compound at a given wavelength and temperature
d = length of the light path through the sample
c = concentration of the sample

Since d is constant and is determined by the size of the sample container (cuvette) and since e is established by measurement with a known standard, the concentration c is proportional to the left-hand term log I/I_o, which is called the *absorbance* or *optical density*. Note that concentration is proportional not to the light absorption but to the *logarithm* of light absorption. Visualize the sample as subdivided into successive layers of liquid. The first increment may absorb 50 percent of the available light; the next increment receives only one-half of the illumination and, in turn, absorbs 50 percent of that, or 25 percent of the total available light. The next layer absorbs 12.5 percent and so on. This is an exponential decay—hence the logarithmic relationship between light path and light attenuation. Of course, having twice the light path or twice the concentration is exactly equivalent, because in either case the light interacts with twice the number of molecules. Beer's law states this relationship mathematically. So that concentration may be read directly from the instrument, the readout meter will be marked off in logarithmic absorbance units; or a logarithmizing circuit will be included in the instrument.

A typical photoelectric colorimeter comprises (1) a stabilized incandescent light bulb, (2) simple optics (mirror, lens) to form a defined beam, (3) filters to isolate a relatively narrow wavelength domain, (4) a sample holder, and (5) a photometer to measure light. The procedure is quite simple—a standard is measured, followed by the sample or a succession of samples. The concentrations are proportional to the meter readings and are computed by simple proportion or are read from a calibration chart or graph.

As contrasted with more costly spectrophotometers, the photoelectric colorimeter is restricted to samples of known character and to tight analytical techniques. In practice, the majority of tasks in industrial quality control and clinical procedures are of a predictable, repetitive type, and thus the colorimeter is quite suitable and enjoys the advantages of easy use, low maintenance, and low cost. The fixed filters of the colorimeter are virtually foolproof as contrasted with the difficulty an inexperienced operator may encounter with the wavelength flexibility offered in even the lowest-cost spectrophotometers.

—Gabor B. Levy, *Photovolt Corporation*

Colorimetric Radiation Detector A simple, inexpensive, rugged, but rather inaccurate radiation detector. The accuracy may be adequate for civilian-defense purposes against an atomic-bomb blast. The principle of operation is based on the color change of an aqueous dye solution due to the acid evolved from chloroform when subjected to radiation. Some use as radiation dosimeters in commercial radiation processing has been found for the device. Discoloration or degradation of plastic films when exposed to radiation also falls into this category of radiation-detection device.

Combustion/Combustibles Analyzer Measurement of combustible gases is important to control their concentrations within safe limits below the lower explosive limit, for optimizing concentrations for efficient reactivity, and to detect process-system upsets. The principal method used for combustible gas or vapor analysis employs self-heated "hot-wire" detectors of a material such as platinum. These are mounted in a Wheatstone-bridge circuit. The hot catalytic detector causes the combustibles in the sample to burn in the presence of oxygen. If the sample does not contain an excess of oxygen, oxygen or air must be added in controlled amounts to provide an excess such that the reaction will be limited by the amount of combustibles present.

The combustion of the mixture releases heat, which is detected as a temperature rise in the catalytic-detector spiral. Since the spiral's resistance is a function of its temperature, an electrical unbalance occurs in the Wheatstone bridge proportional to the concentration of the combustibles. Measurement of this electrical signal provides a measure of the combustible concentration.

This type of analyzer is nonspecific; that is, it cannot differentiate between different combustibles but gives a combined output for the mixture in question. Since the output is a function of the rate of combustion and the heat of reaction, the analyzer is often calibrated as "percent combustibles expressed as percent hydrogen," although it may be calibrated for another specific combustible when it is known that the specific combustible will predominate.

The reference detector may be exposed either to a suitable reference gas such as air or to the sample with the catalytic characteristic destroyed by poisoning or coating. The latter method compensates for thermal-conductivity changes in the sample caused by changing sample compositions, although these are generally minor changes.

Another type of combustibles analyzer measures the heat released by burning the combustibles sample in a small pilot flame using a thermocouple as a sensor. When combustibles are present, the temperature of the flame increases proportional to the combustibles concentration. This method provides freedom from poisoning of the catalytic property of the platinum with such substances as silicone vapors, which may seriously limit the conventional "hot-wire" analyzer.

Combustibles analyzers are available in portable combinations with oxygen analyzers. These devices are valuable for testing combustion in furnaces and various process combustion apparatus and are thus termed combustion analyzers. Both portable and on-stream analyzers are available.

—Lyle K. Barnes, *The Hays Corporation—A Milton Roy Company*

Command (See **Set point.**)

Command Control (See **Display-command operator/ machine interface.**)

Commercial Computer The commercial computer is generally considered to be a digital computer which includes in its instruction set special functions for operating on decimal data. Some of the functions provided are arithmetic operations on decimal data and the editing of data fields. The edit function provides the capability of inserting $ signs, commas, and other forms of punctuation in a field of data. Although these functions could be obtained without special instructions by means of subroutines, the digital computer with these instructions performs the required functions more efficiently and, typically, at a higher speed. A commercial computer also is designed to manipulate efficiently text information needed to print documents or reports from input data.

—W. W. Welz, *IBM Corporation*

Common-mode Rejection Ratio (CMRR) The common-mode rejection ratio is a parameter used to express the ability of a differential subsystem or instrument to reject the effect of a common-mode voltage applied at its input terminals (see **Common-mode voltage**). The definition of the ratio is based on the idealized model of Fig. C-32.

The voltage V_{cm} is the common-mode voltage, V_{in} is the equivalent differential-input voltage due to V_{cm} (see **Differential-mode voltage**), V_o is the output voltage (or output value in the case of a digital measurement subsystem or instrument), and G is the gain or conversion factor of the differential subsystem or instrument.

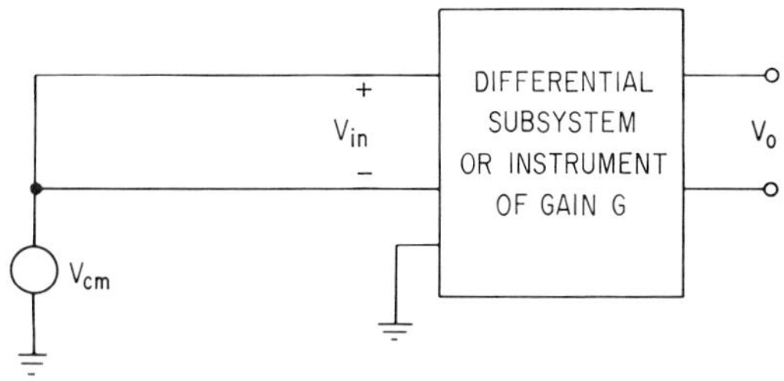

Fig. C-32. Common-mode rejection ratio model.

The common-mode rejection ratio is defined as CMRR $= V_{cm}/V_{in}$. Since it is difficult to measure V_{in} directly, the equivalent definition CMRR $= V_{cm}G/V_o$ is used as the basis for measurement of common-mode rejection ratios. Even if it were possible to measure V_{in}, this second form of the definition is more indicative of system performance, since some common-mode to normal-mode conversion generally occurs internal to the subsystem and is not reflected as an actual input voltage.

If the common-mode voltage is ac, the measurement of the CMRR is based on the ac component in the output. In the case of an instrument or subsystem having a digital output, this is the increase in the repeatability or spread in output values (see **Repeatability**). Consistent units must be used; that is, if peak-to-peak values are used for V_{cm}, then the peak-to-peak increase in the spread of output readings must be used.

The CMRR is often expressed in decibel units. The definition for this form is that CMRR (db) = 20 log (CMRR).

Typical values for the CMRR of an instrumentation subsystem range from 100 to 140 db or, when expressed as a ratio, from 10^5:1 to 10^7:1, with 120 db being the most common. A specification of 10^6:1 means that every volt of common-mode voltage results in the equivalent of 1 μV of normal-mode voltage at the input to the subsystem. Since the CMRR of most subsystems depends on the resistive source unbalance and, in some cases, on the value of the common-mode voltage (see **Error, common-mode**), a CMRR specification should include these parameters. An example of a complete specification is CMRR = 10^6:1 at up to 1,000 Ω source unbalance and 200 V dc.

—Thomas J. Harrison, *IBM Corporation*

Common-mode Voltage The common-mode voltage is that voltage which is common to both inputs of a differential subsystem or instrument when measured with respect to a system reference point (ground) (compare **Differential-mode voltage**). The concept is illustrated in Fig. C-33, which shows a differential subsystem being used to measure a signal voltage V_s. The ground reference point for the measurement is the subsystem ground G_1; the signal voltage, however, is referenced to G_2, which may or may not be at the same potential as G_1. The voltages at the input terminals A and B are $V_A = V_s + V_{cm}$ and $V_B = V_{cm}$; the voltage common to both A and B, the common-mode voltage, is V_{cm}. Another definition which is sometimes used for the common-mode voltage is the average value of V_a and V_b; that is, $(V_A + V_B)/2$. Since common-mode voltages are generally of concern only when $V_{cm} >> V_s$, the difference between the two definitions is usually negligible.

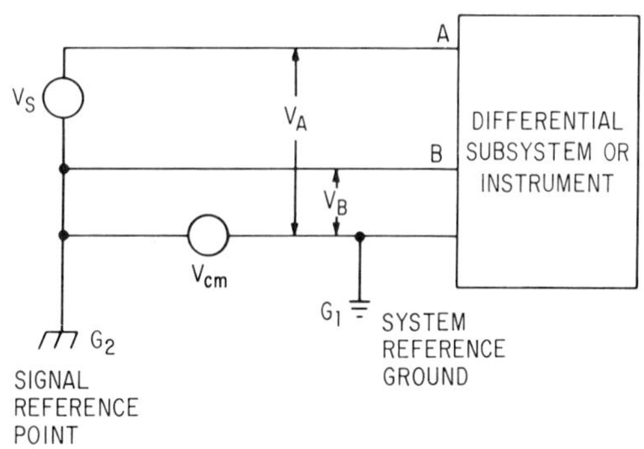

Fig. C-33. Common-mode voltage.

Common-mode sources are of concern in differential-measurement systems because they can cause measurement error due to common- to normal-mode (or differential-mode) conversion (see **Differential-mode voltage**). If the differential-measurement subsystem or instrument were perfect, it would measure only the value of the signal voltage V_s with no contribution due to the presence of the common-mode voltage V_{cm}. In practice, however, imperfections in the subsystem cause some of the common-mode voltage to appear as a signal voltage.

One way in which this conversion can occur is illustrated in Fig. C-34. The resistances R_1 and R_2 represent the input-line resistance and the source impedance of V_S. The resistances R_3 and R_4 are leakage resistances to ground in both the input lines and the differential-measurement subsystem. The resistances R_1 and R_3 form a voltage divider, as do resistances R_2 and R_4.

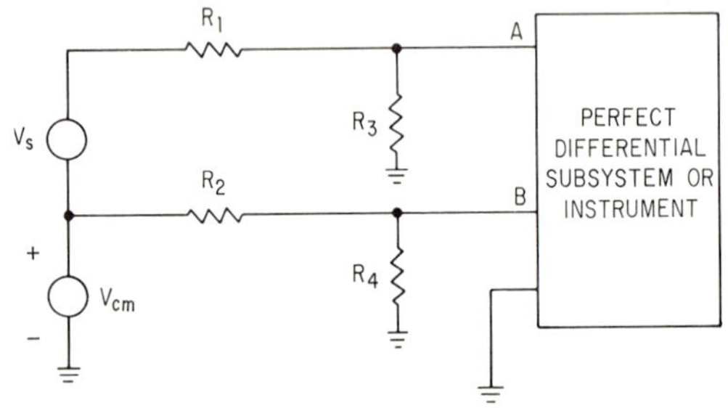

Fig. C-34. Common- to normal-mode conversion model.

Neglecting the input impedance of the instrument, the differential- or normal-mode voltage $V_A - V_B$ is given by

$$V_A - V_B = V_s \frac{R_3}{R_1 + R_3} + V_{cm}\left(\frac{R_3}{R_1 + R_3} - \frac{R_4}{R_2 + R_4}\right)$$

With the assumption that $R_3 \approx R_4 >> R_1$, R_2, which is a reasonable case, this expression reduces to

$$V_A - V_B \approx V_s + V_{cm}\frac{R_2 - R_1}{R_3}$$

Since the desired result of the measurement is the value of V_s, the second term represents the error due to common- to normal-mode conversion, known as the common-mode error (see **Error, common-mode**). The inverse of the coefficient of V_{cm}, $|R_3/(R_2 - R_1)|$, is the common-mode rejection ratio, a measure of the ability of the system to reject the effect of the common-mode voltage [see **Common-mode rejection ratio (CMRR)**].

The model of Fig. C-34 represents only one way in which common-mode voltage can be converted into a normal-mode signal. For example, if the common-mode source is ac, the resistances R_3 and R_4 can usually be neglected, but the leakage capacitance from input terminals A and B to ground must be considered. The model is an idealization of the actual phenomenon in that it assumes lumped parameters, whereas in a real case the resistances and capacitances are distributed along the length of the input lines.

The common-mode voltage may be the result of the manner in which the system is used, or it may be due to unintentional causes. For example, if a thermocouple is bonded to a current-carrying conductor or if the subsystem is used to measure the voltage drop across an ungrounded resistor in a circuit, the common-mode voltage is due to the way in which the system is being used. If, however, the signal source is grounded but the ground potential differs from that at the measurement point, as implied in Fig. C-33, an unintentional source of common-mode voltage has been added to the measurement circuit.

—Thomas J. Harrison, *IBM Corporation*

Communications (See **Classified Index.**)

Commutator (See **Multiplexer.**)

Comparator (See **Amplifier, comparator.**)

Compass An instrument used to point direction. In its simplest form it comprises a magnetized needle that rests on a pivot so that the needle is free to turn with a minimum of friction when subjected to magnetic force. Although the history of the compass cannot be traced with certainty, the instrument dates at least as far back as 2,000 years before Christianity—to the Chinese who were familiar with the magnetic properties of lodestone. The instrument was introduced to Europe during the twelfth century.

Principal uses for the compass are navigation and surveying. In early navigational instruments, the dial always was horizontal and in the form of a full circle, marked off with the familiar cardinal points. North is considered 360°, east is 90° (going clockwise around the dial), south is 180°, and west is 270°. The surveyor's compass differs in that directions are measured clockwise from the south so that a bearing of 270°, for example, indicates east.

The traditionally familiar *magnetic compass* points in the direction of the earth's north magnetic pole which, of course, is *not* true north. The simple magnetic compass is not too satisfactory. Corrections must be made to obtain true directions because of the difference between magnetic and true north and because of the influence of local materials and structures on the action of the instrument. Thus, modern, practical compasses use the properties of either the gyroscope or magnetic induction. See **Flight-control systems**; and **Navigation systems.**

Compensation (See **Ambient conditions, compensation for**; and **Reference-junction compensation.**)

Compiler A program which translates (or compiles) a program written in a compiler language into a program in a language closer to direct machine code, for example, an assembler language.

There are two basic techniques used in compilers. The first is for the compiler output (object code) to be a series of instructions which, with some relatively minor further processing (e.g., assembling and/or link editing), may be directly executed by the computer. The second technique is for the object code to be input to a program, sometimes called an interpreter, which uses the object code as data for controlling which portions of the interpreter should be executed. In this case, the compiler simply performs a syntax check on its source-code input and a relatively minor translation of the source code to produce the object code. Most practical compilers use a mixture of both techniques, the mix depending on such things as the characteristics of the compiler language and the intended type of user.

Compiler Language: A computer language characterized by a one-to-many relationship between the statements written by the programmer and the actual machine instructions executed (compare **Assembler**). The programmer typically has little control over the number of machine instructions executed to perform a particular function and is dependent on the particular compiler implementation. Typically the language is very nearly machine-independent and is biased in its statements and features to a particular group of users with similar problems. These languages are therefore sometimes called problem-oriented languages, or POLs. The various slightly different implementations of a given language are frequently referred to as "dialects." Some compiler languages in wide usage are FORTRAN, BASIC, COBOL, ALGOL, PL/I, and JOVIAL.

—A. J. Arthur, *IBM Corporation*

Compression Testing [See **Testing (materials), destructive.**]

Compton Effect (See **Gamma-ray spectroscopy.**)

Computer, Air-data [See **Air-data systems (aircraft).**]

Computer, Analog (See **Analog computer.**)

Computer, Digital (See **Digital computer.**)

Computer, Fluidic (See **Fluidics.**)

Computer, Mechanical Generally, a digital device designed to perform arithmetic operations on numbers supplied by an operator. Essential elements consist of a selection register that works in conjunction with an accumulator—beyond which a counter, storage, other accumulator registers, and controls for automatic operational sequences are obtainable in a variety of configurations. Mechanical computers generally are restricted to the operations of addition, subtraction, multiplication, and division, although automatic extraction of square roots now is obtainable.

Computer Process Control (See **Direct digital control**; and **Process optimization.**)

Computerized Drafting (See **Drafting system, automatic.**)

Conductivity Analyzer (See **Electrical-conductivity chemical analyzer.**)

Cone, Pyrometric (See **Pyrometic cone.**)

Consistency, Measurement of Consistency is the term used in the paper industry, as well as some food-processing and mining operations, to designate the concentration of weight of air-dry pulp (i.e., of total solids) in any combination of pulp and vehicle, the vehicle usually being water. This can be expressed by

$$\frac{\text{Air-dry weight of solids, lb}}{\text{Weight solids plus water, lb}} \times 100 = \text{consistency, \%}$$

Papermakers occasionally use the terms bone-dry, oven-dry, and moisture-free with reference to pulp fibers or paper from which all water is removed by evaporation. Since such drying seldom is practiced in the paper mill, the term bone-dry is primarily used in laboratory work and in process-equipment design. Air-dry pulp is considered to contain 10 percent moisture, so that

$$\text{Bone-dry consistency, \%} = 0.9 \times \text{air-dry consistency, \%}$$

Uniform consistency is important to the operation of process machinery and to the control of end-product quality.

Through the years, the term consistency has acquired several connotations. Although consistency is a real property of nonnewtonian fluids, consistency measurement is usually relative to arbitrary standards. Thus, the measurement can be stated in no definitive units, such as those of absolute viscosity, temperature, and pressure. Measuring devices infer a value of consistency from measurements of other fluid properties.

Substances which undergo continuous deformation when subjected to shear stress are said to exhibit fluid behavior. The resistance offered by the substance to this deformation is termed *consistency*. The consistency is constant for gases and newtonian fluids if static temperature and pressure are fixed. When referring to such fluids, the consistency is called viscosity. See **Viscosity, measurement of.** The consistency of nonnewtonian fluids is not constant but is a function of the applied shear stress and, in some cases, may vary with time. The term *apparent viscosity* frequently is applied with reference to the consistency of nonnewtonian liquids.

The apparent viscosity of a suspension of certain fibrous materials is related to the percentage of fibers in the suspension. This relationship provides the operating principle for several consistency-regulating devices; that is, these devices employ measurement of apparent viscosity as the base for control of fiber concentration. This principle has, through constant usage, caused the term consistency to acquire the connotation of solids content and has caused its true scientific definition to be disregarded in some industries. The consistency of a suspension of fibrous materials varies primarily with the percentage of fibers in the slurry. As long as the nature of the suspended material is not greatly changed, the use of apparent viscosity as the control point gives proper consistency control. In respect to the operation of any regulator, the question of exact variation of viscosity of the suspension is mathematically unimportant, as may be noted from a consideration of the viscosity-consistency curve of Fig. C-35. Note that the maintenance of a fixed viscosity will involve a corresponding consistency regardless of whether the relationship is linear or not. It is evident that the usefulness of any apparatus based upon the relation between consistency and apparent viscosity of a paper stock, for example, is limited to the range in which the viscosity-vs.-consistency curve may show considerable slope.

The term consistency also has a "flowability" connotation, that is, the ease with which a substance may flow from one container to the next. In the case of some food products, such as tomato paste and catsup, this quality is related to "palatability." Printers refer to this property as "tackiness" in connection with ink; paints sometimes are referred to as being "thick" or "thin."

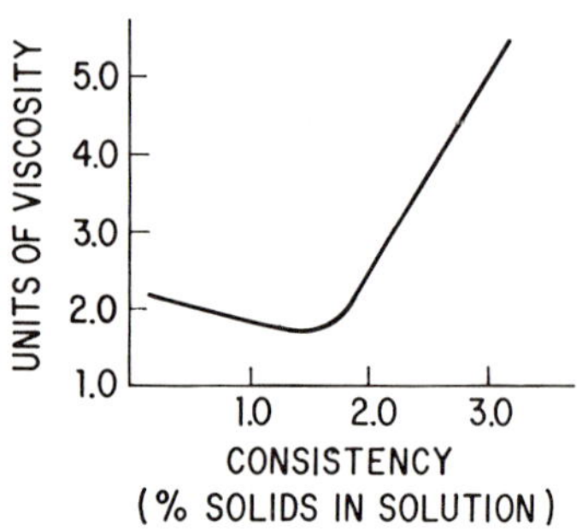

Fig. C-35. Relation of apparent viscosity and consistency for paper stock.

Instruments for the measurement of consistency may be classified as (1) those which determine apparent viscosity, both mechanically and hydraulically; (2) those which measure drainage rate; (3) electrical-resonance devices; and (4) optical devices. Instruments for determining apparent viscosity by mechanical means employ feelers in such forms as paddles, cylinders, egg-shaped spindles, and modified propellers. Variation in consistency of the material is indicated as a variation in the torque required to drive the feeler, usually at constant speed. In the hydraulic measurement of apparent viscosity, the sensing element may take the form of a friction cylinder, tube, or in some cases a relatively simple sluice. Under conditions of constant flow through the device, variation of apparent viscosity produces variations in head of fluid. Change in head may be measured by movement of a float, air-bubbler differential pressure, actual weight of the friction cylinder, or action of the slurry on a loaded flap in a discharge sluice. Most of these devices have been applied principally to paper-stock consistency control.

REFERENCES

Green, H.: "Industrial Rheology and Rheological Structures," John Wiley & Sons, Inc., New York, 1949.

Considine, D. M. (ed.): "Process Instruments and Controls Handbook," McGraw-Hill Book Company, New York, 1957.

Considine, D. M., and S. D. Ross (eds.): "Handbook of Applied Instrumentation," McGraw-Hill Book Company, New York, 1964.

Continuous-path Control System (See **Numerical control;** and **Position measurement and control systems.**)

Contouring Control (See **Numerical control.**)

Control, Electric (See **Electric controller.**)

Control, Hydraulic (See **Hydraulic controller.**)

Control, Pneumatic (See **Pneumatic controller.**)

Control Action In a controller or controlling system, the nature of the change of the output as effected by the input. The output may be a signal or the value of a manipulated variable. The input may be the control-loop feedback signal when the set point is constant, an actuating error signal, or the output of another controller. Various control actions are described elsewhere. See **Adaptive control; Cascade control; Feedback control; Feedforward control; High-limiting control; Integral (reset) control; Optimization control; Proportional control; Proportional-input control; Proportional-plus-derivative control; Proportional-plus-integral control; Proportional-plus-integral-plus-derivative control; Ratio control;** and **Split-range control.**

REFERENCES

Principles of Automatic Process Control, *ISA Publ.* PAPC, 1969.

Tucker, G. K., and D. M. Wills: "A Simplified Technique of Control System Engineering," Honeywell Inc., Fort Washington, Pa., 1958.

Considine, D. M. (ed.): "Process Instruments and Controls Handbook," McGraw-Hill Book Company, New York, 1957.

"Manual on Process Instrumentation and Control," 2d ed., American Petroleum Institute, Division of Refining, API RP 550, New York, March, 1965.

Control Mode A specific type of control action, such as proportional, integral, or derivative. See **Control action.** Generalizations pertaining to the matching of control mode to process application are given in Table C-2.

Control Point (See **Set point.**)

Control System A large portion of this encyclopedia is devoted to descriptions of the various elements that make up a control system. The prime objective of this entry is to classify the many facets of control-system technology to aid the reader to locate information on specific topics. Reference to the alphabetical index also is suggested. In particular, also see **Automatic control.**

Tersely, a control system may be defined[1] as *"a system in which deliberate guidance or manipulation is used to achieve a prescribed value of a variable."* In practice, of course, and outside of rigorous technical circles, the term *control system* takes on a broader meaning, but not necessarily accurate technically. Generally speaking, the term *control system* tends to connote the means required to operate automatically a machine, process, or other assemblage of devices and equipments that, when taken together, perform some desirable function and produce some desired result with a minimum of human intervention. Thus, there are *machine* control systems; *aircraft* control systems; *communication-network* control systems; distillation, drying, and other kinds of *process* control systems; *nuclear-reactor* control systems; *electrical-power-distribution* control systems; and in the handling of data, such as documents, mail, records, and accounts, *information* control systems. The foregoing are examples and are not meant to comprise an exhaustive list of control-system uses.

[1] Adapted from Measurement and Control Terminology, proposed SAMA Standard PMC20, with the permission of the publisher, the Process Measurement and Control Section, SAMA.

TABLE C-2. *Control Mode vs. Application*

Control Mode	Process Reaction Rate	Load Changes		Applications
		Size	Speed	
On-off. Two-position with differential gap	Slow	Any	Any	Large-capacity temperature and level installations. Storage tanks, hot-water supply tanks, room heating
Floating. Single-speed with adjustable neutral zone	Fast	Any	Small	Processes with small dead time. Industrial furnaces and air conditioning
Proportional	Slow to moderate	Small	Moderate	Pressure, temperature, and level where offset is not objectionable. Kettle reboiler level, drying-oven temperature, pressure-reducing stations
Proportional-plus-derivative (rate)	Moderate	Small	Any	Where increased stability with minimum offset and lack of reset wind-up is required. Compressor discharge pressure, paper-strip edge guiding
Proportional-plus-integral (reset)	Any	Large	Slow to moderate	Most applications, including flow. Not suitable for batch operations unless overpeaking is allowed
Proportional-plus-integral-plus-derivative	Any	Large	Fast	Batch control; processes with sudden upsets

Although the mathematics and theory of automatic control systems tend toward the complex, this is not the prime cause of topical complexity. Rather, the complexity derives from the seemingly limitless ways in which a control-engineering problem can be approached and solved. There are a dozen or more major variables that can be measured, manipulated, and controlled, and for each variable, one finds a half dozen or more ways to measure that variable. One need only think of the several ways in which temperature can be measured satisfactorily. The information flowing within a control system can be analog or digital or a combination of the two schemes, thus giving rise to analog-to-digital converters and digital-to-analog converters. The machine or process being controlled can provide information that results from a controlled change or correction—as in a *closed-loop* system—in the form of *feedback;* or a control system—as in an *open-loop* system—can be preprogrammed for control action without feedback. The medium for transferring information within the control system commonly can be pneumatic or electric. The portion of a controller that computes the error signal commonly can be fluidic, pneumatic, or electric. The medium used to power the final controlling element, such as a valve, damper, motor, or heating-element switch, and so on, may be electric, hydraulic, or pneumatic.

The control mode, that is, the type of control action which the controller takes in directing the final controlling element, may vary from a simple on-off action to the more complex proportional-plus-integral-plus-derivative action. Where a large computer may take part in a control system, the computer may function "on-line" or "off-line," terms which are self-evident. The control engineer has the responsibility of matching the many and variable characteristics of the machine or process to be controlled with the most effective control hardware available, the characteristics and specifications of which comprise a long list for each item of equipment.

To satisfy fully the meaning of *total system,* a control system can be visualized best as comprising two subsystems: (1) the controll*ed* system (sometimes referred to as the *process*) and (2) the control*ling* system. This distinction is evident from Fig. C-36. The controll*ed* system in this case is a liquid-heating tank or heat exchanger in which cool liquid enters the tank near the top and exits from the tank near the bottom (right-hand side). A hot liquid flows upward through a heating coil within the tank. As shown, a temperature-sensing element measures the tank-liquid temperature and by way of a pneu-

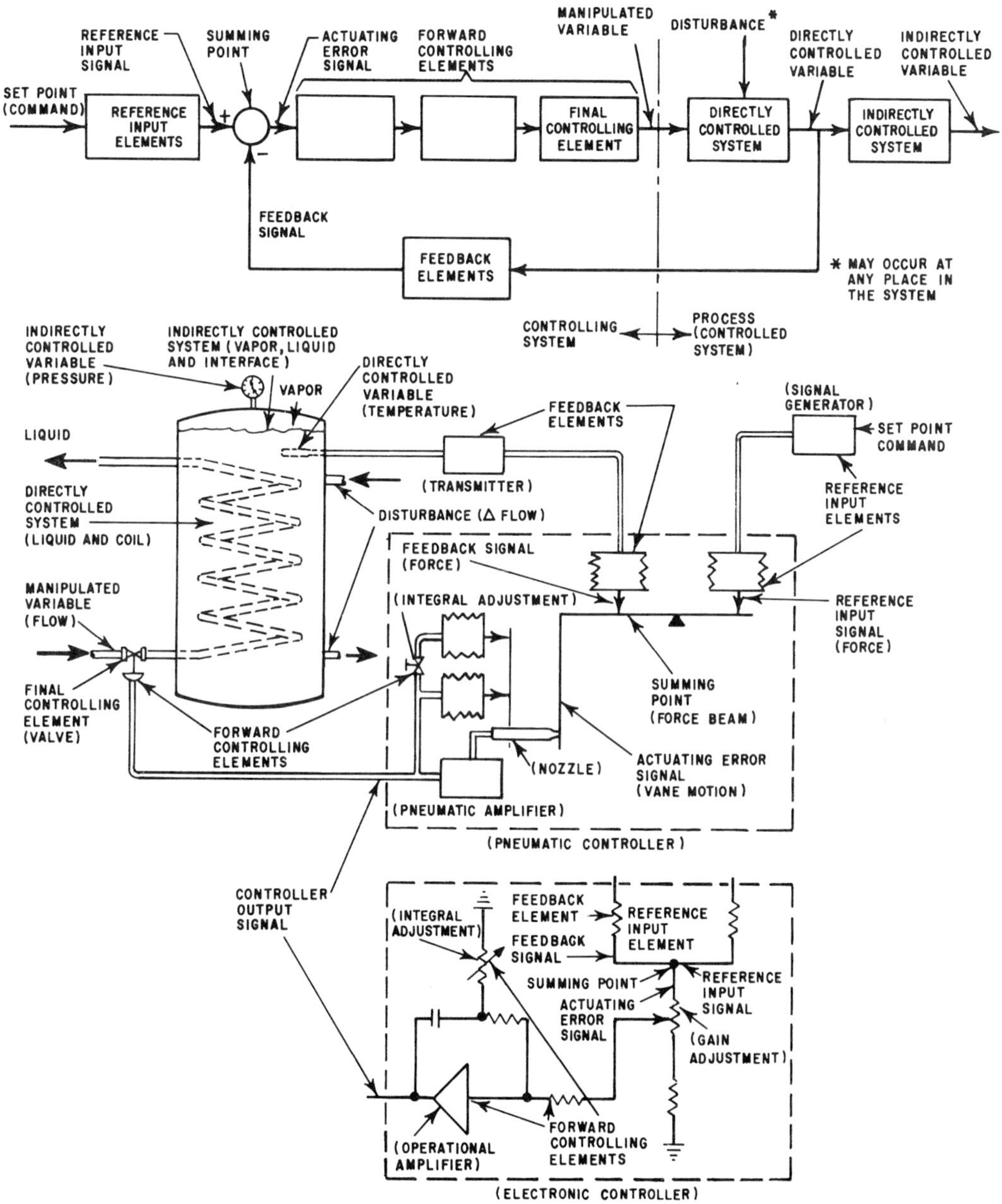

Fig. C-36. Major elements of a control system. *(From Measurement and Control Terminology, proposed SAMA Standard PMC 20, with permission of the publisher, The Process Measurement and Control Section, SAMA.)*

matic or electronic controller adjusts the valve on the incoming hot-liquid line. The major elements that are named on this diagram are described elsewhere and also are included in the topical classification which follows. The controll*ing* system comprises all the hardware involved in taking measurements, making control calculations, and taking control actions. But, it is in the controll*ed* system where the control actions, in effect, are implemented—and in the case of a closed-loop system, the results of this implementation are fed back to the controll*ing* system. The two interacting subsystems are stressed to show that control-system hardware does not act in isolation. Both subsystems have their own characteristics—which must be matched if the total control system is to be effective.

Topical classifications of control-system terms found elsewhere in this encyclopedia are given in Tables C-3 to C-9. Examination of the references listed here also is suggested for automatic control theory and control-system technology.

TABLE C-3. *Terms Related to the Controlled System (or Process)*

See:	
Dead time, process	Process reaction rate
Disturbance	Process variable
Lag, controlled-system	Self-regulation, process
Load change	Set point
Process	Variable

TABLE C-4. *Terms Related Primarily to the Sensing Element and Measuring Instrument*

See:	
Accuracy	Reproducibility
Accuracy, digital-data-acquisition system	Resolution
Backlash	Resolution, digital-data-acquisition system
Calibration curve	Resonant frequency, transducer
Detent	Response time, transducer
Deviation	Ripple
Frequency range, transducer	Rise time, transducer
Hysteresis	Sensitivity
Linearity	Sensor
Range, instrument	Transducer
Repeatability	
Repeatability, digital-data-acquisition system	

TABLE C-5. *Terms Related Primarily to the Controller Portion of the Controlling System*

CONTROL MEDIUM USED

See:

- Electric controller
- Fluidics
- Hydraulic controller
- Pneumatic controller
- Self-actuated regulators

CONTROL MODE

See:	
Control action	Proportional-input control
Control mode	Proportional-plus-derivative control
Derivative (rate) control	Proportional-plus-integral control
Floating control	Proportional-plus-integral-plus-derivative control
High-limiting control	Velocity-limiting control
Integral (reset) control	
Proportional control	

TABLE C-6. *Terms Related to the Overall Configuration of the Controlling System*

See:	
Adaptive control	On-line
Automatic control	Open-loop control
Cascade control	Override control
Closed-loop control	Process optimization control
Direct digital control	Ratio control
Feedback control	Remote control
Feedforward control	Sampling theorem
Negative feedback	Shared-time control
Numerical control	Split-range control
Off-line	Supervisory control
	Timing controllers

TABLE C-7. *Terms Related to the Characteristics of the Overall Controlling System*

See:	
Algorithm	Response, steady-state
Gain (magnitude ratio)	Response, transient
Gain, loop	Settling time
Linear control system	Summing point
Offset (droop)	Time constant
Response, ramp	Transfer function, theoretical

TABLE C-8. *Terms Related to the Information Flow (Signals) in a Control System*

See:

Analog input	Digital-to-analog converter
Analog output	Distortion
Analog-to-digital converter	Error, aliasing
Common-mode rejection ratio (CMRR)	Error, common-mode
Common-mode voltage	Error signal
Data communications	Ground loop
Decoding network, digital-to-analog converter	Grounding
Delay time, transducer	Guard shield
Differential-mode voltage	Interference
	Modulation
	Noise
	Signal conditioning
	Signal-to-noise ratio
	Transmission, electric
	Transmission, pneumatic

TABLE C-9. *Terms Related to the Final Controlling Element*

See:

Actuator, electric	Feeder, volumetric
Actuator, electro-hydraulic	Feeder, weigh
Actuator, electro-pneumatic	Final controlling element
Actuator, hydraulic	Hysteresis
Actuator, pneumatic	Manipulator
Batching systems, automatic-weight	Metering pump
Butterfly valve	Reducing valve
Control valve	Robot
Damper	Saunders patent valve
Electric-motor drives, control of	Servo
Feeder, vibratory	Servomechanism
	Three-way valve
	Valve positioner, pneumatic

REFERENCES

Considine, D. M. (ed.): "Process Instruments and Controls Handbook," McGraw-Hill Book Company, New York, 1957.

Considine, D. M., and S. D. Ross (eds.): "Handbook of Applied Instrumentation," McGraw-Hill Book Company, New York, 1964.

Burdett, G. A. T. (ed.): "Automatic Control Handbook," George Newnes Limited, London, 1962.

Raven, Francis H.: "Automatic Control Engineering," 2d ed., McGraw-Hill Book Company, New York, 1968.

Murphy, Gordon J.: "Basic Automatic Control Theory," 2d ed., D. Van Nostrand Company, Inc., Princeton, N.J., 1966.

Bibbero, Robert J.: "Dictionary of Automatic Control," Reinhold Publishing Corporation, New York, 1960.

Klir, J., and M. Valach: "Cybernetic Modelling," D. Van Nostrand Company, Inc., Princeton, N.J., 1967.

De Roy, B. E.: "Automatic Control Theory," John Wiley & Sons, Inc., New York, 1966.

Grabbe, E. M., S. Ramo, and D. E. Wooldridge: "Handbook of Automation, Computation and Control," 3 vols., John Wiley & Sons, Inc., New York, 1958–1961.

Davies, P. O. A. L.: "Introduction to Dynamic Analysis and Automatic Control," John Wiley & Sons, Inc., New York, 1966.

Kelley, C. R.: "Manual and Automatic Control," John Wiley & Sons, Inc., New York, 1968.

Tomovic, R.: "Introduction to Nonlinear Automatic Control Systems," John Wiley & Sons, Inc., 1966.

Bower, J. L., and P. M. Schultheiss: "Introduction to the Design of Servomechanisms," John Wiley & Sons, Inc., New York, 1967.

Chestnut, H., and R. W. Mayer: "Servomechanism and Regulating System Design," John Wiley & Sons, Inc., New York, 1959.

Mesarovic, Mihajlo D.: "The Control of Multivariable Systems," The M.I.T. Press, Cambridge, Mass., 1968.

Aizerman, M. A., and F. R. Gantmacher: "Absolute Stability of Regulator Systems," Holden-Day, Inc., Publisher, San Francisco, 1964.

Ahrendt and Taplin: "Automatic Feedback Control," McGraw-Hill Book Company, New York, 1951.

Doebelin, E. O.: "Dynamic Analysis and Feedback Control," McGraw-Hill Book Company, New York, 1962.

Smith, O. J. M.: "Feedback Control Systems," McGraw-Hill Book Company, New York, 1958.

Truxal, John G.: "Control Engineers' Handbook," McGraw-Hill Book Company, New York, 1958.

Gibson, J. E.: "Nonlinear Automatic Control," McGraw-Hill Book Company, New York, 1963.

Harriott, P.: "Process Control," McGraw-Hill Book Company, New York, 1964.

—Douglas M. Considine, *Editor-in-Chief*

Control System, Closed-loop or Feedback (See Closed-loop control.)

Control System, Digital-drafting (See Drafting system, automatic.)

Control System, Linear (See Linear control system.)

Control System, Open-loop (See Open-loop control.)

Control Valve In the automatic control of fluids, the control valve is the final control element which directly changes the value of the controlled variable by varying a rate of flow. The control valve has been defined as a continuously variable orifice in a fluid-flow line. The complete control valve consists of an actuator and a valve body. The actuator provides the power to vary

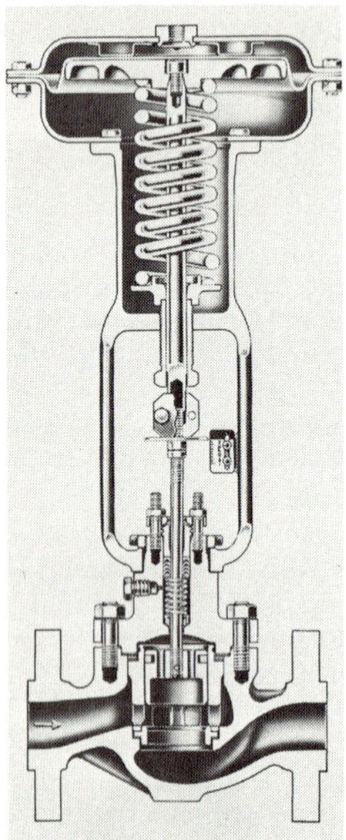

Fig. C-37. Sectional view of diaphragm-actuated control valve with cage-guided valve plug. *(Fisher Controls.)*

the port opening in the body. The valve-body assembly physically consists of a pressure-tight fitting that is screwed, flanged, or welded into a pipeline (see Figs. C-37 and C-38).

Many different control valves have been developed to meet the needs of industry. Various valve-body types and valve flow characteristics are discussed in this entry. Some special valve forms are discussed elsewhere. See **Butterfly valve; Damper; Reducing valve; Saunders patent valve;** and **Solenoid valve.** Sizing of valves is described in a separate entry. See **Control valve, sizing of.** The various forms of valve actuation are described elsewhere. See **Valve positioner, pneumatic.**

Valve-body Designs: It is the function of the valve body to regulate the rate of flow of the controlled fluid in accordance with a predetermined pattern as the valve stem or rotary shaft is moved by the actuator. To fulfill this function properly, the body must contain the fluid without leaking,

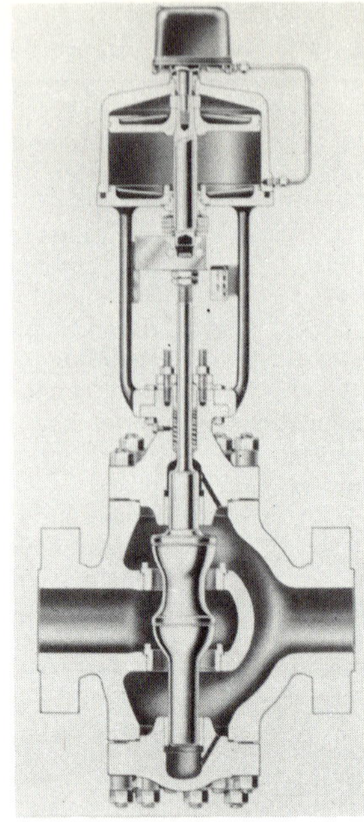

Fig. C-38. Sectional view of piston actuator on high-pressure double-port control-valve body. *(Fisher Controls.)*

must have adequate capacity or flow area, must resist erosion and corrosion, and must be fitted with a movable inner member (generally a valve plug) that can be positioned accurately in accordance with the dictates of the actuator. The predetermined pattern of flow change vs. valve-plug travel is known as the *valve characteristic.*

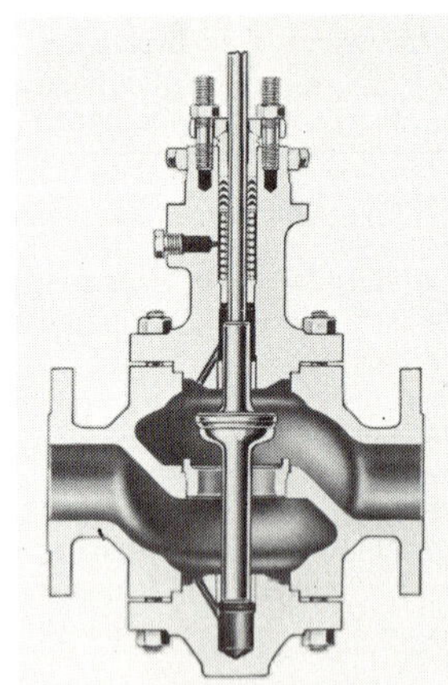

Fig. C-39. Single-port globe body with top- and bottom-guided valve plug. *(Fisher Controls.)*

When the term *automatic control valve* is mentioned, many instrument engineers immediately visualize the reversible-globe type, either single- or double-port. The term "globe body" comes from the more or less global shape of this design. By reversible is meant that the movable inner member can be "push down to close" or inverted to "push down to open." Although this body design was the standard and widely used (and still is), other designs are emerging. For example, another type of globe body with heavy-duty cage-guided valve plug and "quick-change" trim is rapidly becoming well established in industrial-process applications. A single-port globe body is shown in Fig. C-39. A control-valve body with cage-guided valve plug is shown in Fig. C-40.

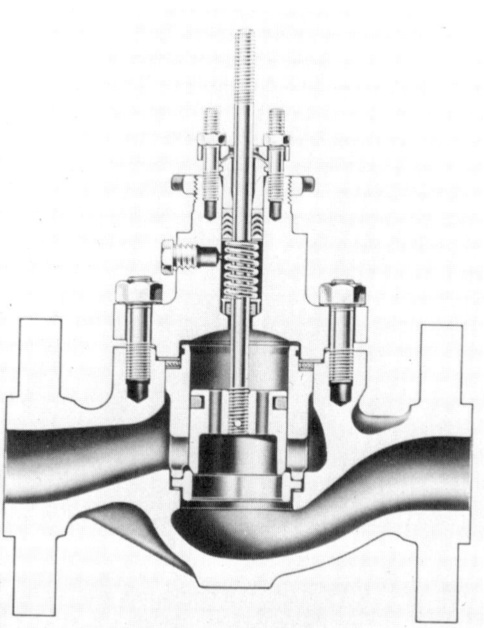

Fig. C-40. Balanced control-valve body with cage-guided valve plug. *(Fisher Controls.)*

A globe valve body with "quick-change" trim is shown in Fig. C-41.

Angle valve bodies generally are made in the single-port style and often are used where a self-draining body is required or where they facilitate piping of the valve into the system. An angle valve for high-pressure service is shown in Fig. C-42. Split-construction bodies and globe styles which feature "quick-change" trim are widely used. Separating the body halves in the split construction exposes internal body parts for inspection and replacement. Other "quick-change" trim bodies feature minimum time required to remove the trim from the body—all accomplished without taking the body out of the pipeline. Forged valve bodies for extremely high-pressure service—up to 50,000 psi—are available. However, because of the pressures involved, these valves generally have a port size of less than 1 in.

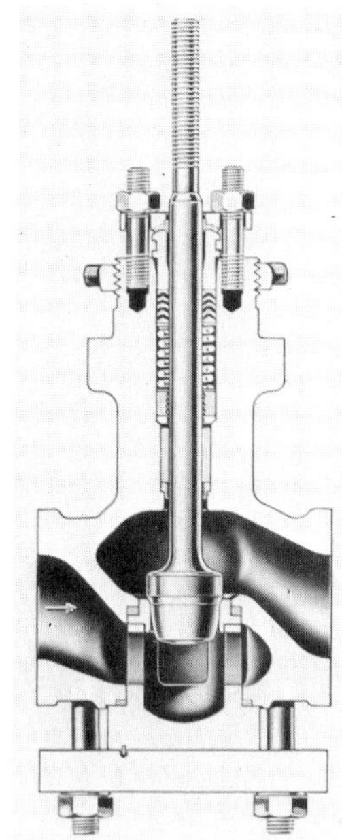

Fig. C-41. Globe-valve body with "quick-change" trim. *(Fisher Controls.)*

Three-way valves, adaptations of single- and double-port globe bodies, provide flow-splitting or -blending service. See **Three-way valve.** Butterfly valves with their rotating disk fill a need for a compact, lightweight control-valve body. Ball valves with a V-notched ball, as shown in Fig. C-43, are widely used as throttling control-valve bodies. These valves exhibit exceptionally high capacity and excellent control and rangeability because of the ball configuration. Ball valves have found wide application in the pulp and paper industry, chemical plants, and for general-processing applications.

Valve-plug Types: Valve plugs for control-valve bodies fall into one of the following categories: (1) quick-opening, (2) linear, or (3) equal-

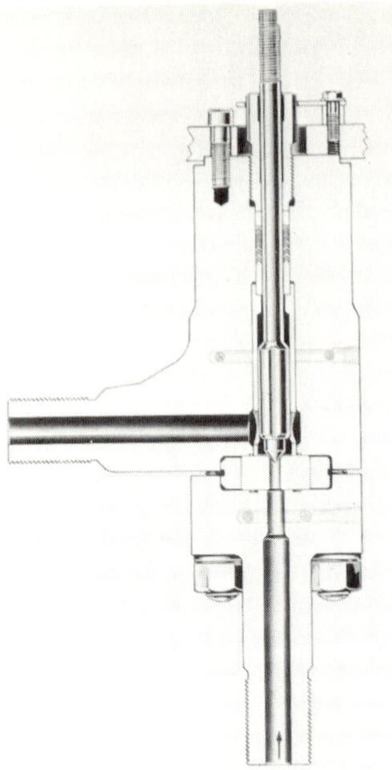

Fig. C-42. High-pressure angle control-valve body. *(Fisher Controls.)*

percentage. These terms apply basically to the flow characteristic obtained when a particular style of valve plug is installed. Valve plugs may be contoured, ported, or fluted to obtain the desired characteristic. Plugs also may be guided in the valve body with top and bottom guides, by a cage, or by skirts. Metal seating or composition seating is available on most styles. Material selection depends on the service, but for the majority of applications, austenitic chrome-nickel stainless steels have proved adequate. Valve bodies with rotating disk or ball generally have a flow characteristic that approaches equal percentage.

Valve-body End Connections: There are three common methods of installing control valves in pipelines: (1) screwed pipe threads, (2) bolted gasketed flanges, and (3) welding end connections. Screwed ends are normally female National Pipe Thread (NPT), and they generally are limited to valves not larger than 2 in. Flanged end connections on valve bodies conform to existing standards published by the American National Standards Institute (ANSI). Common ratings are 125, 200, 250, 300, 600, 900, 1,500, and 2,500 lb. Flange facings vary from a flat-face style to a raised-face or ring-type joint style. Welding end connections are either socket or butt-welding styles—with the socket style usually furnished in small sizes through 2 in. and the butt styles in sizes of 2½ in. and larger. Ultrahigh-pressure bodies utilize self-energizing end connections, such as a cone ring, lens ring, or wave ring, which becomes tighter as the pressure increases.

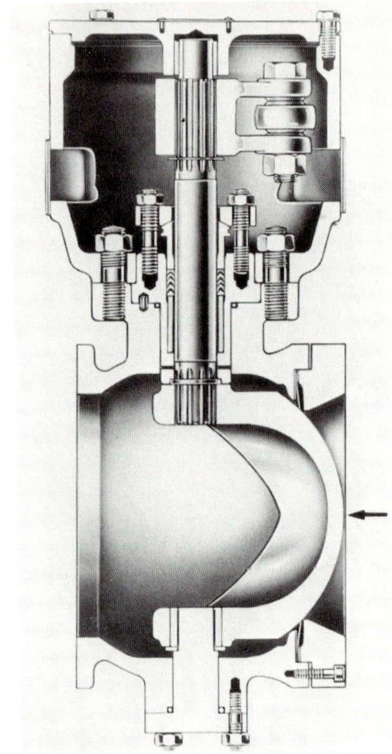

Fig. C-43. Cutaway view of V-notched valve. *(Fisher Controls.)*

Valve-body Bonnet: The bonnet of a control-valve body is that part of the body assembly through which the valve-plug stem moves. The bonnet provides a means for mounting the actuator, it houses the packing box, and it closes off one end of the body (the bottom flange is the closing member for the other end). The bonnet is available as a plain style (standard for most applications), as an extension (for cold- or hot-service temperatures); or as a bellows-seal style which provides a positive seal against leakage of the controlled fluid. Most bonnets are sealed with a gasket, but screwed bonnet joints and pressure-sealed joints have been used successfully. Control-valve body packings generally are one of the following types: (1) molded TFE *V* rings, (2) TFE-impregnated asbestos, (3) graphited asbestos, and (4) semimetallic (for high-temperature service).

Valve-body Materials: Since the majority of control-valve applications are relatively noncorrosive and at reasonable pressures and temperatures, cast iron and cast carbon steel are the most common body materials. Virtually any castable material can be used, and plastics with suitable mechanical properties or bar-stock materials are available when needed for specialized applications. Each material has its particular properties and advantages and disadvantages. The allowable pressure and temperature ratings for various valve bodies depend upon the material and construction. Most valve bodies conform to American National Standards Institute pressure-temperature ratings.

Valve Flow Characteristics: The flow characteristic of a control valve is the relationship between the flow through the valve and the valve travel as the travel is varied from 0 to 100 percent. The characteristic usually is shown as a plot of valve travel vs. flow rate—each expressed as a percent of maximum. The typical flow-characteristic curves for control valves in common use are shown in Fig. C-44. The curves represent what is known as "inherent flow characteristic," that is, the characteristic obtained with a designated constant pressure drop maintained across the valve. Inherent flow characteristic is the basic flow characteristic that is built into a given valve and is the only parameter convenient for the manufacturers to publish. Deviations from these characteristics should be expected in actual service when changes in pressure drop and other conditions are encountered. The term given to the characteristic obtained in service is "installed flow characteristic." Many factors may affect this actual relationship between valve travel and flow, with the result that the difference between inherent and installed characteristic can become quite substantial and always must be considered in the complete analysis of any flow-control system.

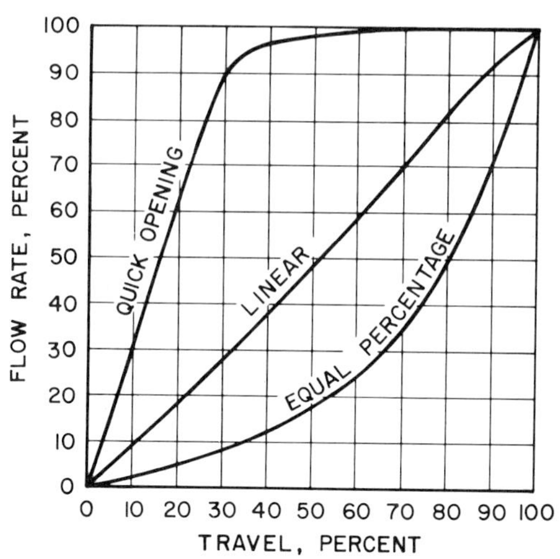

Fig. C-44. Typical flow-characteristic curves for control valves. *(Fisher Controls.)*

Valve flow charactersitics are produced in a control valve from the physical design of the valve plug itself, or from shaped openings in a cage in which the valve plug guides. A wide variety of valve-plug constructions is available, many of which exhibit the same overall flow characteristic. For example, a valve plug with a linear flow characteristic can be a contoured, lathe-turned plug; or it may be a cast plug with rectangular slots or ports in the skirt of the valve plug. A simple, needle-point valve plug for single-port valves also can be machined to provide a linear characteristic (see Figs. C-45 and C-46).

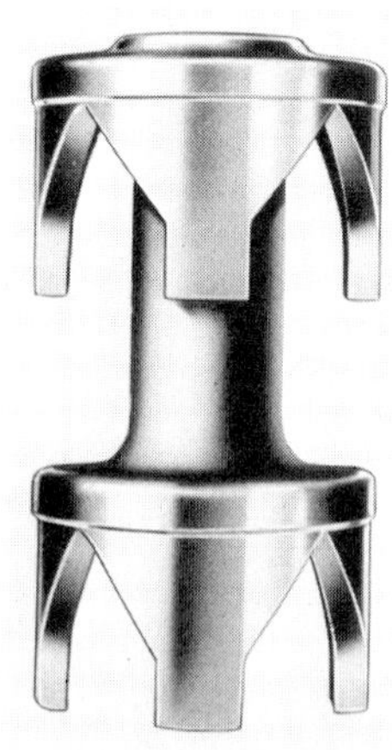

Fig. C-45. Ported double-port valve plug. *(Fisher Controls.)*

The purpose of characterizing control valves is to provide a relatively uniform control-loop stability over the expected range of system operating conditions. To establish the exact flow characteristic needed to "match" a given system requires a dynamic analysis of the control loop. But, since analyses of common processes have been performed, useful guidelines for the selection of the proper flow characteristic can be established.

The quick-opening flow characteristic provides for maximum change in flow rate at low valve travels with a fairly linear relationship, as shown in Fig. C-44. Additional increases in valve travel give sharply reduced changes in flow rate, and when the valve plug nears the wide-open position, the change in flow rate approaches zero. This characteristic cannot be defined mathematically by a first-order differential equation. In control-valve applications, the quick-opening characteristic is used primarily for on-off service, but it also is suitable for many installations where a linear characteristic normally would be specified.

Fig. C-46. Contoured double port valve plug. *(Fisher Controls.)*

The linear flow-characteristic curves show that the flow rate is directly proportional to valve travel; it plots as a straight line. This proportional relationship produces a characteristic with a constant slope, so that with constant pressure drop, the valve gain will be the same at all flows. (Valve gain is the ratio of an incremental change in flow rate to an incremental change in valve-plug position. Gain is a function of valve size and configuration, system operating conditions, and valve flow characteristic.) Expressed mathematically, the linear characteristic is

$$Q = kL$$

where Q = flow rate
L = valve travel
k = constant of proportionality

A linear flow characteristic is commonly specified for liquid-level control, and for certain flow-control applications requiring constant gain.

In the equal-percentage flow characteristic, equal increments of valve travel produce equal percentage changes in existing flow. The change in flow rate is always proportional to the flow rate existing just before the change in valve-plug position is made. When the valve plug is near its seat and the flow is small, the change in flow rate will be small; with a large flow, the change in flow rate will be large. The equal-percentage characteristic may be expressed mathematically by the following equation:

$$Q = Q_0 e^{nl}$$

where Q = flow rate
L = valve travel
e = base of natural logarithms (2.718)
Q_0 = minimum controllable flow
n = constant

Control valves with an equal-percentage flow characteristic are generally used on pressure-control applications, and on other applications where a large percentage of the pressure drop is normally absorbed by the system itself, with only a relatively small percentage available at the control valve. Equal-percentage valve plugs should also be considered where highly varying pressure conditions can be expected.

Arguments on the selection of proper operating characteristics for a control valve are far from being well defined, but guidelines can be established to help in the selection. Remember, however, that there are exceptions to any "rule of thumb" and that a positive recommendation of a particular characteristic is possible only by means of a complete dynamic analysis of the system. It should be noted that where a linear characteristic could also be used, and while the controller might have to operate on a wider proportional-band setting, the same degree of control accuracy can be expected. Here are the guidelines:

Liquid-level Control

1. With constant pressure drop, use a linear characteristic.
2. When the pressure drop decreases with an increase in load, the linear characteristic

is still the best selection except where the full-load pressure drop is less than 20 percent of the no-load pressure drop. For this exception, use the equal-percentage flow characteristic.

3. When the pressure drop increases with an increase in load, again use the linear characteristic except where this increase is greater than 2:1. For this exception, use of a quick-opening characteristic will be helpful.

Pressure Control

1. For liquid applications, use an equal-percentage flow characteristic.
2. For compressible-fluid applications, use an equal-percentage flow characteristic if the system has less than 10 ft of pipe downstream of the control valve.
3. For compressible-fluid applications, use a linear flow characteristic if the downstream system involves a receiver, distribution system, or transmission line exceeding 100 ft. If the pressure drop varies more than 5:1, an equal-percentage characteristic would be a better selection.

Flow Control

1. Choose the proper flow characteristic based on the following tabulation:

Flow Controller Measuring Element	Location of Control Valve	Flow Characteristic	
		Wide Flow Range	Small Flow Range†
With square-root extractor	In series	Linear	Equal percentage
With square-root extractor	In bypass*	Linear	Equal percentage
Without square-root extractor	In series	Linear	Equal percentage
Without square-root extractor	In bypass*	Equal percentage	Equal percentage

*When control valve closes, flow rate increases in measuring element.

†Large changes in pressure drop occur at the valve.

Other Processes

1. Exceptions are more common than the rule.
2. It is best to select an equal-percentage flow characteristic where no guidelines are available and a dynamic analysis cannot be justified.

In practice, few control loops require an optimum level of performance, and consequently, satisfactory operation can generally be obtained even with a serious mismatch between the process and the valve. Where the system has a dominant time lag, or where a wideband proportional-plus-reset controller is used, it is nearly always possible to obtain good *steady-state* control with *any* valve flow characteristic.

—Robert J. Lasher, *Fisher Controls Company*

Control Valve, Sizing of The importance of correctly sizing control valves always must be emphasized. From an economic viewpoint, an undersized valve cannot do the job for which it is intended and must be replaced. A valve that is too large costs more initially and has high maintenance costs, since the seating surfaces wear rapidly. As for operation, an oversized valve provides poor control and can cause system instability. The most expensive, sensitive, and accurate controller is of little value if the control valve cannot correct the flow properly to maintain the control point.

The basic theory of control-valve sizing may be expressed—"The flow rate of the process fluid is mathematically converted to an equivalent flow rate of a reference fluid. Then a valve size is selected which is known by test to be capable of flowing that equivalent quantity of the reference fluid at the process pressure conditions specified." For liquid flow, the reference fluid is water; for gas and vapor flow, the reference fluid is air at standard conditions of temperature and pressure.

Most manufacturers use a valve flow coefficient C_v to provide a comparatively simple method of sizing which is applicable to a wide variety of valve constructions, valve sizes, and field services. By definition, the valve flow coefficient is "the number of U.S. gallons of water per minute passing through a restriction with a pressure drop of 1 psi." This is universally accepted as a measure of valve capacity, obtained by actual flow tests conducted in accordance with accepted practices. Many valve manufacturers employ only the C_v coefficient as their capacity index for all fluids by use of suitable constants which convert the coefficient to convenient units for gases and steam. One manufacturer uses C_v coefficients for liquids only, and separate coefficients C_g and C_s for gases and steam, respectively. The reasoning is based on extensive airflow tests which have indicated that control valves of different design with the same C_v rating can have widely different airflow rates under the same pressure conditions. C_g values are obtained from actual air tests, and the C_s values are determined mathematically from the test C_g.

Flow coefficients published by manufacturers are used in formulas or in conjunction with sizing slide rules and nomographs to obtain the capacity rating of the valve, or the required valve size for a given application. Most engineers prefer the use of the nomographs or slide rule for solving sizing problems. Nomographs consist of a series of flow-data scales arranged so that—by a series of straight-line intersections—the valve size or the flow coefficient can be obtained. By adjusting the location of data scales and spacing of scale markings, nomographs can include various corrections that generally are impractical in sizing formulas. Slide rules solve the various sizing formulas directly, and they too contain corrections for specific gravity, temperature, and superheat.

—Robert J. Lasher, *Fisher Controls Company*

Controlled-volume Pump (See **Metering pump.**)

Controller Setting (See **Process reaction rate**; and **Response, transient.**)

Conversion Time In general, the conversion time is the length of time between the initiation and completion of a single analog-to-digital or digital-to-analog conversion operation (see **Analog-to-digital converter; Digital-to-analog converter**). It is therefore the reciprocal of the conversion rate. In popular usage, however, the term is usually used only in reference to analog-to-digital converters or digital voltmeters. In an analog-to-digital converter, the conversion time consists of the time necessary to reset and condition the logic, a delay to allow for settling time of the input buffer amplifier, a polarity-determination time, the actual analog-to-digital conversion operation, and any time necessary to transfer the digital result into an output register. Depending on the method of analog-to-digital conversion, one or more of these operations may not be included. For example, in a unipolar analog-to-digital converter, there is no polarity-determination time.

The term conversion time is also used in reference to data-acquisition or analog-input subsystems for process control computers (see **Analog input**). A more precise term for this application is *measurement time*. The measurement rate may not be the reciprocal of the measurement time, since some of the operations performed may be overlapped with other operations. For example, when some types of analog-to-digital converters are used, it is possible to select the next multiplexer point during the time that the previous value is being converted from analog to digital form.

In addition to the functions performed during the conversion time as defined in the first paragraph, the conversion time (or measurement time) for an analog-input subsystem may include some of the following operations: multiplexer address decoding, multiplexer switch-selection and settling time, amplifier settling time, range-selection and settling time, and the time necessary to deactivate the multiplexer switches and allow the subsystem to return to an initial state.

—Thomas J. Harrison, *IBM Corporation*

Converter, Analog-to-Digital (See **Analog-to-digital converter.**)

Converter, Digital-to-Analog (See **Digital-to-analog converter.**)

Conveyor Scale (See **Scale, industrial.**)

Coordinate-measuring Machine (See **Measuring machine.**)

Copper Resistance Thermometer (See **Resistance thermometry.**)

Core memory A computer storage system in which thousands of tiny doughnut-shaped magnetic cores each store one bit of information. Current pulses are sent through wires threading through the cores to record or read out data.

Coronagraph For observation of the sun's inner corona, the solar image is projected onto an occulting disk. The light from the corona is less than the solar light by a factor of 10^6, so that scattered light must be very well suppressed. This has resulted in the design shown in Fig C-47, which uses a single lens as an objective and uses stops and a conical shape for the disk to eliminate as much scattered light as possible. The strong emission of certain spectral lines allows the instrument to be used monochromatically. Coronagraphs are mounted at high-elevation sites

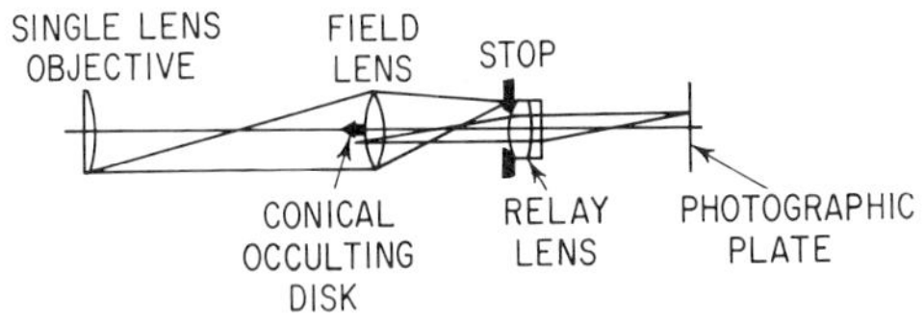

Fig. C-47. Coronagraph.

to minimize atmospheric scattering, and quite special techniques are employed to produce extremely smooth optical surfaces for these instruments.

Also see **Solar telescope;** and **Spectroheliograph.**

—Joost H. Kiewiet de Jonge, *University of Pittsburgh,* and Lloyd M. Geertz, *Owens-Illinois*

Coudé Telescope (See **Telescope, reflecting.**)

Coulometer A device (also coulombmeter) for the measurement of electric current. The *silver coulometer,* originally developed by the National Bureau of Standards in 1916, comprises a small platinum vessel (which acts as the cathode) into which is immersed a pure silver anode that is surrounded by a porous cup. The electrolyte is a high-purity aqueous solution of silver nitrate (15 percent $AgNO_3$ by weight). The quantity of silver deposited is weighed and the time required to deposit the silver is duly noted. From these measurements, the average current strength can be calculated. *A coulomb,* named after the French physicist C. A. Coulomb (1736–1806), generally is defined as one ampere flowing for one second. Prior to 1950, the *international coulomb* was the legal standard in the United States. It was defined as the quantity of electric current which, under specified environmental and apparatus conditions, would deposit 0.00111800 g of silver from an aqueous solution of silver nitrate. Since 1950, the legal U.S. standard has been the *absolute coulomb,* which can be defined as the quantity of electrical charge that crosses a surface in a period of one second if a steady electric current of one absolute ampere is flowing across the surface. One international coulomb is equivalent to 0.999835 absolute coulomb. A *copper coulometer,* comprising copper electrodes in a bath of copper sulfate, also is used. One coulomb deposits 0.000329 g of copper. Both the silver and copper coulometers described are known as *weight coulometers.* In a *volume coulometer,* the volume of a gas, such as hydrogen or oxygen liberated by electrolysis, is measured, and from it the current can be calculated. The effect of current flow on electrolyte concentration also can be determined by titrating the electrolyte after electrolysis. This type of device is known as a *titration coulometer.* In some cases, a coulometer may be referred to as a voltameter.

Coulometry A method of chemical analysis performed by determining the amount of a material released during electrolysis by measuring the coulombs required. The method is very useful for analysis of the following ions and with a possible sensitivity of 1 ppb: Cl_2, Br_2, I_2, OH^-, H^+, Ce^{4+}, Fe^{++}, Cr^{++}, Sn^{++}, Ag^+, Cu^+, and ferrocyanide. The upper concentration range of the method is limited by solution heating and gas evolution. Identification of unknowns is made by choice of potential applied to the electrodes to cause oxidation or reduction of the material sought. Quantity is determined by the amount of current required to support electrode reaction. In a coulometric analyzer or titrator, the oxidizable or reducible material sought is reacted with a coulometrically generated reagent to the end point (either batch or continuously) controlled by an amperometric or potentiometric detector. The amount of current required for generation of the reagent is related directly to the amount of oxidizable or reducible material in the sample.

Countdown The step-by-step process of preparing a missile or spacecraft for launching. In a broad sense, any methodical, sequential procedure for passively checking out a system prior to actual operation of the equipment.

Counter An instrument that is used to count functions or events occurring in a machine or system to which it is connected and to display the accumulated count.

Mechanical Counter: These devices are termed *mechanical* because of the means, such as a shaft, linkages, gears, or belts, used to connect the counter with the system or apparatus that is being counted or controlled. These counters may have electrical outputs to trigger control functions, but their inputs always are mechanical.

Although there are variants, the typical mechanical counter consists of a row of small wheels, each painted with the numbers 0 through 9, mounted on a common shaft and connected together by mutilated-pinion gears or geneva wheels—or other intermittent-motion mechanisms—as shown in Fig. C-48. The first, or lowest-order wheel, is driven by an input shaft or equivalent. As the first wheel completes the last tenth of its first turn, a single gear tooth on its output side engages the pinion gear between the first and second wheels. The pinion is engaged with a continuous gear on the input side of the second wheel. Thus, the second wheel follows the first—at a one-to-one speed ratio—for 36°, or one-tenth of a revolution. Eventually, the first two wheels pick up the third—then the first three wheels pick up the fourth, and so on—as the counter goes from 99 to 100, from 999 to 1,000, and so on.

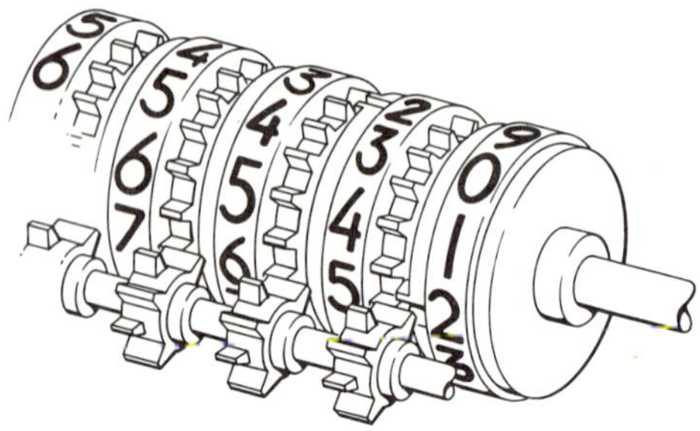

Fig. C-48. Mechanical counter showing wheels and intermittent gearing.

Mechanical counters are designed to accept one of several basic types of input motion. The most common is continuous rotation produced when the counter is connected directly or through gears or some like mechanism to a rotating shaft, or when it is connected by friction roller to a translating surface, such as a slide or belt. Most *rotation counters* are bidirectional; that is, they add counts if rotated in one direction and subtract counts if the rotation is reversed. These devices are termed *direct-drive counters.* One rotation of the input shaft can produce a change of 1 count (one-tenth of a revolution of the first wheel) or 10 counts (full revolution) in the counter. A partial revolution will produce a partial change in either case. The first wheel of a direct-drive counter thus is really an analog indicator. Succeeding wheels move intermittently and so have better claim to the term *digital.*

The second most common input is a reciprocating linkage or lever which drives a ratchet mechanism inside the counter to move the display wheels in a single direction. Bidirectional ratchet counters can be built, but standard catalog counters will only count up or only count down. The first wheel in a ratchet-driven counter advances one full unit (usually one count) for each cycle of the input linkage.

Another common type of ratchet counter looks like a direct-drive counter, but with a shaft instead of a lever for the input. Each rotation of the shaft drives a ratchet mechanism inside the counter and advances the first wheel by one full unit. These devices are termed *rotary-ratchet* counters. Rotation of the shaft in either direction causes the counter to count in one direction (up or down). Another, less common mechanical counter has a detented rotary input to which usually is connected a *star wheel.* The latter appears as a wagon wheel without a rim. The counter is placed near a conveyor belt that may be carrying large objects, such as bottles or boxes. As these objects hit the spokes of the rimless star wheel, they move the counter's input shaft from one detented position to the next. Reversing the flow will reverse the direction of count from up to down.

Mechanical counters are obtainable in a wide range of sizes and shapes. Generally, the smallest available are five-digit counters less than an inch long. The largest counters that are made in any quantity are the familiar counters used on gasoline pumps. A variety of count speeds are available: 0 to 1,000 counts per minute is a common rating for a ratchet counter—and up to 0 to 30,000 counts per minute is possible with a direct drive. The highest-speed counters have a geneva wheel instead of a pinion gear between the lowest-order wheels to reduce impact shocks. Life of mechanical counters is a function of the speed at which they are driven, but they can achieve tens or hundreds of millions of counts if their speed rating is not exceeded. Mechanical counters generally have 4, 5, or 6 digits with count capacities of 9,999; 99,999; and 999,999, respectively. Once the counter reaches its top limit, it simply will start all over from zero. The torque required to drive a mechanical counter varies from a fraction of an inch-ounce for small and military counters to one inch-pound or more for very large counters with built-in control functions.

Counters can be calibrated in a number of ways. They can count by fives, by ones, by tens, by twos, and so on. The first wheel can transfer one unit to the second wheel after 10 counts (most common), or after 12 counts, or 5 counts, or 8 counts, and so on. In various forms, therefore, the counter can be used to keep track of degrees, hours, minutes, dollars, bullets, pounds sterling, or boxes of thumbtacks.

Mechanical counters are available with a number of special features. Certain designs will produce an electrical signal after they have reached a preselected count. The operator selects this count by manually positioning the counter wheels (or a group of secondary wheels) before starting the machine or system to which the counter is connected. When the machine has gone the desired distance, so to speak, the counter will turn on a light, sound a bell, operate a recorder, or shut off the machine. Such devices are called *preset* or *predetermining counters.*

Some counters can be read electrically—periodically or continuously. Usually this is accomplished by mounting a small electrical brush on each counter wheel and running this brush over a printed circuit that is sandwiched between each pair of wheels. These devices are called *readout counters* and can be obtained with binary or decimal outputs. Most mechanical counters can be obtained in *reset* or *nonreset* versions. A reset counter can be stopped and returned to zero whenever the operator wishes. A nonreset device must be counted ahead or back down to zero (direct-drive types) or forward to zero (ratchet types). Mechanical counters also are available with wheels that print instead of display the accumulated count. Generally, these types are used as an integral part of office machinery.

Mechanical counters are used primarily to perform control or accounting functions. They add or subtract, integrate, remember, accumulate, record, calculate, and display.

Electrical Counters: These devices are driven by electrical pulses instead of by rotating shafts or gears. However, the internal mechanical counter elements still are used. Electrical counters can produce electrical outputs for controls of other devices.

Several drive systems are used for electrical counters. The electromechanically driven ratchet mechanism is one inexpensive and widely used drive. An electromechanical solenoid or relay moves a lever on which is mounted a spring-loaded pawl. This engages a ratchet wheel on the first wheel of the counter and thus indexes the counter 36° for every electrical pulse delivered to the counter. A similar, but more widely used form of electrical counter is shown in Fig. C-49. A solenoid or relay (relay shown in illustration) converts the electrical input pulse to mechanical motion. In this case, an inverse-escapement mechanism is used instead of a ratchet mechanism to produce the desired intermittent motion. This drive will index the first counter wheel 18° when the electrical pulse is received and the relay driver is turned on. It will index the wheel a second 18° when the pulse is turned off and a spring returns the escapement lever to the rest position. The inverse escapement is widely used because this is a simple drive mechanism and eliminates the needs for nonoverthrow and no-back pawls required by most ratchet drives. As Fig. C-49 indicates, there is no difference between electrical and mechanical counters except the driving means.

A more sophisticated, but much faster type of electrical counter uses an electromagnetic stepping motor as the input drive means. As shown in Table C-10, such counters typically operate at four or five times the speed of ratchet or inverse-escapement counters. An advantage of the stepping motor is elimination of the high impact forces and accelerations commonly encountered with cruder intermittent-motion mechanisms, such as ratchets and escapements. Thus, stepping-motor counters have a longer useful life than the other, less costly designs.

A less popular type of high-performance electrical counter uses a synchronous motor and high-speed clutch as the driving means. The motor runs continuously. Each input count signal triggers the clutch, which usually is a one-turn or half-turn device connected to the first counter wheel through suitable reduction gearing. The first wheel turns 36° each time the clutch is actuated.

Electrical counters can be obtained with most of the features previously described for mechanical counters. Add-subtract counters generally have two separate relay-drive mechanisms, one of which is energized to add counts, the other to subtract counts. In electrical reset counters, an additional relay-powered mechanism or reset motor is used to return the wheels to zero.

Pulses used to drive electrical counters can be generated by several kinds of devices, including switches, photoheads, magnetic pickups, relays, and other electronic-circuitry components. Pulse shapes can vary quite drastically without seriously affecting the performance of the counter. The relatively slow motion mechanical-drive

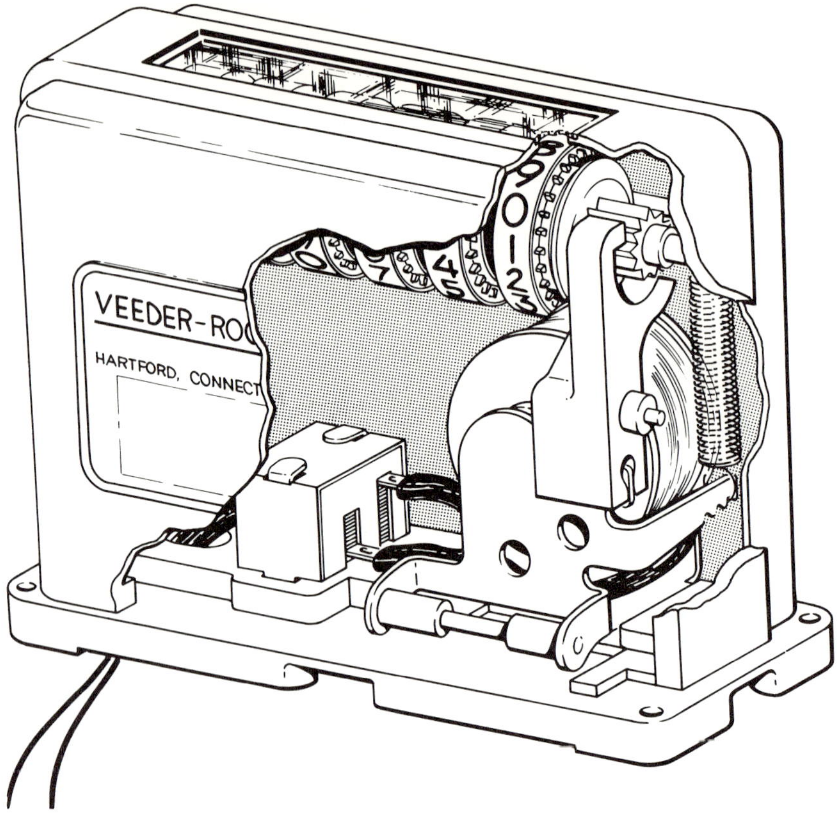

Fig. C-49. Inverse-escapement drive for an electrical counter.

system acts as a mechanical filter to ignore high-frequency noise, ripple, contact bounce, and most other system noise. Thus, fairly crude drive circuits or switches can be used for inputting an electrical counter.

Electrical counters require a fairly high power input as compared with electronic counters. The smallest electrical counter will consume 1 or 2 V-A; the largest 8 V-A or more. Since solenoids, ratchets, and motors all are inductive devices, some care must be taken to protect switches from inductively produced arcs and back spikes in counter-drive circuits. Circuits commonly used are identical to those used with electrical relays. *RC* networks, diodes, zener diodes, and neon bulbs are commonly used. These circuits are designed to slow down the dissipation of energy stored in the counter coil when the circuit is opened. However, this reduces the maximum counting speed obtainable. In critical systems, forcing circuits can be used to obtain some increase in counter speed.

Electrical, Single-wheel Counter: This is the most modern and versatile type of electrical counter and differs from the electrical counters just described in that each count wheel has its own electrical driver and produces an electrical drive pulse for succeeding wheels. Most single-wheel counters are provided with electrical readout and electrical reset. Many also are provided with mechanical preset. A pushbutton mechanism thus can be used to position the wheel in a desired position at the beginning of the count cycle. The counter thus will produce an electrical signal each time it reaches this preselected position.

Because each wheel in a multidigit single-wheel

TABLE C-10. *Electrical-counter Characteristics*

Drive Type	No. of Wheels	Approx. Lengths Available, in.	Max Speeds, counts per min	Estimated Max Life, millions of counts
Ratchet.	3–7	1½ by 5	3,000	4–200
Inverse escapement	3–7	1½ by 5	6,000	4–200
Step motor.	6	3 by 5	30,000	500–1,000
Motor and clutch. .	6	7	15,000	50–100

counter is a separate and self-contained counter, it will perform many functions not possible with other multiwheel electrical counters. For example, the wheels of a single-wheel counter system can be driven in parallel. This is very useful in money counting, for example, where a penny can be counted by delivering one pulse to the first digit, a dime counted by sending one pulse to the second digit (rather than 10 pulses to the first digit), and a dollar by sending one pulse to the third digit (rather than 100 pulses to the first). Another advantage—the wheels can be reset in parallel. Because each single wheel is furnished with a readout system, the wheel can be separately homed to a position that has been selected by a remote-switch means, such as decade switches, punched cards, tapes, and computer inputs.

Many logic and control-circuit functions also can be performed with single-wheel counters wherein they function somewhat like stepping switches in logic systems. Shift registers, parallel adders, add-subtract systems, dwell-cycle control systems, and integrators—to name a few—all have been built using this versatile counter. These systems are not as fast as their electronic counterparts, but they are less costly, they continuously indicate their status, and they remember their condition even during power failure. Because of these characteristics, these counters are finding growing application in digital and numerical control systems. The construction details of a typical single-wheel counter are shown in Fig. C-50.

Electronic Counters: These devices use electronic circuits instead of mechanical wheels and gears to count and to remember a series of events. They perform the same basic functions as those of the counters previously described but also are frequently packaged with other electronic circuits to perform related, but noncounting functions, such as timing, gating, and computing. But, in their basic configuration, they simply count.

Electronic counters are more expensive than mechanical or electrical types, but they are able to count at much higher speeds and for all practical purposes do not wear out. Speeds in the tens of thousands of counts per second are easily achieved (compared with 500 counts per second for the best electrical counters) with count life between servicing in the billions of counts. Thus, they are widely used for computer circuits, for counting atomic events, and for very high-speed packaging and conveyor counting applications.

Since it is more costly to count electronically than electromechanically, a number of counter manufacturers package one or more electronic counting decades with a standard five-digit electrical counter and appropriate interface circuits. The electronic circuits handle the least significant digits at high speed. For every 10, 100, or 1,000 counts, the electronic counter feeds one pulse (stretched) to the electrical counter, which operates in the conventional manner at slower speeds.

Generally, electronic counters do not have a built-in count display capability as do mechanical counters where the mechanism used to count (wheels and gears) easily can be marked to display the accumulated count. Although some electronic counters are used without a display means of any sort, the generally available catalog counter uses an electromechanical or electronic indicator of some type. The neon-filled 10-cathode glow-discharge tube is commonly used. It should be noted that the indicator does not have to follow the counter accurately so long as it appears to be counting when it is connected and will rapidly lock onto the correct count whenever the count rate drops to a speed low enough to be read by the operator (between 200 and 300 counts per minute). Thus, relatively slow-speed electromechanical devices can be used as indicators for very high-speed electronic counters.

An exception is the neon-bulb ring counter in

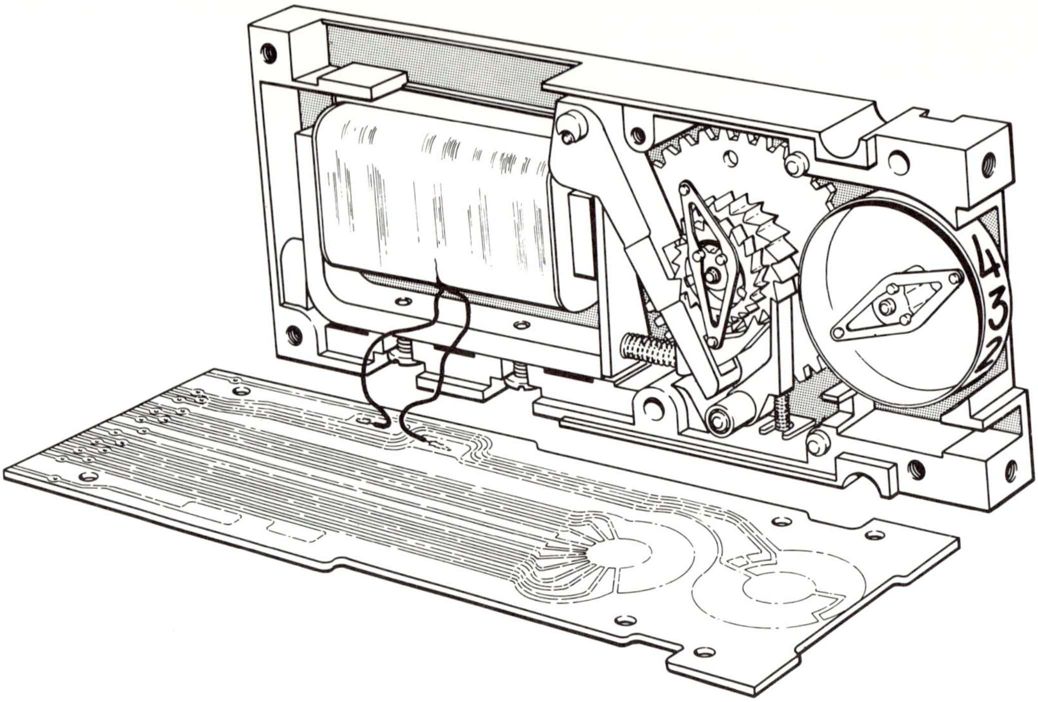

Fig. C-50. Single-wheel electrical counter.

which the same component acts as a memory and display device. This earlier form of electronic counter has lost popularity because of its relatively slow speed (1,000 counts per second) and because multicathode glow-discharge tubes are easier to read. In a neon-ring counter, 10 small neon bulbs act as latching relays and indicators for 10 flip-flop circuits connected in a continuous series or ring. As the first pulse is fed to the counter, the neon bulb associated with 1 lights. The second pulse extinguishes bulb 1 and lights 2 as the first and second flip-flops respond to the count. Every 10 pulses, a single pulse is fed to a succeeding ring—and so on.

Electronic counters are driven by electrical pulses. The pulse requirements differ widely depending upon the amplification and/or pulse-shaping circuits provided by the manufacturer. Low-amplitude and extremely short pulse times (micro or nanoseconds) are acceptable inputs. However, the electronic counter is quite intolerant of electrical noise. Because it is capable of counting at high speeds, it easily can count spikes, grass, ripple, and the like. Thus, switches, pulse generators, and associated transducers must be selected carefully. Any contact bounce or equivalent must be suppressed or filtered before the pulse train is fed to the counter. Also, it is necessary to limit the rise time of the input circuit to the maximum useful count speed to avoid extraneous counts.

Common pulse sources include photoelectric devices of all types, high-speed relays, a wide variety of electronic circuits, magnetic transducers, and capacitative transducers—with photocells the most common source. One frequently overlooked advantage of electronic counters is their noiseless operation, a definite advantage for installation in office and laboratory environments. A disadvantage of electronic counters is memory failure when electrical power is lost. Neither the counting circuits nor the indicators commonly used have any way to remember their condition prior to power failure unless, of course, sophisticated magnetic-core memories are used.

Unlike other counters, the electronic counter often is designed to produce electrical outputs at more than one predetermined count—with 1, 2, and 4 outputs being common. Presetting is done by manual adjustment of 10-position switches in most of the standard catalog counters, but punched cards, punched tapes, and commutators can be used. For example, in the manufacture of customized transformers, the coil-winding machine can be controlled (number of turns to first tap, and so on) with a punched card or computer in conjunction with an electronic counter.

Resetting is accomplished by feeding the counter a single pulse. Reset and count pulses, of course, enter the counter at different points. Reset usually is accomplished in the time required for one count so that an electronic counter (unlike electromechanical counters) can be reset while counting at full speed—without internal damage and often without missing a count.

Electronic counters often are packaged with other circuits to perform related functions. For example, gated counters will produce a train of counted pulses, shutting off the pulses at a preselected count. Usually, the pulses can be produced at varying rates; the counters contain a variable-rate pulse generator, in other words. Another common circuit develops a clock or timing pulse train, one pulse precisely every second or millisecond or microsecond. Counters with this type of circuit serve as accurate digital clocks and again can be preset to produce electrical outputs or stop the pulse train at a predetermined count.

REFERENCES

Burton, Allan L.: Counters and Digital Indicating Devices, in D. M. Considine and S. D. Ross (eds.), "Handbook of Applied Instrumentation," pp. 11–103 to 11–117, McGraw-Hill Book Company, New York, 1964.

Woodson, Wesley E.: "Human Engineering Guide for Equipment Designers," University of California Press, Berkeley, Calif., 1954.

Mate, George: Counters Applied to Measurement and Control, *Process Control and Automation*, 1961.

Temple, L. D.: How to Pick the Right Counter, *Prod. Eng.*, Apr. 17, 1961.

Strong, John: "Procedures in Experimental Physics," chap. 7 (Geiger Counters), Prentice-Hall, Inc., Englewood Cliffs, N.J., 1939.

Frank, M. E., and S. T. Schy: Counting on a Magnetic Drum, *Control Eng.*, vol. 8, no. 10, pp. 75–79, October, 1961.

Giulie, J. D.: Counting to Control Weight Fill, *Automation*, vol. 9, no. 9, pp. 85–88, September, 1962.

Ponstingl, J. C.: Timing Devices for Automatic Control, *Automation*, vol. 7, no. 11, pp. 77–81, November, 1960.

Everett, J. E.: Counting for High Speed Control, *Automation*, vol. 7, no. 6, pp. 92–96, June, 1960.

Tanzman, H. D.: High Accuracy Time Interval Measurements, *Electron. Ind.*, January, 1959, pp. 62–67.

McCready, E. A.: Electronics Does the Counting, *Radio-Electron.*, vol. 30, no. 10, pp. 102–104, October, 1959.

Russell, R. W.: Applying Electronic Counters, *Automation*, vol. 7, no. 4, p. 80, April, 1960.

Bickford, J. H.: Economic Numerical Control, *Instr. Control Systems*, December, 1966.

Bickford, J. H.: Mechanisms for Intermittent Rotary Motion, *Machine Design*, Dec. 23, 1965.

Gomez, R. S.: Contact Protection Studies for Miniature Dry Reed Switches, 13th Annual National Relay Conference and 2d International Conference on Electromagnetic Relays, Oklahoma State University, April, 1965.

Dawirs, Harvel N., Thomas M. Fisher, and Donald M. Palmer: Study to Establish Transient Suppression Techniques and Criteria, *RECON Tech. Rept.* NS22, Marshall Space Flight Center, Huntsville, Ala., August, 1965.

—John H. Bickford and William Kes,
Veeder-Root Company

Counter (Computer Storage) A counter is a storage element capable of maintaining numeric values which can be incremented or decremented by the value of another number.

A counter may be a location in storage (see **Storage, computer**) which is incremented or decremented under the control of the program (see **Program, computer**). This type of counter can be used to record the number of times a program loop has been executed, for example. The counter location is set to the value of the number of times the sequence of instructions is to be performed (see **Instruction**). At the end of the sequence of instructions, the counter is decremented by 1 and then tested for zero. If the counter is nonzero, the sequence of instructions is repeated. If the counter is zero, the program advances to the next instruction.

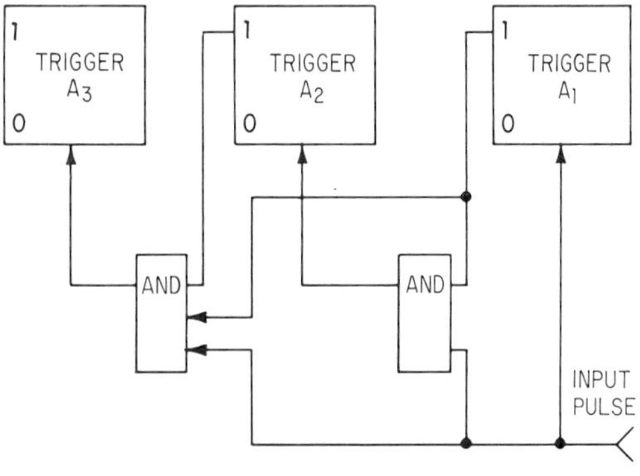

Fig. C-51. Binary counter.

A counter may also be a circuit that records the number of times an event takes place, such as the counter in Fig. C-51, which counts according to the binary number system (see **Number systems**). This binary counter is a three-stage counter capable of counting up to eight pulses. Each trigger (see **Flip-flop**) changes state when a pulse is gated to its input (see **Gate**). In the case of trigger A_2, it receives an input pulse only when trigger A_1 and the input are both 1 and, subsequently, trigger A_3 changes its state only when triggers A_1 and A_2 and the input pulse are all 1s. The truth table in Fig. C-52 shows the value of each trigger after each input pulse. At the eighth pulse, the counter resets to zero.

	A_3	A_2	A_1
P1	0	0	1
P2	0	1	0
P3	0	1	1
P4	1	0	0
P5	1	0	1
P6	1	1	0
P7	1	1	1
P8	0	0	0

Fig. C-52. Truth table for binary counter.

Counters may be used to accumulate the number of times an external event takes place. In this case the counter, which may be a storage location or a counter circuit, is incremented as the result of an external stimulus.

—W. W. Welz, *IBM Corporation*

Counter, Radiation [See **Gas-ionization tube; Geiger counter; Ionization chamber; Proportional counter; Scintillation counter;** and **Radiation (nuclear), detection of.**]

Counting Scale (See **Scale, counting.**)

CPU [See **Central processing unit (CPU).**]

Crane Scale (See **Scale, industrial.**)

Crayon-type Temperature Indicator [See **Temperature-sensitive materials (indicators).**]

Creep Testing [See **Testing (materials), destructive.**]

Critical-edge Spectrometer, X-ray (See **X-ray analysis.**)

Crossbar Switches These are matrix-switching devices used to connect any one of a group of inputs to any one of a group of outputs (see Fig. C-53). This is done by energizing two magnets; one selects an input row, and the second selects an intersecting output column. By this action, a set of contacts is closed, connecting input to output. While these switches are used principally in telephone switching, they are adaptable for many testing and control functions. They do the same tasks as those done by rotary switches or step-by-step switches, but their advantage lies in being able to select any input-to-output connection with only two magnet operations.

Crystal Oscillator The most precise *mechanical* resonator and frequency generator available today is the crystal oscillator. The quest for a stable, accurate, and low-cost frequency generator for the precise control of commercial radio and other higher communications frequencies led to the development of piezoelectric resonators—notably quartz crystals for a wide variety of applications. See **Frequency generator.** Also, the quartz crystal has become highly developed as a frequency standard for timekeeping and as a time-signal generator. See **Time and time measurement.**

Piezoelectricity was discovered by the Curie brothers in 1880. The term *piezo* is derived from the Greek word meaning "to press." The effect causes a crystal to exhibit electrical polarity when the crystal is subjected to mechanical pressure. Conversely, the crystal is physically deformed when subjected to an electrical potential. Specifically, piezoelectricity is a property of nonconducting solids that have a crystal lattice which does not have a center of symmetry.

Quartz, tourmaline, and rochelle salts and such synthetic crystals as ethylene diamine tartrate (EDT), dipotassium tartrate (DKT), and ammonium dihydrogen phosphate (ADP) have varying suitability as piezoelectric elements. Tourmaline, an expensive material used mainly in hydrostatic pressure-measuring devices, is more durable than quartz and, for a given frequency, normally is more rugged than quartz. Rochelle salt has a greater piezoelectric effect than any other crystal but has the disadvantage of a greater sensitivity to temperature change than quartz. EDT has an advantage over quartz when

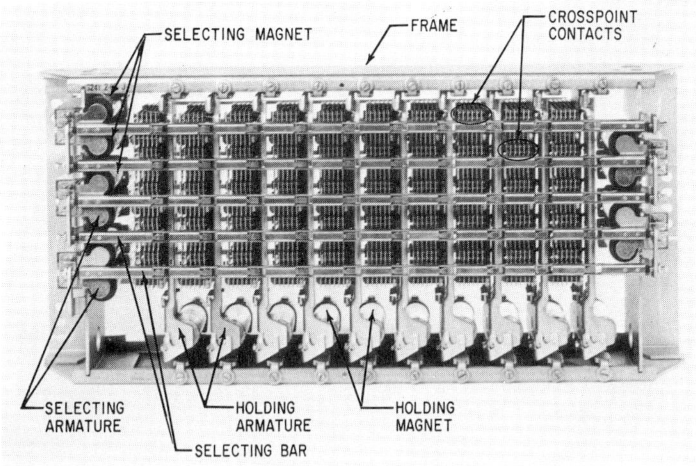

Fig. C-53. Crossbar switch.

used in frequency-modulated oscillators because of the wide gap between its resonant and anti-resonant frequencies.

All piezoelectric crystals should have a good temperature coefficient, that is, should show as little change in resonant frequency as possible under large variations in temperature. Ideally, the piezoelectric constant of proportionality between the mechanical and electrical variables must be the same for both direct (pressure-to-electricity) and converse effects.

Natural quartz crystal and new synthetic quartz crystal have been found best to meet the properties required of a piezoelectric element. Quartz is hard (7 on Mohs' scale as compared with a diamond, which is 10) and is relatively abundant and stable. Because of the anisotropic structure of quartz, cutting a crystal in different orientations makes it possible to obtain crystals for the widest range of applications, including frequency control, filters, resonators, and electromechanical transducers. Each of these uses depends on the orientation of the crystal cut with respect to the crystallographic axes. As shown in Fig. C-54, the principal axes in quartz are identified as the optic *(Z)*, the mechanical *(Y)*, and the electrical *(X)* axes. By use of polarized light and x-ray diffraction techniques, the various axes may be properly located, permitting a crystal plate to be cut from the quartz with the performance characteristics desired. After cutting and extensive processing, the crystal plate is mounted at the nodal points of its normal vibration. These points, which also serve as electrical connecting points, allow the crystal to vibrate freely with a minimum of damping. Finally, the mounted crystal is hermetically sealed in a dry inert atmosphere within a crystal holder.

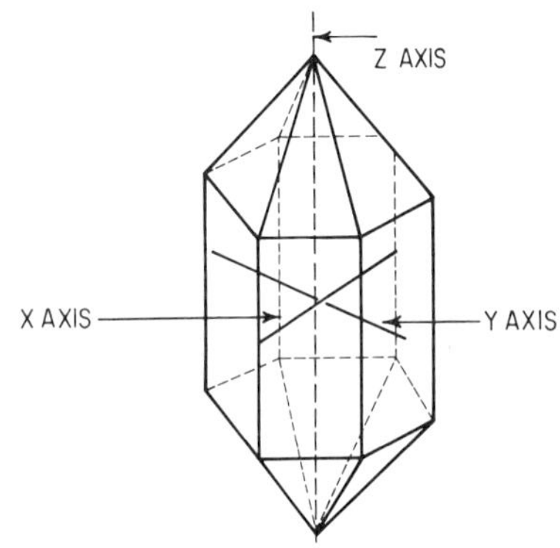

Fig. C-54. Principal axes in a quartz crystal.

A stability of 1 part in a billion is routinely possible by the use of proportional ovens that can regulate the crystal temperature within $\pm 0.001°$ C. By employing multiplying or binary techniques, frequency-generating devices in the range of from less than 1 Hz to more than 600 MHz are available. In the temperature-compensated crystal oscillator (TCXO), stability over a wide range of temperatures is obtained by a compensating network that shifts the frequency of the crystal by an amount approximately equal and opposite to the shift of frequency caused by the temperature change. This network eliminates the need for an oven. TCXOs are available with temperature stabilities of ±0.5 ppm over a temperature range of minus 40 to plus 70° C.

REFERENCES

Mason, W. P.: "Quartz Crystals for Electric Circuits," D. Van Nostrand Company, Inc., Princeton, N.J., 1946.

Edson, W.: "Vacuum-tube Oscillators," John Wiley & Sons, Inc., New York, 1953.

Firth, D.: "Quartz-crystal Oscillator Circuit Design Handbook," Department of the Army, Contract DA 36-039-AMC-00043E, 1965.

—William O. Bennett, John J. Carpenter, Frank Dostal, and E. Van Haaften, *Bulova Watch Company, Inc.*

Crystallography, X-ray (See X-ray analysis.)

Cup Anemometer (See Anemometer.)

Current (Electric), Measurement of (See Ammeter; Coulometer; and Electrical quantities, measurement of.)

Current Divider (See Bridge circuits, electric.)

Current Meter, Ocean Measuring ocean currents is one of the most basic measurements in the field of oceanography. There is a wide assortment of methods, ranging from simple drift bottles to highly sophisticated instruments. Current meters are valuable in the study of rivers, lakes, estuaries, and, in fact, any body of water where water movement is of interest. Other uses for current meters include determining the velocity of ships and submersibles.

Two parameters normally are measured—*current speed,* read in knots or meters per second; and *current direction,* read in degrees.

Accurate current measurements are made with instruments using an impeller or savonious rotor. The rotor speed is determined by the number of times that uniformly spaced magnets, located on the rotor, pass a magnetic-reed switch. The electrical-contact closures are recorded as a function of time. Accuracies of savonious-rotor current meters are within ±1 to ±3 percent of full-scale reading. Current meters are designed for depths in excess of 10,000 ft. Direction sensors normally include a magnetic compass as part of their construction so that readings are relative to magnetic north.

Current meters are designed as *remote-reading* instruments using cables from the sensors to readout modules; or as *in situ* instruments incorporating self-contained data-recording systems capable of recording unattended for long periods of time. More complex systems include analog-to-digital conversion which provides digital outputs in standard or complex formats for data processing (see Figs. C-55 and C-56).

Antifouling paints are used to protect the exposed parts of current meters because of the

Fig. C-55. Remote-reading ocean-current meter showing current-direction and current-speed receiver/recorders in upper portion of photograph. Direction sensor is shown at left; current-speed sensor is shown at right. *(Hydro Products.)*

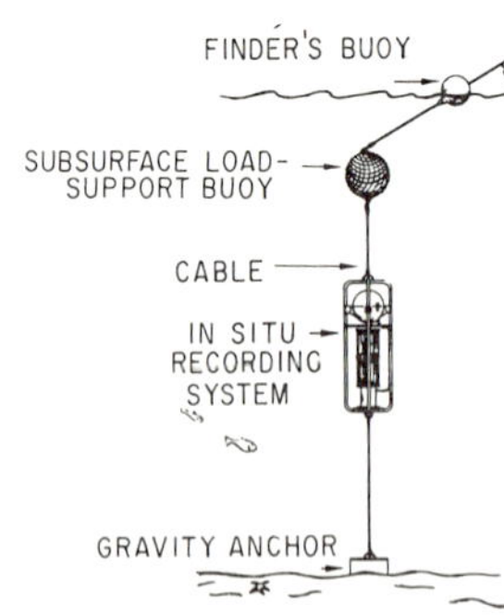

Fig. C-56. Typical installation of an in situ ocean-current meter. *(Hydro Products.)*

effects of the marine environment, particularly in warm tropical waters where marine organisms can completely encrust and inhibit moving parts without protection. Trends in the design of current meters include the development of instruments without moving parts or parts which are less susceptible to marine organisms, and the use of digital outputs via acoustical links to data-processing stations. Also see **Oceanographic instrumentation.**

—Leon J. Corcoran, *Hydro Products Division, Dillingham Corporation*

Curve Follower (See **Drafting system, automatic;** and **Numerical control.**)

Curve-tracing Osilloscope (See **Oscilloscope.**)

C_v **Factor** (See **Control valve, sizing of.**)

Cybernetics A comparative study of the methods of automatic control and communications that are common to man and machines. The science is useful in analyzing and improving the efficiency of communications systems, information-handling systems, and feedback control systems.

Cycle Controller (See **Timing controllers.**)

Cycle Steal The term cycle steal is generally used to describe a particular type of channel [See **Channel (CPU)**] which has the ability to access main storage (see **Storage, computer**) for data while the arithmetic unit is performing an operation. If the arithmetic unit is performing an operation which does not require a storage access at the time of the channel storage-access request, the operation continues while the data from the storage are transmitted to the channel. If both the channel and the arithmetic unit request a storage access at the same time, the arithmetic operation is delayed until the channel storage request is serviced. The delay in the arithmetic operations can be eliminated in a computer by providing separate storage elements for the channels and arithmetic units [see **Arithmetic and logical unit (ALU)**]. This function is also described as the ability to overlap input or output operations with computing operations. Both serial and multiplex channels may have this capability.

—W. W. Welz, *IBM Corporation*

Cyclotron (See **Accelerator.**)

Cylinder Actuator (See **Actuator, pneumatic.**)

D

Dall Tube A primary flowmetering device which, because of its special hydraulic shape, magnifies the differential pressure produced as a fluid flows through it. The Dall tube (Fig. D-1) senses a higher-than-normal apparent inlet static pressure and a lower-than-normal throat static pressure. The basic flow formula (see **Flowmetering devices, primary**) is applicable for this flow tube.

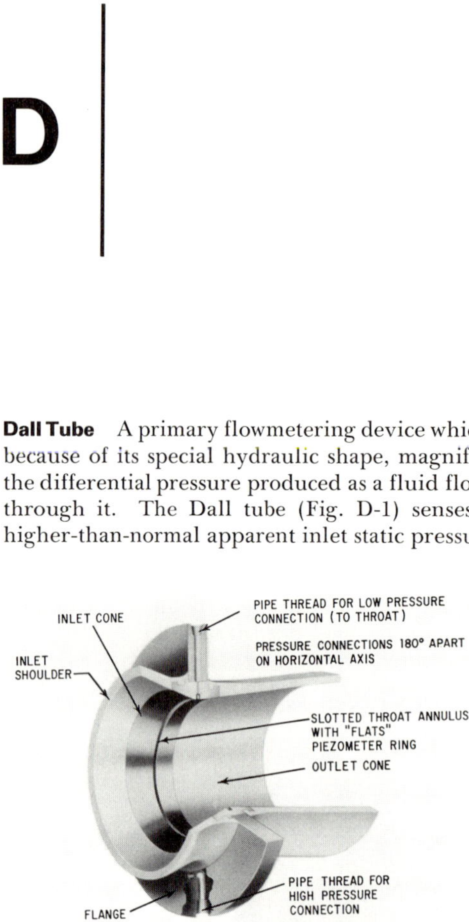

Fig. D-1. Dall tube in the form of a plastic insert—center flanged. The plastic insert also is available with flange on end. Cast-iron Dall tubes or fabricated (welded) construction are also available. *(B I F)*

Discharge Coefficient: The magnitude of the coefficient is between 0.72 and 0.64, and it is a function of beta ratio. The coefficient becomes constant over certain Reynolds numbers but is different for the different beta ratios. The Dall tube, at the same rate of flow and the same beta ratio, produces about twice the differential that can be expected from the equivalent venturi tube. Thus, the tube is the most efficient signal producer available. Typical head-recovery properties are:

Percent of Differential at a	*Beta Ratio* of
3	0.75
5	0.55
9	0.4

The sensitivity of the coefficient of the Dall tube to improper flow patterns has been investigated by the manufacturer (B I F), with published data available. A tolerance level of ±1 percent has been established.

In general, tubes of this design require longer

straight pipe lengths than venturi tubes to eliminate the effect of bad flow pattern on the coefficient of the tube. Dall tubes do not require any downstream piping. The tube also can be used for metering compressible fluids because the Y factor has been established by tests.

—Edmond B. Lynch, *B I F, Unit of General Signal Corporation*

Damper The term *damper* generally refers to a louver that controls the flow of combustion air into the fire box of a boiler, the movable obstruction in the flue of a furnace, or the vanes that regulate the airflow produced by the fans of a forced-draft cooling tower. Butterfly valves often are used for these applications, and when so used, they may be referred to as dampers. However, since butterfly valves are used in so many varied applications, they should not be routinely designated as dampers.

Damping The progressive reduction or suppression of the oscillation of a system. When the time response to an abrupt stimulus is as fast as possible without overshoot, the response is said to be *critically damped;* when overshoot occurs, *underdamped;* and when response is slower than critical, *overdamped. Relative damping,* for an underdamped system, is a number expressing the quotient of the actual damping of a second-order linear system or element by its critical damping. For any system whose transfer function includes a quadratic factor $s^2 + 2z\omega_n s + \omega_n^2$, relative damping is the value z, since $z = 1$ for critical damping. Such a factor has a root $-\sigma + j\omega$ in the complex s plane, from which

$$z = \frac{\sigma}{\omega_n} = \frac{\sigma}{\sqrt{\sigma^2 + \omega^2}}$$

where s = complex variable
ω_n = natural frequency, rad/s
z = damping ratio
σ = real part of the complex variable s
j = $\sqrt{-1}$
ω = frequency, rad/s

Damping factor, for the free oscillation of a second-order linear system, is a measure of damping, expressed (without sign) as the quotient of the greater by the lesser of a pair of consecutive swings of the output (in opposite directions) about an ultimate steady-state value (see Fig. D-2). Also see **Closed-loop control; Resonant frequency, transducer;** and **Scale, damping of.**

Dashpot (See **Scale, damping of.**)

Data-acquisition Computer The data-acquisition computer is a digital computer which operates in real time collecting physical data. The instruction repertoire of the data-acquisition computer and the scientific computer is essentially the same, since once the data have been accepted into the system, the computation requirements are basically the same. The data-acquisition computer is characterized by its ability to attach a wide variety of input or output devices, such as telemetering devices and flying-spot digitizers which are unique to the particular application. It can also handle random demands from devices which are asynchronous with respect to other program activity. Data-acquisition input-output devices may have very high peak data rates, although the average data rates are more reasonable. An example of this type of device is a flying-spot digitizer, which transmits data to the computer at an average rate of 100,000 bytes per second (see **Byte**), but the peak rate may reach 1,000,000 bytes per second.

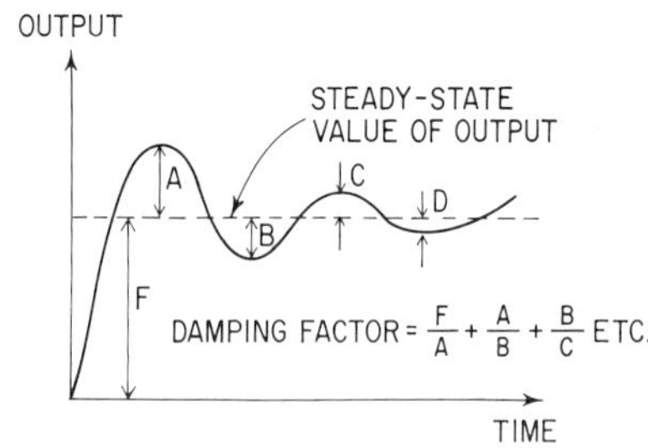

Fig. D-2. Underdamped response of a system with second-order lag.

Since the external devices generally transmit data randomly, the data-acquisition computer is designed to have the ability to respond very quickly to an external "interrupt" signal so that the necessary action can be taken by the control program.

—W. W. Welz, *IBM Corporation*

Data Communications With few exceptions, the communication of instrumentation data was relatively simple a few decades ago. The problem of extending the instrument's effectiveness over relatively short distances often was solved by using longer leadwires, as in the case of thermocouple or resistance-thermometer instruments, or by lengthening the run of capillary tubing in the case of filled-system thermometers. In other words, the instrument per se was stretched without the insertion of a data link into the system. Thus the science of communications was not really considered a part of instrumentation technology.

During and after World War II, considerable stress was placed upon remote control, which required that the instruments per se no longer could be stretched but that ways had to be developed to transmit instrument readings several hundred feet from a remote process location to a central control house. Pneumatic and electric transmitters and receivers were developed to solve these problems, but even then the art of transmitting such information from the instruments—or commands to locally mounted instruments or valves from the control house—was really not considered by many instrument engineers as lying within the province of communications. An early bringing together of instrumentation and control with communications did occur in connection with the control of pipelines and electric-power-distribution systems because of the much larger distances involved. However, the real mingling of the instrumentation and communications sciences was brought about by three principal forces—all of which are continuing to accelerate at a high rate today—to the point where instrumentation and communications technology are so intertwined that it is becoming difficult to differentiate one from the other. The aforementioned forces were (1) the appearance of sophisticated airborne and space systems which require the telemetering of information for a multitude of reasons over very long distances with practically all the information instrument-originated or -related in some fashion; (2) the advent of the practical digital computer and associated data-processing equipment; and (3) closely following (2), the stress on management-information systems. Thus, communications is an important element of many modern instrumentation systems. Of course, aside from the use of communications systems for the handling of instrumentation data, instruments per se play a vital role in making a communications system operate properly.

Communications information will be found spread throughout this encyclopedia, and reference for specific items can be found in the alphabetical index. Reference to the **Classified Index** under Communications also is recommended.

The characteristics of principal electronic communications systems are summarized in Table D-1. The time span and cost rates for some of these systems are compared in Table D-2.

Objectives of Data-communications Systems: These may be summarized as follows:

1. To reduce the time, effort, and expense required to convert information to a form suitable for entry into a data-processing system. This is done by capturing the data at their source.

2. To increase the performance of a system by the more rapid movement of information from its source to the processing point.

3. To provide accurate and timely reports for management information and decision making.

These three objectives may be realized by using data communications to improve existing systems and to open previously impractical application areas to the advantages of modern electronic data-processing equipment.

With all the technological advances, the basic considerations in a modern communications system are not too different from those in a primitive society using drums or smoke signals to send messages between related villages. Some of the important factors which must be considered in system design include (1) the purpose of the system, (2) the number and size of the messages, (3) the urgency of the messages, and (4) the accuracy and cost of the system.

Networks: The three networks shown in Fig. D-3 are the basic building blocks of any communications network. The circle indicates a terminal that can send messages, receive messages, or both. The connecting lines indicate the communications channel or circuit, which can be wire,

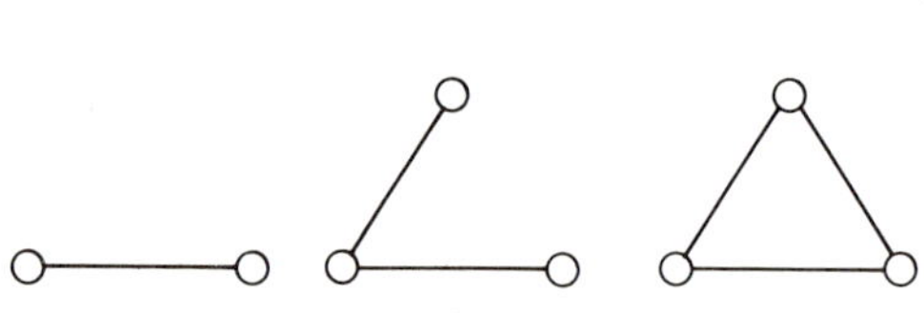

Fig. D-3. Basic communications networks.

cable, radio, microwave, millimeter-wave, or laser. Network *A* is considered a point-to-point network and can represent, for example, two computers or a card reader and punch connected by a channel. Network *B* illustrates a situation in which two terminals can communicate over the channels only by passing through a third terminal. Network *C* indicates a three-terminal system in which each terminal is able to exchange information with one or both of the others.

TABLE D-1. *Characteristics of Electronic Communications Systems*

TELEVISION

Data Content: Visual, in motion. Audio included or separate

Sensing: TV camera tube changes light pattern into analogous electical pattern. Latter is scanned by electron beam to produce a series of picture elements along successive horizontal lines for transmission in sequence to a receiver

Receiver: Cathode-ray display tube reconverts electronic pattern to moving electron beam which activates phosphor on face of tube. Broadcast TV presents 30 complete pictures per second

Transmission: Via coaxial cable or microwave. Essentially instantaneous

Channel Bandwidth Required: Industrial color, 15,000 kHz; broadcast standard, 6,000 kHz; typical closed-circuit TV, 10,000 kHz.

SLOW-SCAN TELEVISION

Data Content: Visual, nonmoving. Graphic and alphanumeric. Can include audio

Transmission: Via coaxial cable, microwave, or telephone voice circuit. Average 10-in.^{2} picture in 5 s

Channel Bandwidth Required: Standard, 8 kHz; minimum, 5 kHz; military special-purpose, 3 kHz

FACSIMILE

Data Content: Visual, nonmoving. Graphic and alphanumeric. Data to be sent must be on a two-dimensional surface

Sensing: Electromechanical scanner. Information moves past fixed detector for photoelectric scan

Receiver: Printer. Electrostatic or photoelectric printout on moving sheet of sensitive paper

Transmission: Via coaxial cable, microwave, or telephone voice circuit. Average 14.7-in.^{2} surface per minute. High-speed, 3 pages (8½ by 11) per second

Channel Bandwidth Required: Slow-speed, 3 kHz; high-speed, 4,000 kHz

TWO-WAY RADIO

Data Content: Audio or code

Sensing: Microphone

Receiver: Speaker

Transmission: Wireless. Essentially instantaneous

Channel Bandwidth Required: Average, 5 kHz; minimum, 1.5 kHz

TELEPHONE

Transmission: Telephone circuit or microwave. Essentially instantaneous

Channel Bandwidth Required: Long distance, 3 kHz

TELEGRAPH

Data Content: Alphanumeric only

Sensing: Mechanical-electric. Typewriter or tape punch converts letters and numbers to electronic code

Receiver: Electromechanical typewriter decodes signal and activates typewriter or punches tape for computer input

Transmission: Telegraph circuit or microwave. Transmission time, standard 65, 75, 100 words per minute

Channel Bandwidth Required: 0.030–0.15 kHz

TELEWRITER

Data Content: Records pen movements to transmit written or sketched information

Sensing: Position of pen on transmitter surface recorded in X and Y coordinates and translates into electronic code for transmission

Receiver: Electronic signal reconverted into position coordinates for reproduction on paper at receiver

Transmission: Radio or telephone circuit. Transmission time is coincident with speed of handwriting or drawing

Channel Bandwidth Required: 0.018–2.4 kHz

Signals: A data-communications system usually provides for two types of signals: (1) control signals and (2) text signals. The control signals indicate that a message is about to begin or has just ended, or other control functions. The text signals carry the actual message. In order to ensure that the message was received properly, an answer-back or acknowledgment signal must be provided by the receiving terminal.

Volume: The volume requirements of a data-communications system include the number and length of the messages. Different message lengths, their origin, direction, and destination, and the peak-loading times are all factors in message traffic-load determination.

Speed: Communications facilities are relatively fast or slow. A wide microwave channel, for example, can carry a much greater amount of information at a higher rate of speed than a single telegraph channel. Other facilities, such as telephone circuits, are available to provide a wide selection of speeds.

In order to use existing networks for data transmission, data-processing equipment must prepare

TABLE D-2. *Time Spans and Cost Rates for Some Communications Systems*

System	Time	Comparative Cost (Telephone Call = 1)
Telegram (15-word)....	40 min	13.3
Day letter (50-word)....	1½ hr	5.7
Night letter (50-word)..	8 to 10 A.M. next day	3.4
Telex (100–150 words in 3 min)	Essentially immediate	0.7–1.4
TWX (3-min message)	Up to a few minutes	1.1–1.7
Facsimile (4–200 words)	1 to 2 hr	1.7
Telephone (3 min person-to-person, 300 words)	Essentially immediate	1

suitable input for network terminals or be connected to the network. In addition, data-processing codes and signals must be converted to signals acceptable to the circuits or channels being used.

Communications Channels: A channel or circuit is a path for electrical transmission between two or more terminals. The channel may be a group of wires or a special part of the radio-frequency spectrum. The purpose of the channel is to carry information from one point to another. Circuit is used conversationally as a synonym for channel, but technically a circuit can contain several communications channels.

The three basic channel types are (1) simplex, (2) half-duplex, and (3) duplex. A simplex circuit, like a one-way street, can carry information (either textual data or control signals) in one direction. A half-duplex circuit can carry information in either direction, but only one direction at a time. A duplex circuit, also known as a full-duplex circuit, can carry information in both directions at the same time. A network can consist of any combination of these, according to the requirements.

In a half-duplex telephone circuit, reversing the flow of data requires a certain amount of time known as "turnaround time," nominally 200 ms. In preliminary planning for a network, consideration must be given to the time required for answer-back signals that can occur before, during, and after each message as a response from the receiving station indicating that it is ready to receive or that the last block of data was received correctly.

Simplex, half-duplex, and duplex circuits are *types* of circuits. The names indicate only the directional capability of the channel. The *grades* of channels are broadband, voice-grade, subvoice-grade, and telegraph. Circuits are graded on their basic line speed, expressed in characters per second, bits per second, or words per minute. Since these different grades of circuits use different types of signaling, it is necessary to understand code structure to evaluate their potential. For various code terms, see the alphabetical index.

Conceivably, if the sending and receiving terminals were controlled by the same timing pulse, start and stop bits could be eliminated. Characters would be detected on the basis of so many bit times. Transmission without start and stop signals is termed *synchronous.* In synchronous signaling, a special pattern of bits is sent periodically to keep the terminals in unison. These are automatically generated and sent as required by the system.

In addition to the serial start-stop and synchronous serial methods of transmission, Fig. D-4 shows how the same characters can be sent over five channels at the same time, allowing one channel for each bit in the code structure. This is termed *parallel transmission.*

In summary, a network may include simplex, half-duplex, or duplex channels. The transmission mode may be either start-stop serial, synchronous serial, or parallel. Within the circuit or channel, the code structure may be indicated by the presence or absence of a signal, or by changes in the signal itself.

Carrier Systems: These systems permit many conversations or messages on a single circuit. Modulation techniques used in connection with carrier systems are illustrated in Fig. D-5. Also see **Amplitude modulation; Frequency modulation; Pulse-code modulation; Pulse-duration telemetry; Pulse-position modulation;** and **Radio telemetry.**

Grades of Channels: The agreement between the communications common carrier and the customer is basically an arrangement for the use of a certain type and grade of communications channel. The channels are classified by type, such as half-duplex or full-duplex, and by grade, such as the capacity of the channel to carry information. The capacity can be expressed in

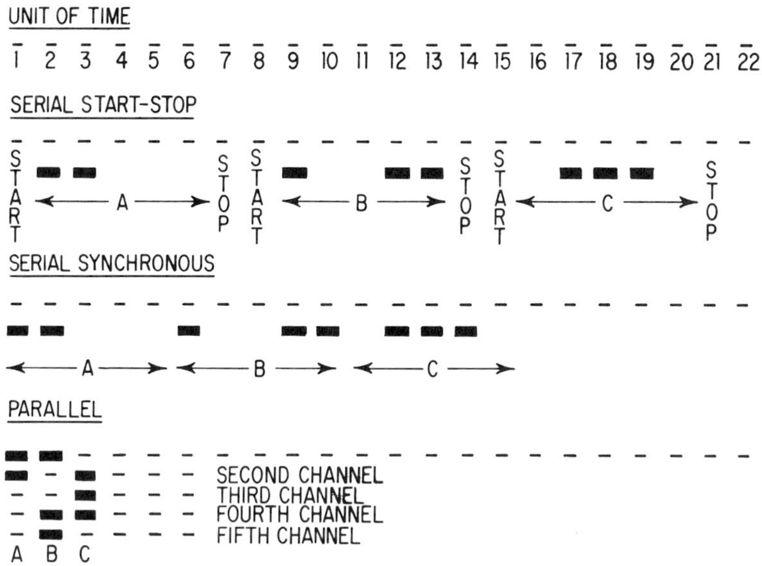

Fig. D-4. Transmission modes.

bits per second or a similar signal-time rate. The physical circuit or technique used, whether a voice-frequency circuit, a wire-carrier channel, or a portion of a microwave band, depends on the facilities available in each geographic location. The customer, for example, is not necessarily concerned with the fact that a voice-grade channel uses a carrier frequency of 15 or 500 kHz,

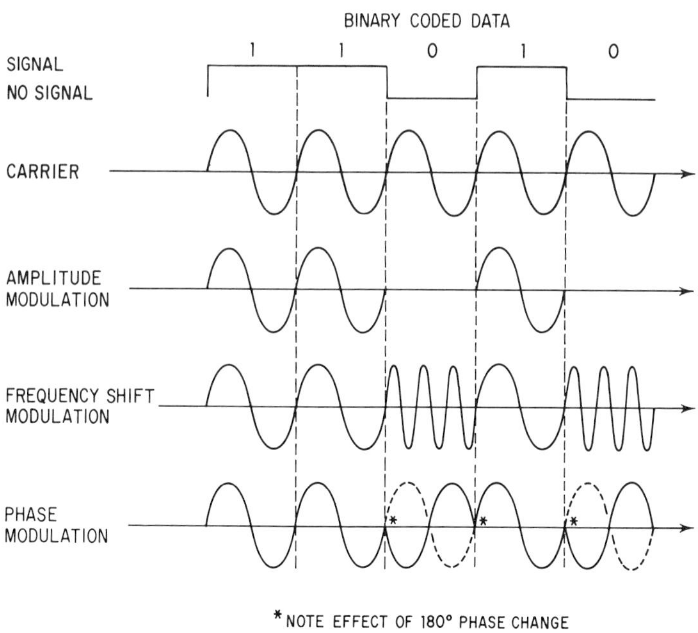

Fig. D-5. Modulation techniques used in communications systems.

but only that the available bandwidth is capable of carrying the required volume of information. The line speed of a channel, measured in bits per second, is directly related to the bandwidth of the channel. A high bit-per-second rate requires a wider channel or more cycles per second. The communications carriers have found that, from a practical point of view considering cost and performance, the bandwidth should be about twice as wide as the number of bits per second passing through the channel. A rate of 2,000 bits per second, for example, would require a bandwidth of about 4,000 Hz, as contained in a frequency range of 200 to 4,200 Hz. Technological advances, however, have made it possible to pass 2,000 bits per second over a 3-kHz bandwidth.

Broadband Channels: Microwave or radio-relay communications systems use the superhigh frequencies above 4 million kilocycles per second, i.e., 4 M Hz. Radio waves at this frequency level tend to travel in a straight line, requiring repeater stations with a dishlike antenna on high buildings or towers every 20 to 35 miles. Carrier, modulation, and channelizing techniques are used. The speed of transmission is limited only by the bandwidth available and the terminals used.

Voice-grade Channels: These channels have a range or bandwidth of approximately 3,000 Hz. When used for voice communication, the channel can carry one conversation. In data communications, a line speed of approximately 2,400 effective bits per second is available. Considerably higher speeds are possible but are dependent upon the commercial availability of suitable channels and terminal equipment. The common carrier can chanelize a voice-grade line into smaller bands of 150 to 200 Hz, each suitable for subvoice or telegraph communication. The cost of a leased voice-grade line used for data is determined by the length and type of channel. Dialed calls for data transmission are usually charged for at the same rate as ordinary voice telephone calls. A telephone and data set is required at each terminal. A telephone connection is established between two terminals either by direct dialing or through an operator. Each terminal attendant then presses a data key on his telephone to connect the data set to the line and disconnect the telephone. Pressing the key establishes a direct line for data transmission between the two terminals and prevents normal use of the telephone.

Subvoice-grade Channels: These circuits have a lower speed than voice-grade channels. Although there are no specific limits, subvoice-grade channels usually are considered to be those within the range of 150 to 600 bits per second.

Telegraph-grade Channels: These circuits have line speeds in the range of approximately 45 to 75 bits per second. Information is transmitted serially. Historically, the telegraph was the first effective means of cross-country electrical communications. One of the characteristics of early telegraph was the use of regenerative repeater stations along the circuit. These repeaters were necessary because of signal weakening or distortion over long distances. Regenerative repeaters, which are code-sensitive and thus restrict transmission to the code for which the repeater was designed, may be replaced or removed.

Data Sets: Also known as line adapters, "subsets," and "modems," these pieces of equipment are designed to provide the necessary compatability between data-processing equipment and communications facilities. Data sets are used at each termination of the communications channel at the interface with the data-processing equipment. Their basic function is modulation and demodulation for data transmission.

Communications-system Control: The requirements for the control of a communications system are shown by the network diagram of Fig. D-6. The control terminal may or may not be adjacent to the central data-processing system. The circles numbered two through eight indicate terminals on the network. This diagram will be used to review message switching, circuit switching, selective calling, polling, line control, and a complete communications control system.

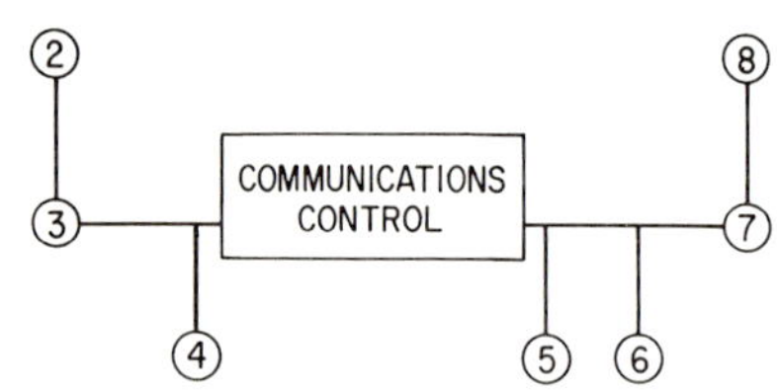

Fig. D-6. Communications network.

Message Switching: This is used in a leased telegraph-wire system with multiple circuits to transfer messages from one circuit to another. In Fig. D-6, a message from terminal 4 to terminal 6 would pass through the communications center and be switched to the circuit connected to terminal 6. In a message-switching system, the entire message is received at the communications center and retransmitted along the appropriate circuit. This is sometimes called "store and forward." The methods described here are manual torn tape, semiautomatic pushbutton, and fully automatic switching.

Manual Torn-tape Switching: In a leased-line system employing manual torn-tape switching, a station wishing to send a message or data to any other station in the network will transmit the message to the communications center with the appropriate message origin and destination codes. The communications center receives the message in perforated-paper-tape form. An operator manually tears the message off the receiving console and transfers it to a tape-storage rack. Sending operators interpret the destination codes and manually insert this same tape in the proper outgoing transmitter connected to the circuit, and the message is directed to its final destination.

Semiautomatic Tape Switching: A leased-line system employing semiautomatic switching is a continuous, rather than a torn-tape operation. An abbreviation of the destination city in the heading of the message is all that is required for an operator to send a message to another station. Upon receipt of a message at the communications center, the operator in the switching aisle reads the destination station abbreviated on the perforated paper tape and depresses a pushbutton corresponding to the desired destination station, and the message is forwarded. At the end of each message, the transmitter stops automatically and signals the operator so that another message can be switched.

Automatic Tape Switching: Several sizes and types of fully automatic switching systems are commercially available. A private-wire system of this type places a greater responsibility upon the operator at the originating station, in that a precise message heading and ending format must be used. The switching is executed on a fully automatic basis. The sequence of codes, entered by the originating operator, instructs the communications center as to the proper destination. In essence, the message is received in the communications center in perforated-paper-tape form. The tape feeds automatically from the tape punch, up over the sensing pins of the tape reader, which interprests the codes designating the destination station, checks these codes against a directory, establishes the proper circuit connection, and assigns the message to an outgoing line.

Circuit Switching: As opposed to message switching, circuit or line switching does not require the receipt and retransmittal of messages. The communications center provides a connection between the terminals. The actual switching may be done by an operator or through the use of dial equipment. The dial-telephone system is an example of circuit switching.

When two or more terminals are operating on the same channel, all stations will receive all messages and data, whether meant for them or not, unless something is done to prevent this. Conversely, if all points are trying to transmit, there will be contention for the circuit. The first problem is solved through the use of a selective-calling system, the second by a polling service. These systems, which vary in design and complexity, ensure that only the proper station receives the message. One such system is designed so that when the sending (calling) station precedes its message with a two-character code, only the machine assigned that code is set up to receive. In this way, the selecting device prevents stations on a multistation line from getting unwanted traffic.

One system used for solving the contention problem (several stations trying to send at the same time) incorporates a device which will automatically permit each station to send in turn. This is known as an automatic polling device. With this device, the communications center sends a code to start the first station. If the station has a message ready, transmission is begun. If there is none, the communications center sends a code to test the next station. Each station in the network is checked in this manner. Polling may be on either a sequential or a priority basis. In sequential polling, the communications center tests the transmitters in sequence. After the last station has been polled, the cycle is started again. The polling sequence can be modified to interrogate active stations more often than others.

Line Control: Polling and selective-calling systems use certain signals and procedures to ensure the orderly and efficient use of the data-transmission system. This is line control. Systems require different types of controls, dependent upon the complexity of the network and the volume and type of traffic. In forwarding data from one terminal to another, the receiving terminal must be tested to determine whether it is ready to accept a message. If the receiving terminal is ready, the data are forwarded along with checking signals. During transmission, the receiving station signals whether or not the data are being received properly. At the end of the message, the circuit must be left clear for other traffic. Performing these functions involves addressing, terminal status, text signals, and error control.

Addressing: A transmission or message consists of the addressing portion in control mode and the data in text mode. A mode-change character is used to differentiate between them. The addressing portion is used to direct the message to one or more terminals, and in some instances to a specific component, such as a page printer or card punch linked to the terminal.

Status: Addressing is done in control mode. The terminals addressed respond in the same mode and indicate their status. The addressed terminal can signal that it is ready to receive, ready to send, not ready to receive, or not ready to send, or it may not respond at all. If the power is not turned on at a terminal, or the terminal is otherwise unable to respond, the sending station does not receive an answer and starts a "time out" or waiting period. Dependent upon the system, this delay can be 1 s to 2 min long. The time out in polling is usually very short to prevent blocking traffic. In the middle of a transmission, however, the waiting period would be longer to permit the terminal attendant to overcome the cause of the delay by such actions as clearing a card jam or replenishing the forms in a printer.

Text: Text signals are classified as data-transmission characters and text "answer-back" signals. The data characters from the sending terminal are used to convey the data, and include end-of-message signals, end-of-block signals, and characters for redundancy checks. The text answer-back signals from the receiving station include positive and negative signals indicating whether or not the preceding message or block of data was received properly. If an answer-back is not received, a time-out delay is initiated.

Error Control: In some systems, a longitudinal redundancy character (LRC) is sent after each block of information. This is checked by the receiving station, and the appropriate answer-back signals, indicating correct or incorrect receipt, are returned. An incorrect transmission signal can be used to initiate the retransmittal of a block of data from the sending terminal.

Line Speeds: The line speeds of communications channels are usually expressed in bits per second, characters per second, or words per minute. In order to compare line speeds, these expressions must be reduced to a common form, such as bits per second. This requires a knowledge of the number of bits in a character for common codes, and the meaning of a "word." A word is considered to be six characters in length, usually five characters and a space. The following formula can be used for conversion:

$$\frac{\text{Words per minute} \times \text{characters per word (6)} \times \text{bits per character}}{\text{60 seconds}} = \text{bits per second}$$

Data Terminals: There is a rapidly growing profusion of data-terminal equipment and suppliers. To provide a representative review of such equipment, the following brief descriptions and condensed specifications are given.[1]

System A: Send-receive terminal. Paper-tape system. 120 characters per second. Unattended operation. 5- or 8-level paper tape generated is computer-compatible. Checks parity as data are received. Error detection stops terminal and instructs transmitter to back up and retransmit portion of data received in error. Off-line tape duplication and editing available. Interfaces—punched-card input, typewriter input-output, and magnetic-tape conversion.

System B: High-speed paper-tape reader and receiving set. Operates at 105 characters per second. Any 5-, 6-, 7-, or 8-level code, including USASCII, may be used. Receiving set automatically answers calls placed to it by sending set. Transmission may be halted at any time to permit voice communication over same line—transmission resumed by action of receiving-set operator. Reverse channel facility available to ensure that if line is interrupted the transmitter will halt until connection is resumed. Other terminals in family have error-detection and correction capabilities by automatic retransmission of data blocks that have been received in error.

System C: Combination of high-speed printer and paper-tape reader. Differs from usual paper-tape terminals in that it is designed to produce directly a hard-copy printout. Transmission is via 5- or 8-level punched paper tape. Up to 300 characters per second, dependent upon modem used. Data may be received and printed at up to 300 lines per minute (equivalent to 600 characters per second) with up to six carbon copies. Four error checks, including horizontal and vertical parity. Error correction is by automatic retransmission.

System D: Magnetic-tape send-receive terminal. Up to 300 characters per second. Higher speeds

[1] Systems *A* and *I*, Tally Corp.; *B*, Teletype Corp.; *C* and *D*, Digitronics Corp.; *E*, Mohawk Data Sciences Corp.; *F*, Communitype Corp.; *G*, Honeywell Inc.; *H*, Infotec, Inc.; and *J*, IBM Corp.

possible if broadband facilities are available. Tapes used may be 7- or 9-track and are optionally computer-compatible. Storage buffer—capable of storing 1,020 characters. Four error checks, including vertical and horizontal parity. Error correction is by automatic retransmission. Transmission is halted if three attempts at retransmission fail to correct error. Block counter protects against loss of or duplication of transmission blocks. Options include fixed or variable code translation and paper-tape input and output.

System E: Basically a keyboard to magnetic-tape recorder. Transmission rate over voice-grade lines is about 100 characters per second, dependent upon message format. Data recorded from keyboard onto a 7-channel, 200-bits-per-inch tape system using standard ½-in. computer-compatible magnetic tape. Nine-channel, 800-bits-per-inch version is available. Other models in family available with punched-card or paper-tape input, and typewriter or line-printer output.

System F: Computer-generated magnetic tapes may be received or transmitted. Can operate asynchronously at up to 200 characters per second or synchronously at 100 characters per second. Unit may be used on private or dial-up lines. Record gaps and parity are generated internally, and parity is double-checked in both read and write modes. Parity error causes retransmission of incorrect block. Tape format is standard 7-track NRZI format with IBM-compatible code. Another model in family is data-communication system for direct typing onto magnetic tape which can later be transmitted.

System G: Combination of communications and special-purpose input-output capabilities in series of magnetic-tape terminals. Basically a 64-character keyboard to magnetic-tape device, units will transmit or receive up to 120 characters per second over voice-grade lines. 7- or 9-channel tape units also available. Unit may be used to transmit all the data on a tape or, by using a program-controlled mode, data may be formatted for transmission, eliminating the need to transmit duplicate fields and blanks. Error detection includes parity check, block-length check, and character format check. Available with a card reader, line printer, or adding machine.

System H: Unit incorporates a 7- or 9-track magnetic-tape system and a small (4,000-character, 12-bit) programmable computer. Terminal will transmit or receive data at up to 105 characters per second asynchronously. Computer may be used for code conversion, automatic formatting, or addition of fixed data before transmission or after reception; compression of data or reconstruction of compressed data and other functions by programming in a simple assembly language.

System I: Incremental magnetic-tape transport is basic unit. System can transmit digital data over ordinary dial-up telephone lines at up to 120 characters per second, asynchronous. Up to 1,600 characters per second continuous may be achieved with private communications facilities. Operates unattended. Automatically retransmits data received in error. Standard interface allows typewriter, punched card, paper tape, or high-speed computer input-output. ¼-in. audio-type tape on a 3-in. reel. Other systems in family available for conversion and transmission of paper tape or ½-in. computer-compatible magnetic tape.

System J: Unit operates by reading data from a tape unit, record by record, into a 200-word storage buffer for conversion into a serial-transmission code. Core storage may be expanded up to 2,400 characters to accommodate larger data records. When receiving, the unit converts the serial-transmission code to binary-coded decimal and places the data in their buffer for synchronization with the tape unit. Operational speeds—150 to 5,100 characters per second. Four separate error checks, including vertical and horizontal redundancy. Errors corrected by automatic retransmission. Unit is designed to interface with a magnetic-tape unit with a communications service.

Unfortunately, space does not permit more detailed descriptions of the foregoing units, or even brief descriptions of many other similar devices. Brief descriptions are included only to acquaint the reader in an overall way with obtainable data-terminal hardware.

For additional information related to data communications, telemetering, and associated communications topics, consult the **Classified Index.**

Note: Portions of this text are from the IBM Systems Reference Library, File Number T-P-OO, Form Number E20-8158,[1] IBM Technical Publications Department, White Plains, N.Y. Brief descriptions of data terminals were extracted from Louis Waltz, What's Available for High-speed Data Transmission (*Control Eng.*, vol. 15, no. 10, October, 1968).

REFERENCES

James, M. L., G. M. Smith, and J. C. Wolford: "Applied Numerical Methods for Digital Computation with FORTRAN," International Textbook Company, Scranton, Pa., 1967.

Oakford, R. V.: "Introduction to Data Processing Equipment," McGraw-Hill Book Company, New York, 1962.

Hein, Leonard W.: "An Introduction to Electronic Data Processing for Business," D. Van Nostrand Company, Inc., Princeton, N.J., 1961.

Brandon, D. H.: "Management Standards for Data Processing," D. Van Nostrand Company, Inc., Princeton, N.J. 1963.

[1] Reprinted by permission from "Data Communications Concepts and Communications Facilities," copyright by International Business Machines Corporation.

Data-conversion Codes Digital data are encoded in a variety of ways for processing and transmittal within data-processing and communications systems. Some of these coding systems are binary-coded decimal (BCD), extended binary-coded decimal interchange code (EBCDIC), and the U.S.A. Standard Code for Information Interchange (USASCII). These various coding schemes have different emphases in their structures. For example, an 8-bit code takes one-third longer to transmit a given message on a given line than a 6-bit code but allows a wider range of characters and superior error recovery. Table D-3 lists the binary equivalents of a range of characters for some common codes.

—A. J. Arthur, *IBM Corporation*

TABLE D-3. *Data-conversion Codes*

Decimal and Graphics	BCD	EBCDIC	USASCII
0	000000	11110000	0110000
1	000001	11110001	0110001
2	000010	11110010	0110010
3	000011	11110011	0110011
4	000100	11110100	0110100
5	00101	11110101	0110101
6	000110	11110110	0110110
7	000111	11110111	0110111
8	001000	11111000	0111000
9	001001	11111001	0111001
A	110001	11000001	1000001
B	110010	11000010	1000010
C	110011	11000011	1000011
D	110100	11000100	1000100
E	110101	11000101	1000101
F	110110	11000110	1000110
$	101011	01011011	0100100
@	111110	01111100	1000000
CR	101101	00001101	0001101
•	111011	01001011	0101110

Data Plotter (See **Drafting system, automatic.**)

Dead Band (See **Hysteresis**; and **Sensitivity.**)

Dead Time, Process Any delay between two related actions in a control system can be defined as *dead time*. An example of dead time is illustrated in Fig. D-7. The pH electrode assembly must be installed downstream from the point of caustic addition to allow time for mixing and the chemical reaction. If the fluid flows at 5 ft/s and the distance is 20 ft, the dead time is 4 s. Therefore, dead time is the interval of time between initiation of an input change or stimulus and the start of the resulting response. The presence of dead time in a process dictates a proportional-band setting of over 100 percent to dampen cycling. The effect of dead time in a process under automatic control is to increase substantially the complexity of control. With proportional control only, the presence of dead time dictates a wide proportional band, which results in large offsets with load changes. Proportional-plus-integral (reset) controllers eliminate offset due to load changes but reduce the speed of response in returning the measured variable to a constant value after a load change.

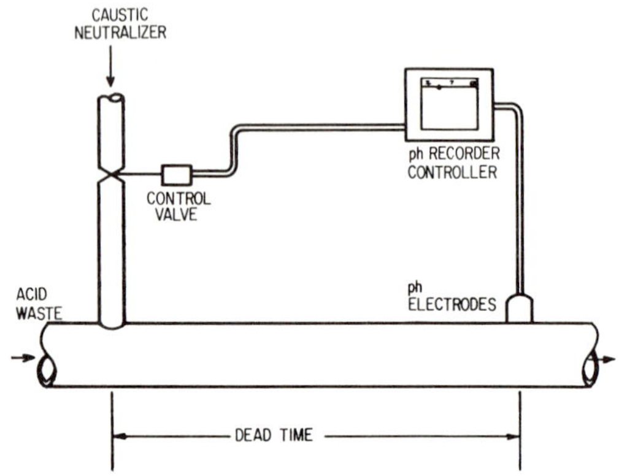

Fig. D-7. pH-control process with dead time.

—Richard H. Kennedy, *Bristol Division, American Chain & Cable Company, Inc.*

Dead-weight Tester An instrument used to measure primary pressures in terms of the fundamental units of *force* and *area* in accordance with $P = F/A$. Pressure ranges are actual or absolute up to 100,000 psi with accuracies as high as one-hundredth of 1 percent. In principle, the device is a piston of known area—inserted into a close-fitting cylinder. Weights of known mass atop the piston are supported by hydraulic or pneumatic pressure that is applied to the other end of the piston. Accuracies generally are referenced to a standard at the National Bureau of Standards to which the dead-weight tester has been compared and calibrated. This device is used to calibrate all types of pressure-indicating or -controlling instruments—gages, switches, transmitters, transducers, and differential-pressure units. Calibration over the pressure range of the instrument is by direct attachment to the dead-weight tester and by applying pressure and weights to obtain a balanced position for the pressure increments desired.

REFERENCES

Cross, J. L.: Reduction of Data for Piston Gauge Pressure Measurements, *Natl. Bur. Std. (U.S.) Monograph* 65, August, 1964.

Johnson, D. P., J. L. Cross, J. D. Hill, and H. A. Bowman: Elastic Distortion Error in Dead Weight Piston Gauge, *Ind. Eng. Chem.*, vol. 49, p. 2046, December, 1957.

Newall, Don H.: A New High Pressure Technique, *Ind. Eng. Chem.*, vol. 49, p. 1993, December, 1957.

Johnson, D. P., and D. H. Newall: The Piston Gauge as a Precise Measuring Instrument, *Trans. ASME*, vol. 75, p. 301, 1953.

—George H. Krell, *Mansfield and Green Div., AMETEK, Inc.*

Decoding Network, Digital-to-Analog Converter A decoding network, when used in reference to digital-to-analog converters, is that portion of the digital-to-analog converter which converts the state of the digital-input analog switches into an attenuation factor (see **Digital-to-analog converter**).

A digital-to-analog converter requires the proper combination of the decoder network, analog switches, a reference source, an operational amplifier (in some cases), and appropriate logic-control circuitry. The reference source is selectively switched into the decoding network by means of the analog switches to provide an output signal which is proportional to the digital-input data.

Although an ac decoder network is possible, the networks used in digital-to-analog converters are generally designed for dc use using precision resistors. The two most popular decoder networks, illustrated in Figs D-8 and D-9, are the weighted-resistor network and the resistor-ladder network. Although the illustrated networks are n-bit binary decoders, other codes can be realized by appropriate modifications. For example, in a weighted-resistor binary-coded-decimal decoder (see **Binary-coded decimal**), the configuration is identical but the resistors are R, $2R$, $4R$, . . .; in the ladder decoder, the resistors are R and $2R$, but an attenuation network providing an attenuation factor of 0.8 is inserted between each 3-bit section of the ladder decoder to provide the binary-coded-decimal output.

The weighted-resistor network of Fig. D-8 is used with a voltage-reference source. Each of the

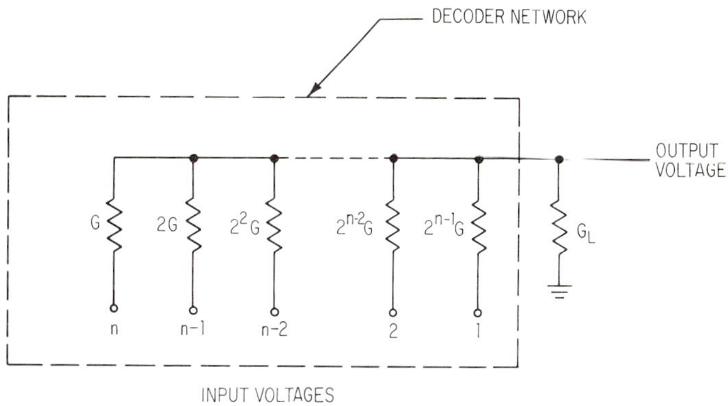

Fig. D-8. Weighted-resistor decoder network.

terminals 1, 2, . . . n either is connected to the reference voltage V_r or it is grounded. If the kth node is connected to V_r and the other (n-1) nodes are grounded, the output voltage V_o is

$$V_o = \frac{V_r}{2^n - 1 + G_L/G} \sum_{k=1}^{n} A_k\, 2^{n-k} \qquad A_k = 1, 0$$

where A_k is an index which is 1 when the k-n terminal is connected to the voltage source and 0 when it is grounded. If the conductance $G_L = G$, the expression reduces to

$$V_o = V_r \sum_{k=1}^{n} A_k\, 2^{-k} \qquad A_k = 1, 0$$

which shows that the output voltage is binary and is determined by the conditions at the input terminals. The input conductance at the kth terminal when G_L is assumed equal to G is

$$G_{\text{in}} = 2^{n-k}(1 - 2^{-k})G$$

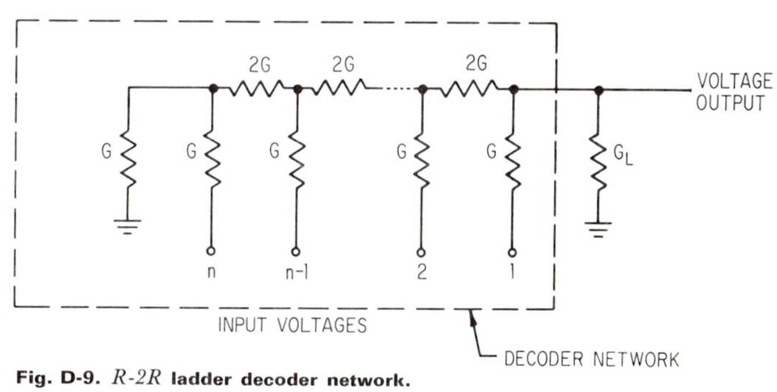

Fig. D-9. *R-2R* **ladder decoder network.**

and the output conductance of the network is $2^n G$, a constant. Although this network is designed for use with a voltage-reference source, a similar network can be designed for use with current-reference sources.

The resistor network of Fig. D-9 is popularly known as the *R*-2*R* ladder network. As in the previous case, either a reference voltage is connected to the *k*th terminal or the terminal is grounded. The expression for the output voltage is

$$V_o = \frac{V_r}{1 + G_L/2G} \sum_{k=1}^{n} A_k \, 2^{-k} \qquad A_k = 1, 0$$

where, as before, $A_k = 1$ when the *k*th terminal is connected to the voltage V_r and 0 if the terminal is grounded. If the load resistance G_L is made equal to 0, the expression becomes

$$V_o = V_r \sum_{k=1}^{n} A_k \, 2^{-k} \qquad A_k = 1, 0$$

which shows the binary contribution of each input. The output conductance is constant and is equal to $2G$ if $G_L = 0$. The input resistance is also constant and is equal to $2G/3$. As in the previous case, a similar network can be used for current sources.

Both networks provide the same basic operation, but there are some distinct differences. The weighted-resistor network uses a minimum number of resistances. However, the ratio of the largest to the smallest resistor is 2^{n-1} (or 512 for $n = 10$) compared with only 2 for the ladder network. This wide spread of resistor values can reduce the choices of network-fabrication techniques, since the spread of obtainable resistance values is limited with some techniques. In addition, the different resistance values may cause the time response of the network to be dependent on the pattern of inputs. The constant input resistance of the *R*-2*R* ladder network also simplifies the design of the reference supply and switches.

The type of resistor used in the network depends primarily on performance limitations and cost. Discrete wirewound resistors are commonly used for high-precision networks having a very low temperature coefficient (see **Temperature coefficient**). Thin- and thick-film resistors can be used in networks of less precision or less stringent temperature-coefficient requirements. The capacitance and inductance of the wirewound resistors generally limit their response to 500 ns or more. Film-resistor networks with faster response time are available.

Independent of the type of resistor used, the networks are generally described by overall specifications rather than by specifications on the individual resistors. As noted from the equations for the output voltage, the important network parameters are the transfer ratios from each input to the output, not the absolute value of the resistors. In addition, the error contribution of each resistor varies depending on its position in the network. By specifying the network in terms of ratio accuracies and temperature coefficient, the network manufacturer can select matched assemblies of stable resistors to satisfy the overall network specification. This method of describing the network usually results in a substantially lower cost than when specifications are imposed on each individual resistor in the network.

Digital-to-analog converter decoder networks are combined with a reference source, analog switches, control logic, and in some cases, an operational amplifier to form a complete digital-to-analog converter. In addition, they are used in conjunction with reference sources and analog switches in the successive-approximation analog-to-digital converter to generate comparison voltages used in the conversion process (see **Analog-to-digital converter, successive-approximation**).

—Thomas J. Harrison, *IBM Corporation*

Defibrillator (See **Cardiovascular instrumentation.**)

Delay Line Delay lines are used in control systems to hold a signal for a discrete period and then retransmit the signal with a minimum of distortion. *Magnetostrictive* delay lines convert electrical pulses into acoustical pulses for transmission along a wire and then reconvert to electrical pulses after a delay dependent upon the length of the wire. The device is limited to about 50 μs, since this length of delay occupies a 15-in. package and longer delays become unwieldly. Maximum signal frequency is about 4 MHz, and the delay is about 5 μs/in. of transmission material. Other delay-system configurations include the use of a tape or drum recorder to delay a signal purposely. In an ultrasonic delay line, the delay is determined by the length of the path taken by acoustic waves through the medium.

Delay Time, Transducer The interval of time between a changing signal and its exact repetition for some specified duration at a downstream point of the signal path. Delays in the response to the initiation of signals are typical of transducers, sensors, and instrumentation systems which incorporate (1) transmission lines, (2) multi-time-constant filters or stabilizing networks, (3) zones of no response or dead bands, (4) backlash, and (5) hysteresis-producing mechanisms. Delay or dead time thus is caused by the finite velocity of signal propagation in association with the spatial separation between input and output, or by what is known as distance-velocity lag. Alternately, delay or dead time is caused by the momentary absence of an effective connection between input and output. The distance-velocity type of delay generally can be distinguished from the hysteresis type of delay by the independence of the former from the waveshape and amplitude of the input signal. The hysteresis type of delay results in significant waveform distortion. In effect, the response is a degraded reproduction of the input or stimulus. Delay also may be generated by employing recording and compatible reproducing mechanisms. The delay can be controlled easily, thus making this technique useful in the analog-computer simulation of large-scale distributed systems. Also see **Dead time, process.**

—Otto Muller-Girard, *Consulting Engineer, Rochester, N.Y.*

Demand Meter For satisfactory service, the electrical energy for a community or region must be generated at a rate that will fully satisfy the need at any given instant of time—for peak loads rather than for the average demand. This means that the electric-utility supplier must install additional capacity to meet the maximum rather than the average demand. So that revenue may be produced that will compensate for the generating equipment that may be used only part of the time, the practice of demand billing was conceived. In other words, the charge to a customer for electric service is based upon the customer's maximum requirements in addition to the usual charges for electric-power consumption. This practice, of course, persuades the customer to diversify his load so that his energy consumption will be more uniform. Obviously, momentary overloads do not require reserve-equipment capacity. Thus, in determining charges that are based upon demand, the system must take into consideration the amount of energy consumption over an appreciable time interval. Demand billings generally are limited to industrial rather than domestic users. The time intervals or *demand intervals* most commonly used in metering practice are 15, 30, and 60 min, the quarter-hour period being the most common. Thus demand meters must relate energy consumption with time.

There are three basic kinds of demand meters: (1) the recording wattmeter, (2) the integrating-demand meter, and (3) the lagged-demand meter. Even though the chart record from a recording wattmeter can be used for estimating demand charges, the amount of manual scanning and computation is large. Thus, this approach is not used frequently. Recording wattmeters have the advantage of providing an exact record of when the demand loads occurred. Curve-drawing wattmeters also are referred to as Class I demand meters.

The *integrated-demand meter* usually is connected with a watthour meter so that the energy consumption measured by the meter is registered from time to time and the maximum demand may be determined from the record. The connection with the watthour meter may be either by way of gearing or by electrical impulses received from contacts made when the meter has performed a given number of revolutions. In the "block-interval" type, the registration is executed at arbitrarily selected moments terminating the selected time interval, thus at 8:00; 8:30; 9:00, and so on. The readings may be registered on a tape or recorded on a chart. In other designs, an indicating pointer is carried upward over a graduated scale as the watthour meter rotates and is left at the maximum position when, at the end of the time interval, the registering mechanism is momentarily thrown out of gear and returns to a zero position. In another design, equal blocks of energy are recorded on a tape or chart beside a separate record of time. Thus, it is obvious that the maximum demand occurred where the greatest number of energy blocks are found within the specified time interval. The time of the beginning of the interval is not arbitrarily fixed in this type of instrument. Integrated-demand meters are also referred to as Class II demand meters.

A *lagged-demand meter* (referred to as a Class III demand meter) is constructed so as to require a certain time interval to elapse before the indicating element can reach the point corresponding to the value of the load. This action is found in meters that utilize heating effects as their operating principle, such as in the Lincoln demand meter. The lagging device usually is so adjusted that the instrument attains 90 percent of the full value of a steady load suddenly thrown on—in the specified time interval.

Densitometer When light passes through a thin material or film, the light is attenuated by the partial opacity of the material. This characteristic is called *optical density* ($\log_{10}$ of opacity) and is measured by a transmission densitometer. When light falls on a surface and is reflected, the light can be attenuated, giving rise to shades of color. See **Color, measurement of.** This property also is quantified in terms of optical density, and a reflection densitometer is used. Parenthetically, spectrophotometry, used to determine the concentration of a dissolved ion or molecular species, is akin. There, too, a logarithmic unit is proportional to the concentration and the term absorbance is used rather than optical density. However, in photographic films, the attenuation is due to scattered as well as absorbed light, and optical density is a more appropriate expression. Of course, when there is no scattering, the two methods converge, but densitometry is the term always used with solid samples, and the instruments differ greatly because films rather than liquid samples are measured. Since the nature of the samples and the purpose of evaluation differ, there are several distinct classes of densitometers.

Transmission Densitometers (Graphic Arts): The optical layout of a transmission densitometer is shown in Fig. D-10. Light from a stabilized source is collimated (or focused) onto the sample, and the transmitted light is picked up by a phototube and read out on a logarithmic scale of a photometer. An essential element is the diffuser that is used to pick up light scattered by the sample rather than only the absorbed portion. The simplest integrating device is an opal glass in contact with the sample. This is one of the methods accepted by the American National Standards Institute (Ref. 1). Other types of integrating devices, such as spheres, are used in some densitometers and densitometric attachments to spectrophotometers.

Transmission densitometers are used routinely in evaluating photographic materials and are invaluable in the control of photographic and printing processes. The usual diffuse-density definition relates to a broad spectral domain in the visible (luminosity curve). However, color films are evaluated by special spectral filters.

Microdensitometers are scanning-transmission densitometers which incorporate magnification to evaluate microscopic areas of films. These instruments cannot evaluate diffuse density because of the narrow angle of acceptance implied in magnification. However, the devices are capable of great resolution, and much attention is given to the linear-drive mechanism by which the films are scanned. An important application of these instruments is the evaluation of spectrograms.

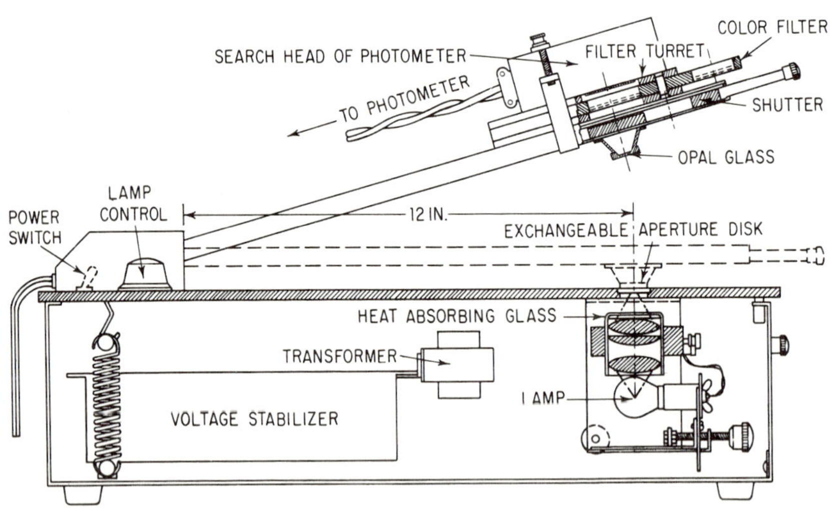

Fig. D-10. Optical layout of a transmission densitometer. *(Photovolt.)*

An instrument for this purpose is shown in Fig. D-11.

Reflection densitometers are similar to transmission densitometers except that the beam of

Fig. D-11. Microdensitometer for scanning spectrograms. *(Baird-Atomic.)*

light is focused on the surface of the sample, such as a printing-press sheet. The diffusely reflected light is picked up at 45° from incidence and quantified by a photometer. These instruments have specific narrow spectral capability, achieved by the insertion of special filters. The characteristics of these filters vary, dependent upon the material to be evaluated. A reflection densitometer has become an essential instrument in the pressroom.

A *scanning densitometer* is a transmission densitometer which is used to evaluate electrophoresis substrates used in clinical chemistry, thin-layer chromatograms, ultracentrifuge patterns, autoradiograms, and the like. The instruments always yield a linear scan of the sample. Spectral discrimination is important, and specular densities rather than diffuse densities are measured. However, the density is not always a simple linear function of the concentration. For this reason, these instruments are provided with variable transfer functions. This feature can be accomplished by mechanical cams or, more elegantly, by an electrical network (Ref. 2). With this arrangement, the ratio of albumin and globulins can be obtained directly from a stained strip of serum electrophoresis material. When scanning cleared cellulose acetate electrophoresis substrates, quantification will be obtained with a precise logarithmic transformation (optical density or absorbance). With paper substrates, the low densities are suppressed and the high densities augmented to give a similar quantitative result. A densitometer of this type shown in Fig. D-12 illustrates a type of device that is widely used in clinical practice. Also see **Electrophoresis apparatus.**

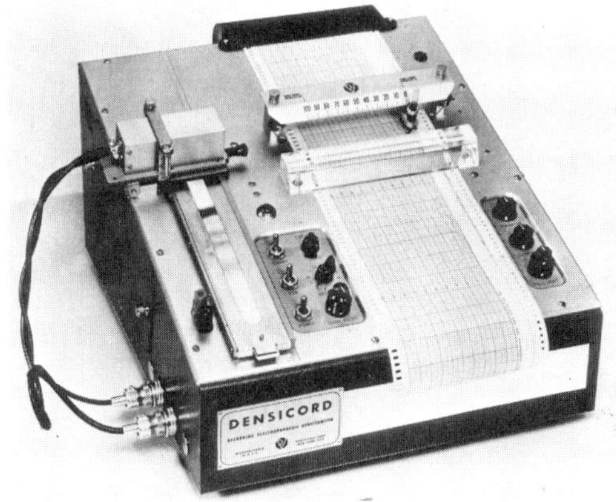

Fig. D-12. Automatic scanning densitometer with variable recording functions. *(Photovolt.)*

REFERENCES

1. American National Standard PH2.19–1959, Diffuse Transmission Density.
2. Levy, Gabor B.: "Instrumental Methods of Experimental Biology," chap. 15, p. 524, The Macmillan Company, New York, 1964.

—Gabor B. Levy, *Photovolt Corporation*

Density (Fluid), Measurement of (See **Specific gravity, measurement of.**)

Density Meter, Radiation-type Nuclear density gages measure and control slurries and liquids moving in process through pipelines. Uses are found in the chemical, cement, petroleum, plastics, paper, sewage, mining, steel, and food industries. Accuracy is as good as ±0.03 percent. The gages can be mounted on pipes from 1 through 30 in. in diameter.

As shown in Fig. D-13, a radioisotope source in a lead-shielded shutter-controlled holder is clamped to one side of the pipe with a detector cell clamped to the other side of the pipe and in line with the radioisotope source. Transmitted radiation is in proportion to the specific gravity of the material in the pipeline. The detector cell converts transmitted radiation into minute electric current. Current is amplified for the readout and to operate control functions. Standard transducers can be used. Span and repeatability are determined by pipe size. The final control servo or the indicating panel may be several hundred feet away from the measuring point.

An ac or dc amplifier may be used. An ac amplifier is used for accurate density determination with very narrow spans. A dc amplifier is used with larger spans and where less precision is required. Both types have industrially accepted drift rates and spans. Zero suppression is supplied by either compensating cell or an electrical zero. The compensating cell virtually eliminates zero drift rate caused by radioisotope-source decay. The radioisotope, with the same half-life as the measuring source, is mounted integrally with the cell. Opposing current from the compensating cell decreases at the same rate as the measuring current. The electrical-zero suppression uses a zener-diode-regulated voltage and is best suited where long-time stability and span requirements are less strict.

Predetermined calibration is accomplished on most applications by determining zero and span settings from calculations of known absorption values of a given process material. Periodic rechecking is done to standardize the unit for source decay by emptying the pipe or by filling the process pipe with water (known factor of 1.0). No access to the gage head is required to recheck zero and span. Equivalent absorbers are used for on-site calibration of known parameters. Absorbers accurately represent the absorption of radiation by process material. This method is more accurate than predetermined calibration, since absorbers are made for specific measurement. Absorbers are used for restandardizing the gage

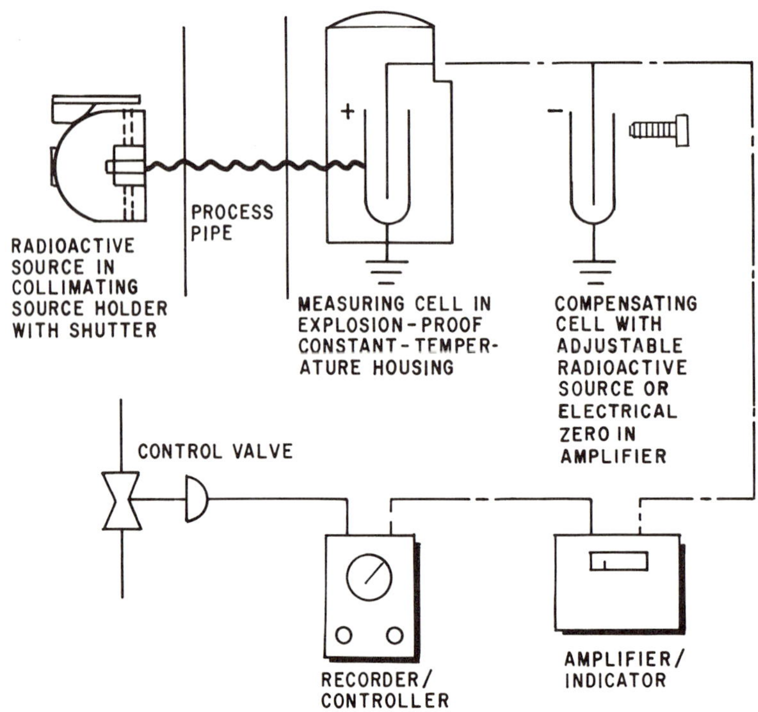

Fig. D-13. Density-control system of the radiation type. *(Ohmart.)*

when the pipe cannot be emptied for long periods.

—Gordon W. Hickenlooper,
The Ohmart Corporation

Depth Measurement, Oceanographic Typical systems utilize standard bourdon-tube pressure transducers or precision absolute-pressure strain gages. Excitation and signal-conditioning circuitry provide the user with analog or digital information in visual form or with outputs for data recording. Systems are available for measuring depth in the deepest parts of the ocean, although most research is performed in depths of less than 6,000 m. Normal accuracies range from ±0.1 to ±3 percent of reading. Basic uses of depth systems include monitoring on submersible vehicles, lowered instrument frames, and instrument packages while being towed or tethered at specific depths. Also see **Oceanographic instrumentation.**

Readout circuitry normally consists of dc amplifiers which provide in the simplest case a dc output that is proportional to the depth of the transducer. More complex systems include analog-to-digital converting circuitry and provide digital outputs in standard or complex formats as required. Most digital systems utilize photographic or magnetic-tape recording so that large amounts of depth data can be processed in conjunction with other parameters, such as temperature and salinity.

Bourdon-tube pressure-measuring systems are not used where accuracy is a primary requirement, but they are less expensive, especially the transducer, and are less complex than strain-gage types. See **Bourdon tubes.** Strain-gage transducers are used for measuring depth to accuracies on the order of 0.1 percent of reading, which can be accomplished with proper signal-conditioning circuitry.

Each of the systems described requires a cable from the transducer to the surface readout unit. The cable normally is the major factor affecting the reliability of the system. Acoustic systems, although generally more complex and expensive, do provide greater accuracy and reliability.

—Charles L. Strickland,
Hydro Products Division,
Dillingham Corporation

Derivative (Rate) Control That component of control action for which the output is proportional to the rate of change of the input. Derivative, rate, and anticipatory control are synonymous terms. Derivative-control action is shown by

$$m = K_c T_d e' + M$$

where m = manipulated variable
K_c = proportional sensitivity
T_d = derivative time
e = deviation
M = constant
e' = first derivative of e with respect to time de/dt

Derivative control does not exist by itself, but only in combination with proportional control. Also see **Electric controller;** and **Pneumatic controller.**

—Richard H. Kennedy, *Bristol Division,*
American Chain & Cable Company, Inc.

Detector A term sometimes used instead of the terms *sensor* or *primary element* to denote that part of a measurement system which is exposed to and affected by the variable being measured—as the thermocouple in a temperature-measurement system, or a strain-gage load cell in a force-measurement system. In electronic circuitry, detector denotes that stage of a receiver in which demodulation occurs. In the case of a superheterodyne receiver, this is termed second detector—sometimes termed demodulator. Also see **Sensor;** and **Transducer.**

Detector, Threshold A threshold detector is a circuit which provides a digital indication that a signal input to the circuit exceeds a predetermined magnitude. For example, if the predetermined threshold value is 5 V, an input of 5.5 V results in an output representing a binary 1 and an input of 4.5 V produces a digital output representing a binary 0.

The threshold detector is a form of comparator (see **Amplifier, comparator**), and the terms are often used interchangeably. Sometimes, however, the term comparator refers only to a circuit which detects whether or not an input signal has changed polarity.

In any practical threshold detector, a band of uncertainty exists around the predetermined threshold value. An input signal whose value falls in this "dead band" produces an output which is indeterminant; that is, it cannot be interpreted as either a 1 or a 0. The magnitude of the dead band can be reduced by increasing the gain of the threshold circuit. There are limitations on gain, however, imposed by stability and overload considerations.

Threshold detectors are commonly used in instrumentation systems and in data-acquisition and process-control computer systems. In digital systems, they are sometimes provided as a feature under a name such as "voltage sense" or "level sense." They are often used to provide alarm indications in processes. For example, the output of a thermocouple transmitter monitoring a critical voltage might be connected to a threshold detector whose threshold value corresponds to the maximum allowable temperature. If the temperature exceeds this value, the detector output changes from one binary state to the other, and this triggers an audible/visual alarm indicator or provides an interrupt to the process-control computer system.

—Thomas J. Harrison, *IBM Corporation*

Detent A portion of a mechanism that facilitates manual positioning or automatic indexing is known as a detent. A common manual application of a detent is for an adjusting knob with preselected positions (as compared with a continuous adjustment), such as for range positions on a voltmeter or the channel selector on a television set. Where the number of necessary positions is limited, detents provide a positive and easy method of exactly repeating the desired knob positions. One common method is shown in Fig. D-14, where the grooves in the knob base are called detents. There is a large variety of detent mechanisms used for automatic indexing. Some form almost always is found in devices where there is some "discontinuous" motion. Detents also are used to facilitate automatic detection, switching, counting, and pulsing, and to prevent backlash in drive systems where the driving forces are applied intermittently.

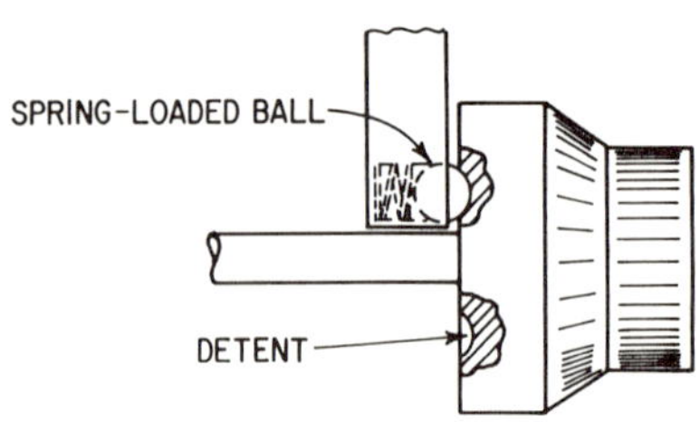

Fig. D-14. One form of detent mechanism.

—Robert L. Wilson, *Honeywell Inc.*

Development Chromatography (See **Chromatography.**)

Deviation In an instrument or control system, any departure from a desired or expected value or pattern.

Dew Point, Measurement of Dew point may be defined as the temperature of liquid water or ice in vapor-pressure equilibrium with an atmosphere. It is common practice to use dew point as a generic term regardless of whether it is ice or liquid water. More precise usage distinguishes between dew point and frost point, which is the temperature of ice (frost crystals) in vapor-pressure equilibrium with an atmosphere. Below 32° F, water can exist as ice or as supercooled water, and the vapor pressures are not the same. Hence, in measuring dew points below 32° F, it is necessary to note which phase is present and to report results accordingly.

Manually Operated Dew Cup: The simplest form of dew-point apparatus consists of a polished-metal cup filled with ether or other volatile material. Air is bubbled through the ether, causing cooling by evaporation. When the observer notes the presence of condensation (dew) on the surface of the cup, he reads the temperature of the ether (hence of the cup surface). Such instruments have the advantage of simplicity and low cost and, if well built and carefully operated, are capable of high accuracy. In another version of the manually operated instrument, the gas whose moisture content is to be measured is passed at a slow rate over a polished-metal target that is cooled by the expansion of liquid propane, butane, or carbon dioxide. When condensate appears on the target, the temperature is read. With these instruments, under laboratory conditions, an accuracy of 0.2° F is obtainable. Under field conditions, errors of 5° F are not uncommon and 1° F accuracy is considered very good.

McMahon Frost-point Tester: This is a special-purpose dew-point instrument that operates in the range −40 to −76° F. It uses bottled carbon dioxide for refrigerant with a back-pressure valve to hold a steady pressure in the cooling system, thus maintaining the target temperature steady at any desired level. The appearance of frost is observed optically. Temperature is measured by a vapor-pressure thermometer whose bulb forms an integral part of the target structure.

Fog Chamber: A version of the Wilson cloud chamber containing a sample of the gas under pressure is suddenly vented to the atmosphere. The initial pressure is adjusted by repeated trials until the cooling due to expansion is just sufficient to cause condensation in the form of a fog in the chamber. Self-contained lighting of the chamber is arranged to make the tiniest fog particles readily visible using the Tyndall effect. Dew point is computed from the ratio of expansion and the initial temperature.

Mechanized Dew-point Meters: These devices are made to give continuous recording. A mechanical refrigeration system can be used to provide cooling, a photocell views the condensate and controls the refrigeration, and a temperature recorder, usually of the thermocouple type, provides a record. The target is alternately cooled and heated to cause the condensate to appear and disappear on a cyclic basis, the record appearing as a sawtooth line or series of dots. The devices are costly and somewhat complex.

Vapor-equilibrium System: In this approach to continuous recording of dew point, the inherent complications of refrigeration are avoided by use of the properties of a hygroscopic salt. The temperature at which a saturated solution of a hygroscopic salt, such as lithium chloride, is in vapor equilibrium with the atmosphere is measured. In like manner, the classical dew-point method may be regarded as a measurement of the temperature at which pure water (condensate on the target) is in vapor equilibrium with the atmosphere. The temperature for the salt solution, being much higher than the temperature for pure water, is reached by electric heating rather than refrigeration. Structurally, a tube containing a temperature-measuring element is wrapped with a glass-fiber wick saturated with the hygroscopic solution. Two conductors are wrapped around the assembly in contact with the wick and are supplied with low-voltage alternating current. Current flow through the salt solution generates heat, raising the temperature. When the temperature of vapor equilibrium is reached, water evaporates, reducing current flow and heat input. The temperature cannot go higher because all the water would evaporate and heat input would cease. It cannot fall below because all the salt would then be in solution and too much heat input would be generated. Hence, equilibrium is reached with a portion of the hygroscopic salt in solution and conductive and a portion dry and nonconducting, thus balancing heat input to heat losses. Also see **Psychrometer.**

REFERENCES

Considine, D. M., and S. D. Ross (eds.): "Handbook of Applied Instrumentation," sec. 7 (Humidity and Dew Point), McGraw-Hill Book Company, New York, 1964.

Considine, D. M. (ed.): "Process Instruments and Controls Handbook," sec. 7 (Psychrometers, Hygrometers, and Dew-point Meters), McGraw-Hill Book Company, New York, 1957.

Dial Scale (See **Scale, pendulum.**)

Diaphragm Actuator (See **Actuator, pneumatic.**)

Diaphragm Motor Valve (See **Actuator, pneumatic;** and **Control valve.**)

Diaphragm Valve (Saunders) (See **Saunders patent valve.**)

Dichroic Absorption (See **Photometers.**)

Dielectric Transducer (See **Capacitance transducer**)

Differential Electrical Instrument (See **Electrical quantities, measurement of.**)

Differential-mode Voltage The differential- (or normal-) mode voltage is that voltage which appears between two terminals or other points in a circuit, neither of which is necessarily at the system reference potential which is normally designated as "ground." (Compare **Common-mode voltage.**) In terms of differential-measurement instruments, amplifiers, or data-acquisition subsystems, the differential-mode voltage is the voltage appearing between the two input terminals. This is illustrated in Fig. D-15, where $V_A - V_B$ is the differential input voltage. If the differential system is perfect, the measurement reflects only the value of the differential signal voltage V_s. In practice, however, unbalances and other deficiencies in the system and measurement

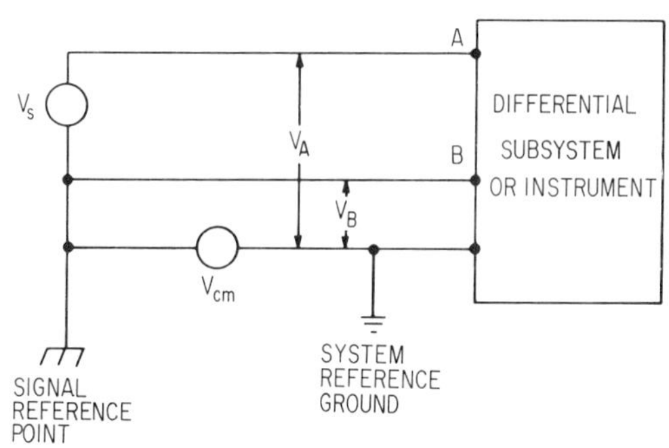

Fig. D-15. Differential-mode voltage.

setup result in some of the common-mode voltage being converted to a differential-mode signal. (See **Error, common-mode.**)

—Thomas J. Harrison, *IBM Corporation*

Differential-pressure Transmitter A widely used process-control instrument, particularly in connection with variable-head flowmeters. The device, sometimes referred to as d/p (differential-pressure) cell or a delta-P cell, also is used in liquid-level measurement and any other kind of measurement wherein differential pressure is the prime criterion of measurement. There are two major types: (1) electric and (2) pneumatic. An electric differential-pressure transmitter is shown in Fig. D-16. Primary beam (1) is fastened to the end of the flowmeter-body force shaft. Differential pressure across the measuring element in the meter body results in a torque being applied to the shaft. This torque attempts to rotate the primary beam in a counterclockwise direction, but as it does so, the air gap of the detector (2) is increased. The detector consists of two pieces of ferrite, one on the primary beam (3), the other mounted on the base. The stationary piece contains an induction coil which is connected to an oscillator circuit. When the air gap increases, the inductance decreases and the output current of the oscillator is increased. This output is fed back to a magnet unit (4), which exerts a force through a strap on the end of the secondary beam (5). The torque generated by this force is transmitted to the primary beam

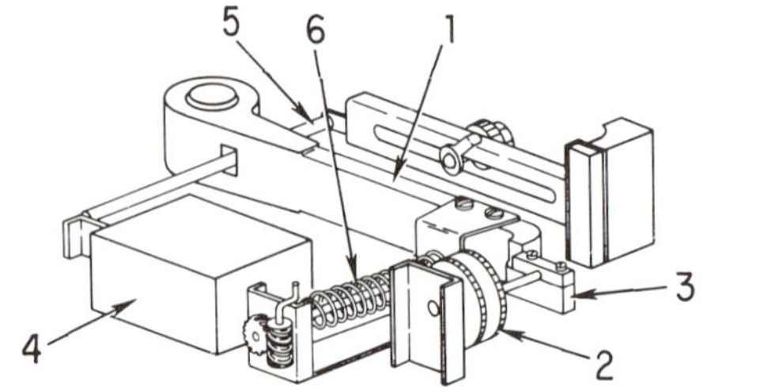

Fig. D-16. Electric differential-pressure transmitter. *(Honeywell.)*

through the adjustable span rider and rebalances the system. Suppression or elevation force is applied through the zero spring (6) to the primary beam. Three different zero springs are available to cover the complete range of suppressions and elevations. The zero spring is attached to this beam, and it carries at its end one of the pieces of ferrite of the detector assembly. Differential pressures are converted into a proportional electric signal of 4 to 20 mA dc. Standard differential ranges are: 0–5 to 0–25 in. water; 0–20 to 0–250 in. water; and 0–100 to 0–1000 in. water.

A pneumatic differential-pressure transmitter is shown in Fig. D-17. The primary beam (1) is fastened to the end of the meter-body shaft, and differential pressure across the measuring

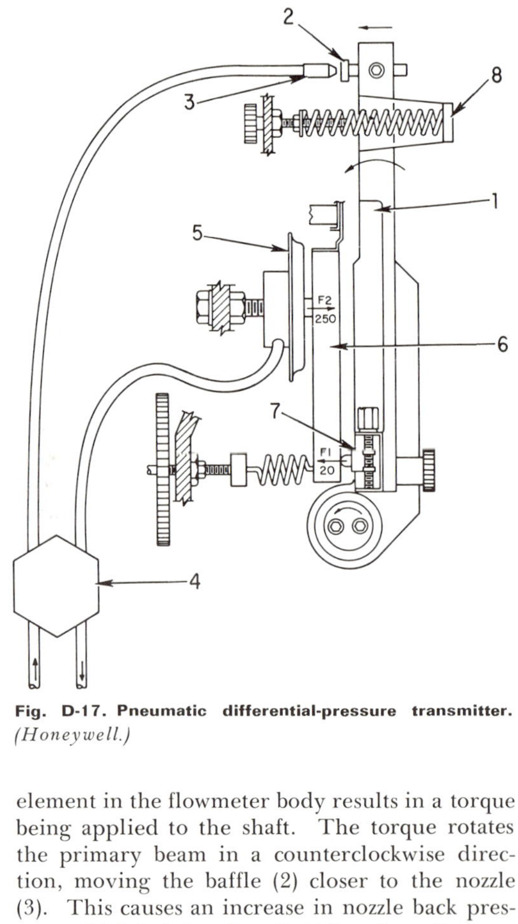

Fig. D-17. Pneumatic differential-pressure transmitter. *(Honeywell.)*

element in the flowmeter body results in a torque being applied to the shaft. The torque rotates the primary beam in a counterclockwise direction, moving the baffle (2) closer to the nozzle (3). This causes an increase in nozzle back pressure, which is fed to the pilot (4) and results in an increased output. The output is fed back into the rebalancing capsule (5) which exerts a force on the secondary beam (6). The torque generated by this force is transmitted to the primary beam through the adjustable span rider (7) and rebalances the system. When a suppression or elevation spring (8) is used, it is adjusted to apply the required amount of suppressing force to the primary beam. The device converts differential pressure into a proportional pneumatic output of 3 to 15 psi.

Also see **Liquid level, measurement of; Manometer;** and **Pneumatic controller.**

—Russell D. Irwin, *Honeywell Inc.*

Differential Producer, Flow (See **Flowmetering devices, primary.**)

Differential Transformer In the parlance of circuitry, a differential transformer combines two or more signal sources to a common transmission line. In terms of instrumentation, the differential transformer is a transducer wherein movement of the iron core of a transformer alters the output voltage that is across two series-opposing secondary windings. See **Eddy-current transducer;** and **Linear-displacement transducer.**

Diffraction Grating (See **Measuring machine;** and **Optical-emission spectrochemical analysis.**)

Diffractometer, X-ray (See **X-ray analysis.**)

Digital-to-Analog Converter A digital-to-analog converter is an apparatus for generating an analog voltage or current which is proportional to the value of a digital-input word. The digital-to-analog converter is often included in the data-acquisition or process-control computer system as a means of generating analog signals under control of the computer system (see **Analog output**). These analog signals may be used to interface with analog devices in the process or experiment being controlled by the computer. Examples of such devices are positioning mechanisms such as valve positioners and graphic-display devices such as *X-Y* recorders, strip-chart recorders, and oscilloscope displays. Digital-to-analog converters may also be used independent of a computer system to provide voltages which are proportional to the settings of input switches.

Most digital-to-analog converters consist of a digital-to-analog converter decoder network, an analog switch for each digital-input bit, a buffer amplifier, and necessary control logic in a configuration such as that shown in Fig. D-18. The decoder network is an electronically controllable attenuation network whose attenuation factor is proportional to the position of the input switches. (See **Decoding network, digital-to-analog converter.**) The analog switches, under the control of the logic, either connect the reference source to the decoder-network input terminal or ground the terminal (see **Switch, analog**). As discussed in the entry for decoder networks, this results in an output voltage proportional to the setting of the switches.

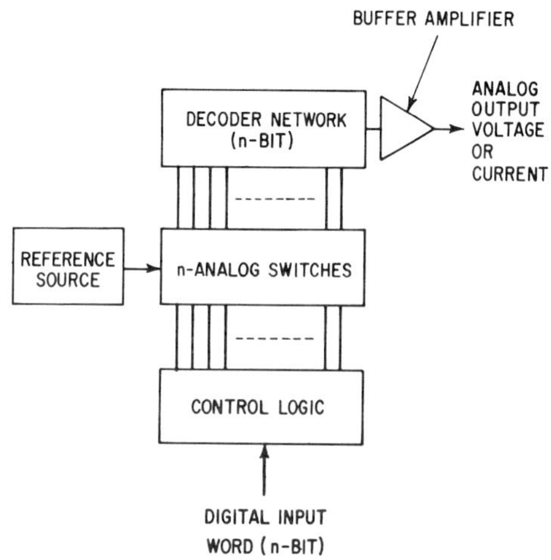

Fig. D-18. Digital-to-analog converter configuration.

The control logic provides digital storage of the input-data word and control of the switch position. In addition, it may provide decoding of addresses and the digital multiplexing of digital data to one of several digital-to-analog converters included in the subsystem.

The buffer amplifier provides impedance-buffering and current-driving capability. It is generally required because the output impedance of the decoder network is relatively high. Driving loads directly from the decoder network can therefore result in loading errors which may be intolerable in a high-accuracy system.

Other digital-to-analog converter configurations are possible. As an example, a digital-to-analog converter consisting of control logic, a decoder network, a reference source, and an operational amplifier is shown in Fig. D-19.

—Thomas J. Harrison, *IBM Corporation*

Digital Computer A digital computer is a device which is capable of carrying out a sequence of operations on information represented by combinations of discrete or discontinuous data, as compared with an analog computer which operates on continuous data. A digital computer has the capability of performing stored sequences of arithmetic and logical operations, not only on data but also on its own program.

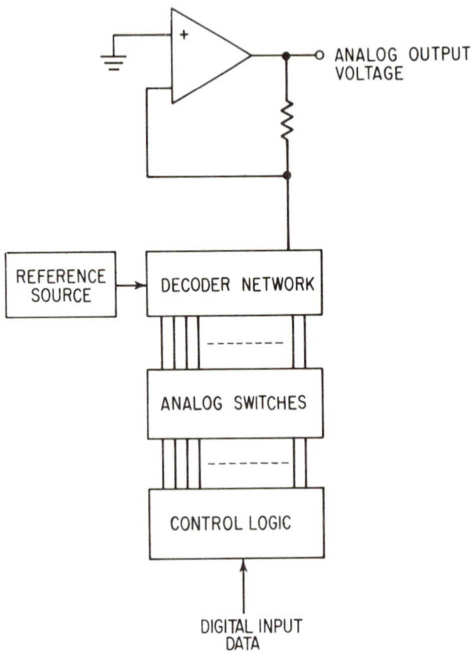

Fig. D-19. Operational-amplifier digital-to-analog converter.

The digital computer generally consists of a central processing unit [see **Central processing Unit (CPU)**] which interprets and executes the instructions of the stored program (see **Program, computer**), the input devices which provide the data to be operated on, and the output devices (see **Input/output devices**) which receive the results of calculations. Also see **Instrument data processing** in the **Classified Index.**

REFERENCES

Miller, W. E.: Digital Computer Applications to Process Control, *Proc. IFAC/IFIP 2d Intern. Conf., ISA Publ. DCA2,* Menton, France, 1967.

Miller, W. E.: Digital Computer Applications to Process Control, *Proc. IFAC/IFIP 1st Intern. Conf., ISA Publ. DCA1,* Stockholm, 1964.

Maurer, Douglas: "Programming: An Introduction to Computer Language and Techniques," Holden-Day, Inc., Publisher, San Francisco, 1968.

Reilly, E. D., Jr.: "The Elements of Digital Computer Programming," Holden-Day, Inc., Publisher, San Francisco, 1968.

Kovach, Ladis D.: "Computer-oriented Mathematics," Holden-Day, Inc., Publisher, San Francisco, 1964.

Kuo, F. F., and J. F. Kaiser: "System Analysis by Digital Computer," John Wiley & Sons, Inc., New York, 1966.

Lee, T. H., G. Adams, and W. M. Gaines: "Computer Process Control," John Wiley & Sons, Inc., New York, 1968.

Nashelsky, L.: "Digital Computer Theory," John Wiley & Sons, Inc., New York, 1966.

Lapidus, L.: "Digital Computation for Chemical Engineers," McGraw-Hill Book Company, New York, 1962.

Ledley, R. S.: "Digital Computer and Control Engineering," McGraw-Hill Book Company, New York, 1960.

Chu, Y.: "Digital Computer Design Fundamentals," McGraw-Hill Book Company, New York, 1962.

Bartee, T. C., I. Lebow, and I. S. Reed: "Theory and Design of Digital Machines," McGraw-Hill Book Company, New York, 1962.

—W. W. Welz, *IBM Corporation*

Digital-control, Process (See **Direct digital control.**)

Digital-control, Drafting (See **Drafting system, automatic.**)

Digital Counter (See **Counter.**)

Digital Electric Instrument (See **Electrical quantities, measurement of.**)

Digital Input The digital-input features of a process-control computer system are devices which permit the reading of external digital information represented by contact closures or voltage levels.

Both points of a contact to be sensed are connected to the digital-input device of the computer. One point is connected to a common-voltage bus, and the other point is connected to a digital-input terminal. If the contact is closed, the voltage is recognized by the circuits connected to the digital-input terminal as a binary 1 (See **Binary state**). If the contact is open, the absence of the voltage at the digital-input terminal is interpreted as a binary 0.

In reading digital information by means of voltage levels, one voltage level is interpreted as a binary 1 and another voltage level as a binary 0. For example, a voltage in the range of 0 to + 20 V may be read as a binary 1, and a voltage in the range of -10 to -30 V may be interpreted as a binary 0.

The digital-input signals are generally arranged in a group and are read under the control of the computer program. The execution of the computer program (see **Program, computer**) which reads the digital-input data may be initiated as a response to an external "interrupt" signal (see **Interrupt**). The digital-input data can be in various formats, depending upon the devices to which they are attached. The data may be bit-significant (that is, each bit signifies a unique condition or event unrelated to the other bits in the group), or the entire group could be connected to the output of a counter or a digital voltmeter and therefore represent a numerical value.

—W. W. Welz, *IBM Corporation*

Digital Motor (See **Stepping motor.**)

Digital Output The digital-output features of a process-control computer are devices which permit the transmission of digital signals to external devices. The digital-output information may take the form of a contact-operate signal or a voltage-level signal.

Under the control of the computer program (see **Program, computer**), the digital information is transferred from the central processing unit to a digital-output register (see **Register, computer**). The output of the register may be connected directly to the digital-output signal lines to provide the voltage-level signal. To provide the contact-operate signals, the output of the register is connected to a driver circuit which is connected to the digital-output signal lines. When a register postion is turned on, the drive circuit provides the current which actuates an external device such as a relay. The information remains in the register until another computer command to the same group of digital-output points is executed, or the register may be reset as the result of a signal from the external device or after a fixed time interval has elapsed.

—W. W. Welz, *IBM Corporation*

Digital Voltmeter (See **Electrical quantities, measurement of.**)

Digitizer (See **Drafting system, automatic.**)

Dimension, Measurement of The following measurands may be involved when considering dimension: (1) length, (2) thickness, (3) diameter, (4) distance, (5) angle, and (6) volume. Numerous bench tools, transducers, instruments, and systems are available to sense and to measure these variables. The important devices are described in this volume. Particularly, see **Gaging, pneumatic**; **Linear-displacement transducer**; **Linear transformer**; **Measuring machine**; **Metrology**; and **Thickness gaging**. Also consult alphebetical index at the end of this volume.

Direct Digital Control A process-control technique—"direct" in the sense that the process loops are tied in directly with computers— and "digital" because the output signals, as compared with conventional analog controllers, are digitally calculated by a computer. A typical example of direct-digital-control loops is shown in Fig. D-20. Also see **Electric controller.**

Raw material is preheated in a furnace and then introduced into a reactor. The main part of the furnace fuel is supplied by purge gas or waste gas from other parts of the plant. When the supply is insufficient, the difference is made up by liquid-oil fuel from storage tanks. Since the percentage of reaction (A001) is a function of the reactor-feed temperature (T001), the control objective is to regulate T001 in accordance with the desired target of A001 by throttling control valves FV001 and FV003.

The signals from measuring devices (pneumatic in this example) enter the input section, where signals are read into the computer through the conversion of an analog-to-digital converter. This usually is done on a time-sharing basis. In this particular case, a multiport pneumatic scanning valve is used. The same conditions exist for the output section, where computed output signals are read out periodically through digital-to-analog converters on a time-sharing basis.

The algorithm used in direct digital control varies in accordance with the applications. For illustrative purposes, it may be assumed that an algorithm equivalent to the conventional three-mode analog control is to be used. This may be expressed mathematically by

$$m(t) = \frac{100}{p}\left(e + D\frac{de}{dt} + \frac{1}{R}\int e\,dt\right) \qquad (1)$$

where m = manipulated variable
e = error
t = time
p = proportional band, percent
D = derivative time
R = reset time

In actuality, Eq. (1) must be converted into a difference equation so that the relationship can be handled by a digital computer,

$$M = \frac{100}{p}\left(E + D\frac{\Delta E}{\Delta T} + \frac{1}{R}E\,\Delta T\right) \qquad (2)$$

Therefore, a change in the manipulated variable (the real object of interest) will be

$$M = M_n - M_{n-1} \qquad (3)$$

$$M = \frac{100}{p}\left\{(E_n - E_{n-1}) + D\left[\left(\frac{\Delta E}{\Delta T}\right)_n - \left(\frac{\Delta E}{\Delta T}\right)_{n-1}\right] + \frac{1}{R}(E_n - E_{n-1})\Delta T\right\} \qquad (4)$$

$$M = \frac{100}{p}\left[\Delta E_n + \frac{D}{\Delta T}\Delta(\Delta E)_n + \frac{1}{R}\Delta E_n \Delta T\right] \qquad (5)$$

where $T =$ sampling interval of process-control loops

$$\Delta(\Delta E)_n = \Delta E_n - \Delta E_{n-1}$$
$$= (E_n - E_{n-1}) - (E_{n-1} - E_{n-2})$$
$$= E_{n-2}\ E_{n-1} + E_{n-2}$$

The foregoing example, given for purpose of illustration, may not be all-inclusive. There may be some problems with regard to the interactions in frequency domain among proportional, derivative, and reset terms while tuning; or overshoots when either set point or measured variable experiences a substantial step change; and other factors. Equation (1) can be modified to include these functions. However, no attempt is made here to go further into control algorithms, since these depend very much upon specific applications and cannot be generalized as such.

A computer can be programmed to handle various types of control, such as cascade, ratio, feedforward, adaptive, and process-optimization—which conventional controllers may be incapable of or unqualified to handle. The system shown in Fig. D-20 is a good example of feedforward direct digital control. Reactor-feed temperature T001 can be calculated according to a mathematical model, simulating the chemical reaction taking place in the reactor. This temperature then is used to predict the fuel requirement according to heat-balance equations.

In addition to what has been mentioned previously, the digital computer may be programmed to perform other functions, such as alarm scanning, logging, sequencing, and logic operations. Although these functions usually are furnished as part of a direct-digital-control package, these auxiliary functions normally are not considered direct digital control per se. Direct digital control implies only those control functions which, after accepting signals from process sensors, direct the involved processes to the desired target points by sending appropriate control signals to the final control elements.

A major concern of direct digital control users

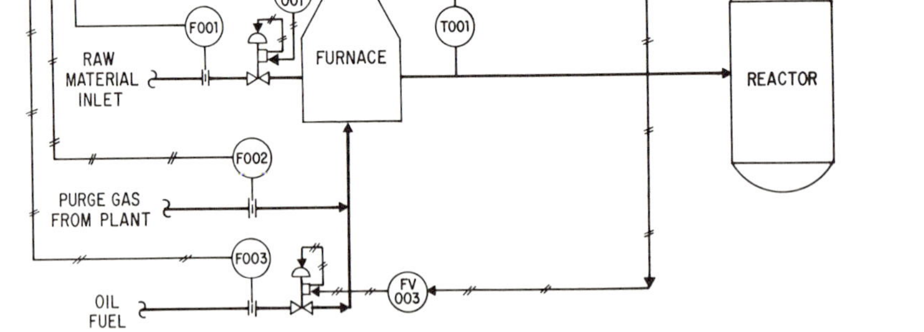

Fig. D-20. Direct-digital-control system. *(Bristol.)*

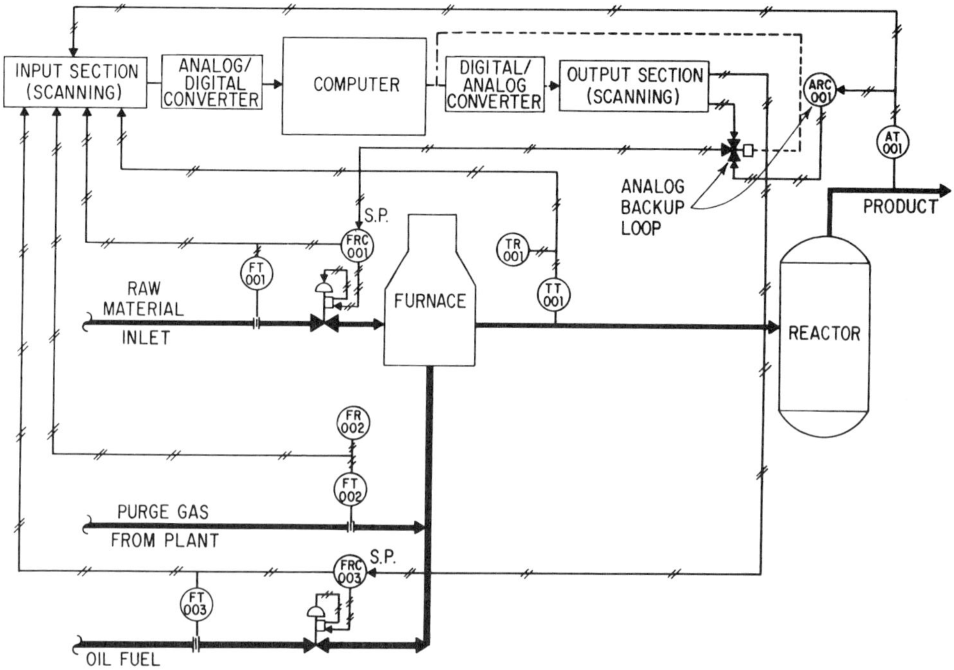

Fig. D-21. Hybrid control system—a combination of direct digital control and conventional analog controllers. *(Bristol.)*

is that of computer reliability. The conservative solution is a hybrid system, of which Fig. D-21 is a typical example. The prime difference between Figs. D-20 and D-21 is the presence of conventional analog controllers in Fig. D-21. For high reliability, the hybrid approach, particularly for critical loops, may be required for quite a long time—at least until some solution to complete reliability of the digital system is found other than costly redundancy.

It is fairly common practice to run a standby computer side by side with the main control computer. The two computers usually share the same input and output bus, and both computers are updated in each scanning cycle so that at the moment of failure one computer can take over from the other.

REFERENCES

DeBolt, R. R., and B. E. Power: A "Natural" 3-mode Controller Algorithm for DDC, *ISA J.*, September, 1966.

Partain, C. L., and R. K. Adams: The Conventional Three-mode Algorithm for Multi-loop DDC Using a Small Digital Computer, ISA Annual Conference, New York, 1968.

Fertik, H. A., and C. W. Ross: Direct Digital Control Algorithm with Anti-windup Feature, ISA Annual Conference, Chicago, 1967.

Woodley, G. V.: Building-block Approach to DDC Algorithm Implementation, ISA Annual Conference, Chicago, 1967.

Lombardo, J. M.: The Place of Digital Backup in the Direct Digital Control System, Spring Joint Computer Conference, Atlantic City, N. J., 1967.

—Anthony Y. Chiang, *Bristol Division, American Chain & Cable Company, Inc.*

Disappearing-filament Optical Pyrometer (See **Optical pyrometer.**)

Discharge Gage (See **Vacuum gaging.**)

Disk File (See **Disk storage.**)

Disk Storage A computer storage in which data are stored as small magnetized spots arranged in circular tracks in magnetic coatings of rotating disks. A disk pack is a set of magnetic disks that can be removed from a disk storage as a unit.

Displacement, Measurement of Instrumental techniques are available for the measurement of displacements as small as a wavelength of light to several feet. Generally, displacement is thought of in terms of a motion of a few millimeters or less. Frequently, a measurement of displacement is made to relate to some other measurand, and hence displacement transducers are fundamental components of many instrumentation systems. Examples would include the measurement of motion of the free end of a bourdon, the movement of a scale beam, and deflection in an accelerometer. Displacement obviously is associated closely with motion and position. Displacement implies motion, that is, from one point to another, and it also implies position, that is, a change from one position to the next. Displacement generally implies the establishment of a new position as related to a stable or norm position.

Displacement can be measured by a number of transducers—capacitance, electronic, inductance, magnetoelectric, photoelectric, piezoelectric, radioactive, and resistive. Several practical examples of displacement transducers are described elsewhere and are listed in the detailed alphabetical index at the end of this volume. Particularly, see **Linear-displacement transducer; Linear transformer; Position, measurement of; Resolver;** and **Synchro.**

Displacement Chromatography (See Chromatography.)

Displacement Transducer, Eddy-current (See Eddy-current transducer.)

Displacement-type Flowmeter (See Flowmeter, positive-displacement.)

Display-command Operator/Machine Interface The communications threshold that exists between the machine or process side of a system and the human operator may be termed the display-command interface. Information flows in and out of both sides of this interface (see Fig. D-22). The machine or process communicates with the operator by means of displays. The operator communicates with the machine or process by means of command hardware. For any given application, the display-command interface must meet one prime criterion, namely, that of providing a reliable, easily understood and manipulated, convenient, and adequate, but not necessarily sophisticated, communications link.

Display-command Objectives: Hardware for display and command functions ranges from a single digital readout or analog record and a knob or toggle switch or two as in the case of a simple instrument that may be used for indicating or recording measurements in connection with a laboratory experiment (where no control commands are involved)—to a scheme where scores of variables may be interrelated in some complex chemical process with great amounts of data involved and important command decisions to be made and transmitted to the process.

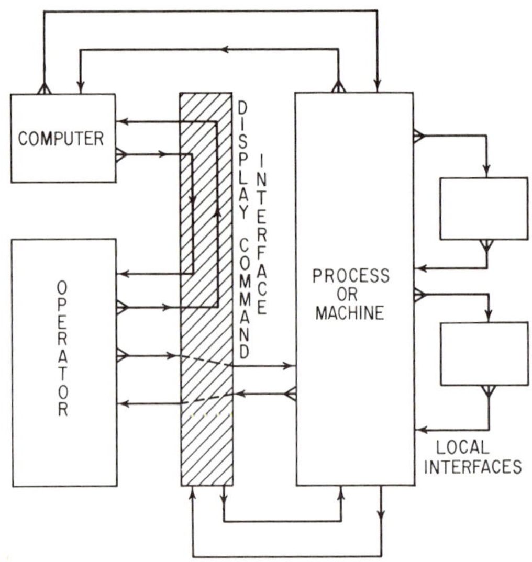

Fig. D-22. Major information flows demonstrating importance of the display-command interface.

The end use of the information involved will affect the choice of display-command hardware. End uses include (1) operation of a machine or process where data presentation will be continuous and where the operator can put new settings into the system rapidly and accurately, (2) monitoring of variables of secondary importance, (3) recording or tabulating information for cost accounting and other historical performance reasons, and (4) gathering data for a research project. This is but a partial list of possible functional end uses of information. In connection with data upon which process or machine-operating commands will be based, a further breakdown of uses would include (1) needs that apply only during periods of start-up and shutdown; (2) means for changing throughputs and production rates; (3) emergency-situation requirements; and (4) equipment-switchover situations where manual intervention is required.

The end-use objectives will determine, in turn, the priorities of information, that is, relative importance of some information with other data that will be brought to the display-command interface. Priority will determine the scheduling of information collection and presentation. Timing may range from continuous to regular periodic sampling to displaying only when certain information falls outside of what may be considered nominal parameters. The importance or lack thereof of storing information also is an important consideration in the selection of hardware.

Location of data presentation also will affect the selection of the display-command format. Local mounting (as with most machines) is simpler and less costly than remote-control panels. Too often the foregoing inventory of display-command needs is not carefully analyzed. Over-engineering of the interface occurs quite often. Usually a second-generation machine or process will incorporate a much simplified version of the original interface. Possibly the greatest source of waste in the field of instrumentation and control is the collection of information that really is not needed. On the other hand, far too little attention is paid to the application of human-factors engineering in the design of the interface so that the human operator literally becomes a part of the total system rather than playing the role of an appendage to the system.

The operator's reaction is greatly influenced by two major factors: (1) the kind of information he is given and (2) the way that information is presented to him. The information he is given is a direct function of his involvement with the controlled process, and this varies widely from industry to industry and process to process. For example, the operator is very much involved in a continuous process that is under semiautomatic control since he has to adjust set points and monitor output variations. In a fully computer-controlled process, however, the operator is involved only when something goes wrong or during the start-up phase. The method of presenting the information to the operator depends primarily on what he is presently accustomed to, modified by what he willing to adapt to. For example, if he is used to judging a meter reading by the position of a needle on a scale, presenting the reading in digital form on a numeric readout can completely confuse him. The display designer has to know the practices that are current in the industry for which he is designing before he finalizes an approach. Often, illuminated displays are not the accepted practice, analog display instrumentation being sufficient.

Continuous Processes: Although computer control is making great inroads in the continuous-processing industries, the majority of processes are under semiautomatic control—where control loops hold the process variables to set points that the operator manually establishes from a console or panel. What is often overlooked is that the operators live with the process for at least 8 hours a day, 5 days a week. Once they know their process, the operators really do not require a display that graphically informs them of every stage in the process (see Fig. D-23). Generally, indicating controllers that are grouped in functional blocks are sufficient (see Fig. D-24). A prime consideration in designing a display is to keep the operator happy—and informed. Operators generally react affirmatively to recorders, because the operator is happier if he sees a little wiggle in the trace so that he knows that something is going on. This philosophy of letting the operator know that action is occurring is important to any display, even though the process may be under computer control.

Batch Processes: Typical batch processes are blending ingredients in the food industry, charging blast and basic oxygen furnaces in the steel industry, and compounding resins in the plastics industry. These processes have a commonality in that their interaction with the operator is periodic. In systems of this type, the operator is responsible for setting into the control unit the quantities and types of material that are to be combined, and for monitoring the status of the weighing and selecting the equipment to ensure that the correct mix is obtained. In some applications, he also must monitor the conditions in the mixing vessels to verify that controlled variables are maintained at the correct level and for the correct time. The display panels for these systems tend to be more complex than for a continuous process, mainly to aid operator recall of the sequence of operations. Graphic panels are useful here because by illuminated indicators, either mounted onto a schematic panel or incorporated in a backlighted panel, the status of each piece of equipment in relation to the others can be observed.

Semicontinuous Processes: The primary difference between a semicontinuous process and

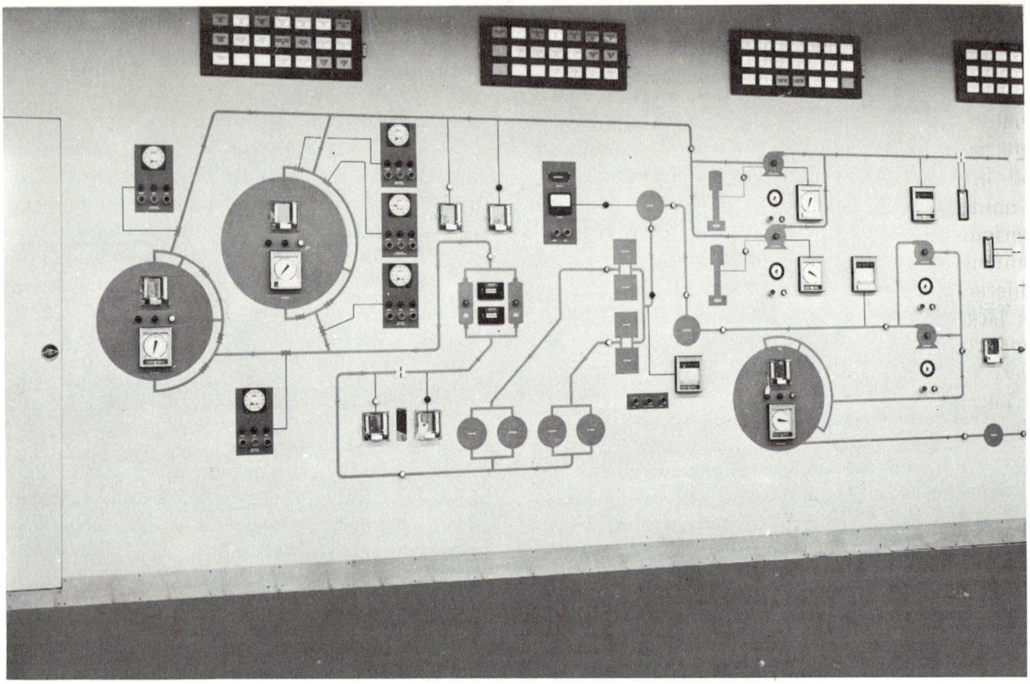

Fig. D-23. Graphic panel in which miniature instruments are inset into a diagram of the process being controlled. Graphic panels are particularly useful for displaying batch-type processes where operating schedules and cycles are complex and difficult for the operator to commit reliably to memory.

a batch process is that the material input to the former process is already "mixed." Typical semicontinuous processes in the metals industry are hot-strip mills, plate mills, and tinning lines. The operator's involvement in a semicontinous process is first and foremost to ensure that the input material is correct for the setting of the processing equipment. If it is not, he must either change the setup or reject the material. A plate mill is a good example. The operator must know the dimensions of the slab and its temperature and metallurgical characteristics before he can adjust the roll gap and decide how many passes are needed to reduce the slab to the required plate dimensions. With material tracking by computer being introduced into steel mills, the operator's job is simplified since he is given a rolling schedule that is automatically synchronized with the slab reaching his station. The information may be printed out on a teletypewriter or in alphanumeric form on a cathode-ray tube.

Digital readouts are commonly used in rolling-mill control panels because mill-setup data are conveniently expressed in digital form. A typical example is roll-gap setting. A numerical readout mounted above thumbwheel switches (for changing the setting) displays the actual roll gap. The operator can quickly compare the setting he requested with what it actually is.

In most semicontinuous processes, graphic panels are not too common because there usually is only one machine involved, the process being a multistage operation. Multistand rolling mills are an obvious exception, but in these the graphic panel is not used because the material moves too quickly through the stands for the operator to track it visually. Continuous-casting machines are an exception, and graphic panels often are used. In this case, the material moves relatively slowly, and there is the added factor that it is almost impossible to track the material by eye because of the length of the strand and the vapors and steam in the atmosphere during the casting operation.

Process Start-up: Although almost every controlled process has a start-up phase, in some systems the operator's primary job is to turn on machines, open and close valves or breakers, fire boilers, run up turbines, and the like. Examples are electrical generating stations and building environment-control centers. Equipment-status reporting and alarm logging are major requirements for the display panels in these systems. The key design objective is to display what will give the operator a good feel for the overall status of the system without visually overwhelming him when something goes wrong. Acceptance of cathode-ray tubes is growing in the area of alarm reporting, because all alarms can be centralized in one compact area on the control panel. The urgency of an alarm can be signaled by causing the indication on the cathode-ray-tube screen to "wink" or change color.

Predictive displays are an important part of any start-up system. Predictive displays require some form of computational capability in the control system, since they must be able to extrapolate from existing conditions. Numeric readouts can be used to display the predicted values, but a line-drawing cathode-ray tube can give an analog display of the expected progress of the process. Typical messages that a predictive display can produce are the speed that a turbine will reach by a certain time or the fact that the safe operating temperature of a boiler will be exceeded if the firing continues at its present level.

The displays for a modern computer-controlled start-up of a 540-MW once-through boiler-turbine combination include:

Pushbutton Control Stations: All functions, such as start-stop, manual-automatic, raise-lower, and open-close are initiated by pushbuttons. Related functions are grouped together under a common bezel. The pushbutton arrangement facilitates operator understanding and reduces the posibility of errors. The pushbuttons are backlighted and color-coded to confirm the operating mode.

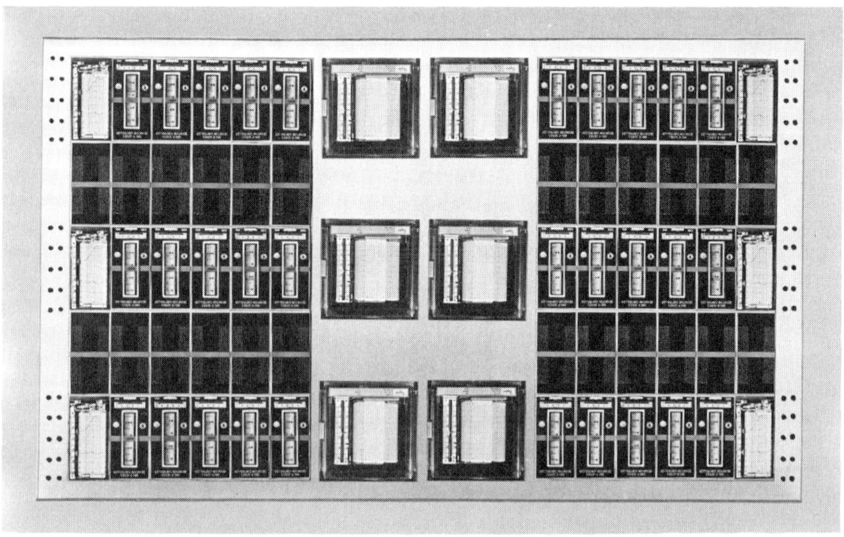

Fig. D-24. Functional groups of indicator controllers with vertical scales are popular for continuous-process control and display. Set points are positioned so that they all align with a horizontal bar on the face of the panel. The drift of any variable can be quickly noted. *(Taylor.)*

Vertical-scale Indicators: Critical variables are displayed on 6-in. vertical-scale indicators mounted integrally with pushbuttons so that the information is displayed at the point of use.

Closed-circuit Television: Four television monitors display furnace conditions and area surveillance for security purposes.

Window Annunciators: Conventional backlighted annunciator panels are provided for monitoring critical variables.

Trend Recorders: Four conventional strip-chart recorders give trend information of such parameters as fuel flow, air flow, oxygen, combustibles, and steam temperature to assist the operator in making split-second decisions regarding furnace conditions. Two other trend recorders are controlled by the computer to give trend information on operator request.

Cathode-ray-tube Displays: One cathode-ray tube displays alarm information as detected by the computer. Any variable found to be abnormal is displayed as a line of alphanumeric information, which includes the variable in engineering units, its identification, and the engineering-unit alarm limits. The variable is continuously updated as long as it is off-normal. The most recent variable is displayed as the top line on the cathode-ray tube. All lines are stepped down as new alarms are detected. A line is cleared from the cathode-ray tube when the variable returns to normal. Alarm information also is typed on a typewriter to provide a hard-copy record. A second cathode-ray tube displays information requested by the operator, who may request a display of any single variable or can request entire groups of related information by depressing a single pushbutton. Critical variables can be displayed by depressing specially engraved pushbuttons. Some examples are the hottest superheater-tube temperature, the highest turbine-bearing vibration, the hottest turbine-bearing temperature, and the hottest boiler-metal temperature.

Environmental Systems: Controlling a building environment differs from a process that is making a product. The time response of the control system in the former case is much longer and less critical. This is reflected in the design of the display system, since indicators are commonly time-shared among many loops. In environmental systems, slide projectors are commonly used to project schematic diagrams of the various loops. These schematics usually have address points printed on them. The points can be accessed by the operator keying them into a keyboard on his console. Security-surveillance displays often are incorporated in the environment-control panel so that one operator can keep an entire building or building complex under complete control. Logging typewriters are commonly used in building control instead of banks of annunciators. These typewriters print out the time, location, and values of the variables generating the alarm.

Display-command Interface Hardware: Several specific types of displays and switching methods are described elsewhere in this encyclopedia. Reference to the **Classified Index** under Display is suggested. The following descriptions are confined to the highlights of a few major categories of display-command instrumentation.

Miniature Indicating Electronic Controllers: These devices now comprise the garden-variety instrumentation format for process-control panels. Most have a similar outward appearance to the casual observer—a small frontal area and great depth. Instruments of this type are available from a dozen or more major makers and are generally designed for high-density arrangements on panels or for placing within process diagrams on graphic panels—hence the emphasis on small frontal area of rectangular shape, ranging from approximately 3 by 6 in. to 6 by 6 in. to 3 or 4 by 9 in. All the instruments have basic indicating and switching functions on the front panel, display of important loop variables, and transfer between automatic and manual outputs. This general type of instrument is diagramed in Fig. D-25. The controller receives a standard input signal from a transmitter that monitors the process variable. The device determines the deviation between the process signal and an internally generated set-point signal.

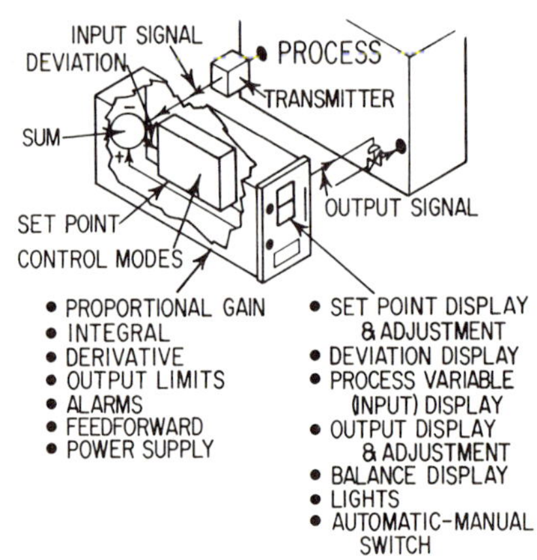

Fig. D-25. Major display-command elements of a miniature electronic controller.

Control-mode hardware then modifes the deviation signal by some combination of amplification, integration, and differentiation to produce a standard electrical output signal. The output signal ultimately goes to a valve or other final controlling element designed to affect the process in a way that reduces the deviation. Besides having its own set-point signal, the controller may receive a remote set-point signal generated by another controller (cascade control) or by a computer.

The operator gets four basic indications from the front panel: (1) set point, (2) value of process variable, (3) deviation, and (4) controller output. In some cases, the process value is implied from indications of set point and deviation. Most controllers, however, indicate the process variable only if the deviation is within a certain range. The operator also commands certain adjusting and switching functions. He can adjust manual output and the set-point value from the front; and he can switch from manual to automatic without abruptly changing the controller output. For switching from automatic to manual, the operator generally needs a fifth kind of indication, that is, balance. This allows him to adjust the manual output to equal automatic output before transfer.

Set-point indication in the controllers takes three basic forms: (1) centered expanded scale, (2) a meter, or (3) a calibrated dial. The expanded scale is the most common. The set-point scale is a mechanical arrangement that consists of a relatively long tape or drum ranging from 6 to 10 in. The scale travels behind a window that shows only a portion of the scale length (see Fig. D-26*a*). The operator adjusts the position of the scale until the desired set-point value rests at the center of the window, usually indexed by a hairline or green band. The mechanical movement of the drum or tape takes part in generating an electrical set-point signal. Commonly, the mechanism for scale movement also drives a multiturn potentiometer. In some designs, a wiper attached to the drive contacts a large fixed circular slidewire connected as a voltage divider. The resulting electrical set point then

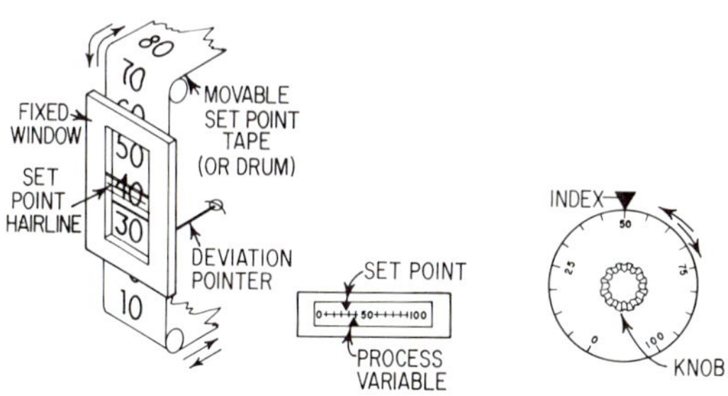

Fig. D-26. Types of set-point display; (a) centered expanded scale; (b) meter; and (c) calibrated dial.

is compared with the process variable to determine the deviation.

The expanded set-point scale serves as a backdrop for a deviation pointer that travels the length of the window. The meter movement matches the visible portion of the set-point scale. For example, if the window shows 40 percent of the set-point scale, the deviation movement is ±20 percent. In this manner, the deviation pointer also indicates the process value as long as this value remains on scale. For zero deviation, the pointer rests at the center of the window at the set point (control point). In another design format, the set point is indicated simply by displaying its electrical value on a galvanometer as shown in Fig. D-26*b*. The calibrated dial is the third way to show set-point value, in which case a calibrated scale attached to the set-point adjustment knob mechanically indicates the knob's rotary position against a fixed index or pointer (see Fig. D-26*c*). The knob attaches to the shaft of a potentiometer.

Those controllers with tape or drum set-point scales offer relatively high readability and resolution, but the actual deviation must be within the range of the meter for the operator to read process values. If the deviation is not on scale, the operator must hunt for the process value by changing the set point, or he must plug in an auxiliary display.

Most controllers have a separate meter to indicate controller output. These meters usually are 2 in. long. Depending on the transfer-switch position, the meters indicate either manual or automatic output. Usually, interchangeable tags at the side of the meter show the control-valve position, open or closed, for the extremes of controller output.

The controllers also have provisions for transferring between automatic and manual operation and for adjusting manual output. Going from manual to automatic is important during plant start-up, and automatic-to-manual procedures come into play during emergencies and plant shutdown. Switch types include lever, toggle, rotary, and pushbuttons. The manual and automatic signal outputs originate in different circuits. Usually the operator wants to make them equal before he transfers in one direction or the other. If the two signals are not equal, the switching operation will cause a step change in controller output and will "bump" the process.

In the manual mode of most controllers, the automatic output signal tracks the manual output signal so that the operator need not balance the two signals before transferring from manual to automatic. Usually, the integral (reset) function of the controller does the tracking; this control mode must be present if automatic balancing is desired. Even though the output signal does not jump after a transfer directly to automatic, the integral function will immediately begin to drive the controller output if a deviation exists between the process value and the set point. This driving action may cause an undesirable process bump, especially for a large deviation. Thus, regardless of output-signal balancing, the operator must carefully match the set point to the process variable if he wishes to avoid a bump caused by integral action. The operator must balance the two controller output signals if he wishes to transfer in the direction of automatic to manual without bumping the signal from the controller. To balance signal outputs, the operator must turn to a "balance" position on the transfer switch and adjust the manual output until a meter indicates equal manual and automatic outputs.

The actual indication of balance takes three basic forms: (1) In the nulling approach (see Fig. D-27*a*), each signal feeds one side of a null meter when the transfer switch is in the balance position—and then the operator adjusts the manual output until the meter reads at midscale. This method offers high sensitivity for accurate balancing, especially if the null meter is a large deviation meter. (2) Other designs indicate a null between the automatic and manual output signals on the controller's output meter when the transfer switch is in balance, as shown by Fig. D-27*b*. In still another design, the difference between the two signals is continuously displayed on a small null meter, as shown in Fig. D-27*c*. (3) In another form of balance indication (Fig. D-28), the operator presses a button or lever that momentarily connects the manual output signal to the controller output meter. The operator then adjusts the manual output knob while he repeatedly presses the button. At balance, the pointer remains motionless regardless of the pushbutton's position. With this method, the transfer switch does not need a balance position; the pressed button represents the balancing state.

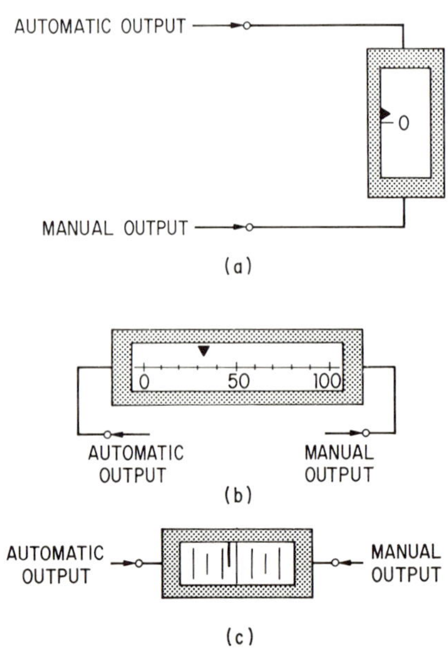

Fig. D-27. Forms of balance indication for making transfer from automatic to manual control: (a) deviation meter; (b) modified output meter; and (c) continuous balance indication.

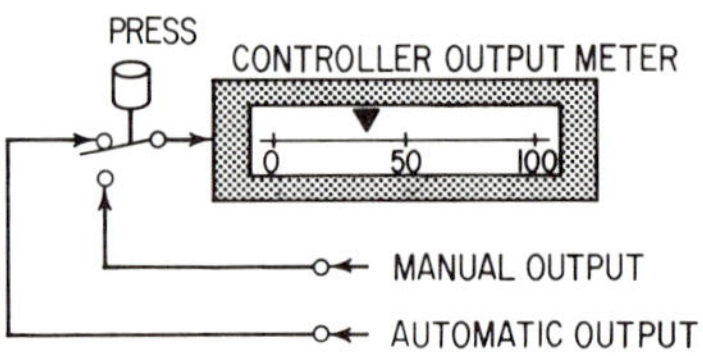

Fig. D-28. To balance automatic and manual outputs with the method illustrated, the operator repeatedly presses the button and adjusts the manual output until the pointer stops moving.

The method shown in Fig. D-29 differs from the others. The switch has the shape of a T. In manual, the control operational amplifier is the simple integrating circuit shown. The manual output remains constant so long as the switch remains in the center top position as shown in Fig. D-29. The operator moves the switch to the left to decrease manual output and to the right to increase it. The top of the T has five positions: one for off (no change) and two each for fast and slow rates of increasing or decreasing manual output. The operator may ramp the manual output from minimum to maximum in either 6 or 100 s. This controller requires no balancing for switching from either automatic to manual or vice versa. The output will remain constant immediately before and after the transfer operation. But, if a deviation exists between process value and set point before transfer from manual to automatic, the same considerations for process bumping as described earlier will apply.

Other features sometimes placed on the front panel of these controllers include lights, legend plates, and communication jacks. Panel lights can indicate alarms, remote set points, and transfer-switch positions. The legend plate identifies the loop under control. Besides loop identification, the legend plates sometimes serve as handles or catch releases for sliding the controller in and out of the case. Communication jacks allow operators in the control room to talk with men in the plant area.

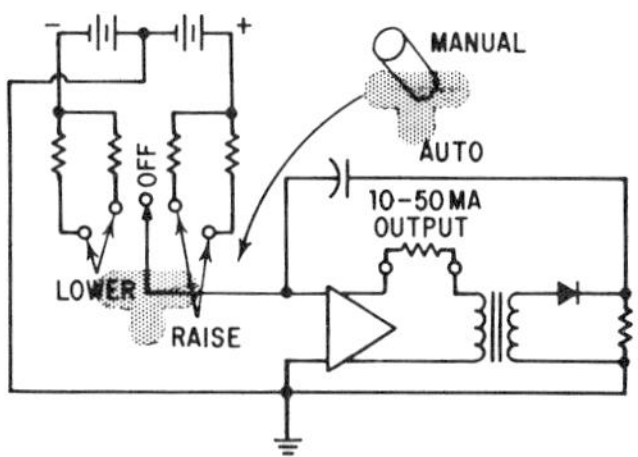

Fig. D-29. To balance automatic and manual outputs with the method illustrated, the operator jogs manual output to higher or lower values by momentarily moving the T switch to the right or left.

Controller size is a factor in fabricating panels, but mounting design is equally important. A few makers require individual cutouts for each station; others can mount standard individual cases side by side horizontally in a single cutout. The "pack" approach also is available wherein the case width is a multiple of two or more controller units. Some suppliers tend to "pack" in the vertical direction by two or more rows of intruments. Some makers supply the single housing, as shown in Fig. D-30*a;* others provide the arrangements shown in Fig. D-30*b* and *c*. High-density installations save panel space between the horizontal rows and cost less by reducing the number of cutouts.

Some makers offer recorder-controllers in the same case, but the usual approach is separation. In other words, the recorder and the electronic controller are individual plug-in stations. The recorder is an element in the input loop of the controller or—in the case of trend recordings—is switched into many inputs. The maximum number of pens ranges up to three. Charts generally are 4 in. wide and run either vertically or horizontally.

Cathode-ray-tube Displays: Much of the cathode-ray-tube display technology evolved from military command and control systems. Among the main applications for cathode-ray tubes is data storage and retrieval. Displays for this can be the relatively simple inquiry-response desk-top terminals, large numbers of which can communicate with one central data processor. Primarily, they give instant visual access to computer-stored records. A popular use of remote-terminal displays which communicate over telephone lines is stock-market quotations. The unit can receive data from all major exchanges, including past performance charts, transactions, financial news, and other related data.

Display systems grow more complex when graphic- and visual-analysis capabilities are required. For many tasks, a graphic input to the computer is the best or possibly the only way to communicate, particularly in computer-aided design or simulation. In chemical research, new compounds are studied visually; molecular structures are computer-generated with certain parameters controlled by the operator. Studies can be conducted in three dimensions; a new-car design can be drawn on the cathode-ray tube and then rotated to be looked at from all sides. Generating visual data requires a great deal of

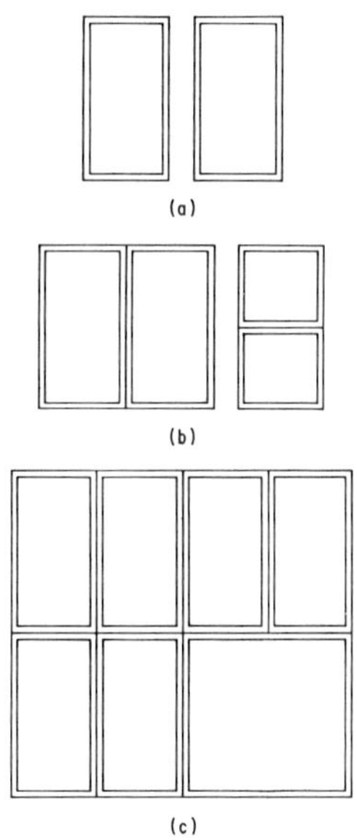

Fig. D-30. Miniature electronic controllers and other instruments can be panel-mounted in three ways: (a) single mounting; (b) multiple mounting, one axis; and (c) multiple mounting, two axes.

software and hardware. One way to ease the computer load and also provide a more realistic picture is to project slide or film images onto the face of a cathode-ray tube and superimpose computer-generated information. This technique is widespread in military command and control systems which, for example, display terrain maps and superimpose troop concentrations and equipment locations. The technique also is useful in connection with aircraft-flight-plan analysis.

A generalized block diagram for a typical cathode-ray-tube display system is shown in Fig. D-31. Entry into the system is at the operator's console, which contains an alphanumeric keyboard for writing instructions or entering data, a function selector to make inquiries or select operating modes, and a communications device common in graphic displays—the light pen. With this device, the operator can point to information on the screen, designate a location at which information should appear, enter information, or even draw a line.

The operator can communicate with a digital computer that is part of his console, one that is located remotely, or with none at all. Simple data-retrieval systems require only a memory bank from which data are extracted, edited, and reentered. Computer information enters the display console through an interface unit which conditions the input to ensure compatible logic levels and data word lengths. Additional signals are provided by the computer for sequencing computer signals with display-generation circuits. Timing and mode control ensures the proper positioning of characters and symbols on the screen.

Display generators are typically required to present three types of visual data: (1) point plotting, (2) alphanumeric characters, and (3) graphic or pictorial views. The simplest function generation is point plotting. This requires binary *X* and *Y* addressing to position the cathode-ray-tube beam. Two output functions are needed to represent the plot adequately: position generation and intensity generation (blanking). The blanking function turns on the beam when the programmed position is reached. Alphanumeric displays are more complex. In addition to beam positioning and blanking, some method is required to generate symbols. This function is handled by a character generator which contains the circuitry to convert digital-coded signals

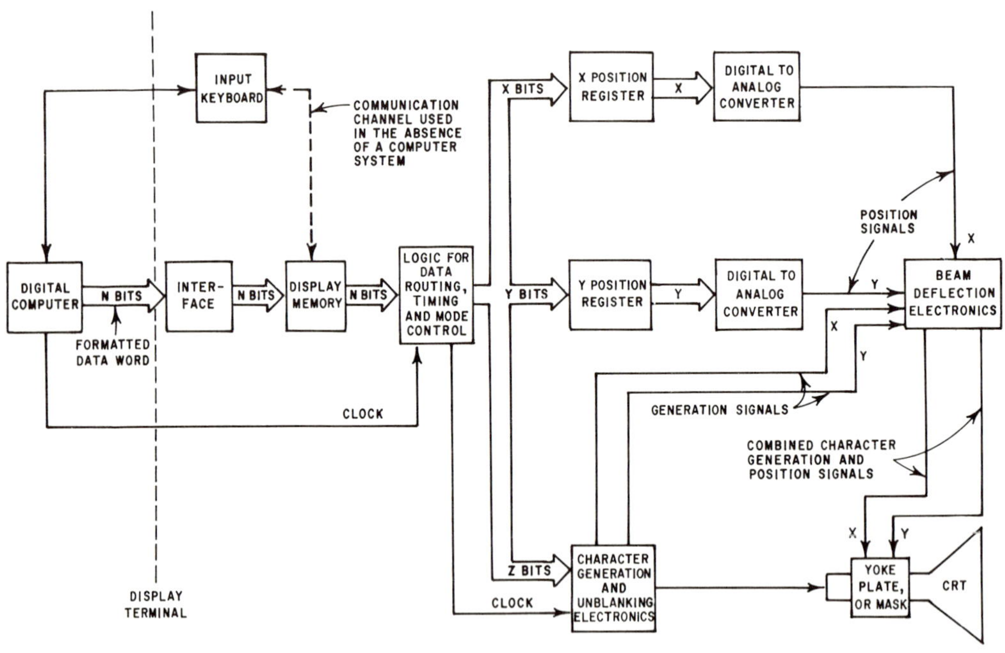

Fig. D-31. Block diagram of typical cathode-ray-tube display system.

into analog deflection voltages to form the symbol. Character generation also can be self-contained within the cathode-ray tube.

To present graphic or pictorial data and limit the increased software load, additional function generators are needed. Vector generators form a line defining the start and end points of the line and join these points with a continuous solid line or a series of closely spaced dots. Circle generators require that the center of the circle and the radius be specified to deflect the beam along the desired path.

The output end of the display is the cathode-ray tube, whose image may be projected onto a large screen or viewed directly. Cathode-ray tubes may be deflected electrostatically, magnetically, or by a combination of both. Electrostatic deflection is fast but requires high driver voltages on the order of 200 V/in. Magnetic-field deflection is slower but produces a smaller spot size and brighter image. One system uses magnetic deflection for positioning and electrostatic deflection for fast character writing. Also see **Cathode-ray tubes.**

Raster systems use techniques similar to television and are less expensive than the positioning technique, but raster systems are more difficult to program, are not as visually appealing, and complicate the use of a light pen.

Displays may be refreshed from the central computer or from a buffer. Introducing a buffer increases the cost. If the terminal is located a considerable distance from the central computer, a buffer is almost mandatory; but if central computer time is available, it may be practical to use it for display refresh. A trade-off is required between a high refresh rate or use of long-storage phosphors. Storage phosphors are not as bright or as effecient and are not easily protected against damage, but there are applications for long-storage phosphors that are practical. Color can be advantageous in numerous applications, such as computer-aided design, but the cost is high and present color sets are limited to three primary colors plus white. Convergence circuits when available will produce the full spectrum of colors.

Acknowledgment: Parts of the foregoing material were excerpted from *Control Engineering,* June, 1968; and *Instrumentation Technology,* October, 1968.

Display Tubes For some instrumentation systems, it is either more effective or more convenient to provide information to an operator on the face of a cathode-ray tube (CRT) than to use other forms of indication, such as a moving pointer or moving dial. See **Indicating instruments.** The choice of a CRT for the information interface with the operator will be determined by several factors, including the rapidity with which the information may change, the complexity of the information, the size of display area required (affecting readability and resolution), whether or not historical information needs to be displayed together with current data, and to a degree, the familiarity of the operator with certain forms of display. For example, CRT displays have not been used very extensively for conveying process-control information, and the use of a CRT in the process-control room represents a marked departure in the operator/system interface. On the other hand, the use of a CRT in the cockpit of an aircraft for navigation or fire control is routine.

Character-generation Tube: In this type of tube, alphabetic, numeric, alphanumeric letters, and various forms of symbols are generated by passing the electron beam through an aperture of appropriate shape. Three basic methods are used to generate character information: (1) The *raster-scan* technique involves controlling the intensity of the electron beam during the sweep. The process is similar to facsimile recording where the characters are generated in segments. (2) The *Lissajous* technique where the electron beam serves as a pencil. (3) The *shaped-beam* technique where the tube incorporates a number of stencil-type openings that are used to shape the electron beam.

Direct-view Storage Tube: This is a CRT in which secondary emission of electrons from a storage grid is used to provide a very bright display over long and controllable periods of time. The tube includes one or more writing guns, a flooding gun, and selective erasing capability. In one use, the image can be viewed on an aircraft instrument panel without a hood in direct sunlight. Aircraft applications include fire control and weather radar. Writing speeds up to 300,000 in./s are possible. A tube of this type also is used in storage-type oscilloscopes where traces can be stored for a period varying from several seconds to many hours or even days. See **Oscilloscope.**

Scan-conversion Tube: Even though this tube is not a display tube per se, it serves an important link in a display system. A scan-conversion tube will convert radar blips (see **Radar**) to a television signal for viewing on a TV screen. The scan-conversion tubes enables information to be put in at one rate and taken out at another rate, thus providing some storage. The technique is particularly useful for retaining aircraft locations in a air-traffic-control instrument so that the path of an aircraft can appear as a dotted line.

Advancements in color, flat, and other forms of cathode-ray tubes are contributing and will continue to contribute to greater utility and flexibility of dynamic display systems. Also see **Cathode-ray tubes; Data communications;** and **Oscilloscope.**

REFERENCES

Barker, Richard A.: Techniques of Dynamic Display, part I, Cathode Ray Tubes, *Control Eng.*, vol. 7, no. 2, February, 1960.

Stevens, Roger T.: Large-scale Displays, *Sci. Technol.*, November, 1968.

Boulding, R. S. H.: "Principles and Practice of Radar," 7th ed., D. Van Nostrand Company, Inc., Princeton, N.J., 1963.

Berkowitz, R. S.: "Modern Radar: Analysis, Evaluation, and System Design," John Wiley & Sons, Inc., New York, 1965.

Dissolved Oxygen, Measurement of (See **Oxygen analyzer.**)

Distance-Velocity Lag (See **Dead time, process;** and **Delay time, transducer.**)

Distortion An undesired change in waveshape. Also see **Interference; Noise;** and **Signal conditioning.**

Disturbance An undesired change in a variable applied to a system which tends to affect adversely the value of a controlled variable. The means to counter the effect of a disturbance on a process is the prime object of an automatic process-control system. An example of *disturbance* is shown in the figure that accompanies the description of a control system. See **Control system.**

Dither (See **Oscillation, measuring-system.**)

Diverting Valve (See **Three-way valve.**)

Divider, Current (See **Bridge circuits, electric.**)

Divider, Voltage (See **Bridge circuits, electric.**)

Doppler Radar (See **Navigation systems.**)

Doppler Spectrograph (See **Astronomical instrumentation.**)

Dosimeter, Chemical Various chemical dosimeters have been developed based on the ability of ionizing radiation to induce chemical reactions. Among these are the ferrous-ferric dosimeters and the cerous-ceric dosimeters. The latter are used for very high radiation doses. The ferrous dosimeter is the most thoroughly studied of these chemical reactions. The ferrous sulfate dosimeter involves the conversion of ferrous ions to ferric ions in an $0.8N$ sulfuric acid solution by radiation-induced oxidation. This detector is most useful for high radiation levels, principally gamma-radiation sources. The amount of ferric ion is determined by comparing the transmission of an irradiated sample with that of a nonirradiated sample on a spectrophotometer.

Draft Gage (See **Manometer.**)

Draft Weigher (See **Scale, industrial.**)

Drafting System, Automatic An automatic drafting system is useful for applications which require (1) precision unattainable by a human draftsman, (2) a large volume of noncreative drafting work, (3) finished drawings from computer-calculated parameters, or (4) rapid graphic display of program information or drawings made up of many redundant components. The graphic output is limited only by the user's ability to describe information in terms of coordinates lying in a plane.

A rather simple graphic-display system can be used to draw cutter paths directly from a numerical-control machine-tool tape. Using an automatic drafting system in this way not only reduces the number of parts which must be scrapped because of errors in the numerical-control tape but also prevents the machine tool itself from being destroyed by an improper command. Simple systems also can produce finished engineering drawings using information from a digitized rough drawing. Detail drawings can be accomplished by scaling up a portion of a drawing and adding the necessary detail information.

Working in conjunction with a large computer, an automatic drafting system can give a hard copy of information described mathematically. As an example, approximately 2,000 equations are required to describe the design of a tire. This information needs to be calculated for only half the width of the tire because the mirror-image feature can be used to draw the other half. Modifications to the basic design for other tire sizes also can be made easily and the modified design drawn. By storing recurring symbols as drawing subroutines, automatic drafting is useful in producing such items as logic diagrams, piping layouts, electrical schematics, flow charts, PERT (program-evaluation-review technique) charts, building layouts, or subdivision plans.

When extremely precise linework is required, such as in the printed-circuit or integrated-circuit technologies, the automatic drafting system is the only practical method of artwork production. Digitized information to form such things as masks for photoetching and chemical milling in the disciplines mentioned may come from a drawing or be generated directly by a computer program. An example of such a program is one which determines the optimum conductor paths for interconnecting given pads on a printed-circuit board (see Figs. D-32 to D-34).

A designer may use a computer with a cathode-ray tube/light pen interface to do creative work. When a pleasing design is achieved, the information stored in the computer can be used to produce a large-scale drawing on an automatic drafting machine. With appropriate software and a large computer, engineering drawings can be produced directly from design specifications for general mechanical forms, such as gears or springs.

Applications in development include mapping and periodic updating of weather or strategic

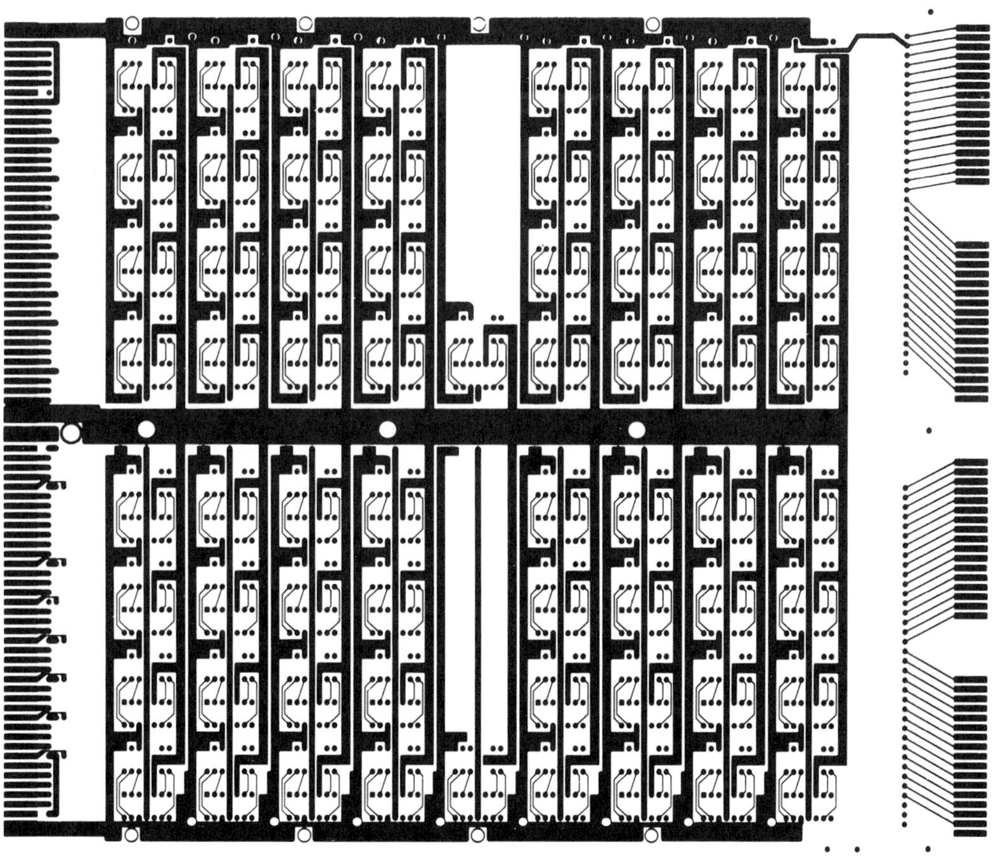

Fig. D-32. Printed-circuit master as prepared by an automatic drafting system.

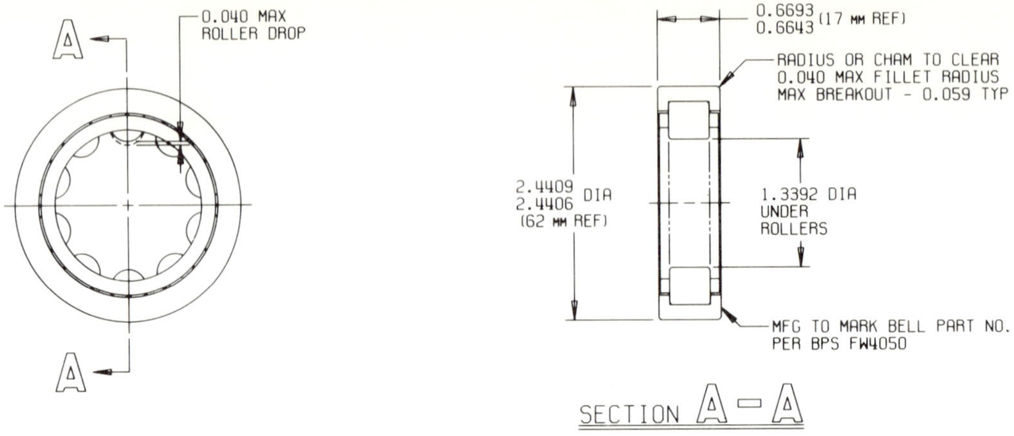

Fig. D-33. Part drawing as prepared by an automatic drafting system.

military information directly as it is received from satellite observatories. The variety of tasks for automatic drafting will increase as further computer systems and software programs are developed.

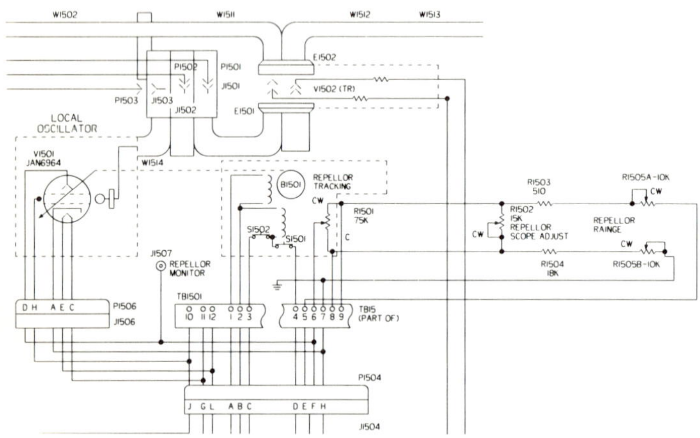

Fig. D-34. Part of an electronic schematic as prepared by an automatic drafting system.

Basic Elements of System: An automatic drafting system rapidly produces highly accurate drawings of quality linework from a program language, cathode-ray tube/light pen sketch, another drawing, or computer calculations. A programmed description is used by a control to move the drawing mechanism on a drafting table to produce a drawing. This description consists of a series of *X* and *Y* coordinates which describe the paths of lines to be drawn and miscellaneous instructions concerning the operation of the drafting table and drawing tools. A large-scale automatic drafting system is shown in Fig. D-35.

A simple drawing consisting of straight lines or circles easily can be described in program form. More commonly, however, the desired result is a complex piece of work. In this case, the programmed description can be prepared by a computer, as in a program to determine the conductor paths for a printed-circuit board from a list of the pads to be connected; or it can be prepared by digitizing an existing drawing. Digitizing refers to the process by which graphic information is described with coordinate data. Complete automatic drafting systems are composed of (1) a digitizer, (2) a drafting table, (3) a control system, and (4) accessories.

Digitizer: A digitizer permits graphical data sources, such as engineering drawings, charts, or maps, to be reduced to digital form as data points in an orthogonal-coordinate system which describe the line paths. The system shown in Fig. D-36 is a manual digitizer and consists of a cabinet which contains solid-state circuitry, coordinate displays, and operator controls; and a large 48 by 58-in. tiltable board fitted with a movable carriage and a reading head. The graphic information that is being digitized is held firmly in place by a vacuum system.

The standard reading head, which is mounted on the *Y* carriage, has crosshairs which are positioned manually over the point to be read. The *Y* carriage moves up and down the *X* carriage, and the latter moves back and forth across the

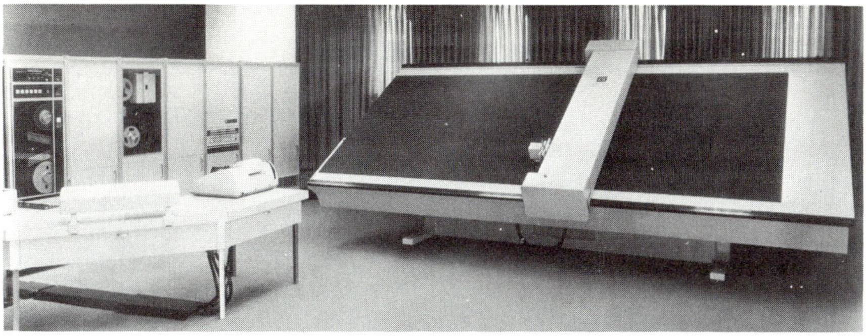

Fig. D-35. A large-scale automatic drafting system consisting of a control (left), an input/output unit (center), and a table (right). *(Gerber Scientific Instrument.)*

board. The position of the crosshairs is monitored by the system circuitry, which uses an up-down counter for each axis. These count output pulses from photoelectric encoders driven through precision gearing by movement of the carriages along mechanically exact racks and ways. The counters count up or down by adding or subtracting the incoming pulses, dependent upon the direction of movement. The total at any instant indicates the absolute position of the reading head with reference to a selected zero point. The output can be selected to be either the absolute position or the difference in each axis direction between the present position and the one at which the last reading was taken. These outputs are referred to, respectively, as being in absolute or incremental form. The resolution of the system can be as fine as 0.001 in. Also see **Numerical control.**

When a reading is taken in the standard system, the coordinates are stored as absolute or incremental data in an output buffer and read serially to a card punch. An optional feature allows the coordinates to be scaled automatically to quarter or half the actual value. Optional outputs record on punched paper tape, typewriter, or magnetic tape. The output is chosen to appear as EIA (Electronic Industries Association) Standard RS-244A character code. A complete program for immediate use with a graphic-display system can be obtained by complementing the coordinate information with the necessary drafting-machine commands (such as pen position) through use of data switches or a carriage-mounted keyboard.

When such information as printed- or integrated-circuit artwork, which consists of many parallel lines, is being digitized, an optional grid-alignment feature will automatically move the point being read to the nearest intersection of two grid lines. The grid size is incrementally selectable from eight sizes ranging from 1 to 0.001 in. This feature enhances the speed with which the information can be digitized by permitting the operator to be less exacting in the placement of the reading head.

Table: The drafting table is the component of a graphic-display system which converts command codes from a control unit into meaningful linework, producing a finished product. The table is called upon to move from an existing position to a new one along a path directed by the control, to produce a line and to draw or print symbols and alphanumeric characters at given

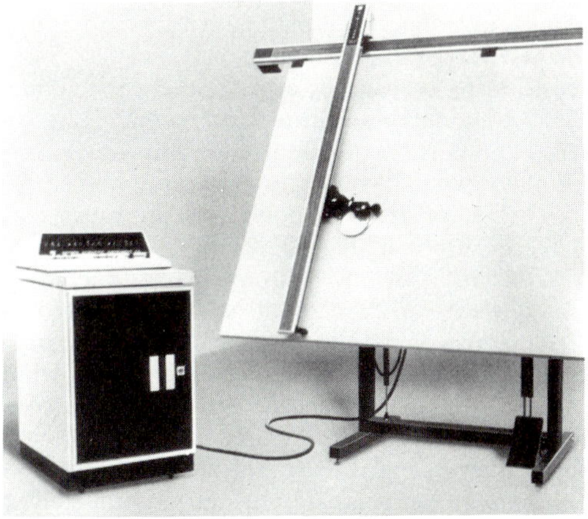

Fig. D-36. Manual, large-area coordinate digitizer used in automatic drafting system. *(Gerber Scientific Instrument.)*

locations. General-purpose automatic-drafting tables consist of a flat bed mounted on a pedestal base which is electrically tiltable to any angle from horizontal to vertical. The bed is constructed of cast aluminum or a precision aluminum plate fastened to box and rib framework which is secured to two rigid trunnion points. The bed of a high-precision table is made of steel plate with waffle-pattern support webbing and is fixed in a horizontal position. The drawing surface is covered by a Neoprene-rubber platen with small holes for manifolding into a vacuum system to hold drawing material securely in place.

The most accurate drafting systems are completely digital because the error inherent in any device required to operate in a linear region is eliminated. Such errors can result from drift in the operating point due to aging, temperature excursions, variations in a reference parameter, and noise sources. Digital systems operate on the basis of variables which have only two discrete levels which define their presence or absence. Since these levels can be widely spaced and each state defined with broad tolerances, outside influences, such as drift characteristics of system electronic components and nominal noise levels, cannot cause erroneous operation.

The carriage in digital systems is positioned by a step-motor drive mechanism which drives each carriage to a new position using digital drive pulses supplied by solid-state amplifiers directly connected to the output pulse trains of the control. Positive high-torque steps are made instantly and without slip or chatter. Each step causes a corresponding movement of the head. The *X*-axis drive is either by a rack and pinion or by two parallel precision ball screws. *Y*-axis positioning is either by a preloaded recirculating ball screw or by a helical lead screw. The carriage can move at speeds up to 750 in./min. Standard drawing heads hold ball-point pens, wet-ink pens, or scribing tools. A high-precision table is shown in Fig. D-37.

Control System: The primary function of the system is to read coded coordinate and miscellaneous command information from an input source and convert it into electrical signals which move the drafting-machine drawing head in the desired manner from an existing position to a new location specified by the given coordinates. The source of information usually is punched paper tape but may be magnetic tape, a teletypewriter, punched cards, or a computer. The most versatile controls consist of a small computer and solid-state logic circuitry which channels the output signals according to a recognized input code. The standard mode of movement is in a straight line between the present position and the new coordinates, called *linear interpolation.* When the control receives new coordinate data, it computes the axial distances to be moved and meters pulses simultaneously to the *X* and *Y* digital step motors at rates such that the result is a straight line. Movement of the head actually occurs in discrete steps, but since the step size is so small (as fine as 0.0001 in.), the steps are not discernible even with line widths as fine as 0.002 in.

In some cases, it is desired to draw a path between two points—the path not being a straight line. Circular, parabolic, and higher-order modes of interpolation are available. Two sets of coordinates are required for *circular interpolation,* an *X-Y* set to locate the end point on the line segment and an *i-j* set to define the center point of the circle. With this information, the equation of the circle intersecting the present position and the line segment end point with its center at the required location is computed. The incremental commands necessary to move the head along this path are then generated. For *parabolic interpolation*—by accepting two sets of coordinate points and using the coordinates of the present position, the equation of the parabola which passes through these points is computed and the incremental commands are generated which will move the machine along this path.

Regulation of the carriage speed is another function of the control. When a line is being drawn, the carriage attempts to accelerate to a speed manually selected by the operator from 10 to 100 percent of the maximum speed attainable by a particular table configuration. If maximum speed is reached, the speed will continue at that rate until deceleration is initiated by proximity to the end point. The processes by which the carriage is accelerated and decelerated are referred to as up-ramp and down-ramp. When a line segment is short, the carriage is up-ramped to the midpoint and then down-ramped to the end point.

The control system also must be able automatically to scale a drawing decimally from zero to a maximum of 999.99999 times actual size and to provide a mirror image of the drawing along either or both axes. The control system also can select and plot on a plane determined by any two

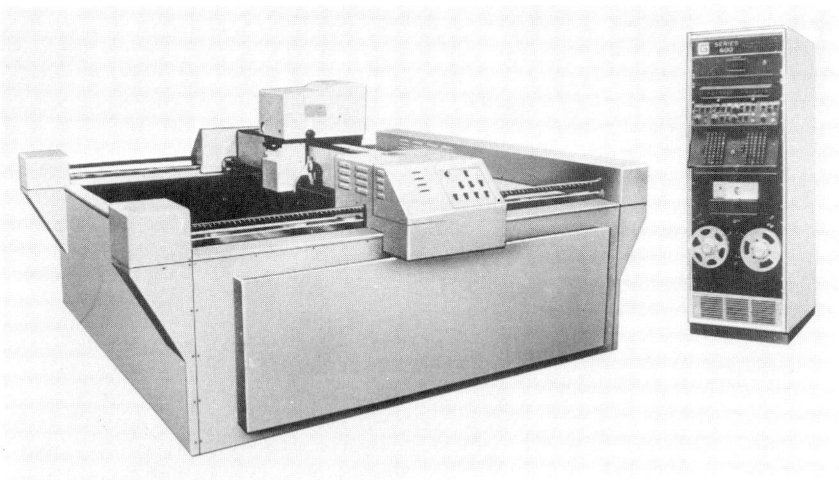

Fig. D-37. High-precision automatic drafting table. *(Gerber Scientific Instrument.)*

of up to five axes present in the input information. Manual controls permit the operator to slew the carriage to any position and to set the reference-zero position. Still other control functions include automatic dashed-line generation, display of carriage position, display of information contained in the input, and sequence-number search and display. The latter also can be used to locate a particular drawing sequence in a program so that only a section of a drawing may be drawn.

When alphanumeric characters or symbols such as those contained in electrical schematics, logic diagrams, PERT charts, plumbing diagrams, or maps are frequently used, they may be digitized and stored in the control system under a code name, or mnemonic. Thereafter, each time the symbol is required, the control will refer to the stored information and draw the symbol before returning to the input program for further instructions.

There are two methods by which the functions described can be accomplished: (1) by hardware in a "hard-wired" system and (2) by software. In the "hard-wired" system, circuits are designed and constructed to recognize and respond to only one set of input conditions. One of these is required for every function to be performed, each of which is defined by a unique set of input conditions. If it is desired to modify the function in any way, the circuit must be removed and rewired. A software-oriented system consists of an array of logic circuits which may be set to respond to unique sets of input conditions simply by setting the required initial conditions with a machine-program tape. To modify the response of any function in such a system, it is only necessary to change the machine program.

Three types of control systems are available: (1) A *computer interface* to permit time-sharing control of an automatic-drafting table by a large computer. The computer supplies the coordinate information and the tool-control commands. All other program interpolations are provided by the control. The time-sharing mode of operation is accomplished by a three-word routine in which the control notifies the computer that it is READY for data, the computer responds to the control when in a DATA READY condition, and the control reads the information and responds with an ACKNOWLEDGE signal. (2) An *incremental, linear-interpolating control* which provides an economical method of converting digital information into a drawing without special postprocessing or data conversion. Internal-velocity computation, adequate manual control, console-monitoring features, scaling, speed selection, and symmetry abilities allow this system to handle straightforward drawing requirements directly, or difficult tasks while operating on-line with a large computer. (3) A *software-oriented system* which allows the user complete freedom of input source and format. Because of the growing application of digital computers to production control, manufacturing, and scientific research, the control system is required to perform conversions and generate drafting-machine-control instructions from a wide variety of rapidly changing computer languages; and it must be compatible with an ever-growing number of computer systems.

A feature of the latter type of control system is the attainment of higher average drawing speeds through a "look-ahead" mode in which the control monitors information 10 blocks ahead of the data currently being used and optimizes the speed as a function of slope changes. Other stored programs include such items as instructions for drawing symbols or alphanumeric characters and making the computations necessary for linear, circular, and parabolic interpolations, and infinite scaling. In a large system of this type a general-purpose computer with an 8,000-word core memory is used.

Drafting Heads: An automatic drafting system usually is only as valuable as the quality and accuracy of its graphic output. The accuracy of the finished product depends primarily on the refinement of the control and drafting table. The required quality of the linework, however, depends on the application. The wide range of types of linework required has led to the adoption of the most versatile and inexpensive head as a standard and the provision of optional accessories for producing special linework.

The standard head is a toolholder which accepts a ball-point pen, a wet-ink pen, or a scribing tool. The head provides micrometer adjustment of height for various drafting-material thicknesses and setscrew adjustment of the tool excursion. A ball-point pen drawing on paper is useful for applications where rapid display of general information is desired. Such an application is verifying tool paths for numerical-control machine tapes. A wet-ink pen produces quality drawings on frosted Mylar or high-quality drafting paper with line widths ranging from 0.010 to 0.050 in. A scribe is used with standard scribe coat to produce lines from 0.003 to 0.015 in. in width when extremely fine lines and high accuracy are required. A felt-tip pen may be used when it is important to display information distinctly and quickly as received. To provide greater flexibility, optional heads for use with the aforementioned tools are two- or six-position toolholders. A symbol-printing and drawing head makes it possible to print up to 72 symbols and characters as well as to draw with either of two drawing tools.

The most precise linework is obtainable with an optical-exposure head. This device produces lines by exposing a photographic emulsion on a base material, such as Mylar or glass. The device can form lines with a uniform width (±0.0002 in.) ranging from 0.002 to 0.187 in. by the selection of an aperture from the 24 available. The quality of the linework is ensured by automatically controlling the light intensity during acceleration and deceleration of the head. Any form of symbol, up to 0.187 in. in size, can be produced by flash exposure as in a normal photographic process.

Automatic Digitizer: The most sophisticated automatic drafting systems utilize an automatic digitizer. This consists of a closed-circuit television, an electronic scan sensor, manual-positioning controls, and a software package. To digitize a piece of graphic information using this system, the operator simply aims a direction vector to designate the direction of travel to get to the line to be digitized and then actuates an accelerator control. The head moves in the given direction until it senses the line. The head then locks onto the line and outputs an initial point and follows it, outputting data as often as necessary. The frequency of output data is determined by a software program which monitors the axis encoders and calculates the linear distance traveled, slope change, internal angle increase and the area defined by the curve followed, and a straight line from the last data point to the present position. It produces an output-data block when a comparison check shows any of these to be greater than the operator-selected magnitudes. These features may be overridden by a manual readout command.

If the digitizer should leave the line while under automatic control for any reason, such as poor line quality, the computer is signaled to stop taking data, the carriage is halted, and the input/output typewriter prints an appropriate message. The operator may return the head to the last valid point simply by pushing a button on the control. The operator then either manually steers the head across the problem area or returns it to automatic control. The automatic digitizer also may be instructed to stop when it experiences a line intersection or to always go left, right, or straight ahead. If the digitizer stops, the operator must be instructed to decide which of the lines is to be followed, and the automatic system then will again take control.

Programming: The graphic-display system controls described accept coordinate data in word-address format in accordance with EIA Standard RS-274B. Character codes and axis and motion nomenclature are in accordance with EIA Standards RS-244A and RS-267A, respectively. The standard input is from 1 in. wide, eight-channel binary-coded-decimal punched control tape. The coordinate data may be either incremental or absolute, depending on the control mode. In the incremental mode, the control moves the drawing head a distance from the present position in each axis direction equal to the coordinate data. A control operating in the absolute mode positions the drawing tool to the position described by the coordinate data with respect to the selected zero-reference point.

The standard-format detail is as follows:

$$x + 13y + 13z + 13dzm2^*$$

The control recognizes this block as a set of words containing coordinates in the positive *X, Y,* and *Z* directions consisting of four integers, the first of which is a whole number and three of which are decimal digits, a tool-command designated by a *d* followed by up to two digits, and a miscellaneous function command provided by an *m* followed by up to two digits. The asterisk is the "end of block" signal and must be included for each data block. The data block may be expanded to accommodate optional features by preceding the above with an *n* with up to three digits and a *g* with up to two digits, in that order. Any information included in the data block other than that specified by the input format will be ignored by the control except to display such data when this feature is specified. All commands are modal; that is, the command remains in effect until negated by another command. Also, any command not required need not be included in the data block.

To enhance speed, the programmer may omit leading or trailing zeros from the input coordinates. Leading or trailing zeros refer to those which precede or follow meaningful data, respectively. This allows more information to be included in a given length of tape, which results in more information being read in less time. The control must be notified when this is done either by setting a switch to the proper position on a "hard-wired" system or by the setup program on a software-oriented system. The control then automatically positions the coordinate data properly with respect to the decimal point.

The *d* nomenclature is used to control pen position, to accomplish automatic dashed-line sequences, to produce symbols from a symbol generator or a print head, or to operate the shutter and select the aperture in an optical-exposure head. The *g* nomenclature is a preparatory command which indicates the type of interpolation required (linear, circular, parabolic), symbol-print-head selection and mode of operation (single-symbol print or printing with automatic spacing to the right), symbol-generator selection and orientation (horizontal or vertical), and tool selection when a six-pen turret is installed, and designates a photoexposure. Miscellaneous functions, controlled by the *m* command, are scaling factors, fixed or optional program stops, or a code to signal the end of a magnetic tape for automatic rewind.

(EDITOR'S NOTE: Considerably more space is devoted to this description of the automatic drafting machine than is normally devoted to systems in the encyclopedia because the type of technical and systems approach described will be useful for many kinds of machines where any kind of tool can be positioned just as the pen is positioned in a drafting machine.)

—David Rager, *Algorex Data Corporation*

Drag-type Tachometer (See **Speed, measurement of.**)

Drift Drift is an undesired change in output over a period of time that is unrelated to changes in input, operating conditions, or load.

For a measurement or control system, drift can be determined for any specific device, or element within the device, or for the control loop itself such as changes in offset (or deviation between the set point and process variable) as a function of time.

Under laboratory conditions, drift of an element is usually determined in one of two ways:

1. By maintaining exact operating and load conditions and monitoring output variations for a fixed input signal as a function of time. This is called "point drift."

2. By maintaining input signal, operating conditions, and load approximately constant and by comparing calibration curves at the beginning and at specified intervals of time. This is called "calibration drift."

Drift for a measuring device can either be systematic (i.e., approximately predictable in direction and magnitude as a function of time), random (nonpredictable), or some combination of the two. For this reason, a drift specification of 1 percent for 6 months is usually not intended to imply that drift for the second 6 months will be an additional 1 percent.

For most devices, drift is measured and specified as a percentage of output span. Drift for a controller (open-loop), however, is necessarily referred to the input. Since the output will drift to one end of the scale in a relatively short time, the actual drift is determined by the change in input magnitude necessary to restore the original output level.

—Robert L. Wilson, *Honeywell Inc.*

Drives, Control of (See **Electric-motor drives, control of.**)

Droop (See **Offset, droop.**)

Drop-weight Tester [See **Testing (materials), destructive.**]

Drum Storage A magnetic storage consisting of a cylinder coated with a magnetic material and driven by an electric motor. Data are recorded on, and read back from, the surface of this cylinder by individual heads for each track, staggered around the circumference of the drum.

Dynamometer An instrument for measuring force or power. Generally, dynamometers are of two types: (1) absorption dynamometer in which the power is absorbed by friction and dissipated as heat; and (2) transmission dynamometer in which the power measured is passed on and very little is wasted on friction. The general ranges of application and accuracies of dynamometers are listed in Table D-4.

TABLE D-4. *Characteristics of Dynamometers*

Type	Approx. Speed Limit, rpm	Usual hp Limit, hp	Probable Error, %
Prony brakes:			
Block	2,000	10	1–5
Band	1,000	5	1
Wooden cleats on bands	1,000	200	$\frac{1}{2}$–1
Rope ($\frac{3}{4}$ in.) with cleats	1,000	50	1
Fluid-friction dynamometers:			
Froude (ordinary water brake)	10,000	25,000	$\frac{1}{5}$–$\frac{1}{2}$
Turbine	4,000	5,000	$\frac{1}{10}$–$\frac{1}{5}$
Alden	1,000	5,000	$\frac{1}{2}$–1
Fan brake	2,000	200	1–5
Electric dynamometers:			
Eddy-current brake	6,000	300	$\frac{1}{5}$–$\frac{1}{2}$
Electric generator	750–4,000	30,000	$\frac{1}{10}$–$\frac{1}{5}$
Transmission dynamometers:			
Torsion	1,000–3,000	50,000	1–5
Kenerson	1,500	100	2

Note: This table is from Marks' "Standard Handbook for Mechanical Engineers," 7th ed., McGraw-Hill Book Company, New York, 1967. Speed limits are approximately the highest available commercially. Horsepower limits refer to largest sizes obtainable. Probable error is very approximate and refers to apparatus in fair adjustment.

Dynamometer, Electric A device or system used for testing and determining the power of engines, motors, and other rotating machinery. In one form, the dynamometer essentially is an electric generator that is driven by the engine or motor. The power developed is dissipated into heat or may be channeled into the power line. The eddy-current dynamometer, consists of one or more conducting disks rotating in a magnetic field.

The power developed by the engine or motor under test is determined by measuring the torque applied to the dynamometer and multiplying by the speed. Since 1 hp = 33,000 ft-lb/min,

$$\text{hp} = \frac{2\pi L \times F \times \text{rpm}}{33{,}000} \qquad (1)$$

where hp = horsepower
L = radius (length of arm), ft
F = force, lb
rpm = revolutions per minute

Since most dynamometer arms are measured in inches, for L in inches,

$$hp = \frac{F \times L \times rpm}{63{,}025} \tag{2}$$

To simplify calculations, L generally is established as 63,025 divided by some round number, such as 2,000 ($L = 31.513$ in.), 3,000 ($L = 21.008$ in.), or 6,000 ($L = 10.504$ in.). The round number then is the "dynamometer constant," and the formula becomes

$$hp = \frac{F \times rpm}{\text{dynamometer constant}} \tag{3}$$

To determine the rating for a dynamometer scale (or other force-measuring device), the formula may be rewritten as

$$F = \frac{hp \times \text{dynamometer constant}}{rpm} \tag{4}$$

Assume a dynamometer with a continuous rating of 200 hp with 150 percent (300 hp) rating for 15 min, 2,500 to 6,000 rpm, with a torque arm of 21.008 in. (dynamometer constant = 3,000),

$$F = \frac{300 \times 3{,}000}{2{,}500} = 360 \text{ lb}$$

A further simplification, common in laboratories with a number of dynamometers of different sizes, is to calibrate the scales in arbitrary units so that for each dynamometer

$$hp = \frac{\text{scale reading in arbitrary units} \times rpm}{1{,}000} \tag{5}$$

In each case, the pounds pull per arbitrary unit is 63.025 in. divided by the actual length of the torque arm in inches.

In most cases, the dynamometer will be called upon to measure torque in either direction of rotation. When mechanical scales, either spring- or pendulum-type, are used, a "reversing linkage" as shown in Fig. D-38 is required. An actuating bar with two knife-edges in opposite directions is bolted to the dynamometer frame. If the direction of rotation is clockwise, the left linkage, in tension, applies an upward force to its pivot in the lever, and a downward force is exerted on the scale. The right linkage is free. If the direction is counterclockwise, the right linkage, in compression, applies a downward force to its pivot in the lever, and a downward force is exerted on the scale. The left linkage is free. The "push-pull" fulcrum pivot is located midway between the tension and compression pivots in the lever so that the scale reads the same in either direction.

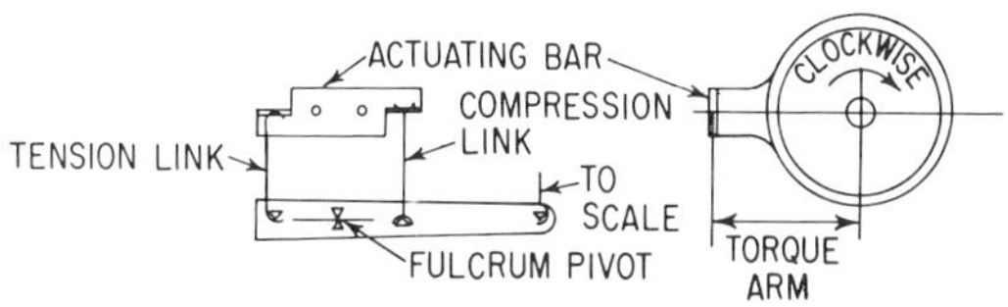

Fig. D-38. Dynamometer connection to a scale.

Use of Strain Gage: Torque from the dynamometer also can be measured by coupling a "universal" (calibrated in tension and compression) strain-gage load cell to the arm, or by coupling two compression-type hydraulic or pneumatic load cells, one above and one below the arm. Or, the torque can be measured by mounting four strain gages, as shown in Fig. D-39, on a calibrated section of shaft and inserting this unit between the engine or motor and the dynamometer. Slip rings are used to carry the leads from the conventional bridge circuit.

To minimize errors in torque measurement, the electrical leads to the dynamometer should be very flexible in the direction of rotation and the shaft bearings should have the least friction possible. On some large precision dynamometers, the outer races of the shaft bearings are rotated so that the bearing friction is always "rolling" and "breakaway" friction never is encountered.

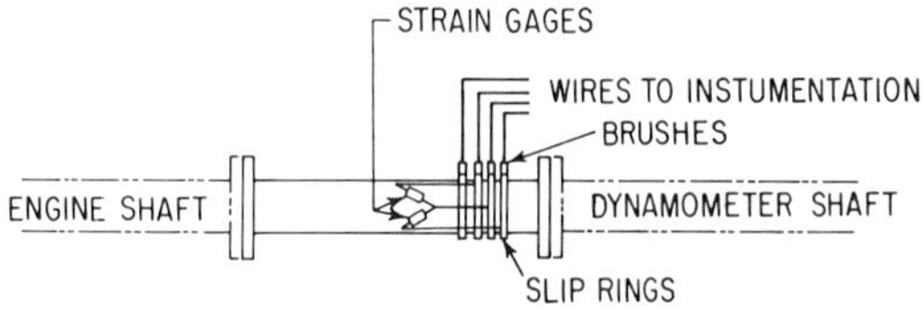

Fig. D-39. Strain gage used for torque measurement.

—Donald B. Kendall, *Toledo Scale Company, Division of Reliance Electric Co.*

E

Ear, Instrumental Measurements of The most common measurement of hearing function is the pure-tone audiogram in which frequency from 125 to 8,000 Hz is plotted against hearing loss in decibels. The audiogram displays the ability of the ear to hear a pure sine-wave tone at a given frequency compared with a "normal" ear. The unit of loudness used is the decibel, which is defined as $10 \times \log_{10} (P_1/P_2)$, where P_1 is the power of the sound being applied and P_2 is the just audible power required at the given frequency for the "normal" ear to hear. The standard audiometer contains a frequency-selection knob, an attenuator calibrated in 5-db increments, and a key which connects the output of the instrument to the earphones placed on the subject's head. The procedure is to increase the amplitude slowly while depressing the key in short pulses until the subject reports that he can just hear the sound. In this way, the audiogram can be plotted.

In addition to pure tones, speech sounds are also used as test signals. Using + 9 db (referred to 0.0002 dyne/cm^2) as a 0-db threshold level, it is possible to determine the extent of the hearing loss for speech using specially selected two-syllable words having approximately equal stress on each syllable (called "spondaic" words).

The equipment used for this measurement consists of a microphone, audio amplifier, and a pair of headsets, the system having a flat frequency response between 125 Hz and 8 kHz. Sensitivity, or gain, of the amplifier is controlled by a step attenuator calibrated in 1-db steps, and the output is arranged to go into either ear separately or both ears simultaneously.

A more sophisticated version of the pure-tone audiometer was developed by von Békésy (Ref. 3), in which the amplitude control is run up and down by a motor while the subject operates a key. The amplitude is slowly increased until the subject hears the sound, which reverses the motor once more. At the same time, a curve is plotted automatically. The frequency is similarly increased slowly and automatically. The resulting curve is somewhat sawtooth in form and more accurately brackets the threshold values. This instrument is more precise than the conventional audiometer previously described, but because of its cost and complexity, the device is used for research rather than clinical evaluation.

In designing an audiometer for any other sound-presentation equipment for hearing studies, great care must be given to the elimination of background noise and hum, the generation of pure sine waves (if this is the waveshape being used), and the design of the switch or other device that turns the sound on and off. If more than one tone is presented at a time, "masking effects" may occur, giving different results than would be obtained with each sound separately. Thus, the instrumentation in this field is dependent upon the psychology and physiology of the hearing mechanism, meaning that "obvious" assumptions must be made with care. Also see **Hearing, sensory aids for.**

REFERENCES

1. Davis, Hallowell: "Hearing and Deafness," Holt, Rinehart and Winston, Inc., New York, 1964.
2. Hearing Aids and Their Components, *Zenith Hearing Aid Sales Corp. Publ.* HA-831.
3. von Békésy, Georg: "Experiments in Hearing," pp. 81–91, McGraw-Hill Book Company, New York, 1960.

—J. Malvern Benjamin, Jr., *Bonic Instruments, Inc.*

EBCDIC Extended BCD Interchange Code is an 8-bit binary-coded-decimal code for up to 256 alphabetic, numeric, graphic, and control symbols. It is used for data prepresentation and control in punched cards, on magnetic tape and disk, and as a transmission code. In some character-oriented machines, the code is used for internal data representation and manipulation. EBCDIC is used extensively in computers manufactured by International Business Machines Corporation. The code representation is shown in Fig. E-1 on page 224. (Compare **USASCII;** see **Binary-coded decimal.**)

—A. J. Arthur, *IBM Corporation*

Eddy-current Dynamometer (See **Dynamometer, electric.**

Eddy-current Tachometer (See **Speed, measurement of.**)

Eddy-current Testing (See **Nondestructive inspection.**)

Eddy-current Transducer If a nonmagnetic, electrically conductive object is placed in the magnetic field of a coil, the effective inductance of the coil is decreased and its resistance is increased—because the field sets up eddy currents in the object that circulate so as to oppose the field that creates them. Moving the object closer to the coil increases the effect. Hence, coil characteristics are a function of spacing between coil and object. Transducers that utilize this effect to sense the position of a specific object or "target" are called eddy-current transducers.

Electric Actuator (See **Actuator, electric.**)

Electric Controller Electricity is one of our common media used in automatic-control systems—the other three forms being pneumatic-control, hydraulic-control, and mechanical-control systems (usually in the form of self-contained regulators).

The term *electric-control system* generally connotes an assembly with the following characteristics or functions:

1. The *input signal* from a sensing device or transmitter is *electrically* (*a*) *measured,* (*b*) sometimes *amplified,* and (*c*) *compared with a reference* or set-point (desired) value.

2. The *output signal* from the controller to the actuator or final control element is *electrical.*

3. The final control element is *actuated electrically*—as by a solenoid, electric motor, or electropneumatic converter; or where the flow of electrical energy is modulated—as by a relay, silicon controlled rectifier, or saturable-core reacter.

Electric controllers range from simple on-off household thermostats which may use a bimetallic or mercury switch—to complex industrial electric controllers that may have proportional, reset, and rate-control actions and that, in turn, may be preprogrammed or be part of a computer-directed system. The descriptions are confined here to electric controllers for industrial applications. For information on other electric controllers, refer to the alphabetical index.

EXTENDED BCD INTERCHANGE CODE

Bits 4 5 6 7	Second Hex Digit	Digit Punches																	
			00				01				10				11				Bits 0, 1
			00	01	10	11	00	01	10	11	00	01	10	11	00	01	10	11	Bits 2, 3
			0	1	2	3	4	5	6	7	8	9	A	B	C	D	E	F	First hexadecimal digit
			12				12	12		12	12	12		12	12				Zone punches
				11				11	11	11		11	11	11		11			Zone punches
					0		0		0	0	0		0	0			0		Zone punches
			9	9	9	9	9	9	9										Digit punches
0000	0	8-1	① NUT	② DLE	③ DS	④	⑤ SP	⑥	⑦ -	⑧					⑨	⑩	⑪	⑫ 0	8-1
0001	1	1	SOH	DC1	SOS				⑬ /		a	j	~		A	J	⑭	1	1
0010	2	2	STX	DC2	FS	SYN					b	k	s		B	K	S	2	2
0011	3	3	EXT	TM							c	l	t		C	L	T	3	3
0100	4	4	PF	RES	BYP	PN					d	m	u		D	M	U	4	4
0101	5	5	HT	NL	LF	RS					e	n	v		E	N	V	5	5
0110	6	6	LC	BS	ETB	UC					f	o	w		F	O	W	6	6
0111	7	7	DEL	IL	ESC	EOT					g	p	x		G	P	X	7	7
1000	8	8		CAN							h	q	y		H	Q	Y	8	8
1001	9	8-1	RLF	EM							i	r	z		I	R	Z	9	9
1010	A	8-2	SMM	CC	SM		c	!	⑮ ¦	:								LVM	8-2
1011	B	8-3	VT	CU1	CU2	CU3	·	$	,	#									8-3
1100	C	8-4	FF	IFS		DC4	<	*	%	@					⑀		⑁		8-4
1101	D	8-5	CR	IGS	ENQ	NAK	(	)	_	'									8-5
1110	E	8-6	SO	IRS	ACK		+	;	>	=					⑂				8-6
1111	F	8-7	SI	IUS	BEL	SUB	|	¬	?	"								EO	8-7
			12				12				12	12		12	12	12		12	Zone punches
				11				11				11	11	11		11	11	11	Zone punches
					0				0		0		0	0	0		0	0	Zone punches
			9	9	9	9									9	9	9	9	Zone punches

CARD HOLE PATTERNS

① 12-0-9-8-1
② 12-11-9-8-1
③ 11-0-9-8-1
④ 12-11-0-9-8-1
⑤ NO PUNCHES
⑥ 12
⑦ 11
⑧ 12-11-0
⑨ 12-0
⑩ 11-0
⑪ 0-8-2
⑫ 0
⑬ 0-1
⑭ 11-0-9-1
⑮ 12-11

Fig. E-1. Extended BCD Interchange Code (EBCDIC).

Input Signals: In addition to providing for many kinds of control with a broad array of actuators, electric controllers must match a broad range of input signals. There are three major signal-input categories: (1) direct-connected, (2) recorder-actuated, and (3) standard-signal. The type of input signal greatly affects the circuitry and physical configuration of the electric controller.

Direct-connected controllers can be connected to sensors, such as radiation pyrometers, resistance thermometers, and thermocouples for temperature measurement; to tachometers for speed measurement; to thermal converters for ac power measurement; to shunts and volt boxes for dc power measurement; to various transducers for frequency measurement; and to process analyzers for chemical-composition determinations of liquid and gas streams. Millivoltmeters and electronic controllers fall into this category.

Recorder-actuated controllers derive their inputs from recording or indicating instruments. These controllers utilize the pen or pointer drive mechanism to convert and amplify the input to the controller. This package offers economy, flexibility, and simplicity where control is required along with the need for recording and a large, bold indication.

Standard-signal controllers actually are receivers, since these devices operate from milliampere signals put out by transmitters. Dependent upon instrument brand, typical signals are 1 to 5 mA, 4 to 20 mA, and 10 to 50 mA. Transmitters are available to measure all the commonly encountered process variables and provide one of the foregoing milliampere inputs to the controller. The standard-signal electric controller offers the desired flexibility for processes with a large number of control loops.

Basic Elements of an Electric Controller: Each controller consists of three principal parts: (1) an *operator's station,* (2) a *computing and control* portion, and (3) an *output section.* The operator's station provides a set-point knob, often calibrated to read directly in units of the process variable, to set the desired value. Usually, the set point is operated manually. For set-point programming, for supplementary control, and for remote operation, the set point may be operated electrically. The operator's station also may include (1) a selector for choosing automatic or manual control, (2) a means to change the position of the actuator and hence the value of the process variable when under manual control, and (3) various indicators to display the value of the variable, the actuator position, and other loop and instrument conditions.

Electric-control Forms: At least five control forms or modes are available in electric controllers. For fundamental descriptions of these modes, see **Automatic control,** and consult the alphabetical index. Electric-control forms include:

1. *On-off or two-position control:* A simple electromechanical-comparison device may sense only the direction of control error and initiate control action by varying the actuator. This kind of control allows the actuator to assume one of two positions which usually turns the process regulator (final control element) *full on* or *full off.*
2. *Multiposition control:* Variations of the foregoing device are available which permit the actuator to assume several positions. This control form requires that the comparison device sense not only direction but also the extent of control error in several discrete steps.
3. *Floating control:* This control form allows the actuator to assume an infinite number of positions. A series of spaced, timed pulses are directed to the actuator by a comparison device similar to that used for multiposition control. For the more stringent process requirements, an infinite number of actuator positions usually is required.
4. *Proportional control:* Positioning of the actuator is based on direction and extent of the error. As the controlled variable moves through a range, termed the proportional band, the final control element is proportionately adjusted from minimum to maximum input.
5. *Proportional-plus-reset control:* Positioning of the actuator is based on the direction, extent, and *duration* of the error.
6. *Proportional-plus-reset-plus-rate control:* Positioning of the actuator combines the actions of type 5 with an additional positioning action that is based on the rate of change of the error. Adjustments are provided in these controllers for setting the amount of each action (proportional, reset, and rate) so that controller settings may be optimized for the process.

Electric-controller models are available which provide useful combinations of two or three actions with various ranges of adjustments. In these controllers, an electrical signal from the output section of the actuator is returned to the controller input to improve control response and stability. This controller feedback is in addition to feedback from the process variable which is an inherent characteristic of all process-control loops.

Mechanically Operated Switches: On-off or two-position, multiposition, and floating control usually are performed with mechanically actuated switches. One arrangement, with a double-level cam, is operated by a change in relative position of the switch and cam elements. One element is coupled to the input and the other to the set point so as to compare the two. The span of motion must be less than 360° on a circular cam to prevent reversing control action on large unbalances.

Another arrangement combines a notched single-level cam with a two-level toggle switch. This requires a change in relative direction along with a position change to operate the switch. With this arrangement, the span of motion can exceed 360° on a circular cam, thereby providing more precise control settings for a given cam diameter. For floating control, a synchronous-motor-driven cam and switch generate time pulses. These three switches and variations thereof are used in many combinations and circuits to provide a multitude of control arrangements.

Other Mechanically Operated Devices: Variable resistors (sometimes called potentiometers, rheostats, or slidewires), capacitors, and inductors are used to derive the control-error signal for some proportional-action controllers, including types with reset and rate action. The movable and stationary elements of these devices are coupled to the input and set point so as to compare the two and provide an electrical signal suitable for the electronic circuit which, in turn, provides proportional, reset, and rate-control actions.

Electric Circuits: The variable resistor is used in a Wheatstone-bridge circuit with an electronic-detector relay connected to the control actuator so as to restore balance when the input signal changes. Manually adjustable variable resistors often are provided to adjust the proportional band and initially adjust the actuator position. The latter is the equivalent of reset in a two-action controller. A typical circuit is shown in Fig. E-2.

Automatic reset may be combined with proportional action by an auxiliary bridge and integrating circuit which accumulates, stores, and inserts an auxiliary voltage into the proportional bridge circuit so as to reposition the actuator based on reset-control action. Rate action similarly is combined with proportional and reset actions by using a differentiating circuit which introduces a temporary actuator offset during intervals of rapid process change, but not during a set-point change. For specific process requirements, the mechanically actuated switches are combined with proportional, reset, and rate-action controllers to perform auxiliary control operations, such as safety interlocks.

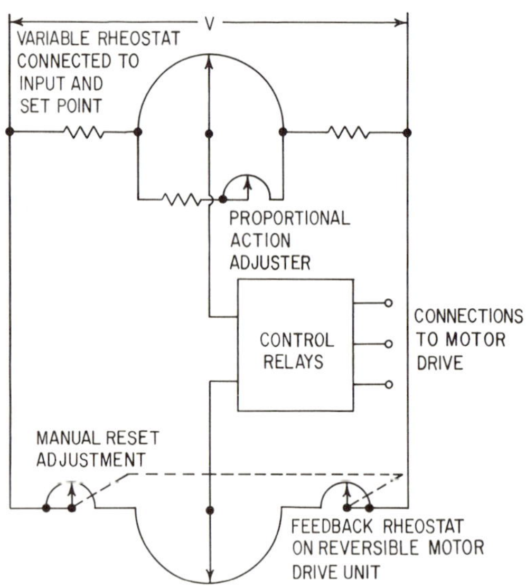

Fig. E-2. Circuit for proportional-action electric controller.

Millivoltmeter Controller: An instrument of the type shown in Fig. E-3 is a direct-connected device; that is, the input electrical signal from the sensor is fed to the millivoltmeter movement, causing movement of the indicating pointer. Excursions of the pointer from the set point or desired value are converted into controller action by a comparison device as previously described.

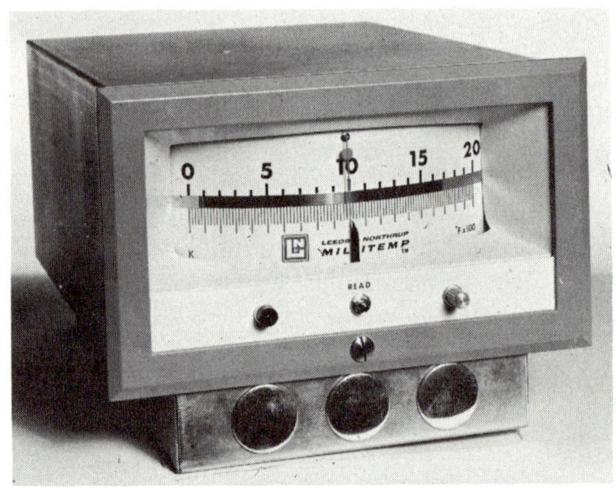

Fig. E-3. Millivoltmeter temperature indicator and controller. Indicating and set-point scale both are about 5 in. long. Ranges must begin at near zero. Speed of response is from 15 to 20 s for full-scale deflection. Accuracy is ±1 percent of span; sensitivity is ±½ percent of span. Power requirements are 20 V-A at 120 V, 50/60 Hz. *(Leeds & Northrup.)*

The millivoltmeter is used widely for automatic temperature control of laboratory and industtrial furnaces, ovens, and related types of batch-process industrial-heating apparatus. The instrument normally is used with a thermocouple, and the overall cost is comparatively low. The millivoltmeter must be adjusted for the thermocouple and leadwire resistances of a particular installation. The maximum tolerable resistance value is about 25 Ω on most commercial models. The sensitivity of the meter movement limits the useful range to a minimum span of 300° for base-metal thermocouples (iron-constantan or chromel-alumel) and 1200°C for noble-metal thermocouples (platinum–platinum 10 percent rhodium). A millivoltmeter should not be used where vibration is excessive or where angle mounting is required. Advantages of the millivoltmeter include a large scale and broad range of temperature indication. Models are available for two-position, multiposition, proportional, and proportional-plus-reset control—and for use with a variety of actuators. In addition to temperature measurement, the millivoltmeter can be used to control other variables provided that the input signal is appropriate and the process dynamics are compatible with the characteristics of the controller.

Direct-connected Electronic Controller: Although essentially the functional equivalent of the millivoltmeter, controllers of this type normally do not indicate or record the process variable, but as shown in Fig. E-4, they do include a display of a limited range of deviation from the set point. Advantages of these controllers include: (1) reasonable levels of vibration do not impair their operation; (2) high external resistances can be tolerated; (3) angle mountings usually are possible with proper allowances for the deviation indicator and relay operation; and (4) the controllers are adaptable to high-speed processes.

Fig. E.5 Recorder-actuated controller for two-position electric control. Indicating and set-point scales are 6 to 11 in. long. Ranges may be suppressed and begin well above zero. Accuracy typically is ±0.3 percent and sensitivity is ±0.1 percent of span. Speed of response is 1 to 5 s for full-scale change. Power requirements are 50 V-A at 120 V, 50/60 Hz. Control forms available are on-off, two-position, and multiposition or floating. *(Leeds & Northrup.)*

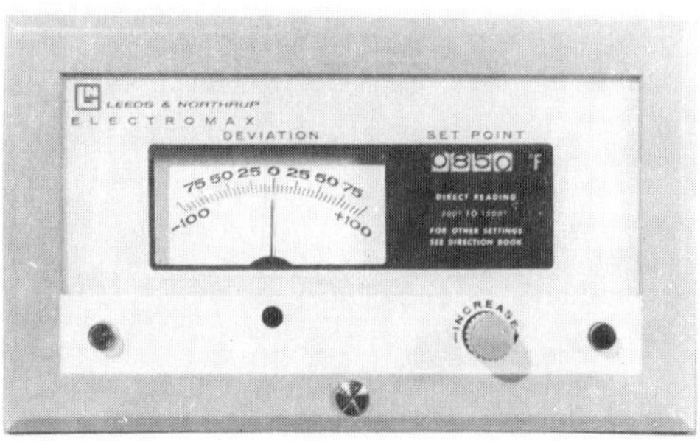

Fig. E-4. Direct-connected electronic controller with deviation indicator. Analog scales usually are 4 to 5 in. long. Digital setters usually have 1,000 to 2,000 counts. Ranges usually begin at zero. Span of deviation meter shown is 200° F. Speed of response typically is less than 0.1 s. Accuracy is ±3 percent of span; settability is 0.5° and sensitivity is 0.2° on high-performance models. Power requirements are 25 V-A at 120 V, 50/60 Hz. Interference rejection varies widely among models. *(Leeds & Northrup.)*

The circuitry of the direct-connected electronic controller includes a filter and an amplifier capable of detecting and amplifying small dc microvolt signals proportional to the extent and direction of control error. This is an ac amplifier with an input dc-to-ac modulator, an output ac-to-dc demodulator, and feedback to provide the necessary stability of zero and gain. The output of this amplifier is an error signal suitable for operating a detector relay or proportional, reset, and rate-action control circuits with a variety of control outputs. All the latter circuits normally are included as an integral part of the direct-acting controller and are similar to curcuits used in standard-signal electric controllers.

Recorder-actuated Electric Controllers: These instruments are so termed because essentially they are recording instruments in which electric-control action has been incorporated (see Fig. E-5). The controller elements of these instruments may be partially or completely contained within the recorder case. If partially contained, an auxiliary case containing the controller elements is often attached to the recorder as shown in Fig. E-6. All common control forms or modes and actuators can be used with appropriate models of recorder-actuated controllers. A variety of sensors can be accommodated because of the flexibility provided by the servoactuated recorders with which the controllers are associated. Instruments of this type are used to control temperature (particularly high temperatures), kilowatts, power-line frequency, pH, and chemical concentration, such as a constituent in a gas stream.

Fig. E-6. Recorder-actuated electric controller with control elements located in attached case. Control forms usually are proportional, proportional-plus-reset, and proportional-plus-reset-plus-rate. Controller outputs are available to position motors with modulated currents and to modulate the ratio of "on" to "off" time of relays or solenoids. Other features are similar to those of instrument in Fig. E-5. *(Leeds & Northrup.)*

Standard-signal Electric Controllers: This type of instrument accepts as its input signal a current change from a transmitter. This signal received varies with the process variable being measured. Standard transmitter-signal ranges have been stated previously. Because standard-signal controllers usually are used on processes with large numbers of control loops and with more stringent control requirements, the devices usually are offered only with proportional and reset action; or with proportional-plus-reset-plus-rate actions—and with output sections adaptable to infinite-position actuators.

Controllers of this type are adaptable to complex processes and are preferred for control loops that require (1) control based on more than one process variable, (2) operation of a large number of actuators, (3) controller interconnection to minimize loop interaction, and (4) special control operations, such as ratioing, cascading, sequencing, and arithmetic computations. Various physical configurations of standard-signal electric controllers are shown and described in Figs. E-7 through E-9.

Electric-controller Output: The type of output section varies with the type of actuator to be used:

Output section	*Actuator*
Single-relay output section	Relay, solenoid, and other two-position actuators
Current-output section and internal-controller feedback	Variable-speed electric motor drive; electro-pneumatic converter; silicon controlled rectifier; saturable-core reactor
Dual-relay output section usually including provisions for a direct position feedback from the drive	Reversible electric-motor drive

Abbreviated sketches of typical electric-controller output connections are shown in Fig. E-10. There are hundreds of variations of these arrangements to meet specific process needs.

The switch of the single-relay output section may be mechanically or electrically operated in an on-off controller. It may be double-throw with or without "dead band" between the open and closed circuits. The switch may be operated by a proportional-plus-reset action

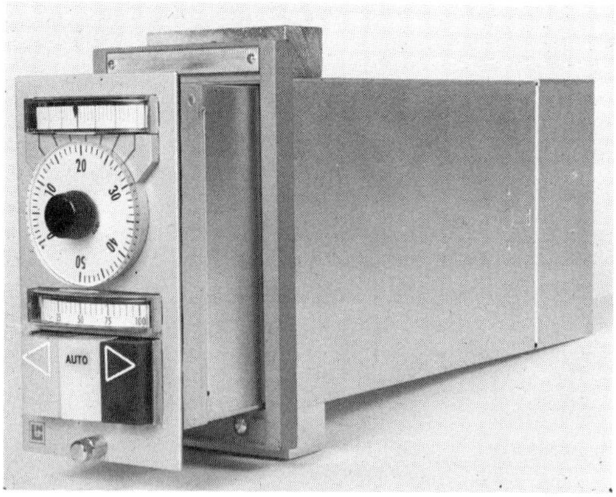

Fig. E-7. Standard-signal electric controller with operator's station attached and with manual set-point dial. Meter at top displays deviation from set point of the process variable. Calibrated circular dial is the set-point scale. The small knob must be rotated several times for full travel of dial. Meter below displays the position of the actuator. Switches at bottom are to select automatic or manual control and to adjust the actuator when on manual control. *(Leeds & Northrup.)*

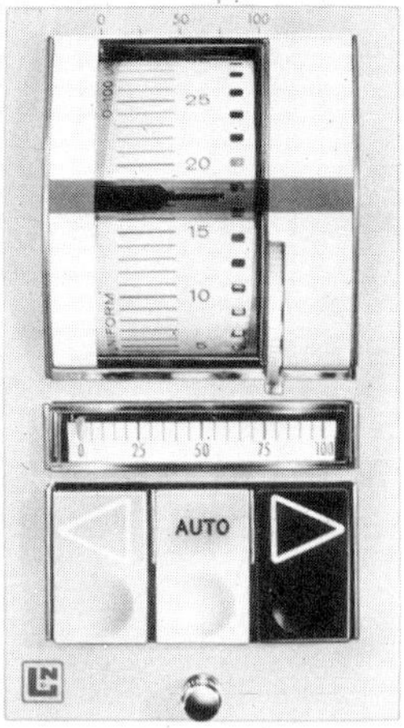

Fig. E-8. Operator's station for a standard-signal electric controller with motor-driven set point for digital-directed analog control. The motor drive is a convenient way to retain the set point when transferring to and from the computer-directed mode. *(Leeds & Northrup.)*

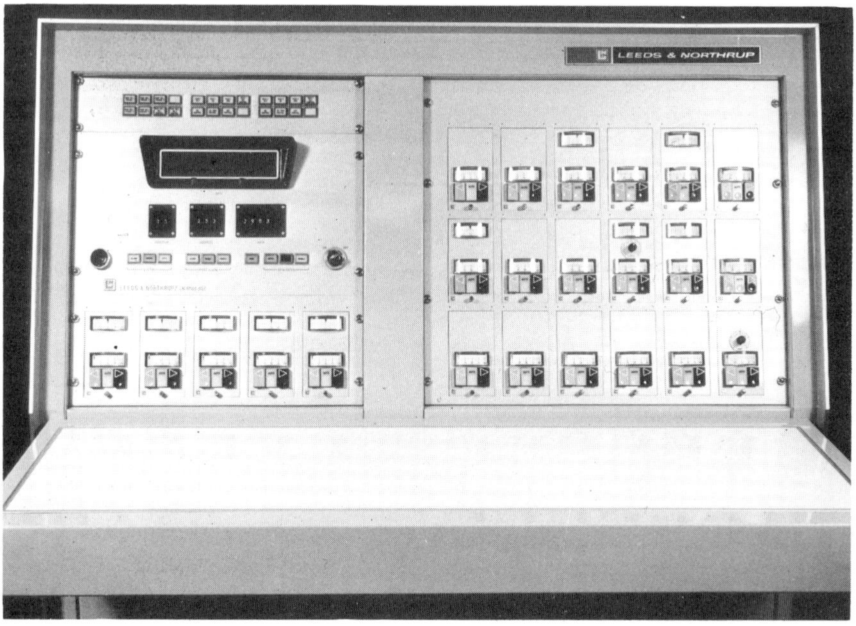

Fig. E-9. Standard-signal electric controllers mounted on central instrument panel for a process on direct digital control. Each instrument displays the process variable and the actuator position as well as providing switches for automatic-manual selection and manual operation. Set-point dials and knobs are not required because the computer performs the control. The two instruments with dials and knobs (on the right-hand portion of panel) are for biasing parallel actuators. *(Leeds & Northrup.)*

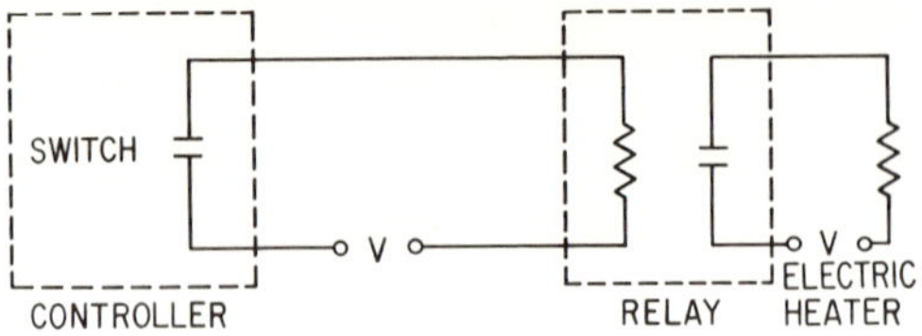

(a) SINGLE-RELAY OUTPUT SECTION

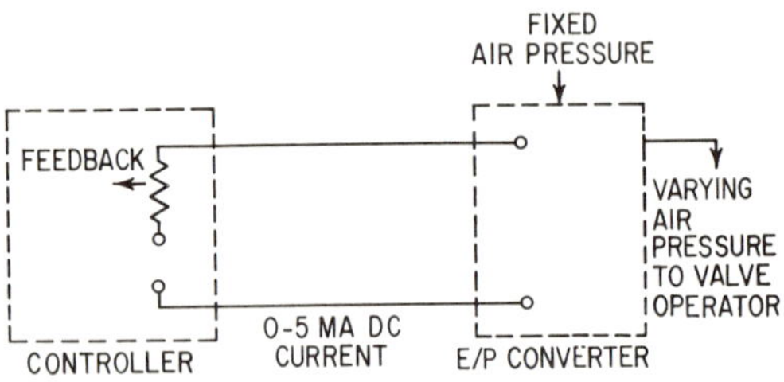

(b) CURRENT OUTPUT SECTION

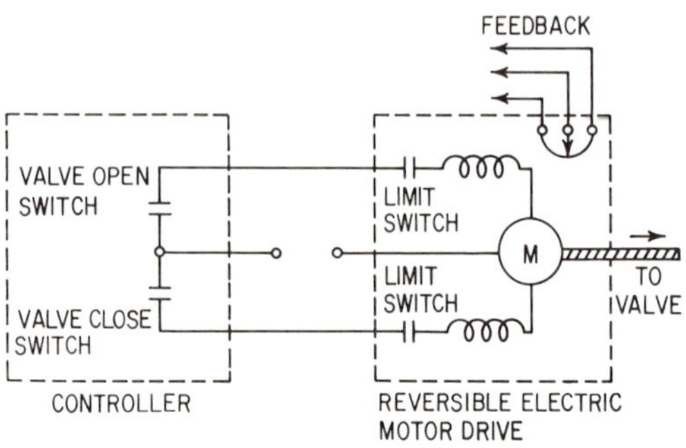

(c) DUAL RELAY OUTPUT SECTION

Fig. E-10. Electric-controller output sections: (a) single-relay; (b) current output; and (c) dual-relay output section.

Fig. E-11. Reversible electric-motor actuator with medium torque. This actuator is suitable for operating valves, dampers, autotransformers, and similar lever-operated process regulators. Models are available with torques ranging from 5 to 75 lb-ft and with full-stroke speeds ranging from 10 to 60 s. Operating voltage usually is 120 V, single-phase, 50/60 Hz—with maximum running current of 1 A. A handwheel is provided for local manual operation. Internal worm and worm gear prevents back drive by unbalanced loads. The linkage may be arranged for characterizing the process regulator. Extra limit switches and feedback transmitters can be provided for complex control. *(Leeds & Northrup.)*

controller, varying the ratio of open to closed time of a continuously pulsed relay. Because loss of controller power opens the relay, an open relay must correspond to a safe-process condition. If restoration to automatic control on return of power is undesirable, a lock-out relay can be provided.

The current-output section provides a direct current proportional to the required output. This current is derived from a proportional-plus-reset-action controller. The controller also may include rate action. On loss of controller power, the valve will move to the position corresponding to zero current; so this must correspond to a safe-process condition.

In the dual-relay output section, the controller switches are interlocked to prevent simultaneously energizing both motor fields. They may be mechanically or electrically operated in an on-off controller for two-position control. A time-pulse switch is in the common lead to the motor for floating control The switches may be in a proportional action or proportional-plus-reset and rate controller. On loss of controller power, the motor will remain in its last position.

Fig. E-12. Reversible electric-motor actuator with high torque. This actuator is similar to Fig. E-10 except that it is designed with torques ranging from 150 to 4,000 lb-ft—for very large valves and dampers. Full-stroke speeds range from 30 to 300 s. Operating voltages for larger sizes are 208/240/480 V, three-phase, 50/60 Hz. Power consumption at full-running load ranges from 0.5 to 5.0 kVA. *(Leeds & Northrup.)*

Fig. E-13. Electronic solid-state power switch used with electric controller. The device uses silicon-controlled-rectifier switches for running, reversing, and breaking high-torque reversible electric-motor actuator. *(Leeds & Northrup.)*

Fig. E-14. Saturable-core reactor used with electric controller to modulate power input to an electrically heated process. Models are available for single-phase and three-phase power in sizes ranging from a few to several thousand kilovolt-amperes. *(Leeds & Northrup.)*

Control Actuators: Several types of actuators, as shown and described in Figs. E-11 through E-15, can be used with electric controllers. Actuators fall into two basic categories:

1. *Inherently two-position actuators*
   Solenoid
   Relay
2. *Inherently infinite-position actuators*
   Reversible electric-motor drive
   Saturable-core reactor
   Silicon controlled rectifier
   Variable-speed electric-motor drive
   Electropneumatic converter

In some situations, an infinite-position actuator is capable of two-position actuation.

Applications of the various control actuators include:

Application	*Actuator*
Dampers, feeders, governors, pumps, valves, and other motion-actuated process regulators to modulate the flow of gases, liquids, and solids	Solenoid, reversible electric-motor drive, variable-speed electric motor drive, electro-pneumatic converter
Modulation of flow of electric energy for heating or magnetic excitation	Relay; silicon controlled rectifier; saturable-core reactor

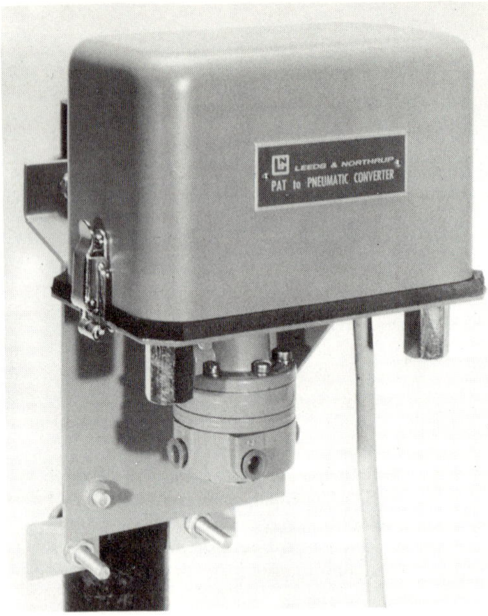

Fig. E-15. Electropneumatic converter used with electric controllers to change electrical output from controller to a proportional 3- to 15-psi standard air signal for operating pneumatic-valve actuators. *(Leeds & Northrup.)*

Applications of Electric Controllers: Generally, these instruments are best suited for the following uses: (1) for controlling high temperatures and the flow of electric power; (2) for regulating electrically heated processes; (3) for installations where there are long distances between sensor, controller, and actuator; (4) where an electric power supply is the most practical and economical; (5) for complex processes with a large number of interrelated controlled variables and process regulators; and (6) where safety and reliability demand that the control fail-safe in its prevailing state. Also see **Intrinsic safety.** See summary in Table E-1.

REFERENCES

Eckman, Donald P.: "Automatic Process Control," John Wiley & Sons, Inc., New York, 1958.

LaJoy, Millard H.: "Industrial Automatic Controls," Prentice-Hall, Inc., Englewood Cliffs, N.J., 1954.

Considine, D. M., and S. D. Ross (eds.): "Handbook of Applied Instrumentation," McGraw-Hill Book Company, New York, 1964.

Grabbe, E. M., Simon Ramo, and Dean E. Wooldridge (eds.): "Handbook of Automation, Computation, and Control," 3 vols., John Wiley & Sons, Inc., New York, 1958–1961.

—W. S. Bloor, *Leeds & Northrup Company*

TABLE E-1. *Characteristics that Determine the Selection of Electric Controllers*

CONTROL FORM OR MODE
On-off or two-position
Multiposition
Floating
Proportional
Proportional + reset
Proportional + reset + rate
INPUT SIGNAL
Standard signal
Recorder-actuated
Direct-connected electronic
Direct-connected millivoltmeter
CONTROL-ACTUATOR POSITIONS
Two-position
Relay
Solenoid
Reversible electric-motor drive
Multiposition
Reversible electric-motor drive
Infinite-position
Silicon controlled rectifier
Saturable-core reactor
Electropneumatic converter

Electric-motor Drives, Control of A classification of electric-motor drives is shown in Fig. E-16.

Constant-speed Drives: Induction motors and synchronous motors traditionally have been used for constant-speed applications since alternating current is the more readily available source of industrial power. As shown in Fig. E-17, these ac motors can be started by full-voltage starters or by reduced-inrush starters. The latter method reduces the initial drain on the power system. The wound-rotor motor, along with its secondary control, also is suitable for starting and accelerating high-inertia loads when low inrush currents are a requirement.

Adjustable-speed Drives: Both ac and dc motors can be used on applications that require adjustable speed. Semiconductors, such as the silicon controlled rectfier (SCR), have made antiquated such technologies as the thyratron rectifier and the saturable-reactor rectifier as a means for converting alternating current to adjustable-voltage direct current. Dc motors, as shown in Fig. E-18, can be connected to large, constant-

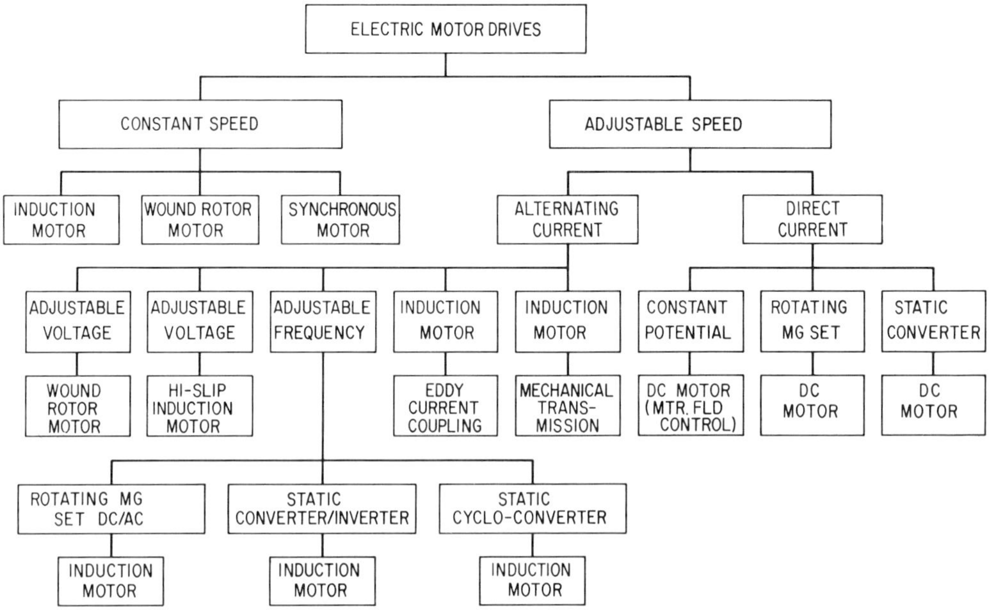

Fig. E-16. Classification of electric-motor drives.

TYPE	BLOCK DIAGRAM	AVAILABLE STANDARD DRIVE SYSTEMS	
		HORSEPOWER *	RPM †
INDUCTION MOTORS CONSTANT SPEED MULTI SPEED	FULL-VOLTAGE STARTER  	1/4 TO 5,000	3,600 1,800 1,200 900 720 600
INDUCTION MOTOR WOUND ROTOR	FULL-VOLTAGE STARTER 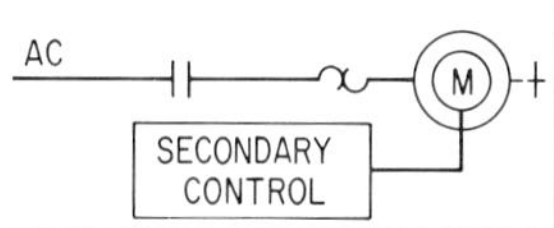 	2 TO 5,000	1,800 1,200 900 720 600
SYNCHRONOUS MOTOR	FULL-VOLTAGE STARTER 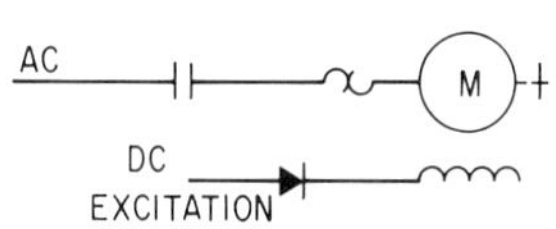 	20 TO 5,000	1,800 1,200 900 720 600 514
AC MOTOR REDUCED IN RUSH STARTER	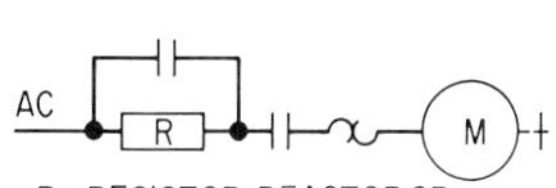 R= RESISTOR, REACTOR OR AUTOTRANSFORMER	5 TO 1,500	3,600 1,800 1,200 900 720 600

Fig. E-17. Constant-speed drive systems. *Larger horsepower available (special). †60-Hz synchronous speed.

TYPE	BLOCK DIAGRAM	AVAILABLE STANDARD DRIVE SYSTEMS	
		HORSEPOWER *	SPEED RANGE
CONSTANT POTENTIAL DIRECT CURRENT	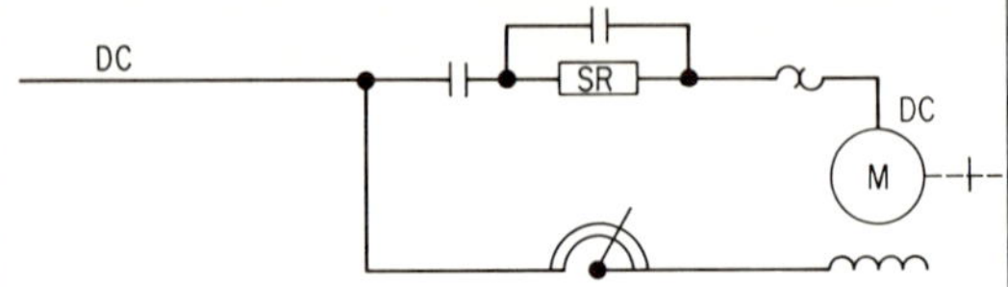	25 TO 500	UP TO 6:1
MOTOR GENERATOR POWER CONVERSION	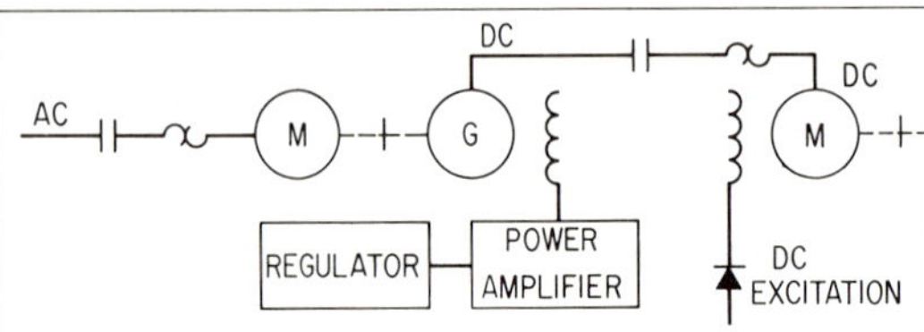	1 TO 5,000	UP TO 100:1
SILICON CONTROLLED RECTIFIER POWER CONVERSION	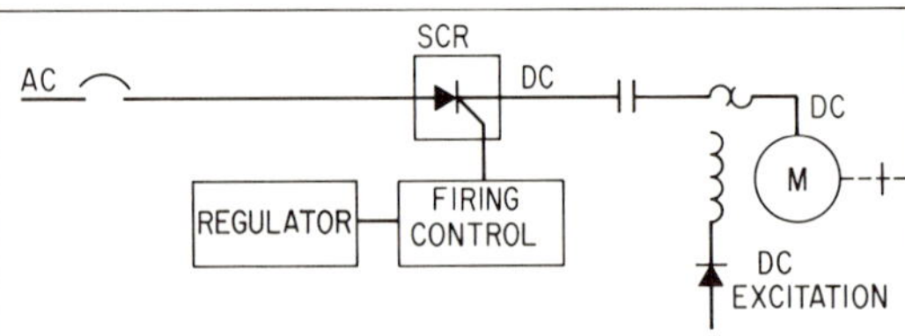	1/4 TO 2,000	UP TO 20:1

Fig. E-18. Adjustable-speed dc drive systems. *Larger horsepower available (special).

TYPE	BLOCK DIAGRAM	AVAILABLE STANDARD DRIVE SYSTEMS	
		HORSEPOWER *	SPEED RANGE
DC/AC ROTATING MOTOR GENERATOR POWER CONVERSION	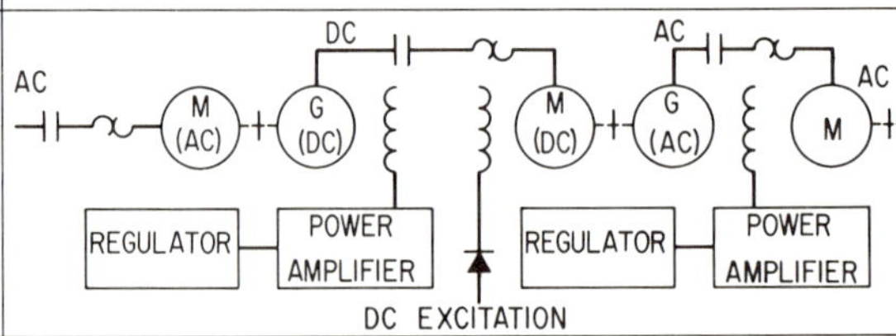	1/4 TO 5,000	UP TO 5:1
STATIC SCR CONVERTER INVERTER POWER CONVERSION	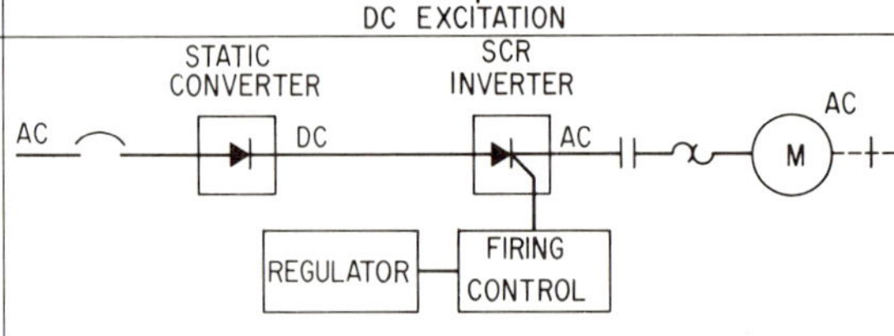	1/4 TO 500	UP TO 20:1
STATIC SCR CYCLO-CONVERTER POWER CONVERSION	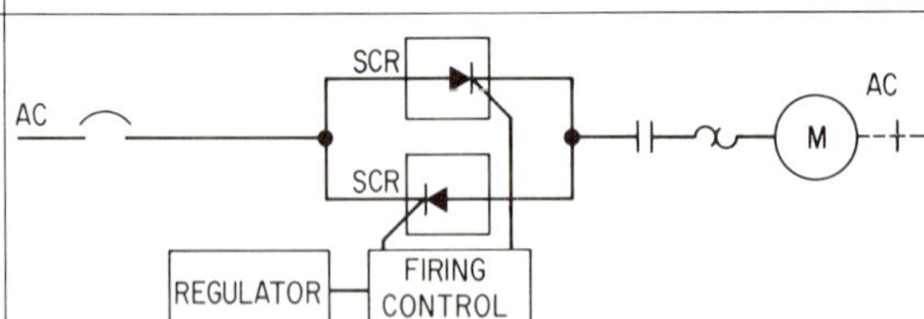	50 TO 5,000	UP TO 10:1

Fig. E-19. Adjustable-frequency ac drive systems. *Larger horsepower available (special).

potential dc sources and controlled over a limited speed range by motor-field control. Very wide speed ranges can be achieved with rotating motor-generator sets or static (SCR) conversion technologies capable of providing adjustable armature voltage. Adjustable-speed ac drive systems are gaining acceptance because of the rapid switching times available with semiconductors. This development has opened up the field of adjustable frequency, as shown in Fig. E-19.

The silicon controlled rectifier led to the development of the adjustable-voltage high-slip induction-motor drive and the wound-rotor induction-motor drive with primary or secondary control, or a combination of both. Two other constant-frequency ac drive systems are shown in Fig. E-20. Both the adjustable mechanical transmission and the eddy-current coupling provide economical choices.

Power-conversion Technologies: Traditionally, *rotating motor-generator* sets have been used to convert constant-frequency ac power to adjustable-voltage dc or adjustable-frequency ac power—primarily because of simplicity and flexibility. Maintenance personnel have an excellent understanding of motor-generator sets. As shown in Figs. E-18 and E-19, the motor-generater set is the only form of power conversion where regeneration is an inherent characteristic. This flexibility permits operation in all four speed-torque quadrants without the need for auxiliary control or power components.

Static (SCR) Converters-Inverters: These devices now are competitive with motor-generator sets. Since the silicon controlled rectifier is capable of passing current in only one direction, the basic system is capable of operation in only the first speed-torque quadrant. However,

TYPE	BLOCK DIAGRAM	AVAILABLE STANDARD DRIVE SYSTEMS	
		HORSEPOWER *	SPEED RANGE *
MECHANICAL TRANSMISSION	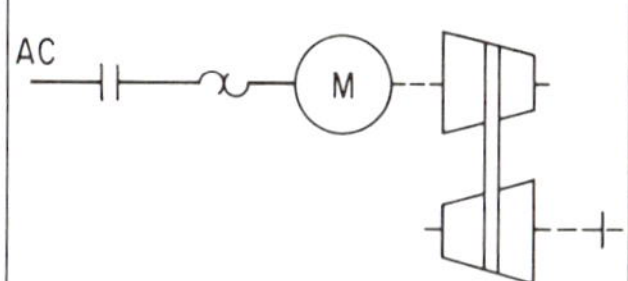	1/4 TO 75	UP TO 10:1
EDDY CURRENT COUPLING	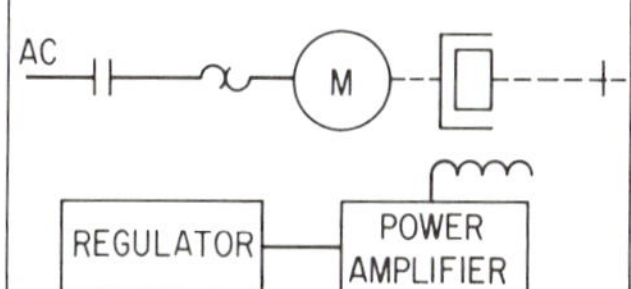	1/4 TO 750	UP TO 20:1
WOUND ROTOR MOTOR	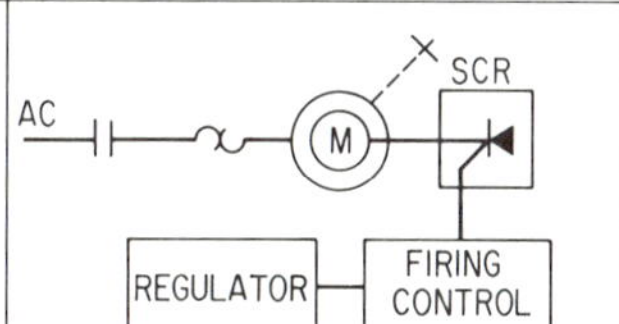	2 TO 5000	UP TO 4:1
HI-SLIP INDUCTION MOTOR	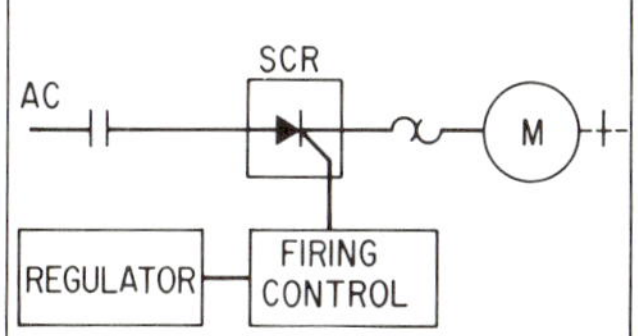	1 TO 150	UP TO 1 1/2:1

Fig. E-20. Adjustable-speed ac drive systems (constant-frequency). *Larger horsepower and wider speed ranges available (special).

it is economically feasible to add power and control components to provide four-quadrant capability. Since there are many drive applications in industry that do not require the ability to operate in all four speed-torque quadrants, the flexibility of this design makes it possible to tailor static-conversion techniques more closely to the driven machine. The SCR also has made possible the static converter-inverter and the cycloconverter. These developments have made the adjustable-frequency ac motor drive possible. In the case of the *converter-inverter* system, shown in Fig. E-19, constant-frequency ac power first is converted to constant-voltage dc power by means of a simple static rectifier. This dc power, in turn, is inverted into an adjustable-frequency power supply by means of static logic circuitry. The *cycloconverter,* as shown in Fig. E-19, converts constant-frequency ac power directly into adjustable-frequency ac power by switching silicon-controlled-rectifier components.

Regulators: Analog regulators of all forms are depicted in Fig. E-21. Feedback control systems are becoming an important part in trends toward automation. Although only the motor-generator set and SCR technologies are shown in dotted lines, these same regulators also can be used with eddy-current couplings as well as with adjustable-voltage and adjustable-frequency ac drive systems (see Figs. E-18 to E-20).

The operational amplifier is the heart of the analog feedback regulator. The amplifier output can be connected to *A,* a power amplifier which, in turn, is used to excite dc generator fields, dc motor fields, or eddy-current coupling fields; or to *A′*, the firing control associated with SCR converters and inverters. The reference *B′* is compared with any number of feedbacks—*D,* voltage; *F,* current; *G,* speed; and *H, I,* and *J,* position—thereby providing voltage, current, speed, and position regulators.

Timed acceleration can be provided by making the connections indicated at *B* and *B′*, whereas current-limit acceleration can be provided by contacting *B* to *C* and *C′* to *B′*, thereby bypassing the timing function and making up the connection at *E* in order to insert the current-limit function. This single motor or master regulator can be converted to a follower regulator by disconnecting points *B* and *B′* and making up the connections at *C* and *C′*. This becomes the basic building block for all multimotor coordinated-drive systems.

Regulator performance becomes a function

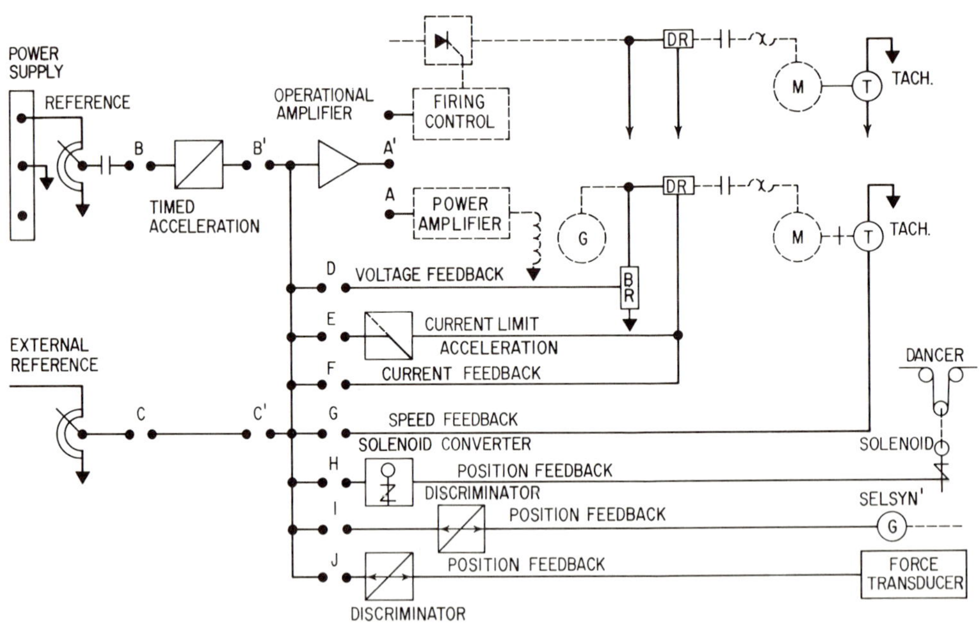

Fig. E-21. Analog regulators—composite block diagram.

of the hardware employed. Standard regulators can be improved by using high-performance power supplies, operational amplifiers, and tachometer generators. Also see **Tachometer generator.**

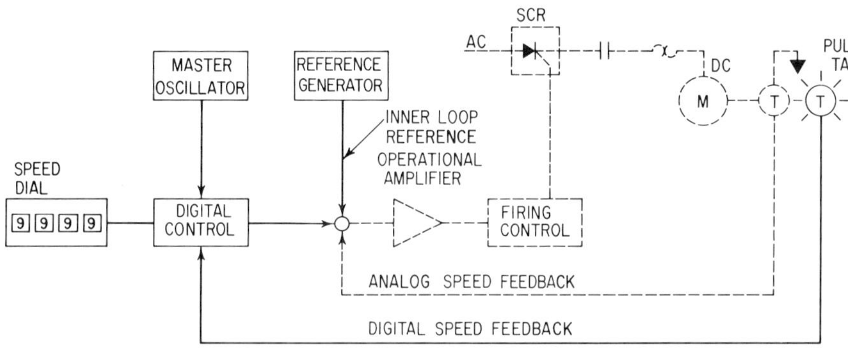

Fig. E-22. Digital speed regulator.

Digital Regulators: The component that limits the performance of analog speed regulators is the tachometer generator. The most accurate tachometer generator is incapable of consistently regulating speed to better than 0.1 percent of top speed. To improve performance, digital techniques can be used. A typical digital speed regulator is shown in Fig. E-22. The digital regulator employs a basic analog regulator with a digital outer loop which is designed to reduce the variation in speed with respect to changes in load to nearly zero. The digital regulator compares pulses from the digital tachometer with the reference pulses generated by a

TABLE E-2. *Functional-logic-control Features of Electric-motor Drives*

STARTING AND STOPPING
Timed acceleration and deceleration
Current-limit acceleration and deceleration
Coast stop
Electrical braking (dynamic)
Mechanical braking (electrically operated)
DIRECTION OF ROTATION
Run forward
Run reverse
Jog forward (inch—creep)
Jog reverse (inch—creep)
Thread
SPEED CONTROL
Manual
Remote preset
Automatic
Process instrumentation
Magnetic or punched tape
Digital computer
COORDINATED SYSTEMS
Master drive
Voltage follower
Speed follower
Torque follower (current)
Position follower

TABLE E-3. *Typical Industrial Applications for Adjustable-speed Electric-motor Drives*

Single-motor Drives	Multimotor Drives
CHEMICAL	ALUMINUM AND
Pump drives	STEEL
GLASS	Annealing lines
Glass drawing	Cut-to-length lines
Lehrs	Foil mills
LUMBER	Metal-decorating lines
Log carriages	Rolling mills
Planer feeds	Slitting lines
Veneer lathes	PAPER
MACHINE TOOL	Corrugators
Broaches	Off-machine coaters
Feeds	Pulp and paper
Planers	machines
Spindles	Winders
MATERIALS	PRINTING
HANDLING	Presses
Cargo winches	RUBBER AND
Conveyors	PLASTICS
Cranes and hoists	Blown film
Missile hoists	Calendars and trains
Mooring winches	Hot-stretch lines
TESTING	Tread conveyors
Aircraft generators	TEXTILE
Dynamometers	Bleaching and dyeing
Gear testers	ranges
	Slashers
	Spinning machines

crystal oscillator. These pulse trains are compared in an electronic up-down counter. When a speed error is detected, corrective action is initiated through the error register. See **Crystal oscillator.**

Logic Controls: In addition to the power-conversion and regulating-system techniques associated with adjustable-speed electric-motor drives, it is necessary to provide logical direction to the drive system in the form of manual or automatic control. The most important control features required by industrial applications are given in Table E-2. Starting, stopping, direction of rotation, speed control, and the coordination of several drives into a multimotor system all must be considered when selecting electric-motor drives.

A tabular summary of typical industrial applications for adjustable-speed electric-motor drives is given in Table E-3.

—Robert C. Berger, *General Electric Company*

Electric Transmission (See Transmission, electric.)

Electrical-conductivity Chemical Analyzer Electrical conductivity provides a useful measure of ion concentrations in water solutions and, where applicable, constitutes a simple and low-cost analytical method. Water per se is a very poor conductor. The very purest water has a conductivity approaching very closely the theoretical minimum of approximately 1/20 μmho/cm at 25° C, which is due to the dissociation products of water itself (hydronium and hydroxyl ions). The conductivity of a water solution is, in practice, almost exclusively due to a dissolved electrolyte rather than to the solvent (water) ions, and thus a criterion for electrolyte concentration can be established.

Solutions of strong electrolytes follow a rather uniform pattern of change of conductance with concentration which is almost linear at low concentrations, rising more gradually to a maximum usually around 20 to 30 percent by weight—then falling off as the concentration rises further (see Fig. E-23). The range of practical conductivity measurement extends from a fraction of a micro mho to more than 1 mho. Although electrolytic conductivity measurements are most numerous in the low-concentration region, where they are almost uniquely sensitive, many applications have been established in medium- and high-concentration regions. Since water is so poorly ionized, addition of a trace of electrolytic material causes a large increase in conductance. For example, distilled water will double its conductance on addition of 1 ppm of a typical salt—and 1 ppm of a strong acid will increase the conductance as much as 500 percent.

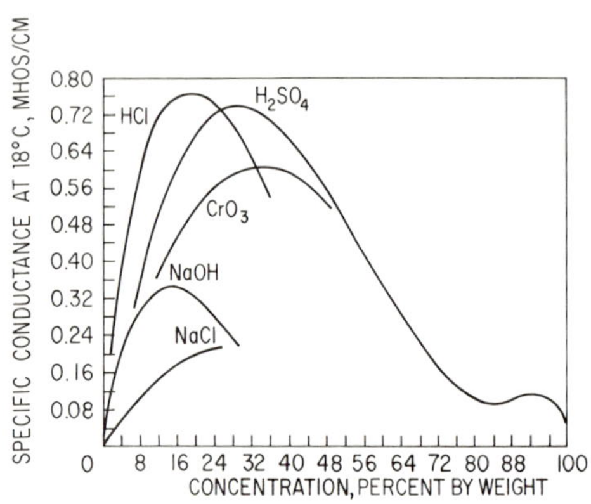

Fig. E-23. Conductivity-concentration curves for certain electrolytes.

At the high end of the range, because of the steep slope of the conductivity-concentration curve, electrical-conductivity measurements in such systems as concentrated sulfuric, hydrofluoric, and nitric acids have proved to be exceedingly sensitive to small concentrations of water. Differences in acid concentrations of 0.050 percent in the concentration span 95 to 100 percent are easily measured by these means.

Major applications for electrical-conductivity measurements include (1) concentrations of salts and acids and bases in simple water solutions; (2) gaging the quality of pure waters (such as distilled or demineralized water as well as boiler steam); (3) measuring the extent of reactions—as in neutralization, precipitation, and washing of soluble electrolytes from insoluble materials; and (4) detecting contamination—as from leaking heat exchangers, condenser coils, and steam coils. Saltwater intrusion in freshwater wells and streams can be detected as well as breaks in steam-condenser tubes.

The determination of specific electrical conductance basically consists of measuring the ac resistance of a column of solution. With alternating current, concentration changes in the electrolyte and polarization effects can be reduced. A typical ac Wheatstone-bridge system is shown in Fig. E-24. In the diagram, *D* represents a suitable ac voltage-sensitive device. Another type of circuit incorporates a crossed-coil ac electrodynamometer. A third system utilizes a simple ohmmeter circuit. A meter, transformer secondary, and a conductivity cell are connected in series so that the current is a function of the cell resistance. The meter may be calibrated in resistance or conductance terms. Although such a system is inexpensive and indicates continuously, accuracy suffers seriously from sensitivity to voltage variations.

The primary element in electrical-conductivity measurement is the conductivity cell. These cells are simple in basic structure, consisting typically of two metal plates or electrodes firmly spaced within an insulating chamber. Dip cells are designed for dipping or immersing into open vessels. For permanent installation in pipelines and tanks, cells provided with threaded fittings are used. Cells also are available that permit removal of the element without closing down or depressurizing the line in which they are installed.

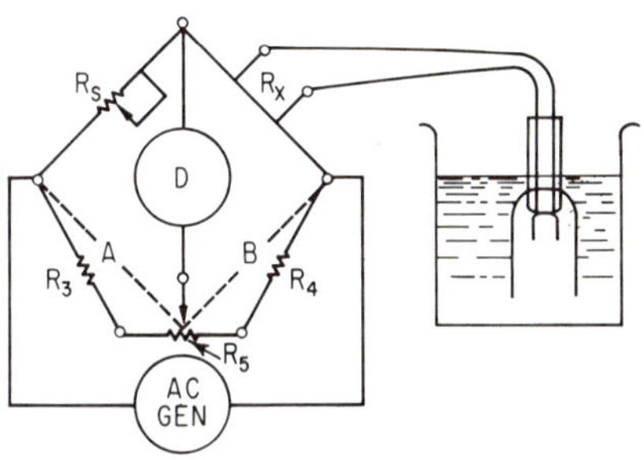

Fig. E-24. Ac Wheatstone bridge as used in some electrical-conductivity chemical analyzers.

Theory indicates and experience confirms that there are practical limits of measured electrolytic resistance for any desired accuracy and sensitivity. The optimum appears to be in the vicinity of 500 to 10,000 Ω measured resistance. The first criterion is to select a cell of such cell constant that the measured resistance of the solution under test will fall within these limits.

When measuring high electrolytic resistances, capacitive impedances in series with the conductivity cell are low in comparison and without significant effect on either the accuracy of the resistance reading or the sharpness of the null-point balance. However, if shunt capacities are present, as in long lengths of lead wires, these impedances do affect the bridge reading and also broaden the null balance. This effect is greater at higher bridge source frequencies. On the other hand, when measuring low electrolytic resistances, capacitive impedance in parallel with the conductivity cell has negligible effect on bridge readings. However, any series capacitive impedance impairs the sharpness of bridge balance. Since this impedance varies inversely with frequency, high bridge frequencies are desirable when measuring low resistances.

To check the cell constant by independent means, it is necessary to measure the resistance of the conductivity cell immersed in a carefully prepared solution of known specific conductance, usually 0.001, 0.01, or 0.1*N* postassium chloride in water. The choice of standard solution to be used for checking a conductivity cell depends largely upon the magnitude of the cell constant, since it is desirable to hold the measured resistance as close as possible to the optimum range of 500 to 10,000 Ω. The values of Table E-4 are suggested. The values in the table do not include the conductivity of the water in making the solutions. Thus as high-resistance water as possible should be used in making up the standard solution, particularly 0.001*N* KCl. The cell constant K is calculated by

$$K = R_x \frac{\text{S.C. KCl} + \text{S.C. H}_2\text{O}}{1{,}000{,}000}$$

where R_x = measured resistance, Ω

S.C. KCl = specific conductance of KCl as in Table E-4 at temperature of measurement

S.C. H_2O = specific conductance of distilled water used in making up solution at same temperature

When measuring electrical conductivity of solutions, the fact that conductance varies with temperature must be borne in mind. Conductance for most electrolytes increases fairly uniformly with temperature—at a rate of about 2½ percent/° C. For continuous and repeatable measurements, some form of temperature compensation should be considered. The most common means is the inclusion in one of the bridge arms of an adjustable resistor calibrated in temperature units. Sample-cooling systems sometimes are used prior to the measurement system so as to even out erratic sample-temperature variations. Automatic temperature compensation is available wherein a second conductivity cell is dipped into an isolated sample of the liquid—which is in thermal, but not electrical con-

TABLE E-4. *Standard Reference Solutions for Electrical-conductivity Measurements*

Approx. Normality	Composition KCl, g/liter	Temp., °C	Specific Conductance, μmhos/cm	Suitable for Cell Constants, *K*	Approx. Cell Resistance, Ω
0.001	0.0744	25	147	0.01–0.1	70–700
0.01	0.7440	{ 18 25	{ 1,221 1,409	0.1–10	70–7,000
0.1	7.437	{ 18 25	{ 11,170 12,860	10	8,000

tact with the flowing stream. Changes in temperature will affect the resistance of both cells to the same extent, thus canceling out this effect.

Electrical Energy, Measurement of (See **Watthour meter.**)

Electrical Integrator (See **Integrator.**)

Electrical Quantities (Variables) A high percentage of modern instrumentation, control, and data-processing systems fundamentally are electrical in nature. Even instruments and systems that may be described as predominantly mechanical, pneumatic, hydraulic, optical, or chemical commonly utilize electrical energy in some fashion—with the required electrical and electronic components, circuits, servos, and subsystems. The broad manner in which electrical phenomena permeate the overall instrumentation/control sciences has given rise to whole subfamilies of devices and technology, as described by electromechanical, electrohydraulic, electropneumatic, electrochemical, electro-optical, and so on.

Electrical quantities or variables are measured and controlled to accomplish a number of purposes: (1) the design, manufacture, and testing of systems; (2) the routine operation and maintenance of systems; (3) fundamental research and development of nearly all kinds of scientific phenomena in the laboratory; and (4) the generation, distribution, and consumption of electrical energy.

Traceability: All electrical instruments are calibrated in terms of the *prime standards* of the quantity measured. These prime standards are established and maintained by the National Bureau of Standards as the legal standards for use in the United States. The method of tracing the accuracy of voltmeters, ammeters, and wattmeters to these standards is shown in Fig. E-25. Also see **Standards, measurement.**

Manufactors of standard cells and precision electrical instruments and systems usually maintain their own standards laboratories which contain *local standards* that are certified or recertified on a periodic scheduled basis by the National Bureau of Standards. Laboratory secondary standards and the working standards used in various stages of manufacturing and assembly are, in turn, calibrated against these local standards. Also see **Standard cell.**

Various types of ratio devices are used to extend the standard-cell voltage to cover the many ranges of commercial instruments and in terms of current in amperes as well as voltage through the use of standard resistors. Extension of voltage values is accomplished through the use of what is commonly referred to as a volt box. Effectively, this is a high-valued resistor, accurately tapped at an exact proportion of its total value. If the total voltage is applied to the total resistance, then the value at the tapped points is low enough to compare accurately with the standard-cell voltage.

Extension of current values is accomplished through the use of very low-valued resistors; passing the unknown current through them results in voltage which, in turn, can be compared with the standard-cell voltage, and through the use of Ohm's law and the resistance value, the value of the current can be established. Such low-valued resistors are called shunts, and standard units of this kind are maintained with much care. These shunts can be certified by the National Bureau of Standards.

The direct voltages and currents established at this level then are used to calibrate dc voltmeters, ammeters, and ac-dc voltmeters, ammeters, and electrodynamic wattmeters. Also shown in Fig. E-25 are ac-dc difference or transfer standards. These are high-accuracy instruments, such as the thermal- and electrodynamic-type voltmeters, ammeters, and wattmeters which are sent to the National Bureau of Standards for certification. In this case, the NBS does not perform an accuracy check but rather determines the difference between the ac calibration at var-

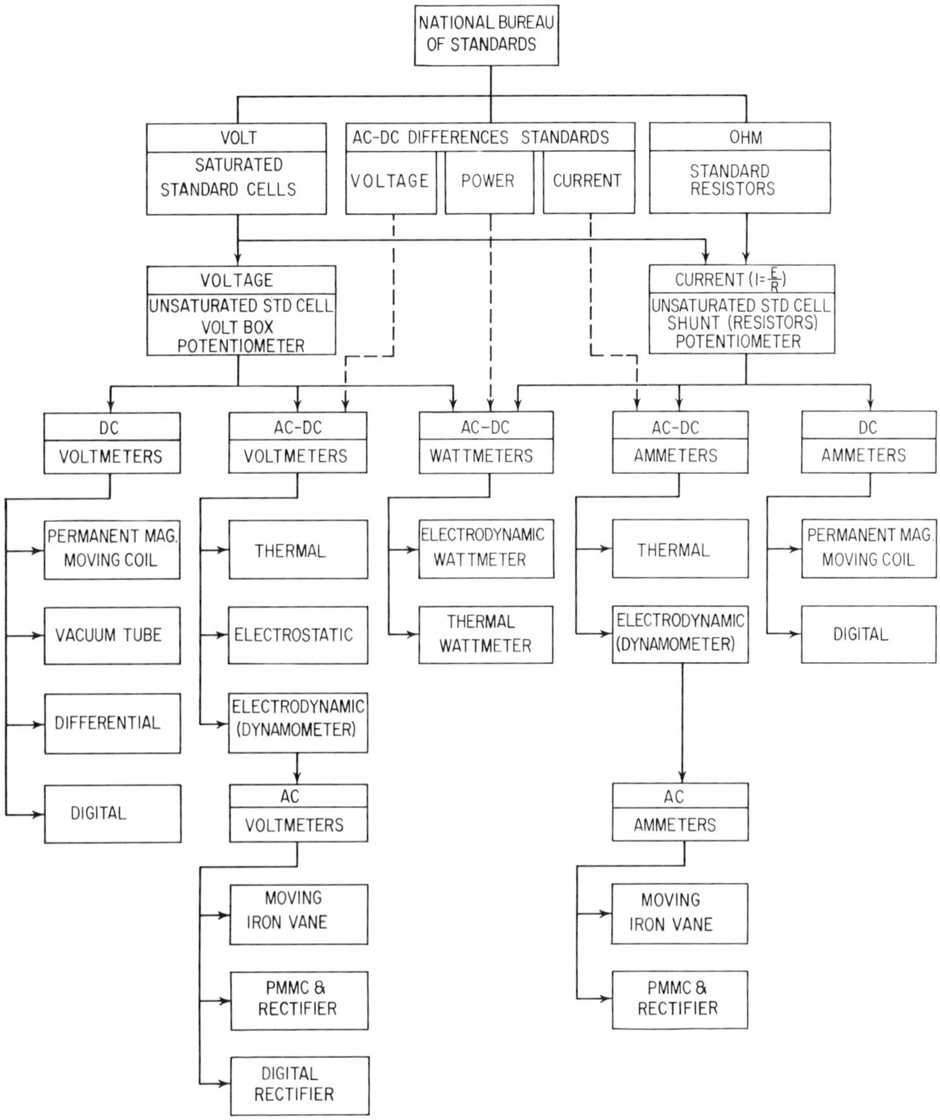

Fig. E-25. Traceability of electrical instruments back to prime standards maintained by the National Bureau of Standards. *("Instrument Sketch Book," Weston Instruments, Inc.)*

ious frequencies and that on direct current. These standards then are used for determining the performance of ac-dc instruments when used on alternating current.

Alternating-current voltmeters and ammeters usually are calibrated against electrodynamic voltmeters and ammeters which have been calibrated on direct current and which have a known accuracy on alternating current as determined by the ac-dc difference test.

—Adapted from "The Instrument Sketch Book," *Weston Instruments, Inc.*

Electrical Quantities, Measurement of A substantial percentage of modern instruments and controls essentially are electrical or electronic in their operation. Literally hundreds of different kinds of transducers are available for the detection of a score or more variables, such as temperature, pressure, flow, and level. These transducers produce electrical outputs which possess a definite relationship with the *non*-electrical variable being measured. Such electrical outputs are relatively easy to convert into means for indication, recording, and control. These are *electrical-type instruments,* and although the circuitry and components of these instruments often are quite similar, they are not to be confused with a whole family of instruments that are designed to measure electrical quantities per se; that is, the end objective is to determine the value of an electric current, or voltage, or capacitance, and so on. The latter instruments are referred to here as *electrical instruments.* Also see **Magnetic quantities (Variables).**

Universal Nature of Electrical Instruments: As is evident from Ohm's and other fundamental electrical laws, electrical quantities are closely interrelated, thus giving the electrical-instrument user much more flexibility in the application of such devices than is true in the case of most other kinds of instruments. For example, it is not easy to use a pyrometer to measure viscosity, but a voltmeter can be rigged to measure current or resistance as well as the voltage for which it was designed. Or, the value of an electrical unknown can be calculated from the known value of other electrical characteristics (values) to which it relates. It follows that many electrical instruments, even though they may be designed for the measurement of a specific electrical variable, will incorporate electrical principles, circuits, and components that may be quite similar to other electrical instruments that are designed to measure some other electrical quantity. This apparent universality of electrical instruments makes it difficult to classify them cleanly and to describe comprehensively the various electrical instruments on the basis of electrical quantities being measured without a lot of duplication. Hence the basic appraoch used here, wherever it applies, is that of describing electrical instruments by operating principle rather than by end application.

Analog and Digital Instruments: Electrical instruments have been undergoing continuous development for at least 150 years. During most of that time, the principal effort of making instruments react to electricity without human intervention, often referred to as direct-acting or direct-reading instruments, was aimed at the perfection of pointer-deflecting instruments. In these devices, the deflection angle of the pointer is a function of, and thus analogous to the value of the electrical quantity measured. Therefore, the name *analog instrument* has been coined to distinguish these devices from a completely different type in which the value of the quantity measured is displayed in numerals, usually in the form of two or more digits. These devices are aptly termed *digital instruments.* See Figs. E-26 to E-28.

Fig. E-26. Conventional portable analog voltmeter. *(Weston.)*

The basic electrical analog instrument can be traced to Oersted's discovery (1819) of the relation between current and magnetism. Faraday (1821) learned that a current-carrying conductor would rotate in a magnetic field. Ampère (1821) demonstrated that a current in one conductor attracted or repelled another current-carrying conductor. Sturgeon (1836) wound the current-carrying wire into a coil and suspended it in a magnetic field. Kelvin (1867) placed a soft-iron core in the center of the coil, shortening the air gap, increasing the sensitivity, and improving the scale characteristics of the device. D'Arsonval (1881) perfected and patented an instrument of this type. In 1888, Weston discovered the factors required to produce a permanent-magnet system, added soft-iron pole pieces, devised current-carrying control springs, and produced the first commercial double-pivoted permanent-magnet moving-coil instruments as such. From Oersted's and Kelvin's work also stem the principles on which polarized iron-vane and electrodynamic types of instruments evolved through the work of Ayrton and Perry (1881), Bruger (1886), Weston (1889), and many others.

Digital electrical instruments appeared much more recently and owe their existence to the art of electronics and the ability of electronic devices to create electrical pulses, to perform

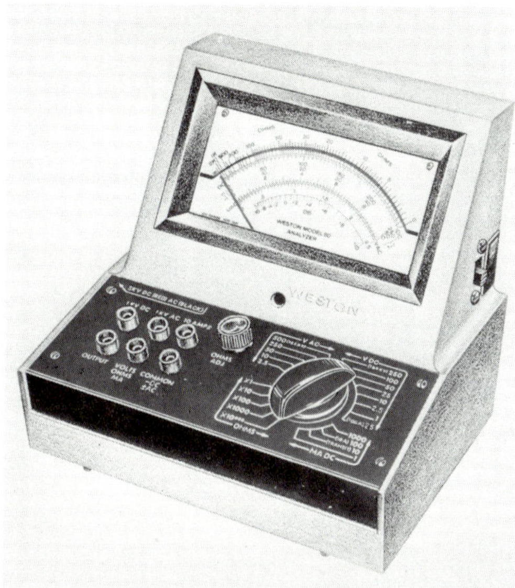

Fig. E-27. Multiple-range tabletop instrument containing a permanent-magnet moving-coil instrument with a variety of circuits and transducers for measuring many electrical functions in one modular housing. Instruments of this type often are termed multimeters. *(Weston.)*

rapid switching operations, to produce precisely predictable time functions of voltage, to make voltage comparisons, to amplify, and to perform the fundamental operations of addition, multiplication, differentiation, and integration.

Modern digital and analog electrical instruments are discussed in more detail later in this section.

Differential Instruments: Related to analog instruments by the functions of direct-acting galvanometers—and to digital instruments by the use of numerical presentations of the measured results—are the *differential instruments.* In these devices, the measurement is achieved by the manipulation of knobs or dials until a condition of balance is obtained and indicated by the galvanometer. Many modern versions

Fig. E-28. Digital voltmeter which converts the analog input into digital logic and displays it in decimal form. *(Weston.)*

of these instruments contain highly stable solid-state reference-voltage sources, such as zener diodes (see Fig. E-29).

Basic Mechanisms Used in Electrical Instruments: Several fundamental mechanisms are used (or are of historical interest) in electrical instruments for the measurement of current, voltage, resistance. and related electrical

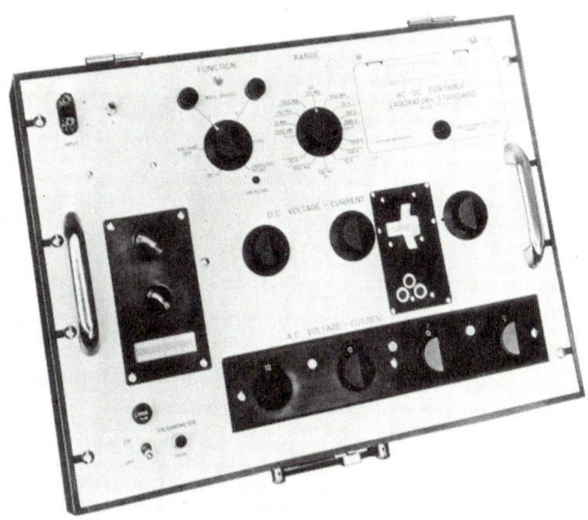

Fig. E-29. Differential instrument in which precision decade dividers, stable solid-state reference voltage, shunts, transformers, and special thermocouples provide both ac and dc measurements of high accuracy. *(Weston.)*

quantities: (1) the early polarized iron vane; (2) the moving iron vane; (3) the electrodynamic mechanism; (4) the electrostatic mechanism; (5) the permanent-magnet moving coil; (6) thermoelements; (7) the thermal watt converter; and (8) the hot-wire mechanism (now obsolete).

Polarized-iron-vane Mechanism: This was the first current-measuring mechanism to be reduced to practice. Originally, as shown in Fig. E-30*a*, the device was dependent upon the earth's magnetic field to provide the control torque. Later, it was made more sensitive and convenient to use by employing a permanent magnet as shown in Fig. E-30*b* through *e*. The mechanism shown in Fig. E-30*e* came into use in large quantities as an automobile dashboard ammeter. Better suited to mass production in relatively crude form than to greater refinement, this mechanism has lost its position in the field of electrical instruments and has, instead, made a place for itself in the "indicator" field where precision is not demanded.

Moving-iron-vane Mechanism: As shown in Fig. E-31, if two similar adjacent iron bars are similarly magnetized, a repelling force is developed between them which tends to move them apart. In the moving-iron-vane mechanism, this principle is used by fixing one bar in space and pivoting the second so that it will tend to rotate when the magnetizing current flows. A spring attached to the moving vane opposes the motion of the vane and permits the scale to be calibrated in terms of the current flowing. An early, suction-type mechanism is shown in Fig. E-32, a concentric-vane mechanism in Fig. E-33, and a radial-vane mechanism in Fig. E-34

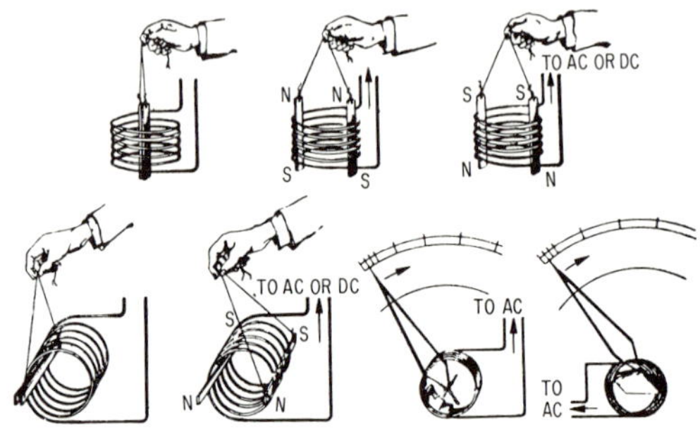

Fig. E-31. Principle of moving-iron-vane mechanism.

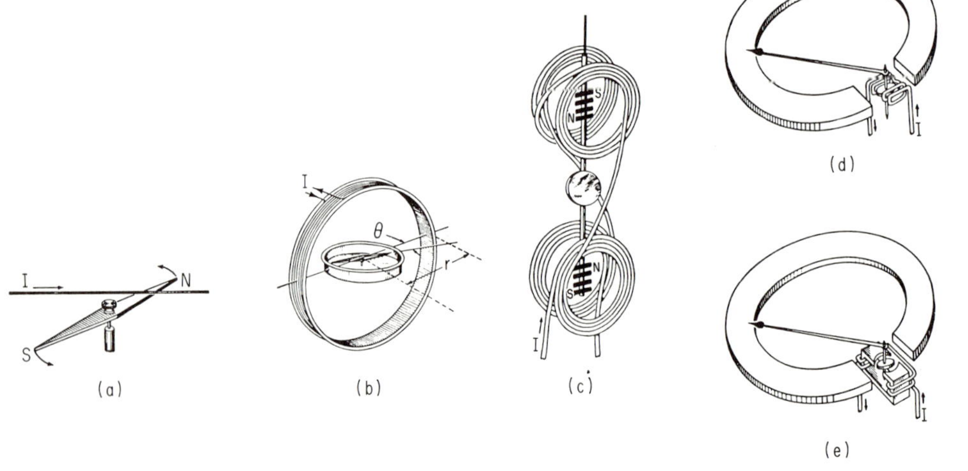

Fig. E-30. Forms of polarized-iron-vane mechanism: (a) original electrical indicator (Oersted, 1819); (b) tangent galvanometer — $I = 10Hr/2\pi N \tan\theta$ A; **(c) early-type astatic galvanometer (Kelvin, 1858); (d) simple polarized iron-vane mechanism; (e) complete polarized iron-vane mechanism where soft-iron core adds sensitivity, requires fewer turns of wire, and improves scale characteristics.**

Electrodynamic Mechanism: These devices are the most fundamental of all the indicating devices now used. Form-factor variations in the waveform of the signal do not affect their readings because of the complete absence of magnetic materials, such as iron, and the indications of current or voltage are mean values of their squared amounts and thus permit calibration in terms of true rms values. The mech-

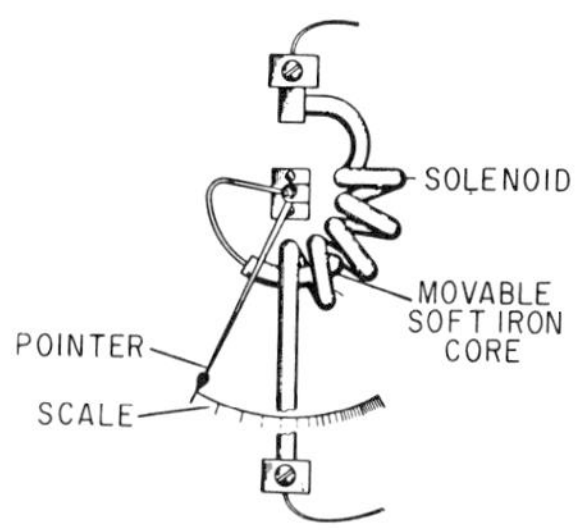

Fig. E-32. Early magnetic-vane mechanism of the suction type.

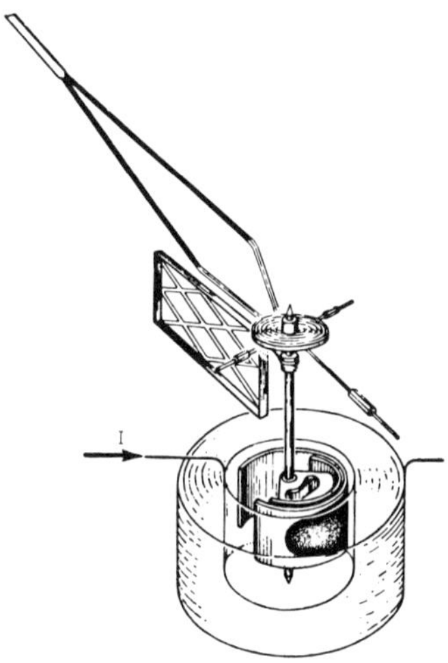

Fig. E-33. Concentric-vane mechanism wherein the vanes slip laterally under repulsion. The device is only moderately sensitive and has square-law characteristics. The short magnetic vanes provide small dc reversal and hence residual magnetism errors. With this mechanism, it is possible to shape the vanes to secure special characteristics, thereby opening the scale where needed.

anism is current-sensitive—the pointer moves because of current flowing through turns of wire. It is the most versatile of all the basic mechanisms, since the single-coil movement, as shown in Fig. E-35, can be used to indicate current, voltage, or power—ac or dc, while crossed-coil and crossed-field modifications can be used for power factor,

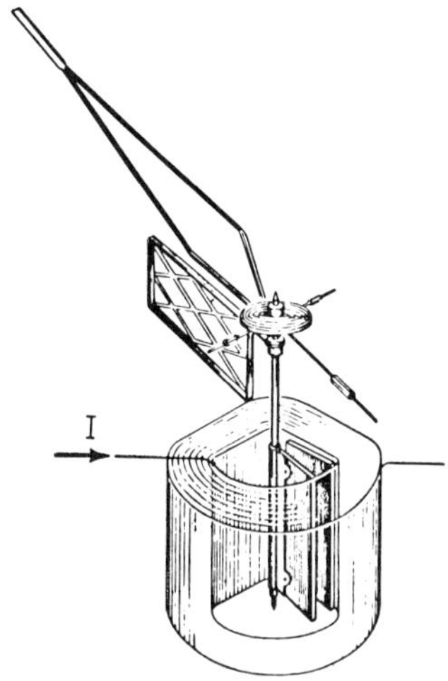

Fig. E-34. Radial-vane mechanism. The device opens up like a book under repulsion. It is the most sensitive and has the most linear scale of the various moving-iron-vane mechanisms. An aluminum damping vane, attached to the shaft just below the pointer, rotates in a close-fitting chamber to bring the pointer to rest quickly.

phase angle, frequency, and capacity measurements. Arrangements for these are shown in Fig. E-36 (for single-element types) and in Fig. E-37 (for double-element types).

Perhaps the most important use of this mechanism is as a transfer instrument between the basic standards of *E, I,* and *R*—all of which are defined for direct current only—and alternating current, in which form most of the electrical power is generated, distributed, and used. The torque produced in the moving coil of an electrodynamic instrument is proportional to the product of the in-phase components of the currents in field and moving coils. This mechanism measures products; and thus currents and

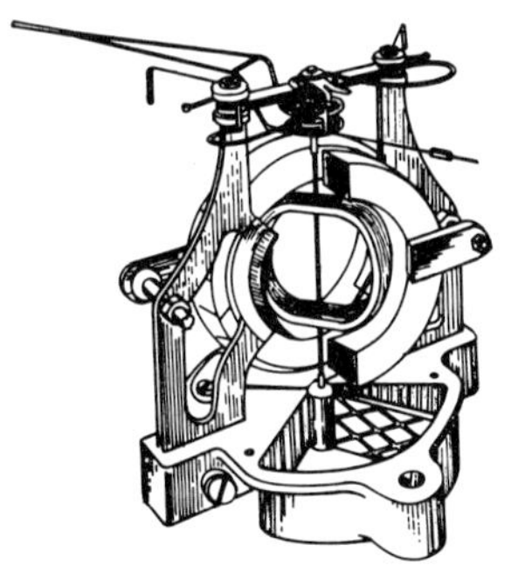

Fig. E-35. Electrodynamic mechanism. *(Weston.)*

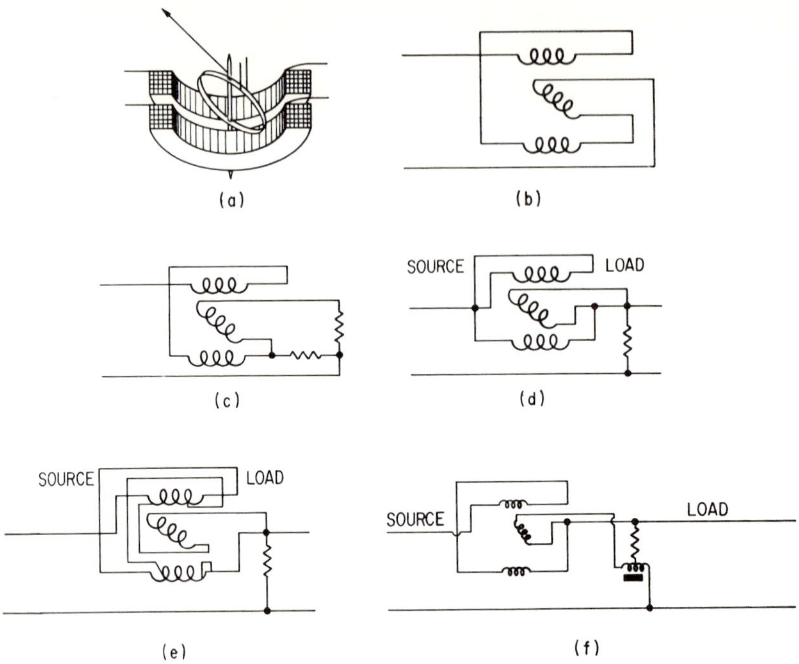

Fig. E-36. Coil arrangements in single-element electrodynamic mechanisms. *(a) Single-element type. (b) Milliammeter-voltmeter.* Fixed and moving coils in series for milliammeter; combination in series with a resistor unit for voltmeter. *(c) Ammeter.* Fixed coils in series with a resistor which is provided as a shunt for the moving coil. *(d) Wattmeter.* Fixed coils in series with the load; moving coil with suitable series resistor connected across the load or source. Across the load, the reading includes moving-coil circuit power; across the source, the reading includes fixed-coil circuit power. *(e) Wattmeter (compensated).* Fixed coils incorporate two windings—a heavy winding connected in series with the load; a fine winding connected in series with the moving coil to cancel the effect of the moving-coil current, thus resulting in an indication of true power. *(f) Single-phase varmeter.* Resembles single-phase wattmeter of *(d)*, but current in moving coil is rotated 90° in time phase.

voltages are squared, and power—which is a product by definition—is measured directly. The mechanism shown in Fig. E-35 therefore is used in wattmeters, voltmeters, and ammeters. With a longer pointer and other structural refinements, it is the mechanism used in a laboratory standard. With further modification of crossed moving coils, shown schematically in Fig. E-37*f*, fundamentally the same mechanism is used for measurements of ratios, such as power factor, phase angle, and capacity. In this form—or with crossed-field coils—this mechanism measures a ratio by balancing two torques. Damping vanes similar to those used in moving-iron instruments (Figs. E-33 and E-34) rotate in close-fitting chambers and provide proper damping for the pointer motion.

If two complete field-coil systems are arranged one above the other, each including and acting upon its own moving coil—and if the two moving coils are attached to a common shaft which carries the pointer—the mechanism can be used to measure the total power in a polyphase ac system (see Fig. E-37*a* through *e*).

Electrostatic Mechanism: This device resembles a variable condenser. Of all the mechanisms used for electrical indications, it is the only one that measures voltage directly rather than by the effect of current. The torque resulting from the attraction between fixed and movable plates is a function of the voltage between the plates, the plate area, and inversely, the distance between the plates. For greater sensitivity, this distance must be reduced, clearances permitting, or the plate area (and thus the weight) must be increased. Greater weight, in turn, calls for still greater plate area if sufficient torque is to be developed to overcome pivot friction in most industrial applications. This limits the use of the instrument to certain special applications, particularly in ac circuits of relatively high voltage, where the current taken by other mechanisms would result in erroneous indications. A protective resistor is used in series with the instrument to limit current flow in the event of a short between plates.

A gold-leaf *electroscope* is shown in Fig. E-38. Like charges on the ends of the leaf cause a separation of the ends. This is a very sensitive indicator, but not an instrument.

A form of the attracted-disk electroscope as devised by Sir William Snow Harris (1834) is shown in Fig. E-39. A guard ring around the disk serves to prevent nonuniformity in the electrostatic lines of force. A very large attracted-disk electrometer, mounted in a shielded cage and

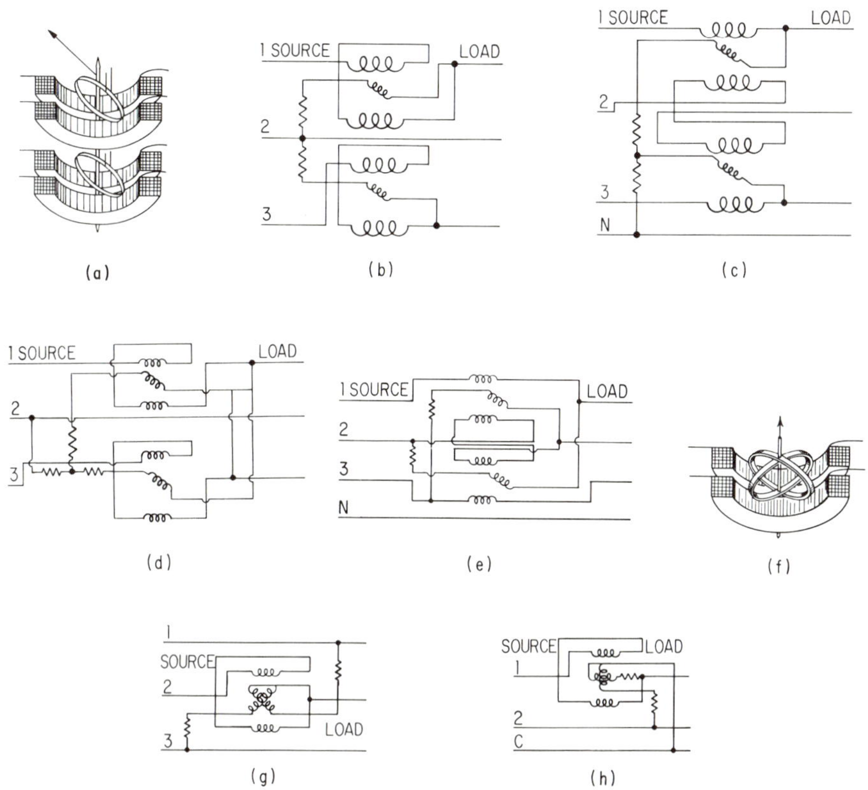

Fig. E-37. Coil arrangements in double-element and crossed-coil-type electrodynamic mechanisms. *(a) Double-element type. (b) Polyphase wattmeter (two-element).* Blondel's theorem states that true power can be measured by one less wattmeter element than the number of wires of the system, provided that one wire can be made common to all element potential circuits. Connections are suitable for single-phase, three-wire; three-phase, three-wire. *(c) Polyphase wattmeter* ($2\frac{1}{2}$ *element*). A compromise on Blondel's theorem for measurement of three-phase, four-wire system power. This system will indicate true power provided that voltages are balanced. This is termed $2\frac{1}{2}$ element because of split connection of fixed coils. Current in line 2 passes through both the upper and lower element, in reverse direction, but is displaced in phase by 60°. The effect is thus half. Summation of two elements equals third element required. *(d) Polyphase varmeter (two-element) three-phase, three-wire system.* This and the four-wire version of *(e)* resemble the polyphase wattmeters of *(b)* and *(c)* except that moving-coil connections produce currents 90° rotated in time phase. *(e) Polyphase varmeter for three-phase, four-wire-system* ($2\frac{1}{2}$ *element*). *(f) Crossed-coil type. (g) Power-factor meter (three-phase).* Crossed moving coils are connected to opposite legs of a three-phase system. Fixed coils connected in series with line used as common for moving-coil connection. Moving system assumes a position dependent upon power factor of circuit. Correct for balanced loads only. *(h) Power-factor meter (two-phase).* Essentially similar to three-phase type except pointer is mounted in line with inner coil. Fixed coils not in the common line. For a single-phase power-factor meter, the double moving coil again is used, with a reactance network to give the required phase difference in the currents in the moving system.

using quartz pillar supports, is used by the National Bureau of Standards (Washington) for voltage standardization up to 300,000 V. Using this high-voltage electrometer, the ratio of trans-

Fig. E-38. Gold-leaf electroscope.

formation of high-voltage potential transformers now can be checked by an independent method.

The principle of an early electrostatic mechanism, devised by Lord Kelvin (1887), is shown in

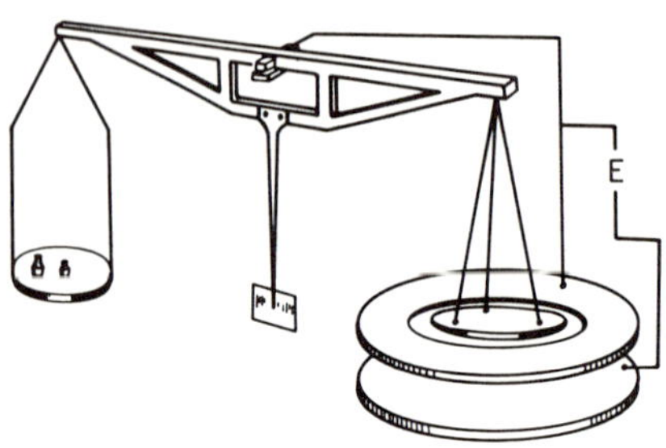

Fig. E-39. Attracted-disk electroscope devised by Sir William Snow Harris (1834). Guard ring around the disk served to prevent nonuniformity in the electrostatic lines of force.

Fig. E-40. Moving and fixed plates attract each other, causing the moving plate to rotate against the torque of a control spring (not shown). The position of balance thus becomes a measure of the

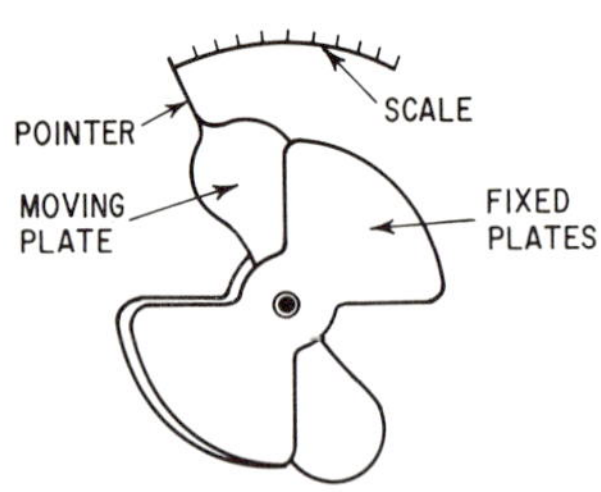

Fig. E-40. Early electrostatic mechanism devised by Lord Kelvin (1887).

potential applied. In modern format, such as the 3½-in. panel-size device shown in Fig. E-41, a multiple set of plates—in ranges from 150 to 3,500 V—is used.

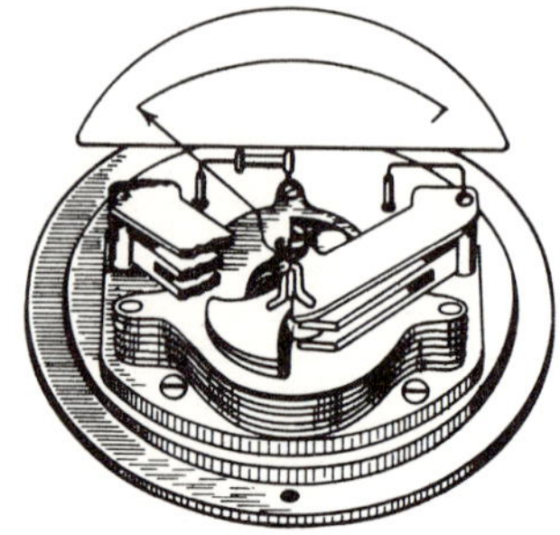

Fig. E-41. Modern form of electrostatic mechanism.

Permanent-magnet Moving-coil Mechanism: The suspension galvanometer, shown in Fig. E-42, is an early type of moving-coil instrument. A coil of wire is suspended in a magnetic field and will rotate when it carries an electric current. The fine-filament suspension of the coil serves to carry current to and from it, and its elasticity sets up a moderate torque in opposition to the rotation of the coil. The coil will deflect until its electromagnetic torque balances the mechanical torque of the suspension. A mirror attached to the coil deflects a beam of light, causing a spot of light to travel on a scale at some

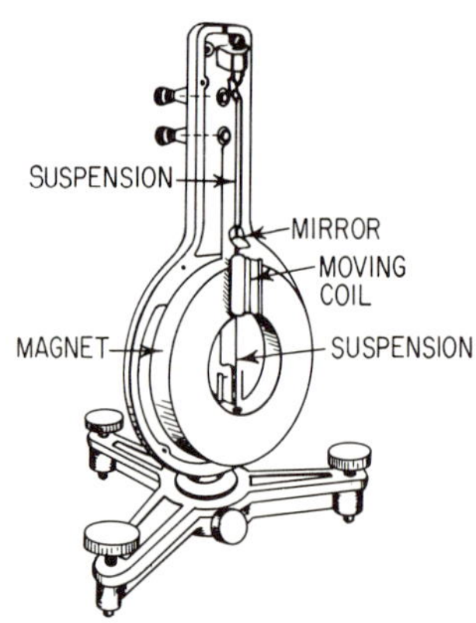

Fig. E-42. Suspension galvanometer, an early type of moving-coil instrument.

distance from the instrument. The effect is that of a pointer of greater length, but of no mass. With modern refinements, the suspension galvanometer still is widely used for certain laboratory work where extreme sensitivity is required and where the delicacy of the instrument is not objectionable.

The principle of operation of the permanent-magnet moving-coil mechanism is represented in Fig. E-43. The reaction of permanent-magnet flux with the current establishes the torque. The induced electromagnetic torque is balanced by the mechanical torque provided by control springs attached to the movable coil. This permits the angular position of the movable coil

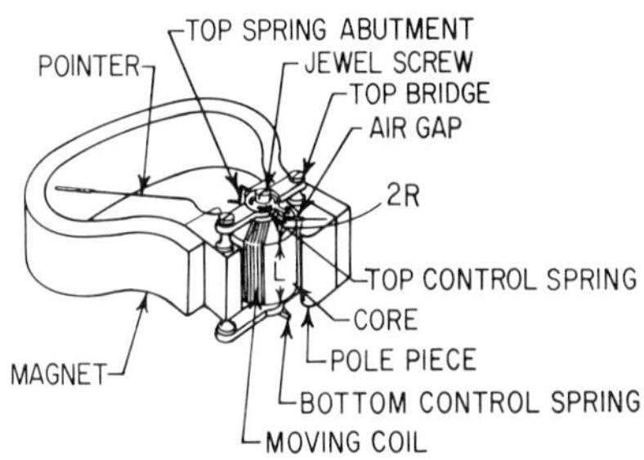

Fig. E-43. Permanent-magnet moving-coil mechanism with external magnet. *(Weston.)*

to be indicated by a pointer against a fixed scalar reference. Torque can be defined as

$$T = \frac{B \times A \times I \times N}{10} \tag{1}$$

where T = torque, dyne-cm
B = flux density in air gap, gauss
A = effective coil area, cm^2 = 2 × radius of coil side from pivot center × length of active coil side
I = movable-coil current, A
N = turns of wire on coil

The torque equation shows that, to measure a given current, the designer may vary only the value of the control-spring torque and the number of turns on the moving coil—since both flux density and effective area of the movable coil are fixed parameters for a given instrument mechanism. The practical coil area generally ranges from approximately 0.5 to 2.5 cm^2; flux density usually ranges from 1,500 to 5,000 gauss. Thus, a wide choice of mechanisms is available to the designer for meeting the needs of numerous applications. A typical panel instrument, with a 3½-in. case and a 1-mA range and scale for 100° deflection, would have the following nominal characteristics: A = 1.75 cm^2; B = 2,000 gauss; N = 84 turns; T = 29.2 dyne-cm. The resistance of the coil would by 88 Ω; the power dissipated would be 99 μW.

Response of Permanent-magnet Moving-coil Mechanism: The response of the dc instrument to a sudden current change is governed by the basic laws that cover the response of any moving system characterized by both mass and stiffness. The torque equation expressing this relation is

$$T = K\frac{d^2\theta}{dt^2} + D\frac{d\theta}{dt} + S\theta \tag{2}$$

where K = moment of inertia of movable system, g-cm^2
D = damping coefficient, dyne-cm/(rad)(s)
S = spring constant of control springs, dyne-cm/rad
T = actuating torque, dyne-cm
θ = angular deflection in time t, rad

The first term of the equation is the torque required to overcome the inertia; the second term is the torque required to overcome any damping (whether air, friction, or magnetic damping); and the third term is the torque required to overcome the suspension-spring restoring moment.

Damping: Three specific solutions to the torque equation can be obtained, dependent upon the relative magnitudes of the damping coefficient, the spring constant, and the moment of inertia. These solutions are equations expressing angular deflection as a function of time and thus define the motion of the movable system. These motions are known as the *underdamped* case, in which the pointer attached to the movable coil will oscillate back and forth several times—above and below its final rest position; the *critically damped* case, in which the pointer will be observed to move promptly up to its final rest position with no overshoot; and the *overdamped* case, in which the pointer will approach the rest position asymptotically.

The *specific damping coefficient n* is used to designate which of the three motions best describes the moving-system characteristics when current changes suddenly. It is expressed as

$$n = \frac{D}{D_0} \tag{3}$$

where D_0 = damping coefficient that causes critically damped motion
D = actual damping coefficient

The possible solutions are $n < 1$ for underdamped motion, $n = 1$ for critically damped motion, and $n > 1$ for overdamped motion.

Use of External Magnet: The permanent-magnetic moving-coil mechanism shown in Fig. E-43 offers the largest magnet in a given space and is used where maximum flux in the air gap is required in order to provide an instrument of lowest possible power consuption. Common practice is to grind one side of an Alnico V block of magnetic material and solder or cement it to soft-iron pole pieces to form a U-shaped magnetic structure.

Use of Core Magnet: Alnico and other improved magnetic materials make it feasible to design a magnetic system in which the magnet serves as a core. Such magnets operate at their highest-energy product with minimum lengths, thereby making the core-magnet mechanism a practical reality. These mechanisms have the obvious advantage of being relatively resistant to external magnetic fields, therefore allowing for the elimination of magnetic shunting effects in a panel, or of the need for magnetic shielding in the form of iron cases (see Fig. E-44*a* and *b*). The advantage of the self-shielded feature makes the core-magnet mechanism particularly useful for aircraft and aerospace applications, especially where a multiplicity of mechanisms must be mounted in close proximity to each other as exemplified by cross-pointer indicators wherein as many as five mechanisms are mounted in one case to form a unified display. The weight reduction also is an advantage.

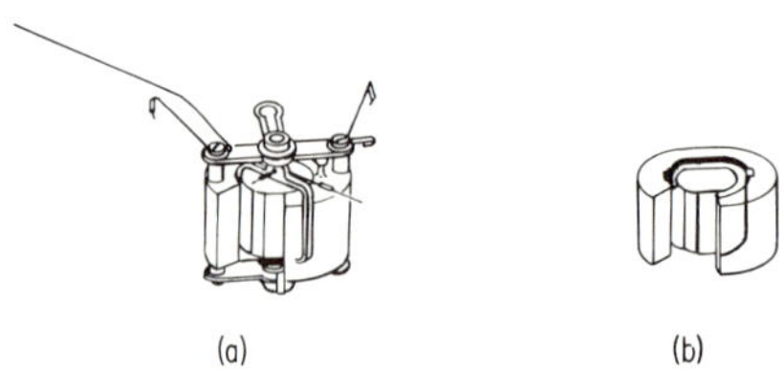

Fig. E-44. Permanent-magnet moving-coil mechanism with core magnet: (a) assembled mechanism; and (b) core magnet.

End-pivoted Permanent-magnet Moving-coil Mechanism: This is a variation of the more common center-pivoted design. In this arrangement, the coil rotates in a single air gap, allowing a full-scale deflection as high as 270°. A modern self-shielded version with a scale deflection of 250° is shown in Fig. E-45. The off-center coil type also is used extensively in edgewise panel meters (Fig. E-46) and in aircraft meters (Fig. E-47). For the latter instruments, the deflection sometimes is limited to 60° or less with the magnetic flux concentrated over the smaller angle, thus offering exceptionally high torque to permit long, heavy pointers to be moved with high speed and good damping.

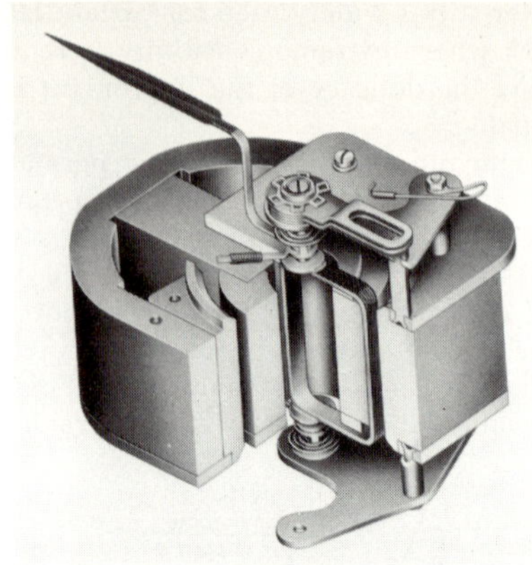

Fig. E-45. Modern version of an end-pivoted or off-center permanent-magnet moving-coil mechanism. *(Weston.)*

Direct-current Ratio-indicating Mechanism: A mechanism in which two independent

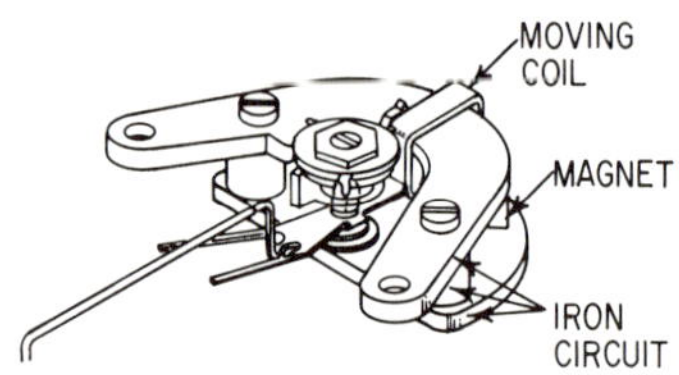

Fig. E-46. Typical edgewise meter mechanism of the permanent-magnet moving-coil type.

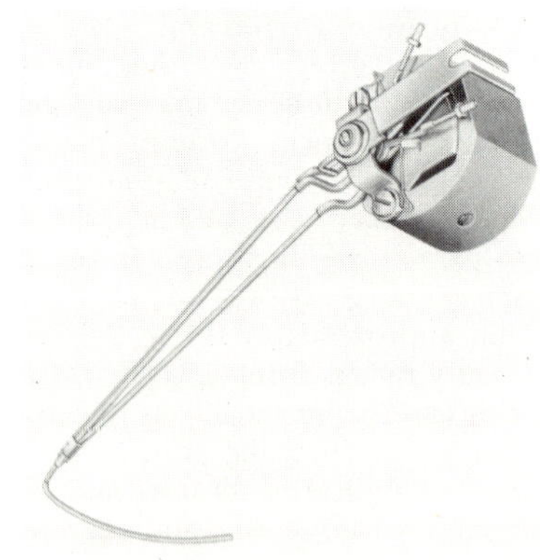

Fig. E-47. Typical aircraft mechanism of the permanent-magnet moving-coil type. *(Weston.)*

coils rigidly mounted on opposite sides of a common axis move in an air gap formed by a core mounted eccentrically between the cylindrical pole faces of a permanent magnet is shown in Fig. E-48. No control springs are attached to the moving system, the current being fed to the coils by fine filaments. The ampere-turns (or flux) developed by each coil will react with the flux in its part of the air gap, causing the pointer to move until the product of ampere-turns times the gap flux will be equal on each side so that balance is reached.

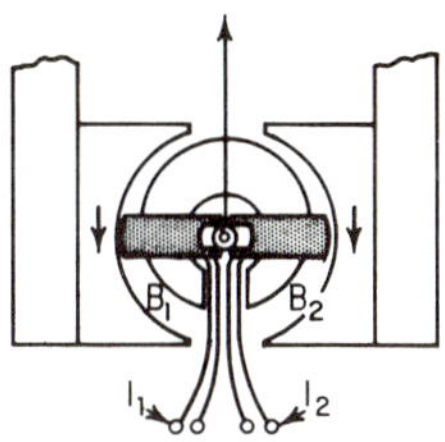

Fig. E-48. Direct-current ratio-indicating mechanism.

Except for a multiplying factor to take care of turns ratio and coil areas, the situation can be written

$$I_1B_1 = I_2B_2 \tag{4}$$

or

$$\frac{I_1}{I_2} = \frac{B_2}{B_1} \tag{5}$$

where I_1 and I_2 = currents in the two coils
B_1 and B_2 = flux densities in the air gap at the points where the coils rest in the state of balance

Equation (5) expresses a ratio of two currents in terms of a ratio of two flux densities. In view of the nonuniform and reversed flux distribution in the two sides of the air gap, a given ratio of flux densities occurs only once within the total angle of moving-system rotation and is therefore associated with only one point on the scale. Each point on the scale thus represents only one current ratio, and the scale can be so calibrated. Thus, the instrument is a ratio meter capable of measuring any condition which can be expressed as a ratio of two currents. An example is the measurement of resistance as a ratio of voltage to current ($R = E/I$).

Taut-band Instruments: The suspension-type mechanism has been known for years. For a long time, the device was confined to laboratory equipment where high sensitivities were required and available torques were low so that it was desirable to eliminate even the low-friction pivots and jewels. The device had to be used in the upright position since the sag in the very low torque ligaments caused the moving system to come in contact with stationary members of the mechanism in any other position. The taut-band instrument provides the advantage of the absence of friction of ribbon suspension while also eliminating the sag by placing the ribbons under sufficient tension. Tension is provided by a tension spring so that the instrument can be used in any position. Generally, taut-band-suspension instruments can be made with higher sensitivities than those using pivots and jewels and can be utilized in virtually every application which presently is served by pivoted instruments. One form of taut-band suspension is shown in Fig. E-49.

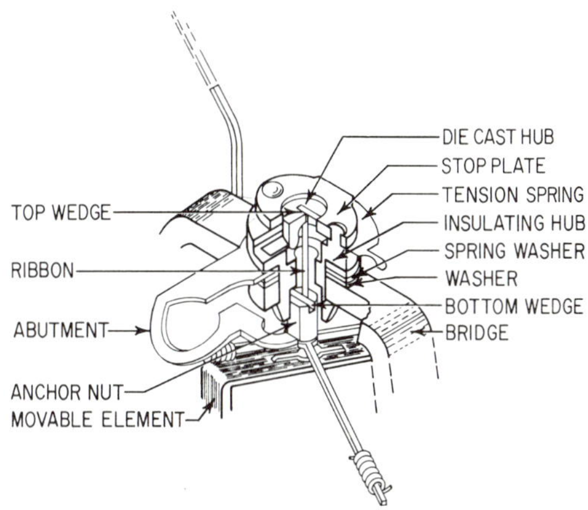

Fig. E-49. Taut-band suspension used in permanent-magnet moving-coil mechanism.

Temperature Compensation: A permanent-magnet moving-coil mechanism is not insensitive to temperature by itself but may be made so that by appropriate use of proper series and shunt resistors of copper and manganin it is temperature-compensated. The changes in the magnet and the copper tend to make the pointer read low on fixed voltage impressed, while the spring change tends to cause the pointer to read high. The effects are not identical, however, with the result that an uncompensated mechanism tends to read low by approximately 0.2 percent/° C. For purposes of specification, an instrument is considered compensated when the change in accuracy due to 10° C change in temperature is not more than one-fourth of the total allowable error.

Special Panel Meters: The equivalent scale length of a 7- by 9-in. pointer-type panel meter is provided by the projected-moving-scale instrument, shown in Fig. E-50, in a 2½- by 3⅛-in. panel space. This is accomplished by mounting a scale of photographic film to the moving coil of a core-magnet mechanism in place of the usual pointer. A lamp, lens system, and mirror projects this moving scale onto a coated window. The optical system expands the scale ten times its original value, and the portion to be read appears on both sides of a hairline on the coated window. This design automatically avoids problems of parallax in reading. Further, while long scale lengths are available in projection, a shortened deflection angle of the mechanism is permitted, allowing high sensitivity.

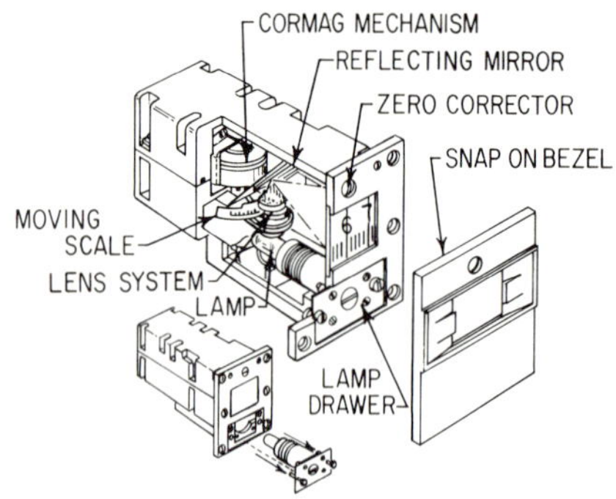

Fig. E-50. Special panel meter in which moving scale is optically projected. *(Weston.)*

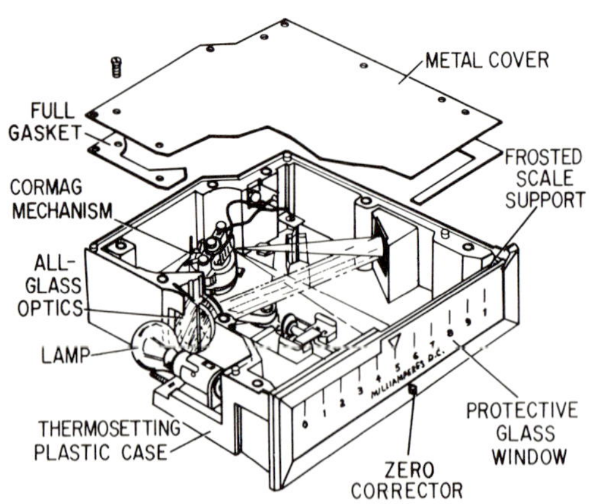

Fig. E-51. Special panel meter in which moving pointer is optically projected. *(Weston.)*

The application of light beams to panel and switchboard instruments further is demonstrated by the projected-moving-pointer instrument shown in Fig. E-51. The scale is fixed, and a pointer of light moves along the scale. The scale markings are printed on the front surface of the translucent scale plate, which is also the light-diffusing surface.

Thermoelements: Thermo instruments measure the effective value of the current flowing. They can be used on dc, ac, audio, or radio frequency. They are used extensively where waveforms or frequencies encountered are such as to cause errors in other types of indicators. No standard ampere at radio frequency being available, thermo instruments generally are calibrated at 60 Hz, sine wave. Instruments giving excellent performance up to 100 MHz are available.

A basic type of thermoelement and permanent-magnet moving-coil mechanism for the indication of dc, ac, and rf is shown in Fig. E-52. A potential is generated in the junction *E* of the dissimilar metals *ED* and *EC* as the current heats the junction. This potential causes a direct current to flow through the indicating mechanism. The current is a function of the heating current in *AB*. This arrangement provides no compensation for ambient temperature changes.

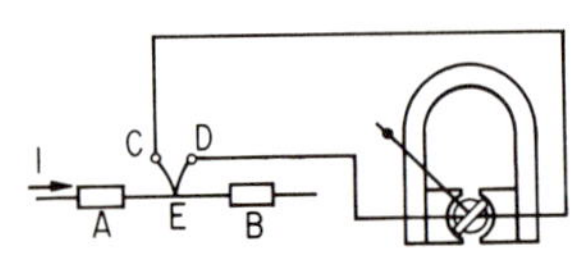

Fig. E-52. Basic type of thermoelement and permanent-magnet moving-coil mechanism.

A compensated thermoelement, shown in Fig. E-53, produces a thermoelectric voltage in the couple *CED* which is a function of the current *I* flowing through the circuit *AB* and causes a temperature rise at point *E*. Since the couple

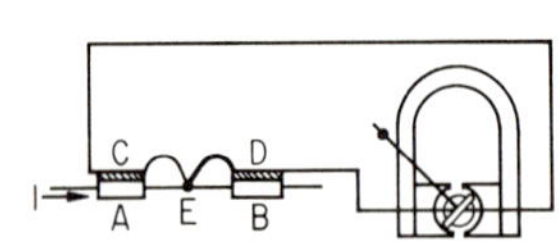

Fig. E-53. A compensated thermoelement. *(Weston.)*

voltage developed is a function of the temperature difference of its hot and cold ends, this temperature difference must be caused by the current being measured. For accurate measurements, points *C* and *D* must be at the mean temperature of points A and B. This is accomplished by attaching the couple ends C and D to the center of separate copper strips whose ends are thermally in contact with A and B, but which are electrically insulated from them.

The scale of the indicator can be calibrated with evenly spaced divisions in terms of I^2 or with a square-law arrangement of divisions—cramped at the left—in terms of I in AB, since heating is a function of the square of the current producing the heat. More nearly evenly divided scales can be obtained by using specially shaped pole pieces. Instruments with the compensation just described are used commercially for 0.5 to 20 A in self-contained designs. Higher ranges are of similar construction except that the heating element is supplied external to the indicator and the connecting lugs are in line with the heating wire or element. Ranges over 60 A are provided with air-cooling fins. The heating wire (0.5 to 3 A) is solid. Above 3 A, it is tubular to reduce skin-effect errors.

A bridge-type thermoelement, as shown in Fig. E-54 is used for currents from 0.1 to 0.75 A. In this arrangement, the voltage generated by a number of junctions is used to drive current through the indicator. Ranges from 0.0015 to 0.1 A use simple heating elements in small vacuum containers.

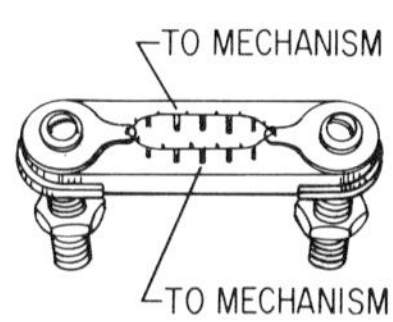

Fig. E-54. Bridge-type thermoelement.

Thermal Watt Converter: A thermocouple arrangement distantly related to the bridge-type heating element is used in the thermal watt converter. This device permits the measurement of power by thermoelectric means. From the principles of geometry, it can be shown that the sum S and the difference D of two vectors e and i which enclose the angle θ are

$$S^2 = e^2 + i^2 + 2\,ei\,\cos\theta \tag{6}$$

$$D^2 = e^2 + i^2 + 2\,ei\,\cos\theta \tag{7}$$

so that

$$S^2 - D^2 = 4\,ei\,\cos\theta \tag{8}$$

where e and i = momentary values of voltage and current, respectively

θ = phase angle

Thus, a circuit arrangement capable of measuring $S^2 - D^2$ can be used to measure a quantity proportional to $ei \cos\theta$ which represents power. Such a circuit is shown in Fig. E-55. If momentarily the direction of current from the current transformer secondary is that of the plain arrows and, simultaneously, the direction of current in the potential circuit is that of the flagged arrows, it follows that in the insulated heater of one thermocouple (A) these currents aid each other (sum S); while in the other (B) they oppose each other (difference D).

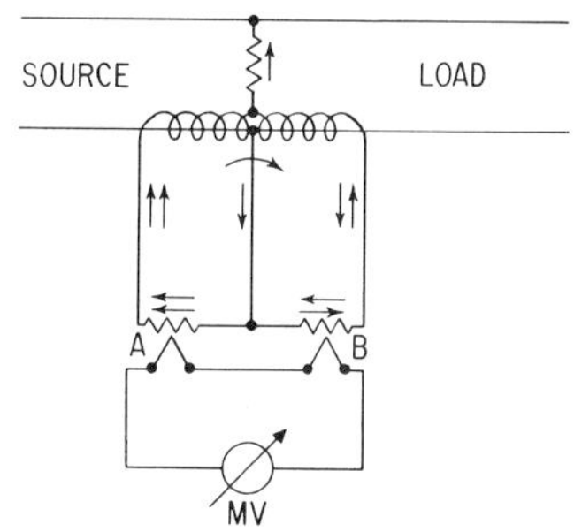

Fig. E-55. Circuit of a thermal watt converter.

The heat generated in the thermoelement heaters and, through the proper design, the temperature differences between hot and cold junctions and, thus, the electromotive forces of the couples are proportional to the squares of currents (S and D, respectively) in the heaters. The outputs of elements A and B are connected in opposition—hence the total emf developed will be proportional to $S^2 - D^2$, which represents power. In practice, chains of couples are used instead of single couples in order to obtain greater emf; also, the couples are self-heated similar to those of bridge-type heating elements.

Hot-wire Mechanism: These devices now are obsolete, having been replaced by the more sensitive, more accurate, and better-compensated heating-element permanent-magnet moving-coil combinations previously described. Of historical interest, in the hot-wire ammeter, the current being measured caused a wire to heat and thus to expand approximately in proportion to I^2. The change in length of the wire was arranged to drive a pointer. Instability due to wire stretch and lack of ambient-temperature compensation made the mechanism commercially unsatisfactory.

Precision Voltage and Current Measurements: Measurements to an accuracy of better than 0.05 percent are beyond the readability of standard indicating instruments. These measurements can be made by modern equipment employing proved methods using voltage dividers, a galvanometer, a precision voltage reference, and thermoelements.

Alternating voltages are measured using a thermal voltage standard as shown in basic form in Fig. E-56. The measuring element is a vacuum thermoelement *A*, which consists of a heater of high resistance alloy and a thermocouple thermally attached to the heater by means of an insulating bead. The output of thermocouple *A* is compared with a similar thermocouple *B*, the heater of which is connected to a zener reference. Both the measuring and reference thermoelements have similar temperature dependence, which cancels the effects of ambient-temperature changes. Also, both couples are mounted in a temperature-lag box to reduce further the effects of ambient temperatures.

they are adjusted for a null indication of the galvanometer.

Decade resistors and thermoelements are designed for wide frequency response such as 50 kHz; both ac and dc readings can be made. The difference between the ac and dc readings (ac/dc difference) is checked on all instruments against known standards calibrated at the National Bureau of Standards. The zener reference contains zener diodes especially aged and tested for long-term stability. The line input is rectified, filtered, and regulated. Temperature-sensitive components are mounted in a temperature-controlled oven.

Current measurements are made in a similar manner by replacing the series decade resistors with decade shunts for the basic range of 10 to 100 mA, with current transformers for the higher ranges. This type of arrangement is shown in Fig. E-57. The circuits of Figs. E-56 and E-57,

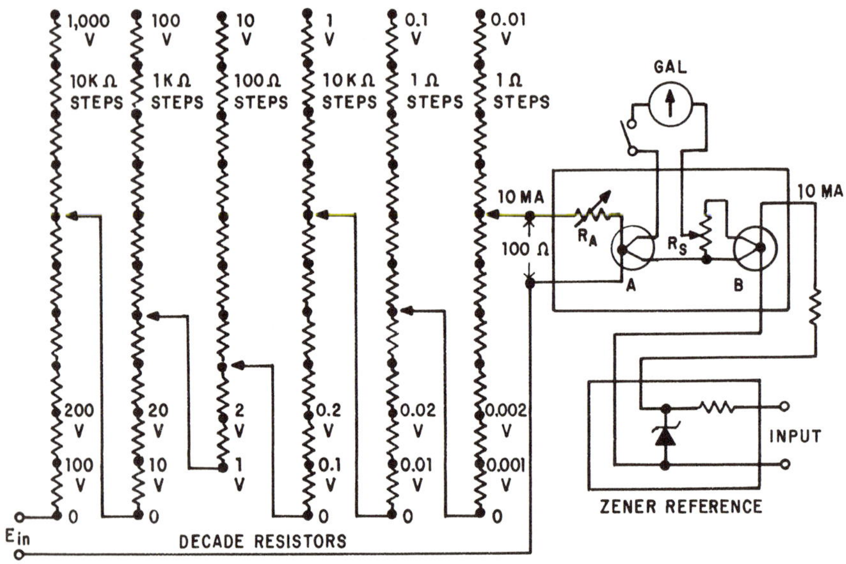

Fig. E-56. Basic circuit format for precision measurement of alternating current.

In use, resistor R_A is adjusted so that its resistance plus the heater of the thermoelement *A* equals 100 Ω. The outputs of both thermocouples are next made equal by adjusting R_B for a null reading on the galvanometer with 10 mA from the zener reference flowing through both heaters in series. The decade resistors are selected to provide 10 mA to the heater of thermoelement *A* when connected as shown in Fig. E-56, and the switches will indicate the input voltage E_{in} as

plus a voltage divider, shunts, and potentiometer, have been combined into one instrument to form a complete ac and dc precision standard. In its simplest form the dc portion is shown in Fig. E-58. The input voltage to the voltage divider or the input current to shunts provides a voltage drop of 0 to 150 mV which is balanced by a voltage from a potentiometer consisting of a decade and slidewire. This voltage is obtained by passing constant current from the zener refer-

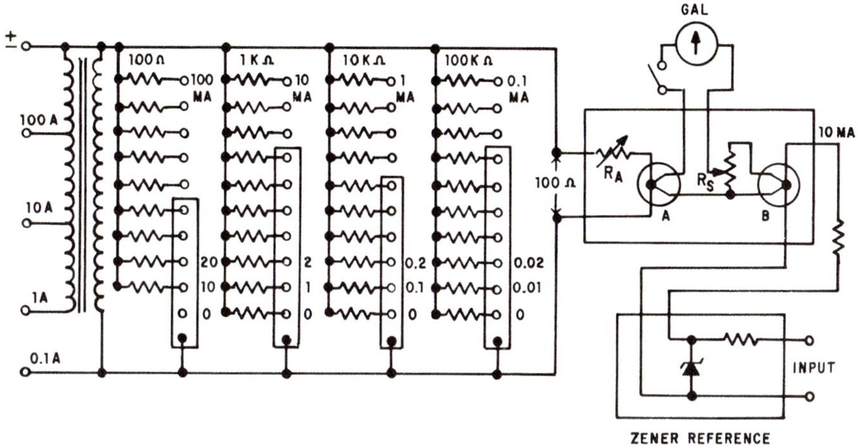

Fig. E-57. Basic circuit format for precision measurement of current.

ence through the slidewire and decade resistors. In its final form, the entire instrument is protected by overload circuits, fuses, and relays to prevent burnout of resistors and thermoelements.

Differential voltmeters are basically a special form of potentiometer whereby an unknown voltage E_{in} is applied against one that is accurately known E_1 and a high-sensitivity meter M_1 (null detector) is used to measure the difference between the two. The basic circuitry of a typical differential voltmeter is shown in Fig. E-59.

If the known voltage E_1 is adjusted through the use of a Kelvin-Varley divider to the exact potential of the unknown voltage E_{in}, the null detector will indicate zero and the value of the unknown voltage may be read from the position of the dials associated with the divider. In the example shown in Fig. E-59, the unknown voltage is 2.7 V and the standard voltage is 10 V. The null detector will indicate zero when $E_2 = 2.7$ V or $E_2 = 2.7 = R_{out}/R_{in} \times E_{in}$.

Since $R_{in} = 10$ kΩ (note this is constant for any setting of the switches), this will occur when $R_{out} = 2.7$ kΩ. Examination of Fig. E-59 shows that this will occur when switch (1) is placed in its second position and when switch (2) is moved to its seventh position. Only two decades are shown. Others can be added whereby the total resistance of each added decade must equal the resistance of two steps of the preceding one. For

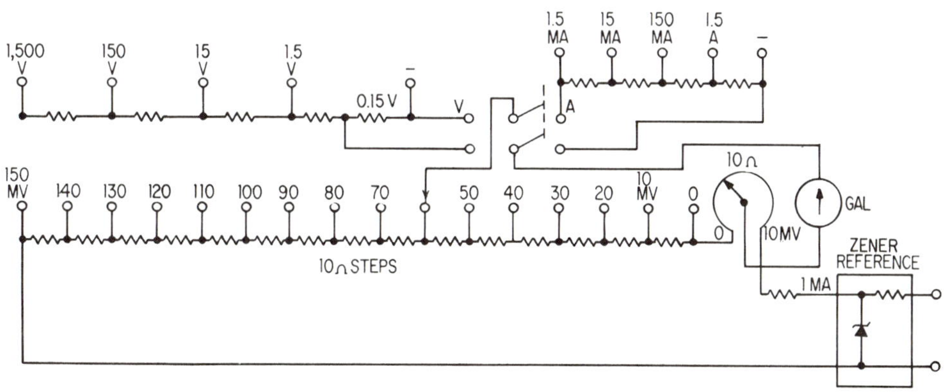

Fig. E-58. Direct-current portion of a complete ac and dc precision standard. Shown highly simplified here, the circuits of Figs. E-56 and E-57, plus a voltage divider, shunts, and a potentiometer, have been combined into one instrument.

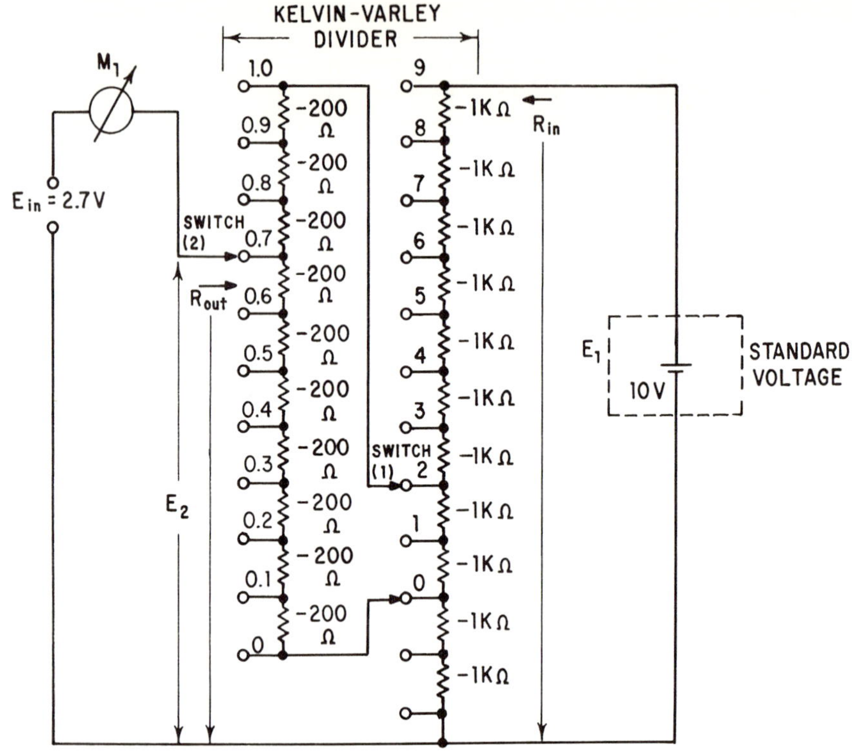

Fig. E-59. Basic circuitry of a typical differential voltmeter.

high-voltage measurements by differential voltmeters, the value of the standard voltage must be greater than the unknown. This may be overcome by inserting a voltage divider between the unknown and the null detector.

Electronic Digital Instruments: These devices employ electrical and electronic means to convert electrical quantities into digital outputs and readings. Almost always the input quantity is a voltage, other quantities being easily converted into voltage before being processed by the instrument. A digital output is more than just a visible reading on a display. The output can be a system of coded signals which can be transmitted over distances or which can actuate printers and typewriters. Such outputs also can be inputted directly into computers.

The electrical means of conversion employed in these instruments essentially are the same passive circuit elements also found in analog instruments, that is, resistors, capacitors, inductors, transformers, and so on. Along with these passive elements are active electronic elements, such as transistors and diodes. The latter elements are not commonly found in conventional pointer-deflecting instruments. The combination of active and passive elements is used to build circuits which, in themselves, perform certain basic functions. These circuits include the following:

1. The *amplifier*—to provide an amplified version of the input.

2. The *pulse generator*—a circuit which produces electrical pulses at its output terminals, either continuously or upon receipt of a command.

3. The *gate*—a circuit which has at least two inputs and one output. One input controls whether or not a signal from the second input is fed through to the output.

4. The *counter*—a circuit whose output is indicative of the number of events that occurred on its input. Frequently coupled with this circuit is a readout circuit which decodes the output of the counter circuit and displays it as a number. This number is either the lighting of a lamp labeled with the appropriate number, or the glowing of an element of a gas tube representing the number.

5. The *clock*—a circuit which provides elec-

trical impulses at its output, at a fixed, never-changing rate.

Digital Voltmeters: Several approaches are used in digital-voltmeter design. Typical means include (1) the principle of successive approximations, (2) dual-slope integration, (3) ramp and counter, and (4) voltage-to-frequency conversion.

Successive Approximations: In this technique, the unknown voltage is successively compared with standardized voltages, generated in the instrument, so that each voltage increment is one-half the amount of its predecessor. Whenever a standardized voltage exceeds the unknown voltage, the measurement is disregarded. Whenever the standardized voltage is less than the unknown, the former is "remembered." All the remembered voltages are added up until the difference between this sum and the unknown voltage can be neglected. The sum is then displayed.

Dual-slope Technique: This requires a fixed-time integration interval during which the voltage to be measured first is integrated. This may be accomplished by charging a capacitor. The capacitor then is discharged by applying an opposed fixed reference voltage greater than the unknown. The time required to discharge the capacitor until its voltage reaches zero is used as a measure of the unknown voltage. During this discharge time, the output of a pulse generator is gated to a counter. At the end of the time, the counter reads the voltage to be measured and the cycle repeats itself.

Ramp and Counter: In this type of instrument, a linear rise in voltage is generated and compared with the unknown voltage. A counter which receives clock pulses is started at the beginning of the process and stopped when the two voltages are equal. The time that this takes is proportional to the unknown voltage. Therefore, the number of pulses counted during this interval is a measure of the unknown voltage.

Voltage-to-Frequency Conversion: In this method, the input voltage determines the rate or frequency at which a series of pulses is generated. This pulse train is gated to a counter for a prescribed, fixed time interval and the number of pulses counted during this time is a measure of the unknown voltage.

An important element in the conversion of voltage to a frequency is the operational integrator. This is an amplifier with a capacitor connected between output and input. In front of the input is a resistor (see Fig. E-60). The amplifier is assumed to have a high gain. As the result of the action of this configuration as an integrator, the application of a *step* input results in a linear *ramp* output in which the slope of the ramp is proportional to the input-voltage step.

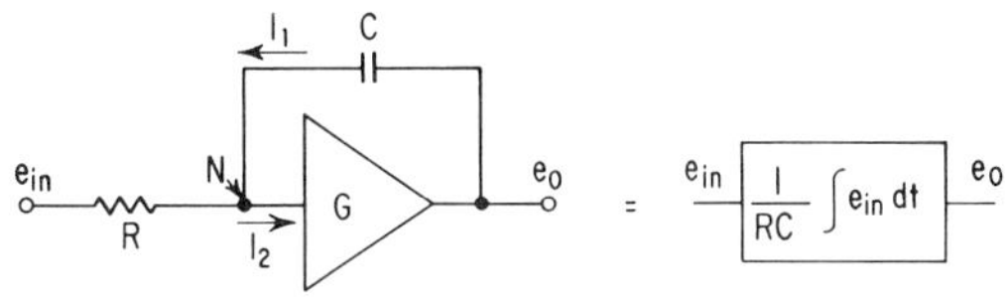

Fig. E-60. Operational-amplifier integrator.

To generate a quantitative measure of the input voltage E_{in}, a relay or gate (solid state) is energized by the output voltage E_o when the slope of the output voltage crosses a preset threshold level. This relay, when closed, permits a square voltage pulse of fixed amplitude and duration to be applied to the input to the integrator. The pulse is produced precisely by the pulse generator when triggered by the clock circuit. The pulse, being of opposite polarity to the input and greater in amplitude than the input, results in a reversal of the integrator output as shown in Fig. E-61. In addition to triggering the pulse generator, the clock frequency controls the width of the reset pulses. The first clock pulse passing through turns the pulse generator on; the next clock pulse gates it off.

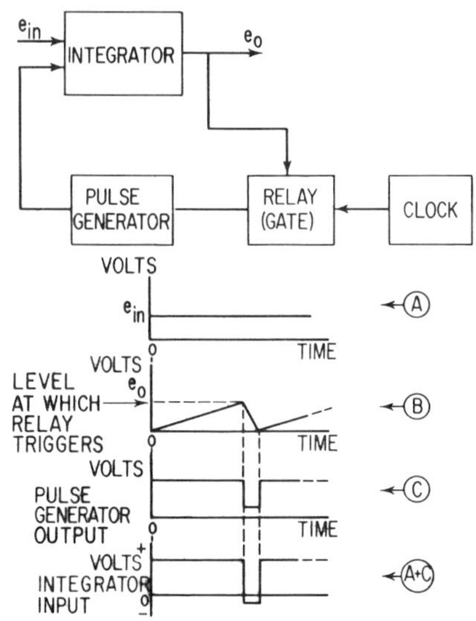

Fig. E-61. Cyclic integrator.

Reviewing the sequence of events, at time zero, the unknown input voltage is applied to the integrator, and the output begins to rise at a uniform rate. As soon as the output voltage reaches a fixed level, the gate is opened, thus permitting a clock pulse to trigger the pulse generator. The pulse generator then supplies a well-defined pulse of voltage to the input of the integrator. The sum of this pulse and the unknown input voltage results in the integrator's seeing, at its input, a potential which has reversed sign. In order to maintain a balanced-current condition at the summing node, the integrator now begins to decrease its output at the required rate. When the pulse has ended, the integrator output resumes its uniform rate of rise, and the cycle of events just described is repeated. This loop is a voltage-to-frequency converter, since the number of pulses requested by the integrator in a given time interval, in order to maintain a balanced condition at its input, is directly proportional to the magnitude of the input voltage.

Some refinements of this basic scheme are necessary, because as just described, a high-gain operational amplifier with its associated large input-summing resistors will tend to degrade (lengthen) the rise time of the pulses supplied by the feedback loop. But, if as shown in Fig. E-62, a dc amplifier is added ahead of the integrator and the summation of the input voltage and the voltage of the feedback pulses is made at this point, the requirement for perfect linearity of output voltage as a function of time is removed from the integrator. This summation is made so that the average difference between the two voltages appears at the output of the first amplifier.

Because the integrator now sees only an error signal as the output of the first amplifier and operates on this error (rather than the full input), the original stringent requirements placed upon it are greatly relaxed. Without the requirement for accurate transformation, the integrator needs only to gate a clock pulse whenever the integral of the error signal exceeds a preset level. The pulse-generator output then resets the integrator as previously described.

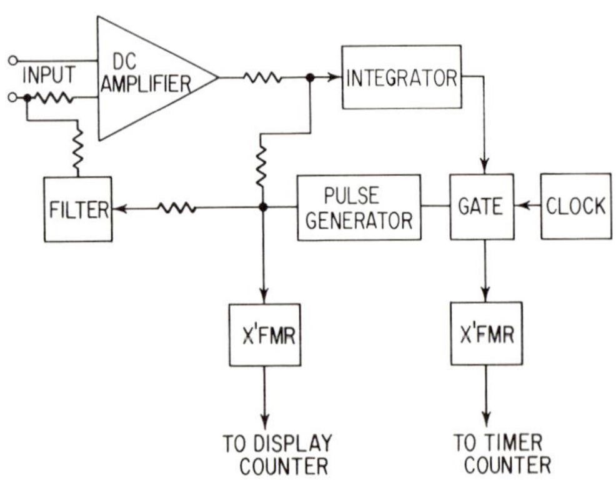

Fig. E-62. Voltage-to-frequency converter.

Accuracy is dependent primarily upon the amplitude and width stability of the pulses which are smoothed and fed back through a potential divider for comparison with the input. Any difference between the average pulse level and the input results in an error signal which adjusts the rate at which pulses are generated, thus maintaining a null balance.

Counting and Display: With a train of pulses available, it remains only to totalize these pulses for a fixed interval of time and to display this total as a direct measure of voltage. Because the timing or totalizing interval is controlled by the same clock which gates the pulse generator in the voltage-to-frequency converter system and which controls the width of the pulses, no degradation of the measurement occurs if the clock is somewhat imprecise.

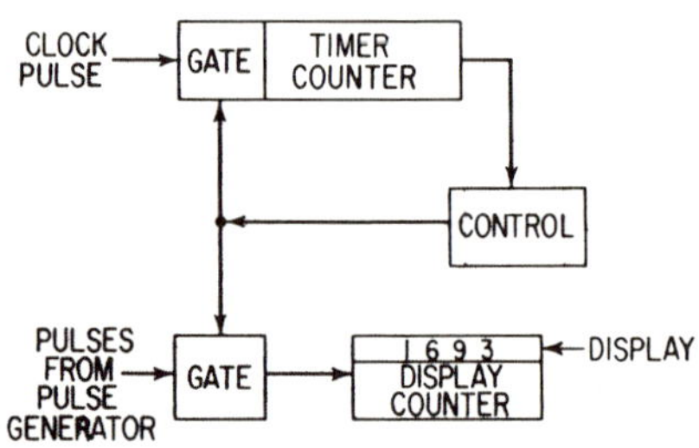

Fig. E-63. Counting and display of pulse train.

The sampling interval normally is 1/60 s. This time corresponds to 1 Hz of 60-Hz voltage and permits the influence of any 60-Hz voltage component at the input to integrate to zero.

The pulse train is measured and displayed as shown in Fig. E-63. With both input gates open initially, the clock pulses and the pulse-generator output are permitted to pass into the timer counter and the display counter, respectively. When the proper number of clock pulses have been accumulated in the timer counter, the timer counter passes a signal to the control section, which closes the input gates. This action prevents further pulses from entering the counters. The number

TABLE E-5. *Major Characteristics of Digital and Analog Electrical Instruments*

Digital	Analog
Power requirements	
Derive operating power from a separate power supply and draw only negligible power from the source to be measured. Digital instruments cannot operate without auxiliary power	Power usually must be furnished by source of same voltage or current to be measured. Use on very low power. Sources may falsify measurements. Most analog instruments function without auxiliary power
Accuracy	
Fifty to one hundred times more accurate than analog instruments	The finest pointer-on-scale instruments are rated to be accurate to within ±0.1% of full scale
Resolution	
One part in several hundred thousand	One part in several hundred
Speed of measurement	
Hundreds or even thousands of measurements per second	Close approximations of their final readings in $\frac{1}{10}$ s
Rate of change	
Rate or the rapidity of change of the quantity measured can be judged more easily when observing the motion of a pointer than when observing a change in the value of a digital readout. Therefore, analog pointer-on-scale instruments are preferred when a visual indication of rate is important.	
Reaction to environment	
Relatively complex and made of large numbers of parts and components whose reaction to environmental conditions varies. Can be made entirely without moving parts	Relatively simple in construction and can be made to perform even under very unfavorable conditions. But, they do have moving parts
Size and portability	
Usually designed for stationary installation. Size and weight are comparable with stereo hi-fi tuners	Can be made about as small as a coin, but where accuracy and resolution are important, the minimum size is about that of a fair-sized dictionary. The more accurate they are—the less portable

of pulses which have been accumulated in the display counter during this totalizing interval establishes the output status of the display counter, and this status is gated to the display. The control section then resets the counters and reopens the gates to permit a fresh measurement to be made and the display to be updated.

Digital vs. Analog Electrical Instruments: When it is necessary to select between digital instruments and pointer-on-scale (analog) instruments, a number of factors must be considered. These are summarized in Table E-5.

REFERENCES

Fink, D. G., and J. M. Carroll: "Standard Handbook for Electrical Engineers," McGraw-Hill Book Company, New York, 1968.

Harris, Forest K.: "Electrical Measurements," John Wiley & Sons, Inc., New York, 1952.

—Adapted from "The Instrument Sketch Book," *Weston Instruments, Inc.*

Electrocardiography (See Cardiography.)

Electrode, pH, and Redox (See pH, measurement of; and Oxidation-reduction potential, measurement of.)

Electrodynamic Electrical Instrument (See Electrical quantities, measurement of.)

Electroencephalogram (See **Central nervous system, instrumental analysis of.**)

Electrohydraulic Actuator (See **Actuator, electrohydraulic.**)

Electrolytic Tranducer A device whose electrolytic resistance is caused to change when exposed to the variable being measured. Like other electrolytic cells, the devices are subject to polarization—hence are not widely used.

Electromagnetic Transducer A large number of transducers associated with the measurement of many different variables operate on electromagnetic principles. Generally, these devices fall into the following major categories: (1) differential transformers; (2) Hall-effect magnetic transducers; (3) inductance transducers; (4) induction transducers; (5) magnetoresistive transducers; (6) magnetostrictive or magnetoelastic transducers; and (7) saturable reactors—although the latter are used principally for controlling rather than sensing purposes. See **Differential transformer; Hall-effect transducer;** and **Saturable reactor.**

Inductance Transducers: Inductors usually take the form of a single turn or several turns of wire. The following relationship applies:

$$L = Kn^2\mu \tag{1}$$

where L = inductance, henrys
K = constant determined by the coil's geometry
n = number of turns
μ = permeability of medium inside and about the coil

If two inductors are connected serially, the total inductance is

$$L = L_1 + L_2 \pm 2M \tag{2}$$

where L_1 = inductance, coil 1
L_2 = inductance, coil 2
M = mutual inductance between coils 1 and 2

If the windings are wound in the same sense, the plus sign is used; if in the opposite sense, the minus sign is applied.

Further, the mutual inductance of two inductors is

$$M = k\sqrt{L_1L_2} \tag{3}$$

where k = coupling coefficient as determined by the geometry of the two coils.

From the foregoing, it is obvious that inductance can be varied by (1) changing the number of turns, (2) changing the geometry, or (3) changing the permeability of the surrounding medium. These are the properties of which advantage is taken in designing an appropriate transducer.

To vary the number of turns physically is difficult, but it is an easy matter to vary the effective number of turns. Potentiometric-type transducers employ this principle by utilizing a wiper contact that moves along the coil. Thus, if linear displacement, for example, is being measured, the wiper is caused to move, which in turn, varies the inductance between the wiper and a fixed contact at one end of the inductor.

It is impractical to vary the geometry of a single inductor, but it is easy to change the geometry of two coils in relation to each other. Essentially, this is what is done in the differential transformer. The variable-mutual-inductance transducer is another example.

The most convenient way to vary inductance is to vary the permeability of the surrounding medium. Variable-reluctance and variable-permeance devices are the same, inasmuch as reluctance is the reciprocal of permeance. Practical examples of these devices are described elsewhere and are listed in the detailed alphabetical index at the end of this volume.

General advantages of variable-inductance devices include (1) high signal levels often with no amplification required, (2) low impedance levels, and (3) small power losses. Bridge techniques can be used to improve the sensitivity of the devices.

Induction Transducers: The principle behind these devices is Faraday's induction law, when simplified is

$$e_o = -n\frac{d\phi}{dt} \tag{4}$$

where e_o = induced output voltage of a coil
n = number of turns in the coil
$\frac{d\phi}{dt}$ = time derivative or change of magnetic flux passing through coil

This relationship is the basis of a number of kinds of velocity, acceleration, and displacement transducers and, of course, is the principle behind many electric meters. Since the induction transducer essentially is a generator, the device produces an electrical signal without the need of an external energy supply. The principle is particularly well adapted to the measurement of rotary motion. See **Tachometer generator.** Other practical examples of induction transducers are described elsewhere and are listed in the detailed alphabetical index at the end of this volume.

Magnetoresistive Transducers: Advantage is taken in some transducers of the fact that metals, when cooled to low temperatures, show a change of resistivity when they are subject to an applied magnetic field. Bismuth, in particular, is quite sensitive in this respect. In practice, these devices are severely limited because of their high sensitivity to ambient-temperature changes.

Magnetostrictive Transducers: Also sometimes referred to as magnetoelastic transducers, these devices operate on the Joule effect, namely, that a ferromagnetic material is dimensionally altered when subjected to a magnetic field. The reverse situation, known as the Villari effect, also occurs; namely, when a ferromagnetic material is stressed, resulting in strain, the material's permeability is changed. Thus, most magnetostrictive transducers are used for force measurement. They have a high-frequency response—of the order of several 10^4 Hz. Some of the materials that exhibit magnetostriction include Monel metal, Permalloy, Cekas, Alfer, and a number of other nickel-iron alloys not previously mentioned. Disadvantages of these transducers include (1) the fact that excessive stress and aging may cause permanent changes, (2) zero drift and sensitivity changes due to temperature sensitivity, and (3) hysteresis errors.

Electrometer (See **Voltmeter.**)

Electromotive Force, Measurement of (See **Voltmeter.**)

Electromyogram (See **Medical instrumentation.**)

Electron Microprobe (See **X-ray analyzer, electronmicroprobe.**)

Electron Paramagnetic-resonance Chemical Analyzer (EPR) An instrument system, as shown in Fig. E-64, based upon the observation of the electronic paramagnetic susceptibilities of substances. Analysis is accomplished by observing the resonance between the Larmor precession frequency of an ensemble of unpaired electrons and an externally applied oscillating magnetic field as a function of an externally applied steady magnetic field. The relation between precession frequency and the steady magnetic field for a free electron is linear and given by the Larmor equation $\nu = \gamma H_0$, where ν denotes the frequency and H_0 the field. γ is the gyromagnetic ratio and has the value of 2.8024 MHz/gauss for a "free" electron (intrinsic spin = ½ and no orbital angular momentum).

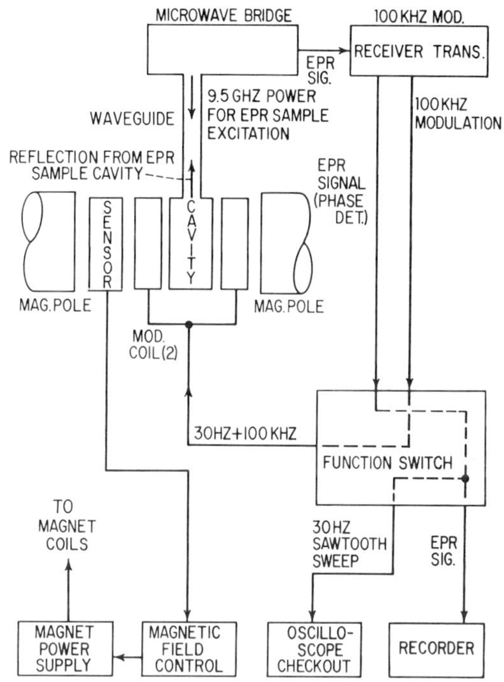

Fig. E-64. Block diagram of electron paramagnetic-resonance chemical analyzer (EPR). *(Varian.)*

This ideal relation normally does not hold, because electrons are not free. Deviations are caused by neighboring nuclei with nonzero nuclear spin (hyperfine structure) and by surrounding electric fields (fine structure) and field gradients. Thus, it is possible to use the unpaired electrons as probes for analysis of their immediate microscopic surrounding. Observation of EPR is contingent upon the presence of *unpaired electrons.* The most important examples are transition-metal ions, free radicals, and impurities in dielectrics.

Applications include problems of structural analysis in medicine, biology, chemistry, and physics; determinations of chemical reaction rates and of impurities in dielectrics; and the identification of free radicals.

EPR spectrometers consist of three main parts: (1) a microwave bridge capable of producing an oscillating magnetic field of the desired frequency at the sample position; (2) a magnet with power supply to provide the steady magnetic field; and (3) the associated signal detection, amplification, and display systems. The sample is contained in a microwave cavity that is capable of supporting the oscillating magnetic field. Usually the instruments operate at a fixed frequency of approximately 10 GHz. The associated magnet systems should be capable of producing fields up to 6,000 gauss or more.

Most EPR spectrometers are multipurpose instruments but are supplied with accessories designed for specific purposes. These accessories may include (1) instruments to control sample temperatures, (2) goniometers for varying sample orientation relative to the steady magnetic field, (3) sample holders for aqueous solutions and tissues, and (4) light sources for optical irradiation and observations of the sample during the EPR experiment. Electron-nuclear double-resonance and electron-electron double-resonance accessories are useful for improving the effective resolution of EPR spectrometers.

Typical instrument sensitivity is such that the order of 10^{-11} mole of a paramagnetic species can be detected. Very high concentrations of paramagnetic species may decrease the apparent instrument sensitivity. Typically, concentrations of 1 percent or less result in the most favorable conditions for observation. The method is *nondestructive.* Minimum lifetimes of the paramagnetic species for observation are of the order of 1 ms. Because of the presence of high-frequency electric fields, dielectric losses of samples can place serious constraints on usable sample volumes. The spectrometer output can be fed directly into on-line or time-shared computers for time averaging, data processing, and data reduction.

—Peter Schindler, *Varian Associates*

Electronic Data Processing (See Data communications.)

Electrophoresis Apparatus When influenced by an electrical field, particles will move through a fluid, a phenomenon termed *electrophoresis* and sometimes referred to as *cataphoresis.* Advantage of this principle is taken to separate and to measure down to microgram quantities a number of substances, such as nucleoproteins that may be difficult to analyze by other methods.

Electropneumatic Actuator (See Actuator, electropneumatic.)

Electroscope (See Electrical quantities, measurement of.)

Elongation, Measurement of [See Testing (Materials), destructive.]

Elution Chromatography (See Chromatography.)

Emission Spectroscopy (See Optical-emission spectrochemical analysis.)

Emissivity (See Radiation pyrometer.)

Emulator An emulator is a computer feature enabling the computer to execute instructions written for a different type of computer. This is a particularly common feature where the different computers are of a similar but not identical type. For example, if the instruction set of a smaller machine has a close equivalent for each instruction in the set for a bigger machine and the word or instruction size is smaller, an emulator would normally be offered by the manufacturer. In this way, those customers who replaced the smaller machine by the larger would not have to rewrite their existing programs immediately for use on the new machine.

An emulator is normally implemented either by special logic circuits or by microcode. The same functions can be implemented as a special "simulator" program at the expense of memory and performance.

The average life span of individual programs is fairly short—it is quite rare for a three-year-old application program to be still in use which has not been modified almost to the point of being unrecognizable—so emulation is not normally required for more than the initial 2 or 3 years of a new computer installation. Emulation is only one of the many techniques used for converting from one computer system to another. Much of the effort expended on compiler languages has been to standardize them to the point that, if the problem program runs successfully on one machine, it may be recompiled for a different machine and will again run successfully. Emulation would then not be required. These efforts, although quite successful, have not yet been fully so, and normally some conversion effort must be undertaken. Also, these efforts have done very little for assembler-language programs; so either such a program must be rewritten or the original machine must be emulated on the new one.

—A. J. Arthur, *IBM Corporation*

Encoder In its purest sense, the term encoder refers to any device or subsystem which accepts an input and produces an output in coded form. As such, it includes digital-logic configurations which convert a digital input word in one code into an output word in another code. An example is a decimal-to-octal or a decimal-to-binary decoding circuit.

In general usage, however, encoder more often refers to a device or subsystem which converts an analog quantity into a digital representation by some quantization technique (see **Quantize**). In this usage, the term is equivalent to analog-to-digital converter (see **Analog-to-digital converter**). Both electronic analog-to-digital converters and electromechanical analog-to-digital converters such as shaft-to-digital converters can be referred to as encoders. The growing trend, however, is to reserve the term analog-to-digital converter for electronic converters and use encoders to refer to electromechanical converters, such as in "shaft-to-digital encoder."

—Thomas J. Harrison, *IBM Corporation*

Energy (Electrical), Measurement of (See **Watt-hourmeter**.)

Engine Controls, Aircraft (See **Aircraft-engine controls**.)

Engine-speed Control (See **Governor**.)

Environmental-test Instrumentation The complexity of instrumentation and control equipment, coupled with more stringent application requirements—particularly in the defense and aerospace field—gave rise a number of years ago to a new subscience of instrumentation, namely, the measurements and controls required to test systems in advance in environments approaching as closely as possible the conditions which the equipment will encounter in actual use. Although a number of special techniques have been developed for this field, environmental-test instrumentation relies predominantly on instruments and controls that also are used for many other industrial and commercial purposes.

Equal-percentage Valve (See **Control valve**.)

Error, Aliasing Aliasing error is an error which can be introduced into sampled data in a digital-data-acquisition system as the result of violation of the basic sampling theorem. The theorem, often called Shannon's sampling theorem, states that in order to reconstruct a signal containing frequency components in the spectrum 0 to f_m Hz from sampled data, uniformly spaced samples must be taken at a rate of at least 2 f_m samples per second. Failure to obtain data at this rate converts high-frequency components into low-frequency components so that the reconstructed signal contains low-frequency energy not present in the original signal. This phenomenon is illustrated in Fig. E-65, which shows a 5-Hz signal sampled at a 4 samples per second rate. The reconstructed signal has a frequency of 1 Hz. A familiar example of aliasing is the slow-frequency rotation of wagon wheels in the western movie caused by the stroboscopic sampling of the motion-picture camera.

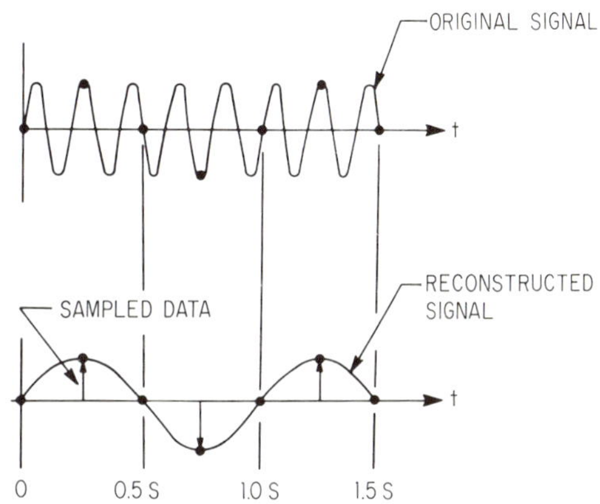

Fig. E-65. Aliasing-error example.

The adjective "aliasing" results from the interpretation that the high-frequency components take the "alias" of a lower-frequency component. An equivalent term, folding error, arises from the interpretation that the frequency spectrum of the signal is "folded" such that high-frequency components appear in a lower-frequency spectrum.

The basis of the sampling theorem and the generation of the aliasing error are easily explained by considering the sampling operation as a modulation process. Sampling a continuous signal is equivalent to modulating the signal with a series of uniformly spaced impulses as shown in Fig. E-66. If a signal containing frequency components up to a frequency f_m is modulated (sampled) by a carrier of frequency f_s, the frequency spectrum of the resulting sampled signal consists of the original spectrum plus harmonic spectra centered on the sampling frequency and its harmonics as illustrated in the figure. This is precisely the same effect as the modulation of an rf carrier with an audio signal in amplitude-modulated radio, except that there are an infinite number of carrier frequencies because the frequency spectrum of a train of equally spaced impulses (the modulating signal) consists of an infinite number of frequency components at f_s and its harmonics.

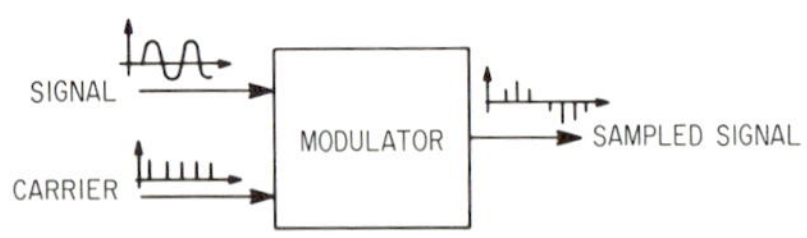

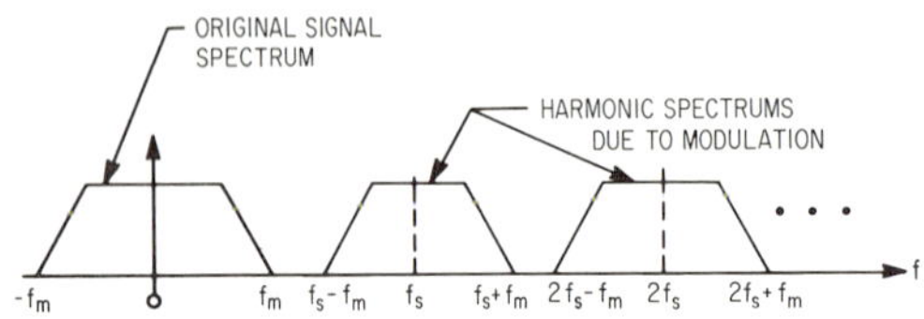

Fig. E-66. Frequency-domain sampled spectra.

Examination of the figure shows that if the sampling frequency f_s is less than $2\,f_m$, the lower sideband of the first-harmonic spectrum overlaps the original signal spectrum. Since mathematical signal-reconstruction methods are equivalent to low-pass filtering in the signal domain, it is obvious that the signal energy from the first-harmonic spectrum which overlaps the original spectrum will be passed by the low-pass filter and therefore will affect the nature of the reconstructed signal.

In the statement of the sampling theorem and in the above explanation, the signal was assumed to have no frequency components above the frequency f_m. In practical applications, signals are not definitively band-limited and high-frequency components are present in any real signal. These may be due, for example, to noise impulses induced in instrumentation lines. In most process-control applications, however, the signals are nominally band-limited by the low-pass characteristics of the process or the signal transducers. For such signals, the sampling theorem is often stated so as to require sampling at twice the highest *significant* signal frequency. In practical applications, the selection of the sampling frequency depends on the interpretation of the word significant. An often-quoted rule of thumb is that the sampling rate should be five to ten times the highest frequency of interest. Although technically incorrect, application of this rule is usually successful, since signals are usually band-limited to reject frequencies above those of interest. The 5 or 10 multiplier provides a safety factor to account for the finite rolloff rate of most low-pass systems. If accurate reconstruction of the sampled signal is required, however, an estimate of the aliasing error should be made.

The aliasing error can be estimated by using a graphical technique based on the frequency-spectrum diagram of Fig. E-66. The frequency spectrum of the original signal and the spectra of the modulation harmonics are first drawn as shown in the figure. Then, at each frequency in the bandpass of the reconstruction filter or its mathematical equivalent, the amplitude of the original frequency components and all contributions from aliased harmonics are summed. This sum is an estimate of the magnitude of this frequency component in the reconstructed signal. This represents a conservative estimate, since the true contribution is represented by the vector sum of the components whereas the estimate is based on the algebraic sum.

—Thomas J. Harrison, *IBM Corporation*

Error, Common-mode The common-mode error is a noise or interference signal appearing at the input of a differential instrument or data-acquisition subsystem as the result of an applied common-mode voltage. This error signal is the result of the common-mode voltage being converted to a normal-mode signal (see **Differential-mode voltage**) because of circuit unbalances or other phenomena in the differential instrument, the instrumentation wiring, or the transducer. A specific example of common-to-normal-mode conversion is described in the entry for **Common-mode voltage**.

Since the common-mode rejection ratio (CMRR) of most differential instruments is relatively high, generally greater than 80 db, the common-mode error cannot be easily measured in most applications. It can, however, be inferred by using the measurement technique described for determining the common-mode rejection ratio [see **Common-mode rejection ratio (CMRR)**]. This test consists of applying a known common-mode voltage to the input of the instrument or subsystem and observing the change in the output. If the CMRR of the instrument or amplifier is known, the common-mode error resulting from a common-mode voltage of V_{cm} is given by V_{cm}/CMRR. This follows directly from the definition of the CMRR.

It is important to note that the common-mode error is a differential signal and cannot be easily distinguished from a valid signal derived from the differential signal source being measured. If the valid signal is dc and the common-mode voltage is ac, the resulting ac common-mode error can be recognized as the ac component in the output signal or digital value. Similarly, a dc common-mode error on an ac signal can be detected as an offset in the output data. In sampled-data systems in which the sampling time is small compared with the period of the signal and common-mode voltage, however, the contribution of the common-mode error to the output data cannot be determined by examination of a single output reading. If many readings are taken, the effect of ac common mode on a dc signal can be effectively eliminated by averaging data values. Similarly, the dc offset error caused by a dc common-mode signal mixed with an ac signal can be subtracted from the data if sufficient data points are available.

—Thomas J. Harrison, *IBM Corporation*

Error, Folding (See **Error, aliasing.**)

Error, Measurement The error of measurement (Ref. 1) is the numerical difference between the true value of a quantity and its value as obtained by measurement. Repeated measurements of the same quantity give different indications. Errors traditionally have been described as of three general types: (1) *mistakes,* (2) *systematic errors,* and (3) *accidental* or *random errors.*

Mistakes (Ref. 2) are errors due to careless reading, mistakes in recording observations, or incorrect application of a *correction.*

Systematic errors (Ref. 2), sometimes referred to as "bias," influence all measurements of a quantity alike. Such errors arise in scaling an instrument where the scale as made does not, throughout its range, fully conform to the measured quantity or where, in adjusting the linkage in a mechanical instrument, the indications deviate from the desired value on a fixed scale. There are numerous other examples of systematic errors. *Scale error* is a term used to express the systematic error for any particular measured quantity. The scale error is the algebraic difference between the average indication and the true value of the measured quantity. The scale error is the quantity which, algebraically subtracted from the average indication, gives the true value (see Fig. E-67). A positive scale error denotes that the indication of the instrument is greater than the true value. *True value = average indication − scale error.*

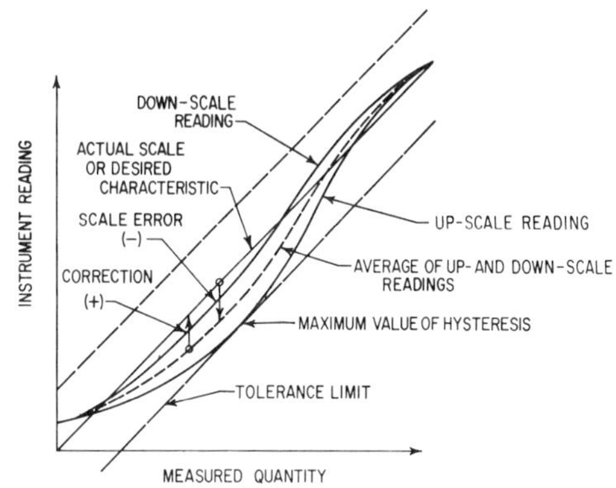

Fig. E-67. Diagram of scale error. The variations shown are exaggerated.

Correction is also the algebraic difference between the true value and the average indication of the measured quantity. Correction is the quantity which, algebraically added to the average indication, gives the true value (see Fig. E-67). A positive correction denotes that the indication of the instrument is less than the true value. Scale error and correction for the same measurement are of opposite sign. *True value = average indication + correction.*

Accidental errors (Ref. 2), sometimes referred to as *random errors,* describe the closeness with which a series of readings of the same quantity repeat. This is the concept of precision, the value of which is expressed as an imprecision. The cause of such errors is unknown or is not determinable in the ordinary process of making measurements. Such errors normally are small and follow the laws of chance. Random errors thus may be treated mathematically according to the laws of probability.

Accuracy of Measurement: In describing the accuracy of an instrument, it is necessary to combine the effects of systematic and random errors and to state specifically the *operating conditions* under which the accuracy is given.

Measurement accuracy (Ref. 3)[1] is a number which defines the limit that errors or uncertainties will not exceed when the device is used under specified operating conditions. The broken lines in Fig. E-67 represent limits within which the observed readings occur. These lines also may represent the measurement accuracy tolerance specified for an instrument. The limit of errors is in reference to the desired characteristic, which is the median line in the figure. The limit of error thus is plus or minus—as, for example, ±2° F. Measurement accuracy frequently is expressed in absolute units as in the foregoing example, or in percent of full scale (±0.5 percent of full scale) where the instrument range begins at zero and full scale represents the upper-range value.

Other Errors: In addition to scale error, other errors in instruments frequently are identified and include (1) *hysteretic error* (Ref. 3); (2) *dead band* (Ref. 3); (3) *hysteresis* (Ref. 3); (4) *operating conditions;* (5) *operating influence,* including ambient-temperature error, immersion error (in a temperature instrument), and barometric-pressure error; and (6) *observational errors,* including estimation error and parallax error.

See **Barometric-pressure error; Dead band;** and **Hysteresis.**

Operating conditions (Ref. 3) are those conditions which may influence the reading of a measuring instrument and involving quantities other than those which the instrument is designed to measure. When operating conditions have an influence on the reading of an instrument under the normal range of operations, the manufacturer should state the specific conditions for which the measurement accuracy applies. When a change in an operating condition occurs which influences the reading of an instrument, such change is referred to as an *operating influence* and may be expressed as a limit or as a coefficient. Operating influence may be defined as a change in a performance characteristic caused by a change in the specified operating condition from some standard or reference operating condition, all other conditions being held steady at specified standard conditions.

Estimation errors are variations of reading an instrument for the same observer in repeated observations or between observers. If the effect of parallax is ruled out, the errors of estimation for a steady indication will vary accidentally or randomly around a particular value, which has been described by Dr. Ku (Ref. 7) as the "limiting mean." A good rule of thumb where the width of the graduation line is less than one-fifth the graduation interval and where the graduation interval is greater than 1 mm is that a careful observer will repeat readings to one-tenth the graduation interval.

[1] Adapted from Ref. 3, with the permission of the publisher, the Process Measurement and Control Section, SAMA.

REFERENCES

1. Scarborough, J. B.: "Numerical Mathematical Analysis," chap. 1, p. 3, The Johns Hopkins Press, Baltimore, 1930.
2. Ref. 1, chap. 14, p. 284.
3. SAMA Standard PMC 20, Measurement and Control Terminology, Process Measurement and Control Section, Scientific Apparatus Makers Association, New York, 1969.
4. Ref. 3, adapted from Fig. 11, p. 22.
5. Swindells, J. F.: Calibration of Liquid-in-Glass Thermometers, *Natl. Bur. Std. (U.S.) Monograph* 90, p. 12, 1965.
6. Ref. 5, p. 21.
7. Ku, H. H.: "Handbook of Industrial Metrology," chap. 2, ASTME, Prentice-Hall, Inc., Englewood Cliffs, N.J., 1967.
8. ASTM Standard E77, Test and Standardization of Thermometers, p. 1054, 1963.

—Wallace Dean Wood, *Taylor Instrument Process Control Division, Sybron Corporation*

Error Procedures, Computer Error procedures are those procedures executed by a computer on the occurrence of various errors. The basic philosophy behind most error procedures is to record the error and attempt recovery. If recovery connot be made, then the recovery failure is recorded—normally by logging it to the operator or programmer. The system capability is then degraded in some predefined and concise manner.

Computer errors fall into three main categories, central-processor errors, peripheral device errors, and program errors. The last are by far the most common but frequently may appear to be processor or peripheral errors. If the error cannot be duplicated using other programs, then it is almost certainly a program error.

For program errors, the normal error procedure is to document the error to the programmer and for minor errors to continue with the program. For major errors, the program causing the error is aborted along with the related job, and documentation of this action is provided to the operator.

Peripheral-device error procedures normally consist of retrying the operation a number of times when possible and logging the failures. If the failure persists, the operator is informed and the computer may wait on operator intervention or abort the current job and logically disconnect the device. A technique coming into common usage is to document the error to the operator and then to return to the current program with a special indicator to allow the application programmer to decide the error procedure to follow. This solves the problem of two different programs running in the same machine concurrently (see **Multiprogramming**) and having different error-procedure requirements.

Central-processor error procedures are much more drastic than those in the other two classes, as it is difficult to justify any action depending on current central processing unit (CPU) resources if the CPU has an error indication, e.g., parity error. The error may have been caused by a noise spike (in which case special hardware techniques are generally used to recover), or it may be a fault in the logic circuitry rather than the actual memory stack. Generally, therefore, a "warm start" is attempted. That is, all current jobs are aborted; all of the core is initialized, if possible; and a restart from a previously defined state is attempted. If the warm start fails, then another will normally be attempted. The computer continues to cycle in this mode until the operator intervenes. It is unlikely that a machine which cannot be warm-started can be safely used without some operator intervention.

—A. J. Arthur, *IBM Corporation*

Error Signal In a closed loop, the signal resulting from subtracting a particular return signal from its corresponding input signal. The *actuating error signal* is the reference-input signal minus the feedback signal.

Event Recorder A recorder that plots on-off information against time to indicate when events start, how long they last, and how often they recur. In one type, the writing pen is displaced a fixed amount to one side when energized. In another type, the pen makes a line only when voltage is applied so that off time is indicated by the absence of a line. These devices can be either circular or strip chart. The chart-drive mechanism is similar to other recorders. The event-writing pen or stylus can be actuated by a solenoid or electromagnet.

Exclusive OR Circuit The exclusive OR circuit is a logical element which has the properties that if either of the inputs is a binary 1, then the output is a binary 1 (see **Binary state**). If both the inputs are a binary 1 or 0, the output is a binary 0. In terms of Boolean algebra (see **Boolean algebra**) this function is generally represented as $F = AB' + BA'$, where the prime denotes the NOT function.

With reference to the transistor exculsive OR circuit in Fig. E-68, the output is positive when either transistor is in saturation. When input A is positive and B is negative, transistor T_2 is in saturation; and when B is positive and A is negative, transistor T_1 is in saturation. When A and B are either both positive or both negative, both transistors are cut off and the output F is negative.

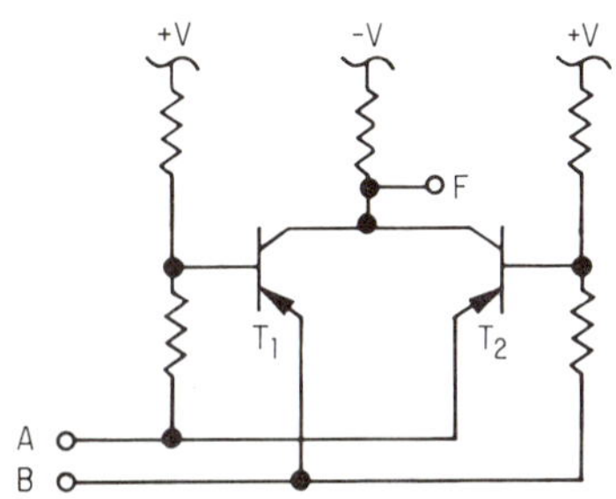

Fig. E-68. Transistor exclusive OR circuit.

—W. W. Welz, *IBM Corporation*

Executive Program The executive is a program which controls a computer system. Alternate names for this program are the "director," "supervisor," "monitor," or "scheduler." These names are sometimes used for completely different types of programs or for modules performing subsets of the executive's functions.

The fundamental purpose of the executive program is to allocate the resources of the system among the jobs contending for these resources. These resources include input/output channels and devices, main memory, the central processor, and even programmed functions. The executive is thus an extremely important program and, particularly if multiprogramming is being used, fundamentally determines the overall performance of the system.

The degree of complexity of the executive depends on the type of algorithm used for resource allocation and also on how many different algorithms are used. The requirement for different algorithms arises from the widely varying nature of the resources involved. For instance, for a resource which is used infrequently and where use takes very little time, it may well be satisfactory to halt or "mask" all other programs in the system, use the resource, and then "unmask." An example of this is the changing of one word or byte in main memory. On the other hand, for a resource which is used heavily and is relatively slow, it may be necessary to provide an elaborate operating procedure to handle its use. For example, the printing of messages on typewriters is normally done using a buffering and queuing scheme so as to run the typewriters at maximum speed.

A common form of allocation of central processing unit usage for control and data-acquisition computers is by "preemptive priority levels." This technique allows the user to structure his application program set on a priority basis in the knowledge that if program *A* is of higher priority than program *B*, then program *A* can preempt and retain control of the central processing unit until program *A* has completed execution.

Various types of queuing algorithms are used for input/output device management, the simplest being a rejection of any other request for the device until the current request is satisfied. A common queuing algorithm is to have a FIFO (first in, first out) queue for the device, but some quite sophisticated algorithms are currently available. One, for example, searches the queue for the "best" request to do next.

Similar concepts can be used to handle requests for program resources. For instance, a large subroutine may be included in only one partition in a multiprogramming system but may be accessed by many programs to save memory space. In this case, some form of resource allocation must be done to decide which partition(s) may currently use the subroutine. The executive program thus contains all the coding (including error-recovery procedures) which must be main-memory-resident to perform resource allocation. It has access to those modules which can be backup-store-resident and contains all necessary data and workspace.

—A. J. Arthur, *IBM Corporation*

Explosion Proofing (See **Intrinsic safety.**)

Exposure Meter (See **Photometers.**)

Extended Range (See **Range, instrument.**)

Extensometer (See **Strain gage.**)

Eye, Instrumental Measurements of Measurements associated with the eye consist of electroretinography—which is best done by the evoked-potential technique (see **Central nervous system, instrumental analysis of**); eye-motion measurement; and measurement of variations in pupil diameter and the internal pressure of the eyeball.

Electroretinography (Refs. 1 and 2): Eye motion can be measured by placing one electrode at the bridge of the nose and another near the temple; the electrical signal produced by the muscles that move the eye in a horizontal direction will then be picked up. This simple technique is used for qualitative indications in such work as sleep research and target-tracking experiments.

If more precise measurement is required, the image of a small spot of infrared radiation focused on the eyeball is detected with a position-sensing photodetector—either an infrared-sensing TV camera or a motor-operated servotracking device. The simpler devices that track eye motion require immobilization of the head; however, other devices clamp directly on the head. These devices are also used for measuring nystagmus—the small, involuntary motion of the eye when it is fixing on a stationary spot.

Pupillometry (Ref. 1): Involuntary dilation of the pupil can be used as a measure of emotional stress and is not under the subject's conscious control. The pupillometer is a device which automatically and continuously measures the pupil diameter. An infrared image is reflected from the region of the eye pupil onto a vidicon tube, and the maximum size of the black spot, which is the pupil, is measured automatically as the image is scanned with a TV-type raster.

Also see **Vision, sensory aids for.**

REFERENCES

1. Allemand, R. J.: Biomedical Measurements, *Measurements and Data News,* vol. 3, no. 1, p. 58, January–February, 1969.
2. Eye Motion: Methods for Measurement and Their Significance, *Proc. 2d Natl. Biomedical Sciences Symposium,* session II, pp. 43–108, Plenum Press, Plenum Publishing Corporation, New York, 1964.

—J. Malvern Benjamin, Jr.,
Bionic Instruments, Inc.

F

Fabry-Perot Interferometer This instrument produces bright circular fringes of high resolution on a dark field by means of multireflection in the air film between two plane plates thinly silvered (or aluminized) on the inner surfaces. The high fringe resolution and associated high chromatic resolving power make the Fabry-Perot interferometer especially suitable for spectroscopic work. The instrument also can be used to determine accurately the relative wavelengths of several spectral lines. The device is a valuable tool for measuring small distances with a high degree of precision. See **Measuring machine.** Also, since optical-path differences can be determined accurately, material indices of refraction can be found with relative ease. See **Refractometer.**

Facsimile A communication system in which graphic material (photographs, maps, etc.) is scanned and the information converted for transmission by wire or radio to a facsimile receiver. Sometimes termed telephoto or wirephoto. In the receiver, the image may be produced by a modulated light beam focused on photographic printing paper mounted on a rotating drum, by sending electric current through special paper mounted on a drum, or by using a pressure stylus with appropriate pressure-sensitive paper. Also see **Recorder, graphic.** In most facsimile transmitters, the copy to be reproduced is placed on a rotating drum driven by a synchronous motor for scanning by a photoelectric system mounted on a carriage that moves along the length of the drum. Alternatively, the rotating drum may move axially past a stationary photoelectric system. Also see **Data communications.**

—From John Markus, "Electronics and Nucleonics Dictionary," 1966, *used with permission of the McGraw-Hill Book Company, New York*

Fahrenheit Temperature Scale A temperature scale named for its inventor, Daniel Gabriel Fahrenheit (1686–1736). The fundamental interval, ice point to steam point, is 180 degrees. Fahrenheit used the ice point, designated 32°, and the body temperature of a healthy man, designated 96°, as fixed points in making scales for his thermometers.

In reporting on his invention in 1724, he referred, incorrectly, to a third temperature as a fixed point designated 0°, attained with a mixture of ice, water, and sal ammoniac or sea salt. The Fahrenheit scale is used extensively in English-speaking countries for meteorological and industrial purposes. Temperatures are denoted by the term degrees Fahrenheit and by the symbol °F.

—Robert D. Thompson, *Taylor Instrument Process Control Division, Sybron Corporation*

Fan-brake Dynamometer (See **Dynamometer.**)

Fathometer A sonar-type instrument used to measure ocean depth and locate underwater objects. A sound pulse is transmitted vertically downward by a piezoelectric or magnetostrictive transducer mounted on a vessel. The interval required for the pulse to return after reflecting from the sea bottom or an intervening object is determined electronically and converted into an indication of depth or distance. A device of this type sometimes is termed a depth finder or echo depth sounder. Also see **Depth measurement, oceanographic.**

Fatigue, Measurement of [See **Testing (materials), destructive.**]

Fault Detection (Electric) (See **Bridge circuits, electric.**)

Feedback Control Control action in which a measured variable is compared with its desired value to produce an actuating error signal which is acted upon in such a way as to reduce the magnitude of the error. See **Closed-loop control.** See Fig. F-1.

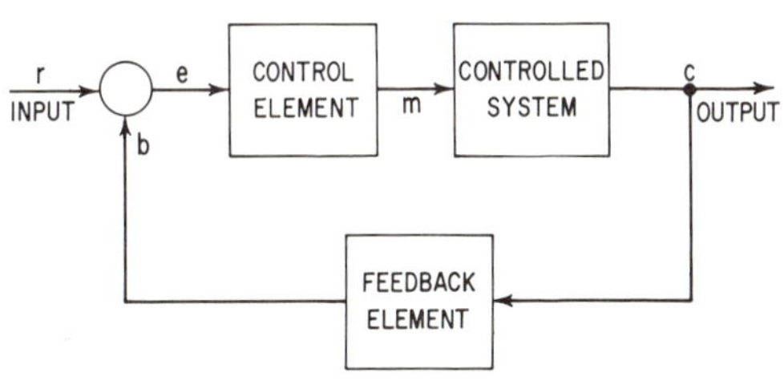

Fig. F-1. Feedback control system. $e = r + b$, where r = reference, e = error, b = feedback (+ or −), m = manipulated variable, c = controlled variable.

Feedback Loop (See **Closed-loop control.**)

Feeder, Gravimetric (See **Feeder, weigh.**)

Feeder, Vibratory A mechanical device used to provide controlled bulk-material flow from storage to processes or to and from various processes. Generally, these feeders can be driven hydraulically, electromechanically, pneumatically, or electromagnetically. An electromagnetic feeder creates vibratory feeding at a steady 3,600 vibrations per minute from half-wave rectified 60-Hz alternating current. Because the feeders have no rotating parts, bearings, eccentrics, or sliding joints, the need for lubrication is eliminated and wear is minimized.

Feeder, Volumetric A device for metering a particulate solid by volume. Functionally, volumetric feeders for solids are analogous to volumetric flowmeters for fluids and are important elements of many process-control loops. Basic characteristics of a volumetric feeding device are (1) a determinable geometry that, in essence, becomes the basic unit of flow measurement and (2) a controlled rate of transfer of material from the feeder to the process. Types of volumetric feeders include the rotary-lock feeder, the helix feeder, the roll-type feeder, the disk/table feeder; the vibrating-tray feeder; and the volumetric belt feeder.

Feeder, Weigh A device for metering a particulate solid by weight. The two main classifications are:

1. Gravimetric feeders
   a. Belt-type devices
   b. Loss-in-weight devices
2. Nongravimetric weigh feeders

Gravimetric Belt Feeder: Most commercially available gravimetric feeders are of this type and comprise a moving endless conveyor belt which actuates a weighing system as the material to be fed passes over the belt. Material is delivered to the belt by a volumetric prefeed device, such as a rotary-lock, helix, or belt volumetric feeder. See **Feeder, volumetric.** Feed rate can be controlled by modulating either the speed of the volumetric prefeed device or the speed of the belt. Gravimetric belt feeders are of two subcategories: (1) *inferential* and (2) *noninferential.*

In the noninferential feeder shown in Fig. F-2, the rate of material delivery is the product of the belt loading and the belt speed. Belt speed is monitored by a tachometer generator; belt load

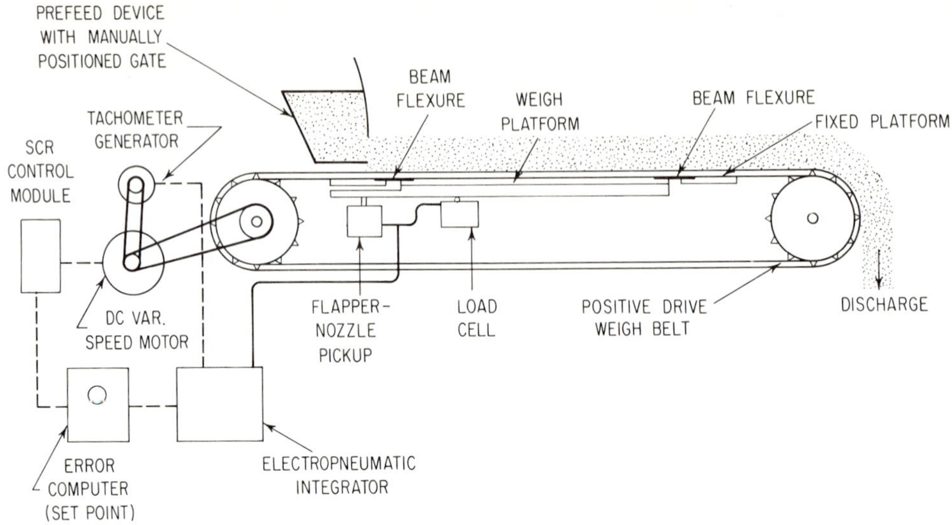

Fig. F-2. Belt-type gravimetric feeder (noninferential type) with fixed prefeed and with belt speed adjusted electrically in response to belt load. *(B I F.)*

is determined by a pneumatic load cell. The product is determined by an integrator, the output of which is compared with a set-point input signal in a computer controller to determine the error. The error signal, in turn, modulates the speed of a dc motor which drives the weigh belt. The feeder gate, which in conjunction with the belt constitutes a volumetric feeding device, is at a constant setting. Thus, the volumetric loading of the belt essentially is constant. The feeder modulates belt speed to correct for density variation and changes in material condition to provide a controlled flow by weight. At low belt speeds, the computer controller takes into account the distance/time for the material to travel from the point of load sensing to the point of discharge by purposely introducing a lag into the system. A feeder of this type provides optimum response, although it is more costly than an inferential feeder.

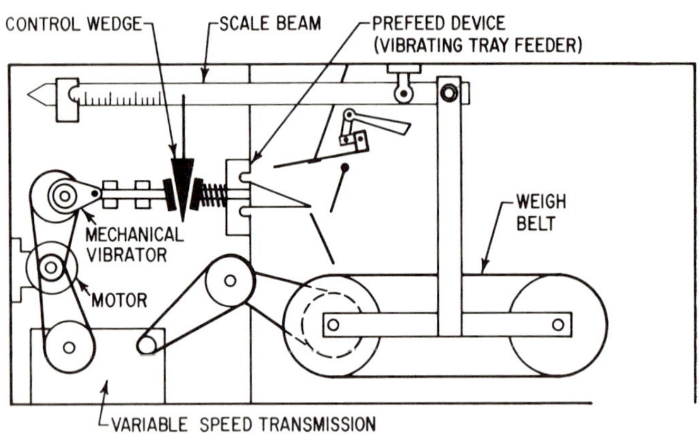

Fig. F-3. Belt-type gravimetric feeder (inferential type) with constant-belt-load feed mechanism adjusted mechanically in response to belt-load variation. *(B I F.)*

A typical inferential device, wherein the belt loading is considered to be constant and where the material delivery or rate is a function of belt speed, is shown in Fig. F-3. The entire belt assembly, plus the material, is balanced against a preset poise weight. Deviations in belt loading are transmitted mechanically to the prefeed volumetric dry feeder in order to modulate the material flow. A device of this type cannot provide a flow-measurement signal other than belt speed. The device is inferential in that the weight of material on the belt is considered to be constant. Within narrow limits, the belt load essentially is constant. The limits can be determined by simple limit switches that actuate alarms in the event that the belt load deviates from the desired weight value. In the arrangement shown, control is accomplished by means of a tapered wedge that is entered between two jaws, one jaw reciprocating at the output of the variable-speed transmission, the other jaw integral with the volumetric-tray prefeed device. The degree of insertion of the wedge between the jaws varies the amount of motion transmitted to the tray—hence varies the tray-feeder amplitude.

In another inferential device shown in Fig. F-4, the belt speed is considered constant because a synchronous motor positively drives the belt. Deflections of the belt under material load are sensed by a pneumatic flapper-and-nozzle pickup. The pickup generates a pressure in the range of 3 to 15 psi in a differential-diaphragm-type load cell. Any pressure in the load cell between 3 and 15 psi is a direct measure of belt load. Conventional pneumatic controls are used to adjust a gate positioner pneumatically, the effects of which are to modulate the thickness of the *layer* of material on the belt—hence return the belt load to the set point or desired value. Feed-rate-error correction is made by adjusting the delivery of the prefeed device. An integrator when used with a machine of this type is driven directly from the feeder belt drive.

Gravimetric Loss-in-Weight Devices: A machine of this type is a novel form of noninferential feeder and comprises a volumetric feeding device and one or more hoppers, all of which are suspended from a scale lever system. The rate of change of hopper mass (loss in weight) can be controlled closely by modulating the delivery of the volumetric feeding mechanism in response to minute deflections of the scale lever system. The loss-in-weight feeder is the most accurate of gravimetric devices because feed-rate correction is made directly at the discharge point (see Fig. F-5.) The machine is not subject to the lag time, with momentary over- and underfeeding, that is experienced with slowly moving weigh belts which must travel over a weigh span at a distance from the prefeed device. The loss-in-weight feeder must be batch-loaded, whereas a belt feeder can operate from a continuous supply of material.

In operation, the loss-in-weight hopper and feeding mechanism are balanced by a counterpoise weight which is set manually at the time the batch of material is loaded. This weight is retracted at a predetermined rate during feeding by an accurate timing motor. A volumetric helix feeding mechanism discharges material at a rate which maintains the scale beam in equilibrium. When the scale beam is disturbed from its equilibrium position, a photocell sensor generates a signal that is proportional to the displacement of the beam. This signal is used to modulate the speed of a dc motor that operates the helix feeding mechanism.

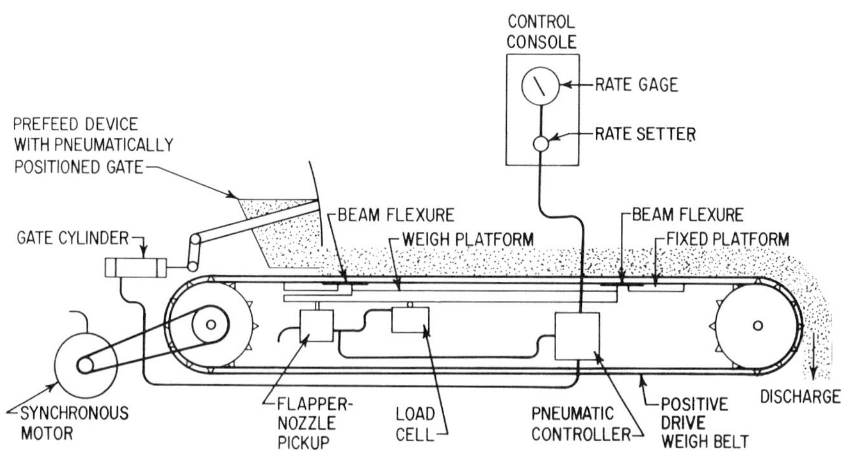

Fig. F-4. Belt-type gravimetric feeder (inferential type) with constant belt speed and with feed gate adjusted pneumatically in response to belt-load variation. *(B I F.)*

Nongravimetric Feeders: These are devices that utilize techniques other than the measurement of gravitational pull to determine mass. The principal device in this category uses the degree of absorption of radiation by the material being fed as a measure of the material mass. The principle can be used with belt or screw conveyors. See **Weighing, radiation methods for.**

Weigh-feeder Applications: Since weigh

feeders include some form of volumetric feeding device, they are susceptible to the same factors as volumetric feeders. See **Feeder, volumetric.** Normally, the more consistent the material, the smaller the variations for which the feeder must

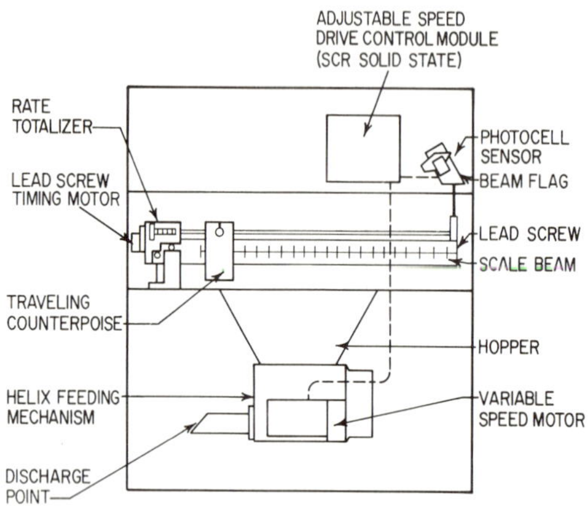

Fig. F-5. Loss-in-weight-type gravimetric feeder. *(B I F.)*

correct—and the more accurate the end result. Broadly speaking, the loss-in-weight feeder is more suitable and accurate for use over extended periods of time. Where short-term accuracy is required, the noninferential variable-speed belt feeder, with its high response, will provide the best results.

REFERENCES

Lee, C. A.: Solids Feeders, *Chem. Eng.,* Feb. 17, 1964, pp. 119–134.

Zilly, Robert G.: Solids Transport and Storage, in Robert H. Perry, "Chemical Engineers' Handbook," 4th ed., sec. 7, McGraw-Hill Book Company, New York, 1963.

Considine, Douglas M.: Process Weighing, *Chem. Eng.,* Aug. 17, 1964; Sept. 14, 1964.

—Norman J. P. Taylor, *B I F , Unit of General Signal Corporation*

Feedforward Control Feedforward control implies the approach to the control problem by means of predicting manipulated signals to make the necessary process adjustments before controlled variables are affected by the disturbances. The predictions usually are obtained by calculations in accordance with some mathematical model and the information collected from measurements of the various process streams involved. A typical feedforward control loop is shown in Fig. F-6.

Unlike feedback control which solves the control problem by "trial and error," feedforward control requires reasonable knowledge of the process being handled in order to set up a mathematical model. The more accurate a feedforward control is desired to be, the closer must the model simulate the controlled process. Depending upon the degree of complexity of the mathematical model, the computing device required may be a group of relays, an analog computer, or a digital computer.

In many instances, feedforward control will succeed where feedback control has failed. This is especially true for difficult processes, such as those which require wide proportional band and long reset time, those processes which are subject to frequent load changes, and those processes involving optimal or adaptive controls. See the example of feed-forward control of pH described under **Automatic control.** The success of feedforward control is almost entirely contingent upon accuracy. Any errors that may result from measurements, calculations, and negligence of

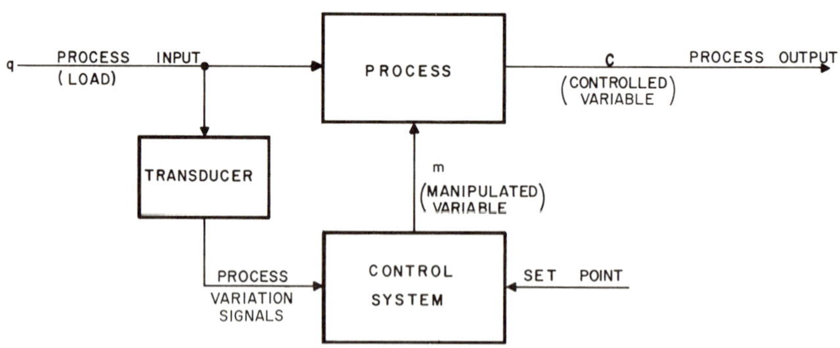

Fig. F-6. Feedforward control system.

some significant factors in setting the mathematical model will cause offset. In such cases, the only cure is to apply feedback (manual or automatic) in order to correct for accumulated errors. This situation is illustrated in Fig. F-7, where a level-control feedback loop is used to reset the feedforward loop. The process is further delineated in Fig. F-8.

In conclusion, feedforward control is by no means a panacea for all the control problems. For one thing a non-self-regulating process is not controllable by feedforward alone. But it indeed is a powerful tool and provides substantial improvement over feedback in response to the load changes and other disturbances.

—Anthony Y. Chiang, *Bristol Division, American Chain & Cable Company*

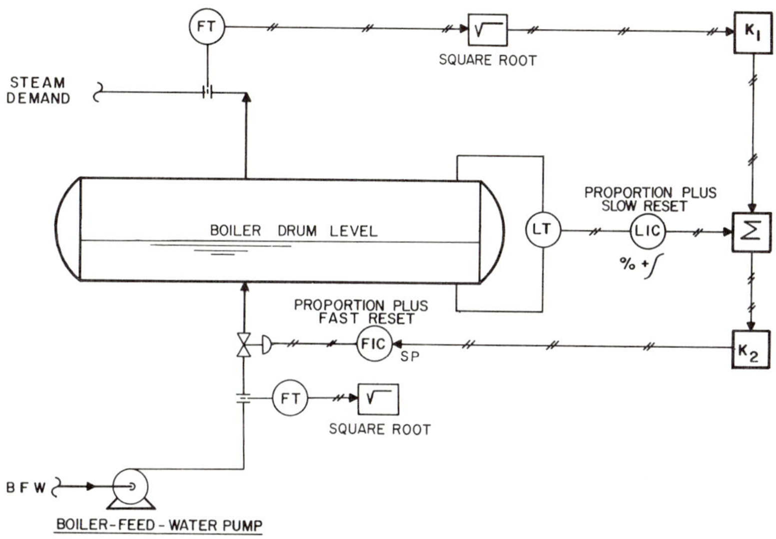

Fig. F-7. Feedforward control of steam-generation system in which a level-control feedback loop is used to reset the feedforward loop.

Fiber Optics The technique of transmitting light through long, thin flexible fibers of glass, plastic, or other transparent materials. Bundles of parallel fibers can be used to transmit complete images. The technique is particularly

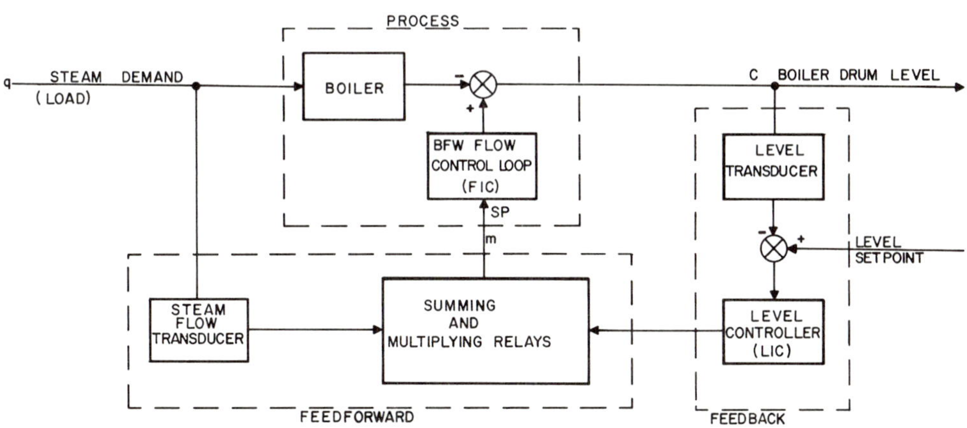

Fig. F-8. Block diagram of steam-generation control system shown in Fig. F-7.

useful in the design of sensing elements where space is tightly limited.

Fiber Testing [See Testing (materials), destructive.]

Filled-system Thermometer (See Thermometer, filled-system.)

Film Detector, Radiation Photographic film is the standard detector for medical x-ray and industrial radiography. Similar films are used for radiation personnel monitoring in the form of film badges. In addition, photographic emulsions have fulfilled a major detection role in high-energy physics and cosmic-ray research.

Final Controlling Element That element of a control system which directly changes the value of the manipulated variable. Common forms include valves, dampers, and louvers. In the broad sense, an electric motor that finally positions some element of a machine or process also can be referred to as a final controlling element.

Fixed Point A method of storing numeric data in a computer such that the data are all stored in integer form (or all in fractional form) and the user postulates a radix point between a certain pair of digits. As an example, consider a computer whose basic arithmetic is in decimal and in which each computer word consists of 7 decimal digits in integer form. Then, if we wish to add 3.141463 to 2.808380, we store them in the computer as 3141463 and 2808380, which when added together, provides the answer 5949843. Externally, we recall that we have postulated a decimal point between digits 1 and 2. The result, therefore, represents 5.949843 (see **Number systems**; compare **Floating point**). For convenience, input and output conversion routines are normally provided which can add or delete the radix point in the external representation and align the data as required internally.

Since the arithmetic operations are normally the basic arithmetic operations of the computer, fixed-point operations are quite fast and are therefore to be preferred to floating-point operations. The major drawback is that the magnitude of the numbers involved must be much better known than for floating-point numbers, since the absolute magnitude is limited by word size and the availability of double-length operations. However, for many applications, fixed-point calculations are quite feasible and the increase in speed is useful.

—A. J. Arthur, *IBM Corporation*

Flame Photometer-spectrophotometer The gas flame was the earliest source of light used for spectrochemical analysis. Earlier investigators had recognized that certain metals imparted colors to flames, but Bunsen and Kirchhoff in 1860 first studied flames through a spectrometer, measured the lines emitted by the alkalies, and demonstrated the possibility of spectrochemical analysis. Ludegardh developed a burner in the late 1920s which atomized the sample but trapped out solution droplets and passed on to the flame only a mist of saturated air. This produced a hotter flame, which he photographed with a medium quartz spectrograph. In the 1930s, this system was employed to give good sensitivities and quantitative results with a repeatability of 3 to 5 percent for the alkalies, alkaline earths, and many of the transition metals. By the 1940s, simple flame photometers had been introduced for the measurement particularly of Na and K, but also for Li and Ca. These consisted of a burner, filter, barrier-layer photocell, and a galvanometer or meter. These instruments quickly displaced chemical methods for the analysis of Na, K, and Li. A typical current instrument is illustrated in Fig. F-9.

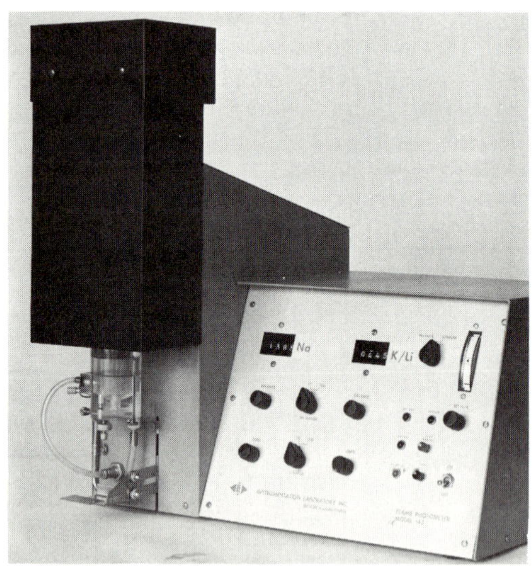

Fig. F-9. Flame photometer used extensively in clinical laboratories for determination of Na and K. *(Instrumentation Laboratory.)*

By the 1950s, spectrophotometers were being employed to secure higher resolution in isolating the lines, and photomultipliers and high-gain amplifiers were providing means of measuring much lower light levels. This made it possible for the first time to identify and measure, quickly and quantitatively, many common metallic elements in solution when aspirated into a flame. A widely used instrument is shown in Fig. F-10.

The spectrophotometer can be employed as a scanning spectrometer to produce a trace of the flame emission with wavelength. Gilbert (Ref. 1) has published such charts showing the emission of some pure elements. The same article includes a table of conditions used to measure 64 elements either directly or indirectly through the formation of band systems, usually of oxides, and the measurement of band heads. Data for flames of air-hydrogen, oxyhydrogen, oxyacetylene, and oxycyanogen are included. The hotter oxyacetylene and oxycyanogen flames are used to excite more energetically the elements with high excitation energies, such as Zn. Another helpful technique, developed extensively by Dean (Ref. 2), is the use of organic solvents in place of aqueous solutions, to produce higher temperatures and more complete excitation, sometimes increasing the sensitivity by a factor of as much as 100 times.

In 1951, Walsh proposed that the concentration of an element in a solution could be determined with greater sensitivity by measuring the light that the element absorbs in a flame rather than the light that it emits. He pointed out that the number of atoms in a flame that are in the ground state, ready to absorb, is many times ($100\times$ to $10^{14}\times$) the number in the lowest excited state, capable of emitting. Also, he pointed out that absorption measurements should not be as affected by variations in flame temperature as emission. Atomic-absorption techniques were developed in Australia and in Europe during the 1950s but were introduced into the United States only in the early 1960s. Since that time atomic-absorption spectroscopy has expanded very rapidly, so that now either flame-emission or atomic-absorption spectrophotometry is certainly the most rapid, frequently the most sensitive, and occasionally the most accurate and reproducible way of measuring some 68 metallic elements, which are given in Table F-1, with their limits of detection.

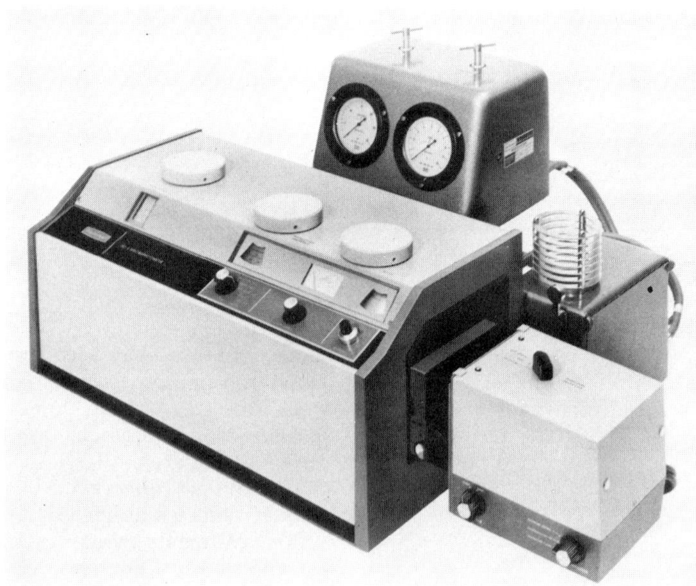

Fig. F-10. An early flame spectrophotometer, still widely used, for measuring many elements in solution. *(Beckman.)*

The components of a typical atomic-absorption spectrophotometer are shown schematically in Fig. F-11, which shows:

(1) A "cool" light source to produce the strong lines of the element to be analyzed with negligi-

TABLE F-1. *Comparison of Estimated Detection Limits,* μg/ml, *for Atomic-absorption Spectrophotometer*

Element	Emission	Absorption	Element	Emission	Absorption	Element	Emission	Absorption
Ag. . . .	0.06	0.02	Hf. . . .	75	5	Rb. . . .	0.002	0.005
Al.	0.2	0.1	Ho. . . .	0.1	0.3	Rh. . . .	0.3	0.3
Au. . . .	4	0.1	Hg. . . .	2	0.5	Ru. . . .	0.3	0.3
As.	2.2	0.5	In.	0.03	0.5	Sb.	1	0.2
Ba.	0.03	1	Ir.	110		Sm. . . .	0.6	5
Be.	0.08	0.003	K.	0.003	0.005	Sc.	0.07	0.1
Bi.	6	0.2	La.	1	20	Si.	5	2
B.	0.1*	15	Li.	0.000003	0.005	Sr.	0.004	0.02
Cd. . . .	0.5	0.01	Lu. . . .	0.2		Ta. . . .	20	4
Ca.	0.005	0.01	Mg. . . .	0.04	0.003	Te. . . .	2	0.5
Ce.	10		Mn. . . .	0.01	0.01	Tb. . . .	1	2
Cs.	0.005	0.05	Mo. . . .	0.03	0.1	Tl.	0.09	0.2
Cr.	0.01	0.01	Na. . . .	0.0001	0.005	Tm. . . .	0.2	
Co. . . .	0.2	0.05	Nd. . . .	1	2	Sn.	0.5	0.1
Cu. . . .	0.03	0.01	Ni.	0.6	0.02	Th. . . .	150	
Dy. . . .	0.1	0.2	Nb. . . .	1	2	Ti.	0.5	0.1
Er.	0.3	0.2	Pb.	1	0.05	U.	10	30
Eu. . . .	0.003	0.2	Pd. . . .	0.1	1	W.	4	2
Fe.	0.1	0.05	P.	1†		V.	0.3	0.2
Gd. . . .	2	4	Pr.	2	10	Yb. . . .	0.05	0.04
Ge. . . .	0.6	0.8	Pt.	10	0.5	Y.	0.3	0.3
Ga. . . .	0.07	1	Os.	10		Zn	16	0.005
			Re. . . .	1	0.8	Zr.	50	5

*With kind permission from V. A. Fassel, presented at the Pittsburgh Conference on Analytical Chemistry and Applied Spectroscopy, 1966.

†Molecular bands.

ble doppler or Stark broadening, so that the radiation emitted will have a narrower width than the absorbing line width in the flame. Hollow-cathode lamps have been used most frequently, but microwave-excited lamps are also suitable. The light source requires a regulated power supply, not diagramed.

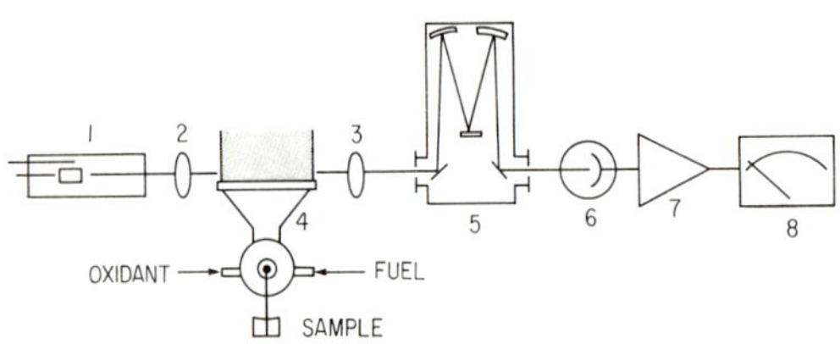

Fig. F-11. Components of a typical atomic-absorption spectrophotometer.

(2) and (3) Condensing lenses to confine the beam to a narrow width through the flame and to illuminate the slit of the spectrometer.

(4) The flame, shown from a "laminar-flow" burner, which has a long path length, 5 or 10 cm, to increase absorption. A stream of compressed air or nitrous oxide across a capillary that dips into the sample both aspirates the liquid and serves as the oxidant or "support" gas for the flame. Fuel, usually acetylene or hydrogen, is fed into the burner through a second connection. Pressure regulators and flowmeters are required to control and indicate the flow rate of both gases. Droplets that are too large to be transported by the support gas collect at the bottom of the mixing chamber and are drained out.

(5) A grating or prism spectrometer, preferably capable of isolating lines separated by only a few angstroms, because good sensitivity and a straight-line response are achieved only if the absorbing line is resolved from all other lines (*a*) of the analytical element, (*b*) of the other elements that a "multielement" lamp may emit, and (*c*) of the argon or neon used as filler gas.

(6) A photomultiplier sensitive over the range 1937 Å (arsenic) to 7800 Å (rubidium). (Usually a special detector will be required for cesium at 8521 Å.) A regulated high-voltage power supply and potentiometer is required for the dynodes of the photomultiplier and is used to set the full-scale deflection.

(7) An ac amplifier measures the hollow-cathode output reduced by the light that is absorbed by the flame. To avoid measuring the light that

is emitted by the flame, particularly at higher concentrations, a rotating-sector disk may be inserted at lenses 2 or 3 to chop the light from the hollow cathode, or the latter's power supply can be pulsed so that the light is emitted as an ac signal. The amplifier then will not respond to the dc signal of the light emitted by the flame but will measure the light from the hollow cathode that remains after absorption by the flame. As explained later, a supplementary logarithmic amplifier and scale expansion will make measurements more convenient.

(8) The output is read on a simple microammeter, a strip-chart recorder, or a digital voltmeter.

Atomic absorption, like molecular absorption, is governed by the law

$$I = I_0 e^{-kcx} \quad \text{or} \quad \log \frac{I_0}{I} = kcx$$

where I_0 = initial intensity of the beam with a "blank" being aspirated

I = intensity transmitted when the sample is being aspirated

k = constant dependent on the element, the instrumental parameters, and the operating conditions

c = concentration of the element in the sample

x = absorbing path length, which, of course, is held constant and so could be included in k

Since log (I_0/I) is absorbance A and x is held constant, we see that $c = kA$. Thus, if a logarithmic amplifier is provided to measure absorbance or log (I_0/I), readings will be linear in concentration. The analytical curve will, however, normally flatten out before the absorbance reaches a value of 1.0.

A typical set of strip-chart readings and the corresponding analytical curve are shown in Fig. F-12.

At very low concentrations, with a stable light source, well warmed up, and a well-adjusted aspirator and flame conditions, and with a time constant for the amplifier set at 5 s or more, the noise level of the signal is so low that the limit of detection is determined by the ability to read the meter. Accordingly, it is desirable to have a "scale-expansion" adjustment. To obtain a 10× scale expansion, the gain of the amplifier is increased, so that the reading for I_0, which was 100, would now be 1,000. If I originally read 90, it would now read 900. At the same time that the gain is increased, the zero is offset electronically by 900 divisions to bring I back to 0. I_0 will then read 100. The absorption deflection has been increased by 10 times. Thus, Fe, shown in Fig. F-13, gives 1 percent absorption at 0.1 ppm. With 10× scale expansion, 0.1 ppm produces an apparent 10 percent absorption, and 0.01 ppm will read 1 percent.

The concentration at which an element produces an absorption of 1 percent is defined as

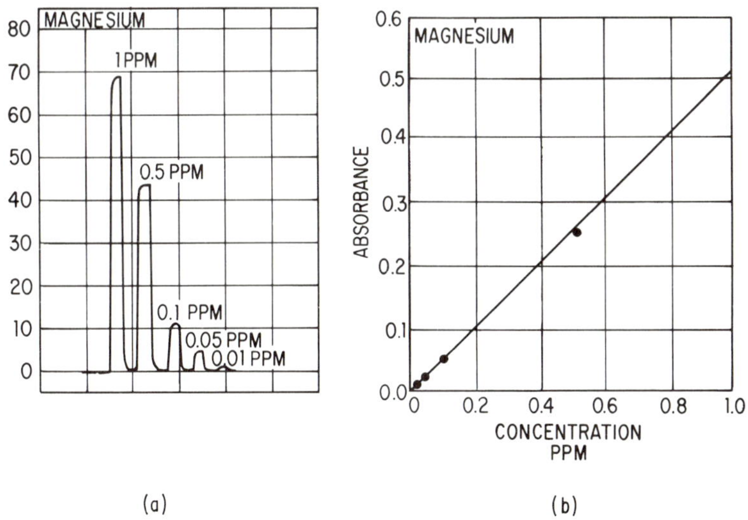

Fig. F-12. Analysis of magnesium: (a) typical strip-chart readings; and (b) corresponding analytical curve.

the "sensitivity" of the element, usually expressed in parts per million (ppm) or micrograms per milliliter (μg/ml), for aqueous solutions. The concentration at which the scale-expanded signal is twice the noise level of the blank is called the "limit of detection," similarly expressed.

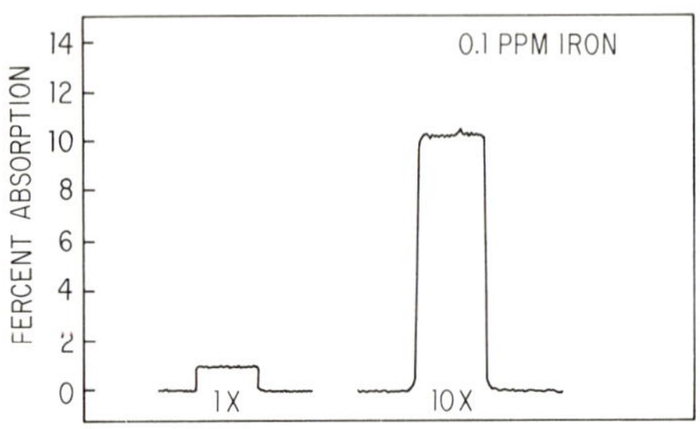

Fig. F-13. Scale expansion of ten times in the measurement of Fe.

Since flame-emission spectroscopy is still a highly useful technique, as illustrated by Table F-1, many instruments are combination flame-emission–atomic-absorption spectrophotometers. Such equipment is illustrated in Fig. F-14.

Atomic-absorption analytical methods are similar in a general way to those developed for flame emission. Many are described in the literature, and several manufacturers of equipment provide analytical methods for a wide variety of elements and matrices.

REFERENCES

1. Gilbert, P. T.: Analytical Flame Photometry, Symposium on Spectroscopy, *ASTM Spec. Tech. Publ.* 268, 1959.
2. Dean, John A.: "Flame Photometry," McGraw-Hill Book Company, New York, 1960.

—Richard F. Jarrell, *Jarrell-Ash Division, Fisher Scientific Company*

Flapper-nozzle Systems (See **Pneumatic controller;** and **Transmission, pneumatic.**)

Flapper Valve (See **Hydraulic relay.**)

Flight-control Systems The control of aircraft ranges from a human pilot who is provided with sufficient instrument information to operate control surfaces and throttles to control the aircraft in flight to the completely automatic flight-control system which can perform automatic landings with man serving only as a monitor. Pilots become hopelessly confused when visual contact with some reference to true vertical is lost. Thus, some instrumentation to assist the aircraft pilot in the problem of attitude sensing is essential. Accelerations, both translational and due to turns, combine with gravity to give a resultant known as *dynamic vertical.* A human being can fairly well sense dynamic vertical but is poor at sensing rate of turn or at remembering true vertical. The gyroscope, which is the main attitude-sensing instrument, provides a means of

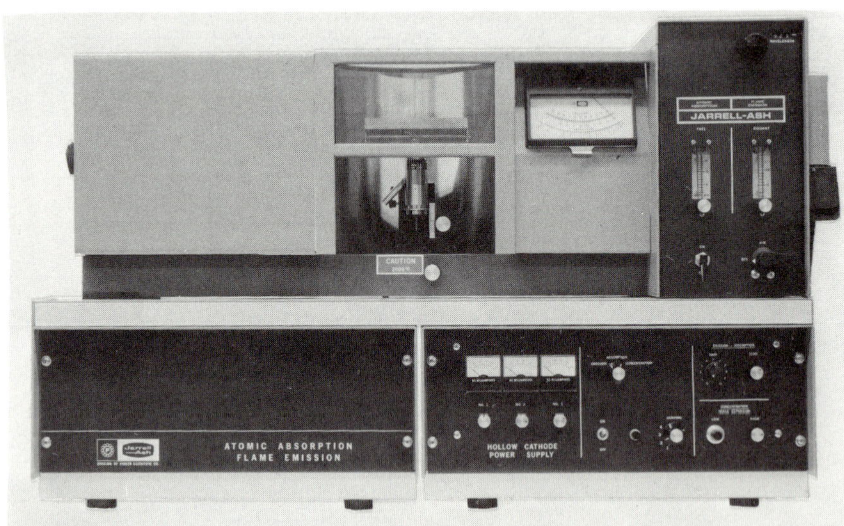

Fig. F-14. A flame-emission-atomic-absorption spectrophotometer. *(Atomsorb, Jarrell-Ash.)*

getting large values of effective inertia in compact form. Thus, gyroscopic instruments are among the most important elements of a flight instrumentation and control system—whether for assisting a manual pilot or for providing inputs to a fully automatic flight-control system.

Gyros:[1] A wide variety of gyroscopic instruments are in daily use, but all have one common, basic element—a spinning mass known as the gyro rotor. Gyros depend upon one of two closely related properties of a spinning mass—either (1) the tendency of the rotor's spin axis to remain fixed in its angular relationship to inertial space or (2) *precession* (see Fig. F-15). The spin axis is basic to the rotating mass. The *input axis* is any axis perpendicular to the spin axis; the *output axis* is an axis perpendicular to both the spin and the selected input axis. If the spinning mass is forced to rotate at an angular velocity ω_{IA} about

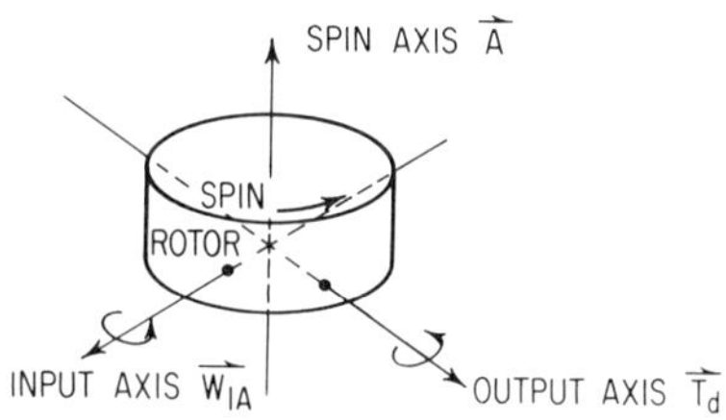

Fig. F-15. Axes of a gyroscope.

the selected input axis, a torque $\vec{T}_d$ appears around the output axis. Conversely, if a torque is applied, an angular velocity will result in the spin axis. This is known as precession. The magnitude and direction are given by

$$\vec{T}_d = H\omega_{IA} \tag{1}$$

where H is the angular momentum of the spinning rotor.

Despite the great diversity in gyroscopic instruments, the basic differences lie in four major areas: (1) the means by which the gyro spin rotor is supported with respect to the case of the instrument; (2) the restraints—intentional and unintentional—placed on angular movements of the rotor with respect to the case; (3) the torques—intentional and unintentional—which are applied to the rotor; and (4) the means by which the angular position of the rotor with respect to the case is exhibited visually or reported to a control system for which the instrument serves as a sensor.

[1] Information on gyros prepared by Henry A. Dinter, Jr., Honeywell Inc.

Generally, the rotor-support system consists of one or more gimbals or cardans, although there are notable exceptions, such as the very high performance electrically suspended gyro in which the rotor is supported in a high vacuum between three sets of charged electrodes. If the support system permits the rotor to move with respect to the case about only a single one of all the possible axes perpendicular to the spin axis, the instrument is said to be a *single-degree-of-freedom*

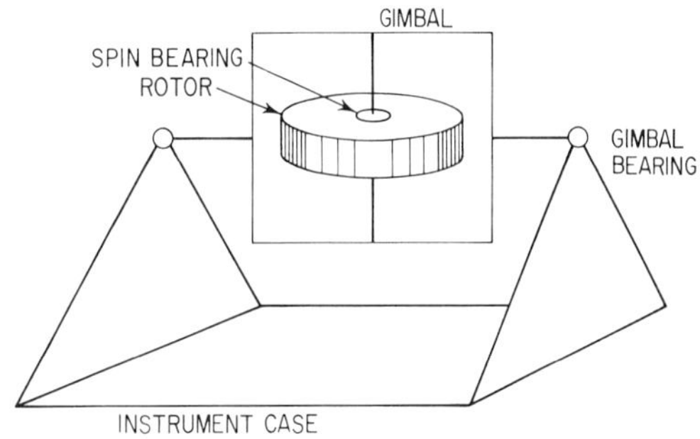

Fig. F-16. Typical gimbal system for a single-degree-of-freedom gyro.

gyro. If the case can move with respect to the rotor about two axes perpendicular to each other and to the spin axis without forcing the spin axis to rotate with respect to inertial space, the gyro is said to be a *two-degree-of-freedom* gyro. Gimbal systems for a single-degree-of-freedom instrument are shown in Fig. F-16, for a two-degree-of-freedom gyro in Fig. F-17.

The angular motion to be sensed determines what intentional restraints will be applied to the gyro. For example, if the instrument is to sense

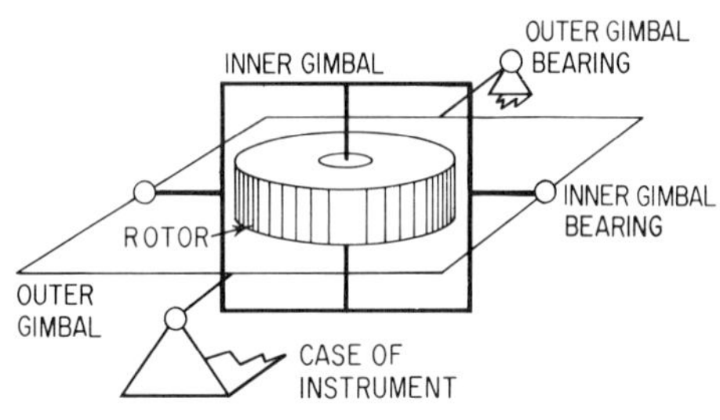

Fig. F-17. Typical gimbal system for a two-degree-of-freedom gyro.

angular velocity, typically a single-degree-of-freedom gyro will be selected and intentional spring and viscous restraints will be applied to the motion of the single gimbal with respect to the case. If angular displacements are to be sensed, a two-degree-of-freedom gyro may be selected with no intentional restraints applied. Intentional torques may be applied about the appropriate gimbal axes to cause the gyro's spin axis to precess toward and to retain some preferred orientation, such as parallel to local vertical. A gyro of this configuration is said to be *captured* or *slaved* and, in this example, to local vertical.

The readout system for an aircraft flight instrument usually is vertical. If the instrument is to be used as a sensor for a control system, a voltage proportional to displacement of the case with respect to the rotor usually is the desired output. The element which produces this voltage is termed a *pickoff*. Electromagnetic pickoffs, such as synchros or differential transformers, are the most widely used, although electro-optical and capacitive pickoffs also are important.

Rate Gyro:[1] The rate gyroscope is a device designed to measure the instantaneous angular-velocity components of a body with respect to inertial space. Typical applications include autopilot damping, rate-of-turn indication and limiting, antenna stabilization, and telemetry instrumentation. The general configuration, as shown in Fig. F-18, consists of a high-angular-velocity spin rotor with constant angular momentum, constrained in a single-degree-of-freedom gimbal whose angular precession is restrained by a mechanical spring.

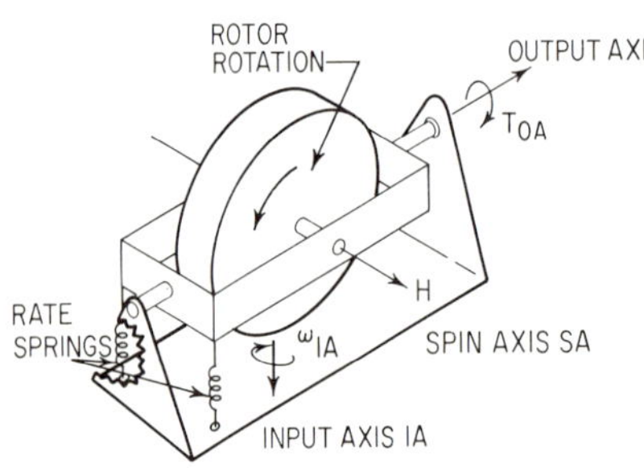

Fig. F-18. Spring-restrained rate gyro.

Forced angular velocity of the instrument case about the nominal input axis is accompanied by a proportional gyroscopic torque about the output axis in accordance with momentum-conservation principles. This behavior is expressed vectorially by the relationship

$$\vec{T}_{OA} = \vec{H}_{SA}\,\vec{\omega}_{IA} \tag{2}$$

where $\vec{T}_{OA}$ = developed precessional torque about the output axis
$\vec{H}_{SA}$ = angular momentum about the spin axis
$\vec{\omega}_{IA}$ = forcing angular velocity about the input axis

A restraining torque is provided by the mechanical spring and is proportional to the relative angular displacement of the gimbal with respect to the gyro case. It may be expressed by

$$T_{SOA} = K_{SOA}\ \theta_{OA} \tag{3}$$

where T_{SOA} = spring-restraint torque about the output axis
K_{SOA} = angular spring rate
θ_{OA} = angular deflection of the gimbal with respect to the case

Substituting, in the steady-state condition where $T_{OA} = T_{SOA}$, and rearranging yields

$$\theta_{OA} = \frac{H_{SA}}{K_{SOA}}\,\omega_{IA} \tag{4}$$

Therefore, the angular deflection of the gimbal is proportional to the angular velocity about the input axis. The addition of an angular-displacement pickoff will provide a signal proportional to angular velocity about the input axis. Since the device is a spring-mass system about the output axis, the addition of a viscous damper is necessary to achieve usable frequency response. Dynamic response, in the ideal case, will conform with second-order low-pass-filter behavior.

A typical arrangement of a spring-restrained rate gyro is shown in Fig. F-19. Constant angular momentum is developed by the rotor of a hysteresis synchronous motor. This rotating mass generally is placed outside the nonrotating stator. The entire spin motor is enclosed in a hermetically sealed gimbal supported at one end by a jewel-pivot bearing and at the other end by a torsion-bar spring. The entire case is filled with silicone fluid surrounding and partially floating the sealed gimbal. Damping is achieved by viscous shear of the fluid. The major portion of this restraint occurs in the small gap between the outer case and a plastic cylinder (balance wheel in Fig. F-19) attached to the gimbal. Variations in this gap due to differences in thermal-expansion rates

[1] Information on rate gyro prepared by George M. Aries, Honeywell Inc.

for the steel case and the plastic cylinder partially compensate fluid-viscosity changes with temperature. Variations of this damping arrangement include (1) gas-damped (piston and cylinder) fluidless gyros, (2) bellows-actuated valves that

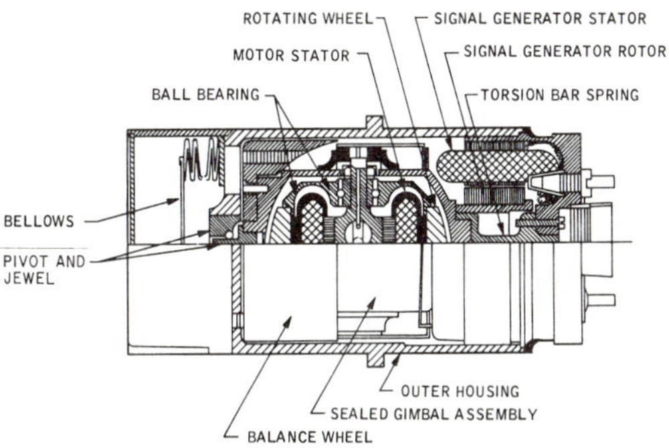

Fig. F-19. Spring-restrained rate gyro employing a hysteresis synchronous motor. *(Honeywell.)*

control the pressure drop through their orifices of the fluid that is displaced or "pumped" by a paddle arrangement on the gimbal, (3) magnetic damping, and (4) "fluidless gyros" that retain small amounts of fluid in shear gaps by capillary action.

The signal generator is an angular variable-differential transformer. The 8-pole "moving-iron" rotor is attached to the gimbal and rotates with it, while the 16-pole stator is attached to the case. Variations on this setup include 4-pole differential transformers, such as the Microsyn, potentiometers, and "E-core" transformers.

Typical performance ranges for two categories of rate gyros are given in Table F-2.

Rate-integrating Gyro:[1] The rate-integrating or floated gyro can be designed as either a single- or two-axis device. The single-axis rate-integrating gyro has wider application and is described here (see Fig. F-20). The spinning wheel is assembled in a cylindrical gimbal or float, and the normal two-axis precessional freedom of this assembly is constrained by jewels and pivots to provide a single output axis. The rate-integrating gyro is designed so that fluid damping between gimbal and housing is the predominant torque which balances the input-rate precessional torque. Torques resulting from elastic restraint (flex leads conducting power to the gimbal), and coulomb friction (jewels and pivots) are made relatively small. Flex-lead torques are reduced by minimizing the flex-lead cross section within the current carrying requirement (approximately

[1] Information on rate-integrating gyros prepared by William N. Johnson, Honeywell Inc.

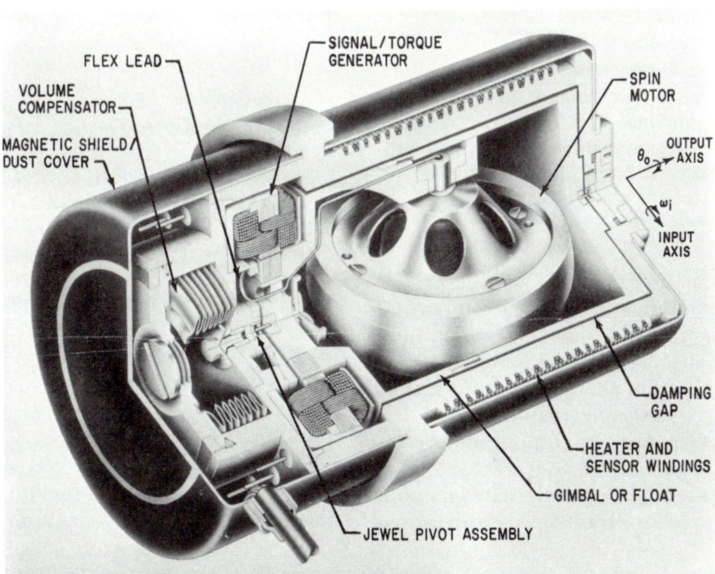

Fig. F-20. Rate-integrating gyro. *(Honeywell.)*

TABLE F-2. *Typical Performance Parameters of Rate Gyros**

Parameter	Category I, Low-cost, Usually Fluidless	Category II, High-performance Devices
Operating range, deg/s	±5– ±1,200	±5–±1,200
Angular momentum, g-cm²/s	50×10^3–3×10^6	5×10^3–1×10^6
Full-scale output, V	5–30	5–10
Threshold and resolution, deg/s	<0.1	<0.01
Hysteresis, % of full scale	<1.0	<0.2
Acceleration sensitivity, deg/s/*g*	<0.1	<0.1
Linearity, % of full scale	1.0–2-0	0.5–2.0
Power requirements:		
Spin motor	28 V dc; 26 V 400 Hz, 1, 2, 3 phase	26 V, 400 Hz 1, 2, 3 phase
Pickoff	28 V dc; 26 V 400 Hz, 1 phase	26 V, 400 Hz 1 phase
Total power consumption, W	5–15	3–10
Size	2 in. diam × 4 in. long — 4 × 4 × 4 in.	1 in. diam × 2.5 in. long—2 in. diam × 4 in. long
Weight, lb	0.5–2.0	0.2–1.0
Dynamic response:		
Damping ratio at 72°F	0.2–5.0	0.2–5.0
Damping-ratio variation from —65 to 200°F of one extreme to the other	2*x*–20*x*	0.4*x*–15*x*
Natural frequency, Hz	2–50	15–100

*Standard Gyro Terminology, *Aerospace Industries Association, EETC Rept.* 5.

0.0003 by 0.004 in.) and by assembling these in a symmetrical array. Coulomb-friction torque is reduced by floating the gimbal. Low-level spin-motor vibration produces dither, which futher reduces friction. Under the condition where precessional torque is balanced primarily by viscous-damping torque, the equation relating steady-state input rate and output or gimbal motion reduces to

$$\theta_o = \frac{H}{C}\int \omega_i \, dt \tag{5}$$

Gimbal displacement about the output axis is proportional to the integral of input rate, where $H =$ angular momentum and $C =$ damping constant. A signal generator is provided to transform mechanical motion about the output axis to electrical signals. A torque generator is included to supply bias or equivalent input-rate precessional torque.

In any practical gyro, the steady-state equation must include a drift-producing error-torque term

$$\frac{H}{C}\int \omega_i \, dt + \frac{1}{C}\int T_e \, dt \tag{6}$$

The major error torques developed about the output axis are separated and identified as follows:

g-insensitive .... Flex-lead restraint
Electromagnetic coupling between the spin motor and external fields
Suspension friction

g-sensitive .... Gimbal-mass unbalance, mismatch between center of gravity and center of buoyancy

g^2-sensitive .... Anisoelasticity, mismatch of spring rates of the structure supporting the spinning wheel

Reduction of these error torques is a primary design goal for a high-performance gyro. Many of these torques vary with temperature, and it is necessary, in most applications, to operate the gyro at a closely controlled temperature. In addition, the viscosity of the flotation fluid or the damping constant varies at approximately 3 percent/° F. A heater and sensor are assembled about the gyro housing to facilitate the required temperature control.

Typical rate-integrating-gyro parameters and

gyro drift rates, based upon an angular momentum of 1×10^5 g-cm²/s, are given in Table F-3.

TABLE F-3. *Typical Rate-integrating-gyro Parameters*

Parameters:	
Angular momentum H	100,000 g-cm²/s
Damping constant c	20,000 g-cm²/s
Gimbal inertia I	100 g-cm²
Signal-generator sensitivity	25 V/rad
Torque-generator sensitivity	50 dyne-cm/mA (100°/hr) mA
Operating temperature	130–180°F range
Drift rates:	
g-sensitive	0.5°/hr
g-sensitive	1.0°/hr/g
g^2-sensitive	0.03°/hr/g^2

Vertical Gyro:[1] A vertical gyroscope is a two-degree-of-freedom device with provisions for maintaining the spin reference axis at local vertical. The unit provides a measure of angular displacement of the case about the two nominally orthogonal, horizontal axes. Typical applications include autopilot stabilization, reference, and control; radar and antenna stabilization; flight indicators; telemetry instrumentation; and a number of other areas including railroad surveys, road grading, and oil-well exploration where need for a vertical reference is required.

[1] Information on vertical gyro prepared by Gary D. Moon, Honeywell Inc.

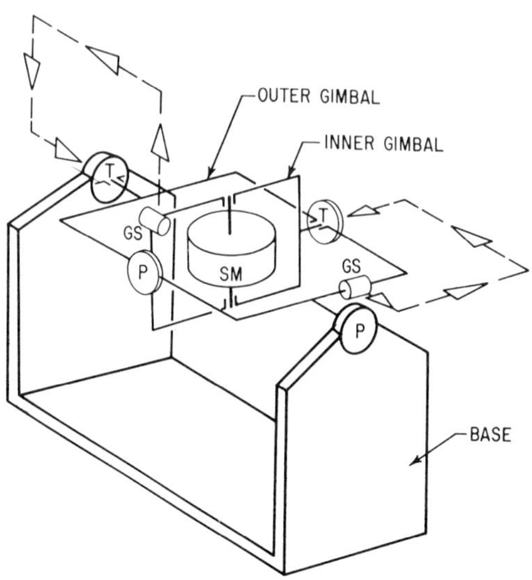

Fig. F-21. Block diagram of vertical gyro.

The basic gyro often is used, without the erection means, to provide a short-term two-axis reference about other axes. The general configuration, as shown in the block diagram of Fig. F-21, consists of a high-angular-momentum spin motor supported within two orthogonal gimbals. Output is provided by pickoffs which sense angular displacement between the inner gimbal and outer gimbal and between the outer gimbal and the case. Electrical torquers slaved to gravity sensors are used to correct for gyro drift and maintain the spin axis vertical. Slip rings normally are used to conduct electrical power to the pickoffs and motors.

The rotating member of the spin motor normally is designed to be external to the stator in order to obtain maximum angular momentum H for a given total rotor and stator spin-motor mass. The spin motor, which may be either a synchronous or induction type, is enclosed in a light case or envelope which forms the inner gimbal. Inner-gimbal freedom normally is limited to ±85° in order to prevent gimbal lock, which occurs when the spin axis is coincident with the outer gimbal axis. The outer gimbal surrounds the inner gimbal and is constructed similarly. Outer-gimbal freedom normally is unlimited. A number of different pickoffs have been used on vertical gyros and predominantly include synchros, resolvers, and potentiometers. Less frequently used are inductive, capacitive, and photoelectric pickoffs. In the normal orientation of a vertical gyro, the pickoffs are used to measure roll with the outer gimbal and pitch with the inner gimbal. Vertical erection normally is controlled through the use of gravity-sensor accelerometers, which may be in the form of mercury switches, electrolytic sensors, or pendulum-type switches. The sensors control electrical torquer motors which provide the necessary torque to maintain the gyro spin axis at vertical by correcting for gyro errors caused by drift or by erection to false vertical.

Gyro error or the departure of the spin axis from vertical is caused primarily by:

1. Apparent earth's rate since the spin axis is stable in space or inertial coordinates rather than earth-base coordinates. Rate of departure is defined by

$$\omega_E = 0.727 \times 10^{-4} \cos \phi \qquad (7)$$

where ω_E = earth's rate, rad/s at latitude ϕ
ϕ = latitude, usually stated in degrees north or south

2. Torques about the gyroscope axes which include both acceleration-sensitive and non-acceleration-sensitive and cause drift at a rate defined by

$$\omega = \frac{T}{H} \tag{8}$$

where ω = drift rate, rad/s
T = torque, g-cm
H = angular momentum, g-cm^2/s

3. Vibrational anisoelastic torques which are due to mechanical rectification caused by the unequal bearing compliance of the spin motor and supporting gimbal structure.

4. Error due to the difference between apparent vertical and true vertical which is caused by various lateral or longitudinal accelerations acting on the instrument. This error is minimized by use of erection-cutout schemes.

An initial erection or mechanical-caging scheme sometimes is utilized to orient the gyro gimbals rapidly to a predetermined position relative to the base. This may be in the form of extremely high torquer erection rates, or a mechanical latching scheme independent of the erection torquers.

Typical performance parameters and characteristics of a vertical gyro are given in Table F-4.

TABLE F-4. *Typical Vertical-gyro Performance Parameters*

Size	4.0 in. diam × 4.25 in. long
Weight	3.5 lb
Spin motor	Synchronous or induction Angular momentum—2 × 10^6 g-cm^2/s Speed—24,000 rpm
Excitations	Spin motor—26 or 115 V, 400 Hz, 1, 2, or 3 phase Torque motors—26 or 115 V, 400 Hz, 1 or 2 phase Pickoffs (synchros or resolvers) —26 V, 400 Hz, 1 phase; (potentiometers)—2 W max ac or dc
Power requirements	Spin motor—11 W start; 8 W run Torque motors—13 W Pickoffs—2 W
Gimbal freedom	Outer axis—360° Inner axis—±85°
Performance	Free drift—0.5°/min Verticality—±0.15° Torquing rate—2–6°/min

Electrically Suspended Gyro:[1] Fundamentally, the electrically suspended gyro (ESG) is a free two-axis gyro. The device essentially consists of the inertial element, a metallic sphere, surrounded by an orthogonal array of electrodes embedded on the inner surface of a ceramic envelope. The rotor is suspended by servocontrolled electric-field forces that act between the rotor and the electrodes. The rotor is spun to operating speed by a set of spin coils external to the ceramic housing and then allowed to coast. Readout of the spin-axis angular position is photoelectric. A high vacuum is maintained in the rotor-electrode cavity to prevent voltage breakdown between electrodes and rotor and to minimize gas-drag effects on rotor speed. The basic elements used in the mechanization of an ESG are shown in Fig. F-22.

The rotor-electrode structure is electrically equivalent to a number of capacitor plates having one common plate, the rotor. The rotor is positioned and maintained equidistant from all electrodes with the attractive electric-field forces developed by servocontrolled potentials applied to each of the six electrodes (*E*1 through *E*6). Rotor-position error signals for the servocontroller are obtained from the capacitive differences between pairs of diametrically opposite electrodes. The acceleration capability of an ESC is limited by the minimum allowable rotor mass, for a given performance level, and the maximum allowable voltage gradient between electrodes and rotor without electrical breakdown. ESGs have been made capable of withstanding greater than 30 *g* of acceleration.

After initial levitation, the rotor is spun up to operating speed by a rotating magnetic field produced by orthogonal spin coils (*A* and *B*) somewhat similar to those of an induction motor. As shown, the rotor has a preferred spin axis by internal geometry. During spin-up, a rotor nutation usually occurs because the initial attitude of the rotor-preferred spin axis is random. This nutation is damped to a suitable level by a magnetic field produced by applying direct current to another coil. The preferred spin axis of the rotor is erected to approximate coincidence with the dc field established by the damping coil. This defines the gyro-spin-axis starting position. If some other rotor position is desired, voltages of

[1] Information on the electrically suspended gyro prepared by Robert C. Staats, Honeywell Inc.

the proper magnitude and phase are applied to the spin coils to torque the rotor.

When the rotor has attained operating speed, spin power is removed and the rotor is allowed to coast. The rotor is slowed primarily by residual gas molecules and magnetic fields. Since the rotor runs in a vacuum of approximately 10^{-8} torr and since the external case of the gyro provides a magnetic shield which attenuates earth's field and other fields to an acceptable level, the rotor rundown rate is very low and no further spin power is required.

Readout of the ESG rotor position is accomplished by optical pickoffs which view patterns on the rotor surface. The pickoff system preserves the integrity of the ESG by exerting no torques on the rotor. Any one of several systems can be used, dependent upon whether the gyro is intended for gimbaled-platform or strap-down-system application.

In the gimbaled application, only two pickoffs are required. One views a pattern on the end of the spin axis; the other views a time-reference pattern on the equator. The first pickoff generates an analog signal proportional to the displacement of the pickoff optic axis from the rotor spin axis. The second pickoff provides phase sense to the analog voltage. The ESG is mounted in a conventional gimbaled platform in which the gimbal servomotors drive the gimbals to a pick-off null. Accurate position information is obtained from the gimbal-angle readout device.

In the strap-down application, the spin axis is fixed in inertial space while the case is free to rotate with the vehicle. Three orthogonal sensors (*PO*1, *PO*2, and *PO*3) are used in the strap-down mode. Outputs from two of these pickoffs are used at any one time to determine the location of the spin axis. A great-circle pattern is applied to the rotor as shown in Fig. F-22. The pickoff optical system directs a light beam to the rotor surface. When the nonreflective patterns pass through the light spot, the light returning to the photosensor is modulated and an output pulse signal is generated. The time interval between the pulses is measured accurately and is proportional to the rotor position beneath the pickoff. Measurement of the time interval of line crossings under any two of the three pickoffs allows computation of the rotor spin-axis position with respect to the gyro case.

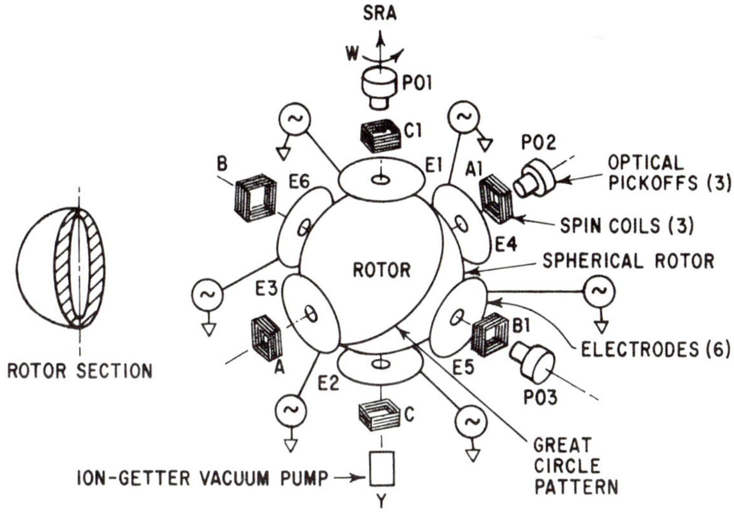

Fig. F-22. Electrically suspended gyro in strapped-down configuration. *(Honeywell.)*

To maintain vacuum levels required within the gyro, a combination ion gage–getter pump is used.

Because of the geometric and mechanical simplicity of the device, there are only four basic torque effects in the ESG: (1) rotor-mass unbalance torque; (2) magnetic-field torque; (3) gas-drag torque; and (4) electric torque. The major advantages of the ESG are: (1) drift sources are predictable; (2) reliability is high; (3) airborne and missile versions are capable of operation under severe dynamic environments; (4) two-axis all-attitude readout capability is obtained from one sensor; and (5) the system is flexible in application—in either a gimbaled or strap-down configuration.

Turn-and-Bank Indicator:[1] This instrument, sometimes called a turn-and-slip indicator, is the simplest form of instrument employing the gyroscope. The device contains a rate gyro mounted with its spin axis athwartship and its spring-restrained gimbal axis aligned fore and aft. Deflection of this axis from its neutral position is translated to motion of a pointer *A* (see Fig. F-23), calibrated in rate of turn. Below the pointer is a ball bank indicator *B* which acts as a damped pendulum showing the dynamic vertical component lying in the plane perpendicular to the fore-and-aft axis. If both pointer and ball are centered, the pilot knows that he is flying straight and that the aircraft's wings are horizontal. The turn-and-bank indicator also is used in making desired rates of turn with proper bank. Both air-driven and electric (ac or dc) gyros are in common use.

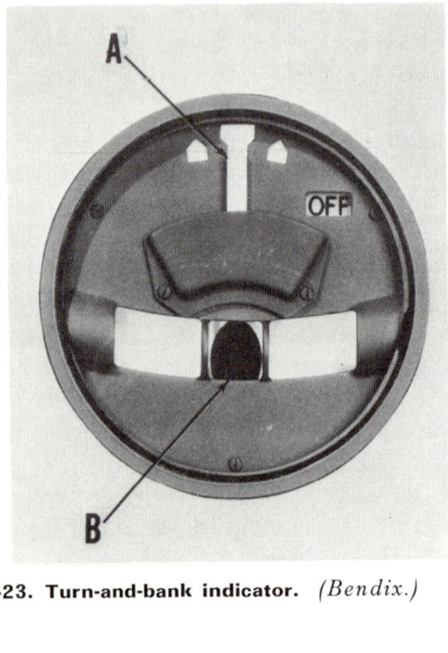

Fig. F-23. Turn-and-bank indicator. *(Bendix.)*

In a properly designed turn-and-bank indicator, the direction of rotation, moment of inertia, and speed of the gyro wheel are so matched to the centering spring that the gyro spin axis remains nearly horizontal as the plane makes a turn with proper bank at its normal cruising speed. Instead of a rotating gyro wheel, other sensing elements have been used, such as a vibrating reed or a stream of liquid. Instead of driving a pointer, the gyro sensor may mechanically perform some function in a computer or may drive a potentiometer, synchro, switch, or similar device giving an electrical function of rate of turn.

Automatic Flight-control Systems[2] These systems augment the basic aircraft characteristics and also provide pilot-relief modes. When signals from rate gyros and accelerometers are fed back to appropriate control surfaces to augment the basic aircraft characteristics, such a system is referred to as a *stability-augmentation system*. For aircraft operating over a wide range of speeds and altitudes, it is a normal requirement to change some of the control parameters to correspond with the changing aircraft characteristics. This may be done by using inputs from total and static-pressure sensors on an open-loop basis or by deriving a measure of aircraft sensitivity by "self-adaptive" techniques and adjusting control parameters accordingly. In addition, feedback signals normally are shaped by suitable filter networks. Such frequency shaping not only serves to provide the desired aircraft response characteristics but also deemphasizes undesirable aircraft structural modes that contribute to instability. Hydraulic actuators normally are used to provide the necessary power amplification to actuate the control surfaces. Control-surface motions caused by the stability-augmentation system do not manifest themselves at the pilot's controls. Force (or motion) transducers on the pilot's controls sometimes are included to provide command signals for the feedback loops. A system of this type is known as a *command-augmentation system*.

In view of the vital role played by such augmentation systems in providing the required vehicle-handling qualities, redundancy techniques are used to ensure undergraded performance even after a component failure. Monitoring also is provided to apprise the pilot of the status (levels of redundancy remaining) of the system. Such monitoring features also are useful for directing maintenance actions.

Automatic pilots traditionally have provided

[1] Information on turn-and-bank indicator prepared by Paul F. Bechberger, The Bendix Corporation.

[2] Information on automatic flight-control systems prepared by David L. Mellen, Honeywell Inc.

a pilot-relief function. Signals proportional to aircraft attitude, altitude, and heading are used to hold the aircraft at the engaged attitude or altitude and heading. In addition, the capability to preselect a desired altitude and heading often is provided. Commands may be accepted by the autopilot from external subsystems such as navigation computers and data-link systems.

Most transport autopilots provide some form of automatic instrument-landing capability. The autopilot receives a signal from a glide slope and a localizer beam for vertical and lateral control, respectively. Automatic "capture" of the beam is provided, and the aircraft is automatically controlled to the beam.

Earlier systems provided what is referred to as a category I capability, that is, the aircraft is controlled to an altitude of 200 ft, at which point the landing is completed manually. In view of this, category I operation is limited to a runway visual range (RVR) of 2,400 ft. More modern installations provide what is known as category II capability. In this case, the autopilot must be capable of providing safe automatic control to an altitude of 100 ft, and an RVR of 1,200 ft is specified. The category III capability will provide automatic control to touchdown and ultimately automatic ground rollout. Obviously, the lower the altitude for which automatic control is provided, the greater the need for failure protection. Category III systems require a "fail-operational" capability; that is, the system must be capable of sustaining any single failure without degradation of performance. Many portions of earlier-type

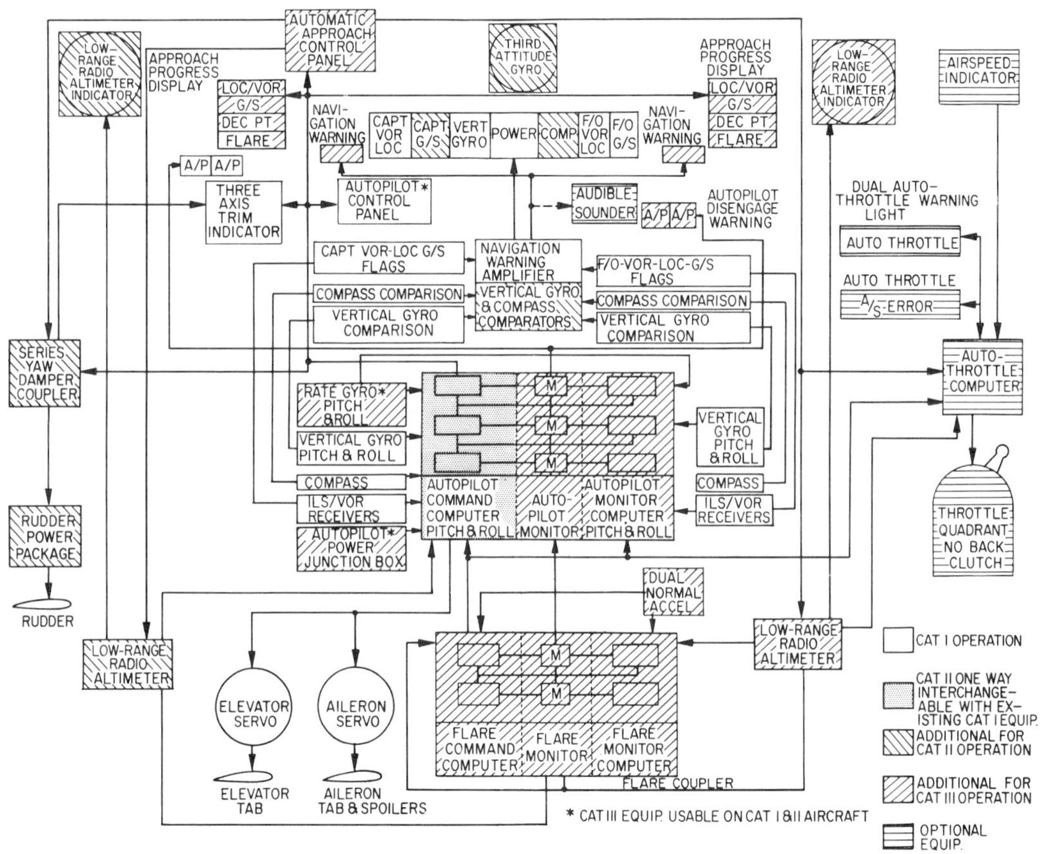

Fig. F-24. Automatic approach and landing system for low-weather minimums. LOC, localizer; VOR, visual omnirange; G/S, glide slope; A/P, auto pilot; DEC PT, decision point; CAPT, captain; F/O, flight officer; COMP, computer; ILS, instrument landing system; M, monitor comparitor; A/S, airspeed. *(Bendix)*

systems must be duplicated for redundancy and other controls added to provide this capability (see Fig. F-24). Also see **Air-traffic control.**

Operation of Automatic Pilot:[1] All automatic flight-control systems (the automatic pilot plus the stability-augmentation system) are composed of three distinct functions: (1) sensing, (2) computing, and (3) actuating. For example, a simple yaw damper, which is one axis of the stability-augmentation system, is shown in Fig. F-25. The yaw-rate gyro senses the yaw rate of the aircraft and transmits a signal, proportional to yaw rate, to the electronic computer. The computer determines the desired rudder positions to damp the yaw-rate motion. In damping, the motion is opposed with sufficient leading effect to slow the motion down, but not to oppose it directly. Sinusoidal displacements would be opposed with a sinusoidal force occurring 90° ahead of the displacement. The actuator accepts the electronic command from the computer and positions the rudder. In most modern aircraft, hydraulic actuators are used because of the large forces and rapid motion required.

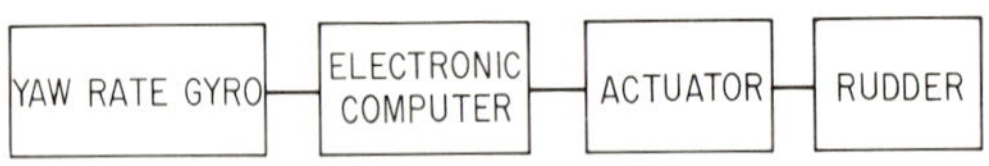

Fig. F-25. Block diagram of yaw-rate gyro system.

[1] Information on operation of the automatic pilot prepared by David L. Mellen, Honeywell Inc.

The sensors in a typical automatic flight-control system will include rate gyros, accelerometers, attitude and heading gyros, barometric-altitude and Mach-number sensors, and guidance-beam displacement sensors—and develop electronic commands to the actuators that drive the rudder, ailerons, elevator, and the throttles. In some of the more extensive systems, the piloting task is reduced to control of the takeoff and the final flare in landing, and monitoring of the automatic system.

A typical autopilot pitch axis is shown in Fig. F-26. Alternatives replace the vertical gyro with a rate gyro or a normal accelerometer. The two alternative systems control the rate of change of flight path and allow control in situations where the aircraft has maneuvered beyond the limits of control of the vertical gyro. These arrangements are typical for fighter aircraft or aerobatics, whereas a vertical gyro is adequate for passenger service and bombers.

A typical autopilot roll axis works in the same way except that (1) the other pickoff on the vertical gyro is used, (2) a rate gyro is oriented to be sensitive to the roll axis, (3) the actuator is the aileron drive servo, and (4) the other axis of the hand controller is used for manual inputs.

An aircraft banks to turn. Since lift forces generated by the wing are perpendicular to the wing planform, wing lift causes the aircraft to turn. In order to prevent loss of altitude during the turn, the vertical component of lift must be maintained. This is accomplished by inserting a command calling for up-elevator which, in turn,

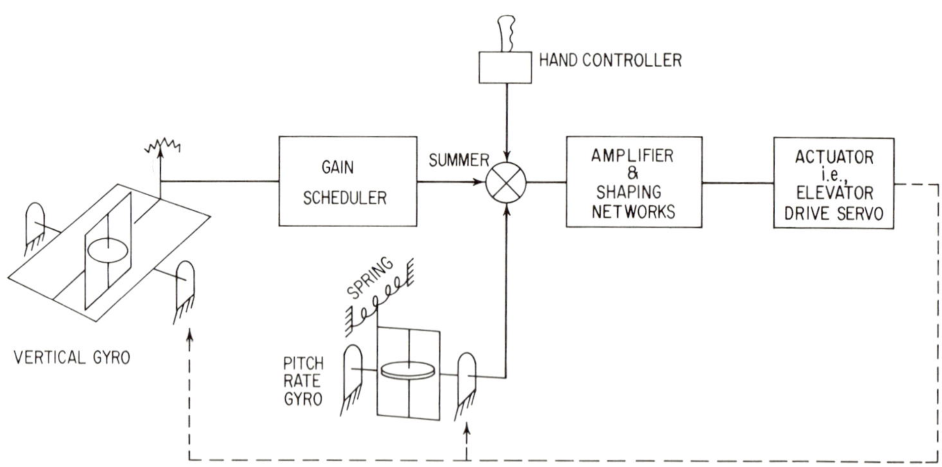

Fig. F-26. Typical autopilot pitch-axis control system. *(Honeywell.)*

results in increased wing angle of attack. The force diagram of Fig. F-27 shows the relationship for a representation of roll angle. The turning rate in a horizontal plane is

$$\omega = \frac{a}{v} \text{ rad/s} \tag{9}$$

$$\omega = g \frac{\tan B_A}{v} \tag{10}$$

$$\omega = \frac{32.2}{v} \tan B_A \tag{11}$$

where ω = turning rate
v = velocity
a = acceleration
B_A = bank angle
g = acceleration due to gravity

The basic function of the yaw axis is to retain zero lateral acceleration in order to (1) make the above turn equations be true, (2) keep occupants from being thrown against the side of the aircraft, and (3) keep the aerodynamic side loads on the aircraft small.

The yaw axis can be mechanized with a lateral accelerometer or sideslip sensor as the basic sensor, but more often a rate gyro is used. During turns, the yaw axis requires turning-rate commands from the vertical gyro—similar to the pitch-axis requirement for up-elevator commands.

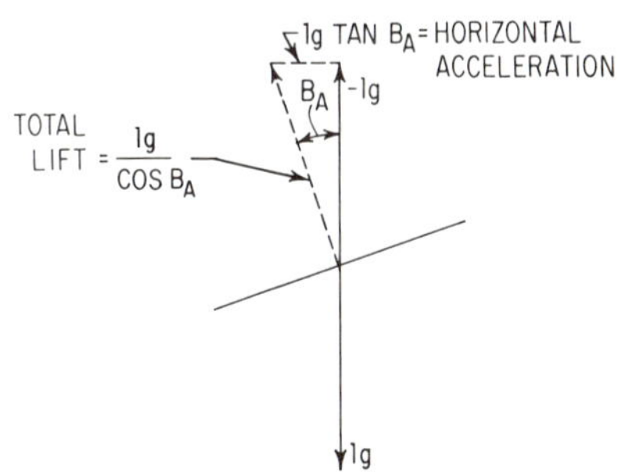

Fig. F-27. Force diagram showing relationships governing roll angle.

In the foregoing situations, the signals are combined so as to maintain a stable, self-restoring system. This is accomplished through servomechanism practice wherein (1) gains are adjusted for stability; (2) rate signals or networks are used for stabilization; and (3) nullifying or augmenting signals are used to alter the apparent characteristics of the aircraft. Typically, the autopilot also will provide trim settings and methods of turning the automatic functions on and off gradually so as to avoid sudden changes. Also see **Aircraft, instrumentation of; Air-traffic control;** and **Navigation systems.**

Flip-flop The flip-flop is a bistable device whose output assumes one of two stable states depending on the state of the most recently applied input signals. The flip-flop is capable of storing information and is used to provide gating. See **Gate;** and **Register, computer.**

Figure F-28 illustrates a direct-coupled trigger or latch constructed of AND and OR logical elements. In this configuration, the output is fed back to the input. A set pulse initially turns on the output, and the output then provides its own input even though the set pulse is removed. The circuit remains "latched" in this condition until a reset pulse breaks the feedback loop.

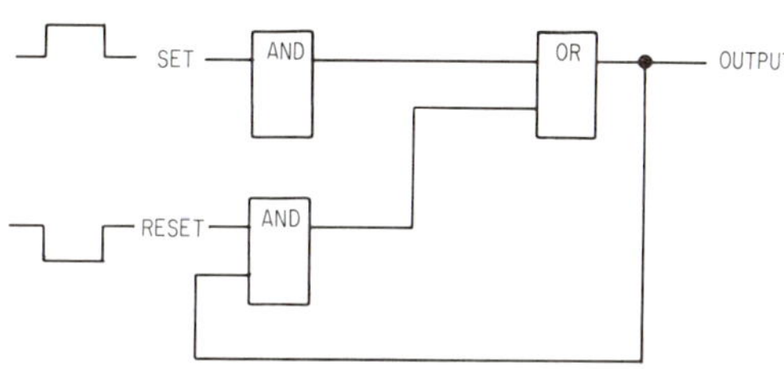

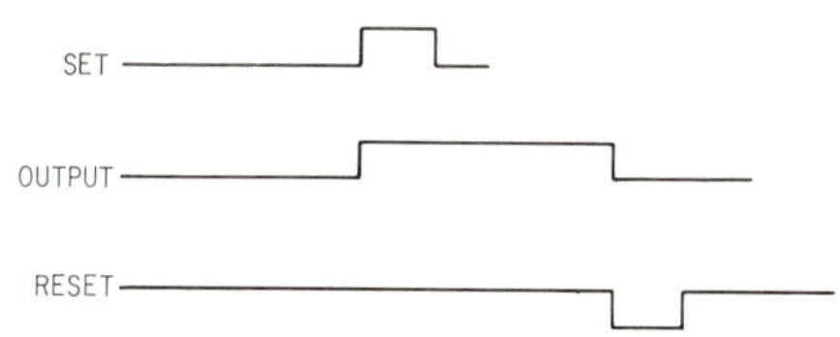

Fig. F-28. Direct-coupled trigger or latch.

The schematic for a transistor trigger is shown in Fig. F-29. When the trigger is in the "off" condition, transistor T_4 is "off" and the "off" output F is at $+V$, representing a binary 1 state. Transistor T_1 is "on" and "on" output E is at 0 V, representing a binary 0 state. Transistor T_3 is in a state of low conduction, and its emitter voltage is not sufficient to forward-bias T_4; however, T_2 is in a state of heavy conduction and its emitter voltage forward-biases T_1 and holds it "on."

To change the output state of the trigger, a conditioning gate of 0 V is applied to input C while the signal on input D is still positive. When the set pulse at input D goes to 0 V, a negative shift is seen at the emitter of T_3 and T_3 immediately goes into heavy conduction. The reduced collector voltage causes T_2 to go into a state of low conduction. The voltage at the emitter of T_2 is reduced and biases T_1 "off." When the ac transient has receded, the T_3 emitter voltage holds T_4 "on." The trigger can be turned "off" by applying gate and pulse inputs to points A and B, respectively.

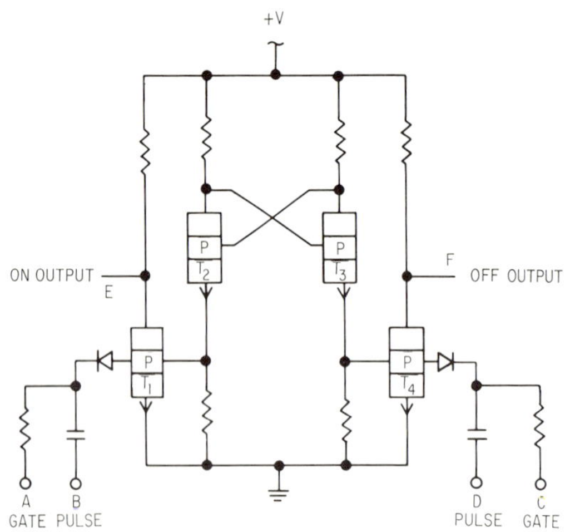

Fig. F-29. Gated direct-coupled trigger.

The trigger may be connected for binary operation by connecting A to F and C to E, and applying the input pluse to both B and D. Since the gate must be applied before the pulse arrives, the trigger alternately changes state whenever a pulse is applied.

The trigger may be used to build a shift register by connecting the "off" output of the previous stage to gate C, connecting the "on" output of the previous stage to gate A, and applying a pulse to inputs B and D.

—W. W. Welz, *IBM Corporation*

Float-level Gage (See **Liquid level, measurement.**)

Floating Control A control system in which the rate of change of the manipulated variable is a continuous (or at least a piecewise continuous) function of the actuating signal.[1] The manipulated variable can remain at any value in its operating range when the actuating signal is zero and constant. Hence, the manipulated variable is said to "float." The final controlling element moves relatively slowly toward either one of its two extreme positions, depending on whether the control variable is above or below the set point. The three common types of floating controls are described as follows.

Single-speed Floating Control: The final controlling element moves at a constant rate regardless of the amount of deviation of the measured variable from set point. The single-contact on-off controller can produce single-speed floating control if used with a slow-running, reversible motor-driven operator. If it is a temperature controller, the final control element will move toward its closed position at a constant speed when the temperature is at or above the set point. When the temperature is below the set point, the final controlling element will move toward the open position at the same single speed.

Single-speed Floating Control with Neutral Zone: When the value of the control variable is in the neutral zone, no contact is made and the final controlling element remains motionless. This type of control action can produce almost an exact correction for any load condition.

Multispeed Floating Control: The final controlling element is moved at two or more rates, each rate corresponding to a definite range of values of deviation. A final controlling element moves at several speeds depending upon the amount of deviation.

—Richard H. Kennedy, *Bristol Division, American Chain & Cable Company, Inc.*

[1] Extracted from American National Standard Terminology for Automatic Control, ANS CE85.1–1963, with permission of the publisher, The American Society of Mechanical Engineers, New York.

Floating Point A form of numeric notation that allows automatic location of the radix point during arithmetic operations. Floating-point notation is generally used by digital computers when performing engineering- and scientific-application-type arithmetic operations, since it does not require that the relative magnitude of operands be precisely known (compare **Fixed point**).

Floating-point Format: A floating-point number consists of two parts which are multiplied together. The first part is the signed mantissa, which denotes the value of the floating-point number. The second part consists of the radix raised to an integer exponent, which denotes the magnitude of the number. Following is an example of a decimal floating-point number with its parts labeled.

$$-12.345 = -0.12345 \times 10^2$$

radix point — signed mantissa — radix raised to an integer exponent

Other examples of floating-point numbers of radix 2 and 10 are

$$(1.01)_2 = +0.101 \times 2^1$$
$$(0.00546)_{10} = +0.546 \times 10^{-2}$$

Floating-point Arithmetic: Digital computers perform floating-point arithmetic in two ways. The first way is via software simulation (see **Software**). This method provides the programmer with a series of subroutines that may be called when he desires to perform some floating-point operation such as the addition of two floating-point numbers. The second method is via a floating-point-hardware arithmetic unit. When floating-point hardware is a feature of the digital computer, the programmer is able to issue floating-point instructions which activate the hardware to perform the floating-point operation. While floating-point arithmetic units perform floating-point operations much faster than the software-simulation approach, the cost of floating-point hardware can be significant. Whether the hardware arithmetic unit or the software simulation approach is better for a given control installation depends entirely on the amount of floating-point computation required for that installation.

An example of the floating-point addition of two numbers is

$$\begin{aligned} 2.41 \times 10^2 + 1.11 \times 10^1 &= 24.1 \times 10^1 + 1.11 \times 10^1 \\ &= (24.1 + 1.11) \times 10^1 \\ &= 25.21 \times 10^1 \\ &= 2.521 \times 10^2 \end{aligned}$$

An example of a floating-point multiplication operation is

$$\begin{aligned} (2.41 \times 10^2) \times (1.11 \times 10^1) &= (2.41 \times 1.11) \times (10^2 \times 10^1) \\ &= (2.6751) \times (10^3) \\ &= 2.6751 \times 10^3 \end{aligned}$$

—B. W. Landeck, *IBM Corporation*

Flow Chart, Programming A method of graphically representing the procedures required to solve a problem on a digital computer. Flow charts consist of function blocks with connectors between those blocks. A given function box may represent an input/output operation, a numerical computation, or a logic decision. Figure F-30 is a flow chart of a program that reads values from the process, converts those values to engineering units, limit-checks the converted values, and if there is a violation, prints an alarm message on the process operator's typewriter.

Flow-chart Usage: There are two primary reasons for flow-charting a digital-computer program. The first reason is for ease of initial program design. By initially flow-charting the procedures involved in the program and then coding from the resultant flow chart, instead of coding without any preliminary design, the programmer saves a significant amount of effort in developing the program.

The second primary reason for flow charts is for program documentation. Once a program has reached an operational status, it is necessary that it be properly documented. The reason for

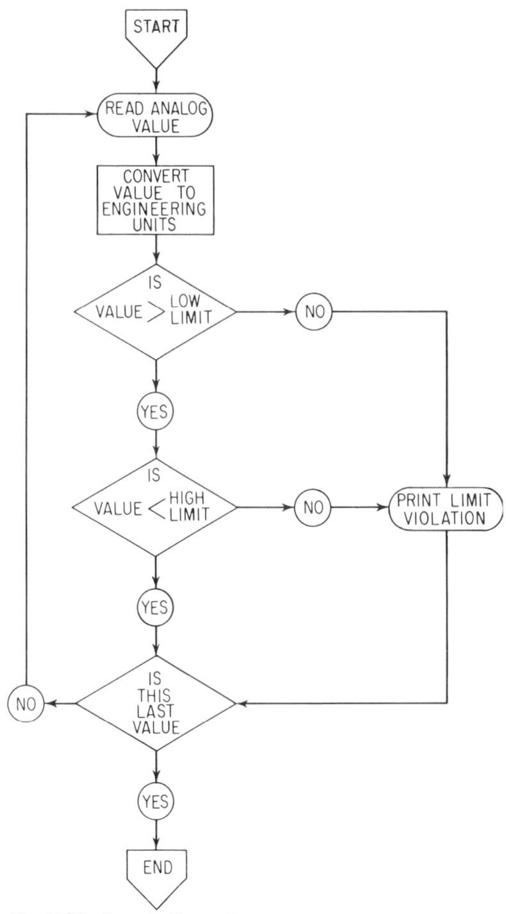

Fig. F-30. Sample flow chart.

this is that in the future it may be necessary to correct a bug in the program or to make some modification to the functional nature of the program. Since the original programmer responsible for the program may not be available at the later date when it is necessary to make the correction or modification, it is necessary that that program be documented in such a way that someone else may readily understand the nature of the program. An effective way of transmitting the understanding of the program is via flow charts of the program.

Flow-chart Levels: Flow charts are produced at various levels of detail depending upon their purpose. A functional block in a low-level flow chart may represent only a few computer instructions, whereas a functional block in a high-level flow chart may represent many computer instructions. High-level flow charts are used primarily for initial program design and as a means of informing a nonprogrammer of what the program does. Low-level flow charts are usually the last stage of flow charting before a program is actually coded and are used as documentation for programmers who in the future may modify or correct a given program.

—B. W. Landeck, *IBM Corporation*

Flow Measurement The most common method of moving a fluid from one point to another is to force the fluid through a closed conduit of known shape and dimension. Conduit (pipe) of circular cross section and constant internal diameter is used most frequently because of its mechanical strength and large area per unit of wall surface. Fluids of all types (liquids, vapors, and gases) are transported in this manner. Advantage is taken of several physical phenomena in measuring directly or inferentially the quantity of material flowing through a pipe. Parameters of the major methods are tabulated in Table F-5. The relative advantages and limitations of these methods are given in Table F-6. Major flowmeters and primary devices also are described elsewhere in this encyclopedia. See **Dall tube; Flowmeter, electromagnetic; Flowmeter, mass, Flowmeter, positive-displacement; Flowmeter, turbine-type; Flowmeter, variable-area; Flowmeter, variable-head; Flowmetering devices, primary; Flow nozzle; Gentile tube (Foster); Metering pump; Orifice plate; Piezometer ring; Pitot tube; Rotameter;** and **Venturi tube.**

Another less frequently used but familiar method of transporting liquids is by means of a suitable hydraulic gradient through an open channel. The cross-section profile of the channel may be irregular over most of its course except in the area where flow measurement is accomplished. In this area, a constriction of known geometric shape and dimension must be used to produce the measurement. Also see Tables F-5 and F-6 and **Flume;** and **Weir.**

TABLE F-5. *Measurement Characteristics of Flowmeters**

Type	Pipe Size, in.	Volume Range	Accuracy	Remarks
Volumetric meters:				
Oscillating-piston positive-displacement	½–2	3–120 gpm	±0.20%	Repeatability =0.015%
Lobed-rotor	1½–24	8 gpm to 25,000 bbl/hr	±0.20%	Repeatability =0.015%
Nutating-disk	½–2	2–160 gpm	±1.00%	Repeatability =0.100%
Metering pump (proportioning)	Off-line	Adjustable 4–100% of range	±1.00%	Response =20 s full scale
Variable-area meters:				
Plugged and ported cylinder	1–4	....	±0.25%	Mainly for highly viscous fluids
Tapered-tube rotameter with free bob	½–4	....	±0.5–2.0%	Generally superseded by guided float-type
Tapered-tube rotameter with center guide	½–4	....	±0.5–2.0%	Inductance bridge often used for transmission
Tapered-tube rotameter with beaded tube	Up to 4	....	±0.5–2%	Enables viscosity-immune floats to be used without center guide

TABLE F-5. *Measurement Characteristics of Flowmeters* (Continued)*

Type	Pipe Size, in.	Volume Range	Accuracy	Remarks
Tapered-tube rotameter, metallic	....	....	2% of max bypass flow rate	Measures bypass flow, hence can be used on any pipe size containing orifice or other differential-pressure-producing primary flow device
Velocity and current meters:				
Vortex meters, liquid flow	1–10	6–60 gpm to 700–6,800 gpm	±0.5%	Repeatability = 0.1% Response: 1- and 2-in. meters = 300 ms 3- and 4-in. meters = 400 ms 6-, 8-, and 10-in. meters = 500 ms
Vortex meters, gas........	1–10	750 std ft^3/hr at 20 psig to 150,000 std ft^3/hr at 1,500 psig to 1.2 million std ft^3/hr at 20 psig to 25 million std ft^3/hr at 1,500 psig	±1.0%	Response: 1- and 2-in. meters = 300 ms 3- and 4-in. meters = 400 ms 6-, 8-, and 10-in. meters = 500 ms
Vortex mass flowmeter, gas	1–10	5–25,000 lb/min	±0.1% over 5:1 range ±0.3% over 25:1 range	Response = 1 s
Turbine meter with magnetic signal pickup...........	1–10	6–7,200 gpm	±0.25%	Repeatability = ±0.02%
Turbine meter (viscosity-compensated)...........	6–16	600–15,000 bbl/hr	±0.15%	Repeatability = ±0.015%
Electromagnetic meter.....		0.1–250 million gal/day	±1.0%	Responds almost instantaneously to changes in flow rate
Current meter...........	2–36	....	±2%	Subtracts reverse flow

Mechanical indicators and recorders for variable-head meters†

Element type	Typical range	Min span	Accuracy	Remarks
Orifice plate	0–10 in. water to 0–400 in. water	5–10 in. water	±0.25 to ±0.5% of full scale for recorders ±0.5 to ±2.0% of full scale for indicators	Usually include adjustable damping Response to a step upset is from a few seconds to minutes
Bellows type..............	0–10 in. water to 0–50 psi	10 in. water	±0.25 to ±0.5% of full scale for recorders ±0.5 to ±2.0% of full scale for indicators	

TABLE F-5. *Measurement Characteristics of Flowmeters (Continued)*

Mechanical indicators and recorders for variable-head meters†				
Element type	Typical range	Min span	Accuracy	Remarks
Float and cable.	0–5 in. water to 0–36 in. water	5 in. water	±1.0% of full scale	Relatively slow response because of mechanical nature of system

Primary devices for variable-head meters

Flow in conduits		Flow in open channels‡		
Type	Size, in.	Type	Channel width, ft	Range of head, in.
Orifice plate.	½–72			
Flow nozzle	3–24			
Flow tube	3–48	Weir	4–17	2–18
Venturi tube.	3–48	Flume	¼–50	4–80
Pitot tube.	3–48			

Remarks: For information pertaining to constancy of coefficients, accuracy, and Reynolds-number range, see ASME, "Fluid Meters Handbook," 5th ed., p. 188, Table 17.

*From Douglas M. Considine, Process Instrumentation, *Chemical Engineering,* Feb. 12, 1968. Used with permission of McGraw-Hill, Inc.

†Motion-balance transmitters used with the same primary devices have approximately the same characteristics as those listed above.

‡Accuracy varies from 2 to 3 percent.

TABLE F-6. *Flow-measurement Systems—Advantages and Limitations**

Advantages	Limitations
Nutating-disk meter	
Relatively low cost	Fair accuracy
Several materials of construction available	Moderate pipe sizes and capacities
Applicable to automatic liquid-batching systems	
Moderate pressure loss	
Oscillating-piston positive-displacement meter	
Good accuracy, especially for low flow rates	Limited to sizes 2 in. or less
Easy to install, maintain	Limited power for driving mechanical accessories
Easily applied to automatic liquid-batching systems	
Good repeatability	
Moderate cost	
Metering (proportioning) pump	
Combines pump and high-accuracy metering	No proof of metered flow
Capable of high-pressure operation	Introduces pulsations into system
Provides zero-leakage measuring device	
Controllable from any signal (electric or pneumatic)	
Orifice plate	
Low cost	Relatively high permanent pressure loss
Available in numerous materials	Tend to clog; not useful for slurries, or entrained particles
Can be used in wide range of pipe sizes (½ to 72 in.)	Flow rangeability about 3:1 for a given plate
Characteristics are well known and predictable	Square-root characteristics
Years of applicational experience	

TABLE F-6. *Flow-measurement Systems—Advantages and Limitations (Continued)*

Advantages	Limitations
Orifice plate	
Can be used with differential-pressure devices	Straightening vanes upstream often required Connecting piping problems (freezing, condensation) unless orifice is integral with transmitter Characteristics change because of erosion, corrosion, and sealing Accuracy dependent upon care during installation
Flow nozzle	
Permanent pressure loss lower than orifice plate Available in numerous materials Can be used with differential-pressure devices Good for fluids containing solids that settle Widely accepted for high-pressure/temperature steam flow	Higher cost than orifice plate Limited to moderate pipe sizes
Flow tube	
Negligible permanent pressure loss Shorter and easier to install than venturi tube Flow turbulence has little effect on measurement accuracy	Higher cost than orifice plate or flow nozzle Limited to moderate pipe sizes Solids that settle interfere with operation
Dall tube	
Low head loss Characteristics well investigated Short laying length Available in numerous materials of construction No upper line-size limit	For given flow and differential, has lower capacity potential than venturi types
Venturi tube	
Very small permanent pressure loss Handles suspended solids Available in very large pipe sizes—widely used for large flows Available in numerous materials of construction Characteristics well established Years of applicational experience Best accuracy tolerance of all venturi-type head meters	High cost Generally not useful below 3 in. pipe diam
Weir	
Low cost, considering volume measured Usually constructed on location Several characteristics (notch designs) possible Not easily damaged	Applicable only to open-channel measurements Requires field calibration Accuracy not over 2 to 3%
Flume	
Handles greater flows than weirs Can be constructed on location Free flow at higher velocities, permitting use in flows having entrained solids without periodic cleaning All sections are plane surfaces, making for easy construction Dimensions not easily altered to cause incorrect measurements Relative absence of harmful effects to measuring data from physical changes downstream	Cost greater than weir Requires calibration on location Accuracy not over 2 to 3%

TABLE F-6. *Flow-measurement Systems—Advantages and Limitations (Continued)*

Advantages	Limitations
Rotameter	
Easily equipped with magnetic, electronic, induction or mercury-switch alarms, or transmitting devices Viscosity-immune bobs available Several rotameters mounted side by side provide convenient, visible flow comparisons Relatively low cost Handles wide variety of corrosives Can be installed without regard to fittings or straight pipe preceding or following meter Good for small flows	Must be mounted vertically Limited to small pipe sizes and capacities unless put in a bypass Relatively low temperature limitations
Variable-area meter (plugged and ported cylinder)	
Good for liquids of high viscosity Good accuracy	Limited size range (1–4 in.) Relatively high cost
Lobed-rotor meter	
Can perform at relatively high temperatures and pressures Low permanent pressure loss Numerous materials of construction available Applicable to gases and wide range of light to viscous fluids, including asphalts Good accuracy and repeatability, especially at high rates	Susceptible to damage from entrained vapors Tends to be bulky and heavy in larger sizes Relatively high cost Slip increases at low flow rates; best used at high rates Not available above 24 in.
Vortex meter	
Handles wide variety of chemicals, including slurries, liquids with entrained particles, and viscous materials up to 500 centistokes No tight clearances between parts Size, weight, and capacity advantage over positive-displacement meters Good response speed Available (with densitometer) for gas mass-flow measurement Very low pressure drop	Relatively high cost Not available above 10 in. Upper temperature limit is 400°F
Turbine meter	
Viscosity compensation available Weight advantage over positive-displacement meter Easy to install, maintain Good accuracy, excellent repeatability Good temperature and pressure ratings Fairly low pressure drop	Relatively high cost Limited range of viscous materials can be handled Not available above 16 in.
Current meter	
Low pressure drop Subtracts reverse flow Can withstand high temporary overloads Installed as unit	Limited to liquids Relatively high cost
Electromagnetic meter	
No obstruction in pipe Available in several materials of construction Large pipe sizes and capacities Can handle slurries and greasy materials	Relatively high cost Fluids measured must be at least slightly conductive

*From Douglas M. Considine, Process Instrumentation, *Chemical Engineering,* Feb. 12, 1968. Used with permission of McGraw-Hill, Inc.

Flowability (See Consistency, measurement of.)

Flowmeter, Electromagnetic Operation of an electromagnetic flowmeter is based upon Faraday's law of electromagnetic induction, where the voltage E induced in a conductor of dimension D moving through a magnetic field H is proportional to the velocity V of the conductor. The voltage is generated in a direction mutually perpendicular to both the velocity plane of the conductor and the plane of the magnetic field. The mathematical expression is $E = CHDV$, where C is a dimensional constant.

In a given size of flowmeter (transmitter), volume flow rate Q equals transmitter cross-sectional area A times fluid velocity V, or $Q = AV$. The original expression may be rearranged to express velocity as $V = E/CHD$. By substitution, $Q = AE/CHD$. For a given size of transmitter, A, C, H, and D become constants. This permits the expression of $Q = KE$, or the generated voltage E is directly proportional and linear with volumetric flow rate Q.

The transmitter consists of a nonmagnetic (metal or fiber glass) tube through which the liquid passes, an insulating liner (metal tube only) on the inside of the tube, an electromagnet to induce the field, and two metallic electrodes that contact the flowing liquid (see Fig. F-31).

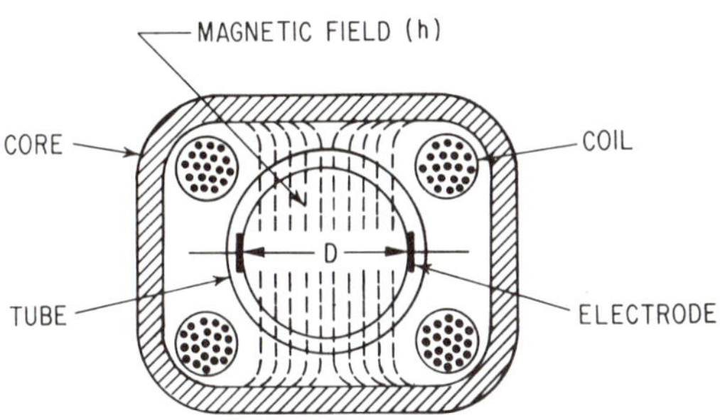

Fig. F-31. Cross section of electromagnetic-flowmeter transmitter.

A uniform magnetic field is induced through a portion of the transmitter tube. The flowing liquid is the conductor. Consider the liquid in cross section to the flow axis as a series of flat disks passing through the tube. As each disk passes through the magnetic field, a voltage is generated. The pair of electrodes mounted in the tube wall detects this voltage. The liquid is the conductor, and the length D is the distance between the electrodes and equals the transmitter-tube diameter. Continuous flow represents a steady progression of disks through the magnetic field, thus yielding a continuous voltage signal.

The specific conductance of the liquid must be above a minimum value, dependent upon the type of electronic-signal converter used, transmitter size, signal-lead length (from transmitter to amplifier), and the accuracy required. Transmitter output is linear with volumetric flow and is unaffected by turbulence and flow profile—hence is unaffected by piping configuration and fluid viscosity. The flow tube becomes a part of the process piping with a variety of possible end connections, such as ANSI flanges, sanitary-type fittings, and plain ends for slip-on coupling. Transmitter liners are available in a number of corrosion- and abrasion-resistant materials.

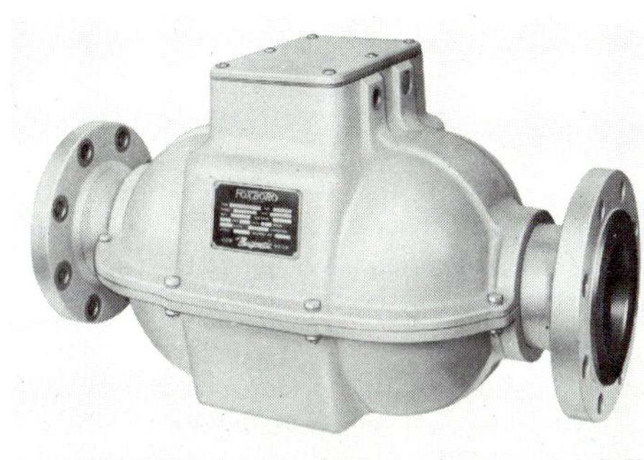

Fig. F-32. Electromagnetic-flowmeter transmitter. *(Foxboro.)*

The unobstructed flow passageway makes the electromagnetic flowmeter especially suited for measurement of solids-contaminated liquids and heavy slurries. Full-scale flow rates range from as low as 0.07 to many thousand gpm. Shown in Fig. F-32 is a 4-in. transmitter with 410 gpm maximum full-scale.

—John S. Boynton, *The Foxboro Company*

Flowmeter, Heat-addition Type (See Anemometer.)

Flowmeter, Mass The output of a true mass flowmeter is in *mass units* and consequently requires no corrections for pressure, specific gravity, temperature, and other variables that affect the readings of volumetric meters. This direct approach toward obtaining flow information in the form that is ultimately required by the user, namely, pounds of material (for production, accounting, and selling and procurement purposes) is the prime advantage of this type of flowmeter. The natural-gas industry in the United States still prefers to measure gas in standard cubic feet, but the new ethylene industry, because the compressibility factor for ethylene is not clearly defined in most tables, has set the pound as the unit of measurement for ethylene. Other materials now being metered in mass units by many users include helium, oxygen, carbon dioxide, nitrogen and air—and liquids, such as gasoline, propane, butane, methane, kerosene, light oils, and ethylene oxide.

An angular-momentum type of mass flowmeter is shown in Fig. F-33. An angular momentum is imparted to the stream of gas or liquid being metered. This is accomplished by using an impeller turning at a constant speed. The angular momentum of the stream of fluid is removed by a turbine, located just downstream of the impeller. This turbine is restrained and cannot rotate. The reaction torque produced by the turbine as it removes the angular momentum of the fluid becomes the meter output. The torque

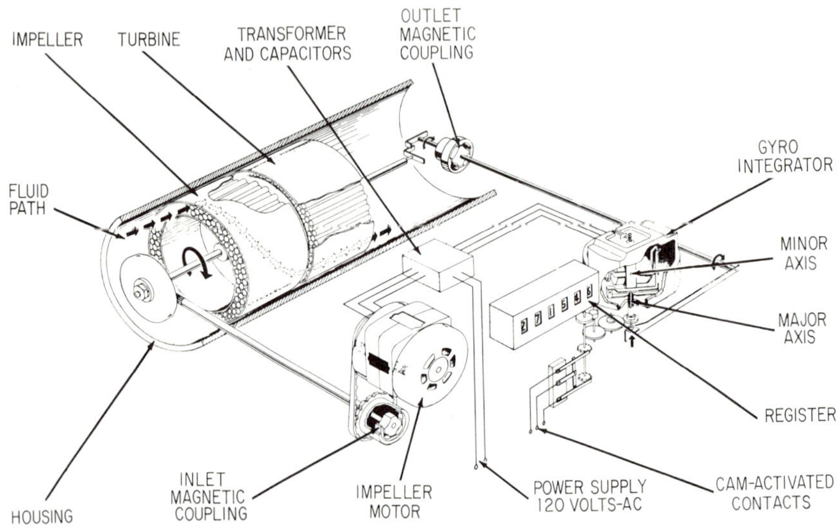

Fig. F-33. Functional diagram of gyro-integrating mass flowmeter. *(G-E; Black Sivalls & Bryson.)*

TABLE F-7. *Characteristics of Angular-momentum-type Mass Flowmeter*

Specification	For Gases	For Liquids
Flow-rate range	250–53,3000 lb/hr	2,000–320,000 lb/hr
Density limits	0.10–30.0 lb/ft^3	30.0–90.0 lb/ft^3
Accurate operating range	10–133% of nominal flow rating	10–133% of nominal flow rating
Accuracy, ±1% of point	20–133% of nominal flow rating	20–133% of nominal flow rating
Operating-pressure rating	Up to 1,500 psig	Up to 1,500 psig
Ambient-temperature range	−20 to +125°F	−20 to +125°F
Ambient-temperature influence	−0.2%/100°	−0.2%/100°F
Fluid-temperature range	+0 to 165°F	0 to 165°F
Fluid-temperature influence	+0.3%/100°F	+0.3%/100°F
Power requirements	120 V ±10% 60 Hz ±3 Hz 55 W	
Explosion-proof features	Meets class 1, group C and D Underwriters Code	

is directly proportional to the mass flow rate. The flow rate is totalized by applying the torque produced to the minor axis of a gyroscope. The gyroscope rotates about its major axis at a rate proportional to the applied force. This rotation is geared to a counter which can be calibrated to the cumulative total pounds flowing through the meter. Meter characteristics are summarized in Table F-7. Mass flowmeters produce their best accuracy in relatively clean streams. Strainers and filters are required on most applications.

—L. F. Holder, *Black Sivalls & Bryson, Inc.*

Flowmeter, Positive-displacement Positive displacement flowmeters represent a broad class of measurement devices in which the quantity of material (gas or liquid) flowing through a pipe is determined (1) by temporarily separating the flowing material into finite volumetric increments and (2) by counting and totalizing these volumetric increments. If the number of counts is related to time, the meter also can be used as a *rate-of-flow* as well as *total-flow* instrument. The term *positive* stems from the fact that direct volumetric measurement is involved—much like determining the amount of liquid in a tank by dipping the liquid out gallon by gallon and keeping track of how many gallon containers were filled during the dipping process. This positive action is contrasted with observation of a secondary phenomenon, such as inferring rate of flow from measuring the differential pressure created across a restriction (such as an orifice plate) in a pipeline. The term *displacement* derives from the fact that the fluid flowing through the metering device actually displaces (replaces) the preceding volumetric increment of fluid that passed through the meter—in such fashion that no quantum of fluid can be counted twice.

This class of flowmeters, in terms of quantities of devices used, by far represents the most common form of flow measurement. Examples include the common household water and gas meters, which also find wide application in industry, particularly for cost accounting, billing, and inventorying purposes. Pressure in the pipeline supplies the energy for operation of these meters. Hence, with the exception of additional energy that may be required for operation of peripheral equipment, such as transmitters and recorders, the positive-displacement meter requires no external source of power.

Classification of Positive-displacement Flowmeters: Generally, these meters are identified by the means used to partition the flowing fluid into volumetric increments. Major types include:

Bellows Gas Meter (Fig. F-34): This type contains four measuring compartments which operate simultaneously. Some of the compartments are filling while others are emptying but all are always conforming to set conditions, thereby ensuring a uniform delivery of gas. The mechanism is constructed so that the number of times each measuring chamber is filled and emptied is registered, thus indicating the total volume in cubic feet on the index. The register or meter is operated from a crank that is rotated by movement of the diaphragms. The mechanism used generally is referred to as a standard Glover-type mechanism. Synthetic-rubber diaphragms are used to ensure that displacement is directly proportional to stroke. Two valves of the D-slide type are actuated by a central crank through a suitable linkage. Motion of the meter mechanism cannot occur unless a pressure differential (hence gas flow) takes place. Normally a pressure differential of 0.10 in. of water will commence actuation of the meter. Accuracy requirements

Fig. F-34. Bellows-type gas meter. *(American Meter.)*

for these meters are quite rigid over a wide range —from full capacity to practically zero. Maximum capacities range from 150 to 17,000 ft³/hr.

Nutating-piston Meter (Fig. F-35): This type of meter commonly is used for measuring water but can be used in many other applications and, when equipped with suitable peripheral equipment, can be used effectively in liquid-blending systems. The liquid enters the meter through the inlet and passes upward through a strainer into the top of the main casing. The liquid then moves downward through the measuring chamber into the base of the meter and discharges through the meter outlet. When the liquid enters the measuring chamber, the liquid drives the single measuring piston, which nutates. The roller or positive-displacement cam, against which the piston bears, controls piston movement and makes a complete nutation at each movement. The position of the piston divides the chamber into two compartments which are successively filled and emptied, each compartment holding a definite volume. The movements of the piston are transmitted by a train of gears or magnetic couplings to the meter register. Meters of this type are relatively low-cost and are limited to moderate pipe sizes and capacities—from about 1 to 1,000 gpm. Several materials of construction are available, and the pressure loss is moderate.

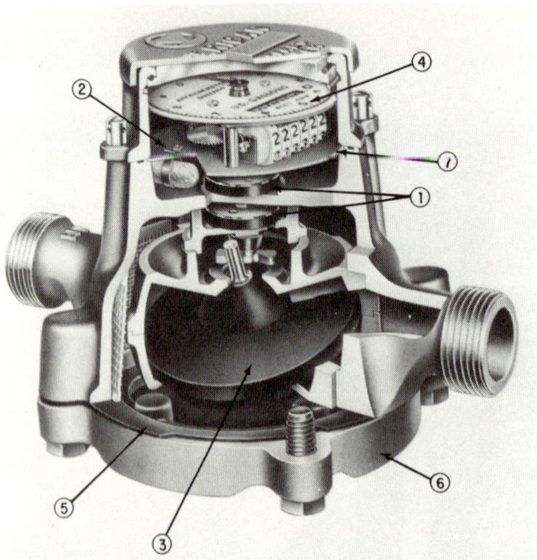

Fig. F-35. Positive-displacement water meter of the nutating-piston type. (1) Magnetic drive; (2) hermetically sealed register; (3) web; (4) dial—registers can be reset to zero; (5) liner to prevent corrosion from galvanic action; (6) frost bottom to prevent internal damage caused by freezing; (7) change gears. *(American Meter.)*

Oscillating-piston Meter (Fig. F-36): This meter is similar to the nutating-piston or -disk type except that mechanical motion takes place in one plane only. The measuring sequence starts with Fig. F-36*d*, where the piston is in the neutral position. Pressure of the inlet fluid forces the piston to the left and causes it to "roll" downward as shown in Fig. F-36*a*. At this point, most of the piston surface, both inside and out, is under the influence of the incoming fluid and is driven vertically downward (Fig. F-36*b*) . In this position, the piston outside surface has full-line input pressure on its left side vs. full-line discharge pressure on its right side. The net effect is a clockwise rolling motion "uphill" to the "neutral" position of Fig. F-36*d*. Generally, oscillating meters are available in sizes ½ through 6 in. and are best suited for clean fluids. Temperature and pressure ranges are below 200° F and 250 psi.

Rotating Meters: These meters are available in many varieties. Vanes are moved radially as cam followers (Fig. F-37) to form the measuring chamber or are articulated in the retracting type as shown in Fig. F-38. The vanes themselves may rotate as lobes (Fig. F-39). Mechanical operation of a sliding-vane rotary meter is shown in Fig. F-37. This unit contains a cylindrical rotor that revolves on ball bearings around a central shaft and stationary cam. As liquid flows against an extended blade, the resulting rotation of the

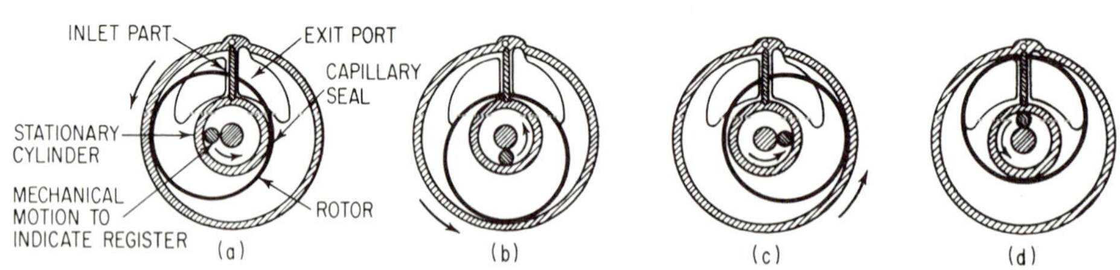

Fig. F-36. Oscillating-piston meter similar to nutating-disk meter except that motion takes place in one plane.

rotor and the action of the cam cause the blades to act as cam followers, creating measuring chambers that accurately measure fluid throughput. Capillary action of the metered fluid effectively seals the blades to form the measuring cavities. These meters are quite accurate and are available in sizes to 16 in. Liquids should be clean and below 400° F and 900 psi.

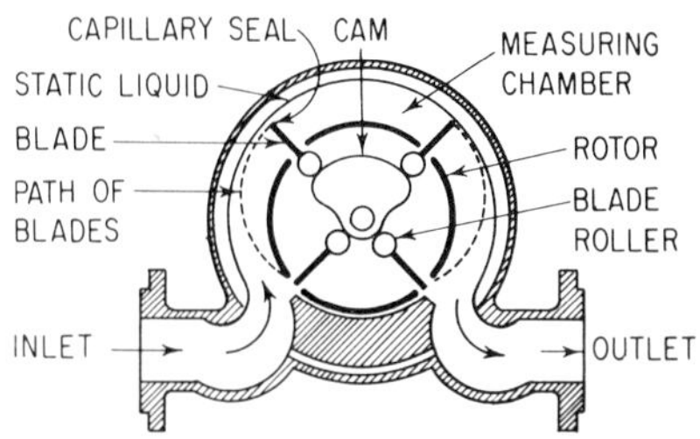

Fig. F-37. Sliding-vane rotary meter.

In the retracting-vane rotary meter (Fig. F-38), fluid entering the meter is deflected downward against the extended blade, causing rotation of the measuring element. Sealing of the metering chamber is accomplished by incorporating features of the piston-ring seal. The vanes or blades, coated with a resilient material, are spring-loaded to provide the wiping action of a piston ring. Unlike the cam action of the sliding-vane rotary

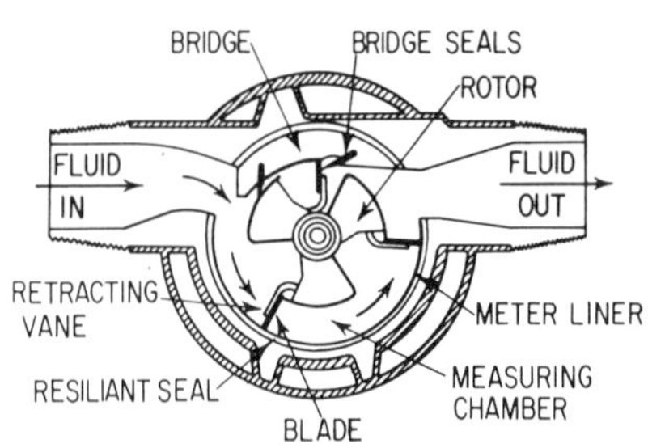

Fig. F-38. Retracting-vane rotary meter.

meter, the vanes of the retracting-vane meter are positioned not only by the configuration of the meter internals but also by foreign materials inside the meter. This feature makes the meter suitable for handling highly viscous (100,000 SSU) or slurry fluids. The meter is available in pipe size from 1 to 4 in. and can handle fluids up to 400° F and 500 psi.

The action of the lobed-impeller flowmeter, which consists of a shaped chamber and a pair of impellers kept in proper relationship to each other through an external gearing arrangement, is shown in Fig. F-39. In some units, the lobes are in the form of eccentric gears, thus avoiding any need for external gearing to maintain the

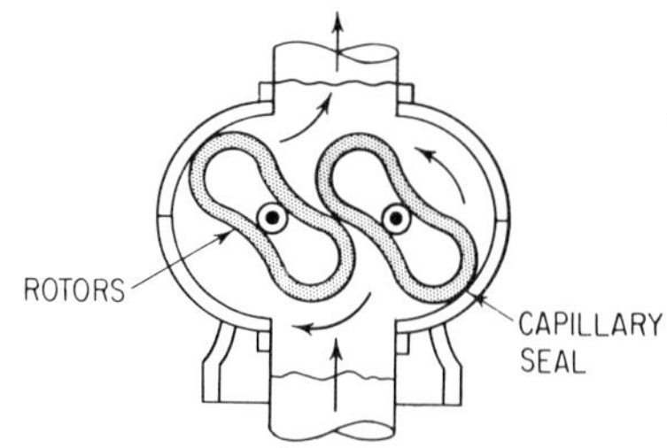

Fig. F-39. Lobed-impeller flowmeter.

proper relationship between the lobe positions and the metering chamber. Lobed meters are best suited for clean fluids and are available in sizes ½ to 8 in. and for handling fluids over the temperature range of − 65 to + 450° F and at pressures up to 500 psi.

—Information on oscillating-piston meter and rotating meters adapted from E. F. Spolidoro, Comparing Positive-displacement Meters, *Chem. Eng.*, June 3, 1968

Flowmeter, Target-type In this device, fluid-flow force is applied on a "target" that is suspended in the fluid stream, as shown in Fig. F-40. Flow is measured as the square root of this force. The device may be installed directly in the flow line, thus eliminating the need for pressure-tap connections. A shaped metal target is attached to a beam, or "force bar," which is positioned in the center of and perpendicular to the flowing stream. Thus, flow occurs through the annular orifice

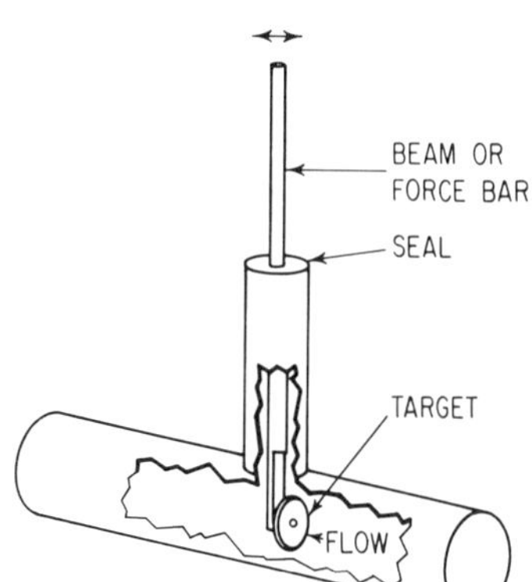

Fig. F-40. Schematic of target-type flowmeter.

around the target. The force bar is extended through a seal flexure to a transmitter. Meters of this type are suitable for dirty hydrocarbon or viscous streams, although the meter also is applicable to other liquids and to gases and vapors. Typical accuracy is ±2 percent of full scale within flow design range. Other operating parameters are pressures up to 1,500 psi and temperatures to 750° F. The device can be installed in either horizontal or vertical lines.

Flowmeter, Turbine-type A flowmetering and transmitting device that derives its name from the operating principle used. A turbine wheel (rotor) is set in the path of a fluid stream. The flowing fluid impinges on the turbine blades, imparting a force to the blade surface and setting the rotor in motion. When a steady rotational speed has been reached, the speed is proportional to fluid velocity—hence to volume flow rate. Although there are distinct design variations between manufacturers, the fundamental concept remains the same.

Fig. F-41. Ball-bearing-type turbine flowmeter with interchangeable shaft-support section and flanged housing. *(Foxboro.)*

A multiblade rotor is mounted at right angles to the axis of the flowing fluid (see Fig. F-41). The rotor is supported by ball or sleeve bearings on a shaft which is retained in the flowmeter housing by a shaft-support section. This section may support the rotor shaft at one end only (commonly referred to as cantilevered or single-ended construction); or it may be in two sections, thereby supporting both ends of the shaft (double-ended construction). Other variations have the rotor affixed to the shaft with the bearing(s) mounted in the shaft support. In these variations, the shaft-support section also commonly acts as a flow straightener (straightening vanes). An exception to the right-angle orientation of the rotor is the "Pelton wheel" design, used in some low-flow turbine flowmeters. This concept is analogous to the overshot or undershot water wheel with the rotor axis at right angles to the flow axis.

In each case, a count of the number of turbine-wheel revolutions per unit time will be a measure of flow rate, and a count of total revolutions will determine total volume flow.

Rotor speed may be transmitted through the meter housing by a mechanical shaft with a magnetic coupling to an external shaft or through a suitable gland in the housing. These mechanical-output types are scaled (adjusted) to desired rotation vs. flow ratios by use of selected gear trains.

Another widely used method of signal generation is a magnetic-pickup coil, consisting of a permanent magnet with coil windings, mounted in close proximity to the rotor but external to the fluid channel. The passage of each rotor blade past the base of the pickup coil changes the total flux through the coil, inducing 1 Hz of voltage. The frequency of the generated pulses is a measure of flow rate, and the total number of pulses a measure of total flow.

Normally, the coefficient of each transmitter is determined by actual flow calibration prior to shipment by the manufacturer. This coefficient, commonly referred to as calibration factor or K factor, is expressed by the equation

$$K = \frac{T_k f}{Q}$$

where $f =$ frequency, Hz
$Q =$ a volumetric flow rate (such as gpm)
$T_k =$ a time constant such as 60 for gpm)
$K =$ pulses per volume unit (such as pulses per gallon)

Turbine-type flowmeters provide very accurate flow measurement over wide flow ranges. The meters are effective in aerospace and airborne applications for energy-fuel and cryogenic flow measurements. Applications in process plants are increasing. Primary uses include flow totalizing for inventory control and custody transfer, precise automatic batching for loading and batch mixing, and automatic flow control using digital electronic techniques. Secondary instrumentation includes signal amplifiers, frequency-to-analog converters, totalizers, totalizer-batchers, and frequency (flow-rate) indicators. Most commonly used sizes range from reduced-bore meters (usually ½-in. nominal size) for fractional flow rates to 12-in. nominal size meters for 8,000 to 9,000 gpm, full-scale flow. End connections include NPT, male flared tube, flanged, and plain for slip-on couplings.

—John S. Boynton, *The Foxboro Company*

Flowmeter, Variable-area Variable-area flowmeters are available in a number of basic design configurations as shown in Fig. F-42. In the variable-area flowmeter, there is a *constant differential* across a *variable orifice* and flow rate is displayed as a function of the *area of the orifice,* which in turn, is a function of the position of a float. This is to be contrasted with an orifice meter, which has a fixed-area orifice plate and where flow rate is displayed as a function of the differential pressure across the orifice.

Essentially, the variable-area flowmeter consists of (1) a chamber that contains (2) a weight or "float" which is free to move in a straight line (usually vertical) away from a rest point against (3) a restoring force. Usually, the force is gravity, but if insensitivity to attitude or position of use is important, spring-loaded units may be used (see Fig. F-42*b*). The chamber is arranged so that the area of an orifice or port is caused to increase with increasing distance of the float from the rest point. Since the restoring force on the float is constant or nearly so, the area of the port will vary directly with the flow rate. The configuration of the port usually is such that the float movement away from the rest point is proportional to the area uncovered.

The varying area can be accomplished in several ways:

1. A bob or float in a tapered or conical section tube (see Fig. F-42*a*). This is the most common method and is that used in rotameters. See **Rotameter.**

2. A float with a tapered stem, the smaller diameter of which is at the bottom. The tapered stem is positioned in a fixed orifice inside the meter chamber, and vertical movement of the float causes the inside diameter of the annular flow port to change (see Fig. F-42*c*).

3. A piston in a straight-bore chamber with slots or holes arranged so that piston movement away from the rest point uncovers a port area proportional to the linear displacement of the piston. An outer chamber carries the fluid to the outlet pipe connection (see Fig. F-42*d*).

Characteristics: Because the volume of flow through a variable-area meter is proportional to area and because area essentially is proportional to float travel, most area meters have *linear scales*

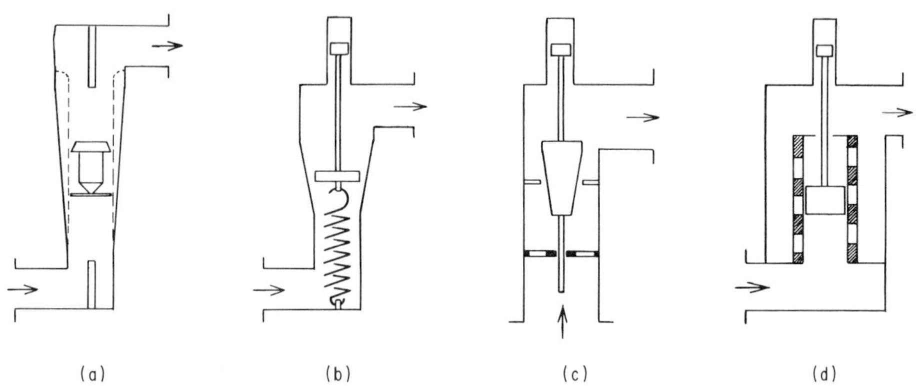

Fig. F-42. Variable-area flowmeters. *(a) Tapered-tube rotameter* with free bob or float. The earliest and most widely used variable-area flowmeter wherein the upward flow causes the float to seek an equilibrium position proportional to flow in order to balance the gravitational force on the float (see **Rotameter**). *(b) Spring-loaded*-type variable-area meter is used to replace the gravitational force on units requiring insensitivity to attitude. *(c) Tapered-plug orifice* is a variation of the variable-area principle used in many commercially available meters using all-metal construction instead of the glass tube commonly used in rotameters. *(d) Piston-and-port*-type device uses a tight-fitting piston in a straight cylinder whose walls have a series of holes or slots. Vertical piston motion uncovers a port area proportional to vertical position. The flow rate then is directed by an outer shell to the outlet connection.

with *equal-scale* increments. The ratio of maximum-flow reading to minimum-flow reading normally is 10:1. This is to be contrasted with the orifice meter whose square-root relationship makes it impractical to obtain readings below 20 percent of full scale. Variable-area meters are simple in design, having only one moving part, which essentially provides freedom from wear and enhances reliability. Variable-area meters are easily equipped to provide secondary functions, including alarms, totalization, recording and control, and remote transmission of readings.

Formulas: The following is not a rigorous analysis, but rather it is an approximation designed to show the effects of important variables

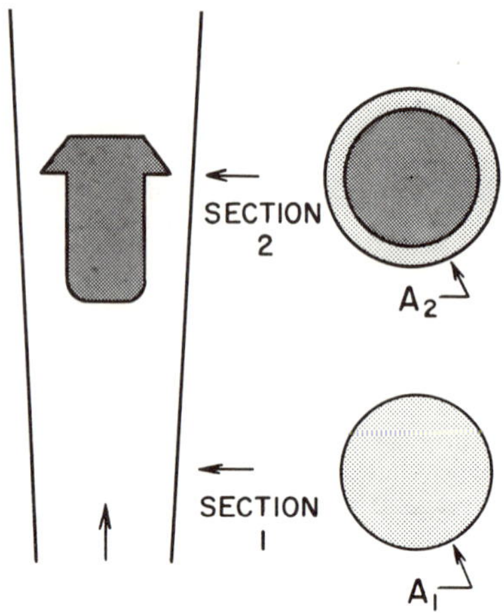

Fig. F-43. Application of Bernoulli's theorem to the study of variable-area flowmeters.

on calculations for variable-area flowmeters. The resulting general equation applies equally well for liquids or gases. Assuming steady flow and no change in heat energy, Bernoulli's theorem can be used to express the kinetic-energy relationship at points 1 and 2 shown in Fig. F-43.

$$V_2 - V_1 = \sqrt{2g(h_1 - h_2)} \qquad (1)$$

where V_1 = velocity of the fluid at section 1, ft/s
V_2 = velocity of the fluid at section 2, ft/s
g = acceleration due to gravity, ft/s^2
h_1 = hydraulic head at section 1, ft of fluid
h_2 = hydraulic head at section 2, ft of fluid

The difference in head at the two points can be expressed as

$$h_1 - h_2 = \frac{P_1 - P_2}{\gamma} \qquad (2)$$

where P_1 = pressure at section 1, lb/ft^2
P_2 = pressure at section 2, lb/ft^2
γ = specific weight of fluid, lb/ft^3

V_1 is dropped because it is very small compared with V_2. Equation (1) then becomes

$$V_2 = \sqrt{2g\frac{P_1 - P_2}{\gamma}} \qquad (3)$$

Introducing flow volume as a function of velocity and orifice area, $V_2 = q/A_2$, where q is volume flow rate in cubic feet per second and A_2 is the area of the annular orifice at section 2 in square feet.

After transposing, Eq. (3) becomes

$$q = CA_2\sqrt{2g\frac{P_1 - P_2}{\gamma}} \qquad (4)$$

Note that a factor C (coefficient of discharge) has been introduced to allow for the dropping of V_1 and for other factors not considered in this derivation. C usually is determined empirically.

To express the flow in gravimetric units (since $q = w/\gamma$), where w = gravimetric flow rate in pounds per second,

$$w = CA_2\sqrt{2g(P_1 - P_2)\gamma} \qquad (5)$$

If Eqs. (4) and (5) are to be used for gas service, it is necessary to use specific weight of the gas or vapor at operating conditions. This can be calculated using Boyle's and Charles' laws for perfect gases but must be obtained from thermodynamic properties tables in the case of vapors and imperfect gases.

—Charles E. Fees, *Fischer & Porter Company*

Flowmeter, Variable-head Variable-head or head flowmeters operate by measurement of the pressure differential or head across a suitable restriction to flow in the pipeline. The flowmeter comprises two main elements: (1) the primary flowmetering device—some form of restriction in the flow line to induce the head; and (2) a means to measure the differential head and interpret that measurement in terms of fluid flowing through the pipeline. The principles of flow are described under **Flow, measurement of**; and the major kinds of restrictions are described under **Flowmetering devices, primary**. Prior to the development and refinement of electromagnetic, mass, target-type, turbine-type, and variable-area flowmeters, the variable-head flowmeter was by far the most common type of flow instrumentation. However, the variable-head flowmeter remains as a widely used approach for many industrial applications.

Variable-head flowmeters operate on the principle of converting energy from one form to another. In the case of liquid flows, only two forms of energy are considered, namely, kinetic energy and the energy due to static pressure. Orifice meters, venturi meters, etc., depend upon the conversion of energy in the form of static pressure into the kinetic energy of velocity. In the case of the pitot tube, the conversion is from kinetic energy to pressure at the impact tube. See **Pitot tube.** In the case of gas or vapor, a third form of energy is involved in the interchange, namely, the internal energy of the compressed fluid. A close approach to an exact mathematical analysis, based on the laws of thermodynamics, may be made for venturi tubes and flow nozzles; but for orifice plates and other primary devices, in which the throat or jet area is indefinite, very little is gained over use of the hydraulic formula with experimentally determined correction factors.

Some of the greatest difficulties encountered in variable-head flowmeter measurements involve the means of connecting the primary devices to the measurement system. Contributing factors include (1) gases or vapors in liquid- or steam-meter lead lines; (2) differences in temperature or composition of the liquid in the two piping connections; (3) leaks, condensate, hydrates, and similar conditions in gas-meter lead lines. Such difficulties have contributed to the use of electric and pneumatic transmission with a transmitter located as close to the level of the pressure taps as possible.

Condensing Chambers: When steam flow is measured, the lead lines and meter body usually are filled with condensate. This condensate must overflow into the pipe at the same level on the upstream and downstream side, or a difference in liquid head results which adds to or subtracts from the differential pressure created by the orifice. Two types of condensing chambers are used. One type condenses the steam continuously and overflows the excess back into the pipe. Intermittent condensing chambers are used to maintain an equal head of cool water on both legs of the meter.

Purges and Seals: Continuous purges of inert gas or air have been used successfully on steam-meter piping. The inert gas or air at a higher pressure than that of the steam line is bled into the piping system through a capillary or small-orifice restriction at a rate such that the friction drop in passing from the point of entrance into the meter piping to the pressure taphole is negligible. Thus, steam is kept purged from the piping system and the meter piping is kept free of condensate.

Pulsation and Unsteady Flow: The head flowmeter is basically a steady-rate measuring device. Because of the square-root relationship between differential and rate of flow, the average of a fluctuating differential does not give a true measure of flow; that is, the square root of the integral of the head does not equal the integral of the square root of the head except when the head is constant. On liquid flows, the usual remedy is to install a cushion chamber partially filled with gas or vapor between the source of the pulsation and the meter. On gases or vapors, the errors due to pulsation are found to be minimized by the use of high operating differentials and high orifice to pipe diameter ratios.

Differential-pressure Measurement: Means for measurement of the pressure differential produced are described elsewhere, inasmuch as these same forms of instrumentation are used to measure pressure differentials in a process per se as well as differentials associated with other variables, including liquid level. See **Differential-pressure transmitter, Ledoux bell;** and **Manometer.**

Total flow is obtained only by integrating the rate over a period of time. The integration may be done automatically by an integrator (see **Integrator**) which is part of the meter, or it may be done by computation of the chart record wherein a planimeter is used. Semiautomatic planimeter devices are available.

Flowmeter, Velocity-type (See Anemometer.)

Flowmetering Devices, Primary Venturi tubes, flow nozzles, Dall tubes, Gentile (Foster) tubes, and orifice plates belong to the family of flow-rate meters of the head class as defined and described by the American Society of Mechanical Engineers (ASME). Generically, these devices often are termed *primary flowmetering devices.* The devices all have the common feature that they produce a pressure difference (head difference) when flow is passing through them. The square root of this head difference is proportional to the rate of flow. Because of this common feature, the devices also are widely known as *differential producers.*

In selecting a primary flow device, it is important to examine the following properties:

1. *Discharge coefficient*—its magnitude, tolerance, and behavior for different Reynolds numbers
2. *Energy-recovery* properties (head recovery)

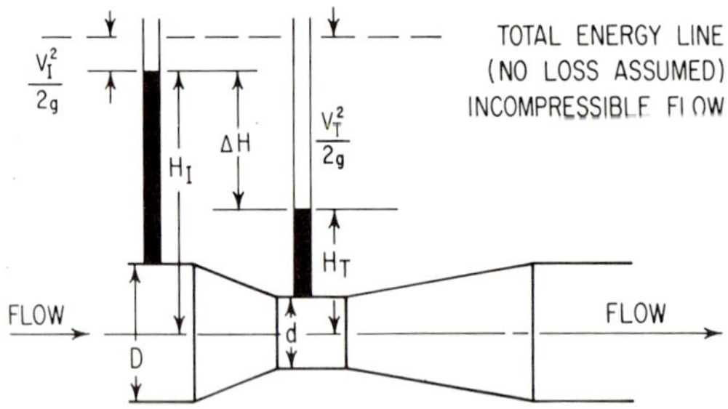

Based on the diagram, the following equations can be derived:

Theoretical throat velocity =

$$\sqrt{2g\Delta H} \times \frac{1}{\sqrt{1-\beta^4}} \quad (1)$$

Theoretical cubic feet per second =

$$\underbrace{d^2 \frac{\pi}{4}}_{\text{Throat area}} \underbrace{\sqrt{2g\Delta H} \times \frac{1}{\sqrt{1-\beta^4}}}_{\text{Theoretical throat velocity}} \quad (2)$$

Actual cubic feet per second =

$$d^2 \frac{\pi}{4} \sqrt{2g\Delta H} \times \frac{1}{\sqrt{1-\beta^4}} \times C \times Y \quad (3)$$

where V_I = inlet average velocity, ft/s
V_T = throat average velocity, ft/s
g = acceleration due to gravity, ft/s^2
H_I = inlet static pressure, ft of line fluid
H_T = throat static pressure, ft of line fluid
D = diameter at inlet pressure tap, ft
d = diameter at throat pressure tap, ft
ΔH = differential pressure, ft of line fluid
β = d/D ratio
C = discharge coefficient = actual flow rate/theoretical flow rate
Y = correction factor accounting for the effect of compressibility in case of gas measurement

Fig. F-44. Derivation of basic flow formula pertaining to primary flowmetering devices that produce a head difference.

3. *Sensitivity to changes in flow patterns* (disturbances caused by upstream and downstream piping configurations)
4. Applicability to *compressible fluids*
5. *Confidence level* of performance

All primary flowmetering devices use the same basic flow equation, derivation of which is given in Fig. F-44. Equation (2) is the theoretical flow formula for incompressible flow. To make this formula useful for actual flow calculations, two factors should be added to it, as in Eq. (3). Factor C is the discharge coefficient, established by hydraulic tests, and Y (*calculated* for venturi tubes and nozzles, and *tested* for orifice plates and Dall tubes) is a factor to account for the effect of compressibility. In the case of compressible flow measurements, such as gases, steam, and air, the Y factor makes the basic flow equation for incompressible fluids usable for compressible flowmetering.

For descriptions of specific primary flowmetering devices, see **Dall tube; Flow nozzle; Gentile tube (Foster); Orifice plate;** and **Venturi tube.** Also see **Piezometer ring.**

—Edmond B. Lynch, *B I F, Unit of General Signal Corporation*

Flow Nozzle A primary flowmetering device, the fundamental principle of which is described under **Flowmetering devices, primary.** Flow nozzles are of two types: (1) nozzles with recovery cones, such as venturi flow nozzles; and (2) nozzles *without* recovery cones. Most popular in the United States are American Society of Mechanical Engineers (ASME) nozzles, which usually do not have recovery cones. Commercially available nozzles have (1) an elliptical approach section, (2) fairly long cylindrical throat section, and (3) the recovery cone. A sectional view of an ASME long-radius flow nozzle of the low-beta series is shown in Fig. F-45. A welded-in-line venturi nozzle is shown in Fig. F-46.

Some flow nozzles sense true static pressure at inlet and throat. Others have a corner-type inlet tap that senses an increased apparent inlet static pressure, and some of these nozzles may pick up a somewhat influenced throat static pressure because of their throat design. The coefficients of these nozzles differ according to the way in which they sense the static pressure. Nozzles that sense true inlet and throat static pressures have a coefficient of about 0.98 to 0.99. Nozzles that have a long throat, but with a corner tap at the inlet, will have a coefficient between 0.98 to

0.99 up to a beta ratio of 0.5. For larger beta ratios, this coefficient will decrease according to the hydraulic shape of the nozzle.

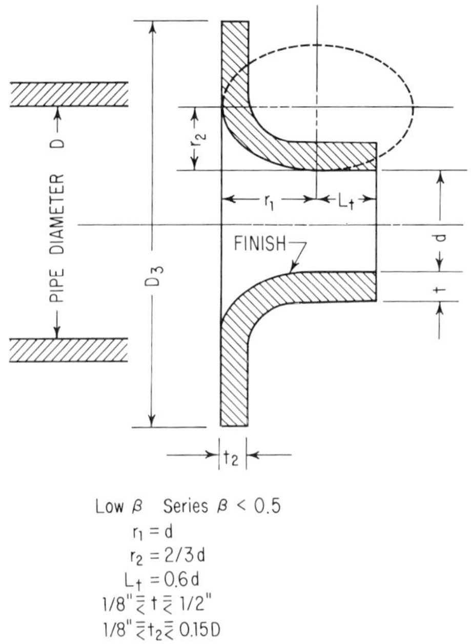

Fig. F-45. Dimensional relations of the ASME long-radius flow nozzle of the low-β series.

The head losses of flow nozzles are basically different, dependent upon whether or not they have recovery cones. Typical head losses are:

	Percent of Differential	
Beta Ratio	With Recovery Cones	Without Recovery Cones
0.75	11	35
0.55	14	57
0.3	24	85

As regards the sensitivity of these nozzles to improper flow patterns, nozzles working with true static-pressure pickup are less sensitive than those which use an inlet corner tap and shorter throats. Furthermore, nozzles having recovery cones are largely unaffected by downstream piping, whereas nozzles with no recovery cones are so affected. Piping recommendations for ASME nozzles are given in the ASME "Fluid Meters Committee Handbook." Manufacturers' recommendations should be followed in other cases.

Flow nozzles are applicable for gas metering because the *Y* factor has been established by calculation; or where not established, the *Y* factor can be determined by test. Tolerances on the coefficients of ASME nozzles have been established by ASME at a value of ±0.95 percent (subject to revision). While the degree of confidence in accuracy of the nozzles produced by various manufacturers depends on the thoroughness with which each manufacturer establishes his coefficient, the tolerance of a high-quality nozzle will be ±1 percent.

Also see **Dall tube** and **Venturi tube.**

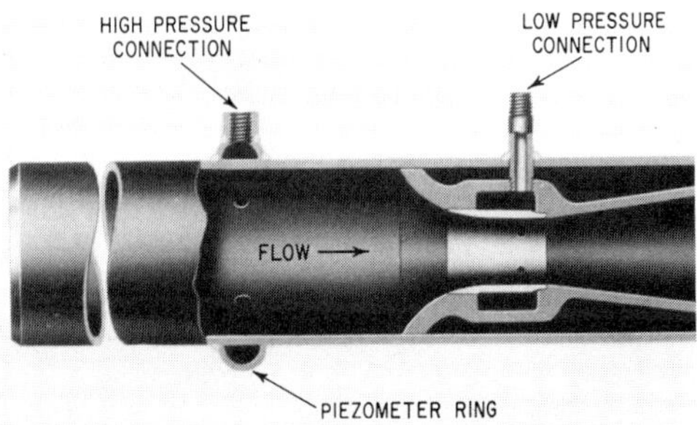

Fig. F-46. A welded-in-line venturi nozzle. Insert nozzle construction also is available. In some cases, the entire nozzle is constructed of cast steel in one piece. *(B I F.)*

—Edmond B. Lynch, *B I F, Unit of General Signal Corporation*

Fluidics As recommended by the National Fluid Power Association, *fluidics* may be defined as "the technology associated with sensing, control, information processing, and/or actuation functions performed solely through utilization of fluid dynamic phenomena." Devices and circuits function without moving parts and may be fabricated from plastic, metal, or ceramics; by molding, machining, chemical milling, or photoetching.

The great promise of fluidics is high reliability at low cost in normal, hazardous, or extreme environments. The lack of moving parts, combined with the fact that devices and circuits may be fabricated by diffusion-bonding a multiple of plates of a single material to the degree that the interface between plates no longer exists, imply reliability. The absence of electricity and mechanical friction is favorable in hazardous areas, the flexibility in choice of material permits operation in extreme (chemical, thermal, nuclear, etc.) environments, and the short free-flow paths of

TABLE F-8. *Common Designs of Fluidic Devices*

FLIP - FLOP

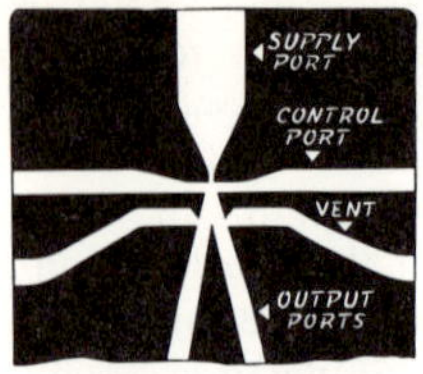

SILHOUETTE OF A WALL-ATTACHMENT BISTABLE DEVICE. Fluid enters at the supply port. The tendency of a high velocity jet to entrain fluid is thwarted by the walls, and a low-pressure area develops. The pressure differential across the jet forces it to the wall, and through the output port on that side. A control signal on that side satisfies entrainment, and the jet swings to the other wall.

BISTABLE FLIP-FLOP
LOAD INSENSITIVE

SYMBOL

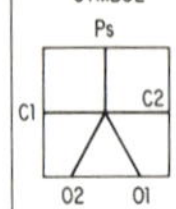

OPERATING PARAMETERS	
POWER NOZZLE PRESSURE	1.5 – 20 psi
SWITCH PRESSURE	5 – 15% OF Ps
PRESSURE RECOVERY	33% OF Ps (MAX)
FREQUENCY RESPONSE	> 1,000 Hz
FAN OUT	4 SIMILAR DEVICES

SWITCH CURVE

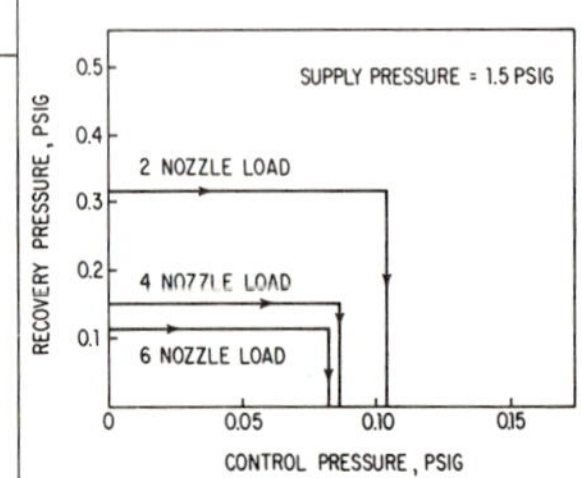

Shown are typical characteristics of a bistable wall-attachment device. The number of nozzle loads refer to the number of similar devices being driven.

OR-NOR GATE

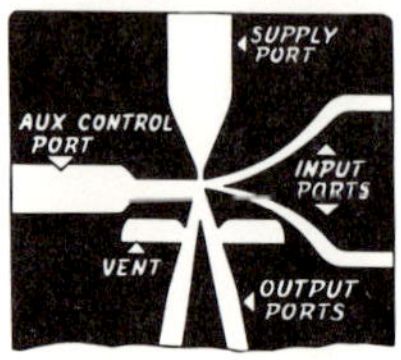

SILHOUETTE OF A WALL-ATTACHMENT-TYPE MONOSTABLE DEVICE. The auxillary control is large enough to permit entrainment, and the jet is always forced to the side with the control ports. The jet will swing to the opposite wall when sufficient fluid is introduced through either or both of the control ports, remaining as long as the signal is present. The vents contribute to stability over a wide range of loading conditions.

OR-NOR GATE

SYMBOL

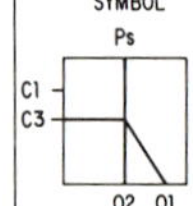

OPERATING PARAMETERS	
POWER NOZZLE PRESSURE	1.5 – 20 psi
SWITCH PRESSURE	5 – 15% OF Ps
PRESSURE RECOVERY	35% OF Ps (MAX)
FREQUENCY RESPONSE	> 1,000 Hz
FAN OUT	4 SIMILAR DEVICES

SWITCH CURVE

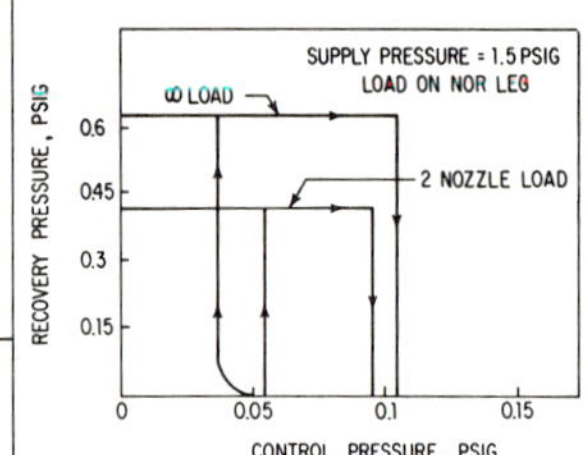

These are typical characteristics of a monostable wall-attachment device. These devices are compatible with each other, and all others in a given line.

TURBULENCE AMPLIFIER

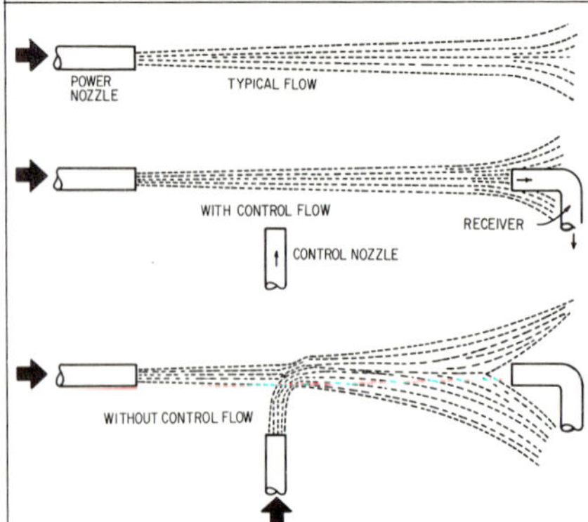

SKETCH OF A TURBULENCE AMPLIFIER. A monostable device switched by driving a laminar jet turbulent. Fluid enters through the power nozzle at low pressure, to ensure laminar flow. The receiver is placed in the laminar area and much of the flow is recovered. A control signal forces the jet into turbulence, drastically reducing the amount of fluid energy entering the receiver. When the signal is removed, the jet returns to the laminar condition.

PROPORTIONAL AMPLIFIER

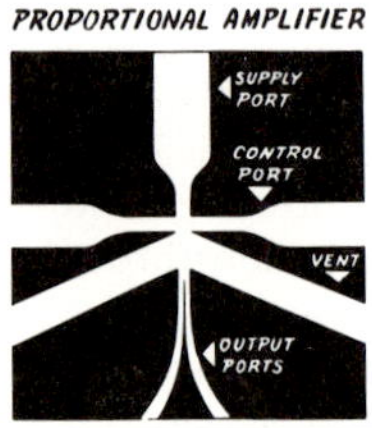

SILHOUETTE OF MOMENTUM-EXCHANGE PROPORTIONAL DEVICE. Replacement of the attachment walls by vents permits entrainment, and the jet is divided between the outputs. A differential in power level of the control ports causes the jet to deflect toward the lower power control, and the output on that side receives an increased percentage of the flow. When cascaded, gains in the thousands are achieved.

PROPORTIONAL DEVICES

SYMBOL

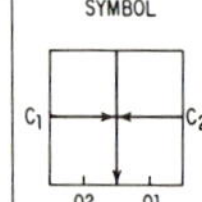

OPERATING PARAMETERS PROPORTIONAL	
POWER NOZZLE PRESSURE	0.5 – 20 psi
PRESSURE RECOVERY	40% Ps
MAX PRESSURE GAIN	7
FREQUENCY RESPONSE	> 1,000 Hz

OUTPUT CHARACTERISTICS
PROPORTIONAL

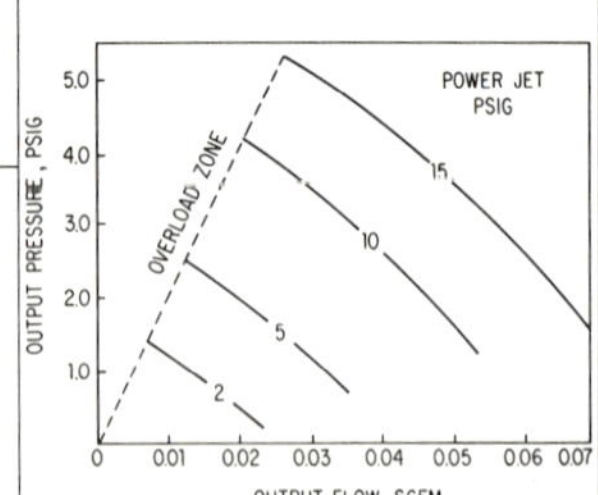

These are typical characteristics for jet interaction-type devices. The overload zone has been eliminated in later models, and operation at pressures in excess of 100 psig is now possible.

IMPACT MODULATOR

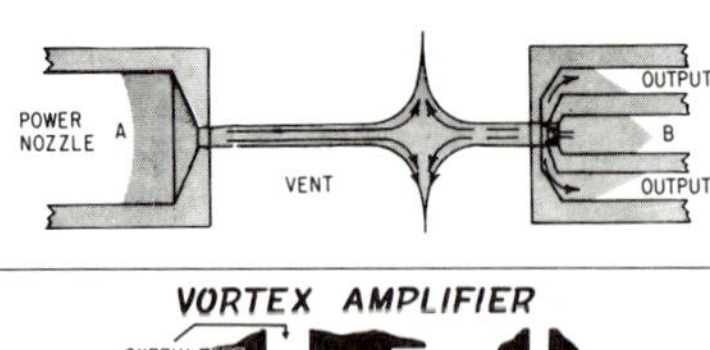

SKETCH OF IMPACT MODULATOR-TYPE PROPORTIONAL DEVICE. This is another momentum-exchange device in which two submerged fluid jets, originating from opposite directions are impacted against each other. The point of impact is a function of the relative momentum of the two jets, and minor changes in this relationship result in large changes in pressure at the outputs.

VORTEX AMPLIFIER

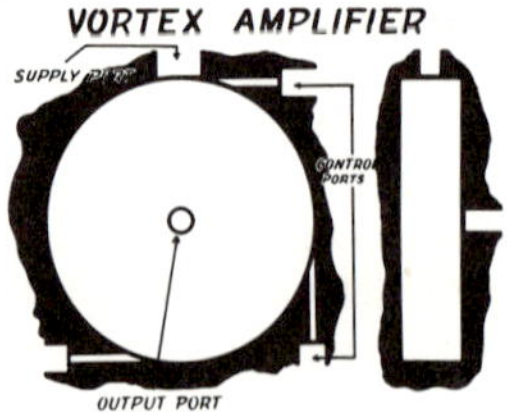

SKETCH OF VORTEX-TYPE PROPORTIONAL AMPLIFIER. In this device, fluid introduced through the supply port, in the absence of control flow, passes directly through the output port. With flow at any of the controls, a vortex forms, and as the flow spirals inward, the velocity increases, shear stresses increase, and the pressure drops.

the small jets render them relatively insensitive to shock and vibration. Although the preponderance of the work done to date has utilized air as the fluid, operation may be either pneumatic or hydraulic. As a rule of thumb, if a nozzle-exit Reynolds number $\rho b V_e/\mu$ of 1,500 can be achieved at realistic temperature and pressure values, the fluid in question is suitable. On the negative side, changes in geometry or flow pattern resulting from contaminant buildup affect operating characteristics, and relative to electronics, signal propagation is slow (the speed of sound to the speed of light).

Table F-8 presents a brief description of the more commonly used designs, and Table F-9 a list of many of the devices and functions currently available as off-the-shelf items. Devices are classified by power-nozzle dimensions, which are generally 0.005, 0.010, or 0.020 in. wide with an aspect ratio of 1, 2, or 4. There are exceptions to this, with some flow diverters dimensioned in feet.

Figures F-47 through F-49 show different interconnection techniques. The breadboard (Fig. F-47) is used for circuit development. Manifold mounting (Fig. F-48) is used to test out the circuit on line or in cases where the total volume (<10) does not justify the cost of integration. The integrated circuit (Fig. F-49) is used for proved circuits in moderate to high volume.

TABLE F-9. *Available Fluidic Components**

Devices:	Multigate functions:
Flip-flop	Schmitt trigger
OR-NOR gate	Binary counter
AND-NAND gate	3-stage proportional
One-shot multivibrator	Summing junction
Inhibited OR gate	Shift register
Digital amplifier	Binary-to-decimal converter
Proportional amplifier	Pulse generator
Preference flip-flop	Educational kit
Rectifier	
Linear resistors	Circuits:
Passive AND gate	Liquid-level control
Passive OR gate	Press-safety control
Flow diode	Sequence controller
Capacitors	Bottle-filler control
	Transfer control

*There is also a significant number of sensors and interface devices.

Although the number of successful applica-

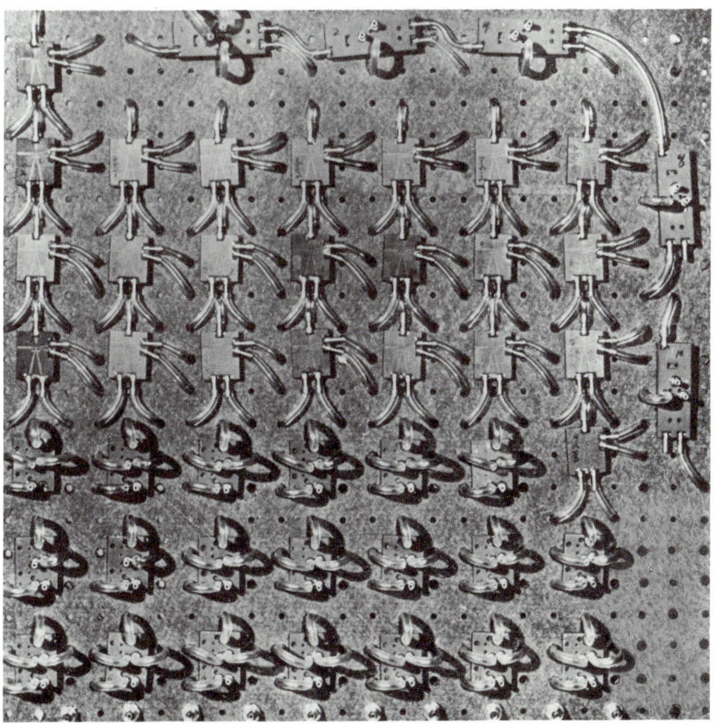

Fig. F-47. Typical fluidic breadboard circuit.

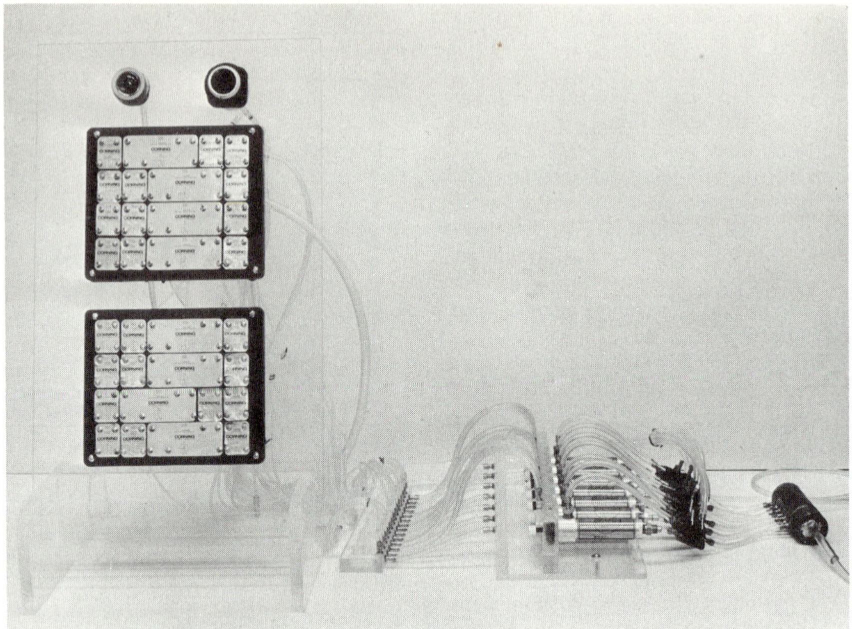

Fig. F-48. Typical fluidic manifold circuit.

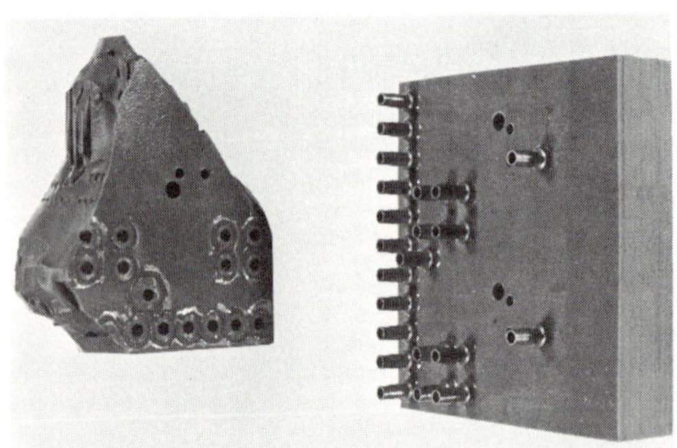

Fig. F-49. Typical fluidic integrated circuit. Holes in piece at left are on ½-in. centers.

Fig. F-50. Fluidic-controlled air-gaging machine. *(Bendix.)*

tions, such as the gaging machine shown in Fig. F-50, is steadily increasing, the technology is still in a stage of development. Active projects in industrial, military, and academic facilities around the world number in thousands, a few of which are presented in Table F-10 for reference.

TABLE F-10. *Typical Fluidics References*

U.S.A.	Meyer, C. H., and R. S. Gluskin: A Fluid Operated Diesel Locomotive Transition Control Unit, *Proc. Fluid Amplifier Symposium,* Harry Diamond Laboratories, 1965. Clayton, B. J., and T. G. Wetheral: Application of Pure Fluid Components to Missile Control System, presented to SAE Aerospace and Fluid Power Systems and Equipment Committee, 1964. Fellows, J. W.: Fluidics in Weighing Systems Control, ASME Paper 66-MD58, 1966.
U.K.	Ward, J. E., et al.: A Lathe Control System Incorporating Fluidics, *Proc. 3d Cranfield Fluidics Conference,* 1968.
Italy.	Galli de Paratesi, S., et al.: Development of Fluidic Systems for Nuclear Plant Control, *Proc. 3d Cranfield Fluidics Conference,* 1968.
Sweden	Ekstrom, U., et al.: *An All Fluidic Gyrostabilized Platform, Proc. 3d Cranfield Fluidics Conference,* 1968.
France.	Champagnon, J. P., and A. A. Thiney: Fluidic Thrust Reversing Control System, *Proc. 3d Cranfield Fluidics Conference,* 1968.

REFERENCES
Kirshner, J. M.: "Fluid Amplifiers," McGraw-Hill Book Company, New York, 1966.
Humphrey, E. F., and D. H. Tarumato: "Fluidics," Fluid Amplifier Associates, 1965.
British Hydromechanics Research Association, *Proc. 3d Cranfield Fluidics Conference,* 1968.
"Fluidic Systems Design Guide," Fluidonics Division, Imperial-Eastman Corporation, 1966.

—R. W. Van Tilburg, *Corning Glass Works*

Flume Flumes are classified by the American Society of Mechanical Engineers (ASME) Fluid Meters Committee as head-area meters. In head-area meters, both the area of the stream and the head are variable, but not independently—the area being a function of head. These meters also are distinguished from other classes of rate meters in that they are used in open conduits or in conduits which do not flow full, that is, in which there is a free surface. Consequently, flumes are applicable to the metering of liquids only. Various types of flumes have been investigated, their proportions established, and flow characteristics published. These include (1) the venturi flume, (2) the Palmer Bowler flume, and (3) the Parshall flume (which is an improved venturi flume). Of these, the Parshall flume perhaps is the most commonly used, since it is used widely in the irrigation field for proportioning of water (see Fig. F-51). Also see **Weir.**

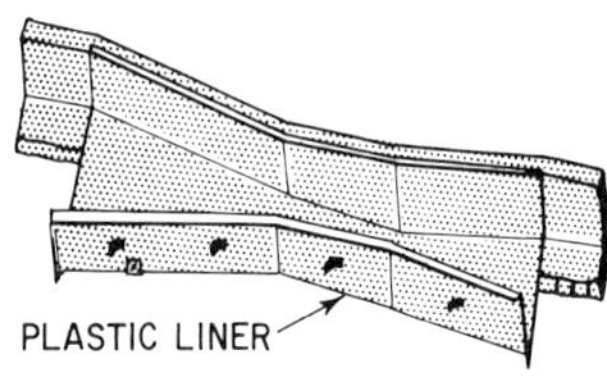

Fig. F-51. Parshall-flume liner constructed of plastic for placement in open conduit. *(B I F.)*

REFERENCES
"Water Measurement Manual," 2d ed., U.S. Department of Interior, Washington, D.C., 1958.
"Fluid Meters, Their Theory and Application," ASME, New York, 1959.
Considine, D. M. (ed.): "Process Instruments and Controls Handbook," pp. 4–82 to 4–85, McGraw-Hill Book Company, New York, 1957.

—Edmond B. Lynch, *B I F, Unit of General Signal Corporation*

Fluorometry An instrumental technique for identifying compounds and determining concentrations through measurement of the fluorescence properties of materials. Phosphorimeters which perform similar functions do so on the basis of the phosphorescence properties of substances. These two related properties can be designated as luminescence.

Luminescence may be defined as the emission of light by materials at lower-than-incandescent temperatures as a result of chemical action, electrical action, physiological processes, exposure to certain types of radiation, or other nonthermal processes. Fluorescence may be defined as the emission of light or other electromagnetic radiation by a material exposed to another type of radiation or to a beam of particles, with the luminescence increasing within about 10^{-8} s after irradiation is stopped. Certain minerals fluoresce in characteristic colors during exposure to ultraviolet radiation. Phosphorescence is a form of luminescence in which the emission of light continues more than 10^{-8} s after excitation by radiation having a shorter wavelength, such as by electrons, ultraviolet light, or x-rays.

Instruments based upon these principles are useful for clinical research, particularly in the areas of cancer and mental illness; for the quantitative assay of pharmacological compounds, metal chelates, pesticides, botulinum toxin, and hydrocarbon studies for air-pollution research.

Fluxgate Magnetometer (See **Magnetometer, airborne.**)

Fluxmeter An instrument for measuring magnetic flux, usually calibrated in maxwells. In one form, the device is a specially designed ballistic galvanometer where (1) the galvanometer is heavily overdamped and (2) the period of the galvanometer is long. These characteristics are necessary so that flux density, regardless of the impulse time involved, may be accomplished without appreciable error. In practice, deflection of the fluxmeter is a measure of changes in flux linkages between a magnetic field and a test coil that is connected to the instrument terminals. The test coil is used under two conditions: (1) where the coil is stationary in a changing magnetic field and (2) where the coil is moved so as to alter the flux which it links in a constant field. In either condition, flux-linkage changes induce an emf in the circuit of the fluxmeter. In some instances, a substantial interval of time may be involved, particularly if the test coil is stationary in a slowly changing field. Thus, an ideal fluxmeter becomes a ballistic galvanometer wherein the restoring couple of the suspension effectively is reduced to zero. Also see **Magnetic quantities (variables).**

REFERENCES

Fink, D. G., and J. M. Carroll: "Standard Handbook for Electrical Engineers," McGraw-Hill Book Company, New York, 1968.

Harris, Forest K.: "Electrical Measurements," John Wiley & Sons, Inc., New York, 1952.

Footcandle Meter (See **Photometers.**)

Force, Measurement of Force as a measured variable may be moment, shock, torque, positive pressure, or vacuum. Some of the major entries in this encyclopedia on these topics are **Bourdon tubes; Load cell; Pressure gage; Shock, measurement of; Strain gage; Tension, measurement and control of; Testing (materials), destructive;** and **Vacuum gaging.** Also consult **Classified Index** under Force.

Force, Standard (See **Standards, measurement.**)

Force-balance System (See **Valve positioner, pneumatic.**)

Force Transducer, Proving-ring Proving-ring deflection is picked up by a linear variable-differential transformer, resulting in a proportional voltage. The unit shown in Fig. F-52 is available in ranges from 10 to 5,000 lb. External amplifiers

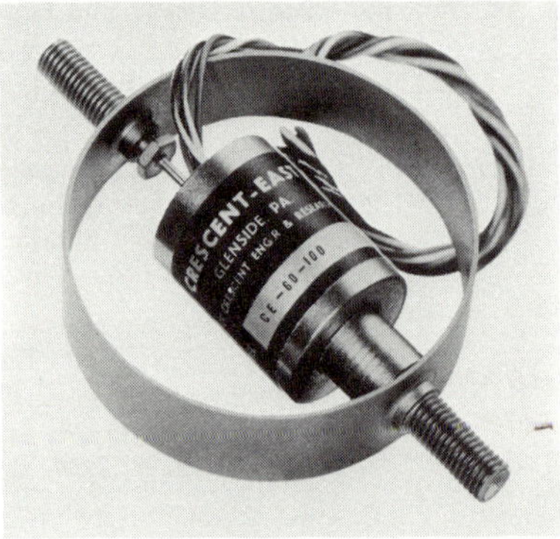

Fig. F-52. Proving-ring-type force transducer in which ring deflection is picked up by a linear variable-differential transformer. *(C-E Electronics.)*

and demodulators are available to provide dc voltage or current to drive conventional recorders or indicators. The transducer can be used in various force-, weight-, and pressure-measurement systems. Also see **Proving ring.**

—Frank G. Daveler, *C-E Electronics, a division of Ametek, Inc.*

FORTRAN A higher-level computer-programming language oriented toward engineering and scientific applications which require a large amount of numeric computation. FORTRAN (FORmula TRANslation) allows arithmetic operations to be expressed in a form very similar to standard algebraic notation. A programmer coding in FORTRAN does not normally need to know the internal architecture of the digital computer upon which the program is to be executed. (Compare **Assembler;** see **Compiler.**)

FORTRAN Source Program: When a programmer codes a problem in FORTRAN, he produces what is known as a FORTRAN source program, which consists of a set of statements. A given statement may express an algebraic equation to be solved or a logic decision to be performed. There are five types of statements in FORTRAN: (1) Arithmetic statements express a series of arithmetic operations to be performed in algebraic notation. (2) Input/output statements allow the programmer to govern the transfer of data between the CPU and the outside world. (3) Control statements allow the programmer to control the sequence in which the statements are to be executed; this control may be exercised on a logical-decision basis. (4) Specification statements allow the programmer to define the nature of the data being manipulated. (5) Subprogram statements allow the user to define the subroutines which may be called at any point within his program. Figure F-53 is a listing of a FORTRAN source program which reads values, computes results from those values, and prints out the results. Statements preceded by a C are comments which do not affect the execution of the program.

FORTRAN Compiler: A FORTRAN compiler is a program for a specific computer which accepts as input the FORTRAN source program and translates it into a machine-language object program. The machine-language object program may then be loaded into the computer and executed to produce the results required. FORTRAN compilers vary radically in the efficiency of their translation operations (see also **Translator**). Some compilers give a very rapid translation, but the object code produced is inefficient in terms of core requirements and execution speed. Other compilers require more time for the translation procedure, but the object code is much more efficient in both core and execution time required. In a process-control environment where the user is coding his control algorithms in FORTRAN (see **Algorithm**), it is normal practice to have those programs compiled only once or twice but executed many times over. Therefore, one of the prime considerations when evaluating a digital computer and its associated FORTRAN compiler for a control application is the efficiency of the machine-language object code generated by the compiler.

REFERENCES

McCracken, D. D., and W. S. Dorn: "Numerical Methods and FORTRAN Programming," John Wiley & Sons, Inc., New York, 1964.

Golden, J. T.: "FORTRAN IV Programming and Computing," Prentice-Hall, Inc., Englewood Cliffs, N.J., 1965.

```
C  SAMPLE FORTRAN PROGRAM
C  READ A AND B FROM A CARD READER (DEVICE
   #2)
C  USING FORMAT 2
C  FORMAT 2 DEFINES 2 FLOATING POINT NUM-
   BER
C  FIELDS OF 10 COLUMNS EACH
     1 READ (2,2)A,B
     2 FORMAT (2F10.2)
C  COMPUTE VALUE FOR C; THE DOUBLE AS-
   TERISK**
C  DENOTES EXPONENTIATION
     C = (A**3/B) + B
C  PRINT C ON DEVICE #3 USING FORMAT 3 WHICH
   DEFINES AN ALPHANUMERIC FIELD ( C = )
   AND
C  A 10 COLUMN FLOATING POINT FIELD
     WRITE (3,3)C
     3 FORMAT (' C = 'F10,4)
C  GO TO BEGINNING AND READ ANOTHER CARD
     GO TO 1
     END
```

(*Note:* Statements preceded by a "C" are comment statements which do not affect the execution of the program.)

Fig. F-53. Example of FORTRAN program.

—B. W. Landeck, *IBM Corporation*

Foster Flow Tube [See **Gentile tube (Foster).**]

Freeness (See **Consistency, measurement of.**)

Freezing-point Determination An instrumental technique for determining the purity of many liquid organic and inorganic substances that solidify at temperatures that can be readily measured. Although an impurity cannot be positively identified by this method, many pure substances are specified by their freezing point, although not uniquely. The amount of impurity is related to depression of the freezing point and is determined by plotting sample temperature vs. time as the ambient temperature is lowered. The freezing point is identified by the discontinuity in this plot. This is essentially a laboratory method. A 5- to 25-ml sample usually is adequate. A batch determination requires from 10 to 30 min. Probability of error is ±1 percent.

Frequency (Electrical), Measurement of Frequency may be defined as the number of periodic phenomena that occur in a unit of time. Frequency and time are related by

$$f = \frac{1}{t} \quad (1)$$

where f = frequency, Hz
t = time, s

The most widely accepted standard of frequency is the period of rotation of the earth. Because this rotation is not absolutely constant, very precise methods have been devised to establish correction factors, which then are applied to what is termed the *mean solar day*. A basic unit of time then is defined to be the *second*,[1] which is equal to 1/86,400 of a mean solar day. Also see **Time and time measurement.**

Electrical frequency is the number of periodic swings or cycles of voltage or current in a unit of time. Measurement can be accomplished by (1) counting the number of cycles in a given time, (2) measuring the time in which a given number of cycles occurs, (3) balancing an impedance bridge in which the impedances of the arms are known functions of frequency, (4) tuning electric circuits to a condition of resonance, (5) resonating tuned reeds magnetically, (6) using a deflection-type frequency meter, and (7) comparing the unknown frequency with a known frequency.

Counting requires electronic means, including sources of pulse trains, pulse counters, and electronic switches (gates). Very wide ranges of frequency up to many millions of hertz are measurable in this manner. Accurate time measurements require similar instrumentation. Methods 3 and 4 require manual adjustment of capacitors or other circuit elements and the observation of a balance or resonance detector in a bridge circuit. Tuned reeds are limited to a fairly narrow range of electromagnetically produced mechanical oscillations. Instruments based upon this method usually extend little beyond the common power frequencies. Most deflection-type frequency meters employ a rotatable iron vane whose position is determined by the superimposition of two stationary magnetic fields whose resultant changes its orientation as a function of frequency. These fields are produced by two coils, one of which is connected into a resistive circuit, the other into an inductive circuit. As the frequency increases, the current in the latter decreases. Such instruments are widely used in electrical-power installations and are calibrated for the range of frequencies encountered in that field. (See Electronic Frequency Meters, below.)

Oscilloscope Frequency Comparison: A common method of oscilloscope frequency comparison involves the generation of Lissajous figures on the face of a cathode-ray tube. A source of known frequencies is connected to one set of deflection plates; the unknown frequency is connected to the other set. If one frequency is a harmonic or integral multiple of the other, a stable and recognizable pattern will be displayed. The equipment arrangement wherein the unknown frequency is connected to the vertical input and the known frequency to the horizontal input is shown in Fig. F-54. The horizontal time-base generator can be switched to observe the waveform of the unknown signal.

Fig. F-54. Equipment arrangement for observation of Lissajous figures in frequency measurements.

The unknown frequency can be estimated to within about ±5 percent by setting the horizontal input to the internal sweep position and measuring the period of the unknown waveform. Then, the known frequency is applied to the horizontal input and the sensitivity and gain controls for both the vertical and horizontal amplifiers are adjusted to obtain equal vertical and horizontal deflections. If the two applied frequencies are adjusted, various Lissajous patterns will be formed to indicate the ratio between the two frequencies. The most often used frequency ratios are illustrated in Fig. F-55. As the phase between the two applied signals is varied, the Lissajous pattern shifts. A specific unknown frequency can be measured by the Lissajous method if an interpolation oscillator is used with a frequency standard to provide continuous adjustment of the standard frequency. Sources of error that may be encountered include (1) instability in the applied frequency signals which will result in phase changes with resultant pattern rotation, (2) jitter in the input signals, and (3) observation errors in determining the more complex ratio patterns. Also see **Crystal oscillator;** and **Signal (electric) generator.**

[1] The International System of Units defines the second as the duration of 9,192,631,770 cycles of the radiation associated with a specified transition of the cesium atom.

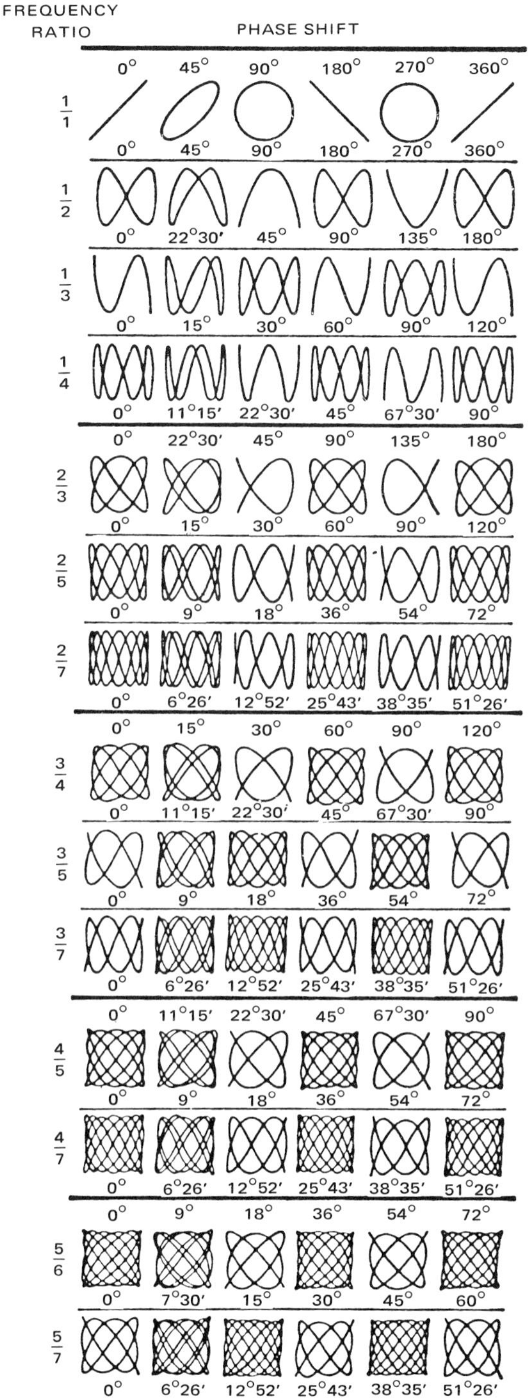

Fig. F-55. Lissajous figures for the most frequently used frequency ratios.

Electronic Frequency Meters: A typical circuit is shown in Fig. F-56. The input signal is amplified and converted to a square wave which charges a capacitor through a diode. A second diode discharges the capacitor through the meter circuit. The current through the meter is proportional to the rate of the charging pulses; hence it is proportional to the frequency of the input signal. Another form of instrument is shown in Fig. F-57. The input waveform drives a trigger circuit which generates a series of negative trigger pulses. These actuate a constant-current source and a phantastron (linear-timing) circuit. The phantastron output turns off the current source, thus determining the width of the current pulse delivered to the meter circuit (and establishing the meter "range"). One such stable pulse is passed through the meter circuit for each input cycle. The meter averages the pulses it receives and presents an indication that is proportional to the average frequency.

Heterodyne Frequency Meter: This instrument consists essentially of a frequency-calibrated stable oscillator, a mixer, and an indicator. Am amplifier for driving the beat indicator also is included, as shown in Fig. F-58. The output signal may be used to drive external indicators, such as headphones, oscilloscopes, or recorders. With the unknown frequency applied to the mixer, the oscillator usually is tuned for a zero beat, which is indicated by a minimum or zero output from the mixer. The oscillator frequency then will be the same as, or a subharmonic of, the frequency of the applied signal. Heterodyne techniques also are used where the oscillator is tuned for a difference frequency in a specified range of frequencies rather than for a zero beat. The difference frequency then is determined by a separate frequency-measuring circuit to obtain the desired information. By using oscillator harmonics to zero-beat against the unknown, the basic range of the instrument may be extended by 50 to 100 times the highest fundamental frequency available from the oscillator. Many heterodyne frequency meters also include a crystal calibrator which is used to check accuracy at various points on the oscillator frequency-control dial. A heterodyne frequency meter with crystal calibrator can provide measurement accuracy of up to 1 part in 10^6 or better.

Electronic Counters: As shown in Fig. F-59, an electronic counter consists of a time-base generator, a signal gate, and decade-counting units. Frequency is measured by counting the number of input cycles over a precisely controlled period of time. Also see **Counter.**

The time-base generator develops control signals which are applied to the signal gate. When the first, or "start," signal is received, the signal gate opens to pass input pulses from the unknown frequency source to the decade-counting units. When the second, or "stop," signal is received, the signal gate closes to prevent further input pulses from reaching the decade-counting units. Totalization of input pulses by the decade-counting units during the interval when the gate was open is a measure of the input-signal frequency. The standard gate time (interval between start and stop signals) may be adjusted for various degrees of resolution. For example, when measuring 1 MHz on a counter, the counter usually indicates the unit of frequency displayed. To decrease the measuring time, a 1-s gate could be used with a resulting display of 01000.000. Note that, as the speed of measurement is increased, the number of significant digits is decreased.

Frequency-measurement accuracy of an electronic counter is ± 1 count ± the accuracy of the time-base generator. Accuracy depends upon stability and on the technique used to calibrate the time-base generator. The error of ± 1 count is inherent in the gating methods used. By using accessory heterodyne frequency converters and transfer oscillators, the basic frequency range of an electronic counter can be extended. In one standard arrangement, frequencies up to 500 MHz or more are measured with a 10-MHz counter and a frequency converter. With a transfer oscillator or other harmonic generator-mixer arrangements, the range is further extended to at least 40 kMHz. At lower frequencies, the accuracy of measurement is the basic accuracy of the electronic counter. Phase-locking systems

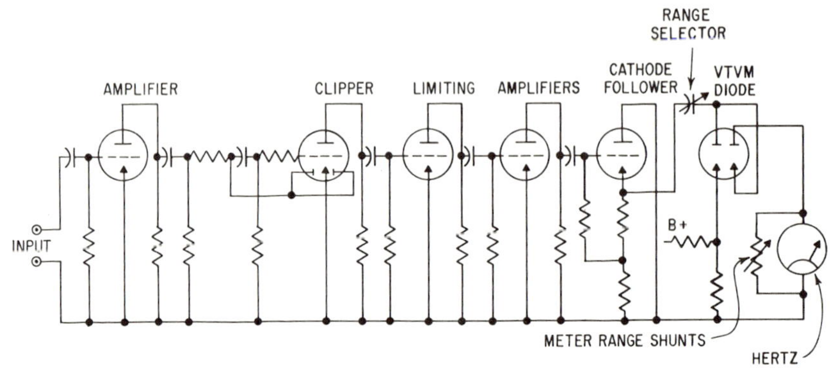

Fig. F-56. Electronic frequency meter.

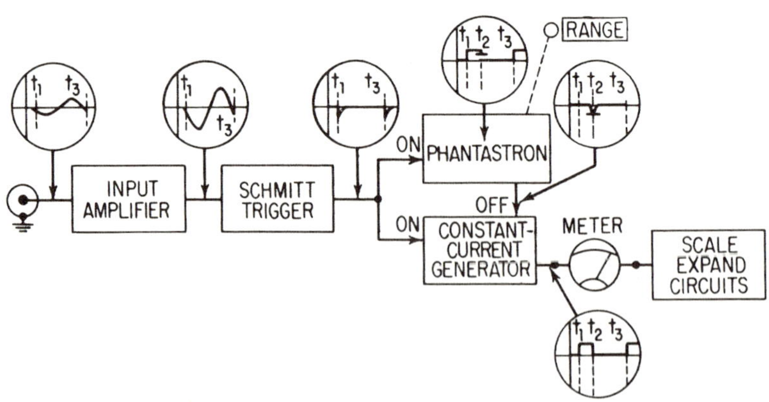

Fig. F-57. Electronic frequency meter incorporating a linear-timing circuit.

allow full use of the inherent accuracy of the counter on continuous-wave signals up to the top ranges covered by some transfer-oscillator systems.

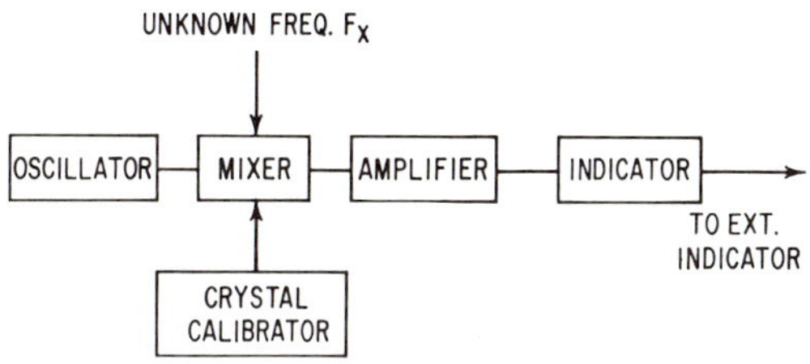

Fig. F-58. Heterodyne frequency meter.

Slotted-line Measurement of Frequency: An electronic counter and transfer oscillator can be used for very accurate measurements of frequency in the uhf and microwave bands. For some measurements, where an accuracy of the

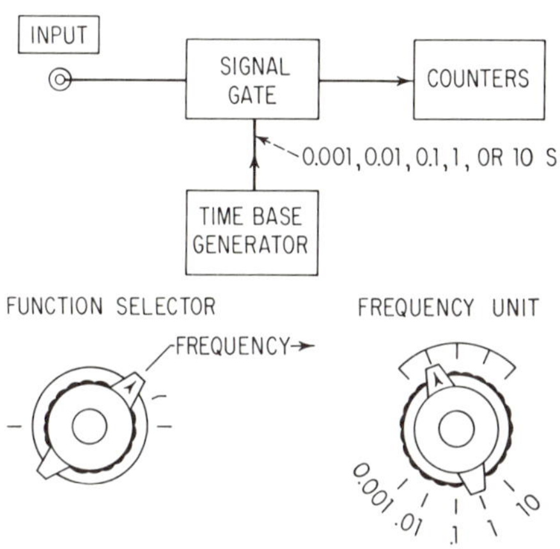

Fig. F-59. Basic electronic counter for frequency measurement.

order of ±0.5 percent is sufficient, the frequency may be determined with a slotted line or slotted section by observing the standing-wave pattern. Figure F-60 shows the standing-wave pattern obtained with a measuring arrangement of the type shown in Fig. F-61. Adjacent minima are one-half wavelength apart. In free space, where the velocity of propagation of electromagnetic waves is approximately 3×10^{10} cm/s,

$$\lambda = \frac{3 \times 10^{10}}{f} \qquad (2)$$

where λ = wavelength, cm
f = frequency, Hz

This equation may be used directly with reasonable accuracy in an air-dielectric coaxial slotted-line measuring system where velocity of propagation is approximately that of free space. In a waveguide system, where velocity of propagation is somewhat less, the measured distance between standing-wave minima corresponds to the guide wavelength, which in turn, depends

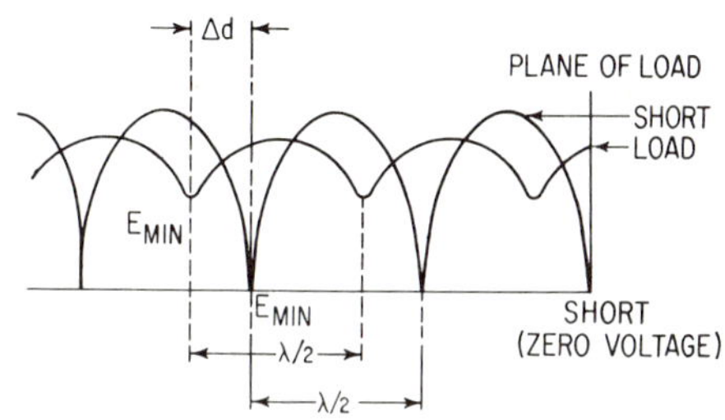

Fig. F-60. Standing-wave patterns.

upon the dimensions of the waveguide. Actual wavelength can be determined from

$$\lambda_g = \frac{\lambda}{\sqrt{1 - (\lambda/\lambda_c)^2}} \qquad (3)$$

where λ_g = guide wavelength, cm
λ_c = cutoff wavelength for the particular mode

In an air-filled rectangular waveguide of width a, operating in the usual TE_{10} mode, $\lambda_c = 2a$.

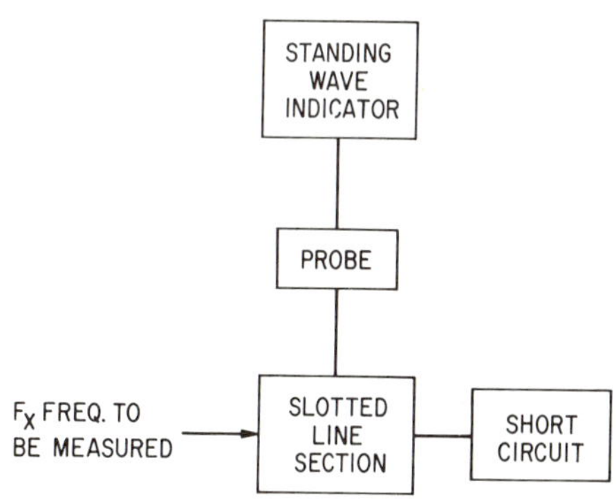

Fig. F-61. Slotted-line measurement of frequency.

Lumped-constant Wavemeters: These are used up to frequencies of 1,200 MHz. As shown in Fig. F-62, a crystal detector is coupled to a resonant LC circuit. For frequencies up to 100 MHz, a fixed inductor and variable capacitor are used to form the resonant circuit. Plug-in coils of various inductance values are used to cover the frequency range. The capacitor dial is calibrated to read frequency directly. When the wavemeter coil is in the presence of the field of the unknown signal, the circuit is tuned to resonate at the same frequency as the unknown signal. The output of the crystal detector is indicated by the meter circuit. In wavemeters operating above 100 MHz, a butterfly resonant circuit may be used. This element varies both the inductance and capacitance simultaneously. The accuracy of a lumped-constant wavemeter depends upon the Q of the resonant circuit. In general, accuracies between ¼ and 1 percent may be obtained.

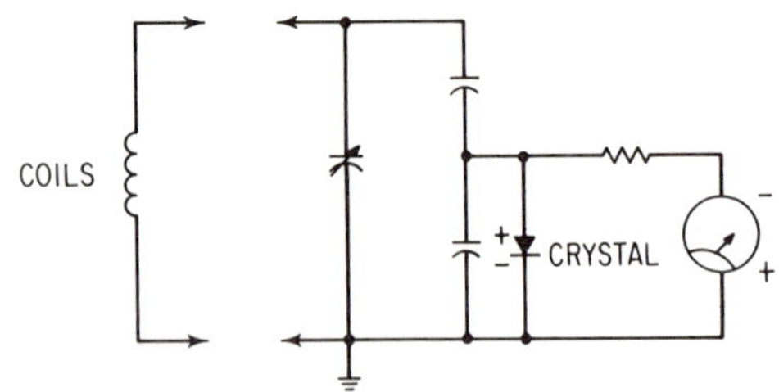

Fig. F-62. Lumped-constant wavemeter.

Cavity-type Wavemeters: Frequency in a waveguide system can be determined easily with a cavity-type wavemeter. Typically, a cylindrical cavity and piston are used as shown in Fig. F-63. As the position of the wavemeter piston is varied, the distributed inductance and capacitance of the cavity, and hence the resonant frequency, are changed. In a reaction-type wavemeter, there is

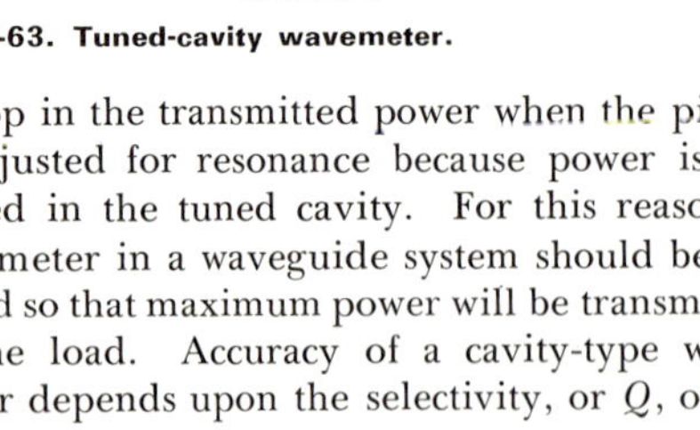

Fig. F-63. Tuned-cavity wavemeter.

a drop in the transmitted power when the piston is adjusted for resonance because power is absorbed in the tuned cavity. For this reason, a wavemeter in a waveguide system should be detuned so that maximum power will be transmitted to the load. Accuracy of a cavity-type wavemeter depends upon the selectivity, or Q, of the resonant cavity. Commercial wavemeters are available with accuracies of 0.1 to 0.01 percent.

Low-frequency Measurement: An interesting scheme is described in Ref. 1.

REFERENCES

1. West, James L.: Two-step Process Speeds Low-frequency Measurements, *Electronics,* May 27, 1968.
2. Harris, Forest K.: "Electrical Measurements," John Wiley & Sons, Inc., New York, 1952.
3. Considine, D. M., and S. D. Ross (eds.): "Handbook of Applied Instrumentation," pp. 15–82 to 15–90, McGraw-Hill Book Company, New York, 1964.

—Extensive editorial guidance rendered by Henry Berring, *Weston Instruments, Inc.*

Frequency, Standard (See **Standards, measurement.**)

Freqency Analysis, Voice (See **Voice-recognition systems.**)

Frequency Generator [See **Signal (electric) generator.**]

Frequency Modulation Modulation (FM) in which the instantaneous frequency of the modulated wave differs from the carrier frequency by an amount proportional to the instantaneous value of the modulating wave. The amplitude of the modulated wave is constant.

An FM-FM telemetry system has the advantage of simplicity and flexibility in the transmission of several analog channels over a common rf link. The high-frequency response is good and the accuracy is quite satisfactory for many applications. An FM system permits the designer considerable latitude in exchanging spectrum utilization and data bandwidth for noise immunity. Since a frequency-division multiplexing format is used, the system enjoys the additional advantage of simplicity in adding or deleting data channels, since all that is required is to add or remove a modular subcarrier oscillator of the appropriate frequency to revise the system data capacity. FM-FM systems normally are used when the following performance levels are desired: *accuracy,* 1 to 5 percent; *data bandwidth,* to about 10 kHz maximum; and *number of channels,* fewer than 25.

Frequency Range, Transducer The range of frequencies, stated as a lower frequency to an upper frequency, over which a transducer or sensor can be expected to produce an output which is proportional to within a stated degree to a sinusoidal variation of input or measurand. The stated degree of proportionality may be specified in decibels, as a percentage, or on a per-unit basis. Also see **Resonant frequency, transducer.**

For many transducers or sensors, the frequency range starts at 0 Hz. Such devices have a static sensitivity to the measurand or input. Often only the upper frequency of the frequency range is stated for such devices. Other transducers, notably those which involve the piezoelectric effect, are devoid of a static response, and the specification of a lower-frequency performance boundary is important. The applicable frequency range for all transducers generally is determined from a plot or data set of the frequency response of the transducer or sensor. The frequency-response plot, or Bode diagram, displays the normalized ratio of output to input or amplitude ratio vs. frequency. Both amplitude ratio and frequency generally are plotted on logarithmic scales, although the amplitude ratio also may be expressed in decibels (20 log e_o/e_i) and then plotted on a linear scale. When such a diagram is prepared, the data points will appear evenly spaced when each test frequency is related to the next higher or next lower test frequency by a constant ratio. A ratio of 1.585, which is the fifth root of 10, thus will produce five data points per decade of frequency.

The specification of the deviation of the proportionality as a function of frequency generally is stated in decibels when the device is of an acoustic, telephone, radio, or related nature. This type of specification, however, is too coarse for most other types of transducers and sensors. Thus, the decibel specification substantially has been replaced by the percentage or per-unit specification. The percentage or per-unit value so given is the difference between the output amplitude and the equivalent input amplitude and can be translated into a limiting amplitude ratio by dividing the percentage by 100 and adding the result—with due respect for algebraic sign—to unity. Thus, if the frequency range is specified as "within $\pm$ 2.5 percent from 30 Hz to 3 kHz," it is understood that the amplitude ratio for the device will be greater than 0.975 and will be less than 1.025 over the stated frequency range of from 30 to 3,000 Hz.

The frequency range over which a certain level of performance can be obtained is for many transducers and sensors a function of the signal or measurand amplitude. Power-handling limitations in the output stage or stages of the transducer or sensor generally are responsible for this. When this condition prevails, the frequency range becomes inversely proportional to the amplitude of the measurand. Device specifications should make a note of this condition. Alternately, it may be more appropriate to specify the maximum rate of output rise of which the device is capable or to give the slewing rate for the transducer or sensor.

—Otto Muller-Girard,
Consulting Engineer, Rochester, N.Y.

Frequency Synthesizer [See **Signal (electric) generator.**]

Frost Point, Measurement of (See **Dew point, measurement of.**)

Fuel Control, Engine (See **Aircraft-engine controls;** and **Governor.**)

Fuel Optimal Control (See **Adaptive control.**)

G

Gage Glass A transparent columnar enclosure which, when connected to a liquid storage tank or pressure vessel through tubular connecting pieces at either end, permits a visual observation of an interface and its position relative to the storage tank or pressure vessel (see Fig. G-1). Gage glasses are of two general categories: (1) tubular and (2) flat or armored.

Tubular Gage Glass: This device consists of a transparent tube (glass or plastic) connected to shutoff valves at each end through stuffing boxes. Where gages are subject to mechanical shock loadings, plastic tubes generally are used to reduce the possibility of gage failure. Tubular gages are inexpensive and are used in noncorrosive applications ranging from low-pressure industrial boilers and liquid-storage tanks to domestic boilers, coffee urns, and other similar apparatus. Principal characteristics are summarized in Table G-1.

Armored Gage Glass: The trend toward higher pressures and temperatures, coupled with the need for greater safety, led to development of the flat or armored gage glass. An armored gage consists of flat pieces of glass clamped to a metal chamber by forged steel covers and gaskets. The metal chamber is connected to shutoff valves by threaded or welded pipe or by flanges (see Fig. G-2). Armored gages are of two types: (1) *transparent* and (2) *reflex*. In the transparent type, a piece of plain flat glass is clamped to each side of the metal chamber to facilitate observation of the liquid color, condition, or interface from one side of the gage while a light source is applied

Fig. G-1. Principle of gage glass. The gage glass and its lower connection to the vessel can be considered as one leg of a U-tube manometer, while the vessel forms the other leg. Since the upper connection between the gage glass and the vessel subjects each leg of the manometer to the same gas pressure, the liquid in the gage glass seeks the same elevation as the liquid in the vessel.

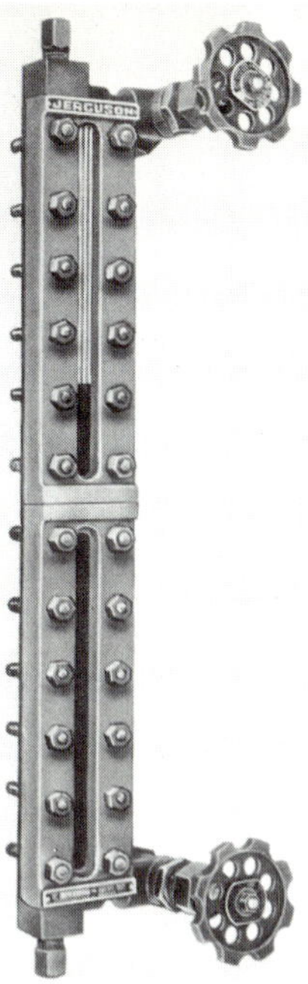

Fig. G-2. Multisection reflex-type armored gage glass. *(Jerguson.)*

at the other side. These gages may be used with liquids corrosive to glass by adding a thin sheet of transparent protective material (mica or plastic) between the glass and liquid (see Table G-1). In the reflex gage, the flat piece of glass is clamped on only one side of the metal chamber. The glass has reflecting prisms molded into one face. The prisms reflect ambient light in the gas space of the gage, creating a silvery mirror effect, while the ambient light in the liquid space of the gage passes through to the back of the chamber, producing a black appearance. Reflex gages are used only where the interface is to be observed and where liquids are not corrosive to glass (see Table G-1).

Magnetic Indicating Liquid-level Gage: A device of this type is shown in Figs. G-3 and G-4. As in a gage glass, the liquid rises or falls within an externally mounted tube. The tube is not of glass but is constructed from a 2-in. type 304 stainless-steel pipe or other nonmagnetic material. A guided float (6 to 20 in. long) made from type 321 stainless-steel tubing moves vertically with changes in liquid level. A permanent magnet is attached to this float, and an indicating mechanism is clamped to the side of the level tube. The coupling between float and indicating mechanism is magnetic. The mechanism comprises a series of small, hinged wafers (vertically spaced on 0.4-in. centers) capable of turning 180°. Each wafer is black on one side (to indicate liquid) and yellow on the other side (to indicate space in the vessel above the liquid line). As the float rides up and down, it flips the magnetized wafers

TABLE G-1. *Principal Characteristics of Gage Glasses*

Characteristic	Tubular Gage Glasses		Armored Gage Glasses: Transparent		Armored Gage Glasses: Reflex		Magnetic Level Indicator
Temperature rating to . . .	150°F	425°F	100°F	750°F	100°F	750°F	450°F
Pressure rating to	600 psi*	350 psi*	5,000 psi	2,800 psi	4,000 psi	2,300 psi	400 psi
Maximum length of tube or of indication	48 in.†	24 in.†	Nine increments—from $3\frac{3}{4}$ to $12\frac{5}{8}$ in.‡				6, 12, 18, 24, 36 in.§
Corrosion resistance and other properties	Limited to noncorrosives. Susceptible to breakage. Usually protected by guard rods		Materials include carbon and stainless steels, nickel, brass, aluminum, and copper-nickel. Carbon-steel chambers may be lined or coated with plastics, lead, and rubber				Types 304 and 321 stainless steels¶

*Operating pressures of plastic tubes (polycarbonate or Kel-F are limited to 300 psi (300°F).

†Glass-tube outside diameter is $\frac{5}{8}$ or $\frac{3}{4}$ in. Wall thickness ranges from $\frac{1}{16}$ to $\frac{3}{16}$ in.

‡Multiple window sections are used to produce gage glasses in visible lengths up to 72 in., with longer lengths possible where required.

§For intermediate ranges and lengths over 36 in., standard sizes can be combined to give continuous indication.

¶This design is particularly useful where glass, gaskets, and standard gage-glass components will not withstand corrosion—where toxic gases and dangerous fluids can be contained within a rigid metal pipe—and for reading interface of similarly colored liquids.

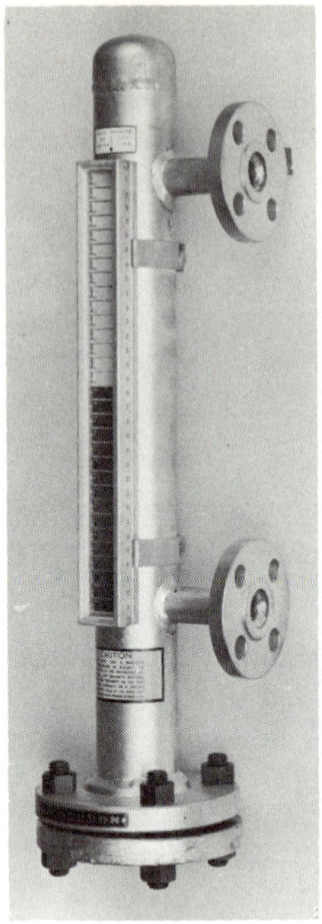

Fig. G-3. Magnetic indicating liquid-level gage. *(Jerguson.)*

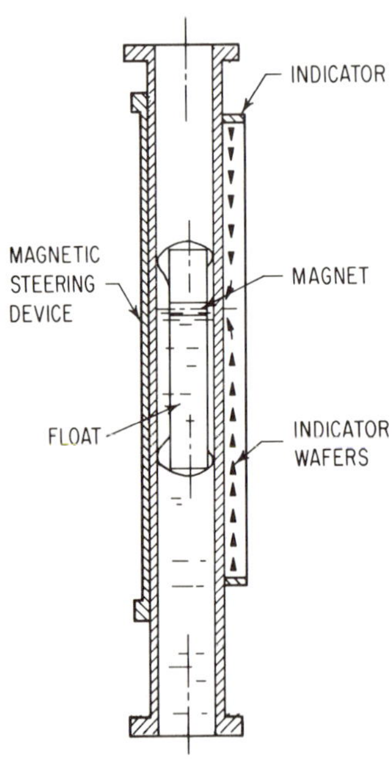

Fig. G-4. Sectional view of magnetic indicating liquid-level gage. *(Jerguson.)*

to either black or yellow. The intersection of the two colors indicates level, and the effect for the operator is the same as a gage glass (see Table G-1).

Gage-glass Variations and Accessories: These include (1) models for low-temperature liquids where a transparent frost-preventing unit is used around the vision slot, (2) heating or cooling chambers to prevent liquid solidification or boiling, (3) pads on the gage-glass assembly to permit welding the unit directly to the vessel wall, (4) use of an internal deflecting vane to indicate the presence or absence of flow in a pipeline, and (5) specially calibrated gage glasses for use as volumetric measuring devices.

REFERENCES

Van Ham, Ernest E.: Some Design Considerations of Liquid Level Gage Glass Design, *ISA Paper,* Apr. 5, 1955.

Van Ham, Ernest E.: True Boiler Water Level Indication, *ASME Paper,* Feb. 23, 1958.

Considine, Douglas M.: Liquid Level Measurement Systems: Their Evaluation and Selection, *Chem. Eng.,* Feb. 12, 1968.

Fribance, Austin E.: "Industrial Instrumentation Fundamentals," McGraw-Hill Book Company, New York, 1962.

—R. A. Pontefract, *Jerguson Gage & Valve Co.*

Gaging, Pneumatic A pneumatic gaging system consists essentially of components that provide a constant-pressure air supply, indicating means, and a metering orifice. The principle of operation is based on the effects of varying the flow of air from the metering orifice. For example, as flow from the orifice is obstructed, pressure in the system will build up to the regulated value, and flow through the system will drop. Over a significant range of values in such a gaging system, there exists a linear relationship between flow or pressure and the size of the escapement orifice. In many gaging applications, the linear relationship is equated to the clearance that separates a sized metering orifice and an obstruction. The indicating component shows the change in flow or pressure as a linear measurement.

The orifice obstruction can be either the workpiece being gaged or an integral part of the gage spindle. An example of the first is found in the gaging of a bore by inserting a gaging spindle, containing two diametrically opposed orifices through which air is flowing. Here, there is no contact between the gaging spindle and the part being measured; the closeness of fit serves as the

TABLE G-2. *Basic Applications of Pneumatic Gaging*

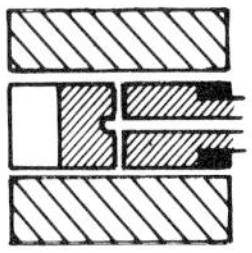
Two diametrically opposed open jets check true diameter of holes having tolerances of 0.005 in. or less.

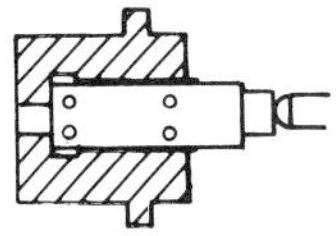
Contact-type gaging head is used in measuring diameters for interrupted bores and bores having a keyway

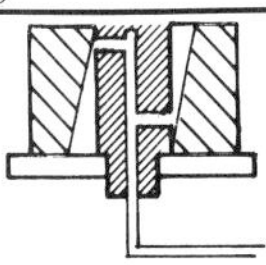
Spindle with opposed open orifices spaced longitudinally checks squareness of a bore axis with a face.

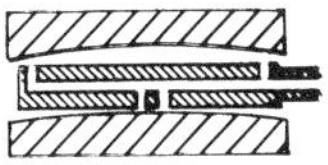
Camber or straightness of hole is checked by rotating through 180° a spindle having four jets

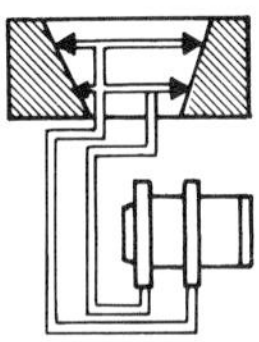
Taper is indicated as any spindle is passed through a bore

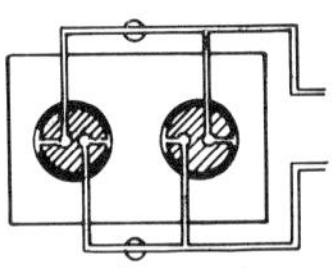
Center distance between holes can be checked by a fixture with two spindles, each having two opposed jets

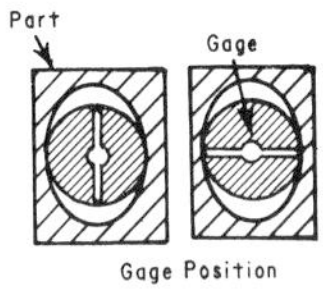

Out-of-round is indicated when any spindle is rotated through 90°

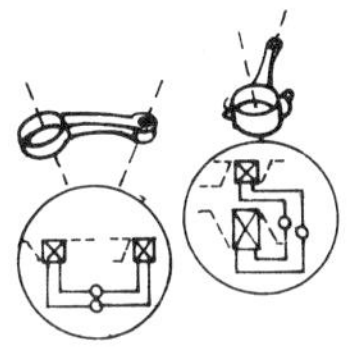
Parallelism of holes can be indicated by combining two spindles of type used in squareness checking

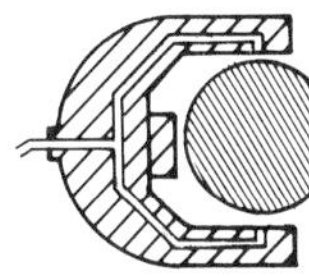
Outside diameter can be checked by two opposed standard jets; tolerance should be 0.002 in. or less

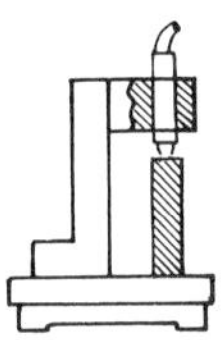
Height, width, or depth can be measured by a suitably fixtured contact-type gaging head

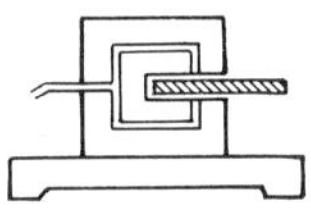
Thickness gage for thin parts and items having close tolerance incorporates two opposed standard jets

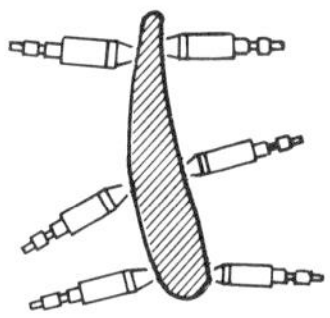
Multiple-contact-type gage heads mounted in a suitable fixture can check contour

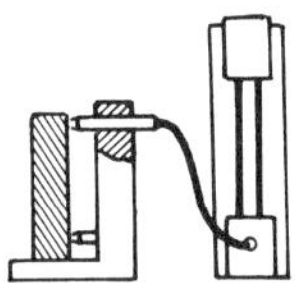
Squareness of surfaces for close-tolerance parts having good surface finish can be checked by a fixtured spindle

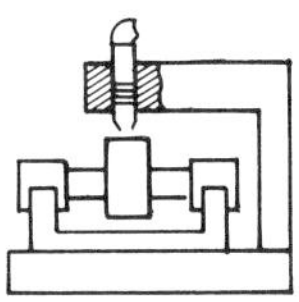
Concentricity can be indicated by one or more contact-type spindles mounted in a suitable gage fixture

obstruction. In the second kind of gage, the gage spindle includes a component that mechanically contacts the work being measured. Movement of this component relative to a sized orifice affects airflow and establishes the basis for measurement.

Basic applications of pneumatic gages are shown in Table G-2. Industrial installations range from the simple application of a single gaging spindle, for checking a single dimension, to a multiple sensing application in which, for example, forty or more dimensions may be checked simultaneously. Pneumatic gages are well suited to postprocess gaging, in which parts are checked against specifications after machining. Pneumatic units also are applied to *in-process* gaging and, in some cases, are tied into complex computer control systems. Air-to-electric transducers frequently are used in automated systems.

Combinations of various gaging devices are used to provide signals and controls for a variety of applications, including (1) adjusting movement of cutting or forming tools, (2) signaling replacement of worn tools, (3) stopping machines, (4) warning when part-size trend approaches minimum or maximum limits, (5) segregating parts at various stages of production, (6) weighing and checking weight relationships, (7) connecting or straightening parts automatically, (8) matching parts for selective assembly, (9) inspecting finished parts, (10) classifying and matching parts, and (11) selectively packaging parts.

Gain (Magnitude Ratio) For a linear system or element, the ratio of the magnitude (amplitude) of a steady-state sinusoidal output relative to the causal input; the length of a phasor from the origin to a point of the transfer locus in a complex plane.

The quantity may be separated into two factors: (1) a proportional amplification often denoted as K which is frequency-independent and associated with a dimensioned scale factor relating the units of input and output; and (2) a dimensionless factor often denoted as $G(j\omega)$ which is frequency-dependent. Frequency, conditions of operation, and conditions of measurement must be specified. A loop-gain characteristic is a plot of log gain vs. log frequency. In nonlinear systems, gains are often amplitude-dependent.

Gain, Loop The absolute magnitude or ratio of the Laplace transform of the return signal (controlled variable) to the Laplace transform of its corresponding error signal (deviation). The loop gain must be less than 1.0 to obtain stability of control. The controller proportional-band setting which produces a loop gain of 1.0 is called the "ultimate proportional band" and is characterized by cycling at constant amplitude and frequency of the controlled variable. Loop gains less than 1.0 result in cycling gradually dying out; with loop gains greater than 1.0, the cycling amplitude will increase. Constant loop gain is a characteristic of a linear control loop where the same damping is applied to all disturbances. If gain varies with the amplitude of oscillation, the loop is nonlinear—with the loop gain varying either directly or inversely with amplitude.

An objective of control-system design is to maintain a constant loop gain and ensure a constant response of the system to disturbances under all conditions of operation. For example, as described by Shinskey (Ref. 1), in a heat exchange process which is a single capacity plus dead time process, the dynamic gain is inversely proportional to product flow. Compensation is accomplished by use of an equal-percentage valve.

Loop gain is the product of the gains of all the elements of the loop including the process. Normally, the gain of only one of the elements is adjustable, that is, the controller. The other elements—final control element, transmitter, process, and so on—contribute their gains as determined by their design. The element gains may not be constant as in the case of an equal-percentage valve, which intentionally is designed to compensate for process-gain variations, such as pressure loss in piping and fittings.

REFERENCE

1. Shinskey, F. G.: "Process-control Systems," McGraw-Hill Book Company, New York, 1967.

—Richard H. Kennedy, *Bristol Division, American Chain & Cable Company, Inc.*

Galvanometer An instrument for indicating or measuring the directional flow of a small electric current, or a function of the current, by means of a mechanical motion derived from electromagnetic or electrodynamic forces which are set up as a result of the current. The basic mechanism is used as a component in the measuring systems of many types of instruments. (See **Electrical quantities, measurement of.**) In some cases, the instruments are, in themselves, termed galvanometers as found in connection with laboratory-type galvanometers.

d'Arsonval Galvanometer: As shown by Fig. G-5, this device comprises a lightweight coil of fine wire suspended in the field of a permanent magnet. An iron cylinder is added inside the coil to intensify the field, and a mirror is mounted

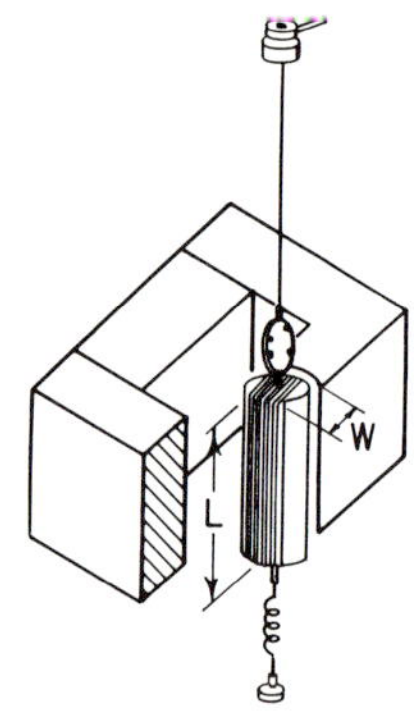

Fig. G-5. d'Arsonval galvanometer.

on the rotating system to indicate motion. The device has the following characteristics:

N = number of turns in coil
W = width of coil, cm
L = length of coil, cm
B = field strength in which coil operates, gauss

The force on a conductor 1 cm long carrying 1 abampere (10 amperes) in and normal to a magnetic field of 1 gauss or 1 line/cm² is 1 dyne at right angles to both the conductor and the field. The total force on one side of the coil, when it is carrying a current of I' abamperes, is

$$F = I'BNL \tag{1}$$

Since the force is applied at a radius of $W/2$ cm, the turning moment, or torque in dyne-centimeters, is

$$T = \frac{I'BNLW}{2} \tag{2}$$

Since both sides of the coil are working, the total torque is twice that of Eq. (2), or

$$T = I'BNLW \tag{3}$$

The product of the width and length of the coil W is equal to the area. Therefore, A can be substituted for LW. Also, it is possible to write I (amperes)/10 $= I'$ (abamperes). The result is

$$T = \frac{IBNA}{10} \qquad (4)$$

Equation (4) gives the torque in dyne-centimeters in terms of the dimensions of a moving coil, the ampere-turns in it, and the magnetic field through which it moves. This is the fundamental equation for all moving-coil electromagnetic instruments. Equation (4) applies to highly sensitive suspended galvanometers as well as to conventional millivoltmeters, voltmeters, and ammeters. Also Eq. (4) may be considered to apply to electrodynamometer wattmeters, in which case the magnetic field is that produced by the fixed-field coils rather than by a permanent magnet.

The shape of the moving coil depends upon the final intended use. If sensitivity to low currents is the only criterion, the coil should be square. However, when arranged for a high sensitivity with low restoring force and high flux density, the square coil responds very slowly. This is because it has a relatively high moment of inertia and a high degree of damping. The most sensitive galvanometers generally have a coil so proportioned that the length is three to four times the width. Indicating instruments, where ample energy is available, will be found to use coils more nearly square.

Magnetic-system Characteristics: Since the current sensitivity is proportional to the magnetic field, consideration of the permanent-magnet system which produces the field is next in order. Referring to Fig. G-6, the dimensions of the magnetic system are:

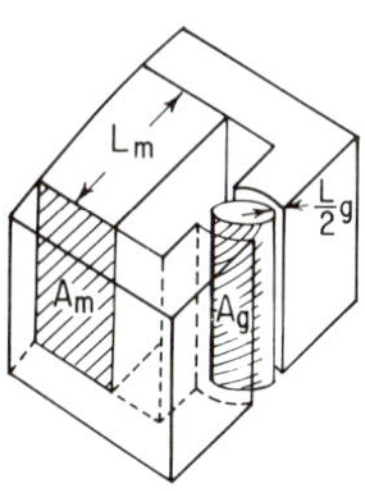

Fig. G-6. Magnetic system for a galvanometer with important dimensions indicated.

L_m = length of magnet
L_g = length of air gap
A_m = cross-sectional area of magnet
A_g = cross-sectional area of air gap

In designing a moving-coil instrument, whether galvanometer or ammeter, a certain flux density B_g is needed. This is the flux density of working air gap which has an area A_g and a total air-gap length L_g (the sum of two gaps in series). This is to be obtained by means of a suitable magnet material, the magnet having a length L_m and a cross section A_m. A permanent-magnet material can be described by (1) its demagnetization curve, which is that portion of the BH (gauss vs. oersteds) curve in the second quadrant; and (2) an associated external-energy curve comprising the product of the BH values. A typical curve for Alnico V is shown in Fig. G-7. The BH product curve varies with the BH ratio and is important to magnetic-system design, since it is desired that the magnet operate at or close to its point of maximum energy.

Assuming that it is desired to use this particular magnet material in a moving-coil instrument, it is necessary to solve for the dimensions of the magnet by means of the following equations:

$$\text{Magnet length } L_m = \frac{B_g L_g}{H} \qquad (5)$$

$$\text{Magnet area } A_m = \frac{B_g A_g}{B} \qquad (6)$$

where L_m = minimum length of magnet, cm
B_g = flux density in air gap, gauss
L_g = length of total air gap, cm
H = magnetizing force at operating point
A_m = cross section of magnet, cm^2
A_g = cross-section area of air gap, cm^2
B = flux density in magnet, gauss

Thus, it is possible to obtain values for the magnet cross section and length. If the shape of the air-gap system is such that it is necessary to place the permanent magnet rather far away, more flux will be lost in leakage than would be the case with some other conformation where the magnet itself could be closely associated with the air-gap system. Thus, flux leakage may vary from as little as 10 to as much as 50 percent. Additional magnet cross section is required to furnish this leakage flux. The magnet length from Eq. (5) is a minimum for stable operation. Slight excess in length is desirable.

Equation of Motion: The moving system of

a galvanometer has mass, inertia, a restoring force, and an electrical winding. The balancing of all these factors, together with the magnet system, indicating means, and a housing, is the designer's problem. Since galvanometers are used for many purposes, many different designs result. In a spring-controlled moving system where the controlling torque is proportioned to the angular deflection and where the damping associated with the system is proportional to the velocity of motion, the oscillation resulting is a harmonic motion. This motion was first represented in the form of a differential equation by Gauss and Weber:

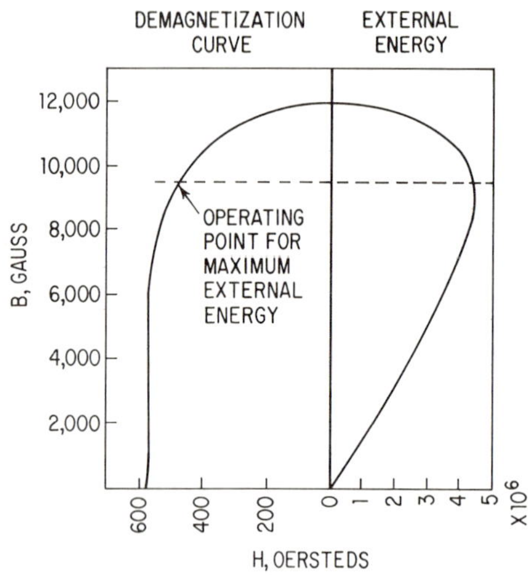

Fig. G-7. Demagnetization and external-energy curve for Alnico V.

$$K\frac{d^2\theta}{dt^2} + G\frac{d\theta}{dt} + S\theta = 0 \qquad (7)$$

where θ = angular deflection from initial position, rad
t = time from initiation of motion, s
K = moment of inertia of system around axis, g-cm^2
G = damping coefficient, g-cm/(rad)(s)
S = spring or control constant, g-cm/rad

When such a system is critically damped, however, that is, when its character of motion lies at the border line between just oscillatory and just overdamped, the following relation holds:

$$G_0 = 2\sqrt{KS} \qquad (8)$$

It is desirable to term the ratio of the actual damping coefficient G to the critical damping coefficient G_0 a new factor n. Thus, Eq. (8) can be reduced to a simple dimensionless form based upon two parameters. Expressed mathematically, the value of the damping coefficient for any damping is

$$G = nG_0 = 2n\sqrt{KS} \qquad (9)$$

Also, for damping less than critical,

$$n = \frac{\delta}{\sqrt{\delta^2 + \pi^2}} \quad \text{and} \quad \delta = \log_e k \qquad (10)$$

where δ = logarithmic decrement of motion
k = damping factor (defined as the ratio of a deflection or swing to the successive swing in the opposite direction)

The factor k also may be considered as the reciprocal of the percent overthrow when an impulse is applied to the instrument. This value n is less than 1 for less than critical damping, is equal to 1 when damping is critical, and is greater than 1 for the overdamped condition. The term n is variable, since other factors being maintained the same, it *varies directly* as the square of the total flux through the coil, the coil area, and the number of turns of wire in the coil—and it *varies inversely* as the coil circuit resistance, and the square root of the moment of inertia, the spring constant.

Damping associated with the moving system usually is obtained by induced currents in the moving-coil winding which, for example, flow through (1) the external closed circuit of a thermocouple or (2) a shunt resistance connected internally in the instrument which may or may not involve a more elaborate network for temperature compensation. The moving coil also may be wound on an aluminum frame which contributes to mechanical strength and secures further damping arising from eddy currents. With high flux density and many turns in the moving coil, an overdamped condition frequency exists, causing an unduly long time delay for the instrument pointer to reach its final position. In some instances, the circuit resistance arbitrarily is increased or the flux reduced in order to obtain less damping—even at the sacrifice to some degree of absolute torque.

Ballistic Galvanometer: This device usually is used to measure the quantity of electricity (coulombs) and generally is of the moving-coil type. In ballistic applications, it is desirable that the time of current flow in the galvanometer be short as compared with the period of the galvanometer, since it only is under these conditions that accurate indication will take place. Thus, it is desirable to make the galvanometer period long. When impulses being measured have a prolonged time duration, it is important that the instrument be very highly overdamped if high accuracy in integrating the value of the impulse is required. This is true, for example, in a fluxmeter.

Galvanometer Recorder (See Oscillograph.)

Gamma-ray Spectroscopy Gamma rays are high-energy electromagnetic radiation with very short wavelengths (10^{-18} to 10^{-11} cm). The rays penetrate matter deeply—on the average much more deeply than do alpha and beta rays, which are charged particles. Deep penetration makes gamma rays useful in the laboratory and industry in much the same way as are x-rays. Gamma rays usually are detected by observing the effects that they produce in matter when the rays encounter an atom. Important among these effects are (1) the photoelectric effect and (2) the Compton effect. The photoelectric effect occurs when the gamma ray strikes one of the orbital electrons of the atom, transferring its energy to the electron. This produces a free electron and an ionized atom. The Compton effect arises where the gamma ray strikes an orbital electron without imparting all its energy to the electron. The electron is detached from the atom but receives only part of the gamma energy. The remaining energy persists as a scattered gamma ray with lower energy than the initial ray. This scattered ray may further collide with one or more other atoms, freeing other electrons. These types of interactions occur variously in nuclear-radiation detectors. In each detector type, some observable reaction results and in one manner or other produces an electrical-output charge suitable as an input for an electronic measuring system.

Measurements of gamma radiation are made chiefly in two ways: (1) a record is made of the number of counts as a function of energy, in which case a gamma-ray spectrum is obtained; and (2) time relations are observed, in which case several types of information may be desired. Commonly used detectors include scintillation, semiconductor, and gas proportional detectors. The scintillation detector often is preferred where high efficiency is more important than resolution. Efficiency is defined as a measure of the probability that an incident gamma ray will interact with the material in the detector. Semiconductor types are used increasingly, particularly where high resolution is required.

—Tracy S. Storer, *Hewlett-Packard*

Gas Analysis (See **Chromatography; Combustion/combustibles analyzer; Infrared process analyzers; Oxygen analyzer; Thermal-conductivity gas analyzer;** and **Ultraviolet spectrophotometry.**)

Gas Density, Measurement of (See **Specific gravity, measurement of.**)

Gas-ionization Tube One of the simplest devices for radiation detection (see Fig. G-8). Consider a gas-filled cylinder with a fine axial wire and an ionizing radiation particle traversing the tube volume. As the particle traverses the tube, the gas is ionized. This ionization consists of positive ions and electrons. As the fine central wire (anode) is at positive potential with respect to the cylinder (cathode), the electrons move to the wire in a region of increasing field strength and the

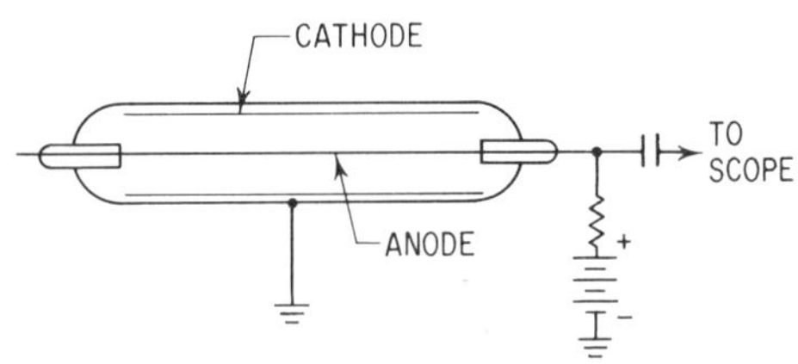

Fig. G-8. Gas-filled diode.

positive ions move outward to the cathode. The electrons are collected in a short period of time while the positive ions have moved only a short distance out toward the cathode. At low voltages (ion-chamber region), no additional ions are created by collision and the number arriving at the central wire will be essentially the number produced in the initial *ionizing* event (neglecting recombination). The proportional region is reached when the counter voltage is raised to that point where secondary electrons are first formed by collisions of the primary electron with gas molecules. As the voltage on the counter is increased further, this multiplication of electrons increases until the phenomenon known as the *Townsend avalanche* occurs. This is the cumulative process in which the initial electrons and their progeny also produce more electrons. The pulse on a proportional counter is given by

$$V = \frac{Ane}{C}$$

where V = pulse size, volts
A = gas amplification
n = number of electrons in initial ionizing event
e = charge on electron, coulombs
C = distributed capacity of central wire system, farads

The gas amplification A is defined as the total number of all electrons that are produced (including the original electron) as a result of each initial electron traveling from its origin to the central wire ($A = 1$ for an ion chamber). The proportional region is characterized by amplifications greater than 1, up to values of 10^5 or 10^6. In this region, each discharge is a single avalanche, originates from a primary ion pair, and is limited to a small distance along the wire. The output-voltage pulse is proportional to the number of ions formed in the initial ionizing event. As the voltage is increased further, each avalance produces new avalanches until the discharge spreads along the whole length of the central wire. The amplification increases until it is no longer dependent on the amount of primary ionization and all discharge pulses become equal in amplitude. This is the beginning (threshold) of the Geiger-counting region (see Fig. G-9).

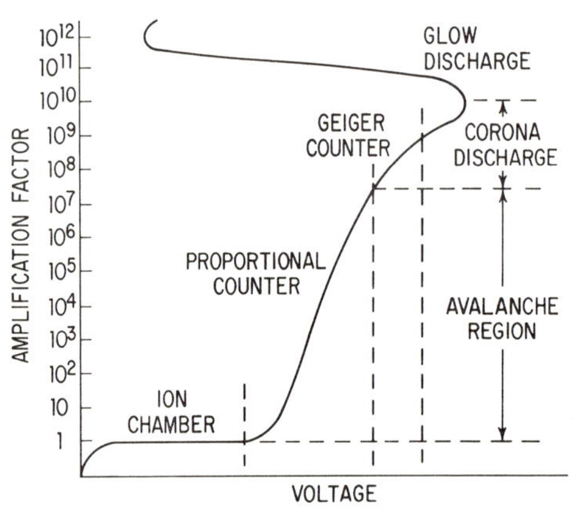

Fig. G-9. Relation of voltage and amplification in gas-filled diode.

Gas Meter (See **Flowmeter, positive-displacement.**)

Gas Thermometer (See **Thermometer, filled-system.**)

Gasometer Sometimes referred to as a *meter prover,* the gasometer is a laboratory-type device used for calibrating gas meters, particularly bellows-type gas meters used for measuring fuel and other gases. The apparatus consists of two cylindrical tanks, each with one open end, of dimensions such that the smaller will have a diameter from 3 to 6 in. less than the larger (see Fig. G-10). If the larger tank is filled with water to within about 6 in. from its upper end and if the smaller tank is inverted, with its open end downward, and

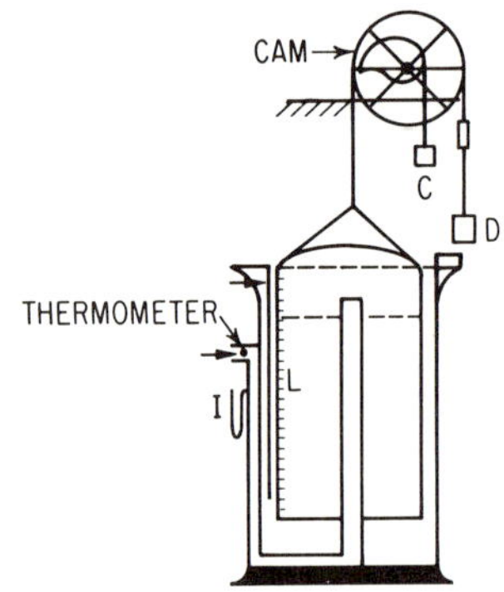

Fig. G-10. Gasometer or meter prover.

introduced into the first tank, the result essentially will be a gasholder. The balancing mechanism to compensate for the weight of the tank consists of a weight D to balance in whole or in part the weight of the upper tank. Weight C compensates for the depth of immersion or buoyant effect of the bell or upper tank—hence the special design of cycloidal cam giving a variable effective radius. Readings of gas temperature and pressure within the bell and a means to measure the vertical linear displacement of the tank are required. A thermometer, introduced as shown, measures the temperature. A manometer I, as connected, determines the pressure of the bell. A scale L, reading in cubic feet and decimals, is fixed vertically to the moving tank. A stationary indicating finger on the fixed tank indicates the vertical distance moved or volume displaced. Thus, the apparatus serves as an accurate means for volumetrically determining the gas that has passed through a meter under test. The instrumentation provided allows for calculating all measurements in terms of standard conditions.

Gate A gate is a signal which enables an AND circuit to pass a signal. The gate signal usually is of longer duration than the signal to ensure that coincidence occurs. For example, in conditioning the set pulse of a flip-flop, the gate must precede the set signal so that the negative shift will be recognized by the transistor.

The term gate is sometimes used to describe a circuit having a binary output which is completely determined by the binary state of its input signals such as in the AND and OR gate circuits.

—W. W. Welz, *IBM Corporation*

Gaussmeter A form of direct-reading magnetometer particularly suited for the measurement of field in the air gaps of instrument magnets as well as magnetic structures that present difficult access problems. In one version, a small magnet of cylindrical shape is placed at the lower end of a long, thin shaft, which is part of the moving system of the instrument. Jewel bearings support the moving system and a spiral spring provides the restraining torque. A pointer is carried by the upper end of the shaft and moves over a small instrument dial that is calibrated in gauss. Shaft and magnet are enclosed within a walled tube that also carries the lower pivot jewel. Thus, the instrument is configured as a probe with diameters ranging from 0.052 to 0.090 in. Probe lengths vary from 1.25 to 5 in. Measurement ranges are available from 100 to 5,000 gauss. In use, the probe simply is placed in the magnetic field; then the instrument case is rotated until the pointer reaches zero on the scale. Inasmuch as the probe magnet is in alignment with the external field, there is no torque on the moving system. Thus, the pointer indicates the direction of the field. The instrument case then is rotated until the pointer reaches a maximum up-scale reading. When in this position, the probe magnet is at a right angle with reference to the field being measured. The scale at this time reads field strength directly in gauss. Other forms of gaussmeters include the Grassot fluxmeter and the Hall-effect gaussmeter. See **Magnetic quantities (variables).**

Geiger Counter A radiation counter that uses a Geiger-counter tube. See **Geiger-Mueller tube;** and **Gas-ionization tube.** The Geiger region for counting begins at that point where all discharge pulses attain the same amplitude. This pulse size is independent of the primary ionization. The Geiger threshold is best located by use of an oscilloscope to determine the lowest voltage at which all pulses become equal in size.

Geiger counters usually are constructed with a cylinder and a fine axial wire; the cylinder is operated at ground potential and the wire at positive high voltage. Geiger counters may be classified into two categories as characterized by their filling gases: (1) the non-self-quenching counter filled with the rare gases or air or some mixture of them and (2) the self-quenching counter usually filled with a mixture consisting of a rare gas and a small percentage of some polyatomic organic quenching vapor (or halogen gas-filled).

Geiger-Mueller Tube A radiation-counter tube operated in the Geiger region and the primary component of a Geiger counter. The tube usually consists of a gas-filled cylindrical metal chamber containing a fine-wire anode as its axis. See **Geiger counter.**

General-purpose Computer A stored-program digital computer which is designed to solve a large variety of problems is usually considered a general-purpose computer, as compared with the special-purpose computer whose design is optimized toward solving a particular problem. A particular general-purpose computer may operate more or less efficiently in some applications than another general-purpose computer. For this reason, general-purpose computers are sometimes divided by application into commercial, scientific, process-control, and data-acquisition computers (see also **Commercial computer; Direct digital control; Scientific computer; Data-acquisition computer**).

—W. W. Welz, *IBM Corporation*

Gentile Tube (Foster) A primary flowmetering device that operates on a principle quite different from other devices in this general category (see Fig. G-11). As with other primary flowmetering devices, the Gentile tube accelerates the flow in the throat. Unlike the others, however, the Gentile tube does not sense the static-pressure difference between the inlet and throat created by this acceleration, but rather the device compares

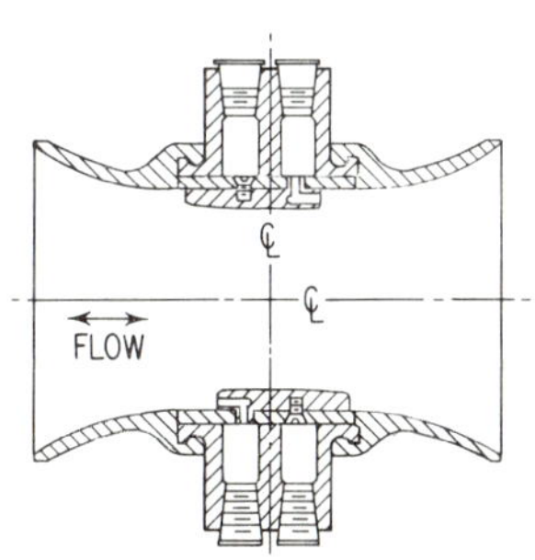

Fig. G-11. Sectional view of Gentile tube.

the total upstream and downstream pressures in the high-velocity region to produce a differential pressure.

The discharge coefficients of these tubes are not published. An ASME paper by Prof. Hooper (No. 49-A-88) presents the coefficients of a few tubes, indicating coefficient values of 0.85 to 1.30. As reported in this paper, the coefficients of these tubes do not obtain a constant value at any Reynolds number. According to the inventor's note, the power of the differential in the flow formula has to be changed in order to get a constant coefficient. The uncalibrated tolerance on this discharge coefficient is claimed to be ±1 percent.

Energy-recovery properties, although somewhat poorer, are similar to Dall tubes. The Gentile tube is sensitive to flow-pattern changes and is affected by downstream piping. Advantages of the tube include short laying length and a capability to measure flow from both directions. Since the *Y* factor has not been established, accurate metering of compressible fluids cannot be assured.

—Edmond B. Lynch, *B I F, Unit of General Signal Corporation*

Geophysical Instruments Instruments for finding oil, gas, and mineral deposits have evolved from such classic devices as (1) the pocket transit, (2) the clinometer (for measuring angles of slope and dip in earth strata), (3) the hand-carried aneroid barometer (for course measurements of altitude), (4) the protractor, and (5) the surveyor's Jacob's staff. In a general sense, instrument development has coincided with expansion of the field of geophysics proper—from basic surface exploration to more sophisticated subsurface exploratory techniques.

Classes of Measurements: Instrumentation is determined by the basic divisions into which geophysical exploration falls: (1) *naturally occurring phenomena,* such as gravity and electrical and magnetic fields; and (2) *induced effects,* typified by deliberate explosions to induce seismic waves, sound impulses, or the generation of electric currents in the earth. Further, data may be gathered (1) at the earth's surface, (2) by airborne instruments, (3) atop or at the bottoms of inland and coastal waters, and (4) in boreholes through the earth's crust.

Gravity Methods: Though mass, size, and shape of the earth itself are prime determinants of gravitational acceleration, rock masses in a given environment also are important factors in variations of gravity. Gravity increases from the equator to the poles because of polar flattening and equatorial bulge. Gravity varies with elevation at any latitude and with topography in the vicinity of an observation point. Hence, the basis of gravity exploration lies in adjustment of observed gravity values in direct relation to the effects of local geology. This is most often true in petroleum exploration, although metallic-mineral prospectors utilize the technique in view of density differences between metallic ores and surrounding earth. Instrument design is classified by methods of obtaining readings: (1) torsion balances, (2) pendulums, and (3) gravimeters most often are used.

Earlier days of prospecting were associated with both the Eötvös torsion balance for determining gravity gradient and pendulum-type instruments. Also see **Gravity, measurement of; Gravimeter.** Such instruments needed hours to settle into a reading, accounting for gravimeters' supplanting the older instruments. Gravimeters do not read gravity directly. Rather, their detection of differences in gravitational attraction depends upon a weighted boom pivoting about a hinge point (see Fig. G-12). Linked to a spring system, the boom provides a state of relative unstability—hence sensitivity to changes up to 1 part in 100,000,000.

Boom deflections from null position are measured by the observation of changes in spring tension required to return the boom to null. Graduated dials show the necessary readings. Two calibration methods are typical (for required

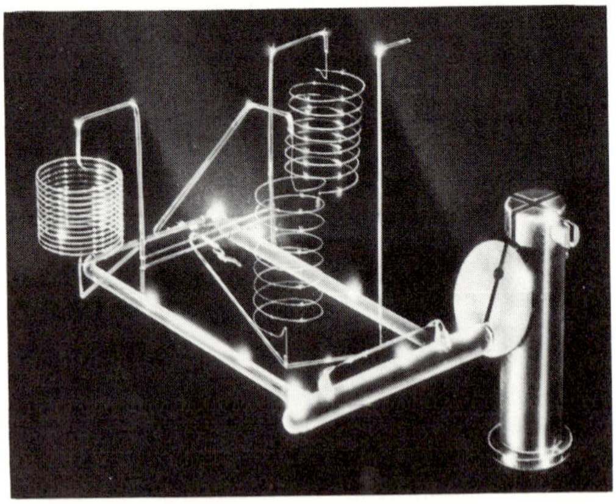

Fig. G-12. Gravity-sensing element in the Worden gravity meter. The weighted boom, pivots, and springs are made of fused quartz.

gravitational significance): (1) one method follows a series of *comparative readings* at sites at which the change in gravity is known from pendulum readings; and (2) the other method involves *tilting the instrument* through a known angle, with readings being taken for simulated gravity changes and calculating "apparent" decreases in gravity that vary as the cosine of the angle of tilt.

Whatever instruments are chosen for gravity calculations, "densities and disposition" of underlying formations are the variables which are detected and evaluated in terms of gravity differences. For oil and gas exploration, significant geologic structures associated with such deposits (salt domes, anticlines, buried ridges) produce significant readings which lead to relatively accurate exploratory drilling.

With watertight housings, self-leveling gravimeters are used, with remote electronic logging, for underwater work.

Concentration on development of petroleum resources led to the development of seismic exploration techniques. The first field work in the United States began in the mid-1920s, using analog instruments. In 1957, a joint effort by two oil companies, a geophysical-service company, and an electronics company resulted in the development of digital systems which rapidly supplanted the older analog techniques. The primary justification for the development of digital seismic systems was to allow a far greater exploitation of the potential information which may be obtained from seismic data utilizing statistical communication theory.

Typically, the complete seismic-data program consists of three distinct phases: (1) *acquisition* of data in the field (either land or marine environments), (2) *processing* of the data by a computer system, and (3) *display* of the data. In field operations on land, a rugged vehicle to facilitate exploration in all types of terrain is required. Processed data are made into a seismic cross section by use of a multitrace, oscillographic, X-Y recorder or a cathode-ray oscilloscope. Evaluation and interpretation are done by geophysicists to locate those conditions most favorable to the presence of oil.

Magnetic Techniques: For subsurface exploration (both mineral and petroleum), magnetic techniques depend upon local field variations caused by differing magnetization of surface and near-surface rocks. With the exception of magnetite (Fe_3O_4), most ore bodies display only slight magnetism. Usually magnetic explorations are relative only; hence measurements are made over a uniform grid pattern. A vertical field balance often is used for surface work. Essentially, this consists of a magnet system in which a turning moment (arising from the magnetic field) about a horizontal axis is compensated by gravity moment that results from an off-center weight. An instrument of this type measures the vertical component of magnetic intensity. Similar instruments measure the horizontal components of magnetic fields at points on the grid. *Universal* magnetometers can measure the intensity of all magnetic components, plus their direction in space, but they are relatively cumbersome.

Airborne magnetometers, available since the mid-1940s, have been used extensively in mineral exploration. The more recent "bucket of water" magnetometer has come into wide use. This instrument provides absolute measurements of magnetic strengths through a count of the magnetic properties of protons. In use, a field transverse to and larger than the field to be measured is applied. When suddenly removed, the applied field allows precession of the proton gyromagnets to a new distribution that is characteristic of the field being measured. Frequency of such precession is a function of the field intensity.

Sonic (also continuous velocity, acoustical) *logging* automatically records variations in acoustical properties of strata penetrated by a borehole. Instrumentation consists of a sound transducer with frequencies ranging from 1 to 10 Hz, plus one or two detectors. Kept parallel to the bore's axis by a "centralizer," the instrument's impulses travel through the drilling mud, through a short section of the rock stratum, and back across the mud column to the detector(s). At the surface, a sonic log records transit time (time required for sound to travel 1 ft through rock formation) in the form of curves similar to electrical-wave log readouts. Varying porosities of discrete strata show different transit times, hence variation in the recorded log. Such measurements are often correlated with self-potential electric logs or gamma-ray logs to facilitate accurate interpretation of subsurface conditions. Primarily for oil and gas exploration, sonic logging is little utilized for detection of minerals.

Because subsurface geologic structures associated with natural gas are closely analogous to those for petroleum deposits, instruments for natural-gas exploration include all those typically used for petroleum detection—and include gravimetric, magnetic, electrical, and seismic means. Also see **Seismology.**

REFERENCES

Runcorn, S. K.: "Methods and Techniques in Geophysics," vol. 1 (1960), vol. 2 (1966), John Wiley & Sons, Inc., New York.

Worzel, J. L.: "Pendulum Gravity Measurement at Sea," John Wiley & Sons, Inc., New York, 1965.

White, J. E.: "Seismic Waves," McGraw-Hill Book Company, New York, 1965.

Chow, V. T.: "Handbook of Applied Hydrology," McGraw-Hill Book Company, New York, 1964.

—Anthony F. Amante, *Texas Instruments Incorporated*

Glass Electrode (See **pH, Measurement of.**)

Glass Thermometer (See **Thermometer, liquid-in-glass.**)

Glossmeter Gloss is the light reflecting property of a surface. In precise terms, this is *specular gloss,* or mirrorlike reflection. Gloss is sensed as shininess, sheen, and brilliance of highlights. A glossmeter serves to evaluate this visual impression in a quantitative way (Ref. 1). The physics of the phenomenon are clear. Light can be considered as a stream of particles (photons). When photons impinge on a surface, some of them are bounced back at the interface. Some photons may penetrate the material and be absorbed or scattered, giving rise to color appearance. See **Color, measurement of.** The gloss component represents only those photons which do *not* penetrate but which are *reflected.* The angle of reflectance is equal to, and opposite the angle of incidence normal to the surface.

A typical glossmeter consists of a light source and simple optics to direct a defined beam onto the sample. In the opposite direction, there is a light detector (photocell or phototube) connected to a readout meter. The instrument has to be set up with respect to the angle of incidence (and reflectance). Various industries have reached agreement in this respect. Gloss of paper and paperboard is determined at an angle of 75° (Ref. 2). The surfaces of metals and porcelain enamel are measured at 45° (Ref. 3). The gloss of paints, lacquers, and related products usually is evaluated at 60° (Ref. 4); and at 85° for assessing the "sheen" of flat paints, or alternatively, at both these angles and at 20° according to some methods (Ref. 5). The internal arrangement of a typical glossmeter is shown in Fig. G-13. Instruments also are available that can be set to a variety of angles to accommodate the diverse test methods prescribed for various materials. In most routine applications, glossmeters of this general type are used. The instrument first is set to read the value of the standard. Then, the gloss of one or of a series of samples simply is read. More complex and costly instruments are available for research applications.

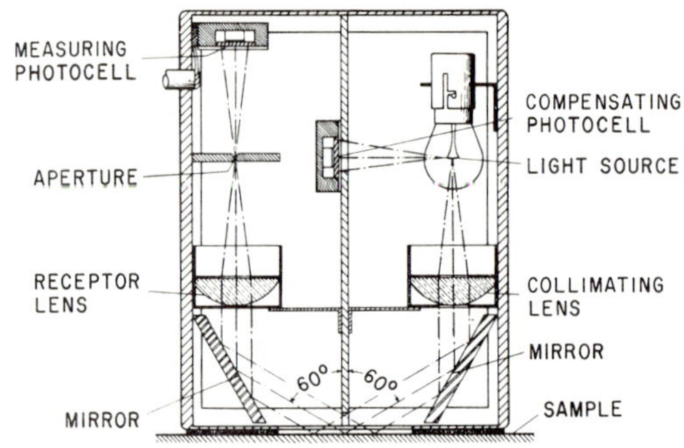

Fig. G-13. Optical scheme of a typical 60° glossmeter. *(Photovolt.)*

The *unit* or *degree of gloss* is defined as 1,000 for a theoretically perfect mirror. A physical standard, the gloss of which is known, is used to calibrate the instrument. For instance, a black glass of refractive index 1.567 is assigned a gloss value of 100 (Ref. 5).

Goniophotometers scan and map reflectance through almost a complete arc. Thus, the diffusely reflected light, as well as the specular component, is included. This information is valuable in research work. The instrument will pick up significant differences on *pearlescent* materials and, in general, will put subtle differences in reflecting characteristics into a quantitative form by way of a plot of intensity of reflected light vs. angle (Ref. 6).

Distinctness-of-image Glossmeter: This is a specialized instrument (Ref. 7) which is particularly applicable to evaluation of consumer goods, such as automotive finishes and chrome-plated flat surfaces. While a standard glossmeter will quantitate the average gloss impression as seen by the eye, the instrument will give no clue to distortions or haziness which modify the subjective visual impact. This can be done by more sophisticated instruments which evaluate the distinctness of reflection. In essence, an instrument of this type is a dynamic glossmeter in which a rotating slit scans the reflected beam.

All glossmeters require exact positioning of the sample against the instrument. The least vibration, with resultant change of geometry, will destroy the measurement. Surfaces which are not absolutely flat or which have scratches or are rough cannot be measured accurately. Some glossmeters furnish an electrical analog to the reading, and this signal can be used for printout or computer manipulation.

REFERENCES
1. Judd, D. B., and G. Wyszecki: "Color in Business, Science and Industry," John Wiley & Sons, Inc., New York, 1963.
2. TAPPI Standard T480ts-65.
3. ASTM Method C 346–59.
4. Federal Test Method Standard No. 141/6101.
5. ASTM Method D 523–67.
6. ASTM Method E 167–63.
7. Hunter, R. S.: *ASTM Bull.* 186:48.

—Gabor B. Levy, *Photovolt Corporation*

Glover-type Gas Meter (See **Flowmeter, positive-displacement.**)

Golay Detector A small (3-mm circle) radiation-transparent cell that contains a gas, the pressure of which is a measure of the incident radiation. The cell contains a thin film that absorbs radiation and hence affects the gas temperature and pressure. Time constant of the cell is 0.015 s and frequency of measurement is 10 Hz.

Goniometer An angular measuring device essentially similar to a circular table or a dividing head with the exception that autocollimating telescopes are used to establish the datum values for the readings that are taken from the divided circle—either directly or through a micrometer eyepiece. The instrument is used for determining circular divisions and utilizes the basic principle of the divided circle (see Fig. G-14).

The instrument consists of a table support that rotates about a vertical axis and carries a divided circle and reading system. One, two, or four microscopes may be used to read the divided circle. In addition to the table and divided-circle system, the instrument normally will utilize two autocollimator telescopes mounted to the instrument. Usually one telescope is fixed; the second is allowed to rotate about a horizontal axis. Therefore, a datum can be established for use of the fixed telescope, and the angular values can be read through use of the rotatable telescope.

The goniometer permits rapid inspection of plane surfaces, such as prisms or parallel plates. After one face is viewed, the telescope can be turned with the divided circle so that a second face of a prism can be viewed—the angle through which the telescope is turned is the measurable angle between the two faces. Some precision goniometers are capable of establishing values to 0.1 s of arc. The device has been used to establish parallel faces and Brewster angles of laser rods and for production inspection of prisms and optical components.

Fig. G-14. Precision goniometer. *(Hilger-Watts.)*

—A. W. Young, *Rank Scherr-Tumico*

Goniophotometer (See **Glossmeter.**)

Governor An automatic controller used to maintain the rotative speed of a machine at a desired value. The governor measures the speed, compares the measured value with the desired value, and acts to correct any error between the two values—usually by adjusting the flow of energy to the machine. Governors may be divided into two main types: (1) a device where the speed-sensing element operates the energy-metering device directly and (2) a device which uses one or more stages of power amplification between the speed-sensing element and the energy-control device. There is a natural distinction between these two types arising from the fact that the first type usually will give stable control on an engine or other prime mover, while the second type requires the presence of some stabilizing factor to prevent continual oscillation of the speed, or *hunting*.

A centrifugal governor of the direct-acting type is shown in Fig. G-15. The operator sets the desired speed by varying the compression of the reference spring. When the engine exceeds this speed, the centrifugal force of the flyweights overcomes the spring force and lifts the fuel-metering rod to decrease the engine-fuel flow. Should the speed drop below the desired value, or should the operator select a still higher speed, the spring overcomes the flyweight force and increases the fuel flow.

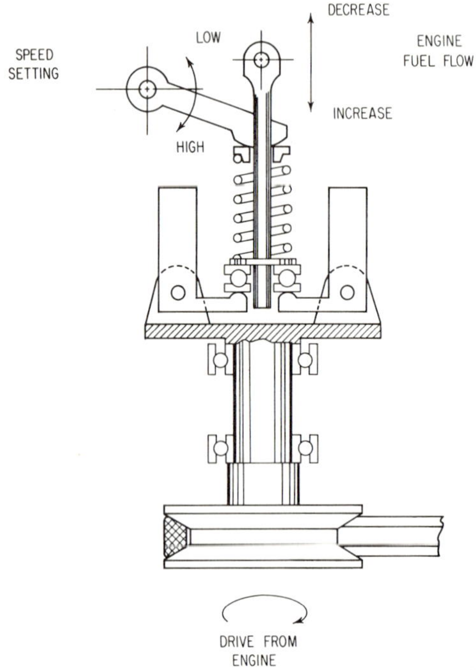

Fig. G-15. Centrifugal governor of the direct-acting type. *(Woodward Governor.)*

Governors of this type have been used for centuries to control the speed of water and wind mills by varying the pressure between the millstones. To this day, the little friction governor in your telephone dial operates on this principle. James Watt adapted this mechanism to control the flow of steam to his engine, and the word "governor" still calls to mind the whirling flyballs of the steam-engine governor.

In designing a governor such as this, it must be kept in mind that the centrifugal force changes with flyweight position as well as with speed. Unless the reference-spring rate exceeds the "rate" of the flyweights, a snap action will result, which may be desirable in an overspeed trip but not in a regulating governor.

Other examples of direct-acting governors are the vane governors used on small engines, matching the force of the cooling-fan blast against a spring, and the "velocity" governors on automotive engines, weighing the impact of the induction-air stream against a spring-loaded plate.

All these governors require a change in engine speed to obtain the force necessary to overcome friction in the fuel-metering mechanism. This causes some hysteresis and inaccuracy at best, and where large forces are required, such governors cannot be used. These governors also have inherent speed droop, since a lower steady-state engine speed is required in order to maintain a greater fuel-flow rate. Thus the governed speed of the engine droops off as the load is increased.

With the advent of the hydraulic turbine, the need arose for a governor powerful enough to operate the wicket gates against the force of the water and the high friction forces. Most of the early power-amplifier governors were mechanical devices, the flyweights controlling the power (derived from the turbine wheel) through clutches or differential gearing of various ingenious designs. Today, most governors use hydraulic power amplification because of its simplicity and the ease with which the power may be modulated.

A single-stage power-amplified governor is shown in Fig. G-16. The speed-sensing centrifugal element operates a small pilot valve which controls the flow of pressurized fluid to or from the power piston. The accuracy of this governor is not appreciably affected by friction or other forces on the device controlling the flow of energy to the prime mover. Note also that the drooping characteristic of the direct-acting governor is not present here. This governor will maintain the speed for which it is set, i.e., the speed at which the pilot valve is centered, regardless of the load on the prime mover and the corresponding position demanded of the power piston. Governors of this type are commonly used to control the speed of propeller-type aircraft power plants. Propeller-type loads are inherently stable because of the increased torque required by them when the speed increases, and no stabilizing device is required in the governor.

In other applications, such stabilizing factors are not present in the engine or its load, and the governor in Fig. G-16 will not control the system stably. The principle involved here may be understood, at least qualitatively, from the following.

When the load on an engine changes, the governor should ideally move the throttle instantly to

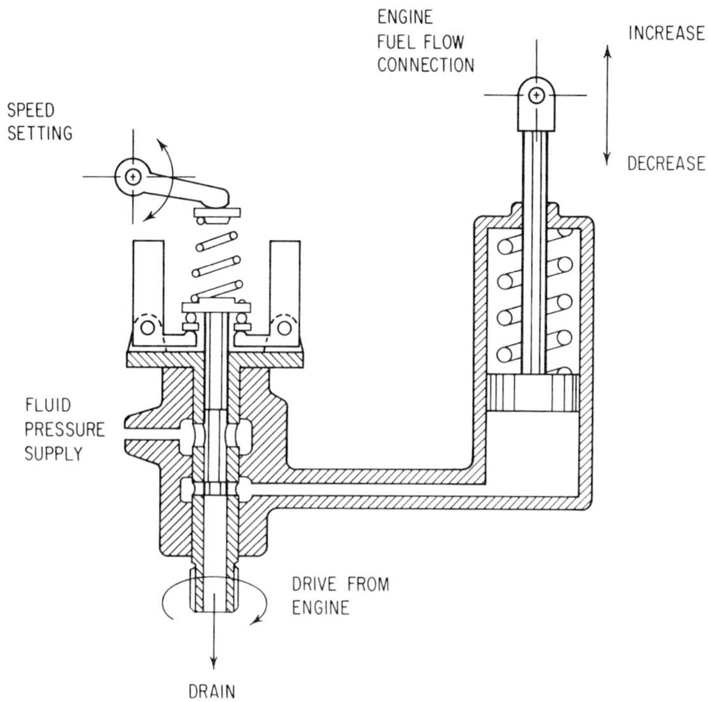

Fig. G-16. Single-stage power-amplified governor. *(Woodward Governor.)*

the required new position and the engine speed should instantly follow the throttle position. Neither of these conditions obtains. The response of the governor to a speed error is a *velocity* of the power piston *toward* the required position, and the response of the engine to a new throttle position is an *acceleration toward* the required speed. Now, position is the time integral of velocity, and speed is the time integral of acceleration. Once the system is disturbed and starts to hunt, with the speed describing a sine curve about the equilibrium value, the governor power piston will follow a cosine curve, the integral of the sine-wave speed curve, and lagging it by 90°. The engine-speed response to a cosine-wave throttle movement will be a sine wave, lagging by another 90° and thus in phase with the original disturbance. There is nothing to damp out the oscillation, and the system continues to hunt.

Adding feedback from the throttle position to the speed-sensing element will short-cut the engine lag and reduce the system phase shift to something less than 180°, and stability will result. A governor with such a connection is shown in Fig. G-17. Note, however, that we have introduced the drooping characteristic of the direct-acting governor so that the only gain is the increased power and accuracy this governor delivers.

Since the droop is required only to stabilize the system during an upset, it may be allowed to dissipate at a rate which will not again disturb the system. In the governor shown in Fig. G-18, the pressure oil flowing to or from the power piston displaces the buffer piston against its springs. The resulting pressure differential appears across the compensating piston on the pilot valve, adding to, or subtracting from, the reference-spring force to cause speed droop. After the system is stabilized, the compensating pressure equalizes again by leakage through the reset needle valve, and the governor maintains a constant speed regardless of the load on the engine. Since such a governor, when used on an engine-generator set, will maintain constant frequency and clock time, it is called isochronous.

Today, governors are expected to minimize speed overshoots and time of return to equilibrium as well as to provide accurate, stable speed

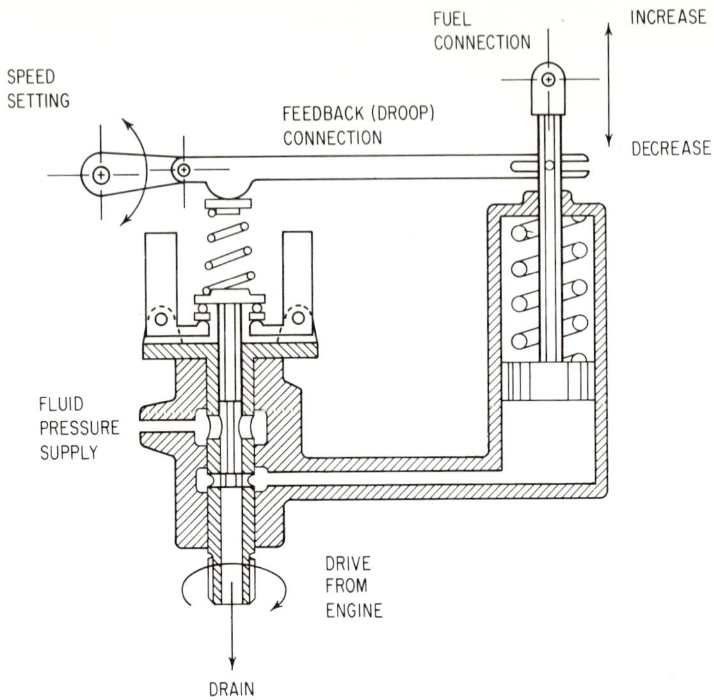

Fig. G-17. Governor in which feedback from the throttling position is added to the speed-sensing element. *(Woodward Governor.)*

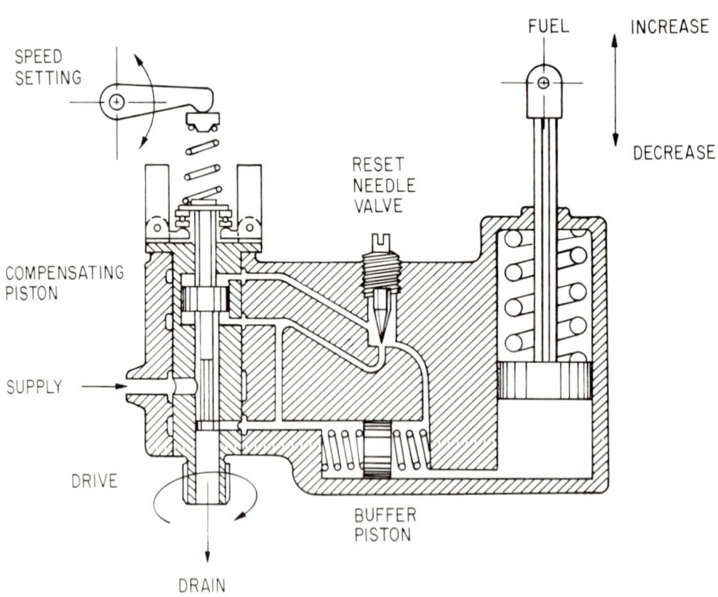

Fig. G-18. Governor with automatic pressure compensation. *(Woodward Governor.)*

control. This is especially true of generator sets supplying electronic gear requiring close frequency control. In these cases, an electrical signal proportional to the load is used to actuate the governor power piston before the speed-sensing part of the governor can act. Depending on the engine characteristics, a five- or tenfold reduction in speed overshoot and in time to return may be achieved by such governors.

Speed may also be sensed electrically, in some cases more simply and better than mechanically. Electronic circuits also facilitate combining speed, load, and other input signals in an optimum manner, and feedback signals, derivative, proportional, and integral, in any combination, are readily obtained electrically. Hydraulic power amplification, for the final stages, remains superior in rigidity, smoothness, and energy-storage capability. Thus, the modern high-performance governor is likely to be a combination of electronic brains and hydraulic muscle.

—Urban Floor, *Woodward Governor Company*

Graphic Panel (See **Display-command operator/machine interface.**)

Graphics, Computer A graphics device is a computer input or output device which utilizes a cathode-ray tube to provide a visual representation of data from the computer, in the form of either characters or line drawings. When used in conjunction with a keyboard or light pen, the graphics device can be used for inputting data into the system.

An application of a graphic device is the solution of an engineering-analysis problem which requires many iterations of the program. That is, the input variables are changed based on the results of the previous interation. The advantage of using graphic equipment in these applications lies in the fact that faster turnaround is achieved at the graphic console (see Fig. G-19).

The digital computer positions information on the graphics display area by specifying the horizontal and vertical coordinates of a virtual square grid composed of possible electron-beam deflection and points. The grid of addressable coordinates for a device is called its raster. Each set of positioning data addresses one coordinate to which the beam is to be repositioned, and the beam is positioned to this coordinate from its previous position. The digital information from the X and Y deflection registers is gated to a digital-to-analog converter (see **Digital-to-analog converter**) which converts the data into weighted current steps which are applied to the deflection yoke of the cathode-ray tube. Since the glow caused by the electron beam hitting the phosphor coating of the tube decays with time, the display must be redrawn continuously at such a rate that it appears steady and stationary to the observer.

Fig. G-19. IBM 2250 display unit.

The light pen is a light-sensing device used to alter the display pattern. It accomplishes this by causing an interrupt when the electron beam intersects the position to which the pen is pointing. This interrupt causes a branch to a program which reads the coordinates at which the pen is pointing.

—W. W. Welz, *IBM Corporation*

Grassot Fluxmeter [See **Magnetic quantities (variables).**]

Gravimeter Any instrument capable of performing relative measurements of gravity may be described as a "gravimeter." Instrumentation available today includes types capable of both static and dynamic measurements.

The demands of exploration geophysics have prompted the development of precise, portable gravimeters, capable of resolution to 0.01 mgal. Fundamentally they are designed to balance the force on a fixed mass against elastic stresses of a system of springs and/or torsion fibers. The majority of portable gravimeters in use today utilize fused quartz in the fabrication of the mass-tension balancing system, which is usually housed in a partial vacuum to protect it from contamination as well as to eliminate errors due to change in barometric pressure.

Minimum total operating ranges are usually specified from 5,200 to 7,000 mgal. Errors in the use of such instruments may arise from errors in connecting fundamental reference stations to the world base at Potsdam or from inaccurate calibration of the gravimeter. Because calibration lines have been extended over a large portion of the earth, intercontinental ties are numerous, and measurements with gravimeters can be made accurately and with great reliability. Modern, portable gravimeters are available with long-term drift rates less than 7 mgal a month.

—Howard Cannon, *Texas Instruments Incorporated*

Gravimetric Feeder (See **Feeder, weigh.**)

Gravitometer (See **Specific gravity, measurement of.**)

Gravity, Measurement of Gravitation is the mutual attraction between all masses and particles of matter in the universe. In gravimetric and geodetic studies of the earth, consideration of this universal force alone is not enough. Also to be considered is the effect of the centrifugal force of the earth's diurnal rotation. Gravity g is gravitation minus the effect of the centrifugal force. This denotes the acceleration of gravity at the surface of a planet or other celestial body, and while commonly called gravity, it is understood to have the dimensions of acceleration. A rotating planet is flattened at the poles to a form such that the observed acceleration of gravity at any point—i.e., the planet's attraction diminished by the centrifugal acceleration of the rotation—is perpendicular to the surface.

The acceleration of gravity g can be measured by either *dynamic* or *static* devices. The most common dynamic method measures the period of a swinging body under the attraction of the gravity force by means of the pendulum, the reversible pendulum, or the gravity variometer. More recently, measurement of the velocity of a falling body has been used. The static group consists of the common spring-balance and gas-pressure gravimeters. Gravity measurements are either *absolute* or *relative*.

Earlier relative-gravity observations were made mostly by the pendulum method, which is described here; gravimeters are now used almost exclusively. See **Gravimeter.**

Pendulums of Invar, a nickel-steel alloy, replaced even earlier brass pendulums because the length of the Invar instrument is almost independent of temperature variations. On the other hand, the crystalline structure of Invar, like that of alloys in general, is variable and can cause small changes in pendulum length. Invar is also sensitive to magnetic effects, so that a Helmholtz coil must be used to protect the pendulum against them.

The single-pendulum apparatus has been outmoded for some time, because the sway of the device's platform cannot be easily eliminated, and changes in pendulum length cannot be estimated at all. However, when two pendulums swing in the same plane and have almost the same length, the effect of sway can be eliminated. Changes in length can be determined, except in the rare circumstance that both pendulums are changed the same amount and in the same direction.

In four-pendulum devices, two pairs of pendulums swing in different planes, 90° from one another. The accuracy of this method is about 1 mgal. However, Cambridge University and the Gulf Oil Corporation have obtained accuracies of 0.3 to 0.5 mgal; 1 gal is equivalent to an acceleration of 1 cm/s^2, and 1 mgal to 0.001 cm/s^2. The Sterneck type of ½-s pendulum has been frequently used because it is relatively short, 0.25 m in length. It swings in a high vacuum to minimize air-induced errors. In operation, the apparatus is a high-vacuum flask. Two pendulums swing in the same plane to eliminate the sway of the stand.

Since absolute measurements are difficult and time-consuming, relative methods are most used. However, absolute gravity is very important in several sciences. In these measurements, the gravity g itself is measured directly. For this purpose, H. Kater's reversing pendulum is often used. Since the obtained accuracy is not yet satisfactory, a number of institutions around the world have made absolute-gravity observations, with the measurements of some of these still in progress. Gallileo's classic falling-body method for absolute-gravity measurement has been improved, and several others invented. Among these is the reversible pendulum already noted, photography of a freely falling graduated scale, free fall of an emulsion-coated rod, motion of two reversible pendulums in opposite phase, photoelectric timing of the free rise and fall of a body, free fall of an interferometer mirror, and motion of a short and a long (200-m) wire pendulum in a shaft.

The absolute-gravity values obtained have been

tied directly or indirectly to the location of Potsdam, where the first accurate absolute measurements were made in 1898–1903. In succeeding years, the following corrections to the Potsdam value were obtained: from Washington, D.C., in 1936—16.2 mgal; from Teddington, England, in 1939—13.0; from Leningrad in 1956—12.1; from Canada in 1960—14.4; from Sevres, France, in 1961—12.7; and from Princeton, N.J., in 1963—14.6, for an average correction of 13.3 mgal.

Thus, the absolute value of 981,274 mgal measured in Potsdam is 13 to 14 mgal too high. The negative correction of the equatorial value—4 mgal, added to this gives a total correction of 18 mgal.

In relative measurements, the ratio between the gravity g_0, measured at the gravity base station, is obtained. The equation of the relative measurements is the identity of $g = g_0 g/g_0$. If g_0 is known, the ratio g/g_0 obtained from the observation gives the gravity g.

In the relative observations, the length l, which is difficult to measure, disappears, and only the swinging time T and T_0 are needed, as can be seen in the formulas $g = T^2 l/\pi^2$, $g_0 = T_0{}^2 l/\pi^2$, $g/g_0 = T_0{}^2/T^2$.

Practical measurements of relative gravity are made with a spring balance such as the *Worden* gravity meter. These units measure the apparent change in mass due to gravity variations by means of fused-quartz linkages and springs. See **Geophysical instruments.**

—J. B. Damrel, Jr., *Texas Instruments Incorporated*

Gregorian Telescope (See **Telescope, reflecting.**)

Ground Loop This is an imprecisely defined term which is generally used to describe a current path which exists between two signal reference points (grounds) in a circuit or subsystem. Such current flow is unintentional, since it is generally assumed that the two reference points are at the same potential. As the result of such current flow, noise can be introduced into an instrumentation or digital-data-acquisition subsystem.

Figure G-20 illustrates the existence of a ground loop in a thermocouple measurement circuit. The thermocouple and shield are grounded at a reference point G_1 in the process being measured, whereas the amplifier input is grounded at the location of the subsystem. If a potential difference exists between the reference points G_1 and G_2, a current I flows through the shield. Furthermore, if the shield has a small but finite resistance R_3 which is substantially less than the resistance R_2 of the thermocouple lead wire, the input signal seen by the amplifier is the sum of the thermocouple voltage V_s and the voltage $R_3 I$ in the shield. Measurement of the input signal, therefore, results in an error equal to the

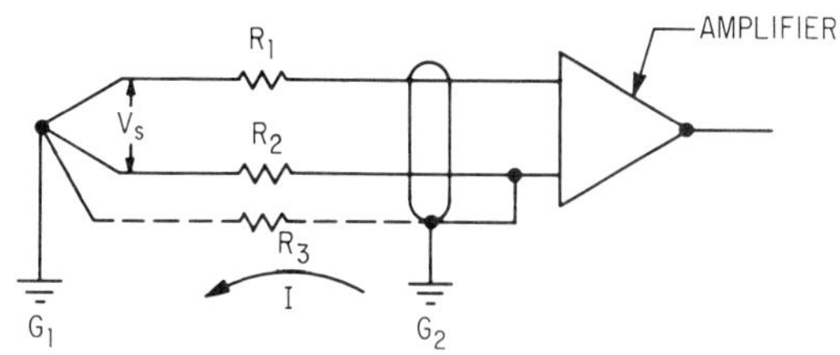

Fig. G-20. Illustration of a ground-loop problem.

voltage drop across the shield resistance. Since the potential difference between the signal reference points can easily exceed the thermocouple output voltage, the error may completely obscure the true signal value.

—Thomas J. Harrison, *IBM Corporation*

Ground Radar (See **Air-traffic control.**)

Grounding An electrical ground is a point in a circuit which is used as the reference point from which other circuit potentials are measured. It is possible for more than one ground to exist in a subsystem, and these multiple grounds may or may not be at the same potential. In some cases, a distinction is made between a ground as the circuit reference point and the potential of the physical earth. The terms "earth" and "earthing" are not, however, in common use in the United States.

Grounding practices are an important consideration in the design and installation of high-accuracy digital-data-acquisition and instrumentation systems. Difference between ground potentials can result in measurement errors due to both normal and common-mode interference potentials (see **Common-mode voltage; Differential-mode voltage**).

Most instrumentation and digital-data-acquisition systems have three identifiable grounding systems. The first of these, the safety ground system, is established for the protection of personnel and equipment in the event of a power-system fault. It may also provide a path for unbalanced current flow, although this is usually unintentional. For compliance to most national and local electrical codes, all exposed, non-current-carrying conductors must be connected to a safety ground. This includes items such as metallic panels, conduit, equipment cabinets, and building structural members.

The power-supply ground functions as the return circuit for dc supply sources in the instrumentation or digital-computer subsystem. It may or may not be the reference point for digital and analog signals in the subsystem. The third grounding system, the signal ground, provides the reference point for information signals. The power-supply ground is sometimes used as the signal ground. Usually, however, a separate grounding system is provided for information signals. This is particularly true in the case of low-level analog signals.

The three grounding systems are usually commoned at a single point in the instrumentation or data-acquisition subsystem. This common point is connected to a ground conductor which ultimately is connected to earth, either through the primary power wiring or directly through a driven ground rod. It is important that the connection between the subsystem ground and earth be a very low impedance path if noise-interference problems are to be avoided in low-level signal processing.

The interference between ground currents in an improperly designed grounding system can be illustrated with the model of Fig. G-21. This shows an amplifier with a ground point A and an analog-to-digital converter with a ground point B (see **Analog-to-digital converter**). The impedances Z_1, Z_2, and Z_3 represent the impedances of the ground conductors in the subsystem.

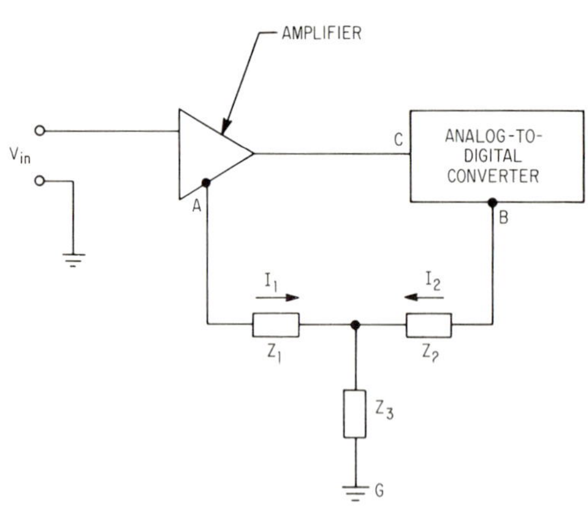

Fig. G-21. Grounding model (improper).

Since the analog-to-digital converter is a single-ended measuring instrument, it measures the potential appearing between points B and C. It is easily seen that ground currents I_1 and I_2 flowing through the ground impedances alter the potential between these two points. In addition, the current I_2 may contain transients due to the digital impulses present in the analog-to-digital converter, and this causes noise to be coupled into the analog-to-digital converter input, causing inconsistencies in the digital output. Similarly, the current I_1 flowing in the amplifier and common ground leads produces an offset at the input of the analog-to-digital converter. Since the current flow in the amplifier ground is generally proportional to the input signal, this offset appears as a gain error in the system characteristic.

A similar situation applies to the amplifier. The amplifier amplifies the signal appearing between the input terminal and the amplifier reference point A. Since this path contains the impedances Z_1 and Z_3, current flow from the amplifier and analog-to-digital converter produces a voltage which is added to the input signal and amplified.

The errors produced by the improper grounding illustrated in the figure can be eliminated by connecting all ground points directly to the subsystem ground reference point, thereby avoiding current flow in common impedances. This concept, often called single-point grounding, is fundamental if interference problems are to be avoided in a low-level data system. In an actual instrumentation or digital-data-acquisition application, it is difficult to provide a single-point grounding system because of the physical dimensions of the total measurement system. For example, in a process-control application, the sensors may be located a mile or more from the amplifier and analog-to-digital converter. In such cases, separate grounds may be established and then connected to the system reference ground by a single conductor. It is extremely important, however, that the impedance of this conductor be very low at all frequencies in the bandpass of the measurement system.

A second effective method of avoiding errors due to grounding problems is to utilize differential measurement systems. These systems measure a voltage impressed between two terminals, neither of which is ground, without respect to the potential existing between the terminals and ground. If the potential between the terminals and the reference ground (the common-mode voltage) is significant and depending on the common-mode rejection of the differential system, there may still be some interference due to the ground potential (see **Common-mode voltage; Error, common-mode**). It is possible, however, to minimize this interference.

Even within the confines of an instrument, improper grounding practices can result in errors due to current flow in common impedances. It is generally easier to establish a single-point grounding structure in a small instrument, however, and most interference problems can be eliminated with careful design. Systems involving both analog and digital signals are particularly susceptible to grounding problems because maintaining a low-impedance ground for narrow pulses is difficult. It is particularly important to design the grounding systems specifically to meet the requirements of these situations. Grounding is often termed an "art." The phenomena involved, however, obey the fundamental laws of electricity, and careful analysis can usually explain performance related to the ground structure.

—Thomas J. Harrison, *IBM Corporation*

Guard Shield Guard shielding is an electrostatic-shielding technique used in precision-measurement instrumentation and some types of differential amplifiers to enhance the common-mode rejection ratio [see **Common-mode rejection ratio (CMRR)**]. The shield surrounds all, or a portion of the instrument, and it is floating with respect to the instrument ground. When the instrument is used, the guard shield is connected to the common-mode source which is, in turn, connected to one side of the instrument input. The shield provides a low-impedance path for the flow of common-mode currents and, as a result, minimizes common-mode current flow through the signal source and lead impedances.

The guard-shield concept is illustrated in Fig. G-22 for an isolated differential amplifier being used to measure a thermocouple signal. The guard shield of the amplifier is connected to the shield surrounding the input leads which, in turn, is connected to the common-mode source V_{cm}. The capacitances C_1 and C_2 represent leakage capacitances between the input lines and shield, and the shield and ground reference point, respectively. A common-mode current I flows from the common-mode source through the shield and the capacitance C_2. Since the potential across C_1 is essentially zero, very little current flows from the common-mode source through the input line resistance and C_1. Thus, no offset or error voltage is generated across the line resistance R. In essence, the guard shield has shorted out the line resistance.

If the guard shield is not present, a common-mode current flows from the common-mode source, through the line resistance R and the input to ground capacitance. The voltage generated across the line resistance R by this current is in series with the signal voltage and is indistinguishable from a valid signal. Therefore, it represents an error in the signal and contributes to the total measurement error of the subsystem.

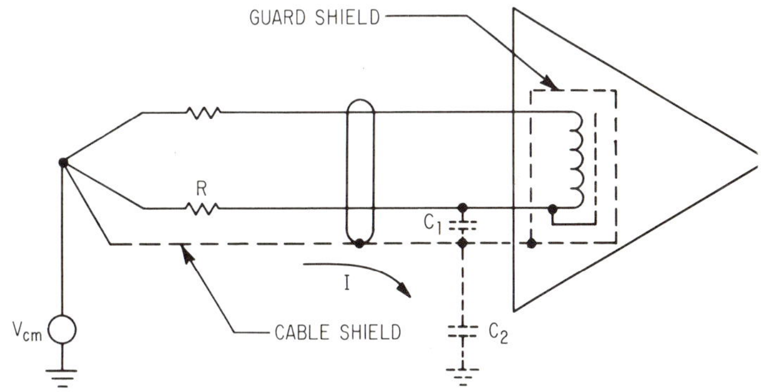

Fig. G-22. Guard-shield example.

—Thomas J. Harrison, *IBM Corporation*

Guidance Systems (See Navigation systems, and **Space and missile-systems instrumentation.**)

Gyro (See **Flight-control systems** and **Space and missile-systems instrumentation.**)

Gyroscopic Accelerometer (See Accelerometer.)

H

Hail, Measurement of (See **Precipitation gage.**)

Hair Hygrometer (See Hygrometer.)

Hall-effect Transducer First noted by Hall at Johns Hopkins University in 1879, this physical principle was not incorporated into practical transducer designs until relatively recently. Known as the Hall effect, the phenomenon takes place when a transverse magnetic field is applied to a current-carrying conductor. The transverse electric field resulting produces a voltage drop across the conductor. This is because the applied magnetic field causes the electrons to deflect within the bounds of the conductor (see Fig. H-1).

The Hall coefficient is a ratio of the transverse electric field and the product of the magnetic field times the longitudinal current density. The coefficient varies with various materials but is much higher for semiconductors than for most metals. Hence, semiconductors are preferred for Hall-effect transducers. These devices generally are characterized by high speed of response, excellent temperature stability (although usually requiring individual calibration), and small, mechanically simple packaging.

The effect is nonreactive and can be realized at impedance levels of a few ohms at milliwatt

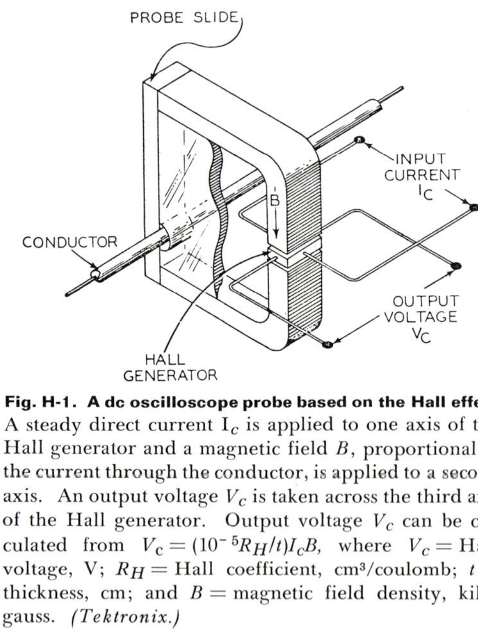

Fig. H-1. A dc oscilloscope probe based on the Hall effect. A steady direct current I_c is applied to one axis of the Hall generator and a magnetic field B, proportional to the current through the conductor, is applied to a second axis. An output voltage V_c is taken across the third axis of the Hall generator. Output voltage V_c can be calculated from $V_c = (10^{-5} R_H/t) I_c B$, where V_c = Hall voltage, V; R_H = Hall coefficient, cm³/coulomb; t = thickness, cm; and B = magnetic field density, kilogauss. *(Tektronix.)*

power levels. Thus, the devices are compatible with intrinsically safe instrumentation. The devices may be used to measure magnetic fields directly of the order of 10,000 gauss, or they may be used to indicate induced magnetic fields in connection with other devices. Hall-effect devices are incorporated in a number of transducers for displacement, speed, torque, shaft horsepower, and flow measurement.

Hardness, Measurement of After years of experience in testing materials, particularly metals, workers in the field have reached the general conclusion that hardness is not a basic property of a material but rather a combination of properties that may include modulus of elasticity, true tensile strength, work hardening, and yield strength. If hardness (the measured variable in this case) is not an absolute quantity, it then follows that attempts to measure hardness must be relative in nature. This does not mean that hardness measurements are not of value to metallurgists in specifying and checking materials, or to the machinist, or to the product designer who can relate hardness to resistance to abrasion, wear, and yielding. For practical purposes, hardness can be considered the property of a material to resist local penetration and scratching—hence abrasion and wear. The reader is referred to the American Society for Testing and Materials for more detailed definitions.

Hardness-testing instrumentation falls into two broad categories: (1) scratch hardness and (2) indentation hardness. A possible third classification might be termed "rebound" hardness, although it can be argued that the latter method is a form of indentation-hardness measurement.

Scratch hardness is of particular interest to geologists for the identification of minerals and to metallurgists for certain limited testing where a microscope must be used to aid in evaluating the test results. The method revolves around Mohs' scale of hardness, which simply lists 10 materials in order of increasing hardness as given in Table H-1. The principle states that, for example, a diamond will scratch quartz, but not vice versa—hence the diamond is the harder material; or that quartz will scratch feldspar, but not vice versa, and so on.

TABLE H-1. *Mohs' Scale of Hardness of Materials*
(Hardness increases with number designation)

1. Talc	5. Apatite	8. Topaz
2. Gypsum	6. Feldspar	9. Sapphire
3. Calcitespar	7. Quartz	10. Diamond
4. Fluorspar		

Indentation-hardness Testing: This method is used widely in industry and essentially involves a measurement of the indentation (or penetration) resistance of a material when the material is subjected to an instrumental "indentor" of exacting materials and dimensional specifications and under carefully controlled forces. In *Brinell* hardness testing, a hardened sphere under a known loading is forced into the surface of the material. The diameter of the resulting indentation becomes the measure of Brinell hardness. The Brinell number becomes

$$B_r = \frac{\text{load, kg}}{\text{surface area of indentation, mm}^2}$$

Hardened steel and especially treated steel balls or jewels are used as the indentors. The standard ball size is 10 mm diameter; the standard test load is 3,000 kg for steels, 500 kg for softer nonferrous materials, and 100 kg for very soft materials.

Rockwell hardness testing involves measuring the depth of indentation. A steel ball of specified diameter or a spherical-tipped conical diamond (120° angle and 0.2 mm tip radius) may be used as the indentor. The latter type is called a "Brale." In operation of a Rockwell hardness-testing instrument, a minor load of 10 kg first is applied. This causes an initial penetration and holds the indentor in place. Then, the depth-measuring scale is set to an arbitrary maximum value of 130 if balls are used, and to 100 if the Brale is used. Then a major load of 60, 100, or 150 kg is applied under dashpot control. The major load is removed and a return made to the minor load of 10 kg. The hardness then is read directly from a scale calibrated in penetration. The scale is arranged such that soft materials with deep penetrations yield low hardness numbers. The foregoing description simply describes the basic priniciple; there are numerous variations.

In the *rebound hardness tester,* a diamond-tipped hammer is dropped, under carefully controlled conditions, onto the test specimen, and the height of rebound is measured. The higher the rebound, the harder the material. An instrument employing this principle, the Scleroscope, is widely used industrially—with tabular information published along with the various Brinell, Rockwell, and Vickers data. In particular, the Scleroscope is used for testing roll hardness.

In the *Vickers* method of hardness testing, which is similar to the Brinell method, the result is expressed in terms of the pressure under the indentor. The same units of measurement (kg/mm^2) are used. A diamond indentor in the form of a square pyramid with an apical angle of 136° is used. The loads are lighter, ranging from 2 to 120 kg. The impression is measured microscopically, and the following formula is used:

$$V = \frac{P}{0.5393d^2}$$

where V = Vickers hardness number
P = imposed load, kg
d = diagonal of indentation, mm

A fully nondestructive method for approximating hardness uses eddy-current techniques. See **Nondestructive inspection.**

REFERENCES

Baumeister, Theodore, and L. S. Marks: "Standard Handbood for Mechanical Engineers," 7th ed., McGraw-Hill Book Company, New York, 1967.

Le Grand, Rupert: "The New American Machinists' Handbook," McGraw-Hill Book Company, New York, 1955.

American Society for Testing and Materials, Philadelphia, Pa., various publications.

Hardware, Computer A digital computer consists of physical equipment for performing the operations and a program (see **Program, Computer**) to control the sequence in which the operations are performed. The physical equipment, such as electronic circuits, devices, or peripheral equipment which make up a computer, is referred to as the hardware, in contrast to the program, which is generally called software.

—W. W. Welz, *IBM Corporation*

Hay Bridge (See **Bridge circuits, electric.**)

Haze Detection (See **Turbidimeter.**)

Hearing, Sensory Aids for Hearing losses (sensorineural) are generally greatest at the higher frequencies (above 2 kHz). Thus, the sibilants and percussives in speech are the most difficult to perceive. In addition, there usually is a loss in dynamic range so that not only must sound be louder to be perceived, but the range between threshold of perception and threshold of pain is much narrower. Thus, the ear cannot accommodate in many cases to an undue amount of amplification. This condition is known as "recruitment." Any hearing aid should take these two factors into account and hence should not be a simple audio amplifier if best results are to be achieved. Furthermore, stereophonic sound location is lost with a single-channel hearing aid. Thus, some patients find that binaural hearing aids are of real value. They seem to help the wearer to discriminate a desired conversation from background noise in a rather dramatic fashion.

As a further aid in helping the profoundly deaf, whose hearing may be cut off as low as 1,000 Hz, a device known as a transposer has been developed (see reference) which translates sounds at the upper frequencies made by sibilants and percussives into the lower-frequency region where they can be amplified sufficiently to be audible. There are several variations on the basic scheme, but in principle the devices contain filters which detect only the upper frequencies and then trigger sound generators below 1,000 Hz which indicate to the user the presence of the high frequencies. These instruments have been under evaluation in Sweden, Denmark, England, and the United States.

Various other interesting instruments (see reference) have been developed for use in schools for the deaf to help teach speech production. These instruments monitor the speech of the deaf person and then display production of the "s" and "th" sounds, pitch, and rhythm on an oscilloscope screen and meters.

REFERENCE

Pickett, James M.: Recent Research on Speech-analyzing Aids for the Deaf, *IEEE Trans. Audio Electroacoustics,* vol. AU-16, no. 2, pp. 227–234, June, 1968.

—J. Malvern Benjamin, Jr., *Bionic Instruments, Inc.*

Heart, Measurements of (See **Cardiography**; and **Cardiovascular instrumentation.**)

Heaviside Bridge (See **Bridge circuits, electric.**)

Helix (See **Bourdon tubes.**)

Heterodyne Frequency Meter [See **Frequency (electrical), measurement of.**]

Hexadecimal (See **Number systems.**)

Heydweiller Bridge (See **Bridge circuits, electric.**)

High-limiting Control Control action in which the output never exceeds a predetermined high-limit value. Also see **Split-range control.**

Histogram (See **Laser, instrumentation of.**)

Hollerith Code A system of encoding alphanumeric data on cards via the punching of those cards (see **Alphanumeric code**). Hollerith code was first developed by Herman Hollerith in 1890 for the U.S. Census Bureau. Figure H-2 is an example of a Hollerith punched card showing the codes for the various characters. A single character is punched in each column of the card.

—B. W. Landeck, *IBM Corporation*

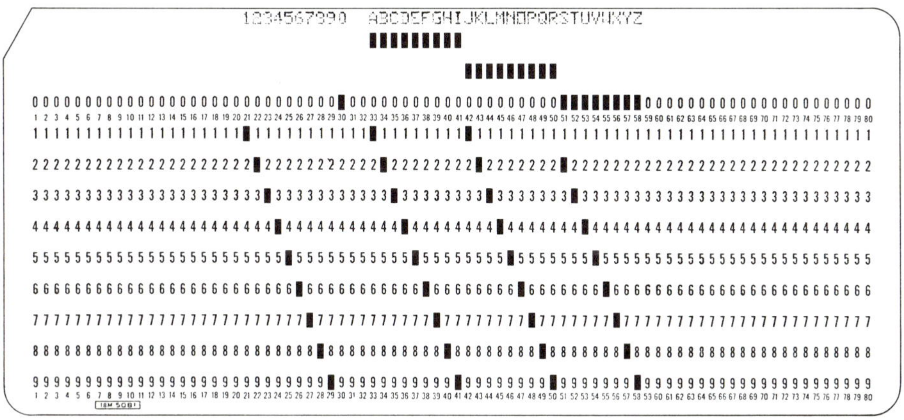

Fig. H-2. Hollerith punched card.

Holography (See **Laser, instrumentation of.**)

Hopper Scale (See **Scale, industrial.**)

Hospital, Monitoring of (See **Medical instrumentation.**)

Hot-filament Vacuum Gage (See **Vacuum gaging.**)

Hot-wire Anemometer (See **Anemometer.**)

Humidity, Measurement of Humidity is the presence of water vapor in the atmosphere. The process of measuring the humidity or the amount of moisture is called hygrometry.

Relative humidity is the ratio, usually expressed in percent, of the pressure of water vapor in the gas to the saturation pressure of water vapor at the temperature of the gas; or relative humidity is defined as the ratio of the mass of moisture actually in a given volume of air at any time to the mass that the given volume could hold if saturated at that temperature.

Absolute humidity of a gas mixture may be defined as the pressure of water vapor or as the weight per unit volume of water vapor. Specific humidity is the weight of water vapor in air per unit weight of dry air.

Dew point is the temperature to which water vapor must be reduced in order to obtain a relative humidity of 100 percent, that is, to obtain saturation vapor pressure.

Instruments for measuring humidity include (1) dry- and wet-bulb thermometers (see **Psychrometer**), (2) hygrometers (see **Hygrometer**), and (3) dew-point indicators (see **Dew point, measurement of**).

Humidity of the atmosphere, along with air temperature and motion, affects human comfort, and hence humidity measurement and control are important elements of comfort air-conditioning systems. Humidity also affects certain manufacturing processes, often critically. Because of the influence of humidity on natural and on some synthetic fibers, this variable is extensively controlled in the textile industry. The effect of humidity changes on paper requires that humidity be controlled in paper manufacture as well as during the printing process. Humidity control also is important to the operation of certain types of electrical equipment; in the preservation of materials during shipment and storage—to prevent rotting, cracking, and other forms of deterioration; in combustion processes where the moisture content of air affects operating efficiencies; and in natural-gas transportation and storage.

Physical Effects Used in Humidity Measurements: Several basic principles are utilized in humidity instrumentation, including:

1. *Condensation:* A surface in contact with the atmosphere is cooled until condensate or dew appears. Dew-point indicators are designed around this principle.
2. *Dimensional Change:* Humidity variations cause changes in the physical dimensions of materials, usually organic. The hair hygrometer operates on this principle.
3. *Thermodynamic Equilibrium:* The bulb of a thermometer is surrounded or wetted by a water-soaked wick and exposed to the atmosphere until equilibrium is achieved. By comparing the wet-bulb temperature with that of a dry-bulb thermometer in the same atmosphere, the relative humidity can be determined. This is the principle of the psychorometer.
4. *Gravimetric Absorption:* A water-absorbing material, such as phosphorus pentoxide, absorbs moisture from the atmosphere and thus gains weight. The volume of air passed through the absorbent must be established in order to determine moisture content quantitatively.
5. *Conductivity of an Absorbent:* A material, such as a hygroscopic salt, will vary its ability to conduct electricity when placed between two electrodes. When corrected for temperature effects, the conductivity becomes a measure of relative humidity.
6. *Electrolysis:* Moisture from the air is absorbed by a material such as phosphorus pentoxide, the absorbed water content of which then is decomposed electrolytically. The electrolysis current is directly proportional to the rate of decomposition of water and thus to the moisture content of the air. The volume of air passed through the absorbent must be established.
7. *Heat of Absorption:* The temperature of a solid is measured to determine the amount of heat released when a solid absorbs water vapor. A measurement of temperature change when water vapor is alternately adsorbed and desorbed is interpreted as moisture content.
8. *Vapor Equilibrium:* A saturated solution of a hygroscopic salt is maintained at the temperature at which it is in vapor-pressure equilibrium with the atmosphere. The temperature of the salt converts directly to dew-point temperature.
9. *Infrared Absorption:* This takes advantage of the fact that water vapor absorbs electromagnetic radiation in certain portions of the infrared region. A measurement of this absorption can be interpreted as moisture content.

REFERENCES

Wexler, Arnold, and W. G. Brombacher: Methods of Measuring Humidity and Testing Hygrometers, *Natl. Bur. Std. (U.S.) Circ.* 512, Sept. 28, 1951.

Considine, D. M., and S. D. Ross (eds.): "Handbook of Applied Instrumentation," sec. 7 (Humidity and Dew Point), McGraw-Hill Book Company, New York, 1964.

Ruskin, R. E.: Principles and Methods of Measuring Humidity in Gases, *ISA Publ.* HMPI, 1963.

Amdur, E. J.: Applications of Humidity and Moisture Measurements in Science and Industry, *ISA Publ.* HMPII, 1963.

Wexler, A., and W. A. Wildhack: Fundamentals and Standards of Humidity and Moisture Measurement, *ISA Publ.* HMPIII, 1963.

Hunting (See **Oscillation, measuring-system.**)

Hybrid Computer A hybrid computing system is a combination of an analog computer with either a digital computer or a digital logic array, interfaced for compatible intercommunication as shown in Fig. H-3. Hybrid computers can play a critical role in digital-control schemes when they are used in off-line simulation of the process before attempting on-line control. Pointed out here are the advantages of hybrid simulation—with three examples of how analog-digital combinations of computers are applied in the process and aerospace industries. These examples show how substantial savings can be made without sacrificing accuracy.

Design of the interface is critical in obtaining the most powerful combination. Under some interfacing conditions, the combination may be synergistic, exhibiting greater problem-handling power than a simple linear linkage of computing capability would indicate. The relative computing roles of the analog and digital portions will vary widely with the problem being solved. At one end of the scale, all computation may be performed on the analog portion while the digital

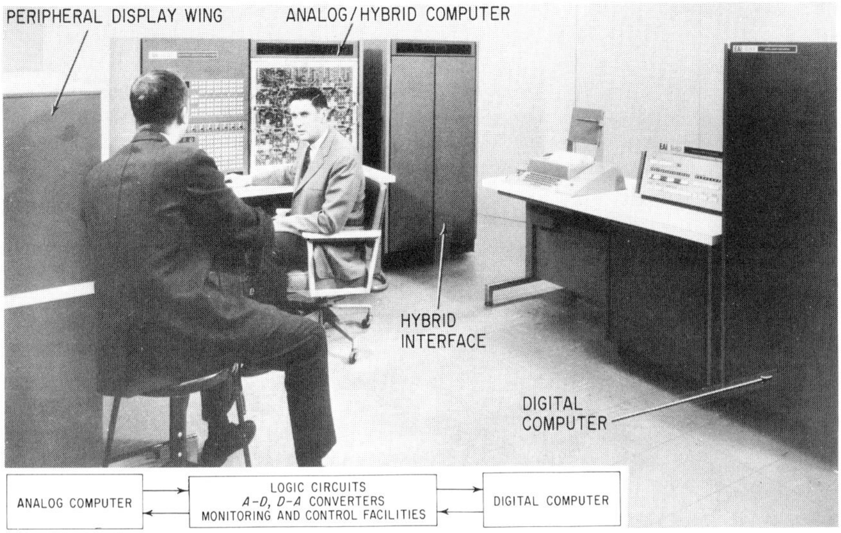

Fig. H-3. A hybrid system showing how the analog and digital computers are interconnected through a specially designed interface that contains logic circuits, digital-to-analog and analog-to-digital multichannel converters, and monitoring and control facilities. Such arrangements are effective in research and development laboratories and in industrial-control teaching as well as in off-line preparation for on-line digital process control. *(Electronic Associates.)*

portion stores and documents. At the other end, all the problem flow and control and most of the calculations may be performed on the digital portion, while the analog portion is used for data reduction, interpolation, and display.

Optimum savings accrue when a balanced, integrated hybrid exhibits the greatest number of natural computing advantages of both analog and digital techniques. In the process industry, the flexibility, speed, and economy of such integrated hybrids is widely recognized, as the following examples illustrate. A third example, dealing with missile-flight control, emphasizes the economic advantages of hybrids.

Tubular Chemical Reactor: In this example, the problem is to determine the most advantageous controller settings for a tubular chemical reactor under various load disturbances (see Refs. 1 and 2). The physical equipment consists of a tubular reactor, separator, and control loop—as shown in Fig. H-4. The three reactants A, B, and C combine in the reactor to form the waste product D and the desired product E. All five components travel to the separator where A and C are recycled to the reactor. B is returned to tankage, D is discarded, and E is monitored and stored.

Reactant B is fed to the reactor in large excess at a constant flow rate B_0. The recycled reactants A and C combine with the makeup streams A_i and C_i to form the input rates A_0 and C_0. The control system adjusts the makeup stream C_i to maintain the flow rate of product E from the separator relatively constant when subjected to load fluctuations in the reactant A makeup stream. The basic assumptions are:

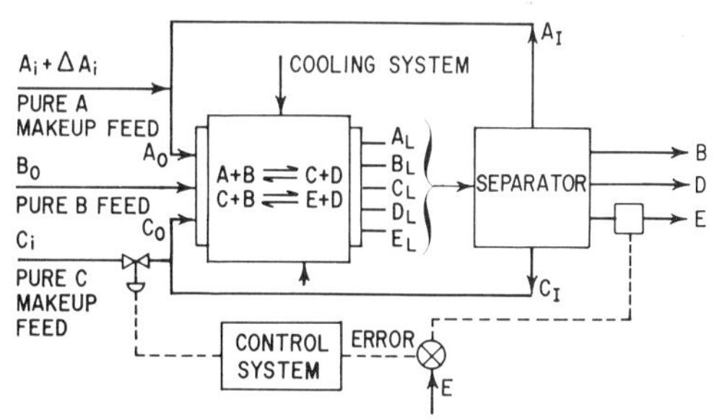

Fig. H-4. Block diagram of a tubular chemical reactor with a product separator and control system. The problem is to find the most advantageous controller settings for the reactor under various load disturbances.

Velocity through reactor is fairly constant because flow rate B is large compared with flow rates of other reactants.

Reaction-rate coefficients are constant because the process is isothermal.

Density and average molecular weight of the reaction mixture are substantially constant.

Fluid movement is plug flow with no back mixing.

Partial differential equations that appear in Ref. 2 comprise the mathematical model that is implemented in the analog computer. This set of equations expresses the concentrations as a function of time and distance along the axis of the reactor. First-order lag equations represent the separator, which is assumed to be 100 percent efficient. The proposed control system is proportional-plus-integral-plus-derivative.

The relative roles of each computer are shown in Fig. H-5. The reactor model was reduced to ordinary differential equations by mathematical manipulation. The analog computer implemented them in a time-shared loop, solving segments of the equations on each pass. The loop operated faster than real time to provide a continuous solution. The separator and control equations, in transfer-function form, were solved in real time. The digital computer served as a storage and playback device to synchronize the iterative fast-time solutions of reactor equations to the real-time handling of separator and control transforms in the analog computer.

Analog	Interface	Digital
High-speed solution of reactor equations	A — D	Time delay to synchronize high-speed solution with real-time solution
Real-time simulation of separator and controller using transfer functions	D — A	
Control of repetitive operations	Logic	Mass storage
Control of storage and playback		
Output		

Fig. H-5. Relative roles of the analog and digital computers for the chemical-reactor example. Note how the digital computer assisted in keeping the high-speed and real-time activities of the analog computer in full synchronization.

The results, obtained in easily interpretable graphic form, proved that the separator would filter out high-frequency disturbances. The results also established the open-loop stability of the system and determined the best controller setting for proportional band, reset rate, and derivative time. The advantages of simulation over empirical approaches has been recognized, but the hybrid method offers increased speed and efficiency over either analog or digital pure simulation. In this problem, the advantages of the analog portion were individual integrator-mode control, integrator speed, analog repetitive and iterative operation, and analog logic. On the digital side, the advantages were the compute function, storage, playback, and data and logic transmission between machines.

Reacting Distillation Column: Simulating the dynamics of a reacting distillation column is more difficult with pure analog or digital techniques than by hybrid methods (Refs. 1 and 3). Figure H-6 shows a 25-plate column with partial and total overhead condenser, reboiler, and controls. Reactants A and B feed the column at plates 21 and 20. Reactant A's flow rate is controlled by sensing plate 3 temperature, while B's flow rate is maintained constant during the process.

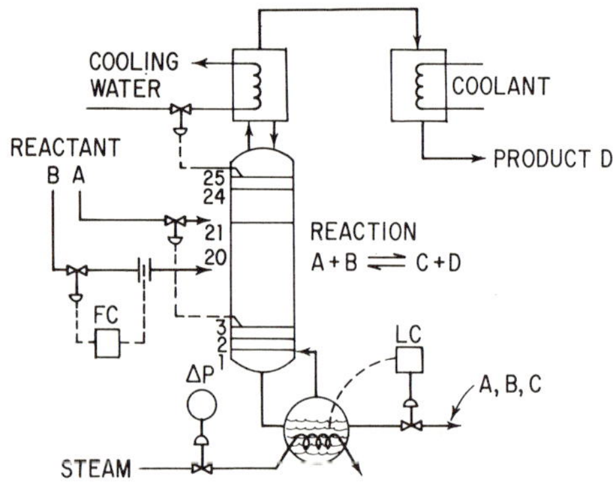

Fig. H-6. Schematic of a reacting distillation column to illustrate how a difficult situation often may be handled more readily with a hybrid setup than with a pure analog or digital program.

Reactant A is added in large excess, and A and B combine in a reversible relation to form products C and D. This exothermic reaction is assumed to have forward and backward rate constants that obey the Arrhenius relationship. Product D is so relatively volatile that the overhead product is substantially pure D. The bottom product consists of excess A, product C, small amounts of unreacted B, and very little D. The mathematical model for this system is derived by performing mass and enthalpy balances around the overhead system, an arbitrary plate, and the reboiler. A set of first-order, nonlinear ordinary differential equations describes the rate of change of each component's composition and the total enthalpy as a function of time. The model also contains a vapor-liquid equilibrium relationship, plate-efficiency correlation, a pressure-drop expression, and equations for reaction kinetics.

A pure analog simulation of each component on each tray would require an enormous amount of computing hardware. In pure digital simulation, the limiting factor would be computing time. But hybrid computation is able to take greater advantage of physical similarities among process modules. The high-speed iterative capability of the analog computer is employed in conjunction with digital-computer support. An analog representation of a single plate is used as a subroutine to perform plate-to-plate calculations up the column in fast time, while solving the condenser and reboiler equations in slow time. The single-plate analog circuit acts like a numerical integration in pure digital simulation, but at a much faster rate and without numerical-instability problems.

For example, the differential equations representing a plate are integrated over a short period of time, holding the flow rates, equilibrium values, and other quantities constant over the interval. At the end of the interval, they are updated. The values of composition, enthalpy, flow rates, and other quantities at the end of the interval serve as initial conditions at the beginning of the next interval as the computation recycles. The maximum allowable integration-step size is limited by the time constant of the system, and not by numerical instability or truncation error as in pure digital simulation. With larger integration-step sizes, the advantage offered by the superior speed of hybrid becomes more apparent.

Two important features of hybrid computation illustrated in this example are the mass-storage facility of the digital portion and dynamic rescaling. The mass-storage facility makes it possible to store plate compositions, enthalpies, flow rates, and pressures from one time interval to the next. As to dynamic rescaling, compositions of components may often vary from top to bottom of a large column by several orders of magnitude. When key trace components are critical, dynamic ranges of 10^4 and 10^5 are common.

Pure analog computation would require rescaling at fixed plates, a method likely to be inadequate for severe transients. In hybrid computation, the digital computer rescales automatically at any point in the column where overranging is likely to occur. Using the digital computer also takes full advantage of its multiplication and division accuracy. Relative computer roles for this example are presented in Fig. H-7. Graphic displays provided almost instantaneous solutions for flow rates and temperature control. Thus, it is possible to work interactively with the hybrid system, gaining insight into the dynamics of the control system under synthesis. The hybrid simulation ran 120 times faster than real time, making 1 min of computer time equivalent to 2 hr of plant time. If the digital portion had not been used, analog hardware for the simulation would have been increased by a factor of 30.

Analog	Interface	Digital
High-speed solution of tray dynamics	A — D	Storage of tray parameters
Slow-speed solution of condenser and reboiler dynamics		Nonlinear function generation and high-accuracy multiplication
Control of repetitive operations	D — A	Dynamic rescaling
Control of storage and playback of tray parameters	Logic	Control of data transfer between top tray and condenser, and between bottom tray and reboiler
↓ Output		↓ Output

Fig. H-7. Relative computer roles for the reacting-distillation-column simulation.

Missile-flight Control: This example describes a hybrid simulation of a six-degree-of-freedom air-to-ground missile-control system. The system previously had been programmed digitally using coarser approximations and with considerably less complexity. The factors of time, cost, and accuracy are compared. Since cost per solution almost always is determined by run time on computers, the task of making a comparison may be reduced largely to one of converting run time into dollars.

A rate of $450 per hour is set for a hybrid system of the type needed; a comparable digital machine would be a large general-purpose unit whose time would be commercially available for a similar amount, or possibly more. To remove any question pertaining to the comparison, an hourly cost of $250 for the digital computer is assumed.

This example concerns a 5-min missile flight. One flight run on the digital computer takes 4 hr 10 min, or 4.17 hr. Multiplying by the hourly rate, this yields $1,042.50 per flight. When run on a hybrid computing system, the flight would take 5 min (0.083 hr), or $37.35 per flight.

Time is the next consideration: (1) run time, (2) turnaround time, and (3) setup time. The advantage of run time has been described. The hybrid ran at real time, the digital at one-fiftieth real time. The hybrid could have run at twice real time, or 100 times as fast instead of 50 times as fast. It is important to note that if the problem takes an hour on digital, the advantage of doing it in 36 s may not seem great. But if the problem will take 100 hr on digital, doing it in 1 hr on hybrid may have tremendous appeal.

In this example, digital batch operation dictated a schedule of one flight per day with 5 to 7 hr turnaround. Taking an arbitrary time per missile flight of 5 min, time to examine results of 10 such flights would be about 10 days using the digital operation. This delay makes it difficult for engineers to receive answers to their problems on a timely basis. On the other hand, the hybrid system may be directly operated by the engineer with some laboratory assistance, and the entire 10 flights usually can be run in less than 1 hr.

Setup time on a hybrid system is much less time-consuming than it once was. Modern hybrids are substantially automated. Digitally controlled attenuators and comprehensive hybrid software have reduced setup time from hours to minutes. Setup time for this example averaged 15 min. A cost of $65 is typical for hybrid, compared with $20 per run for digital simulation. Since hybrid runs are usually made in continuous strings, whereas digital runs are usually made one at a time, the setup charge per flight will average less for hybrid than for digital simulations.

In this example, hybrid simulation yielded target-miss distance as accurately as digital. Repeatability of miss distance, the yardstick of stability in the majority of missile-flight simulation, was well within specified limits. It is concluded that hybrid computation is a powerful tool when added to the tools available for control-system simulation.

REFERENCES

1. Distefano, G. P.: "Hybrid Computation in the Process Industries," Electronic Associates, Inc., Princeton, N.J., 1968.
2. Coughanower, D. R., R. Vichnevetsky, J. P. Landauer, and A. M. Carlson: "EAI Hybrid Computation Applications Study 6.4.2h," Electronic Associates, Inc., Princeton, N.J., 1965.
3. Ruszkay, R., and E. E. L. Mitchell: Hybrid Simulation of a Reacting Distillation Column, *Proc. Joint Computer Conf.*, 1966.
4. Lewis, D.: "Hybrid Application Notes," Electronic Associates, Inc., Los Angeles Computation Center, 1968.

Note: This entry is based largely upon Hybrid Computers Aid Process Control, *Control Engineering*, August, 1968, and is reproduced here with permission of the copyright owner, The Reuben H. Donnelley Corporation.

—Edgar B. Sharpe, *Electronic Associates, Inc.*

Hydraulic Actuator (See **Actuator, hydraulic.**)

Hydraulic Controller A device that uses a liquid control medium to provide an output signal which is a function of an input error signal. The error signal is the difference between a measured variable signal and a reference or set-point signal. The significant difference between the hydraulic controller and other types of controllers is the use of a liquid as the control medium as opposed to air in the case of pneumatic controllers or electricity in the case of electric controllers. Various liquid control media can be used, but the choice usually is between a petroleum oil or a fire-resistant fluid. If fire hazard is not a major concern, petroleum oil is preferred because of its superior lubricating and corrosive-protecting features.

Advantages of Hydraulic Controllers: These include: (1) *high speed of response*— the liquid control medium, being effectively incompressible, makes it possible for a load, such as a work cylinder, to respond very quickly to output changes from the hydraulic controller; (2) *high power gain*—since liquids can be converted readily to high pressures or flows through the use of various types of pumps, hydraulic controllers can be built to pilot this high-energy fluid; (3) *simplicity of final actuator*—most hydraulic controller outputs are two hydraulic lines that can be tied directly to a straight-type cylinder to provide a linear mechanical output; and (4) *long life*—the self-lubricating properties of most hydraulic controllers, particularly when used with petroleum oils, are conducive to long, useful life.

Some Limitations of Hydraulic Controllers: These include: (1) *maintenance of the hydraulic fluids*—dependent upon the types of hydraulic hardware used, filtration is required to keep the fluid clean; use of fire-resistant fluids, because of poor lubrication and corrosive effects, requires careful maintenance; (2) *leakage*—care must be taken with seals and connections to prevent leakage of the hydraulic fluid; and (3) *power supply*—new hydraulic power supplies usually must accompany new hydraulic-controller installations. This is contrasted with pneumatic and electric systems where such power normally is available.

Elements of Hydraulic Controllers: The major elements include (1) a hydraulic relay or amplifier, (2) an error detector, and (3) a signal-sensing section.

As in all control-loop applications, a measured variable must be sensed, compared with a set point, and the difference or error signal manipulated by the controller to cause some final control element to correct the process. Dependent upon the particular construction of the hydraulic controller, often these functions can be incorporated physically into the hydraulic controller per se.

The hydraulic relay is common to all hydraulic controllers. (See **Hydraulic relay.**) The relay allows a small mechanical displacement to control the output from a hydraulic power supply in a fashion that will drive a work load. The jet pipe and spool valve are examples of these relays. If the process variable is being sensed directly, such as pressure by means of a bourdon tube or differential pressure with a diaphragm, the signal sensing and error detection can become an integral part of the hydraulic controller. As shown in Fig. H-8, a force is generated from the signal system and is compared with a set-point force operated by the spring, and the resulting mechanical displacement (error signal) moves the hydraulic relay (in this case, a jet pipe) to control the work cylinder.

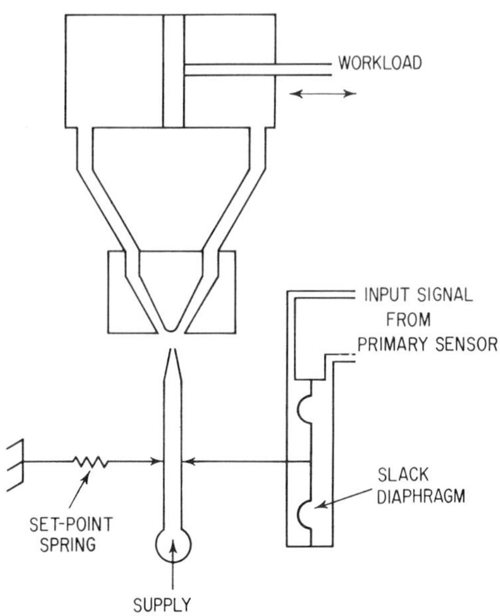

Fig. H-8. Hydraulic controller with integral signal sensing.

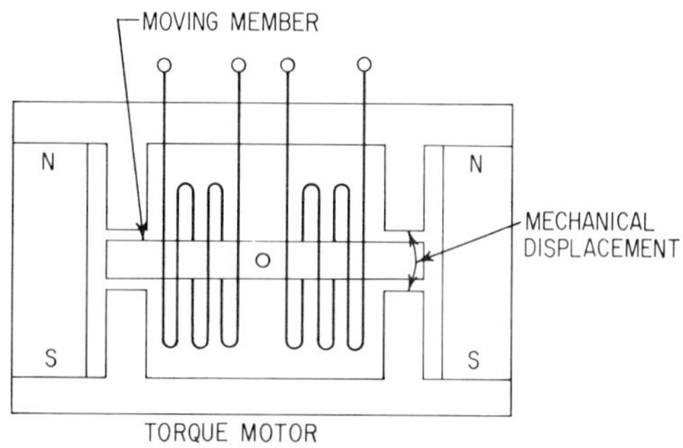

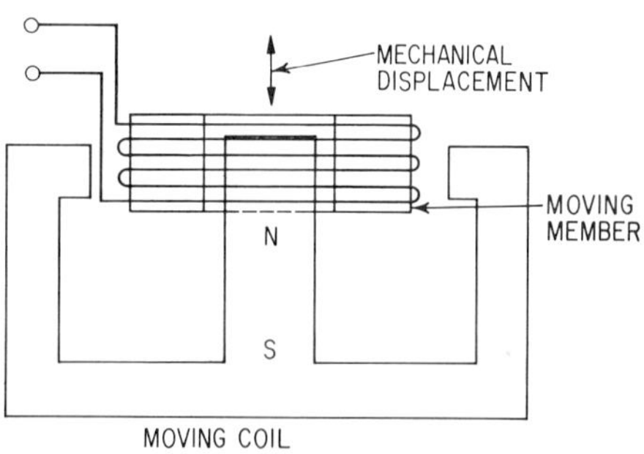

Fig. H-9. Electromechanical transducers used with hydraulic controllers.

Frequently, the process variable will be converted to an electrical signal by an appropriate transducer. In these cases, the generation of the set-point and error detection are done electrically and the error signal which is fed to the hydraulic relay is an electrical signal. When this is done, an electromechanical transducer is required to generate the mechanical displacement. Two of the more popular schemes are shown in Fig. H-9—the moving coil and the torque motor. Electromechanical transducers that are mechanically packaged together with a hydraulic relay commonly are referred to as servovalves.

Applications: Hydraulic controllers perform very well where both high power and fast speed of response are needed. Edge-guiding control systems are examples. It is not uncommon for a hydraulic edge-guide control system to control the edge of a moving steel strip (typical strip velocity of 1,000 ft/min) to $\pm\frac{1}{64}$ in. and do this by shifting a coil of steel weighing up to 50,000 lb. Hydraulic controllers also are widely used as liquid pipeline-pressure controllers. Here a pipeline control valve is operated to control against sudden pressure surges of another pumping station in the line. Generally, hydraulic control systems will be selected based on more demanding control requirements rather than on lower initial costs.

—John R. Huebner, *Singer, GPE Controls, Inc.*

Hydraulic Relay A device that allows a small mechanical displacement to control the output from a hydraulic power supply in a fashion that will drive a work load. The three major types of hydraulic relays are (1) jet-pipe valve, (2) flapper valve, and (3) spool valve.

Jet-pipe Valve: By pivoting a jet pipe, as shown in Fig. H-10, a fluid jet can be directed from one recovery port to another. The fluid energy is converted entirely into a velocity head as it leaves the jet-pipe tip and then is reconverted into a pressure head as it is recovered by the recovery ports. The relationship between jet-pipe

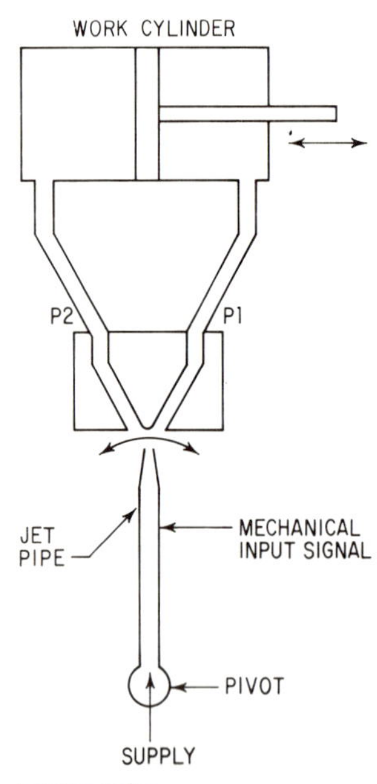

Fig. H-10. Jet-pipe valve.

motion and recovery pressure is shown in Fig. H-11. Although the jet pipe can be used at higher pressures, most applications are less than 800 psi. The proportional operation of the jet pipe makes it very useful in proportional-speed float-

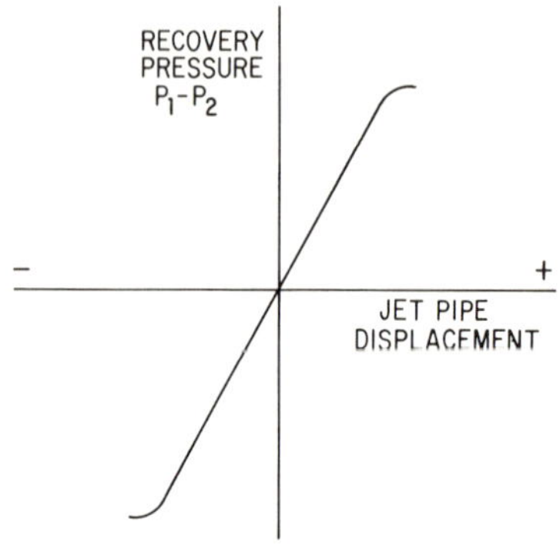

Fig. H-11. Relationship between jet-pipe motion and recovery pressure.

ing systems (integral control), as shown by Fig. H-12*a*.

Position feedback can be provided by rebalancing the jet pipe from the work cylinder as shown by Fig. H-12*b*. A proportional-plus-reset

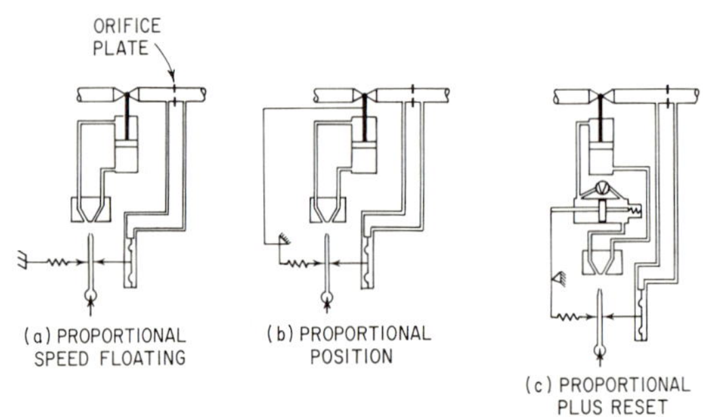

Fig. H-12. Hydraulic controllers arranged in different control modes.

arrangement is shown by Fig. H-12*c*. In the last case, the proportional feedback is reduced to zero as the oil bleeds through the needle valve. The hydraulic flow obtainable from a jet pipe is a function of the pressure drop across the jet pipe. Typical flows for jet pipes are shown in Fig. H-13.

Flapper Valve: This device makes use of two orifices in series, one of which is fixed, the other variable. The variable orifice consists of a flapper and nozzle. The nozzle restriction is changed as the flapper is positioned closer to or farther

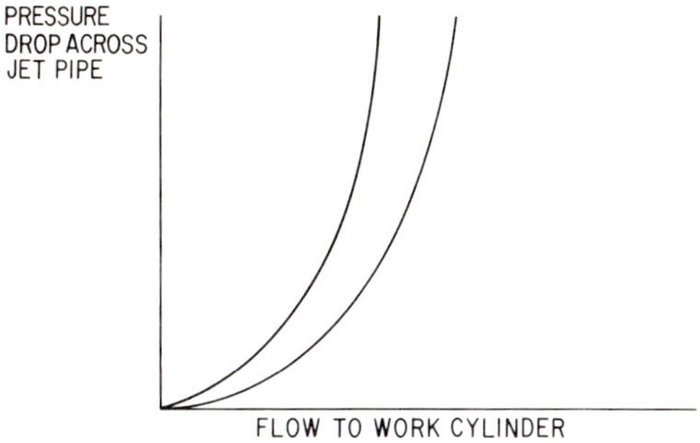

Fig. H-13. Typical flow for jet pipes. P_1 and P_2 are identified in Fig. H-10.

from the nozzle. If the variable restriction is changed, the pressure drop across the fixed orifice changes, thus producing a variable output pressure with flapper position. Single- and double-flapper arrangements are available, as diagramed in Figs. H-14 and H-15.

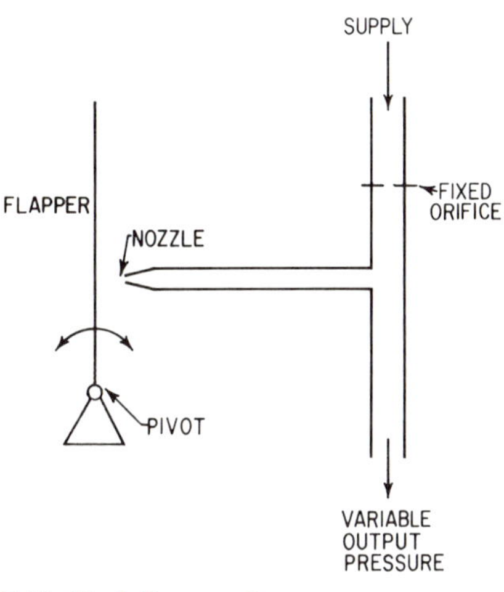

Fig. H-14. Single-flapper valve.

Spool Valve: This device (see Fig. H-16) when used as a hydraulic relay generally is constructed in either a three-way (Fig. H-17) or a four-way (Fig. H-18) valve-porting arrangement. In both cases, the mechanical displacement of the spools allows the hydraulic-pressure supply to be ported in a fashion that will displace the work cylinder in either direction dependent upon the spool displacement. The four-way valve can put full supply pressure on one side of the work cylinder and drain the other, or vice versa, while the three-way valve requires one side of the work cylinder to be

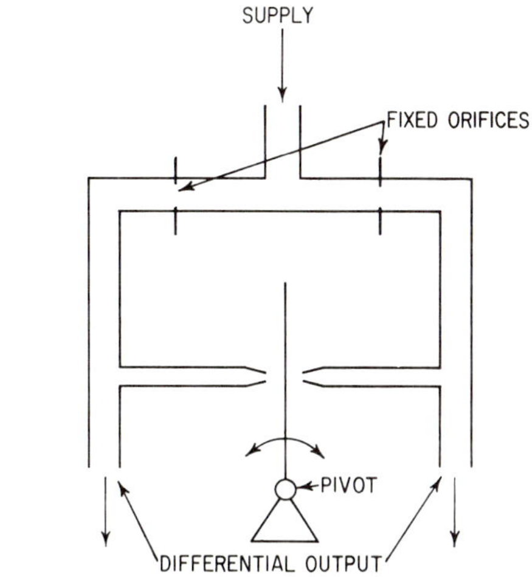

Fig. H-15. Double-flapper valve.

pressurized from a constant-pressure source. If the width of the spool lands is greater than that of their respective ports, the valve is overlapped and is referred to as a closed-center valve. An underlapped valve is referred to as an open-center valve. A line-to-line valve is one in which the spool lands are the same width as the porting. Flow through the spool valve is proportional to the square root of the differential pressure across the valve, as is the case with the jet pipe.

Two-stage Valves: When higher flows and pressures are required, a two-stage valve may be

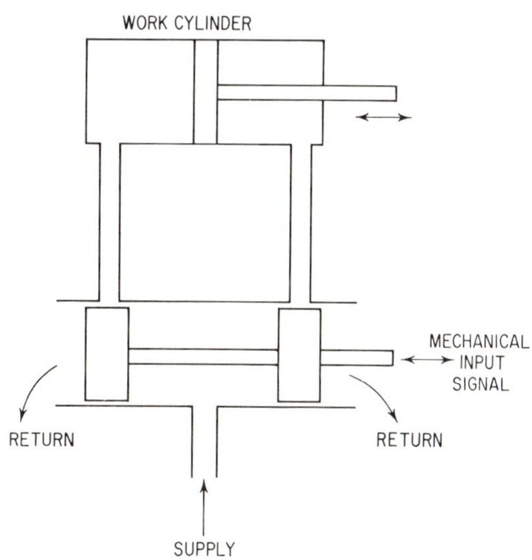

Fig. H-16. Spool valve.

used. In these cases, the second stage usually is a spool valve while the first stage may be either a jet pipe, a spool valve, or a flapper valve. Some different combinations of two-stage valves are

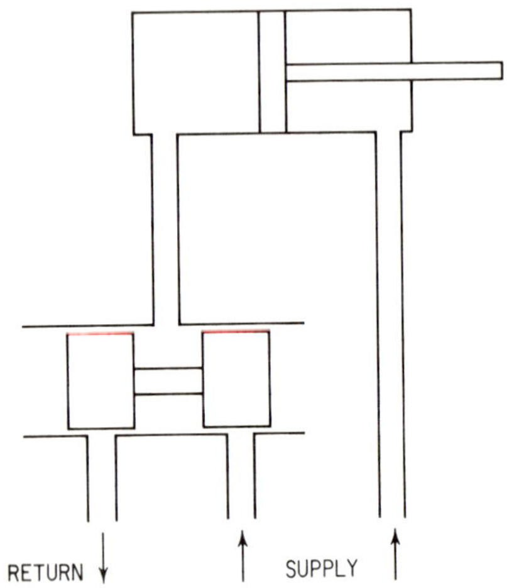

Fig. H-17. Three-way spool valve.

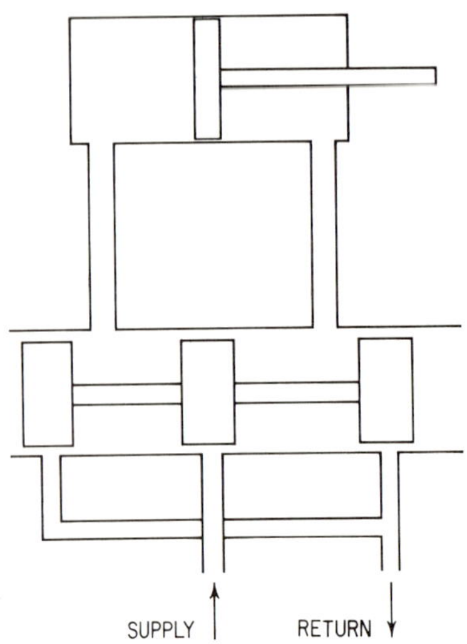

Fig. H-18. Four-way spool valve.

shown in Fig. H-19. In two-stage valves, the second stage must be either positioned, such as with springs, or fed back to the first stage as required. Feedback may be mechanical, electrical, or hydraulic.

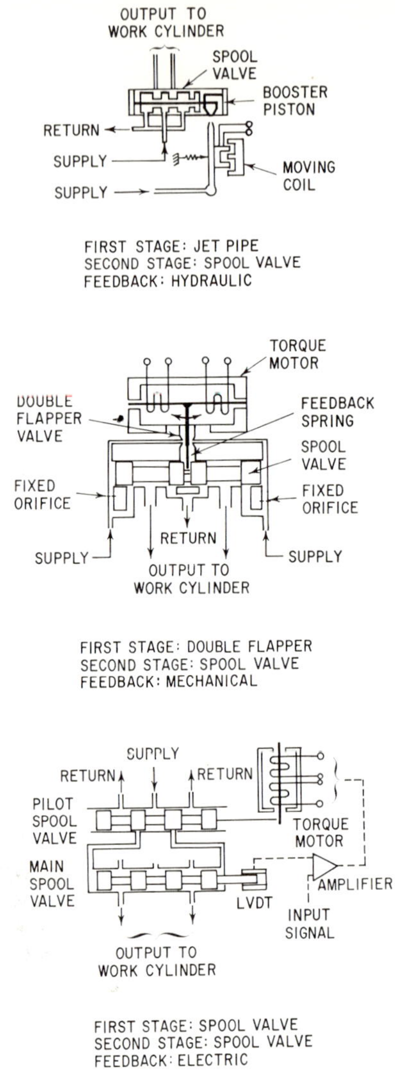

Fig. H-19. Various combinations of two-stage valves.

REFERENCES

Considine, D. M. (ed.): "Process Instruments and Controls Handbook," pp. 9-35 to 9-47, McGraw-Hill Book Company, New York, 1957.

Dushkes, S. Z., and S. L. Cahn: Analysis of Some Hydraulic Components Used in Regulators and Servomechanisms, *ASME Paper* 41-A22, Atlantic City Meeting, 1951.

Ziebolz, H. W.: Designing Hydraulic Servos, *Mach. Design*, vol. 19, no. 7, pp. 123–126, July, 1947.

Tess, G. F.: Direction-control and Servovalves, *Mach. Design,* vol. 39, no. 22, pp. 46–55, September, 1966.

—John R. Huebner, *Singer, GPE Controls, Inc.*

Hydrogen Electrode (See **Oxidation-reduction potential, measurement of.**)

Hydrogen-ion Concentration (See **pH, measurement of.**)

Hydrometer (See **Specific gravity, measurement of.**)

Hydrophone An electroacoustic transducer that responds to waterborne sound waves and delivers essentially equivalent electric waves. The hydrophone is used mainly to detect the approach of submarines and other watercraft. Examples include moving-coil hydrophones and moving-conductor hydrophones. Also see **Acoustic instrumentation.**

Hygrometer An instrument used to measure the amount of moisture in the atmosphere. The most commonly used types are wet- and dry-bulb thermometer instruments called psychrometers, and direct indicators of relative humidity called hygrometers. Also see **Hygrothermograph** and **Psychrometer.**

Hair hygrometers are used widely to indicate relative humidity directly. These instruments are actuated by a bundle of hairs, under tension, connected to a simple linear system and a pointer which operates over a graduated scale calibrated in relative humidity. These instruments operate reliably over a wide range of temperatures. Their time constant at room temperature is on the order of several minutes. The time constant increases as the temperature is decreased, approaching ten times greater at 10°C. The accuracy at room temperature approaches 3 percent relative humidity when the device is exposed sufficiently long to reach equilibrium. Calibration accuracy is best between 15 and 90 percent relative humidity. If this range is exceeded, the unit should be cycled a few times in its normal range to restore the original calibration.

Many other hygroscopic materials have been substituted for hair, such as nylon, wood fibers, cotton filaments, and goldbeater's skin. However, none of these materials is as widely accepted as properly treated, selected human hair.

REFERENCE

Wexler, Arnold, and W. G. Brombacher: Methods of Measuring Humidity and Testing Hygrometers, *Natl. Bur. Std. (U.S.) Circ.* 512, Sept. 28, 1951.

—James R. Cosby, *Environmental Science Division, The Bendix Corporation*

Hygrothermograph An instrument for recording ambient temperature and relative humidity. The temperature-responsive element is a bourdon tube. This specially shaped and formed tube is filled with alcohol. As the temperature varies, the alcohol expands and contracts, forcing the tube to bend. The bending motion of the bourdon tube operates a pen arm, through a mechanical linkage, to scribe a trace on a calibrated temperature chart. See **Bourdon tubes;** and **Thermometer, filled-system.**

The humidity element is a specially treated, multiple-strand hygroscopic element of human hair. Variations in humidity cause the hair to lengthen or shrink, depending on whether the humidity increases or decreases. The motion thus developed is utilized through a mechanical linkage to operate a pen arm and scribe a trace on a calibrated chart. Units of this type generally use a single chart, mounted on a clock-driven drum, containing both the temperature and humidity scales (see Fig. H-20). The spring-driven clock mechanism operates approximately 8 days on one winding. Time-scale gears can be changed

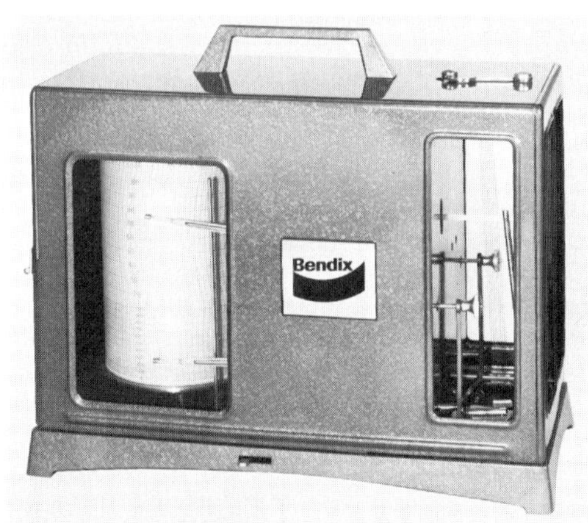

Fig. H-20. Hygrothermograph for recording relative humidity and air temperature. *(Bendix.)*

as desired to provide a full-chart recording period of one day or a full week.

—James R. Cosby, *Environmental Science Division, The Bendix Corporation*

Hysteresis When used as a performance specification, hysteresis is the maximum difference for the same measured quantity between the upscale and downscale readings during a full-range traverse in each direction (see Figs. H-21 and H-22).

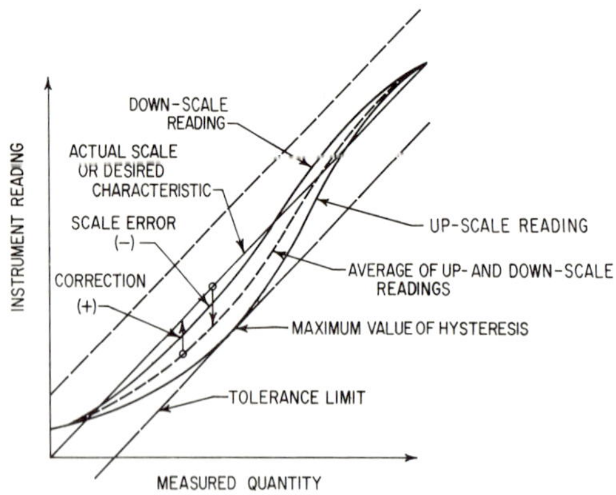

Fig. H-21. Typical hysteresis curve. Variations shown are exaggerated.

This common-usage definition—very useful in practice—includes *hysteretic error* and *dead band.* That portion of the difference which is dependent upon the history of prior excursion is hysteretic error, while that portion due to dead band may be determined by a conventional dead-band test. Some difference of the reading will occur on any small reversal of the measured quantity if an instrument exhibits hysteretic error without dead band.

Hysteretic error sometimes is referred to as elastic hysteresis. The hysteretic error is that portion of hysteresis due to energy absorption in the elements of a measuring instrument. The hysteretic error is obtained by subtracting the value of dead band from the corresponding value of hysteresis for a given value of the measured quantity (see Fig. H-22). The energy absorbed is conceived as produced by molecular friction and appears as heat in dynamic cycling when cyclic mechanical force is applied to a spring, or cyclic magnetizing force to a magnetic material.

Dead band is the range through which the measured quantity can be varied without initiating response (see Fig. H-22). Dead band usually is expressed in percent of span. In a mechanical instrument, the dead band is made up of two components: (1) friction and (2) play. Friction is caused by the rubbing of elements together. Play is caused by looseness in bearings or gaps in the mesh of gears, as in a gear-driven indicator. In Fig. H-22, the diagram represents a situation where the friction and the play are the same over the range of the instrument. Tapping will reduce the effect of the friction, and the change noted when tapping is done is referred to as "tapping error." However, when play is involved, the relative position of elements is random—unless there are forces, such as produced by springs or gravity, to influence the play. Also see **Magnetic quantities (variables).**

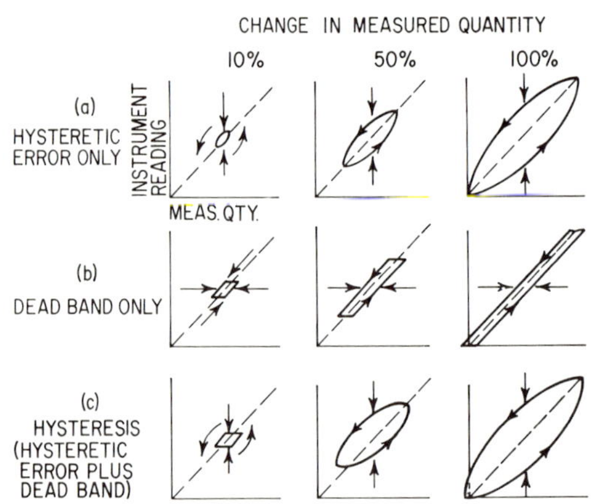

Fig. H-22. Hysteretic error, dead band, and hysteresis. The errors shown are exaggerated.

REFERENCES

SAMA Standard PMC 20, Measurement and Control Terminology, Process Measurement and Control Section, Scientific Apparatus Makers Association, New York, 1969.

Ku, H. H.: "Handbook of Industrial Metrology," chap. 2, ASTME, Prentice-Hall, Inc., Englewood Cliffs, N.J., 1967.

—Wallace Dean Wood, *Taylor Instrument Process Control Division, Sybron Corporation*

I

Ice Detector [See **Air-data systems (Aircraft).**]

Illuminometer (See **Photometers.**)

Ils (See **Flight-control systems.**)

Image Detection (See **Pattern recognition.**)

Impact, Measurement of [See **Shock, measurement of;** and **Testing (materials), destructive.**]

Impedance Measurement (See **Bridge circuits, electric.**

Inclined Manometer (See **Manometer.**)

Incremental Control System (See **Drafting system, automatic;** and **Numerical control.**)

Index Register The index register is a register (see **Register, computer**) whose contents is generally used to modify the data address of the instruction as the instruction (see **Instruction**) is being read from storage (see **Storage, computer**). The modified address is referred to as the effective data address. A particular index register is addressed by a specified field in the format of the instruction.

Index registers provide the ability to modify the program efficiently as it is being executed. For example, if the operation to be performed is a repetitive sequence of instructions on a table of data, it is necessary to change only the index-register value instead of the data-address portion of each instruction. The data address of the instructions would contain the address of the required data with reference to the start of the table. All instructions which reference table data are indexed by the specified index register which contains the address of the start of the table. When the program is to perform these operations on another table of data, the value in the index register is changed to the start address of the new data table. This effectively modifies all the indexed instructions in the sequence.

Index registers may be fixed locations in main storage, or they may be implemented in logic using flip-flops or triggers. In the case of storage-resident index registers, the index-register address contains only the number of bits required to specify the register number uniquely, and a fixed prefix is supplied by the system logic to provide the actual storage address. This technique minimizes the length of the computer instruction, since most computers have only a few index registers. Thus, only two or three address bits are required in the instruction even though a complete storage address may require 10 or more bits.

The index registers are also used as counters by the program, and the same register may be used as both a counter and an address-modification value to step through tables. That is, the index register is initialized to contain the number of factors in the table to be operated on, and the instructions to be performed contain the table-start address as a data address which is to be indexed by the same index register. Each time the sequence of instructions is performed, the index register is decremented by one and tested for zero value. If the result of the test is nonzero, the sequence of instructions is repeated, but since the index-register value has been reduced by one, the effective data address references the next lower table value. If the test result is zero, all the factors have been operated on, and the program steps to the next sequential operation.

—W. W. Welz, *IBM Corporation*

Indicating Instruments Much ingenuity has gone into creating different schemes for displaying instrumental data. Only the fundamental methods are described here. The use of cathode-ray tubes as data indicators is discussed elsewhere—see **Display tubes.**

Generally, an indicator consists of an arrangement of parts for producing and controlling (1) the motion of a pointer in relation to a stationary scale or (2) the motion of a scale in relation to a fixed point of reference.

Movable-pointer Indicators: Conventional fixed-scale movable-pointer indicators vary in appearance, form of scale, and the plane wherein the indicating pointer moves. In all cases, it is essential that the scale be graduated and mounted so as to minimize reading errors. To avoid parallax errors, precision indicating instruments, which normally are read at close range, often employ a mirror mounted beneath the pointer and adjacent to the scale graduations. In another arrangement, the graduated portion of the scale is raised to the level of the pointer tip, thus enabling accurate readings to be made at greater angles of observation. On instruments designed to be read at close range, the index may be in the form of a thin edgewise strip, a hairline scratched on each side of a transparent member, or a sharp arrow-shaped tip. Indicators which are intended for distant reading require scales having fewer graduations and pointers with proportionately larger tip areas.

Several types of movable-pointer indicators are shown in Fig. I-1. Frequently, the scales shown are used with two indicating pointers. The second pointer (shown in broken line) indicates values of a second variable, or it may function as a manually set control index or reference spotter.

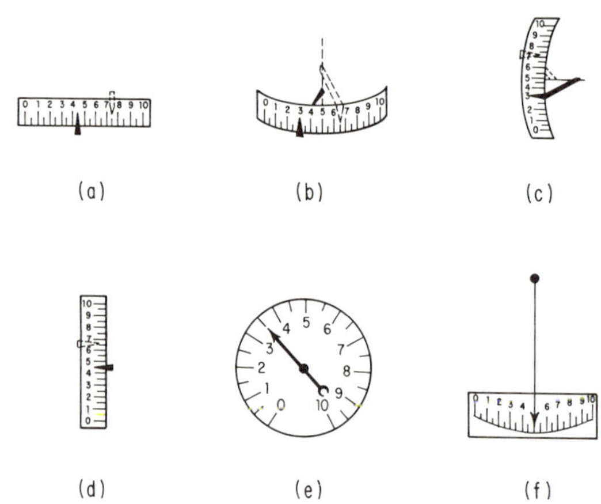

Fig. I-1. Types of movable-pointer indicators: (a) straight scale; (b) arc scale; (c) arc scale (vertical); (d) straight scale (vertical); (e) circular scale; and (f) segmental scale.

Movable-scale Indicators: Three movable-scale (fixed-pointer) arrangements are shown in Fig. I-2. A flat-disk scale is shown in Fig. I-2*a*. In practice, the angle of rotation seldom exceeds 340° with a maximum calibrated scale length of about 30 in. Figure I-2*b* shows an arrangement sometimes referred to as a precision indicator and is a variation of Fig. I-2*a*, the chief difference being that the scale is finely divided (from 300 to 600 divisions in a 24-in. scale length). The stationary index generally is in the form of a sharp tip or hairline. Figure I-2*c* shows a horizontally mounted rotating-drum type scale. The angle of rotation, scale length, and graduations are similar to those of the rotating-dial-type scale. A number indicator in the form of several vertically mounted, numbered drums is shown in Fig. I-2*d*. The drums are geared together to form a number measurement. Both rotating-drum and dial-type indicators, as a general rule, are servo-operated. Drum- and dial-type indicators permit close observations for relatively precise readings. Another type of movable-scale indicator uses a

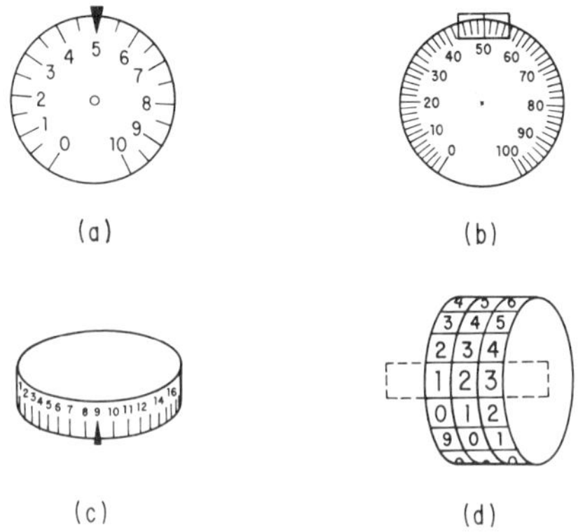

Fig. I-2. Types of movable-scale indicators: (a) dial; (b) precision indicator; (c) drum; and (d) drum counter.

two-color painted metal strip which is moved in a vertical plane. Often, this device is applied to liquid-level measurement—with the dividing line representing the top of the liquid.

Multipoint Indicators: A multiple-point indicator should not be confused with a multiple-pointer instrument. The multipoint indicator consists of a single indicating instrument used in conjunction with a point selector, which may be either manual or automatic. Selection is accomplished by switching electric circuits. Such systems usually are confined to measurement of variables which can be converted to electrical signals. A variety of buttons and switches, including dial systems, are used for point selection.

Multirange Indicators: The utility of an instrument can be increased by incorporating calibration for several ranges in one instrument, each range having a separate scale. Correctly identifying the range selected with its associated scale is made more certain by use of a scale-range pointer that is automatically tied to the range-selector switch. In an alternate arrangement, a mask is used to cover all scales but the one selected.

Inductance, Measurement of (See **Bridge circuits, electric;** and **Electrical quantities, measurement of.**)

Inductance Transducer (See **Electromagnetic transducer.**)

Inductive-bridge Transducer (See **Position-measurement and control systems.**)

Information Processing (See **Character-recognition systems; Code readers and systems; Computers; Data communications; Pattern recognition; Voice-recognition systems**—as well as the **Classified Index.**)

Information Theory The mathematical theory concerned with information rate, channel width, distortion, noise, and other factors affecting the transmission of information. Also see **Data communications;** and **Telemetry.**

Infrared Detector, Thermistor A detector that uses a temperature-sensitive resistor as its sensing element. Infrared radiation causes the temperature of the resistor to change. The thermistor element is made of polycrystalline manganese, nickel, and cobalt oxides. A bias voltage or current permits the change in resistance to cause an equivalent change in output voltage. The thermistor is a true power detector. The output voltage is proportional to the radiant power falling on the detector. These detectors are widely accepted for use in military, scientific, industrial, and space-exploration applications.

The infrared radiometer or infrared pyrometer collects and measures radiation from a target to provide the temperature of the target without touching it. Special versions are used to detect fires, plot ocean-surface temperatures from aircraft, detect overheated bearings on moving railroad cars, and provide remote temperature inspection of transformers and transmission lines. Military applications of the radiometer include heat-seeking missiles, aircraft and missile trackers, aiming systems for gunfire control, and detection devices for location of vehicles and personnel in the dark. Astronautical radiometers track the position of the sun, moon, and stars. Long-range weather forecasting depends on infrared radiometer surveillance of the earth's cloud coverage from satellites. The horizon sensor is a special radiometer used to establish a stable vertical reference for attitude control of missiles and satellites in space.

Thermal Imaging: Infrared scanning instruments or image-forming cameras produce thermal pictures of their targets. Medical applications use infrared instruments for remote thermal mapping of the human-body surface temperatures. Areas that exhibit abnormally high or abnormally low temperatures are readily revealed and may be analyzed for body malfunction or disease.

Standard-thermistor Detector: This device is mounted so that one surface is exposed to radiation while the other surface is separated from a thermal heat sink by a thin film of thermal insulation. The heat capacity H of the thermistor and the thermal resistance of the insulator Z govern τ, the thermal time constant of the detector, as expressed by $\tau = HZ$. The thermistor is enclosed in an evacuated housing with an infrared window. The device exhibits high sensitivity over the spectral region from 2500 angstroms to 35 microns and beyond.

Immersed-thermistor Detector: This device is optically attached to a lens which serves as the optical element and heat sink. The lens makes possible the use of a smaller-area thermistor, and this results in increased radiation-detecting ability. A germanium lens increases the detecting ability 3.5× over the standard or unimmersed thermistor. A gain of approximately 3× results with a silicon lens. The spectral range of the immersed thermistor is governed by the transmission characteristics of its lens. The immersed thermistor is mounted in an evacuated housing so that the thermistor is protected in the vacuum environment while the lens is exposed to radiation.

Circuitry: As shown in Fig. I-3, the thermistor may be connected as a single resistor to a constant-current circuit. The output voltage or signal then is derived as voltage variations taken directly across the thermistor. This circuit does not compensate for ambient-temperature variation, changes of which produce a signal voltage.

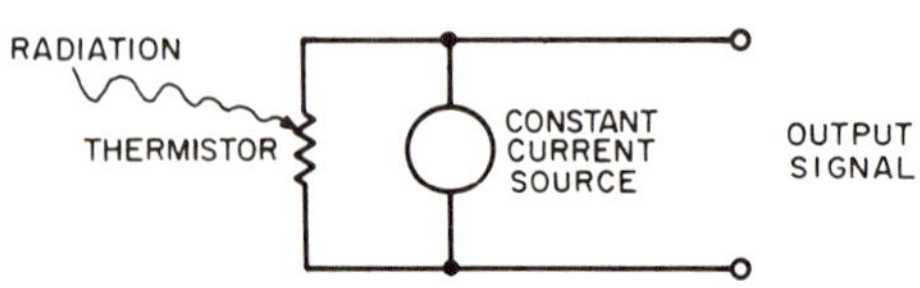

Fig. I-3. Single thermistor—constant-current operation.

The common circuit for the thermistor requires that two thermistors be connected as a balanced bridge to equal bias voltages of opposite polarity with the common bias junction at ground potential (see Fig. I-4). The signal voltage then is obtained between the junction of the two thermistors and ground. The two thermistors essentially are equal in resistance at any temperature. Thus, the bridge remains balanced as ambient temperature varies. In operation, one thermistor is exposed to radiation while the second thermistor is shielded permanently from radiation. This compensated circuit produces only one-half the signal voltage of an equivalent single-thermistor constant-current circuit.

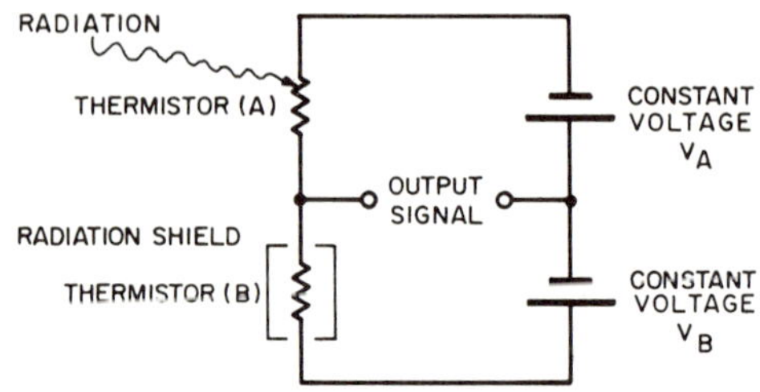

Fig. I-4. Two-thermistor bridge circuit—constant-voltage operation.

Thermistor Design: Dimensions of the thermistor and the focal length of the optical system determine the angular field of view of the instrument. Thermistor size is affected by the smallest target area that must be distinguished and its distance from the instrument. The target must be traced through the optical system to determine the size of the image in the focal plane. A detector size should be selected that will be completely covered by the target image while maintaining the desired field of view. For point targets, a field lens can be used to disperse target radiation uniformly over the entire detector area, thus achieving equal signal output regardless of target position in the field of view.

Responsivity: This is the measure of the detector's ability to convert radiation power into volts of output signal. A decrease in detector area increases responsivity, since the latter varies inversely as the square root of the area. Minimum responsivity required is (1) the smallest change in detector output voltage that can be processed by the associated electronics and (2) the smallest change in incident radiant power that the thermistor must detect.

The thermal time constant of the thermistor causes a time lag between a radiation change and the resulting output-voltage variation. Because of thermal lag, fast radiation chopping does not permit detector output voltage to reach its maximum amplitude and thus produces an effective decrease in responsivity (see Fig. I-5). For practical purposes, the time constant is the time required for the detector to produce 63 percent of

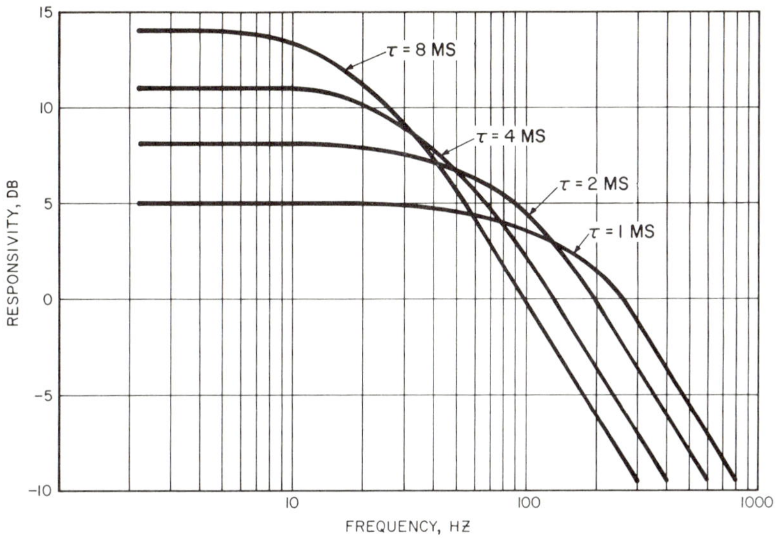

Fig. I-5. Effect of time constant on responsivity of thermistor. *(Barnes Engineering.)*

its full output from a radiation step-function input. The longest detector time constant τ, in seconds, that should be used for any chopping frequency f, in hertz, is

$$\tau = \frac{1}{2\pi f}$$

Thermistors normally are made to a standard thickness of 10 microns. Two types of material are available: At 25° C, material 1 has a resistance of 2.5 MΩ for a square of any size; material 2 exhibits 0.25 MΩ/square. The aspect ratio of the thermistor is a major factor in determining resistance.

—Seymour Weiner, *Barnes Engineering Company*

Infrared Moisture Meter (See Moisture, measurement of.)

Infrared Process Analyzers When one considers the tremendous variety of chemical compounds exhibiting infrared absorption, it is not surprising that infrared is one of the most widely applied analytical techniques in industrial-process control. About the only materials which do *not* absorb at some infrared wavelengths are O_2, N_2, H_2, halogens, and the rare gases. Infrared has been applied in many forms to process control; for example, the field of temperature-sensing radiation pyrometry is based largely on near infrared emission. The thickness of one layer in a multilayer plastic film can be monitored at high speeds if it has a unique infrared absorption spectrum. Water in sheet paper can be continuously monitored by selective infrared reflectance. Stack gases can be remotely identified by their characteristic emission. The majority of infrared process analyzers, however, utilize selective infrared absorption by the component of interest in a representative sample that continuously flows through a windowed cell in the analyzer.

Traditionally, infrared analyzers have been classified in two general groups, dispersive and nondispersive, depending on the means of wavelength selection. The recent development of the narrow-bandpass interference filter (single or selectable wavelength) creates a third category, the optical-interference filter analyzer.

Several arrangements fall under the description "nondispersive infrared analyzer." The most widely used, however, is the so-called positive filter, or Luft (Ref. 1) analyzer, depicted in Fig. I-6. This arrangement utilizes two matched sources S_1 and S_2 operating at approximately 800° C having a blackbody-emission continuum spectrum. An opaque (usually) chopper modulator blocks both sources simultaneously with equal on-off periods at frequencies as low as 2 Hz. The sample cell C_1 is a highly polished tube that conducts the radiant energy through it by multiple reflection. The gaseous sample from the process flows continuously through this cell. The reference cell C_2 is similar but is usually used merely to transmit the radiant signal and is simply filled with an inert gas like air. For applications where there is no interfering absorption by other sample components, the filter cells F_1 and F_2 would not be used. The radiant energy then passes into the pneumatic microphone detector D, which is the heart of this instrument. The detector is a sealed chamber with a very thin flexible diaphragm separating the two halves exposed to each beam. The chamber contains the same gas as the component to be measured in the sample stream, plus an inert pressurizing gas for optimizing sensitivity. This gas sensitizes the analyzer to the desired component by absorbing radiation at only those wavelengths where the sample absorbs. An electrical output signal results when the sample absorbs radiation from one beam, thus changing the relative pressure in the two chambers, deflecting the flexible diaphragm which forms a capacitor plate in an electric circuit.

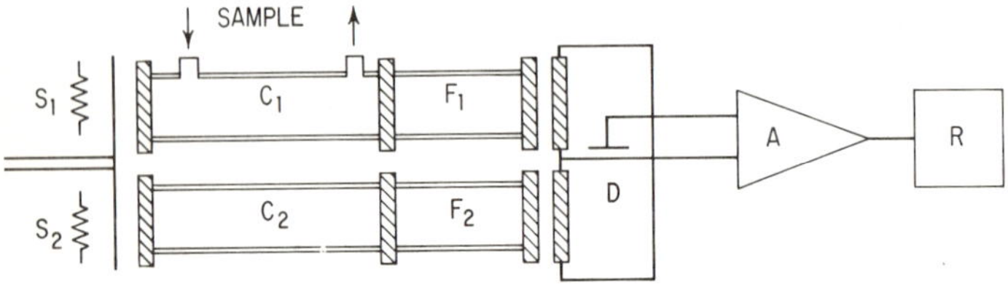

Fig. I-6. Typical positive-filter analyzer.

A unique feature of the gas in this detector is the very high spectral resolution provided by the vibration-rotation absorption bands of the gas. This resolution is better than the best laboratory grating instrument and accounts for the very high sensitivity of this technique. Analysis in the 0 to 100 ppm range with 6-in. cells is common. For complex samples, absorption bands of other components may partly overlap those of the gas in the detector and hence interfere. This can be eliminated by removing those wavelengths by putting the interfering gas in the filter cells F_1 and F_2 or by using optical filters. The signal from the detector is amplified (A) and recorded (R) on a scale calibrated in concentration of the measured component. Occasionally liquid samples are also measured where a suitable sensitizing gas can be found.

Almost all dispersive infrared process analyzers in use today are of the "homemade" variety; hence a wide assortment of designs is evident. Figure I-7 shows a dispersive analyzer utilizing the two-wavelength ratio-recording technique (Ref. 2). This uses a grating monochromator with two entrance slits to yield two wavelengths, one at the measuring wavelength of the desired component in the liquid (in this case) sample and one at an appropriate reference wavelength. The flow-through sample has a secondary cell as a protective device in case of leaks. The detector is a radiation thermocouple as used in laboratory instruments, but there the similarity ends. The

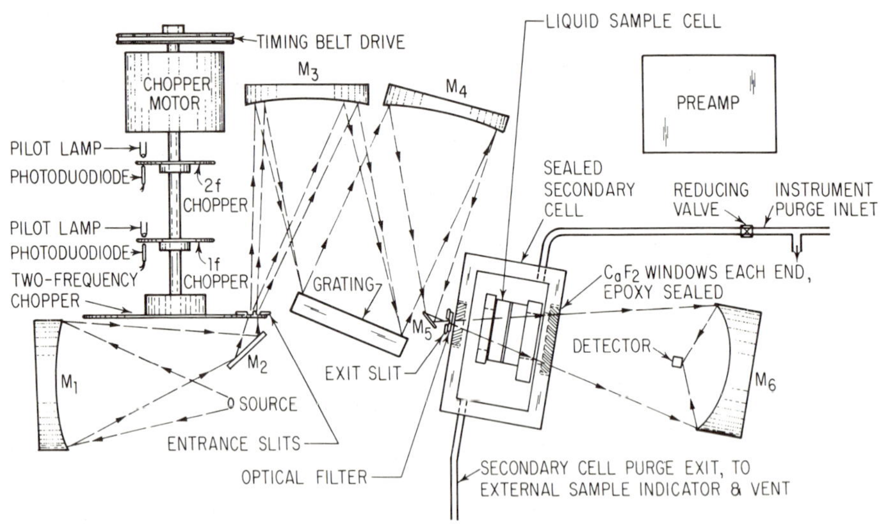

Fig. I-7. Infrared analyzer with secondary safety cell.

process analyzer is far more rugged and designed for long-term reliable quantitative analysis with very little versatility compared with a laboratory instrument.

The bandpass optical filter analyzer has been used for some time in the near infrared and can now be built to operate in the entire infrared range. The variable-wavelength wedge-type interference filter analyzer simplifies the monochromator portion of a "dispersive"-type analyzer drastically. The filter replaces the entire monochromator in its function of isolating the operating wavelength. Analyzers using bandpass or variable-wavelength filters in the longer infrared wavelengths (beyond 5 microns) are not generally available but presumably will be in the future.

The hardware costs of an infrared analyzer may vary from ≈$3,000 to ≈$10,000 without a sample system. The sample system may equal the cost of the hardware but is an absolutely necessary expenditure, for no analyzer can operate without a clean, bubble-free sample at a reasonable temperature and pressure.

REFERENCES

1. Luft, K. F.: *Z. Tech. Physik,* vol. 5, p. 97, 1943.
2. Bartz, A. M., and H. D. Ruhl: *Chem. Eng. Progr.,* vol. 64, no. 1108, pp. 45–49, 1968.

—A. M. Bartz and H. D. Ruhl,
The Dow Chemical Company

Infrared Spectroscopy The infrared region of the electromagnetic spectrum lies between the visible region and the submillimeter, or microwave, region and extends from approximately 0.78 to 300 microns. It is of scientific importance because molecular vibration and rotation occur in this frequency range. Analysis of infrared spectra provides basic information on molecular structure in terms of geometry and interatomic forces and distances, but the widest industrial application is probably the use of vibrational spectra for qualitative and quantitative analysis. Each molecule is characterized by a unique set of vibrational absorptions which serve as a set of identifying "fingerprints," and the depth of absorption bands serves as a measure of the number of molecules present. Sample-size requirements vary from milligrams for routine studies to as little as a fraction of a microgram when special accessories are used in conjunction with the basic spectrophotometer.

Single-beam spectrometers are used in many laboratories, both for control and for certain types of specialized physical measurement, but the overwhelming majority of spectrophotometers manufactured today are double-beam instruments. A typical research-type spectrophotometer is shown in Fig. I-8, and its optical schematic is given in Fig. I-9.

Infrared radiation from the source is separated as shown into two beams—a sample beam and a reference beam. After passing through the sample compartment, these two beams are subsequently recombined by a rotating semicircular chopper-mirror (*M*7) which passes first one, then the other, and so on. Mirror *M*9 then focuses the sample (or reference) beam onto the monochromator entrance slits *S*1. The beam is collimated (by *M*11) and travels to the diffraction grating *G*1, where it is dispersed. Wavelength (or frequency) is scanned by rotating the grating about an axis *C* perpendicular to the plane of the figure. As the grating is rotated, successive frequencies are swept across the exit slits *S*2 and focused (by *M*14) onto the thermocouple detector target.

In use, a sample is placed in the sample beam. If it absorbs energy at a given frequency, then the energy content of sample and reference beams will differ, and as they are passed alternately through the system by the chopper-mirror, the thermocouple will generate a fluctuating voltage corresponding to this difference. This voltage is used to drive a motor, which in turn moves an optical attenuator (shown in front of mirror *M*6) into the reference beam until the attenuator removes energy from the reference beam equal to that which the sample removes from the sample beam. It is this nulling out of the optical difference between sample and reference beams which gives the system its name as an optical-null spectrophotometer. The recorder pen, which shows percent transmission of the sample, is linked to the optical attenuator.

Chart readout usually is in linear percent transmission vs. linear frequency. The infrared spectroscopist, however, usually defines spectral location not in terms of wavelength or frequency but as a unit, the cm^{-1}, which is the reciprocal of wavelength (expressed in centimeters) and is therefore proportional to frequency.

The unusually wide choice of commercial instrumentation makes it extremely difficult to give a concise summary of characteristics and performance. Somewhat arbitrary characterization can be made in terms of cost, wavelength range, resolution, photometric accuracy, and auxiliary

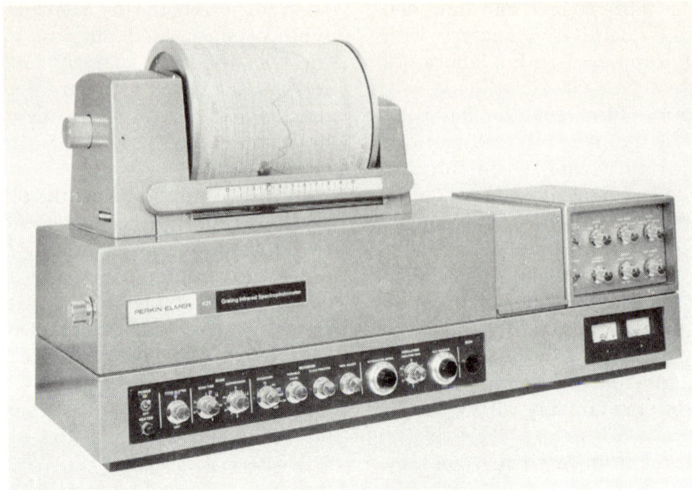

Fig. I-8. Typical research-type double-beam infrared filter-grating spectrophotometer. *(Perkin-Elmer.)*

controls. Resolution is related to information density, wavelength range together with resolution determines total available data, and photometric accuracy defines the error magnitude of each data point.

Small, inexpensive instruments ($3,000 to $6,000 selling price) usually have scanning ranges of 4,000 to approximately 600 cm^{-1}. Resolution at 1,000 cm^{-1} (midrange) is typically 3 to 5 cm^{-1}, and photometric accuracy ranges from 1 to 3 percent.

Medium-performance instruments ($6,000 to $10,000) often provide extended spectral range, e.g., 4,000 to 250 cm^{-1}, but many scan over the narrower 4,000 to 600-cm^{-1} range, which is still adequate for many analyses. Best resolution is of the order of 1 cm^{-1} at 1,000 cm^{-1}, although survey scans are usually made at resolutions of 2 to 4 cm^{-1}. Photometric accuracy of 1 percent is common, and auxiliary controls often provide a choice of slit programs and scan speeds.

Research-type spectrophotometers ($15,000 to $30,000) naturally incorporate the maximum in performance and versatility. While some instruments are available with restricted frequency ranges, the top-of-the-line instruments usually have a range of 4,000 to 200 cm^{-1} and require dry-air purging to reach the lowest frequencies. Best resolution varies from 0.15 to 0.3 cm^{-1} and photometric accuracy from 0.5 to 1 percent, depending on the instrument. Auxiliary features usually include continuously variable slit program and scan speeds, linear-absorbance presentation, ordinate and abscissa expansion and com-

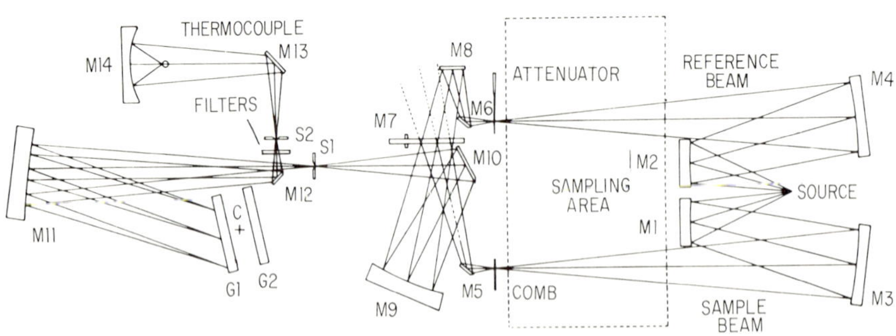

Fig. I-9. Optical schematic of double-beam optical-null spectrophotometer.

pression, scan-speed suppression, recycling, and automatic energy control.

Specialized research-type instruments are available for use in the far-infrared region from 200 to 33 cm^{-1} and sometimes further. The near-infrared region above 4,000 cm^{-1} (specifically from 0.78 to 2 or 2.5 microns) is not scanned by most infrared spectrophotometers, since it is included in the range of many of the larger ultraviolet-visible spectrophotometers.

REFERENCES

Alpert, N. L.: Instruments, in H. Szymanski, "IR Theory and Practice of Infrared Spectroscopy," pp. 8–87, Plenum Press, Plenum Publishing Corporation, New York, 1964.

Whetsel, Kermit: Infrared Spectroscopy, *Chem. Eng. News,* vol. 46, no. 82, p. 96, 1968.

Spectroscopy Nomenclature, *Anal. Chem,* vol. 39, no. 14, p. 1943, December, 1967.

—W. D. Perkins, *The Perkin-Elmer Corporation*

Infrared Thermometer (See **Infrared detector, thermistor**; and **Photometers.**)

Input/Output Devices A device by means of which information is read into a computer is called an input device, and a device which accepts information from a computer is called an output device. The information which is transferred to or from the computer is in either digital or analog form.

Some examples of input and output media used for digital data are punched paper tape, punched cards, electronic signals from keyboards or graphic devices, and contact points.

The *paper-tape reader* is a device which senses holes punched in a strip of paper tape and transmits the information into the computer. The information may have been punched in the tape by a *paper-tape punch* as output from a computer or other electrical device, or by a manually operated punch (see **Paper-tape reader/punch**).

Information contained in punched cards is read into the computer by a *card reader*. The punched card may be the output of a computer using a *card punch* or the output of a keypunch device (see **Card reader/punch**).

A keyboard may be used to input data to a computer. In this case, when the operator depresses a key, the keyboard-entry device translates the key position into a unique digital value which is read into the computer. The keyboard may be a part of a graphic-display console (see **Graphics, computer**) or a separate input device. The light pen may also be used as an input device on a graphic-display console. In this case, the computer reads the coordinates of the display tube at which the light pen points. By moving the light pen from one set of coordinate points to another set, a digital representation of a line can be generated for use in the computer. By means of a graphics console, digital data may be displayed on a screen and modified by a light pen or keyboard, and the information in the computer can be updated immediately.

The digital-input device (see **Digital input**) or feature provides the means of reading digital information represented by contact closures or binary voltage levels. The contacts may be a part of a piece of manufacturing equipment, for example, and the closure may signify that a particular parameter is outside of acceptable limits. The digital-output feature (see **Digital output**) provides the ability to operate a contact external to the computer. This output could be used to operate a relay which would cause an out-of-limit unit to be placed in a reject hopper, for example.

In many applications, a substantial portion of the data to be processed by a computer is the result of the measurement of some physical characteristic such as pressure, temperature, electric current or voltage, or weight. Data in this form must be converted into digital values before they can be processed in a digital computer. The information may be accepted in analog form by the computer via the analog input (see **Analog input**) device and translated to a digital value in an analog-to-digital converter (see **Analog-to-digital converter**). The data may also be read into the computer via digital input if the analog-to-digital conversion takes place external to the computer.

Analog signals for the control of a process or device can be obtained from the digital computer by means of the analog-output (see **Analog output**) subsystem. The analog-output subsystem receives digital data from the central processing unit and translates them into the required analog form by means of a digital-to-analog converter (see **Digital-to-analog converter**).

Printers and plotters are output devices which provide a means of obtaining a hard copy of alphanumeric or graphic output from a digital computer.

—W. W. Welz, *IBM Corporation*

Instruction A series of characters which defines a single digital-computer operation. An instruction might, for example, command a computer to load a register (see **Register, computer**), add two numbers together, or move a data field.

An instruction consists of two parts, an operator and an operand. The operator specifies the function to be performed, such as addition. The operand specifies the address of the data to be used or the data themselves, such as the value to be added to the accumulator contents. The operand portion of an instruction is optional depending upon the nature of the operator. Each type of computer has its own instruction set. Usually these instruction sets are not compatible between different types of computers.

—B. W. Landeck, *IBM Corporation*

Instruction Counter The instruction counter is a counter register (see **Register, computer; Counter**) which contains the address of the instruction being accessed in storage (see **Storage, computer**). Each time an instruction (see **Instruction**) is executed, the register is incremented such that, at the completion of the operation, the instruction counter is able to address the next instruction.

When a program is interrupted (see **Interrupt**), the instruction-counter address must be saved so that the program may resume at the point of interruption when the interrupt program is finished. If a "branch on condition" instruction is executed and the branch is taken, the contents of the instruction counter is replaced by the "branch to" address.

—W. W. Welz, *IBM Corporation*

Instrument Lexicons generally define an instrument as a medium or means through which or by which something is effected or achieved. Secondarily, an instrument usually is defined as an implement, a tool, or a utensil. Other meanings include a device for creating musical sounds, an agent for communicating or accomplishing some deed, and a document, such as a legal instrument.

In this encyclopedia, we are concerned with instruments as tools and, more specifically, tools that measure. Measurement involves comparing—hence instrument in the context of this volume means measuring instruments that compare whatever factor or variable is sensed with some form of standard. By means of prior calibration—when we read temperature from a thermometer, we are in essence obtaining data from the instrument that relate the temperature to which the bulb presently is exposed to the ice or steam points (or whatever other standard, reproducible points) against which the instrument was calibrated. Thus, the measuring instrument always is the bridge between some standard (including empirical standards) and the unknown. With a hand instrument, the comparison is made manually. Most of the instruments described in this volume accomplish this comparison automatically. Many automatically record as well as indicate. Still others also perform an automatic controlling function that is closely coupled to the measuring function. Thus, there are indicating, recording, and controlling instruments in all combinations.

Some instruments essentially are fully mechanical in nature—thus mechanical instruments. Similarly, there are electrical, electronic, pneumatic, fluidic, and hydraulic instruments. Frequently, an instrument will be identified by the application which it serves—thus, aircraft instruments, process instruments, meteorological instruments, medical instruments, etc.

Instrumentation in the context of this encyclopedia simply means the science of putting measuring instruments to constructive use.

Insulation Tester (See **Megger**; and **Ohmmeter.**)

Integral (Reset) Control The control action in which the output is proportional to the time integral of the input; that is, the rate of change of output is proportional to the input.[1] The control variable is changed at a rate that is proportional to the error or deviation from set point. If the deviation were tripled over a previous value, the final controlling element would be moved three times as fast. When the measured variable is at the set point, the final controlling element remains stationary. Integral control is expressed by

$$m = \frac{1}{T_i} \int e \, dt + M \qquad (1)$$

where m = manipulated variable
T_i = integral time
e = deviation of measured variable from set point
M = constant of integration

The integral time T_i is the time of change of the

[1] Extracted from American National Standard CE85.1-1963, American Standard Terminology for Automatic Control, with permission of the publisher, The American Society of Mechanical Engineers, New York.

manipulated variable caused by a unit change of deviation.

Response of an integral controller to a step input is shown in Fig. I-10. In this example, 2 min is required for the output to change an

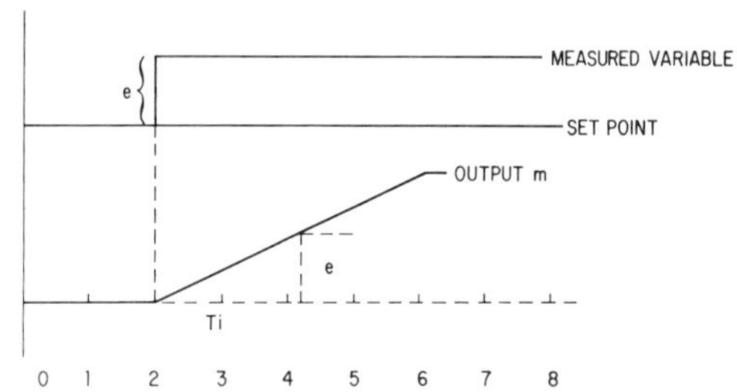

Fig. I-10. Response of an integral controller to a step change in input.

amount e equal to the step change in input. Therefore, the integral time is 2 min per repeat, or 0.5 repeats per minute. Figure I-11 shows the response of an integral controller to a sinusoidal input. The output exhibits a phase lag of 90°

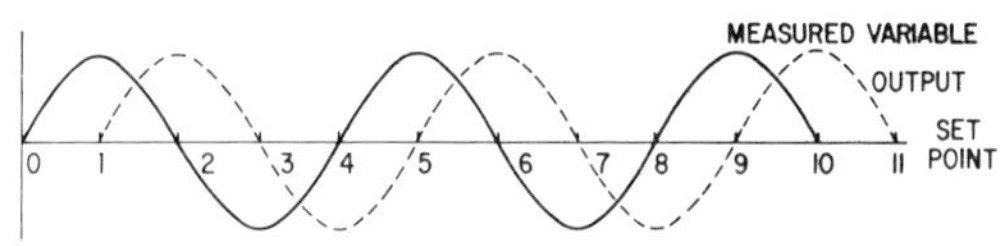

Fig. I-11. Response of an integral controller to a sinusoidal input.

regardless of the period of the cycling. This can be shown as follows:

Let $e = \sin wt$ and substitute in Eq. (1),

$$m = \frac{1}{Ti}\int \sin \omega t \, dt + M$$

$$m = -\frac{1}{\omega T_i} \cos \omega t + M$$

where $\omega = 2\pi f$, rad/s

$f =$ frequency, Hz

Since the output m is given by a negative cosine function, the phase lag is seen to be 90°. Also see **Electric controller**; and **Pneumatic controller.**

—Richard H. Kennedy, *Bristol Division, American Chain & Cable Company, Inc.*

Integrating Accelerometer (See Accelerometer.)

Integrator In instrumentation, an integrator is a means or technique to totalize or integrate a measured quantity, such as electric power, fluid flow, sunlight, and temperature (heat). Integration normally is made with respect to time, and the dimensions of the integral typically are kilowatthours, ampere-hours, gallons, and cubic feet. The means used may be (1) mechanical, (2) electrical, or (3) a combination. The integrator assembly consists of (1) an input developed by or resulting from the measured quantity, (2) a device for totalizing this quantity proportional to the time integral of its input signal, and (3) an output readout device, such as a numerical or an electromechanical counter. The counter may be arranged to run continuously and the interval or period of integration determined by observation of the counter reading. Or, the interval may be recorded periodically and the counter automatically reset to zero—normally done at some specified interval, such as one hour or one day.

Mechanical Integrators: The device shown in Fig. I-12 rather typifies the principles of a mechanical integrator. A synchronous motor drives a disk F at a constant speed. A ring G is held against disk F by an output shaft H coupled to a numerical counter. (Also see **Counter.**) Ring G also is coupled mechanically to the instrument measuring the input variable and is positioned proportionally along disk F from the center to the circumference of the disk. Dependent upon the position of ring G, the output of shaft H (recorded on the counter) varies from the center (zero output) to the circumference of the disk (100 percent output). Thus, as the instrument moves to record the measured variable, ring G moves to adjust the speed of shaft H in accordance with the magnitude of the measured quantity. The total count

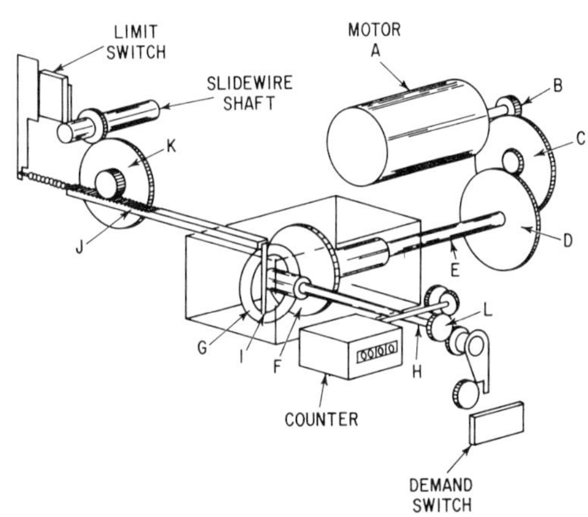

Fig. I-12. Mechanical integrator.

over a given period of time is an integral or summation with respect to time.

Other mechanical integrators employ similar versions of a constantly rotating force, either varied continuously—as with the ring technique—or interrupted by ratcheting devices in an intermittent pulsing operation, to drive the counter proportionally to the input signal.

Electrical Integrators: Typically, these are solid-state electrical operational amplifiers that accept a dc input signal and develop a proportional pulse-type output frequency. The output is capable of driving an optional electromechanical counter. Also see **Amplifier.** The basic circuitry which utilizes the integrating properties of an operational amplifier with capacitance feedback is shown in Fig. I-13. Analysis of the circuit shows that the voltage appearing at the output of

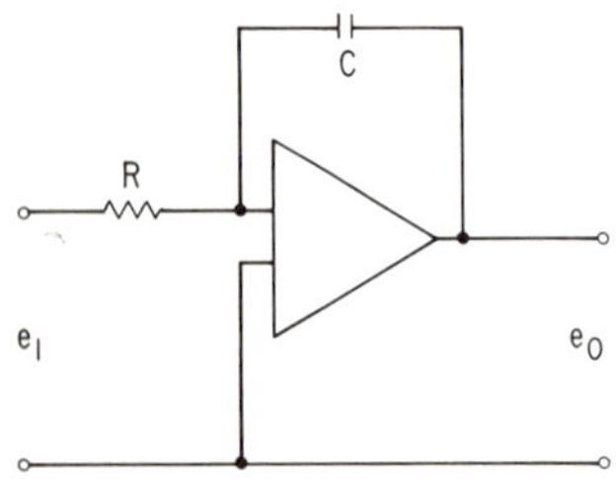

Fig. I-13. Operational amplifier with capacitance feedback.

the amplifier e_o is proportional to the integral of the input voltage e_i, or

$$e_o = \frac{1}{RC}\int_{t_0}^{t_1} e_1 dt$$

A typical medium-level pulse-type continuous integrator, designed to integrate a slowly varying analog signal, is shown in Fig. I-14. A simplified circuit diagram is given in Fig. I-15. Two medium-gain operational amplifiers are used to supply input voltages to a third amplifier connected as a conventional analog integrator. A fourth amplifier functions as a comparator and provides the drive voltage to operate the switching circuitry that reverses the polarity of the integrating amplifier. This provides a ramp-function output signal whose slope alternates positively and negatively at designated output frequencies—which frequencies are proportional to the magnitude of the input signal. Full-scale frequency output can be varied by changing circuit constants of the integrating capacitor or input resistance.

Fig. I-14. Medium-level pulse-type continuous integrator. *(Leeds & Northrup.)*

In Fig. I-16, the block diagram of another type of continuous integrator, a dc reference (applied) voltage is compared with a dc feedback voltage generated by a dc tachometer generator. An error voltage which is the difference between the reference and feedback voltages forms the input for an amplifier which drives a two-phase motor mechanically coupled to the tachometer generator. For any given value of reference voltage, the motor is made to operate at a speed such that the input and feedback voltages are equal within a small limit of error. The result, assuming that the output of the tachometer generator is linear with respect to speed, is a speed-control system in which the speed of the motor-generator combination is proportional to the dc reference (input) voltage. A revolution counter, mechanically coupled to the output shaft of the motor-generator combination, provides an integral of the dc reference voltage with respect to time.

Flow-measurement Integration: A typical flow-computing system begins with the detection of actual flow by one of several primary devices. The resulting input rate signal, which is a measure of the rate of flow, may be in the form of milliamperes dc (4 to 20, 1 to 5, 0 to 4) or dc voltages (1 to 5 or 0 to 4). The signal is applied to an electronic integrator designed to accept the signal and to provide an output count which is proportional to the integral (or the integral of the square root) of the input. The input signal may be linear or follow a square law, which is the case when an orifice meter is used. The output signal provides a count which is totalized on a conventional electromechanical counter, an electronic counter, or counter printers. The frequency or count rate is proportional to the input signal so that with a higher flow rate, the input-signal magni-

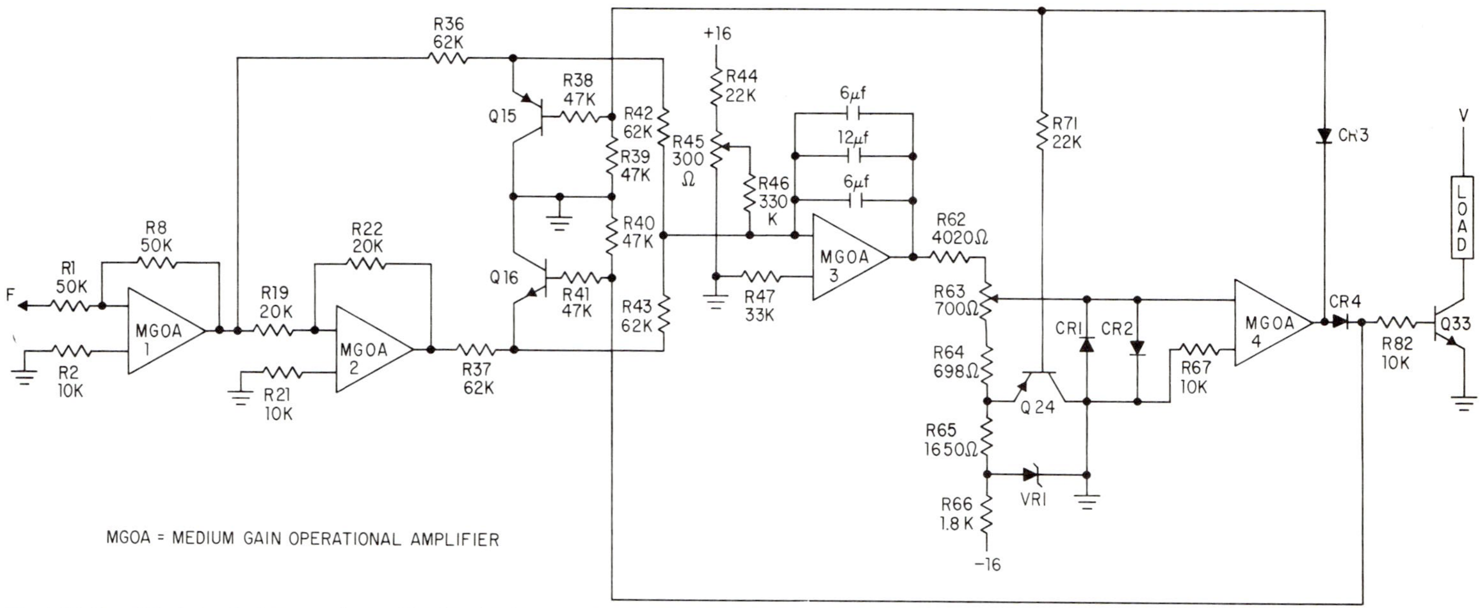

Fig. I-15. Simplified circuit diagram of a medium-level pulse-type continuous integrator.

tude is higher and consequently the count rate is higher.

Weight-measurement Integration: This can be accomplished in several ways, including lever systems; counterbalancing systems, such as

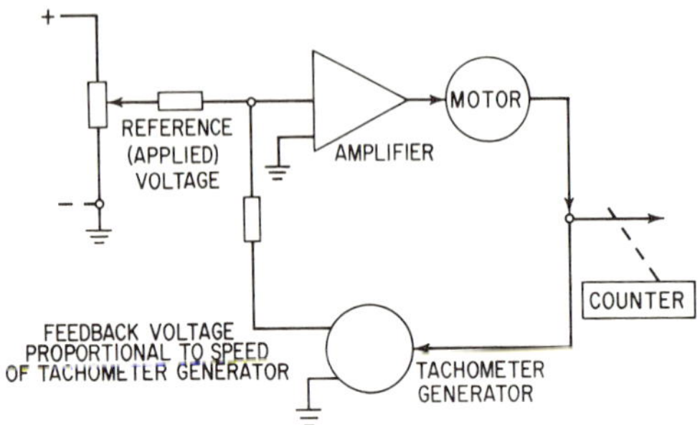

Fig. I-16. Block diagram of continuous integrator.

beams and poises; pendulum systems, or springs; hydraulic, pneumatic, or electronic systems. Loads to be measured may be batch, intermittent, or continuous. See **Batching systems, automatic-weight;** and **Weight, measurement of.** A typical continuous system may involve a continuous material-handling belt conveyor in which a section of the idler-support system is independently supported—with electrical strain gages attached to the supporting structural members. The strain gages are connected to form an electrical bridge so that, with a definite input voltage, an output voltage proportional to the load (weight) is developed. The rate (belt speed) can be determined by various means, one of which is an electrical tachometer coupled to the belt drive and developing a electrical output proportional to the speed of the belt. The two outputs, if desired, can be measured independently by two recorders and by means of transmitting slidewires in the recorders have their outputs combined as $e_w \times e_t$ and this combined output integrated by an electronic integrator. The two outputs also can be integrated without intervening recording. Signal levels can be made compatible by means of an intermediate amplifier if necessary. For example, the strain-gage output circuit may require amplification to equal the tachometer output. The amplified output may be introduced into a multiplier/divider-type operational amplifier designed on the pulse-height–pulse-width multiplication principle to produce a dc voltage output proportional to the product of two unipolar analog voltages. This output, in turn, can be integrated as previously described.

—Robert A. McCarron, *Leeds & Northrup Co.*

Interface, System The set of signals which form a boundary between parts of a system is referred to as an interface. Sometimes the set of rules controlling communications across such a boundary is termed an interface.

The storage interface consists of data bit (see **Bit**) signals into and out of storage (see **Storage, computer**), the address bits into storage, and the necessary read and write control signals.

The input-output interface consists of the communication lines between the central processing unit [see **Central processing unit (CPU)**], channel [see **Channel (CPU)**], and the peripheral devices (see **Peripheral equipment, computer**) attached to the system. The control signals in this interface permit the selection of the device and the synchronization of the data transfer between the device and the central processing unit.

An interface may be a serial interface in which the signals are propagated through each device attached to it or a parallel interface in which the signals are distributed to each device in parallel.

—W. W. Welz, *IBM Corporation*

Interference Any undesired energy that tends to adversely affect the reception of desired signals. *Common-mode interference* is a form of interference which appears between measuring-circuit terminals and ground. Sometimes this type of interference is termed longitudinal interference. *Normal-mode interference* is a form of interference which appears between measuring-circuit terminals. This type of interference also may be referred to as differential interference or transverse interference.

Interference also may refer to the systematic alternate reinforcement and attenuation of two or more coherent waves when they are superimposed. Also see **Common-mode voltage; Differential-mode voltage;** and **Noise.**

Interferometer A precision device which uses the interference of light waves as the basic of measurement. The instrument usually is composed of a combination of mirrors which are silvered alternately to transmit and reflect light. The optics of an interferometer are designed so that the variance of known wavelengths and path lengths within the instrument permits accurate measurement of distances, the resolution of spectral lines, the measurement of diameters of distant stars, and other kinds of precision work. Also see **Fabry-Perot interferometer;** and **Laser, instrumentation of.**

International System of Units (See Standards, measurement.)

International Temperature Scale A temperature scale selected by international agreement to reproduce the Kelvin Thermodynamic Scale as closely as possible within the limits of the state of the art. Over the years there have been a number of international temperature scales.

The normal thermometric scale (1887) was based upon the constant-volume hydrogen thermometer with a fundamental interval of 100 degrees between the fixed points melting ice (ice point) at 0° C and condensing water vapor (steam point) at 100° C. The scale, commonly referred to as the International Hydrogen Scale, was reproduced for international comparisons by means of four liquid-in-glass thermometers.

The International Temperature Scale (1927), developed to extend the range and accuracy of measurement, was based upon six fixed and reproducible equilibrium temperatures to which numerical values were assigned and upon the indications of interpolation instruments calibrated according to specified procedures at the fixed temperatures. The fixed points, all at 1 standard atmosphere of pressure, were:

Oxygen	—182.97°C
Ice	0.000°C
Steam	100.000°C
Sulfur	444.60°C
Silver	960.5°C
Gold	1063°C

The interpolation instruments and usage ranges were:

Platinum resistance thermometer	—190 to + 660°C
Platinum vs. platinum 10% rhodium thermocouple	+660 to + 1063°C

The scale was revised in 1948 in accord with advance in knowledge to make it more uniform and reproducible. At this time the designation degree Celsius was adopted in place of degree centigrade. The scale was named the international temperature scale of 1948.

Up to 1954, the Kelvin Thermodynamic Scale was defined by two fixed points, ice point and steam point. In 1954, the scale was redefined by means of a single fixed point; the triple point of water was chosen, and the value assigned was 273.16° C. A text revision of the International Scale was made in 1960 in which the ice point was replaced by the triple point. This is the only fixed point common to the Kelvin and International Scales. The name of the scale was changed at that time to International Practical Temperature Scale of 1948. The fixed points, all at 1 standard atmosphere of pressure except for the triple point of water, are:

Oxygen	—182.97°C
Triple point of water	+0.01°C
Steam	100°C
Sulfur	444.6°C
Silver	960.8°C
Gold	1063°C

The interpolation instruments remained the same as in 1927, but the usage ranges were changed slightly:

Platinum resistance thermometer	—182.97 to +630.5°C
Platinum vs. platinum 10% rhodium thermocouple	+630.5 to + 1063°C

The International Conference in 1960 recommended replacement of the sulfur point (444.6° C) by the zinc point (419.505° C) as being more reproducible.

In order to cover the range from absolute zero to the oxygen point, the various national laboratories developed scales such as the 1958 He^4 Temperature Scale, the 1962 He^3 Temperature Scale, and the NBS provisional scale 2-20 (1965). At the 1967 International Conference on Weights and Measures, authorization was given to revise the international practical scale to incorporate extension of the lower part of the range as well as to make other agreed-upon changes.

The 1967 International Conference changed the unit of thermodynamic temperature from degree Kelvin (symbol: ° K) to kelvin (symbol: K). The name and symbol are used to express temperature intervals rather than the terms degrees Kelvin or deg K. Temperature intervals may also be expressed in degrees Celsius.

—Robert D. Thompson, *Taylor Instrument Process Control Division, Sybron Corporation*

Interpolation, Geometric (See Drafting system, automatic.)

Interpreter An executive routine which translates a stored program coded in a pseudo computer language into actual computer instructions at execution time. A program executed under an interpreter normally takes longer to run than a machine-language version of that program. However, the one-time-only program may take less total computer time when executed under an interpreter than if it is first compiled and then executed. Some interpretive systems do a partial compilation of the user's source language such that, when the program is executed under the interpreter program, its efficiency is increased. Even with partial compilation, however, interpretive systems are less efficient at execution time than the actual computer-instruction execution approach.

If a program is to be executed many times without change, as in the case of a process-control environment, the interpreter approach can be very inefficient. This is because a compiler need only perform the translation once and the translated code can then be executed many times. (See also **Translator.**)

—B. W. Landeck, *IBM Corporation*

Interrupt An interrupt is a signal which causes the central processing unit [see **Central processing unit (CPU)**] to change state as a result of a specified condition. In terms of programming, this change of state is the temporary discontinuance of normal program execution. The interrupt may be caused by an external condition, an input or output device (see **Input/output device**), or the program (see **Program, computer**) currently being processed in the central processing unit itself.

When an interrupt is recognized, the program currently being executed is stopped and a new program takes control of the central processing unit (CPU) to provide the proper response. When the new program has been completed, the control of the CPU is returned to the interrupted program at the point at which it was discontinued. If another interrupt is recognized at this time, its program is executed before control of the CPU is returned to the program which was initially interrupted.

In a system in which there is more than one interrupt condition, a priority must be established to determine in which sequence the requested servicing programs are to be executed. This priority may be established by either hardware logic or programming. If the priority is established by programming, the interrupt condition causes the system to transfer control to an interrupt-service subroutine. This routine determines the specific cause of the interrupt and, based on the assigned priority of the interrupt condition, schedules the execution of the interrupt-service routine in the correct sequence. If a hardware priority-interrupt structure is implemented, the hardware is designed to prevent any interrupt from being recognized which is of lower priority than the program or task currently being executed. If the interrupt is of higher priority, however, it causes the transfer to control to the higher-priority interrupt-service routine.

An external interrupt condition may be a cyclic condition such that at a specific time interval sensor information is to be scanned and transmitted to the computer. Or, it may be random and the interrupt signal could be generated by a unique set of conditions in the process being controlled. An input- or output-device interrupt is used to signal that the device has completed its operation and that it is ready to receive a new job or set of commands. An interrupt generated within the CPU itself may be the result of a counter overflow or the addition of two numbers which has exceeded the capacity of the space allocated to the result.

In general, an interrupt is the means by which the system program is alerted to a special condition. The system program then generates the required response in order to satisfy the condition.

—W. W. Welz, *IBM Corporation*

Interrupt Response Time The amount of time required to recognize an interrupt once an event has occurred in the process and to transfer control of the digital computer to the service routine for that interrupt. The interrupt response time of a digital computer is one of the basic parameters evaluated when determining which computer is most desirable for a particular control application.

There are two quantities which must be evaluated before the response time of a digital computer can be determined. The first is the maximum mask time (see **Mask**) that the programming system contains. The maximum mask time determines the worst-case delay in the recognition of an event occurring in the process. The second quantity to be determined is the overhead time required to transfer control from the interrupted program to the interrupt-service program once the interrupt has been recognized. The overhead involves such things as saving the registers of the interrupted program, initializing the registers for the interrupt-service program, and branching to the interrupt-service program. Once the maximum interrupt response time has been determined for a given digital-computer system, the process-control engineer can determine whether that digital-computer system can respond to the process in the time allowed.

—B. W. Landeck, *IBM Corporation*

Intrinsic Safety To be intrinsically safe, equipment and associated wiring must be incapable of releasing sufficient electrical energy, under normal or abnormal conditions, to cause ignition of a hazardous atmosphere. Abnormal conditions include failure of electrical components, application of overvoltage, damage to wiring and insulation, or other damage that may occur during routine maintenance and adjustment. Where power levels are high, as in the case of lighting and power lines, the concept is not applicable. Where power levels are low and with proper circuit design, the concept is applicable, as pioneered in England and Canada, for indicating, recording, and controlling instruments—and for alarm circuits. Where practical, intrinsically safe equipment is much less costly than surrounding apparatus with explosion-proof housings of sufficient mass to contain an explosion or to isolate potential ignition sources from explosive atmospheres by immersion in oil or by purging with a nonexplosive medium.

REFERENCES

McKinney, A. H.: Electrical Safety Abstracts and References, *ISA Publ. ESAR,* 3d ed., 1967.

Electrical Safety Practices, *ISA Monograph* no. 110, 1965.

Electrical Safety Practices, *ISA Monograph* no. 111, 1967.

Electrical Safety Practices, *ISA Monograph* no. 112, 1969.

Inverter Circuit The inverter is a circuit which provides a logical NOT of the input signal. That is, if the input signal is a binary 1, the output is a binary 0 (see **Binary state**). If the input signal is in the 0 state, the output is in the 1 state. With reference to Fig. I-17, if A is positive, the output F is at 0 V since the transistor is biased into conduction; if A is at 0 V, the output is at +V because the transistor is cut off. The Boolean-algebra expression (see **Boolean algebra**) can be represented as $F = A'$, where the prime denotes the NOT function.

—W. W. Welz, *IBM Corporation*

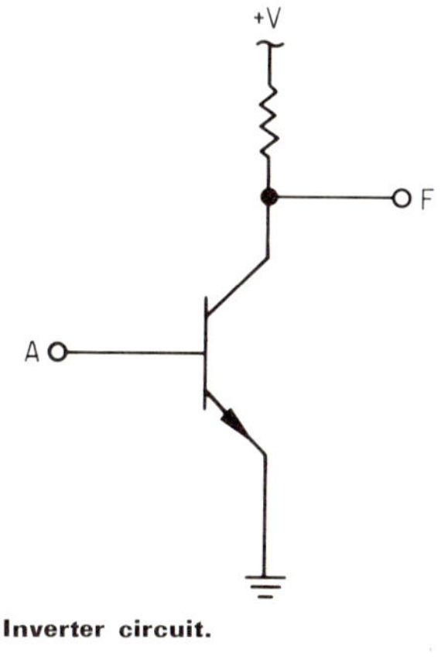

Fig. I-17. Inverter circuit.

Ionization Chamber The ion-chamber detector is commonly used for industrial-radiation measurement with beta-ray and gamma-ray gages. The device consists of an enclosure containing two oppositely charged electrodes in a gas. When the chamber is exposed to nuclear radiation, the gas is ionized and each ion is drawn to the electrode of opposite polarity. The resulting current through the chamber is proportional to the intensity of the radiation. An ion chamber may be used as a counting device registering each individual ionizing event or as an integrating device measuring the total current output due to the charge collected on its electrodes from the internal ionization. Counting chambers are used where the number of ionizing events is small. For large amounts of radiation, the integrating type is used. Ion chambers may be constructed with parallel plates or coaxial cylinders and have been made from many materials—metals, plastics, and waxes.

Ionization-type Vacuum Gage (See **Vacuum gaging.**)

Iron-vane Electrical Meter (See **Electrical quantities, measurement of.**)

Irradiance Meter (See **Oceanographic instrumentation.**)

Isotopes, Radioactive (See **Radioisotope-tracer instrumentation.**)

Iterative Array An array of a large number of interconnected identical processing modules, used with appropriate driver and control circuits in a computer to permit a large number of simultaneous parallel operations. The technique permits the use of slower, less expensive circuits for individual operations while still providing the required fast total execution time for a problem.

Iterative Process A mathematical process for calculating a desired result by means of a repeating cycle of operations that yields a result closer and closer to the desired result.

J

Jet-engine Controls (See **Aircraft-engine controls.**)

Jet-pipe Valve (See **Hydraulic relay.**)

Job A set of related sequential tasks that are to be performed on a digital computer. The job may consist of one or more tasks (see **Task**). An example of a job is a FORTRAN compilation followed by the execution of the program compiled. This is a two-task job.

—B. W. Landeck, *IBM Corporation*

Joule, Standard (See **Standards, measurement.**)

K

Katharometer (See Thermal-conductivity gas analyzer.)

Kelvin, Standard (See Standards, measurement.)

Kelvin Bridge (See Bridge circuits, electric.)

Kelvin Temperature Scale A temperature scale, named in honor of Lord Kelvin (Sir William Thomson, 1824–1907), that involves the interrelation of temperature, volume, and pressure of a perfect gas and is therefore independent of the material used. In practice it is most commonly reproduced by the constant-volume gas thermometer.

The scale is defined by means of a single fixed point, the triple point of water, assigned value 273.16° K. Temperatures are denoted by the term kelvin and the symbol K.

—Robert D. Thompson, *Taylor Instrument Process Control Division, Sybron Corporation*

Kymograph (See Cardiovascular instrumentation.)

L

Ladder Network (See **Decoding network, Digital-to-analog converter.**)

Lag, Control-system Any retardation of an output with respect to the input. The lags in a process determine the difficulty of control. A signal that travels through a closed-loop path can change in only two ways—either in magnitude or in time. The effect of any portion of a feedback loop on the relationship between input and output is known as the transfer function, which expresses the signal size and signal time effects in mathematical form. See **Transfer function, theoretical.**

Process response to a step change in input is a combination of a few basic responses, including *first-order lag* (single-capacity process), *second-order lag* (two-capacity process), dead time, and integration. A second-order lag is one which results from changes in energy storage at two separate points in the system. A second-order lag can be underdamped, overdamped, or critically damped. For processes with an underdamped second-order lag response—after an increasing step change in input—the output will start to change gradually and then rapidly approach and pass the final steady-state value. The output cycles into this final value with successively smaller cycles.

For a process with an overdamped second-order lag response—after an increasing step change in input—the output does not change immediately. After a short period of time, the output will exhibit an S-shaped response which levels off at the final steady-state value without overshoot. Most processes exhibit this response. Examples include (1) two resistance-capacitance networks series connected, (2) a thermocouple in a well, and (3) two pressure tanks separated by a resistance loop.

For a process with a critically damped second-order lag response—after an increasing step change in input—the output responds similarly to an overdamped second-order process, but the rate of increase to the final steady-state value is the fastest obtainable without overshoot. See **Response, steady-state;** and **Response, transient.**

—Richard H. Kennedy, *Bristol Division, American Chain & Cable Company, Inc.*

Landing System (See **Flight-control systems.**)

Language, Computer A language for transmitting information between people and digital computers. A programmer describes a problem and how the problem is to be solved in the computer language so that the digital computer can provide the results required. A computer language consists of a well-defined set of characters and words and a series of rules (termed syntax) for combining them into computer instructions or statements. Computer languages vary greatly in their flexibility and ease of use. In general, the more flexibility and freedom of expression allowed in a computer language, the more complex it is for a programmer to use. Computer languages vary from those which provide complete flexibility in use (actual machine language) to those which provide little flexibility but great ease of use (problem-oriented languages). One way of describing the hierarchy of computer languages is that on the first level are machine-oriented languages, on the second level are procedure-oriented languages, and on the third level are problem-oriented languages.

Machine-oriented Languages: Machine-oriented languages are the most flexible of all computer languages. Anything that the computer is capable of doing can be achieved through programming in a machine-oriented language. There are two problems that are associated with the great flexibility of machine-oriented languages. First, the flexibility requires that a programmer have extensive training in the specific machine-oriented language before he is able to make efficient use of it. Second, each type of digital computer has its own machine-oriented language; the language for one type of digital computer is generally incompatible with the language on another computer. Therefore, once a program has been written for a digital computer in a machine-oriented language, it must be recoded before that same program can run on a different type of computer.

Machine-oriented languages can be broken down into two subclasses. The first subclass is the actual machine language of the computer. Each type of digital computer has only one machine language and does not require any prior translation before the computer can execute the steps defined in the program. The second subclass of machine-oriented languages is symbolic languages. Symbolic-language instructions generally have a one-to-one correspondence to the machine-language instructions of a digital computer, but instead of using the numbers that the digital computer uses for operation codes, mnemonics (see **Mnemonic**) are used to represent operation codes. Symbolic languages are easier to use than the machine language, but they require translation into machine language before they can be executed (see **Assembler; Macroassembler**).

Procedure-oriented Languages: Procedure-oriented languages are designed to allow the programmer to specify the process by which a problem is to be solved independent of the machine upon which the problem is to be executed. Examples of procedure-oriented languages in common use are ALGOL, COBOL, FORTRAN, and PL/I. FORTRAN and ALGOL are procedure-oriented languages aimed at mathematical problems. They allow mathematical procedures to be coded via algebraic expressions. COBOL is a business-oriented language for problems requiring the manipulation of large amounts of data. PL/I is a general procedure-oriented language for both scientific and business applications (see **ALGOL; COBOL; FORTRAN**).

Problem-oriented Languages: Problem-oriented languages allow the user to specify the problem to be solved but not the procedure to be followed in solving that problem. Problem-oriented languages are the easiest to use but the least flexible. An example of a problem-oriented language is PROSPRO/1800 (process supervisory program for the IBM 1800 data-acquisition and control system). PROSPRO/1800 is a problem-oriented language for the control of continuous processes. The user specifies through the PROSPRO/1800 language the general organization of his process and the various control actions to be taken. The PROSPRO compiler then converts the user's specifications into actual machine-language programs to be run on the IBM 1800 data-acquisition and control system for the control of his process.

—B. W. Landeck, *IBM Corporation*

Laser, Instrumentation of Three laser properties are of primary interest for most applications: (1) monochromaticity (spectral purity), (2) collimation (ability of radiation to be within a small cone angle), and (3) spatial coherence (the phase relationship of the wave). Three principal types of lasers, categorized by the nature of the active lasing material, are (1) optically pumped solid, (2) gaseous, and (3) semiconductor. Each type has its particular characteristics of emission wavelength, conversion efficiency, operating-temperature range, method of excitation, time sequencing of output, modulation, and output power level.

Optically Pumped Solid Lasers: A fluorescent solid normally is used as the active laser

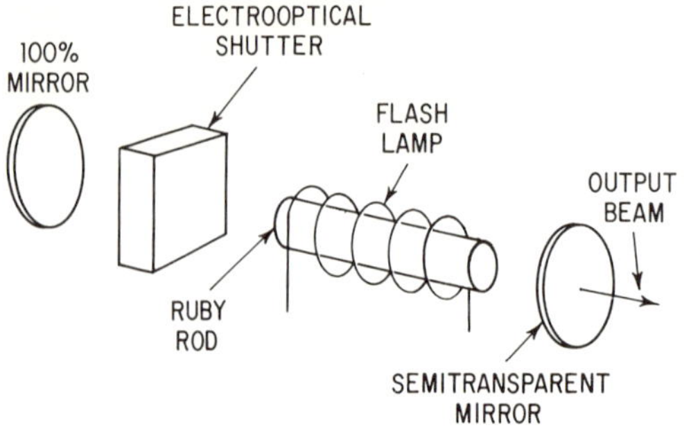

Fig. L-1. Giant pulsed ruby laser.

medium (see Fig. L-1). A synthetic ruby crystal which emits a visible red beam of 0.69-micron light was used in the first laser. Pulsed ruby lasers can emit hundreds of joules of energy in a pulse of a millisecond duration. When the pulse is shortened to a few nanoseconds, by a technique termed "Q switching," thousands of megawatts of power can be obtained. Such radiation focused into a spot a few microns in diameter will produce a power density billions of times greater than that of the sun's surface. Other optically pumped laser materials are glass or YAG (yttrium-aluminum-garnet) doped with neodymium. Both types emit in the near infrared (1.06 microns). With YAG, continuous (CW) power of 200 W or pulsed operation at repetition rates of hundreds of pulses per second can be achieved. Solid laser materials usually permit beam angles of a few milliradians and efficiencies of less than 5.0 percent.

Gas Lasers: The noble gas mixture of helium-neon that operates by excitation through electron collision is the most significant gas laser (see Fig. L-2). Continuous (CW) output of a few milliwatts can be achieved in a low-cost, small, rugged unit. Ionized noble gases (argon, krypton, and xenon) produce emission of several laser wavelengths simultaneously in the blue, green, and red regions. Power output up to 10 W with beam angles of less than a milliradian can be achieved. Molecular gas lasers, with carbon dioxide as the lasing material, can emit thousands of watts of continuous 10.6 micron radiation with efficiencies up to 15 percent.

Semiconductor Lasers: Excitation is provided directly by electron flow across a semiconductor junction. Gallium arsenide is the most popular type. Several watts of 0.84- to 0.9-micron radiation can be emitted continuously with efficiencies up to 35 percent. The beam angle usually is several hundred milliradians (see Fig. L-3).

Laser Applications: Each laser type offers characteristics which must be evaluated for a

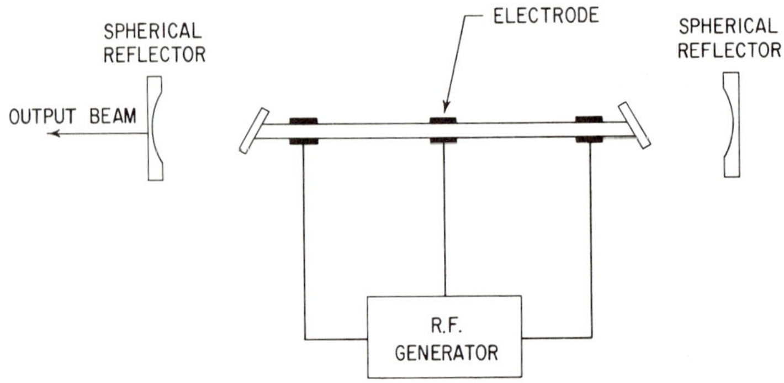

Fig. L-2. He-Ne laser with external spherical reflectors (curvature of mirrors is exaggerated).

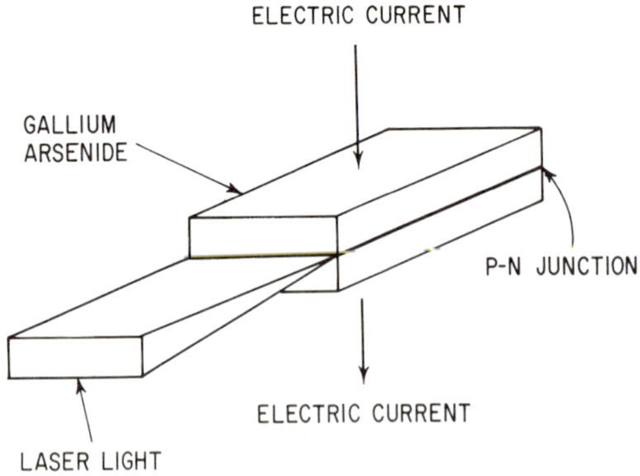

Fig. L-3. Gallium arsenide laser.

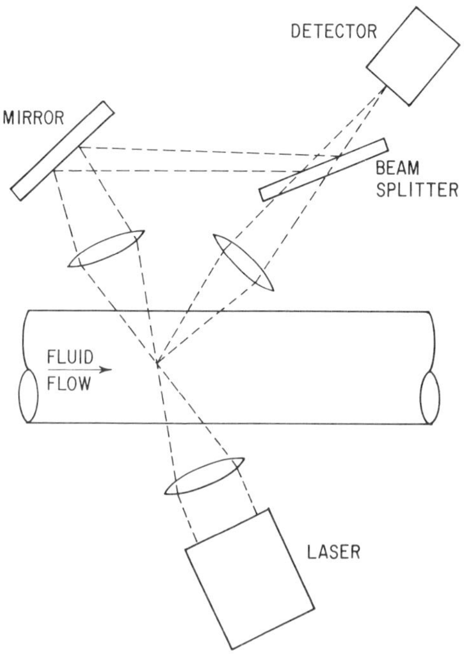

Fig. L-4. Laser doppler flowmeter.

particular application. They include (1) power density, (2) monochromaticity, (3) spectral brightness, (4) beam angle, and (5) pulsed or continuous operation—or precision timing of the laser pulse.

A laser doppler flowmeter is shown in Fig. L-4. Fluid flow is determined by measuring the doppler shift in laser radiation scattered from particles in the moving fluid stream. No sensor or transducer is necessary in the moving fluid stream. The laser radiation focal point can be moved across the flow tube to measure velocity profiles. Fluid flows from 0.01 to 5,000 in./s have been measured. Contaminants, such as smoke, may have to be added to gases to provide scattering centers for the laser beam.

Material velocity, such as moving metal, plastic, or paper sheets, can be determined from laser energy reradiated from the moving material. High accuracy and noncontact operation provide advantages over mechanical counters. The technique also can be used to measure the stretch in moving materials, such as plastic and paper.

Laser Interferometer: For linear measurement of lengths up to 200 in., a laser interferometer can provide accuracies of better than 20 millionths of an inch. The detection system determines the precise difference in optical path length between a known length and the one to be measured. This technique is used to measure the height of ocean waves or contours of machine parts and earth masses (see Fig. L-5).

Surveyor's Transit: A laser, attached to a surveyor's transit, will create a visible line of sight which is especially useful across inaccessible areas. The system can be used at night to measure long distances. Laser seismographs permit determination of movements along earth faults of less than one-millionth of an inch. The technique is insensitive to temperature and humidity and has excellent frequency response.

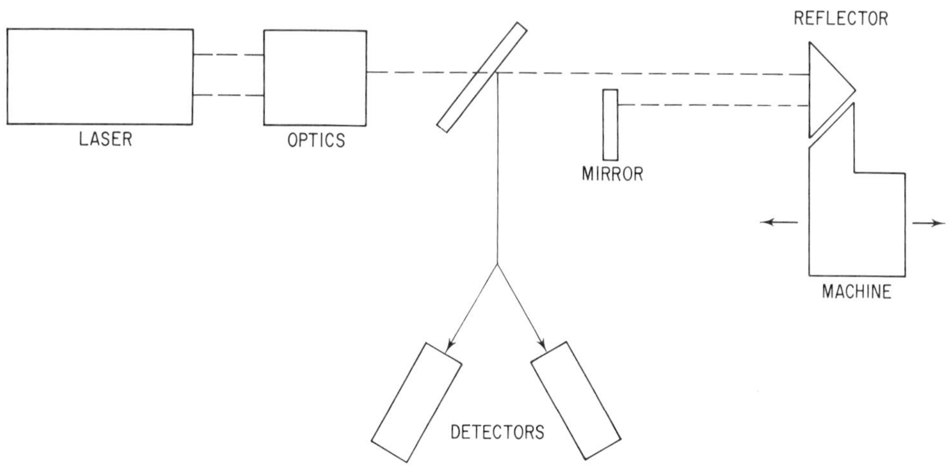

Fig. L-5. Laser interferometer.

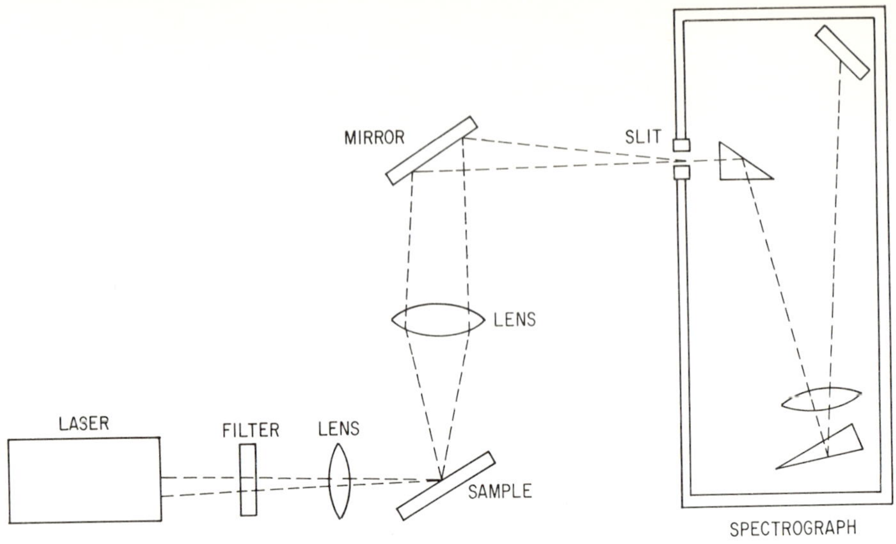

Fig. L-6. Laser-probe emission spectrograph.

Particle size and, to some extent, the composition of the pollutant in air-pollution studies can be determined from light reflected from smoke or haze.

Schlieren Photography: Variations in refractive index or density in a transparent medium, as shock waves around a projectile, can be illuminated to determine the changes in the interference pattern. This phenomenon is termed "schlieren." When the millions of watts of illumination as well as the precision timing capability of the laser are used for schlieren photography, much higher resolution is obtained. Improved illumination in undersea exploration is possible with green lasers.

Laser-probe Emission Spectrograph: Focused laser beams vaporize minute samples of material for analysis in mass spectrometers, spectrographs, and gas chromatographs. The laser provides temperatures higher than those generated by conventional carbon-arc sources and reduces contamination. Only a small amount of the material is consumed. Selective layers of thin films can be removed for analysis without changing or disturbing the other layers (see Fig. L-6).

Holography: Conventional photographs record only brightness variations in a two-dimensional record of the intensity of light reflected from a three-dimensional object. Regardless of distance from the camera, all objects that reflect the same amount of light appear on film in equal shades of gray. Holography, or wavefront reconstruction, records the light diffracted from an object as wave-interference patterns. From this record of wave patterns of an object, it is possible to recreate a true three-dimensional image of the object. The laser light illuminates both the object and a mirror so that the film received reflects light from both the object and the mirror. There is no lens. Holograms made with gas lasers require lengthy exposure times, but ruby lasers make holograms of transient subjects, such as moving objects and flowing fluids. After the film is developed, a He-Ne or argon laser is used to reconstruct the true three-dimensional image in space.

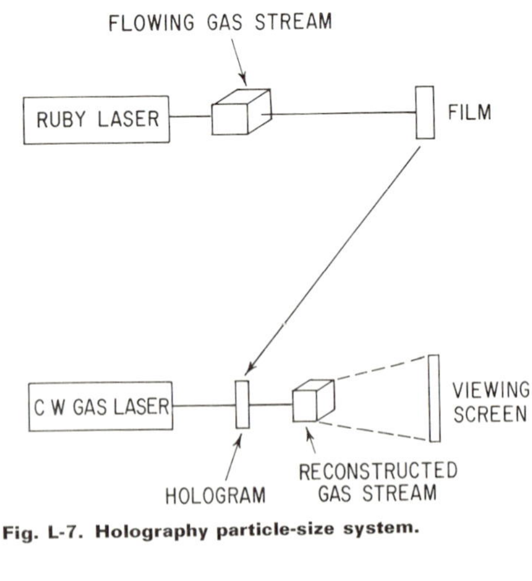

Fig. L-7. Holography particle-size system.

Applications of holography include making histograms of aerosol particle size, evaluation of dynamic aerosols, and fluid-mechanics studies. One system measures particle sizes as small as 4 microns with a resolution of 0.3 micron. Advantages are that the moving fluid is not disturbed, no samples are taken, and the information can be obtained in almost real time through use of well-designed electronics (see Fig. L-7).

The term *laser* stands for light amplification by stimulated emission of radiation. Laser and maser often are used interchangeably because their ranges of output wavelengths overlap. Generally, laser is used for wavelengths below 15 microns, in the visible and near-infrared light ranges. Maser is used for far-infrared operation, approaching the limit of the millimeter-wave spectrum.

REFERENCES

Lengyel, B. A.: "Introduction to Laser Physics," John Wiley & Sons, Inc., New York, 1966.

Marshall, S. L.: "Laser Technology and Applications," McGraw-Hill Book Company, New York, 1968.

Smith, W. V., and P. P. Sorokin: "The Laser," McGraw-Hill Book Company, New York, 1966..

—A. M. Johnson, *Apollo Lasers, Inc.*

Laser Cane (See **Vision, sensory aids for.**)

Latch (See **Flip-flop.**)

Lead Screw A rotating screw that is principally used for conversion of rotary motion to linear motion (but not the reverse) and finds application in the positional control of machine slide members. Characteristically, a lead screw is placed between two pillow bearings with end or thrust pressure applied against the free end of the screw. A lead nut operates on the screw threads as shown in Fig. L-8. The nut generally is suspended in a cage that is attached to the machine element undergoing resultant linear motion.

Machine lead screws may have one of the following configurations:

1. Ball-nut screw—for minimum friction
2. Ground screw, bronze nut—medium friction
3. Smooth cylinder, disk or ball friction nut —low friction

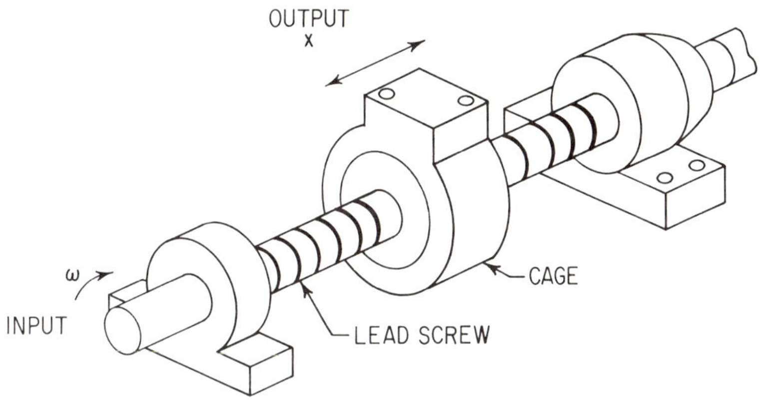

Fig. L-3. A lead screw–nut combination used to convert rotary motion to linear motion.

The specification for a lead screw–nut combination must describe the lead or advance of the nut per revolution of the screw in inches per revolution, the cumulative error per unit length in thousandths of an inch per foot, and also must include the backlash or dead band expressed in thousandths of an inch per radian or thousandths of an inch per milliradian. Also see **Position-measurement and control systems;** and **Rack and pinion.**

—James R. Walker, *De Vlieg Machine Co.*

Leadwire and Connections (See **Resistance thermometry; Thermocouple;** and the **Classified Index.**)

Leak, Calibrated (See **Vacuum gaging.**)

Leak Detector An instrument for sensing the escape of a gas under pressure from a system that is closed and that should be pressure-tight (hermetically sealed)—hence a device for checking on the integrity of new equipment, such as refrigeration and air-conditioning systems as they come off the production line; or as a means for checking, for reasons of safety and efficiency, equipment that has been installed for some time. Leak detectors can be designed to be sensitive to specific gases or chemical radicals.

Learning Mechanism (See Adaptive control; and Robot.)

Ledoux Bell A device used for measuring differential pressure that uses a parabolically shaped bell in the measurement chamber to extract the square root of the pressure differential, thus providing uniform scale readings. Largely of historical interest.

Length, Standard of (See Standards, measurement.)

Level Gage, Radiation-type A nuclear level gage can measure and control the level or interface position of liquids and slurries in process vessels, tanks, and pipes—and of solids in silos, hoppers, and bins. The gage does not require physical contact with the material—hence there are no problems of fouling, gumming, or clogging of components. Gages are available with recording only, recording with simple alarms, high- and low-level alarms, and full proportional control. Installations range from indicating high- and low-level alarm on vessels as large as 50 ft in diameter in which gamma switches are used—to continuous measurements of heights as great as 20 ft with a repeatability of ±1 percent.

As shown in Fig. L-9, a gamma-emitting radioactive source is mounted in a lead-shielded holder on one side or chord of the vessel and a suitable detector is mounted on the opposite side. Absorption of gamma radiation by any material is a function of the mass of material lying between the radiation source and the detector. As the material level changes, varying amounts of radiation are detected, thus causing a change in detector output. A minute electric current is generated which is amplified for readout, alarm, or control. For vessels greater than 15 ft in diameter, the radioactive source may be mounted in a well that projects into the vessel.

The amplifiers used are ac stabilized type for high accuracy. Where less accuracy is required, dc amplifiers can be used. Zero suppression is electrical for normal stability, or a compensating cell can be used for long-term stability. The gamma switch uses a Geiger-Mueller counter tube and provides high or low "on-off" alarms and signals. The range is 3 in. vertically, ½ in. when mounted horizontally. Multiple installations on the same vessel can provide alarm or pump control. Continuous level measurement uses either a single-point radioactive source or a patented strip source with a stack of Ohmart cells as detectors.

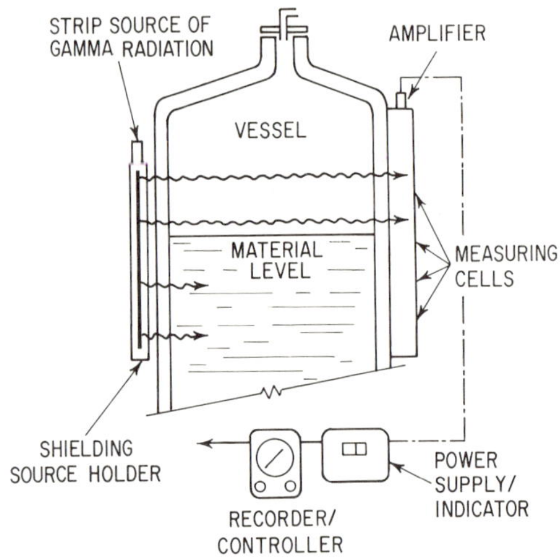

Fig. L-9. Level-gaging system of the radiation type for use in measuring the level of liquid or solid materials. *(Ohmart.)*

Greatest accuracy can be achieved with a motor-driven level gage. The radiation source and detector cell move up and down in unison as controlled by a motor-driven drum. Potentiometers provide output voltages; one indicates full range in feet; a second potentiometer indicates inches within range. A digital readout also may be provided.

—Gordon W. Hickenlooper, *The Ohmart Corp.*

Level Instrumentation, Bulk Solids The chemical, food, basic metals, and mining industries utilize numerous bulk solids in the forms of flakes, powders, granules, and lumps. The storage and processing of bulk solids normally involves the use of vessels, such as bins, silos, and tanks. The mechanized systems for loading and discharging the vessels require monitoring of the bulk-solids level. The principal objective of the monitoring system is to prevent overfill and total runout and to provide data for inventory auditing. An automated bulk-solids-handling system requires that signals be provided for each critical level change. Generally, these signals are electrical and actuate alarms, audible or visual, start or stop motors, or actuate solenoid valves.

Several electromechanical, electronic, and nucleonic level controls are available. Although the highest-cost device may approach universality in application, often a low- to medium-cost device will satisfy the immediate requirements. Selection of the most suitable level control will be governed by such factors as (1) minimum/maximum temperature, (2) minimum/maximum air or gas pressure, (3) vibration, (4) moisture content of the bulk solid, (5) compatibility of control components with the bulk material, and (6) properties of the bulk material, including density, flowability, particle shape, and acoustic and dielectric characteristics.

Pressure-sensitive Sensors: The presence or absence of bulk solids can be determined by a pressure-sensitive instrument, such as by the flexing of a diaphragm or the deflection of a pendant-like component. In both cases, a linear motion is derived for purposes of actuating or deactuating an electrical control-circuit switch. The sensitivity of the device will be determined by (1) the size of the diaphragm, (2) spring tension or compression, (3) counterweights, (4) length of pendant stem, (5) size of the pendant. Normal changes in the bulk-solid properties will not alter performance. These devices, along with other types of bulk-solids level sensors, are shown very schematically in Fig. L-10.

These devices normally can be used over the temperature range of −30 to +800° F for medium-low to very high bulk densities, for powders to medium-large lump sizes, and at atmospheric pressure. They are low-cost devices and can be mounted internally or externally with suitable exposure of the sensing element to the bulk solids. Service is infrequent and does not require special skills.

Rotating Paddle: The operation of this device centers around a low-torque slow-speed motor. The motor either turns the paddle in the absence of the bulk material or turns itself to actuate a control-circuit switch when paddle rotation is stopped by the presence of bulk material. The sensitivity of the device is determined by the configuration of the paddle, the characteristics of the electrical switch, and the switch-actuation technique. Normal fluctuations in bulk-solid properties do not alter the performance.

In normal use, the device may be used at temperatures from −30 to 800° F for all bulk densities and all common particle sizes and for pressure differentials up to 200 psi. They are low-cost devices and may be mounted internally or externally with suitable exposure of the paddle to the bulk materials. Service is infrequent and does not require special skills.

Weighing Methods: Mechanical, pneumatic, hydraulic, and electronic force-sensing systems are utilized to indicate vessel gross-weight changes. The vessel may be suspended from or supported by the force-sensitive components, often involving some form of lever system. Relays or switches are included to provide electrical signals at specific weights (hence levels). Of course, the true mass and not the actual bulk-solids level will determine when the signals will be given. Normal use requires only that the vessel weight can be suitably isolated from its surroundings. The system is relatively high in cost and requires scheduled maintenance by skilled technicians.

Capacitance-bridge Sensor: This device comprises an electrode (probe), bridge circuitry, and control-circuit relay. Changes in the amount of bulk solids in the immediate area of the electrode will alter the capacitance. The units are adjusted to respond to specific capacitance fluctuations and are capable of indicating all level changes on a vertical probe. Signal repeatability may be influenced by changes in the density and moisture content of the bulk material and any retention of solids on the electrode. The electrode may be mounted internally or externally with suitable exposure to the bulk solids. In normal usage, the device may be applied over a temperature range of −200 to +1000° F to all bulk densities and all common particle sizes, and for pressure differentials up to 1,000 psig. These are medium-cost devices and usually require scheduled maintenance by skilled technicians.

Ultrasonic Sensor: This device consists of a beam or wave transmitter, detector, and control-circuit relay. Two different schemes are used. In one scheme, the detector responds to the presence or absence of sound-wave shielding as provided by bulk solids when either on the transmitter or on the detector exposed surfaces. Normal fluctuations in bulk-solid properties or normal environmental changes do not affect the signal repeatability. This device can be used over a temperature range from −250 to +400° F and for pressures up to 1,000 psig. The device is in the medium-cost category.

A second ultrasonic scheme involves a combined transmitter and detector mounted to *look*

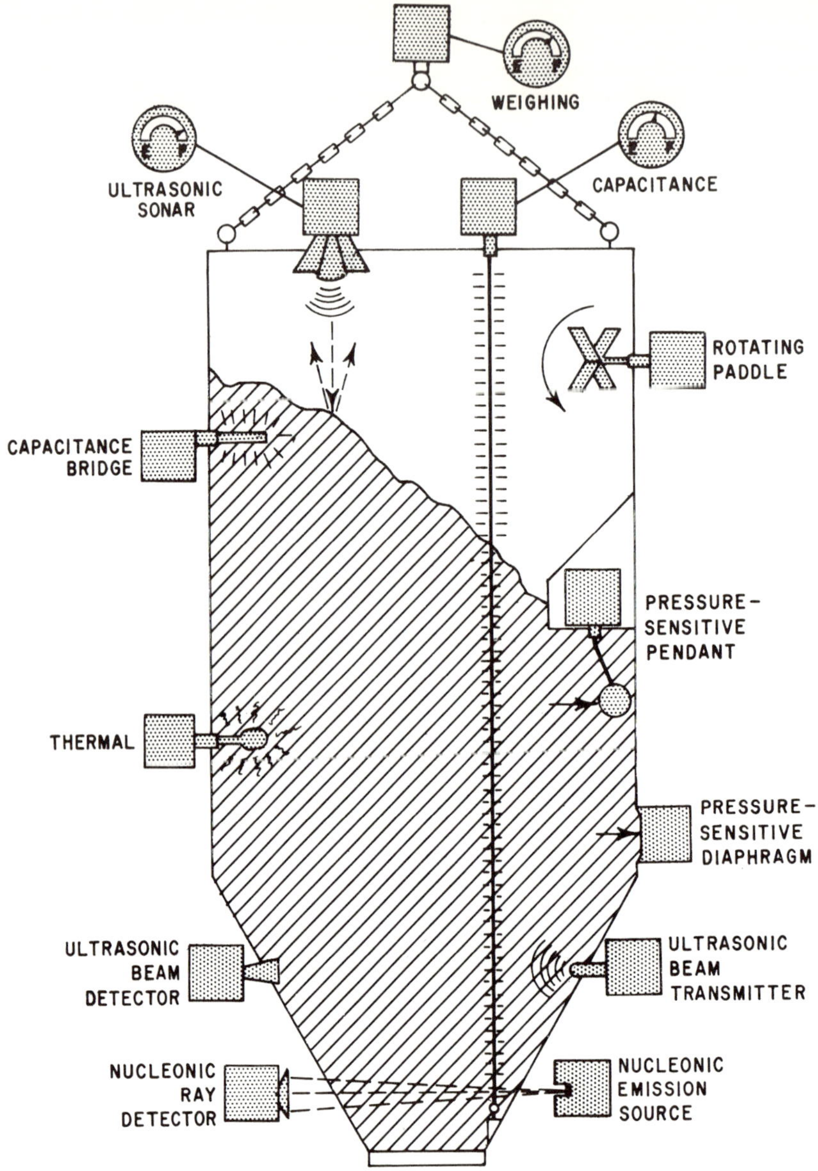

Fig. L-10. Various instrumental schemes for measuring the level of bulk-solid materials.

down over the bulk-solids level. A pulsed or intermittent sound beam is emitted. The elapsed time for the sound wave to rebound (echo) from the surface of the bulk solids provides a means of indicating and recording the vertical distance to the solids below. Fluctuations in temperature, airborne dust, and constituents in the air above the solids affect the speed of the sound wave and hence the accuracy of the signal. Normally, this system can be used in the temperature range from −40 to +250° F for a depth range of solids up to 100 ft and for pressures up to 200 psig. The system is in the medium- to high-cost category. The electronic components of the ultrasonic devices usually require scheduled maintenance by skilled technicians.

Thermal Devices: These sensors employ dc circuits with low-heat sources at the bulk-solids point of contact. The bulk material must provide a greater heat-transfer coefficient as compared with the air above. When the presence of the bulk material causes a lowering of the temperature of the contact element, the resistivity varies, thus unbalancing a bridge circuit and energizing a control-circuit relay. These are medium-cost devices primarily used where environmental conditions are at controlled constants.

Nucleonic Sensors: These bulk-solids level controls utilize a gamma or x-ray emission source, detector, and control-circuit relay. See also **Level gage, radiation-type.**

REFERENCE

Considine, D. M., and S. D. Ross (eds.): "Handbook of Applied Instrumentation," pp. 5-79 to 5-84, McGraw-Hill Book Company, New York, 1964.

—Harry A. Downer, *The Bin-Dicator Company*

Level Measurement, Liquid (See **Liquid level, measurement of.**)

Lie Detector (See **Polygraph.**)

Light Meter (See **Photometers.**)

Limb (Artificial) Servosystems (See **Medical instrumentation.**)

Limit Alarm (See **Alarms, control-system.**)

Limit Control (See **Velocity-limiting control.**)

Limit Switch An enclosed electromechanical device (switch) which alters electric circuits when actuated by an external force—as by a machine member (lever, cam, dog, and so on) or other object when a selected position of travel or "limit" is reached. The circuits that are switched may be for (1) safety stop, (2) position indicating, or (3) an interlocking part of a total sequence of automatic or semiautomatic operations. The limit switch is an important element of many automated systems as exemplified by machine tools, conveyors, and materials-handling equipment. Unlike photoelectric and proximity switches, the limit switch as described here requires actual physical contact for actuation. Limit switches are expected to operate reliably for tens of millions of actuations in industrial environments. The switches are available in numerous current and voltage ratings—ranging from the switching needs of solid-state devices to motor-controlling relays and large solenoids. Voltage requirements may be as high as 600 V.

Line-following System (See **Drafting system, automatic.**)

Line-scan Recorder (See **Oscillograph.**)

Linear Acceleration (See **Accelerator.**)

Linear Control System A linear system or element is one whose time response to several simultaneous inputs is the sum of their independent time responses (Ref. 1). A linear system can be represented by linear algebraic equations, difference equations, or differential equations. The assumption of linearity usually is valid only over a narrow range of operation, as nothing in nature behaves in a strictly linear fashion.

The most important feature of a linear system is the applicability of the *principle of superposition;* namely, the effect of simultaneous input functions is equivalent to the sum of their individual effects. Fig. L-11 illustrates a system with more than one input—where the output $V(t)$ depends upon all the inputs $X(t)$, $Y(t)$, and $Z(t)$. If the system is linear, the total output can be taken as the sum of the three output components. The system, therefore, can be represented as shown in Fig. L-12, where the effect of each input on the

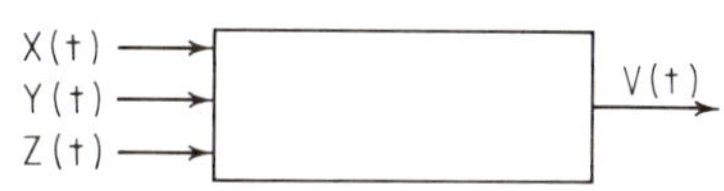

Fig. L-11. Block representation of a linear system.

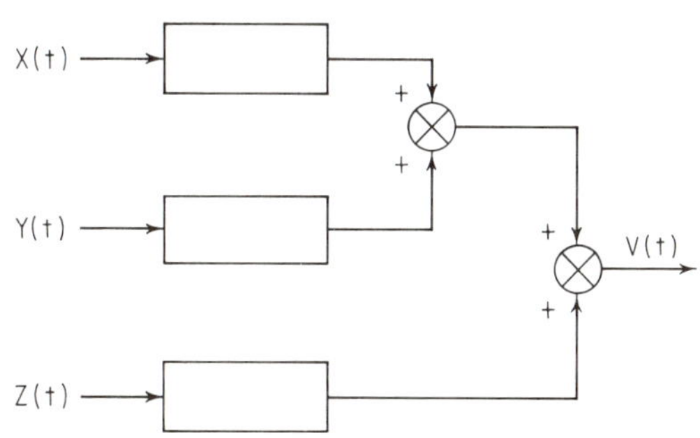

Fig. L-12. Linear system illustrating the superposition principle, which applies to linear systems only.

output may be analyzed separately and then these effects may be added to obtain the total output.

Examples of linear systems are resistors, capacitors, and inductors, since these devices have the same values of resistance, capacitance, and inductance regardless of applied current or voltage as long as the current and voltage ratings are not exceeded. Examples of nonlinear elements are transistors and diodes.

Automatic controllers are linear systems with the output proportional to some dynamic function of the input(s). This includes controllers with proportional, reset, derivative, or any time function of a linear variable. As in any element or system, controllers are linear over a specific range which is the maximum and minimum design outputs, usually 3 to 15 psig for a pneumatic controller and 4 to 20 mA dc for an electronic controller.

Fig. L-13 illustrates a linear control system where the flow transmitter measures the differential pressure developed across the orifice and provides an output signal proportional to the square of the flow. Since output is not linear with input, a square-root function generator is introduced to accomplish this. Otherwise, the control loop will not perform consistently at different rates of flow. If the proportional band of the controller were adjusted for optimum control at 50 percent flow, the system would be unstable at 100 percent flow and sluggish near zero flow—without the square-root function generator. The flow-control valve of Fig. L-13 is provided with a positioner to eliminate hysteresis, a nonlinear phenomenon; and to characterize the valve so that

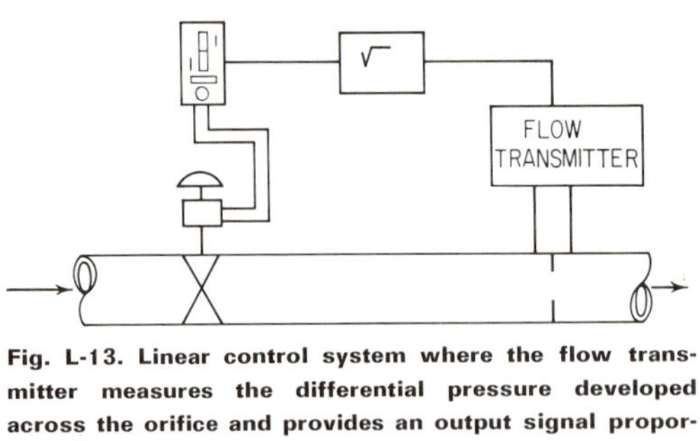

Fig. L-13. Linear control system where the flow transmitter measures the differential pressure developed across the orifice and provides an output signal proportional to the square of the flow.

valve stroke is proportional to the valve opening, a second nonlinear function.

Because most processes are nonlinear, it is necessary to consider control operation to be over a narrow range so that linear analysis may be done by the several methods available. These include methods such as the Bode diagram, the Nyquist diagram, the Nichols chart, the root-locus technique, and the analytic-design method suggested by James, Nichols, and Phillips (Ref. 1) and developed by Newton, Gould, and Kaiser (Ref. 2). There are no general methods for the analysis and synthesis of nonlinear control systems. Each method of analysis developed for nonlinear systems is specific for the type and order of the differential equations, the number of nonlinearities, and other factors (Refs. 3 and 4).

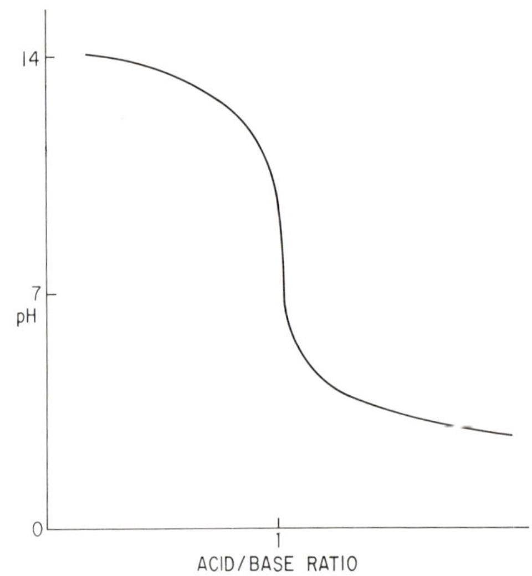

Fig. L-14. pH neutralization is an example of a nonlinear process.

An example of a nonlinear process is pH neutralization, where the relationship between pH and the acid/base ratio is extremely nonlinear. The curve of Fig. L-14 is exponential with the slope, changing as much as 1,000 : 1 at a set point of 7 pH, the steepest region of the curve.

REFERENCES

1. James, H. M., N. B. Nichols, and R. S. Phillips: "Theory of Servomechanisms," MIT Radiation Laboratory Series, vol. 25, McGraw-Hill Book Company, New York, 1947.
2. Newton, G. C., Jr., L. A. Gould, and J. F. Kaiser: "Analytic Design of Linear Feedback Controls," John Wiley & Sons, Inc., New York, 1957.
3. Rajko, Tomovic: "Introduction of Nonlinear Automatic Control Systems," John Wiley & Sons, Inc., New York, 1966.

4. Gibson, John E.: "Nonlinear Automatic Control," McGraw-Hill Book Company, New York, 1963.

—Richard H. Kennedy, *Bristol Division, American Chain & Cable Company, Inc.*

Linear-displacement Transducer, Eddy-current (See **Eddy-current transducer.**)

Linear-displacement Tranducer, Potentiometric Linear potentiometers for the measurement of position and displacement take numerous forms dependent upon intended application. The simplest, least costly form is a single length of wire along which a slider or other form of moving device contacts the wire. The position of the slider determines the effective length of the conductor. Hence a change in electrical resistance or voltage drop is related to the position or displacement of the slider. This simple device is useful for laboratory demonstration but seldom is used for military and industrial applications. The latter make use of wirewound, thin-film, or printed circuits because these techniques make possible a considerably greater length of resistor within the same linear length. In summary, the voltage drop per unit length is larger, thus producing stronger, more useful signals.

One example of a linear-displacement transducer of the wirewound potentiometric type used for the high environmental and reliability requirements of aircraft and missiles is shown in Fig. L-15. The basic design is cylindrical because the device generally is used in connection with hydraulic actuators. The device comprises the following major elements: two resistance elements *j* which are mounted (molded) along with slide bars *d* is the element block *s* which, in turn, is contained within the outer case *n*. Wiper assemblies *h* are attached to wiper carrier *e* and are aligned to coincide with the resistance elements and contact bars immediately opposite each other. See the sectional view. The wiper carrier is secured to the actuating shaft *a* in such manner as to eliminate backlash and yet allow the shaft to rotate freely 360° when installed. This is necessary in many cases because the end fitting *f* provides a thread for attachment to a moving member and free rotation of shaft is necessary to avoid deflection of the wiper system or dislocation of the wipers from the element and contact bars. The forward end of the transducer provides a bushing *b* into which are inserted a pair of *O*-ring seals, providing adequate sealing against intrusion of salt spray, sand, dust, and moisture. These transducers are ½ and ¾ in. in diameter, dependent upon application—the larger size usually used where motion is in excess of 6 in.

Applications of these transducers include aircraft-control-surface indication, such as rudder,

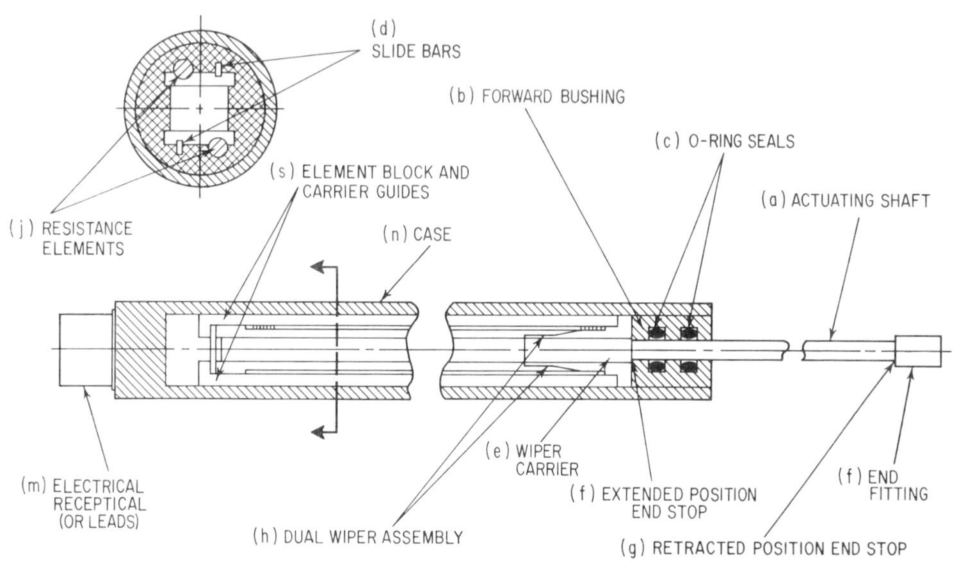

Fig. L-15. Cylindrical wirewound potentiometric-type linear-displacement transducer. *(Gulton.)*

aileron, and elevator—either combined into a screw jack or separated from it—landing-gear-retractor systems, missile-stage separation, and industrial uses, including machine tools and valve indicators.

—Richard N. Clarke, *Servonic Division, Gulton Industries, Inc.*

Linear-displacement Transducer, Variable-reluctance These devices, often abbreviated LVRT, are useful for laboratory measurements of growth, stress, thickness, vibration, shock, and contour and are used to provide electrical feedback for valves and actuators in servocontrol systems for missiles, aircraft, and industrial machinery. As shown in Fig. L-16, the variable-reluctance unit forms half of a four-arm bridge. For many servo applications, a center-tapped transformer is used to form the reference half of the bridge. The output signal then is produced between the transformer center tap and the transducer center pin.

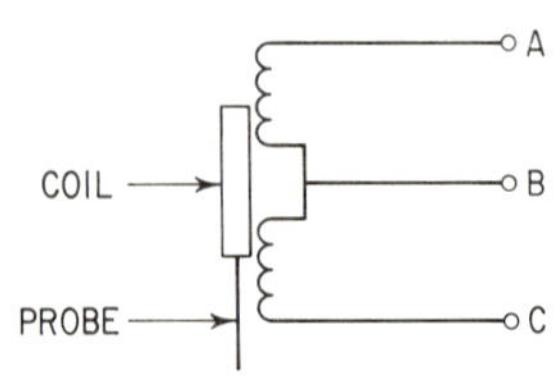

Fig. L-16. Circuit of linear variable-reluctance transducer (LVRT). *(Gulton.)*

The signal phase changes 180° at midrange position. A variable resistive ratio divider or a variable inductive ratio divider may be used to complete the bridge. The null position is adjustable to any point within the transducer range.

The winding is continuous over the length of a spool-type bobbin. There are no intermediate partitions that divide the bobbin segments, as is the case in linear variable-differential transformers. The advantages of the LVRT include excellent linearity over long strokes—up to 24 in.; and good temperature stability. A disadvantage—the overall length must be double the total stroke. The devices can withstand rapid temperature changes from − 160 to + 450° F. Encapsulation of the coils in an inert epoxy resin hermetically seals the unit and protects it against vibration and shock. The working portion of the probe is made of magnetic stainless steel; nonmagnetic stainless steel is used for the extension. Standard probe diametric clearance is 0.001 to 0.020 in. Carrier frequences from 1,000 to 20,000 Hz may be used. Optimum results are obtained with a carrier frequency of 3,000 to 5,000 Hz. Operation with other frequencies results in slight changes in the maximum excitation voltage, phase shift, and linearity. The size of the devices (less probe extension, but including connector) ranges from ¼ by 2.2 in. to ½ by 10.2 in. in standard configurations. Linearity is ± 1 percent or better.

—Frank G. Daveler, *C-E Electronics, a division of Ametek, Inc.*

Linear Interpolation (See **Drafting system, automatic.**)

Linear Programming A mathematical technique for finding the optimum solution to a particular problem, in terms of lowest cost, least effort, shortest time, least equipment, or other requirements.

Linear Transformer A specialized synchro, sometimes referred to as an LT, consisting of a salient-pole rotor and a single-phase stator, distributively wound. The winding on the stator is designed to produce an output voltage that varies linearly with rotor position. This linear function is valid only within a restricted band about the zero position—generally ±50° or ±85°. This band is known as the *excursion range*. Past the excursion range, the output voltage no longer is linear and tends to become sinusoidal.

The linear transformer is used in many cases as a replacement for a potentiometer. The LT is an infinite-resolution device. The LT also has the advantage of being constructed in the same manner as other servosystem components. Thus, service behavior of the LT will be the same as that of the other elements in the system. Also see **Resolver; Synchro;** and **Transolver.**

—Joseph P. Spetzer, Jr., *Clifton, Division of Litton Industries*

Linear Variable-differential Transformer (LVDT) An electromechanical transducer that produces an electrical output proportional to the *displacement* of a separate movable core. As shown by the cross-section view of Fig. L-17, three coils are symmetrically spaced on a cylindrical coil form. A rod-shaped magnetic core, positioned axially inside this coil assembly, provides a path for magnetic flux linking the coils.

When the primary or center coil is energized with alternating current, voltages are induced in the two outer or secondary coils. The secondary coils are connected in series opposition so that the two voltages in the secondary circuit are opposite in phase. The net output of the transformer is then the difference between the voltages of the two secondary windings. With the core position approximately in the center of the transformer, the output voltage will be zero. This is termed the balance point or null position.

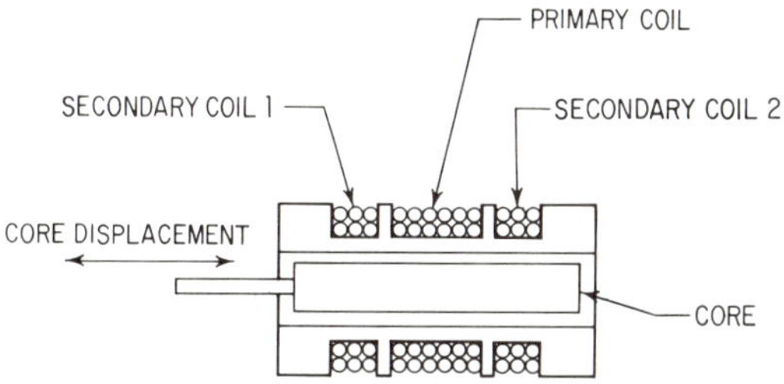

Fig. L-17. Sectional view of a linear variable-differential transformer (LVDT). *(Gulton.)*

When the core is moved from the null position, the voltage induced in the secondary coil—toward which the core is moved—increases. Simultaneously, the voltage induced in the other secondary coil decreases. This produces a differential voltage output which varies linearly with change in core position. Motion of the core in the opposite direction beyond the null position produces a similar linear voltage characteristic with the phase shifted 180°. A continuous plot of voltage output vs. core position appears as a straight line through the origin provided that opposite algebraic signs are used to indicate opposite phases, as shown by Fig. L-18. Actually, the output voltage does not quite become zero at the null position of the core because of small residual

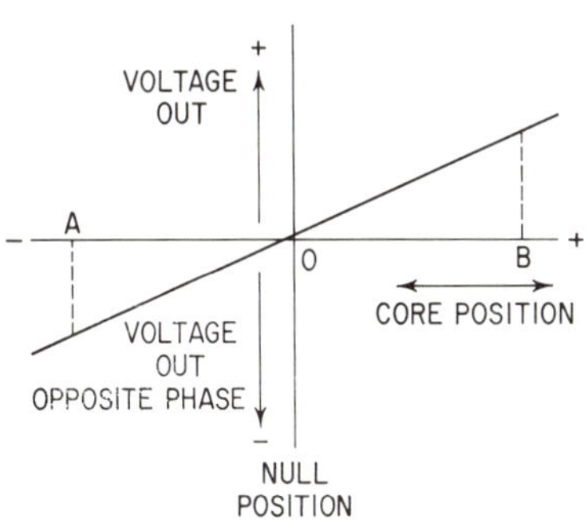

Fig. L-18. Output voltage as a function of core position of a linear variable-differential transformer.

voltages that do not cancel. By proper design, this null voltage output can be reduced to an insignificant quantity. A schematic of a typical linear variable-differential transformer is shown in Fig. L-19.

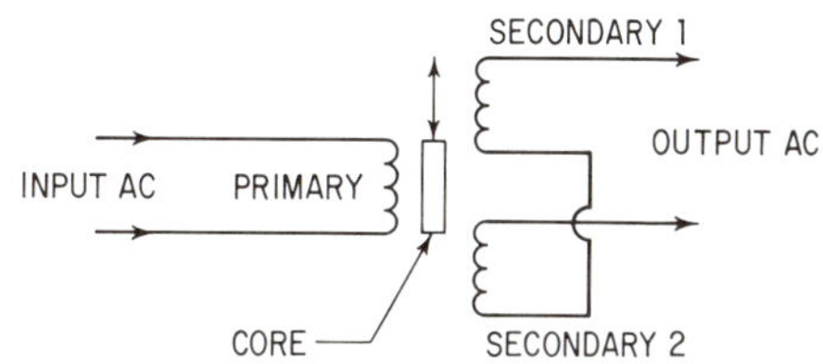

Fig. L-19. Schematic diagram of linear variable-differential transformer. *(Gulton.)*

Desirable features of the LVDT include (1) simplicity of design, (2) availability of a wide variety of linear ranges, (3) ease of installation, (4) negligible actuating force required, (5) absence of friction between parts, (6) inherent ruggedness, (7) ability to operate over a wide range of temperatures, (8) noncritical alignment of the moving element, and (9) mechanical isolation of

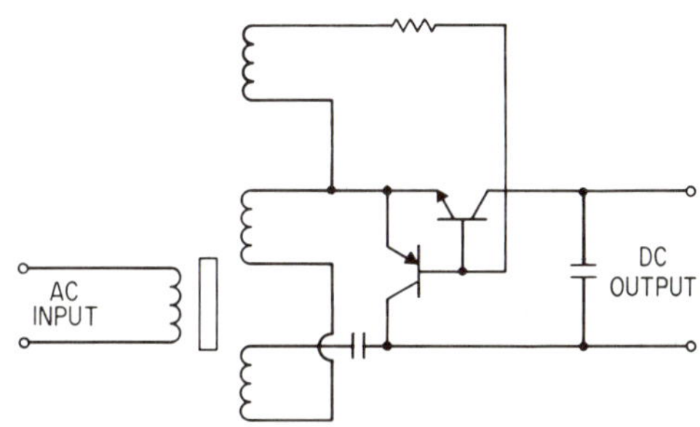

Fig. L-20. Linear variable-differential transformer with demodulator.

the moving element from other parts. Electrical features include: (1) a voltage output linearly proportional to core displacement, (2) high sensitivity, (3) stepless variation in output, (4) low output impedance, and (5) stable zero output. Conversion of the ac output voltage to a dc voltage is basic and simple. Demodulation can be accomplished using simple diode networks or may utilize transistors as switching elements as shown in Fig. L-20. Typical specifications for linear variable-differential transformers are given in Table L-1.

Although the LVDT is used to measure electronically linear displacement, such as actuator or mechanism positioning, the device also is used

TABLE L-1. *Characteristics of Linear Variable-differential Transformer (LVDT)*

Electrical stroke (linear range)	± inches from null
Mechanical stroke (limit of overstroke)	± inches minimum from null
Excitation voltage	1.0–115 V rms
Excitation frequency	60–20,000 Hz
Excitation power	Normally less than 1V-A
Insulation resistance	Greater than 50 MΩ at 500 V dc
Dielectric strength	As high as 1,500 V rms
Sensitivity	Volts out/volts in/inch displacement
Linearity	From ±0.25% full-scale
Output accuracy	From ±1%
Repeatability	Normally less than 0.01% full-scale
Null voltage	Normally less than 30 mV rms
Output impedance	Normally less than 1,000 Ω
Temperature range	Normally from −65° to +350° F, but can be designed to operate from cryogenic temperatures to over 1,000° F

in conjunction with pressure-sensing devices. In the latter case, the pressure sensor should deflect linearly proportional with pressure, thus providing a linear motion that can be sensed by an LVDT. The LVDT also can be used in an acceleration-measuring device where displacement of a seismic mass will produce a linear displacement of a magnetic core within the windings.

Comparatively high outputs can be obtained (up to 0.05 V/V) for motions as low as 0.005 in., thus making the LVDT useful in ac applications where temperature or cost limitations prevent the use of electronics for signal amplification. LVDTs generally are found for linear-displacement measurements of mechanisms or linkages where the motion usually varies between 0 to 0.05 in. and 0 to 6.0 in. The long life and high reliability of the LVDT make the device particularly suitable for remote positioning, as exemplified by aircraft and missile control surfaces. The extremely fine resolution, high accuracy, and good stability also make the LVDT particularly suitable for short-stroke position devices as found in the instrumentation field.

—Richard N. Clarke, *Servonic Division, Gulton Industries, Inc.*

Linearity The closeness to which a curve approximates a straight line. Linearity generally is expressed as a deviation from linear or as a *nonlinearity*. This is the maximum deviation of a curve or data set from a corresponding point on a specified straight line. Unless otherwise specified, the nonlinearity is expressed in output terms—very frequently in percent of full-scale output. The nature of the straight line which is used in the establishment of the deviation from linear must be established if comparisons are to be meaningful.

When the straight line passes through the lower and upper end points of the data, the resulting linearity is called *end-point linearity* or *terminal-based linearity*. Similarly, if the straight line passes through the output-data point corresponding to zero input and that line is inclined so as to minimize positive and negative deviations of the data from that line, the resulting linearity is the *zero-based linearity*. In the case of *independent* linearity, the straight line is placed to minimize the maximum positive and negative deviations from the straight line. The *least-squares* linearity is obtained when the straight line is arranged to be located through the data set such that the sum of the squares of the deviations from that straight line is minimized. This latter linearity finds some favor when data are routinely reduced with the aid of a digital computer. When the line is located to pass through the theoretical end points, the linearity is referred to as the *theoretical-slope* linearity. This latter linearity is independent of the character of a given data set and thus forms a firmer basis of comparison than the previously described types of linearity.

—Otto Muller-Girard, *Consulting Engineer, Rochester, N.Y.*

Linearity, Differential The differential linearity of a quantizing system is a measure of the consistency of the quantizing interval (see **Quantize**). The differential linearity in percent is defined as

$$\text{Differential linearity} = 100\,\frac{Q_a - Q_t}{Q_t}$$

where Q_a is the actual quantizing interval and Q_t is the theoretical quantizing interval. In the case of a binary system with equal quantizing intervals, the theoretical interval is $V_{\text{f.s.}}/2^n$, where n is the resolution of the system in bits (see **Resolution**) and $V_{\text{f.s.}}$ is the full-scale value. Thus, differential linearity as a percent of full scale is

$$\text{Differential linearity} = 100\,\frac{2^n Q_a - V_{\text{f.s.}}}{V_{\text{f.s.}}}$$

The differential linearity usually depends on the level of the signal at which the measurement is made. For this reason, the maximum differential linearity is generally accepted as the specification over the entire range. This convention also requires that a testing procedure include examination of all quantizing intervals. An exhaustive test is quite extensive, therefore, since a high-resolution system has many thousands of quantizing intervals. In a 14-bit system, for example, there are over 16,000 quantizing intervals.

Differential linearity is of importance when the distribution of the quantized data is of significance. For example, in a nuclear-scattering experiment, the distribution of the energy levels of particles striking a detector may be of importance. If the differential linearity of the analog-to-digital converter being used to measure the level is poor, the distribution may be altered significantly.

In general, analog-to-digital converters which utilize a continuous-comparison technique have better differential linearity than those which perform conversion through a series of comparison steps (see **Analog-to-digital converter**). Thus, the ramp-type converters exhibit better differential linearity than the successive-approximation or parallel converters. A differential specification of 1 percent or better is considered quite good in most applications. To appreciate the significance of such a specification, this means that the quantizing levels in a 5-V 14-bit analog-to-digital converter vary less than $\pm 3\ \mu V$.

—Thomas J. Harrison, *IBM Corporation*

Liquid Chromatography (See **Chromatography.**)

Liquid Level, Measurement of Methods range from simple mechanical devices to very sophisticated systems, dependent upon performance required and environment of use. Level measurement, by definition, always is relative to some dimensional reference. Liquid level usually is (1) the interface between a gas and a liquid or (2) the interface between two immiscible liquids of unequal density. Liquid-level measuring systems may be (1) *continuous,* that is, indicate the level continuously over the full range of level or (2) *point or incremental*, wherein the level is measured at only one or a series of discrete interface locations. Liquid-level sensors may be connected to locally mounted or remote indicators, recorders, and controllers (electric or pneumatic). All the many variations of instrumentation are applicable to liquid-level sensors as with other sensors.

Criteria useful for evaluating liquid-level systems include (1) accuracy, (2) cost, (3) reliability, (4) use with corrosive fluids or vapors, (5) use with explosive vapors, (6) use in pressurized vessels, (7) output independent of pressure, (8) output independent of fluid density, (9) output independent of fluid or ambient temperature, (10) capable of remote indication, (11) capable of control, (12) ease of characterization for near-linear output, (13) easy zero and span adjustment, (14) easy servicing or replacement, (15) ease of providing alarm signals, (16) suitable for liquid interface, and (17) use under vibration and impact conditions.

The fundamental approaches to liquid-level measurement, with comment on normal accuracy and span capability, are summarized in Table L-2. A qualitative comparison, listing some of the advantages and disadvantages, of the more common level-measurement systems is given in Table L-3. This information is intended only for initial selection guidance.

Float-operated Systems: A float arm, operating through a bellows-type seal to drive a pointer mechanism, is shown in Fig. L-21. Remote electrical indication can be achieved by replacing the pointer mechanism with a potentiometer. An even simpler approach uses a pivoted float arm to drive a permanent magnet which, in turn, positions a permanent-magnet pointer through a seal wall, thus eliminating the bellows seal.

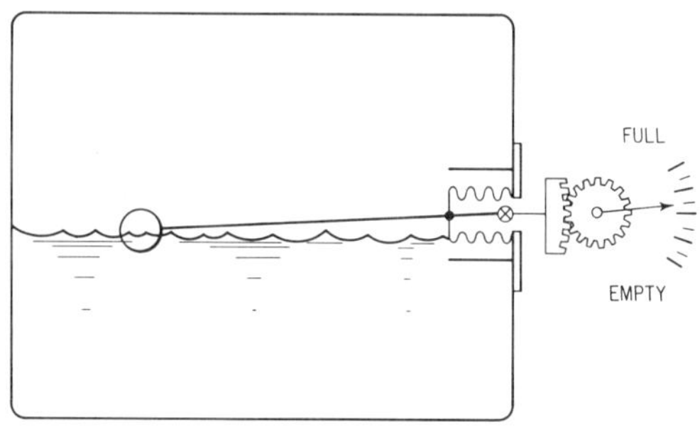

Fig. L-21. Float arm, operating through bellows-type seal, for measuring liquid level.

Liquid level in high-rise tanks can be measured to better than $\pm\frac{1}{8}$ in. with tape or cable-type designs. A direct-reading system with the tape read at eye level outside the tank is shown in Fig. L-22. A weight maintains tension in the tape as the float follows level changes. More com-

TABLE L-2. *Liquid-level System Characteristics*

Type	Practical Span		Normal Accuracy	Remarks
	Max	Min		
Float-operated:				
Float arm, mechanical.	10 ft	6 in.	1–2% span	Usually under 25 psi
Float arm, hydraulic.	10 ft	1 ft	1–2% span	Self-powered, remote indication
Float arm, magnetic pointer.	6 ft	3 in.	2–5% span	Low-cost, direct-reading
Float arm, potentiometer.	10 ft	6 in.	2% span	Remote electrical indication
Float, tape or cable.	75 ft	. . . .	1/16 in.	On-site and remote indication
Hydrostatic head:				
Direct-pressure gage	Tank height	Gage sensitivity	1% span	Vented tanks only
U-tube manometer	400 in. H_2O	15 in. H_2O	1% span	Pressures up to 800 psi
Mercury sealed, bell.	70 in. H_2O	12 in. H_2O	1% span	Pressures up to 3,500 psi
Gas bubbler tube	Tank height	Gage sensitivity	1% span	Vented tanks only
Isolating diaphragm, pneumatic	Tank height	Gage sensitivity	1% span	Used where bubbler tube is not acceptable
Buoyant-element gage	Tank height	. . . .	1% span	
Differential-pressure transmitter. Force-balance and open-loop systems	1,000 in. H_2O	1 in. H_2O	0.5% span	25,500, 2,000, and 6,000 psi pressure ratings. Response = 0.2–2 s
Sight gages.	Tank height	3 in.	0.02 in.	Direct reading of level
Radiation:				
Point sensing	Single point	. . . .	0.1 in.	Response 0.1–10 s
Continuous	20 ft	1 in.	1–3% span	Response 0.1–10 s
Capacitance:				
Nonconducting fluids	Tank height	6 in.	0.5% span	Response 0.5 s
Conducting fluids	Tank height	3 in.	0.5% span	Response 0.5 s
Point sensing.	Single point	. . . .	0.1 in.	Response 0.05 s
Thermal transfer:				
Thermistor, thermocouple, or semiconductor point sensor	Single point	. . . .	0.1 in.	Response slow
Sonic:				
Point sensing	Single point	. . . .	0.1 in.	Response 0.02 s
Continuous.	100 ft	3 ft	1% span	Digital and analog systems
Electrode probe	Single point	. . . .	0.1 in.	Response 0.1–0.5 s
Electrical-resistance types:				
Potentiometer, magnetic float	12 ft	6 in.	0.5–2% span	Response 0.1 s
Variable resistance in pressure-sensitive sheath	200 ft	2 ft	1 in.	2–3 in. H_2O head required to actuate
Weighing.	Tank size	. . . .	±0.1% span	Response may be slow

Note: Much of the information in this table is excerpted by special permission from *Chemical Engineering,* Feb. 12, 1968.

plex systems store the tape or cable on a drum that is driven by a negator spring or electric drive. These systems will operate in pressurized tanks and many have remote-indicating capability.

Remote indication is a feature of the self-powered float-operated hydraulic system of Fig. L-23. A two-bellows transmitter is positioned by a float arm-linkage mechanism, the float-arm motion compressing one bellows while extending the other. A similar two-bellows receiver element drives a pointer-indicating mechanism. Transmitter and receiver bellows are cross-connected with metal tubing and completely filled with hydraulic fluid.

Hydrostatic Head Systems: A pressure gage of the proper sensitivity, located at the bottom

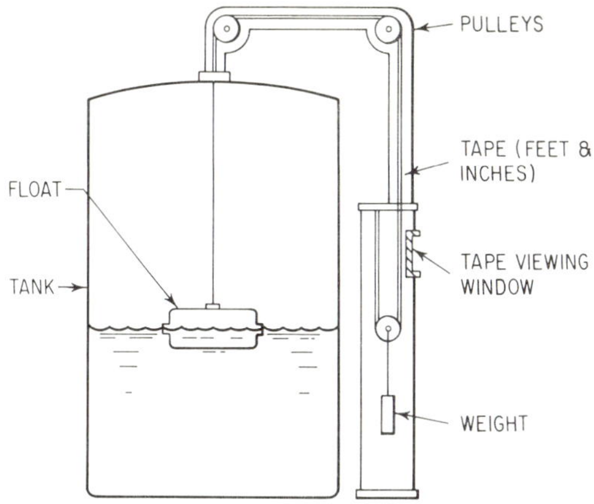

Fig. L-22. Tape-type direct-reading level-measurement system.

of a vented tank, will indicate liquid level by sensing the hydrostatic head. If the tank is pressurized, a differential-pressure gage is needed. A U-tube manometer with a fluid of known density is an excellent standard for pressure or for differential-pressure measurement. The mercury manometer and mercury-sealed-bell differential transmitter largely have been replaced by bellows- or diaphragm-type transmitters.

Use of a force-balance transmitter in a pressurized tank is shown in Fig. L-24. This pneumatic unit has a 3 to 15 psi output. In addition to a pneumatic output signal, force-balance differential-pressure (d/p) and pressure-transmitter designs are available with a dc output (either 10 to 50 mA or the more popular 4 to 20 mA). These units have adjustable span and zero. Most units have a 10-to-1 span adjustment; some have adjustable damping. Diaphragm and body materials, such as 316 stainless steel and Hastelloy C, are available where corrosion is a problem. Models are available for flange mounting at the bottom of a tank such that the diaphragm is flush with the tank wall to give better performance with settling or coagulating fluids. Open-loop nonfeedback designs also are available with various pressure ratings and spans. The latter are self-powered units with pointer indication of level or head. (See **Self-actuated regulators.**)

A problem in the application of a differential-pressure transmitter (as in Fig. L-24) involves location of the transmitter and the effect of con-

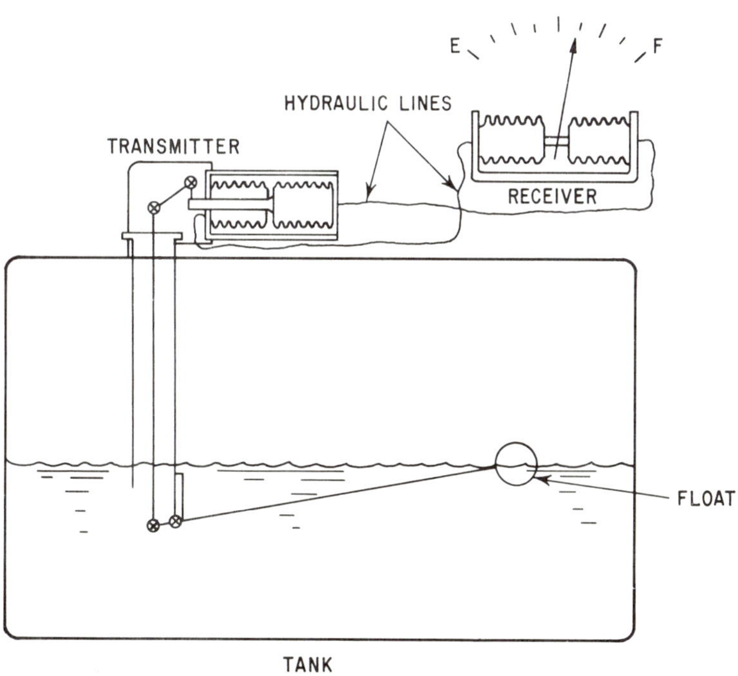

Fig. L-23. Self-powered float-operated level-measurement system.

TABLE L-3. *Liquid-level Measurement-systems Characteristics*

Relative Advantages	Relative Limitations
Gage glass	
Direct reading Special designs for use up to 600°F and 10,000 psi "Glassless" designs available in numerous materials for corrosion resistance	Overlapping gages needed for long, level spans Remote readout somewhat difficult Some fluids tend to coat and scale glass Limited to relatively still liquid surfaces Accuracy and readability depend on cleanliness of glass and fluid
Diaphragm-type pneumatic-pressure gage	
Low-cost diaphragm seal Good for use with slurries Air does not contact measured fluid Applicable to pressurized tanks Performs well with turbulent fluids Good for batch food processing	Less accurate at high static pressure Air supply to diaphragm must be higher than tank or vessel pressure Limited because of flexibility (fragility) of diaphragm material Output a function of fluid density
Gas bubbler tube	
Good in corrosive fluids Low cost Good for use with slurries and viscous fluids with large tube Performs well with turbulent fluids	Tank cannot be sealed Not applicable where measured fluid will coat equipment Output a function of fluid density
Buoyant-element gage	
Adaptable to many fluid densities Numerous materials of construction for corrosion resistance Tank pressure not a problem Good for condensing fluids	Should not be used with slurries or with adherent fluids Not good with turbulent fluids Output a function of fluid density
Manometer methods	
Good for interface level measurement Available for relatively high temperature and pressure applications	Best when measured fluid is clean and not severely corrosive Output a function of fluid density
Electrode probe	
Easy to install and maintain (throwaway probe) Good for use with slurries Good accuracy and response speed Applicable to level measurement of moist bulk solids No moving parts exposed to fluid	Fluids measured must be electrically conductive Cannot be used with adherent fluids Susceptible to electrolytic corrosion Continuous measurement difficult
Electrical capacitance probe	
No moving parts exposed to fluid Suitable for continuous indication and/or control Remote adjustment of span and zero is possible Models for conducting or nonconducting fluids Probe materials for most corrosive fluids Good for use with slurries	Measured fluid must have proper dielectric qualities Usually requires recalibration if measured material changes in composition or moisture content Questionable performance with adherent fluids Probe length and mounting must suit tank

TABLE L-3. *Liquid-level Measurement-systems Characteristics (Continued)*

Relative Advantages	Relative Limitations
Sonic gage	
Intrinsically safe or explosion-proof No moving parts Useful at high temperatures/pressures Good accuracy and response speed Not affected by foam, froth, or droplets	Operation with adherent fluids not fully proved Continuous level-reading systems expensive
Nuclear-radiation gage	
No physical contact with liquid Suitable for molten metals as well as liquids of all types (corrosive, abrasive, highly viscous, adherent) Useful at very high temperatures/pressures Good accuracy and response speed No moving parts (except servogage design)	Fear (greatly exaggerated) of using radioactive source Affected by density change of liquid Radiation-source holders may weigh several pounds Relatively high cost
Differential-pressure gage (force feedback)	
Pneumatic or electric outputs suitable for remote indication and control Withstand high pressure and overpressure Span adjustable over 10-to-1 range Outside of tank mounting favors servicing Diaphragm-flush models for slurry or coagulating fluids	Difficult to characterize output Output a function of fluid density Response time relatively slow Possible condensation problems in the transmitting lines
Potentiometer-magnetic float	
Output easily characterized Electrical span and zero adjustment Output suitable for remote indication and control Intrinsically safe or explosion-proof Large choice of float and transmitter materials High and low alarms adjustable 0–100%	Present designs have limited pressure rating Adherent or very dirty fluids troublesome Multiple units required for high tanks Transmitter length and mounting must suit tank Interface use not fully proved
Float-type direct-reading gages	
Low cost, reliable designs Operate over large temperature range Choice of corrosion-resistant materials	Questionable impact and vibration resistance Tailored to tank geometry Normally limited to moderate pressures

Note: Much of the information in this table is excerpted by special permission from *Chemical Engineering,* Feb. 12, 1968.

densation in the transmitting lines. The reference line (and often the high line as well) may be liquid-filled with a large isolating flexible diaphragm or bellows-located at the tank wall to accommodate thermal expansion of the trapped liquid. By this means, the reference leg can be controlled and condensation problems eliminated, and (within limitations) the transmitter may be located at various levels with respect to the tank.

The *gas bubbler-tube system,* shown in Fig. L-25, is a common head-sensing system. A regulated air supply pushes air through a restrictor and into the bubbler tube, whose open end is near the bottom of the tank, thus making gage pressure equal head pressure. If the fluid density remains constant, the gage pressure will be a good measure of liquid level in the tank. The tank must be vented to allow the supply air to continue to flow. The vent can be a pressure-relief valve so long as

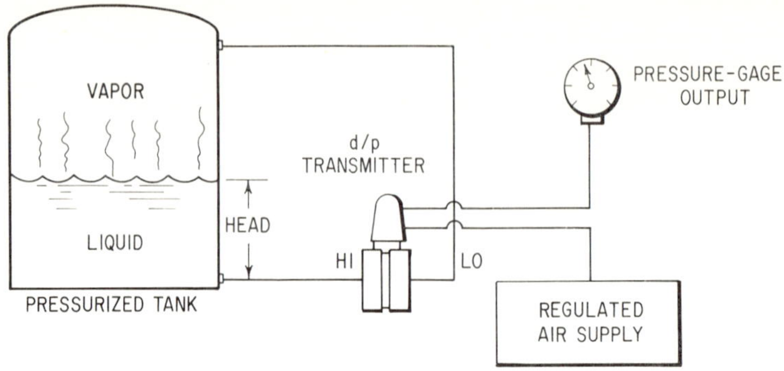

Fig. L-24. Force-balance transmitter for measuring level in a pressurized tank.

the supply pressure always exceeds maximum tank pressure. In this situation, the indicator should be a differential-pressure gage to eliminate static pressure from the reading. With the continuous air supply of Fig. L-25, continuous level indication is provided. A lower-cost approach, appropriate for remote tanks, uses a hand pump built into the pressure indicator to pump pressure up to head pressure for a temporary reading of level.

Where it is undesirable to bubble air through a fluid, an isolating diaphragm may be used as shown in Fig. L-26. The head-measuring pneumatic-pressure transmitter is mounted at the bottom of the tank. A flexible diaphragm isolates tank fluid from the supply air as well as acting as a flapper to control bleed through a nozzle to atmosphere to make air pressure equal the head pressure. A restrictor in the air-supply line absorbs the pressure difference between supply and output pressure. Unlike the gas bubbler tube, this system can be used to gage level in a sealed, pressurized tank.

Archimedes' *buoyant-force principle* sometimes is used to indicate or control level. Most designs use a buoyant cylinder that is heavier than the liquid displaced by it. The cylinder is supported by a spring such that cylinder position

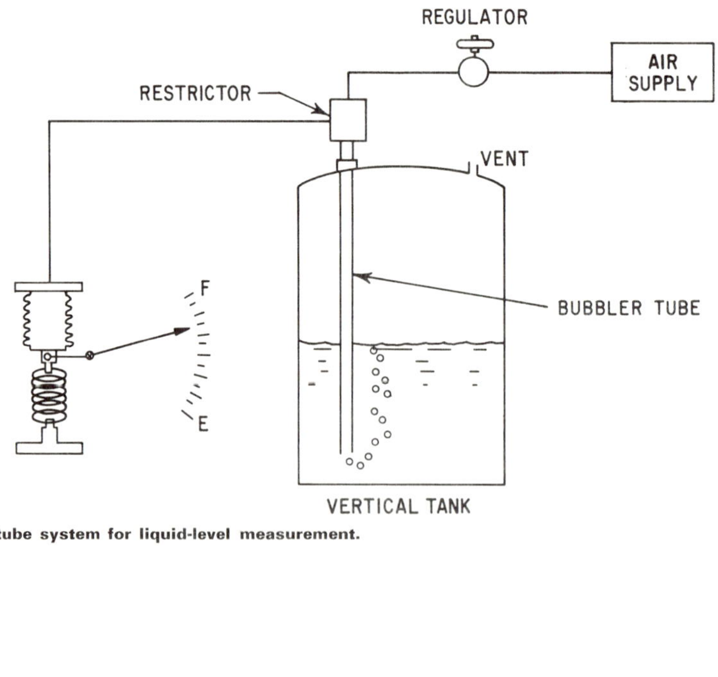

Fig. L-25. Bubbler-tube system for liquid-level measurement.

is a function of liquid level and density. A coil spring or torque tube may be used. Position is converted to a pneumatic or electrical signal. Force-balance designs operate on the same principle with little motion at the seal flexure. On-off controllers often use this principle to operate electric switches to control valves or pumps. Such controllers often are self-contained for mounting outside a process tank with two pipe connections that bring tank fluid to the controller.

Capacitance gaging systems use the variation in electrical capacity of a probe with change in tank level to indicate level. In aircraft fuel tanks, probes often are characterized to match tank geometry. With nonconducting fluids, the sensed capacity may be between a single tube and the tank wall. With conducting fluids, an insulated cylinder commonly is used wherein capacitance of the insulated sensor varies with the level.

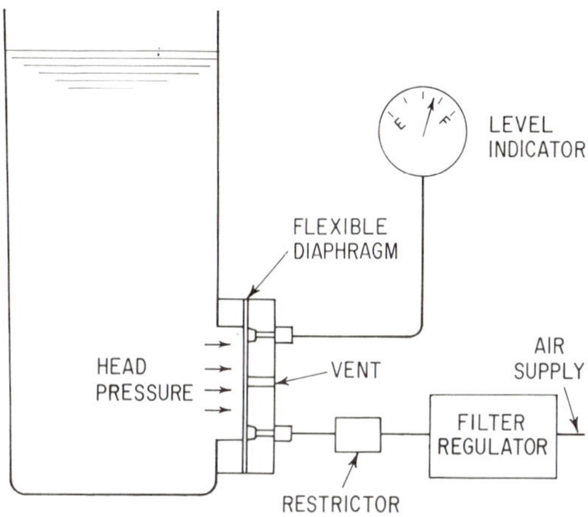

Fig. L-26. Diaphragm-type level-sensing element.

The dielectric constant involved with nonconducting fluids is that of the fluid; with conducting fluids, it is that of the insulating coating. Electrical span and zero adjustment in the electric circuitry make it possible for capacitance systems to be conveniently adjusted for accurate, continuous measurement of liquid level. Capacitance probes also are used as point-sensing alarm and control elements.

Electrode-probe systems can be used as single-point sensors to operate electrical relays for alarm and control purposes. Electrode potentials depend on the relay used and the sensitivity required for the liquid. Though ac voltages are preferred to minimize electrolysis, dc voltages are selected for liquids that have poor conductivity. These systems require a relay with very high sensitivity.

Electric-resistance Systems: There are a number of potentiometric-type level transmitters. A simple open rotating potentiometer,

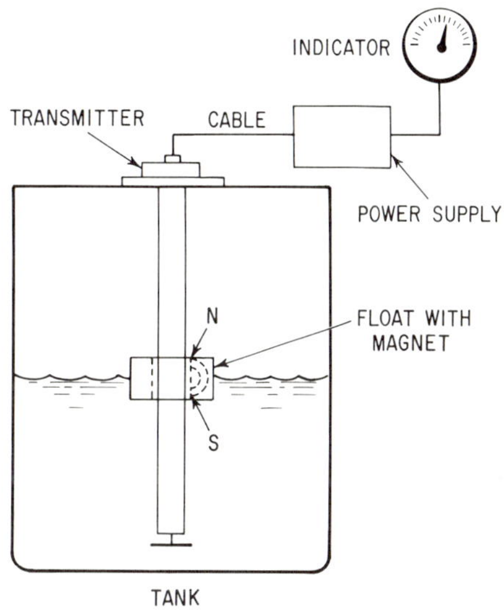

Fig. L-27. Potentiometric transmitter with a magnetic float that actuates the moving point of a potentiometer.

driven by a float arm in a nonconducting fluid, is commonly found in automobile fuel tanks. A more elaborate potentiometric transmitter is shown in Fig. L-27 with a magnetic float that actuates the moving point of the potentiometer. The resistance element is located within a sealed tube which also guides the vertical motion of the float. In another transmitter design, a series of reed switches and resistors within the tube sense magnetic-float position. A potentiometric system with associated electronics to excite the transmitter and convert input voltage to dc output proportional to level is shown in Fig. L-28. Span and zero can be adjusted at the receiver to give 10 to 50 mA or other desired range of current output. The system is well suited to remote indication and control.

A two-terminal variable-resistance level sensor[1] is available for use in liquids, slurries, and granular solids. The system packages a resistance strip within a pressure-sensitive sleeving such that the weight of flowable materials shorts out that

[1] Metritape.

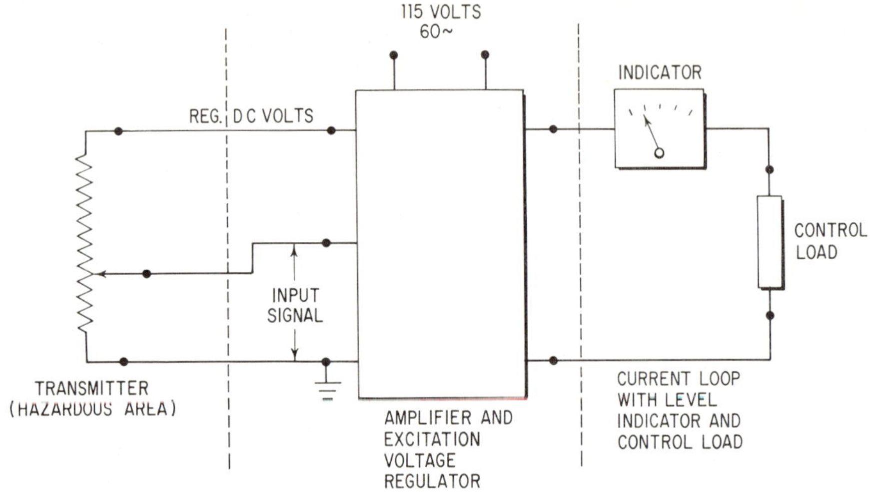

Fig. L-28. Potentiometric system with associated electronics to excite the transmitter and convert input voltage to direct current output proportional to level.

part of the resistance element that is slightly below level. These tranducers are made to specified lengths, from 2 to 200 ft, in coilable-strip form so that they may be inserted easily through a hole in the top of a tank. A breathing bag at the top of the sensor allows for gas expansion and isolates the electrical system from the fluid environment. The sensor may be ac or dc excited—with electric analog or digital output. In addition to liquids, the sensor is suitable for the continuous level measurement of dry granular solids in large tanks and silos.

Also see **Level gage, radiation-type; Level instrumentation, bulk solids;** and **Gage glass.**

REFERENCES

Considine, D. M.: Liquid Level Measurement Systems: Their Evaluation and Selection, *Chem. Eng.* Feb. 12, 1968, pp. 31–38.

Akeley, L. T.: Eight Ways to Measure Liquid Level, *Control Eng.* July, 1968, pp. 72–75.

Lawford, V. N.: Remote Tank Gauging Using Liquid-filled Capillaries, *ISA Publ.* P16-PHYMMID-67, 1967.

—L. T. Akeley, *Beckman Instruments, Inc.*

Lissajous Figure [See **Frequency (electrical), measurement of.**]

Load Cell, Hydraulic One form of hydraulic load cell is shown in Fig. L-29. Capacities range from 0 to 6,000 lb to 0 to 1 million lb. The diaphragm of the sensing element has a U-shaped annulus that is held in place by the load-cell housing. When the diaphragm makes a stroke, the unsupported portion of the flexible, U-shaped outer edge moves hydraulically with a rolling motion. Although the diaphragm develops a piston-like

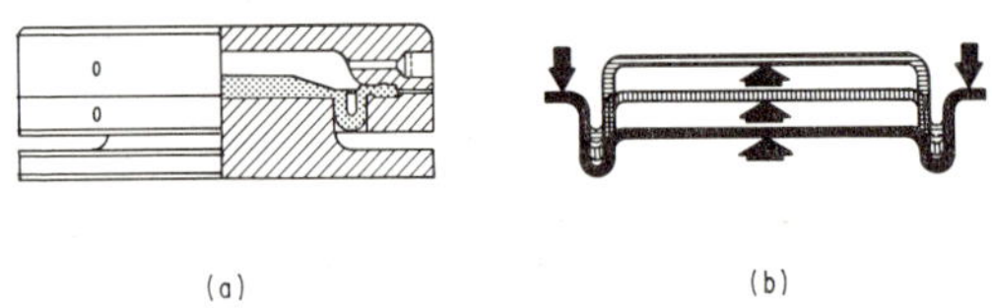

(a) (b)

Fig. L-29. Hydraulic load cell: (a) cutaway view showing patented fluid-seal diaphragm; and (b) view of mobile diaphragm in three positions. *(Martin-Decker.)*

stroke, it is so engineered that one component cannot rub against another. The dial indicator normally used with a cell of this type can be located up to 100 ft from the cell—by means of hydraulic fluid-filled hose. Full-scale accuracy of 0.2 of 1 percent is claimed. Cells of this type are used in crane scales, as cable-tension indicators, lift-truck-load indicators—where materials can be weighted as they are lifted, for test gages, and in most other applications where force and weight must be measured. Also see **Testing (materials), destructive.**

Load Cell, Pneumatic A device wherein a force

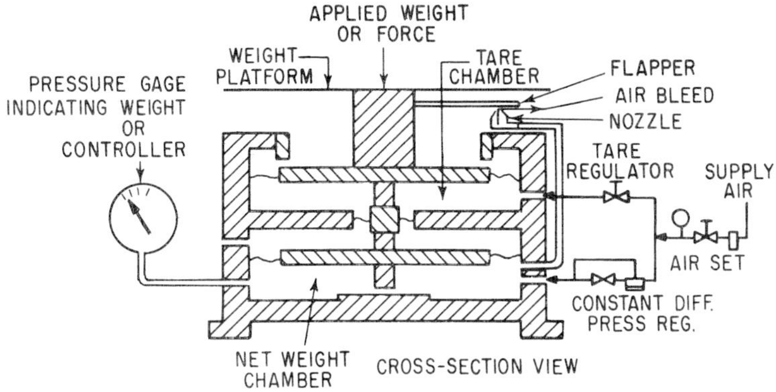

Fig. L-30. Pneumatic load cell used in connection with a hopper scale. Note means for incorporating compensation for tare weight.

or weight is countered by air pressure—the pressure required for balance being a measure of the unknown force. A pneumatic load cell used in connection with a hopper scale is shown in Fig. L-30. The pneumatic load cell is not to be confused with pneumatic position detectors that often are used with mechanical lever scales wherein the weighing device has a pneumatic output signal.

Load Cell, Strain-gage An electromechanical transducer which translates changes in force or weight into changes in voltage. The change in voltage can be calibrated directly in terms of the load applied to the cell. Simply constructed, strain-gage load cells take advantage of all the characteristics of bonded resistance strain gages. Hermetically sealed within the load cell are one or more sets of matched strain gages bonded to a high-strength steel element that is machined to close tolerances. The strain gages are connected electrically to form a balanced Wheatstone bridge. Additional compensation resistors are added to the circuit for maintaining the accuracy of the bridge over a wide range of temperatures. The principle of operation depends upon the deflection of the element. Deflection creates a change in gage resistance, thereby unbalancing the bridge circuit. The result—for a given input voltage, the output voltage of the bridge varies proportionally with the load and thus is indicated, recorded, or used for control by way of appropriate instrumentation. See **Bridge circuits (electric)**; and **Strain gage.**

Typical strain-gage-type load cells are available for force or weight measurement in the range from 1 to 12 million lb. The general accuracy of the devices is ±0.1 percent of full-scale output—with sensitivities of 2 or 3 mV output per volt of input. Calibrated electronic-weighing-system accuracies of 0.05 percent have been achieved, and units with greater than normal sensitivity are available as manufactured with semiconductor-type strain gages.

—Leon J. Weymouth *BLH Electronics, Inc., a subsidiary of Baldwin-Lima-Hamilton Corporation*

Load Change A load change is any factor or

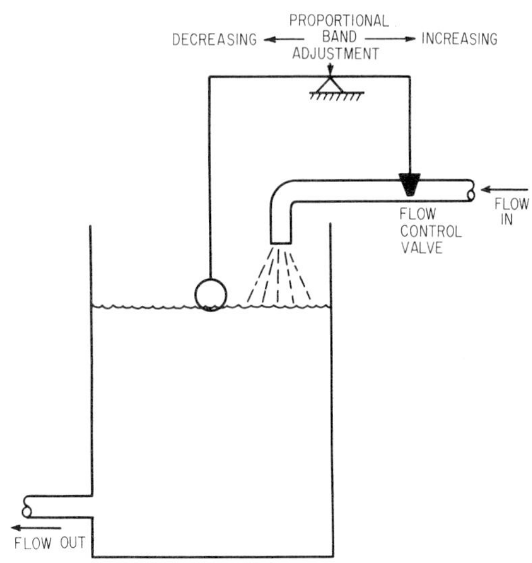

Fig. L-31. Tank level with proportional control.

condition that changes the balance between input and output of a control loop and usually is the most important factor affecting the quality of control. The main function of the controller is to change its output to manipulate the measured variable so as to balance the load. An example of a load change is shown in Fig. L-31. A level is measured, causing the input flow valve to operate by means of an automatic controller which adjusts the flow into the tank. The purpose of the system is to maintain a constant level despite variations in flow out (load).

—Richard H. Kennedy, *Bristol Division, American Chain & Cable Company, Inc.*

Lockout (See Mask.)

Logger A recorder that automatically scans measured quantities at specified times and records or logs their values on a chart.

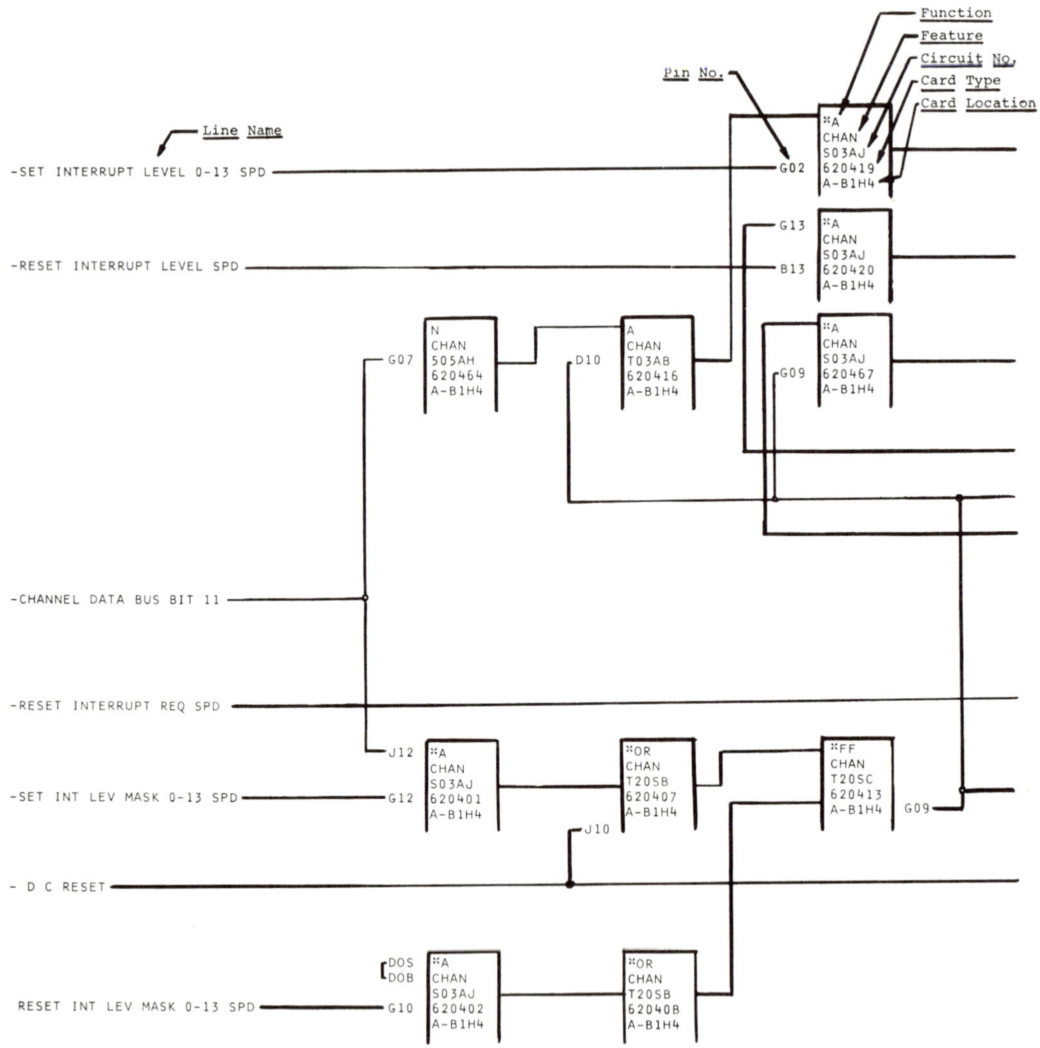

Fig. L-32. Computer logic diagram.

Logic, Computer Computer logic is a term used to describe the circuits which perform the arithmetic and control operations in a digital computer. In designing a digital computer, the designer applies the principles of Boolean algebra (see **Boolean algebra**) and combines the basic logical elements of AND, OR, INVERTER, EXCLUSIVE OR, NOR, NAND, etc., to perform the specified function. Each of the logical elements is implemented as an electronic circuit which is connected to other circuits to achieve the desired result.

—W. W. Welz, *IBM Corporation*

Logic Diagram, Computer A logic diagram is a drawing which shows the interconnection of the individual logic elements in the computer. The logic diagram contains all the information required for the wiring of a computer. With reference to Fig. L-32, the logic diagram consists of logic blocks and electrical interconnections between them.

The logic block in the diagram contains the name of the function performed by it, such as AND, OR, and FLIP-FLOP. It also contains the physical location of this circuit within the computer. The input and output signal connections for this function are designated by the logic block. The logic diagram also shows the connection of this logic block to the other logic block within the system.

The logic diagram, which is initially drawn by the designer, is used as a source of input information to a digital-computer program (see **Program, computer**). The computer provides wiring lists for the machine and a printed logic diagram. The computer can also be programmed to check for deviations from the design rules such as the overloading of circuits or invalid connections of circuits. The computer also provides other outputs necessary for the manufacture and design of the system such as bills of material and lists of unused circuits and locations.

Since the logic diagram represents a logical description of the system, it is used as an aid in troubleshooting the system to isolate malfunctioning parts.

—W. W. Welz, *IBM Corporation*

Longitudinal Voltage (See **Common-mode voltage.**)

Loop A loop generally can be defined as a complete control circuit from system input to output. In the case of *open-loop* control, a controller and a controlled element, as shown in Fig. L-33, would be included. In the case of *closed-loop* control, in addition to controller and controlled element, there would be a feedback circuit consisting of a measuring element and transducer as shown in Fig. L-34. In the process industries, the controlled element usually would be a control valve or controlled-volume pump, whereas in the field of servomechanisms, the controlled element may be a mechanical linkage, a slidewire, or a stepper motor.

Fig. L-33. Open-loop control.

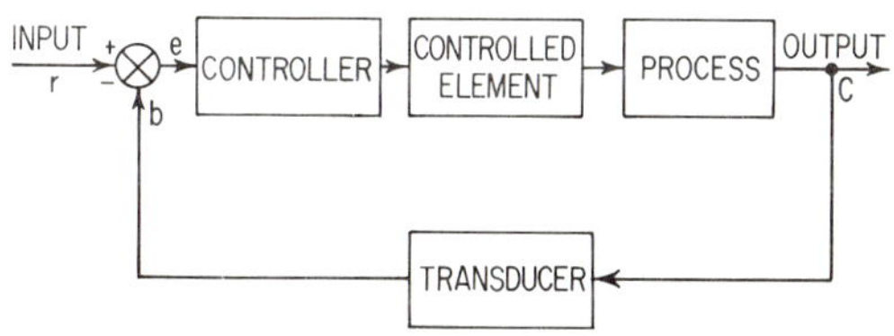

Fig. L-34. Closed-loop control.

—Anthony Y. Chiang, *Bristol Division, American Chain & Cable Company, Inc.*

Loop Gain (See **Gain, loop.**)

Loran (See **Navigation systems.**)

Loss-in-weight Feeder (See **Feeder, weigh.**)

Luminescence, Measurement of (See **Fluorometry.**)

Luminous Intensity, Standard (See **Standards, measurement.**)

Lumped-constant Wavemeter [See **Frequency (electrical), measurement of.**]

Lungs, Measurements of (See **Pulmonary system, instrumental analysis of.**)

M

Mach Number [See **Air-data systems (aircraft)**.]

Machine Control, Adaptive (See **Adaptive control**.)

Machine Control, Digital (See **Drafting system, automatic**; and **Numerical control**.)

Machine-speed Control (See **Governor**.)

Machine-tool Control (See **Numerical control**; and **Position-measurement and control systems**.)

McLeod Gage (See **Vacuum gaging**.)

Macroassembler A type of assembler which allows the user to define pseudo computer instructions which may generate multiple computer instructions when assembled (see **Assembler**). Source statements which may generate multiple computer instructions are called macrostatements (see **Statement, programming**). A macroassembler on a process-control digital computer can prove to be a very powerful tool when properly used by a process-control engineer. By properly defining a set of macrostatements, a process-control engineer can define a process-control programming language specifically oriented toward his process.

Macroassembler Statements for Process Control: One of the problems in installing a digital-computer control system is that the process engineer must become familiar with the programming languages used on the machine. In fact, he normally must become a competent programmer to install the system. Many digital-computer systems provide higher-level languages like FORTRAN (see **FORTRAN**) which the process-control engineer may use. A limitation with languages like FORTRAN, however, is that they were not specifically designed for process-control applications and as such can be rather cumbersome to use. A powerful macroassembler essentially allows a process engineer to define a computer-programming language which is specifically oriented toward his process. Once the process engineer has defined the macrostatements he requires, he is able to complete the installation of a digital-computer control system much more rapidly. Therefore, when investigating various digital process-control computers, it is necessary to determine if they have macroassemblers and then to evaluate how powerful and flexible those macroassemblers are.

Macrolibraries: A macrolibrary consists of a set of defined macrostatements and the associated computer instructions. When a process-control engineer defines a language for his system, he creates a macrolibrary. The usefulness of a macroassembler depends greatly upon the ease of creating, manipulating, modifying, and linking various macrolibraries. Therefore, when evaluating various macroassemblers, the user should investigate the macrolibrary capability and flexibility.

—B. W. Landeck, *IBM Corporation*

Magnetic Amplifier A device, available in several configurations, for amplifying dc or ac outputs—ranging from a fractional watt to many kilowatts. Amplification is obtained by using the magnetic-core saturation properties of the control winding (see Fig. M-1). The signal is applied to control winding C, which controls the current flow (from ac voltage source S) through load resistor L_R. Gate windings X and Y operate as switches that are closed for a fraction of the ac cycle, dependent upon the amplitude of the applied signal. Magnetic amplifiers differ in circuit details depending upon whether the requirement is for an ac or a dc output, for half-wave or full-wave output, or for operation on single-phase or polyphase ac supplies.

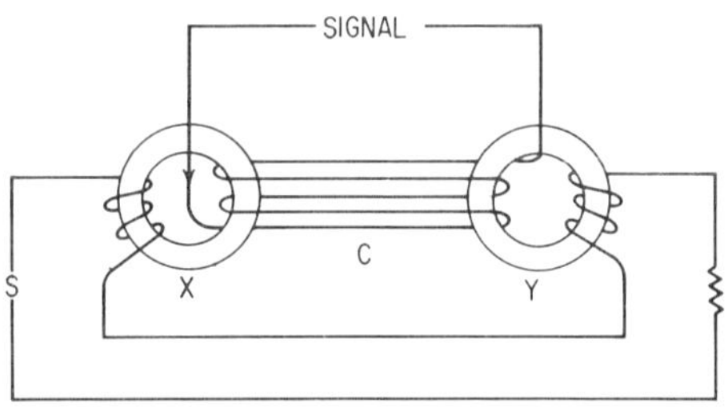

Fig. M-1. Schematic of very basic magnetic amplifier. X, Y are gate windings on separate cores, each with same number of turns; C is control winding; L_R is load resistor; and S is ac voltage source.

Magnetic-ink Character Recognition (MICR) MICR is the common machine language for check handling, the development of which dates back to the early 1950s, at which time a Technical Subcommittee on Mechanization of Check Handling of the American Bankers Association commenced writing specifications for an appropriate system and began discussions with equipment suppliers. The system now is employed by virtually every banking institution in the United States. The number of checks processed by United States banks is now approaching 22 billion annually. MICR, however, is not confined to banking use but finds application in other areas where tremendous quantities of paper information must be processed. MICR is one of two fundamental character-recognition schemes—the other being optical character recognition (OCR). See **Character-recognition systems; Optical character recognition (OCR);** and **Pattern recognition.** Bar and other types of codes also are used for information processing. See **Code readers and systems.**

In MICR, special character shapes are imprinted on the documents in question, such as checks, with a special magnetic ink (not ordinary printer's ink) so that when the characters are scanned by a reading head each character will set up its own distinctive wave pattern. When this pattern is compared by the machine with patterns stored in its memory, the characters or symbols are recognized and so recorded. The fundamental operating principle is evident from Fig. M-2, which illustrates the form of the character or symbol as printed on the document and the corresponding waveform which that character or symbol generates when it passes through the reader. One of the early design problems was that of coming up with waveforms that would be highly distinguishable, one from the other, electronically and still conform reasonably well with the shape of the character or symbol as man sees and recognizes it. As will be noted from Fig. M-2, some changes in conventional shape or format, particularly in the weighting of certain areas of various symbols, were required. Until the layman becomes fully accustomed to the new shapes, they tend to appear odd and distorted, but it must be admitted that considering the complexity of the problem, a good likeness between the "electronic" characters and conventional characters has been achieved.

Much progress has been made in other areas of character and code recognition since the adoption of MICR by the banking industry. Because new schemes are constantly being created and because of the expansion of need for faster automatic reading equipment, it is of interest to review the main factors considered by the bankers' committee when they had the problem of selecting one of the following several systems available or feasible at that time: (1) fluorescent inks, (2) binary or bar

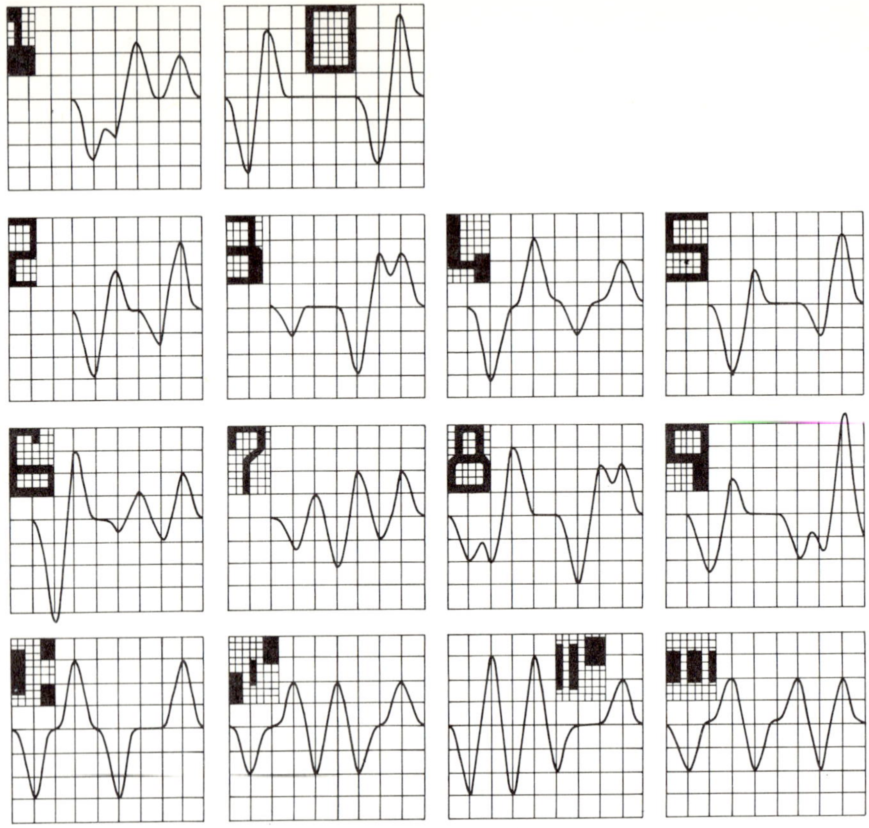

Fig. M-2. Magnetic-ink characters and associated electronic waveforms. Note how these numerals are modified slightly from their usual format. These modifications were adopted by the American Bankers Association in order to improve their waveform for more precise machine reading.

codes, (3) spot code (decimal system), (4) magnetic ink, (5) magnetic bar code—as well as perforations, notches, etc. Factors considered in the choice included (1) accuracy, (2) tolerances, (3) printing, (4) customer acceptance, (5) verification, (6) cost, (7) format, and (8) mutilation and obliteration.

Magnetic Quantities (Variables) Magnetic quantities, like electrical quantities described elsewhere [see **Electrical quantities (variables)**], are extremely important in the fields of instrumentation and control. There are several basic instrument designs for the measurement of magnetic phenomena, including magnetometers and fluxmeters, which find wide application in a variety of areas, such as determinations of the earth's magnetic field and of other bodies in outer space; in the design and operation of rotating electrical equipment, such as motors and generators; and in the design and testing of instrument and control-system circuits and components, the wide range of which may be exemplified by galvanometers, paramagnetic oxygen meters, solenoid valves, ferrite-core memories in computers, and latching relays.

In their characteristics, magnetic quantities closely parallel electrical quantities. A magnetic circuit may be defined as a closed path for magnetic lines of force. This circuit will have a reluctance (similar to resistance or impedance) that will limit the amount of magnetic flux (similar to electric current) that can be sent through the circuit by the magnetomotive force (similar to emf). Magnetic circuits do differ from electric circuits in two fundamental respects, namely, that there is nothing similar to the transport of charge that occurs in an electric circuit, and there is no equivalent of insulation. In an electric circuit, there usually are precise boundaries between conducting and insulating media. Even though materials do have more or less magnetic permeability, generally magnetic flux cannot be regarded as being confined within precise limits.

Magnetic quantities are summarized as follows:

Magnetomotive Force: The total magnetizing force that acts around a completely closed magnetic circuit. If derived from the current flowing in a coil, the mmf is proportional to ampere-turns. The *gilbert,* which equals about 0.8 ampere-turn, is the cgs unit of mmf.

Flux: May be defined as that property of a magnetic circuit which induces an emf in conductors which may be linked with it. The induced emf is proportional to the time rate of change of flux linkages. The *maxwell,* which equals one gauss per square centimeter or one magnetic line of force, is the cgs unit of flux.

Reluctance: The measure of opposition to magnetic flux in a magnetic circuit and, hence, the reciprocal of permeance. Also, it is magnetomotive force/magnetic flux and usually is expressed in this manner rather than by a special symbol.

Magnetizing Force: Also called magnetic-field strength, this is the gradient of the mmf at any point in a magnetic circuit. Usually, it is expressed in terms of the *oersted* or ampere-turn/meter.

Induction: Also called flux density, this is the number of magnetic lines of force per unit area at right angles to the lines of force. Usually, it is expressed in *gauss.*

Permeability: Magnetic induction (in gauss) divided by magnetizing force (in oersteds) and a measure of how a given material compares with air as a path for magnetic lines of force.

Magnetic Measurements: These are of two kinds: (1) those in which the *strength of a magnetic field* is determined (such as the earth's field, the field due to a conductor carrying a current, or the field in the air gap of a magnet) and (2) those made to determine the *properties of a magnetic material.*

Field-strength Measurements: These may be made by (1) induction methods, (2) with an oscillating bar magnet, (3) with a bismuth spiral, (4) with a magnetron, and (5) with a Hall-effect fluxmeter.

In the *induction method,* a coil of known turns and area is arranged so that it can be made to cut the field in the desired region in a direction perpendicular to the field. The emf generated in the coil—and hence the field producing it—is determined from the quantity of electricity discharged through a *ballistic galvanometer* connected to the coil terminals. In measuring the field strength in air gaps, the coil is arranged so that one side can be moved quickly across the entire gap, or through a definite distance, by means of a spring or weight that is suddenly released. When there is sufficient space, a more accurate and convenient method is to rotate the coil through 180°. Very weak fields, such as the earth's field and that due to conductors, may be measured in this way by using a sufficiently large coil of many turns.

More convenient than the ballistic galvanometer is a galvanometer whose moving coil deflects through a definite angle for each flux increment in a search coil connected to it. The total deflection then is the sum of all previous deflections. The instrument should have very low restoring torque and low moment of inertia. The instrument then will act as if it were critically damped, and the moving coil will remain for a long time in each position assumed. Thus, if preset at zero and then having the search coil "dipped" into a magnetic field and held there, the instrument will read the final field strength. Upon removal of the search coil from the field, the pointer will return to zero. In general terms, the deflection will be a measure of $\int e\,dt$, with e being the emf induced in the search coil by the rate of change of the flux linkages during the "dipping." Although faster "dipping" increases e, the time of the dipping motion decreases in proportion so that the integral $\int e\,dt$ remains the same for the same total change in flux linkages. Within a wide range of search-coil velocities, the instrument will read correctly on a scale calibrated in maxwell or gauss. An instrument of this type is known as the *Grassot fluxmeter.* See Ref. 3. See **Gaussmeter.**

In the *oscillating-magnet method,* a small, simple bar magnet is suspended by untwisted silk fibers. The magnet is set to vibrating through an angle of about 5°, and the period of oscillation is determined. The average of at least three observations should be taken. The method is suitable only for weak fields.

The *bismuth-spiral method* takes advantage of the fact that the resistance of bismuth wire increases when placed in a magnetic field. This property is utilized by noting the increase in resistance of a flat spiral coil of bismuth wire when placed in the field to be measured, the lead-in wires being arranged noninductively. The device is calibrated with known field strengths. It is particularly suited for exploring small air gaps, such as those in motors and generators. The considerable resistance-temperature coefficient of bismuth must be taken into account.

When a *magnetron,* which is a two-electrode electron tube in which the flow of electrons to the anode is controlled by a combination of crossed steady electric and magnetic fields in such a way as to produce ac power output, is placed in a magnetic field of which the direction is parallel to the axial cathode, the paths of the electrons become curvilinear. If the field is of sufficient intensity, the radius of curvature of the paths becomes so small that the electrons miss the anode altogether and return to the cathode. The plate current, which hitherto had remained practically constant as the field intensity increased, now falls abruptly to zero.

A *Hall-effect fluxmeter,* usually calibrated in gauss to measure magnetic induction or flux density, utilizes the distortion of the lines of flow of an electric current through a conductor under the influence of a magnetic field. Specifically, with the field vector and the direction of current flow at right angles to each other, electric charges are displaced along a third axis perpendicular to both. In certain semiconductors in which the conditions for the Hall effect are particularly favorable, a measurable voltage develops across the third axis. For constant current, this voltage is proportional to the flux density of the magnetic field in which the Hall element finds itself. The small size of these elements permits their insertion into the narrow flux gaps of indicating electrical instruments and similar devices. The Hall voltage is measured with an appropriately calibrated voltmeter. Also see **Hall-effect transducer.**

Magnetic-property Measurement: The magnetic properties of iron and steel which are of most commercial importance are (1) normal induction or permeability, (2) hysteresis loss, and (3) total losses (core loss) with alternating magnetizing forces of commercial frequencies. Instruments for measuring magnetic permeability are termed *permeameters.* See **Permeameter.**

The *hysteresis curve* of a material is a curve plotted between magnetic induction or flux density and the intensity of a magnetic field—for various values of magnetic-field intensity in the negative direction and back again, or through a complete cycle of values. The ends of a hysteresis loop will lie in the normal induction curve. The *ballistic step-by-step method* is most commonly used for determining hysteresis. This involves first adjusting the magnetizing circuit to a value corresponding to the maximum flux density. It then is reversed a few times to get the specimen in a cyclic state. Through use of appropriate switches, some resistance suddenly is cut into the circuit, thus reducing the magnetic-field intensity to a new value that is carefully noted. The corresponding change in flux density is determined with a ballistic galvanometer in the usual manner. The process is continued, step by step, until zero current is reached.

After this, the circuit is reversed and the resistance is cut out step by step until the same maximum value of magnetic-field intensity is reached in the opposite direction. The entire process then is repeated, and the other side of the loop is obtained. Errors are accumulative, but this is overcome in a second method in which the magnetizing current is brought back each time to the original maximum field intensity. Then it is a simple matter to ensure that the conditions remain constant by occasionally checking the value of flux density corresponding to the maximum field intensity.

A cathode-ray oscillograph can furnish hysteresis cyclograms with a deflecting coil carrying the magnetizing current while the deflecting plates are furnished with a voltage that is proportional to flux density. The hysteresis curve of sheet steel may be computed from waveforms of induced voltage and magnetizing current of a transformer in which the core is the material under test.

Other instrumental methods for determining hysteresis, hystereis loss, and core loss include the Robinson method, Epstein apparatus, standard solenoids, and Helmholtz-coil arrangements but are too detailed for description here. See Ref. 1 and 2. Several instrumental methods for measuring the foregoing magnetic quantities are described elsewhere in this volume. Particularly, see **Bridge circuits, electric; Fluxmeter; Galvanometer; Gaussmeter; Magnetometer;** and **Permeameter.**

REFERENCES

1. Fink, D. G., and J. M. Carroll: "Standard Handbook for Electrical Engineers," McGraw-Hill Book Company, New York, 1968.
2. Harris, Forest K.: "Electrical Measurements," John Wiley & Sons, Inc., New York, 1952.
3. Drysdale and Jolley: "Electrical Measuring Instruments," part II, pp. 453–454, Ernest Bonn, Ltd., London, 1924.

Magnetic-tape Storage Audio-frequency and other signals can be recorded or stored on plastic, paper, or metal tape that is coated or impregnated with magnetizable iron oxide particles. Recorded information in the form of magnetic variations can be removed from the tape by passing the tape through a strong, constant magnetic field (dc erase) or through a high-frequency alternating magnetic field (ac erase). Several tracks can be recorded on one length of tape.

Magnetic Transducer (See **Electromagnetic transducer.**)

Magneto (See **Tachometer generator.**)

Magnetoelastic Transducer (See **Electromagnetic transducer.**)

Magnetometer An instrument for measuring the magnitude and sometimes also the direction of a magnetic field, such as the earth's magnetic field.

Magnetometer, Airborne In order to use the magnetometer for airborne studies, special equipment is required to locate the airplane in space. This includes altimeters, either barometric or radar, to determine the altitude of the aircraft; cameras to photograph continuously the path of the aircraft over the ground, or electronic navigational equipment to record the plane's position relative to transmitting stations; and the means to correlate all these records. In making aeromagnetic surveys, the airplane is directed along a regular pattern of parallel lines at an elevation and spacing chosen to sample adequately the field to be measured. A series of cross or base lines are usually flown so that corrections can be made for the effect of diurnal variation and instrumental drift.

Aeromagnetic surveying was made possible by the development of self-orienting and direction-insensitive electronic magnetometers. When coupled with the means for determining the position of the airplane in space, magnetic maps can be made to study the magnetic field and its spatial variations and to find out how this field is distorted by the magnetic properties of nearby geologic materials. Airborne measurements can be made more rapidly and cheaply than surface measurements and can be made in areas where surface measurements are difficult or impossible.

Airborne magnetometers are of two types: self-orienting fluxgate instruments, which measure the variation in the magnetic intensity as the magnetometer is moved through space, and nuclear magnetometers, which measure the absolute total magnetic intensity.

Fluxgate magnetometers have a detecting element that is a small rod of Permalloy, a very easily magnetized metal, surrounded by a coil of wire in which an electronic oscillator impresses an alternating current. This is distorted by the effect of the external magnetic field. Electronic circuits amplify and measure these distortions, which are then recorded on a continuous paper chart calibrated in magnetic units. Three mutually perpendicular detecting elements are used, with two elements connected to motors so arranged that these two elements are continuously oriented at right angles to the magnetic field. The third element is then aligned parallel to the magnetic field, and its output represents variations in the total magnetic intensity. These variations are all that is required to study the local effect of the magnetic materials on the magnetic field.

Measurement is accomplished by supporting the entire detector assembly and associated orienting system as a pendulum. Synchrotransmitters geared to two orienting motors are used to measure directly the inclination and declination of the earth's field with reference to the heading of the aircraft. When relative to a celestial body, this heading is determined with a sextant which is geared like a synchrotransmitter. Fluctuations in angular measurements are removed by a circuit which automatically averages the observed values over any period, usually about 100 s.

Nuclear Magnetometers: These instruments have been developed for airborne measurements by utilizing the magnetic properties of atomic structures. The two types, proton precession and alkali vapor, both have the advantage of being insensitive to the direction of the field and of producing absolute measurements in terms of frequency, the physical quantity most easily measured to high accuracy.

The *proton-precession magnetometer* utilizes a large magnetic field to orient the magnetic fields of hydrogen protons in water or hydrocarbons. When this field is cut off, the protons precess about the direction of the field to be measured. The frequency of their precession depends on the intensity of the magnetic field. By measuring this frequency, the absolute magnetic intensity can be determined.

Alkali-vapor magnetometers make use of the fact that if circularly polarized light is shone through vapor of an alkali metal which is being excited by a varying radio-frequency magnetic field, light is absorbed at certain frequencies which are related to the energy levels in the atoms of a steady external magnetic field. By measuring the absorption frequencies, it is possible to determine continuously the external magnetic field.

The magnetic intensities measured during the field survey, whatever the type of instrument used, must be compiled—usually as a map. The flight path of the airplane is plotted on a map; the magnetic records are adjusted to a common magnetic datum and reduced to profiles at the chosen horizontal and vertical scales. Magnetic values are read from these profiles and are plotted on the map of the flight path. Lines connecting all points of the same magnetic intensity are drawn so that the final map is a contoured representation of the magnetic intensity—either total or that of one of the components, depending on what data were measured.

The data collected in an aeromagnetic survey are the same as those obtained by ground methods; therefore, interpretation of maps and profiles involves the same fundamental theories as have been applied to the results of ground surveys. The aim of the interpretation is to detect irregularities or anomalies in the magnetic field as mapped and to determine, if possible, the depth and quality of the geologic feature responsible for each anomaly.

Airborne magnetometers are ideally suited for making reconnaissance surveys of large, relatively flat areas which are not easily accessible by surface transportation. Good maps and photographs must be available, or the survey must be located where electronic navigation can be easily used. Unless a helicopter is employed, the airborne magnetometer can seldom be applied economically to projects covering less than 25 sq miles, or for projects requiring locational accuracies better than plus or minus 50 ft. It is of limited usefulness for low-level surveys in mountainous regions, or for surveys of complex and detailed anomalies produced by shallow magnetic deposits.

—F. E. Romberg, *Texas Instruments Incorporated*

Magnetometer, Metastable-helium Developed for spaceborne instrumentation where requirements involved measurements of changes in planetary magnetic fields accurate to 1 part in 5 million and sensitivity of 0.01 gamma, the first metastable-helium magnetometer was produced in 1960. It represented improvement of an order of magnitude over previously available instrumentation for this application.

Installed aboard the Mariner IV in its historic, picture-taking flight to Mars, the metastable-helium magnetometer was responsible for the determination that, unlike earth, Mars is a solid-core planet.

In operation, the system measures the absorption of infrared radiation by helium in a metastable state. Mechanically, a helium spectral-emission lamp and a helium-filled absorption cell are excited by a radio-frequency source. A sensitive infrared detector then measures the amount of infrared radiation absorbed by the metastable helium. The amount of absorption is proportional to variations in planetary magnetic fields, thus providing an accurate measurement.

Current applications of the metastable-helium magnetometer, in addition to space survey, include military projects such as antisubmarine-warfare detection systems.

—Howard Cannon, *Texas Instruments Incorporated*

Magnetometer, Rubidium-vapor Classified as an absolute-type magnetometer, rubidium-vapor instruments were developed in 1958 by United States government scientists. Included in the development project were both the Coast and Geodetic Survey and the National Bureau of Standards.

In operation, the rubidium-vapor magnetometer utilizes a rubidium lamp, which is mounted in the tank coil of a radio-frequency oscillator. After collimating and filtering, the rubidium light is circularly polarized and passed through a rubidium-vapor cell, then focused on a sensitive photocell. By a pumping process, the rubidium in the vapor cell is brought to the $m=+3$ level. Application of a weak alternating magnetic field perpendicular to the ambient field will result in a redistribution of atoms to all the m sublevels. Once out of the $m=+3$ level, the atoms are again capable of absorbing D_1 light. During the pumping process, the cell containing the rubidium vapor loses a substantial amount of its transparency. The transmission changes are observed by focusing the light passed by the cell upon a photodetector. By a process of modulation, amplification, and phase shift, the process becomes self-exciting, and the frequency of oscillation will change as the ambient field changes.

Various combinations of amplifier parameters and rubidium isotopes allow considerable range in the measurement of ambient magnetic fields. Since the total world range is from 15,000 to 80,000 gammas, any system capable of this span is usable anywhere on the earth.

—Howard Cannon, *Texas Instruments Incorporated*

Magnetoresistive Transducer (See **Electromagnetic transducer.**)

Magnetostrictive Delay Line (See **Delay line.**)

Magnetron [See **Magnetic quantities (variables).**]

Main Frame [See **Central processing unit (CPU).**]

Manipulator A device or mechanism designed to perform useful work as directed by a human operator. The distance from the operator to the work area may vary from directly adjacent to any desired distance. Work capacity may range from microscopic objects to tasks requiring superhuman strength. The mechanism may vary from a simple linkage—to complex servocontrol with *sense-of-feel* feedback to the operator—to fully remote control by electrical or radio link with closed-circuit television.

A manipulator with a single arm and hand, using direct linkage, can perform a large variety of individual tasks under control of an operator. For work with radioactive materials or hazardous chemicals at a distance from the operator, a manipulator may consist of a pair of mechanical arms and hands connected electrically to a harness worn by the operator. Hydraulically powered actuators, responding to electrical signals, cause the arm and hand to carry out precisely the same motions as those made by the operator. By concentrating on the manipulation task, the operator observes the actions of the mechanical arms and hands as if they were his own. In this way, an operator can remain protected while performing operations in a hostile environment.

A submarine manipulator, consisting of an electrically powered, articulated arm and hand mounted on an undersea craft with work area viewed by closed-circuit television, may perform salvage or research operations at great depths or distances from the operator. In the medical field, a harness worn by a patient who has suffered paralysis or loss of muscle control permits amplification of weak motions still available to control a power device for performance of specific tasks, such as propelling a wheelchair.

Use of feedback information about position and force by the operator—to provide a *sense of feel*—makes possible manipulators with great power under safe control with the ability to move loads rapidly with great dexterity. Such machines represent a considerable potential economy in the handling of materials and offer the prospect for performing tasks that were previously considered impossible.

When programmed for automatic performance, without an operator, of a task consisting of specific repeated operations, a manipulator becomes a robot. See also **Robot.**

—Harvey A. Senior, *Unimation, Inc.*

Manometer An instrument for *directly* measuring positive pressure, vacuum, and differential pressure; and *indirectly* for measuring flow. Reportedly the oldest known technique for air-pressure measurement, the manometer still is universally recognized as a standard for the calibration of other highly accurate pressure-measurement instruments. Manometers are inherently simple in design and require no mechanical linkage to impair accuracy or to detract from reliability.

The manometer operates on the fundamental principle of displacing a liquid column by the unknown pressure force to be measured. The liquid column is of three main configurations: (1) a U tube as shown in Fig. M-3, (2) a vertical tube, and (3) an inclined tube as shown in Fig. M-4.

U-tube Manometer: When no differential pressure is applied to the device, the two columns of liquid are of equal height and rest at the zero point on the scale. A differential pressure will cause the column on the high-pressure side to fall, the column on the low-pressure side to rise. The value of the differential pressure is the algebraic sum of the negative reading of one column and the positive reading of the other column. U-tube manometers are available with (1) a glass or plastic indicating tube, (2) a dustproof steel case, (3) a white epoxy-coated nonglare scale, (4) a micrometer-type zero adjustment, and (5) a standard pressure rating for 70 psig, or optional readings to 1,200 psig.

Vertical Manometer: This is the most popular style of manometer for measurements above

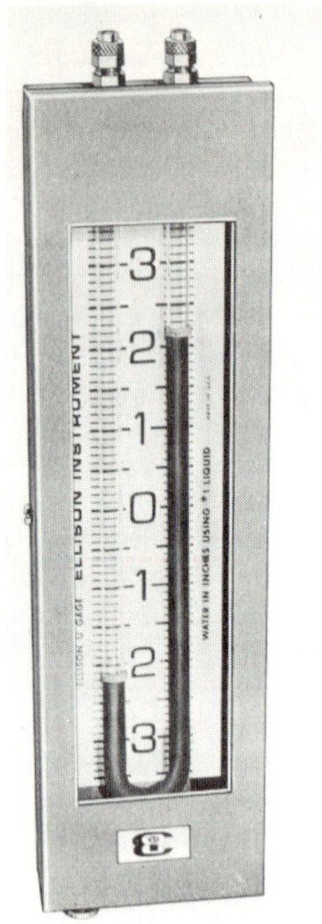

Fig. M-3. U-tube manometer. *(Ellison.)*

12 in. of water. A direct-reading scale facilitates fast, accurate readings because comparison of two column heights is not required. A wide variety of indicating liquids is available for single-source or differential-pressure readings up to 49 psi.

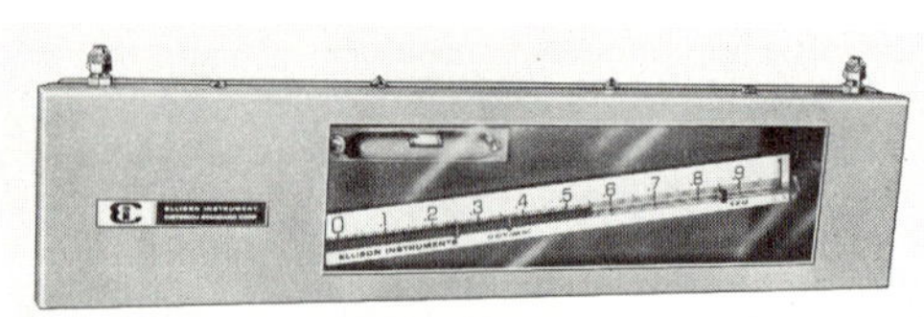

Fig. M-4. Inclined manometer. *(Ellison.)*

High-pressure vertical manometers also are available up to 1,200 psi. Convenience factors include (1) square-root scales for direct reading of flow measurements, (2) movable indicating markers, and (3) multiple tube and scale assemblies for reading several associated pressures or flows. Well-engineered vertical manometers will have a consistent accuracy of from ±½ to 1 percent of full scale.

Inclined Manometer: This device also is called a micromanometer, slant-tube gage, or draft gage. The inclined manometer features an expanded scale of fine readings down to 0.001 in. of water and an accuracy within ±½ to 1 percent full scale. The instruments usually are selected for single-source or differential-pressure readings below 12 in. of water. They are available in a wide variety of scale lengths with numerous mounting styles. Convenience features include (1) special leveling points—factory calibrated, (2) a spirit level for fast leveling, (3) means for easy field cleaning or replacement of glass tubes without disconnecting the instruments, (4) adjustable limit markers, and (5) micrometer-type zero-adjustment knob.

Manometer with Photoelectric Pickup: To enable remote indication and alarming of measured pressures and flows, glass-tube vertical and inclined manometers can be equipped with a

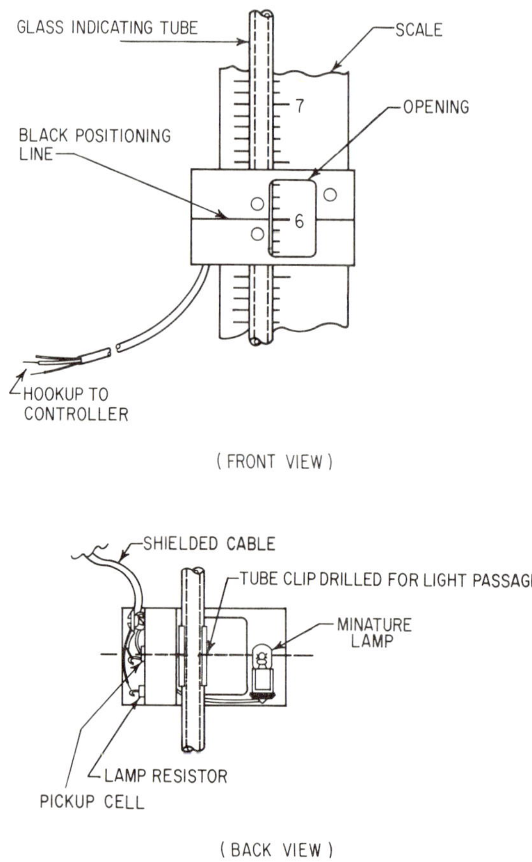

Fig. M-5. Photoelectric pickup used with vertical and inclined manometers. *(Ellison.)*

"snap-on" pickup of the type shown in Fig. M-5. The pickup may be positioned at any point throughout the scale range of the manometer by simply sliding it on the tube. The pickup will sense all indicating liquids provided that they are slightly colored.

—P. D. Dieterich, *Ellison Instrument Division, Dieterich Standard Corporation*

Manual-Automatic Transfer (See Display-command operator/machine interface.)

Mask A digital-computer feature controlled by the digital-computer program which allows the servicing of an interrupt to be delayed until such time as it is convenient (See **Interrupt; Interrupt response time**). Process-control computers require an interrupt feature in order to be able to recognize certain phenomena occurring in the process. There are times, however, when it is desirable to delay the recognition of a process interrupt by the digital computer. A mask instruction in the digital computer allows the recognition of specific interrupts to be delayed until it is convenient to service them.

—B. W. Landeck, *IBM Corporation*

Mass, Standard of (See Standards, measurement.)

Mass Flowmeter (See Flowmeter, mass.)

Mass Spectrometer A general quantitative and qualitative analyzer for most components in all types of samples—gas, liquid, or solid, but with some volatility limitations. A complete analysis is obtained from submicrogram samples in a few seconds or minutes. The range is from ppm or ppb to 100 percent purity; accuracy is ±1 percent; the specificity is good. The instruments are moderately complex and costly.

Operating Principle: With reference to Fig. M-6, the sample to be studied is introduced into an evacuated area (ion source), where it is ionized, accelerated by an electrostatic field, and separated according to mass, and the various masses are collected and measured.

Production of Ions: Several methods are used: (1) bombardment with electrons from a heated filament; (2) direct emission of ions from a solid sample that is deposited on a heated filament (surface ionization); (3) vaporization from a crucible and subsequent electron bombardment (e.g., Knudsen cell for high-temperature studies of solids); and (4) radio-frequency spark bombardment of sample for ppb elemental analysis of solids, as encountered in metallurgical, semiconductor, ceramics, and geological studies. Ions also are produced by photoionization, laser ionization, and electrostatic-field ionization.

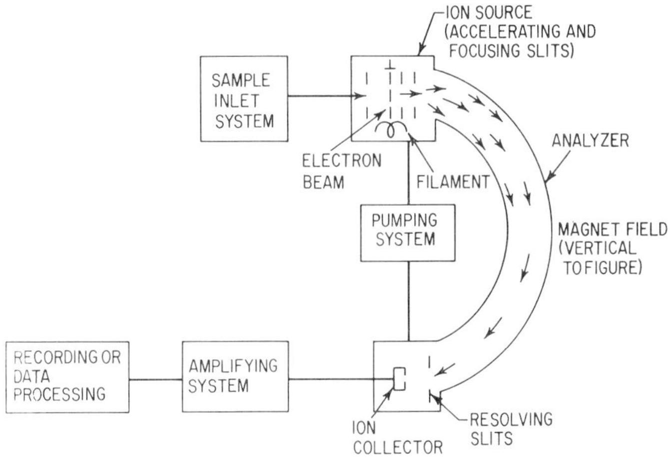

Fig. M-6. Basic elements of mass spectrometer.

Fragmentation: Ionization usually is accompanied by partial fragmentation of the molecule. The fragmentation pattern is constant for a specific molecule and operating conditions. Fragmentation complicates computation but permits distinguishing between isomers and gives molecular-structure information.

Ion Separation: After acceleration, the ions are focused and separated according to mass. The most common separating means is a magnetic field which causes the ions to follow curved paths of radii proportional to their masses. Many different geometries are used. The masses may be scanned by varying the accelerating voltage or magnetic field. Other separation means include (1) combinations of electrostatic and magnetic fields (double focusing and cycloidal focusing), (2) crossed alternating electrical fields (quadrupole mass filter), and (3) use of a field-free drift tube combined with pulsed ion source and gated detection (time of flight). Still additional means are omegation, radio frequency, and cyclotron resonance.

Detection: Commonly used detection means are: (1) *Electrical*—Ion beams are successively scanned across a collector where they pick up electrons. The resulting current is amplified with an electron multiplier and/or electrometer and recorded or computer-processed. (2) *Photographic plate*—Ions strike a photographic plate, activating the emulsion and thus giving a line for each mass after development. Line intensity is proportional to ion abundance. All masses are recorded simultaneously. Automatic data-processing equipment is available for interpreting the photoplate information.

Sample Introduction: A variety of inlet systems are available for gases, liquids, and solids: (1) *Heated batch inlet*—The sample is expanded into a volume at approximately 50 microns pressure and bled into the ion source through a molecular or viscous leak. This method will handle gases, liquids, and solids with vapor pressures above approximately 1 torr at 350° C. (2) *Direct introduction system*—The sample is inserted directly into the ion source through a vacuum lock on a heatable probe. Liquids and gases with vapor pressures above approximately 10^{-7} torr at 350° C can be handled. (3) *Direct insertion by venting*—or through a vacuum lock (spark or surface ionization). This method is well suited for involatile samples. (4) *Continuous capillary inlet*—very useful for monitoring processes or identifying gas-chromatograph effluents.

Mass-spectrometer Characteristics: Several of the common types of mass spectrometers are listed, together with their major characteristics, in Table M-1.

TABLE M-1. *Types and Characteristics of Mass Spectrometers*

Generic Class	Type	Major Uses	Cost, Approx thousands of dollars
Leak detector.	Magnetic analyzer (helium only)	Detects small leaks, using helium-gas tracer	4
Residual-gas analyzer	Magnetic or quadrupole analyzer	Analysis of gases in vacuum systems	5–8
Low-resolution (approx 1/50 to 1/250)	Magnetic, cycloidal, or quadrupole analyzer	Analysis of gases or light liquids	7–20
Medium-resolution (1/2,500)	Magnetic analyzer	Identification, molecular-structure studies, mixture analysis, isotope ratios	35–50
High-resolution (1/50,000)	Magnetic and electric (mass and energy) focusing	Molecular-structure determination, ppb solids analysis, isotope ratios, high-temperature chemistry	120

Uses and Fields of Application: Common uses include (1) compound identification, (2) elemental-formula determination, (3) molecular-structure determination, (4) quantitative mixture analysis, (5) ppb solids elemental analysis, (6) isotope-ratio determination, (7) leak detection (helium tracer), (8) residual-gas analysis in vacuum systems, and (9) age dating. Mass spectrometers are widely used for studies and determinations in organic chemistry; petroleum and petrochemistry; medical, pharmaceutical, and biological laboratories; nuclear investigations; geochemistry and cosmochemistry; metallurgical, semiconductor, and ceramics investigations; pollution control; space programs; agricultural and pesticide research and manufacturing; flavors and fragrances chemistry; and as a basic research tool in studies of ion-molecule reactions, high-temperature chemistry, kinetics, free radicals, and thin films.

A high-resolution mass spectrometer is shown in Fig. M-7.

REFERENCES

Ahearn, A. J.: "Mass Spectrometric Analysis of Solids," Elsevier Publishing Company, Amsterdam, 1966 (spark source).

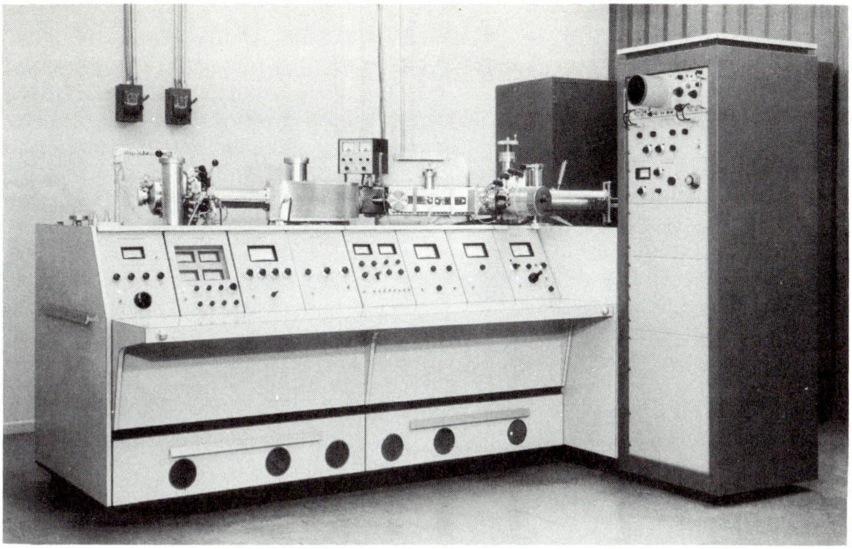

Fig. M-7. High-resolution double-focusing mass spectrometer.

ASTM Committee E-14, Proceedings of Annual Conferences.

Biemann, K.: "Mass Spectrometry—Organic Chemical Applications," McGraw-Hill Book Company, New York, 1962.

Beynon, J. H.: "Mass Spectrometry and Its Applications to Organic Chemistry," Elsevier Publishing Company, Amsterdam, 1960 (comprehensive text).

Kiser, R.: "Introduction to Mass Spectrometry," Prentice-Hall, Inc., Englewood Cliffs, N.J., 1965.

McLafferty, F. W.: "Interpretation of Mass Spectra," W. A. Benjamin, Inc., New York, 1966.

—E. E. Escher, *Instrument Products Division, E. I. du Pont de Nemours & Company, Inc.*

Maxwell Bridge (See Bridge circuits, electric.)

Maxwell-commutator Bridge (See Bridge circuits, electric.)

Measuring Machine A highly instrumented device used to measure and record linear dimensions and *X-Y-Z* coordinates of holes and surfaces in parts and tools. Various styles of these machines operate under manual, motorized, or numerical control. The mode of measurement may be direct, by deviation, or a combination of these two methods. Direct-measuring digital-readout machines are the more common. These machines provide direct or absolute measurement of probe position or movement in three axes to 0.001 or 0.000050 in. revolution. Computer-aided measuring machines generate tapes for use on numerically controlled machine tools. Coordinate data from a prototype model measured on the machine or fed into the machine from a drawing are combined with numerical tool-code information and a numerical-control tape produced by the input/output accessory to the measuring machine.

A typical measuring system consists of three basic components for each axis of measurement: (1) a steel grating, (2) a reading head, and (3) the readout. The grating and reading head are mounted so that the short glass grating in the reading head is superimposed over the steel grating at a slight angle. A beam of light is passed through the glass grating, reflecting back from the steel grating onto four photocells. The relative movement of the two gratings produces optical patterns (moire patterns) or fringes. As these fringes move, they interface with the beam of light being reflected from the steel grating. Photocells in the reading head detect the interference fringes and convert the changes in light intensity into electrical signals, thus counting the fringes and indicating the precise amount of movement by digital display.

Mechanical Integrator (See **Integrator.**)

Medical Instrumentation Many conventional instruments that are designed for general use find wide application in medical and biological research and to some extent in clinical applications. Descriptions of these general kinds of devices will be found throughout this encyclopedia. There are scores of special medical instruments, only some of which can be described in this volume. Some of these devices are described here. Others will be found under **Cardiography; Cardiovascular instrumentation; Central nervous system, instrumental analysis of; Ear, instrumental measurements of; Eye, instrumental measurements of; Hearing, sensory aids for; Pulmonary system, instrumental analysis of; Ultrasonics, medical instrumental use of;** and **Vision, sensory aids for.**

Patient Monitoring (Ref. 1): In the "intensive-care units" or "coronary-care units" of most hospitals, monitoring installations typically display and record blood pressure, respiration rate, electrocardiogram, and temperature. System outputs vary from simple displays on lighted boards at the nursing station to paper-tape and magnetic-tape recordings—and on to automatic computer analysis of variations in the parameters being measured. Though there is general agreement that patient monitoring is useful, agreement does not exist as regards which parameters should be recorded, the precision and speed with which the desired parameters should be observed, or the method of presentation and storage of information.

Prosthetics: A prosthetic device replaces a missing body part in either appearance, function, or both. The field thus deals with artificial limbs and organs. As examples of prosthetic devices, this description discusses artificial powering and control of limbs.

To date no effort has been made to develop external powering for legs, because large body muscles are usually adequate and the amount of external power required would be prohibitive. However, much work has gone into the external powering of artificial arms because of the lack of readily available muscles for "natural" powering and because of the desirability of automatic control of arm and hand positioning, which in turn, requires control of the power source.

Arm amputations are classified as below-elbow, above-elbow, and shoulder disarticulation. Unfortunately, the higher the level of amputation, the greater the need for external powering and control, the more power is required, and the more degrees of freedom for control are necessary.

Electric, pneumatic, and hydraulic power have all been employed, with no clear advantage emerging. Electric power is most amenable to electric control, but the transducers are relatively bulky. Pneumatic power uses pistons smaller than solenoids as actuators, but the tubes connecting them with the controls and the power source are larger than wires, and the energy storage per pound is not as efficient with compressed air as it is with batteries. Hydraulic control has the advantages of higher speed and quieter operation than the other methods, but the high pressure required often produces leaks.

The minimum desirable motions to control are thumb-to-forefinger grasp, rotation at the wrist joint, flexion and extension at the elbow joint, and several motions at the shoulder joint.

Because of the large amount of power required and the difficulty of control, the shoulder joint has not yet been powered, but a minimum of six control points is required to produce the other motions listed above.

One control method uses switches placed at various sites around the neck and shoulders which the subject can learn to operate without too much attention. However, the process does not become fully automatic; the subject must always put close attention on operating his prosthetic device.

Because of these problems, electrical signals emanating from controlling muscles (electromyograms, or EMG) are often used for control. Electromyographic voltages are on the order of 20 μV to 5 mV and contain frequency components from 10 to 100 Hz. They can thus be handled by the same electronic techniques described in connection with the EEG and the ECG. See **Cardiography.** With proper electrode placement, the hum-pickup level can be kept low enough to be unobjectionable. The problem then becomes one of electrode placement in several adjoining locations so that there will be no cross talk between the adjacent sites (Refs. 2 and 3). Several prostheses requiring only one or two degrees of freedom have been built using electromyographic control. One group (Ref. 4) has explored in detail the different frequency contents of electromyograms issuing from closely adjoining muscle bundles and has developed electrical filtering techniques for separating them by computer analysis.

Orthotics: Orthotic devices aid still existent but defective body parts. A brace is an example. Patients who because of paralysis are confined to beds or wheelchairs are in need of a series of aids to let them achieve some measure of independence. Many interesting types of wheelchairs have been developed in the attempt to solve the problems of going up and down stairs and crossing streets, but most of them are heavy, expensive, and too elaborate to be practical.

Various individuals throughout the world have developed many ingenious devices for typing, turning pages, drinking liquids, dialing and answering telephones, and the like—by means of eye blinking, operating switches with a little finger or a toe, or blowing through a tube. Some of these efforts are reported (Ref. 5), but little organized effort has been directed toward solving these problems in a systematic way.

Sensory Aids: Sensory aids include hearing aids, guidance and reading devices for the blind, and special tactile-replacement devices for sensorially paralyzed limbs. Also see **Hearing, sensory aids for**; and **Vision, sensory aids for.**

REFERENCES

1. *Proc. Natl. Conf. Coronary Care Units,* Government Printing Office, Washington, March, 1968.
2. Godden, A. K.: The Techniques of Myoelectric Control of Prostheses and the Prospects of This Type of Control for Thalidomide Casualties, *University of Oxford Rept.* 1048-68, 1968.
3. Bottomley, A. K.: Design Considerations for the Prosthetic Prehension Device, *Proc. Intern. Symp. External Control of Human Extremities,* Dubrovnik, 1966.
4. Wirta, R. W., and D. Taylor, Jr.: Development of a Multiple-axis Myoelectrically-controlled Prosthetic Arm, *Proc. 3d Intern. Symp. External Control of Human Extremities,* Dubrovnik, August, 1969.
5. Rehabilitation Gazette, Box 149, Chagrin Falls, Ohio 44022
6. Alt, F.: Advances in Bioengineering and Instrumentation, *ISA Publ.* AB11, 1966.
7. Allison, R. D.: Dynamic Bioinstrumentation, *ISA Publ.* DB, 1968.
8. Weber, T. B., and J. Poyer: Instrumentation Methods for Predictive Medicine, *ISA Publ.* IMPM, 1966.
9. Thomas, F. D.: Biomedical Sciences Instrumentation, vol. 6, *Proc. 7th Natl. ISA Biomedical Sciences Instrumentation Symp.,* Ann Arbor, Mich., *ISA Publ.* BM69, 1969.
10. *Proc. 20th Ann. Conf. Engineering Medicine and Biology,* Boston, Mass., *ISA Publ.* EMB67, 1967.

—J. Malvern Benjamin, Jr., *Bionic Instruments, Inc.*

Megger[1] A high-range ohmmeter having a hand-driven dc generator as its voltage source and used chiefly for measuring insulation-resistance values in megohms.

Melting-point Determination An instrumental technique for determining the purity of salts, polymers, waxes, resins, and other materials that melt without decomposition. Essentially, this is a batch method. The identity of impurity in a sample cannot be determined. However, many pure materials are specified by their melting point. The range of usefulness generally is from 0 to 5 percent impurities. The amount of impurity is related to the deviation of the melting point from that obtained with the pure material. Determination is made by measuring the sample temperature with an embedded temperature-sensing element as the ambient temperature is raised and observing temperature discontinuities or sample appearance to detect melting. Determinations usually require from 10 to 30 min with a probable error of ± 2 percent.

Memory, Analog An analog-memory circuit is a circuit which is capable of maintaining an output voltage or current proportional to an input signal even after the input signal is removed. It is analogous to digital-memory circuits such as the latch or flip-flop except that the output value may be maintained at any level in a predefined range. Although a sample-and-hold amplifier is a form of analog memory (see **Amplifier, sample-and-hold**), the term analog memory is often restricted to devices which retain an output value for long periods of time, usually greater than several minutes, whereas most sample-and-hold amplifiers are designed to hold a value accurately for only a fraction of a second.

Analog-memory circuits can be fabricated from a number of different devices. The stepping motor is an electromechanical device which can be used as an analog memory when combined with a potentiometer. The armature of a stepping motor is positioned and controlled in discrete steps by applying pulses of varying duration to the motor. Thus, analog information in the form of pulse-duration signals can be converted into shaft-position information. Once the armature has been positioned, it remains fixed until another pulse command is received. By coupling the armature to a potentiometer shaft, the shaft

[1] Trademark of James G. Biddle Co.

position is converted into an electrical quantity, specifically a resistance. The value of the resistance can be sensed and used to control an analog-output voltage or current. Stepping-motor control of set point is provided on several commercially available analog controllers. The stepping motor is controlled by pulses generated by a digital process-control computer or control unit.

An electronic integrator circuit can be used as an all-electronic analog-memory circuit. The integrator may consist of a standard operational-amplifier integrator configuration. When a signal is applied to the integrator, in the form of either pulse-width-modulated signals or an analog voltage, the integrator output increases or decreases depending on the polarity of the signal. When the input signal is removed, the integrator output remains constant at its last value. This analog signal may be used directly, or it may drive a voltage-to-current converter circuit to provide a current-output signal for controlling process actuators.

Another electronic integrator circuit suitable for use an an analog memory is shown in Fig. M-8. The circuit consists of an input multiplexing switch S_1, a capacitor, and a field-effect transistor (FET). A buffer or voltage-to-current amplifier provides the output signal. In operation, an analog voltage signal, often derived from a

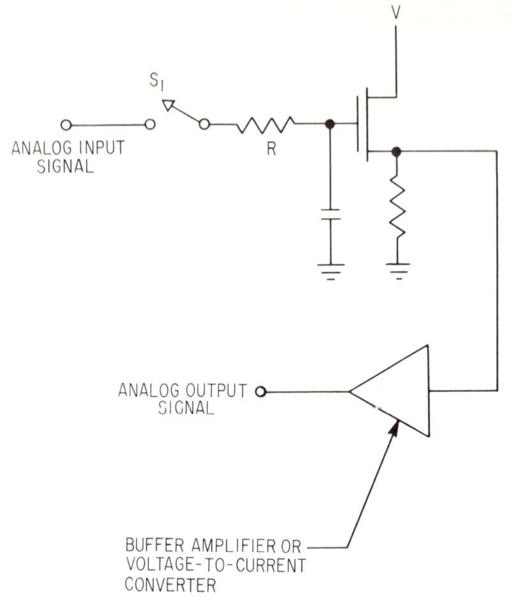

Fig. M-8. FET-capacitor analog memory.

digital-to-analog converter, is applied to the input (see **Digital-to-analog converter**). The switch S_1 is closed under control of the digital process-control computer or control unit, and this allows the capacitor to charge or discharge until the voltage across the capacitor equals the input volt-

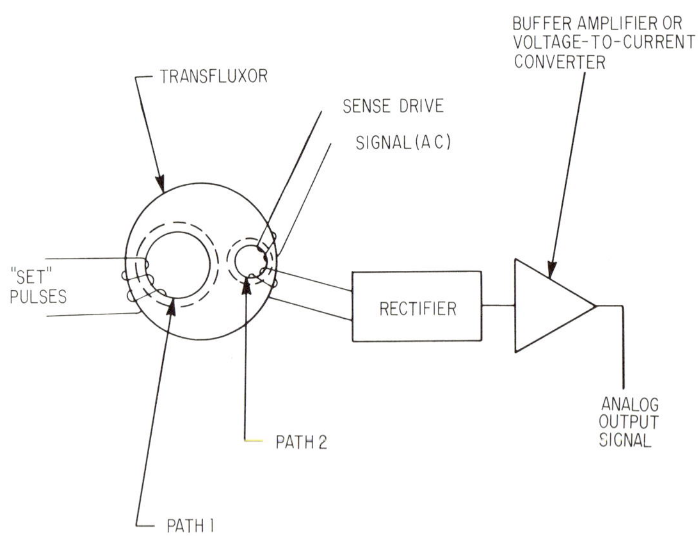

Fig. M-9. Transfluxor analog memory.

age. The switch S_1 is then opened. Since the input inpedance of the FET stage is extremely high, usually greater than $10^9\ \Omega$, the capacitor remains charged. The voltage across the capacitor is sensed by the FET circuit to provide an input to the buffer amplifier or voltage-to-current converter. Some charge leaks off the capacitor because of the finite FET gate current and other leakage resistances which exist in any physical circuit. However, with careful design and construction techniques, this can be reduced to a very small value. Circuits of this type have been built which exhibit less than 0.1 percent/hr change in output voltage even under relatively high humidity conditions.

An analog-memory circuit based on the use of a magnetic core is illustrated in Fig. M-9. The multiaperture core, called a transfluxor, is wound with three windings. When control pulses are applied to the "set" winding, the inside perimeter of the large aperture is magnetized in one direction. Since this perimeter includes the leg between the large and the small aperture, this magnetization controls the reluctance of the magnetic path around the smaller aperture. If a sense-drive signal is applied to a winding wound through the smaller aperture, the voltage induced in another winding through the same aperture depends on the magnetization of path 1. Thus, the output voltage is a function of the pulses applied to the "set" winding. The output voltage is rectified and buffered by a buffer amplifier or voltage-to-current converter circuit to provide an analog-output signal.

An electronically controlled resistor is another example of an element which can be used for analog-memory circuits. An example is the Memistor (registered trademark, Memistor Corp.) device (Fig. M-10), which essentially consists of a screened resistive substrate immersed in a solution of copper sulfate. A copper electrode is also immersed in this solution. By applying a dc potential between the resistive substrate and the copper electrode, a thin layer of copper is electrolytically plated onto the substrate, thereby changing its resistance. By reversing the polarity of the plating voltage, the substrate can be "deplated"; that is, the operation is reversible. If the digital computer or control unit is used to control the switch S_1, the resistance of the Memistor can be made proportional to a digital value stored in the computer. The value of the resistance is sensed by means of an ac voltage source to prevent the sensing current from altering the value of the resistor. The output voltage, derived from a current-sensing resistor in the output circuit,

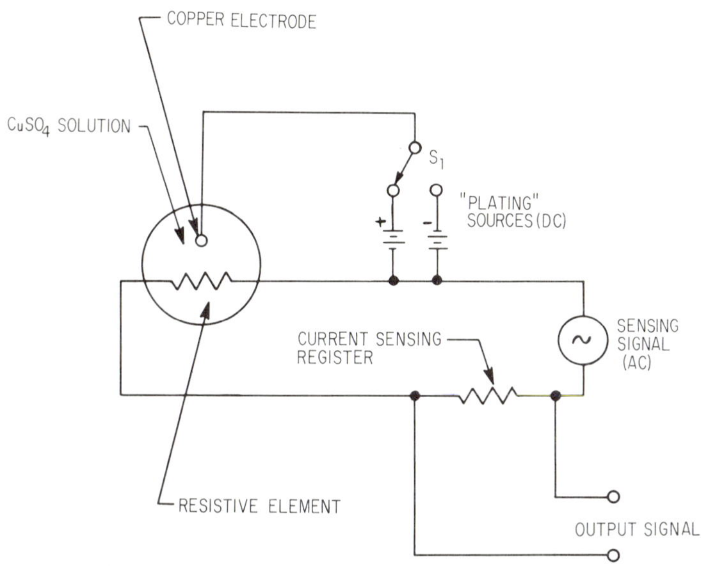

Fig. M-10. Memistor analog-memory element.

can be rectified and used to drive a buffer amplifier or voltage-to-current converter.

Analog-memory circuits are used in the analog-output subsystem of process-control computers to eliminate the need for digital-to-analog conversion equipment for each analog output (see **Analog output**). The memory circuits are controlled either by the digital-output function provided by the computer or by a single digital-to-analog converter which is time-shared among the analog-memory circuits. Although the frequency response of the analog output is reduced by this time sharing, many applications require only limited speed of response, and the use of analog-memory circuits can appreciably reduce the overall system cost.

—Thomas J. Harrison, *IBM Corporation*

Memory-access Time (See **Access time, computer.**)

Memory-cycle Time Memory- or storage-cycle time is the interval of time required for a complete memory operation. More explicitly, it is the minimum allowable time interval between two successive accesses to storage. (See **Storage, computer**; compare **Access time, computer.**)

In magnetic-core storage, cycle time consists of the time required to read the data from the core and the time required to regenerate or write the data back into the core. In a nondestructive readout storage, such as monolithic storage (see **Monolithic memory**), it is either the time required to read the selected location or the time required to write into the selected location.

—W. W. Welz, *IBM Corporation*

Mercury Barometer (See **Barometer.**)

Mercury Switch A mechanically operated switching device in which two or more electrodes are bridged by a pool of mercury to effect circuit closure. Usually the electrodes are sealed in a glass or metal enclosure. In the latter instance, the enclosure may act as one electrode; but in either instance the sealed construction makes the mercury switch valuable in applications involving severe environmental conditions. Relatively immune to ambient conditions, mercury switches may be used from −35 to over 400° F and within pressure ranges from high vacuum to several hundred pounds per square inch. The switches normally are filled with hydrogen gas to keep the mercury pool from oxidizing and to increase the load-handling capability.

Metastable-helium Magnetometer (See **Magnetometer, metastable-helium.**)

Meteorological Instrumentation Meteorology is the science of the atmosphere. There are many subdivisions, including (1) dynamic meteorology, (2) physical meteorology, (3) climatology, (4) synoptic meteorology, (5) aeronautical meteorology, (6) maritime meteorology, (7) agricultural meteorology, (8) hygrometeorology, (9) medical meteorology, and (10) aerology. Although there are many special instruments designed specifically for meteorological use, in most cases these instruments utilize the same basic principles used in instruments for other applications. A number of meteorological instruments are described elsewhere in this volume. Particularly see **Anemometer; Aneroid barometer; Barograph; Hygrometer; Hygrothermograph; Precipitation gage; Psychrometer; Radiosonde; Rawin set;** and **Weather balloon.** Also consult the alphabetical index.

Meter, Electrical An instrument for measuring one or more of the several fundamental electrical quantities, such as current and voltage. Operating characteristics, ranges, and other specifications are given under **Electrical quantities, measurement of.** See also specific meter and instrument types, such as **Ammeter, Voltmeter,** and **Wattmeter.**

An electrical meter is a *self-contained* device for measurement as contrasted with an array of devices and instruments that may be required for making measurements in the laboratory with greater precision. The electrical meter is designed for the day-to-day requirements of industrial and commercial service with ruggedness and long life as major design parameters. Electrical meters are designed for panelboard or wall mounting to maximize the convenience of reading. Also available are standard lines of portable meters that are useful as basic tools for field testing, multimeters for service-shop practice, and laboratory standards and substandards for checking meters of lower accuracy.

Meter Formats: The majority of electrical meters in use are of the analog type rather than digital, although the preference for digital displays is increasing. Analog-type meters tend to be considerably lower in cost and also offer other

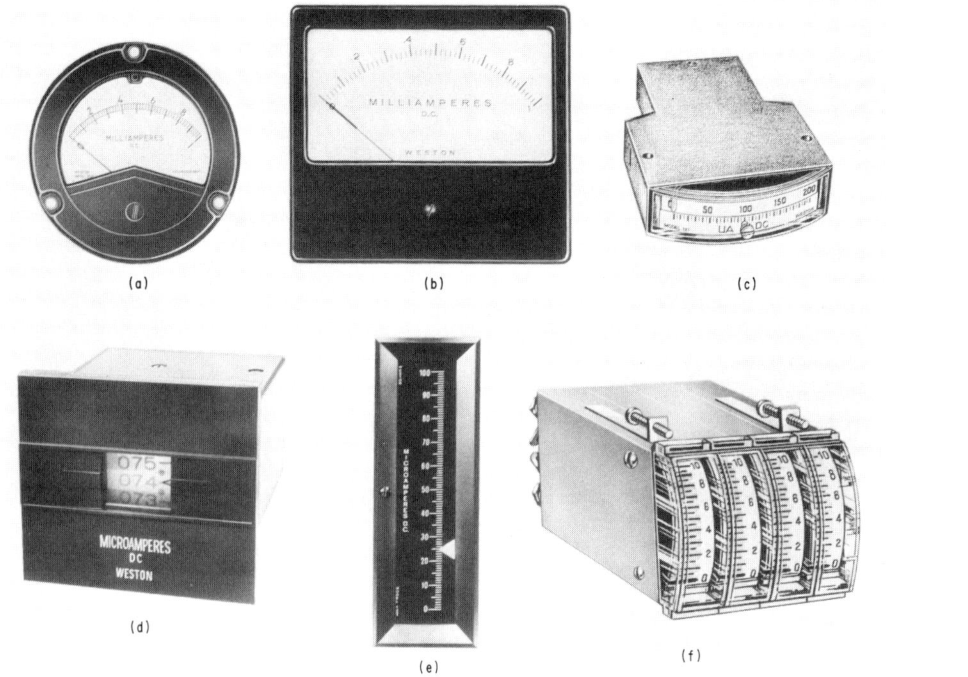

Fig. M-11. Format of standard electrical meters for panel and switchboard mounting: (a) round case with $3\frac{1}{2}$-in. scale; (b) rectangular case with $3\frac{1}{2}$-in. scale; (c) horizontal indicator with $2\frac{1}{2}$-in. scale (panel cutout is approximately $\frac{1}{2}$ by $1\frac{3}{4}$ in.); (d) square case in which moving scale (equivalent to 30 in.) is projected optically in a $3\frac{3}{4}$-in. coated window with digital readout; (e) rectangular case in which moving pointer is projected optically over a $4\frac{1}{2}$-in. scale; and (f) vertical-scale moving-pointer meters ($1\frac{1}{4}$-in. scale) mounted side by side to provide ease for comparative readings. *(Weston.)*

advantages, including an ability to display drift and other measurement changes better. Largely dependent upon accuracy and readability requirements, electrical meters for panel and switchboard use will range from a useful scale of $1\frac{1}{2}$ to $7\frac{1}{2}$ in. As shown in Fig. M-11, the meter and indicating mechanism may be obtained in several forms. Dependent upon the number and complexity of functions to be performed, portable meters for testing and laboratory use will vary in size from about 5 by 5 by 3 in. (with carrying handle) to perhaps 20 by 10 by 10 in. (for table mounting, but still considered portable). Meter weights will range from a few pounds up to about 40 lb. As shown in Fig. M-13, a "clamp-on" meter is also available.

The majority of meters for panel and switchboard mounting incorporate a single calibrated scale. On the other hand, portable meters most frequently feature multiscales for one measured quantity or for multifunctional measurements,

Fig. M-12. Multiple scales of an analyzer-volt-ohm-milliammeter (multimeter). *(Weston.)*

Fig. M-13. Representative portable electrical meters for industrial testing and laboratory applications: (a) industrial-class portable ammeter (5¼ by 3¼ in.) with 4.04-in. scale length; permanent-magnetic moving-coil type for dc measurements—self shielded with ½ percent accuracy; movable-iron type for ac measurements with ¾ percent accuracy; (b) laboratory-class portable standard voltmeter (8¾ by 7¾ by 4+ in.) with 6.1-in. scale length and 100 scale divisions; 1 percent accuracy for dc and ½ percent accuracy for ac measurements; sensitivity 1,000 Ω/V; (c) ac/dc transfer ammeter standard (8½ by 10⅜ by 5¾ in.) with 5¼-in. scale length; ¼ percent accuracy. (d) ac industrial analyzer (18⅞ by 10⅞ by 6⅞ in.) with 3½-in. scale length (per meter); for measuring instantaneous current surges, such as motor-starting circuits; (e) multirange analyzer with 4.9-in. scale length calibrated in volts, ohms, and milliamperes; a single switch controls both functions and ranges; (f) d-c bench tester; (g) ac clamp ammeter and volt-ammeter (13⅝ by 4⅜ by 2½ in.) with 2.36-in. scale length; accuracy is within 3 percent; can be hung on a conductor, leaving operator's hands free. All accuracies stated in the foregoing are percent of full scale. *(Weston.)*

such as a combination ammeter and voltmeter. Meters of this kind are termed *multimeters.* The scale of an analyzer-volt-ohm-milliammeter is shown in Fig. M-12.

It is beyond the scope here to delineate fully the many kinds of electrical meters commercially available. A cross section of meters is given in Fig. M-13.

—Adapted from "Instrument Sketch Book,"
Weston Instruments, Inc.

Meter Prover (See Gasometer.)

Metering Pump A positive-displacement pump that provides (1) high repetitive volumetric accuracy, (2) high-pressure capability, and (3) a means for convenient adjustment of capacity—both manually and automatically. Metering pumps are used in the feeding of boiler-water-treatment chemicals, corrosion inhibitors, filtration aids, sterilizing agents, and in many other industrial applications where it is desired to add a small quantity of one or more liquids to a system. Where it is essential that the quantity of material added be maintained at a precise value, the metering or proportioning pump frequently is found to be the most convenient and economical method. The three basic types of metering pumps are (1) packed plunger, (2) mechanically actuated diaphragm, and (3) hydraulically actuated diaphragm (see Table M-2).

TABLE M-2. *Metering-pump Types and Characteristics*

Characteristics	Packed-plunger Pump	Mechanically Actuated Diaphragm Pump	Hydraulically Actuated Diaphragm Pump
Pressure range, psig	Up to 7,000	Up to 125	Up to 5,000
Capacity range, gph	Up to 1,000/cylinder	Up to 25/cylinder	Up to 300/cylinder
Capacity controls available	Manual—stopped, manual—running, pneumatic, electric	Manual—running, pneumatic (limited), electric (limited)	Manual—running, pneumatic, electric
Max viscosity, centipoise	3,000	100	3,000
Accuracy	±1%	±3%	±1%
Range	10:1	10:1	15:1
Advantages	Low cost	Leakproof,enclosed mechanism, low cost	Leakproof, enclosed mechanism, good suction lift
Disadvantages	Packing leakage, packing adjustment, exposed mechanism (in some designs), poor suction lift	Limited in pressure and capacity	More complex design, temperature limitations

Packed-plunger Pump: This device consists of a plunger which is reciprocated within a cylinder (see Fig. M-14). The plunger is linked by a connecting rod to a crank mounted on a rotating shaft. The rotary shaft motion imparts a reciprocating motion to the plunger. As the plunger is moved, liquid is forced out through the discharge check valve. As the plunger is retracted, the cylinder is filled with liquid entering from the suction line. A conventional stuffing box is used to prevent uncontrolled external leakage past the plunger. Some leakage past the plunger is necessary to prevent overheating of the packing and possible scoring of the plunger. Where such leakage poses problems, such as safety hazards, a different form of pump must be used. The packed-plunger pump is termed a *controlled-leakage* pump in contrast to leakproof designs. Generally, these pumps find greatest usage where the liquid being pumped is not extremely corrosive or hazardous. Because of design simplicity, the packed-plunger pump usually is the lowest-cost metering pump.

Mechanically Actuated Diaphragm Pump: In this design, a flexible diaphragm is used to isolate the pumped liquid from the operating mechanism. The diaphragm is mechanically attached to a pushrod which is caused to reciprocate. The diaphragm generally is circular and shaped to allow flexibility while at the same time providing a means of attachment to the pushrod. Usually a spring-return mechanism is used to return the diaphragm from its extended position. Since the diaphragm is unsupported on one side, this type of pump is limited to low-discharge-pressure use. Prime usage is found in applications at pressures not exceeding 125 psig and at low capacities (see Fig. M-15).

Hydraulically Actuated Diaphragm Pump: This device uses a diaphragm supported on both sides to isolate the pumping liquid from the operating parts. In general, the design represents a combination of a plunger pump and a

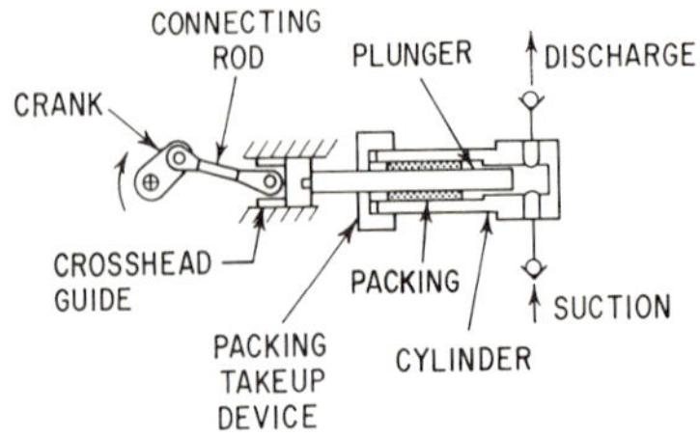

Fig. M-14. Packed-plunger metering pump.

diaphragm pump in which the plunger pump is used as a source of pressurized liquid which is applied to the reverse side of the diaphragm—the diaphragm serving to pump the process liquid. Neglecting minor flow losses, the diaphragm is said to be balanced—allowing use of the pump at pressures many times higher than is possible with the mechanically actuated diaphragm pump.

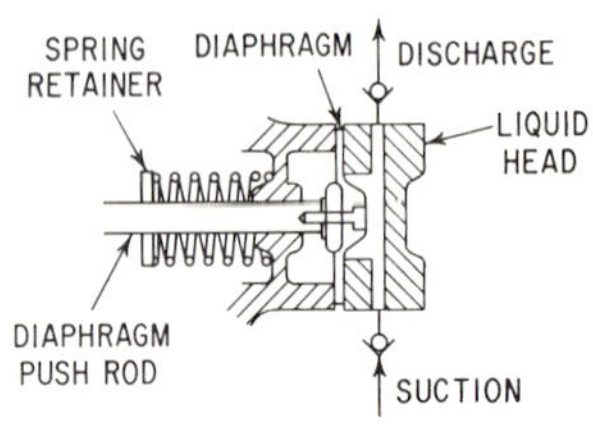

Fig. M-15. Mechanically actuated diaphragm-type metering pump.

Since the plunger can be enclosed in a noncorrosive liquid, the stuffing box is eliminated (see Fig. M-16).

The plunger is caused to reciprocate in a chamber filled with hydraulic fluid. The displacement of this fluid, in turn, causes the diaphragm to be flexed, alternately discharging and recharging the pumping chamber. Capacity control is accomplished by varying the volume of hydraulic fluid displaced by the plunger—either (1) by changing the stroke length of the plunger or (2) by bypassing a portion of the fluid displaced by the plunger. Hydraulically actuated diaphragm pumps have a clearance volume which is large relative to displacement per stroke. Thus, there results a change in delivery with changes in discharge pressure because of the compressibility of liquids. This is a significant factor in the selection of a pump for a high-pressure application.

Pump-capacity Control: Although design details vary with the brand of pump, available capacity controls generally are similar in function.

Manual Adjustment While Stopped: This control generally is found on packed-plunger pumps and provides a means to adjust pump capacity only when the pump is stopped. Usually a change in the effective length of the crank arm is accomplished by loosening a nut that attaches the connecting rod to the crank and repositioning the crankpin at a different radius from the center of the rotating shaft.

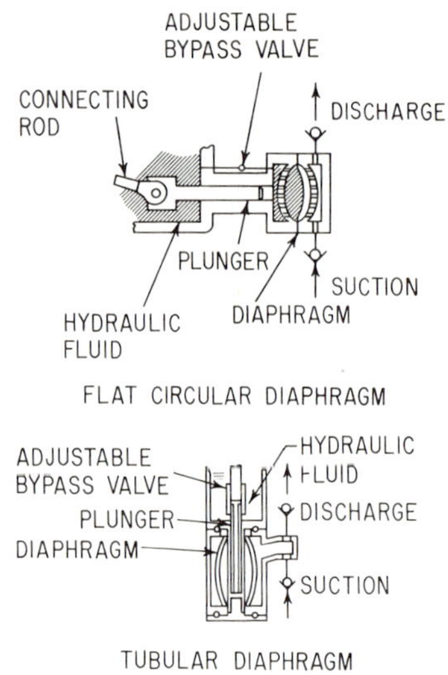

Fig. M-16. Hydraulically actuated diaphragm-type metering pump.

Manual Adjustment While Running: On mechanically actuated diaphragm pumps, this usually consists of a micrometer-screw control which limits the retracted position of the diaphragm. On adjustable-stroke-length hydraulically actuated diaphragm pumps and packed-plunger pumps, adjustment of stroke while running may be accomplished by use of a variable-angle tiltplate design. On fixed-stroke-length hydraulically actuated diaphragm pumps, capacity control is accomplished by adjustable valving to change the amount of displaced fluid bypassing each stroke.

Pneumatic Control: This form of control generally is available on hydraulically actuated diaphragm pumps but finds limited usage on packed-plunger and mechanically actuated diaphragm pumps. The most common form accepts a 3 to

15 psig air signal and provides zero to maximum delivery proportional to the air signal. See **Pneumatic controller.** Operation on signal ranges other than 3 to 15 psig can be accomplished by use of a separate reversing air relay. Ratio control is available in some designs of pneumatic controls. If this is not provided as part of the control, ratioing can be accomplished by use of a separate ratio relay. The ability to adjust ratio on this type of control permits adjustment of dosage on automatic-control applications.

Electric Control: This form of control usually is available on hydraulically actuated diaphragm pumps and on some packed-plunger pumps. Pump delivery is automatically adjusted in response to an electrical signal. See **Electric controller.**

Variable-speed Drive: This form of control usually is available on packed-plunger and on hydraulically actuated diaphragm pumps. Most commercially available variable-speed drives can be adapted as prime movers on these pumps. The use of a variable-speed drive, together with a pneumatic or electric capacity control, provides means for automatically adjusting pump delivery from two independent variables. Flow-proportional response can be accomplished by the variable-speed drive, along with automatic dosage control by means of a pneumatic or electrical signal.

Installation Factors: Proper installation of metering pumps is important to obtaining reliable, accurate pump operation. Pump manufacturers usually specify the minimum net-positive suction head (NPSH) required at the pump suction connection to overcome the fluid-flow losses through the suction check valve and pump-cylinder porting. The user must calculate the head losses for a particular installation to ensure that the required minimum will be available under most adverse conditions. Applicable formulas are given in manufacturers' literature. Since metering pumps are positive-displacement pumps, provisions must be made to prevent damage due to excessive pressures resulting from pump operation with a closed discharge line. Without some pressure-relieving device, operation against a closed line will result in a constant increase in pressure until either the driving motor stalls, the pump fails mechanically, or the system piping ruptures. All metering-pump installations should include a relief valve between the pump and the first downstream stop valve on the discharge. High-viscosity liquids can present problems in accurate pumping because of high flow losses in the suction piping and impeding of proper check-valve action. Manufacturers normally specify viscosity limitations. Steam jacketing and other means to apply external heating frequently are used to reduce viscosities.

—Theodore F. Masse, *B I F, Unit of General Signal Corporation*

Metrology The science of weights and measures. For example, in the metalworking industry, metrology involves standards, instruments, environments, and methods for measuring dimensions, part geometry, surface finish, and numerous other conditions and relationships. Metrology involves national and international cooperation among governmental agencies, industry, trade, and technical societies.

In the United States, the National Bureau of Standards (Washington, D.C.) has the principal responsibility for basic and primary standards for length, mass, time, temperature, electrical quantities, and luminous intensity. The Institute for Basic Standards of the NBS provides the central basis for a complete and consistent system of physical measurement, coordinates that system with the measurement systems of other nations, and furnishes essential services leading to accurate and uniform physical measurements throughout the nation's scientific community, industry, and commerce.

Industrial metrology laboratories (see Fig. M-17) supplement the services of NBS on a private basis. With instrumentation and environment comparable with the NBS, to which all standards trace, the well-equipped industrial-standards laboratory provides measurement and inspection services to private industry and others in doing such things as calibrating and certifying gage blocks and end measures, as well as plain and taper plug and ring-gage calibration, thread-gage calibration, hardness testing, and surface-finish analysis. The degree to which the environment is controlled is one of the main differences between the average industrial-gage laboratory and a metrology laboratory. For example, interferometric measurement of gage blocks should be performed at $68°F \pm \frac{1}{20}°$ and at a relative humidity of 50 percent or less.

Environment is a composite of five factors: (1) temperature, (2) barometric pressure, (3) humidity (vapor pressure), (4) elevation above sea level, and (5) latitude as it affects the gravitational constants. Each factor must be measured accurately and appropriate compensations made when

Fig. M-17. Portion of the Eli Whitney Metrology Laboratory of the Bendix Automation and Measurement Division, Dayton, Ohio.

making metrological observations on the order of six decimal places. A well-equipped metrology laboratory will contain instruments and machines for making interferometric and mechanical measurements of gage blocks, curved surfaces, and balls, as well as internal diameters and holes, flatness, and surface finish.

For other information related to metrology, see **Dimension, measurement of; Gaging, pneumatic; Goniometer; Measuring machine; Standards, measurement; Theodolite;** and **Thickness gaging.**

REFERENCES

Advances in Metrology, *ISA Publ.* 68C3, 1968.

Young, A. W.: New Dimensions in Metrology, *Iron Age,* Feb. 24, 1966, pp. 55–62.

Sterns, Robert B.: Digital Logic Improves Photoelectric Length Measurement, *Control Eng.,* December, 1960, pp. 114–115.

Wiles, A. P.: Keeping Stock on Length, *Control Eng.,* September, 1962, pp. 117–119.

Young, A. W.: Recent Developments in the Use of Optics as an Engineering Tool, Semiannual meeting of the American Society of Tool Engineers, Los Angeles, October, 1958.

Fink, Tony: Miniature Measurement, *Autom. Machining,* June, 1967, pp. 54–55.

Mollet, P.: "Optics in Metrology," Pergamon Press, New York, 1960.

—Henry L. Boppel, *Automation and Measurement Division, The Bendix Corporation*

Microcode (See **Microprogram.**)

Microphone An electroacoustic transducer that responds to sound waves and delivers essentially equivalent electric waves. Microphones are of two basic types: (1) *pressure-sensing,* used to observe the pressure waveform of sound and (2) *velocity-sensing,* for observing the velocity waveform. For measurement purposes, the pressure-sensing device usually is preferred because of its superior frequency response and dynamic range. These devices operate on the capacitive, piezoelectric, or moving-coil principle.

Microprobe Analyzer (See **X-ray analyzer, electron-microprobe.**)

Microprogram A program coded using elementary logical and control commands whose purpose is to simulate a higher-level instruction set for a digital computer. The term is best understood by comparing it with other types of instruction sets. In many computers, the basic machine-oriented instruction set (see **Language, computer**) consists of commands such as add, divide, subtract, and multiply which are executed directly by the hardware. The hardware which provides these functions, however, actually implements each function as a combination of elementary logical functions such as AND, OR, and EXCLUSIVE OR. The exact implementation is usually not of concern to the programmer. Compared with the elementary logical functions, the add, subtract, multiply, and divide commands are a higher-level language set in the same sense that a macrostatement (see **Macroassembler**) is a higher-level instruction set when compared with the machine-oriented language-instruction set.

In a microprogrammed computer, the executable (micro) instructions which may be used by the (micro) programmer consist of a logical functions such as AND and OR and some elementary control functions such as shift and branch. The (micro) programmer then defines microprograms which implement an instruction set analogous to the machine-oriented language-instruction set in terms of these microinstructions. Using this derived instruction set, the systems/application programmer writes programs for the solution of a problem. When the program is executed, each derived instruction is executed by transferring control to a microprogram. The microprogram is then executed in terms of the microinstructions of the computer. After execution of the microprogram, control is returned to the program written in the derived instruction set.

Many recent digital computers provide microprogramming capability to some degree. The advantage is in the increased flexibility which can be realized by tailoring the derived, machine-oriented instruction set to a particular application. For example, microprogramming capability allows the programmer to implement a function such as square root directly without the use of subroutines or macrostatements. This can result in significant programming and execution efficiency if the square-root function is commonly required in the application area of interest. This is the case in a computer programmed for use in aircraft navigation, for example.

Another application is that of providing different derived instruction sets for the same computer system. For example, the instruction set of a character-oriented computer can be implemented on a word-oriented machine if an adequate microprogramming feature is available.

—B. W. Landeck, *IBM Corporation*

Microwave Moisture Meter (See **Moisture, measurement of.**)

Microwave Spectroscopy (See **Molecular rotational-resonance spectroscopy.**)

Millivoltmeter Controller (See **Electric controller.**)

Missile Systems (See **Space and missile-systems instrumentation.**)

Mixing Valve (See **Three-way valve.**)

Mnemonic An abbreviated notation which assists in remembering the full term. Mnemonic codes are used extensively in digital-computer programming. The reason for this is that instructions such as addition or division are represented by binary numbers in a digital computer. Since those numbers can be very difficult to remember, a mnemonic code such as ADD or DIV is substituted for the numbers required to instruct the computer to do an addition or division. A programmer, therefore, writes his program using mnemonic operation codes and produces what is known as a source program. The source program is then inputted to the digital computer's assembler (see **Assembler**), which translates the mnemonic operation codes into the actual machine code of the computer.

—B. W. Landeck, *IBM Corporation*

Mode, Control (See **Control mode.**)

Modulation The process or result of the process whereby some characteristic of one wave is varied in accordance with some characteristic of another wave.

Moisture, Measurement of The measurement of moisture in air and gases is described under **Dew point, measurement of,** and **Humidity, measurement of.** Also see **Hygrometer; Hygrothermograph;** and **Psychrometer.**

Moisture content has a number of synonymous terms, several of which are specific to certain materials and applications. The water content of solid, granular, or liquid materials usually is referred to as moisture content on either the *wet basis* or *dry basis.* For industrial measurements, the wet basis is most common. Wet basis refers to the quantity of water per unit weight or volume of the wet material. The textile industry uses the dry basis for moisture content of textile fibers. Often referred to as the *regain* moisture content, the dry basis refers to the quantity of water in a material expressed as a percentage of the weight of the bone-dry material. The relationship between the wet and dry moisture-content basis is shown in Fig. M-18.

Instruments for moisture measurement may be typed as *periodic* or *continuous.* Periodic instruments generally are automated versions of conventional laboratory moisture-analysis procedures—with a measurement typically requiring

2 min or longer and sometimes up to 20 min. Thus, periodic instruments are quite impractical for automatic-control systems. Moisture-measuring instruments also may be classified by the operating principle used. Those instruments using electrical conductivity (dc or ac), absorption of electromagnetic energy (radio-frequency regions), electrical capacitance (dielectric-constant change), and infrared-energy radiation are most adaptable to continuous measurement because of the comparatively fast speed of response. Instruments that use automatic oven drying, chemical titrations (Karl Fischer technique), equilibrium hygrometric methods, and distillation methods usually are of the intermittent periodic type.

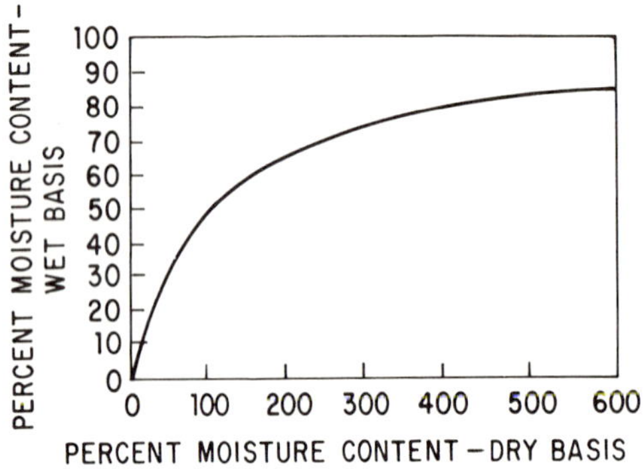

Fig. M-18. Relation of dry vs. wet moisture-content basis.

Electrical-conductivity Methods: The relationship between dc resistance and moisture content of such materials as wood, textiles, paper, and grain is the physical basis of measurement. Specific resistance when plotted against moisture content results in an approximate straight line up to the moisture-saturation point. Beyond saturation, where all the cells and intermediate spaces in the material are saturated with free water, electrical-conductivity methods are not reliable. This point varies from about 12 to 25 percent moisture content, dependent upon the type of material. The sample under test is applied to suitable electrodes in the form of needlepoints for penetration into the material. Granular or fibrous materials may make use of electrodes in the form of a cup or clamp arrangement to confine the material to a fixed volume. The electrodes and sample under test comprise one arm of a Wheatstone bridge as shown in Fig. M-19. The high sensitivity required of the detector dictates the use of electronic amplifiers. The range of resistance values corresponding to normal moisture contents varies from less than 1 to 10,000 MΩ or higher, dependent upon material, electrode design, and moisture content. Increasing moisture content results in decreasing resistance values.

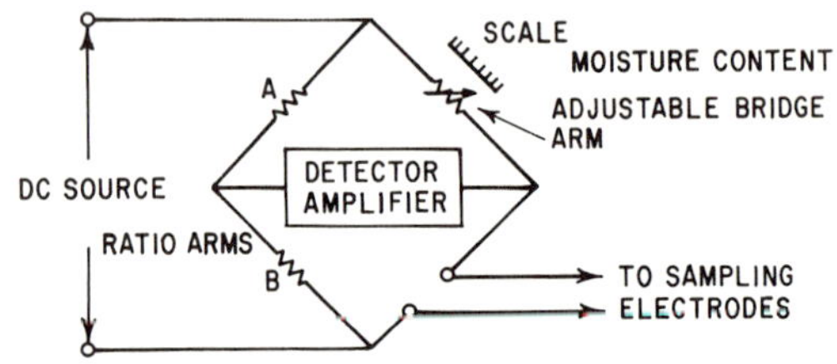

Fig. M-19. Basic bridge circuit for conductivity method of moisture-content measurement.

Electrical-capacitance Methods: The change in dielectric constant between dry and moist conditions of a material is the physical principle used. The dielectric constant of most vegetable organic materials is 2 to 5 when dry. Water has a dielectric constant of 80. Therefore, the addition of small amounts of moisture to such materials causes a considerable increase in dielectric constant. The material being measured forms part of a capacitance bridge which has radio-frequency power applied from an electronic oscillator as shown in Fig. M-20. Electronic detectors measure bridge unbalance or frequency change, dependent upon the method employed. Electrode design varies with the type of material

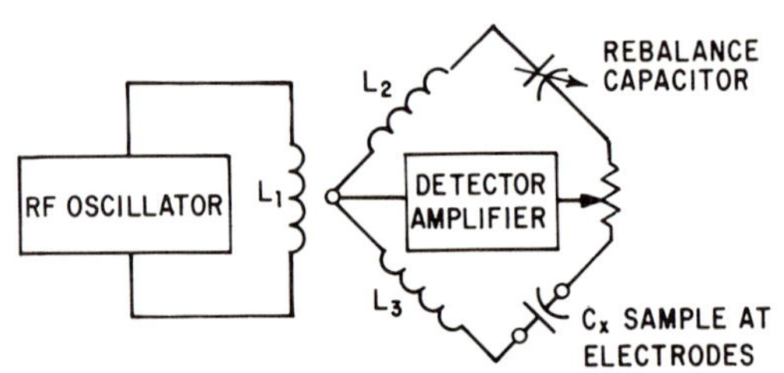

Fig. M-20. Basic bridge circuit for capacitance method of moisture-content measurement.

under test. Parallel-plate electrodes are used for sheet materials, whereas cylindrical electrodes usually are used for liquids or powders. Moisture content measurable by electrical-capacitance methods ranges from 2 to 3 percent to 15 to 20 percent, dependent upon the product involved.

Radio-frequency Absorption: The attenuation of electromagnetic energy when passed through a material is the physical principle used. Radio-frequency devices are operated at frequencies below 10 MHz, and for best results, the products being measured must be comprised of polar materials. The rf energy is passed through the polar material, and the water molecules absorb some of the energy in the form of molecular motion. The block diagram of a typical rf instrument is shown in Fig. M-21. By means of suitable electrodes for the sampling technique, the instrument can be applied to solid or sheet materials. Granular materials can be placed into cell-type electrodes. The test results will be

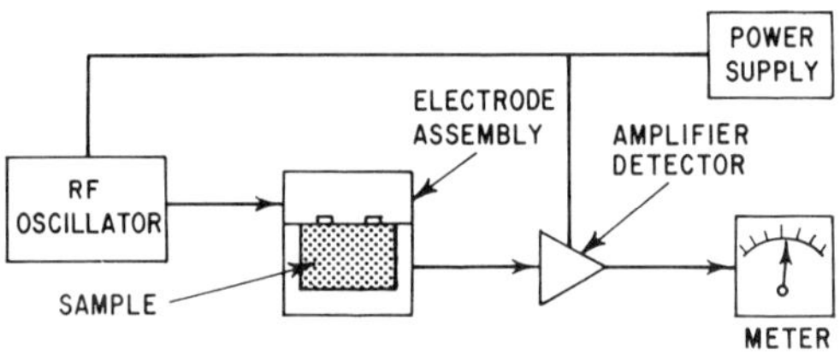

Fig. M-21. Radio-frequency absorption-type moisture indicator.

affected by any polar material in the sample. However, water of hydration in a product will not be detected. Instruments of this type must be calibrated to the particular product under test. Although these instruments usually are available as periodic devices, they can be adapted for continuous measurements on some products. The moisture-content range is dependent upon the material and is from 0.1 to 60 percent. For certain materials, the range is quite narrow.

Microwave Absorption: The frequency of electromagnetic energy used in these devices is in the region of 1,000 MHz and higher. Frequencies of 2.45 GHz (S band), 8.9 to 10.68 GHz (X band), and 20.3 to 22.3 GHz (K band) all have been used. Water molecules greatly attenuate the transmitted signal with respect to other molecules in the material in the case of S- and X-band frequencies. In the case of K-band frequencies, the water molecule produces molecular resonance. There are no other molecules that respond to this particular resonant frequency. Thus, the frequency is most specific to the moisture (free-water) content in paper products. The wavelengths of the K-band frequencies are approximately 1.35 to 1.5 cm. At these very short wavelengths, the energy may be guided or transmitted by waveguides. The block diagram of a microwave moisture instrument is shown in Fig. M-22. The detection of energy absorption can be made by attenuation (loss) or by phase-shift methods, or both measurements may be made simultaneously. Useful

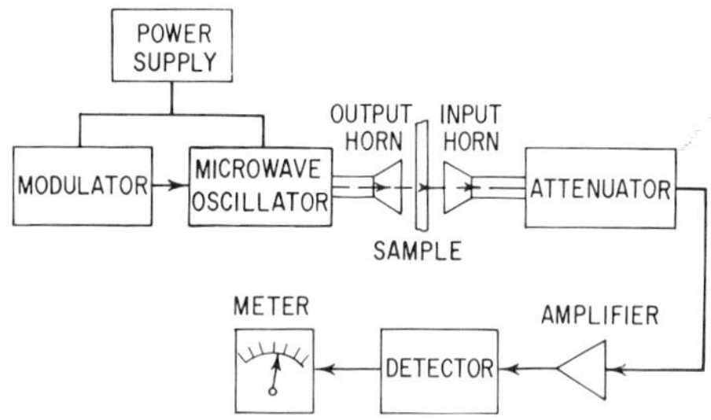

Fig. M-22. Microwave absorption-type moisture indicator.

range of these devices is from less than 1 to 70 percent moisture and higher. At low moisture contents, the S- and X-band devices require a minimum mass of material for practical operation. Accuracy of these instruments is within ±0.5 percent of indicated value—up to approximately 15 percent moisture content.

Infrared Absorption: The water molecule becomes resonant at certain infrared frequencies, and thus the amount of energy absorbed by the water-absorption band is a measure of moisture content. One device for measuring the moisture content of sheet paper makes use of infrared energy at two different wavelengths by measuring the differential absorption. Infrared radiation is passed through a rotating-disk chopper that contains two filters. One filter passes a wavelength of 1.94 microns which covers the water-absorption band; the second filter passes a band of 1.80 microns which is not significantly affected by moisture content. A block diagram of the instrument is shown in Fig. M-23. The paper sheet is exposed to the dual-frequency narrow-band radiation, and the reflected radiation is collected in an integrating sphere where it activates a lead sulfide detector. The signals are amplified, and the ratio of the two narrow-band signals is read out directly as moisture content. The system is quite immune to the effect of other paper variables, such as basis weight, coatings, density, and composition. The useful range is from 1 to 12 percent.

An infrared instrument useful for gas and liquid measurements operates on the principle of selective energy absorption of the sample when compared with a reference material having little or no absorption. The instrument contains two infrared sources—one source is directed through a cell that contains the sample material; the second source is directed through the reference cell. In the case of gas analysis, the reference

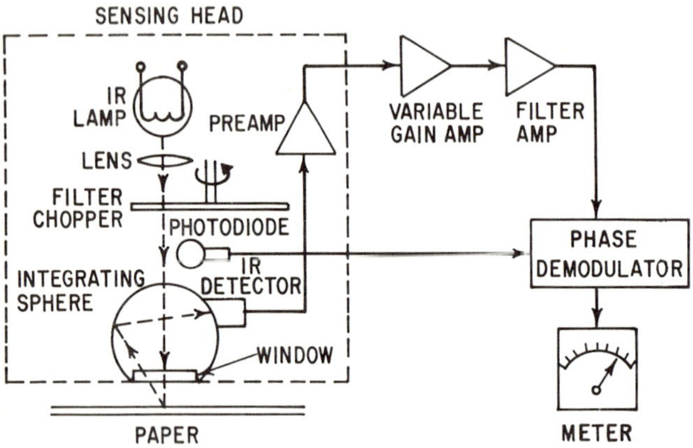

Fig. M-23. Infrared-backscatter moisture instrument.

cell is filled with an inert gas or air. A detector unit consisting of two cavities separated by a metallic diaphragm, which acts as a variable capacitor in respect to a fixed plate within the cavity, is used. The major components of the instrument are shown in Fig. M-24. Operation consists of chopping the infrared energy at 10 Hz to the sample and reference cells. The gas or fluid in the sample cell absorbs the wavelengths which are characteristic of the material and proportional to concentration. The reference cell

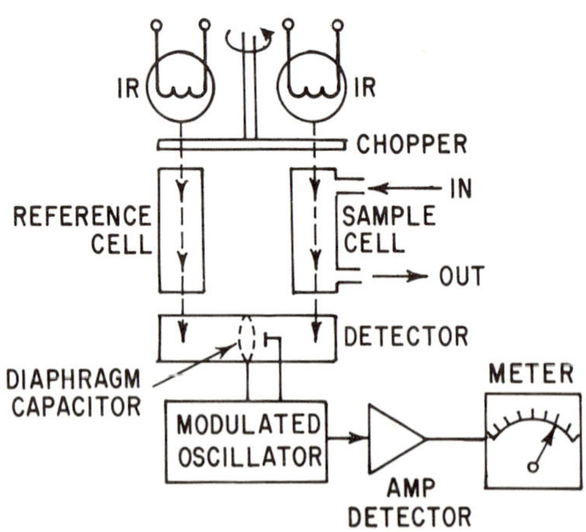

Fig. M-24. Infrared-absorption moisture instrument.

does not absorb any appreciable infrared energy. The infrared energy passes through the cells and into the detector, which causes the temperature of the gas in the detector to rise. However, if some of the energy has been absorbed by the sample cell, the sample side of the detector will receive less energy than the reference half of the detector. The temperature on either side of the diaphragm causes a definite pressure—with the diaphragm moving in a direction that will tend to equalize the two pressures. This movement, in turn, causes the capacitance between the diaphragm and its sensing plate to change. This change is used to modulate an oscillator. The chopper alternately blocks and then passes the energy from each infrared source. Each time the energy is blocked, the pressures equalize in the detector compartments. Then, alternately, the energy is permitted to enter the detector cavities. The modulating frequency is 10 Hz, and the magnitude of capacitance change is related directly to the amount of energy absorbed in the sample cell. The signal from the oscillator is amplified and demodulated for meter indication or is applied to a recorder. The range of moisture in gases is from 1,000 ppm to 100 percent water vapor; for fluids, from 6 to 100 percent water. The response speed is 0.5 s for 90 percent of full-scale output.

Equilibrium Methods: The equilibrium moisture content of the air at the surface of a material is representative of the moisture within the material. This is particularly true of hygroscopic granular and fibrous materials. The instrumentation involves the use of humidity-measuring devices.

Karl Fischer Technique: This method offers extreme sensitivity with good accuracy and covers a wide range of materials. Small samples can be analyzed readily. End points can be determined either visually or electrometrically. The Karl Fischer reagent is added in small increments to a glass flask containing the sample until the color changes from yellow to brown or until a change in potential is observed at the end point. Dark-colored solutions require electrical endpoint measurement. The method does not differentiate between free and combined water. Other methods involving chemical reactions with the free water in liquids or solids include (1) the evolution of free acidic or basic compounds, (2) the evolution of an inert gas, and (3) the formation of an insoluble precipitate. Generally, these methods fail to distinguish active hydrogen from water in many materials and are reliable only under laboratory control.

Distillation Methods: The water is separated from the liquid sample by virtue of the differences in boiling points of the other liquids and water. The water fraction is collected carefully and measured volumetrically.

Oven Drying: The oldest and still a very common analytical method is that of determining moisture content by heating the sample to ensure complete drying. Moisture is calculated on the basis of loss in weight between the original and the dried sample. Semiautomatic drying and weighing ovens are available where the moisture content is indicated directly by a weighing scale built into the oven. Other volatile materials interfere with the method and must be accounted for to determine how much weight loss is due specifically to water evaporation.

REFERENCES

Considine, Douglas M. (ed.): "Process Instruments and Controls Handbook," sec. 7, pp. 150–163, McGraw-Hill Book Company, New York, 1957.

Considine, Douglas M., and S. D. Ross (eds.): "Handbook of Applied Instrumentation," sec. 7-3, McGraw-Hill Book Company, New York, 1964.

Amdur, E. J. : Applications of Humidity and Moisture Measurements in Science and Industry, *ISA Publ.* HMPII, 1963.

Wexler, A., and W. A. Wildhack: Fundamentals and Standards of Humidity and Moisture Measurement, *ISA Publ.* HMPIII, 1963.

—Lee E. Cuckler, *Robertshaw Controls Co.*

Moisture (Atmospheric), Measurement of (See **Hygrometer; Hygrothermograph;** and **Psychrometer.**)

Molecular Rotational Resonance Spectroscopy Molecular rotational resonance (MRR) spectroscopy is a form of spectroscopy occurring in the microwave region. This is generally taken to be from 3,000 to 300,000 MHz (1 MHz equal 10^6 Hz) or in units of wavelength from 10 to 0.1 cm. Absorptions of energy by molecules in this region result primarily from transitions between rotational-energy levels. The general features of the spectrum can be predicted if the molecular rotational constants are known. After the spectra of the molecule and a sufficient number of its isotopic species are assigned, the molecular structure (i.e., bond distances, bond angles) can be determined. Once assigned, the microwave spectrum of a particular molecule becomes a fingerprint of that molecule. The absorptions are not changed by any impurities in the sample.

—Stuart Armstrong, *Hewlett-Packard*

Monitor An instrument used to measure continuously or at intervals a condition that must be kept within prescribed limits, such as radioactivity at some point in a nuclear reactor, the image picked up by a television camera, the sound picked up by a microphone at a radio or television studio, a variable quantity in an automatic process-control system, the transmissions in a communication channel or band, or the position of an aircraft in flight. Also see **Alarms, control-system.**

Monitor, Hospital (See **Medical instrumentation.**)

Monitor Program (See **Executive program.**)

Monitoring, Radiation [See **Radiation (nuclear), detection of.**]

Monochromator A form of spectrometer which emits light of a single wavelength from any portion of the spectrum in the source. This single wavelength (or a narrow band of wavelengths) is isolated by means of a prism or grating. A grating is the most convenient form to use because it offers constant dispersion throughout the entire range covered by most monochromators.

As shown in Fig. M-25, light from a lamp passes through a quartz-entrance condensing and col-

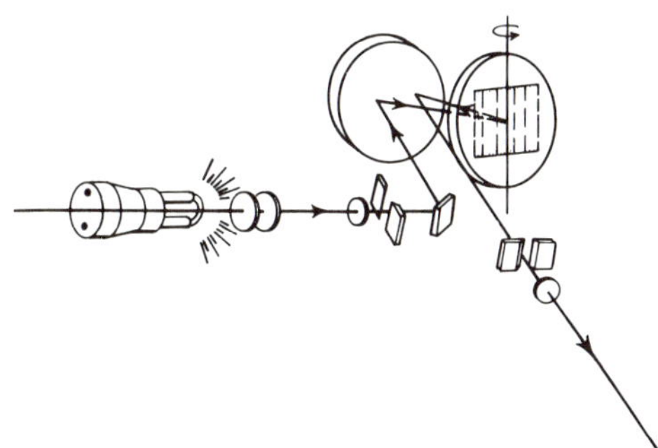

Fig. M-25. Optical system of monochromator with precision grating. *(Bausch & Lomb.)*

lective lens system and illuminates the entrance slit. From the illuminated slit, the light is directed by a plane-reflecting diffraction grating. The dispersed light is incident on the same concave mirror which forms an image of the entrance slit on the exit slit. Since the mirror acts in a dual capacity as collimator and telescope, it is necessarily larger than the parallel beam. This ensures that the full aperture is obtained in both the entering and dispersed light beams. In the device shown, a quartz collective lens at the exit slit forms a 12.5-mm-square image of the grating at a point about 60 mm in front of the monochromator housing. An achromatic condenser lens is available which will form an image of the slit instead of the grating. This lens also eliminates the need to refocus for different wavelengths.

Monochromators can readily be combined with other equipment for photomicroscopy, spectrophotography, and spectroradiometry. Gratings in commercially available instruments range from 300 to 1,200 grooves/mm.

Several light sources are used in monochromators, including:

1. *Air-cooled deuterium arc*—spectral range from below 2000 up to 4000 Å. Used for work requiring continuous spectrum illumination in the ultraviolet.

2. *High-pressure mercury source*—for use in ultraviolet microscopy and other work requiring line spectra.

3. *Coil-filament (quartz iodine) source*—provides a continuous spectrum with a minimum of filament image for use in the near ultraviolet, visible, and infrared regions.

4. *Superpressure mercury source*—for use where a line spectrum of high intensity is desired.

5. *Xenon light source*—provides useful transmission from about 2000 Å to 3.5 microns. Light in the visible region has a color temperature of about 5000°K.

—*Bausch & Lomb Corporation*

Monolithic Memory Monolithic storage is a high-speed nondestructive readout storage device used for the storage of binary information. Its basic storage element is a direct-coupled transistor flip-flop (see **Flip-flop**). By means of monolithic semiconductor technology, a large number of these circuits can be contained on a single silicon chip.

Monolithic-storage devices have the capability of achieving faster access times than are possible with magnetic-core-storage units. Since information is not erased when a bit position is read out, it is not necessary to write the information back into the location following a read operation. Therefore, the storage-cycle time (see **Memory-cycle time**) is the same as the read or write access time (see **Access time, computer**). Another advantage of monolithic storage is that it is feasible to design the storage device to use the same voltage levels as the logic circuits. Therefore, the cost of the smaller-sized storage units can be minimized. Since the basic storage element of the monolithic-storage device is a direct-coupled transistor flip-flop, information in storage is destroyed when power is turned off. Unlike magnetic-core storage, which can retain its information when power is properly sequenced off, the monolithic-storage unit must be reloaded with information when power is turned on.

On a read operation, the selected bit is gated to the sense amplifier, and amplitude-discrimination logic in the sense amplifier interprets the output as a 0 or a 1. On a write operation, the selected storage element is set to a 0 or 1 depending on the data on the input lines.

—W. W. Welz, *IBM Corporation*

Monte Carlo Method A method of solving a group of physical problems by making a series of statistical experiments. Accuracy of results depends upon the number of trials. The method usually is practical only when set up to be performed by a computer.

Motion, Measurement of One or more of four measurands may be involved when considering motion. Displacement, velocity, and acceleration may be linear or angular. A fourth measurand, *jerk,* may be defined as the rate of change of acceleration. The need to measure and to control motion variables is encountered widely in a variety of systems for a host of applications. For example, vibration and shock measurements are concerned with the motion variables, notably acceleration, and constitute a major topic of instrumentation. These and many other applications are discussed eleswhere in this volume. Reference to the detailed alphabetical index at the end of this volume is suggested. Key words to consult include *displacement, velocity,* and *acceleration.*

Motion-balance System (See **Valve positioner, pneumatic.**)

Motor Controls (See **Electric-motor drives, control of.**)

Motor-operated Valve (See Actuator, electric.)

Mueller Bridge (See Bridge circuits, electric.)

Multimeter, Electric (See Electrical quantities, measurement of; Meter, electrical; and Voltmeter.)

Multiple Range (See Range, instrument.)

Multiplexer, Analog An analog multiplexer is an array of analog switches which is used to select one of several signals for transmission to subsequent equipment (see **Switch, analog**). Analog multiplexers are commonly used to select analog-input signals for amplification and conversion to digital form in time-shared data-acquisition subsystems. They are also used in time-shared analog-output subsystems for the distribution of voltage signals to analog-memory circuits (see **Analog output**). Figure M-26 illustrates the use of an analog multiplexer in the analog-input subsystem of a data-acquisition system.

These examples relate specifically to analog multiplexers used in data-acquisition and control systems; in the most general sense, the term covers any switching array used to select one of several analog signals. Thus, for example, the audio-switching networks used in a central telephone exchange are a form of analog multiplexer.

Structurally, the multiplexer consists of a number of switches which are controlled by a control unit as indicated in Fig. M-27. If the multiplexer is differential, two switches are provided for each input signal; if the multiplexer is single-ended, only one switch per input is provided and a common ground return completes the circuit for all input signals.

The functions provided by the control unit vary widely with the particular application. Its basic function, however, is to provide the signals which select or address the various switches in the multiplexer. The particular address to be selected may be obtained from a computer, or it may be determined by logic in the control unit. For example, the switches may be selected in sequential order under the control of a ring counter circuit in the control unit. Similarly, the selection of each address may require a data word received from the computer. In this case, the control unit decodes the digital-input word and provides the switch-selection signal.

In addition to providing the switch-control signals, the control unit may provide timing for the multiplexer and other associated equipment. In the case of most switches, for example, the analog-to-digital converter cannot begin a conversion immediately after the multiplexer switch is selected (see **Settling time**). A time delay must

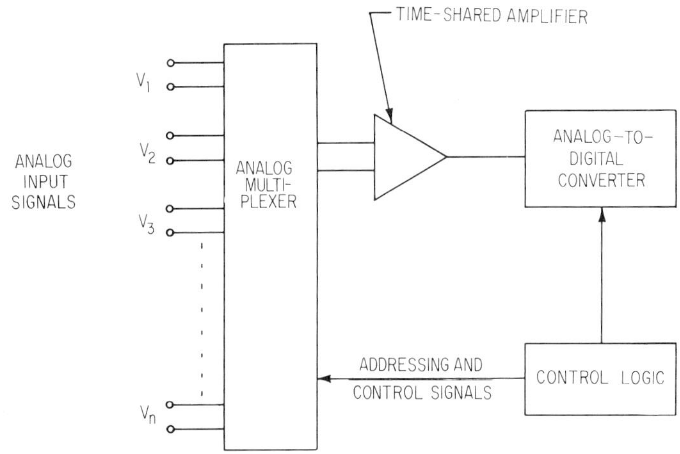

Fig. M-26. Analog-input subsystem configuration.

be provided to allow for the actuation time of the switch, for amplifier settling, and for other transients to die out. The control unit often provides these timing signals through the use of counters or delay circuits.

The analog switches used in the multiplexer may be any one of several types of electromechanical or solid-state switches (see **Switch, analog**). Examples of electromechanical switches include dry-reed relays, mercury-wetted contact relays, and crossbar switch assemblies. Commonly used solid-state switches include bipolar transistors and field-effect transistors, although other devices such as diodes and silicon controlled rectifiers are used in specialized applications. Each of these devices has its particular characteristics which largely determine the overall performance of the multiplexer.

The important performance parameters for an analog multiplexer are the sampling speed, accuracy, noise, and in the case of a differential multiplexer, the common-mode rejection ratio (CMRR) [see **Accuracy; Common-mode rejection ratio (CMRR); Noise**]. Sampling speed is primarily determined by the characteristics of the multiplexer switching device. Electromechanical devices such as dry-reed and mercury-wetted contact relays are generally limited to about 250-samples-per-second sampling speeds. Solid-state switches such as bipolar or field-effect transistors can be used at sampling speeds in excess of 100,000 samples per second. In addition to the basic switching speed of the multiplexer switch, noise and settling-time characteristics are important characteristics in determining the overall multiplexer speed. Even though the switch may close

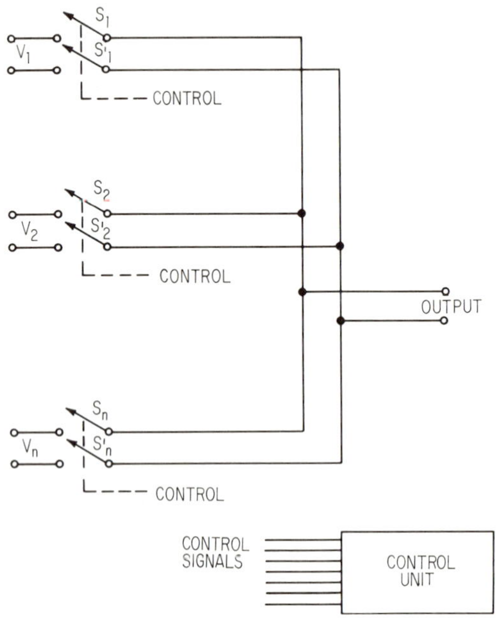

Fig. M-27. Analog-multiplexer configuration.

quickly, the switch transient must be allowed to die out before the output signal of the multiplexer can be converted to digital form. For example, the actual switch-closing time of a mercury-wetted relay is on the order of 1 ms, and there is a noise transient which persists at levels in excess of 10 μV for 5 ms or more. Therefore, in a low-level multiplexing application where this level of noise is objectionable, a delay must be provided between the selection of the switch and the initiation of a conversion. Even though the switch can theoretically be operated at more than 400 samples per second when only the actuation and drop time is considered, the noise consideration limits the practical sampling speed to about 250 samples per second. Similar noise considerations apply to solid-state switches, except, of course, that the duration of the noise is generally less and the attainable speed is greater.

The multiplexer characteristics can contribute errors to the operation of a digital-data-acquisition subsystem. In the case of solid-state switches, the leakage currents associated with the "off" switches flow through the source impedance of the channel being sampled by the "on" switch and also through the multiplexer load impedance. These currents can cause an offset in the voltage being sampled. Although the leakage current from a single switch is small, much less than a microampere, the cumulative effect of all the switches connected to the common multiplexer output bus can result in an appreciable error. In the case of bipolar transistor switches, there is also a direct offset error due to the "on" transistors. This voltage is a result of the bipolar transistor characteristics and may amount to 1 or 2 mV. By using special switch configurations, such as the Bright switch, this offset voltage may be reduced to a few hundred microvolts or less (see **Switch, Bright**). It is this error source which usually restricts the use of bipolar switches to multiplexers for signals greater than 1 V.

The field-effect transistor (FET) does not exhibit an offset voltage in the "on" condition, and for this reason it is suitable for multiplexing low-level signals (see **Switch, FET**). Some FETs, however, have an appreciable "on" resistance, and this can induce errors due to source loading effects. In addition, the "on" resistance increases the effect of leakage currents discussed above and sometimes increases the output-signal rise time.

In differential multiplexers, the common-mode rejection ratio is an important performance parameter. Leakage impedances between the switch-signal path and the drive circuit contribute to the impedances which determine the common-mode performance of the multiplexer (see **Error, common-mode**). In addition, the breakdown voltage of the switch determines the maximum common-mode voltage which may be applied without damaging the multiplexer. Since the electromechanical devices have high drive-to-signal path and contact-breakdown voltages, they are suitable for multiplexers which can withstand hundreds of volts of common-mode voltage. Solid-state multiplexers, however, are generally limited to less than 20 or 30 V of common mode unless special isolation techniques such as transformers are utilized.

Many of the errors associated with multiplexers are increased by increasing the number of input channels in the multiplexer. For example, the addition of channels decreases the common-mode rejection ratio and increases the offset due to leakage currents. A technique which is often used to minimize these effects is called "submultiplexing" or "block switching." In block switching, two levels of multiplexing are used. The first level is identical to that shown in Fig. M-27 and provides selection of an input signal. A second level of multiplexing is then provided at the output of the first-level multiplexer. For example, a second-level switch may be provided for each group of 16 first-level multiplexing switches. To select an input, both the first-level switches and the associated second-level switch must be actuated. The effect of second-level multiplexing is to isolate each block from the leakage currents and other disturbances which are due to other inputs. For example, in a system of 16 blocks of 16 channels each, the leakage errors are essentially determined by the 15 "off" switches in the same block as the addressed point and the 15 "off" block or second-level switches. Thus the errors are equivalent to those encountered in a 30-channel single-level multiplexer even though the two-level multiplexer services a total of 256 input channels.

—Thomas J. Harrison, *IBM Corporation*

Multiplexer, Digital The digital multiplexer is a device which permits the sharing of a common information path between several groups of input or output digital signals. By means of this device, information is tranferred between the central processing unit [see **Central Processing Unit (CPU)**] and any one of several digital input or output devices (see **Input/output devices**). An example of a digital multiplexer for reading data from groups of contact points is shown in Fig. M-28.

With reference to Fig. M-28, in order to read the status of the contacts in group A, a positive signal is transmitted on the group A select line. If the contact is closed, the positive signal appears on the associated bit line (see **Bit**). The bit lines are sampled while the select line is active, and the status of the group of contacts is stored in the computer. When the operation is complete, the signal on the select line is removed. When group B is desired to be read, a positive signal is placed on the group B select line and the status of the contacts in group B is placed on the associated bit lines. The diodes in series with the contacts are required to prevent back circuits, and the number of groups which can share the same input bit lines is a function of the back resistance of the diode.

—W. W. Welz, *IBM Corporation*

Multiprocessing The simultaneous execution of multiple programs in a single computer system which is composed of two or more individual processors and operated by one control program. The individual systems pool their main memories, peripheral units, and processing power, thus increasing the combined effectiveness of the total system. This allows for dynamic allocation of the computer resources of those programs executing in the individual processors. Also, the individual systems making up the multiprocessor system can be used independently if needed. Reliability can be enhanced by programming the system such that each processor is capable of assuming the function of a processor which fails. Although a failure reduces the capability of the

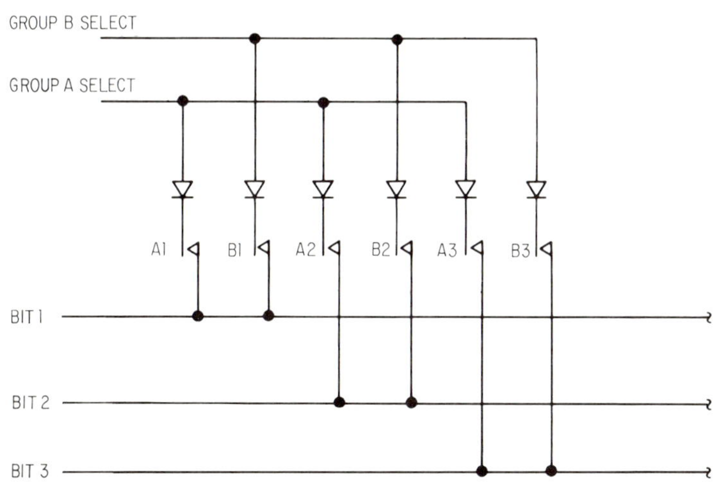

Fig. M-28. Digital multiplexer.

total system when it occurs, the redundancy allows continued operation.

—B. W. Landeck, *IBM Corporation*

Multiprogramming A type of digital-computer operation which allows for several programs to be resident in core storage simultaneously and to have central processing unit (CPU) time-shared among them. Multiprogramming is a method whereby the resources of a computer system may be better utilized. In a computer system where only one program may reside in the core at a time, inefficient use of CPU time results when a program requests data from an input/output device, for example. The CPU must delay its operation until the requested information is in the core because it generally requires the information to continue processing. This unusable input/output delay time in programs becomes a very large portion of the total program-execution time in many cases. Multiprogramming systems allow this input/output delay time to be used by other programs resident in the core to process data which are already available. In a multiprogramming system, therefore, one program may be waiting for data to be entered through an input/output device while another program is processing data which it already has. In this way, no CPU time is lost through input/output delay. Since process-control digital-computer systems involve large amounts of data input and output, multiprogramming systems on process-control digital computers can significantly increase the throughput of the system.

Input/Output Delay-time Control: One of the basic methods of controlling the interplay between multiple programs in a multiprogramming environment is via input/output delay time. In a system where multiprogramming is controlled via input/output delay, the various programs resident in the core are normally structured in a hierarchy. When a given program in the hierarchy initiates an input/output operation, that program is suspended until such time as the input/output operation is complete, and a lower-priority program is allowed to execute during the delay time.

Time-slice Control: Another method of regulating the interplay between the multiple programs resident in the core is via time slicing (see also **Time slicing**). In a time-slice multiprogramming system, each program resident in the core is given a certain percentage of the CPU time. Then, on a round-robin basis, each program is allowed to execute for its proportioned time period. Multiprogramming systems for applications where much more computation is done than input/output operation normally use the time-slice approach.

Multiprocess Applications: One of the capabilities of multiprogramming systems is that they allow multiple functions to be controlled simultaneously by a single process-control digital computer. In the case of laboratory automation, for example, different types of instruments may all be controlled simultaneously from one computer with a multiprogramming operating system. Another example is the use of one computer system to control two or three paper machines in a paper mill. In these examples, each instrument and paper machine has its own program resident in the core with program interplay controlled by input/output delay.

Single-process Applications: Even where only a single process is being controlled by a digital computer, multiprogramming systems allow for greater efficiency. Control of a process involves certain asynchronous functions such as a direct digital control (DDC) program and various data-logging programs. In a single-program system, situations arise where both types of programs (DDC and data logging) may require use of the CPU simultaneously. Since in a single-program system only one of these programs can be in the core at one time, the other program has its execution delayed until such time as the first program has completed operation. A multiprogramming system allows a partition of the core to be dedicated to each type of function required in the control of the process and thus eliminates interference between the various types of functions.

—B. W. Landeck, *IBM Corporation*

Murray Bridge (See **Bridge circuits, electric.**)

N

NAND Circuit The NAND, a contraction of NOT AND (see Fig. N-1), is a logical element which has the characteristic that the output F is 0 if, and only if, all the inputs are 1s (see **Binary state**). Conversely, if any of the input signals A or B or C of the three-input NAND of Fig. N-1 is not a 1, the output F is a binary 1. The NAND function may be obtained by inverting the output of an AND circuit; however, the NAND circuit has the advantage of requiring fewer circuit elements.

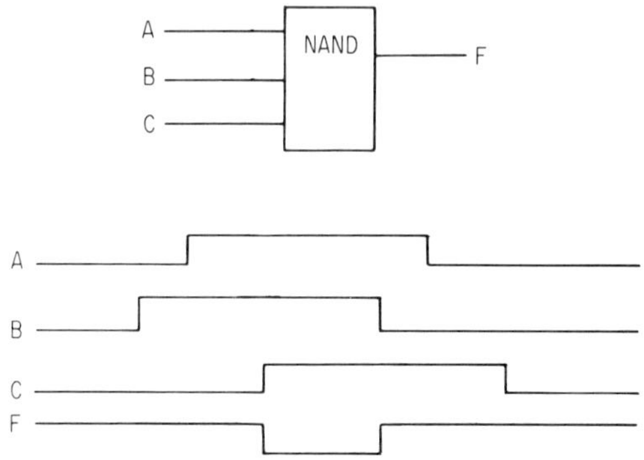

Fig. N-1. NAND circuit.

A two-input transistor NAND circuit is shown in Fig. N-2. The output F is negative only when both transistors are cut off. This occurs when both inputs are positive. The number of inputs, or fan-in, is a function of the components and circuit design.

—W. W. Welz, *IBM Corporation*

Natural Frequency, Transducer (See **Resonant frequency, transducer.**)

Navigation Systems A navigation system functions to answer the age-old questions of all travelers—"Where am I?" "Where am I going?" and "How fast am I going?" Taken as a whole, any navigation system can be considered as an instrumentation system, inasmuch as navigation (1) is based upon taking measurements—some simple and some sophisticated; (2) is involved with data processing, ranging from simple, manual arithmetic calculations to complex computing; and (3) requires a display of information. Further, automatic control, based upon these measurements and calculations, finds applications in a number of modern navigation and vehicular-control systems.

Position and speed are determined with respect to some reference coordinate system, such as latitude, longitude, altitude, and horizontal velocity relative to the earth. Direction is determined

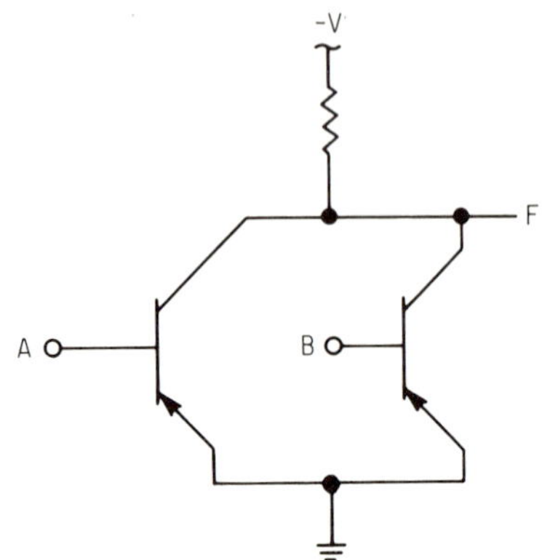

Fig. N-2. Transistor NAND circuit.

with respect to some reference direction, such as the ground-track angle relative to true north. With this basic navigational information and information about a desired path or a destination, a computer (possibly human) can determine deviations from the desired path and send steering signals to a control system to alter the path of the vehicle being guided.

Practically every navigation system can be described as one form or another of *dead reckoning*, such as depicted in Fig. N-3. By integrating appropriate velocity components from a known starting point, one can determine the distance traveled from that starting point and thus present position. The accuracy of dead-reckoning systems tends to degrade with time. Position *fix taking* often is required to update or reset the navigational information. If continuous position fixes are available, then the dead-reckoning extrapolation can be dispensed with or used as an auxiliary mode.

Navigation systems thus are observed to consist of a combination of sensors, computers, and if man is involved, displays. The sensors detect or measure some physical quantity which, by appropriate computation or reference to stored data, enables the determination of some navigation variable, such as time, acceleration, velocity, position, attitude, or heading. The computers can range from human beings with various com-

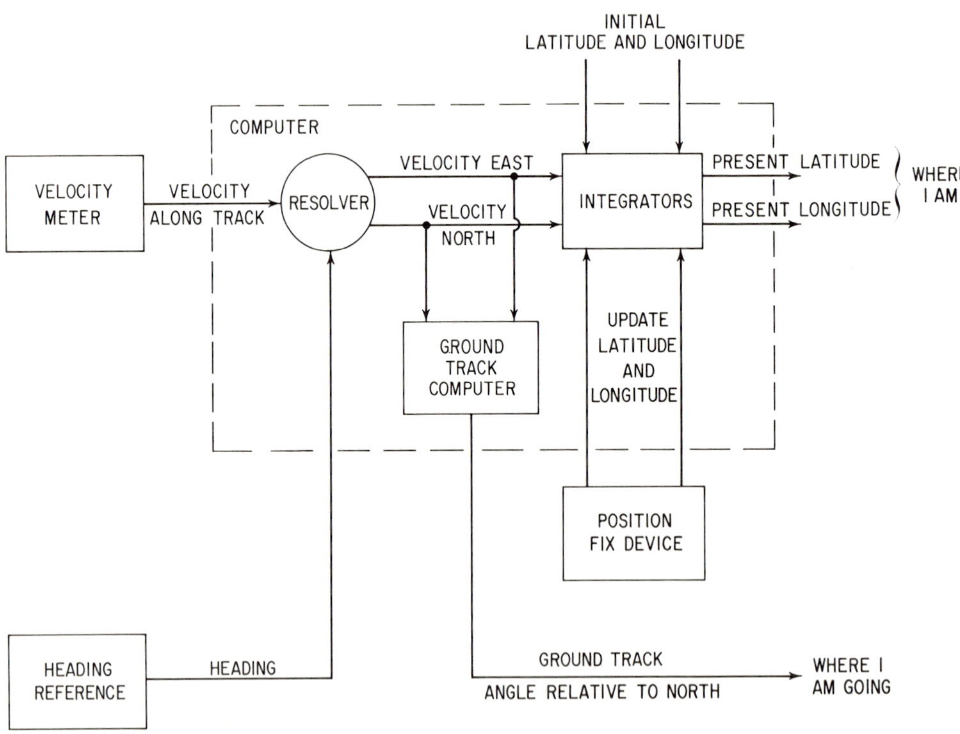

Fig. N-3. Dead-reckoning navigation.

TABLE N-1. *Functional Classification of Navigation Sensors*

Meters	Heading Reference	Position-fix Devices
Velocity meters:	Magnetic compass	Sextant
Pitot log	Gyro compass	Radio aids:
Electromagnetic log	Optical celestial tracker	LORAN
rpm tachometer	Radiometric celestial tracker	OMEGA
Airspeed (pitot) meter plus wind	Inertial systems	SHORAN
Doppler radar		RAYDIST
Laser radar		DECCA
Doppler sonar		VOR/DME
Inertial systems		TACAN
Time meters:		Navigation satellite (TRANSIT)
Chronometer		Landmarks:
Radio-synchronous signals		Optical sighting
Time standards (crystal oscillators)		Radar sighting
Altitude or depth meters:		Seamarks (sonar bottom fixing)
Pressure altimeter		Optical celestial tracker
Pressure-depth meter		Radiometric celestial tracker
Radar altimeter		Radio transponders:
Sonar fathometer		Ground or "mother ship"
Inertial systems		Trackers and data link
Ground or other vehicle		Gravity anomaly, map matching
Trackers and data link		Magnetic anomaly, map matching

puting aids, such as maps, charts, almanacs, plotting boards, and slide rules, to sophisticated automatic high-speed digital computers. Displays or recorders present the navigational data for man's interpretation and use.

Two broad categories of navigation should be noted: (1) *absolute navigation* and (2) *relative navigation*. Absolute navigation is generally defined to be the knowledge of present position in relation to an overall earth-coordinate system. For example, this may be latitude and longitude. Relative navigation is the knowledge of present position relative to some special *local* coordinate or grid system. One can readily navigate from New York City to Philadelphia, for example, without having to know the absolute latitude and longitude of either of the two cities. The accuracy attainable in a relative navigation situation usually is significantly better than the accuracy that can be attained on an absolute basis.

Classification of Navigation Sensors: Primary function of navigation sensors is the basis for the classification of Table N-1. This table includes the more commonly employed sensors in marine and aircraft navigation. Many navigations systems make use of various combinations of these sensors to obtain greater capability or for backup modes.

Recognizing the basic facts that the vehicle being navigated is in some relation to its surroundings (earth, space, or other vehicles), it is possible to construct a navigation-sensor tree, such as that diagramed in Fig. N-4. The major classes of this tree can be theorized from purely logical reasoning. Specific, present-day sensors that fall into the various categories are listed in the figure. A sensor tree of this type can be used as an aid in the synthesis of new navigation devices. The tree also can be employed as a means to determine the vulnerability (for military applications) of the various sensor techniques to detection or jamming or being knocked out by enemy action.

The sensor tree of Fig. N-4 can be subdivided in another manner to indicate the degree of self-containment of a given navigational technique. The possible classes and their definitions are given in Table N-2. Self-contained systems are less vulnerable than the other classes, although they may involve more weight to be carried and it may be dangerous to depend solely upon the on-board equipment.

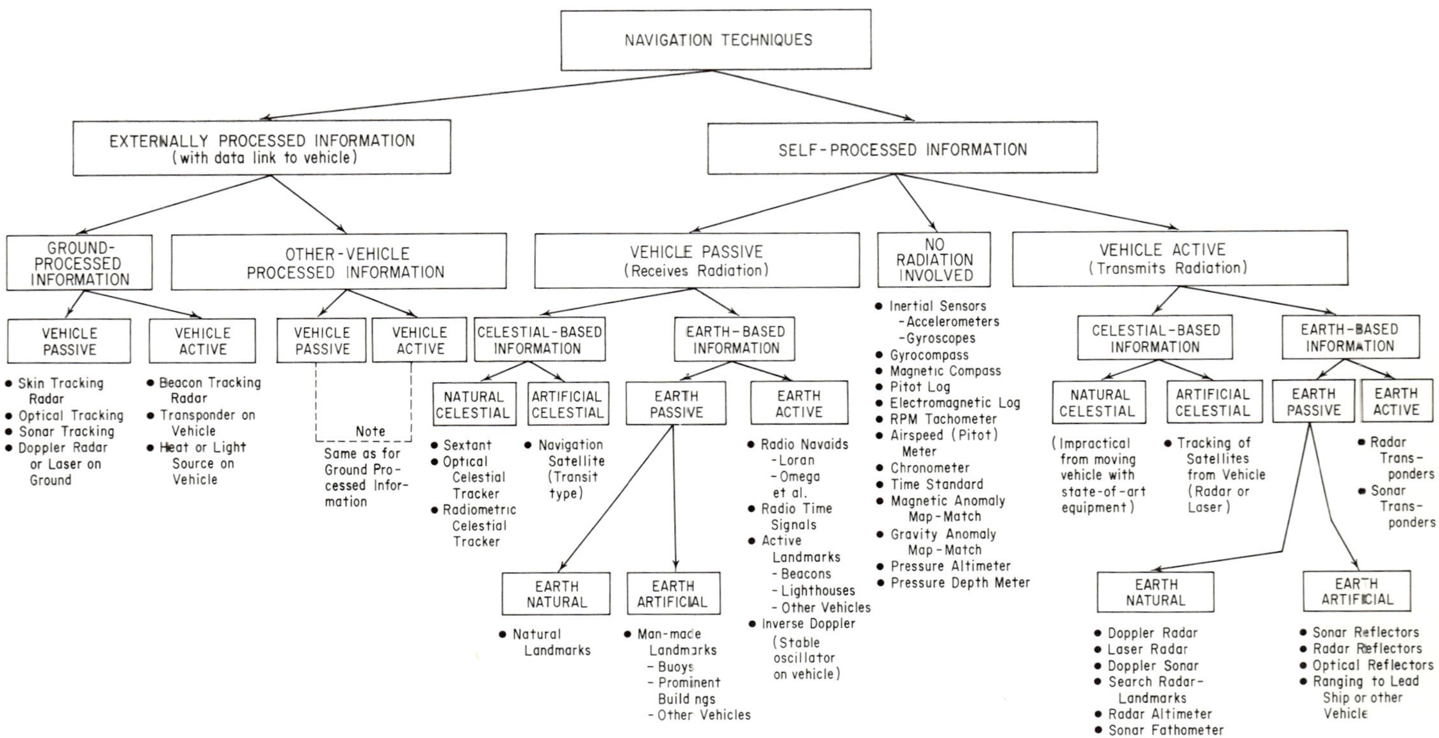

Fig. N-4. Navigation-sensor tree.

TABLE N-2. *Classification of Navigation Systems by the Factor of Self-containment*

1. Self-contained systems
   a. Completely self-contained
      Self-processed information—no radiation involved
      All measurements internal or in local vicinity of vehicle, such as inertial navigation
   b. Self-processed information—vehicle passive—"natural" earth or natural sky
      "Natural" can include man-made landmarks not purposely placed as navigation aids
      Optical star trackers
   c. Self-processed information—vehicle active—"natural" earth or natural sky
      Doppler radar
2. Self-controlled systems
   a. Self-processed information—vehicle passive—earth passive, artificial
      Purposeful navigation landmarks
         Navigation buoys
   b. Self-processed information—vehicle passive—active earth or active sky
         LORAN
   c. Self-processed information—vehicle active—earth passive, artificial or artificial celestial
      Ranging to lead ship
   d. Self-processed information—vehicle active—earth active
      Radar transponders
3. Externally controlled systems
   a. Ground or other vehicle processed information and data link—vehicle passive—external tracker passive
      Theodolite optical tracker
   b. Ground or other vehicle processed information and data link—vehicle passive—external tracker active
      Skin tracking radar
   c. Ground or other vehicle processed information and data link—vehicle active—external tracker passive
      Optical tracking of light on vehicle
   d. Ground or other vehicle processed information and data link—vehicle active—external tracker active
      Beacon tracking radar

Note: The examples given in Table N-2 are simply to identify each classification and are by no means all-inclusive.

There are many other factors that could be employed to categorize navigation sensors. In addition to the normal parameters of any physical equipment, such as accuracy, cost, size, weight, power consumed, reliability, operability, degree of automaticity, interface compatibility, and environmental capabilities, there are certain special factors that may come into play in the navigation field.

Worldwide availability is one factor. For example, LORAN C transmitters presently do not cover the entire world. Magnetic compasses and gyrocompasses (normal-mode) are essentially unusable in the polar regions of the earth. Certain techniques which depend upon prior surveys may not be available worldwide. Also, some sensors perform well under water while others do not.

All-weather operability and day-or-night capabilities are other factors. For example, optical star trackers cannot be employed if a cloud cover obscures the stars. Similarly, laser dopplers cannot be used with clouds below an aircraft. Certain optical devices cannot be used at night, while others cannot be used during daylight (like seeing stars against a bright background). Devices which track the sun or the moon obviously are limited in these respects.

Continuous vs. periodic availability is another factor. For example, because of the present utilization of only a limited number of satellites, a transit position fix is not available at all times, but only at periodic intervals when a satellite comes into radio view, the time interval varying with latitude.

It is rare that a single navigation sensor can satisfy all the desired requirements of a given application. Therefore, almost always, a navigation system will be made up of some special combination of sensors.

Celestial Navigation: Observations of celestial bodies, including the sun (solar), the moon (lunar), and the planets or stars (astral), are made to determine one's position. In 1837, Captain Thomas H. Sumner conceived what is referred to as the *Sumner line,* namely, a *line of position* that could be ascertained by measuring the altitude of a known celestial object. This required the solution of an astronomical triangle, requiring at least three known quantities. The altitude of the celestial body was determined with a sextant. See **Sextant.** Declination of the celestial body was determined with a meridian-circle telescope. The third quantity was the observer's latitude or longitude. But, the coordinates of the latter quantities ironically are what the observer is attempting to determine! By approximating the value of latitude and longitude from dead reckoning (the dead-reckoning position need only be known within 20 miles) and by sighting a celestial object that is nearly due north or south (a longitude change will cause but a small effect in the computed latitude) or an object that is nearly due east or west (a latitude change will cause but a small effect in the computed longitude), the observation can be made accurate within ¼ mile. Determining a line of position in this manner was a standard practice for shipmasters until the early part of the present century.

At any given instant of time, a celestial body is at the zenith for a particular location on the earth's surface. The U.S. Navy terms this location the ground position, or GP. Earth coordinates, of GP, can be given in terms of celestial coordinates. Thus, the *declination* of the object is equal to latitude; the *Greenwich hour angle* is equal to *longitude.* The *Nautical Almanac* and *Air Almanac,* published by the Nautical Almanac Office of the U.S. Naval Observatory, contain these coordinates for the sun, moon, and planets, what are termed *navigator's stars.* Greenwich Civil Time is the base, and sextant observations for altitude are recorded on this time base. As an observer departs from the GP, the object's altitude decreases in an amount equivalent to the distance moved by the observer in *nautical miles.* Plotting could be done on a sphere to represent the earth, but to achieve an accuracy equivalent to the sextant observation, a sphere of very large dimension would be required. Numerous ingenious geometric aids, including Mercator projection, have been developed. See Ref 1. The foregoing observations, while simple in concept, are difficult to accomplish on a moving body, particularly one subject to accelerations, and thus these methods have virtually been abandoned for airborne use. However, the importance of celestial navigation has taken on new significance in connection with certain deep-space flights as well as in certain missile-guidance systems where a telescopic instrument will lock onto a celestial body and give guidance during flight. A star tracker may be optical or radiometric.

Radio Navigation:[1] Radio direction finders were used very early to avoid the difficulties of celestial navigation from an aircraft and for emergencies in bad weather. The device simply uses a directional antenna to locate the direction to several radio stations. A simple triangulation then locates the aircraft with respect to the locations of the stations—usually well known and in the map of reference with which the pilot is familiar. The method is complicated by the aircraft velocity, but not by accelerations, weather, or the availability of tables. The pilot can tune in a station near his destination and merely follow the signal to it. Two deficiencies are apparent: (1) the location of the stations and (2) the errors inherent in a directional antenna. These problems led to the development of improved radio systems.

Hyperbolic Beacon Systems: A pair of radio-beacon transmitters located at known positions transmit signals that are in a known time relationship to each other. Since all radio signals travel at the same velocity, the locus of all points where the signals are received with a constant time delay between them is a hyperbola of possible positions. For each time delay there is one of a family of hyperbolas. A second pair of transmitters located on a different axis produces a family of intersecting hyperbolas. Each intersection defines a location within the range of the transmitters. The aircraft receiver obtains the time delays, and these are converted to a charted position from published data. The receiver display usually is a cathode-ray tube from which the operator makes the time-delay measurements. More elaborate systems have been built that use computers in which the display reads directly in latitude and longitude or in the form of a moving map.

[1] Information on radio navigation was prepared by Paul F. Bechberger of The Bendix Corporation and John W. Miller of Honeywell Inc.

DECCA: This is a continuous-wave hyperbolic radio aid to navigation in which a receiver measures and indicates the relative phase difference between signals received from two or more synchronized ground stations. The system provides differential distance information from which position can be determined. The system is quite accurate (0.03 nautical mile), with a range of about 300 nautical miles.

LORAN: This is a long-distance radio-navigation system for aircraft and ships, utilizing synchronized pulses transmitted simultaneously by widely spaced transmitting stations. Hyperbolic lines of position are determined by measuring the difference in the time of arrival of these pulses. The intersection of two of these lines of position, obtained from either three or four stations, gives a position fix. Standard LORAN operates on frequencies between 1,800 and 2,000 kHz.

LORAN C: This is a widely used version of LORAN that uses pulse signals for more precise time-delay measurement and operates at a frequency of 100 kHz. The range is 2,000 nautical miles.

LORAN D: This is a tactical LORAN system that uses the coordinate converter of low-frequency LORAN C.

OMEGA: This is a long-range system that can provide worldwide coverage with only eight stations. The operating frequency is 10 kHz, and the estimated accuracy is 5 nautical miles with typical receivers. The system was developed originally for totally submerged submarines. The operating principle is similar to *delrac,* a British radio-navigation system designed to provide worldwide coverage by using 21 pairs of master-slave stations, with a 3,000-mile range for each pair of stations. Frequencies used in delrac are in the band from 10 to 14 kHz. DECCA indicating equipment can be used with delrac.

OMNI: This is a radio system that includes the directional information by modulating its radio signal with a simple dot-and-dash code, one code for a position to the left of the beam and another for a position to the right of the beam. The stations are located on airways and at the approaches to airports. With the system, the pilot selects his OMNI way point or destination on the radio and listens for the code that tells him when he is to the left or right of the path. The system has been highly refined with on-board computers and displays. The system also is supplemented with distance-measuring equipment (DME). With the latter, the pilot interrogates the station with his transmitter and receives a distance indication from the station on his receiver. The measurement is made by determining the travel time of the radio wave. DME complicated the airborne equipment considerably but opened the way for easy, continuous navigation by using only a single set of equipment. Use of two stations can provide coverage for all locations within their range and thus free the pilot from flying the designated lines toward the OMNI stations. Very-high-frequency (vhf) omnirange operates in the band from 112 to 118 MHz.

GEE: A vhf radio-navigation system in which three or more ground stations transmit synchronized pulses. Hyperbolic lines of position are determined by measuring the differences in times of arrival of these pulses. The system was developed in Great Britain and is similar to LORAN.

TACAN: An air-navigation system in which a single uhf transmitter sends out signals that actuate airborne equipment to provide range and bearing indications with respect to the transmitter location, when interrogated by a transmitter in the aircraft. Each TACAN station broadcasts a location-identifying Morse-code signal at regular intervals. Also termed tactical air navigation.

SHORAN: A precision short-range position-fixing system using a pulse transmitter and receiver in an aircraft or other vehicle and two transponder beacons at fixed points. A receiver in the aircraft measures the round-trip times of the signals and converts these into distances to the fixed ground stations. Ordinary triangulation on a map then gives position.

There are numerous other radio-navigation systems, many principally of historical interest, but generally they are variations of the concepts and configurations presented by the foregoing systems.

Satellite-navigation Systems: The known position of one or more earth-orbiting satellites can be used as the basis for obtaining a position fix. TRANSIT, an early U.S. Navy satellite system, uses the doppler shift in frequency obtained when the satellite-transmitted signal moves relative to the aircraft. Two pairs of orbital readings are made to obtain a position fix at the intersection of two location hyperbolas. Correction is made for the aircraft travel that occurred between readings. Accuracy of this system is within $\pm\frac{1}{4}$ nautical mile. Proposed systems envision multiple satellites for simplified position fixing and/or global coverage. Some of these systems use satellites that are in synchronous orbit with the earth's rotation. The measurements may be in the form of travel times between signals, phase difference, incidence angles, or detection of swept signals from spinning satellites.

Inertial-navigation Systems: Some large aircraft now are using inertial navigators like those used by missiles and rockets.

Doppler Velocity Measurement: If a narrow-beam radar signal of a given frequency is reflected back from an object and there is a velocity difference between the radar set and the object, the frequency of the return signal will differ from that of the transmitted signal by an amount proportional to the velocity difference. There are sufficient irregularities in the surface of both land and sea to give effective stationary reflecting objects. Thus, an airborne doppler radar can determine the true fore-aft and side drift velocities of an aircraft relative to the earth.

Compass: A *float-type magnetic compass* is shown in Fig. N-5. If a two-pole magnet is pivoted about a vertical axis with its poles in the horizontal plane, the magnet will align itself with the horizontal component of the earth's magnetic field and indicate magnetic north, which differs from true north by the variation angle—which, in turn, varies from place to place on the earth's surface. In aircraft compasses, the magnet is

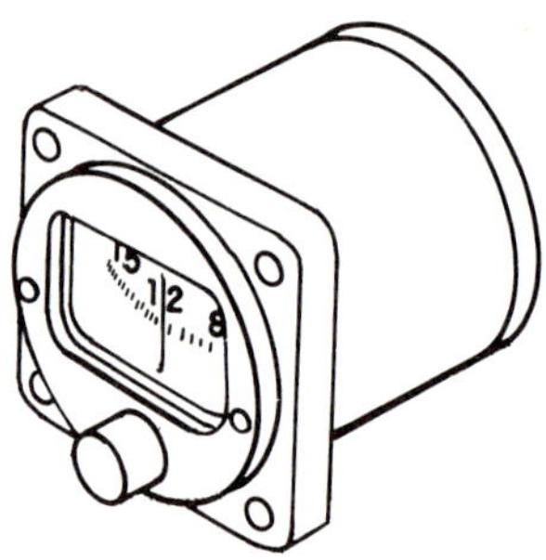

Fig. N-5. Float-type magnetic compass. *(Bendix.)*

attached to a float in an approximately spherical chamber filled with liquid, thus removing most of the weight of the magnet from its pivot to reduce pivot wear and friction. The float provides a reference to dynamic vertical, and the liquid provides damping. Usually there is considerable magnetic material in the aircraft acting as a disturbing influence on the compass. Thus a compensating arrangement of small adjustable magnets usually is provided to balance out this effect.

During any acceleration—as in a turn—the float, in referencing dynamic (not true) vertical, does not align the compass magnet to read the true horizontal component of the earth's field. In fact, in a turn in one direction in high latitudes, it is possible to have the dynamic vertical at a

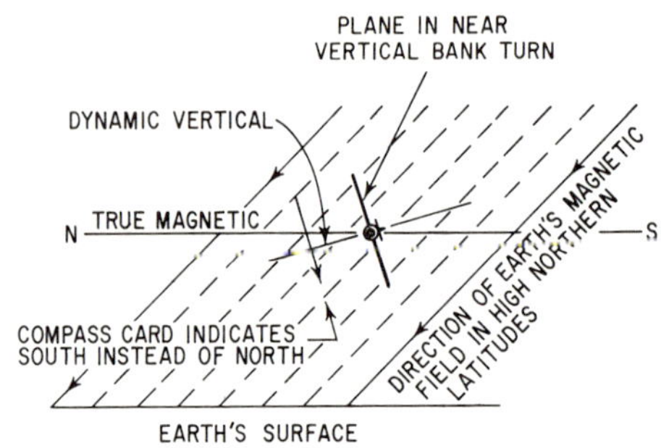

Fig. N-6. Northerly-turning error.

greater angle to the true vertical than is the earth's field and thus have the compass point south instead of north. This effect, shown in Fig. N-6, is known as the northerly-turning error. A signal can be taken from a float or pendulous-type compass and used as a slow correction on a directional gyro, and such a long-time average will be very nearly correct. This is known as a *gyrosyn compass* and is shown in Fig. N-7.

Directional gyros are in use with drift rates of 0.1 to 4° an hour. Correction must be applied for rotation of the earth and aircraft velocity. Such corrections, when properly applied, give a satisfactory heading reference in very high latitudes where magnetic headings are valueless. A device of this type, known as an "all-latitude" or "polar-path" compass, is shown in Fig. N-8. Gyroscopic instruments are described in greater detail under **Flight-control systems.**

REFERENCES

1. "Van Nostrand's Scientific Encyclopedia," 3d ed., D. Van Nostrand Company, Inc., Princeton, N.J., 1958.
2. Markus, John: "Electronics and Nucleonics Dictionary," 3d ed., McGraw-Hill Book Company, New York, 1966.
3. Gruenberg, Elliot L. (ed.): "Handbood of Telemetry and Remote Control," McGraw-Hill Book Company, New York, 1967.
4. "Private Pilot," Aero Products Research, Inc., Culver City, Calif., 1967.
5. Parvin, Richard H.: "Inertial Navigation," D. Van Nostrand Company, Inc., Princeton, N.J., 1962.
6. Mixter, George W.: "Primer of Navigation," D. Van Nostrand Company, Inc., Princeton, N.J., 1967.

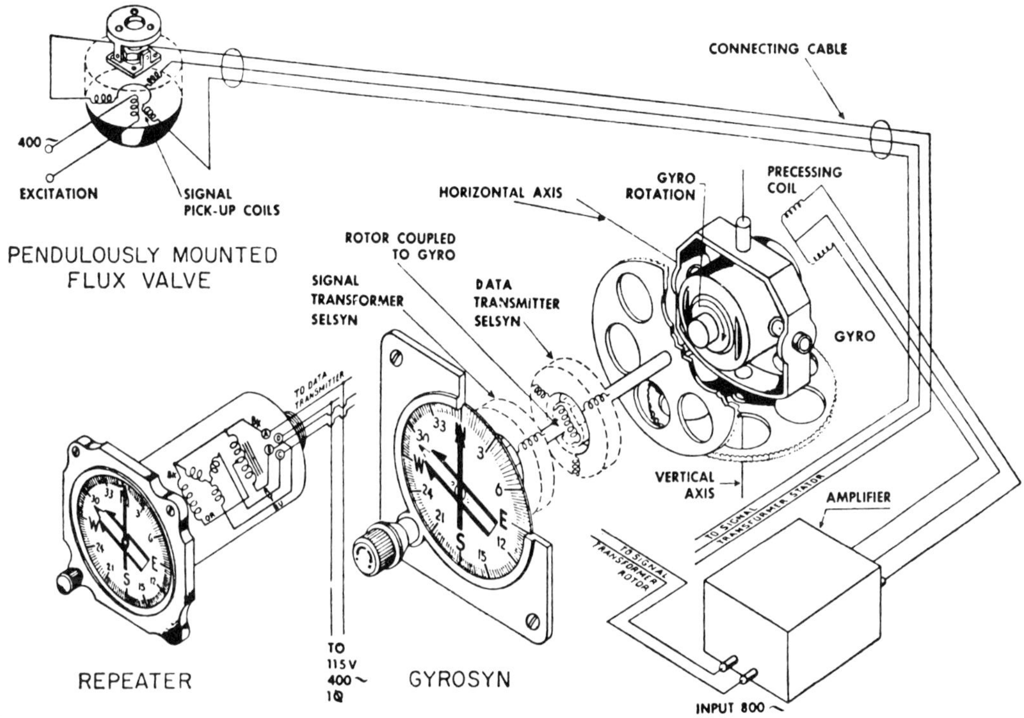

Fig. N-7. Gyrosyn compass. *(Sperry Rand.)*

Negative Feedback Feedback in which a portion of the output of a circuit, device, or machine is fed back 180° out of phase with the input signal. Negative feedback decreases amplification in such a way as to stabilize the amplification with respect to time or frequency, and also serves to reduce distortion and noise. Negative feedback also is termed *degeneration, inverse feedback,* and *stabilized feedback.*

Nephelometer An instrument for determining particle size or particle concentration by measuring the amount of light transmitted or scattered by the suspended particles. Quantitative determinations are made by comparing a given sample with a known standard. Also see **Photometers;** and **Turbidimeter.**

Nervous System, Measurements of (See **Central nervous system, instrumental analysis of.**)

Newton, Standard (See **Standards, measurement.**)

Newtonian Telescope (See **Telescope, reflecting.**)

Nickel Resistance Thermometer (See **Resistance thermometry.**)

Noise Noise is "an unwanted component of a signal or variable which obscures its information content." (SAMA Standard RC20-11-1964.) Although the standard from which this definition is taken defines "interference" as "Any spurious voltage or current arising from external sources and appearing in the circuits of a device," common practice is to use the terms synonymously.

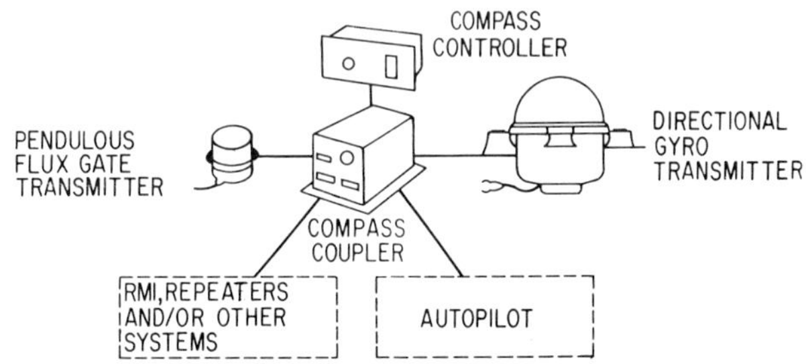

Fig. N-8. Polar-path compass. *(Bendix.)*

In common practice then, noise or interference is s spurious voltage or current signal component which obscures or alters the information content of a signal. The noise may be transient in nature or it may be a well-defined ac or dc signal. The source of the noise either may exist within the subsystem under consideration or may be external to the subsystem.

There are many potential sources of noise in an instrumentation or digital-data-acquisition system. Transient and ac noise may be coupled into measurement circuits via electrostatic and electromagnetic coupling. Dc noise signals may be introduced into a subsystem by current flow through impedances which are common to the grounding structures of two circuits or subsystems or as the result of common-to-normal-mode conversion.

There are two primary means of controlling noise in electronic circuits. The first, and by far the most effective when it is possible, is total elimination of the noise source. This might consist, for example, of eliminating arcing across current-carrying contacts by use of suppression circuits, thereby eliminating spurious electrostatic fields. The second approach is to provide electrical isolation between the noise source and the receiving circuit. This isolation might consist of an electrostatic or electromagnetic shielding enclosure surrounding either the source or the receiving circuit. Merely providing spatial separation between the source and the receiving circuit is often sufficient since electromagnetic- and electrostatic-field intensity decreases rapidly with distance. Thus, sensitive circuits and signal lines should be carefully positioned to avoid sources of electrical noise. If the noise interference is due to currents in common ground impedances, providing separate grounds to a common subsystem or system signal-reference ground is often the required solution (see **Grounding; Ground loop**).

—Thomas J. Harrison, *IBM Corporation*

Noise Analyzer (See **Wave analyzer.**)

Noise Generator A device for generating a random-noise signal for test purposes. Common noise sources are a temperature-limited diode operated at cathode saturation, an electron multiplier, a crystal diode with positive bias, a nonoscillating reflex klystron, and a nonoscillating magnetron.

Noise Level, Sound (See **Acoustic instrumentation.**)

Nondestructive Inspection A body of technology has built around the need for and the instrumental methods developed to examine materials and products without altering them in any significant way. These inspecting and testing methods are to be contrasted with destructive tests in which the specimen suffers physical or chemical damage during the testing procedure. In the usual interpretation, nondestructive inspection—also sometimes referred to as nondestructive testing—is used to detect unsafe or potentially unsafe conditions, such as cracks, voids, holes, inclusions, and other inconsistencies in metal forms, such as sheets, plates, bars, and tubes, in castings, forgings, and weldments. Such defects may arise from faulty manufacturing or from later use of products that may arise from the effects of corrosion, abrasion, vibration, and mishandling and inattention to maintenance procedures, such as lubrication and painting. Although nondestructive inspection was pioneered around metals, many of the techniques can be applied effectively to other kinds of materials, such as plastics, ceramics, wood, in various composition substances. Of course, where nonmetallics are involved, the range of testing procedures becomes more limited.

The major nondestructive-inspection techniques are (1) radiographic methods, (2) magnetic-particle methods, (3) penetrant methods, (4) eddy-current methods, and (5) ultrasonic methods.

Radiography: Radiation from either an x-ray or an isotope source is focused on the test piece, and the amount of radiation that passes through the piece is recorded on film. More or less radiation reaches the film in proportion to the thickness or density variations in the test piece. Thus, a shadow image of light and dark shades is produced. These images reveal flaws or inclusions in the test piece. Large assemblies can be radiographed weld by weld to prove the dependability of an entire structure. Radiation-absorption techniques can be applied over large areas or concentrated over a thin portion of the test object by using narrow-beam radiation-absorption gages. In the beta-backscatter technique, the intensity of radiation that is reflected rather than the radiation that is transmitted is measured. This method is particularly suited to measuring the thickness of coatings or boundary conditions between a thin layer of material that may be mounted on a substrate. X-ray emission or fluoroscopy also is a useful nondestructive-testing method. See **Thickness gaging, nuclear-radiation; Thickness gaging, x-ray;** and **X-ray fluorescence analysis.**

Magnetic-particle Method: In this technique, iron powder is used to reveal the leakage magnetic field created at a flaw or break when any part is magnetized. The familiar horseshoe magnet illustrates the principle. If the magnet is bent into a circle, the field between the ends attracts and holds magnetic iron powder. But, if the magnet is completely closed, the field will be contained entirely within the ring and no iron powder will be attracted. However, if the round magnet is cracked, poles are created at the break and iron powder is attracted to the cracked area to pinpoint the defect. Thus, in use, any crack or defect at or near the surface of the test piece is clearly outlined by the iron powder attracted to the area. Fluorescent paste and black light will reveal minute flaws. Ferrous-iron or steel parts that can be magnetized may be tested. Flaws on the surface or immediately beneath the surface, weld defects, and inclusions are reliably revealed. Tubing, piping of any size or shape, plate, and steel castings and forgings can be tested by this method in either the laboratory or the field. The simplicity of the method makes magnetic-particle testing one of the most popular procedures for locating surface or near-surface flaws in critical parts that can be magnetized.

Penetrant Methods: In this method, dirty parts are washed with a solvent cleaner and dried. Then penetrant is applied freely on the part and allowed to work into tight cracks. Penetrant is removed from all surface areas, and the piece is sprayed with a developer. The developer dries to an even white coating, while the penetrant bleeds up from any flaws through the developer, forming bright red or fluorescence indications on the white surface. The size of the defect is indicated by the richness of color produced, by the speed of bleed-out, and by the dimensions of the indication. Penetrant testing will locate defects open to the surface in welds, cracks in castings; cracks, laps, and seams in forgings and extrusions; and cracks due to heat treating. All metals, ferrous or nonferrous, and many nonmetallic materials can be tested by this method. It should be stressed that the flaw must be open to the surface.

Eddy-current Methods: This testing method reveals any differences in electrical impedance between parts to be tested and a reference sample. Parts to be examined are passed through a coil or explored with a probe, and a trace appears on a cathode-ray tube. Since magnetic and electrical characteristics are related closely to metallurgical properties, a trace position or pattern clearly shows up in variations of metal hardness and composition as well as any defects. Both ferrous and nonferrous parts can be tested, and various coils, probes, and detector tips are available. Eddy-current testing is highly sensitive and reveals flaws and checks hardness within a few points on the Brinell or Rockwell scale. The instruments are rugged and compact and well adapted for electronic sorting and testing up to 4,000 pieces per hour. Also see **Hardness, measurement of.**

Ultrasonic Methods: A pulsed beam of ultrasonic vibrations is passed through the piece to be tested. The beam travels unimpeded through large parts, may be angled for testing sheet stock, and can inspect materials immersed in a liquid. Any flaw reflects vibrations back to the instrument, which indicates the location and size of the discontinuity on a cathode-ray tube. Access is required to only one side of the material being tested. This method is useful for testing all metals and solid nonmetallic materials, such as plastics. The method is widely used in field and laboratory for tests on machinery, ships, aircraft, aerospace vehicles, and refinery equipment. In one interesting application, the device can measure the effects of corrosion (in terms of varying thickness). Also see **Thickness gaging, ultrasonic.**

REFERENCES

Lamble, J. H.: "Principles and Practice of Non-destructive Testing," John Wiley & Sons, Inc., New York, 1963.

Cofield, R. E.: Survey of Nondestructive Test Methods, *Instr. Technol.*, vol. 14, no. 11, pp. 38–44, November, 1967.

Ostrofsky, Bernard: Nondestructive Tests for On- and Offline Inspections, *Chem. Eng.*, May 20, 1968, pp. 174–180.

Kirchner, Russell W.: Equipment Testing, *Chem. Eng.*, Aug. 28, 1967, pp. 107–130.

(EDITOR'S NOTE: Special acknowledgment in the preparation of this summary is made to Automation Industries, Inc.; Krautkramer Ultrasonics, Inc.; and Magnaflux Corporation.)

Nonlinearity (See **Linearity.**)

NOR Circuit The NOR circuit is a logical element which provides a binary 1 output if all the input signals are a binary 0 (see **Binary state**). Another way of expressing this function is that it is the overall NOT of the logical OR operation. The Boolean-algebra representation of this circuit is $F = (AB)' = A' + B'$, where the prime denotes the NOT function.

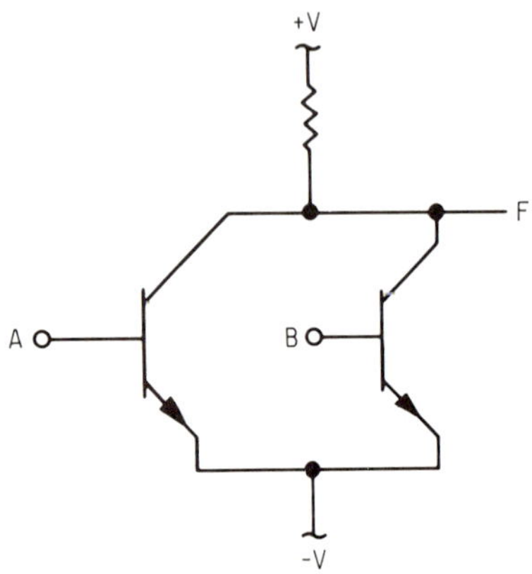

Fig. N-9. Transistor NOR circuit.

Figure N-9 illustrates a two-input transistor circuit of the NOR function. The output F is positive only when both transistors are cut off, and this occurs when both inputs A and B are negative.

—W. W. Welz, *IBM Corporation*

Normal-mode Voltage (See Differential-mode voltage.)

Nuclear Magnetic Resonance Chemical Analyzer (NMR) A spectroscopic technique based upon the principle that nuclei having a magnetic moment, when placed on a homogeneous magnetic field, will precess around the axis of the magnetic field at a rate which is dependent upon the strength of the magnetic field. If, at the same time, these nuclei are brought in contact with a radio-frequency (rf) source of the same frequency as the precessional frequency, the nuclei will absorb energy from the rf source in direct proportion to the number of nuclei present.

Different nuclei will have different precessional frequencies and therefore will absorb energy at different radio frequencies. Also, nuclei of the same nuclear species (such as hydrogen nuclei) will absorb energy at slightly different frequencies, dependent upon their molecular environment. In very homogeneous magnetic fields, fine structure also is present which gives information concerning the number and type of neighboring nuclei.

When a magnetic field is applied, nuclei act like miniature gyros and begin to precess about the axis of the magnetic field. The rate of precession or precessional frequency is called the Larmor frequency and is given by the formula $\omega = \gamma H$, where ω is the Larmor frequency, ω is the gyromagnetic ratio of the nucleus, and H is the strength of the external magnetic field. Magnetic moment is that moment induced by the nuclear spin.

An NMR spectrometer is very useful for determining the type and quantity of various nuclei present in a chemical sample. The technique is especially powerful for determining the chemical structure of an unknown material and for measuring the relative concentrations of mixtures of compounds and isomers with slightly different chemical structures. Common applications include (1) structural determinations of complex organic molecules, (2) quantitative determinations of mixtures, (3) determination of isotopic abundances and the extent of isotopic substitutions, and (4) studies of the structural characteristics of synthetic polymers and bipolymers.

—Jerome L. Holcomb, *Varian Associates*

Nuclear Magnetometer (See Magnetometer, airborne.)

Nuclear-radiation Detection [See Radiation (nuclear), detection of.]

Nucleus Program (See Executive program.)

Null-balance Method A method of measurement in which the measuring circuit is balanced to bring the instrument pointer to zero, as in a Wheatstone bridge, and the settings of the balancing controls are then read—or are used to position automatically a recording or controlling mechanism. Also called balance method and zero method. See **Bridge circuits, electric**; and **Potentiometric instruments.**

Null Detector (See **Bridge circuits, electric**; and **Potentiometric instrument.**)

Number Systems A method of positional notation using numerical symbols for quantity representation. The Arabians invented the numerical symbols and the system of positional notation on which our present decimal system and many other number systems are based. Each of the symbols has a fixed value one higher than that of the symbol before it in the progression from smallest to largest: 0, 1, 2, 3, 4, 5, 6, 7, 8, 9. When several symbols (or digits) are combined, the value of the number depends upon the relative positions of the individual digits, as well as on the digit values. In any system of positional notation, the digit position on the extreme right is the one of the least value, or lowest order, and is called the "least significant digit" (LSD); the digit on the extreme left is the one of the highest value and is called the "most significant digit" (MSD). The increase in value of each digit position depends on the radix (or base) of the number system. Thus, in the decimal system with radix 10, the value of the digit positions to the left of the least significant digit (or unit) increases by a power of 10 for each position. The decimal system has the radix (base) 10 because it has 10 discrete number symbols (0 to 9) available for counting.

As an example of positional notation, consider the decimal number 6,594. Although its value is immediately apparent, the notation 6,594 actually signifies

$$\begin{array}{llllll} & 6 \text{ thousands} & + 5 \text{ hundreds} & + 9 \text{ tens} & + 4 \text{ units} & \\ \text{or} & 6{,}000 & + 500 & + 90 & + 4 & = 6{,}594 \end{array}$$

The positional value of each digit is made even clearer when the number is expressed in powers of 10:

$$6{,}594 = 6 \times 10^3 + 5 \times 10^2 + 9 \times 10^1 + 4 \times 10^0$$

Positional notation is not possible without the zero. Its presence within a number simply means that the power of the base represented by the 0 digit position is not used. Thus, the decimal number 8,003 signifies:

$$\begin{array}{llllll} & 8 \times 10^3 & + 0 \times 10^2 & + 0 \times 10^1 & + 3 \times 10^0 & \\ = & 8 \times 1{,}000 & + 0 \times 100 & + 0 \times 10 & + 3 \times 1 & \\ = & 8{,}000 & & & + 3 & = 8{,}003 \end{array}$$

Fractions and mixed numbers are treated in positional notation in a similar fashion. Each digit position to the *right of the radix point* is assigned a negative power of the radix starting with -1, in ascending sequence. Thus, in the decimal system, the first digit to the right of the decimal point is multiplied by 10^{-1}, the second digit by 10^{-2}, the third by 10^{-3}, and so on. For example, the mixed decimal number 436.578 may be expressed as

$$\begin{array}{lllllll} & 4 \times 10^2 & + 30 \times 10^1 & + 6 \times 10^0 & + 5 \times 10^{-1} & + 7 \times 10^{-2} & + 8 \times 10^{-3} \\ = & 400 & + 30 & + 6 & + 5/10 & + 7/100 & + 8/1{,}000 \\ = & 436.578 & & & & & \end{array}$$

These rules of positional notation are generally applicable to all number systems, regardless of the radix used.

Binary Numbers: The binary (radix 2) number system uses only two distinct symbols, 0 and 1, which signify "no units" and "one unit," respectively. (See also **Binary state.**) In contrast to the decimal system, however, the place value of binary digits to the left of the least significant digit (LSD) increases by a power of 2 each time, rather than by powers of 10. For example, the binary number 101101 signifies

$$\begin{array}{llllllll} 101101 = & 1 \times 2^5 & + 0 \times 2^4 & + 1 \times 2^3 & + 1 \times 2^2 & + 0 \times 2^1 & + 1 \times 2^0 \\ = & (1 \times 32) & + 0 & + (1 \times 8) & + (1 \times 4) & + 0 & + (1 \times 1) \\ = & 32 & + 0 & + 8 & + 4 & + 0 & + 1 \\ = & 45 \text{ (in the decimal system)} & & & & & \end{array}$$

Expressing a binary number in powers of 2, thus, is one way (though not usually the best) of finding its decimal equivalent.

To avoid confusion when several systems of notation are employed, it is customary to enclose each number in parentheses and to write the base as a subscript, in decimal notation. Using the previous example:

$$(101101)_2 = (45)_{10}$$

Fractions are handled in the same way by assigning negative powers of 2 to the right of the *binary point* in ascending sequence. For instance, the binary number 0.1011 means

$$\begin{array}{llllll} (0.1011)_2 = & 1 \times 2^{-1} & + 0 \times 2^{-2} & + 1 \times 2^{-3} & + 1 \times 2^{-4} & \\ = & 1/2 & + 0 & + 1/8 & + 1/16 & = 11/16 \\ = & 0.5 & + 0 & + 0.125 & + 0.0625 & = (0.6875)_{10} \end{array}$$

Again, a literal expansion of the binary number in powers of 2 yields the decimal equivalent. Simpler methods of conversion are described below.

For reference, the first 16 binary numbers and their decimal equivalents are

Decimal	Binary
0	0000
1	0001
2	0010
3	0011
4	0100
5	0101
6	0110
7	0111

Decimal	Binary
8	1000
9	1001
10	1010
11	1011
12	1100
13	1101
14	1110
15	1111

Hexadecimal Numbers: Large binary numbers consist of long strings of zeros and ones which are frequently awkward to interpret and handle. The hexadecimal (radix 16) numbering system is used as a convenient way of representing such large binary numbers. Each hexadecimal digit stands for four binary digits.

Hexadecimal notation requires the use of 16 symbols to represent 16 number values. Since the decimal system provides only 10 number symbols (0 to 9), six additional marks are needed to represent the remaining values. The letters A, B, C, D, E, F have been adopted for this purpose, though any other six marks could have served equally well. The entire list of hexadecimal symbols, thus, consists of 0, 1, 2, 3, 4, 5, 6, 7, 8, 9, A, B, C, D, E, and F, in ascending sequence of value. Table N-3 shows equivalent decimal, hexadecimal, and binary numbers (through decimal 31). Note that upon reaching decimal 16, the hexadecimal symbols are exhausted, and a "1 carry" is placed in front of each hexadecimal symbol during the second cycle, from decimal 16 through decimal 31.

To convert binary numbers to hexadecimal notation, simply divide the number into groups of four binary digits, starting from the right, and replace each group by the corresponding hexadecimal symbol. If the left-hand group is incomplete, fill in zeros as required. For example, the binary number

$$\begin{aligned} 111110011011010011 &= 0011/1110/0110/1101/0011 \\ &= 3 \quad E \quad 6 \quad D \quad 3 \\ &= (3E6D3)_{16} \end{aligned}$$

If the binary number is a fraction or a mixed number, care must be taken to mark off groups of four bits from each side of the binary point position. Thus, the binary number

$$\begin{aligned} 1011001010.1011011 &= 0010/1100/1010.1011/0110 \\ &= 2 \quad C \quad A \quad B \quad 6 \\ &= (2CA.B6)_{16} \end{aligned}$$

Similarly, to convert hexadecimal numbers into binary, substitute the corresponding group of four binary digits for each hexadecimal symbol and drop off any unnecessary zeros. For instance, the hexadecimal number

$$\begin{aligned} (6C4F2E.7B8)_{16} &= 6 \quad C \quad 4 \quad F \quad 2 \quad E \quad 7 \quad B \quad 8 \\ &= 0110/1100/0100/1111/0010/1110.0111/1011/1000 \\ &= (11011000100111100101110.011110111)_2 \end{aligned}$$

The meaning of hexadecimal numbers is made clear by expansion in powers of 16. For example, the hexadecimal number 2CA.B6, when decimals are substituted for hexadecimal symbols, is evaluated as

TABLE N-3. *Decimal, Hexadecimal, and Binary Notation*

Decimal	Hexadecimal	Binary
0	0	0000
1	1	0001
2	2	0010
3	3	0011
4	4	0100
5	5	0101
6	6	0110
7	7	0111
8	8	1000
9	9	1001
10	A	1010
11	B	1011
12	C	1100
13	D	1101
14	E	1110
15	F	1111
16	10	10000
17	11	10001
18	12	10010
19	13	10011
20	14	10100
21	15	10101
22	16	10110
23	17	10111
24	18	11000
25	19	11001
26	1A	11010
27	1B	11011
28	1C	11100
29	1D	11101
30	1E	11110
31	1F	11111

$$\begin{aligned}&2 \times 16^2 + 12 \times 16^1 + 10 \times 16^0 + 11 \times 16^{-1} + 6 \times 16^{-2}\\&= 2 \times 256 + 12 \times 16 + 10 \times 1 + 11/16 + 6/256\\&= 512 + 192 + 10 + 0.6875 + 0.0234375\\&= 714 + 0.7109375\\&= (714.7109375)_{10}\end{aligned}$$

In working out an example of this type, it is best to arrange the products in a vertical column for convenient addition.

Note: This entry is reprinted by permission from "Number Systems," IBM Form C20-1618-1, 1965.

—B. W. Landeck, *IBM Corporation*

Numerical Control In general and historic usage, the term *numerical control* connotes a positioning-control system for machine tools and other kinds of industrial-production machines. Numerical values corresponding to desired positions of tools and symbolic values corresponding to auxiliary functions are recorded on punched paper tape, punched cards, or magnetic tape where the information can be stored indefinitely. Tape or card readers convert the information into signals which operate servomechanisms on each axis of the machine whose motion is to be controlled. Because perforated paper or plastic tape has been used so widely, numerical control sometimes is referred to as "tape control" and sometimes simply "N/C." Although comprising but a small fraction of the total installations, dial-input numerical-control systems also are in use. Numerical control originally was refined for metalworking machinery, such as mills, drills, boring machines, and punch presses, and the largest number of installations is in connection with such equipment. In recent years, N/C has moved out of the bounds of conventional milling-drilling-turning applications and, in the metalworking field, now is used in connection with cutoff machines, broaching machines, grinders of many types, tube benders, gear cutters, ECM machines, flame cutters, and welders. Outside of metalworking, N/C is finding growing acceptance for inspection, automatic-testing systems, simple drafting systems, electronics-assembly machines, step-and-repeat photography, laser cutting and bonding processes; woodworking machinery, and garment-production equipment. The direct coupling of digital computers to machine-tool control systems also is broadening the field of application and the basic definition of numerical control.[1]

Objectives of Numerical Control: The basic objective for the development of numerical control for the metalworking field was that of *reducing the economic lot size* for machined parts production. N/C achieved this objective by reduced machining time, reduced fixture costs, increased cutter life, reduced tooling storage, a lowering of manpower skills required, increased product uniformity, reduced setup times, and a general reduction of other direct costs. Numerical control, like other forms of automation, makes it possible to predict the production process, taking decisions out of the hands of operators. Historically, N/C has permitted closer shop scheduling, more accurate cost prediction, faster cash turnover, higher facilities utilization, and a more rapid return on equipment investment—than manual or other less sophisticated machine control techniques.

Choice of Numerical Commands: One may logically ask, "Why select a control system based upon *numerical* inputs instead of some form of nonsymbolic information?" The other forms of controls that also are used successfully on machines include (1) *tracer* controls which use a form of *mechanical-analog input,* usually a machine template or a three-dimensional model, wherein the analog is scanned and sensed by an electric or hydraulic probe—and where the probe's motions then are repeated by a servo-controlled cutting tool; (2) the "automatic" machines, such as automatic screw machines[2] and large turning machines as found in high-production plants, which use manually preset hard stops to determine the limits of slide motions during cuts; and (3) *graphic-input* systems where an optical line follower can trace or follow curves in graphic form and hence guide machine spindles or, more commonly, flame cutters.

As contrasted with mechanical-analog or graphic inputs, the symbolic inputs of N/C offer the following advantages:

[1] By way of historical comparison, it is interesting to note that the slotted rolls used in the old player piano (a pneumatic system), the arrangements of pins and cams in music boxes, the notched and slotted disks used in nicklodeons and automated calliopes, and early knitting machine and flour-milling machines represented, in essence, open-loop numerical control.

[2] In practicality and certainly in sophistication, N/C can be considered more automatic than the so-called "automatic screw machine." Use of the term *automatic* was preempted many years ago for automatic screw machines and has persisted in usage.

1. The input usually is expressed in binary-coded decimal format and hence can be processed easily by digital-logic controls.

2. A physical model is not required, and often an accurate drawing of the desired part is not needed. Thus skills for preparation are saved. Also it requires less space to store tapes or cards for future production runs than to store models, templates, and so on.

3. Symbolic control lends itself well to the expression of nondimensional commands, such as spindle-speed change, tool change, drill cycle, and coolant on/off.

4. With N/C, the accuracy of the input medium is easy to verify.

It should be stressed, however, that both mechanical-analog-input and graphic-input control systems also have advantages for certain applications. It is interesting to note that N/C machines frequently are used to create the tooling for the other types of machines.

Major Classes of N/C Systems: Numerical-control systems range from very simple to quite complex systems. The more important variations are defined as follows. For additional control-system details, also see **Position-measurement and control systems.**

Point-to-Point Control: This is the simplest form of N/C and is used to guide the machine to particular programmed points, as in drilling, with no particular regard for path or velocity control. The system also is called *positioning* control. Special versions of point-to-point control systems have the ability to control *straight cuts,* usually parallel to machine axes, or at 45° to the axes. Such systems will have velocity control as well as end-point and limited-path control.

Continuous-path Control: This system contains computing elements (interpolators) which permit computation of successive points on desired line segments or curves, starting only from minimal input data, such as end points, radii, or center coordinates. Interpolation may be circular or parabolic. Continuous-path controls also are referred to as *contouring* controls.

Closed-loop Control: These systems use position, and often velocity, feedback to control the dynamic behavior and final position of machine slides. A variety of position transducers of both the analog and digital type is used. See **Position, measurement of.** Analog types include synchros, resolvers, and linear inductive scales. Digital devices include binary encoders, rotary pulsers, linear optical scales, and laser interferometers.

Open-loop Control: In these systems, there is no feedback to the main console. These systems rely upon the drive motor to provide both the actuation and measurement functions. A form of local feedback, integral to the motor, often is used to ensure command execution. Such controls usually use stepping-motor drives (sometimes called digital motors or pulse motors). See **Stepping motor.**

Absolute Control System: This term usually is used to identify controls that use absolute rather than incremental feedback. Absolute feedback connotes assignment of a unique value of each possible position of the machine slide, whereas *incremental feedback* simply gives the control a pulse each time the slide has traveled one increment of resolution (such as 0.0001 in.). Absolute systems do not require rezeroing after shutdown.

Programming: The process of converting the desired tool motions and auxiliary commands to a tape or card input. Programming may require the use of a general-purpose computer with special-purpose software for complex parts. In *absolute programming,* all dimensions are referenced to a single point. In *incremental programming,* each dimension is merely referenced to the preceding point in the program.

Machining Center: This is a new generation of machine tool, made possible by numerical control, and has no manual counterpart. The machine is characterized by an automatic tool-changing mechanism, together with a large magazine of tools. Most machining centers basically are milling-type machines, but turning centers are becoming more common.

Features of N/C Systems: The following terms commonly appear in the literature of controls builders:

Tool-length Offset: A feature that permits on-floor correction for worn or reground tools without changing tapes or presetting tools.

Manual Data Input: This feature permits manual (nontape) entry of dimensions and auxiliary functions into the system.

Position Readout: A display, usually digital, of slide positions.

Cutter-radius Offset: A feature that permits on-floor compensation for different-diameter cutters. In some controls, the same tape can be used to make both male and female parts.

Mirror Image: The ability to change the sign of commands on one axis so that one tape can make a right and a left part.

Tape Search: A feature that permits the operator to find his way to a particular block of information on a tape. This feature often is accompanied by a numerical display of current tape position.

Shared Axes: A feature that permits one data channel to move two or more different machine axes, one at a time. This feature is used to minimize the control cost for a multiaxis machine. The term *time-shared axes* also is used to describe this feature.

Operating Principles of N/C: The most common form of closed-loop numerical control (often called semi-closed-loop control, since the position of slides is only inferred from a rotary device) is shown in Fig. N-10. In this system, the *control-logic unit* includes the operator controls and the tape reader, together with circuits for reading, decoding, and timing commands. The output of the control-logic unit in most cases is a train of pulses that correspond in length and rate to the desired distance and velocity of motion.

Most closed-loop controls employ velocity feedback as well as position feedback for two reasons: (1) to permit precise control of cutting feed rates and (2) to permit use of low-gain servos and thus reducing end-point overshoot, hunting, and the other undesirable effects of high-gain servos. Rotary position transducers normally are used on all but the most precise machines, such as jig borers, inasmuch as rotary devices are far less costly than linear transducers. Rotary transducers, inherently incremental, can be made "absolute" by cascading several of the transducers through gear trains on a single axis.

The actual servomotor may be electric or hydraulic. In some machines, the actuator is not a motor on a lead screw, but a hydraulic postion. The latter approach, however, does exhibit difficulty in dynamically controlling a variable-volume oil column. Because of the necessity of designing the servo around the machine and then tuning it during installation, closed-loop control is appropriate for original equipment application, but less so for retrofitting.

An open-loop numerical-control system is shown in Fig. N-11. The control-logic sector is about the same as used in closed-loop systems. In the open-loop system, however, the drive motor (usually a stepping motor) and its amplifier subsume all the functions of the elements that are enclosed within the dotted box of Fig. N-10. The open-loop system offers equal accuracy and performance, but unlike the closed-loop system, there is no inherent position verification. Both systems depend equally upon the accuracy of the lead screw and other mechanical elements for final workpiece accuracy.

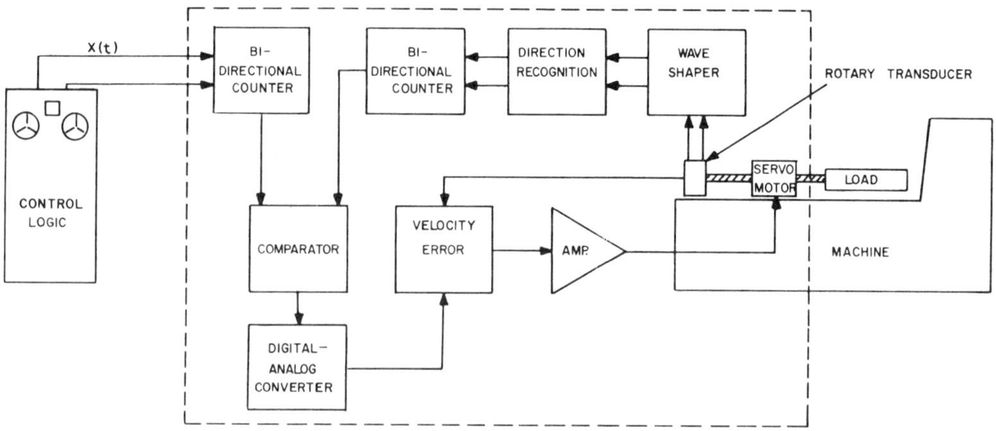

Fig. N-10. Closed-loop numerical-control system.

Unlike the closed-loop system, the open-loop system is relatively indifferent to machine dynamics and resonances over a wide range. The control never is told what the machine is doing; any correction is done within the motor. For this reason, open-loop control is popular for retrofitting where machine dynamics may be unknown. On the other hand, closed-loop control is used extensively on new N/C machine designs.

Mechanical Considerations and Machine Dynamics: In connection with N/C, the same considerations apply in adapting a servosystem to a machine tool as to any other mechanical system. The machine must be designed so that low resonant frequencies are avoided, thus promoting servo stability. Static and dynamic friction must be greatly reduced and eliminated where possible through the use of ball lead screws, roller ways, and related devices—for the dual purpose of achieving servo stability and reducing wear. See **Lead screw.** Positive lubrication is essential.

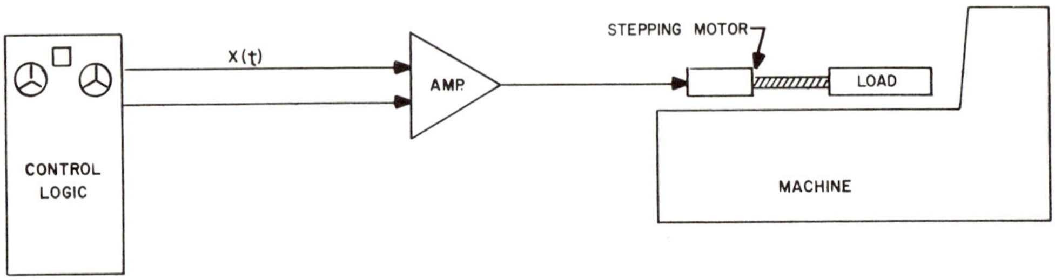

Fig. N-11. Open-loop numerical-control system.

Wherever possible, it is desirable to eliminate gearing in axis drives because gears are a source of inaccuracy, wear, and cost. Direct lead-screw drives are popular for this reason. Several types of electrohydraulic and electric motors can be used in this way. Rotary position transducers present some special mechanical problems. While the simplest solution may be to mount the transducer directly on the motor shaft or lead screw, this is often undesirable because of gear-train windup, lead-screw stretch and windup, and similar mechanical deflections. The preferred solution is to couple the transducer by way of a non-load-bearing linear actuator, such as a precision ball screw or instrument-grade rack.

When the mechanical elements have been fully specified, the servodrive can be designed. The importance of selecting, matching, and tuning the various servosystem components to the servodrive cannot be overstressed.

The open-loop system is substantially easier to implement than the closed-loop system, but in either case a rigid mechanical structure, low friction, and low inertia are always desirable. But, in the open-loop system, dynamic considerations are less critical, no position transducers require mounting, and the drive and amplifier usually are not optimized for the particular slide design.

N/C Systems Progress: Constant progress in N/C systems design and application has resulted from a number of factors, some of which are described briefly:

1. *Performance/cost ratio:* Logic costs have been drastically reduced because of major improvements in solid-state circuitry, such as medium-scale integrated circuits. The consequent reduction in component count has had a beneficial effect on reliability—with further implicit cost reductions. Integrated circuits lower the cost differential between point-to-point and contouring systems. Long-lived solid-state switches are improving the performance/cost ratio as compared with the use of mechanical contactors, mostly relays. Evolutionary improvements in servodrive electronics, drive motors, and feedback elements are a continuing source of design and performance progress. Wide use of stepping-motor drives is just one example. Component and subsystem-design progress is contributing to lower total system cost and, at the same time, making more functions and better performance possible.
2. *Adaptive control:* This concept permits the machining system to adapt to conditions not completely foreseeable by the programmer. An example is feed-rate control whereby the tool is forced to the maximum feed admissible within the bounds of material hardness, cutter sharpness, and other factors. Force transducers are used in a closed-loop mode to achieve maximum productivity via this technique. See **Adaptive control.**
3. *Drives:* Excellent progress in the design of high-performance low-inertia electric motors has altered the earlier dependence upon hydraulic drives. Advantages claimed for electric drives include cleanliness, quiet operation, and lower maintenance. The traditional advantages of the hydraulic approach include high power/weight ratio and cool operation. The latter advantages

will continue to favor hydraulic systems for a large number of applications.

4. *Open-loop vs. closed-loop control:* Although extensively used by European and Japanese machine builders for several years, the acceptance of open-loop control by United States builders was slow. Acceptance is increasing in the United States because of the inherent low cost and high reliability of the open-loop approach. Major improvements in stepping motors are accelerating this acceptance.
5. *Time-shared computer facilities:* The availability of rented-terminal computer services is substantially improving the ability of many firms, especially smaller ones, to utilize numerical control. Improved software packages are leading to smaller programming staffs, shorter programming lead times, easier tape modification, and implementation of "point-to-point contouring" programs for point-to-point machines.
6. *Computer numerical control:* This area is broad, ranging from the direct coupling of a small general-purpose computer to a single machine tool all the way to the direct integration of all plant functions, including axis control, workpiece changing, machine scheduling, and cost control. Computer systems offer many advantages over conventional N/C, including elimination of tape handling, on-line program-revision capability, correction for machine inaccuracies, control of many axes from a single controller, and the capacity for future integration into a sophisticated total manufacturing-control system.
7. *On-site retrofitting:* Improvements in cost and performance of open-loop N/C systems have accelerated retrofitting. Easily installed by the user or by an outside contractor, such controls seldom demand the moving and rebuilding of the entire machine—as was formerly the case with traditional retrofitting. Thus, a major source of cost and downtime is eliminated. Retrofitting makes possible the penetration of N/C into roughing and first-operation work, and very importantly, retrofitting is an attractive route for the firm that is trying N/C for the first time.

REFERENCES

Leone, William C.: "Production Automation and Numerical Control," The Ronald Press Company, New York, 1967.

Childs, James J.: "Principles of Numerical Control," The Industrial Press, New York, 1965.

Various publications, Numerical Control Society, Princeton, N. J.

—Gordon B. Baty, *Icon Corporation*

Nutating Disk Piston Meter (See Flowmeter, positive-displacement.)

O

Observatory (See **Astronomical instrumentation.**)

Oceanographic Instrumentation An identifiable scientific discipline that might be termed *oceanographic instrumentation* has been slow in developing even though oceanographic research and exploration is very instrument-dependent. The situation stems largely from a combination of relatively low funding for developments and the comparatively complex instrumentation problems to be solved. Essentially, oceanographic instrumentation represents adaptations of measurement designs and principles originally conceived for other purposes. Some of the instrument techniques involved (and described elsewhere in this encyclopedia) include analytical tools, such as electrical-conductivity, pH, and salinity meters; chromatographs; infrared analyzers—force-measuring devices, such as accelerometers and pressure transducers; magnetometers, gravitometers; and seismographic equipment—meteorological instruments—computing and data-processing apparatus—and automatic controls and servos, with emphasis on remote, underwater operation. Transmission of data over long distances and battery operation in large numbers of unattended instrument locations characterize a significant portion of oceanographic-instrument requirements. Because of great ocean depths, an ability to operate under high pressures also is a frequent requirement. Gerlach and Nussear (Ref. 1) report on the effects of high pressure on various instrument components. General oceanographic-instrument requirements were analyzed by Gaul and Schule (Ref. 2), including the data shown in Table O-1. The instrument capabilities of a deep-sea instrument capsule were reported by Snodgrass (Ref.3).

Several specific oceanographic instruments are described elsewhere in this encyclopedia. Particularly, see **Current meter, ocean; Depth measurement, oceanographic; Fathometer; Ocean-wave gage; Sonar;** and **Tide and wave recorder.**

Temperature-measuring Systems:[1] Typical water-temperature-measuring systems utilize standard resistance thermistors as transducers or

[1] Information on temperature-measuring systems, relative-irradiance measurements, and transmissometers prepared by Leon J. Corcoran and Charles L. Strickland, Hydro Products, a division of Dillingham Corporation.

precision platinum resistance probes to provide an extremely accurate temperature measurement. Standard temperature systems normally provide direct visual reading of temperature in either analog or digital form. More sophisticated systems provide temperature data for precise analog or digital recordings so that large amounts of temperature data can be collected and stored in an effecient manner.

Temperature systems are used in all phases of oceanography, normally in conjunction with a depth system to provide a temperature vs. depth profile. Temperature systems also are used in conjunction with equipment for measurement of salinity, transmissibility of water, and relative irradiance. Sound propagation under water in many cases is affected greatly by large changes in temperature or thermoclines. Thus, temperature measurement is of prime interest to scientists who are engaged in underwater acoustic research. Operation of oceanographic-temperature systems is similar to that of other applications. See **Resistance thermometry.**

Normal temperature-measurement ranges are from approximately −5 to +35° C. Systems which utilize thermistor probes as the transducer are less costly than the normal platinum-resistance-element systems, but the thermistor systems are less accurate. Thermistor systems measure temperature with a typical reading accuracy of ±1 to 2 percent of full scale. Platinum resistance probes can measure temperature with an accuracy of ±0.05° C or better with the proper signal-conditioning circuitry. Trends in the design of oceanographic temperature-measuring systems include transducers and readout modules with an acoustic link to eliminate the cable now required to the surface readout. An acoustic link can provide temperature data to the user without introducing the reliability problem that normally is encountered when a cable is used. A combination temperature- and depth-measurement system is shown in Fig. O-1.

TABLE O-1. *Typical Oceanographic Variables and Parameters*

Variable	Full-scale Range	Required Absolute	Tolerances, Relative*	Sampling Interval
Air temp., °C	−10 to 40	1.0	0.2	5 min
Sea temp., °C	−2 to 30	0.25	0.02	15 s
Salinity, %	30 to 40	0.05	0.01	15 s
Pressure, psig	0 to 15,000	10	2	5 min
Density, g/cm³	1.00 to 1.04	5×10^{-4}	5×10^{-5}	15 s
Surface profile, ft	0 to 40	....	0.2	0.25 s
Current speed, knots	0.05 to 5.0	0.2	0.1	1 min
Current direction, deg	0 to 360	15	5	1 min
Wind speed, knots	0 to 100	10	5	5 min
Wind direction, deg	0 to 360	20	10	5 min
Absolute humidity, g/cm³	0 to 0.03	0.001	0.0001	5 min

*All values correspond to the middle of the full-scale range. Increase or decrease in tolerance for pressure, surface profile, current speed, and wind speed will be in some manner proportional to the magnitude of the variable.

This table reproduced with permission of the copyright owner, Instrument Society of America, Pittsburgh, Pa. From Roy D. Gaul and J. J. Schule, Jr., *ISA J.*, vol. 9, no. 1, January, 1962.

Relative-irradiance Measurements: Instrumentation is required to provide a means for measuring the ambient irradiance at any chosen depth as a percentage of incident surface irradiance in a body of water. Normally, the self-contained portable instrument is used to measure upwelling or downwelling light flux relative to the ambient light flux at the surface. General applications of the relative-irradiance meter are descriptive oceanography and geophysical applications, photosynthesis, and other biological phenomena. The phytoplankton productivity of various areas and the total standing crop at any location are intimately associated with the light available for photosynthesis.

The major elements of a relative-irradiance meter are (1) the underwater sensor, (2) a gimbaled surface sensor, and (3) the deck readout module. The two sensors are identical except for mounting and suspension. Both sensors collect light in accordance with the cosine law and generate voltages proportional to the total light flux collected. The deck module normally has provisions for balancing via precision potentiometer and a galvanometer null to measure the ratio of light flux collected by the underwater sensor to the light flux similarly collected at the

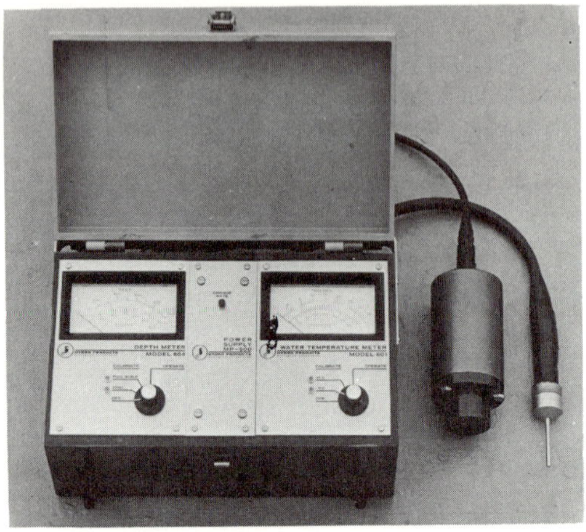

Fig. O-1. Precision bathythermonitor system which incorporates a matched, high-accuracy temperature-measurement system and a depth-measurement system as a combined network with separate readout modules and self-contained power-supply unit for oceanographic work. Operation is from internal batteries or external alternating current, which automatically recharges batteries. The small sensors (at right) can be lowered by hand on a wire, attached to other instruments, or secured to a submerged platform. *(Hydro Products.)*

surface sensor. The relative-irradiance level is read from the potentiometer dial as a percentage of incident surface value. The diffuse attenuation function K can be determined as a function of depth. For this determination, it is common to employ a depth sensor and readout module (see Fig. O-2).

Relative-irradiance meters of this type are easy to operate and do not require a battery or external power source. The photocells normally are self-generating types that power the readout and null-balance the circuit directly. Modifications allow dual subsurface sensors for simultaneously reading upwelling and downwelling light fluxes.

Trends in the design of relative-irradiance meters include analog-to-digital conversion, automatic null-balance circuits, and the automatic selection of filter elements for determining the extinction coefficient, which is wavelength-dependent. Such measurements will be made with narrow-band optical filters over the cells.

Transmissometers: Water clarity is an important parameter of measurement to the oceanographer and, in particular, to scientists and engineers who work in underwater fields that require photography and television documentation. General uses for the transmissometer include measuring water pollution, sediment suspension, plankton concentration, and plotting river outflows. Transmissibility measurements are useful to determine the attenuation coefficient of the water samples for research and evaluation of underwater visibility.

A transmissometer consists of a known light source placed a fixed distance from a calibrated photocell. The fixed distances usually are 1 meter and 10 cm. The 1-meter transmissometer (alpha meter) makes a direct measurement of the

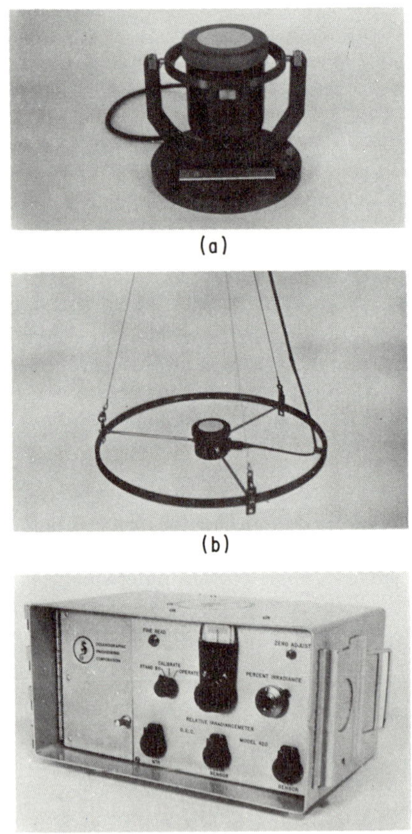

(a)

(b)

(c)

Fig. O-2. Major elements of a relative-irradiance measurement system: (a) gimbaled surface sensor; (b) underwater sensor; and (c) deck readout module. *(Hydro Products.)*

optical-absorption coefficient of an *in situ* water sample; that is, the instrument measures the percentage of light absorbed in traversing a 1-meter path. The transmissometer can be towed behind a ship or boat or lowered over the side or mounted on a submersible. Readout is provided by a

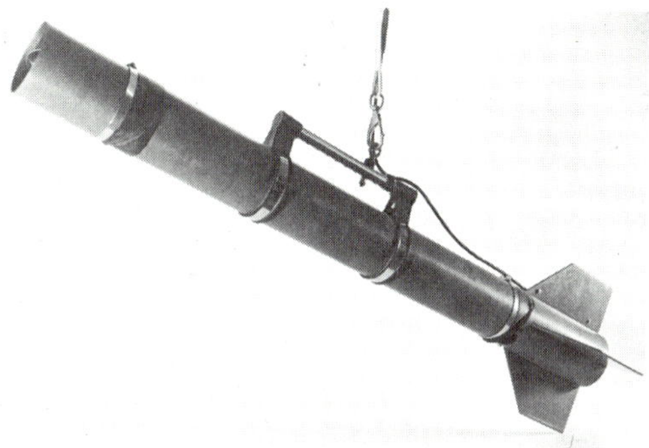

Fig. O-3. Transmissometer (1-meter; alpha meter) being lowered into position for transmissibility measurements underwater. *(Hydro Products.)*

deck unit which contains all the electronic circuitry, controls, and indicating meter. Other instruments can be used in conjunction with the transmissometer, such as charting transmissivity vs. depth or temperature. Provisions are made for recording or plotting. The transmissometer normally is calibrated in air to provide a known reference. Then, when it is placed in water, readings are read as a percentage of the *in-air* calibration. Trends in the design of transmissometers include the incorporation of digital readouts and selectable filters so that light absorption as a function of wavelength for a known column of water can be determined (see Fig. O-3).

REFERENCES

1. Gerlach, Glenn W., and George A. Nussear: High Pressure Electronics, *ISA J.*, vol. 13, no. 2, February, 1966.
2. Gaul, Roy D., and J. J. Schule, Jr.: Oceanographic Requirements vs. Instrumentation Capabilities, *ISA J.*, vol. 9, no. 1, January, 1962.
3. Snodgrass, Frank E.: Deep Sea Instrument Capsule, *Science*, vol. 162, no. 3849, Oct. 4, 1968.
4. Ocean Sciences and Ocean Engineering, *Transactions of the Joint Conference and Exhibit*, sponsored by the Marine Technology Society and American Society of Limnology and Oceanography, Washington, D.C., 1965; vols. 1 and 2 (and subsequent publications) contain considerable information on oceanographic instrumentation.
5. Hatchett, G. L.: "Optimization of Light Sources for Underwater Illumination," Hydro Products, San Diego, Calif., 1969.
6. Kenny, J. E.: Some Design Problems in Wireless Diver Communications, *IEEE Trans. Audio Electro-Acoustics*, vol. AU-14, no. 4, pp. 174–177, December, 1966.
7. Hollien, H.: Underwater Speech Communication, *CSL/ONR Rept.* 3, Communications Sciences Laboratory Department of Speech, University of Florida, Apr. 1, 1967.
8. *Proc. 4th Natl. ISA Marine Sciences Instrumentation Symposium, ISA Publ.* MSD68, Cocoa Beach, Fla.
9. *Proc. 3d Natl. ISA Marine Sciences Instrumentation Symposium, ISA Publ.* MSD65, Miami, Fla.
10. Albers, V. M.: Underwater Acoustics Instrumentation, *ISA Publ.* UA1, 1969.

Ocean-wave Gage Wave measurement is important to oceanographic research. Parameters of interest include wave direction, height, and period—with correlation to current velocity, current direction, and tide level. Visual perception and phase-angle correlation are two commonly used methods for making wave determinations, but both methods have serious disadvantages—the first requiring a visual observer, the second requiring two or more precisely positioned wave-height gages. These limitations are overcome by the three-component balance-wave gage,[1] which measures the force exerted against a fixed, cylindrical probe by each passing wave or combination of waves. The cylindrical probe, acted upon by the forces present when it is submerged, in turn exerts forces on a *Y-Y*, *X-X*, and *Z-Z* axis flexure plate. Each plate (hence each axis) is associated with a force transducer. Mathematical analysis of the transducer data yields wave direction, height, and period. Also see **Oceanographic instrumentation.**

Octal A number system based on a radix of eight (see **Number Systems**). Octal notation is commonly used on binary digital computers. One octal digit corresponds to three binary digits, thus providing easy conversion from one system to the other. The table on page 460 shows the correspondence between numbers in decimal, binary, and octal notation.

[1] Banwell, T. J.: A New Approach to the Analysis and Prediction of Wave Force on Marine Structure, *Aeroject Rept.* S62-249, October, 1962.

Decimal	Binary	Octal
0	0	0
1	1	1
2	10	2
3	11	3
4	100	4
5	101	5
6	110	6
7	111	7
8	1000	10
9	1001	11
10	1010	12
11	1011	13
12	1100	14
13	1101	15
14	1110	16
15	1111	17
16	10000	20

—B. W. Landeck, *IBM Corporation*

Odometer A distance-measuring instrument. A very common and simple example of odometer is the total-mileage and/or trip indicator in an automobile. Also see **Navigation systems.**

Odor Measurement (See **Taste- and odor-measurement instrumentation.**)

Off-line A mode of operation for a process-control digital computer whereby it is not interfacing with the process (compare **On-line**). Process-control digital computers are considered to be in an off-line mode when they are not responding to the process. A process-control digital computer may be off-line because maintenance is being performed on it or the process-control operating system is being generated on it, or for a variety of other reasons.

The term off-line is also applied to peripheral equipment when it is not under the control of the main processing unit.

—B. W. Landeck, *IBM Corporation*

Offset (Droop) The deviation between the actual value and the desired value or set point of a controlled variable that exists as characteristic of proportional control. For any control loop, there is only one load condition where agreement between the measured variable and the set point will occur. For any other load condition, the amount of offset will increase proportionally as the proportional-band setting is increased. The following equation relates the output to the difference between the measured variable and set point of a proportional controller:

$$m = \frac{100}{P} e + b$$

where m = output
P = proportional band
e = error or deviation of measured variable from the set point
b = output bias

For a 4- to 20-mA dc output controller, the bias equals 12 mA dc when there is 0 error. For the output to change from 12 mA dc, an error must exist. It is this error that appears as the difference (or offset) between measured variable and set point.

—Richard H. Kennedy, *Bristol Division, American Chain & Cable Company, Inc.*

Ohm, Standard (See **Standards, measurement.**)

Ohmmeter An electrical instrument arranged to permit the measurement of resistance. An ohmmeter usually is actuated by a self-contained or external source of voltage, such as a battery. In contrast to resistance bridges which permit the comparison of a known with an unknown resistance, direct-reading ohmmeters are milliammeters calibrated in ohms. The two simplest types of ohmmeters are those in which the actuating current is a direct or an inverse function of the resistance connected to the instrument terminals. In both types, the actuating voltage is assumed either to remain constant under all conditions or to be adjustable so that the pointer can be set to a specific mark on the scale with the unknown resistance either open-circuited or short-circuited, depending upon whether the instrument is of the *parallel* or *series* type.

The basic circuit of the so-called "parallel" ohmmeter is shown in Fig. O-4. The unknown resistance is connected in parallel with the milliammeter. The current through the milliam-

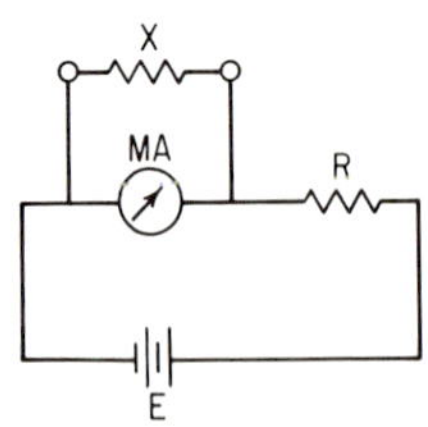

Fig. O-4. "Parallel" ohmmeter.

meter rises with increasing resistance X so that 0 on the meter scale corresponds to 0 resistance and full scale corresponds to infinite resistance. The supply voltage E will remain reasonably constant if a resistance R is inserted to limit the total current which can be drawn from the source.

The circuit of the so-called "series" ohmmeter is shown in Fig. O-5. The unknown resistance X is connected in series with a milliammeter. As the resistance of X rises, the current through the meter diminishes and becomes 0 when X approaches infinity. In order to make full scale on the meter correspond to $X = 0$ without causing the low meter resistance to draw excessive current from the source E when X is low, a resistor R_v is inserted. Usually R_v is made adjustable so that the meter can be set at top mark (0 on the resistance scale) with the terminals for the unknown resistor shorted out, thus compensating for any variation of the supply voltage E.

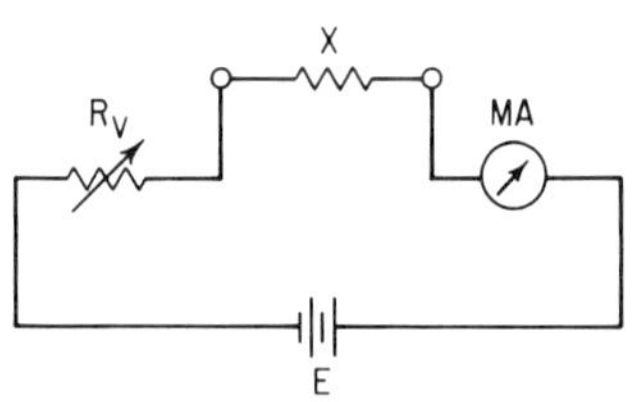

Fig. O-5. "Series" ohmmeter.

The desire to make the readings of a direct-reading ohmmeter independent of fairly wide variations in supply voltage without the need for manual adjustment has been fulfilled by *ratio meters* whose pointer deflection corresponds to the ratio of the voltage across an unknown resistance to the current flowing through it. Instruments of this type usually balance the torques produced by two moving coils attached to the same shaft so that the position of rest represents one specific ratio. See **Electrical quantities, measurement of** (Fig. E-48).

Most ohmmeters are offered in combination with a multiplicity of current and voltage ranges —in the form of multimeters or analyzers that serve a wide variety of purposes. Another popular type using electronic circuitry for extremes in current sensitivity permits the measurement of very high resistance values, such as those encountered in electrical insulation.

REFERENCES

Harris, Forest K.: "Electrical Measurements," John Wiley & Sons, Inc., New York, 1952.

The Instrument Sketch Book, Weston Instruments, Inc., *Circ.* E3G, Newark, N.J., 1966

—Henry Berring, *Weston Instruments, Inc.*

OMEGA (See **Navigation systems.**)

OMNI Systems (See **Navigation systems.**)

On-line The status of a software-hardware real-time system when it is performing its intended function. This state is characterized by the system's being sensitive and responsive to the external stimuli that arrive from the process being controlled or monitored. The state is normally established by a program-initialization function sometimes referred to as "cold start" or initial program load (IPL). This function conditions the system as required to communicate with the external world.

Systems which are examples of on-line operation are airline-reservation systems, rolling-mill control systems, distillation-plant control systems, and data-acquisition and monitoring systems (see **Data-acquisition computer**).

This term is also applied to the state of devices and functions with respect to the host system. Devices are off-line logically when the system has been informed that the device is unavailable for use. The concept of off-line and on-line devices is used in input-output resource-management segments of system programs to allow maintenence of individual units while the remainder of the system is operating.

System functions such as input-stream-preparation devices (e.g., card-to-tape units) and output printing are also termed on-line or off-line depending on whether or not the subject system or a completely independent system performs the function. In an on-line situation where job-stream input is being read, single or multiple job executions are in progress and, concurrently, output is being printed, the term spooling is sometimes used. In systems where these peripheral operations are performed off-line, magnetic tape and disk are the common media used to transport data between the systems. Actual physical handling may be avoided by providing input-output devices that can be electrically switched between the main and peripheral systems.

—L. E. Wise, *IBM Corporation*

On-Off Controller (See **Floating control.**)

On-stream Analyzer (See **Process-stream analyzer.**)

Opacimeter (See **Densitometer;** and **Photometers.**)

Open-loop Control An open loop may be defined as a signal path without feedback (ref.). Any process or machine that is programmed to operate on a time basis regardless of the end results or products which it produces is open-loop. In other words, no measurements are made—no information is fed back to alter the action of the controller.

REFERENCE

Shinsky, F. G.: "Process-control Systems," McGraw-Hill Book Company, New York, 1967.

—Richard H. Kennedy, *Bristol Division, American Chain & Cable Company, Inc.*

Operating System, Computer A group of interrelated programs (see **Program, computer**) to be used on a computer system to increase the utility of the hardware. An operating system may vary in size and complexity from a small batch-processing monitor to a large group of control and service programs controlling a multicomputer complex. It may range in application from commercial and scientific orientation through time sharing (see **Time sharing**), real-time control, and combinations of all these.

Operating systems evolved from the need to obtain the maximum amount of service possible from a computer. Simple monitor systems providing a smooth job-to-job transition were the first step. Operating systems of today contain coordinated programs to control input-output scheduling, task scheduling (see **Task**), error detection and recovery, data management, debugging, multiprogramming, on-line diagnostics, and other services (see **Error procedures, computer**).

The programs in an operating system may be thought of as falling into two broad categories: control programs and processing programs. Within these may be classed all the services provided by a system.

- Control programs
  - Data management—including all input-output
  - Job management
  - Task management
- Processing programs
  - Language translators (see **Language, computer**)
    - FORTRAN, ALGOL
  - Service programs
    - Linking programs
    - Sort/merge
    - Utilities
    - System library routines
  - User-written programs

The *control programs* provide the structure and environment in which work may be accomplished more efficiently. These are the components of the supervisor protion of the system.

Data Management: The movement of data to and from all input-output devices and all storage facilities is the area of concern of data management. This area of the system encompasses the functions referred to in some systems as the input-output control system (IOCS) and is usually segmented into two parts, physical IOCS and logical IOCS.

The physical portion of the system handles device and channel operations, error procedures, queue processing (see **Queue**), and in general, all operations concerned with transmitting physical data segments from storage to external devices.

The logical IOCS portion of the system is concerned with data organization, buffer handling, data-referencing mechanisms, logical-device reference, and device independence. The logical IOCS is the system interface which the problem programmer usually uses, although he may not be restricted from addressing the physical units directly.

Job Management: The movement of control cards or commands through the system input device, their initial interpretation, and the scheduling of jobs (see **Job**) indicated therein is the function of the job-management portion of the system. Job queues and priority scheduling of the jobs are handled in this segment of the system. Job-accounting functions, if provided, are also handled here.

Task Management: The control of the order in which work is performed in the system is the function of the task-management system (see **Task**). Management of the control facilities (central processing unit, storage, input/output channels and devices) according to some task-priority scheme is done by this part of the system. Simple systems may execute only one task at a time requiring simple controls. Larger systems incorporating multiprogramming and real-time response have a much more extensive control structure. (See **Multiprogramming; Real-time computing.**)

The *processing programs* may be thought of as programs that have as a specific objective some function unrelated to controlling the system work. These programs use the services of the supervisory programs rather than operating as a part of them. (See **Executive program.**)

Language Translators: This class of programs exists specifically to allow users to concentrate on solving their logical problem with as little thought as possible to the detailed hardware requirements necessary to provide the problem solution. Higher-level languages, report generators, and special translation programs fall into this category.

Service Programs: The service programs are those programs which are necessary and/or desirable to facilitate system operation, or which provide auxiliary functions. Program-linking functions, sort programs, tape-print and file duplication, and system libraries may be classed in this group.

User-written Programs: This group of programs makes up that portion of the system which is prepared by a user to accomplish his work. Most of the rest of the system which is supplied by the system supplier exists expressly to provide for efficient execution of this part of the system.

REFERENCE

Rosen, S.: "Programming Systems and Languages," part 5F, McGraw-Hill Book Company, New York, 1967.

—L. E. Wise, *IBM Corporation*

Operator Interface (See **Display-command operator/machine interface.**)

Optical Character Recognition (OCR) Three basic reading techniques are used in optical character recognition: (1) spot scanning, (2) columnar sensing, and (3) retinal sensing.

Spot Scanning: The simplest reading mechanisms are spot scanners that cover one small character segment at a time. The most commonly used spot scanner has a rotating disk between the light source and the character—with a pattern of slots that breaks the character into distinguishable elements. The device is mechanically reliable and relatively inexpensive to construct but is limited to a speed of from 300 to 500 characters per second and usually is limited to one highly stylized type font. The device generally is utilized in systems with limited reading requirements.

Two methods of electronically generated spot scanners are used. One method uses a vidicon tube similar to that used in telecasting which picks up the character image. The surface of the tube is scanned by an electron beam that breaks the character down into digital components. In the other electronic method, known as the "flying-spot scanner," a cathode-ray tube generates a beam of light that moves across the character in a scan pattern. A lens system projects the reflected light to photomultipliers that translate black-and-white areas into electrical signals. The electronic methods are faster than mechanically generated spot scanners. Speeds up to 2,000 characters are available, but these units are considerably more expensive and are limited in

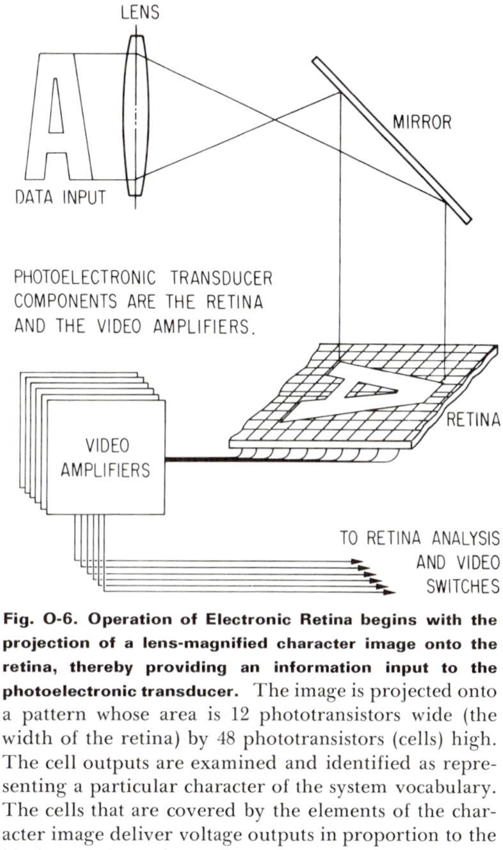

Fig. O-6. Operation of Electronic Retina begins with the projection of a lens-magnified character image onto the retina, thereby providing an information input to the photoelectronic transducer. The image is projected onto a pattern whose area is 12 phototransistors wide (the width of the retina) by 48 phototransistors (cells) high. The cell outputs are examined and identified as representing a particular character of the system vocabulary. The cells that are covered by the elements of the character image deliver voltage outputs in proportion to the blackness of that image. The outputs (analog) can reproduce electronically many shades of gray between pure white and absolute black. The pattern of voltage outputs representing white and black thus completely describes to the examining circuits the image projected onto the retina. *(Recognition Equipment.)*

the variety of type fonts and quality of printing that they can recognize.

Columnar Sensing: This method uses an array of photosensitive devices that sweeps horizontally across the character so that the viewer "sees" the character in vertical segments rather than as individual spots. Earlier implementations of this method also required precise sweeping and timing circuits to maintain the correct spatial relationship among the individual elements of a character and had a reading speed approximating that of mechanically generated spot scanners.

Retinal Sensing: This method[1] uses a two-dimensional matrix of photosensors that senses an entire moving character rather than a segment of it. The retina has a character resolution approaching that of the human eye and reads up to 2,400 characters per second. The device eliminates sweeping and timing circuits and critical adjustments because spatial relationships are fixed by the retina per se. Information regarding the entire character is transmitted simultaneously to the recognition unit in analog form. The other methods described must store information until a complete character has been scanned before attempting to identify it. The retina begins identification instantly and can recognize an infinite range of gray shades which, in effect, adds a "third dimension" to the character. The technique has achieved an ability of causing less than one error in 100,000 characters and is capable of further improvement. The operating principle of the device is shown in Fig. O-6. Essentially, the device is a mosaic-image sensor, or a mosaic of photocells onto which each character to be read by the system is focused. The photocells are physically constrained, and thus their dimensions with respect to each other can be held constant to avoid character distortion. The mosaic, like the scanner, is one character width wide and three character heights high. Behind each photocell in the retina, there is a silicon chip that is sensitive to black, white, and the shades of gray between. These photocells are interconnected so that each single cell "sees" not only its own portion of the character being read but also the portions covered by other cells around it. This arrangement enables the device to judge relationships, dismiss smudges as not being part of characters, and accept even the light portion of the character because the area next to the character is even lighter. In other words, the recognition logic establishes relative values between each cell and those surrounding it.

A later adaptation of this basic technique, utilizing an integrated sensor with 90 photodiodes in a single silicon slice 1½ in. long, overcomes many of the limitations of earlier implementations and reads at speeds up to 3,600 characters per second. By making extensive use of integrated circuits in associated recognition logic and creating what is in effect an "electronic mosaic," the device is used in an OCR system that also has the ability to make relative value judgments.

[1] Electronic Retina, a registered trademark of Recognition Equipment Incorporated.

Fig. O-7. Electronic Retina computing reader. *(Recognition Equipment.)*

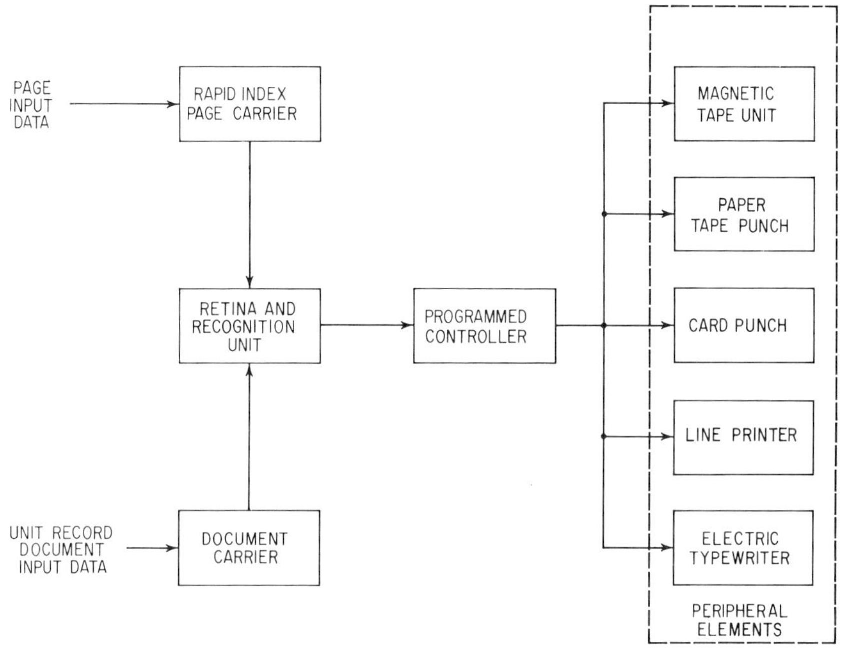

Fig. O-8. Typical block diagram of Electronic Retina computing reader. *(Recognition Equipment.)*

Electronic Retina[1] *Computing Reader:* The retina device just described, in one system configuration, is incorporated in the computing reader shown in Fig. O-7. This system reads ordinary typewritten or printed information in a variety of standard fonts from a range of intermixed paper sizes and weights and translates the information into computer language. Because each system includes a general-purpose computer as a programmed controller, the system can edit and format data during a single read pass. A single system can include more than one paper-handling mechanism. The system shown has a reading rate of 2,400 characters per second and handles pages from 4⅞ to 14 in. width by 3¼ to 14 in. height. The usual output medium for a system of this type is magnetic tape, although punched paper tape, punched cards, and line printers may be used. A typical system configuration is shown in Fig. O-8. The system can be used with a rapid-index page carrier or a document carrier alone or as a dual system using both paper-handling devices. The system can have a vocabulary of up to 360 upper- and lowercase, intermixed, alphabetic, and numeric characters or symbols.

For information on associated subject matter, also see **Character-recognition systems; Code readers and systems; Magnetic-ink character recognition (MICR);** and **Pattern recognition.**

—Israel Sheinberg, *Recognition Equipment Incorporated*

Optical-emission Spectrochemical Analysis A spectrochemical technique which utilizes an optical device to analyze radiation from electrically "excited" sample atoms. The analyzing device provides monochromatic images whose intensities are measured and related to the concentration of the elements within the sample that produces the specific radiation measured. The technique is precise, rapid, and adaptable to solid, powder, or liquid samples. More than 70 elements may be detected by standard procedures. Atomic gases, such as O, N, H, He, A, Ne, Kr, X, and Rd and the halogens are excluded. Nonmetallic substances, such as C, S, Se, require vacuum-path spectrometers for optimum detection and measurement. Analytical ranges may extend from fractional ppm to above 40 percent concentration. Computer-controlled, photoelectric, optical-emission spectrometer systems will output printed percent concentrations for 30 to 50 elements per sample in less than 2 min. This form of analytical instrumentation is widely used in production and quality control as well as research studies that require rapid metallic-content analysis.

Operating Principle: The basic equipment consists of a source unit, a spectrometer (or spectrograph), and a measuring console (or film-reading equipment). Sample atoms may be "excited" by absorbing specific energies from an electric discharge. These atoms, raised to higher-than-usual energy levels, are unstable and revert to their stable states by emitting the absorbed radiation according to the relation

$$E_2 - E_1 = h\nu = \frac{hc}{\lambda}$$

where E = energy, electron volts
h = Planck's constant (6.624×10^{-27} erg s)
ν = frequency, Hz
$\lambda = c/\nu$ = wavelength, Å ($= 10^{-8}$ cm)
C = velocity of light (3×10^{10} cm/s)

Atomic transitions may be expressed in terms of wavelength, and qualitative analysis may be performed by wavelength determination and identification.

The commonly used dispersive device is the diffraction grating, which produces spectra by light interference according to the relation

$$N\lambda = d(\sin\alpha \pm \sin B)$$

where N = an integer
λ = wavelength
d = grating constant (width of single groove)
α = angle of incident light
B = angle of diffracted light

For constant α and the same sin B, integer values of N produce spectra of ½λ, ⅓λ, and so on—called *spectral orders.*

A grating ruled on a spherical surface combines the properties of the diffraction grating with the focusing ability of the optical surface. Such a device, with radius of curvature R, focuses spectra as images of the entrance (primary) slit on the circumference of a circle of diameter R, when the entrance slit is also located on the circumference of the circle. The usual measure of how well a grating separates individual wavelengths is given by the reciprocal linear dispersion (Å/mm) as follows:

$$\text{Å/mm} = \frac{d\cos B}{Nf}$$

Thus, dispersion is governed by the fineness of the grating ruling d and the focal length f of the focusing element.

The concentration C of an irradiating element is related to the intensity I of the emitted spectral line, according to the following relationship:

$$I = kC^n$$

where k and n depend on the excitation conditions employed. Accuracy and precision are improved by use of an internal standard reference line of another element of constant concentration. The relationship becomes

$$\frac{Ix}{Ir} = k_1 Cx^{n_1}$$

where Ix and Ir are the intensities of spectral lines emitted by elements x and r. Cx is the concentration of element x, and k_1 and n_1 are constants depending on the line pair and on the excitation conditions. The relative intensities of lines having different excitation energies depend on the temperature of the spark-discharge column.

The source unit[1] must vaporize and excite a portion of the sample, which is generally used as one of the two electrodes between which the electrical discharge takes place. Table O-2 summarizes the important characteristics of the commonly used spectrochemical source units.

Figure O-9 shows schematically the "multisource" excitation unit which separates the condenser charging cycle from the discharge cycle. Minute variations in the time of ignition have little bearing on the characteristcs of the dis-

[1] No single excitation source is ideally suitable for all applications of emission spectrochemistry. Trace impurities in metals, alloying constituents in high concentrations, biological substances, ceramics, slags, oils, nonconductors, refractories—all may require different excitation techniques and sample-preparation procedures.

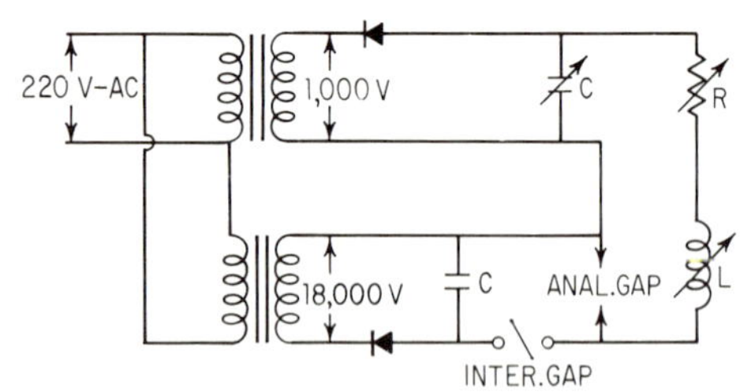

Fig. O-9. Schematic of a "multisource" excitation unit used in optical-emission spectrochemical analysis. *(Applied Research Laboratories.)*

TABLE O-2. *Spectrochemical Excitation-source Units*

Type	Voltage, V	Current	Characteristics
Dc arc.	220	3–30 A	Most sensitive, least reproducible, quantitative analysis and trace-element quantitative analysis
Ac arc.	2,500 5,000	5 A 2.5 A	Good sensitivity, more reproducible, best use for self-electrode metal analysis
High-voltage spark-interrupted auxiliary gap	15–40,000	3–20 rf A	Least sensitive, most precise, ±1% or better. Excites higher-energy lines. Parameter selection allows variations between arclike and sparklike in spectral excitation
Multisource.	1,000	Peak discharge currents from 5 to in excess of 500 A at time constants ranging from 8 to less than 1 ms	Sensitive and precise. Parameter selection allows wide variety of controlled unidirectional and oscillatory discharges variable from arclike to sparklike in spectral excitation

charge, since the condensers were charged to a definite potential during the preceding half cycle. Thus, a definite amount of energy is released in each discharge.

REFERENCES

Harrison, George R., Richard C. Lord, and John R. Loofbourow: "Practical Spectroscopy," Prentice-Hall Inc., Englewood Cliffs, N. J., 1948.

Sawyer, Ralph A.: "Experimental Spectroscopy," Prentice-Hall, Inc., Englewood Cliffs, N.J. 1951.

—Paul W. Flynn, *Applied Research Laboratories*

Optical Pyrometer An instrument for measuring temperatures above 750°C. It finds extensive use in research, control laboratories, and industrial processing. Above 1064.4°C, the instrument defines the international practical temperature scale. With a suitable certificate of calibration, the optical pyrometer can be maintained as a reference standard and used to verify the calibration of other very high temperature pyrometers. Several types of optical pyrometer are available.

Manually Balanced Type: In the manually operated version, the human eye serves as a difference detector and compares the brightness of a standard lamp filament and the object being measured as both are viewed through a red filter. As the temperature of the object increases and as the current through the lamp increases, so does the brightness of each. A knob coupled to the pyrometer scale permits the operator to adjust the current and thus match the filament with the object (see Figs. O-10 and O-11). An electronic control circuit within the pyrometer compares

(a) (b) (c)

Fig. O-10. Contrast of standard lamp filament and object as seen through the telescope of a disappearing-filament optical pyrometer: (a) when pyrometer is set above object temperature, the lamp filament is brighter than the object; (b) when set below, the lamp filament is darker than the object; (c) when at the same or a precalibrated temperature, the lamp filament blends with the object and disappears.

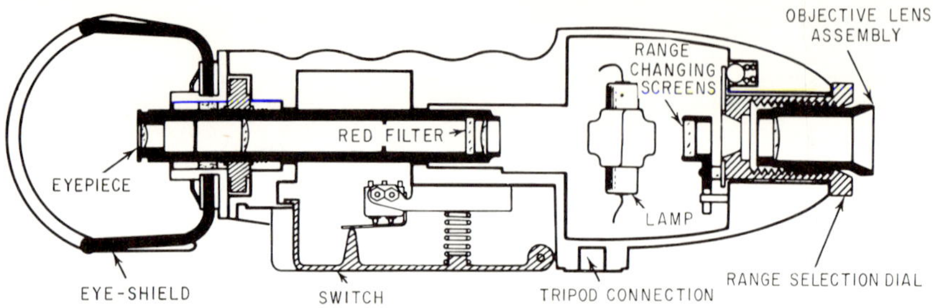

Fig. O-11. Telescope of a manual optical pyrometer showing major parts. Scales and current-adjustment knob are on a separate control box. Models are available with a variety of ranges, generally on two or three scales in the span from 775 to 4200° C, for blackbody conditions. Pyrometers also are available for nonblackbody conditions, including 0.4 emissivity, commonly encountered when measuring molten iron and steel as it is poured from ladles. The objective lens can be changed to accommodate extra-small or close-up objects.

lamp current with an internal-reference standard cell and maintains current precisely proportional to the knob setting for the life of the batteries that supply the lamp current. To extend the useful range of this instrument, absorption screens are placed in the optical path to reduce the apparent brightness of the object.

Another type of manually operated optical pyrometer employs a constant-current lamp and a wedge-shaped absorption screen. The brightness match is made by varying the wedge position while the lamp brightness remains constant. A scale on the wedge-adjustment knob indicates the temperature of the object. A milliammeter and rheostat are provided so that the lamp current can be preset at the proper value before a measurement is made. Although generally not as precise as the disappearing-filament instrument, the varying wedge-type device usually is smaller and lighter in weight.

Automatic Optical Pyrometer: In this device, the human eye is replaced by a photocell which compares the brightness of a standard lamp filament with the object and actuates a control loop within the pyrometer which maintains lamp current at a level that continuously matches filament and object. Precise measurements of lamp current made with a laboratory potentiometer are used to ascertain temperature from a lamp-calibration curve. This instrument eliminates differences in matching by human observers and generally is more precise than manual types. The instrument also is used with null-balance recorders for continuous temperature measurements.

The optical pyrometer is subject to the same adverse effects of emissivity, interfering gases and particles in the sighting path, and surrounding objects as are other radiation pyrometers. See **Radiation pyrometer.** Unless otherwise indicated, the optical pyrometer is calibrated for use under blackbody conditions (emissivity = 1.0).

Two-color Pyrometer: By ratioing two optical-pyrometer measurements made with different color filters, the emissivity effect on targets of some materials can be reduced. An automatic optical pyrometer utilizing this technique is termed a two-color pyrometer. Only modest use has been made of this device generally because of its greater complexity and the few situations where it provides reliable advantages over other optical or radiation pyrometers.

REFERENCES

Dike, P. H., W. T. Gray, and F. K. Schroyer: Optical Pyrometry, Leeds & Northrup Co. *Tech. Pub.* Al. 4000–1966, Philadelphia.

Theory and Methods of Optical Pyrometer, *Natl. Bur. Std. (U.S.) Monograph* 41.

—W. T. Walker, *Leeds & Northrup Company*

Optical Tooling (See **Goniometer;** and **Theodolite.**)

Optimization Control (See **Process optimization.**)

OR Circuit The OR circuit is a logical element which has the characteristics of providing a binary 1 output if any of the input signals are in a binary 1 state (see **Binary state;** compare **Exclusive OR circuit**). The Boolean-algebra representation of this circuit is $F = A + B$.

A diode and a transistor representation of this circuit is shown in Fig. O-12. In the diode OR circuit, if either (or both) input signal A or B is positive, the respective diode is forward-biased and the output F is positive. The number of allowable input signals to the diode OR gate is a function of the back resistance of the diodes. The

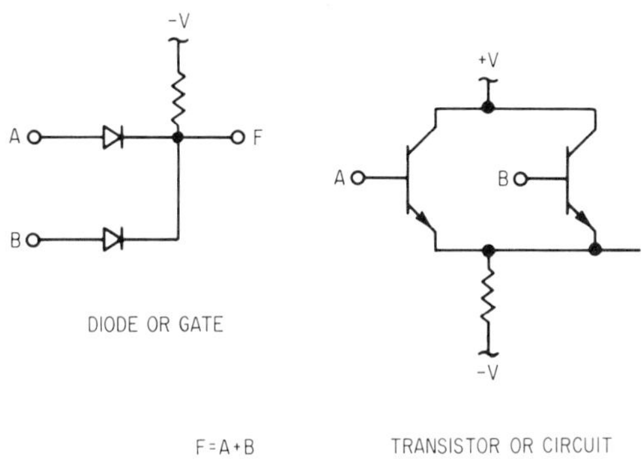

Fig. O-12. OR circuit.

input transistors of the transistor OR circuit are forced into higher conductivity when the respective input signal becomes positive. As a result, the output signal becomes positive when either or both inputs are positive.

—W. W. Welz, *IBM Corporation*

Orbiting Observatory (See **Telescope, orbiting and high-altitude.**)

Orifice Plate A primary flowmetering device that produces a pressure difference when flow is passing through a hole in the plate. The square root of this head difference is proportional to the rate of flow. The working principle and basic flow formula of orifice plates are the same as for other head-difference devices. See **Flowmetering devices, primary.**

The thin-plate square-edge orifice can be constructed of numerous materials. Essentially, the device is a thin plate with an opening, usually concentric (although it can be eccentric or segmental), which restricts the flow and hence causes the pressure drop. Although orifice characteristics are well known and predictable and although duplicate orifices are relatively inexpensive, they do have rather high permanent pressure losses. Orifice plates are applicable for compressible flowmetering because their Y factor has been established by ASME and is published in the ASME "Fluid Meters Committee Handbook" from carefully installed and properly applied orifice plates designed in accordance with ASME specifications. Good metering accuracy can be expected if the orifice plate can be inspected frequently to make certain that its very vulnerable hydraulic shape has not suffered from fouling or damage.

A given orifice plate can be used with the following tap arrangements: (1) flange taps, (2) vena contracta taps, (3) radius taps, (4) pipe taps, and (5) corner taps. These different tap arrangements were introduced to satisfy specific installation requirements. In accordance with the ASME Handbook, the discharge coefficients of orifice plates change with beta ratios, line sizes, and Reynolds numbers. The magnitude of the coefficient depends upon all these components as well

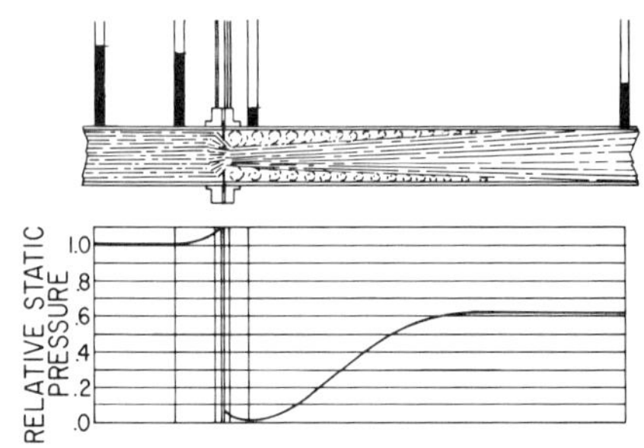

Fig. O-13. Fluid flow through a thin-plate orifice showing positions of pressure taps in common use and the relative static pressure.

as upon the type of pressure tap used. Coefficients range from about 0.62 to 0.85.

With respect to energy recovery, orifice plates are the poorest of the various primary flowmetering devices. Typical head losses are given below. Also see Fig. O-13.

Percent of Differential at a	*Beta Ratio* of
88	0.3
68	0.55
40	0.8

The behavior of discharge coefficients of orifice plates to improper flow patterns has been investigated thoroughly by ASME, and recommendations are published in the ASME Handbook. Orifice plates require the same approach piping as is required for ASME nozzles; also, orifices require downstream piping to protect the discharge coefficient from downstream influence. See **Flow nozzle.**

—Edmond B. Lynch, *B I F, Unit of General Signal Corporation*

Orthotic Devices (See Medical instrumentation.)

Oscillating-piston Meter (See Flowmeter, positive-displacement.)

Oscillation, Measuring-system Oscillations in measuring systems generally are objectionable except when oscillations are applied as test signals or in the form of dither. Oscillatory response delays (or even prevents) the measured or controlled quantity from settling at its steady-state value. Transient oscillations indicate the presence of complex roots in the system equation. Oscillations having constant or time-increasing amplitudes denote system instability. The latter can occur in a closed-loop system that has—at a particular frequency—a phase shift of 360° around the loop and a loop gain of 1 or larger.

There are several classifications or types of oscillations that are commonly referred to, but in a broad sense, any oscillation can be classified in two ways—either *harmonic* or *nonlinear* and either *damped* or *sustained*. The various forms of oscillations that may occur in measuring systems are illustrated and described in Fig. O-14. As indicated, there are a number of types of os-

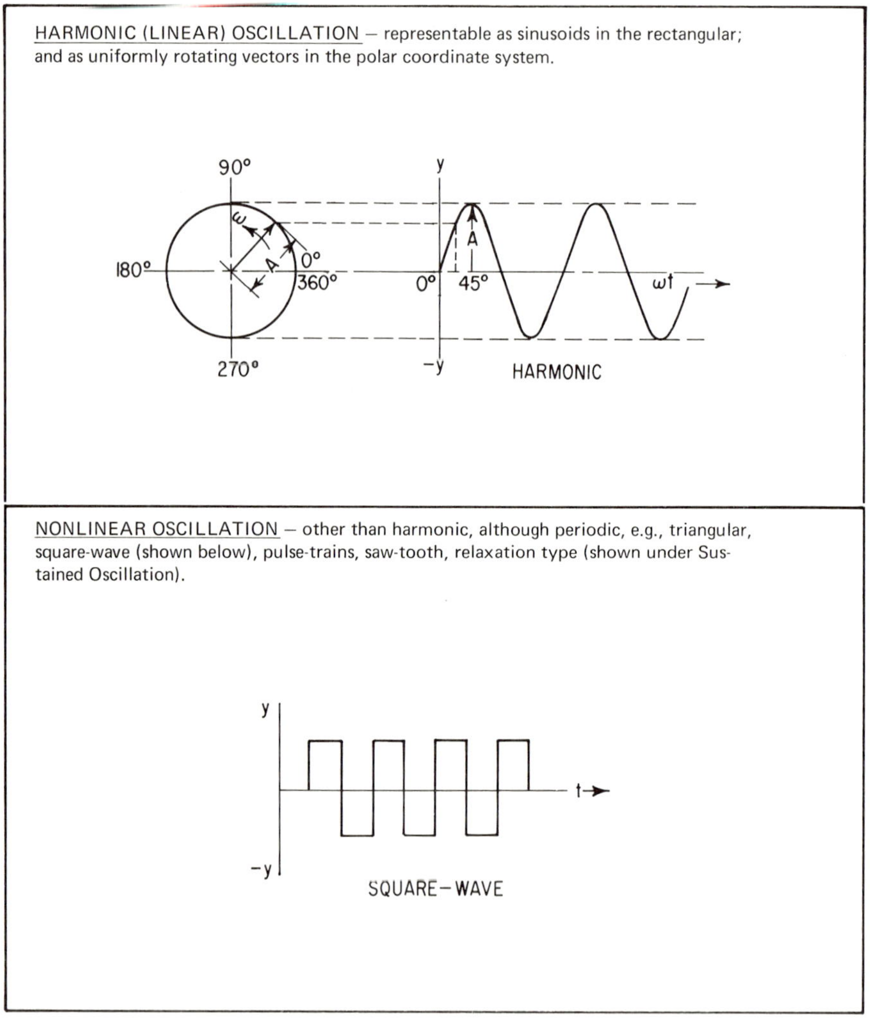

Fig. O-14. Types of oscillations that may occur in measuring systems.

DAMPED OSCILLATION – decaying transient oscillations.

LINEARLY DAMPED – having an exponentially decaying amplitude caused by velocity damping, e.g., viscous friction (as in a galvanometer).

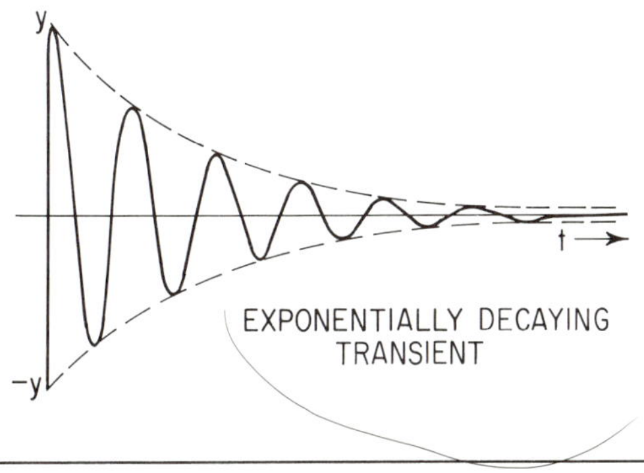

NONLINEARLY DAMPED – e.g., straight-line decaying when friction is present.

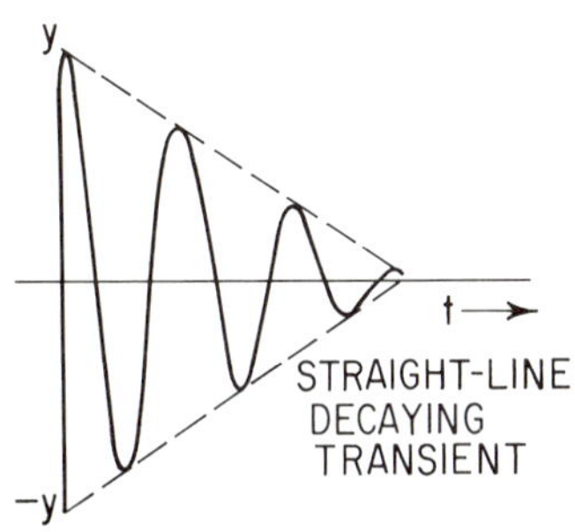

Fig. O-14 (continued). Types of oscillations that may occur in measuring systems.

cillations that are sufficiently common to be given special names. Some of these include:

Limit Cycles: Nonlinear oscillation in a feedback system containing such nonlinearities as hysteresis, saturation, coulomb friction, and dead band.

Buzzing: Audible oscillations (transmitted mechanically) in the approximate frequency region between 100 and 500 Hz.

Motorboating: Audible nonlinear (relaxation-type) oscillations in the approximate 1- to 20-Hz frequency range, and mechanically transmitted oscillations.

Hunting: Low-frequency harmonic oscillation of the controlled variable in a slow-responding process loop or in a regulating system; requires some (soft) nonlinearity.

Dither: Fixed-amplitude oscillations introduced into the system to overcome threshhold levels due to "stiction" or slidewire resolution; usually a fraction of the ac supply voltage.

Vibrations: Environmental mechanical oscillations, harmonic or nonlinear, picked up and amplified by resonance in a mechanical member located in the signal path.

—Robert L. Wilson, *Honeywell Inc.*

SUSTAINED OSCILLATION – harmonic or nonlinear oscillations, generally nondecaying; or decaying to a limit cycle. In each cycle of sustained oscillation, the supplied energy equals the dissipated energy.

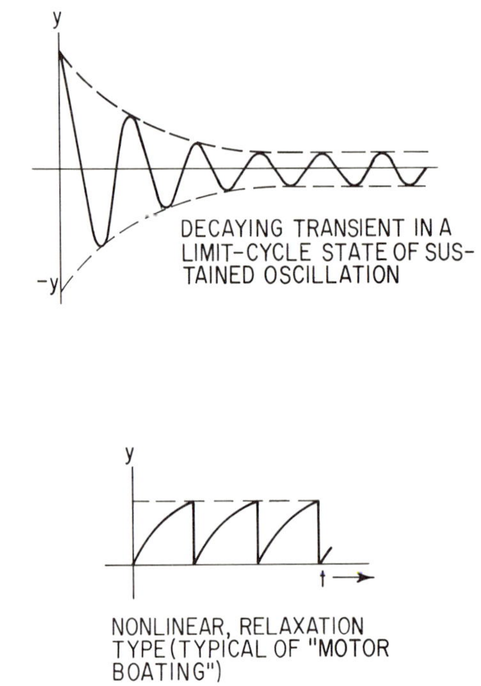

Fig. O-14 (continued). Types of oscillations that may occur in measuring systems.

Oscillograph A device for recording analog or digital information on paper or film. The information may be recorded with inks, powders, electrostatic charge, liquids, electric arc, or light. Technically, all the following recorders may be considered as oscillographs: (1) galvanometer recorders, (2) cathode-ray-tube (CRT) recorders, (3) pen recorders, (4) *X-Y* graphic recorders, and (5) strip-chart recorders. See **Recorder, graphic.**

The main limitation of many types or recorders is frequency response. The wider the bandwidth of a recorder, the greater the amount of information that can be recorded and displayed. Of the foregoing recorders, only the galvanometer and CRT recorders have a bandwidth in excess of 20 kHz. Presently, the term *oscillograph* implies a galvanometer or CRT recorder, that is, a device that uses a beam of light with which to write.

Galvanometer recorders are two-dimensional display-recording devices which contain a light source, mirror galvanometers and optical system, light-sensitive recording media (paper or film), and a system for transporting the record. The basic operation of a galvanometer recorder is shown in Fig. O-15. Part of the light source is reflected from the galvanometer mirror, then reflected off a recording mirror, and passed through a recording lens and onto the recording medium. When the galvanometer is connected to an external source, the electrical signal will conduct through the coil and cause the galvanometer mirror to rotate. Thus, the light beam moves in a straight line across the record, proportional to the amplitude and to the rate of change of the input signal. The movement of the record provides a time history. Thus, the record provides an amplitude (Y-axis) and time (X-axis) recording of the input signal. This configuration is known as *longitudinal recording*.

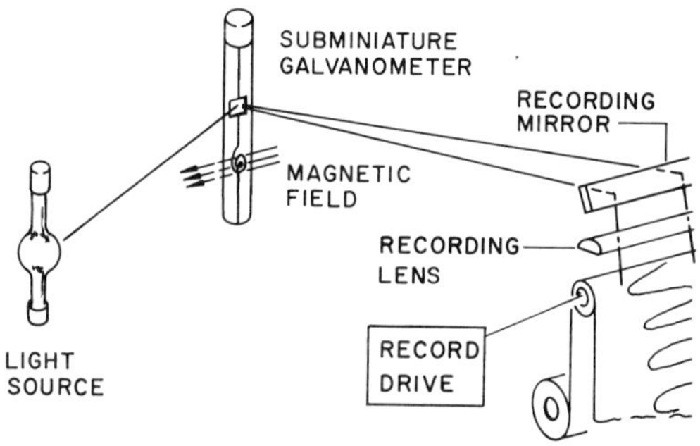

Fig. O-15. Essential elements of a galvanometer recorder.

Figure O-16 shows a model of an oscillograph recorder. This recorder uses a high-pressure mercury-vapor lamp as an ultraviolet-light source. Using an ultraviolet-sensitive paper as the recording medium, an instantly readable and permanent record can be obtained without the use of chemical processing.

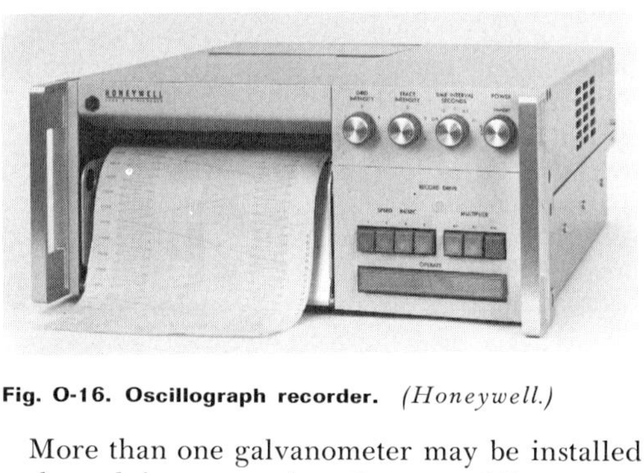

Fig. O-16. Oscillograph recorder. *(Honeywell.)*

More than one galvanometer may be installed and used in a recorder, thus providing many time-coincident channels of recording data. The galvanometers share the same light source and may be electrically or mechanically positioned.

With several galvanometers making simultaneous and overlapping recordings, specific identification of what trace goes with what galvanometer can be difficult. The traces can be identified by a momentary, sequential interruption. The trace interruption is caused by pins located in a moving endless belt which are positioned so that they pass laterally in front of the galvanometers. The pins interrupt the light beam from the recording lamp as it enters the galvanometer lens, and as the pin casts a shadow on a particular galvanometer mirror, the recording trace for that galvanometer is momentarily interrupted. As these breaks occur in the traces on the record, they correspond to the releative positions of the galvanometers in their magnet assemblies, regardless of whether the traces do or do not overlap.

Some recorders identify the trace interruption by printing a series of numbers corresponding to the galvanometer positions on the edge of the record. This system consists of a numbered film-wheel assembly and the optics necessary to transmit a portion of the light from the recording lamp through the wheel to the paper.

Reference grid lines may be recorded on the record by passing a portion of light via mirrors from the recording lamp through a grid-line aperture. This aperture blocks out all but a series of fine bars of light running the same direction as the record travel which are exposed on the record simultaneously with the galvanometer traces.

Reference timing lines may be recorded across the record, perpendicular to the grid lines. The timing lines may be controlled externally or internally. A timing circuit activates a flash tube which directs light to the recording mirror, through the recording lens and into the record.

A typical optical system for the galvanometer trace, grid-line trace, and time lines is shown in Fig. O-17.

Cathode-ray-tube recorders are either three- or four-dimensional display-recording devices. A four-dimensional device simply is a continuously recording oscilloscope while a three-dimensional device is a line-scan recorder.

Recording Oscilloscope: The four axes of a recording oscilloscope are X axis (time base or horizontal), Y axis (amplitude or vertical), Z axis (spot intensity), and Y' axis (movement of recording medium). See Fig. O-18. Also see **Oscilloscope.**

The phosphor on the faceplate of the CRT is selected to have a spectral output which matches that of the recording media. This provides an efficient transfer of energy to the light-sensitive record. The light output from the CRT is fo-

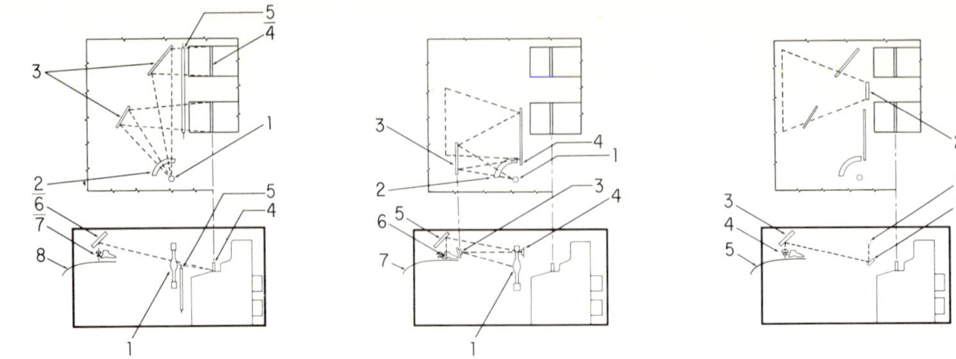

GALVANOMETER TRACE OPTICAL SYSTEM

1. Recording lamp
2. Collector lens
3. Light control mirrors
4. Galvanometer
5. Trace identifier
6. Recording mirror
7. Recording lens
8. Recording paper

GRIDLINE TRACE OPTICAL SYSTEM

1. Recording lamp
2. Collector lens
3. First gridline mirror
4. Second gridline mirror
5. Recording mirror
6. Gridline aperture bar
7. Recording paper

TIME LINE OPTICAL SYSTEM

1. Time line flashtube
2. Time line mirror
3. Recording mirror
4. Recording lens
5. Recording paper

Fig. O-17. Recorder optical system for galvanometer, grid-line, and timing traces.

cused on the record either by an optical-lens system or by providing a fiber-optic faceplate on the CRT and placing the record in intimate contact with the fiber optics.

The four axes provide the capability of recording continuous or transient signals that previously required magnetic-tape or oscilloscope-camera techniques. In the continuous mode, signals may be recorded longitudinally, *X-Y* axes (such as the galvanometer recorder), or transversely, *X-Y-Y'* axes. In the transverse mode, high-frequency data may be recorded without utilizing high record speeds (see Fig. O-19).

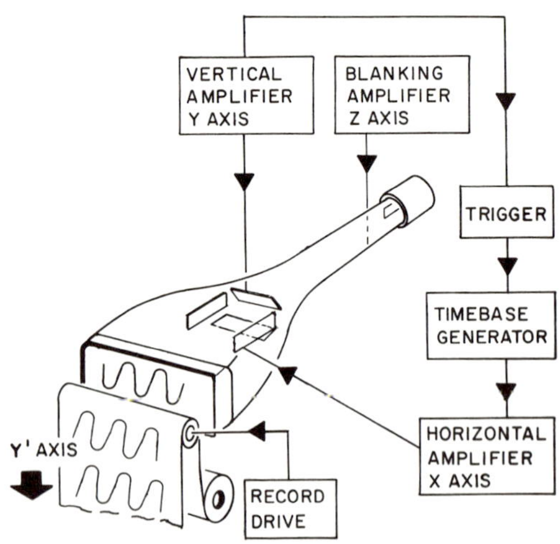

Fig. O-18. Recording-oscilloscope block diagram.

Because of the large combination of horizontal sweep times (typically 2.5 s/cm to 1 μs/cm) and record speeds (typically 0.1 to 250 cm/s) the transverse signal may be skewed, i.e., not perpendicular to the edge of the record. Electrical or mechanical compensation must be provided in order to minimize this skew error.

Reference grid lines in the horizontal and vertical direction are recorded on the record by means of light from the flash tube passing through an aperture. The timing of the flash tube can be controlled externally or internally. Internal control may be from a self-contained timer or synchronized off the record-drive system.

The four axes provide a recorder which is very versatile. Besides transverse and longitudinal recordings, video pictures from television or facsimile may be recorded by use of the *Z-X* axes and either *Y* or *Y'* axis. The quality of the pictures will depend on the sensitivity of the recording media, the bandwidth of the *Z*-axis amplifier, and the spot size that can be recorded.

The recording-media sensitivity and the spot size also determine the writing speed with which one may record data. Characteristically, recording oscilloscopes have been able to record data 100 to 1,000 times faster than galvanometer recorders.

Line-scan recorders are three-axes-recording oscilloscopes. They do not have a vertical amplifier channel and therefore lose the capability of Y-axis recording.

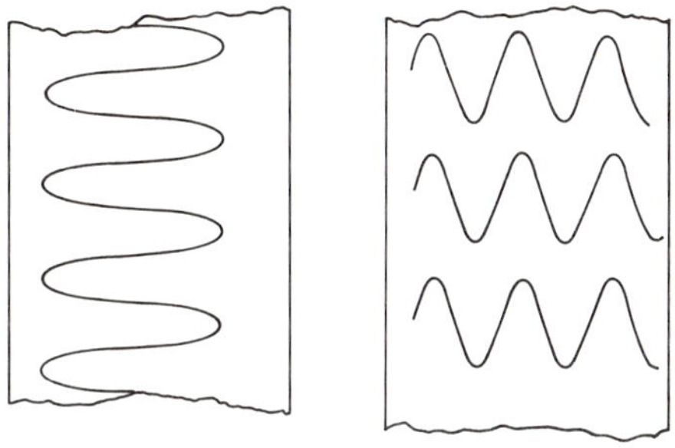

(a) (b)

Fig. O-19. Types of oscillograph records: (a) longitudinal; and (b) transverse.

Alphanumeric characters can be printed out, thus making the recorder a high-speed printer. The characters can be recorded by the use of X-Y-Z axes.

—T. E. Wolk, *Test Instruments Division, Honeywell Inc.*

Oscilloscope The cathode-ray oscilloscope (CRO) is a three-dimensional display device composed of an X axis (time base or horizontal), a Y axis (amplitude or vertical), and a Z axis (brightness range of display). The X and Y axes convey precise quantitative information and usually are specified as time/div and volts/div. The Z axis usually is modulated by blanking or unblanking voltages in order to eliminate *retrace* of presentation. Some displays use Z-axis information to present additional data. A portable oscilloscope is shown in Fig. O-20.

The CRO is a voltage-sensing device. By using suitable transducers ahead of the oscilloscope input, the CRO can convert current, strain, acceleration, pressure, and other physical quantities into voltages suitable for display on a cathode-ray tube (CRT). The CRT has a fluorescent screen which displays waveforms of electrical quantities that vary as a function of time (or as a function of some other quantity in X-Y displays).

To make a measurement, the user counts the number of horizontal divisions and multiplies that value by the time/div control to obtain elapsed time. Voltage differences are obtained by multiplying the number of vertical divisions by the volts/div control. The laboratory oscilloscope basically has an accuracy of ± 3 percent, although greater accuracies are obtainable.

The basic parameters of oscilloscopes are (1) rise time, (2) bandwidth, and (3) deflection factor (sensitivity). Bandwidth and rise time are related by $BW \times RT = 0.35$. Bandwidth often is more meaningful in low-frequency oscilloscopes, while rise time usually is more important in high-frequency oscilloscopes. Generally, an oscilloscope should have from four to ten times the response of the signal under observation. This minimizes attenuation of the higher-frequency information of the display.

The general-purpose oscilloscope has a wide range of calibrated sweeps (typically from 5 s/div to 10 n/div for a 50-MHz version) to allow observation of a full spectrum of events. The CRO generates a sweep ramp internally which moves a spot uniformly from left to right across the CRT display area. This X axis represents time, while the voltage being observed at the Y-axis input is an electrical analog of some phenomenon. The trace varies up or down, dependent upon the instantaneous value of Y for various points of X. A block diagram of a typical oscilloscope is given in Fig. O-21.

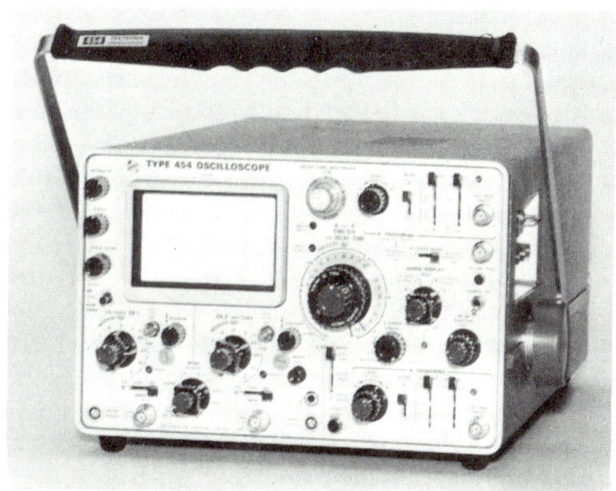

Fig. O-20. Portable oscilloscope with dual trace and delaying sweep. *(Tektronix.)*

Particular attention is given to amplifier-response characteristics to assure faithful reproduction of pulse waveforms as well as sinusoidal response. The vertical portion of the oscilloscope has various deflection factors (10 mV to 10 V/div is typical) to ensure a sensitivity that provides adequate range for the signal under observation. Low-frequency high-sensitivity

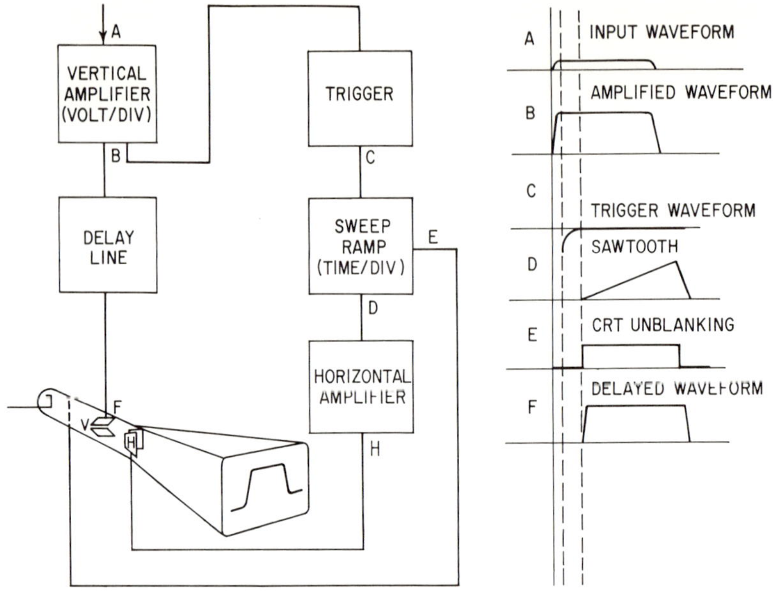

Fig. O-21. Oscilloscope block diagram. A delay line allows all of B to be observed. If a delay line were not inserted, only that part of B from t_2 would be on the CRT screen.

oscilloscopes are available with deflection factors of 10 μV/div.

To maintain a stable display on the CRT screen, each sweep must start at the same point on the waveform being displayed. A portion of the input signal is fed to a trigger circuit, which then initiates the sweep ramp. Delay lines often are used between the vertical amplifier and the CRT vertical deflection plates. By retarding the signal, the delay lines allow the trigger and sweep circuits sufficient time to start before the signal reaches the deflection plates. The entire waveform then may be viewed, even though the leading edge of the waveform is used to trigger the sweep. The rate of rise of the sweep is determined by the calibrated time/div control.

Delaying-sweep Oscilloscope: Additional measurement resolution may be provided by using two linear calibrated sweeps. Sweep 1 (delaying sweep) allows the operator to select any specific delay time by means of a 10-turn calibrated dial. This delay is generated by applying a sweep ramp to a voltage comparator that produces a trigger pulse at a later point in time. Sweep 2 (delayed sweep) starts at the selected time and usually is a decade or two faster than sweep 1. This technique increases resolution, since very fast sweeps may be used for the delayed sweep that begins precisely where desired, regardless of a trigger point. A delaying-sweep oscilloscope acts as a magnifier whose magnifi-

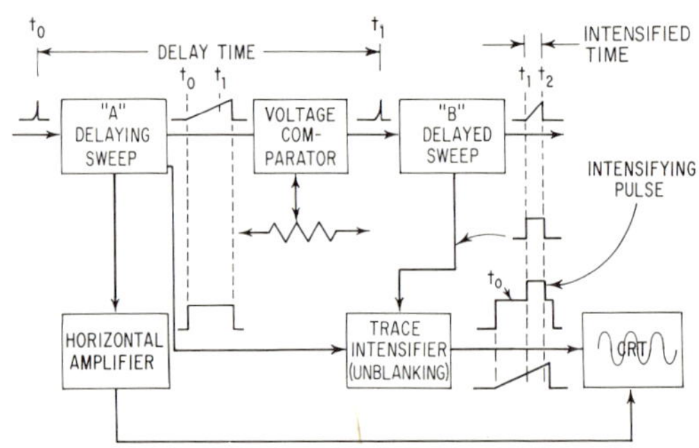

Fig. O-22. Block diagram of delaying-sweep oscilloscope, showing A sweep intensified by B sweep.

cation ratio is equal to delaying sweep rate/delayed sweep rate. Two different unblanking levels are used so that the user can observe the relationship between the two sweeps (see Figs. O-22 and O-23).

Sampling Oscilloscope: This instrument samples many cycles of a signal to reconstruct and to display one cycle. The display (Fig. O-24) is made with each consecutive sample being taken $T + \Delta T$ from the original trigger point. A sampling rate is used to take a very short (25 to 350 ps) sample, and the voltage is measured at that interval. The CRT indicates the amplitude of the input by a vertically deflected spot. The next sample is taken on the next cycle at a progressively later point, as shown in Fig. O-24. The spot moves horizontally slightly and plots the vertical amplitude of this sample. With this technique, a slow-speed waveform is constructed that is a replica of the high-speed input signal. The time required to construct a display varies with the number of samples taken. If a normal sampling rate of 100 kHz is used and 1,000 samples are taken for one display, 10 ms are required to complete a display.

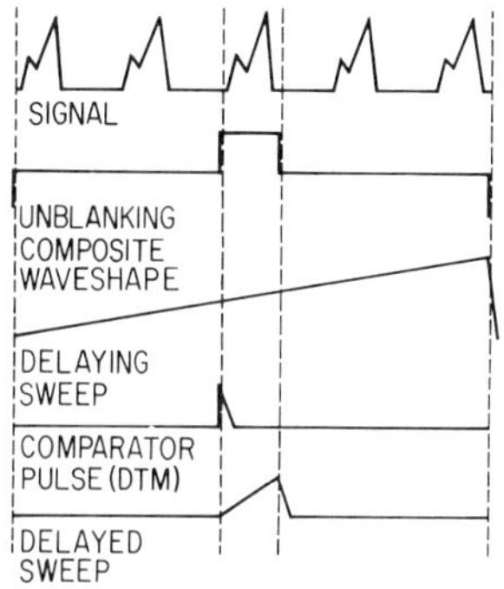

Fig. O-23. Delaying-sweep-oscilloscope circuit relationships.

Compared with conventional oscilloscopes, sampling oscilloscopes offer (1) more bandwidth at identical sensitivity, (2) better overload recovery (at 100-kHz sampling rate, the input amplifier has 10 μs to recover), and (3) recorder outputs (sampling converts a high-speed input signal to a low-speed output signal suitable for *Y-T* recording). Sampling oscilloscopes currently offer rise times of 25 ps (14-GHz equivalent bandwidth). A development in sampling is the sampling-head concept. Sampling heads are miniature modules that determine the input characteristics of a sampling oscilloscope. These heads plug into vertical sampling units and offer the ability to change measurement characteristics quickly and economically.

Digital-readout Oscilloscope: A natural evolution of the sampling oscilloscope is the digital-readout oscilloscope. When the time between samples is calibrated accurately, the samples between two points of interest may be counted to obtain elapsed time. By using a linear voltage ramp whose slope is known, it is possible to obtain counts between two voltage levels and to obtain a voltage-difference reading. Both time and voltage measurements may be read out with an accuracy and resolution that cannot be obtained by the eye. A time measurement made by a typical digital-readout oscilloscope is shown in Fig. O-25. Zero and 100 percent intensified zones are generated that are variable in ½-cm increments. By positioning these zones and the signal delay, the user positions the 0 to 100 percent zones where desired. After the first sweep, amplitudes that correspond to the zones are stored in memory circuits. Changes in amplitude automatically reestablish new 0 and 100 percent memory amplitudes.

In a typical time measurement, digitally selected voltage-divider taps between the 0 and 100

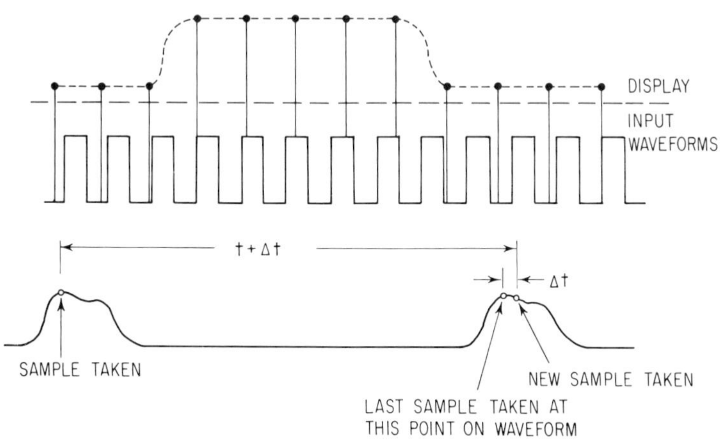

Fig. O-24. Reconstruction of a sampled waveshape.

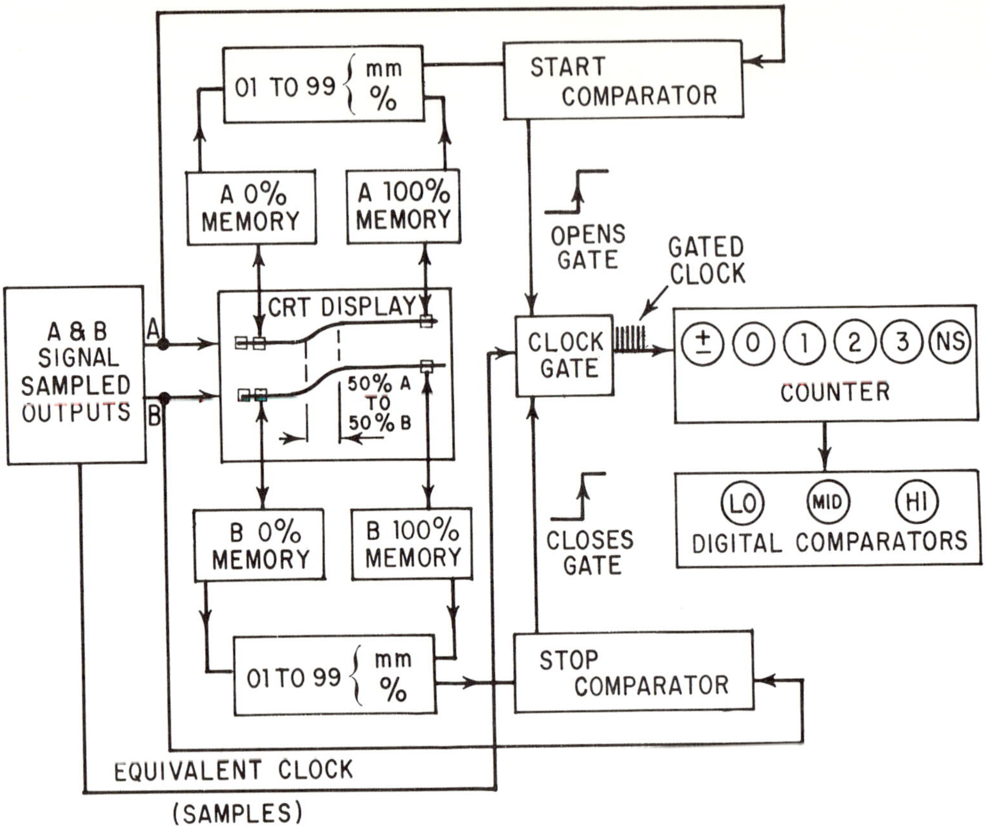

Fig. O-25. Digital readout of typical time-measurement delay from 50 percent A to 50 percent B.

percent memory outputs are set for "start" and "stop" timing in 1 percent or 1-mm increments of either waveform of a dual-trace display. The selected percentage reference levels are compared against the sampled input waveform on the second sweep. Coincidence of the waveform amplitudes with the selected percentage reference amplitudes is sensed by comparators which open and close the gate to the digital counter. The CRT display can be intensified for the duration of the measured interval as a visual reference check. The number of clock pulses is read out digitally with decimal points included. Repetitive pulse measurements, such as pulse voltage, rise time, delay time, storage time, and pulse width, are read out directly with four digits of resolution. Output connectors permit recording of the displayed digital information by the use of such devices as printers, tape punches, and magnetic tape.

External programming of the oscilloscope allows a sequence of measurements to be made without setting up each individual measurement. It is possible to use limit controls (go/no go) to classify and categorize measurements.

Storage Oscilloscope: This instrument retains a waveform on the screen that does not fade at the persistence rate of the CRT phosphor. Waveforms usually can be stored without appreciable degradation from 15 min to 1 hr. High-resolution split-screen storage displays allow simultaneous use of either half of the screen for storage or conventional display. A split-screen display allows storage of a model waveform on one half of the screen while adjusting a circuit to approximate the reference waveform on the other half of the screen. Automatic erase circuitry and variable viewing-time controls are available.

Television Oscilloscope: Vector oscilloscopes are available that display the relative phase and amplitude of television chrominance signals on polar coordinates. In addition, specialized circuitry for making accurate differential-gain and differential-phase measurements is provided. Television *waveform monitors* are designed for signal-level monitoring and the monitoring of vertical-interval test signals. Low-cost television monitors provide measurement capability for educational and closed-circuit television systems. General-purpose oscilloscopes with additional circuitry added permit stable triggering from composite video or composite sync waveforms.

Curve Tracers: Curve-tracing oscilloscopes are valuable for users of semiconductors. Common-base and common-emitter test circuits are included so that dynamic displays of characteristic curves may be obtained. Applying a sweep voltage to the collector and a step voltage or current to the base or emitter provides a dynamic display of the device characteristics. Some transistor curve tracers include readout of sensitivity and step settings, as well as a beta/div readout. This allows display information to be digitized and placed near the CRT.

Oscilloscopes are designed in two basic packages: (1) The non-plug-in oscilloscope combines a limited number of functions in one instrument sufficient to do a given measurement job. (2) The plug-in oscilloscope is an excellent choice where versatility is required and offers a hedge against obsolescence since the oscilloscope characteristics can be changed markedly by changing a plug-in unit. The plug-in concept offers additional flexibility since bandwidth, deflection factor, multitrace operation, differential operation special-purpose functions, and time-base features may be considered separately when choosing an instrument package.

Electronic Switching: Multiple vertical amplifier channels allow comparison between two to four separate sources. Signals may be observed separately or added algebraically, and differences may be displayed. For example, two dual-differential displays can be obtained from a four-trace preamplifier. Multitrace units have two modes of timesharing: (1) the *chopped mode*—where the channels time-share by electronically switching between channels at 100-kHz or 1-MHz rate and (2) the *alternate mode*—where switching occurs from channel to channel at the end of each sweep.

Multitrace configurations have an inherent geometry advantage over multibeam operation since all signals are displayed with the same electron-gun optics and there are fewer correlation problems. Multitrace units are popular for comparing an output signal with an input signal. These units permit the user to observe the effect of an adjustment on various signal inputs and outputs and are useful in time-coincidence applications.

Dual-beam oscilloscopes are used where fast transients are monitored and the switching time between channels obscures a significant portion of the signal information. Two complete electron guns with independent vertical and horizontal deflection systems are used, and thus two transients can be monitored simultaneously. (Some versions have common horizontal deflection plates.) The independent Y deflection systems also allow simultaneous X-Y and Y-T displays. If a spectrum-analyzer plug-in is used, a dual-beam display offers simultaneous viewing of the time domain and frequency domain.

Differential amplifiers measure the instantaneous difference between voltage levels existing at its two input terminals. Common-mode signals are rejected through these amplifiers, since they are common to both terminals. Differential-amplifier characteristics are particularly useful in low-signal-level medical and mechanical applications.

Slide-back differential amplifiers supply a highly accurate dc reference voltage into one input of a wideband differential amplifier. The dc level of an input signal then is compared directly with this reference voltage at a high sensitivity with null-comparison methods. This permits more precise voltage measurements with greatly improved resolution. One version, with a 1 mV/div deflection factor, has an effective screen height of ±11,000 cm and permits a resolution of approximately 0.001 percent or 100 μV/mm.

Spectrum-analyzer plug-ins take advantage of the sweep circuits, power supplies, and wideband amplifiers inherent in a CRO and convert the time base to a frequency base. The plug-in contains the intermediate-frequency circuitry, mixer, and sweeping and fixed local oscillators that are necessary for such a display. A spectrum-analyzer plug-in allows evaluation of the relative amplitudes and frequencies that compose an electrical signal. Plug-in spectrum analyzers are available from 50 Hz to 40 GHz.

Other Plug-ins: Time-domain reflectometry plug-ins offer 140 ps system rise times. Sampling units convert conventional oscilloscopes to scopes with 1 GHz, 2 mV/performance. Carrier-amplifier units are used for mechanical applications, such as stress analysis, vibration studies, and fatigue tests, where it is desired to measure quantities, such as displacement, acceleration, and strain. Carrier-amplifier plug-ins convert resistive, capacitive, and inductive transducer outputs for a standard CRT display.

The *oscilloscope probe* is the interface between the oscilloscope and the signal source. Most oscilloscopes have an input impedance of a 1-MΩ resistive input in parallel with a small value of capacitance (approximately 15 to 50 pF). For many circuits, this input impedance is low enough to permit undesirable loading of the monitored circuit. A wide variety of probes, both active and passive, are available to reduce both the resistive and capacitive loading to the circuit under investigation. The basic passive probe includes a 9-MΩ resistor in series shunted by a variable capacitor. The probe impedance is then 10 MΩ paralleled by approximately 5 to 15 pF.

The loading effect of the oscilloscope can be reduced with a probe. Because high-performance probes are necessary for high-performance oscilloscope operation, the instrument and probes usually are designed together.

—Richard E. Kehrli, *Tektronix, Inc.*

Overlay, Program Program overlay is a segmentation technique to provide for functions which will not all fit into a specified area of the computer memory at one time. To allow for this situation, the program may be prepared in segments (also called phases), each segment containing one or more functions. As each function is required, it is brought into the memory from auxiliary storage (usually disk, tape, or drum) by a special loader component of the system. In this way, it overlays the previous phase. To maintain control of this process, a root segment of the program is core-resident to control the loading and execution of the various segments and provide intersegment communication and coordination as necessary.

The overlay capability in a given system is a part of program-preparation subsystem (link loader, linkage editor, and builder) and requires specific grouping information for the various functions. When the overlay technique is used, it is planned for as a part of the program organization by the programmer.

—L. E. Wise, *IBM Corporation*

Override Control Control action where the output of a number of controllers is selected to control one final controlling element. Gas-compressor control is a common example. Control ordinarily is from a flow transmitter—except that the discharge pressure must not be allowed to exceed a safe limit. Protection against damage to the compressor due to low suction pressure during periods of high demand is provided by a suction-pressure controller. In this system, the discharge-pressure set point is a high limit and the suction-pressure set point is the low limit. So long as these respective variables are on the safe side of their set points, control is from the flow controller. A fourth variable, current to the compressor motor to protect against overloads, sometimes is added to the system.

—Richard H. Kennedy, *Bristol Division, American Chain & Cable Company, Inc.*

Overshoot (See **Closed-loop control.**)

Owen Bridge (See **Bridge circuits, electric.**)

Oxidation-reduction Potential, Measurement of All chemical reactions involving an exchange of electrons are considered to be oxidation-reduction reactions. These reactions produce measurable and predictable potentials. The relative strengths of the oxidants and reductants involved can be measured by determining the oxidation-reduction potential (ORP) prevailing. This can be done by inserting an unattackable electrode, such as platinum, rhodium, or gold, into the solution, together with a reference electrode, and measuring the resulting emf by means of a high-impedance potentiometer or amplifier. The emf or voltage measured is the difference between the individual voltages developed at each electrode. If a hydrogen-gas reference electrode could be used, the potential at the measuring electrode would be the true oxidation-reduction potential of the solution, since the hydrogen-gas reference-electrode potential, by convention, is zero. Generally, the silver-silver chloride or calomel reference electrode is employed instead of the hydrogen, and the potential developed by either of these electrodes requires a correction to the ORP meter reading to relate the measuring electrode potential to the hydrogen "zero" potential.

To understand and properly relate the ORP values and equations found in handbook data and elsewhere, it is necessary to appreciate the rather confusing sign conventions prevailing. Two conventions exist which differ only by being opposite in polarity. The *American convention* expresses the potential with the polarity as it truly exists in the solution *surrounding* the noble measuring electrode. The *European convention* concerns itself with the opposite potential which exists *on* the measuring electrode. Instrument makers and many users are concerned only with the sign of the electrode, since the meter is aware only of this polarity. Thus, most ORP meters and recorders read out in signs followed by the European convention.

Specific oxidation-reduction values for a wide variety of reactions involving an exchange of electrons can be found in handbooks and literature. The values, in volts, typically are referenced to the hydrogen–hydrogen ion couple (having zero potential) at unit activity and at 25° C. These values (for the reaction being measured) can be inserted into the general form of the Nernst equation:

$$E = E_0 - \frac{0.0591}{n} \log \frac{\text{oxidant}}{\text{reductant}}$$

where E = measured emf, volts, opposed to the normal hydrogen electrode whose potential is zero

E_0 = standard emf as found in literature for the reaction involved under the conditions specified

n = number of electrons involved in the reaction

From the foregoing relationship, the expected potential can be predicted, depending upon the ratio of oxidant and reductant. Conversely, by reading the measured potential, the prevailing ratio of oxidant to reductant can be determined.

Limitations of Method: Many types of chemical reactions and processes can be controlled by ORP measurement, but with the following qualifications:

1. The measuring electrode "sees" all oxidants and reductants in a solution. For example, hexavalent chromate within a waste effluent may be the component requiring measurement. However, other oxidants and reductants, such as iron salts and sulfides, may be present, and these can influence the net potential.
2. Variations in the pH of the solution will cause variations in the net potential of the solution.
3. Temperature will influence the potential slightly. The necessary correction usually will be less than 1 mV/° C.

Calibration: If the electrodes and amplification stage are performing properly, the measured potential is a true measurement not requiring calibration. On the other hand, the measuring electrode can become coated (and thereby insensitive) or can be "poisoned" by prior exposure to very strong oxidants or reductants and still retain a "memory" of this exposure. Thus, some users find it opportune to calibrate the system periodically. This can be done by placing the electrodes first in a pH 4 buffer solution containing a few grams of quinhydrone, then placing the electrodes in a pH 7 buffer solution also containing quinhydrone. The solutions should be freshly prepared just prior to usage. Table O-3 lists the millivolt values that can be expected from such solutions at several temperatures and using various reference electrodes. The values are all *plus* and follow the European convention.

TABLE O-3. *ORP of Quinhydrone Solutions*

Electrode	Solution pH and Temp., °C					
	pH = 4			pH = 7		
	20°C	25°C	30°C	20°C	25°C	30°C
Calomel (mercury–mercurous chloride).....	223	218	213	47	41	34
Silver–silver chloride	268	263	258	92	86	79
Hydrogen.....	470	462	454	295	285	275

Applications: Water treatment probably employs the ORP measurement to a greater extent than any other group of processes. ORP is relied upon almost universally, both to monitor and to control the cyanide oxidation and the chromate reduction processes so frequently required in industrial-waste treatment of metal-treating wastes prior to discharge into sewers or streams. Regulatory agencies often monitor freshwater streams with ORP analyzers to determine the relative "health" of receiving waters and to detect unwanted dumpings of strong oxidants or reductants upstream. Some sewage plants employ oxidation-reduction potential to monitor sewage influent and digester sludge and, to a limited extent, to control chlorination for in-plant odor control.

Oxidation-reduction potential is used to control the flow of chlorine or other oxidizing agents in various bleaching processes in the pulp and paper industry. A wide variety of applications exist throughout the chemical industry. Various mine-ore beneficiation processes employ ORP measurement wherever some prescribed oxidizing or reducing level is required for optimum separations.

Measurement of Specific Ions: Another type of ORP measurement involves the use of a metal-metal salt as the measuring electrode rather than a noble metal electrode. As an example, a silver-silver chloride measuring electrode will measure chloride concentrations ranging from a few to over 10,000 ppm. This essentially is an oxidation-reduction-potential measurement. Thus, other oxidants and reductants in the solution being measured can, if in sufficient concentration, introduce a secondary potential, although most applications are not concerned with this aspect. Applications include the monitoring of streams for chlorides as an index of water quality, for salt-intrusion detection, and to detect dumping of high-chloride-containing wastes. Other metal–metal salt combinations, such as silver–silver sulfide to measure sulfide levels in aqueous solutions, have been used as process analyzers.

REFERENCES

Kehoe, T. J., and R. H. Jones: ORP Measurement in Waste Treatment, *Water Sewage Works,* vol. 107, no. 8, August, 1960.

Latimer, W. M.: "The Oxidation States of the Elements and Their Potentials in Aqueous Solutions," Prentice-Hall, Inc., Englewood Cliffs, N.J., 1952.

Lingane, J. J.: "Electroanalytical Chemistry," Interscience Publishers, Inc., New York, 1953.

—Thomas J. Kehoe, *Beckman Instruments, Inc.*

Oxygen Analyzer Methods used for oxygen analysis in both the gaseous and dissolved states are based upon the following characteristics and measurements:

1. Oxygen's high *paramagnetism*—by (*a*) thermal or (*b*) mechanical means
2. Electrical potential or current caused by reactions specific to oxygen concentration or partial pressure in *electrochemical cells*
3. *Thermal energy released* when oxygen is caused by catalytic means to react with an excess of a suitable fuel
4. Sample-volume reduction due to *chemical absorption* of oxygen
5. *Color change* in a sample solution in which the presence or absence of free oxygen causes specific color differences

The most common gaseous-oxygen analyzers employ paramagnetic or electrochemical principles. Most common dissolved-oxygen analyzers use electrochemical methods, while laboratory methods employ color comparison of chemically fixed samples against color standards.

Paramagnetic Oxygen Analyzers: Oxygen's large and almost unique paramagnetism provides an excellent means of analysis. Of common gases, only nitric oxide and nitrogen dioxide are paramagnetic and to a much lesser degree than oxygen. The latter gases seldom are encountered in significant concentrations.

Thermal-paramagnetic Analyzers: These instruments employ oxygen's paramagnetism and the tendency of oxygen to become *less paramagnetic when heated.* These phenomena are used to generate a magnetically induced pressure difference, or "magnetic wind," in the measuring cell. The change in cell sample flow resulting from pressure difference is measured by sensitive thermal detectors in a Wheatstone bridge. The unbalance created in the bridge is proportional to the oxygen concentration or partial pressure. Several analyzer configurations are used to achieve different sensitivity ranges, and tolerances for background-gas variation (see Table O-4). These analyzers generally are used as continuous on-stream units, since they are of rugged and dependable construction.

Mechanical-paramagnetic Analyzers: These instruments detect oxygen partial pressure by measuring the displacement or the force required to prevent displacement of a nonmagnetic test body placed in a nonhomogeneous magnetic field. Typically, the test body is suspended on a fine metal or quartz fiber. When oxygen is present in the gas surrounding the test body, the oxygen tends to displace the test body toward a zone of lesser magnetic strength. By measuring the displacement or the force necessary to prevent displacement, the oxygen partial pressure is determined. This principle is suitable for narrow and wide ranges and suppressed ranges. The method is reasonably independent of background-gas variation except for diamagnetism, which is mainly significant in low ranges. Analyzers of this type generally are less rugged than thermal-paramagnetic instruments, and they are less temperature-sensitive.

Electrochemical Oxygen Analyzers: Several instruments of this type are in use. They are used to measure oxygen in both the gaseous and dissolved states, for trace concentrations, and for moderate and high ranges.

TABLE O-4. *Characteristics of Oxygen Analyzers*

Method	Oxygen Range	Average Full-scale accuracy, %	Remarks
Paramagnetic:			
Thermal	0–1% 0–100%	±1–2	Rugged. Needs close temperature control. Background-gas effect depends on design
Mechanical	0–0.1% 0–100%	±1–2	Accurate. Fragile. Reasonably free of background-gas effect
Electrochemical:			
Membrane	0–5% 0–100%	±5	Low cost. Needs routine regeneration. Sensitive. Slow response
High-temperature fuel cell	0–1% 0–100%	±2	Handles high-temperature samples. Needs close temperature control. Needs reference gas
Combustion-heat rise	0–5% 0–100%	±2	Subject to catalytic poisoning. Requires fuel
Absorption	0–5% 0–100%	±2	Portable. Uses grab sample. Results depend on operator skill
Dissolved-oxygen analyzers:			
Electrochemical	0–20 ppb 0–1,000 ppb	±5	Simple. Fast. Accurate
	0–10 ppm 0–20 ppm	±5	
Thallium cell	0–50 ppb 0–20 ppm	±5	Thallium is poisonous. Needs careful control of conductivity. Medium response
Hydrogen scrubbing	0–20 ppb 0–100 ppb	±5	Complex. Medium response
Colorimetric	0–10 ppb 0–20 ppm	±5	Manual. Laboratory standard

Amperometric-type Analyzer: This analyzer consists of an anode, a cathode, a suitable electrolyte, and an electric-current-measuring load circuit for converting the signal from the cell to an analog of the oxygen concentration. In general, the anode consists of a metal such as zinc or lead which goes into solution as metallic ions while oxygen plus water reacts at an inert cathode, forming hydroxyl ions. These combined half-cell reactions produce a current signal in the measuring circuit which is proportional to the oxygen concentration in the cell electrolyte. The chosen electrode materials and electrolyte may provide adequate cell-polarizing potential, or an external polarizing voltage may be used. The electrolyte may be liquid with gaseous sample sparged into it, it may be a paste or liquid contained within an oxygen-permeable membrane such as a thin layer of Teflon or polyethylene which contacts the sample, or it may be the actual sample in certain dissolved-oxygen analyzers.

Typical cell reaction is as follows:

$$2Me \rightarrow 2Me^{++} + 4e \qquad \text{(anode reaction)}$$

$$O_2 + 2H_2O + 4e \rightarrow 4OH^- \qquad \text{(cathode reaction)}$$

$$2Me + O_2 + 2H_2O \rightarrow 2Me\,(OH)_2 \qquad \text{(combined reaction)}$$

The electrons indicated in the anode and cathode half-cell reactions comprise the current signal of the cell and are transferred from the anode to the cathode through the signal-measuring circuit. The signal may or may not require amplification. It is vital in this type of analyzer that equilibrium be reached according to Henry's law between the gaseous sample and the electrolyte.

Reaction-emf-type Analyzer: This instrument consists of an anode and cathode of a metal, such as platinum plated on opposite surfaces of a ceramic such as calcium-stabilized zirconium oxide. The sample is exposed to one electrode while a reference gas, such as air, is exposed to the other. At high temperatures (550 to 1300° C), the ceramic becomes conductive and the following type of reaction occurs:

$$O_2 + 4e \rightarrow 2O \qquad \text{(anode reaction)}$$
$$2O \rightarrow O_2 + 4e \qquad \text{(cathode reaction)}$$

This reaction produces an emf which is proportional to the difference in oxygen concentration between the sample and reference. The millivolt output signal is a logarithmic function of oxygen concentration. This method may be used to analyze for oxygen at higher temperatures than other methods and is suitable for trace and wide ranges as well as suppressed ranges.

Dissolved-oxygen Analyzers: Determination of dissolved oxygen is important in corrosion control in high-pressure boilers, in water-pollution control, and in food and beverage preservation. In addition to the electrochemical methods described, the oxidation of thallium to thallium hydroxide—as a sample containing dissolved oxygen flows over a thallium bed—produces a change in electrical conductivity of the sample. Measurement of this change provides an analog of the oxygen concentration.

Other Oxygen Analyzers: One method employs thermal detectors to sense the heat rise resulting from combustion of oxygen in the sample in an excess of fuel, such as hydrogen which is metered into the sample. The detector is a heated catalytic platinum spiral of the type used in combustibles analyzers. The signal is measured with a Wheatstone bridge and is an analog of oxygen, since the reaction is limited by the oxygen in an excess of fuel. Chemical absorption is used in the Orsat portable analyzer. An absorbing solution, such as pyrogallic acid, is used to remove oxygen from a measured volume of the sample. The reduction in sample volume is a measure of oxygen concentration. Colorimetric methods are principally laboratory devices for dissolved-oxygen analysis and generally are not adapted to on-stream analysis.

—Lyle K. Barnes, *The Hays Corporation, a Milton Roy Company*

P

Pacemaker (See **Cardiovascular instrumentation.**)

Paint, Temperature-sensitive [See **Temperature-sensitive materials (indicators).**]

Panel Displays (See **Display-command operator/machine interface.**)

Paper Chromatography (See **Chromatography.**)

Paper-tape Reader/Punch Paper tape is an input/output medium which utilizes holes punched in a continuous strip of paper as the information-storage mechanism. Each row of the paper tape consists of an index hole used to locate the row along the length of tape and five, six, seven, or eight hole positions depending upon the code used (see also **Input/output devices**).

In a paper-tape reader, the tape is mechanically fed by means of a sprocket drive wheel or friction rollers through a read station. The index hole provides the timing for the read station and the bit (see **Bit**) holes in the row are sensed by means of brushes or photoelectric sensors. The bits are stored in a buffer [see **Buffer (storage)**] and transmitted to the computer while the next row is being moved into the read station.

In a paper-tape punch, the index hole provides the timing signal for the punch mechanism. When the index hole is sensed, the contents of the

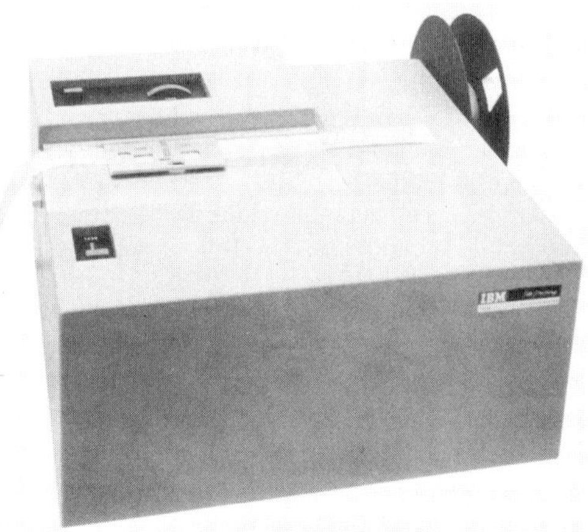

Fig. P-1. IBM 1054 paper-tape reader.

buffer is gated (see **Gate**) to the punch magnets and the selected bits are punched in the tape. While the tape is moved to the next row position, the information to be punched is transferred from the computer to the punch buffer.

Fig. P-2. IBM 1055 paper-tape punch.

Views of the IBM 1054 paper-tape reader and IBM 1055 paper-tape punch are shown in Figs. P-1 and P-2, respectively.

—W. W. Welz, *IBM Corporation*

Parabolic Interpolation (See **Drafting system, automatic.**)

Parallax Error Error of reading when the eye of the observer is not in line and at right angles to the scale at the point of indication. Many instruments have a mirror associated with the scale. When the observer fixes his line of sight so that the reflected image of the indicator is directly behind the indicator, he will be reading the position of the indicator correctly in relation to the scale and will avoid parallax reading error. In a glass thermometer, "The error due to parallax may be eliminated by taking care that the reflection of the scale can be seen in the mercury thread, and by adjusting the line of sight so that the graduation of the scale nearest the meniscus exactly hides its own image; the line of sight will then be normal to the stem at that point." (See reference.)

REFERENCE

ASTM Standard E77, Test and Standardization of Thermometers, p. 1054, 1963.

—Wallace Dean Wood, *Taylor Instrument Process Control Division, Sybron Corporation*

Paramagnetism (See **Oxygen analyzer.**)

Parity As a means of providing error detection in a digital computer, a redundant bit (see **Bit**) is carried with each character or word. This redundant bit is called a parity or check bit.

In an odd parity check system, the value of the parity bit is determined by counting the number of binary 1s (see **Binary state**) in the word, excluding the parity bit. If the number of binary 1s is even, the parity check bit is set to a binary 1; if the number of binary 1s is odd, the parity check bit is set to a binary 0 value. By means of parity bit, a simple error-detection circuit can be designed to detect single-bit changes in the word. Although this type of checking can detect the existence of an error if an odd number of bits are changed, an even number of errors may go undetected.

In an even parity check system, the parity bit is set to a binary value such that the number of binary 1s in the word, including the parity bit, is even. If the absence of any bits in a word is a valid combination, the odd parity check has the advantage in that it protects against the blanking of a complete word.

—W. W. Welz, *IBM Corporation*

Parshall Flume (See **Flume.**)

Partial-pressure Gage (See **Vacuum gaging.**)

Particulates, Measurement of (See **Air-pollution instrumentation.**)

Parts, Dimension Gaging of (See **Gaging, pneumatic.**)

Patient Monitoring (See **Medical intrumentation.**)

Pattern Recognition A pattern for the purpose here may be defined as an *information design.* Information with a design or pattern possesses some form and order even though the data organization and structure in some cases may be very difficult to detect or recognize. Information with a pattern results from some form of rationale at work, and consequently data on all scientific phenomena fall within the patterned information category. The fact that man may not have detected the pattern for such information simply indicates that the phenomenon is not fully understood. In essence, of course, one of man's most effective tools toward understanding is that of attempting to find information patterns. Nonpatterned information may be described as amorphous, nonordered, delirious, and chaotic in nature, that is, information which does not appear to stem from any system of rationale. The term *random* is not used in this connection, because when subjected to thorough scrutiny, many random functions are found to possess some order, and thus were only apparently random.

Pattern recognition is an umbrella-type term that embraces all kinds of information and the instrumental and mathematical means for detecting and analyzing information structure. Because of its great power as a scientific investigatory tool, pattern recognition is a rapidly developing technology. Continued progress is assured because of constantly improving instrumental methods for recording and detecting information patterns and the availability of computers to aid the mathematician in analyzing complex information banks.

Classes of Information Patterns: The most familiar form of information pattern may be the *graphic pattern* in which functions of x and y are represented. Graphic patterns may be analyzed in terms of time sequence, that is, x and y with reference to t. By means of holography (see **Laser, instrumentation of**), patterns in x, y, and z can be determined and analyzed.

In addition to the geometric classification of patterns, there are patterns by plan, that is, information forms that are purposely designed to be detected and interpreted either manually or instrumentally, such as alphabetical and numerical characters and various forms of codes. One also might term these artificially constructed patterns. On the other hand, there are patterns of interest which result from activities that were not conceived or taken with the end purpose of producing a pattern. Terrain, for example, may represent certain patterns, as the delta of a river or a crater or a mountain range, which for navigational purposes one may desire to recognize. The map of a city is a pattern which may have resulted fully or only partially from planning. Thus, there are purposely planned patterns designed to be recognized easily; and there are unplanned patterns whose recognition may be easy or very difficult.

For instrumental detection and recognition, some data must be converted to another form before the information can be recognized and analyzed in a quantitative fashion. Voice recognition is an example where frequencies and amplitudes are first reduced to graphic form prior to instrumental recognition techniques. See **Voice-recognition systems.**

Practical Applications of Pattern Recognition: Some major uses are described briefly as follows:

Radar-signal Detection: Where time delay vs. angle is plotted. See **Radar.**

Character Recognition: Where letters of the alphabet, complete words, numbers, and other symbols are, in essence, read instrumentally. Some fonts (such as the American Bankers Association forms) have been designed specifically with machine reading in mind. In other instances, machines have been designed to read common-type fonts and well-executed, hand-printed letters, numbers, and characters. Cursive script, such as handwriting, introduces many variables and complexities that make the machine-reading approach very costly in terms of practical economics. See **Character-recognition systems; Magnetic-ink character recognition (MICR);** and **Optical character recognition.**

Recognition-prepared Codes: Information may be converted into some form other than characters, which will simplify the instrumental-reading process. Bar codes, punched cards, and tapes fall into this category. See **Code readers and systems;** and **Perforated tape.**

Aerial Photography: Where information on natural and man-made objects is gathered by a camera (visible or infrared) or by radar techniques and made available for analysis in the form of photographic prints or transparencies, or analyzed and stored by electronic means. Closely associated is photogrammetry, which is a technique for constructing a map (particularly a contour map) from a series of aerial photographs. See **Photogrammetry.**

Terrain Recognition: Wherein data pertaining to both natural and man-made objects may be stored electronically and used for navigational and military objectives. Because of the vast variation in both cultural and natural features, each terrain pattern or print is quite specific.

Earth-resources Classification: Where for economic-planning purposes, for example, it may be desired to analyze terrain in terms of various kinds of agricultural output, forests, bodies of water, and in connection with seeking mineral resources, certain types of geologic structures often associated with minerals underground.

Military Reconnaissance: Where it may be desired to ascertain the numbers of enemy trucks, tanks, aircraft, and other equipment in various geographical locations. Visual or ordinary photographic means may not suffice, in which case an electronic recognition system may be used. As may be expected, these systems can become quite complex and costly.

Fingerprint Recognition: In which a lot of effort is going forth to apply instrumental pattern-recognition techniques to detect valley ends and ridge ends in such prints.

Medical Applications: Include the remote transmission and recognition of certain charted data, such as electrocardiograms and electroencephalograms. The examination of tissue slices and smears and the chemical stains produced in certain biomedical applications is being subjected to pattern-recognition technology.

Template matching, in which a model or template of the pattern that is being sought is made and then compared with fresh input information from the subject under study, is a common and the simplest form of pattern-recognition technique. Other methods involving the concept of pattern space become combinations of instrumental techniques and advanced mathematical analysis and are beyond the scope of coverage here. Both coherent and noncoherent optical correlation are used.

REFERENCES

Giuliano, Vincent E.: How We Find Patterns, *Intern. Sci. Technol.,* February, 1967, pp. 40–51.

Stanford Research Institute: Pattern Recognition, *SRI Rept.* 306, Menlo Park, Calif., December, 1966.

Seminar on Image Information Recovery, Society of Photo-optical Instrumentation Engineers, Franklin Institute Research Laboratories, and U.S. Naval Material Command, Philadelphia, Pa., Oct. 24–25, 1968.

Peltier Effect (See **Thermocouple.**)

Pendulous Accelerometer (See **Accelerometer.**)

Perforated Tape A plastic or paper tape in which data may be recorded by means of a pattern of partly or completely punched holes. Tape width varies with the maximum number of holes punched across it. Common widths are five holes plus feed hole, 11⁄16 in.: six or seven holes plus feed hole, 7⁄8 in.; and eight holes plus feed hole, 1 in. Also see **Numerical control.**

Peripheral Equipment, Computer The input or output devices (see **Input/output devices**) which may be connected either on-line or off-line to the central processing unit [see **Central processing unit (CPU)**] are referred to as peripheral equipment.

A peripheral device is said to be on-line when it is under the control of the central processing unit and information reflecting current activity is introduced into the data-processing system as it occurs. A device is off-line when it is not in direct communication with the central processing unit.

Examples of peripheral equipment are printers, tape units, card readers, and card punches. Also see **Data communications.**

—W. W. Welz, *IBM Corporation*

Permeameter An instrument for determining the magnetic permeability of a material by measuring the magnetic flux produced in a specimen of ferromagnetic material under test by a known magnetic intensity. The permeameter represents an improvement over simpler bar or ring tests. There are several configurations.

The Burrows Permeameter: This is a double-bar double-yoke apparatus and is recognized as an instrument of high precision for tests at magnetizing forces up to 300 oersteds. The essential feature is the distribution of the magnetizing winding in sections that permits the independent adjustment of the magnetizing force in various parts of the magnetizing circuit. The method provides dependable results only with magnetic materials of known uniformity, because variations in permeability along the specimen result in magnetic leakage which is not compensated.

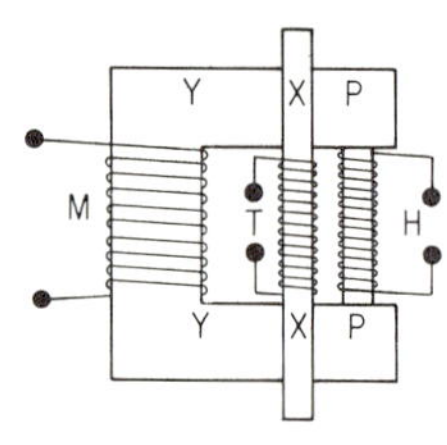

Fig. P-3. Fahy-Simplex permeameter.

The Fahy Permeameter: This instrumental technique determines magnetic-field intensity as the difference of magnetic potential between the ends of the specimen. The "Simplex" form is shown in Fig. P-3. *Y-Y* is a yoke of high-grade laminated steel with a magnetizing coil *M* to give a field intensity up to 200 oersteds. *X-X* is the specimen (usually strips of sheet material or a rod in adapters) surrounded by the test coil T of 100 turns and pressed against Y-Y by two iron pieces P and P which are moved by jackscrews. P and P carry between them a nonmagnetic bar encased throughout its entire length by a coil H consisting of several thousand turns of fine wire. Current in M is adjusted to the appropriate value of field intensity desired and then reversed. Deflection of a ballistic galvanometer connected to T is noted. Then, the galvanometer is connected to H and a second deflection is noted upon reversal of current. Flux density is computed from the first reading using appropriate equations. Field intensity is determined from the second reading.

Other permeameters that cannot be described in detail here include Ewing's isthmus method; the Sanford-Winter M-H permeameter; the high-H permeameter of Sanford and Bennett; and for testing low-permeability materials, the ASTM null method and the Fahy low-mu permeameter. Also, see **Magnetic quantities (variables).**

REFERENCES

Fink, D. G., and J. M. Carroll: "Standard Handbook for Electrical Engineers," McGraw-Hill Book Company, New York, 1968.

Harris, Forest K.: "Electrical Measurements," John Wiley & Sons, Inc., New York, 1952.

Permeance Transducer (See **Electromagnetic transducer.**)

pH, Measurement of pH is a measure of the *effective* acidity or alkalinity of a solution. More accurately, pH is expressed as the negative logarithm of the hydrogen-ion concentration. Pure water has a hydrogen *ion* concentration equal to 10^{-7} mols per liter at standard conditions. The negative logarithm of this quantity is 7. Thus, pure water has a pH value of 7.

Normally, the pH scale is considered to be 0 to 14. If a strong acid completely dissociates (or ionizes) in water, a $1N$ solution of this acid will have a pH value of 0.0. Conversely, a $1N$ base completely ionized in water will have a pH value of 14. Hydrochloric acid and sodium hydroxide come close to meeting these stipulations. Since the pH expression is logarithmic, there is a tenfold change in hydrogen- and hydroxyl-ion concentration per unit change in pH. A slightly acidic solution having a pH of 6 contains ten times as many active hydrogen ions as a solution of pH 7.

The foregoing description stresses the *effective* acidity (or alkalinity) and the hydrogen *ion* content rather than the total hydrogen present. The most important identifying factor of an acid is the *activity* of the hydrogen in the acid. Sulfuric acid and boric acid both contain significant amounts of hydrogen, but essentially all the hydrogen in sulfuric acid *dissociates* in the presence of water to become free hydrogen ions. Boric acid, when added to water, dissociates very little into free hydrogen ions. Instead, the hydrogen remains relatively inert in the undissociated molecules of the acid. Thus, sulfuric acid is called a strong acid, whereas boric acid is a weak acid. The degree of ionization determines the strength or activity of an acid (or base). The pH of a solution of an acid or base at some normality is an indication of the strength or activity of the solution. The pH of $0.1N$ sulfuric acid is about 1.3, while that of $0.1N$ boric acid is 5.3.

Alkalinity Measurement: Water dissociates slightly to form hydrogen ions and hydroxyl ions as indicated by

$$H_2O \rightleftarrows H^+ + OH^-$$

$$(H^+)(OH^-) = Kw$$

In accordance with the law of mass action, at any given temperature the product of the concentration of the hydrogen ion and of the hydroxyl ion is a constant, known as Kw, the dissociation constant of water. Since one hydroxyl ion is formed for each hydrogen ion produced, and since it was noted that the hydrogen-ion concentration of water is 10^{-7} (or pH 7), then the hydroxyl-ion concentration also will be 10^{-7}. The dissociation constant Kw therefore is 10^{-14}. If an alkali is added to water, the hydroxyl-ion concentration increases, depending upon the amount of alkali added and the degree to which it dissociates. Whatever the resulting hydroxyl-ion concentration may be, the hydrogen-ion concentration will automatically decrease proportionately (by combination of hydrogen ions and hydroxyl ions to form undissociated water molecules) so that the product remains 10^{-14}. Thus, for any given hydroxyl-ion concentration, there always is a corresponding hydrogen-ion concentration. Therefore, the measurement of pH indicates equally well the concentration of both alkalies (bases) and acids.

Influence of Temperature on pH: The dissociation constant of water has a substantial temperature coefficient. Thus, the pH value for the neutral point changes with temperature, as indicated by Table P-1.

TABLE P-1. *Effect of Temperature on Dissociation Constant of Water*

Temp., °C	Dissociation Constant of Water Kw*	Hydrogen-ion Concentrations*	Corresponding pH
0	0.114×10^{-14}	0.34×10^{-7}	7.47
25	1.01×10^{-14}	1×10^{-7}	7.0
100	65×10^{-14}	8×10^{-7}	6.10

*Mols per liter.

Buffer Solutions: These are solutions that resist change of pH despite additions of acid or base to the solutions. Buffer action will be exhibited by any solution which contains substantial concentrations of both a weak acid and the salt of that acid. The salt will yield a substantial concentration of anions (A^-), and the weak acid will yield mostly undissociated molecules (HA). Mixtures of weak bases and their salts also act as buffers in alkaline solutions. The hydrogen-ion concentration (H^+) of the solution will be determined by the mass-action law:

$$\frac{(H^+)(A^-)}{(HA)} = K_A \qquad \text{ionization constant}$$

If some strong acid should be added to this solution, the resulting hydrogen-ion concentration will be less than if the acid were added to the same amount of pure water, because the hydrogen-ion concentration cannot increase and the A^- and HA concentration remain unchanged. This would be in violation of the law of mass action. What happens is that the hydrogen ions (H^+) combine with the anions (A^-) to form HA molecules, so that the ion-concentration product will at all times satisfy the law. Conversely, if something should occur, such as dilution or neutralization, to decrease the hydrogen-ion concentration, the HA molecules will dissociate until the law again is satisfied. In the measurement of pH, buffer solutions are employed to calibrate the electrodes and the system to assure accuracy of measurement.

Hydrogen Electrode: Considered to be the standard for the electrometric measurement of pH, the hydrogen electrode was used experimentally to follow changes of acidity as long ago as 1897 (W. Böttger, *Z. Physik. Chem.*, vol. 24, p. 253, 1897). The hydrogen electrode is employed by bubbling hydrogen gas past a wire or foil which is able to catalyze the reaction $H^+ + e \rightarrow \frac{1}{2}H_2\,(g)$ and thus establish an equilibrium between molecular hydrogen and the hydrogen ions. The metal wire or foil usually is platinum which has been pretreated to provide a "platinum-black" catalyzing surface.

Quinhydrone Electrode: This electrode consists of a gold or platinum electrode surrounded by a saturated solution of quinhydrone. Quinhydrone consists of equimolecular amounts of benzoquinone and hydroquinone. These two substances, together with the hydrogen ion to be measured, form a reversible oxidation-reduction system. Thus, the quinhydrone electrode will assume a potential based upon the activities of these three substances. The electrode is limited to pH measurements in the acidic region and hence is impractical for continuous-process measurements.

Antimony Electrode: This electrode simply consists of antimony metal coated with antimonous oxide. The potential of the electrode varies with pH and thus serves as a means of measuring pH. The potential is an oxidation-reduction type developed by the metal-metal salt. The electrode is simple to construct and rugged. However, the potential is influenced by (1) the level of oxygen dissolved in solution, (2) any oxidants or reductants present, (3) a variable temperature coefficient over its 0 to 11 pH span of useful measurement, and (4) the need for frequent standardization, possibly because of change of electrode-surface characteristics with time.

Glass Electrode: It was not until the development of the glass electrode that pH measurement became a simple, reliable tool for all types of users. The possibility that a thin glass membrane of special composition could develop a potential was described as early as 1909 by the German chemist Fritz Haber. However, little progress with this approach was made until the mid-twenties. Developments since have accelerated to the point that both laboratory and continuous-process pH measurements today are almost as common as temperature and pressure measurements. This acceptance primarily is due to the development of the glass electrode together with improved means of extremely stable electronic amplification of the potential developed by the electrode.

The mechanism, whereby certain types of glass develop potentials depending upon the pH levels prevailing, is an extremely complex one. Let it suffice to state that the glass electrode does produce a predictable potential directly related to the hydrogen-ion concentration of the solution in which it is immersed. The glass electrode responds in a predictable fashion throughout the normally accepted 0 to 14 pH range, developing 59.2 mV per pH unit at 25°C, consistent with the classical Nernst equation. Contrary to the earlier types of pH electrodes, the glass electrode is not influenced by oxidants or reductants in the solution. With suitable temperature compensation, pH measurements can be made from 0 to 100°C, and to even higher temperatures under controlled conditions. The glass-electrode pH measurement, by its obvious character, involves high resistances in the measuring loop. This demands that a high-impedance amplifier be employed. Glass electrodes will have resistances varying from a few to a thousand megohms. In recent years, this has ceased to be a significant problem with use of modern electronic circuitry.

Reference Electrode: In pH measurement, a second electrode is required in the solution simply to complete the circuit. A good, stable reference electrode (1) must produce a predictable potential compatible with the glass measuring electrode, (2) must be linear with respect to temperature change, and (3) must be simple to use.

Although various reference electrodes have been employed, only the mercury–mercurous chloride (calomel) and, more recently, the silver–silver chloride reference electrodes are in common use today. The calomel electrode consists of an inner glass tube packed with mercury mixed with mercurous chloride. A hole in the bottom of the tube communicates with a saturated potassium chloride solution, and this, in turn, is contained in a relatively large glass chamber. At the bottom of this chamber one of various types of junctions is employed to permit the potassium chloride to diffuse or leak into the solution being measured. To complete the circuit, a wire is inserted into the packed column and goes to the amplifier along with the shielded cable from the glass electrode. With this arrangement, the potential developed at the calomel/KCl interface is constant at a given temperature since the solution surrounding it remains the same.

The silver–silver chloride electrode can be nearly identical to the calomel electrode. Instead of the packed column, a silver wire need only be employed if coated with a heavy coating of silver chloride. The silver–silver chloride electrode finds broader acceptance, since it remains stable and reproducible at much higher temperatures than the calomel. The latter electrode is limited to about 80°C maximum, particularly for prolonged or continuous use.

Aside from the type of internal employed, the other important parameter of the reference electrode is its type of junction. Various schemes are employed, such as an asbestos fiber embedded in the glass; use of a glass bead of different temperature coefficient from the glass body surrounding it to permit a controlled crack around the bead for liquid communication; a palladium wire embedded into the near-molten glass body with, again, an annular space surrounding the wire resulting after the glass cools; and a ground-glass sleeve mated to a ground-glass portion of the body. All these junctions have their advantages and limitations.

Colorimetric pH Determinations: The color of certain compounds in solution is dependent upon the hydrogen-ion concentration. Generally, these compounds are weak acids and bases and their salts. Color indicators have been found for most ranges of pH. Determinations are made by adding a small quantity of indicator solution to the sample and comparing the color with that of a color standard. Color comparison can be done manually or with the aid of a photometric instrument. Turbid and colored solutions cannot be observed with accuracy, and indicators are not stable in many strongly oxidizing or reducing solutions. See **Colorimeter.**

Importance of pH Measurement: There are more continuous pH analyzers in use than all other continuous analytical-type instruments combined. Continuous pH analyzers can be found in practically every industry which uses water within its processes. Installations of pH analyzers throughout the free world probably are in excess of 25,000. A sizable portion of these form the primary measuring element within closed-loop automatic-control systems. Applications range from industrial water and waste treatment to pH control of flotation processes in mining. Many pH applications can be found in pulp and paper manufacture, metals and metal treating, petroleum refining, synthetic-rubber manufacturing, power-generation plants, pharmaceuticals, chemical-fertilizer production, and a broad spectrum of the chemical industry.

When properly applied, pH measurements offer the following benefits: (1) Many chemical reactions can be controlled better—resulting in greater processing efficiency and better quality of the ultimate product, and sometimes contributing to a safer process. (2) Records are provided to assure management that optimum processing operations prevail. In the case of industrial waste discharge into public streams, lakes, or collection systems, the recorded measurement provides evidence of compliance from a regulatory standpoint. (3) In certain cases, continuous pH measurement has permitted certain processes to be converted from a batch to a continuous basis. Two examples are certain chemical-fertilizer processes and some food-manufacturing processes.

REFERENCES

Bates, R. G.: "Determination of pH Theory and Practice," John Wiley & Sons, Inc., New York, 1964.

Dole, M.: "The Glass Electrode," John Wiley & Sons, Inc., New York, 1941.

Eiserman, George: "Glass Electrodes for Hydrogen and Other Cations; Principles and Practices," Marcel Dekker, Inc., New York, 1967.

Ives, D. J. G., and G. J. Janz: "Reference Electrodes," Academic Press, Inc., New York, 1961.

—Thomas J. Kehoe, *Beckman Instruments, Inc.*

Phase-angle Meter (See **Power-factor meter.**)

Philips Vacuum Gage (See **Vacuum gaging.**)

Phonocardiography (See **Cardiography.**)

Phosphorimeter (See **Fluorometry.**)

Photocell A solid-state photosensitive electron device whose current-voltage characteristic is a function of incident radiation. Examples include photoconductive cells, phototransistors, and photovoltaic cells. Sometimes referred to as an "electric eye." Also known as photoelectric cell. A phototube is not a photocell. Consult the alphabetical index, and also see **Photometers.**

Photoconductive Transducer (See **Photoelectric transducer.**)

Photoelasticity Polariscope (See **Polariscope.**)

Photoelectric Colorimeter (See **Colorimeter.**)

Photoelectric Transducer A device of this type reacts in some way when exposed to visible light. Because a light beam provides a means of sensing a measurand without physical contact and hence is free of friction, and because a light beam is weightless and hence imposes no physical force on the measurement system, a photoelectric transducer has a distinct advantage over other kinds of sensors for many applications. Disadvantages for some applications include the combined costs of the optics and electronics required and temperature sensitivity, although the latter is not always a problem.

Photoelectric transducers fall into three major categories: (1) photoconductive transducers, (2) photoemissive tubes, and (3) photovoltaic cells.

Photoconductive Transducers: The resistivity of most semiconductors, such as selenium, metal sulfides, silicon, and germanium, decreases with exposure to light. Thus, if a potential is applied across a semiconductor, the current flow will be in proportion to the intensity and wavelength of the incident light.

Photoemissive Tubes: Photoemission is the capability of emitting electrons upon exposure to radiation in or near the visible-light region of the spectrum. The three principal types of photoemissive tubes used in transducers are (1) high-vacuum diodes, (2) gas-filled diodes, and (3) photomultipliers.

Photovoltaic Cells: These cells generate an emf internally when they are exposed to incident-light energy and hence require no external power source. Materials used include copper oxide, selenium, germanium, and silicon. The transfer characteristic between the emf generated and the incident illumination is quite nonlinear. The cells also are quite sensitive to temperature changes and, because of a fatigue effect, require frequent recalibration.

Practical applications of photoelectric transducers are described elsewhere and are listed in the detailed alphabetical index at the end of this volume.

Photoemissive-tube Transducer (See **Photoelectric transducer.**)

Photogrammetry A technique for making maps from photographs. In the simplest form, carefully taken aerial photographs (special cameras are required) are matched and a mosaic is prepared from scores or hundreds of such views. Because of numerous variables encountered in the operation of the aircraft and airborne camera, it is difficult to maintain perfect spatial control, and each photograph of the mosaic must be properly scaled. Carefully marked and known ground reference points aid in this scaling and matching process. Because of the growing importance of mapping to both civilian and military requirements, photogrammetry has become quite sophisticated. In particular, survey lenses have been improved. The objectives, such as the Aviotar and Aviogon by L. Bertele, have become standard in high-performance photogrammetric lenses. Computing and automatic-drafting techniques also play a role in photogrammetry. See **Drafting system, automatic.**

REFERENCE

Aerial Photogrammetry Boosted by Survey Lenses, *Space Age News,* May, 1969, p. 51.

Photographic Exposure Meter (See **Photometers.**)

Photometers Photon is the description of an electromagnetic wave in quantized "particle" form. The energy associated with a photon corresponds to the wavelength, and the photon flux to the amplitude. A photometer is an instrument which measures this flux, that is, the intensity of the beam. In common usage, only instruments which respond to the central part of the electromagnetic spectrum, that is, the ultraviolet, visible, and infrared regions, are called photometers. X-rays, gamma rays, and radio waves are excluded.

A photometer consists of a transducer which transforms electromagnetic waves (photons) into an electric current, and a current-measuring readout device. In its simplest form, the instrument could be a voltage-generating photocell connected to a microammeter. Photographic exposure meters and light meters which measure ambient illumination are of this type. The latter are equipped with green filters corresponding to the relative spectral sensitivity of the human eye.

More recently, photoresistors have been used for this purpose. Photoresistors require a voltage source (battery) in the circuit. The microammeter reads out the change in resistance caused by the illumination. These devices are substantially more sensitive but do not lend themselves to high-precision measurements because of the fatigue and "memory" effects of photoresistors.

If more precise work is required, particularly if the measurements must reach into the ultraviolet, phototubes or photomultiplier tubes are the transducers of choice. These components are vacuum tubes equipped with a metal-coated cathode. When light strikes this surface, electrons are emitted and drawn off toward the anode, which is set at a stabilized, positive potential. In multiplier tubes, these electrons are made to cascade over a large number of increasingly more positive plates (dynodes), typically 8 to 14 in number. At each stage, there is a multiplication of the electron flux. This multiplication in photomultipliers can give tremendous internal amplification (many-million-fold). Photomultipliers are the choice for precise measurement of low-light levels. The general principle of a multiplier photometer is the same as previously described, but there is a need for a well-regulated high-voltage supply (instead of a battery). The mechanical construction is more critical because of the need to exclude stray light. The readout amplifier and meter are of matching quality. A photocell-galvanometer-type photometer to measure illumination, a high-quality phototube, and a photomultiplier-tube photometer for scientific laboratory-type applications are shown in Fig. P-4.

Of the transducers mentioned, the photocells and photoresistors have their main application in the visible part of the spectrum and sometimes in the near infrared where they respond. The photomultipliers are useful in the ultraviolet (down to about 200 nm) and up to the infrared. For measurements extending through the infrared, special infrared detectors are needed. These are thermocouples, thermopiles, or pneu-

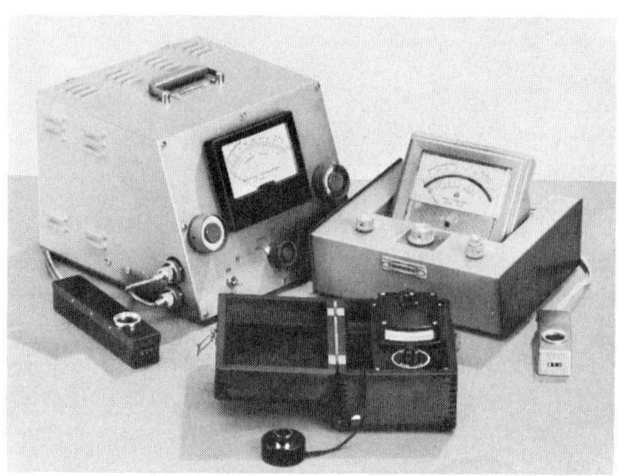

Fig. P-4. Types of photometers. Multiplier photometer shown at left; phototube instrument shown at right; photocell light meter shown in foreground. *(Photovolt.)*

matic detectors. Essentially, these detectors are highly sensitive "thermometers." The synchronous chopping of the beam to be measured is essential here, although the technique also can be used in the visible and ultraviolet region. The reason for chopping is that the detector must distinguish between the "heat" associated with the infrared beam to be measured and the "heat" associated with the equipment, that is, stray radiation. An infrared photometer incorporates a light interrupter (chopper), an infrared transducer, an amplifier tuned to the chopper frequency, and a readout meter. A simple device of this nature is a remote-sensing thermometer as shown in Fig. P-5. Also see **Radiation pyrometer.**

There are many more or less sophisticated devices which serve special purposes and are properly called photometers. For instance, photographic work with flash or strobe light requires photometers which integrate the total light during a flash. In agricultural research, integrating photometers limited to the physiologically active wavelengths are used. In far-ultraviolet work, photon counters are used.

Fig. P-5. Infrared field thermometer. *(Barnes Engineering.)*

In industrial practice, light sensors play an important role. These devices serve as transducers in digital transformation of punch cards; they count containers coming off a line—wherever light and the interruption of light can be used to indicate some event or phenomenon. These applications strictly speaking do not represent *photometers,* since there is no photometric readout involved, but they do represent clever ways of using light in measurement and detecting systems. Some controllers based on turbidity or color discrimination may be considered photometers, particularly when the control function is connected with a strip-chart recorder.

Perhaps more important is the fact that photometers very frequently are part of a larger photometric system. For instance, a colorimeter is simply a light bulb–filter–photometer combination. In many European countries, colorimeters are called "photometers." In English usage, too, spectrophotometers, fluorophotometers, and several other terms include the term "photometer." Actually, most optical measuring devices include photometers as essential parts. Some of these are covered elsewhere in this encyclopedia. The following tabulation, although not totally inclusive, lists the commonly used devices which include a photometer as an essential and integral part of the system.

Atomic-absorption photometers serve a function similar to that of the flame photometer. However, atomic-absorption photometers operate on the very specific spectral absorption of an atomized sample rather than emission. This situation enhances stability and also makes a wider choice of elements possible. The equipment consists of a stabilized hollow-cathode lamp (one for each element to be analyzed), a flame with sample nebulizer, a monochromator, and a photometer. Atomic absorption is one of the most useful of the newer laboratory techniques (Ref. 1). Also see **Flame photometer-spectrophotometer;** and **Optical-emission spectrochemical analysis.**

Brightness meters are special reflection meters to evaluate "paper brightness" by measuring the diffuse reflectance in the blue range of the spectrum. In essence, these meters quantify the yellow characteristics of the paper.

Circular Dichrographs: The purpose and applications of circular-dichrograph instruments are similar to those of spectropolarimetry. However, instead of the change in angle of optical rotation vs. wavelength, they record the difference in dichroic absorption vs. wavelength. The equipment differs in that after the monochromator and polarizer, an electrically driven crystal is disposed in the light path, which gives rise to alternating circular polarization. The photometer circuit is arranged so as to record circular dichrosim directly. The recorder and the monochromator drive are linked, and the instrument yields the final spectral curve (Ref. 2). Also see **Polariscope.**

Color-difference meters are specialized reflection meters used to assess small color variations. See **Color, measurement of.**

Colorimeters are very popular for routine chemical analysis. Compounds or ions which absorb light in the visible part of the spectrum (400 to 800 nm) or which are convertible by specific reagents to such compounds yield to this approach. A typical colorimeter contains an incandescent light bulb as the light source, filters to separate the spectral region, a cuvette to contain the sample solution, and a photometer. See **Colorimeter.**

Densitometers measure the attenuation of a beam passing through, or reflected from the surface of solid samples. See **Densitometer.**

Ellipsometers are used to determine the thickness of very thin films of monomolecular dimensions. These instruments basically are polarization interferometers utilizing a photometer as a readout device.

Fluorimeters, fluorometers, and fluorophotometers are synonymous terms. In these instruments, the sample is excited by a light beam of suitable short wavelength. The reemitted fluorescent light is picked up by a photometer, usually placed at 90° from incidence. A filter or monochromator is provided which excludes the exciting waveband and transmits the fluorescent light (Refs. 3, 4). Also see **Fluorometry.**

Footcandle meters are color-corrected illumination meters calibrated in footcandles. Some instruments are constructed to read correctly at all angles of incidence (eliminating effects of reflectance at the sensor) and are designated as "cosine-corrected."

Glossmeters measure specularly reflected light from the surface of a flat sample. The angle of incidence and the angle of the light pickup are identical and opposed from the normal to the surface. Also see **Glossmeter.**

Hemoglobinometers are specialized colorimeters for the determination of hemoglobin in blood.

Light-scattering Photometers: There are two classes of these instruments. One class is used to determine the amount (or counts) of suspended particles. The arrangement and use essentially are similar to nephelometers used to determine turbidity. Also see **Nephelometer.** The other class is of high precision, uses monochromatic light, and is used to determine the molecular weight of macromolecules dispersed in solution.

Lux meters essentially are footcandle meters except that their calibration is in international lux units (1 footcandle = 10.8 lux).

Microdensitometers are specialized scanning transmission densitometers. See **Densitometer.**

Nephelometers serve to determine turbidity, that is, quantify precipitates in solution by the amount of scattered light. See **Nephelometer.**

Opacimeters are reflection meters specifically designed to evaluate the opacity of thin sheets, such as paper, by measuring the diffuse reflectance over a white and a black surface in turn. Also see **Color, measurement of;** and **Densitometer.**

Polarimeters serve to analyze the concentration of optically active compounds in solution by determining the angle of rotation of plane-polarized light passing through the sample (Ref. 5). See **Polariscope.**

Reflection meters are photometers arranged to pick up diffusely reflected light from the surface of a flat sample. The spectral evaluation of the reflected light permits quantitative color evaluation as seen by the human observer. Also see **Color, measurement of;** and **Densitometer.**

Refractometers serve to determine the refractive index of solutes. As this property is additive, the instruments can be used to measure the concentration in similarly constituted samples. Many of these instruments employ photometric-readout systems. This is invariably true of differential refractometers, which are high-precision instruments used to monitor industrial-product streams and effluents of sample collectors in chromatography. Also see **Refractometer.**

Saccharimeters are polarimeters, calibrated in "sugar degrees" for the purpose of analyzing the concentration or purity of sugar solutions. These instruments usually contain a quartz compensating element which matches the rotary dispersion of sugar. Thus, a saccharimeter operates with white light, obviating the need of a monochromator. A saccharimeter cannot be used as a general-purpose polarimeter, while the reverse always is possible. Modern saccharimeters use photometers coupled with self-balancing servosystems instead of the former visual-manual technique. Also see **Polariscope.**

Spectrofluorimeters are fluorimeters which incorporate two separate monochromators. One serves to scan through the spectrum of the exciting light source; the other scans the emitted, fluorescent light. These instruments are particularly valuable in exploratory or research work. Also see **Fluorometry**.

Spectrophotometers comprise a light source, means of monochromatizing the light, a sample space, and a photometer. The quality and complexity, however, are widely varied, depending upon the analytical task. Spectrophotometers usually serve to determine the concentration of a solute by the measurement of light attenuation, the logarithm of absorption being proportional to concentration. Instruments equipped to operate in the infrared region are especially valuable in qualitative organic analysis because the vibrational-rotational spectra of complex organic molecules are rich in character and serve as "fingerprints." Automatic double-beam scanning spectrophotometers are useful in plotting the absorption spectra. Once it is known, single or multicomponent quantitative analysis is possible. Similar considerations apply to the ultraviolet and visible range. Also see **Spectro terminology**; and consult the alphabetical index.

Spectropolarimeters: In the elucidation of structure and conformation of complex organic molecules, spectropolarimetry is one of the most incisive methods. The method involves the determination of optical rotation as dependent upon wavelength, that is, rotatory dispersion. Instead of the visual polarimeter, the use of a photoelectric instrument is almost mandatory because most of the important information lies in the ultraviolet region where the human eye does not respond. Actually, automatic self-balancing spectropolarimeters usually are employed to eliminate the tedium of taking individual data points (Ref. 6).

Spectroradiometers are photometers equipped with spectrum scanning to evaluate the total spectral intensity of light sources.

Turbidimeters: See **Turbidimeter**.

REFERENCES

1. Slavin, Walter: "Atomic Absorption Spectroscopy," Interscience Publishers, a division of John Wiley & Sons, Inc., New York, 1968.
2. Velluz, Leon, Maurice Legrand, and Marc Grosjean. "Optical Circular Dichroism," Academic Press, Inc., New York, 1965.
3. Udenfriend, S.: "Fluorescence Assay in Biology and Medicine," Academic Press, Inc., New York, 1962.
4. Kallmann, H. P., and G. M. Spruch: "Luminescence of Organic and Inorganic Materials," John Wiley & Sons, Inc., New York, 1962.
5. Levy, Gabor B.: "Standard Methods of Chemical Analysis," vol. IIIA, chap. 15, pp. 258–276, D. Van Nostrand Company, Inc., Princeton, N.J., 1966.
6. Djerassi, C.: "Optical Rotatory Dispersion," McGraw-Hill Book Company, New York, 1960.
7. Phototubes and Photocells, Radio Corp. of America, *Tech. Manual* PT-60, Lancaster, Pa., 1963.
8. Kingslake, R.: "Applied Optics and Optical Engineering," Academic Press, Inc., New York, 1967.
9. Smith, R. A., F. E. Jones, and R. P. Chasmar: "The Detection and Measurement of Infra-red Radiation," Clarendon Press, Oxford, 1957.
10. Newman, D. W.: "Instrumental Methods of Experimental Biology," chaps. 10, 11, 14, and 15, The Macmillan Company, New York, 1964.

—Gabor B. Levy, *Photovolt Corporation*

Photometry, Stellar (See **Astronomical instrumentation.**)

Photomultiplier Tube A phototube with one or more dynodes between its photocathode and the output electrode. The electron stream from the photocathode is reflected off each dynode in turn—with secondary emission adding electrons to the stream at each reflection. See Fig. P-6. Also called electron-multiplier phototube.

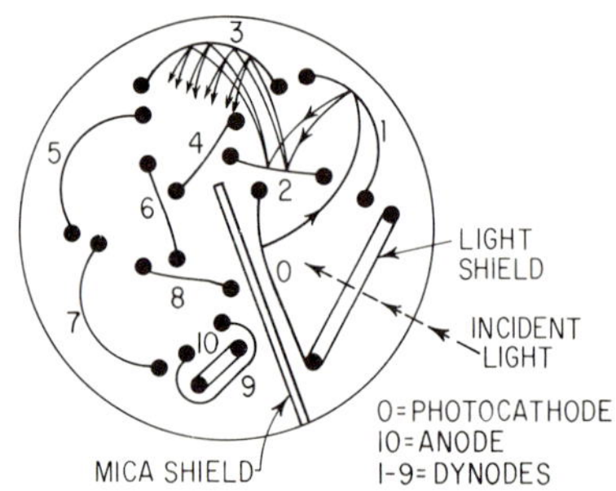

Fig. P-6. Schematic diagram to illustrate operation of a photomultiplier tube.

Photoresistor (See **Photometers.**)

Phototube (See **Photometers.**)

Photovoltaic-cell Transducer (See **Photoelectric transducer.**)

Piezoelectric Transducer Several materials, including quartz, ammonium dihydrogen phosphate (ADP), Rochelle salt, tourmaline, and barium titanates, exhibit a phenomenon known as the piezoelectric effect. These asymmetrical crystalline materials, when subjected to force or stress, develop an electrical-potential difference between certain of their surfaces. This is a property of which advantage is taken in a number of transducers and, in particular, devices for the sensing of motion, position, acceleration, and velocity.

Natural crystals have a very high directional sensitivity because distribution of the asymmetrical charge depends upon the structure and orientation of the crystal. With synthetic crystals, which can be made into almost any desired shape, the desired directional sensitivity can be obtained, or in some cases eliminated. Piezoelectric transducers are relatively simple to design and construct, they are rugged and possess excellent high-frequency characteristics, and they are relatively insensitive to aging. A number of piezoelectric materials are sensitive to humidity and hence require protection. Piezoelectric transducers lack dc response, but this is overcome by use of external ac carrier excitation. Practical examples of piezoelectric transducers are described elsewhere and are listed in the detailed alphabetical index at the end of this volume.

Piezometer Ring A vital element of a primary flowmetering device, such as a venturi or Dall flow tube. The ring provides a point from which accurate static-pressure readings may be taken at critical diameters on a flow tube. A piezometer ring always is provided at the throat section of a primary flow device. Many installations require rings at both inlet and throat sections.

The piezometer ring consists of an annular cavity containing four or more equally spaced vents in the same plane, all of which are open to the bore of the flow tube. See **Dall tube** and **Venturi tube** for illustrations. When velocity distribution is not uniform, slightly different flow pressures can enter the ring at each vent. The total pressure inside the piezometer ring, however, will represent the true average static pressure at that diameter of the flow tube. A tapped hole is provided in the outside diameter of the flow tube, from which a passage leads to the piezometer ring. The port thus made is called the *pressure tap*. Instrument readings taken at the pressure taps on the inlet of the flow tube and the throat section of the piezometer rings sense the true static pressures existing at the inlet and throat diameters.

—Edmond B. Lynch, *B I F, Unit of General Signal Corporation*

Pilot-actuated Regulator (See **Self-actuated regulators.**)

Pirani Gage (See **Vacuum gaging.**)

Piston Actuator (See **Actuator, pneumatic.**)

Pitot Tube In the context of aerospace applications, a pitot device is used for the measurement of undisturbed atmospheric pressure, termed *static* pressure, and often is referred to as a pitot-static tube. See **Air-data systems (aircraft).** The pitot tube also is the basis for flow measurement. Although of relatively minor importance as a primary element for commercial head flowmeters, the pitot tube is an effective tool for laboratory use or for spot checks. The device, shown in Fig. P-7, has a tendency to plug when the flowing fluid contains small amounts of solid matter. The volocity range is limited when used with standard commercial instruments, and the sensi-

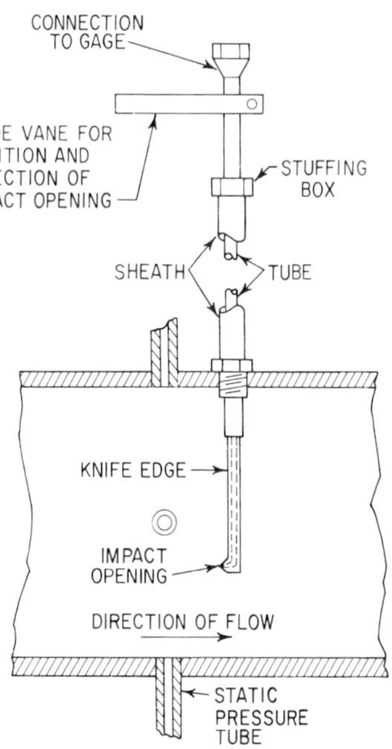

Fig. P-7. Single-opening or elementary type of pitot tube.

tivity to abnormal velocity-distribution effects limits the usefulness of the pitot tube for commercial flow-measurement applications. An elementary pitot tube has two pressure passages, one of which faces into the flowing fluid and, by intercepting a small portion of the flow, reacts to the total pressure of the fluid, that is, static plus kinetic pressure. The other passage faces perpendicularly to the axis of flow and reacts to static pressure only. The difference between these two pressures is a measure of velocity. The energy balance for the device is

$$\frac{p_2}{\gamma} = \frac{p_1}{\gamma} + \frac{V_1^2}{2g}$$

where p_2 = impact pressure (absolute) where velocity $V_2 = 0$
p_1 = static pressure in flowing fluid, absolute
γ = specific weight of flowing fluid
V_1 = velocity of flowing fluid at a point in line with the axis of the impact opening

Solving for velocity and introducing a velocity coefficient,

$$V_1 = C_p \sqrt{\frac{(p_2 - p_1)2g}{\gamma}}$$

which indicates the basic relationship between velocity and pressure differential.

Platinum Electrode (See **Oxidation-reduction potential, measurement of;** and **pH, measurement of.**)

Platinum Resistance Thermometer (See **Resistance thermometry.**)

Plethysmograph (See **Cardiovascular instrumentation;** and **Pulmonary system, instrumental analysis of.**)

Plotter, Curve The plotter is an output device which draws curves of one or more variables as a function of one of more variables. It is used to produce graphical material such as bar charts, flow charts, and engineering drawings. Data from the digital computer (see **Digital computer**) are translated into plotter-actuating signals and converted to incremental plotter movements to produce the drawing.

Figure P-8 is a photograph of an IBM 1627 plotter. The drawing is produced by incremental movement of the pen on the paper surface and/or the paper movement under the pen. The pen is mounted in a carriage that travels horizontally across the paper as viewed from the front of the plotter. The vertical plotting motion is achieved by the rotation of the pin-feed drum which acts as a platen. The drum and the pen carriage are bidirectional; that is, the paper moves up or down and the pen moves right or left. Control is also provided to raise or lower the pen from or to the paper surface.

The information from the computer is decoded into fixed incremental movements of the drum and the pen carriage. That is, each plotter command transmits information which is decoded into the number of pen-carriage increments to be moved (y axis) and the number of drum increments to be moved (x axis). Specified bit combinations also control the raising or lowering of the pen to the paper.

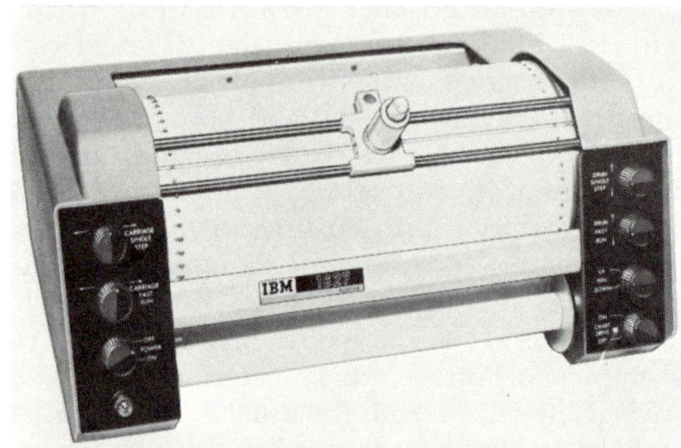

Fig. P-8. IBM 1627 plotter.

—W. W. Welz, *IBM Corporation*

Plotter, Data (See **Drafting system, automatic;** and **Recorder, graphic.**)

Pneumatic Actuator (See **Actuator, pneumatic.**)

Pneumatic Controller The basic functions of a pneumatic controller, shown in Fig. P-9, include (1) *receiving* a measure M of the value of the variable being controlled, (2) *comparing* that value with a reference or desired value R, (3) *determining the magnitude and direction* of any deviation or error E, and (4) *providing an output* O as some function of the deviation. Pneumatic controllers perform basically the same operations as electronic or hydraulic controllers, the primary difference being the penumatic medium with which pneumatic devices operate.

Displacement-type Detectors: While there are many kinds of pneumatic controllers, all employ one of a number of displacement-type pneumatic detectors—the most common of which is variously called a *flapper nozzle,* a *baffle nozzle,* or an *orifice nozzle.* This device, as shown in

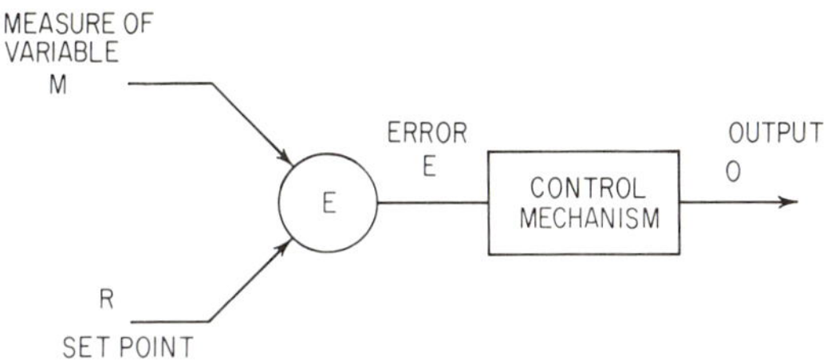

Fig. P-9. Block diagram of basic pneumatic controller.

Fig. P-10, consists of a fixed orifice restriction and a variable nozzle restriction in series. Input motion E is applied to a simple pivoted baffle to change the clearance X between a flat surface on the baffle and the nozzle. The nozzle normally exhausts directly to atmosphere. Under fixed conditions, air flows through the orifice and out of the nozzle through clearance X. As X is increased from zero, the nozzle back pressure P_n

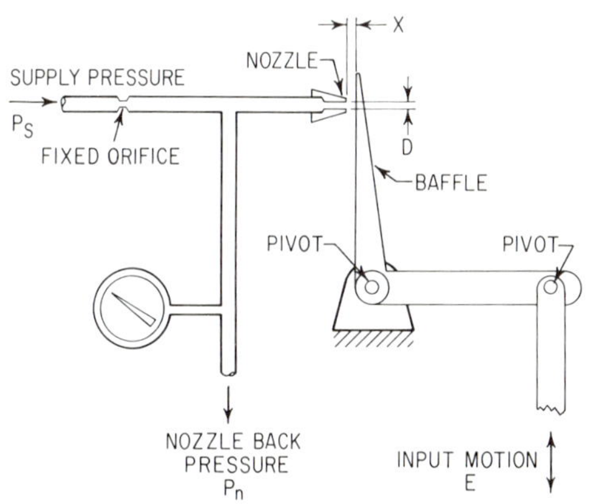

Fig. P-10. Baffle-nozzle detector in pneumatic controller.

decreases as shown in Fig. P-11. The upper limit of the value for nozzle pressure is determined by the air-supply pressure when X equals zero; the lower limit is determined by the ratio of resistance to airflow which is established between the fixed orifice and the nozzle when X is very large. Between these two extremes, various nozzle pressures are established. The slope of the curve at any point is called "nozzle sensitivity." This term also may be called *gain.* The gain is very high for a detector of this type. For the commonly used 0.010-in.-diameter orifice and 0.025-in.-diameter nozzle combination, a motion of the baffle of 0.001 in. causes a change in nozzle pressure greater than 8 psi in the central portion of the curve.

Besides high gain, the nozzle-baffle sensor has two other important characteristics: (1) it develops only a small reactive force—equal to the product of nozzle pressure and the nozzle throat area; and (2) it uses only a small amount of air—less than 0.04 scfm with a 20-psi air supply. The airflow through the orifice is given by

$$Q = kA\sqrt{(P_S - P_n)\,P_n}$$

where the pressures are absolute pressures, where A represents the area of the orifice, and where k

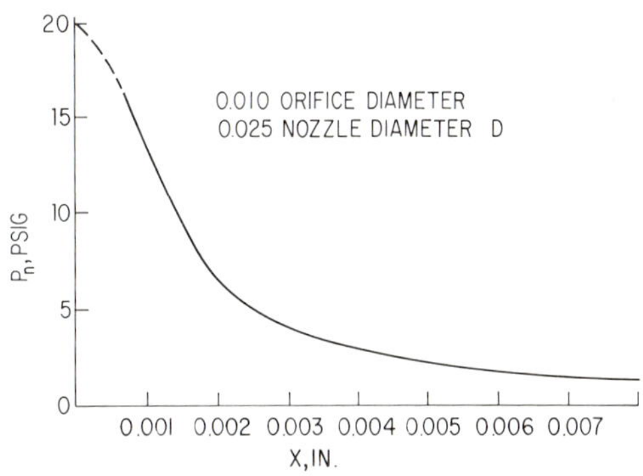

Fig. P-11. Baffle-nozzle characteristic relation between X and P_n for steady state.

is a constant, dependent upon the flow pattern and the condition of the fluid. For air, k equals about 32 $ft^3/(lb)(min)$. Q is given in scfm.

In some designs, the flat baffle of Fig. P-10 is replaced by a ball, as shown in Fig. P-12. This arrangement eliminates the need to align the baffle surface parallel with the lip of the nozzle but may result in slight friction if the ball rubs against its guide. Another form of detector, the "free vane," is shown in Fig. P-13. The two nozzles share a common restriction and have a common centerline. The baffle is moved parallel with the nozzle lips. Sensitivity of this detector is lower than that of a flapper nozzle. Typically, a baffle motion of 0.005 in. is required to change nozzle back pressure about 1 psi.

Still another detector, the *direct-operated pneumatic ball pilot,* is shown in Fig. P-14. As the stem is moved upward, the supply port is opened and the port resistance to airflow from the supply connection is reduced. Simultaneously, the exhaust port is closed and that port's resis-

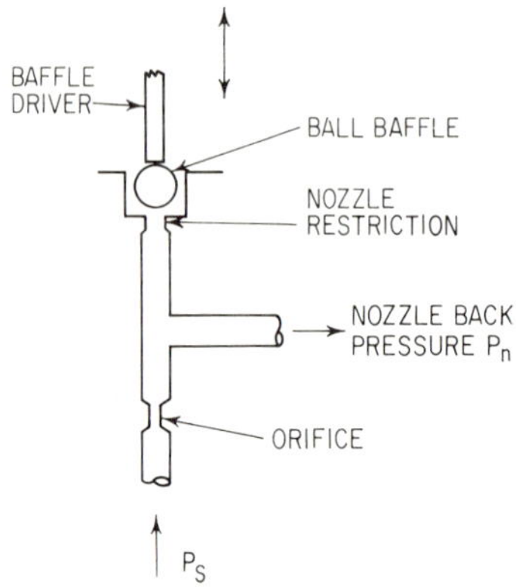

Fig. P-12. Ball nozzle used in pneumatic controller.

tance to airflow from output to atmosphere is increased. Both these resistance changes result in increased air-output pressure. The direct-operated pilot can respond faster than the usual nozzle-baffle system but uses more air because the supply and exhaust ports are large. The usual travel of a ball pilot ranges from 0.003 to 0.010 in. for an output change from 3 to 15 psig. Air consumption often is 0.1 scfm or more.

Pneumatic Relays: To *increase gain,* a pneumatic pilot relay or pilot valve is used in conjunction with a nozzle detector in a controller or transmitter. A relay also increases airflow capacity when a change of output pressure is required. This improves the dynamic response when the pneumatic instrument is connected to a large volume or to a long transmission line. Pneumatic relays differ from pneumatic boosters in that relays have gains higher than 1—usually between 3 and 10. Relays are biased so that no output response occurs until the input pressure from the nozzle has increased somewhat above its minimum. Thus, the nozzle will operate on the central, linear portion of its characteristic curve.

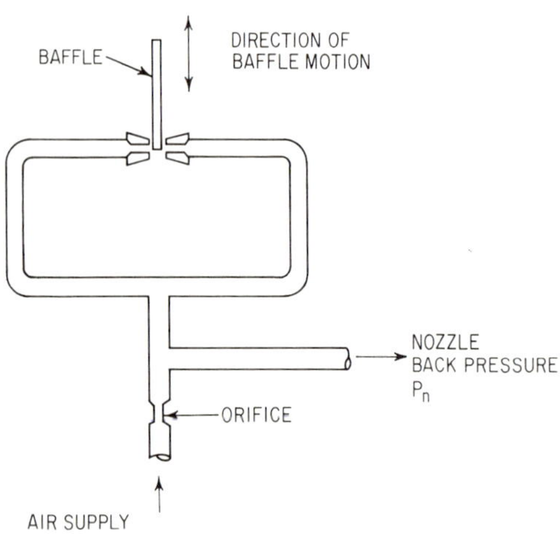

Fig. P-13. Free-vane detector used in pneumatic controller.

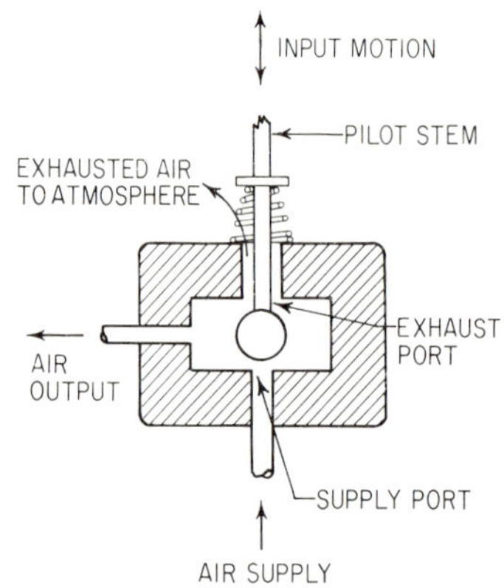

Fig. P-14. Direct-operated pneumatic ball pilot used in pneumatic controller.

A direct-acting relay is shown in Fig. P-15. The metal diaphragm is preformed, holding the supply port fully closed and the exhaust port fully opened—until the nozzle pressure rises to a given value, typically 5 psig. Further increase in nozzle pressure overcomes the diaphragm force and allows the valve stem to move downward—opening the supply port, closing the exhaust port, and increasing the output. At any output between zero and maximum, both supply and exhaust ports are open and considerable air is used. Hence these devices are termed, "bleed-type" relays. A similar relay with reverse action is shown in Fig. P-16. Typical response characteristics for bleed-type relays are given in Fig. P-17. The response is nearly linear in the 3 to 15 psig output range. Linearity of an instrument using such a relay is improved because the characteristic response of a nozzle-baffle detector is quite linear for nozzle back-pressure changes—from about 5 to 15 psig (as shown in Fig. P-11). The bleed-type relay has no feedback; that is, the force developed by nozzle pressure acting on the diaphragm is large compared with the forces of the airflow acting upon the valve. The continuous-bleed relay uses a substantial amount of

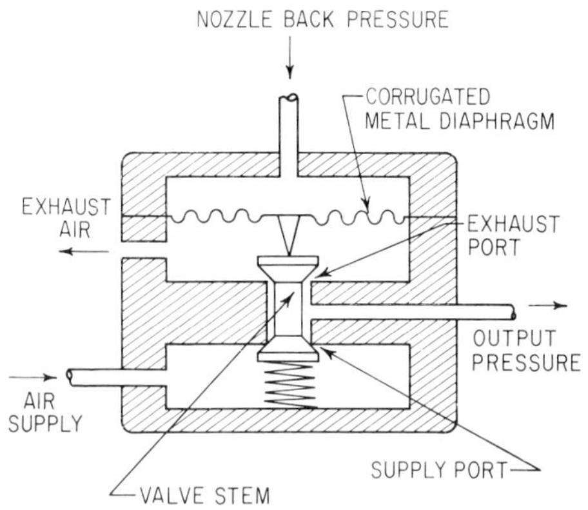

Fig. P-15. Direct-acting pneumatic relay (bleed-type) used in pneumatic controller.

air—frequently as much as 1 scfm—and has maximum air capacity of about twice its consumption.

The nonbleed relay, shown in Fig. P-18, is widely used in pneumatic controllers. This is a force-balance relay with feedback and provides a constant gain determined by the ratio of input area to feedback area. The input (nozzle) pressure applied to a large rubber-covered fabric diaphragm develops a downward force. The output pressure applied to a smaller follow-up area provides an upward force. As the input pressure from the nozzle increases, the diaphragms move downward—causing the valve stem to open the

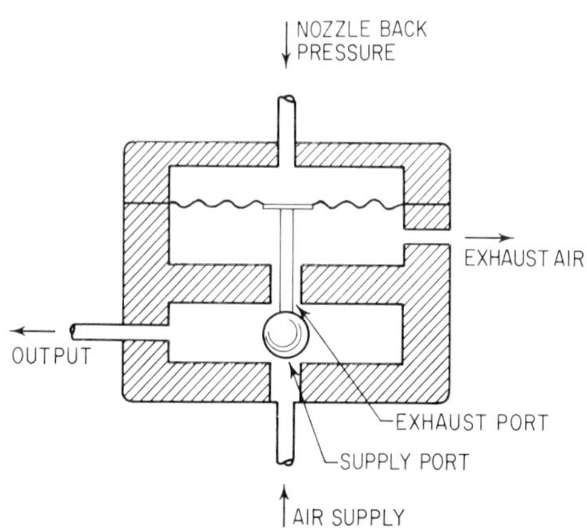

Fig. P-16. Reverse-acting pneumatic relay (bleed-type) used in pneumatic controller.

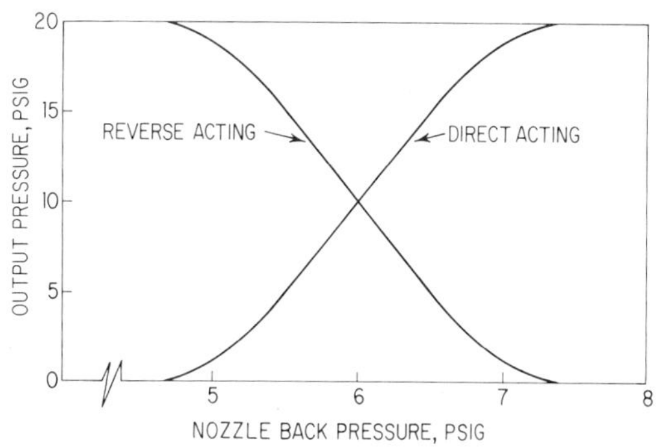

Fig. P-17. Response characteristics of bleed-type relays.

supply port, which admits supply air and increases output pressure until the diaphragm forces are restored to their original balanced condition. Similarly, a reduction in input pressure moves the diaphragm upward—causing the supply port to close and then the exhaust port to open, thereby decreasing output. In balance, both the supply and exhaust ports are closed and no air is used. The ports of the nonbleed relays are large and thus provide high output airflow.

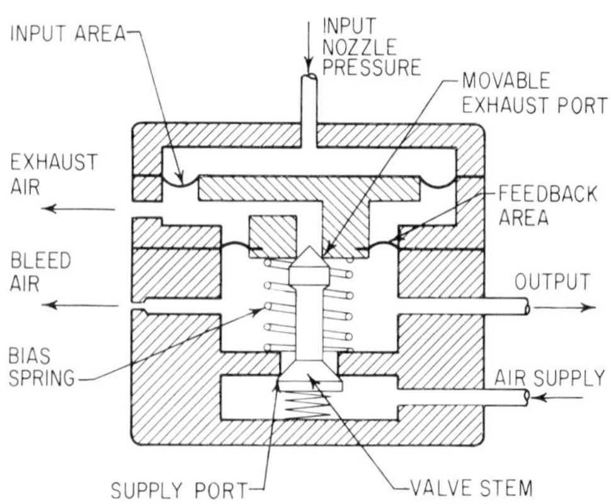

Fig. P-18. Nonbleed relay used in pneumatic controller.

In practice, a "nonbleed" relay often has a small bleed to atmosphere—designed to hold the supply port slightly open. This design ensures that the relay will respond immediately to small changes in input pressure. A bias spring provides upward force to keep the output pressure at zero psig until the nozzle pressure has increased to 5 psig or more—within the linear portion of the nozzle characteristic of Fig. P-11.

Modes of Control: Pneumatic controllers offer several control modes or control responses.

Commonly found are (1) on-off, (2) proportional, (3) integral, and (4) derivative. These various responses are obtained by the choice of the feedback elements and feedback networks, as shown in Fig. P-19. Also see **Automatic control.**

Two-position or On-Off Action: In some forms, on-off is only an approximation. The controller has no feedback, as shown in Fig. P-20. The nozzle detector is a high-gain mechanism, and the relay usually has a gain of 3 or more. Therefore, a small change in input error will cause a large output change. The action of this controller in response to ramp changes in input is shown in

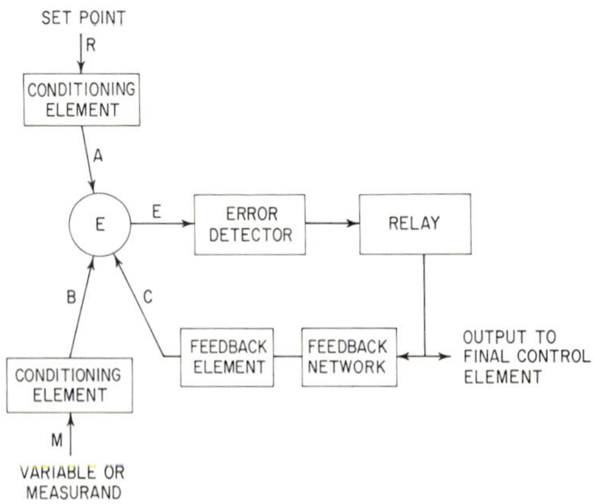

Fig. P-19. Block diagram of pneumatic controller.

Fig. P-21. The narrow band through which the input must change to cause the output to change from the bottom to the top of its range, usually considered to be 3 to 15 psig, is called the *proportional* or *throttling* band and frequently is as small as 1 or 2 percent of input range.

The proportional response sometimes is expressed as gain, which is the percent change in output divided by the percent change in input which produced it—in the proportional mode. On-off controllers of this type are used for simple temperature control where one long time constant dominates the dynamic response of the process. Continuous cycling of the controller output and of the controlled variable may occur—indicating that the controller gain is too high.

On-off action also is provided as *differential-gap action,* shown in Fig. P-22. As the input error increases, the baffle is moved toward the nozzle—increasing the controller output through the direct-acting relay. The output-pressure change is regeneratively applied to the bellows, further closing the nozzle and further increasing output to its maximum. As the input error is reversed (by action of the final control element through the process), the baffle opens slightly, decreasing the output pressure slightly. This decrease in output pressure, acting on the bellows, further opens the nozzle and provides minimum output pressure. The action of this controller is shown in

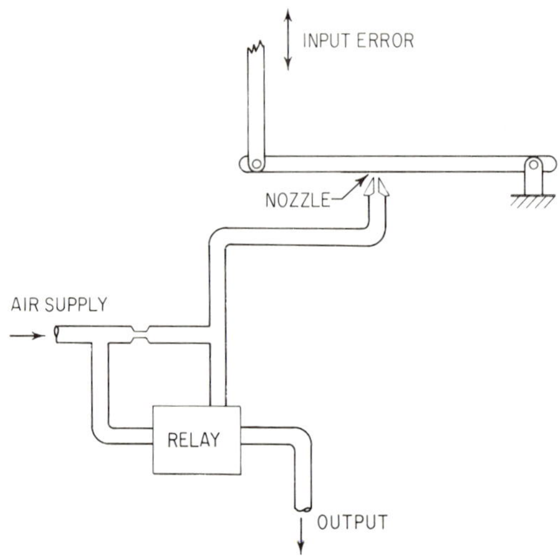

Fig. P-20. On-off pneumatic controller (throttling type).

Fig. P-23. The value of the measured variable which causes the output to rise to maximum is not the same as that value which causes the output to fall to minimum. This difference is called the *differential gap* and is shown to be adjustable in Fig. P-22. Differential-gap controllers hold a controlled variable between two limits without throttling action of the final control element (valve).

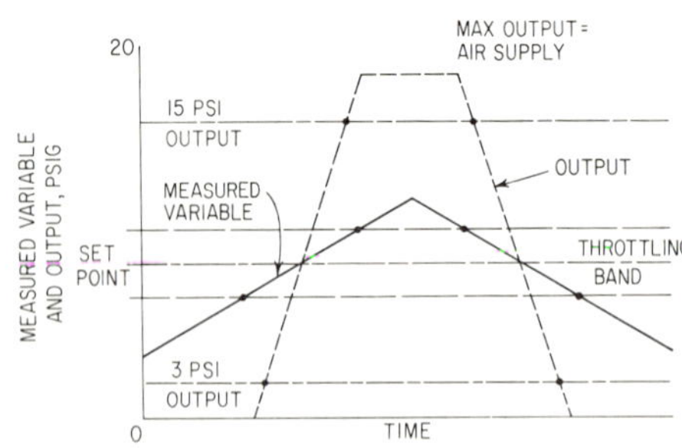

Fig. P-21. Response of a fixed high-gain pneumatic controller.

Proportional Action: An adjustable proportional controller is shown in Fig. P-24. The output of the controller is applied to the feedback bellows to reduce the change in baffle position caused by the initial change in input. Adjustment of pin A to one extreme position virtually eliminates the effect of feedback—producing high gain, or a narrow proportional band. Adjustment of the pin to the other extreme position maximizes the effect of feedback and provides low gain. The gain is continuously adjustable—typically from 1 to 100.

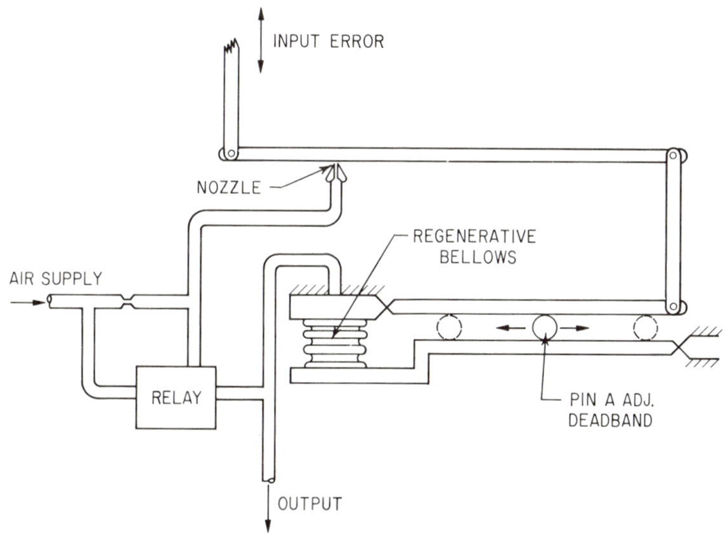

Fig. P-22. On-off pneumatic controller (regenerative type).

Proportional controllers frequently have an adjustment termed *manual reset* or "squaring." This adjustment provides a force which opposes the feedback bellows, making it possible to raise or lower the output to the value required to hold the variable at the set point. In practice the proportional band is adjusted to be as narrow as possible without causing the measured variable to cycle continuously. If the resulting proportional band is wide, objectionable offsets may occur between the set point and the measured value of the variable. To avoid these offsets, *automatic reset* or *integral action* is used.

Reset or Integral Action: Reset gradually raises the gain of the controller from that determined by the proportional-band setting to a higher value if the input error persists for a time. The addition of a reset bellows (see Fig. P-25) in opposition to the follow-up bellows and the addition of an adjustable pneumatic restrictor or needle valve calibrated in time units provide reset or integral response. As the input changes, the output-pressure change is applied to the needle valve as well as to the follow-up bellows. The pressure in the reset bellows changes slowly—as air is passed through the needle valve—to counteract gradually the pressure in the follow-up bellows. This has the effect of removing the follow-up and raising the gain of the controller to the value which would result without any follow-up action. The effect of this higher gain is to reduce the offset between the set point and the measured value of the variable.

Derivative or Rate Response: Some processes with one or more long time constants use derivative or rate response to reduce the effect of these time constants, thereby improving the dynamic

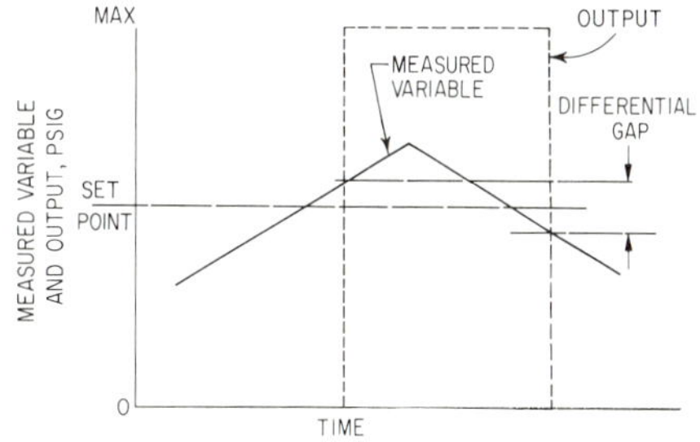

Fig. P-23. Action of a differential-gap pneumatic controller.

response of the control loop. Derivative response is obtained by adding an adjustable restrictor, calibrated in units of time, in series with the follow-up bellows as shown in Fig. P-25. The time constant of the derivative restrictor must be shorter than that of the reset restrictor to avoid regenerative action when both reset and derivative responses are used in the same controller.

When pneumatic transmission systems are used, controllers need not contain large primary measuring elements such as bourdons. Thus, small receiver-controllers are available for use in central control rooms which may be remote from the point of measurement.

proportional response. The minimum gain is determined by the ratio of areas and usually is one-half. The maximum gain is determined by the maximum ratio of R_2/R_1 and typically is 50. Adjustable resistor R_3 controls the reset response. Derivative response is secured with a rate relay with delayed feedback, such as that shown in Fig. P-27. The rate relay may be a separate unit, or it may be housed as part of the controller.

Receiver-controllers: Some of these devices have gain continuously adjustable from 100 down to zero. A controller of this type, employing bellows as pressure elements, is shown in Fig. P-28. Four bellows are arranged symmetrically

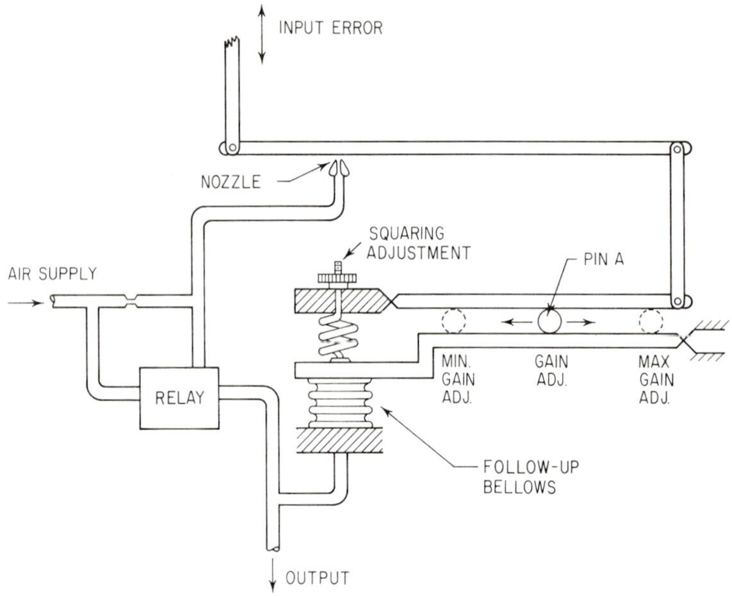

Fig. P-24. Pneumatic controller with adjustable gain.

Stack-type Controllers: One class of these controllers, termed *force-balance* or *diaphragm-stack* controllers, is shown in Fig. P-26. The diaphragms convert pressures received from a pneumatic set-point regulator, from the pneumatic transmitter, and from the feedback networks into coaxial forces. The sum of these forces produces small deflections to control a baffle nozzle. There are no adjustable levers to modify feedback—thus resistor networks are used to adjust the feedback pressures and their resulting forces, allowing adjustable proportional and reset responses. The fixed restrictor R_2 and the adjustable restrictor R_1 act as an airflow divider for

about a motion plate supported by a central tension rod. The bellows which receives set-point pressure is located opposite the bellows which receives transmitted pressure. Similarly, the follow-up bellows is located opposite the reset bellows and in quadrature with the set-point bellows. Any difference between set and transmitted pressures imparts a motion to the central motion pin which moves the baffle. The output-pressure change repositions the motion pin to restore balance. The relative effects of input and follow-up pressures are determined by rotatably positioning the gain-adjustment dial around the motion pin. Integral and rate re-

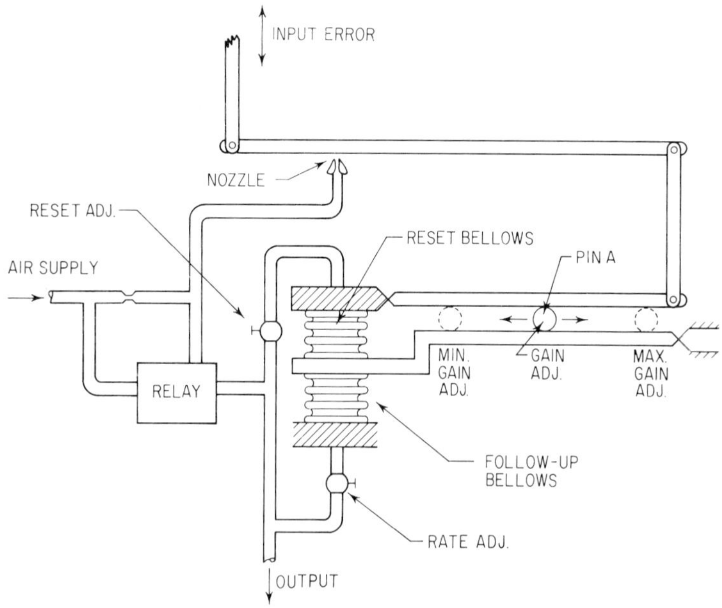

Fig. P-25. Pneumatic controller with reset bellows and adjustable pneumatic restrictor to provide reset integral response.

sponses are adjusted by use of pneumatic restrictors.

Receiver-controllers often mount directly on the back of small panel-mounted indicators or recorders which have provision for manual control. Typically, a pressure regulator—adjustable from the front of the instrument—is connected into the controller output line by valving which isolates the controller output. Thus, the operator can adjust pressure to the final control element while observing the measured value of the variable or its deviation from the set point as shown on an indicator or recorder. Various schemes allow the operator to obtain controller output equal to manual output before switching from manual to the automatic mode.

Cascade Control: At times, control of a variable can be improved by using two controllers—with the output of one controller as the set point for the other. This concept is developed elsewhere. See **Cascade control.**

Response: The response of pneumatic controllers frequently is adjusted on the basis of an experimental technique. The gain is raised—using only proportional response—to the lowest value at which the controlled variable will cycle continuously after a set-point change or other

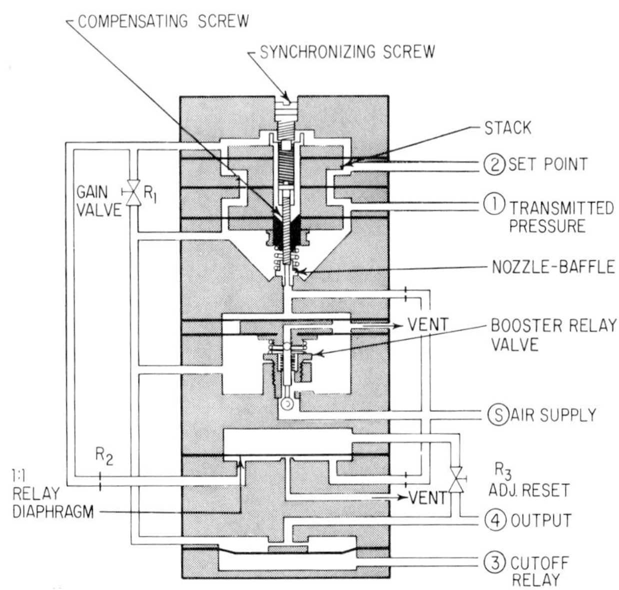

Fig. P-26. Force-balance (stack-type) pneumatic controller. *(Taylor.)*

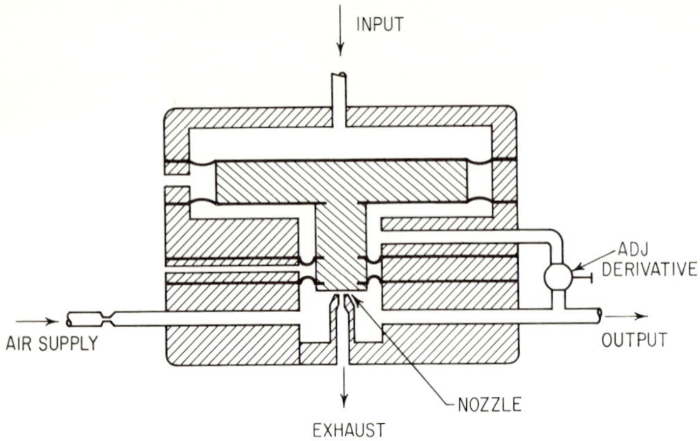

Fig. P-27. Derivative relay used in pneumatic controller.

upset. This value is called the "ultimate gain," or G_u. The period of the cycling is observed, and this value is termed the "ultimate period," or P_u. Then values for the gain, reset, and derivative response settings are established. For a three-response controller, the settings are gain $= G_u/2$, reset $= 2.0/P_u$, and derivative $= P_u/8$. The settings for two-response controllers are slightly less. These settings give acceptable controller response for a wide range of applications.

Styles and Configurations: Pneumatic controllers frequently are designed to be housed within a recorder or indicator case or to be plugged onto the rear of such instruments—using leakless manifolds to complete the required air circuits. Sometimes, controllers may be field-mounted adjacent to the plant equipment and point of measurement. Often, controllers are mounted in centrally located control rooms that enable just a few people to monitor a large number of measured variables. The controllers or the associated indicating and recording instruments may have provision to bypass the automatic controller, allowing the operating personnel to set the position of the final control element directly without the intervention of the automatic controller. Also see **Transmission, pneumatic.**

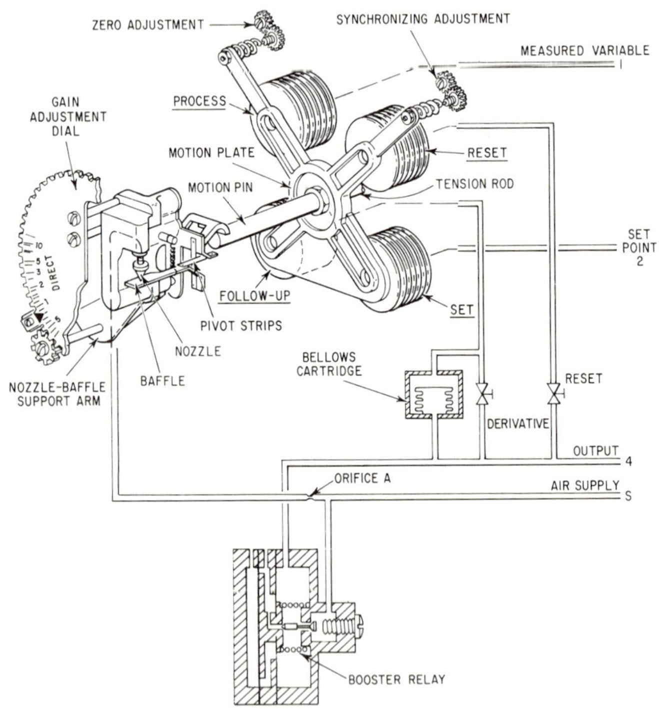

Fig. P-28. Pneumatic receiver-controller. *(Taylor.)*

REFERENCES

Caldwell, Coon, and Zoss: "Frequency Response for Process Control," McGraw-Hill Book Company, New York, 1959.

Eckman, Donald P.: "Automatic Process Control," John Wiley & Sons, Inc., New York, 1958.

Harriott, Peter: "Process Control," McGraw-Hill Book Company, New York, 1964.

Nichols, N. B.: The Linear Properties of Pneumatic Transmission Lines, *ISA Trans.*, vol. 1, no. 1, p. 5, 1962.

Ziegler, J. G., and N. B. Nichols: Optimum Settings for Automatic Controllers, *Trans. ASME*, November, 1942.

—Howard R. Jaquith, *Taylor Instrument Process Control Division, Sybron Corporation*

Pneumatic Load Cell (See **Load cell, pneumatic.**)

Pneumatic Relay (See **Pneumatic controller.**)

Pneumatic Transmission (See **Transmission, pneumatic.**)

Point-to-Point Positioning System (See **Numerical control;** and **Position-measurement and control systems.**)

Polariscope An optical instrument for studying transparent or semitransparent materials in *polarized light.* Basic elements consist of a light source and two polarizing devices: (1) a *polarizer* and (2) an *analyzer.* A system of lenses, quarter-wave plates, and other devices may be part of the system, dependent upon application. Essentially, the polariscope measures the interaction between a sample and polarized light. Three basic operations occur in the polariscope: (1) production of a parallel beam of polarized light, (2) transformation of the state of polarization of the light beam by the sample, and (3) analysis of the new state of polarization of light to determine properties of the sample. The chemical industry, especially that segment involved in organic and carbohydrate chemistry, has made extensive use of the polariscope. The photoelasticity polariscope has been used for many years by engineers to analyze stress distributions in mechanical parts and structures. Electro-optic and magneto-optic phenomena, such as the Faraday effect, the Kerr effect, and the Zeeman effect, can be studied with a polariscope. Typically, the sample placed in the field of the polariscope is immersed in strong electrical or magnetic fields or combinations of these.

—E. Gordon Watson, *Gaertner Scientific Corporation*

Polarography A method of chemical analysis based upon comparative measurements of current-voltage curves obtained during electrolysis of a solution under conditions which ensure concentration polarization of one electrode and constant potential at the other electrode. Ions or molecules in solution may be subjected to identification and measurement by this method if they are susceptible to oxidation or reduction at the polarized indicator electrode by applied potentials in the neighborhood of a few volts. The inherent selectivity and accuracy of the method make it attractive in many circumstances.

Pollution-control Instrumentation (See **Air-pollution instrumentation.**)

Polygraph An instrument that indicates or records one or more functional variables of the human body that may change when a person undergoes the emotional stress associated with a lie. These variables include blood pressure, heart action, and skin resistance. Also called psychointegroammeter and lie detector.

Position, Measurement of The measurement of position is related closely to the measurement of motion and of dimension. Position is a quantity or situation that is related to some other position—often a base reference point or line. Consequently, position usually is expressed in terms of a dimension, a grid, or a vector. Position may be expressed in linear or angular terms.

Position measurement is accomplished in a large variety of ways—each tailored to a specific application. Hence, it is difficult to generalize meaningfully on this broad topic. For example, an optical range finder fixes the position of some target as related to one's present position. The target may or may not be illuminated by an instrumental source of radiation. Radar is used to determine the position (or series of positions—*path*) of a target (usually moving). A surveyor, by taking a series of optical measurements, determines position with reference to a base line or bench mark. In navigation, one determines the position of a ship on a grid, represented by the longitude and latitude of the earth's surface, by determining the position of a distant source of radio-frequency energy, or by sighting the position of the sun, a star, or an artificial satellite—with a simultaneous accurate knowledge of time. In numerical control of machines, the position of the tool may be determined with reference to a base or starting point (absolute system) or with reference to the last stopping point (incremental system).

Most of the key words for seeking information on position measurement elsewhere in this volume have been mentioned in the foregoing description.

Position-measurement and Control Systems The quality of work achieved by machine tools and many other kinds of production machines requires that the moving or operating members of the machine be carefully positioned—sometimes in relationship to each other, but particularly as regards the relationship between machine and workpiece. The degree of positioning accuracy and repeatability required varies with machine and application, as shown by Table P-2. The complexity of the control system will depend upon numerous factors, including the number of axes to be controlled. The speed of response of the control system will be related to the total cycling time of the machine. The environment within which the control system must operate may vary from uncontrolled, outside conditions to carefully monitored and environmentally controlled toolrooms. Thus, a wide selection of measurement and control hardware has developed to meet this broad range of requirements. The final choice of position-measurement and control system will depend upon its reliability characteristics for a given machine application.

Position measurement and control is not limited to machines but finds other applications, such as carefully controlling the movement of antennas in large tracking stations. A generalized machine-control system is shown in Fig. P-29. The control elements of such a system are common to other industrial-control applications and also are described elsewhere in this volume. Particularly see **Numerical control** and consult key words in the alphabetical index.

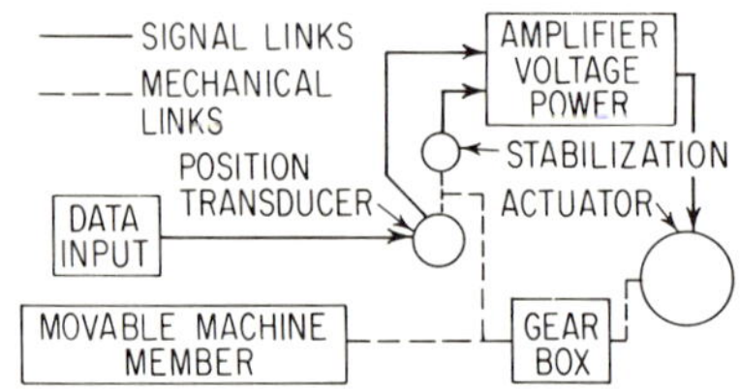

Fig. P-29. Generalized position-control system.

Position Transducers: Machine characteristics and control-system parameters greatly influence the choice of a position transducer:

1. In order that a machine member can be positioned with acceptable accuracy, it is first necessary to establish the extent of the *backlash* or dead-band region for the positioning mechanism used therewith. The measuring transducer and its attendant dead-band characteristics—when employed with the machine member—will determine the amount of dead band or backlash to be included in the total control-system loop. Thus, for all types of repeat-back devices whose mechanical input is provided by rotating a shaft, an important consideration is the means of coupling this device to the positioned machine element.

TABLE P-2. *Positioning Requirements of Typical Machines*

Machine	No. of Controlled Axes	Repeatability, in.	Total Cycling Time
Asphalt batching	6	±0.1% of full-scale reading	2 min
Seat-cushion spring welding	5	±0.015	35 s
Armature insulator assembly	3	±0.005	72 s
Drilling	2	±0.001	*
Boring	6	±0.0001	16 s
Tube bending	5	±0.010	35 s
Frame welding	2	±0.010	Varies
Contour machining	3	±0.001	Varies

*Depends upon number of holes.

In most cases, a gear train is required to reduce member travel to one revolution or less of the transducer shaft. It is necessary to determine whether any backlash in this train is comparable, when expressed as an arc of the transducer shaft, with the positional-accuracy requirement of the machine member. Satisfactory results on the basis of this comparison rest upon the assumption that the electrical and mechanical error factors for the transducer are small (one-tenth or less) compared with the machine-member positioning tolerances. If a transducer must derive its mechanical input from a gear train used for transmission of power to a machine member and if this gear train includes a backlash factor that is large in comparison with the required machine tolerance, a transducer that does not require a mechanical-input shaft should be considered. Where this is impossible, a separate rack and gear train—both exhibiting minimal backlash—should be used to position the mechanical-input shaft of the associated transducer.

2. The controlled member may have undesired movement with respect to the machine base—in a direction transverse to the controlled axis of travel. For example, if the slide and table ways wear nonuniformly, variation in transverse position of a point on the table of a machine may cause a variation in the air gap of a magnetic-slot transducer system. This same machine problem will result in misalignment of optical-transducer systems if the table motion becomes crablike after wear of the slides has progressed.

3. Nondata components of both cyclic and random nature may be superimposed on the true data because of machine-induced vibrations on the transducer.

4. Contaminants may interfere with transducer performance. Iron dust may enter the air-gap region of magnetic-lobe-type transducers, or dust and dirt may partially obscure the sighting path of an optical transducer.

5. Brush position may vary with respect to its segmental disk on a digital-drum-type transducer owing to machine vibration. The result of this variation will be the existence of false pulses of data input which ultimately cause improper machine positioning.

Transducers commonly used for machine applications include the (1) resolver, (2) synchro, (3) inductive plate, (4) optical grating, (5) toothed magnetic drum, (6) digital contactor, (7) digital magnetic drum, (8) magnetic pin, (9) magnetic tooth, (10) multiturn potentiometer, (11) magnetic scale, (12) optical binary scale, and (13) inductive-bridge transducer.

Resolver: Among the several shaft-type mechanical-input transducers available, the resolver is perhaps the most versatile because of the many possibilities available for its electrical input. The resolver shaft is connected to the positioned machine member inside a closed control loop where it will be moved in a direction to reduce its error voltage essentially to zero. This minimal output will generally denote the exact machine position given by the input data. Also see **Resolver.**

The electrical diagram of Fig. P-30 discloses the data input or stator winding of the resolver to be a double winding disposed in the manner of a two-phase motor stator. The electrical characteristic of this device is such that the output

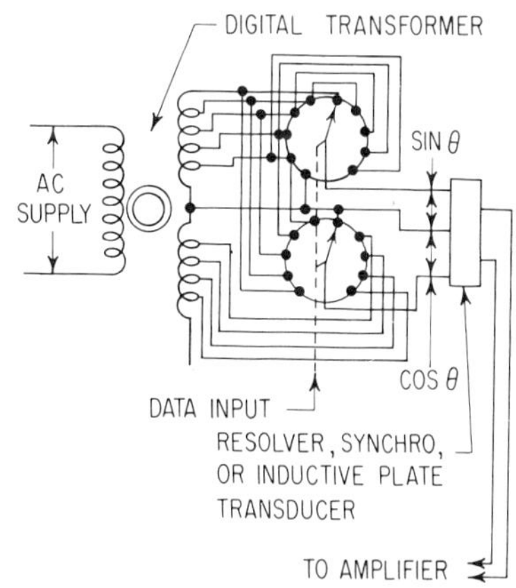

Fig. P-30. Digital transformer circuit for resolvers, synchros, or inductive plates.

voltage of the rotor winding varies in a sinusoidal manner with respect to the data-readout position. Zero voltage will occur at the readout position that corresponds to chosen data. But zero voltage also may occur 180 shaft degrees from this position. Thus, there is the possibility of a false-data region. To eliminate the possibility for false positioning of the machine member, the designer chooses a gear ratio for use between the machine member and its transducer that will permit only 180 mechanical degrees of transducer-shaft rotation during full-member travel, or in some other fashion he designs a nonambiguous data-input system utilizing two 90-degree, related sine and cosine windings.

Synchro: Unlike the resolver, a synchro stator

has a three-phase winding configuration, while the rotor may use a single- or a three-phase winding. See **Synchro.**

Inductive Plates: This device, shown in Fig. P-31, includes an etched stator winding that has been projected upon a dimensionally stable nonconducting surface by a photographic process. The rotor associated with this transducer is constructed in a like manner. Variations in inductor displacement are averaged over a large number of inductors by summing the voltages from a like number of coil sides contained on the rotor plate. Thus, the reproduced rotor and stator inductors need not be printed with a positional accuracy equivalent to that of the final transducer. Either a trigonometrically related transformer amplitude-analog system or a clock-pulse-derived phase-analog system can be used to supply sine and cosine voltages to the two windings contained in the stator plate. This is the same arrangement that is used with rotating resolvers and synchros. Essentially, this transducer is a two-phase synchro or resolver whose windings have been projected onto a linear medium.

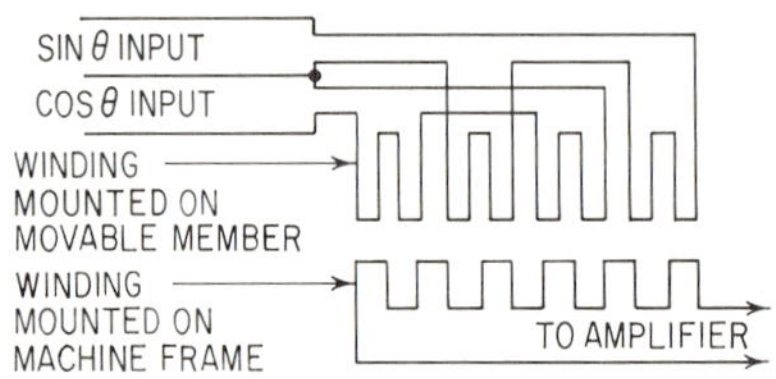

Fig. P-31. Schematic circuit for a position transducer utilizing inductive plates.

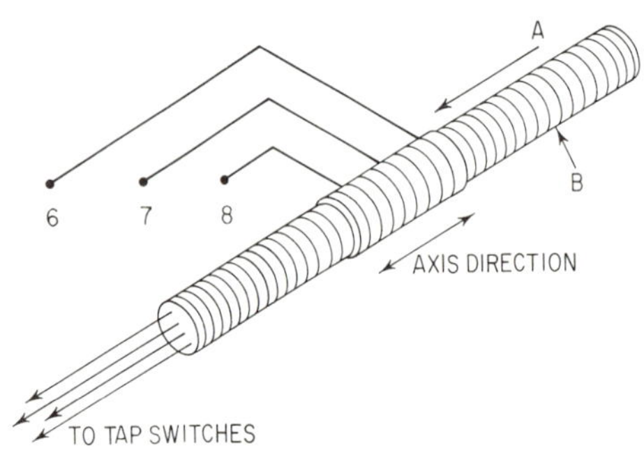

Fig. P-32. Inductive-bridge transducer for position measurement.

The advantage in use of the printed inductive-plate resolver is the elimination of gear backlash between the positioned machine member and the movable element of the transducer. Disadvantages of the device include (1) low sine and cosine winding impedance, largely resistive; (2) low output level—approximately 6 μV/0.0001-in. displacement with 10,000 Hz at ½-V carrier frequency sine and cosine excitation; and (3) low signal-to-noise ratio. The transducer elements must be covered with a shroud that excludes all iron or other material. The accuracy of the printed inductive-plate transducer is equivalent to that of rotary-shaft synchros. Dispersion of data due to quadrature voltages arising from other current-carrying members and by unbalanced impedances in the data transformer is encountered with the transducer. A related type of device uses bifilar windings on a rod and sensing sleeve.

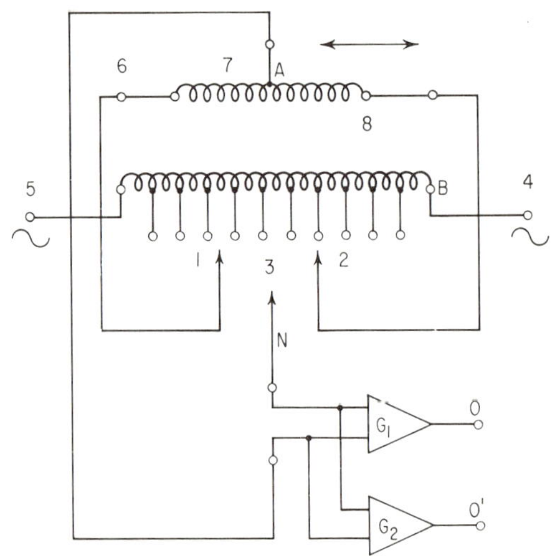

Fig. P-33. Schematic circuit diagram of inductive-bridge transducer for position measurement.

Inductive-bridge Transducer: This device, shown in Figs. P-32 and P-33, is used in high-performance production machines with restricted axis motion. Operation is based upon the use of a fixed inductive member *B* that is slightly longer than the axis to be measured, and a movable member *A* that is approximately half the length of *B*. Selectable taps are placed on *B* in a successive decade with externally located inductors to provide a bridge configuration that may be externally unbalanced by placing *A* (coil) and *N* (point) across a pair of tap points—then moving the coil until equal voltage prevails between the two ends of the coil as evidenced by the existence of a small voltage at *O* and *O'*. A disadvantage of this system is the relatively large number of wires that must be taken from the device through the machine to the control system. An advantage is the high output voltage per unit of displacement (5 mV/0.001 in.). The supply frequency usually is between 400 and 1,500 Hz.

Pulse-type Transducers: A general class of transducer devices used in machine-control systems for measurement of position utilizes pulse-producing elements to convert displacement into pulses that represent units of displacement. Transducers in this class are optical gratings, toothed magnetic drums, digital contactors, digital magnetic drums, single-speed linear resolvers, magnetic scales, and binary scales. Typical circuits of commercial pulse devices are shown in subsequent figures. These transducers function in one of the two following ways:

1. Transducers that require counting of least-dimension units for machine-member positional measurement. An optical grating with a photocell, a toothed magnetic drum with a reluctance pickoff, and a magnetic scale are examples of pulse-counting systems.
2. Transducers that use logic networks for readout of machine-member position. A digital contactor using a geared sequence of printed-circuit drums contacted by multiple brushes and a sequence of magnetic-drum sections with reluctance pickoffs are examples of systems using logic networks to readout positions. A single-speed linear resolver with decade counters and an accumulator register can be used both to drive a visual readout and to position sensitive contacts.

Accuracies for pulse transducers are shown in Table P-3 and are comparable with those for synchros, resolvers, or inductive plates. As in the case of the inductive-plate devices, optical gratings have the advantage that they do not require a gear drive between the transducer and the positioned machine member.

Optical Grating: This transducer, shown in Fig. P-34, requires an amplifier to raise the power level of the pulses from the associated photocell so that position-readout displays or relays used in conjunction with other relay logic elements can

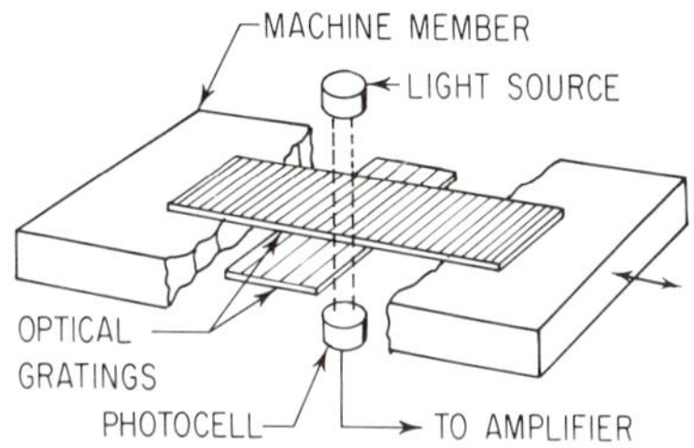

Fig. P-34. Position transducer utilizing optical gratings.

be operated (see Fig. P-35). Also see **Measuring machine.**

Digital Contactors: These transducers do not require any auxiliary relay or amplifier circuit. Digital contactors do have the distinct disadvantage that they must employ at least one printed-circuit drum and brush combination where the drum rotates at 1,000 times the input-shaft speed of the transducer—if the unit is to have positional accuracy comparable with other transducers. In some machines, this has required the high-speed drum to be rotated at 16,000+rpm. Consequently, these units have exhibited short life.

Designers of digital drum-brush transducers

TABLE P-3. *Typical Position-transducer Tolerances*

Transducer	Manufacturer's Tolerance
Synchro (three-phase)	4′ of shaft arc
Resolver	4′ of shaft arc
Optical grating	0.00001 in.
Magnetic toothed drum	15′ of shaft arc
Digital drum	35′ of shaft arc
Magnetic pin or tooth	0.00005 in.
Linear or printed-circuit resolver (single-speed)	0.1000-in. cycle = ±0.00050 in.
Linear or printed-circuit resolver (three-speed)	0.2000-in. fine cycle = ±0.0002 in. 10-in. cycle = ±0.020 in. 400-in. cycle = ±0.800 in.
Rotary printed-circuit resolver	180° cycle 2°/cycle = ±1½ s
Magnetic scale	0.02 mm/cycle = ±0.002 mm
Optical binary scale	1 m/section = ±0.01 mm
Inductive-bridge transducer	30 in. max = ±0.001 in.

have attempted to minimize this difficulty by using a brush-lifting means designed to remove the brush from the high-speed drum during coarse positioning of the transducer shaft. Gen-

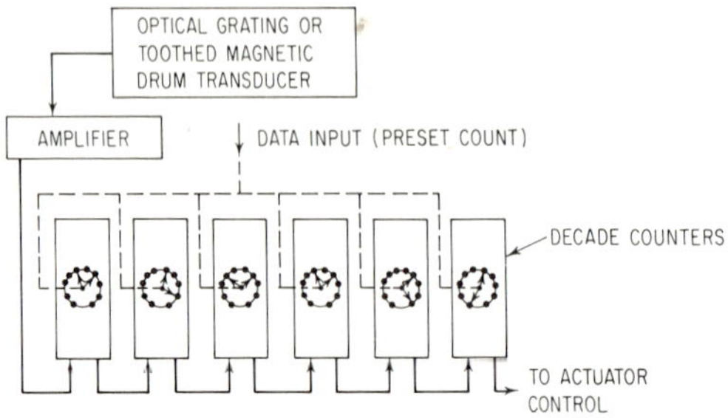

Fig. P-35. Tap-switch circuit for optical gratings or toothed magnetic drums.

erally unacceptable reliability of operation has attended the use of this type of transducer in control systems involving four significant digits or more (see Figs. P-36 to P-38).

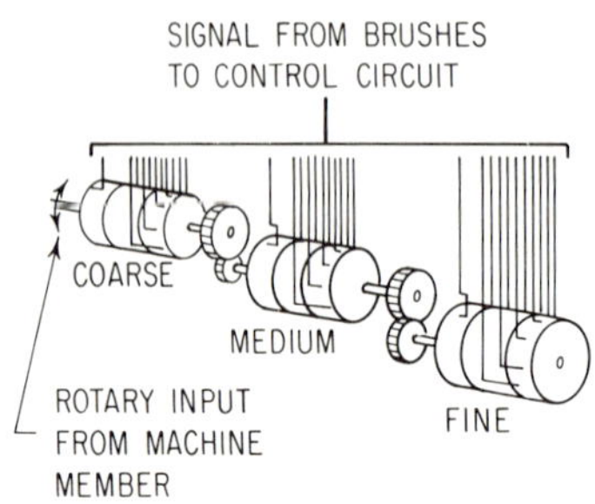

Fig. P-36. Digital-contactor-type position transducer.

Magnetic-pin or -tooth Transducers: These devices are shown in Figs. P-39 and P-40. Comparison of positional resolution for these transducers with other position transducers is given in Table P-3. The magnetic-pin or -tooth transducer has a useful range of ±0.150 in. from the readout point. This transducer must be used with a synchro so that the measuring range can be extended outside the restricted ±0.150-in. range. A switching means must be provided to switch successively from one transducer to another. A

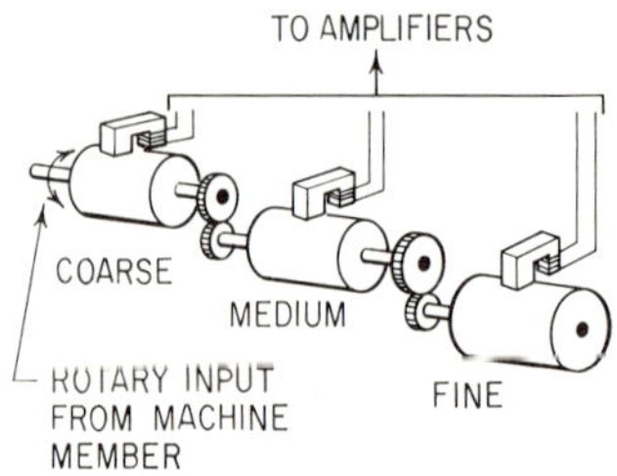

Fig. P-38. Position transducer utilizing digital magnetic drums.

position transducer employing a toothed magnetic drum is shown in Fig. P-41.

Potentiometers: Multiple-turn potentiometers are used in machine-control systems that do

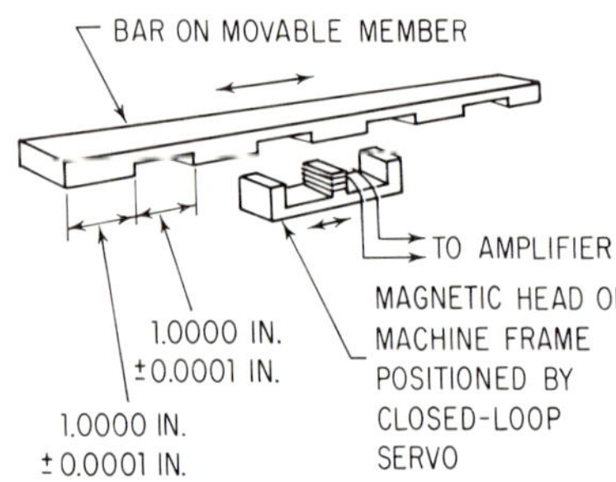

Fig. P-39. Principle of the magnetic tooth as used in a position transducer.

not require an accuracy greater than the positional equivalent of ±0.001 in. Circuitry for such a transducer and an electric circuit commonly employed for inserting digital data into the trans-

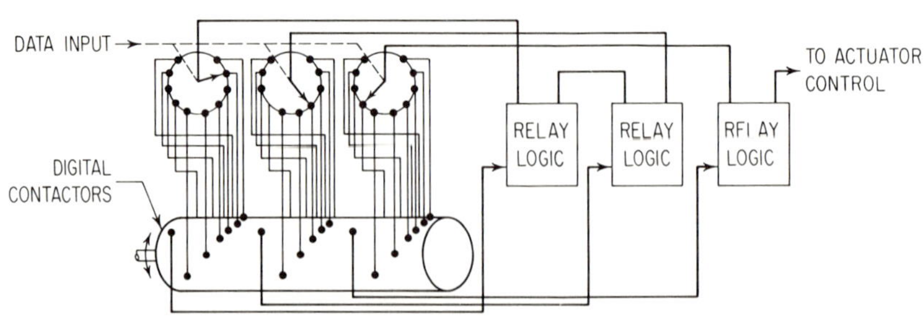

Fig. P-37. Switch and relay logic circuit for digital contactors.

ducer system are shown in Figs. P-42 and P-43. A disadvantage of the potentiometer is unreliability related to frictional wear in the slidewire.

Optical Binary Scale: This is a direct-position readout using the scale shown in Fig. P-44. A binary-coded decimal pattern is deposited on a glass plate that is slightly greater than 1 m in length. This scale is used with a plurality of light-emitting diodes and photodiodes arranged in fixed relation to register the bit condition of at least a 14-bit metric pattern. Accordingly, it is possible to measure a length of 3 m with a three-

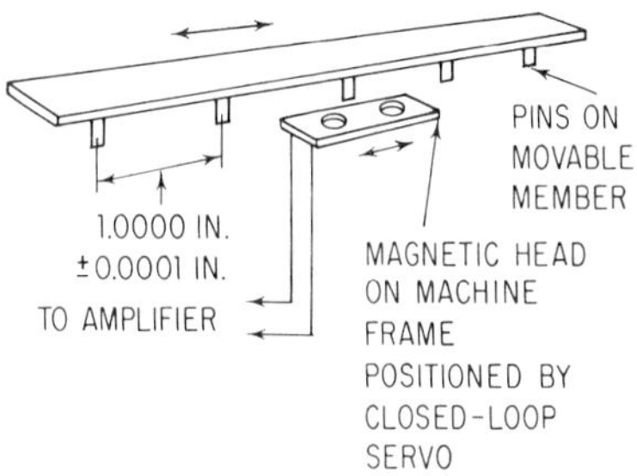

Fig. P-40. Use of magnetic pins in position transducer.

scale array. In combination with this device, a logic network of NAND and NOR gates can be used with appropriate input devices, such as decade preset tap switches, to provide position-coincidence logic for linear position readout and/or position-contact actuation.

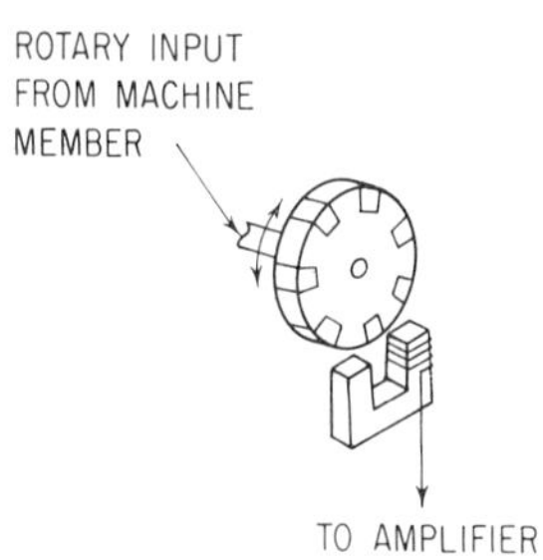

Fig. P-41. Principle of position transducer employing a toothed magnetic drum.

Magnetic Scale: A system for magnetizing the surface of an essentially square flux-current loop ferromagnetic-alloy tape provides alternate magnetized and unmagnetized linear bits that are 0.02 mm in length. The location of a magnetized bit can be determined by "reading" the surface with a dual-pole second-harmonic modulator that gives an output when its poles are subject to a steady field of 0.0005 oersted or greater. A series of decade counters and storage registers can be used to store the bits read per unit of reading-head traverse. One disadvantage of this system

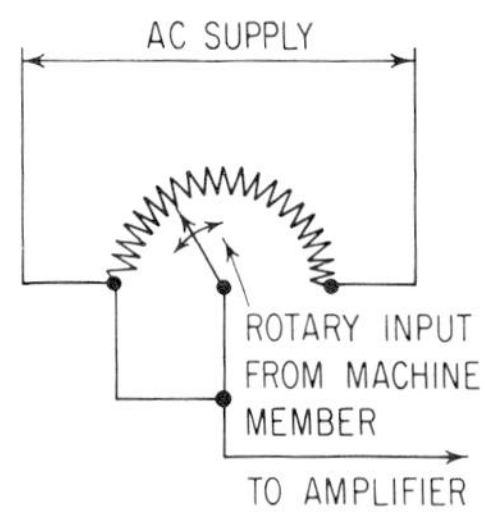

Fig. P-42. Basic circuit of a multiturn potentiometer as used in a position transducer.

is concerned with the necessity for a high-permeability magnetic shield that must be located on three sides of the tape. Also, the tape must be maintained in a prestressed condition in order to

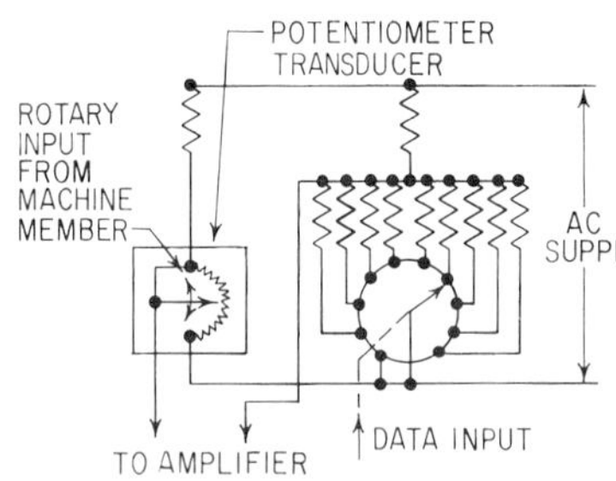

Fig. P-43. Decade switch circuit for multiturn potentiometer.

exhibit acceptable magnetic characteristics (see Fig. P-45).

Translation of Transducer Error Signals: Accuracy in position measurement cannot be expected to be better than the accuracy of the error signal obtained from the position transducer. Accordingly, a machine member cannot be positioned accurately, regardless of transducer used, unless the error signal can be properly translated into movement of the machine.

Transducer Error Amplifiers: The generalized control-system block diagram of Fig. P-29 shows the use of an error amplifier following the transducer in the signal-flow direction. This amplifier increases the voltage and/or power level of the transducer signal to control a higher-power-level amplifier. Generally, the error amplifier should have the following characteristics: (1) high magnitude of voltage gain (10,000 to 100,000), (2) low distortion for control signals lying in the null region (1.5 μV to 2.5 mV), (3) stability of gain and linearity characteristics with changes in ambient temperature, (4) high reliability (capable of continuous operation without failure for at least 10,000 hr), and (5) low internal temperature rise with no requirement for external cooling. The factors of stability and reliability are especially important for machines that must be maintained in a controlled environment in order to maintain an accurate relationship between one machine section and another.

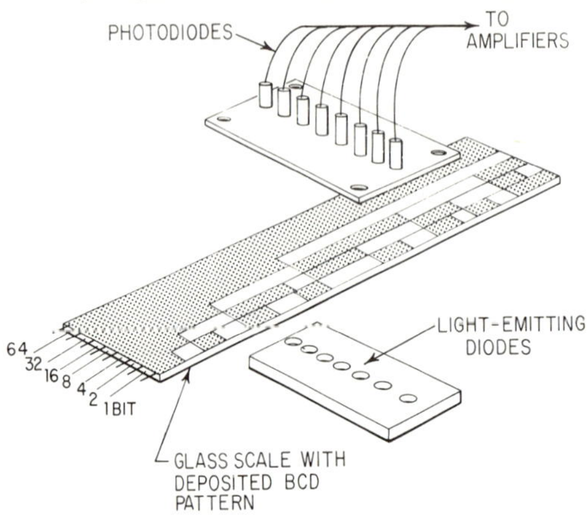

Fig. P-44. Optical binary scale used for position measurement.

Amplifiers used in machine-control systems use control elements that exhibit widely varying characteristics. These amplifiers may be differentiated by their principal control elements: (1) compensated transistor amplifiers, (2) integrated circuit amplifiers, (3) magnetic amplifiers, and (4) fluidic amplifiers (see Table P-4).

Effects of Temperature Changes: Operation of uncompensated transistor amplifiers in the environmental range of 13 to 43° C results in negligible drift or deviation of control characteristics. In magnitude, the drift is comparable with that of a vacuum-tube amplifier operated in a much wider temperature range. To achieve stabilized operation in this restricted thermal environment, it is necessary to utilize temperature-compensating networks or balanced element construction, exhibiting minimal thermal sensitivity within the amplifier configuration. Consideration of unit temperature rise for the three basic amplifiers indicates minimal rise to be experienced with a transistor or magnetic amplifier—probably never greater than 2° C above room ambient for multielement systems. Operation of fluidic components as linear amplifiers and discrete logic elements can be assured in a thermal environment to 200° C. Other amplifier components, excluding magnetic amplifiers, cannot operate in this range. Another difference shown by Table P-4 is that a full-wave high-gain-type magnetic amplifier exhibits a large time lag when operated at a 60-Hz carrier frequency. This inherent lag can approach 25 to 30 Hz of supply frequency when used in a multistage amplifier capable of a voltage gain of 10,000. As a consequence, this amplifier cannot be used in high-gain

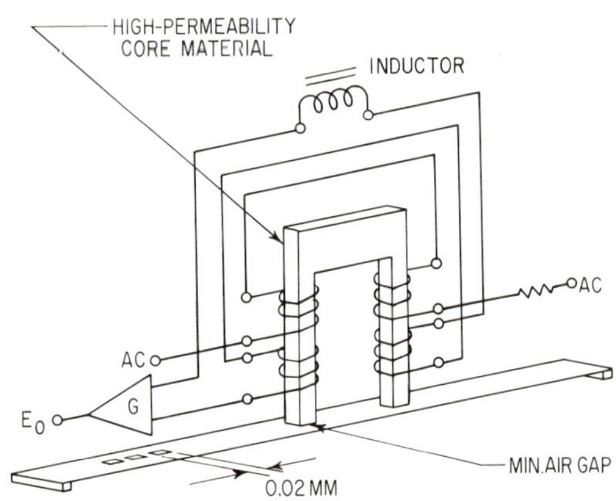

Fig. P-45. Magnetic scale used for position measurement.

TABLE P-4. *Characteristics of Typical Transducer Error Amplifiers*

Amplifier Type	Response Time, s	Voltage Gain	Power Gain	Auxiliary Power Supply	Gain Change (0–70°C), % Drop	Supply Frequency, Hz
Compensated transistor . . .	Carrier rate	10,000	2×10^7	Yes	−1.1	60
Integrated circuit amplifier	Carrier rate	22,000	5×10^{10}	Yes	−0.5	60
Magnetic amplifier	0.075	10,000	2×10^7	No	−4.7	60
Fluidic amplifier	0.005	. . . . .	2×10^7	Yes	None	

control systems to be operated at 60-Hz carrier frequency. If a higher-frequency carrier system can be employed with the machine-control system, such as 400 Hz, then the response time of 25 to 30 Hz is acceptable.

Power Amplifiers and Actuators: Control-system designers sometimes have questioned the necessity for a division of system-amplification means into two sections, i.e., voltage amplifier and power amplifier. Good reasons for this division rest upon the electrical isolation of circuits for high gain at high impedance from sections

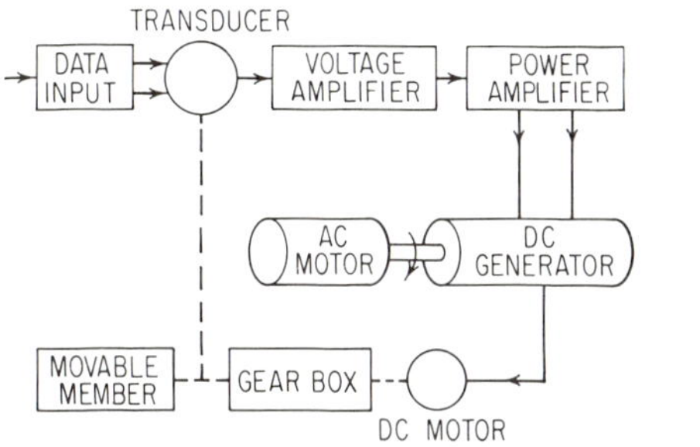

Fig. P-46. Electric power-amplifier system using the position error signal to control the output of a dc generator to a dc motor.

of the amplifier controlling larger power. Such isolation of one unit from the other has eliminated the possibility of unwanted or false electrical-signal components combining with control-data components. A division accordingly has been effected—in the 0.1- to 0.2-W power region—between the output-power level of a voltage amplifier and that of a power amplifier. Each of the

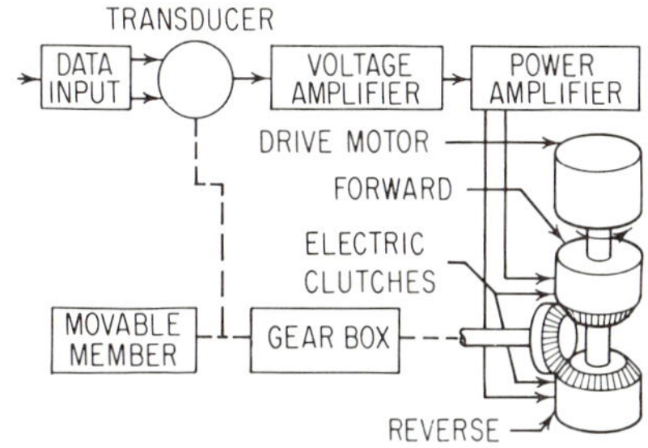

Fig. P-47. Electric power-amplifier system using the error signal to control opposed electric clutches.

power-amplifier systems listed in Table P-5 utilizes a voltage amplifier of the types found in Table P-4. Block diagrams (Figs. P-46 to P-49) show the arrangement of several power-amplifier systems.

A factor to be considered in the choice of a power actuator and amplifier is the "slewing rate," which is a measure of the ability of a closed control loop to position a member, following input data, with a minimum of lag in position and time. This is one of the power-amplifier char-

TABLE P-5. *Typical Power-amplifier System Characteristics*

Components	Transistor Amplifier, SCR, dc Motor	Transistor Amplifier, SCR, or Relay Controlled Constant-speed Motor	Magnetic Amplifier, Amplidyne Generator, dc Motor	Transistor or Magnetic Amplifier, Constant-speed Motor, Double Clutch	Magnetic or Transistor Amplifier, Proportional Valve, Hydraulic Motor or Cylinder	Fluidic or Transistor Amplifier, Proportional Valve, Hydraulic Motor or Cylinder
Rating	3 hp	4 hp	3 hp	3 hp	5 hp	80-lb force
Speed ratio	400 : 1	2 : 1	50 : 1	50 : 1	400 : 1	260 : 1
Time constant, s. . .	0.005	0.27	0.2	0.2	0.035	0.080
Drift factor, in./hr	0.000	Depends on voltage amplifier	Depends on voltage amplifier	±0.00009	±0.00007	±0.00005
Auxiliary power supply	None	None	Drive for generator	None	Hydraulic-power source	Fluidic-power source
Projected life, hr . .	30,000	30,000	10,000	10,000	8,000	8,000
Approach	Velocity servo	Controlled direction	Random servo	Random servo	Random servo	Random servo
Cost factor*	2.9	1.7	2.4	1.9	3.3	1.8

*This is a relative factor. It indicates that the cost of the system in the first column as related to the system in the second column would be on the order of 2.9/1.7.

acteristics known to contribute appreciably to system accuracy and stability, especially in machines continuously positioning movable members according to continuous data command. As shown by Table P-5, the systems that use a proportional valve with a hydraulic motor and a dc motor controlled by a silicon controlled rectifier (SCR) provide the greatest dynamic response.

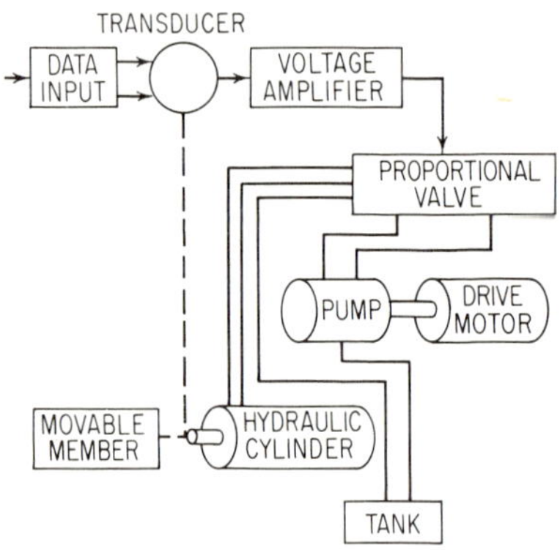

Fig. P-48. Hydraulic power-amplifier system using the error signal to control a proportional valve.

For use in a system such as that for controlling contouring machines, the latter systems offer a greater degree of responsiveness than is useful on most machines—because of inherent machine characteristics. Many discrete cycle-positioning systems, as used in table positioning for boring and drilling machines, jaw and anvil positioning in tube-bending machines, and press-head positioning for a multiple-operation press, do not

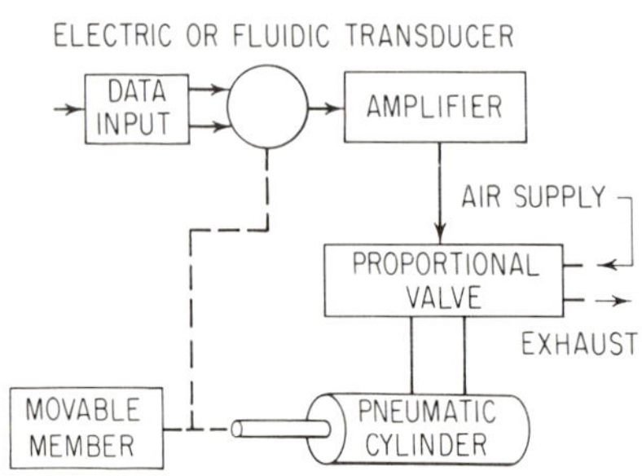

Fig. P-49. Pneumatic power-amplifier system using a proportional valve.

require minimum cycling time for member positioning. A high-response power amplifier is not useful with such systems.

Successful utilization of high-performance power amplifiers in machine-control systems requiring minimum error performance and high reliability is dependent upon the initial conditions of design and adjustment remaining constant during the useful life of the machine, especially if the original system were marginally stable. Highly unstable operation, resulting in oscillation of the positioned member and error in position measurement, can occur when machine inertial or frictional components change during the life of the machine, operating with a low initial margin of stability.

Comparison of Pneumatic, Electric, and Hydraulic Actuators: The responsiveness of the fluid actuator is much the same as that for electric types. A fluidic system requires pressure controls and filters similar to those required with hydraulic control systems. Choice of a power amplifier for use in a machine-control system is based upon the same factors as cited for the control error amplifier, i.e., required stability, reliability, and response time. The choice of a power actuator will involve the same considerations.

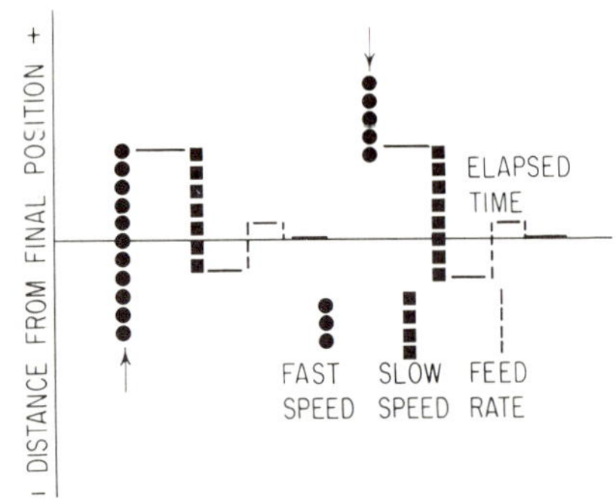

Fig. P-50. Typical response for a system using a limited number of fixed positioning speeds and always approaching the final position from a given direction.

Backlash initially present in the link between the actuator motor and positioned member is another important consideration in the choice of a power amplifier and actuator motor for machine control. If the initial design of the machine does not warrant the use of a zero-backlash system of gears and lead screw in the drive link, then it is necessary to use a controlled positioning of the machine member so that it will always approach its final position from the same direction and at the same speed. If the necessity exists for minimization of positioning time, a lower settling time will result if a servo approach that uses variable speed and direction is made to final member position (see Figs. P-50 and P-51). In the servo approach, the member is positioned at a speed dependent upon error. Both controlled-direction approach and random-servo approach

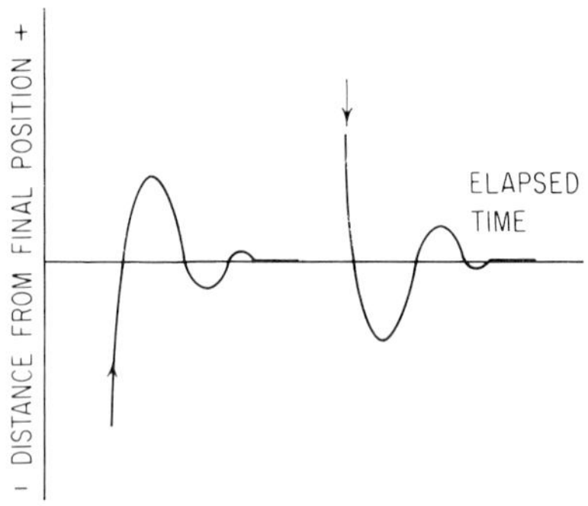

Fig. P-51. Typical response for a system using an infinite number of speeds, with the rate of approach proportional to the error and with random direction of approach to the final position.

positioning cycles are used in point-to-point systems. However, only a refined random-servo approach cycle can be used with continuous-positioning systems.

Positioning-system stability and reliability also are affected by control loops within the principal control loop shown in Fig. P-29. These inner loops or inside control blocks are used for (1) switching from one transducer to another in accordance with the size of control error, (2) anticipating arrival at command point in point-to-point positioning systems, (3) providing constant torque or velocity within the actuator motor loop to maintain positioning accuracy, and (4) supplying stored data according to the requirements of the control loop.

Switching of Transducers: Values of resolution obtainable for various transducers, as shown in Table P-3, are representative of those possible from a single unit operating with a low gear ratio between the positioned machine member and the transducer shaft. Usually an arrangement of transducer and driving member closely geared together will result in ambiguous data transmission from the transducer by virtue of the transducer's rotation through a multiple number of revolutions. Such rotation is permissible if at least one transducer of a multiple-transducer system always is kept within its relevant half-cycle of rotation. It is customary for the control-system designer to use a number of transducers and associated decades with a means suitable for switching from one to another. In this manner, it is possible to ensure final selection of the proper half revolution of the high-speed transducer to achieve desired accuracy. The block diagram of Fig. P-52 shows an arrangement for a multiple-transducer control system to accomplish switching from one transducer control system to another in the basic control loop. The four-resolver system is designed to provide a positioning accuracy of ±0.0001 in. Such a switching system is used in most multiple-transducer systems. Switching

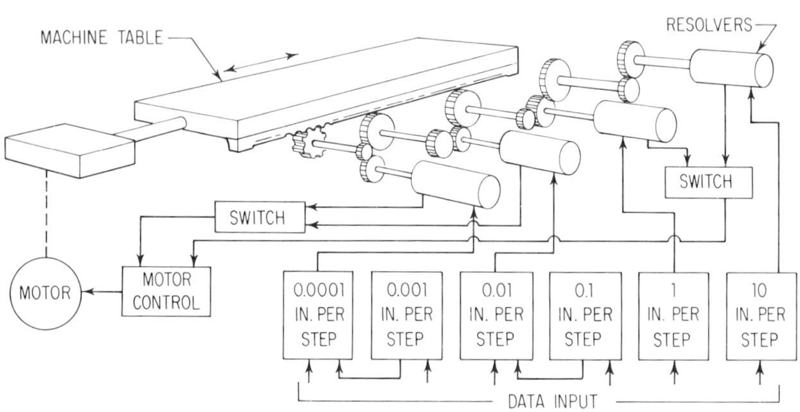

Fig. P-52. Four-resolver position-transducing system designed to give a positioning accuracy of ±0.0001 in. without rotating the coarse resolver more than 180°. By keeping that resolver always within its relevant half-cycle of rotation, this system avoids ambiguity. The control system switches to successively finer resolvers as the command position is approached.

from one transducer to another is based on coarse transducer amplitude and is accomplished with relays or diodes.

A relay switching arrangement is shown in Fig. P-53 and a diode switching circuit in Fig. P-54. The circuit in Fig. P-53 is for use in a point-to-point positioning system, while that of Fig. P-54 is for continuous-positioning systems, such as those used in contouring machines. An advantage of the circuit in Fig. P-53 is that it is incapable of oscillating between coarse and fine control during a closed-loop positioning cycle of a machine member. This is true because once the amplitude of the coarse transducer has fallen below a preset switching-point value, the coarse transducer then is prevented from commanding the positioning system until another overall positioning cycle has been initiated. The switching means of Fig. P-54 allows the coarse transducer to return to command of a positioning system whenever the amplitude exceeds a preset level. For normal functions with a continuous-data system, this type of circuit action is acceptable. The positioned member always will be at or near the command-data position and thus the command always will be given by the fine-data transducer. However, if the positioned member should become temporarily restricted in motion, or if its power is interrupted, then it is possible for the error signal to become large enough so that the command will be taken over by the coarse resolver.

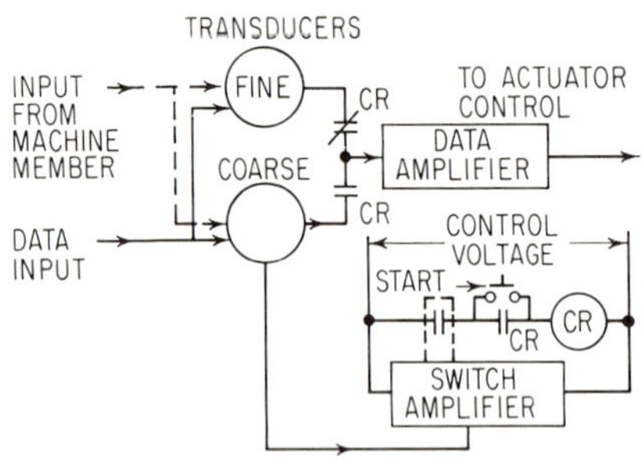

Fig. P-53. Typical switching circuits used to transfer control of a positioning action from transducers for coarse increments to transducers for fine increments. Switching from one transducer to another usually is controlled by the coarse transducer.

In normal operation, the change-over from the coarse to fine resolver results in no system-position instability. However, it is possible, after some change in machine parameters—friction, backlash, and so on—during the life of the machine, that the originally adjusted gain factors for the coarse- and fine-data system may be unsuitable. As a consequence, an unstable condition characterized by switching back and forth from one transducer to the other can arise. This condition can be eliminated by a readjustment of the gains in coarse and fine loops to values corresponding to the new parameters of machine operation.

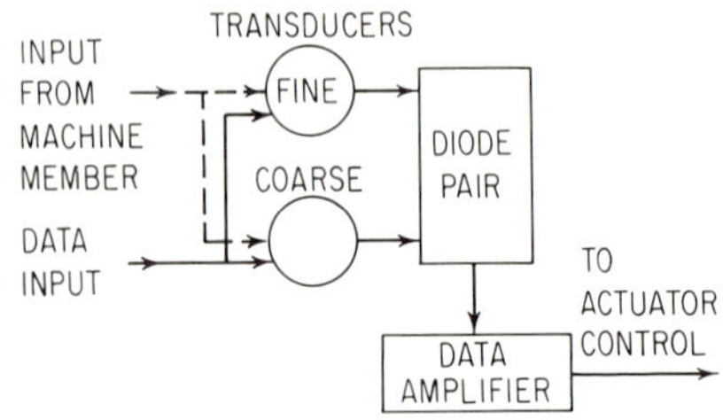

Fig. P-54. Diode switching circuit for continuous positioning.

Use of a transducer in a pulse mode with multiple-decade preset counters and a storage register can make it unnecessary to use multiple transducers and their attendant switching means. The less expensive analog-to-digital components from an all-digital system employing integrated circuits can be appreciated as a distinct cost advantage.

Command-point Anticipation: Systems for point-to-point positioning of machine members, as in drilling, boring, paper cutoff, tube bending, and assembly machines, frequently use two fixed positioning-motor speeds and possibly two traverse gear ratios. Such systems require an anticipation signal to reduce the traverse rate of the positioned member before the command point is reached. Variation of the anticipation point with respect to the command point can result in traverse speed reductions being made too near to the command point, and the final position measurement may be in error. The amount of such variation in anticipation point can change with input data, and dependent upon the data elements being used, this effect can be quite troublesome. Unplanned variation in anticipation point is due principally to analog and transducer error.

Maintenance of Constant Torque: The closeness of actual position to a given command position, whether a point-to-point or a continuous-positioning system, depends upon still another stability factor. This involves the maintenance of constant torque at the output shaft of the actuator motor. When a machine is new, an electric or hydraulic actuator is adjusted to deliver constant torque to the gear train of the machine member being positioned and for the machine parameters existing with this operation. If the frictional-damping component associated with the machine member becomes less as it is subjected to wear, then the loop containing the power amplifier and motor actuator may become unstable and require a reduction of loop gain or an increase of velocity stabilization. The general effect of instability within the power amplifier-motor actuator loop is a reduction of accuracy in final positioning of the associated machine member. The block diagram of Fig. P-55 shows a typical power amplifier–motor actuator circuit. This circuit effectively provides constant torque at any input or command velocity or speed signal from a selected small minimum to a chosen maximum. The overall ratio between maximum velocity and minimum lies between 100:1 and 400:1, depending on the class of control actuator used.

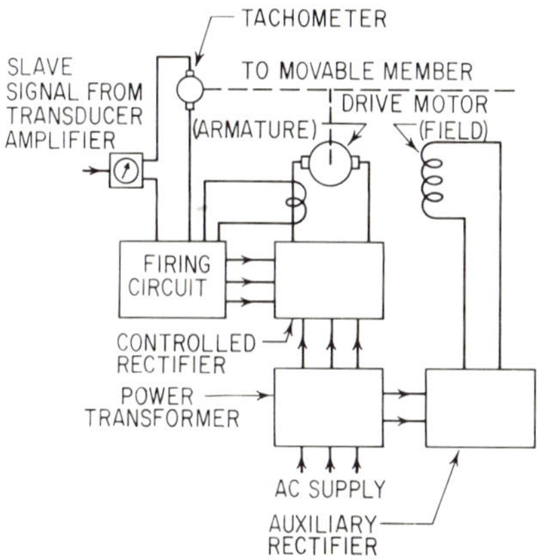

Fig. P-55. Block diagram of a power-amplifier and motor-actuator circuit to provide constant torque output to the movable member. An arrangement of this type is used to maintain positioning accuracy when factors such as frictional-damping change with wear of machine components.

Control-system Costs: The cost of power components for a control system can approach half the cost of the entire system. The relative cost of the actuators listed in Table P-5 varies greatly and can be compared on the basis of two factors: (1) ratio of speed variation and (2) response time. The hydraulic system or silicon controlled rectifier dc motor system can provide a speed ratio of as much as 400:1 with a response time of 5 ms; while a dc motor-generator system can give a speed ratio of 100:1 and a response time of 350 ms. Thus it is seen that a highly responsive control capable of positioning a light machine member at a high rate and over a wide variation of speed will, in turn, cost more than a less responsive control.

Although both controlled-direction and servo approaches can be used in point-to-point positioning systems, there is a cost advantage to be gained by using the controlled-direction approach. The initial cost of the servo approach is much larger because of the higher cost of proportional components, such as hydraulic control valves, pumps, silicon controlled rectifiers, and other similar components, as compared with the cost of nonproportional relays. However, it should be noted that the single-loop approach cycle, which offers a shorter positioning time than that of Fig. P-50, requires the same components as the system of Fig. P-51.

Machine accuracy also determines control-system cost. The position accuracy of a control system is proportional to transducer error and to stability of associated error amplifiers. Investigation of the dependency of machine and control-system cost upon a repeatable accuracy factor indicates that the cost varies approximately as the 2.4 power of accuracy requirements. It is well to consider this so that the accuracy of the machine will not exceed practical requirements.

REFERENCES

Childs, James J.: "Principles of Numerical Control," The Industrial Press, New York, 1965.

Leone, William C.: "Production Automation and Numerical Control," The Ronald Press Company, New York, 1967.

—James R. Walker, *De Vlieg Machine Co.*

Positioner, Valve (See **Valve positioner, pneumatic.**)

Postprocessor A generic term applied to a program which operates in conjunction with another program or programs as a successor. In the field of numerically controlled machine tools, for instance, a postprocessor is a program which takes the output of a general numerical control compiler (see **Compiler**) such as APT (automatically programmed tool) and produces the specific machine-tool-controller command information (usually in the form of a paper or Mylar tape, but it may be a storage file in direct computer control). The APT compiler does the job of generating tool-path information from tool and part geometry descriptions and constraint information supplied by a programmer; the postprocessor adapts this output to a specific machine-control format.

In the aerospace industry, a postprocessor is a data-reduction or -analysis program which is run with recorded information gathered during the actual run or flight. The term "postprocessing of data" is heard in this reference.

—L. E. Wise, *IBM Corporation*

Potentiometer Generally, a component of an electric circuit used mainly as a voltage divider. The common potentiometer (or "pot") is a resistor having a continuously adjustable sliding contact, usually mounted on a rotating shaft. The term potentiometer also may denote an instrument for measuring a voltage by balancing it against a known voltage. See **Voltmeter.**

Potentiometer recorder—an instrument that provides a permanent record of a varying voltage by using a potentiometer to balance the unknown voltage against a known voltage. See **Potentiometric instrument.**

Potentiometer-type transducer—a transducer in which the pressure or other mechanical sensing element moves the contact arm or wiper of a potentiometer. See **Linear-displacement transducer, potentiometric.**

Potentiometer, Linear (See **Linear-displacement transducer, potentiometric.**)

Potentiometric Instrument A potentiometer is a measuring device which operates on the principle of equality of potentials. In the process of measurement, the value of the unknown potential or emf is determined by finding a known potential which is equal to the unknown. When the measurement is made (when the unknown and the known potentials are equal), the current flowing in the measuring circuit essentially is zero. This measuring principle places no load on the unknown source of potential.

The potentiometer may utilize direct or alternating current, and the instrument may employ either a manual or an automatic means to achieve equality of the potentials. A basic potentiometer system consists of the following main elements: (1) the *unknown potential,* that is, the quantity to be measured; (2) a *slidewire*—a linear slidewire with a sliding contact which represents a specific repeatable point on the slidewire; (3) a *reference source*—an electrical source to establish a constant current through the slidewire; (4) a *detector*—a very sensitive and stable device to detect the difference between the known and the unknown potentials; (5) a *measuring circuit*—the wiring and other necessary components required to accomplish the "comparison"; and (6) the *actuating means*—the means by which the measurement is made, manual or automatic.

Manually Balanced Potentiometer: The circuit of one of the original forms of the potentiometer—a direct-current, manually balanced, slidewire type—is shown in Fig. P-56. The circuit consists of (1) a source of emf E_R to supply current through the slidewire, (2) a uniform slidewire R_{SW}, (3) a slidewire contact C which slides along the length of the slidewire, (4) a current detector in the form of galvanometer G, (5) a switch S, and (6) the circuit containing the unknown potential E_{UNK}. In the process of measuring E_{UNK}, the switch S is closed and contactor C is adjusted until the detector G indicates no current flow. At this point, the known potential E_1, as determined by the contact position on the slidewire R_{SW} and the current I through the slidewire, is equal to the unknown potential and balance is achieved; that is, $E_{UNK} = E_1$. By utilizing a known potential E_2 in place of the unknown potential, by determining the new slidewire contact location E_3, and by confirming that I is constant, the unknown potential is determined from the relationship

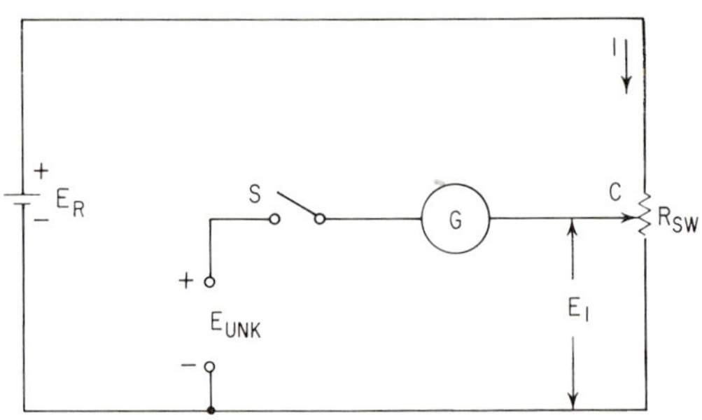

Fig. P-56. Basic potentiometer.

$$\frac{E_{UNK}}{E_1} = \frac{E_2}{E_3} \quad \text{or} \quad E_{UNK} = \frac{E_1 E_2}{E_3}$$

As the potentiometer grew in importance as a measuring tool, the need for a more accurate and stable current source and a more practical approach to standardizing was evidenced. The accuracy of the measurement is directly related to the current source and the ability to maintain the accuracy of that source over long periods of time.

The advent of the standard cell and the standardizing circuit not only increased the accuracy

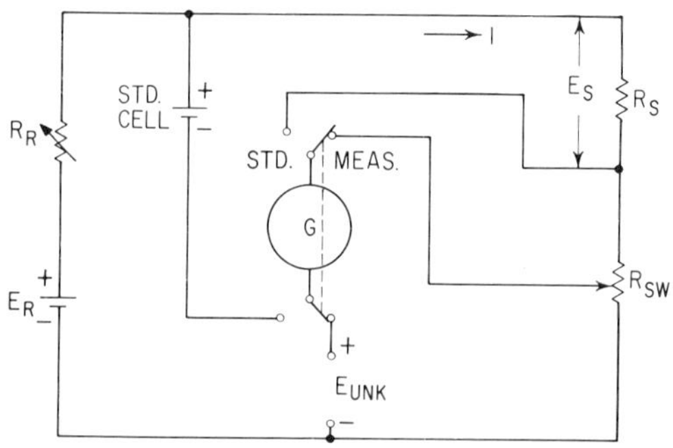

Fig. P-57. Basic potentiometer equipped with standardizing function.

and stability of the potentiometer but greatly broadened the application of the instrument. An instrument equipped with a standardizing circuit is shown in Fig. P-57. With the detector G in the MEAS mode, the circuit and its operation are identical to Fig. P-56. With the detector in the STD mode, the potentiometer principle is applied to the comparison of the potential E_S and the standard cell STD CELL potential. By adjusting R_R until E_S is equal to the standard cell potential, it is possible to maintain a constant current through the slidewire. Many sophisticated versions of this basic arrangement are available in commercial manually balanced potentiometers. The trend over the years has been to increase the possible accuracy of measurement and the speed and convenience of making the measurement.

Automatic Potentiometers: The logical evolution of potentiometer design was that of modifying the manually operated, direct-current, slidewire-type instrument to become an automatic, servo-actuated null-balance, slidewire-type instrument. The latter improvements extended the use of the instrument far beyond the ability to measure low-level potentials accurately. Thus, the modern potentiometric instrument finds wide usage in complex and varied control applications throughout science and industry. One specific design[1] is used here as representative of this broad class of potentiometric instruments. The basic circuit of the instrument is shown in Fig. P-58. An instrument of this type can be used with any sensing element whose dc voltage or current output changes as a function of the measured variable. Examples include thermocouples, tachometer magnetos, pH amplifiers, gas ana-

[1] Bristol Dynamaster.

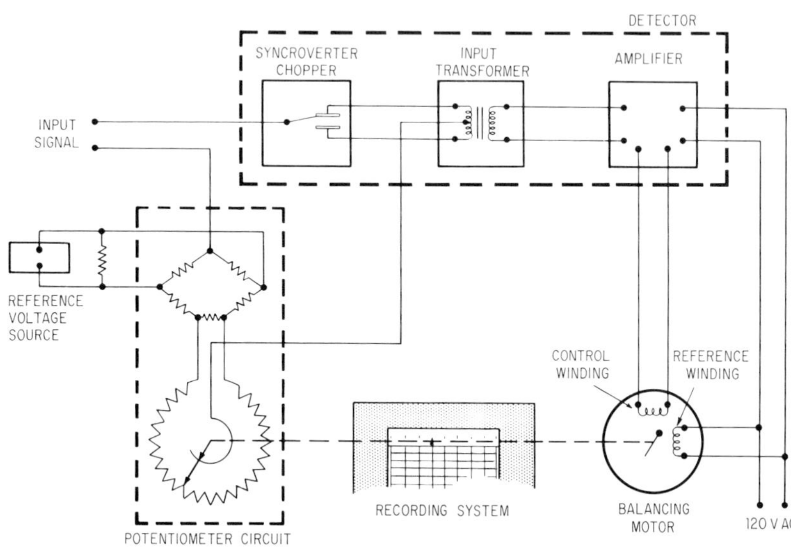

Fig. P-58. Basic circuit of automatic, servo-actuated, null-balance, slidewire potentiometer (Bristol Dynamaster).

lyzers, radiation thickness gages, photoelectric cells, smoke-density cells, and converters of all types with dc voltage or current outputs.

With reference to Fig. P-58, the potentiometer measuring system consists of the following basic components:

1. *Potentiometer circuit*—supplied from a stable and accurate source of dc voltage (the voltage-reference source). The complete circuit consists of the sensing element, the measuring slidewire, and calibration resistors in addition to the voltage reference source.

2. *A converter*—which converts the dc unbalance voltage or error signal from the potentiometer circuit into an ac error signal.

3. *An input transformer*—which increases the error signal applied to the servoamplifier and provides isolation between the measuring circuit and the servoamplifier circuit.

4. *A servoamplifier*—which receives the error signal from the input transformer and amplifies it to a power level sufficient to operate the balance motor.

5. *A servomotor or balance motor*—which is a two-phase motor used to drive the slidewire contact to "balance" the potentiometer circuit. The servomotor has sufficient torque to drive an indicating/recording mechanism and other required auxiliary functions.

6. *An indicating/recording mechanism.*

The balancing action is started by a change in dc voltage output from the sensing element. This change unbalances the potentiometer circuit and produces a dc error signal. This dc error signal is converted to an ac error signal by the converter. The ac output voltage of the converter is increased by the input transformer. The input transformer secondary voltage is applied to the servoamplifier input. The amplifier increases the ac error signal to a power level sufficient to operate the balance motor. The amplifier output voltage drives the balance motor in the direction which will restore the potentiometer circuit to an electrical balance. The indicating pointer and recording mechanism are simultaneously positioned to show the new value of the measured variable. This balancing action is continuous and begins immediately upon the slightest change in the sensing element. When it has been balanced, the entire potentiometer remains motionless until another change in the measured variable occurs. Thus, wear on instrument parts is minimized.

The direction of the balancing action is based upon the phase relation between the ac error signal coming from the converter and the ac power supply voltage applied to the reference winding of the two-phase balancing motor. This phase relation depends upon the direction of unbalance in the potentiometer circuit and, of course, is determined by the direction of the change in emf from the sensing element.

Measuring-circuit Components: Calibration resistors are selected for negligible temperature coefficient and are preaged and impregnated to eliminate any resistance change with time. Each resistor is marked with its precise resistance value. A range card with the fixed resistors mounted so that changing the instrument range consists simply of inserting a new range card is available. The recording charts and indicating scale of the instrument can easily be changed to accommodate the new range. The slidewire is a precisely wound constantan wire with a negligible temperature coefficient. A large number of turns is used to provide a high degree of resolution and sensitivity. Sliding contacts of a special gold-alloy material are used to give long contact life with negligible wear on the slidewire. A zener diode is used as the voltage-reference source. The diode circuitry operates from an isolating transformer powered from the ac power supply, furnishing a precise, stable, and highly isolated dc potential to the potentiometer circuit. This reference source eliminates the need for a battery and standardizing mechanism.

The converter[1] converts the dc error signal into an ac error signal. The phase of the ac error signal produced by the converter has a definite relation to the polarity of the dc unbalance voltage that is fed into the converter. The converter is a single-pole unit utilizing a very low-mass metallic reed. The reed is magnetized by a low alternating voltage derived from the ac power supply and is magnetically driven between two permanently magnetized Alnico pole pieces. The action of the converter in conjunction with a center-tapped input transformer provides an ac error signal to the high-gain electronic servoamplifier. The reed vibrates between the two contacts, across which is connected the full primary of the input transformer. As the reed vibrates between the contacts, the dc error signal is alternately applied to each half of the transformer primary with the reversal of polarity. This results in an alternating current in the secondary of the

[1] Syncroverter.

transformer of a magnitude and phase related to the magnitude and polarity of the dc error signal. Some step-up in the error signal is obtained in the input transformer by the ratio of the secondary-to-primary turns.

The input transformer, in addition to providing isolation between the measuring and amplifier circuits, provides some amplification of the ac error signal. The transformer is carefully shielded to eliminate the effects of stray magnetic fields and is enclosed in a metal case. The primary of the input transformer is center-tapped for use with the single-pole converter previously described. The impedance of the input-transformer primary dictates the input impedance of the basic recorder. The most common input transformer used provides a recorder input impedance "at balance" (at null) of 10 MΩ. The "unbalance" input impedance (100 percent off null) is 10,000 Ω. Other impedances are available.

An electronic servoamplifier is used to amplify the ac error signal to a power level sufficient to operate the balancing motor. Complete potentiometric instruments of this type can be furnished with alarm contacts, various modes of electric and pneumatic control, program control, event markers, retransmitting slidewires for remote indication and recording, and one, two, three, or four pens for recording. Also see **Electric controller; Pneumatic controller;** and **Recorder, graphic.** A bridge-circuit arrangement of essentially this same instrument also is available and is show in Fig. P-59. Also see **Bridge circuits, Electric.**

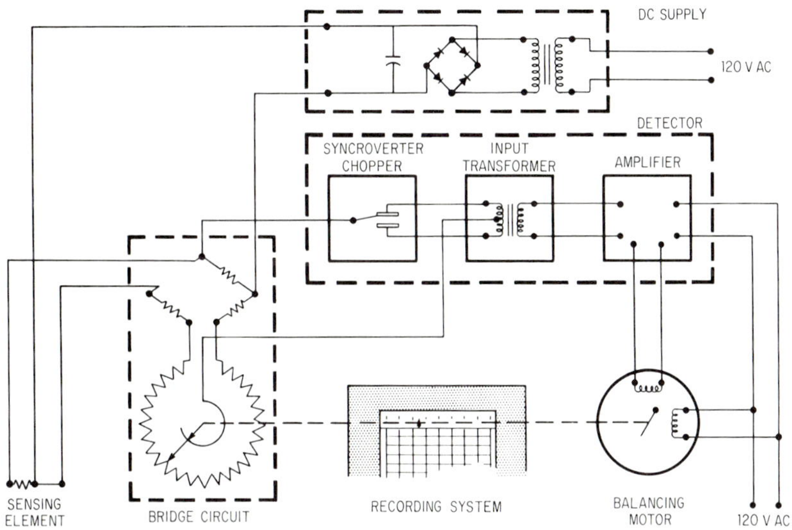

Fig. P-59. Basic circuit of automatic, servo-actuated, self-balancing dc bridge (Bristol Dynamaster).

In summary, a potentiometric instrument utilizes the null-balance method for comparing an unknown potential with a known potential and, of course, is applicable to any sensor or transducer wherein the measured signal can be converted to potential. Null-balance instruments are not confined to potentiometric methods but may involve the application of the null method to comparing other physical quantities, such as electrical capacitance, mechanical force, and pneumatic and hydraulic pressures. Also see **Null-balance method.**

—D. A. Narducci, Jr., *Bristol Division, American Chain & Cable Company, Inc.*

Pour-point Analyzer An important specification of certain materials, such as jet, diesel, and furnace-class fuels, is the pour point. Both laboratory and on-line instruments are available for making a determination of this characteristic. In one instrument, the pour point of individual process streams is automatically recorded. The pour-point temperature, as measured by the analyzer, is equivalent on most stocks to within ± 3° F of the temperature determined by the ASTM pour-point test. The analyzer operates in the −90 to +50° F range. The sample enters a cooling unit at a continuous rate of approximately 1 ml/min. Flowing through a capillary passage, the sample passes through a carefully machined hole in a finned, copper cylinder. The cylinder is cooled by evaporation of liquid CO_2. Cooling continues until a frozen plug of the sample is produced. The system then causes the cylinder to warm—the sample plug melts, dropping into a drain funnel. At that instant, light reflection across the path is lost, causing the recorder-pen drive motor to stop and the recorder chart to drive to advance. A short, vertical pen line indicating the melting point is produced as the chart drives for approximately 10 s. After the 10 s interval, the chart stops and the pen returns to tracking the cylinder temperature until the upper-limit setting is reached. The timer completes the cycle, and a programming switch repeats the analysis.

Power, Measurement of (See **Dynamometer**.)

Power (Electrical), Measurement of The rate at which work is performed or energy transformed is called *power.* One ampere of electric current flowing under the pressure of 1 volt (thus, in a circuit of a resistance of 1 ohm) will do work at the rate of 1 watt (or 1 joule/s). Basically, electrical power in a circuit equals the product of voltage across the circuit times current in the circuit taken at the same point in time. If P is the power, E the voltage, and I the current, the equation $P = EI$ can be written. If the resistance of the circuit is R, by Ohm's law and substituting $E = IR$ for E, it is found that $P = I^2R$.

Electrical power is measured by its effects. One of the basic effects is the same as that observed whenever power is converted, namely, the generation of heat. In a given conductor, power can be converted into heat and measured as a rise in the temperature of the conductor. Taking into account heat loss through conduction, convection, and radiation and assuming the resistance to remain constant, it will be found that the temperature rise closely follows the square of the current. Therefore, it is possible to measure electrical power in terms of temperature. A method by which this can be done and which uses the thermoelectric effect (see **Thermocouple**) is embodied in the *thermal watt converter.*

When an instrument of this type is used on sinusoidal alternating current and voltage, the moving-coil deflection of the instrument will represent the average power as the product of the momentary in-phase components of current and voltage. It also can be shown that this product is the same as that of the rms ("effective") values of current and voltage. By artificially changing the phase of the current in the moving coil by insertion of a large inductive reactance in series with the moving coil and causing the moving coil to be in quadrature with the circuit voltage, the same basic instrument can be used for the measurement of reactive power (vars). Thus, the instrument becomes a *varmeter.*

In polyphase systems, at least two electrodynamic measurements are required for the measurement of the total power of the system. Polyphase wattmeters have two moving coils attached to the same shaft and arranged to operate between two separate sets of field coils. By proper connection of the field coils into at least two of the conductors and the movable coils across the proper conductors, such an instrument will measure either watts or vars depending upon the phase relationships between the currents in the fixed and movable coils.

Circuits of the instruments described can be found in **Electrical quantites, measurement of** and in Refs. 1 and 2. Microwave power measurements are described in Ref. 3.

REFERENCES
1. Harris, Forest K.: "Electrical Measurements," John Wiley & Sons, Inc., New York, 1952.
2. "The Instrument Sketch Book," Weston Instruments, Inc., *Circ.* E3G, Newark, N.J., 1966.
3. Considine, D. M., and S. D. Ross (eds.): "Handbook of Applied Instrumentation," pp. 15-79 to 15-82, McGraw-Hill Book Company, New York, 1964.

—Henry Berring, *Weston Instruments, Inc.*

Power-factor Meter In an alternating current in which voltage and current are sinusoidal, average power is defined as the product of the in-phase components of voltage and current, or $P = EI \cos \phi$, where E and I are the rms or "effective" values of voltage and current, and ϕ is the phase angle between voltage and current. The expression $\cos \phi$ is called *power factor.* This factor can be measured as the ratio P/EI by an arrangement of crossed moving coils in an instrument mechanism otherwise resembling that of a wattmeter. The manner in which the fixed and movable coils of power-factor meters are connected into single-phase, two-phase, and three-phase circuits is described elsewhere. See **Electrical quantities, measurement of,** also Ref. 1. Although power-factor meters indicate $\cos \phi$, they can be calibrated to read directly in ϕ—in which case they are called *phase-angle meters.*

REFERENCES

1. "The Instrument Sketch Book," Weston Instruments, Inc., *Circ.* E3G, Newark, N.J., 1966.
2. Fink, D. G., and J. M. Carroll: "Standard Handbook for Electrical Engineers," McGraw-Hill Book Company, New York, 1968.

—Henry Berring, *Weston Instruments, Inc.*

Power Standard (See **Standards, measurement.**)

Precipitation Gage One of several forms of gages used to measure precipitation in the form of rain, snow, sleet, or hail—over defined periods of time. Major types of precipitation (water) gages are (1) the manual-measurement gage, which essentially is a calibrated measuring stick and bucket; (2) the weighing-type gage, in which the mass of collected water is determined; and (3) the automatic-bucket-type gage, in which the precipitation is measured in 0.1-in. increments. The latter device actuates a remote instrument by sending an electrical pulse for each increment.

Precipitation in the form of snow or sleet is measured by several means. When both snowfall depth and the quantity of water deposited on the earth must be known, two measurements are required. Or, in some instances, the water depth is assumed to be one-tenth that of the snowfall depth. A hollow tube, sometimes called a snow sampler, can be used to cut a core-type sample which is measured for depth and weighed to determine water content. The radioactive-type snow gage consists of a small radioactive source (usually isotope Co-60) mounted at ground surface in a collimating shield. A Geiger-Mueller-type counter is mounted above the maximum snow-depth level. Attenuation in radiation reaching the Geiger-Mueller tube is a measure of the water content of the snow. Reference to a number of exotic snow-measurement schemes, such as the pillow-type gage and the radio-wave-absorption gage, is omitted because of their limited success and application.

Precision (See **Repeatability.**)

Pressure Gage, Elastic-element Three basic kinds of elastic measuring elements are in common use for pressure gages: (1) bourdon tubes, (2) bellows, and (3) diaphragms. For design details of these elements, also see **Bourdon tubes.**

Pressure ranges normally associated with these elements are given in Table P-6. Other forms of elastic elements, not further described here because they are not common, include the slack diaphragm with auxiliary springs, twisted bourdon tubes, and eccentrically drilled rods.

Bourdon-tube Pressure Gages: The bourdon tube, patented by Eugene Bourdon in 1852, is by far the most common of elastic-pressure elements in use even today. It takes three principal geometric forms which are shown in Fig. P-60: (*a*) the common C shape, (*b*) the spiral, and (*c*) the helix. Regardless of form, one end of the bourdon tube is fixed, and it is through this fixed

TABLE P-6. *General Ranges of Elastic-element Pressure Gages*

Element	Application	Min Range (Commonly Supplied)	Max Range (Commonly Supplied)
Bourdon spring	Pressure	0–12 psig	0–100,000 psig
	Vacuum	0–10 in. Hg vacuum	0–30 in. Hg vacuum
	Compound	30 in. Hg, 0–15 psig	30 in. Hg vacuum, 0–300 psig
Bellows	Pressure	0–1 in. Hg	0–100 psig
	Vacuum	0–1 in. Hg vacuum	0–30 in. Hg vacuum
	Compound	Any span within pressure and vacuum ranges, with total span of 10 in. H_2O	Any span within pressure and vacuum ranges, with total span of 10 psi
Metallic diaphragm	Pressure	0–1 in. Hg	0–200 psig
	Vacuum	0–1 in. Hg vacuum	0–30 in. Hg vacuum
	Compound	Any span within pressure and vacuum ranges, with total span of 1 in. Hg	30 in. Hg vacuum, 0–200 psig

end that the pressure to be measured is applied. The other end is free to move in response to the applied pressure. Movement of the free end is termed *tip travel.* Various linkages or motion-sensitive devices are used to amplify the relatively small motion of the free end. Motion is the result of elastic deformation of the tube cross section. The mechanism of a pressure gage employing a C-type bourdon, showing the mechanical linkage and gage pointer, is illustrated in Fig. P-61.

To obtain maximum travel, bourdon tubes usually are designed to operate at or near their elastic limit. This imposes a responsibility on the gage designer and user to be aware of unusual conditions that will affect the stress levels, particularly for high-pressure ranges—in excess of 1,000 psi. Operating conditions, such as high temperature, pulsating pressure, shock-wave propagation, overpressures (even momentary), vibration, and sustained pressures at or near the maximum range of the gage will limit the stress for which the tube can be designed. Because it is highly stressed, corrosive atmospheres, both within and surrounding the tube, are especially detrimental. A wide variety of materials is used for bourdon springs. Nonferrous materials include admiralty and trumpet brass, phosphor bronze, beryllium copper, Monel, and K Monel. Ferrous metals include types 316, 347, 403, and 431 stainless steels, types 4130 and 4650 alloy steels, and Ni-Span C.

Helical and spiral bourdon tubes are special forms of the more common C shape and are used for the purpose of reducing the stress level per unit length of tube. Obviously, for a given amount of tip motion, the longer the tube the lower the stress levels; conversely, for a given stress level, the longer the tube the more motion will be obtained. Helical and spiral springs also allow packing a long length of tube in a small space. The tube may be designed to utilize relatively thin-wall tubing of a very flat cross section which permits winding into a compact spiral form. Helical and spiral tubes frequently are used where it is desirable to eliminate the amplification linkages. By using a very long length

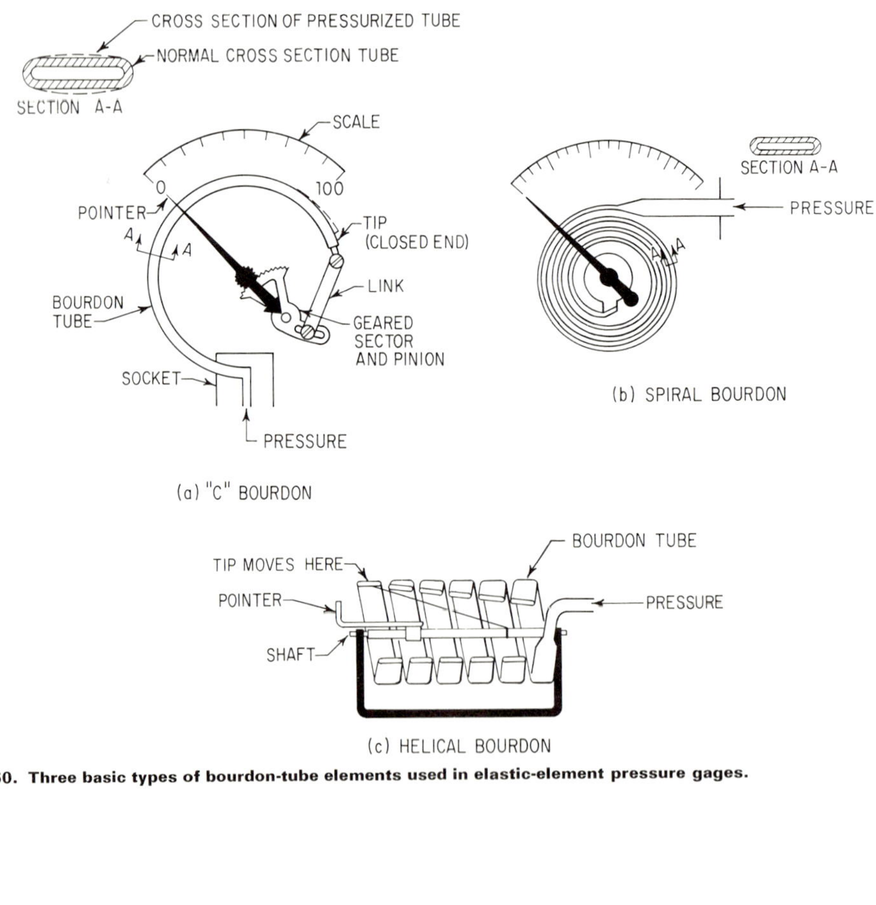

Fig. P-60. Three basic types of bourdon-tube elements used in elastic-element pressure gages.

of tubing, sufficient tip travel can be obtained to indicate directly against a calibrated dial. Thus, wear, friction, and the inertial effects of the linkage are eliminated.

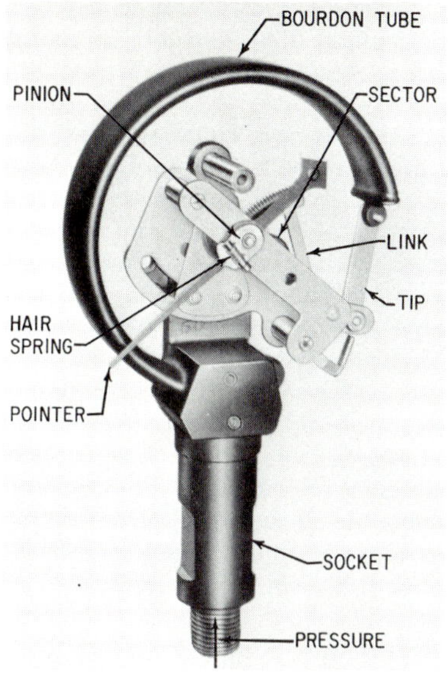

Fig. P-61. Internal mechanism of C-type bourdon-tube pressure gage.

Bellows Pressure Gages: Bellows are more costly than bourdon tubes and are used for low-pressure ranges where it is impractical to design a bourdon tube or where greater energy must be obtained from the measuring element than is afforded by tip movement of a bourdon tube. Prior comments regarding stresses in bourdons also apply to bellows elements—as the latter also are designed to operate close to the elastic limit of the material. Selection of materials for bellows is more limited and the response rate of a bellows is lower than that of a bourdon tube because of the greater volume change required to move the bellows. In many cases, the bellows is used mainly as a separating element and its own spring rate is supplemented by auxiliary coil or leaf springs. A bellows element is shown schematically in Fig. P-62.

Diaphragm Pressure Gages: A diaphragm may be considered as a special form of bellows. A diaphragm is made by corrugating flat disks or plates of the desired material. These plates may be used singly or may be joined at their lips to form a capsule. Capsules may be joined at their center to form a stack.

Diaphragms are used in applications similar to those for bellows but also have advantages over bellows. For example, diaphragms may be designed to have a motion which is nonlinear with pressure—an advantage, for example, in altimeters, where it usually is desirable to have the motion proportional to altitude rather than to pressure. Diaphragms can be made from materials that are difficult or impossible to form into bellows. Generally, diaphragms are used without auxiliary springs. Except for single capsules, diaphragms are more costly than bellows, but because of their inherent advantages they are commonly used in aircraft instrumentation. Diaphragms lend themselves well to use in opposing stacks for differential-pressure and absolute-pressure measurement. A metallic diaphragm element is presented schematically in Fig. P-63.

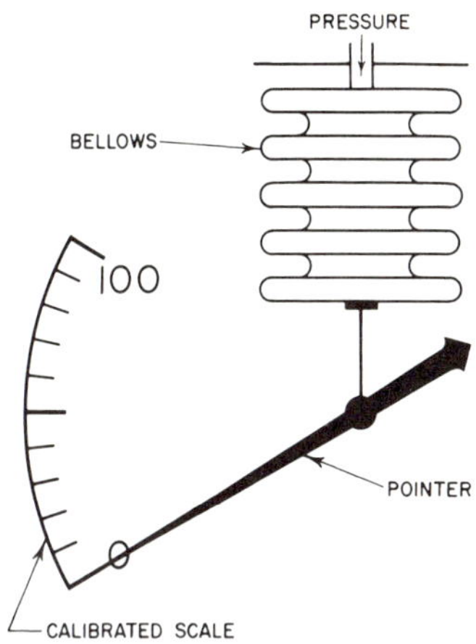

Fig. P-62. Bellows element as used in elastic-element pressure gage.

Pressure-gage Selection: A wide variety of standard elastic-element pressure gages is available. Gages are obtainable in sizes from 1 to 16 in. in diameter and with accuracies ranging from ±5 to ±0.1 percent full scale. Many case styles and case materials are offered—with pressure-connection size and location options. Round-dial gages with concentrically or eccentrically located pointers are widely used. Edge-reading gages are available. The internal movements can be procured in many different materials, with or without metal or plastic bushings, and having varying degrees of adjustment features. Since such a large variety of standard gages is available, a large majority of applications can be satisfied by catalog items permitting maximum economy to be achieved. Reference 1 specifies mounting dimensions, grades of accuracy, and sizes—and includes an excellent gage-specification sheet.

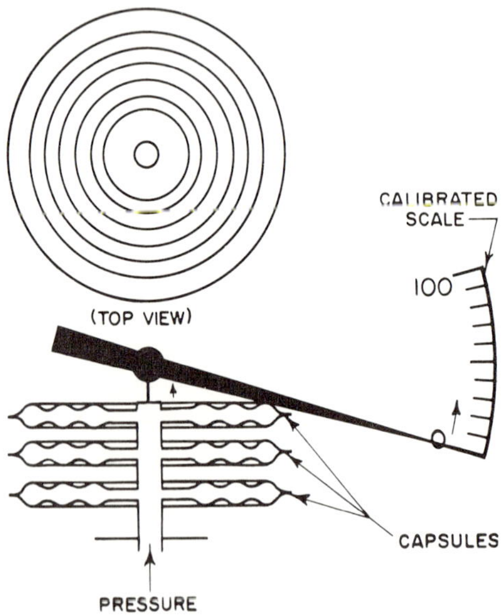

Fig. P-63. Diaphragm element as used in elastic-element pressure gage.

REFERENCES

1. Standard Indicating Pressure and Vacuum Gages, American National Standards Institute, *Publ.* B-40.1, January, 1967.
2. Harland, Philip W. (ed.): "Pressure Gauge Manual," U.S. Gauge Div., AMETEK/U.S. Gauge, 1963.
3. Harland, Philip W.: Pressure Gages, *Machine Design,* Sept. 22, 1966.

—Philip W. Harland, *AMETEK/U.S. Gauge*

Pressure Gage, Manometer (See Manometer.)

Pressure Gage, Piston-type Generally a historic form of pressure-measurement device comparable with a slack-diaphragm gage with the exception that greater friction is present. (see Fig. P-64.) The pressure measured is balanced by a calibrated spring. Variations include weight-loaded- and scale-beam-type balancing methods. The dead-weight tester, still commonly used, is a form of piston-type pressure gage. See **Dead-weight tester.**

Pressure-reducing Valve (See Self-actuated regulators.)

Pressure-relief Disk (See Rupture disk, pressurized-system.)

Pressure Sensor (Eddy-current) (See Eddy-current transducer.)

Pressure Transducer, Vibrating-wire A device that operates on the principle that a taut wire's natural frequency changes as the square root of its tension. In application, one end of the wire is attached to a diaphragm. As pressure deflects the diaphragm, the wire tension and hence the natural frequency of the wire decreases.

This device[1] is used principally in the field of oceanography where the variable-frequency output makes it possible to transmit pressure/depth information over long cables with no degradation of pressure information. The device also has been used in the oil industry to transmit pipeline-pressure information over leased telephone lines. When placed in a constant-temperature environment—in either the deep ocean or a temperature chamber—resolutions of 1 part in 10^7 can be obtained with the device. Two pressure ranges are available—(low) from 0 to 100 psia and (high) from 0 to 15,000 psia. Repeatability is better than ±0.1 percent full scale. The transducer measures approximately 1 in. diameter by 3 in. long and weighs less than 6 oz.

—Malcolm D. Lefcort, *Transducer Division, United Control, a subsidiary of Sundstrand Corporation*

[1] Vibrotron, a registered trademark of United Control.

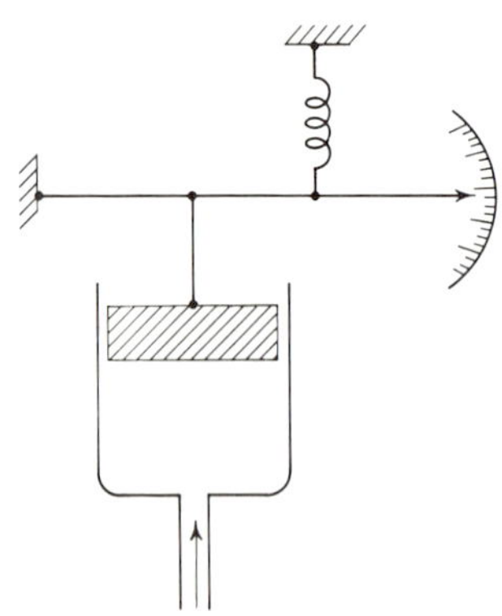

Fig. P-64. Sectional view of piston-type pressure gage.

Primary Element (See Sensor; and Transducer.)

Primary Flowmetering Device (See Flowmetering devices, primary.)

Primary Standard (See Standards, measurement.)

Printer A form of display device used with instrumentation systems wherein the readout is a series of marks or symbols as contrasted with continuous-line recording as effected by pen-type recorders, for example. The kinds of symbols and marks that can be printed are almost limitless. Any symbol that can be printed in any other fashion can be printed instrumentally. Multiple-variable recording instruments often incorporate a print wheel which is advanced automatically as the instrument is switched from one measuring circuit to the next. The symbols printed can be dots or other small geometric forms, or numbers, or letters. See **Recorder, graphic.** The end result is, in essence, a disconnected line which from a slight distance may appear as a continuous line. For additional distinction between records, multicolor printing sometimes is used. In modern instrument systems, tabular printing is more common—such devices resembling in basic function a typewriter. The teletypewriter was one of the earlier forms for line-by-line printed-data transmission. The trend has been toward much higher printing speeds. Computers may print from 600 to 1,100 lines per minute and provide from one to six copies. Printers are available up to 30,000 lines per minute and higher. However, the fastest conventional speed is 1,100 lines per minute. Cathode-ray-tube microfilm systems make very high speeds possible. Ink-jet printing also increases printing speeds. See **Printer, high-speed ink-jet.** On the much slower side, an event recorder essentially is a printing device wherein a mark is made on a circular or strip chart each time a given external event occurs. See **Event recorder.**

Printer, High-speed Ink-jet The printing system[1] shown in Fig. P-65 is a nonimpact output printer that utilizes a stream of controlled ink droplets to form characters on a printing medium. In the printer, ink under pressure is forced through a nozzle which is energized with an ultrasonic signal (66 kHz) to cause velocity modulation of the ink stream. This modulation causes the ink stream to break up into 66,000 discrete droplets per second. The droplets are of uniform size and spacing, both being a function of pressure at the nozzle, viscosity of the ink, nozzle diameter, and the vibration frequency of the nozzle.

[1] Videojet 960 printer, A. B. Dick Company.

Each droplet may be given a precise electrostatic charge as it forms. The voltage present at the time the droplet breaks away from the ink stream induces the charge. Once it breaks, the charge is trapped—staying with the droplet until it reaches its destination. Because the droplets pass between a pair of high-voltage deflection plates, the charge of the droplet determines the precise path of travel and final position of the droplet on the medium being printed. The nozzle is moved in a horizontal plane, driven by a servomechanism-controlled, high-speed, low-inertia motor. The movement of the nozzle provides horizontal deflection of droplets; the high-voltage plates provide vertical deflection. A functional analysis of the system is shown in Fig. P-66. The physical components of the printing system are shown in Fig. P-67.

Fig. P-65. High-speed Videojet 960 printer. *(A. B. Dick Company.)*

A 9 × 11 dot matrix (see Fig. P-68) is used in character generation to determine the charges for droplets. The 99 dots of the 9 × 11 matrix correspond to 99 droplets. If each droplet were

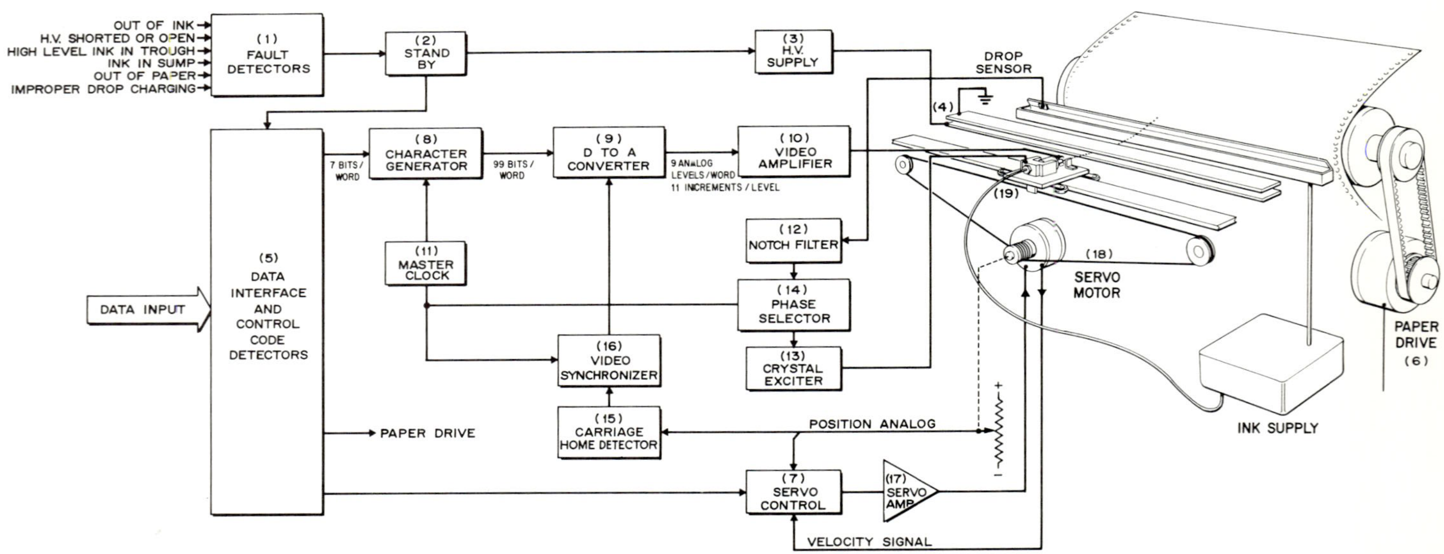

Fig. P-66. Functional analysis of Videojet 960 printer. *(A. B. Dick Company.)*

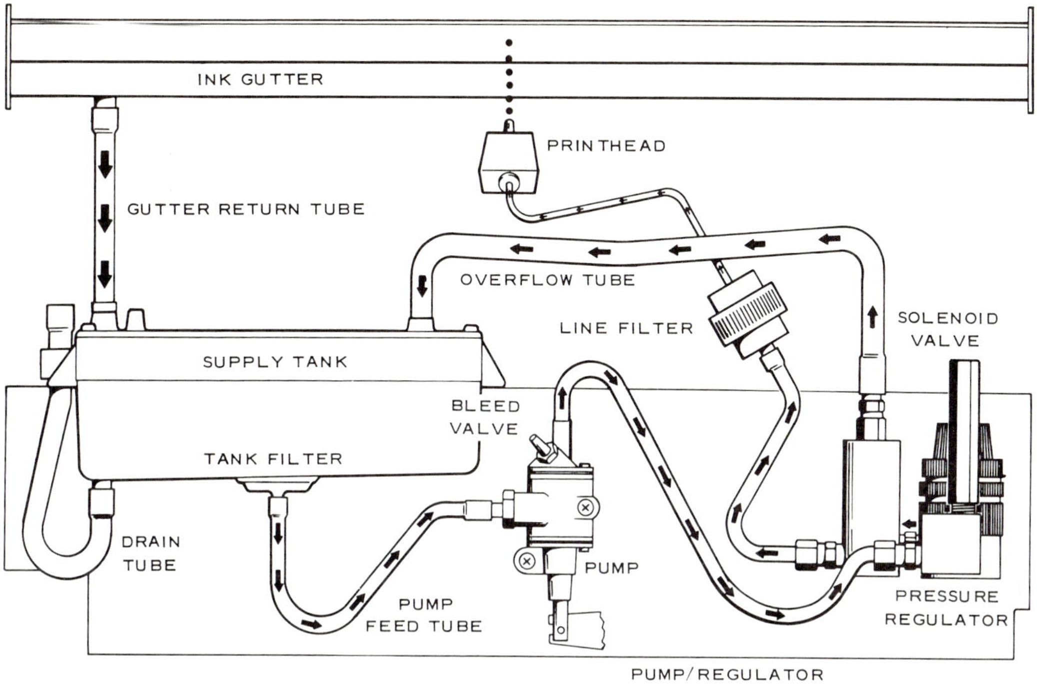

Fig. P-67. Physical components of printing-ink system of Videojet 960 printer. *(A. B. Dick Company.)*

appropriately charged, the resulting character would be a rectangle 9 droplets wide and 11 droplets high. Only those droplets required to form a character are charged. Uncharged droplets are collected and returned to the ink reservoir for reuse. Operating at 250 characters per second, utilizing a 9 × 11 matrix, and allowing for space

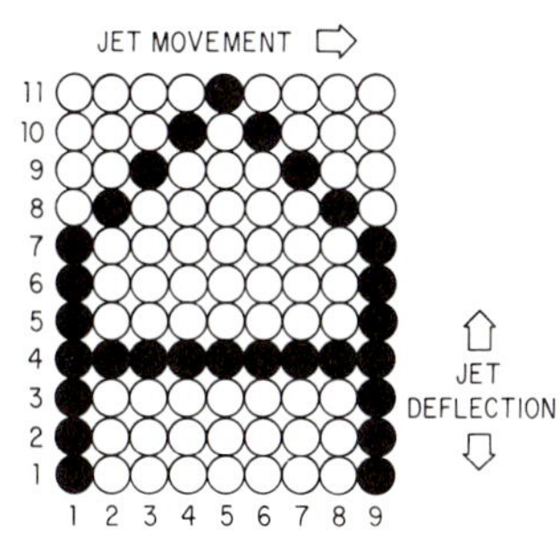

Fig. P-68. 9 × 11 dot matrix used with Videojet 960 printer. *(A. B. Dick Company.)*

between characters requires only 33,000 droplets per second. Droplet formation is at the rate of 66,000 per second in anticipation of expanding transmission capabilities. Operating as a "receive only" device utilizing the USASCII code set, the system will print a question mark in lieu of a character failing parity. The printer also prints an automatic error flag at the end of the line containing a character in question.

Vertical line spacing is six lines per inch with single- or double-line space available. Character spacing ranges from 5 to 15 per inch with nominally 10 characters per inch. The print area is a 13.6-in. print line providing 136 characters at 10 per inch, 204 characters at 15 per inch. A portion of a bank statement printed by the Videojet 960 printer is shown in Fig. P-69 (on page 532).

—Videograph Engineering, A. B. Dick Company

Process A term used in some technical circles to describe the collective functions performed in and by the equipment in which one or more variables are to be controlled. In a control system, the distinction is made between the equipment that is being controlled, that is, the controll*ed* subsystem (or process) and the instrumentation and control hardware required to effect control, namely, the controll*ing* subsystem. See **Control system.**

ACCOUNT NUMBER	NAME	AVERAGE BALANCE	CHECKS	DEPOSITS
145-623-8	JONES	57.4Ø	21	4
145-633-7	ADAMS	234.96	1	2
145-635-2	BRADLEY	89.74	5	Ø
145-641-3	MOSQUE	2.ØØOD	Ø	Ø
145-647-2	LAME	876.4Ø	25	4
145-65Ø-Ø	ALEXANDER	134.6Ø	17	2
145-654-5	FURLONG	86.61	22	3
145-658-6	BERRY	243.8Ø	28	5
145-66Ø-8	MCMILLON	1Ø.18	14	2
145-662-1	STEWART	58.9Ø	17	3
145-663-4	EVANS	162.47	3Ø	4
145-674-8	JONES	347.88	26	2
145-678-3	ARMSTRONG	87.4Ø	11	2
145-683-7	AARON	236.84	23	4
145-683-7	BANKS	976.23	34	4
145-688-5	DONASKI	433.71	2Ø	4
145-688-8	BULLOCK	84.9Ø	14	1
145-69Ø-2	JOHNSON	1.ØØOD	Ø	Ø
145-695-2	YOUNG	136.32	12	4
145-7ØØ-7	WHITE	2Ø1.67	23	2
145-7Ø2-6	WILLIAMS	74.2Ø	7	1
145-72Ø-1	GOLDBLATT	274.6Ø	23	2
145-72Ø-1	SINCLAIR	43.72	6	1

Fig. P-69. Section of bank statement prepared on Videojet 960 printer. *(A. B. Dick Company.)*

Process, Self-regulating (See **Self-regulating process.**)

Process-analysis Instrument [See **Analytical instrumentation, (chemical).**]

Process Control The application of automatic-control theories and hardware to operations found in the process industries. These industries fundamentally handle three forms of materials: (1) fluids, including a vast variety of liquids and gases which are transported in pipes and ducts using pumps, blowers, and gravity as the transportation energy sources and various forms of valves, dampers, and other final controlling elements to control flow; (2) bulk solids which are transported pneumatically or by means of various types of conveyors; and (3) sheeted and webbed materials, such as steel and aluminum sheets, paper, and textile fabrics. The process industries include the chemical, petroleum, metal-producing, and pipeline industries, and portions of other industries, such as various food products. Sometimes a distinction is made between dry and wet processing. The most important variables measured and controlled in the process industries include temperature, pressure/vacuum, flow, liquid-level, bulk-solids level, density and specific gravity, and chemical composition, including pH, oxidation-reduction potential, and many forms of spectrometry and spectrophotometry. From the standpoint of instrumentation and control, the process industries differ rather dramatically from the discrete-piece handling industries, such as metalworking, electrical machinery, electronic components, and apparel manufacturing where the most important variables are dimension, shape, count, speed, position and other physical properties, such as hardness, ductility, and color—with much emphasis on nondestructive inspection.

REFERENCES

Canfield, E. G.: "Electrochemical Control Systems and Devices," John Wiley & Sons, Inc., New York, 1965.

De Roy, B. E.: "Automatic Control Theory," John Wiley & Sons, Inc., New York, 1966.

Grabbe, E. M., S. Ramo, and D. E. Wooldridge: "Handbook of Automation, Computation, and Control," vol. 1, 1958; vol. 2, 1959; vol. 3, 1961, John Wiley & Sons, Inc., New York.

Davies, P. O. A. L.: "Introduction to Dynamic Analysis and Automatic Control," John Wiley & Sons, Inc., New York, 1966.

Kelley, C. R.: "Manual and Automatic Control," John Wiley & Sons, Inc., New York, 1968.

Tomovic, R.: "Introduction to Nonlinear Automatic Control Systems," John Wiley & Sons, Inc., New York, 1966.

Aizerman, M. A., and F. R. Gantmacher: "Absolute Stability of Regulator Systems," Holden-Day, Inc., Publisher, San Francisco, 1964.

Principles of Automatic Process Control, *ISA Publ.* PAPC, 1969.

Tucker, G. K., and D. M. Wills: "A Simplified Technique of Control System Engineering," Honeywell Inc., Fort Washington, Pa., 1958.

Considine, D. M., (ed.): "Process Instruments and Controls Handbook," McGraw-Hill Book Company, New York, 1957.

"Manual on Process Instrumentation and Control," American Petroleum Institute, Division of Refining, *Publ.* API RP 550, 2d ed., New York, March, 1965.

Ahrendt and Taplin: "Automatic Feedback Control," McGraw-Hill Book Company, New York, 1951.

Doeblin, E. O.: "Dynamic Analysis and Feedback Control," McGraw-Hill Book Company, New York, 1962.

Raven, F. H.: "Automatic Control Engineering," 2d ed., McGraw-Hill Book Company, New York, 1968.

Truxal, J. G.: "Automatic Feedback Control System Synthesis," McGraw-Hill Book Company, New York, 1955.

Truxal, J. G.: "Control Engineers' Handbook," McGraw-Hill Book Company, New York, 1958.

Harriott, Peter: "Process Control," McGraw-Hill Book Company, New York, 1964.

Savant, C. J., Jr.: "Control System Design," McGraw-Hill Book Company, New York, 1964.

Leondes, C. T.: "Modern Control Systems Theory," McGraw-Hill Book Company, New York, 1965.

Del Toro, Vincent, and S. R. Parker: "Principles of Control System Engineering," McGraw-Hill Book Company, New York, 1960.

Siskind, C. S.: "Electrical Control Systems in Industry," McGraw-Hill Book Company, New York, 1963.

Process-gas Chromatography (See **Chromatography.**)

Process Optimization This technique is rapidly gaining recognition as a valuable management tool in the operation of a manufacturing plant. A chemical plant, for example, can operate within certain parameters without optimizing the involved processes, but the result can be less product of lower quality which, in turn, is likely to mean less profit. The situation is likened to an automobile engine that will continue to generate power even though not fully tuned for maximum efficiency. In this case, of course, the result is less mileage per gallon of fuel consumed.

The management of a plant must establish the control policy. In the case of a chemical plant, which may involve a large number of interacting variables, optimization of the processes to achieve a maximum yield of products at a minimum cost requires considerable analysis. Further, the problem might be more complicated than a problem of maximizing the products and minimizing the costs. For example, it may be that a particular product of a chemical plant can be made from different kinds of raw materials—as in the case of an ammonia plant. Thus, raw-materials selection becomes a part of the problem. Or, the problem may be one of making different kinds of products from a particular raw material—as often is the case in connection with petrochemical plants. In establishing the best control policy, management must consider product-market values as well as raw-material availability and costs—and often the basic relationships of these two important factors.

For many years, inefficient trial-and-error methods have been used, and they still are commonplace. However, with the advent of process computers and optimization theories, process-optimization procedures have been greatly improved. Perhaps the technique can be explained best by reference to a specific process—a synthesis ammonia plant shown in simplified form in Fig. P-70.

Raw materials are shown entering from the left-hand side of the diagram. After the necessary process preparation and purification procedures, the materials are compressed to a high pressure and fed to the synthesis reactor. The partially reacted gas is introduced intermediately into the second stage of the synthesis gas compressor—goes through the last stage of compression—and finally enters the ammonia separator where all ammonia product condenses and flows to the storage tank.

All process-state variables, such as the flow and composition of raw materials, the pressure and composition of reactor feed, the pressure, temperature, and composition at various points of the synthesis reactor, and other variables are measured and fed as signals to the computer. Output signals from the computer to the final controlling elements include signals for (1) controlling the

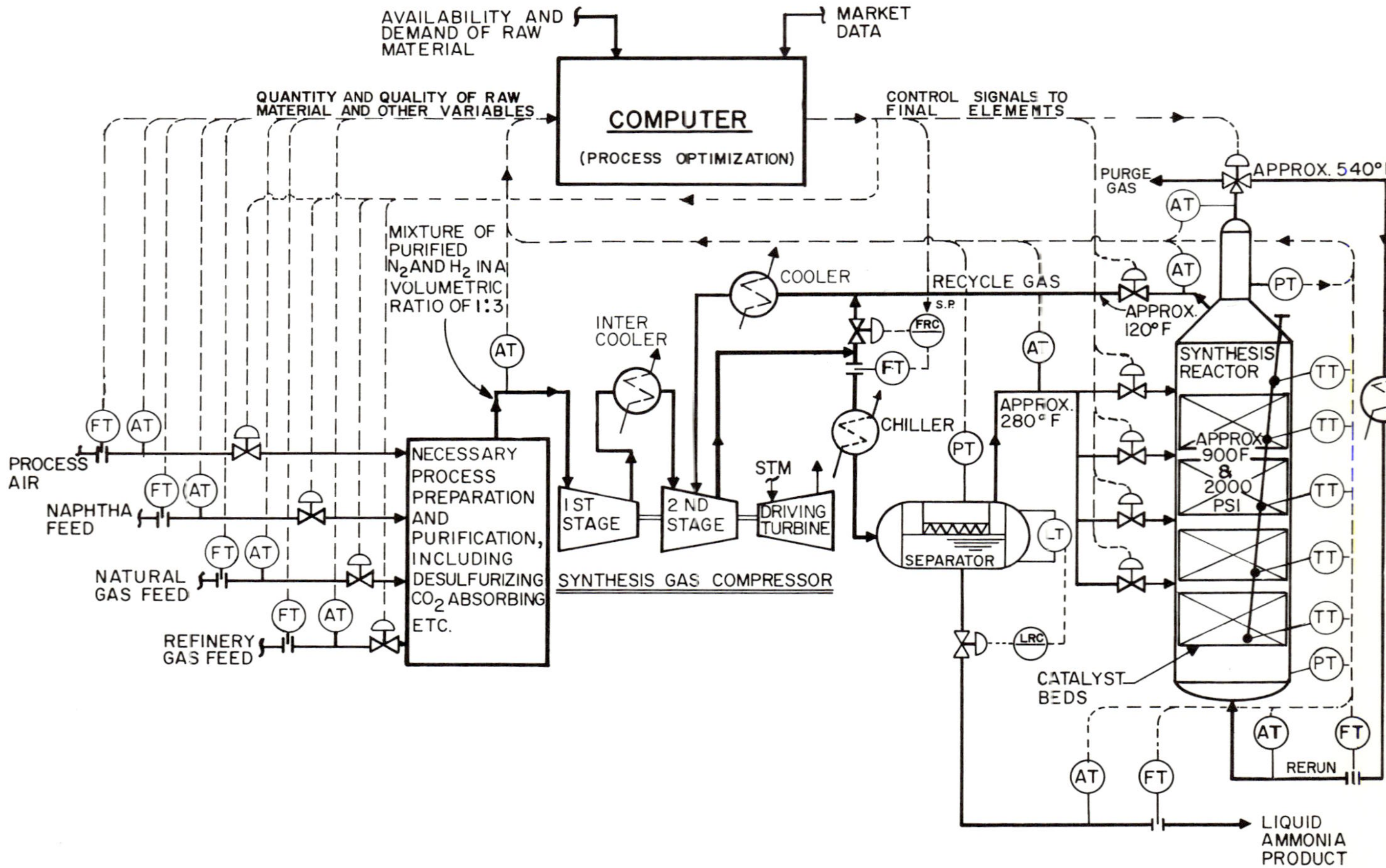

Fig. P-70. Simplified diagram of synthesis ammonia plant.

raw material inlet from battery limits, (2) distribution of the low-temperature synthesis feed gas as quench to various catalyst beds in order to maintain all the catalyst at an *optimum* temperature for *maximum* yield, and (3) the very important control signal for the determination of percentage rerun required in order to maintain a certain product quality for guaranteed market value. The problem involved is that of making decisions so as to maximize plant profits within a certain period of time—say $[0,T]$. Visualize vector x which contains all the process-state variables,

$$x = \begin{bmatrix} x_1 \\ x_2 \\ \cdot \\ \cdot \\ \cdot \\ x_n \end{bmatrix} \qquad \begin{array}{l} x_i(0) = a_i \\ i = 1, 2, \ldots, n \end{array}$$

and another vector m which contains all the control-policy signals,

$$m = \begin{bmatrix} m_1 \\ m_2 \\ \cdot \\ \cdot \\ \cdot \\ m_r \end{bmatrix} \qquad \begin{array}{l} m_j(0) = bj \\ j = 1, 2, \ldots, r \end{array}$$

Then, the problem involved can be described by the following mathematical expression:

$$m = \max_{0 \leq t \leq T} \int_0^T p(x, m, t)\, dt$$

where p represents the profits at a certain time $t \in [0, T]$. Since p is not necessarily a continuous and differentiable function, perhaps it would be better to rewrite the preceding equation as

$$m = \max_{0 \leq t \leq T} \sum_{k=1}^{k=T} p_k(x, m, t)\, \Delta t$$

where the number of stages $k = 1, 2, \ldots, T$.

The equations, simplified as they are, represent the generality of a problem of this kind, which constantly involves multivariable simultaneous equations. The equations usually appear in algebraic, differential, integral, or "difference forms" if numerical approaches are employed, which is usually the case where digital computers are involved. The techniques for solving problems of this kind include the method of variational calculus, the maximum principles of Pontyragin, and the dynamic programming of Bellman.

The variational calculus, or calculus of variations, as it is sometimes called, and the maximum principle of Pontyragin—both belonging to the family of analytical approaches—were developed originally by Euler, Lagrange, Hamilton, and Jacobi and then further expanded in recent years by Russian mathematicians headed by Pontyragin. These analytical methods, powerful as they are and capable of solving most of the problems not treatable by elementary calculus, still are handicapped by the solution of the partial differential equations involved. This is especially true as regards the area involving a large number of variables and is further complicated by the initial conditions and boundary-value problems. For this reason, these analytical methods serve only as a guideline for setting up a program of a plant of moderate complexity, such as that illustrated in Fig. P-70.

Thus, the major burden is left to the third method—dynamic programming, a method developed in recent years by Richard Bellman. This method, simple but powerful, first looks at the last stage of the process and then calculates backward by treating each of the multiple stages individually. This procedure is based upon the concept that a series of optimal policies should result in an optimal policy for the whole process.

The concept may be explained best by the famous statement made by Bellman in 1957—"An optimal policy has the property that whatever the initial state and initial decisions are, the remaining decisions must constitute an optimal policy with respect to the state resulting from the first decision." The method of dynamic programming, with the help of digital computers, has found useful applications in solving the problems of optimization which generally involve multistate, multivariable discrete processes.

—Anthony Y. Chiang, *Bristol Division, American Chain & Cable Company, Inc.*

Process Reaction Rate The speed of response of a control loop to a step change in input. Determination of the process reaction rate is made by opening the control loop just before the final controlling element and introducing a step change to the final controlling element. The shape of the resultant reaction curve, that is, response of the measured variable, provides two factors that are characteristic of the particular process. These factors are reaction rate R and time lag L. With these factors known, controller-mode adjustments can be calculated.

The method of obtaining a process reaction rate R and time lag L from a process-reaction curve is shown in Fig. P-71. The process reaction rate is the slope of the line drawn tangent to the process-reaction curve at the point of inflection. The time lag L is the time in minutes between the

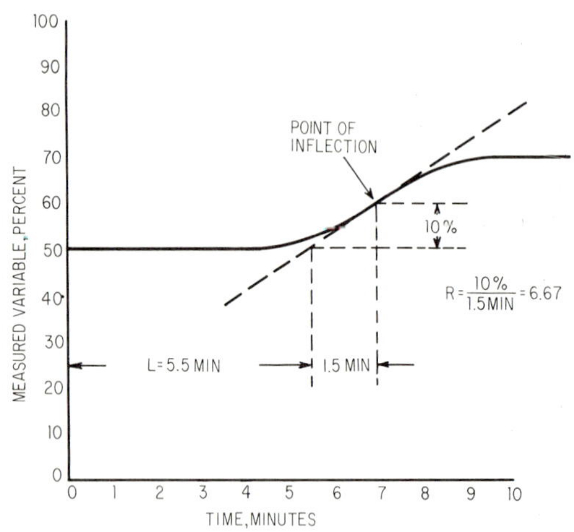

Fig. P-71. Process-reaction curve produced by step change of final controlling element.

step change and the point where the tangent crosses the initial value of the controlled variable. Calculations of control-mode settings are as follows:

Proportional controllers:

$$\text{Proportional band, percent} = \frac{100\,RL}{\Delta P}$$

where ΔP = percent change in final control-element position used to produce the reaction curve

Proportional-plus-integral (reset) controllers:

$$\text{Proportional band, percent} = \frac{110\,RL}{\Delta P}$$

$$\text{Reset rate (repeats per min)} = \frac{0.3}{L}$$

Proportional-plus-integral (reset)-plus-derivative (rate) controllers:

$$\text{Proportional band, percent} = \frac{83\,RL}{\Delta P}$$

$$\text{Reset rate, repeats per min} = \frac{0.5}{L}$$

$$\text{Rate time, min} = 0.5\,L$$

The foregoing procedure cannot always be used. A process that is not self-regulating will not come to rest and cannot be treated by the process-reaction-rate technique. In addition, the technique assumes that the process will react in the same manner at all load and set-point conditions—which does not always hold.

—Richard H. Kennedy, *Bristol Division, American Chain & Cable Company, Inc.*

Process Simulation (See **Hybrid computer.**)

Process-stream Analyzer An instrument for determining the chemical analysis, partially or completely, of a liquid or gas as it flows through a pipeline or piece of processing equipment. A process-stream analyzer generally is assumed to be on-stream, that is, with the sensing element actually interfacing with the process liquid or gas, or where the instrument is very closely coupled to the line. This situation differs from that where "grab" samples are taken from the line, carried to a remote point, such as a control laboratory, and analyzed. Also see **Analytical instrumentation (chemical); Chromatography; Infrared process analyzers; Ultraviolet spectrophotometry;** and consult the alphabetical index.

Process Variable A physical or chemical quantity, the variation of which will affect the operation of a manufacturing process. A list of these quantities is given under **Variable.** Process variables are measured and controlled for a number of reasons, including: (1) *Product quality*—most products are made to meet certain specified physical or chemical properties. As the quality of the product goes up, the tolerances governing deviation from specifications become narrower. As tolerances grow narrower, the need for instrumental control increases. (2) *Material savings*—by the consistent maintenance of product quality, rejects are cut to a minimum, thus effecting savings of raw and in-process materials as well as the productive time of machinery and labor. (3) *Cost accounting*—where instrumental measuring systems provide accurate raw data as contrasted with rough manual estimates and arbitrary cost allocations. (4) *Production inspecting and testing*—where generally instruments can provide faster and more accurate determinations of materials and product quality. (5) *Plant safety*—where instruments detect excessive heat and pressures, noxious vapors, inadequate ventilation, the presence of human limbs in dangerous machine areas, electrical shorts and leakage, flame failure, and numerous other unsafe conditions.

—Douglas M. Considine, *Editor-in-Chief*

Program, Computer A program is a sequence of operations which describes a job to be done by a computer. The operations are either machine instructions which can be implemented by the machine or they are operations which are translated into machine instructions by a compiler or assembler program (see **Compiler; Assembler**). Programs are created by the procedure known as programming (see **Programming, computer**) and ultimately represent a series of machine operations to be executed by the computer. When programs are written in a language other than the fundamental machine language (see **Language, computer**), the collection of programming statements which form the program are called source statements and are collectively referred to as the source program. When the source program has been converted into machine language using an assembler or compiler, the resulting program is called the object program.

Programs are the means by which computers are directed to produce meaningful work. The useful work may consist of mathematical problem solutions, record-keeping operations, process control, and many other operations limited only by the computer-hardware capability.

Groups of interrelated programs called operating systems are usually supplied with a computer system to increase its power and utility. The users of these systems write problem programs to solve specific problems as indicated above. Programs are written in special languages which may be translated into basic machine operations to accomplish the desired result. Examples of languages used are ALGOL, BASIC, COBOL, FORTRAN, and PL/I.

The following is an example of a program to evaluate an arithmetic expression based on values read in from an input device. The results of the computation and the input data are printed at the end of the program. The form shown here is similar to several of the programming languages now in common use.

```
Begin
Read x, A, B, C, D
Let Y = a + x*[B + x*(C + x*D)]
Write x, Y, A, B, C, D
Stop
```

—L. E. Wise, *IBM Corporation*

Program Controller (See **Timing controllers.**)

Programming, Automatic Drafting System (See **Drafting system, automatic.**)

Programming, Computer The process of describing in some computer language (see **Language, computer**) a procedure or sequence of events to be done to accomplish desired work within the capabilities of the hardware system. Programs may be written in machine language (the most basic level), assembly language (a symbolic level providing some assistance in detail chores), or a higher-level problem-oriented language such as ALGOL or COBOL where the actual computer used may not be significant (see **ALGOL; COBOL; FORTRAN**).

Programming may be considered as that group of activities which causes a solution to a problem using the computer as the tool. Steps to accomplish this are:

Problem definition
Solution definition
Design of solution algorithm
Program coding of solution algorithm
Program testing and debugging

Problem Definition: An exact definition of the problem to be solved is essential to be able to allow a feasible solution to be found. This part of the procedure does not necessarily have any bearing on the computer. In the scientific world this may be a specific mathematical statement of the problem. In commercial application this may be a business or record-manipulation problem. Control-system problems may describe a process to be controlled.

Solution Definition: The solution to the defined problem must exist before the computer can be of use. The solution may in fact be finding if a mathematical solution to a problem even exists; the solution in this case, of course, is the procedure used to find if the mathematical solution exists. The problem in such a case would be to determine if a solution to the mathematical problem exists.

Design of Solution Algorithm: This segment of the procedure is really the first step in computer implementation of a problem solution. Here the solution procedure is planned step by step. The problem of precision and accuracy in mathematical problems is investigated to ensure expected results. These procedures or algorithms (see **Algorithm**) describe the logical steps in the solution procedure. The performance of the system in execution of the problem can depend to a large extent on the algorithm used. In scientific work, for instance, functional approximations which are within the desired accuracy range may provide much faster execution than traditional methods of calculations. This is especially important in programs that may require extended amounts of time for execution. Similar considerations apply in other application areas.

Program Coding of Solution Algorithm: This step in the procedure represents the description of the solution algorithm in the language of the computer system. The facilities of the selected language (see **Language, computer**) are used to implement the stated procedure. The performance of the resulting code can be affected by the coding techniques used. Performance may or may not be an important aspect of the problem, however. In systems programming, where the resulting code is essentially overhead for a system, execution speed is, of course, vital. In a one-shot program to generate a specific set of answers or a report, execution time may be of secondary importance to factors such as ease of use or time required to write and test the program.

Program Testing and Debugging: This important final stage of the programming procedure is used to verify that the design and coding of the solution is correct. Normal procedures involve running the program with known input and comparing the actual output with the expected results. Debugging is the isolation and elimination of errors in the program. Program testing can be a very large and extensive part of the overall procedure, especially in programs which comprise real-time response systems and process-control systems. Auxiliary means are often used in these cases, for instance, the use of plotters (see **Plotter, data**) to verify programs which generate numerical control information for machine-tool control. The program-test function is extremely important because of the possible implications of erroneous output from the program.

REFERENCE

Rosen, Saul: "Programming Systems and Languages," parts 1 and 2, McGraw-Hill Book Company, New York, 1967.

—L. E. Wise, *IBM Corporation*

Propeller Anemometer (See **Anemometer.**)

Propeller Flowmeter (See **Flowmeter, turbine-type.**)

Proportional Control Control action in which there is a continuous linear relation between the output and the input. This condition applies when both the output and input are within their normal operating ranges and when operation is at a frequency below a limiting value.[1]

The effects of measured variable deviations from set point upon the control valve with proportional control are shown in Fig. P-72. The valve position changes in exact proportion to the measured-variable excursions from set point. The changes in valve position are proportional to the amount of the deviation of the measured variable from set point and are not affected by the rate or duration of the deviation.

The proportional band or gain is the change in value of the measured variable that is necessary to cause full travel of the final controlling element. In Fig. P-72, a deviation of 30 percent causes a 20 percent change in valve position—which is equivalent to a proportional band of

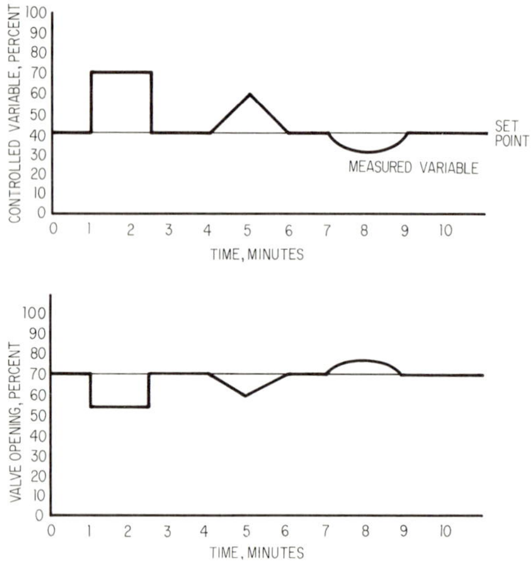

Fig. P-72. Effects of measured-variable deviations from set point upon control-valve position with proportional control.

150 percent, or gain of 0.66 (proportional band = 100/gain). For a proportional band of 100 percent, or gain of 1.0, the valve would move an amount equal to the deviation of the measured variable from set point in percent.

[1] Extracted from American National Standard CE85.1-1963, American Standard Terminology for Automatic Control, with permission of the publisher, The American Society of Mechanical Engineers, New York.

The proportional band usually covers an adjustable range of 1 to 400 percent (gains from 100 to 0.25). Proportional bands greater than 100 percent cannot cause the final control element to travel full scale and therefore cannot completely turn off the supply of energy or material to the process regardless of how far the measured variable deviates from the set point. As shown by the figure, at 2½, 6, and 9 min—when the measured variable returns to its set point—the valve also returns to its original value. Thus, it is obvious that there is no valve correction without a deviation existing between measured variable and set point. From 1 to 2½ min, the measured variable is at a constant value with a constant valve position. The deviation of the measured variable from set point under this condition is called offset (droop) and is a characteristic of all proportional controllers. Proportional controllers can produce an exact correction for only one load condition. With all other loads, there must be some deviation of measured variable from set point. See **Offset (droop).**

It is further shown by Fig. P-72 that when the measured variable rises above the set point, the valve opening is reduced, while when the measured variable deviates below the set point, the valve opening is increased. This situation is known as reverse action. Direct action occurs when an increase in measured variable over the set point causes an increase in valve opening. Most controllers can be changed from direct to reverse action depending upon the process requirements. Also see **Electric controller** and **Pneumatic controller.**

—Richard H. Kennedy, *Bristol Division, American Chain & Cable Company, Inc.*

Proportional Counter A radiation counter that utilizes a proportional-counter tube. The radiation-counter tube is operated at voltages high enough to produce ionization by collision and is adjusted so that the total ionization per count is proportional to the ionization produced by the initial ionizing event. See **Gas-ionization tube; Geiger counter;** and **Ionization chamber.**

The proportional-counting region is distinguished by the fact that the voltage pulse produced is proportional to the primary ionization. This makes it possible to differentiate between alpha and gamma or alpha and beta radiation with the electronic equipment that follows the detector. Slow neutrons are thus counted above a background of gamma radiation in a counter tube filled with boron trifluoride gas. Here the pulses produced by alpha particles from the neutron-boron reaction in the gas are much larger than the pulses produced in the gas from gamma radiation. Beta proportional counters are widely used because of their short resolving time (which permits high-speed counting) and their long life.

When counting beta radiation in the proportional region, the output-voltage pulse may vary from a fraction of a millivolt to a fraction of a volt in the same counter because of the energy distribution of the beta spectrum. If the counter is to have a plateau, a region from 100 to 500 V long where the counting rate rises only 1 to 5 percent with this increase of voltage, the electronic equipment used must be able to accept pulses over a large range in amplitude.

Proportional counters are constructed with a variety of shapes and configurations, of which the most common is the cylindrical shell with a fine axial wire from 0.001 to 0.004 in. in diameter. This fine wire is necessary to obtain field strength near the wire sufficient for electron multiplication. The shell usually is operated at ground potential and the central wire at positive high voltage. Glass-envelope counters (with a thin film of metal deposited on the inner surface) may be used; or metal wall counters may be constructed, using glass-to-metal seals to insulate the central wire. These insulators must be clean, or they will produce high-voltage leakage pulses that are of the same magnitude as the counting pulses. If the counter is evacuated, then filled with gas and sealed, some of the plastics are not acceptable as insulators because of their high vapor pressure.

The proportional counter may be used as a flow counter with the counting gas continuously passing through the counter. This technique is used where solid or liquid radioactive samples are placed in the counter for counting. A radioactive gas also may form part of the fill gas of a counter and thus be counted with no attenuation due to wall thickness. Mixtures, such as argon plus 10 percent methane or argon plus 10 percent carbon dioxide at atmospheric pressure, operate in the proportional region in the neighborhood of 2,000 V for a counter whose axial wire is in the order of 0.001 to 0.002 in. in diameter and whose cathode (shell) is ¾ to 1 in. in diameter. The operating voltage depends on the wire size, cathode size, gas mixture, total gas pressure used, and the electronic counting equipment. Also see **Radiation (nuclear), detection of.**

Proportional-input Control Control action where there is a predetermined relation between the value of the controlled variable and the time average of a two-position final controlling element. This type of control provides proportional average-position action that achieves close to an exact correction of the process input for various conditions of load. This form of control can produce essentially noncyclic (so-called straight-line) control of many processes that are too difficult to permit the use of a simple two-position controller.

Percentage "time on" of the final controlling element is dependent upon the value of the controlled variable in proportion to the deviation of the variable within the control band from the set point. For example, assume a controller with a scale range of 0 to 1000°F and a set point adjusted to 500°F on the scale (see Fig. P-73). The maximum control bandwidth that could be obtained

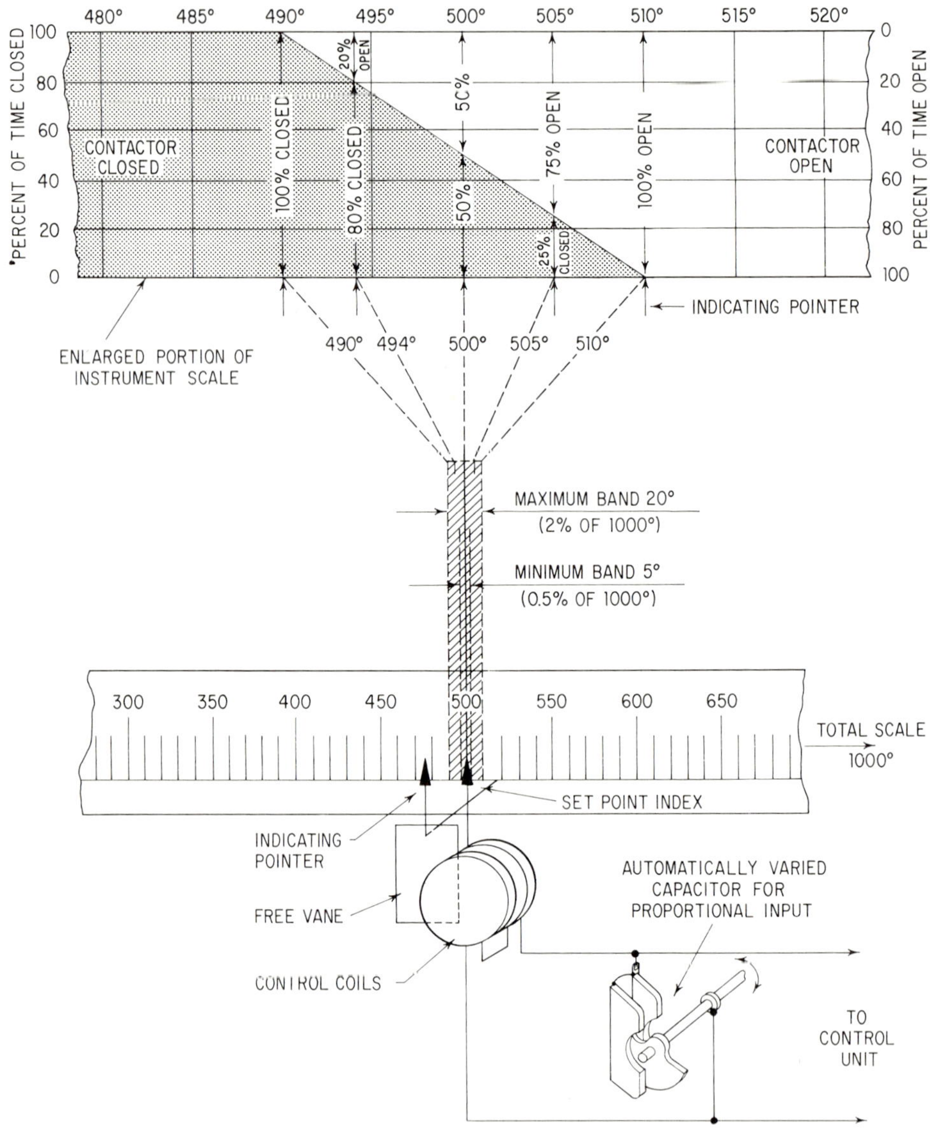

Fig. P-73. Proportional-input control.

would be 2.0 percent of the full-scale range of 1000°F, or 20°F. Half of this control band would be above the set point, half below the set point—so that the band would extend from 490 to 510°F. The minimum control bandwidth that could be obtained would be 0.5 percent of 1000°F, or a control band 5°F wide, extending from 497.5 to 502.5°F. These maximum and minimum bandwidths are shown on the figure by the cross-hatched areas on the scale.

Assume that the controller was regulating a temperature-control process by opening and closing an electric contactor and that the maximum bandwidth of 2 percent was being used. Assume that heat is being added when the contactor is closed. The diagram shown on top of the figure shows the effect of the pointer and vane position on the percentage of time open to time closed of the electric contactor. The shaded portion of the diagram represents contactor-closed time; the white area, open time. When the temperature is at or below 490° F (the lower edge of the band), the contactor will be closed 100 percent of the time. When the temperature is 500° F (set point), the contactor will be open 50 percent and closed 50 percent of the time. When the temperature is at or above 510°F (upper edge of the band), the contactor will be opened 100 percent of the time. When the temperature is at any other value within the control band, the percentage of contactor-open time to contactor-closed time will be proportional to the position of the pointer within the band.

Proportional-input control is suitable for single-capacity processes having slow reaction time, little transfer lag or dead time, and slowly applied load changes. Many electrically heated furnaces and ovens are processes of this type. Also see **Electric controller.**

—Richard H. Kennedy, *Bristol Division, American Chain & Cable Company, Inc.*

Proportional-plus-Derivative Control Control action in which the output is proportional to a linear combination of the input and the time rate of change of input,[1] i.e., where there is a continuous linear relation between rate of change of the controlled variable and position of the final controlling element. Proportional-plus-derivative control action is shown by

$$m = K_c e + K_c T_d e' + M$$

where m = manipulated variable
K_c = proportional sensitivity
e = deviation
e' = rate of change of e with respect to time = de/dt
T_d = derivative time
M = constant

The relationship of the controller output to proportional-only and to proportional-plus-derivative control to a linear or ramp change of input as shown in Fig. P-74. Note that the control output with derivative action leads in time the control output with proportional action only—by the amount of derivative time T_d. Derivative time is the amount of lead, expressed in units of time, that the control action is advanced over the proportional-only control action.

Since derivative control is a function of the rate of change of the error, it responds to noise on the input signal, such as occurs when measuring liquid flow across an orifice. Therefore, to avoid high-frequency instability in the controller, a derivative gain limit is introduced. This usually is on the order of 10 or 20.

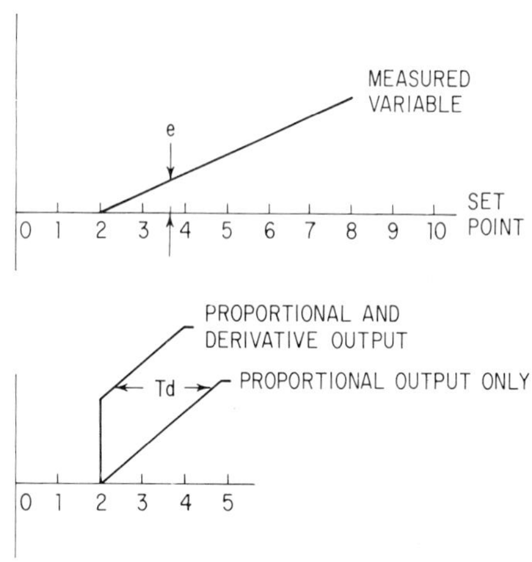

Fig. P-74. Response of a proportional-plus-derivative controller to a ramp change in input.

The response of a proportional-plus-derivative controller to a step change in input with a derivative gain limit of 10 and a proportional band of 50 percent (gain equals 2) is shown in Fig. P-75. Note that for a unit change in measured variable equal to e, the output change is 10 times e and

[1] Extracted from American National Standard CE85.1-1963, American Standard Terminology for Automatic Control, with permission of the publisher, The American Society of Mechanical Engineers, New York.

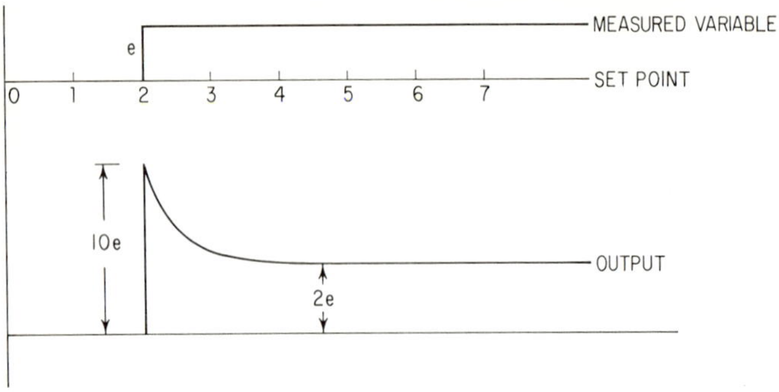

Fig. P-75. Response of a proportional-plus-derivative controller to a step change in input.

then decays to the output dictated by the proportional band of 50 percent or 2 times e. The time constant of the decay from $10e$ to $2e$ is equal to the derivative time T_d.

—Richard H. Kennedy, *Bristol Division, American Chain & Cable Company, Inc.*

Proportional-plus-Integral Control Control where the output is proportional to a linear combination of the input and the time integral of the input.[1] The combination of proportional-plus-integral control combines the advantages of both proportional and integral control. Integral control also is known as reset control. Also see **Integral (reset) control.**

Proportional action provides a continuous linear relation between values of the deviation of the measured variable from the set point and is expressed by

$$m = K_c e + M \tag{1}$$

where m = manipulated variable
K_c = proportional sensitivity (gain)
e = deviation of measured variable from set point
M = constant

The constant M is a bias that provides midscale output of the manipulated variable when e is zero.

Reset or integral control provides that the value of the manipulated variable m is changed at a rate proportional to the deviation of the measured variable from the set point. If the deviation is tripled over a previous value, the final controlling element is moved three times as fast. When the measured variable is at the set point ($e = 0$), the final controlling element remains stationary. Integral control is expressed by

$$m = \frac{1}{T_i}\int e\,dt + M \tag{2}$$

where m = manipulated variable
T_i = integral time
e = deviation of measured variable from set point
M = constant of integration
t = time

The integral time T_i is the time of change of manipulated variable caused by unit change of deviation.

Proportional-integral control combines Eqs. (1) and (2) to give

$$m = \frac{K_c}{T_i}\int e\,dt + K_c e + M \tag{3}$$

where K_c = proportional sensitivity (gain).

In controllers with proportional-plus-reset action, both the gain K_c and integral time T_i are adjustable. The integral time usually is expressed as reset rate "repeats per minute" (inverse of integral time in minutes per repeat) and is defined as the number of times per minute that the proportional part of the response is duplicated.

The top portion of Fig. P-76 shows the effects upon the measured variable by a step change in set point at 1½ and 4 min. The valve action resulting from the step change in set point is shown to be made up of the proportional response plus the reset response. When the step change in set point of 20 percent is made, there is a step change in valve action due to proportional response of 40 percent showing that the gain setting K_c is equal to 2. Since the measured variable is not at the set point at the end of the proportional response, reset action takes place and continues to change the valve opening until such time as the measured variable equals the set point ($e = 0$). Note from the figure that in 1 min the reset response changes the valve opening by 20 percent. Therefore, the integral time T_i is equal to 2 min per repeat, or 0.5 repeats per minute. The same action is shown upon the return of the set point to the original setting at 4 min.

Since the reset action will continue to change the output of the controller until the error signal is zero, the action eliminates the offset (droop), which is characteristic of proportional-only action. See **Offset (droop).**

A serious disadvantage of reset action occurs in

[1] Extracted from American National Standard CE85.1-1963, American Standard Terminology for Automatic Control, with permission of the publisher, The American Society of Mechanical Engineers, New York.

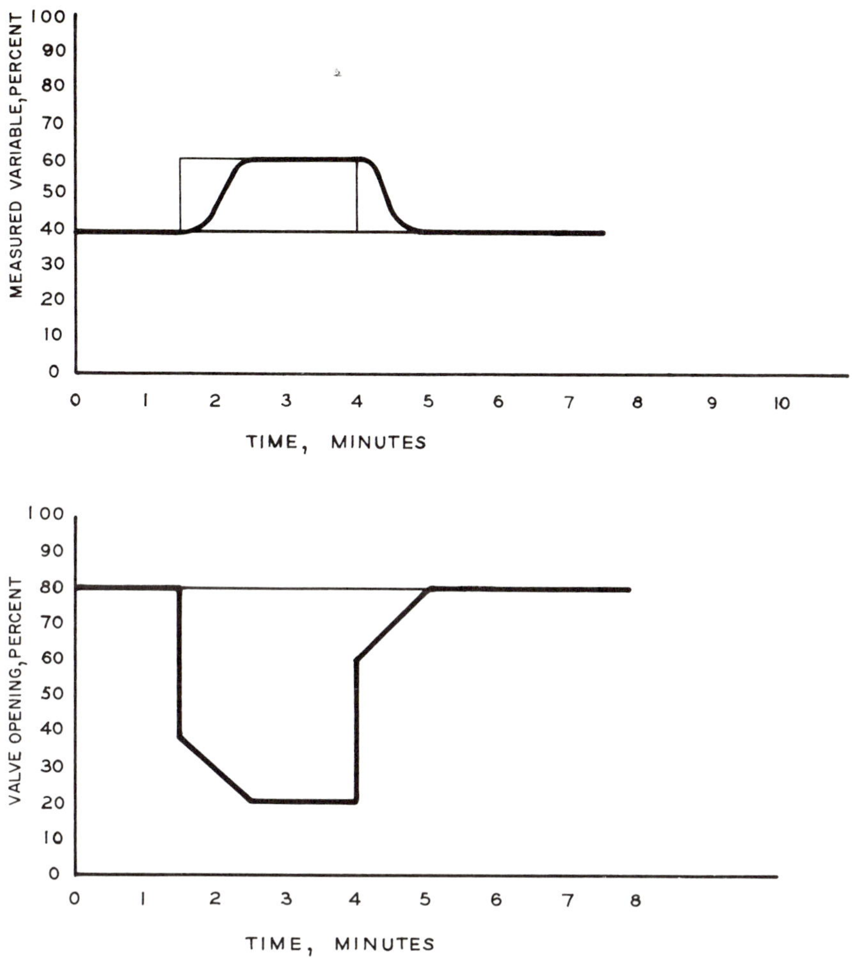

Fig. P-76. Effects of measured-variable deviations from set point on control-valve position with proportional-plus-integral control.

automatic start-up of processes under control. Reset action results in large overshooting above the set point by the control variable—due to reset wind-up or saturation. Under automatic start-up with the measured variable initially at zero, the controller output will be at a maximum with maximum error signal between measured variable and the set point. The output will remain at maximum until the measured variable increases and crosses the set point. Thus, the valve will not begin to close until the measured variable crosses the set point because of reset action. This same action can occur with any large, long-sustained load change where the measured variable deviates from the set point for long periods.

To overcome the aforementioned effects, it is common to add rate or derivative action to the input of the proportional-plus-integral controller. Since the derivative or rate action is an anticipating type, the reset action begins desaturating before the measured variable reaches the set point.

—Richard H. Kennedy, *Bristol Division, American Chain & Cable Company, Inc.*

Proportional-plus-Integral-plus-Derivative Control[1] Control action in which the output is propor-

[1] Extracted from American National Standard CE85.1-1963, American Standard Terminology for Automatic Control, with permission of the publisher, The American Society of Mechanical Engineers, New York.

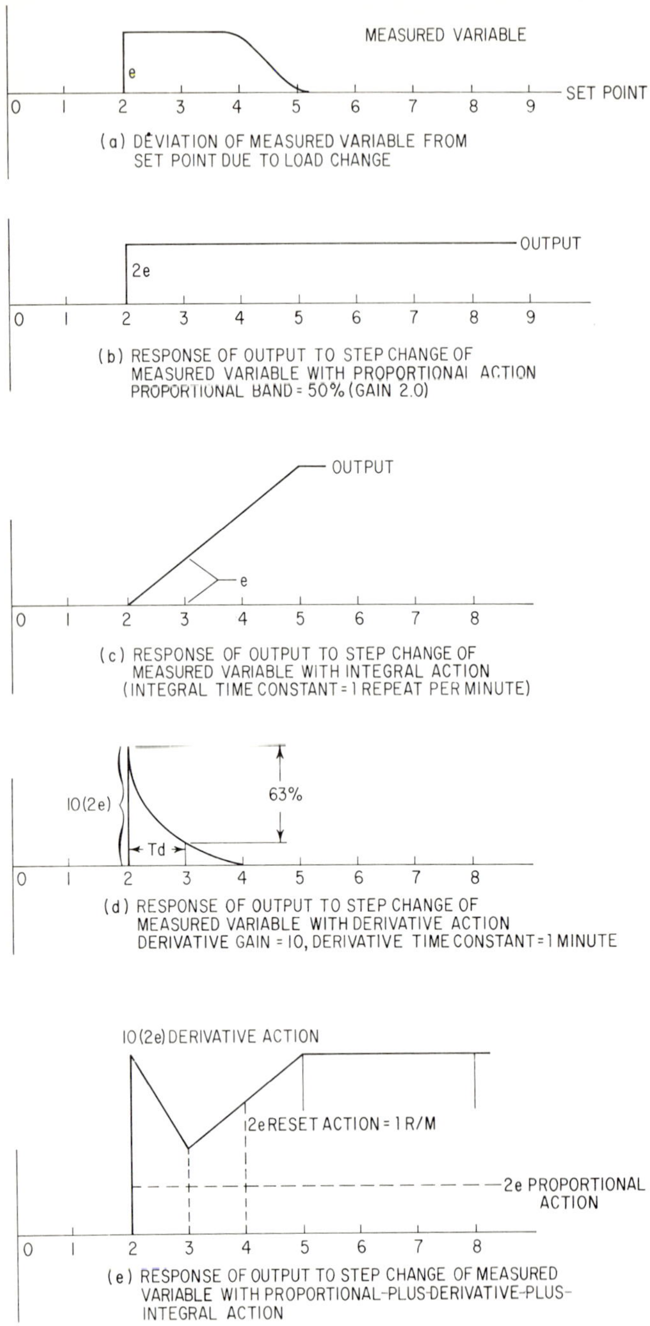

Fig. P-77. Response of input under proportional-plus-derivative-plus-integral control action (e) following a step change in the measured variable (a) with separate responses due to proportional, integral, and derivative actions shown by (b), (c), and (d), respectively.

tional to a linear combination of the input, the time integral of input, and the time rate of change of input. The mode of control is the additive combination of proportional action, integral action, and derivative action and is defined by

$$m = \frac{K_c}{T}\int e\,dt + K_c e + K_c + T_d e' + M$$

where m = manipulated variable
K_c = proportional sensitivity
T_i = integral time
e = deviation
e' = rate of change of deviation with time, de/dt
T_d = derivative time
M = constant

The response of a proportional-plus-integral-plus derivative controller to a step change in measured variable is shown by Fig. P-77. The composite response curve e can be observed by the action of the output indicator on any controller after a step change is made in the set point. Note from the figure that the proportional action changes the output by an amount proportional to the deviation and thereby produces a permanent increased energy or material input to the process. The derivative action corrects the output by an amount proportional to the rate of change of the control variable. Integral action changes the output at a rate proportional to the deviation.

The result of the output, curve e, shows that at first an excess is applied to oppose the change of the variable away from the set point. This excess energy or material to the process then decreases proportional to the derivative time, but since the measured variable still is not at the set point, integral action continues to change the output at the rate proportional to the integral time.

The combination of these actions brings the measured variable back to the set point at 5 min time when the output stops changing and holds steady because the error signal e becomes zero. The foregoing equation and Fig. P-77 assume that there is no interaction between the integral and the derivative action. However, whenever integral and derivative actions operate successively on a signal, the two modes interact—since integral action produces a rate of change of output and the derivative action reacts to a rate of change of output. Also see **Electric controller** and **Pneumatic controller.**

—Richard H. Kennedy, *Bristol Division, American Chain & Cable Company, Inc.*

Proportioning Pump (See **Metering pump.**)

Prosthetic Devices (See **Medical instrumentation.**)

Protecting Tube, Sensor The sensor protecting tube is a broad term covering tubular protection of the sensing element of a wide variety of measuring devices. Protecting tubes range from the thermal well, commonly encountered in the process industries, to perforated tubes that provide mechanical protection to nylon humidity elements. The protecting tube is designed to protect the sensor from one or more of the following hazards and thus assure accurate and economic use of the sensing element: (1) pressure effect, (2) corrosive attack, (3) contamination, (4) impingement effect, and (5) electrical effect.

Protective Devices, Control-system The function of protective instrumentation is to recognize a condition that may become hazardous and to act to eliminate that condition before it can cause damage to equipment or injure personnel. The devices may be installed either to protect unattended equipment, or because detection and correction of a critical condition must be made too quickly for human accomplishment. Some alarms also serve as protective devices. See **Alarms, control-system.**

Alarm and protective devices often are arranged to operate in sequence. The alarm is set to alert operating personnel prior to operation of the protective device. It is advisable to keep protective devices independent of alarms and other instrumentation to provide for maximum and reliable protection and to avoid inadvertent actuation of the protective devices. Full or partial shutdown of a plant due to a faulty protective device (operating when not justified) can be very costly in terms of lost production and subsequent start-up procedures.

Proving Ring An instrument, in the form of a ring of precisely known diameter, used for determining the accuracy of force-measuring and weighing devices. When the ring is subjected to compression or tension loads, the ring changes its diameter—known as "ring deflection." A Morehouse compression proving ring is shown in Fig. P-78. Forces are applied to ring *A* through the integral external bosses *B* and *C*. The resulting deflection is measured with micrometer screw *D* and vibrating reed *E* which are attached to the integral internal bosses *F* and *G*. By means of graduated dial *H* and an index pointer *I*, ring deflection is measured in micrometer dial divisions. Micrometer sensitivity is 0.000004 in. Rings with tension fittings also are available. Proving rings with capacities ranging from 1,000 to 1,200,000 lb are available. Also see **Force transducer, proving-ring.**

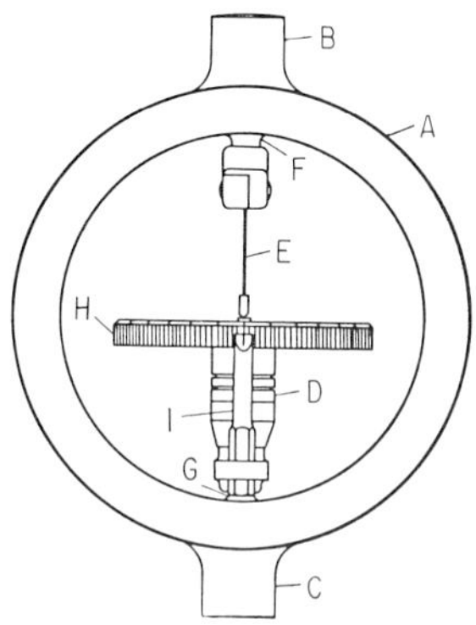

Fig. P-78. Assembled Morehouse proving ring for measuring compressive forces.

Proximity Switch A proximity switch makes or breaks an electric circuit when the presence of an object is sensed. No physical contact between the object and the proximity switch is required. Six operating principles are employed in various designs: (1) the change in inductance of a coil excited with 1,000-Hz or lower-frequency alternating current, which occurs when a ferromagnetic object is brought close to the coil; (2) eddy currents, induced in a metal when it is near the coil excited with a radio frequency, change the loading on the coil; (3) where a metallic plate is excited with radio frequency, the capacitance to ground can be changed by the presence of objects that have a dielectric constant different from air; (4) the field around a permanent magnet can be changed by the presence of a ferromagnetic object near the magnet; (5) the amount of light falling upon a photocell can be altered by any material with optical characteristics different from those of air; and (6) ultrasonic sound can be used much like a light beam.

Psychrometer An instrument used for the precise measurement of relative humidity, dew point, and vapor pressure. These instruments operate on the temperature-differential principle and normally utilize a matched pair of thermometers. The bulb of one thermometer is exposed directly to the air and measures the ambient temperature. The bulb of the second thermometer is encased in a water-wetted wick and is cooled by evaporation. The amount of cooling is a function of the amount of water vapor in the atmosphere and the temperature of the atmosphere at a known airflow rate of 600 ft/min or more. When the wet-bulb temperature and the temperature differential are known, the relative humidity, vapor pressure, and dew point can be computed with a psychrometric slide rule or found by reference to charts supplied with the instrument.

The battery-operated psychrometer, as shown in Fig. P-79, uses a small 4½-V motor-driven aspirator that is powered by three *D*-size flashlight cells to provide a steady, precise airflow to the thermometers. The thermometers normally are calibrated in 1° graduations, either Fahrenheit or Celsius. The minimum readable values are 0.2° F or 0.1° C. Response time of such units is 1½ to 2 min.

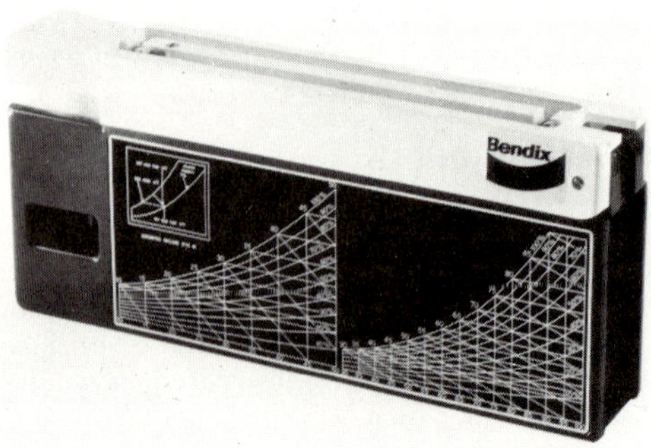

Fig. P-79. Battery-operated psychrometer. *(Bendix.)*

A hand-aspirated psychrometer is shown in Fig. P-80. A rubber hand pump and pressure bellows supply air at a constant rate through a venturi tube located between the thermometers. The psychrometric slide rule is an integral part of this instrument.

Sling Psychrometer: This is a simple yet very reliable instrument for measuring humidity. The device consists of a matched pair of thermometers mounted on a suitable scaled frame attached to a thong or handle. The bulb of one thermometer extends below the other and is enclosed in a small cloth pouch called a wick. When an observation is made, the wet-bulb wick is moistened with water (usually distilled) and the thermometers are whirled rapidly for approximately 20 s and then read quickly. This operation is repeated several times until two successive readings indicate that the temperature of the wet bulb is at its lowest point. The wet- and dry-bulb temperatures, thus observed, make it possible to determine percent humidity from a psychrometric chart. Some sling psychrometers

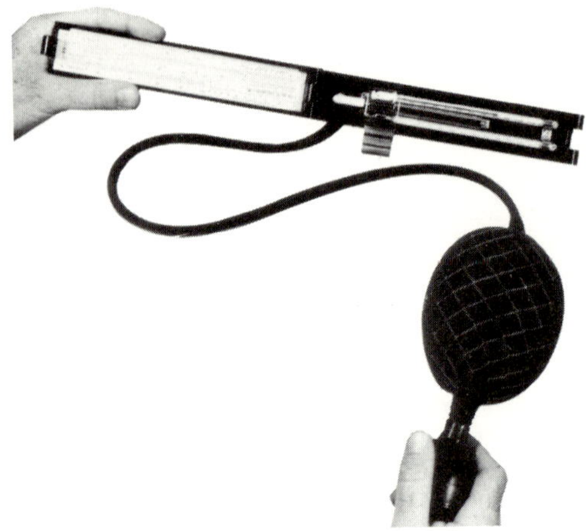

Fig. P-80. Hand-operated psychrometer. *(Bendix.)*

are constructed to utilize a hand crank and gears for whirling the thermometers.

—James R. Cosby, *Environmental Science Division, The Bendix Corporation*

Pulmonary System, Instrumental Analysis of The pulmonary system consists of the lungs and the right heart, which accepts the returned blood from the venous system, pumps it into the lungs, through the pulmonary artery (where it becomes reoxygenated and releases CO_2), and then returns it to the left heart through the pulmonary vein to be pumped by the left heart into the arteries of the body. To study the functioning of this system, one must measure the oxygen level and pH of the blood, the expired CO_2, the amount of oxygen in the lungs, the rate of blood flow through the lungs, and the rate at which air moves into and out of the lungs (inspiration and expiration). pH meters and gas analyzers used in general instrumentation are also used in this field (Ref. 1) and are described elsewhere in this encyclopedia. For blood-flow measurement, see **Cardiovascular instrumentation.**

Some measurements are not easy to make. In the intact animal it is difficult to get a catheter into the pulmonary artery, because this requires threading it through the heart. Volume airflow measurement is made difficult by the need to keep very low air resistance in the measurement line.

Air consumed per unit length of time (known as minute volume) is usually measured by letting the subject breathe through a mouthpiece connected by a 2- to 3-in.-diameter flexible short tube to a storage container known as a spirometer. The spirometer consists of a cylinder closed at the top, the open bottom of which rests in a pan of water the same height as the cylinder. Air is withdrawn from the upper cylinder through a tube that passes through the bottom and rises vertically through the water to open above the water surface. The weight of the upper cylinder is counterbalanced by a weight-and-pulley arrangement attached to the upper cylinder so that the air inside will remain exactly at atmospheric pressure. As the air is withdrawn, the upper cylinder falls, and the distance that it has fallen is a measure of the volume of air that has been used.

The patient's mask contains two check valves made of very flexible rubber, which allows inspired air to come from the spirometer and forces expired air through a different outlet tube. This expired gas can then be routed through a CO_2 analyzer when desired.

In another frequently employed technique a fine-mesh screen is introduced into the airstream, and the pressure drop across it (a few inches of water) is measured with a differential-strain-gage pressure transducer (see direct blood-pressure measurement under **Cardiovascular instrumentation**). The wire-mesh screen is heated slightly to prevent condensation.

For monitoring an acutely ill patient, for studying psychological stress, and for some other research purposes, it is frequently desirable to measure respiration rate. This can be done without recourse to a spirometer by several simple but less reliable techniques.

The same impedance-measuring process used in plethysmography and blood-flow measurement can also be used to infer the amount of air in the lungs (Ref. 2). For this purpose, a pair of electrodes is placed at about the sixth interspace along the left and right mid-axillary lines (about 5 in. below the armpits). A 100-mA 40-kHz signal passed between the electrodes will vary in intensity by about 1 percent with each breath; this is ample for extraction of respiration rate. This procedure is not satisfactory for measurement of the actual volume of air in the lungs because the electric field does not flow uniformly across the lungs but is concentrated by other, higher-conducting structures around the lungs. The system is also subject to motion artifacts, and some computing function must be built into the circuitry to discriminate against coughing, if this is desired.

A quite simple technique utilizes a heated thermistor placed in front of the nostrils and mouth. Temperature of the thermistor will depend upon the rate of airflow past. It will not easily discriminate between inspiration and expiration and is, of course, subject to artifacts caused by general air movement in the vicinity or rapid movement of the patient's head, to which it must be attached.

The third common procedure is to place a band around the chest and note its motion as the chest expands and contracts. The band may be either a mercury strain gage, described under **Cardiovascular instrumentation,** or a thin nylon cord connected to a spring-loaded switch. This technique has the virtues of simplicity and ruggedness, but it is quite susceptible to motion artifact, particularly if the subject is lying in bed.

REFERENCES

1. Dawes, M. J. R., and D. R. Garbe: Criteria of Measuring Tests of Lung Volume and Ventilatory Capacity, *World Medical Electronics,* vol. 3, no. 6, pp. 187–196, September, 1965.
2. Geddes, L. A., and L. E. Baker: "Principles of Applied Biomedical Instrumentation," chap. 10, pp. 150–205, John Wiley & Sons, Inc., New York, 1968.

—J. Malvern Benjamin, Jr., *Bionic Instruments, Inc.*

Pulse-code Modulation Modulation (PCM) in which the peak-to-peak amplitude range of the signal to be transmitted is divided into a number of standard values each having its own three-place code. Each sample of the signal then is transmitted as the code for the nearest standard amplitude.

Information messages of many types, including instrument readings, television pictures, speech, and tables and columns of figures, can be transmitted over a general-purpose communication network in the form of closely spaced pulses instead of as continuous waves which vary in amplitude or frequency. In digital communication, information content is communicated by means of a sequence of numbers. Electrical pulses (termed pulse codes) represent the numbers (termed code words). The code words normally are described in terms of binary numbers and thus in the form of a sequence of 0's and 1's. This, of course, is the same language employed by modern digital computers. Although not limited to binary digits, the latter can be expressed by the simple absence or presence of an electrical pulse. The availability of fast-switching solid-state devices has made PCM techniques practical.

A majority of communications traffic is comprised of speech and television pictures. The originating signals in these cases are analog. Originating instrumentation data signals, of course, may be analog or digital. Where conversion of an analog signal into a digital signal is required, a pulse-code modulator must be used. A device of this type samples and measures the analog-signal voltage. This is done at frequent intervals. The pulse-code modulator also assigns a numerical value to each sample. Provided that the analog signal is sampled at a rate no less than twice the component of highest frequency in the signal, no information will be lost. Insofar as voice signals are concerned, the sampling frequency required approximates 10,000 times per second; for television signals, the rate is approximately 10 million times per second.

In converting analog signals, some rounding off is required. If a sample value of 2.787 V is measured, it may be encoded to the binary equivalent of 2.8. The 0.013-V difference is known as the *quantizing error*. See **Encoder.** The converter changes an analog signal of bandwidth W (pps) into a pulse stream that contains $2mW$ pulses per second. PCM transmission needs approximately $2m$ times greater bandwidth than the original analog signal. Thus, it is observed that PCM requires greater bandwidth than conventional AM and FM modulation systems. However, this disadvantage of PCM is countered by several other advantages that favor digital transmission.

The binary numbers, when they arrive at the receiver, need to be converted back to their original analog format. The train of pulses is passed through an electrical filter. The low frequencies that are below half the sampling frequency are passed by the filter. High frequencies are practically eliminated. Thus, with exception of the quantizing error, the received signal essentially is the same as the original analog signal that was put into the pulse-code modulator.

In the present state of the art, PCM transmission generally is by means of pairs of wires of a telephone cable. The cable per se may embrace up to 900 pairs. The wire pairs present tend to attenuate the digital signal, and thus the system would be confined to rather short distances were repeaters (amplifiers) not used at various locations along the path of the cable so that the signal strength can be restored. Repeaters may be located on telephone poles or underground. The practical distance between repeaters is approximately 6,000 ft. Thus, on a long system, the requirement for components of high reliability, small power consumption, and small size is apparent. In this regard, solid-state component technology has made an invaluable contribution to PCM schemes.

The design of the repeater used with PCM systems, as contrasted with the high-quality amplifier-type repeaters used in conventional AM systems, is simply a circuit that is configured to sense whether or not an electrical impulse is present. Thus, there is no amplification of noise of all types along with the signal, such as may be gathered by the cable, or cross talk or thermal noise. Some noise will not alter the fact that a pulse is present; noise in the absence of a pulse must be quite significant before it may be interpreted as a true pulse. A repeater of this type is termed a *regenerative repeater.* Sometimes a repeater may add an extra pulse or omit a true pulse, but if the error rate does not exceed one error in one million pulses, the signal alteration usually goes beyond detection. An error rate of this magnitude is commonplace in circuits that may contain a thousand or more repeaters in tandem.

To date, PCM schemes have found their major applications in missile and space systems. As pointed out in Ref. 1, the primary advantages of PCM systems include (1) capability for handling both analog and digital signals, (2) flexibility as to numbers of channels and channel sampling rates, (3) capability of transmitting data of high accuracy with little or no degradation of the rf link, and (4) generally superior characteristics of information efficiency and noise immunity in the rf link. Offsetting these advantages is the fact that the complexity of a PCM system varies little with the number of channels; and thus although the complexity of a PCM system compares favorably when the number of input channels is large (> 50), for systems with small numbers of input channels and moderate accuracy requirements, analog systems may be desirable for reasons of simplicity. Also see **Data communications.**

REFERENCES

1. Gruenberg, Elliot L. (ed.): "Handbook of Telemetry

and Remote Control," McGraw-Hill Book Company, New York, 1967.

2. Harman, Willis A.: "Principles of the Statistical Theory of Communication," McGraw-Hill Book Company, New York, 1963.
3. Perkins, F. A.: The Effect of Bit and Group Synchronization on the Reception of PCM/FM Telemetry Signals, *Proc. Natl. Telemetry Conf.*, June, 1964, Los Angeles.
4. McRae, D. D., and F. A. Perkins: PCM Synchronization and Demultiplexing Study—Final Report, George C. Marshal Space Flight Center, Contract NAS 8-5111, Radiation, Inc., June 14, 1963.

Pulse-duration Telemetry Pulse-duration modulation (PDM) is a common form of telemetry used for the transmission of instrumentation data. Transmission may be by wires or by a radio link. The time durations of the received pulses are measured to obtain a form of the original data. Pulses also may be stored on magnetic-tape link.

Creation of the pulses can take several variations. In one industrial transmitter,[1] an opaque, characterized silhouette is photographically reproduced on the surface of an otherwise transparent glass disk revolving at 30 rpm between arms projecting from the input shaft. A light source on one arm beams a sharply focused light through the disk at a photoconductive silicon diode mounted on the opposite arm. The light strikes the diode except when the arms are positioned so that the opaqued area on the disk interrupts it. When the light beam is interrupted, the photodiode exhibits high resistance and no current flows in the transistor amplifying circuit. When the light beam is not interrupted, the photodiode appears as a low resistance to the circuit which passes a flow of current through a coil to "make" the contacts of a hermetically sealed, dry-reed-type switch. Duration of the "make" period, always some portion of the system's 2-s cycle time, is directly proportional to the position of the input shaft and therefore to the measurement.

[1] Foxboro Electronic Telemeter.

Appearing once every 2 s, the pulses are converted at the receiver to a voltage whose magnitude is proportional to pulse duration and therefore to measurement. Basis of this conversion is measurement capacitor C_1, which charges during the period when transmission circuit is energized, then transfers charge through a relay to "memory" capacitor C_2 (see Fig. P-81). Potential on C_2, in turn, is compared with that from a potentiometer operated by the servomotor. The difference in potential (error signal) is fed to a shunt-diode chopper in the input of the transistor amplifier. The resulting 60-Hz square wave is amplified and fed to the phase-sensitive coils of the servomotor, which drives the pen and repositions the potentiometer to balance out the error signal.

Remote Index Adjustment: A common application of PDM in industrial instrumentation is that which permits an operator, located at a central site, to adjust the set point in a telemetering transmitter many miles away. Examples include the pipeline industry, where suction and discharge pressures are transmitted from unattended stations to a central location from which the con-

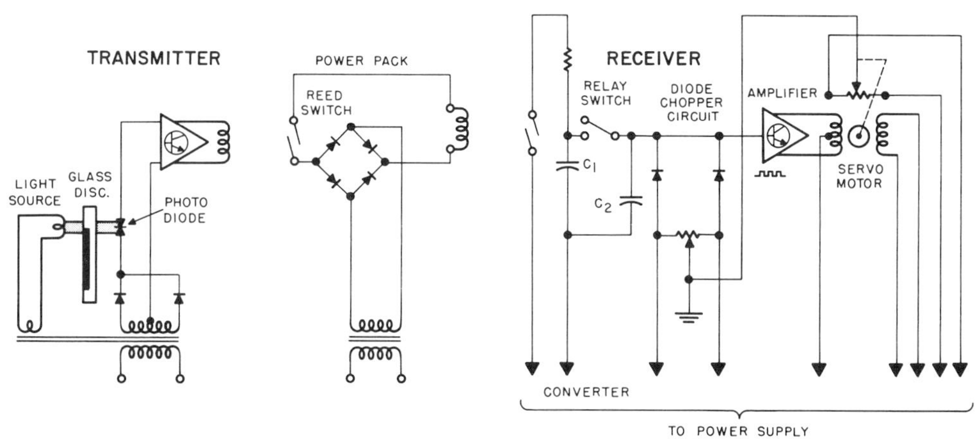

Fig. P-81. Solid-state transmitter/receiver system (Foxboro Electronic Telemeter).

trol indexes are adjusted; or oil production, where gathering systems require adjustments over a considerable geographic area; and gas-distribution systems.

When remote index adjustment is used, the variable at the remote site must be under automatic control. Communications must thus be established in both directions between the remote and central locations. Such communication can be accomplished in several ways:

1. Through separate wire pairs, one for each control index signal. As many paths as required can be used.

2. Through ac/dc duplexing and filters. A variable can be sent via ac signal and its control index adjustment returned via dc signal over the same wire pair.

3. Through latch and reversing relays and time-multiplexing devices. Remote index setting or quantity adjustment may be combined with other variables over a single wire pair.

4. Through tone transmitters and tone receivers. Using one pair (transmitter and receiver) for each signal, up to approximately 20 signals can be transmitted over a single wire pair.

5. If more than 15 signals are necessary, a combination of tone and time multiplexing may be the most economical method of transmitting all signals over one wire pair.

A remote-control index adjustment requires two channels of communication. The first is for the variable being controlled. Here the flow of information is from the remote site to the central location. The second is for the signal which adjusts the control index. This information travels from the central location back to the remote site. Although these schemes are called remote-control index-adjustment systems, they can be used to operate valves remotely or for other related purposes where a position is manually set.

Radio Links: Although applicable to industrial applications, radio links for use with PDM systems have been used most widely in connection with airborne instrumented-flight vehicles. In such applications, the data-transmission rates are many times greater than required by industrial uses. Commutation rates, for example, may be as high as 900 samples per second. As pointed out in Ref. 1, the application of PDM-FM equipment has been strongly influenced by standards developed some years ago based largely upon mechanical commutation limitations. It is reasonable to assume that PDM systems of the future will be extended to include much higher information capacity with many improvements in performance. Following is a condensed version of PDM-FM standards:

Specifications for the Pulse-duration-modulated Signal

No. of samples/frame	30	45	60	90
Frame rate, frames/s	30	20	15	10
Commutation rate, samples/s	900	900	900	900

The number of samples per frame available to carry information is two less than the number indicated because the equivalent of two samples is used in generating the frame-synchronizing pulse. Commutation rate is equal to frame rate multiplied by the number of samples per frame. The commutator speed (or frame rate) shall not vary more than plus 5 to minus 15 percent from nominal. Frame synchronization of the receiving station shall be provided for by leaving a longer than normal gap time in the train of pulses transmitted. This gap shall be the same as that normally occupied by two successive data-channel pulses. A representation of the pulse-train waveform is shown in Fig. P-82.

The information being transmitted in each channel shall determine the duration of the corresponding pulses. The relation between information magnitude and pulse duration should, in general, be linear.

Min pulse duration (zero-level information), μs	90 ± 30
Max pulse duration (max-level information), μs	700 ± 50
Pulse rise and decay time (measured between 10 and 90% levels), μs	10–20*

*Constant to 3 μs for a given transmitting set.

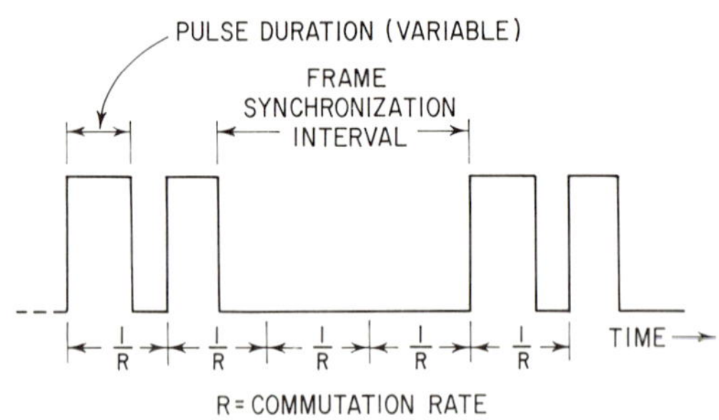

Fig. P-82. Standard Inter-Range Instrumentation Group format and framing method for PDM. *(Elliot L. Gruenberg, "Handbook of Telemetry and Remote Control," McGraw-Hill Book Company, New York, 1967.)*

The time interval between the leading edges of successive pulses within a frame shall be uniform from interval to interval and shall be constant within plus or minus 25 μs. In practice, channel-period scatter is allowed to exceed 150 μs, since this does not affect measuring accuracy and is more typical of electromechanical equipment.

The basic components of an FM link for PDM are shown in Fig. P-83. These include the PDM transmitter, the transmitting antenna, the receiving antenna, and the rf receiving system. In radio telemetry for vehicle instrumentation, the characteristics of the electronic equipment including a radio link are largely controlled by IRIG (Inter-Range Instrumentation Group) standards.

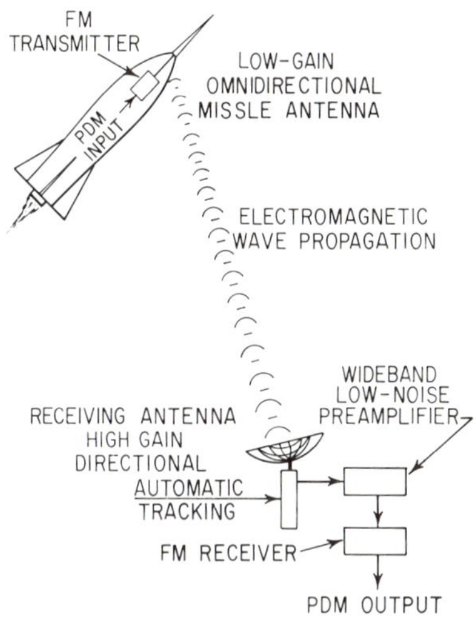

Fig. P-83. The basic FM link for PDM. *(Elliot L. Gruenberg, "Handbook of Telemetry and Remote Control," McGraw-Hill Book Company, New York, 1967.)*

Further, as pointed out by Ref. 1, conventional telemetering links for FM-FM telemetering are sometimes used to handle PDM data. For some applications, compensation is desirable to minimize tilt due to poor response at low frequencies. Since the signal is regenerated in PDM, considerable tolerance on rise time and tilt usually is acceptable. It is preferable to utilize an FM radio link having dc response from the modulator input to the receiver output. Under these conditions, the center frequency corresponds to the 50 percent or nominal slicing level of the wave train. If, on the other hand, the link lacks dc response in the modulator, the effective center frequency of the link will adjust to the average of the waveform. The output-signal center frequency would vary considerably as the channels are simultaneously varied from zero to full-scale condition. It would appear that a somewhat greater bandwidth is required under such conditions or the modulation must be restricted to a lesser portion of total normally available range of the link in order to provide for this shift. Slicing at the 50 percent level can be fixed by clamping the signal at the receiver output.

Major sources of error in PDM systems include impulse noise, random noise, synchronization errors, crosstalk, sampling and interpolation errors, and commutator noise. Means for reducing the effects of such errors is beyond the scope of coverage here. An excellent description of PDM systems for airborne-telemetry applications will be found in Ref. 1.

REFERENCES

1. Gruenberg, Elliot L. (ed.): "Handbook of Telemetry and Remote Control," McGraw-Hill Book Company, New York, 1967.
2. Riblat, H. B., and J. P. Randolph: "Satellite Telemetry and Data Processing," International Telemetering Conference, London, September, 1963.

Pulse Generator In the laboratory, the oscilloscope as the detector and the pulse generator as the signal source largely have replaced the voltmeter and sinusoidal oscillator. A pulse generator is a highly versatile and controllable switch. Two parameters must be controlled: (1) the pulse-repetition frequency, or switching rate; and (2) the pulse duration, or length of time the switch is closed (or open). Rise time or speed of switching also is important. In electronic equipment pulse-repetition rates range from nearly dc to over 100 MHz—and durations from seconds to less than 1 ns. For computation and data-transmission systems, most applications can be served by relatively low power outputs, while radar and certain magnetic data-storage systems may require pulses of extremely high energy. Because of these wide variations, a universal type of pulse generator is not practical.

Pulse Motor (See **Stepping motor.**)

Pulse-position Modulation The PPM telemetry system has been used in the United States mainly by the Navy. The system has not been selected as a test-range standard, but instead, receiving equipment is moved into the test ranges as required. There have been developments of high-capacity PPM-AM telemetry systems which have not been put into use in the United States. The PPM system is very similar to the PDM system (see **Pulse-duration telemetry**) except that instead of using the entire duration of the pulse, only the beginning and the end of the pulse are transmitted as intelligence markers, and a conversion from voltage amplitude to pulse position is used. PPM also is a time-multiplexed system, and equipments of up to 36 primary channels have been developed with a total sampling rate of 50,000 samples per second. Further information on PPM can be found in the listed reference.

REFERENCE

Gruenberg, Elliot L. (ed.): "Handbook of Telemetry and Remote Control," McGraw-Hill Book Company, New York, 1967.

Pump, Metering (See **Metering pump.**)

Pupillometry (See **Eye, instrumental measurements of.**)

Pyrometer Modern usage of the term pyrometer generally signifies a noncontacting temperature-measuring device, such as a radiation pyrometer or optical pyrometer—and no longer implies high-temperature-measuring capabilities only. Originally, a pyrometer was a device or system for measuring high temperatures, i.e., temperatures above the range of liquid-filled thermometers. A few decades ago, a thermocouple and its associated indicating instrument was called a pyrometer. Some portable or special-purpose instruments still retain this designation, i.e., diesel-engine pyrometer, which is a thermocouple-actuated millivoltmeter for measuring engine temperatures. See **Radiation pyrometer.**

—Ernest C. Magison, *Honeywell Inc.*

Pyrometer, Infrared (See **Infrared detector, thermistor.**)

Pyrometric Cone First developed in Europe and introduced to the ceramic industry in the United States in the late 1880s, pyrometric cones are used mainly to follow the temperature rise and to determine the finishing point of ceramic ware being fired in a kiln. These cones are manufactured as slender pyramids 1⅛ or 2½ in. high (base to peak). The cones are composed of raw materials similar to those employed in the manufacture of ceramics. In normal use, one or more pyrometric cones will be placed in the kiln—usually on specially constructed *cone holders.* In the kiln, the cones will be subject to the same heat treatment as the ceramic ware being fired. The softening and deformation of a specified pyrometric cone, observed through a sight hole in the kiln, will signal attainment of the desired heat process.

Although attempts have been made to assign definite temperature calibrations to the several available cones, such calibrations do not have precise meaning because the deformation temperature of a particular cone will depend upon the *rate and duration* of heating. A slow rate of heating will point to a lower temperature than a rapid rate—where the difference can be as much as 100 to 200°F. Although the time-temperature dependence may be disadvantageous for some applications, this is a distinct advantage in the ceramic industry because the physical characteristics of the cone closely parallel the state of the ceramic ware. Sixty-two different cones are available, covering the temperature range of 1100 to 3600°F.

—Elias Kimmel, *Tempil° Division, Big Three Industries, Inc.*

Q

Quantitative-analysis Instrumentation [See **Analytical instrumentation (chemical).**]

Quantize Quantization is "a process in which the range of a variable is divided into a finite number of distinct sub-ranges (called *quanta*), not necessarily equal, each of which is represented by an assigned or 'quantized' value within the sub-range. Example: A person's age is quantized for most purposes with a quantum of one year."[1]

If the widths of the subranges are nominally equal, they are called "quantization units;" if they are intentionally not equal, the subranges are called "quantization steps." An example of a quantization unit is the degree used in temperature measurement. Quantization performed by means of a nonlinear device, such as the quantization of the range of numbers 1 to 10 on a slide rule, results in unequal subranges or quantization steps.

[1] IFIP-ICC Vocabulary, E29, North-Holland Publishing Company, Amsterdam, 1966.

The word "quantize" and its related terms are not widely used in instrumentation and data-acquisition applications. Neither are the essentially synonymous terms "digitize" and "digitalize." These are, however, convenient terms for use in explanations of analog-to-digital conversion concepts and other processes common to data-acquisition applications.

As related to instrumentation and digital data acquisition, the electronic device which performs quantization is the analog-to-digital converter or encoder (see **Analog-to-digital converter**). In the broadest sense of the word, it also applies to the assignment of discrete values to a quantity obtained by reading the position of a pointer on a calibrated scale.

In digital data acquisition, most quantizers or analog-to-digital converters are linear so that the discrete subranges are properly called quantization units. Since much of this equipment is based on the use of binary representations, the term commonly used for the quantization unit is the "bit." Thus, for example, in a 10-bit converter the least significant bit is the quatization unit, which is equal to $V_{f.s.}/2^{10}$, where $V_{f.s.}$ is the full-scale input to the analog-to-digital converter.

—Thomas J. Harrison, *IBM Corporation*

Queue A waiting line or list organized to allow a certain sequence of access to the list elements. Queues are used in programming systems when events may occur at a faster rate than they can be acted upon and a waiting line must be formed. A typical example is an I/O (input/output) channel where requests for channel usage may arrive together and a waiting line is required.

In general, the elements of any queue are blocks of information containing one or two location pointers (addresses) to other elements in the queue. In addition to the queue elements there exists control information to point to a reference point in the queue. Several methods of organizing queues are used, depending on the sequence of processing desired.

A common type of queue is a sequential queue. This is similar to a line waiting at a bus stop. As new elements arrive (are put into the queue), they are put at the end. As elements are processed, they are taken from the front. This type of queue organization is called first in, first out (FIFO).

A second type of queue is the push-down queue. This is like a bus with only one door. The last one in is the first one out. This type is known as last in, first out (LIFO) organization.

A third queue organization is the multipriority queue. This is like the military bus stop where higher rank goes to a corresponding higher place in the line instead of to the end. Elements are always processed from the front of the line. This is really a modified sequential queue containing subsequences and is sometimes referred to as priority in, first out (PIFO).

A fourth type of organization is the "any sequence" organization. Here the elements are linked to (point to) both adjoining elements. This organization is characterized by being scanned for some comparison condition to determine which element to process.

—L. E. Wise, *IBM Corporation*

R

Rack and Pinion A basic mechanical combination that provides conversion of rotary motion to linear motion (or the reverse action) in a machine or intrument. These elements are diversely used—as in cog-railroad cars, machine tools, and pile drivers. Applications in the instrument category are several. The diaphragm or bellows of an altimeter or other pressure-operated device moves a sectored rack that, in turn, meshes with the pinion of a related resolver, synchro, or potentiometer. In certain types of pendulum weighing scales, the rotary motion of the pendulums is caused by the pull on a rack that engages a sector. See **Scale, pendulum.** Other applications occur in machine rack-and-pinion drives for multiple resolver systems, providing an analog-to-electrical control-signal array. Rotary table or platform motion often is accomplished with an internal ring gear or circular rack that is rotated by means of a motor with a shaft-mounted pinion gear. The specifications for a rack-and-pinion mechanism must include the rack-and-pinion tooth geometrical form and data and the dead zone or backlash (see Fig. R-1).

—James R. Walker, *De Vlieg Machine Co.*

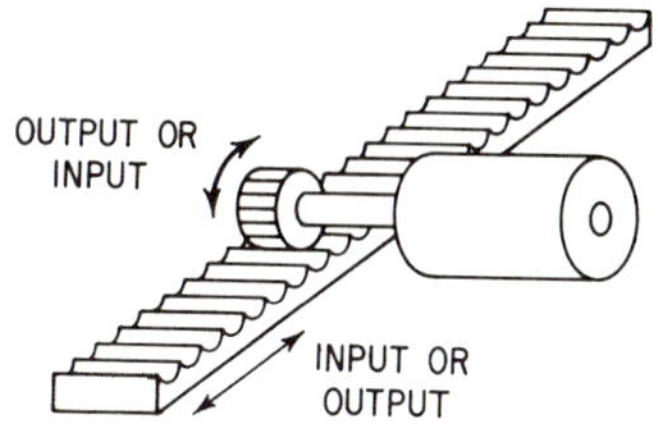

Fig. R-1. Rack and pinion.

Radar An acronym that stands for *radio detecting and ranging.* A radar system uses beamed and reflected radio-frequency energy for detecting and locating objects, particularly airborne; for measuring distance or altitude; for navigating; and for homing, bombing, tracking, and other special purposes. For detecting and ranging applications, the time interval between the transmission of energy and the reception of the reflected energy is a measure of the range of an object in the path of the beam. If the return signal is caused by reflection of the transmitted energy from the target object, this is known as *primary radar.* If the transmitted energy is used to trigger a responder beacon which, in turn, sends back a new signal to the radar set, this is *secondary radar.*

Acquisition Radar: A radar set that detects an approaching target and feeds approximate-position data to a fire-control or missile-guidance radar, which takes over the function of tracking the target.

Airborne Early-warning Radar: A radar system carried by an aircraft wherein the radar signals are relayed from the aircraft to surface stations, or their significance is reported by radio.

Air-intercept Radar: An airborne radar that searches for, acquires, and tracks a target to provide data needed for control of an air-to-air guided missile.

Airport-surveillance Radar: A radar located on or near an airport to provide an indication of the bearing and distance of each aircraft within the terminal area. It is used by itself for air-traffic control and is used with precision-approach radar to form a ground-controlled approach system.

Air-route-surveillance Radar: A long-range (approximately 150-mile) radar used by some governments to control air traffic between terminals. In area-control radar, a system is used to control air traffic over a relatively large area to provide a smooth flow of traffic to the approach-control radar.

Array Radar: A radar having a fixed array of many thousands of antennas, each with an electronically variable phase shifter that permits steering the beam over wide angles without antenna motion.

Radarsonde: An electronic system for automatically measuring and transmitting high-altitude meteorological data from a balloon, kite, or rocket by pulse-modulated radio waves when triggered by a radar signal. Also, a system in which radar techniques are used to determine the range, elevation, and azimuth of a radar target carried aloft by a radiosonde.

Radar Surveying: The use of airborne radar to measure accurately the distance between two ground radio beacons positioned along a baseline, thus eliminating the need to measure the baseline from the ground in rough terrain.

Radar Track-command Guidance: Command guidance that uses two radars external to a missile, one for tracking the target and the other for tracking the missile. The radar-receiver outputs are fed to a computer, and the output of the computer, in turn, is fed into a data transmitter that transmits flight information to the missile for correcting its flight path to the target.

Radar Chart: A special map used in radar navigation. For radar-equipped ships, radar charts emphasize the coastline, hills, large buildings, and other objects that give prominent radar echoes to ships well off shore. For air navigation, radar charts similarly show outlines of cities, rivers, lakes, bridges, railroads, and other objects that appear on radar screens.

Radarscope: The cathode-ray oscilloscope or screen used as an indicator in a radar set. A radar repeater is a cathode-ray indicator used to reproduce the visible intelligence of a radar display at a remote position. When used with a selector switch, the visible intelligence of any one of several radar systems can be reproduced.

Also see **Altimeter; Navigation systems;** and **Radio telescope.**

Radiation (Nuclear), Detection of Radiation detectors fall into two general classes: (1) for the measurement of *intensity* of radiation and (2) for the measurement of the *energy* of radiation. Measurement of intensity of radiation may be subdivided further into *rate* measurements and *integrated radiation-dose* measurements. Common detectors for the measurement of radiation intensity (rate) are (1) the Geiger counter, (2) the ionization chamber, (3) the proportional counter, and (4) the scintillation counter. All these are described in separate encyclopedia entries. For measurement of a radiation dose integrated over a period of time, the common detectors are (1) the electroscope, (2) photographic film, and (3) activation foil. Less commonly used are chemical dosimeters, colorimeters, and other dosimeters based on change of color degradation of material, or mechanical-electric action, such as discharge of the static electricity holding glass beads to an insulated surface. The chemical and color dosimeters, in particular, are used generally for integrated doses from high-intensity radiation fields. The colorimeters are used occasionally to determine the total energy output from a radiation source for absolute standards. The principal detectors for measuring the energy of alpha, beta, and gamma radiation are (1) proportional counters (alpha, beta), (2) ionization chambers (alpha, fission fragments), (3) scintillation detectors (beta, gamma), and (4) photographic film (alpha, *p,* mesons, and heavy nuclei). For special laboratory measurements of high-energy ionizing particles, such as those found in cosmic rays or from high-energy machines, there are very special detectors, some costing up to millions of dollars. This class includes (1) the liquid-hydrogen and liquid-helium bubble chambers, (2) cloud chambers, (3) spark chambers, and (4) the Cerenkov detector. Cloud chambers now are largely superseded by the much higher efficiency bubble chambers. The latter often are very large and very expensive; they are frequently used in conjunction with a high-speed computer. Still competing with these exotic methods is the low-cost photographic emulsion. Spark chambers comprise a series of charged plates which can be made to discharge successively by the traverse of an ionizing particle. They are used chiefly in conjunction with coincidence-counting techniques of high-energy particles.

Types of Radiation Measured: The foregoing detectors are sensors of ionizing radiation, such as alpha particles, beta particles, charged mesons, protons, and heavier charged nuclei. Detection of nonionizing radiation, such as gamma rays and neutrons, requires a conversion to ionizing radiation before detection. Gamma rays are absorbed by the material of the detector (walls, gas, or solid volume), and whatever the interaction mechanism (photoelectric effect, Compton scattering, or pair production), electrons are produced which may be measured by the ionizing-radiation detectors.

Slow-neutron detectors depend upon a nuclear reaction for conversion to ionizing radiation. The most common among these is high-cross-section reaction (high probability of occurrence) of the boron-10 isotope with a thermal (slow) neutron:

$$B^{10} + n \rightarrow Li^7 + \alpha$$

Both the lithium-7 and the alpha particle are released as ionizing radiation. The second common nuclear reaction for detector use is the fission of U^{235}, U^{233}, or Pu^{239}. Again, the reaction has a high cross section, and a thermal neutron will cause the release of the large ionizing energy of the two heavy-fission fragments.

Fast neutrons generally are detected through collisions with hydrogen nuclei. The high-energy transfer in such a collision makes the resultant fast proton readily detected as an ionizing particle. This conversion from fast neutrons to protons may be caused to occur exterior to the detector, inside the detector, or integral with the detector material itself, e.g., organic phosphors for scintillation detectors. Similarly, alpha and beta particles may be detected from a source inside the detector or external to it. In the latter case, a thin window must be provided for entry of the radiation into the detector. The thinness requirement is particularly severe with alpha particles and limits their usefulness in applied radiation problems.

Radiation-measurement Applications: There are two important areas: (1) area and personnel radiation monitoring and (2) industrial-radiation measurements. Radiation monitoring is primarily a measurement of radiation intensity or accumulated radiation dose. Consequently, all the techniques mentioned are used.

The Geiger counter, scintillation counter, and ion chamber are commonly used for area monitoring to determine intensity of radiation. The pencil electroscope and the photographic film are trademarks of every radiation area for personnel monitors where the integrated dose by the day, week, month, or throughout a specific operation is desired. For neutron detection, occasionally activation foils may be employed as area monitors, although they are more commonly used to determine the integrated neutron flux associated with a nuclear-reactor experiment.

When area monitoring includes the analysis of radiation to determine the source of contamination, the energy-sensitive detectors must be employed together with some type of beta or gamma spectrometer. (See **Gamma-ray spectroscopy.**) For alpha-particle determination, the proportional counter, the ion chamber, and the solid-state detector are used. For beta-particle determination, the proportional counter normally is used unless a magnetic-focusing beta spectrometer is used, in which case any beta-sensitive intensity detector may be used. Gamma-ray analysis customarily is made with the scintillation detector, although for precision work at higher energies a "pair spectrometer" and a gamma-sensitive intensity detector are used. The latter, however, is found chiefly in scientific laboratories. The manner in which various types of nuclear-radiation detectors can be tied into allied equipment to achieve a complete recording, alarm, or controlling system is shown in Fig. R-2.

The ion-chamber detector is the prime component of industrial-radiation measurement with beta-ray or gamma-ray gages. The inherent stability, high reliability, and simplicity of the ion chamber make it a favorite for such measurements. The Geiger counter sometimes is used

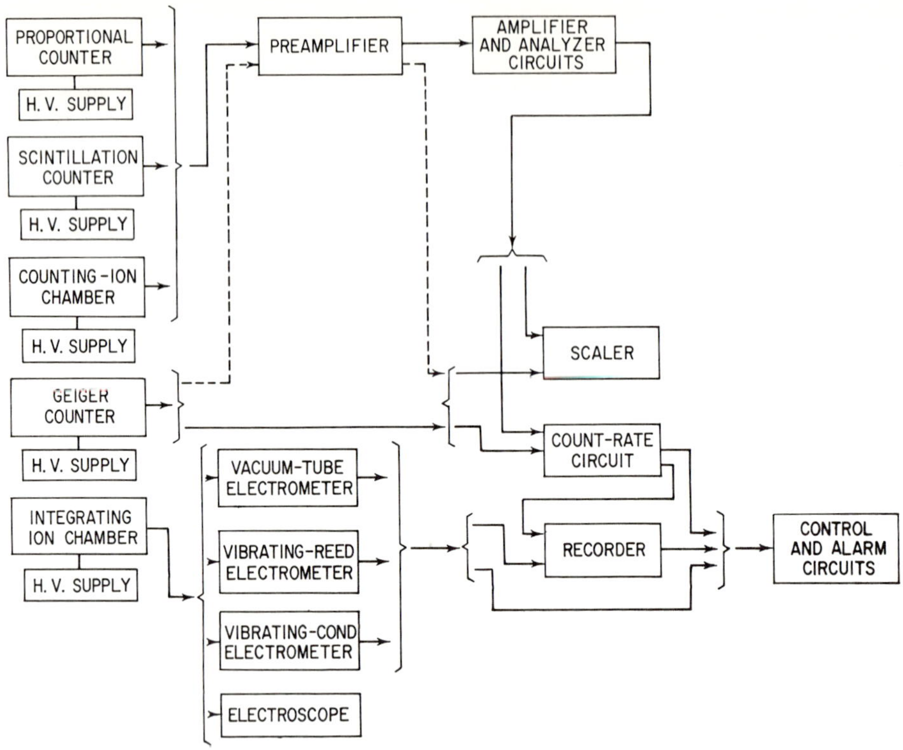

Fig. R-2. Various types of nuclear-radiation detectors and how they can be tied into allied equipment to achieve a complete recording, alarm, or controlling system.

for such purposes where the highest stability is not required and minimum cost is desired. The Geiger detector is not of itself essentially lower in cost than the ion chamber, but the high internal amplification of the Geiger tube reduces the associated electronic equipment required. The scintillation detector is used where the highest detection sensitivity of gamma radiation is required—even at the expense of highest stability. The next most common area for industrial application of radiation is in gamma-ray industrial radiography where the detector is the photographic film. In this work, both area and personnel monitors are used. Other industrial-radiation work, such as tracer studies, commonly employ the Geiger or proportional counters for sample counting. (See **Radioisotope-tracer instrumentation.**) In the area of high-intensity radiation (radiation fields from 10^3 to 10^8 roentgens/hr), the ion chamber, although of very special instruction, is the most accepted detector for intensity measurements. For total-dose measurements on individual processed items, colorimetric- and degradation-type detectors are used in which the radiation produces a color change, e.g., cobalt glass or the darkening of a plastic film. For more precision dose measurements, chemical dosimeters or calorimetric techniques are used.

Solid-state Detector: This may be regarded as a parallel-plate ion chamber in which the plates separate proportionally to the applied voltage. A detector of this type is shown in Fig. R-3. The sensitive volume of the detector is known as the "depletion layer." This is essentially the extent of the region at the junction of the semiconductor materials where a sharp potential gradient exists and consequently where ionization created by an external influence, such as nuclear radiation, will be quickly removed with negligible recombination. The thickness of the depletion layer x is given by $x = 1/3\sqrt{\rho V}$ approximately, where x is in microns, ρ is the conductivity of the semiconductor material in ohm-centimeters, and V is the reverse-bias voltage applied to the junction.

For the measurement of the energy of ionizing particles, particularly alphas and protons, it is necessary that the particle dissipate negligible energy before entering the depletion layer. Two techniques are used. A very thin layer of *p*-doped semiconductor may be superimposed on an *n*-type semiconductor, thus forming a junction very near one outer surface, or the surface-barrier-type diode may be used. In the latter, the discontinuity of space charge within a semiconductor as present at a surface is used to provide a high-potential-gradient region. The surface is coated with a thin layer of gold to provide a conductive contact.

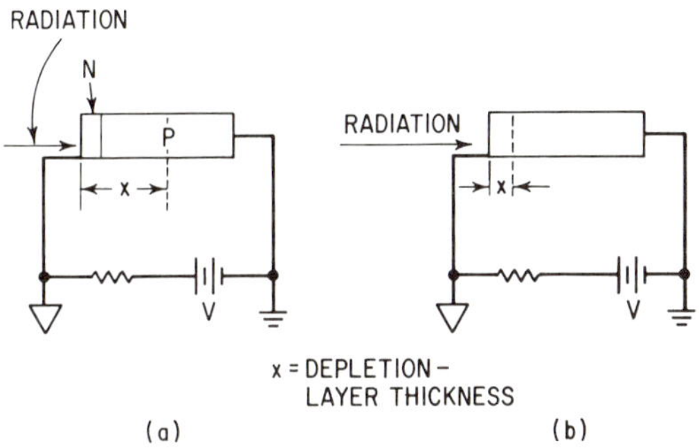

Fig. R-3. (a) Junction-diode solid-state detector; (b) surface-barrier solid-state detector.

The advantages of the solid-state detector are associated with its small size and the high specific ionization for these materials. This specific ionization, the number of ion pairs per electron volt or its reciprocal, the number of electron volts of energy lost by the ionizing particle to create one ion pair, is ten times more favorable for solid-state detectors (~ 3 electron volts per ion pair) than for gas-chamber detectors (~ 30 electron volts per ion pair). Hence, the ionization intensity is an order of magnitude higher in the semiconductor for a given ionizing particle, and as a result, the energy resolution is greater. Solid-state detectors, however, do not appear nearly so favorable when compared with the gas-ionization chamber for integrated-current measurement.

For detector-device descriptions, see **Boron thermocouple; Colorimetric radiation detector; Dosimeter, chemical; Electrical quantities, measurement of; Film detector, radiation; Gamma-ray spectroscopy; Gas-ionization tube; Geiger counter; Ionization chamber; Photomultiplier tubes; Proportional counter; Radioisotope-tracer instrumentation;** and **Scintillation counter.**

Radiation Pyrometer A temperature-measuring instrument for industrial and laboratory applications which measures radiant energy emitted from an object and converts that energy to a corresponding temperature. Characteristics of the instrument include (1) temperature measurement without contact with the object being measured, (2) measurement independent of distance provided the object fills the instrument target area, (3) capability to measure very high temperatures continuously, (4) response to very rapid changes of temperature, and (5) design and positioning flexibility to sight on very large or very small objects. These characteristics make possible the temperature measurement of moving surfaces, inaccessible objects, very small masses, and the temperature averaging of large areas.

A radiation-pyrometer assembly consists of (1) a sensing head, (2) a display instrument, and (3) accessories for mounting, aligning, and protecting the sensing head. As shown in Fig. R-4, the sensing head contains a sensing element and an optical system inside a protective housing. The sensing element may be a thermopile (see **Thermopile**), a resistance element called a bolometer, or a semiconductor device. The display instrument usually is a null-balance-type industrial recorder with a compensator dial to allow adjustment of the pyrometer for the prevailing emissivity. Because of the nonlinear relationship

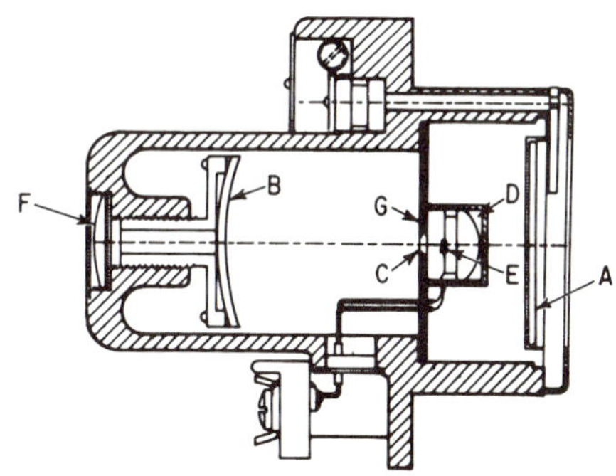

Fig. R-4. Cutaway view of sensing head of a radiation pyrometer. Radiation from the object enters window *A* and is focused by mirror *B* to form a sharply defined image on surface *G*. A small opening *C* in the disk permits a portion of the radiant energy to pass through to mirror *D* which focuses it onto thermopile *E*. Lens *F* and surface *G* are provided to facilitate sighting on the object. Models are available in several ranges from 400°C up. Full-range output will vary from less than 1 to 20 mV. Models are available for object distances from 6 in. to infinity; for target diameters from 1/10 in. to several feet; and with response to 60 percent of any temperature change in 1/10 to 6 s. *(Leeds & Northrup.)*

between radiant energy and temperature, the usual display instruments have scales with expanded divisions at the upper end. For optimum readings, these scales are selected so that the temperature of interest normally will fall in the upper two-thirds of the span.

Accessories are available to adapt the sensing head to many industrial processes. Included are mounting flanges with provisions for keeping the sensing head cool, sighting tubes with open ends, protecting tubes with closed ends, safety shutters to protect the sensing head against protecting-tube failure, and gastight mountings with shutoff valves for protective-atmosphere furnaces (see Fig. R-5).

Most radiation pyrometers are of the *total-radiation type* which receive the maximum available radiant energy at the widest range of wavelengths possible. Another type, called *spectrally selective,* is sensitive to only a portion of the waveband. The latter type is useful because the narrower spectral band may minimize or eliminate adverse effects of emissivity and absorbing gases in the sighting path.

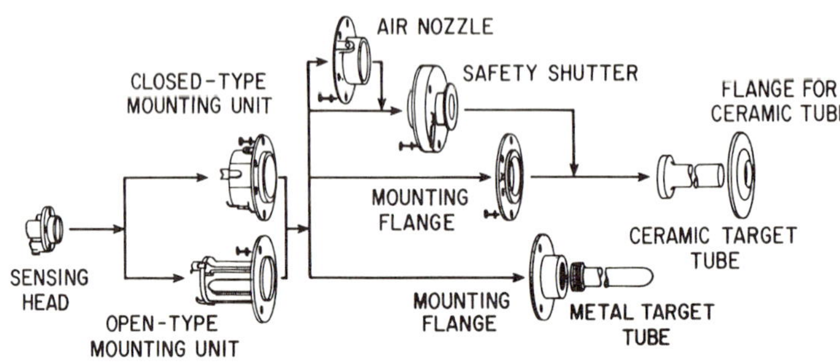

Fig. R-5. Typical accessories used to adapt the sensing head of a radiation pyrometer to a variety of industrial processes. *(Leeds & Northrup.)*

Theoretical Aspects: The basis for radiation-pyrometer measurement is that all objects, above absolute zero, emit radiant energy. This energy is transmitted at wavelengths ranging from 0.01 to 100 microns. As the temperature of the object increases, the energy is carried by progressively shorter waves.

The energy transfer is expressed by the Stefan-Boltzmann equation $W = ek(T^4 - T_1^4)$, in which W is the radiant energy, e is the emissivity, k is the Stefan-Boltzmann constant [5.6697×10^{-12} W/(cm^2)/° K)], T is the temperature of the object being measured, and T_1 is the temperature of the pyrometer sensing head. For temperatures above 400° C, the sensing-head temperature can be ignored and the relationship stated as $W = ekT^4$. For temperatures below 400° C, the pyrometer must be compensated for its sensing-head temperature. This compensation is usually obtained by measuring sensing-head temperature with a thermistor.

The *emissivity* depends on the material of the object being measured, its surface condition and temperature, and the temperature of surrounding objects. The maximum value for emissivity is 1.0. This is the ideal emissivity for temperature measurements. At this value, it is said that blackbody conditions prevail, and factory pyrometer adjustments usually assume blackbody conditions will prevail. For many industrial-process temperature measurements, emissivity is less than 1.0, but meaningful measurements are possible because emissivity remains relatively constant within the temperature range of interest. Minimum values for emissivity approach zero, and at very low values meaningful measurements are not normally possible.

Adverse Factors: Emissivity, interference between the pyrometer and the measured object, and the influence of surrounding objects can adversely affect measurements. If emissivity varies significantly with changing temperature, meaningful measurements are unlikely over a very wide temperature span. Fortunately, most industrial processes operate within a narrow temperature span. In such cases, emissivity changes do not degrade measurements significantly.

If radiation-absorbing gases, luminous particles, smoke, or flames are in the sighting path between the object and the sensing head, these will interfere with the pyrometer. However, their adverse effect can often be counteracted by blowing a stream of air or inert gas down the sighting path. If necessary to sight through a window, it should be a quartz or similar material with low absorption to minimize interference.

The influence of surrounding or nearby objects can be a help or a hindrance. If the object to be measured is inside a furnace of the same temperature as the object, blackbody conditions will prevail on surfaces facing furnace walls irrespective of other factors. However, if the furnace is much hotter or colder than the object, the influence of the furnace walls will cause the pyrometer to read high or low. If the furnace has its door open, the object surface facing this opening will usually appear to have a low temperature because the emissivity of that surface will be that of the object in an open area. Large hot objects standing adjacent to each other in the open may display different emissivity characteristics on the facing and nonfacing surfaces.

Unless essentially blackbody conditions are known to prevail, the radiation pyrometer must be adjusted to read an acceptable temperature based on an established temperature standard. For temperatures above 1200° C, this will usually require on optical pyrometer. For lower temperatures, thermocouples may be temporarily inserted to obtain check readings. Several careful measurements should be made under typical process operating conditions by a capable operator using quality test instruments. The best of these readings should be used to set the compensator dial of the pyrometer. After this adjustment, the pyrometer will read true temperature for most process conditions. If it does not read true temperature because of limitations of the test readings or substantial variations in process conditions, the radiation pyrometer still often provides a temperature guide adequate for process operation.

REFERENCES

Belcher, W. E., Donald Robertson, and W. F. Hickes: "Temperature Measurement," American Society of Metals, Cleveland, Ohio, 1956.

Dike, Paul H.: "Temperature Measurements with Rayotubes," Leeds & Northrup Company, Philadelphia, Pa., 1953.

—W. S. Bloor, *Leeds & Northrup Company*

Radio Navigation (See Navigation systems.)

Radio-frequency Moisture Meter (See Moisture, measurement of.)

Radio Telemetry Radio telemetry of instrumentation data and other intelligence became impor-

TABLE R-1. *Radio-frequency Allocations*

Frequency	Service	Frequency	Service
kHz:		MHz:	
2,500	Radio astronomy	143.6–143.65	Space research (telemetry and tracking)
5,000	Radio astronomy		
10,000	Radio astronomy	149.9–150.05	Radio-navigation satellite
10,003–10,005	Space research	150–153	Radio astronomy
15,000	Radio astronomy	183.1–184.1	Space research
15,762–15,768	Space research	267–272	Space (telemetry)
18,030–18,032	Space research	322–329	Radio astronomy
19,990–20,010	Space research	399.90–400.05	Radio-navigation satellite
20,000	Radio astronomy	400.05–401	Meteorological satellite (maintenance)
20,007 ± 3	Space distress		
25,000	Radio astronomy	401–402	Space (telemetry)
MHz:		404–410	Radio astronomy
30.005–30.010	Space research (satellite identification)	406–410	Radio astronomy
		449.75–450.25	Space telecommand
37.75–38.25	Radio astronomy	460–470	Meteorological satellite
39.986–40.002	Space research	606–614	Radio astronomy
40.68 ± 0.25	Scientific, industrial	608–614	Radio astronomy
73–74.6	Radio astronomy	610–614	Radio astronomy
117.975–132	Commercial satellite	900–960	Space research
132–136	Commercial satellite	1,400–1,427	Radio astronomy
136–137	Space research (telemetry and tracking)	1,427–1,429	Space (telecommand)
		1,525–1,535	Space (telemetry)
137–138	Meteorological satellite	1,535–1,540	Space (telemetry)

tant during World War II and has expanded much during the intervening period. One early use was for the remote radio control of drones, followed closely by the remote control of pilotless missiles. As pointed out in Ref. 1, many functions needed to be measured in the early missile programs—servo control systems, remote-control systems, guidance systems, missile structure, aerodynamics, and propulsion. In addition to the measurement of voltages, these measurements usually were linear motions or shaft positions, or rotations, thrust, acceleration, temperature, and pressure. The present widespread use of radio telemetry in connection with instrumentation data can be gleaned from Table R-1, which summarizes frequency allocations by regulatory agencies.

REFERENCES

1. Gruenberg, Elliot L. (ed.): "Handbook of Telemetry and Remote Control," McGraw-Hill Book Company, New York, 1967.
2. Foster, Le Roy E.: "Telemetry Systems," John Wiley & Sons, Inc., New York, 1965.
3. Nichols, M. H., and L. L. Rauch: "Radio Telemetry," John Wiley & Sons, Inc., New York, 1954.

Radio Telescope Essentially, a sensitive radio receiver used with a large and directional antenna to receive signals from outer space. Also see **Astronomical instrumentation.**

Radio telescopes can be divided into three basic types: (1) interferometers, (2) phased arrays, and (3) reflecting mirrors. Since radio (microwave) wavelengths are at least five orders of magnitude longer than visible-light wavelengths, very large radio telescopes are required to obtain resolution and intensity comparable with that of optical telescopes. See **Telescope.**

Radio telescopes are operated at frequencies from about 10 to 10,000 MHz or wavelengths from 30 m to 3 cm. Connected to the radio telescopes are very sensitive radio receivers, such as the Dieke radiometer, very low noise maser preamplifiers, and radio spectrometers. These receivers are of very wide bandwidth and long integration time. Oddly, *pulsars* (pulsing radio sources) were discovered because the long-period integrating circuits were omitted. Spectrometers are used to measure spectral radio emission such as 21-cm H lines, 18-cm OH lines, and recombination spectra as well as doppler velocities associated with the 21-cm H line. All radio telescopes, as with optical telescopes, have resolution of λ/D and intensity or energy collection proportional to telescope's area D^2, where D is the telescope's diameter and λ the wavelength.

Interferometers, in their simplest form, consist of two antennas spaced many wavelengths apart. The antennas are connected with transmission lines and then to a receiver. Their sensitivity is relatively low, but the resolution is high, given by λ/D, where D is the antenna separation. The interferometer pattern can be steered by varying the phase of one antenna with respect to the other.

Interferometers are limited in baseline (resolution) because of the phase instability of the transmission line connecting the two antennas. Very long baseline (thousands of kilometers) interferometers have been developed by using an atomic clock at each antenna. The two clocks remain in synchronism or are phase-coherent for sufficiently long periods of time to make measurements at each antenna site. These measurements are later correlated, producing an interference pattern of very high resolution, tenths to hundredths of an arc second. These very long baseline interferometers have been used for accurate determination of the position of radio sources as well as for the measurement of the diameters of radio sources, particularly quasi-stellar objects.

Higher gain and a pencil beam can be obtained with unfilled apertures consisting of arrays of antenna elements in the form of crosses (such as Mill's cross), large circular arrays of elements, and other configurations such as T's and Y's. The beams formed by these arrays are steered by phasing the elements. The length of the arrays is of the order of kilometers, and they produce beamwidths (resolution) of several arc minutes.

Large-diameter steerable and fixed parabolic antennas have the advantage that the frequency covered can be much larger than with phased arrays because the feed at the focus of the paraboloid need only be changed. Very low noise maser amplifiers can be used when only a single-element feed is employed. Several large and movable antennas appropriately positioned within a very large area have been used to synthesize a very large aperture.

Examples of steerable parabolic antennas are the 250-ft parabolic antenna at Jodrell Bank and the 210-ft antenna at Parkes, Australia, which has the most precise surface for its diameter of any of the steerable antennas. It may be combined with a 60-ft parabolic antenna movable on tracks for an adjustable-length-baseline interferometer (see Fig. R-6).

Fig. R-6. Radio telescope (210-ft) at Parkes, Australia, with a 60-ft-diameter telescope which can be moved on rails to different positions. The combination of the two telescopes forms an interferometer. *(Australian News and Information Bureau.)*

Large fixed reflectors have been built in which the feed at the focus is movable to steer the beam. An example is the 1,000-ft spherical reflector with movable feed corrected for spherical aberration at Aracibo, Puerto Rico. By combining several large steerable parabolic antennas to form interferometers, together with maser amplifier receivers, both high sensitivity and high resolution may be obtained.

REFERENCES

Pawsey, J. L., and R. N. Bracewell: "Radio Astronomy," Oxford University Press, Fair Lawn, N.J., 1955.

Brown, R. Hanburg, and A. C. B. Lovell: "The Exploration of Space by Radio," John Wiley & Sons, Inc., New York, 1958.

Davies, R. D., and H. P. Palmer: "Radio Studies of the Universe," D. Van Nostrand Company, Inc., Princeton, N.J., 1959.

Shklovskii, I.: "Cosmic Radio Waves," Harvard University Press, Cambridge, Mass., 1960.

Kraus, J. D.: "Radio Astronomy," McGraw-Hill Book Company, New York, 1966.

—Gordon Ferrie Hull, Jr., *Hull Associates*

Radioisotope-tracer Instrumentation In addition to medical and biological applications, radioisotopes find important applications in industry for the measurement of such variables as level, density, thickness, pressure, flow, concentration, and coating thickness. Some definitions are in order.

Isotope—Atomic species of the same atomic number, i.e., belonging to the same element but having different mass numbers, are called isotopes.

Radioisotope—Those isotopes distinguishable from other species of atoms with the same atomic number by radioactive transformation are known as radioactive isotopes or radioisotopes.

Radioisotopes are useful as sources of radiation and as tracers. Their tracer use arises because they exhibit substantially the same chemical behavior as the stable species of isotopes while emitting radiation which permits determining their identity and location. The principal types of radiation emitted by radioisotopes are summarized in Table R-2. X-rays resulting from the filling of nuclear shells and bremsstrahlung generated by the deceleration of beta particles are two additional sources of electromagnetic radiation produced by radioisotopes that find use in instrumentation.

Neutrons—While neutrons are not emitted in the disintegration of radioactive isotopes (except in rare side reactions), they are important to the production of radioisotopes. Neutrons are particles with zero charge and with a rest mass (as compared with relativistic mass) approximately equal to that of the hydrogen nucleus. Neutrons are unstable and decay to a proton and beta particle with a half-life of approximately 12.8 min. In nuclear reactors, neutrons produce radioisotopes by causing fission of fissionable materials, such as uranium-235 or plutonium-239, and by activating substances placed in the reactor through a neutron-absorption reaction. Fission

TABLE R-2. *Principal Types of Radiation Emitted by Radioisotopes*

Type	Symbol	Description	Rest Mass (O = 16)	Charge (electron = 1)	Range in Air	Ion Pairs, per cm in air
Alpha particles	α	Nuclei of helium atoms	4	+2	2–0 cm (for 3–10 meV)	30,000–70,000 (varies with distance from source)
Beta particles	β	Electrons ejected from a nucleus	1/1,840	−1	160–2,000 cm (for 0.5–5 meV)	150–40 (for 0.1–50 meV)
Gamma rays	γ	Electromagnetic radiation produced only in nuclear processes	None	None	15,000 cm ½ value thickness (for 1.5 meV)	1/100 of number of pairs produced by same energy β

meV = million electron volts.

products and activated materials are available for instrumentation applications. Neutrons for instrumentation sources usually are obtained from the reactions

$$\text{Alpha} + Be^9 \rightarrow \text{neutron} + \text{carbon}$$
$$\text{Gamma} + Be^9 \rightarrow \text{neutron} + 2\text{ alphas}$$

Bremsstrahlung—The deceleration of electrons (beta rays) in matter produces x-rays just like those produced in x-ray tubes. When produced by betas, these x-rays are called bremsstrahlung.

Half-life—Radioisotopes are unstable; they decay in accordance with the relation

$$N = N_0 e^{-\lambda t}$$

where N = number of unchanged atoms at time t
N_0 = number of atoms present when $t = 0$
λ = constant characteristic of the particular radioactive species called the decay constant

λ is related to the half-life $t_{1/2}$ of the radioactive isotope by the relation

$$t_{1/2} = \frac{0.693}{\lambda}$$

Half-lives of the known radioisotopes cover the range from a fraction of a second to many years. For example, carbon-10 is a radioisotope with a half-life of 20 s, while carbon-14 has a half-life of 5,100 years.

Detectors for radioisotope radiation are described under **Radiation (nuclear), detection of.** Specific applications are included throughout the encyclopedia where specific measurement variables are described. Highlights of some radioisotope applications are given below.

Medical Uses: Tracer elements are used to study the functioning of particular body organs. If a subject is fed radioactive iodine, the concentration of iodine in the thyroid gland can be measured readily by means of external instrumentation. In outlining the size and shape of the gland, the radioactive iodine will show any nonfunctioning sections of the gland. If cancer of the thyroid has occurred and the tumor of thyroid tissue has become established elsewhere (metastasized), the iodine will concentrate in the new location, indicating where a new cancerous growth may have started. This is only one of several examples of the medical applications for radioisotope instrumentation.

Pressure Measurement: The interaction of alphas with a gas (or solid) results in the formation of positive and negative ions, which can be collected and measured as an electric current. Since the number of ions produced in a gas by alpha particles from a source of fixed size is dependent upon the density and composition of the gas, measurement of ionization can be used as a means of measuring pressure or composition.

Level Measurement: The ability of gamma radiation to penetrate steel walls of pressure vessels and brick linings of cupolas and the absorption or scattering of this gamma radiation by materials present inside these vessels provide phenomena which permit measuring the level of solids or liquids without the use of floats or other internal devices. See **Level gage, radiation-type.**

Thickness and Coverage Measurement: The absorption of radiation by matter is a phenomenon that permits measurement of the mass of material between a source and a receiver. For very thin materials (under 5 mg/cm², 0.0009 in. Al), the absorption of alpha radiation is used to provide a measure of weight per unit area. In the range 0.5 to 1,000 mg/cm^2 (0.00009 to 0.16 in. of Al), beta radiation, such as emitted by Sr^{90}, Kr^{85}, C^{14}, and other radioisotopes, is used for measuring weight per unit area. For thicknesses above 1,000 mg/cm^2, bremsstrahlung or gamma sources are used. See **Thickness gaging, nuclear-radiation;** and **Weighing, radiation methods for.**

Flow Measurement: The three principal techniques involving radioisotopes for measuring flow rate are (1) peak timing, (2) dilution, and (3) total count. The three methods are illustrated in Fig. R-7. In the *peak-timing technique,* a gamma emitter, such as Co^{60} or Sb^{124}, is injected quickly at a point close to the section of the pipe in which the velocity is to be determined. The time of passage of the peak of the tracer wave is determined, using two detectors, such as Geiger-Mueller counters or scintillation detectors, located at a known distance apart and external to the pipe. The detectors usually are connected to a rate meter and recorder. The time interval for the liquid to travel between the two detectors is calculated from the distance on the recorder chart between two radiation-level peaks, or it is determined by having the arrival of the activity at the first detector start a timer which is stopped by the arrival of the radioactive material at the second detector. The *dilution technique* utilizes the fact that the concentration of mixed tracer in a line resulting from the continuous bleeding of a tracer at a known rate into the line is inversely proportional to the relative flow rates of the line and the bleeder—if the tracer uniformly mixes with the flowing liquid. The concentration of the tracer in the line can be determined by measuring either the beta or gamma activity in samples taken from the line or by using a detector outside the line. In the *total-count method,* flow rate is based on measurement of the total counts from a radioactive tracer which has been added to a flowing stream. The total count bears a simple inverse relation to the flow rate. Total count can be measured with a detector in or out of the stream and is useful for measuring flow rates in pipes, open channels, rivers, or any type of flowing system for which a calibration constant can be measured or calculated. The number of counts measured is independent of the distribution of the tracer along the stream, although the tracer must be uniformly distributed *across* the stream so that the final count can be converted into flow rate by use of a determined constant. As the flow increases, the time the radioactive atoms are in the vicinity of the counter decreases. Thus, the number of counts decreases in direct proportion to the increase in average fluid velocity.

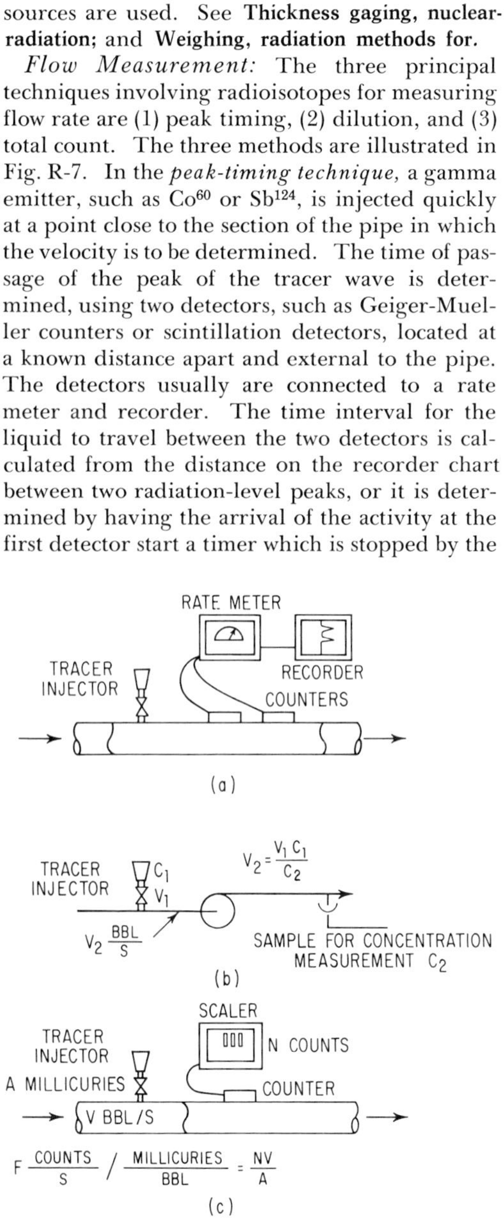

Fig. R-7. Basic radioisotope techniques for measuring flow: (a) peak timing; (b) dilution; and (c) total count.

Interface Location: In oil-product pipelines, a number of different products frequently are handled in one line. To minimize loss of different products at the terminus, an oil-soluble Sb^{124} tagged compound can be introduced into the line at the time a change in product is made. The pumping stations along the route have radiation detectors located ahead of the station which pick up the coming of the interface and automatically operate controls to forewarn operators.

Density and Specific Gravity: Since absorption of beta or gamma radiation is a function of the mass of the material between the source and detector, the density of slurries and other difficult-to-measure fluids and bulk solids can be tackled with radiosotope instrumentation.

Radiometer, Thermistor (See **Infrared detector, thermistor.**)

Radiosonde A device used to obtain meteorological data from the upper atmosphere. Radiosondes are carried to the upper atmosphere by balloons, rockets, or aircraft. The ballon-borne sonde performs its functions as it rises. The rocketsonde normally is ejected at the rocket's apogee and performs its functions during a parachute-controlled descent. Dropsondes, which are special radiosondes, are ejected from aircraft at desired altitudes and perform their functions during parachute-controlled descents. Instruments aboard the sonde transmit radio signals related to pressure, temperature, and humidity to a ground receiving station, known as a *rawin* set.

Rain Gage (See **Precipitation gage.**)

Raman Spectrophotometry An analytical instrument technique utilizing the Raman effect for determining the concentration of various molecules in aqueous or organic systems, separation of geometrical and rotational isomerism, and inter- and intramolecular association where infrared techniques cannot be applied. As an example, if carbon tetrachloride is radiated with light from a mercury arc, a portion of the violet lines in the scattered light will have associated with them a Raman spectrum which is comprised of several lines. The Raman effect is analogous to the Compton effect in the x-ray region.

Random Error (See **Error, measurement.**)

Range, Instrument The range of an instrument designates the limits within which an electrical, mechanical, pneumatic, or other signal may be indicated, recorded, measured, or transmitted. Unless otherwise specified, the input range of the instrument is generally implied and is expressed by stating the lower range value (the lowest quantity of the input signal that the device *is* adjusted to measure) and the upper range value (the highest quantity of the input signal that the device *is* adjusted to measure); i.e., if it is stated that the range is 0 to 15 in. H_2O pressure, it means that the instrument is adjusted to measure a minimum of 0 in. H_2O and a maximum of 15 in. H_2O pressure.

Transmitting instruments (devices that receive information in the form of one physical quality and convert it into information in the form of the same or another physical quality) must be described by stating two ranges, the input and output range. It may be stated: For an input-pressure range of 0 to 15 in. H_2O pressure, the output range will be 4 to 20 mA. The gain of a transmitting instrument then would be equal to output range/input range. An example would be a current-to-pressure transmitter which, for a given range of current input, produces a range of output pressures. A pressure-to-pressure transmitter has a pressure-input range and a pressure-output range.

For multiple-range instruments (devices whose range may be varied by changing switch positions or jumpers, or by other methods), the definition of range applies to the particular range that the instrument is adjusted to measure at any one time. This means that even though the instrument might be capable of measuring beyond its present range if it were adjusted, the "range of the instrument" applies to the limits at which it *is* adjusted.

The "total range," or "rangeability," of an instrument refers to the region between the extreme limits within which the input signal may be measured. It is expressed by stating the lower range limit (the lowest possible quantity of the input signal that the instrument *can be* adjusted to measure) and the upper range limit (the highest possible quantity of the input signal that the instrument *can be* adjusted to measure). When stating the total range or rangeability of an instrument, it is generally assumed that the instrument is capable of receiving inclusively all values of the input signal between the lower range limit and the upper range limit. If the instrument is not capable of receiving all values of the input signal continuously between the lower and upper range limits, this must be noted.

Some instruments, however, may not be capable of meeting the requirements for accuracy over the total range of the instrument. Thus the term "measurement range" of an instrument is employed to denote that part of the total range within which the accuracy requirements for that particular instrument are met.

The "span" of the instrument (input or output) is directly related to the range of the instrument, since it is equal to the algebraic difference of the upper range value and the lower range value. Note that if the span of an instrument changes, the range also must change. However, a change in range may or may not change the span.

One way in which the range may be changed while the span remains constant is by suppressing the "zero value" (that magnitude which is neither positive nor negative, i.e., 0° F, 0 psi, etc.) of the measured signal. The instrument is then referred to as having a "suppressed zero range." This means that the zero value is less than the lower range value and therefore would not be within the span of the instrument. This could be explained by saying that the span of the instrument was biased and that the lower range value and the upper range value are therefore increased above the zero value equal amounts. If, however, the lower range value and the upper range value are shifted unequal amounts above the zero value, or if the lower range value is increased to above the zero value and the upper range value is decreased from its original value, the instrument is still referred to as having a suppressed zero range, but the span has not remained the same. This particular type of instrument range is valuable when measuring an input signal which fluctuates slightly about some value much greater than zero.

TABLE R-3. *Examples of Range and Span Relationships*

Description	Type	Range	Range Valve		Span
			Lower	Upper	
0 100		0 to 100	0	100	100
25 125	Suppressed zero range	25 to 125	25	125	100
—100 0	Elevated zero range	—100 to 0	—100	0	100
—100 —25	Elevated zero range	—100 to —25	—100	— 25	75
— 25 0 +50	Elevated zero range	—25 to 50	— 25	50	75

Conversely, the zero value of the input signal may, in some instruments, be shifted to a value greater than the lower range value, and simultaneously may be greater than, equal to, or less than the upper range value. This instrument would then be referred to as having an "elevated zero range." It can be seen that if the zero value of the measured signal is elevated, the span of the instrument may remain the same, increase, or decrease, depending on the desired range. The purpose of elevating the zero value is to enable negative values of the measured signal to be measured. (See Table R-3.)

Extended-range Instrument: This is a measuring instrument that incorporates the following characteristics:

1. It has a number of input ranges with (*a*) an equal span of each range and (*b*) an overlap of the ranges of 0 to less than 100 percent of span.

2. The total rangeability may include negative, zero, and positive values of the measured variable.

3. All ranges usually have a common output with different scale factors.

The purpose of an extended-range instrument is to retain the resolution and accuracy of a small-span instrument while maintaining a large total rangeability of the measured input.

Extended-range instruments can be automatic or may require manual switching of range to retain the measured signal within the operating range. Automatic extended-range instruments usually contain some overlap of range to make

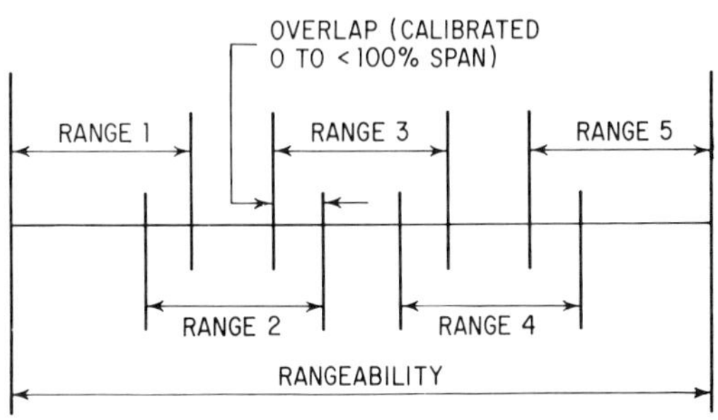

Fig. R-8. Possible arrangement of ranges in an extended-range instrument.

TABLE R-4. *Examples of Multiple-range Instrument Capabilities*

Type	Input Range 1	Input Range 2	Output
Zero suppression.	0–10 mV	10–20 mV	0–100% pen position
Span change.	0–10 mV	0–100 mV	4–20 mA
Zero-suppression span change. .	0–10 V	10–90 V	0–100% pen position
Output-range change	0–10 mV	. . . .	4–20 or 1–5 mA

balance at the switching point possible. Many manual extended-range instruments do not contain overlap. (See Fig. R-8.)

Multiple-range Instrument: An instrument of this type has the capability of changing the upper and/or lower limit of the input or output signal a predetermined amount. This may be accomplished by means of a switch, a jumper connection, or a mechanical linkage, such that at each position, the upper and lower limit and the accuracy are defined. Examples of multiple-range capability are given in Table R-4.

—Robert L. Wilson, *Honeywell Inc.*

Rankine Temperature Scale A temperature scale named in honor of William J. M. Rankine (1820–1872), based upon absolute zero of the Fahrenheit scale, −459.67°F. It is used in engineering in English-speaking countries. Temperatures are denoted by the term degrees Rankine and by the symbol °R.

—Robert D. Thompson, *Taylor Instrument Process Control Division, Sybron Corporation*

Raster (See **Cathode-ray tubes.**)

Rate-of-climb Indicator [See **Air-data systems (aircraft).**]

Rate Gyro (See **Flight-control systems.**)

Rate Response (See **Response time, transducer.**)

Ratio Control A system that maintains two or more variables at a predetermined ratio. An example of ratio control where air flow is controlled in ratio to fuel flow is shown in Fig. R-9. The ratio-control system consists of a fuel-flow transmitter, an air-flow transmitter, a ratio station, and a fuel-flow controller and an air-flow controller. The fuel-flow transmitter sets the set point of the air-flow controller through the ratio station. As the fuel flow changes, the set point of the air controller is changed automatically to a new value so that an exact ratio is maintained between air flow and fuel flow. The system ensures that no matter how the fuel flow changes, the correct amount of air for optimum combustion conditions always will be supplied.

The ratio station simply is a multiplier with the multiplying factor manually selected as dictated by the needs of the process. The normal ratio setting for most applications is in the vicinity of 1.0, which point usually is selected as a midscale position. This condition can be accomplished by sizing the orifices and flowmeters in proportion so that, under normal operation, they will read about the same percentage flow. The range of typical ratio stations is on the order of 0.3 to 3.0 or 0.2 to 4.0.

Since the output of the transmitters commonly is 3 to 15 psig or 4 to 20 mA dc, the ratio station first must remove the zero elevation before performing the multiplication and, in turn, then add in the elevation. The ratio station therefore takes the output of the fuel-flow transmitter and subtracts 3 psi or 4 mA, multiplies by the selected ratio, and then adds the 3 psig or 4 mA back to the signal before sending it to the air-flow controller as the set point.

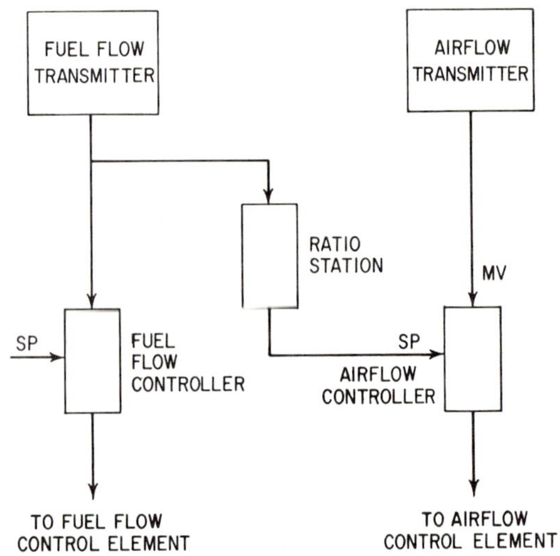

Fig. R-9. A fuel-air ratio control system.

In the example of ratio control shown in Fig. R-10, a master controller sets the flow rates of several ingredients to a process. In this system, it is desired to add each feedstock in a particular ratio to the total of all ingredients. A system of this type usually requires the sum of the ratio settings to be constant—where a percentage increase in one ingredient will be accompanied by the sum of the ratios always equaling 100 percent. A system of this kind is used in preparing the ingredients for a lead smelter prior to sintering. Similar systems find wide application in the food- and feed-processing industries.

In some instances, the ratio station or multiplier is made a physical part of the controller—

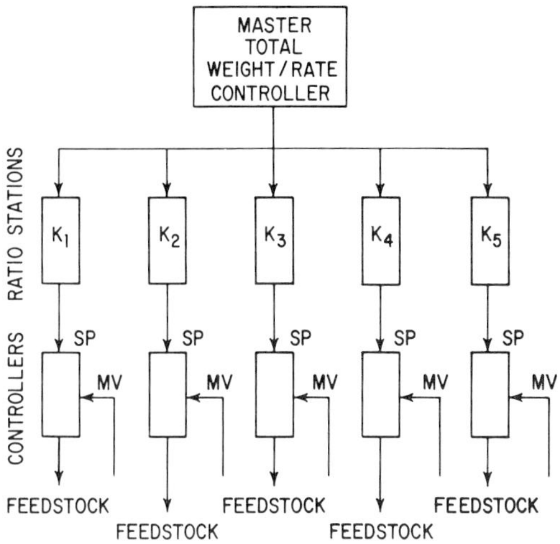

Fig. R-10. Multiple ratio control system.

in which case the device is termed a *ratio controller.*

—Richard H. Kennedy, *Bristol Division, American Chain & Cable Company, Inc.*

Ratio Meter (See **Electrical quantities, measurement of;** and **Ohmmeter.**)

Rawin Set A radio direction finder designed to track radiosondes automatically during flight and to provide azimuth and elevation angle and meteorological data. Radiosondes can be tracked to altitudes of 100,000 m or more and to horizontal distances of 125 miles or more, dependent upon the surrounding terrain and the altitude of the radiosonde. A rawin set consists of an antenna, antenna control, receiver, control-recorder, and pedestal.

Reader [See **Code readers and systems; Magnetic-ink character recognition (MICR); Optical character recognition (OCR);** and **Pattern recognition.**]

Reading Aids (See **Vision, sensory aids for.**)

Reading Error (See **Error, measurement;** and **Parallax error.**)

Readout Output information as presented in some form of display, including moving indicators, dials, lights, and printed or punched tape or cards. Normally associated with information that is visibly seen or "read out," although the term may be used in connection with data that automatically go into some form of storage without a visual interface. See **Indicating instruments; Recorder, graphic.**

Real-time Computing The use of a computer in conjunction with or interconnected with events or processes external to the computer which are time- or event-dependent. Historically, the computer has been used to solve scientific problems and automate record keeping. In these situations, the problem description and solution is presented to the computer in the form of a program and data are supplied on which the program operates (see **Program, computer**). The results of the program runs are reports and similar information normally distributed for use by people. No real-time constraints exist on jobs of this nature other than the time to complete execution.

More recently, however, installations have been created where the computer must respond to external events and perform computations and control functions within specified time limits. Systems of this type are said to be response- or real-time-oriented. The exact time required for response varies with the system function and may be as long as several seconds and as short as several microseconds.

Airline-reservation systems are examples of systems which must respond in a matter of a few seconds with certain information. Here the problem is one of large amounts of data files (reservation schedules), constant changing of the data, and their use and modification by many sources (ticket agents).

Certain forms of data acquisition may require scientific computations on data as they are being received and stored for further analysis. This preliminary analysis may provide the basis for dynamic control or merely provide a first cut at the analysis to guide the later more detailed data reduction. The time constraints of the acquisition rate govern the computing to some degree (see **Data-acquisition computer**).

In control applications, the chief job of the system may be to control information output as a result of sensor information read in. Petrochemical-plant control, engine and transmission production and performance testing, paper-mill control, and gas-distribution control are applications of real-time computing. The computing work performed by the computer varies from simple control algorithm calculation (see **Algorithm**) to complex process optimization and resource scheduling.

One significant factor in real-time computing systems is their response attribute. This facility allows the computing and control power of the computer to be activated by and synchronized with external events to produce more useful work.

REFERENCES

Martin, J.: "Programming Real-time Computer Systems," Prentice-Hall, Inc., Englewood Cliffs, N.J., 1965.

Desmonde, W. H.: "Real-time Data Processing Systems: Introductory Concepts," Prentice-Hall, Inc., Englewood Cliffs, N.J., 1964.

—L. E. Wise, *IBM Corporation*

Réaumur Temperature Scale A temperature scale evolved from the invention made prior to 1730 by René-Antoine Ferchault de Réaumur (1683–1757). The initial concept was to use diluted wine as the thermometric liquid and to designate the ice point as 1000 and the boiling point of the liquid as 1080. In time this was changed so that the fundamental interval, ice point to steam point, was 80 degrees, with the ice point designated 0° and the steam point 80°. At present the scale is little used except in the brewing and liquor industries. Temperatures are denoted by the term degrees Réaumur and by the symbol °R.

—Robert D. Thompson, *Taylor Instrument Process Control Division, Sybron Corporation*

Recognition System, Character [See **Character-recognition systems**; **Magnetic-ink character recognition (MICR)**; **Optical character recognition (OCR)**; **Pattern recognition**; and **Voice-recognition systems.**]

Recorder, Graphic An instrument which automatically plots, often in the form of a continuous curve for immediate or later visual analysis, the values of one or more measured variables generally against time, but in some cases, the values of one or more variables against one or more associated variables.

Graphic recorders are available in a multitude of configurations from numerous manufacturers—at prices ranging frgm considerably less than $100 for a small, portable, single-record instrument with no preamplifier and very modest performance specifications—to several thousand dollars for a high-speed multichannel recording system that is designed for high accuracy, with many options and adjustments designed to maximize applicational flexibility and reliability. Thus, a limited review of this complex topic can only attempt to categorize the subject into its main elements. It should be emphasized that exceptions in the form of special instruments can invalidate some of the generalizations reported here.

Record Format: From the standpoint of the type of graphic record produced, recorders are of two major types: (1) strip-chart recorders (see Fig. R-11) in which the variable being measured is drawn on rectangular coordinates—generally with time as the ordinate and the measured variable as the abscissa; and (2) circular-chart recorders (see Fig. R-12) in which the time graduations fan out from the center much as the pattern that would result were arcs drawn from the center of a clock face to fan out to the time numerals

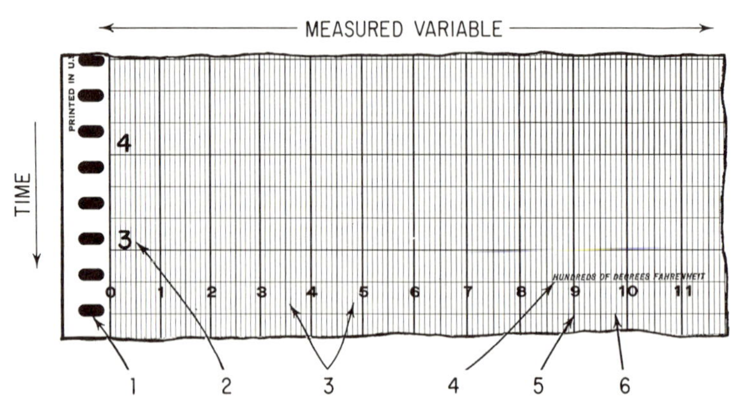

Fig. R-11. Section of an 11-in. strip chart: (1) humidity compensating holes, (2) major time markings, (3) minor graduations, (4) identifying calibration, (5) major graduations, and (6) minor time markings. *(Honeywell.)*

along the circumference. As shown in Fig. R-12, the values of the measured variable generally lie along an arc which depends upon the linkage geometry of the particular recorder.

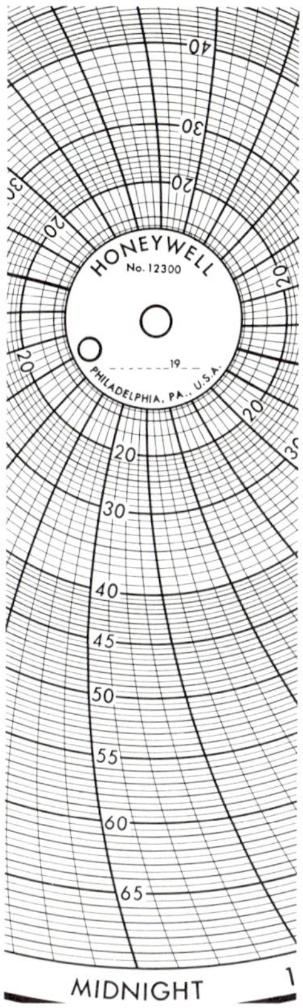

Fig. R-12. Section of a 12-in. circular chart. *(Honeywell.)*

The calibrated width of strip charts on standard, single-channel recorders ranges from as narrow as 2 up to 11 in. The actual width of the chart is greater to accommodate edges for the chart-drive mechanism. Thus, an 11-in. chart actually will be 12$\frac{3}{8}$ in. wide, and so on. Normally, strip charts are furnished in rolls varying from about 30 ft in length up to 400 ft. However, the fan-fold chart in which the strip chart is prefolded in the manner of an accordion has grown in popularity, particularly for the wider charts. The number of calibrated divisions across the width of the chart naturally varies with the chart width, ranging from 10 divisions for narrow charts up to 100 or more divisions for the wider charts. A broad range of chart speeds is available, from several inches per second to a fraction of an inch per hour. Most standard strip-chart recorders are equipped with means to change the chart speed conveniently in fixed increments. For example, the chart speeds available for a given recorder may be 0.1, 0.2, 0.5, 1, 2, and 5 in./min—and 0.1, 0.2, 0.5, 1, 2, and 5 in./s. Chart calibrations and speeds also are expressed in metric units in the United States, the metric specifications commonly being used for high-speed, multichannel recorders. In lower-cost recorders, speed changes often require the physical changing of gears; whereas in more sophisticated instruments, pushbutton speed changing is featured.

The geometry of round charts obviously makes this configuration less flexible and, without experience, more difficult to interpret than the straightforward curve presentation on rectilinear coordinates. However, circular-chart recorders are easy to design and construct and round charts are comparatively easy to store simply by stacking them on pegs. Round charts still are popular for many kinds of industrial use—as contrasted with research and laboratory use—where one chart may serve for a full shift, full day, or even a week or more where the measured variable does not change rapidly. Because the calibrated length (arc) of a round chart is considerably less than half the chart diameter, small chart sizes are impractical for any but the very crudest measurements. The 12-in.-diameter chart with a calibrated width of 4$\frac{5}{8}$ in. is a popular size.

Most chart drives use synchronous electric motors equipped with some form of sprocket that fits into perforations along the edge of strip charts. In the case of round charts, a small hole near the center of the chart fits over a pin that secures the chart to the driving mechanism. In some instances, pneumatic motors have been used. For remote locations where power sources are unavailable or unreliable, a hand-wound spring drive can be used. Or where unregulated alternating current is available or where the recorder must operate during a power failure, a motor-wound spring drive is suitable. Selsyn chart drives also are obtainable where it is desired to proportion chart travel to some quantity other than time.

The drum-type recorder is much less commonplace than either the strip- or circular-chart recorder (see Fig. R-13). The pen scribes its traces on a chart that is mounted on a revolving cylinder so that the measured-variable values appear along an arc. At the end of each recording period, a new piece of chart strip is fastened to the drum.

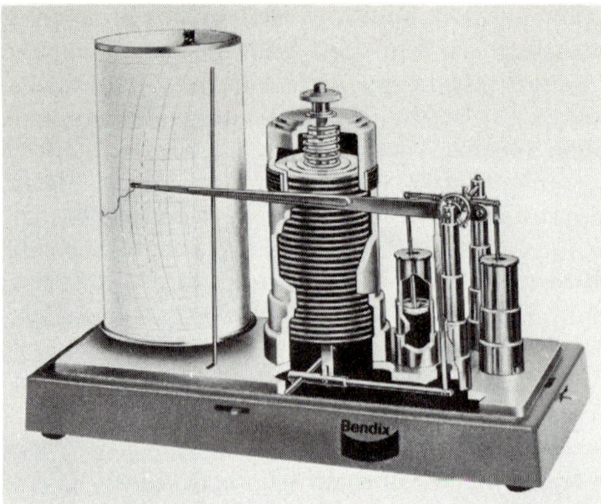

Fig. R-13. Barograph in which a drum- or cylinder-type recorder is used. *(Bendix.)*

Recording Media: The art of graphic recording has progressed much since the very early method of making instrumental scratches on a smoked surface. Some form of paper remains the predominant, permanent-record material, but numerous ways have been developed to mark the paper. One of the oldest and still widely used methods simply is that of pen and ink. There are numerous kinds of pens, including the bucket, the V pen, and the ball-point pen—with various types of capillary feeding systems, both pressurized and gravimetric. Refilling has been made more convenient through the use of ink cartridges or replacement-pen assemblies. For all types of recording the paper must be hygroscopically and dimensionally stable as well as durable for long-term storage. Where wet inks are used, both ink and paper must be matched so that no feathering or blotting of the record will occur. Any treatment of the paper must be such that it will not increase the friction between pen and paper, and ink compositions must be such that they will not clog or discolor over long periods. One form of ink now is offered that does not flow until heated.

Other recording media include the use of solenoid hammers and a carbon ribbon which makes chart markings upon impact; a thermal stylus which makes markings on specially treated papers, some using wax; pressure-sensitive paper upon which permanent markings appear when drawn upon by a stylus under pressure; and Teledeltos paper which reacts to a pulsed electric current to cause markings. Earlier, when recorder pens had to be directly connected to the sensing element through a linkage with no significant amplification, friction with the paper was extremely important. With servo-driven markers, this no longer is so vital, making possible several recording media. Recording on film with a light beam (see **Oscillograph**), of course, is an example of a friction-free "paper/stylus" interface.

Pen- or Stylus-driving Mechanism: Because of the comparatively short distance a pen must travel in a circular-chart recorder—as compared with most strip-chart instruments—the pen can be close-coupled, often with a simple linkage. Pressure and liquid-level gages and pressure-filled thermometers frequently are supplied with close-coupled round-chart recorders. No servo is required—hence the costs are low. In strip-chart recorders, the nature and quality of the pen servosystem largely determine the performance parameters of the instrument—hence the price and applicational flexibility and user satisfaction.

Performance of the pen servo depends, of course, upon the total circuitry of the recording instrument. Most recorders incorporate some form of self-balancing bridge or potentiometer circuit—with pen-position feedback control. Often movement of the pen will be tied in physically with the balancing system, such as movement of a slider along a slidewire. See **Bridge circuits, electric**; and **Potentiometer.** The linearity of the pen servosystem is important and ranges from $\pm\frac{1}{4}$ percent of full scale to 1 percent or more, dependent upon the overall quality of the device. Full-scale pen speeds range from a fraction of a second to several seconds. This characteristic is affected, of course, by the width of the chart that must be traversed as well as the applicational need. The average, rugged industrial-type recorder has a full-scale pen speed of from 2 to 10 s. Pen speed cannot always be equated with recorder quality because, in some cases, rapid pen speed can produce an undesirable record. A quality recorder incorporates some means for damping to prevent overshooting the balance position. The frequency range of a recorder will go from dc up to 200 kHz and higher, again dependent upon need. Most recorder circuits use either a zener diode or mercury-cell reference system.

Recorder Signal Inputs: Recorders fall into two general classes: (1) the recorder that is an intimate part of a total measuring system and where the recorder is simply one instrumental element and (2) general-purpose recorders that are designed to accept a wide variety of input signals.

The close-coupled round-chart recorders previously mentioned fall into category 1. Also in this class will be a large number of industrial indicating-controlling instruments, either electric or pneumatic, where the recording function is but part of the total configuration. General-purpose recorders find wide industrial and particularly extensive laboratory use. Dependent upon quality, complexity, and price range, these recorders will have from a few to several ranges each of which can be switched in or out conveniently. The flexibility of these recorders is enhanced further through use of a series of preamplifiers. As a sampling of what is available, following is a list of the built-in ranges and preamplifiers available with a wide strip-chart recorder.

9–100 mV dc, continuously adjustable	Accuracy ±1% of span
10, 20, 50, 100, 200, and 500 mV and 1, 10, and 100 V dc in 9 fixed steps	Accuracy ±2% of span
10 standard ranges for thermocouple inputs	Accuracy ±1% of span
Ac amperes input	Accuracy ±1% of span
Ac watts input	Accuracy ±1% of span
Frequency input	Accuracy ±2% of span
Power-factor input	Accuracy ±1% of span
Speed input	Accuracy ±1¼% of span

Recorders, often miniaturized, for use in process-control systems are designed to accept the standardized signals of 1 to 5 mA, 4 to 20 mA, and 10 to 50 mA. Also see **Electric controller.**

Recorder-case Configurations: The chart size determines the overall dimensions of a recorder—with case sizes ranging from about 3½ by 5 in. (front) and 6 in. (depth) for a small 2¼-in. strip-chart recorder to 18 by 20 in. (front) and 13 in. (depth) for a large 11-in. strip-chart recorder. A standard circular chart of 12 in. diameter (4⅝-in. calibrated width) will be housed in an instrument case that measures approximately 16½ by 18 in. (front) and 10½ in. (depth). The larger recorders frequently house electronic or pneumatic controlling elements as well as the recording mechanism per se. Recorders are available as battery-operated portable units, for flush and flange mountings on panels, and with numerous accessories all of which affect case size and configuration.

Multipoint or Multichannel Recorders: These instruments are used where it is desired to record simultaneously the values of several measured variables—as, for example, inputs from a number of thermocouples or inputs from a number of different variables, such as temperature and pressure sensors. Normally multiple recorders are used when (1) the relationship between two or more variables is significant or (2) panel space is at a premium. Particularly where the measured variables do not change rapidly, it is less expensive to record several variables by one instrument than to invest in separate instruments—even though multichannel recorders are costly.

The display of multiple points on a single recorder can be handled in several ways: (1) a print wheel can be used instead of a pen—where the print wheel is advanced (much like a date stamp) to a new symbol for each variable or point measured until its full cycle is complete; (2) several pens or styli (usually not over six) can operate off separate servos and operate essentially each in the fashion of a single-point recorder—useful where the resulting records are not too closely spaced, although multiple ink colors assist in differentiation; or (3) the charts can be segmented vertically—with each pen or stylus spanning only a portion of the chart width—in essence an integrated series of narrow strip-chart instruments in one case. Because of the comparative ease of attaining multiple records, the strip-chart recorder offers many more arrangements than the circular-chart recorder in this respect. Where a print wheel may be shared among (up to 48) several points of measurement, the instrument can be designed such that the print wheel will skip a point in sequence should that point have altered (since the last reading) so much that the print wheel carriage has not had time to reach a point of balance. Numerous arrangements are available to cope with this problem.

Space will not be taken here to comment on the application of recorders. It is safe to state that any variable that can be measured can be recorded in one fashion or other. One commonly used although specially designed recorder should be described briefly, namely, the *X-Y recorder.* These instruments are used where it is desired to plot the relationship between two variables [$y = f(x)$] rather than to plot each variable separately as a function of time. The *X-Y* recorder closely resembles and functions similarly to a single-pen recorder with the exception that the chart (*Y* axis) is moved in response to changes in a variable rather than at a uniform time rate. The chart may be driven by a separate servo-actuated measuring element similar to that used in positioning the recorder pen or by a self-synchronous motor in a remote-control system. In most applications for the X-Y recorder, a square-shaped graph sheet is used. Thus, the full-scale chart travel (Y axis) is made equal to the full-scale pen travel (X axis). The addition of a second servo-actuated measuring element and recording pen to an X-Y recorder yields an X-X_1-Y or an X-Y-Y_1 recorder where the relationship among three associated variables can be plotted.

REFERENCE

Considine, D. M. (ed.): "Process Instruments and Controls Handbook," McGraw-Hill Book Company, New York, 1957.

Reducing Valve The term *reducing valve* can be applied to any pressure-regulating device that functions to maintain a preset value of downstream pressure. The reducing valve may be a pressure regulator (operated by the controlled fluid), a relay-operated pressure regulator (operated by the controlled fluid with amplification by a relay), or a control valve used in conjunction with an automatic pressure controller (operated by a separate source of air or gas). In each case, the pressure in the downstream system is sensed or monitored and a corrective signal or pressure change is transmitted to the reducing valve to control the flow and thus maintain the set pressure downstream.

Reentrant A method of program coding (usually used with subroutines or called procedures) which allows the code to be used concurrently by different calling programs. One copy of the code is resident in core and can be used by several programs simultaneously, saving the core that would be required for multiple copies of the code. A multiprogramming atmosphere makes use of this technique (see **Multiprogramming**).

Some properties of code that is written reentrant are: it does not modify itself (is read only); it refers to work areas and parameters through pointers to the calling program or a work area; and it may have a short period of noninterruptibility to set up the reentrant conditions in the case where return-address information may be stored in the core rather than in registers.

—L. E. Wise, *IBM Corporation*

Reference Electrode (See **pH, measurement of;** and **Oxidation-reduction potential, measurement of.**)

Reference-junction Compensation This is the technique whereby variations in the temperature of the reference junction (R/J) of a thermocouple are either manually or automatically taken into account in the determination of the temperature of the measuring junction (M/J) of the thermocouple. See also **Thermocouple.**

The voltage appearing across the terminals of a thermocouple is nonlinearly proportional to the difference in temperature between the measuring junction and the reference junction. The measuring junction is that junction where the two-wire types, comprising a particular thermocouple type, are welded together. The reference junction is that junction where the thermocouple wires are terminated at two terminals made of the same material (usually copper). The circuit is then connected by normal copper wire to the measuring device (see Fig. R-14).

The voltage appearing at the input to the measuring device is proportional to the temperature difference between the M/J and R/J; i.e. $E_{in} = f(T - T_0)$. For a constant measuring-junction temperature, changes in temperature at the reference junction will produce an erroneous voltage input to the measuring device. The influence of a change in reference-junction temperature on the accuracy of the measurement is inversely proportional to the span of the measuring device.

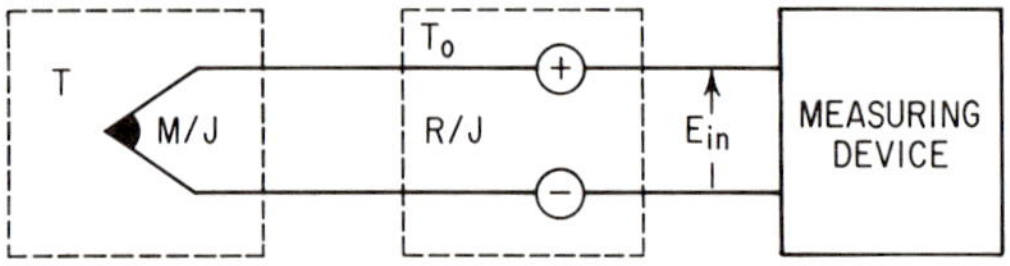

Fig. R-14. Fundamental thermocouple circuit.

Various techniques may be employed to compensate for reference-junction temperature shifts. Among these are:

1. Using a table of correction temperatures
2. Keeping the R/J in a fixed-temperature enclosure (e.g., an ice bath)
3. Passing a temperature-dependent current through a constant resistance in the measuring circuit
4. Passing a constant current through a temperature-dependent resistance in the measuring circuit

Reference-junction compensation may be accomplished manually by measuring the reference-junction temperature, looking up the corresponding emf in the temp/emf tables, adding it to the emf measured across the thermocouple terminals, and looking up the corresponding temperature in the same temp/emf table.

Passing a constant current through a temperature-dependent resistance is a common technique and will be discussed in detail.

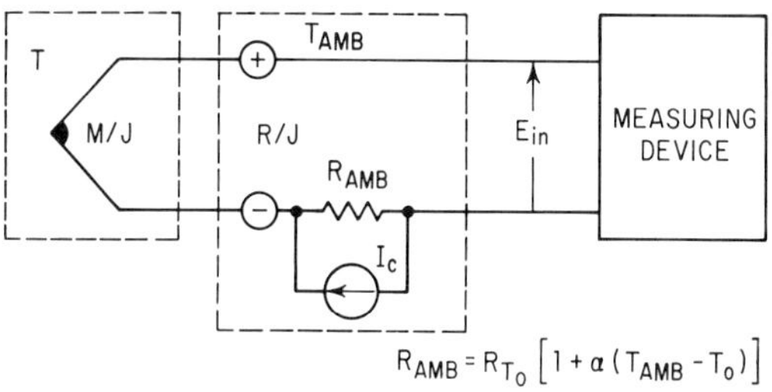

Fig. R-15. Reference-junction-compensation circuit in which a temperature-dependent resistance is used.

A resistor made of copper wire will change its resistance with temperature in a predictable manner. With a constant current flowing through it, the voltage across the resistor will change in a predictable manner.

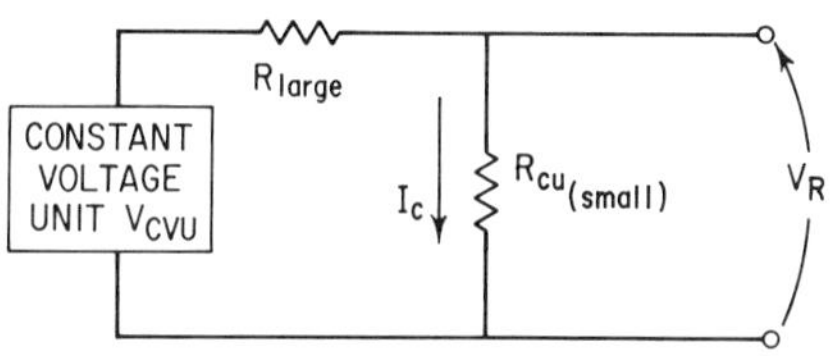

Fig. R-16. Source of constant current in a reference-junction-compensation circuit.

In Fig. R-15, the voltage across the copper resistor R_{amb} is

$$V_R = I_c R_{T_0}\,[1 + \alpha(T_{amb} - T_0)]$$

where R_{T_0} = resistance at temperature T_0

T_{amb} = any ambient temperature

α = temperature coefficient of resistance of copper

I_c = a constant current determined by ambient-temperature limits over which the reference-junction resistor must compensate and by the thermocouple type and other measuring circuit considerations

The constant current may be provided by the circuit configuration shown in Fig. R-16. In this circuit,

$$V_{CVU} = (R_{large} + R_{cu})\, I_c$$

Large changes in R_{cu} will produce negligible changes in I_c. Thus, for practical purposes, a constant-current generator is achieved.

In addition to the temperature-dependent voltage, a temperature-independent voltage has been generated. Expansion of the voltage/current relationship demonstrates this fact.

$$V_R = I_c R_{T_0}\,[1 + \alpha(T_{amb} - T_0)]$$

$$R_{T_0} = R_{cu} \quad \text{at } T = T_0$$

$$V_R = \underbrace{I_c R_{T_0}}_{\text{temp independent}} + \underbrace{I_c R_{T_0}\ \alpha(T_{amb} - T_0)}_{\text{temp dependent}}$$

The temperature-independent voltage can be accounted for by including it in the design of the measuring circuit of the measuring device, or it can be opposed by an equal and opposite voltage, making use of the constant voltage used previously. The circuit in Fig. R-17 shows the entire compensation circuit.

If R_1 and R_{large} are specified as low-temperature-coefficient resistors, E_{in} is now properly compensated so that $E_{in} = f(T_{amb} - T_0)$, independent of the ambient-temperature swing of the reference junction. The only unknown is the $\Delta V/\Delta T$ to be produced by I_c flowing through R_{cu}.

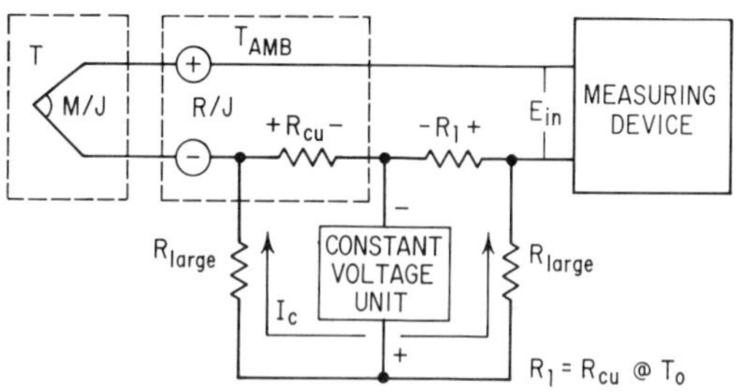

Fig. R-17. Resistance-type reference-junction-compensation circuit.

Figure R-18 shows a typical thermocouple emf vs. temp curve. With the circuit shown in Fig. R-17, a linear compensation voltage will be produced. However, the emf vs. temp relationship is a nonlinear one that has the form

$$e(T) = a_n T^n + a_{n-1} T^{n-1} + \ldots + a_1 T + a_0$$

For small excursions around a given point on the curve, the emf vs. temp curve is approximately linear, and a straight-line compensation voltage is quite accurate. For larger and larger excursions around a given point, the straight-line approximation becomes increasingly more inaccurate.

The problem then becomes one of minimizing average error. In Fig. R-18, the largest errors will occur at T_A, T_0, and T_B. In Fig. R-19, the largest error will occur at T_0. In Fig. R-19, the end points of the ambient-temperature swing are used to determine the compensation line. In Fig. R-18, two intermediate points T_1 and T_2 are chosen such that the error, at T_A and T_B, is approximately equal to the error at T_0. The magnitude of the largest error obtained in Fig. R-18 will be approximately half of the largest error obtained in Fig. R-19.

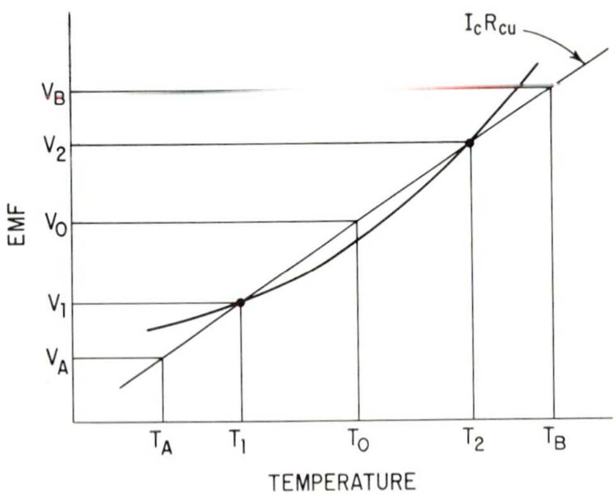

Fig. R-18. Typical thermocouple emf vs. temp curve.

In order to calculate I_c for the smallest maximum error, plot the emf vs. temp relationship for the ambient temperature over which compensation is desired. Determine the T_1 and T_2 which will give the smallest error split around the compensation line. Corresponding to T_1 and T_2 will

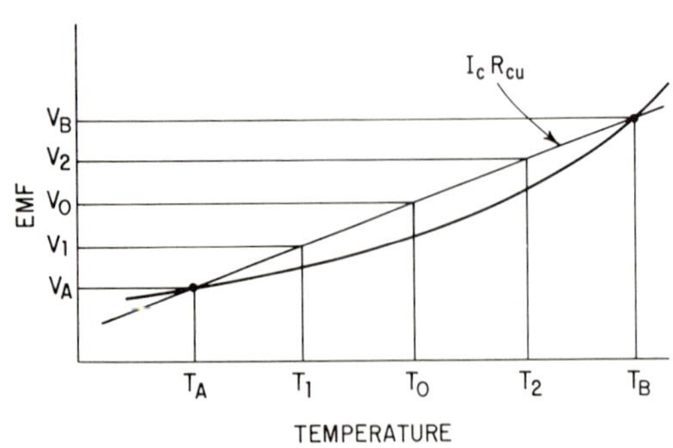

Fig. R-19. The end points of the ambient-temperature swing are used to determine the compensation line.

be two voltages V_1 and V_2. $V_2 - V_1$ will give ΔV. $T_2 - T_1$ will give ΔT. Choose a value for R_{cu} at T_0 (on the order of 10 Ω). By making an appropriate choice of I_c, we wish to make

$$\frac{\Delta V}{\Delta T} = I_c \frac{\Delta R_{cu}}{\Delta T}$$

where $\Delta R_{cu} = R_{T2} - R_{T1}$

$$\Delta R_{cu} = R\,T_0 \alpha \Delta T$$

$$\frac{\Delta R_{cu}}{\Delta T} = \alpha R_{T0}$$

Then
$$\frac{\Delta V}{\Delta T} = I_c \alpha R_{T0}$$

Solving for I_c yields

$$I_c = \frac{\Delta V}{\Delta T} \frac{1}{\alpha R_{T0}}$$

where $\Delta V = V_2 - V_1$ and $\Delta T = T_2 - T_1$.

Or
$$I_c = \frac{V_2 - V_1}{T_2 - T_1} \frac{1}{\alpha R_{T0}}$$

—Robert L. Wilson, *Honeywell Inc.*

Reflectance, Measurement of (See Glossmeter.)

Reflecting Telescope (See Telescope, reflecting.)

Reflection Densitometer (See Densitometer.)

Refractometer An optical instrument for measuring the refractive indices of liquids, solids, or gases. Commercial uses include quality-control applications in the food-processing, brewing, pharmaceutical, chemical, and petroleum industries. Although not widely used in this manner, refractometers have been placed "on line" for continuous recording and control of processes. A precision refractometer will give direct readings to one unit in the fourth decimal place.

Refractometers are of two general types: (1) instruments which utilize the critical angle of total reflection and (2) those which utilize interference techniques to determine the refractive index. The Abbe and Pulfrich refractometers are of the first type; the Jamin and Rayleigh instruments are of the second type.

Regain, Measurement of (See Moisture, measurement of.)

Register, Computer A hardware circuit (see Hardware, computer) which is used to store bits (see Bit) of information is usually called a register.

The term generally implies storage for multibit units of information such as words, bytes, or characters. The register is usually constructed of flip-flops (see **Flip-flop**). The name of the register generally denotes the function for which it is used. A data register is used to store a data word which may be operated on by the program, and an address register (see **Address register**) is used to store an address. A shift register has the capability of moving bits in the word either right or left under control of the computer instruction. An index register (see **Index register**) contains a value which is added to the operand address of an instruction and therefore permits effective instruction modification without changing the instruction itself.

The term register may also apply to a storage location in the immediate-access storage unit.

—W. W. Welz, *IBM Corporation*

Regulator A device that maintains a desired quantity at a predetermined value or varies it according to a predetermined plan. In a broad sense, the term regulator is synonymous with controller. In practice, regulator usually refers to certain kinds of controllers, although the distinction is not sharp. A controller is aptly termed a regulator when it is relatively simple in design, easy to apply, relatively inexpensive, and has comparatively coarse performance characteristics. There are exceptions, of course, and offense should not be taken by makers of more sophisticated controllers who elect to call their products regulators rather than controllers. Usually, a regulator does not incorporate means for transmitting or receiving signals from remote locations, indicating, or recording. Generally, a regulator is very close-coupled to the equipment that it is controlling. Often, regulators are self-contained—self-actuating without requiring external energy sources but depending solely upon the energy derived from the equipment or system that the regulator is controlling.

Regulator, Motor-speed (See **Electric-motor drives, control of**.)

Regulator, Self-actuated (See **Self-actuated regulators**.)

Relative Humidity, Measurement of (See **Hygrometer**; **Hygrothermograph**; and **Psychrometer**.)

Relay A device that is operated by a variation in the conditions in one circuit and serves to make or break one or more connections in the same or another circuit. The circuits may be electrical, pneumatic, or hydraulic. Electromagnetic and thermal relays are common components of electric control systems. A thermal relay is operated by the heat produced by current flow. A latching relay has contacts that lock in either the energized or the deenergized position or both, until reset. The latching action may be produced by a mechanical latch, a holding coil, or a permanent magnet. A stepping relay is a relay whose contact arm may rotate through 360°, but not in one operation. Also see **Electric controller**; **Fluidics**; **Hydraulic relay**; and **Pneumatic controller**.

Relay, Dry-reed The dry-reed relay is an electromechanical, hermetically sealed, magnetically actuated switch used for analog multiplexing and for controlling low-power loads (typically, 3 to 5 VA). The relay capsule consists of overlapped cantilevered contacts hermetically sealed in a glass capsule containing an inert or reducing atmosphere under moderate pressure. This construction is shown in Fig. R-20. Depending on the power level and the application for which the switch is designed, the capsule may be from several inches long to less than an inch with a diameter of 0.25 to 0.1 in. The contacts are a nickel-iron alloy with reasonably high permeability and low retentivity. The contact portion of the reed is often plated with another metal to reduce contact resistance and provide a more reliable contact.

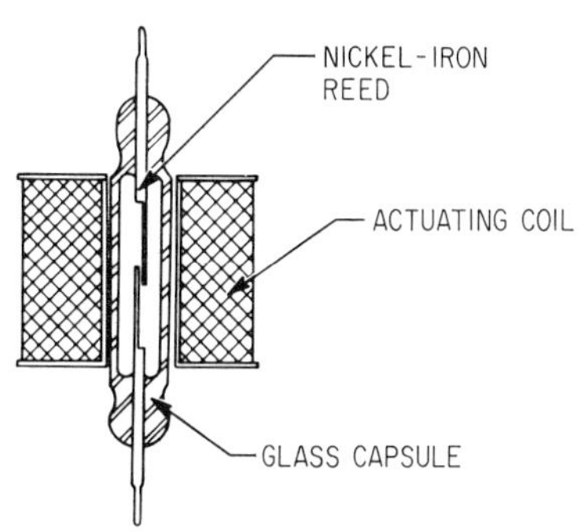

Fig. R-20. Dry-reed-relay construction.

In most applications, the switch is actuated by a magnetic field produced by a coil wrapped around the capsule. The switch is opened by the restoring spring force of the reeds themselves.

Permanent magnets can be used to bias the reeds to produce normally closed, latching, and other contact actions.

The pick time of the relay depends on the construction of the capsule and the characteristics of the drive signal. Typically, pick times are less than 1 ms (including bounce), and drop times may be as small as a few tens of microseconds. Following the closing action, a noise voltage is produced in the reed because of magnetostrictive effects. This noise may persist at significant levels ($>20\ \mu V$) for several milliseconds after the contacts close. The noise and pick-time characteristics usually limit the reed relay to sampling speeds of less than 250 samples per second in high-precision analog multiplexing applications (see **Multiplexer, analog**). Contact life is highly dependent on the switching application, but relays are typically specified for a life in excess of 10^6 operations at full rated resistive load.

—Thomas J. Harrison, *IBM Corporation*

Relay, Hydraulic (See **Hydraulic relay.**)

Relay, Mercury-wetted-contact The mercury-wetted-contact relay is an electromechanical, hermetically sealed, magnetically actuated relay similar in construction and operation to the dry-reed relay (see **Relay, dry-reed**). The difference is in the use of mercury to provide low contact resistance and a reliable contact action. The construction of the mercury-wetted relay capsule is shown in Fig. R-21. The contacts are a nickel-iron alloy which has high permeability and low retentivity. The glass capsule contains an inert or reducing atmosphere under moderate pressure. The base of the armature is immersed in mercury, and a coating of mercury is maintained on the armature and contacts by the surface tension of the mercury. Small capillary lines scribed on the armature surface assist in the capillary action.

Typically, the relay is actuated by a magnetic field produced by a coil which encloses the relay capsule. As in the case of the reed relay, small biasing magnets can be positioned near the capsule to alter its switching characteristics and to provide contact actions such as latching and normally closed. Depending on the construction of the capsule and the drive conditions, the pick time of the mercury-wetted-contact relay is from 0.75 to 6 ms with drop times of up to 2 or 3 ms. As in the case of the reed relay, noise is produced by magnetostriction in the armature. The noise is somewhat less for a capsule of comparable size, however. The decay time of the noise limits the speed of the relay in high-precision multiplexing applications to less than about 300 samples per second.

The mercury-wetted-contact relay is widely used as a low-level multiplexing device in addition to its use in power switching. Its characteristics of low contact resistance, high leakage, high breakdown voltages, and long life make it an excellent multiplexing switch, particularly when high common-mode voltages are encountered (see **Common-mode voltage**). Life is typically specified in excess of 10^9 operations at full rated resistive loads. Although the relay must be operated in a near vertical position, it is a rugged device and easily withstands the vibration and shock found in most industrial environments.

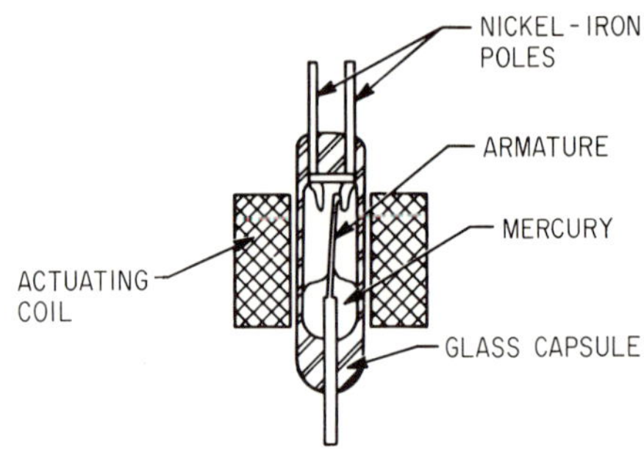

Fig. R-21. Mercury-wetted-contact relay capsule.

—Thomas J. Harrison, *IBM Corporation*

Relocation, Program The process of modifying a unit of code (program, subroutine, procedure) to operate from a different location in core than it was prepared for. This feature is an integral part of virtually every current programming system. For instance, this technique permits a library of subroutines to be maintained in object form and made a part of a program by relocation and appropriate linking.

The process in which a program is prepared for execution by relocating the main routine and any included routines to occupy a certain part of the core on a one-time basis may be referred to as static relocation. The resulting relocated program may be resident in a library and be loaded into the same location in the core each time it is executed.

Some systems have dynamic core-allocation schemes and are set up so that each time a program is loaded, it is relocated to an available space in the core. This is termed dynamic relocation. Some systems, such as time-sharing systems (see **Time sharing**), may also temporarily stop execution of programs and store them on auxiliary storage and at a later time reload them into a different location for continued execution. This is another example of dynamic relocation.

Several means of relocation are available. Software relocation, a commonly used means, is a method whereby a program loader processes all the code of a program as it is loaded and modifies any required portions. Auxiliary information is carried in the code to indicate which parts must be altered. This method can be time-consuming, since all code must be examined.

Hardware relocation utilizes special machine components (such as a base or relocation register) to alter addresses automatically at execution time to give the desired results. This method is fast and desirable in dynamic relocation situations.

Another means of relocation is to use coding techniques which cause the resulting code to be self-relocating. The code executes in any core location into which it is loaded. Two methods are available, depending on the machine capabilities. Index registers may be used to make all references to memory. Double indexing is required to do both the relocation and normal indexing operations. Alternatively, the code may actually modify all addresses as it executes to provide the correct reference. This is slow and requires additional core, but it is useful in some instances.

—L. E. Wise, *IBM Corporation*

Reluctance Transducer (See **Electromagnetic transducer.**)

Remote Control The facility to operate equipment, apparatus, and processes from a distance. A high percentage of industrial control systems fall into this category—where the indicating, recording, and controlling instruments will be located up to several hundred feet from the equipment that is being controlled. In installations of this type, the signals of information flowing from control house to process and back are either electric or pneumatic. Without the use of boosters, pneumatic transmission is distance-limited. See **Transmission, pneumatic**. Remote control has a bearing upon the selection of instrumentation—with certain primary sensing elements being distance-limited. For example, without incorporating an electric or pneumatic transmitting device, a filled-system thermometer must be reasonably closely coupled to the equipment which it is sensing and controlling.

Repeatability For a measuring device, this term may be defined as the closeness of agreement among a number of consecutive measurements of the output for the same value of the input under the same operating conditions, approaching the measurement from the same direction, and for full-range traverses. See Fig. R-22.

It is important to note that this characteristic usually is measured as a *non*repeatability and expressed as repeatability in percent of span. The characteristic does not include hysteresis (SAMA RC20-11). When the measurement of nonrepeatability is made over a specified period of time (preferably a long period) and when hysteresis is included, the characteristic properly is referred to as *reproducibility* (see **Reproducibility**).

In process-control applications, repeatability frequently is the most important operating characteristic of the control loop. This is particularly true in batch processes. In this context, the term relates to the ability of the control loop to repeat a value of the process variable (or end result) for consecutive tries, over a short period of time, and under the same operating conditions. This

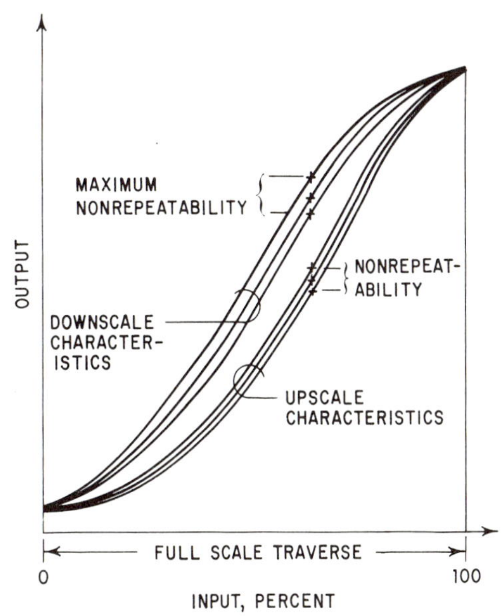

Fig. R-22. Repeatability curves for a measuring system.

characteristic should not be confused with *accuracy* (see **Accuracy**).

For example, assume that a process consists of cutting a continuous run of bar stock into lengths of 4 in. Assume further that, for a given control setting, the resulting bar lengths vary from 4.01 to 4.03 in. around a statistical average of 4.02 in. The accuracy might be stated as 0.03 in. or 0.75 percent, repeatability as 0.02 in. or 0.50 percent, or precision (a term no longer generally used) as the rms deviation of the sample from the average of 4.02 in.

—Robert L. Wilson, *Honeywell Inc.*

Repeatability, Digital-data-acquisition System In digital-data-acquisition systems, repeatability is often defined in terms of the statistical distribution of the reading obtained by reading a stable input many times. A common definition is that the repeatability is the maximum difference between the mean value of the distribution and two boundary points which are determined as the $\pm 3\sigma$ (sigma) limits on the distribution, where σ is the standard deviation of the assumed normal distribution of values (see **Accuracy**). When this definition is used, repeatability is stated as a plus-and-minus quantity (e.g., ± 0.02 percent). Note that this differs from the definition implied by the figure. When specifying the difference between the 3σ limit points, the term "spread" is sometimes used. This term corresponds to the definition of repeatability shown in the figure. Because of the lack of accepted definitions, the user is cautioned to investigate the definition of terms used in specifications when attempting to compare performance parameters.

—Thomas J. Harrison, *IBM Corporation*

Reproducibility The closeness of agreement among repeated measurements of the output for the same value of input made under the same operating conditions over a period of time, approaching from both directions. Reproducibility usually is measured as a *non*reproducibility in percent of span for a specified time period. Normally, this implies a long period of time, but under certain conditions the period may be a short time during which drift may not be included. Factors making up reproducibility are described elsewhere. See **Drift; Hysteresis;** and **Repeatability.**

Residual-gas Analyzer (See **Vacuum gaging.**)

Resistance Thermometry The art of measuring temperature by utilizing the characteristic relationship of electrical resistance to temperature for pure metals expressed by

$$R = R_0(1 + at + bt^2 \cdots ct^n) \quad (1)$$

where R_0 = absolute value of resistance
a, b, c = temperature coefficient of resistance
t = temperature

Only a few of the pure metals having this characteristic relationship can be used in resistance thermometers. Required properties include:

1. The *resistance-temperature relationship* must be highly stable so that neither the absolute value of resistance R_0 nor the coefficients a, b, and c change with repeated heating and cooling within a useful temperature range.
2. *Specific resistance* (ohms per cubic centimeter) must be within limits that permit fabrication of practical size elements.
3. Small resistance changes for *nontemperature effects* such as strains must be minimized.
4. *Change in resistance* with temperature must be relatively *large.*
5. Metal must be *conveniently and repeatedly obtainable* with an essentially identical resistance-temperature relationship.
6. Metal must *not undergo any change of phase or state* within a reasonable useful temperature range.

Platinum, nickel, and copper have a long history of use for resistance thermometry. Additionally, materials such as germanium have been fabricated into temperature-sensing elements which are closely related to metal resistance thermometers and useful for certain temperature measurements. See **Infrared detector, thermistor.**

Platinum Resistance Thermometer: The following properties make platinum the best metal for fabricating sensing elements for resistance thermometers: (1) Platinum is a noble metal, not subject to corrosion and capable of being drawn into very fine wire. (2) The melting point is high and the metal does not volatilize appreciably at temperatures below 1200° C. (3) The metal can be obtained in a very pure state and when pure has a very reproducible temperature-resistance characteristic. While more expensive than nickel or copper, platinum is used in laboratory-type thermometers for highly precise measurements in the range of −183 to +630° C and in industrial-type thermometers in the range of 0 to +650° C.

For industrial types, the materials and fabrication procedures can be controlled with sufficient precision to permit production of thermometers of such accuracy that they can be replaced or interchanged without concern for calibration. With laboratory types, much higher accuracy is obtained by individually checking thermometers at known, fixed temperatures in order to determine their absolute resistance and coefficients precisely.

Check points used are (1) boiling point of oxygen (−182.962° C), (2) triple point of water (0.01° C), (3) freezing point of tin (231.9681° C), (4) boiling point of water (100° C), and (5) freezing point of zinc (419.58° C) [IPTS 68 (International Practical Temperature Scale, 1968)]. These points can be reproduced precisely and with relative ease in a properly equipped standardizing laboratory, and they are appropriately spread over the working-temperature range. In the range 0 to +650° C, the relationship between resistance and the thermodynamic temperature of the platinum resistance thermometer is defined by the *Callendar* equation. The constants can be determined by resistance measurements at the triple point, tin point, and zinc point. The equation can then be used to relate resistance and temperature throughout the useful range of the thermometer.

$$t = \frac{R_t - R_0}{\alpha R_0} + \delta\left(\frac{t}{100} - 1\right)\frac{t}{100} \tag{2}$$

where R_t = resistance at temperature t

R_0 = resistance at 0° C, derived by correcting the triple-point resistance

α and δ = constants derived from resistance measurements made at the tin and zinc points

For temperature measurements below 0° C, the Callendar equation requires an additional term with an additional constant β called the VanDusen constant. This constant β in the expanded equation is determined by checking at the oxygen point.

$$t = \frac{R_t - R_0}{\alpha R_0} + \delta\left(\frac{t}{100} - 1\right)\frac{t}{100} + \beta\left(\frac{t}{100} - 1\right)\frac{t}{100} \tag{3}$$

Laboratory-type thermometers are used to define the international practical temperature scale from −183 to +630° C and for precise measurements in laboratory testing and research, including calorimetry and determining the purity of substances by the freezing-point method.

Nickel Resistance Thermometer: For industrial temperature measurements in the range of −70 to +150° C, nickel sensing elements are commonly used. The low cost of nickel as compared with platinum is the major determining factor. The upper temperature limit is imposed by the materials used in insulating the nickel wire, such as enamel, silk, or cotton. Where the wire is insulated with fiber glass, high temperatures may be measured, but above 300° C, the nickel wire undergoes transformations which make its temperature-resistance curve irregular. Because of nickel's high coefficient of strain, care must be exercised to avoid inducing strains when the thermometer elements are fabricated.

Copper Resistance Thermometer: Copper can be secured commercially in a very pure state, and thus it is not difficult to match an established resistance-temperature table. Like platinum, the temperature coefficient of copper is high. The resistance-temperature curve is straight within narrow limits from about −50 to +200° C. For this reason, copper thermometers frequently are used for temperature-difference measurements. The disadvantages of copper include its tendency to oxidize at higher temperatures and its low resistivity as compared with platinum and nickel. Copper windings must be of fine wire to avoid massiveness, and thus, for the same resistance of a platinum or nickel thermometer, the copper thermometer has a slower speed of response to temperature changes. Thus, copper thermometers frequently have a resistance of 10 Ω at 25° C instead of 25 or 100 Ω, which are the values commonly used for platinum and nickel. The stability of calibration of copper is superior to that of nickel. Copper resistance thermometers are used extensively to monitor winding temperatures in heavy electric-power equipment, such as generators, motors, and transformers, Copper thermometers are useful from −200 to +150° C.

Thermistors: Semiconductor resistance thermometers are called *thermistors.* In contrast to pure metals, semiconductors show a decrease in resistance with temperature rise. This property occurs because of the increased availability of electrons from the breakdown of covalent bonds. The increase in available electrons with increased temperature is very great in semiconductors, thus giving the thermistor a negative coefficient much greater than the pure metal's positive coefficient. Thus, thermistors have greater sensitivity to small temperature changes. Thermistors also have very high room-temperature resistance, which permits the use of only two leads in many applications. The size and shape of thermistors can be varied to provide very fast response with little self-heating.

While not as widely applicable to temperature measurement as the metal resistance thermometers, thermistors are useful when a negative coefficient of resistance is required—as in cryogenic work or in certain temperature-compensating circuits. Thermistors generally are difficult to match exactly or to duplicate when replacement is required. Some form of matching compensator is necessary for accurate measurements.

Resistance-thermometer Displays: For industrial measurements, the temperature is displayed on deflecting-type indicating meters, digital indicators, null-balance indicators, and null-balance recorders. These instruments employ some form of Wheatstone-bridge circuit to measure the thermometer's resistance and can be located a considerable distance from the thermometer. Because of the relatively large change in ohms per degree available from resistance thermometers, display instruments can have range spans as narrow as a few degrees.

Leadwire connections must be given careful consideration to avoid introduction of errors due to leadwire resistance that occur between the thermometer and the display instrument. The use of two leads to connect the thermometer with the display instrument is restricted to thermometers of several hundred ohms and with the display instrument located close to the thermometer. By using three leads, as shown in Fig. R-23, the measurement can be made independent of lead resistance, thus permitting long leads with low-resistance thermometers. In the three-lead method, one lead is placed in each arm of the bridge by running the third lead to one side of the thermometer. For temperature-difference measurements, four-lead loop-type thermometers are used. Two leads are connected at the thermometer to form a loop. Each loop is connected in a bridge arm opposite to its thermometer to provide compensation for the equal amount of lead resistance attached to the thermometer.

Bridges for Laboratory Measurements: For precise laboratory measurements, resistance-measuring bridges are used which have circuits designed to provide the highest possible accuracy. These bridges display the thermometer resistance in ohms, which must be converted to temperature. To minimize lead-resistance effects, the laboratory-type thermometer will have a four-lead potential-type connection. In addition, a reversing switch will be provided so that the leads can be reversed with respect to the bridge and the average of two readings used. The Mueller-type bridge is commonly used with platinum resistance thermometers for laboratory measurements. Other types of bridges also are used to accommodate higher-resistance thermometers or to provide convenience for repetitive measurements in calorimetry or freezing-point determinations. See **Bridge circuits, electric.**

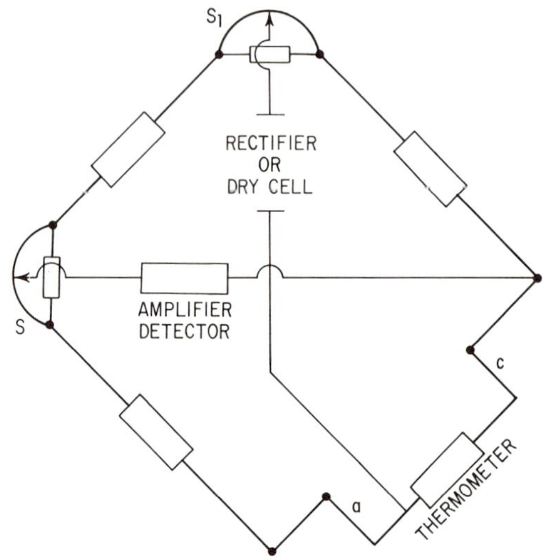

Fig. R-23. Bridge circuit for recorders and controllers used with industrial resistance thermometers with three leads. By mechanically coupling and proportioning S and S_1, the bridge ratio is maintained at 1:1 and effects of lead resistance a and c cancel if these leads are of equal size and length. In industrial thermometers, the resistance of sensitive elements varies from 10 to 500 Ω at 25°C. Resistance changes are 0.4 Ω/°C for 100-Ω nickel, 0.5 Ω/°C for 25-Ω platinum, and 0.04 Ω/°C for 10-Ω copper.

REFERENCES

Stimson, H. F.: The International Temperatures Scale of 1948, *J. Res. Natl. Bur. Std.*, vol. 42, Res. Papers RP 1962, March, 1949.

"Precision Measurements and Calibration," Handbook 77, vol. 2, pp. 1–9, National Bureau of Standards, Washington, D.C., February, 1961.

Stimson, H. F.: "Precision Resistance Thermometry and Fixed Points," Handbook 77, vol. 2, pp. 40–67, National Bureau of Standards, Washington, D.C., February, 1961.

Robertson, D., and K. A. Walch: Calibration Techniques for Precision Platinum Resistance Thermometers,

Leeds & Northrup Co., *Tech. Publ.* A1. 2101, North Wales, Pa., 1961.

Daneman, H. L., and G. C. Mergner: Recent Trends in Resistance Thermometry, Leeds & Northrup Co., *Tech. Publ.* A1. 2101, North Wales, Pa., 1968.

Shirk, W. H., Jr., and R. F. Pistoll: A New High Accuracy Extended Range Resistance Thermometer Bridge, *Proc. 5th Ann. ISA Temp. Meas. Symp., Advances in Metrology,* Paper 68–620, vol. 5, Instrument Society of America, Pittsburgh, Pa., October, 1968.

—W. T. Walker, *Leeds & Northrup Company*

Resolution The least incremental value of input or output that can be detected, caused, or otherwise discriminated by the measuring device, controller, or system. It is often used as an expression of observer error in reading. That is, the resolution of a scale or chart is the minimum error that must be assumed to exist, that can be attributed exclusively to the observation. This is generally considered to be about one-fifth of the smallest scale division. If these least incremental values are considered small, this would be termed fine resolution; if they are considered large, this would be termed coarse resolution.

Many measuring devices effectively have infinite resolution. That is, for a continuously changing input, the smallest additional incremental change in the input—in the same direction—will initiate an additional response on the output. Other devices do not have infinite resolution, and for these, there is a limit in the smallness of an incremental change in the input or output that can be discriminated.

Consider a variable wirewound resistor. The smallest incremental change in resistance value possible (in percent) is the inverse ratio of the number of active turns. In an automatic, null-balancing potentiometer, the feedback element is a slidewire (variable wirewound potentiometer), and for a very slowly changing input, the output will be seen to change in discrete steps with the least incremental change (assuming non-shorting contacts) equal to one over the number of active turns of the slidewire.

—Robert L. Wilson, *Honeywell Inc.*

Resolution, Digital-data-acquisition System In digital-data-acquisition systems, where either the input or output of the subsystem is expressed in digital form, the resolution is determined by the number of digits used to express the numerical value. For example, in a digital-to-analog converter, the output analog signal takes on a finite number of discrete values corresponding to the discrete numerical input. Similarly, the output of an analog-to-digital converter is discrete even though the analog input signal is continuous. (See **Analog-to-digital converter; Digital-to-analog converter.**)

In digital equipment of this type, the resolution is commonly expressed in terms of the number of digits in the input or output digital representation. In the binary system, for example, a common specification is that "resolution is x bits." If V_{fs} is the full-scale input- or output-voltage range, this specification means that the resolution is $V_{fs}/2^x$. If $x = 10$ and $V_{fs} = 5$ V, the resolution is $5/2^{10} = 0.00488$ V. Another common expression of the resolution of a digital system is in terms of "parts." Thus, a 4-digit decimal converter is said to have a resolution of "1 part in 10,000" and a 10-bit binary converter has a resolution of "1 part in 1,024." A third expression is to state the binary resolution as "$\pm\frac{1}{2}$ LSB," where LSB is the least significant bit; in a binary analog-to-digital converter at any given input value, a variation in the input of $\pm\frac{1}{2}$ LSB produces an output change. A similar expression of $\frac{1}{2}$ LSD (least significant digit) is sometimes used in relation to decimal or other nonbinary digital equipment.

—Thomas J. Harrison, *IBM Corporation*

Resolver This device derives its name from its ability to resolve a vector into its coordinate components. The resolver consists of a drum rotor with two windings 90° apart (or in space quadrature) and a stator with two windings also in space quadrature as shown schematically by Fig. R-24. For economy, the low sides of each rotor phase normally are connected together internally and a common connection is brought out. When circuit conditions demand it, the grounds are left separate and four slip rings and brushes are used. The four-ring resolver generally is called a *computing resolver,* the three-ring unit an *electrical resolver.*

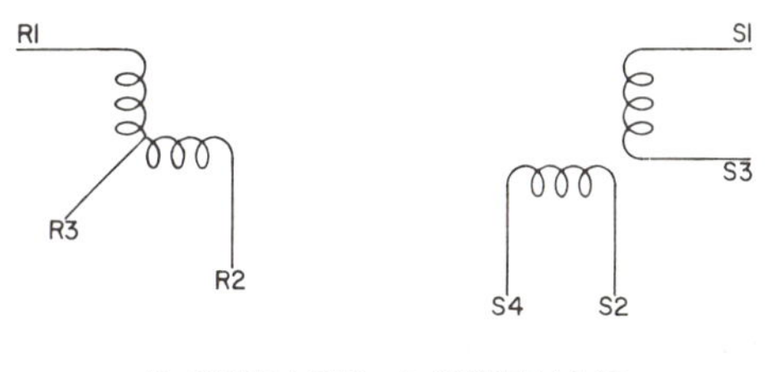

Fig. R-24. Schematic of a resolver.

Consider a resolver with one input phase excited with a voltage V and with its rotor positioned at some angle θ. These two input parameters represent a vector $V\underline{/\theta}$. The resolver output voltages for these input conditions are proportional to $V \sin \theta$ and $V \cos \theta$, which are the rectangular components of the input vector. By use of a simple servosystem, the reverse process also can be generated; that is, given the components as input voltages, the resolver can erect the vector in magnitude and angle.

If both phases of the resolver are excited with two in-phase voltages V_1 and V_2, the output equations will be

$$V_3 = V_1 \cos \theta + V_2 \sin \theta \tag{1}$$

$$V_4 = V_2 \cos \theta - V_1 \sin \theta \tag{2}$$

These are the equations for the transformation from a coordinate system in x and y to another system in x' and y'. The systems have the same origin, but they have been rotated through the angle θ. Thus, a resolver can be used to solve this trigonometric problem. Several resolvers, when properly interconnected, can be used to solve problems of greater complexity, such as spherical triangles. Units used in this manner form what is known as a *resolver chain.*

Often, resolver chains will employ a particular type of unit known as the compensated or feedback resolver (see Fig. R-25). This unit has an additional set of coils, termed the feedback winding, that is wound in the same slots as the stator winding. Practically all the flux generated by the stator links all the turns of the feedback winding, thus making the compensator output essentially equal to the stator input voltage. The compensator voltage then is fed into a booster amplifier. Changes in the stator flux field due to temperature, frequency, or any other service condition are reflected in the compensator output. Through negative feedback, the change in the compensator output restores the stator field to its original condition. The resolver output thus can be maintained at a stable value over a wide range of service conditions, thereby minimizing system errors.

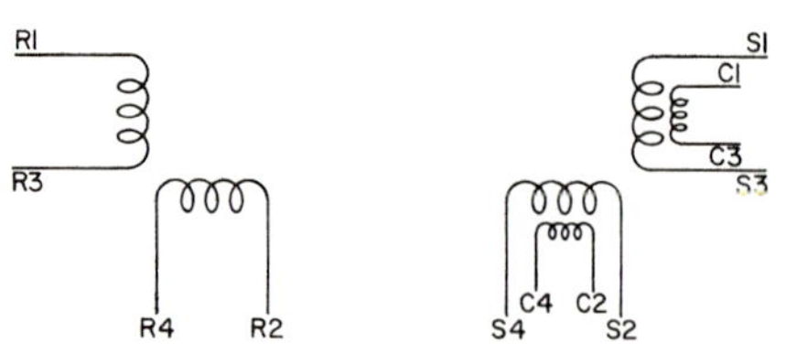

R = ROTOR LEAD S = STATOR LEAD C = COMPENSATOR LEAD

Fig. R-25. Schematic of a feedback resolver.

A resolver also may be used as a phase shifter. When connected in the circuit of Fig. R-26, assuming R is equal to X_c and both are large with respect to the resolver output impedance, the output voltage no longer will vary sinusoidally with rotor-position angle. Rather, the voltage will remain constant, but the time-phase shift between the input and output voltages will vary directly with rotor position. Thus,

R = ROTOR LEAD S = STATOR LEAD C = COMPENSATOR LEAD

Fig. R-26. Schematic of a resolver used as a phase shifter.

$$V_{\text{out}} = kV_{\text{in}}\underline{/\theta} \tag{3}$$

In addition to the foregoing specialized functions, resolvers may be used as control-system components exactly like their three-phase counterparts. In such cases, the units sometimes are indentified by their function in the system, that is, *resolver transmitter, resolver differential,* or *resolver-control transformer.* Resolvers often are used in these kinds of applications because resolvers generally are more accurate than three-phase synchros. The use is not widespread, however, because generally resolvers are more expensive. Resolvers could be designed for use in torque systems, but there would be no advantage. Transmitters and receivers are more than adequate for torque-transmission systems. Also see **Linear transformer; Synchro;** and **Transolver.**

—Joseph P. Spetzer, Jr., *Clifton, Division of Litton Industries*

Resonance Bridge (See **Bridge circuits, electric.**)

Resonant Frequency, Transducer The measured frequency at which a transducer responds with maximum output amplitude. Transducers or sensors may have more than one resonant frequency. Unless stated to the contrary, it is assumed that the given resonant frequency is the lowest frequency of resonance for the particular device. The resonant frequency commonly is expressed in hertz or a multiple thereof. The resonant frequency can be determined readily from the inspection of the frequency-response plot. (See **Frequency range, transducer.**) The term derives its popularity from the convenience with which it can be determined—without a knowledge of the internal workings of a transducer or sensor. However, resonant frequency should not be confused with *natural frequency*.

The natural frequency may be defined as the frequency of a resonant system or device at which the phase of the output lags the input of the forcing function by 90°. The natural frequency of a linear mechanical device is proportional to the square root of the ratio of spring constant or mass. Similarly, for an electrical device, the natural frequency is proportional to the square root of the reciprocal of the product of inductance and capacitance. Thus, the natural frequency is derived most readily from the knowledge of the construction of a transducer or sensor, whereas the resonant frequency is determined most readily from the performance data of the device.

The natural frequency f_n and the resonant frequency f_r are related to each other by way of the damping coefficient or damping ratio h in the following manner:

$$\frac{f_r}{f_n} = \sqrt{1 - 2h^2}$$

The damping ratio h is the ratio of the actual damping to the critical damping, that is, the damping required to make the device nonresonant.

—Otto Muller-Girard, *Consulting Engineer, Rochester, N.Y.*

Respiration, Measurement of (See **Pulmonary system, instrumental analysis of.**)

Response, Ramp The total (transient plus steady-state) time response resulting from a sudden increase in the rate of change in the input from zero to some finite value.

Response, Steady-state That part of the time response of a system or element remaining after transients have expired.[1] The mathematical equations describing the dynamic behavior of a linear control system will be a linear differential equation. The complete solution of a linear differential equation consists of two separate solutions. One of these describes the temporary or transient response of the system from the start of the response until the steady state has been reached. This transient response generally is called the complementary function and represents the solution of the homogeneous equation which is defined as the equation of motion with the forcing function equal to zero.[2] In stable systems, the transient response eventually becomes zero.

The second part of the complete solution is a steady-state solution of response, describing the final or continuing state of the system. This is the solution of what is termed the particular integral of the equation of motion. For a series circuit containing resistance and capacitance, as shown in Fig. R-27*a*, the transient and steady-state responses are shown in Fig. R-27*b*.

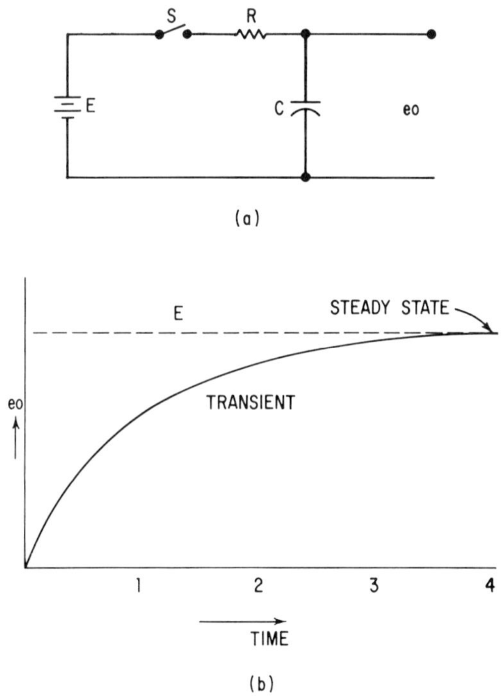

Fig. R-27. Steady-state response of a resistance-capacitance circuit.

[1] Extracted from American National Standard CE-85.1-1963, American Standard Terminology for Automatic Control, with permission of the publisher, The American Society of Mechanical Engineers, New York.

[2] Also see Joseph A. Goldberg, "Automatic Controls and Principles of System Dynamics," pp. 53–59, Allyn and Bacon, Inc., Boston, 1964.

A steady-state response also can be a time-changing response of rate as in the case of a first-order system with a ramp input. In this case, the steady-state output is equal to the input with the displacement output lagging the input because of inertia.

—Richard H. Kennedy, *Bristol Division, American Chain & Cable Company, Inc.*

Response, Step The typical time response of a system to a step increase of input is shown in Fig. R-28. Most of the terms appearing on this diagram also are defined separately elsewhere in the encyclopedia.

If the loop gain of a control system, i.e., controller and process, is equal to 1.0, an input disturbance will never die out but will recirculate around the loop unchanged in size. Since the controller gain (proportional band) is adjustable while that of the process is fixed by its design, the transient response is determined by the controller gain. By reducing the controller gain to a point where system gain is less than 1.0, the cycling effects of a step input or disturbance can be made to die out. For example, in a hot- and cold-water mixing system where the mixed water temperature is controlled by proportioning a valve in the cold-water pipeline, a loop gain of 1.0 results in mixed-water-temperature cycling, as shown in Fig. R-29. The controller gain setting resulting in this type of response is called the "ultimate proportional band." If the controller proportional band were increased (gain decreased) to reduce loop gain below 1.0, the tran-

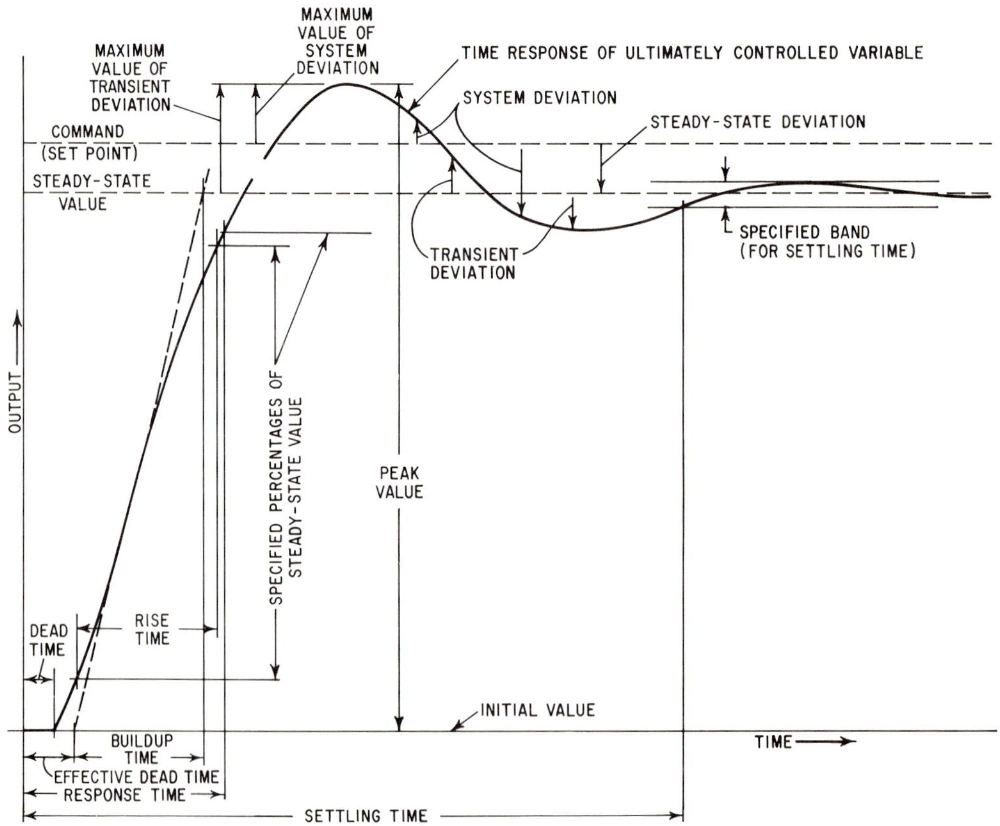

Fig. R-28. Typical time response of a system to a step increase of input.

Response, Transient The time and manner in which the output of a system or element reaches a specified value after the application of a step input or disturbance. Transient response is a measure of system performance, i.e., how it responds to a set point or load change.

sient response would appear as in Fig. R-29*b*, while decreasing the proportional band will increase loop gain above 1.0, which would cause the cycling to increase as shown in Fig. R-29*c*.

Shinskey (Ref. 1) points out that determining the period of oscillation r_0 by finding the ultimate proportional band where constant-amplitude cycling occurs, plus knowing the process dead time r_d, will give an accurate picture of the process.

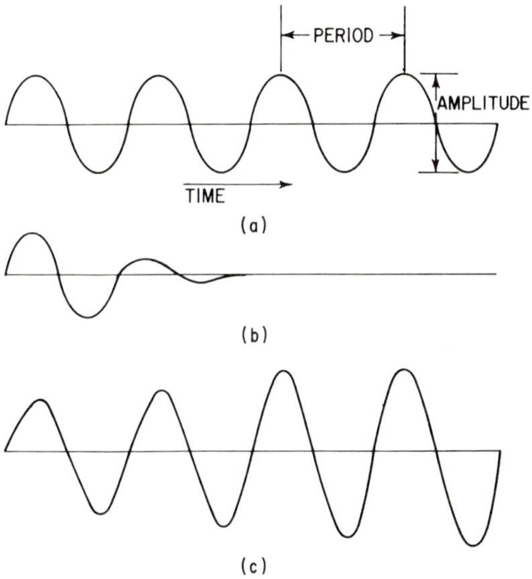

Fig. R-29. Transient response of typical control loop with different loop gain settings. Controlled variable transient response (*a*) with loop gain equal to 1.0; (*b*) with loop gain less than 1.0; and (*c*) with loop gain greater than 1.0.

Ziegler and Nichols (Ref. 2) proposed a method of finding controller settings from the transient response of the loop to a step disturbance. This calls for finding the ultimate proportional band PB_u of the controller where constant cycling occurs and measuring the cycle period P_u. If these two variables are known, the optimum proportional-band, reset, and derivative settings can be determined by the following equations:

For *proportional-only* controllers:

$$S = 0.5\,S_u$$

where S = controller gain setting for stable control

S_u = ultimate gain = $100/PB_u$

For *proportional-plus-reset* controllers:

$$S = 0.45\,S_u$$

$$R = \frac{1.2}{P_u} \quad \text{repeats per minute}$$

where R = reset setting

For *proportional-plus-reset-plus derivative* controllers:

$$S = 0.6\,S_u$$

$$R = \frac{2}{P_u} \quad \text{repeats per minute}$$

$$D = \frac{P_u}{8} \quad \text{min}$$

where D = derivative setting

REFERENCES

1. Shinskey, F. G.: "Process-control Systems," McGraw-Hill Book Company, New York, 1967.
2. Ziegler, J. G., and N. B. Nichols: Optimum Settings for Automatic Controllers, *Trans. ASME,* vol. 64, pp. 759–768, 1942.

—Richard H. Kennedy, *Bristol Division, American Chain & Cable Company, Inc.*

Response Time, Transducer The length of time required for the output of a transducer to rise to a specified percentage of its final value as a result of a step change of measurand. Or an output, expressed as a function of time, resulting from the application of a specified input under specified operating conditions.

The first definition relates response time to a step change of measurand. The second definition leaves the specification of the nature of the measurand change to the user and also reminds the user to specify the operating conditions. This circumstance has merit since the step change, while often simple to produce, also is often very artificial and atypical of the signal shape to which a transducer or sensor is subjected in actual use. Thus, the response time to a ramp, an impulse, a parabolic, a logarithmic, or a sinusoidal signal may be more appropriate. The term *response time* thus should be qualified—as *ramp* response time—to draw attention to the nature of the measurand change.

In the event that the response to the chosen measurand change does not result in a reverse initial reaction or in an overshoot beyond the final value or in oscillations about the final value, then—unless stated to the contrary—it is customary to include the time of the reverse initial reaction but to exclude the time devoted to overshoot or oscillations from the response time. When the response to a specified measurand change is complex, the description of the character of the transducer or sensor often is better served by avoiding a recitation of the *response time.* Other terms such as dead time, frequency range, resonant frequency, and the like may serve better to quantize the nature of the transducer or sensor.

REFERENCES

SAMA Standard RC 20-11-1964, Measurement and Control Terminology, Recorder-Controller Section, Scientific Apparatus Makers Association, New York.

Minnar, Emil J.: "ISA Transducer Compendium," Instrument Society of America, Pittsburgh, Pa., 1963.

Tentative Recommended Practice, RP 37.1, Nomenclature and Specification Terminology for Aerospace Test Transducers with Electrical Output, Instrument Society of America, Pittsburgh, Pa., April, 1963.

American National Standard C85.1-1963, Terminology for Automatic Control, American Society of Mechanical Engineers, New York, 1963.

Considine, D. M., and S. D. Ross (eds.): "Handbook of Applied Instrumentation," McGraw-Hill Book Company, New York, 1964.

American National Standards Institute, New York: C42.25-1957, Industrial Control Equipment; C42.30-1957, Instruments, Meters, and Meter Testing; C42.65-1957, Communication.

Scientific Apparatus Makers Association, Chicago: "Standard Load Cell Terminology and Definitions.

Instrument Society of America, Pittsburgh, Recommended Practices: RP26.1-1957, Dynamic Response Testing of Process Control Instrumentation, Part I—General Recommendations; RP26.2-1960, Part II—Devices with Pneumatic Output Signals; RP26.3-1960, Part III—Devices with Electrical Output Signals; RP26.4-1961, Part IV—Closed Loop Actuators for Final Control Elements; RP31.1-1961, Terminology and Specifications for Turbine-type Flow Transducers (Volumetric); RP37.3-1964, Guide for Specifications and Tests for Strain Gage Pressure Transducers for Aerospace Testing.

—Otto Muller-Girard, *Consulting Engineer, Rochester, N.Y.*

Restart The process of beginning again a computing job, task, or other operation. In a control and acquisition system a restart function may be specified for each of the various individual programs or tasks that make up the component parts of the system. Normally, the restart function is activated by a system failure or other unusual occurrence in the area of the task being executed, such as an attempt to initiate an invalid I/O (input/output) operation. The restart function may be a separate task or program specified by the system designer to recover in the event the desired function cannot continue. A restart function may be specified for the whole system to provide for total system recovery in case of detectable system-wide failures. This function may be referred to as the reload function and may be core-resident or auxiliary-storage-resident.

—L. E. Wise, *IBM Corporation*

Rheological Measurements (See **Consistency, measurement of;** and **Viscosity, measurement of.**)

Richter Scale, Seismic Earthquakes are graded or classified for size by intensity and by magnitude. Intensity is measured by the local effects of the motion, so that a given earthquake will have different intensities at different locations. The magnitude of an earthquake, on the other hand, is characteristic of the earthquake itself and independent, at least in theory, of the spot where it is observed.

The scale of intensity recommended by leading seismologist Charles Richter is the so-called modified Mercalli scale, in which intensity is estimated by the effects of the earthquake on people and objects. Magnitude I means not felt except by instruments. Magnitude V means "felt outdoors . . . sleepers wakened . . . liquids spilled . . . doors swing." Magnitude IX means "General panic . . . damage to foundations . . . conspicuous cracks in ground." The largest intensity XII means "Damage nearly total. Large rock masses displaced. Lines of sight and level destroyed. Objects thrown into the air." In areas where earthquakes occur often, studies of the motion have been made with so-called "strong-motion seismographs." These measure the intensity more quantitatively than the modified Mercalli scale but are not of general usefulness because to date so few of them are in operation.

The scale of magnitudes can, of course, be formed with much better quantitative accuracy than the intensity scale, for the reason that it is based on the readings of carefully calibrated instruments rather than, for example, the effects on household objects or on different classes of masonry. On the other hand, there are obvious theoretical difficulties, since the only variable which can be rationally used in a magnitude scale is the energy, and the energy released in an earthquake can be measured only indirectly. This means that a theoretical model must be constructed in which the release of energy by earthquakes, whether their source be shallow or deep, will give predicted effects as recorded on seismographs. Then in the case of actual earthquakes the observed effects of an earthquake to which a magnitude has been assigned must be consistent for the nearby and distant observatories. The problem is still under active investigation, but a theoretical relation between the magnitude M and the energy E (in ergs) which has been widely used is given by the formula

$$\log E = 11.4 + 1.5M$$

in which the largest value of M assigned to any recent earthquake is 8.9 and in which magnitudes are read to the nearest tenth.

—F. E. Romberg, *Texas Instruments Incorporated*

Ripple The conversion of an ac signal to a dc signal utilizing a rectifier circuit causes pulsations in the output circuitry. The unidirectional output voltage may be considered to be the resultant of a periodic alternating current of any frequency superimposed upon a direct current as shown in Fig. R-30.

The pulsating component of the output voltage is referred to as the ripple voltage. The amount of tolerable ripple voltage depends on specific application; however, in most measuring systems, small ripple voltages will produce undesirable noise and offset.

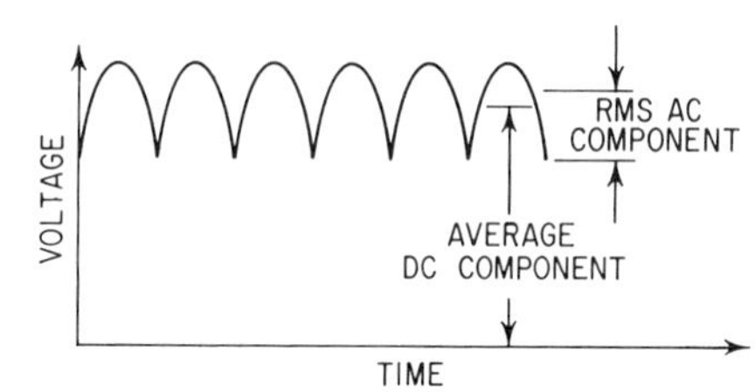

Fig. R-30. Ripple voltage vs. time.

Filters are generally used to reduce ripple. A filter can be as simple as a shunt capacitor which shorts the ac component to ground or as complex as a multiple lattice network consisting of series inductors and shunt capacitors.

Ripple due to rectification is the case most frequently encountered; however, it may result from any demodulation process such as a chopper or dc oscillating amplifier. Summarizing, a general definition expressed as "percent ripple" is the ratio of the rms ac component of the waveform to the dc component multiplied by 100.

—Robert L. Wilson, *Honeywell Inc.*

Rise Time, Transducer The time required for the output of a system (other than first-order) to make the change from a small specified percentage (often 5 or 10 percent) of the steady-state increment of a large specified percentage (often 90 or 95 percent), either before or in the absence of overshoot. Or, the length of time for the output of a transducer to rise from a small specified percentage of its final value to a large specified percentage of its final value.

It is assumed, unless stated differently, that the rise time is the output response of the transducer or sensor to a step change in the measurand or input. The rise time excludes, as partially noted in the definitions, the initial and the final response time. Also, overshoot time is excluded. In the case of transducers or sensors that exhibit a reverse initial response, that response time also is generally excluded. Thus, the rise time is a measure of the maximum rate of rise (stated alternately, of the maximum output velocity) of which a transducer, sensor, or system is capable.

The rise time of a linear transducer or sensor with input amplitude-independent behavior can be computed from the transfer function. The specification *rise time* for such a device adds little to what is already known about the device. However, the specification is useful when the sensor or transducer is subject to internal saturation as a result of a large and sudden input change. The response and the rise time which are a part of the response then may provide data concerning the output element and its capability to change the energy level in the connected load. The load to a transducer or sensor may be one or more other instrumentalities, or the combination of such instrumentalities with transmission lines. Reactive loads, that is, loads of a capacitive or inductive nature, tend to have a significant influence on performance. The rise time is one way to assess that influence.

—Otto Muller-Girard, *Consulting Engineer, Rochester, N.Y.*

Robot A programmed manipulator designed to perform useful work automatically without human attendance. Particularly useful for performance of material-handling functions in industry, the robot not only can increase productivity but also can relieve human operators from tasks that become excessively burdensome, dangerous, routine, or monotonous. The robot can perform these tasks with a consistency and reliability to duty that are beyond the physical capabilities of

TABLE R-5. *Typical Industrial-robot Applications*

Application	Cycles per hour	Tooling Required for Hand	Weight of Part and/or Tools, lb
Unload two die-casting machines, quench and deliver casting to trim press	500	High-temp fingers	25
Operate forging press	150–400	Shaped high-temp. clamp	80
Weld auto frame parts	50–100	Fixture to hold weld-gun	70
Unload plastic-molding machine, paint and box parts	75	Double suction cup on extension rod	23
Spring-clip assembly	1,800	Hopper-fed clip-gun mounting	25
Load and unload two chucking machines alternately	100	Special shaped fingers	25

man. Just a few tasks to which robots are assigned are given in Table R-5.

Unlike special-purpose automation equipment, the commercially available robot is designed with application flexibility and transportability so that it may be moved about and adapted to various tasks. Other advantages of the robot include (1) production-scheduling flexibility, (2) increased product quality and yield, (3) reduced scrap losses, (4) adaptability to existing manufacturing facilities, (5) economy through reduction of labor costs, and (6) multishift operation without concern for the human factors of boredom, carelessness, and fatigue.

While robots have a rather broad potential for industrial applications, they are particularly suited for (1) machine loading and unloading, (2) repetitive stacking, (3) sorting, and (4) positioning and transferring of parts. Use of a robot requires consideration of a number of factors: Are all motions required within the capability of the robot? Are parts presented singly at fixed location and orientation? What methods will be used to coordinate robot action with operation of existing equip-

Fig. R-31. Industrial robot picks up metal stamping on conveyor at left, rotates the stamping 180° while turning, and places stamping on bed of press in background.

ment? How will malfunctions be corrected with minimum production delay?

Robot-operating Principles: Typically, the industrial robot has an articulated mechanical arm and hand having freedom of movement equivalent to a human waist, shoulder, elbow, and wrist. Thus, the robot's hand can be brought to any point within reach and can grasp parts or tools with ease. Application of a robot in a stamping plant is shown in Fig. R-31. Movement of the arm and hand in five degrees of freedom (see Fig. R-32) is adequate for performing most industrial tasks. The hand consists of one or more sets of fingers or grippers under independent open-and-close control; or the hand may be a mounting bracket for a welding gun, spray gun, or other operating device under automatic control. In place of a gripper, the hand may be equipped with suction cups, magnetic pickups, spreaders, or hand tools, such as nut spinners. A servosystem usually is employed for each degree of freedom to provide adequate load and speed abilities. This is backed up by a memory system that can direct the hand to accomplish or repeat a complex pattern of movements.

Robots must be *taught,* that is, programmed to perform the desired task. In the *teach mode,* the robot is directed through various movements in sequence while the necessary information is recorded in the memory. In the *playback mode,* the robot repeats, as desired, the sequence of movements taught by the operator and stored in the memory. Control of motion may be *point-to-point* or *continuous-path.*

Point-to-Point Robot: A robot of this type not only can perform operations as recorded for each step of a program but also can hold for an unspecified time at any desired step of a program. Since an industrial robot almost always works in conjunction with automatic equipment operating on a timed cycle, the point-to-point robot may be synchronized by use of interlocks at any point in a program. The robot may provide command signals for operation of external equipment or may wait for command by external equipment.

In the *programming mode,* teach buttons, as shown in Fig. R-33, control hydraulic power to an actuator for motion of arm-hand assembly to the desired position. The shaft encoder continuously monitors position. The record button is depressed to impress on the memory drum the position data for all five axes of motion. A stepping drive then indexes the memory drum one step. The process is repeated to record all steps in the program. An end-of-the-program signal is recorded for automatic drum return to the first step.

In the *playback mode,* as shown in Fig. R-34, data for the desired position of each axis of motion, as recorded on the memory drum, are transferred by read heads to the comparator. Actual position, as shown by the encoder, is compared with the desired (memory) position. A difference in information activates the servosystem to drive the actuator until the positions coincide. When all five axes of motion are at memorized positions and compliance with additional control and auxiliary functions is obtained, the memory drum is indexed to the next step.

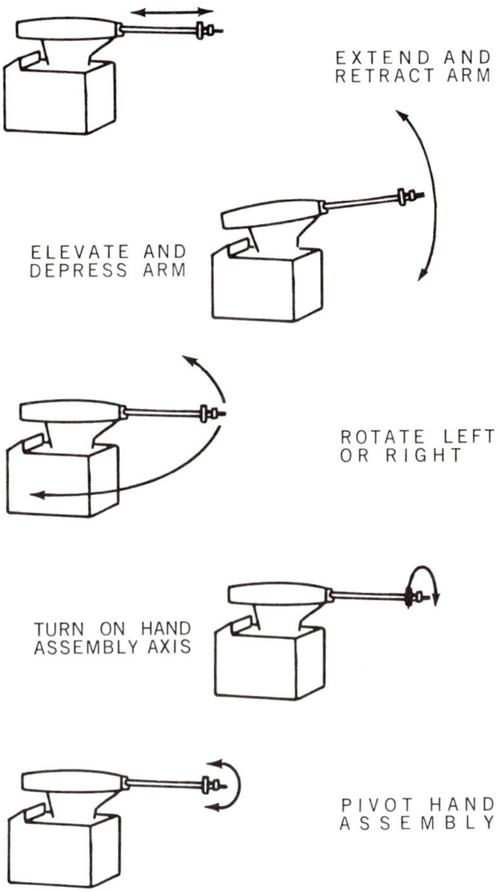

Fig. R-32. Five degrees of freedom designed into industrial robot.

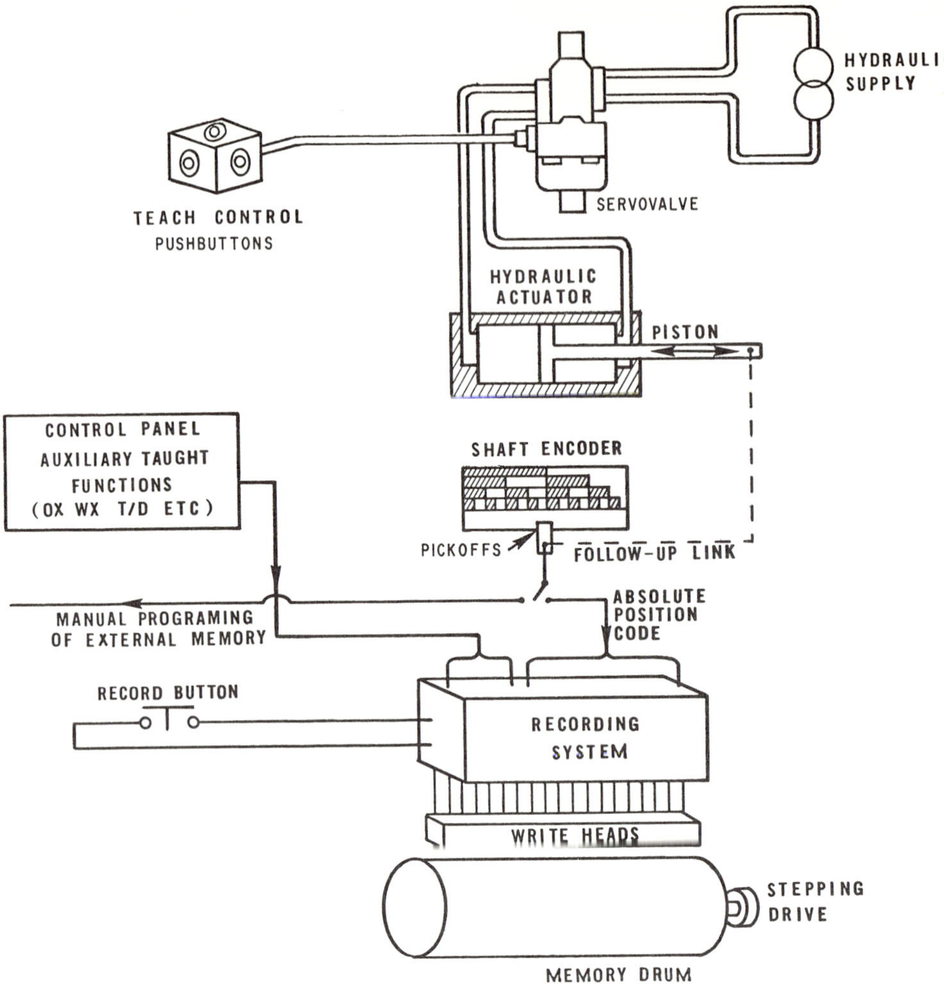

Fig. R-33. Programming operation for a point-to-point robot.

Continuous-path Robot: The arm of a continuous-path robot can move from one point in space to another point in space by any one of a number of paths, each of which may be predetermined and controlled completely. The outer end of the arm describes a smooth compound curve—hence the term continuous path. Complete control of the path greatly increases the information requirements for playback. The length of the program depends upon the memory capacity. With the continuous-path robot, programs are more difficult to revise and synchronization with external equipment is more difficult.

In the *programming mode,* as shown in Fig. R-35, a manual programmer lever directs the hydraulic servosystem for motion of the transfer arm. A resolver continuously monitors the arm position. The operator directs the transfer arm in three degrees of freedom, manually, through the movement pattern to be recorded. The gripper at the end of the arm is actuated, and control and synchronizing signals for external equipment are activated if called for by the program. Continuous position information for three axes of motion and auxiliary signals are recorded simultaneously on two magnetic tapes.

In the *playback mode,* as shown in Fig. R-36, the data for the movement pattern, as recorded on one of two identical magnetic tapes, are transferred to a summing circuit. Actual arm position, as shown by the resolver, is compared with the desired position as recorded. A difference in information activates the servosystem to drive the actuator until the positions coincide. Arm motion in three axes is controlled to duplicate the movement pattern, and auxiliary signals are presented as recorded. At the end of the program, an alternate tape is activated for immediate repeat of the program while the first tape is being rewound. The delay between replays is in milliseconds.

Robot Specifications: In one commercially available robot,[1] the device is equipped with an arm having a 7½-ft reach and a two-finger hand with an adjustable clamping force up to 300 lb. With this arm, the robot is able to reach from 4 in. above the floor to an elevation of slightly less than 8 ft and to position a part anywhere within a 350 ft^3 working area. In place of fingers, the arm can be equipped with other manipulative devices and working tools. The unit can move a 25-lb load at top speed or a 75-lb load at reduced speed. Maximum operating speeds of the standard machine with arm at full radial extension are: radial velocity, 30 in./s; vertical velocity, 50 in./s; rotational velocity, 110°/s; wrist-bend velocity, 110°/s; and wrist-swivel velocity, 110°/s. Other maximum arm-hand movement parameters are: radial extension, 3.5 ft; vertical stroke, 87.5 in.; arm rotation 220°; wrist bend, 220°; and wrist swivel, 180°. The unit is designed for a 40,000-hr

[1] Unimate, Unimation, Inc., Danbury, Conn.

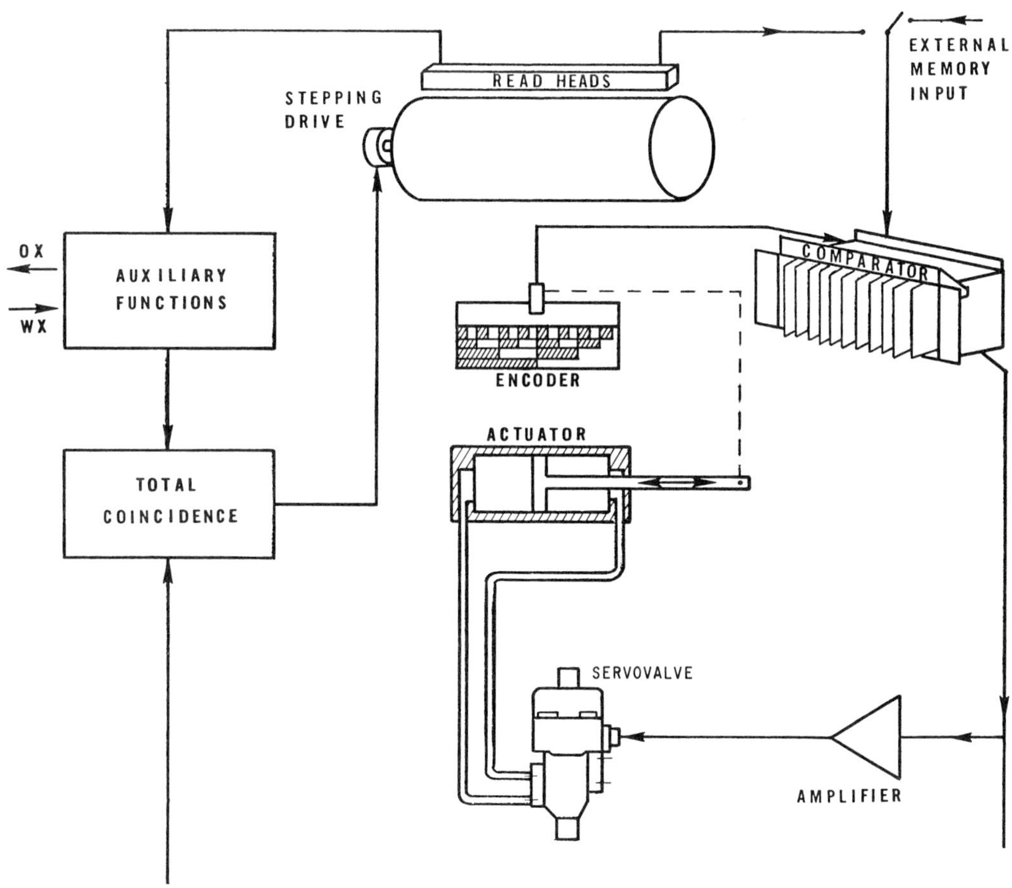

Fig. R-34. Playback operation for point-to-point robot.

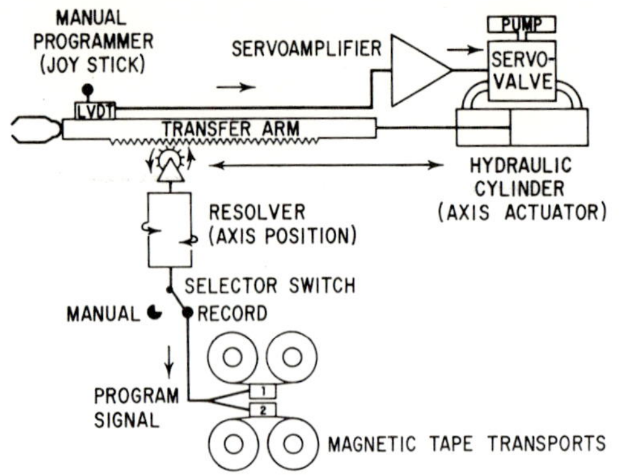

Fig. R-35. Programming operation for continuous-path robot.

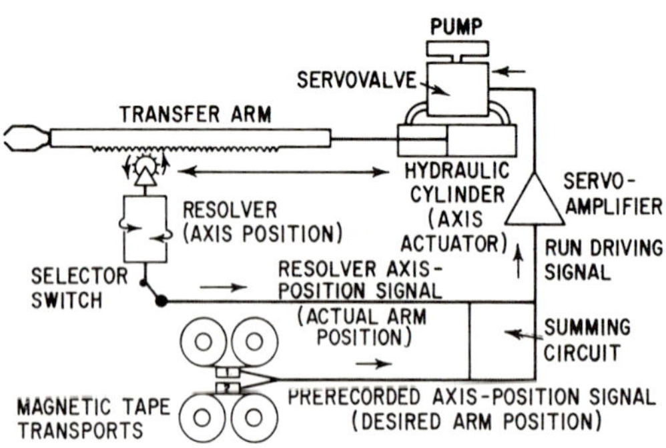

Fig. R-36. Playback operation for continuous-path robot.

operating life in temperature environments extending to 120° F. The device weighs approximately 3,000 lb.

Other industrial robots[2] and the addition of other features to the equipment described here make it possible to obtain a robot for almost all industrial applications. See also **Manipulator.**

—Harvey A. Senior, *Unimation, Inc.*

Roll in/Roll out A function where a portion of the core is temporarily stored on auxiliary storage media to make the core available for a different (usually higher-priority) function and the later reloading of the stored portion for continued execution. This function is often used to allow low-priority tasks or jobs to be run when the main or higher-priority task does not require the core but must have it available on demand. In control and data-acquisition situations this method is used to allow routine computing work to use the computer when the process is idling or waiting. (See **Time sharing.**)

Another use of the technique is for exceptional conditions which require priority use of the computer. Any system function which uses roll in/roll out must be able to tolerate the time required to accomplish the moves to and from the auxiliary storage.

—L E. Wise, *IBM Corporation*

Rotameter The most extensively used form of the variable-area flowmeter. For diagram, see **Flowmeter, variable-area.** The rotameter uses a tapered tube mounted vertically with the smaller (inlet) end at the bottom. The largest diameter of the float is slightly smaller than the internal tube diameter at the bottom of the tube. The float is free to move vertically through the metering tube, and at no-flow rate it rests on a bottom float stop. The float seeks a position in the tube where the area is such that the *upward forces* acting on the float due to pressure loss *balance* the *downward force* of gravity.

The term rotameter had its origin in the fact that in earlier meters of this type, the floats were produced with slots which imparted spinning about their vertical axis to provide spin stablization needed to center the float. Mechanical centering is now common and can be accomplished with three or more longitudinal ribs or beads formed on the inside surface of the glass metering tube. Plain conical glass tubes with no beads often are equipped with a center rod, supported from the ends, on which a float, with a hole through its center, is used. Metal-tube rotameters generally have plain conical tubes and floats equipped with rigidly attached guide rods above and below the metering edge.

Sharp-edged floats, with a certain degree of immunity from viscous-drag effects, are available in ½-in. sizes and larger. Turbulent flow (thus, viscosity immunity) is maintained for viscosities up to approximately 9 centipoise in the ½-in. size and up to nearly 100 centipoise in the 3-in. size. Although the rotameter can be used in the laminar-flow region, a different calibration and flow curve for each different operating condition is required. In practice, meters used for varying viscosities above the viscosity limit are supplied with a family of curves in which scale reading is plotted vs. the actual flow for various viscosities (or temperatures). The viscosity or temperature

[2] Including Versatran, AMF Thermatool, Inc., New Rochelle, N.Y.

must be known at the time the rotameter reading is taken.

Spherical ball floats are standard for all meters ¼ in. and smaller, with $\frac{1}{16}$-in.-diameter balls being the smallest available from most manufacturers. Because of their very small mass, all such ball floats operate in or near the laminar-flow region for all liquids and gases.

Glass-tube General-purpose Rotameter: The most commonly used all-purpose indicating rotameter is the glass-tube type with 5- or 10-in. scale length. While there are exceptions, the majority in the ¼-in. size have 5-in. scales, while those with ½-in. or larger pipe size have a 10-in. scale length. This class of meters generally has an accuracy of ± 2 percent of full scale with a potential accuracy of ± 1 percent of full scale with special calibration. Individual meters typically have a 10:1 ratio between full-scale flow rate and minimum scale reading. The maximum flow rates range from about 0.5 cm^3/min of water or 35 cm^3/min of air to 200 gpm of water or 400 scfm of air. The largest glass-tube meter generally has 3-in. flange connections, although some 4-in. meters are available.

The glass tube may be sealed to the metal-pipe fitting with either a packing gland or *O*-ring seals. A body, generally constructed of metal, maintains rigidity and correct dimensional separation of the two end fittings. Connections on the metal-pipe fittings generally are female national pipe threads (NPT) or 150- or 300-lb ANS flanges. Rotameters can handle almost all fluids because of the availability of glass- or plastic-lined or solid plastic fittings.

Purgemeters: These typically are short-scale (1½- to 3-in.), low-flow (below 0.5 gpm or 2 scfm), small-pipe-size (¼-in. NPT or ⅛-in. NPT) glass-tube rotameters. The name purgemeter stems from the wide use of these meters for indicating the rate of flow of a purging liquid or gas. Because of the short scale, the accuracy generally is ± 5 percent of full scale.

Rotameters for Secondary Functions: In addition to flow-rate indication at the point of measurement, rotameters can be equipped with additional functions, including alarm, pneumatic or electronic transmission, recording, controlling, or totalizing. For these purposes, the industry standard is a 5-in. float travel. The combination of linear output and 10:1 flow range makes the rotameter particularly attractive for transmitting flow signals to central control rooms. These meters must be equipped with a magnet or iron core for performing a secondary function. The lowest practical full-scale flows are of the order of 1 gpm in a ½- or 1-in.-size meter, up to 300 gpm for the 3-in. size.

Precision Rotameters: The use of an expanded scale length of 20 to 24 in. makes possible rotameters that will give accuracies of from ¼ to 1 percent of reading. In some cases, the percent of reading capability is enhanced by the use of logarithmically shaped tubes with a very slight taper at the bottom (inlet) end and a continuously increasing taper angle at increased heights in the tube. The resultant scale has equal-percent-of-reading resolution at all points on the scale—as opposed to the equal-percent-of-full-scale resolution on standard linear-tube meters.

Rotameter Corrections: Since the majority of fluid applications of rotameters are for low-viscosity fluids, the most important corrections are for changes in fluid density or specific weight. These corrections are especially important for gas flows where changes in operating temperatures and pressures cause significant changes in the specific weight of the flowing medium. Equations and tables for making such corrections are readily available from rotameter manufacturers.

—Charles E. Fees, *Fischer & Porter Company*

Rotating-type Flowmeter (See Flowmeter, positive-displacement.)

Rubidium-gas Cell (See Clock, atomic.)

Rubidium-vapor Magnetometer (See Magnetometer, rubidium-vapor.)

Rupture Disk, Pressurized-system An intentionally designed "weak spot" within a pressure system. Thus, the rupture disk is expected to fail before other more valuable equipment is damaged or destroyed, or before an explosive force can be created that will endanger human life.

S

Saccharimeter (See **Polariscope.**)

Safety, Intrinsic (See **Intrinsic safety.**)

Safety Controls (See **Alarms, control-system**; and **Protective devices, control-system.**)

Safety Device, Pressure-relief (See **Rupture disk, pressurized-system.**)

Sampling Theorem This term generally refers to a theorem developed by Shannon which states the conditions under which a continuous signal can be reconstructed if its value is known at particular instants of time. Specifically, the theorem states that a band-limited signal containing no frequency components greater than f Hz can be completely reconstructed from equally spaced samples taken at a rate of $2f$ samples per second. Failure to satisfy the conditions of this theorem in practical applications results in distortion in the reconstructed signal. This distortion is known as "aliasing error" or "folding error." (See **Error, aliasing.**)

In practical situations, the conditions of the theorem are rarely satisfied in that few signals are strictly band-limited and most sampling techniques result in a sample which is representative of the sampled signal over a small, but finite, period of time. Although the conditions for complete reconstruction of the signal under these nonideal conditions can also be derived, the result is more complex. As a result, the theorem is generally accepted as a guideline for the analysis of sampled data. The result of using the theorem in these nonideal cases is rarely of consequence if the duration of the sampling time is small compared with the period of the highest significant frequency and if the signal is essentially band-limited, i.e., if it contains no significant frequency components in excess of f Hz. In critical applications where accurate reconstruction of the signal is essential, however, the characteristics of the sampling process and the sampled signal should be carefully analyzed and an estimate of the aliasing error should be obtained.

—Thomas J. Harrison, *IBM Corporation*

Satellite Instrumentation (See **Space and Missile-system instrumentation.**)

Saturable Reactor The permeability of a magnetic material varies with the amount of applied magnetization. The saturable reactor takes advantage of this principle for control purposes. Its use is limited, however, because of poor high-frequency response. If a control winding is added to an iron-core reactor, the passage of a direct current (variable, adjustable) through this winding will alter the degree of saturation of the core. In turn, this changes the reactance offered by the ac winding to the flow of alternating current. By proper arrangement of external circuits, a saturable reactor can be used as a magnetic amplifier. Also see **Electric controller.**

Saunders Patent Valve In effect, a Saunders patent valve is a simple pinch-clamp valve in which a diaphragm is forced against an internal body weir or bridge to provide a leaktight closure. Also known as diaphragm valves, Saunders valves have been developed to handle corrosive or hazardous substances as well as air, water, normal liquids, gases, and slurry flows. Since the diaphragm isolates the body bonnet and associated parts, toxic, chemical, and flammable substances can be handled with ease and safety. The diaphragm also prevents lubricants, products of decomposition, or atmospheric impurities from entering and contaminating the flow. Thus, diaphragm valves have found wide acceptance in the food-processing industry.

Scale, Beam The simplest beam scale uses an even-arm beam that is suspended at a central fulcrum point. At balance, the beam is horizontal and the weight of material in one pan equals the calibrated weight in the other. Other types of beam scales use a sliding poise of known weight on a graduated beam to change the ratio of the balancing weight. Each position of the poise counterbalances a different weight on the opposite end of the weighing system. Beam scales are of several types and sizes, including the sensitive laboratory balance, the rugged all-purpose portable floor scale, and high-capacity motor-truck and railroad-track scales. (See **Scale, industrial.**) In industrial weighing, the term *beam* refers to the type of weight readout, namely, a graduated bar with sliding poise. Also see **Weight, measurement of.**

Cutoff Controls for Beam Scales: Perhaps the most convenient and commonly used cutoff switch for beam scales is the tilt-type mercury switch as shown in Fig. S-1. The switch unit is held in a clip that is attached to an actuating lever. The lever is mounted on a ball-bearing joint and is tilted by contact with a fork. The fork is attached to a stationary arm mounted on the beam rack. As the preset weight is reached on the scale, movement of the beam moves the fork. Contact

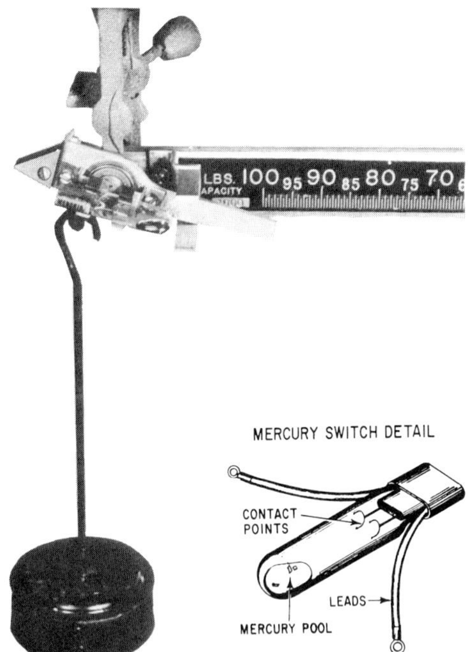

Fig. S-1. Tilt-type mercury switch attached to beam of an industrial scale to provide automatic weight cutoff. *(Howe Richardson.)*

of the fork with the actuating lever tilts the lever and attached switch. As the switch tilts, the mercury moves to the contact points and closes the circuit. Removing the weight from the scale opens the circuit.

—Arthur J. Burke, *Howe Richardson Scale Co.*

Scale, Counting Generally a platform-type scale that is supplied with a set of ratio beams (9:1 and 99:1) and ratio pans. The container of parts to be counted for inventory, shipping, or receiving is set on the scale platform. Small quantities of parts are taken from the container and put in the ratio pans until the scale indicator returns to zero. Parts in the ratio pans, plus any left in the operator's hand, become a part of the count. An automatic counting system eliminates operator computing and balancing out—and gives a printed record of the count almost immediately upon depression of a button. For predetermined weighing, such as packaging identical amounts of small parts, 10:1 and 100:1 ratio beams are supplied. The procedure followed is the same as that just described, but the parts used for balancing out are *not* part of the count.

—Arthur J. Burke, *Howe Richardson Scale Co.*

Scale, Damping of If there were no friction present in a scale or balance, the device would oscillate continuously and would not reach a point of equilibrium where the pointer would come to rest and indicate a reading. Fluid friction originating in a properly designed dashpot (essentially a plunger and cylinder) possesses the desirable property of decreasing as the velocity of the oscillating parts falls off—so that as the parts are about to come to rest, friction approaches zero. Magnetic damping also is suitable for this purpose. In addition to decreasing oscillation, the dashpot absorbs the shock of sudden loading and, in a large scale, should be located near the supporting levers where the forces are relatively large. The ability of the dashpot to absorb shock effectively decreases as its location approaches the counterbalancing mechanism.

Scale, Industrial There are two classes of industrial scales: (1) static (or platform) scales and (2) automatic scales. With a static scale, the load/unload sequence must be done manually. With an automatic scale, the load/unload sequence is programmed and done automatically.

Light-capacity Static (or Platform) Scales: There are several types including:

1. *Bench-type Scales*—used for package filling, low-output manual checkweighing, formulation, parts counting, shipping-room weighing. These scales mount on a bench (or cart for portability). Capacities range from 3 to 2,000 lb.

2. *Portable Floor Scales*—used for larger and heavier loads with capacities ranging up to 3,000 lb. The scales are floor-mounted and provided with wheels.

3. *Self-contained Floor Scales*—used in shipping rooms, freight terminals, and manufacturing departments for larger and heavier loads than can be handled by the foregoing types. These scales are supplied with their own support frame and container to house the lever system and are mounted with their platform flush with the floor. Capacities range up to 8,000 lb. Indication can be by beam, by dial, or digital. Weight printing is available for most types. See **Scale, beam;** and **Scale, pendulum.** See Fig. S-2.

Heavy-capacity Static (or Platform) Scales: Scales in this category include:

1. *Built-in Scales*—used for weighing the heaviest of in-plant loads, such as ingots, coils of steel strip, bundles of bars and tubes, crated machines. These scales require reinforced-concrete pits. Capacities range up to 500 tons.

2. *Motor-truck Scales*—used in shipping and receiving and for checking gross vehicle weights to prevent exceeding legal load limits. These scales require reinforced-concrete pits. The scales have multisection lever systems that can extend up to 170 ft, dependent upon load and maximum truck length. Capacities range up to 200 tons. See Fig. S-3.

Fig. S-2. Self-contained floor scale. *(Howe Richardson.)*

3. *Railroad-track Scales*—used in shipping and receiving by manufacturers and by railroads in weigh stations at freight-classification yards. These scales require reinforced-concrete pits. In freight classification, the speed of weighing is important, as also is the recording of weights, car numbers, freight codes, destination—for use in billing. These scales have multisection lever systems that can extend up to 170 ft. Capacities range up to 500 tons.

4. *Tank Scales*—used primarily in the chemical and food-processing industries for shipping, re-

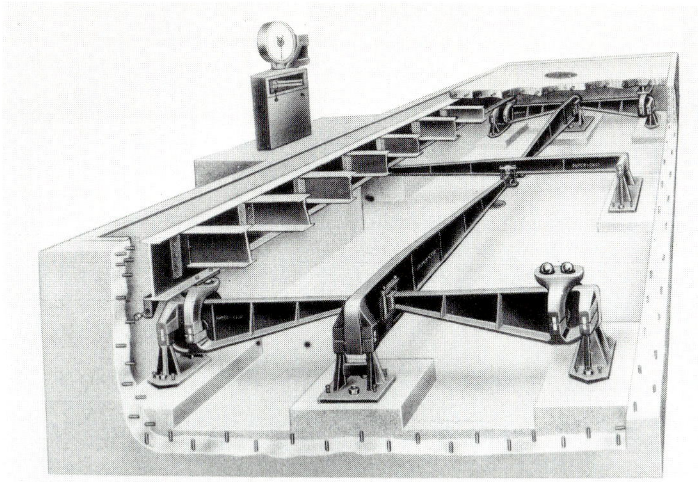

Fig. S-3. Large motor-truck scale. *(Howe Richardson.)*

ceiving, and in-process weighing. These scales usually are mounted on the floor or on reinforced-concrete stands or steelwork. Capacities range from 500 to 400,000 lb. See Fig. S-4.

5. *Bulk-hopper Scales*—used to weigh dry-bulk materials in shipping, receiving, or into processes. These scales usually are floor-mounted, suspended from overhead, or mounted in a separate steelwork frame complete with upper and lower garners. Capacities range up to 125,000 lb.

Indication on all these scales can be by beam, by dial, or digital. Automatic range changing, remote weight indication and recording, and weight printing also are available. Heavy capacity platform scales can use load cells instead of levers as their weight-supporting elements. Load cells also can be used at the output end of a lever system to provide a means for remote indication and recording.

6. *Crane Scales*—used for weighing loads picked up by cranes. These scales can use beam, dial, or load cells (electronic and hydraulic) for weight measurement. Capacities range up to 50 tons. Load cells offer compactness. The hydraulic cell is self-powered, and accuracies often are adequate.

Automatic Scales: These scales are used primarily in shipping, receiving, and in-process weighing and bagging of bulk materials. The scales sequence the feeding, weighing, and discharge of the materials automatically. There are two major types: (1) draft (or increment) weighers and (2) continuous weighers.

Automatic Draft Weighers: These are of three principal types:

1. *Automatic Bulk-hopper Scales*—either (*a*) a postweigher or (*b*) a preweigher. A *postweigher* delivers an approximate preset increment of weight, stops the feed, and then indicates, prints, and totalizes the weight delivered. Variations from preset approximate weight are not significant because the *actual* weight delivered al-

Fig. S-4. Tank scale. *(Howe Richardson.)*

ways is recorded. Postweighers are usually installed with upper and lower garners to permit on-stream, nonstop weighing. Capacities range up to 125,000 lb.

Postweighers depend upon an analog-to-digital converter on the dial indicator to feed weight information to a calculator or printer. The calculator prints gross and residue weights on a tape and also totalizes net weight received or shipped at any time. The printer gives a printed record of gross and residue weights but does not totalize net weights. The printer can be integral with the dial indicator or remotely located. See **Scale, pendulum.** Cycling relays and cutoff switches take the automatic postweigher through its feed-weigh-discharge sequence and impulse the transmission of gross and residue weights after the feed and discharge operations, respectively. Pilot lights indicate cycle progress and also alert the operator to any unprogrammed event.

A *preweigher* delivers a *precise* (to one-fourth of 1 percent or better) increment of weight, stops the feed, and then discharges automatically. A counter tallies the number of discharges. In mechanical preweighers, the weight of material provides the force needed to sequence the events in the weighing cycle. In electrical preweighers, solenoid-controlled air valves sequence the events and give higher speed and accuracy. Preweighers can be made to deliver a preset number of drafts and in a timed sequence if desired. For best accuracy and output, the feeder must be matched to the material handled to assure a uniform feed stream to the scale. Capacities range from 10 to 1,800 lb or more. See Fig. S-5. Also see **Feeder, weigh.**

Preweighers handle the feed-weigh-discharge sequencing by means of a mechanical-linkage system for mechanically operated scales and by means of limit switches or proximity switches for electrically operated scales. Where mechanical scales are supplied with feeders, limit switches actuated by the opening and closing of the scale inlet gate are used. With electrically operated scales, switches are an integral part of the scale, and the addition of feeders, programming of discharges, and control of tonnage delivered is relatively simple. Preweighers indicate the amount of material weighed by means of counters which totalize the number of discharges made. When a given tonnage of material is to be delivered by the scale, a stop counter is used which will stop the scale after a preset number of discharges. Normally, the preweigher will continue to deliver repeated increments of the preset weight unless stopped manually or automatically.

2. *Automatic-bagging Scales and Systems*— These scales are automatic preweighers with an additional bottom hopper or bag spout which funnels the preweighed increment into a bag held on the bottom of the spout manually or by a clamping device (see Fig. S-6). Automatic-bagging scales are used widely to bag animal feeds, fertilizers, salt, flour, and chemicals in 10- to 150-lb bags. Speeds of 20 bags per minute or better and accuracies of ±one-fourth of 1 percent or better are common, dependent upon type of material and weight increment. In addition to the bagging scale, bag conveying, closing, and check-

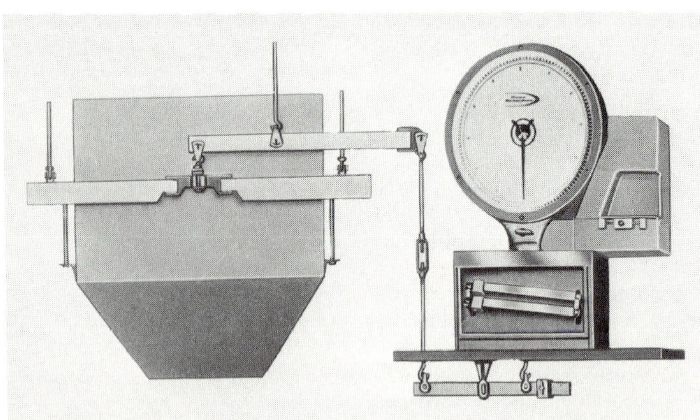

Fig. S-5. Hopper scale. *(Howe Richardson.)*

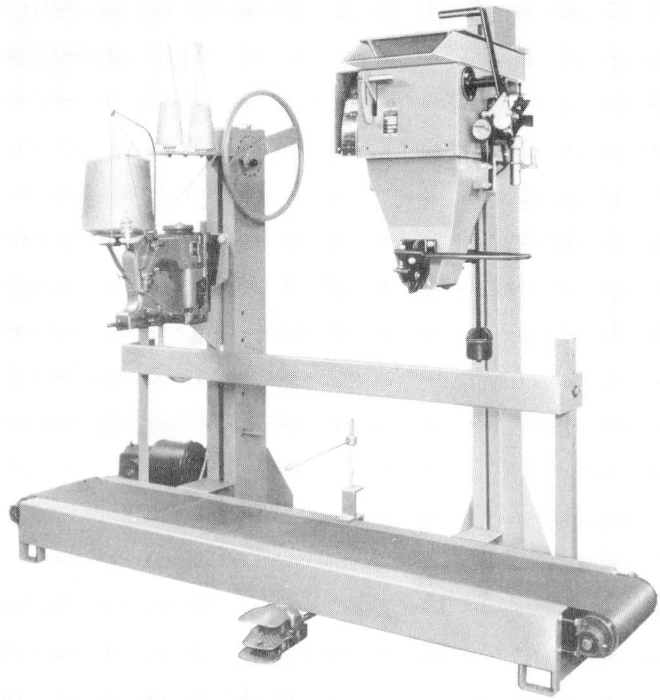

Fig. S-6. Bagging scale and system operable by one man. *(Howe Richardson.)*

weighing equipment are supplied for a complete bagging system. One high-speed high-accuracy automatic bagger is equipped with a weight-feedback system to keep bagging accuracy within preset limits.

3. *Automatic Batching Scales*—See **Batching systems, automatic-weight.**

Automatic Continuous Weighers: Scales of this type also are known as conveyor or belt scales and are used for on-stream continuous weighing into a process, for checking daily production, for monitoring process performance by recording low and peak-load points and hourly flow, and for constant-rate feeder control when feeding a main ingredient and an additive into a process. These scales consist of a suspension section inserted in an existing conveyor which weighs the material passing over the conveyor. The suspension section, or weigh bridge, transmits weight information to the scale's integrator, where it is combined with speed information from the conveyor to provide an indication of flow rate in pounds, or tons per hour, and total pounds or tons weighed. Readout equipment usually includes a flow-rate indicator and a weight totalizer. Optional readout equipment can include remote weight indicators and totalizers, chart recorders, and ticket and tape printers. Several continuous weighers can be operated from one master control unit to give continuous proportioning of ingredients or to feed an additive in a preset ratio to a main ingredient. Continuous weighers offer low-headroom advantages compared with draft weighers and accuracies to ± one-half of 1 percent over a period of delivery. Unlike draft weighers, it is impractical to check-weigh successive weighed increments of material. Capacities range to 400 tons/hr, but with special equipment, capacities can go as high as 12,000 tons/hr. Also see **Weighing, radiation methods for.**

REFERENCES

Considine, D. M.: "Industrial Weighing," Reinhold Publishing Corporation, New York, 1948.

Considine, D. M.: Process Weighing, *Chem. Eng.*, Aug. 17 and Sept. 14, 1964.

Kendall, D. B.: Weight and Weight Rate of Flow, in D. M. Considine and S. D. Ross (eds.), "Handbook of Applied Instrumentation," McGraw-Hill Book Company, New York, 1964.

—Arthur J. Burke, *Howe Richardson Scale Co.*

Scale, Pendulum The pendulum scale (also referred to as dial scale) is said to have self-powered and automatic indication because the force applied against the unknown weight changes proportionally and automatically to maintain a constant state of equilibrium in the scale lever system. Since the displacement of the force opposing counterweight is exactly proportional to the weight applied, the amount of displacement of the balancing device (pendulum) can be converted into the angular displacement of the pointer moving over a graduated dial. As shown in Fig. S-7, as the load is applied to the scale, the two counterbalancing pendulums move out in an arc to restore weight balance.

Dial scales find their greatest use in commercial and industrial applications and are available in all types and sizes. See **Scale, industrial;** and **Weight, measurement of.** These scales eliminate dependence on human judgment in balancing out the load because this function is performed automatically. Where a wide range of industrial loads is to be weighed on one dial scale, the dial can incorporate several ranges which can be made to change automatically to match the range of the applied load. This feature also assures good reading accuracy with each weighing range. Printout on tape or ticket is available, as well as analog-to-digital conversion for digital readout and totalizing.

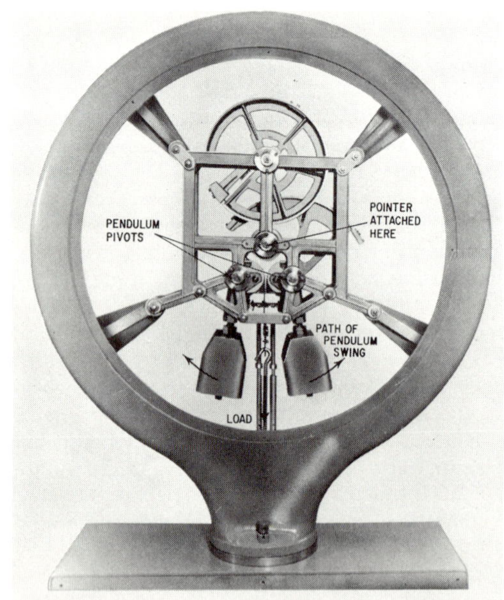

Fig. S-7. View inside the head of a dial scale showing how pendulums swing out automatically to counterbalance applied load. *(Howe Richardson.)*

A variation of the pendulum or dial scale is one that indicates numerically (instead of by pointer) and projects these weight values on a ground-glass screen where the numerals are enlarged and illuminated. Reading is faster and more accurate—and parallax is avoided with this arrangement. In this system, the applied weight pulls a connecting rod upward and swings an optical scale in an arc between a stationary light and a magnifying lens. The figures and graduations of the optical scale are magnified about forty times their original size and reflected to the ground-glass screen. A dashpot dampens oscillations.

Analog-to-Digital Conversion: Frequently, it is necessary to furnish weight values in digital form—for applications such as at-the-site weight printing, remote weight printing, weight totalizing, and scoreboard displays. Analog-to-digital conversion usually is done by mechanical means. The dial indicator is furnished with a star wheel. When a digital reading of the pointer position is taken, a locking pawl moves into a notch of the star wheel. Index fingers then slide into their proper settings on an index cam. When this occurs, the type sectors of the printer and the wiper arms of the printed-circuit switches are positioned automatically and a digital readout is registered. After readout, a clearing lever pulls away the fingers from the cam and the pawl leaves the star wheel, freeing the dial mechanism for the next weighing.

Cutoff Controls for Pendulum Scales: Switches are used to control the flow of materials in repeated weighings of the same amount of material into weigh hoppers, for low-output feeding into processes, and for batching and filling containers. Switches used include (1) the differential transformer which converts mechanical motion into electrical information by the movement of a small iron rod (armature) within an associated transformer; (2) mercury-magnetic switches which depend upon a small external magnet to attract an armature in the glass envelope, thereby moving a contact in and out of a mercury pool; (3) magnetic-reed switches which have two flexible iron strips in a sealed glass cylinder—with their ends overlapping, but not touching until a magnet approaches; and (4) photoelectric switches which have a light source and photocell mounted in one enclosure and a reflecting surface on the indicator pointer to provide cutoff. The armature of the differential transformer is attached to the dial mechanism, which coordinates it with pointer position. As shown in Fig. S-8, the mercury-magnetic, magnetic-reed, and photoelectric switches can be

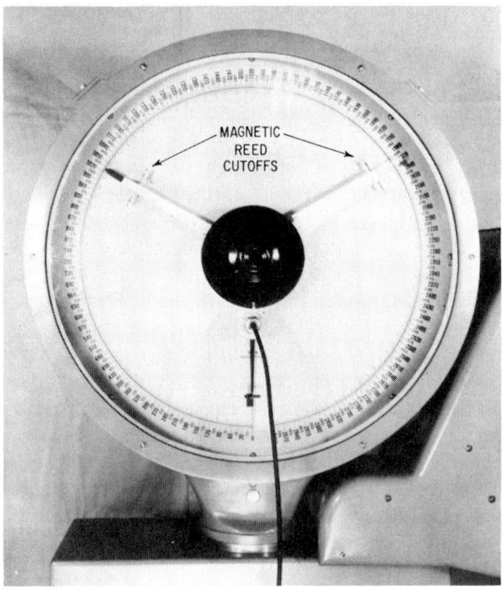

Fig. S-8. Two magnetic-reed cutoffs mounted on dial of a pendulum scale to provide weight cutoff at two values. Magnet to actuate reed switches is mounted on indicating pointer. *(Howe Richardson.)*

mounted in the scale dial head at the required cutoff points; or where some flexiility of positioning is desired, they can be mounted at the end of a movable arm. Also see **Batching systems, automatic-weight.**

—Arthur J. Burke, *Howe Richardson Scale Co.*

Scale, Radiation-type (See **Weighing, radiation methods for.**)

Scale, Spring The common spring scale uses a helical extension spring as the weight-measuring element. Extension of this spring under load is translated into an indication of weight value. Many spring scales are used in retail trade as well as in industrial applications—usually in the low-weight ranges. Scales for industrial applications in the higher-weight ranges (100 to 100,000 lb) use an alloy-steel beam (instead of spring) whose few ten-thousandths of an inch deflection under load is translated into a dial indication of weight. Spring scales generally are compact, self-powered, and low in cost. Spring scales are more likely to be affected by vibration and the extremes of temperature than other types of scales. They also have a lower accuracy potential. High-sensitivity balances, using the torsion of a taut band as the restoring force, also are in the spring-scale category. See **Weight, measurement of.**

—Arthur J. Burke, *Howe Richardson Scale Co.*

Scanner In the broad context of instrumentation, scanner has a number of interpretations. (1) A scanner may be a radar antenna and reflector assembly designed to oscillate back and forth about a center position during search and then move in any required plane during tracking of a target. (2) A scanner may be that part of a facsimile transmitter that systematically translates the densities of the elemental areas of the subject copy into corresponding electric signals. (3) A scanner may be an instrument system that automatically samples, measures, or checks a number of quantities or conditions in sequence, as in process control. (4) A scanner may be a manually operated multipoint instrument. Also see **Multiplexer, analog.**

Scanner, Optical [See **Character-recognition systems;** and **Optical character recognition (OCR).**]

Scheduler A component of a programming system which handles the allocation of system resources according to the system-priority scheme. In actual implementation the scheduling function may be broken into several separate subfunctions such as task or job scheduling and I/O (input/output) scheduling (see **Task; Job**). The scheduler acknowledges the requests for activity, ensures that any requisite conditions are met, and may initiate the requested action or put it in a queue (see **Queue**) to be acted upon at a later time. (See also **Executive program.**)

—L. E. Wise, *IBM Corporation*

Schering Bridge (See **Bridge circuits, electric.**)

Schmidt Camera This device makes use of a spherical primary mirror with the aperture stop placed at the center of curvature of the mirror to obtain a very wide field (although the field is curved). Spherical aberration is corrected by a corrector plate placed at the aperture stop. Speeds usually are fast, and the primary mirror usually is larger than the aperture stop. Figure S-9 is a schematic of the Schmidt camera. These

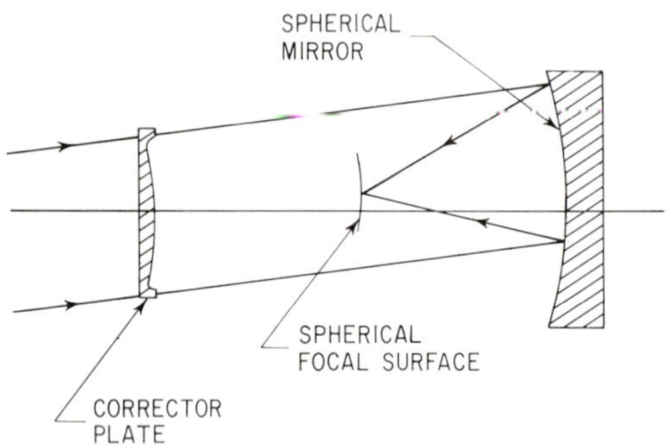

Fig. S-9. Schmidt camera.

instruments are used for photographic sky surveys and nonastrometric mapping.

—Joost H. Kiewiet de Jonge, *University of Pittsburgh,* and Lloyd M. Geertz, *Owens-Illinois*

Scientific Computer A general-purpose digital computer (see **Digital computer**) which is designed with an instruction set biased toward fast arithmetic operations is generally considered a scientific computer. The arithmetic operations are generally performed on binary numbers (see **Binary state**) as opposed to decimal arithmetic in a commercial computer. Another characteristic of the scientific computer is the ability to operate on bit (see **Bit**) significant data by means of shifting operations. In the larger scientific computers, the ability to perform floating-point arithmetic operations efficiently is a requirement. Also see **Data communications.**

—W. W. Welz, *IBM Corporation*

Scintillation Counter One of the most important radiation detectors. A scintillation counter consists of a scintillation phosphor optically coupled to a photomultiplier tube and both enclosed in a lighttight shield. The particle to be detected loses energy in exciting and ionizing the molecules of the crystal. Energy in the form of light is radiated from these molecules, and some of this light falls onto the photocathode of the photomultiplier tube.

Photoelectrons are ejected from the photocathode and are accelerated to the first dynode, where each photoelectron ejects several other electrons by secondary emission. This process of multiplication is repeated at each succeeding dynode by placing the dynode at a higher potential than the preceding dynode. On arrival at the anode or collector, a multiplication in the order of 10^6 may be produced. This charge (initiated by one particle) on arrival at the anode then is used to produce a voltage pulse on the anode-circuit capacitance.

The prerequisite property of a scintillation phosphor is that it be transparent to its own excited radiation. When an atom of any material is raised to an excited state by an ionizing particle, it will emit light in the process of returning to the ground state. With most materials, the atoms are equally capable of reabsorbing the light thus emitted. The peculiar property of phosphors useful for scintillation counting is that they are excited to a particular energy level by ionizing radiation such that the light energies emitted on return to the ground state do not correspond to the resonant absorption energies of that phosphor for light quanta. Thus, the phosphor is transparent to its own excited radiation. This property occurs naturally with certain phosphorescent crystals, particularly those of organic nature. Most of the inorganic crystals require "doping" with another element (notably thallium) to produce this condition.

There are many properties to be considered in the selection of a phosphor for a particular application, including conversion efficiency of energy in to light out, response of a phosphor to different particles and quanta, linearity of a phosphor to a wide range of energies of a particle or quanta, and decay time of light emission. A multicrystalline ZnS (Ag) phosphor screen $\sim$10 mg/cm^2 thickness makes a highly efficient alpha detector. Pulses produced in this layer by beta and gamma radiation can easily be biased out electronically because they are many times smaller than the average alpha pulse. Also see **Radiation (nuclear), detection of.**

Second-order Lag Response (See **Closed-loop control.**)

Seebeck Effect (See Thermocouple.)

Seismic-type Vibration Transducer (See Vibration pickup, velocity.)

Seismograph (See Seismology.)

Seismology The basis of scientific seismology is the observation and analysis of elastic energy as it propagates itself through the earth. When mechanical energy is released in a homogeneous earth, it is propagated outward in waves whose fronts are spherical and whose mechansim is alternating compression and rarefaction of the material through which they pass. These waves, called pressure or *P* waves in seismology, are physically analogous to the sound waves that spread outward from an explosion in air or water. The earth, however, is not generally homogeneous, and though imperfectly elastic, it has rigidity or shear strength which is absent in air or water. This results in a second kind of wave action, in which the material through which the wave passes moves transversely to the direction of wave motion. These shear or *S* waves travel through solid material at a velocity a little more than half that of the *P* waves. In addition to the *P* and *S* types (called body waves) there is also energy propagation along surfaces or interfaces. At the earth-air interface waves similar to surface waves in water are propagated; these are called Rayleigh waves and cause, as in water, circular vertical motion of the material through which they pass, in the plane containing their direction of propagation. In solid material, surface waves like *S* waves also occur, the material moving horizontally in a plane transverse to the direction of motion of the wave; these are called Love waves. If the velocities of the different modes of wave propagation are known, deduction can be made, for example, of the distance between the earthquake and the observation station by measuring the time interval between the arrival of the faster and slower waves.

When earthquake waves cross an interface between earth materials with different elastic properties, their velocity, direction, and phase may be changed, and they will generally give rise to waves of other modes. For example, a *P* wave encountering an interface may give rise to a refracted and a reflected *P* wave and a refracted and a reflected *S* wave. A wave may also be refracted even if it does not cross an interface, because of the change in pressure and density of the earth material as the wave penetrates more deeply into the earth.

Waves recorded at an observatory can be analyzed for their directions of travel, times of arrival, and frequency content. When data from different observatories are compared, it can then be deduced what interfaces they have passed through. Such deductions have led to the discovery that the earth is divided into three main layers, the crust, the mantle, and the core, each with its own set of characteristics with respect to the passage of seismic waves and each with its own subdivisions. The crust or outer layer of the earth reaches to a depth of perhaps 10 miles under the oceans and 30 miles under the continents; it consists of a bottom "basaltic" layer, an intermediate "granitic" layer, and an upper layer containing the sedimentary rocks, the near-surface igneous and metamorphic rocks, and the ocean. Directly under the crust is the mantle, denser than the crust and recognizable by seismic methods, because *P* waves in the upper mantle have everywhere a velocity close to 8 km/s. The discovery of the mantle surface (the Mohorovicic discontinuity) was one of the early achievements of seismology. The central part of the earth, the core, begins at a depth of 2,900 km (1,800 miles); it is far denser than the crust or mantle and is proved to be molten (to lack rigidity) because *S* waves are not transmitted through it.

Seismology has proved to have extensive commercial as well as scientific applications. The principles used to deduce large-scale earth structure have been put to use in mapping small-scale geologic structure with especial emphasis on features such as faults and anticlines which may form traps for the accumulation of oil. Small artificial earthquakes are produced by detonating chemical (and lately nuclear) explosives in shallow boreholes, and the refracted or reflected energy is sensed and analyzed as for natural earthquakes. The chief difference between scientific and commercial seismology is in the scale of energy and instruments. The frequencies of interest are in the 10- to 100-Hz range instead of the 1 to 0.01 range as with natural earthquakes, the seismometers are small and portable, and the receiving stations are yards instead of miles apart. Because of its intensive application, commercial seismology has given rise to important developments in the art of data processing. The effectiveness of sensing earthquake energy is limited by the fact that the earth is never absolutely quiet; small motions caused by distant sources such as ocean waves are always present even if local disturbances caused by wind or traffic are absent. This kind of energy is called, and treated as, noise by the seismologist, and efforts to detect significant seismic signals through the background of noise have resulted in important technological advances. Recording earth motion in digital instead of analog forms permits both a greatly increased dynamic range and highly advanced electronic-computer methods of data processing. The latter involves the application of the most advanced information theory to the interpretation of seismic data. The result has been that data from large arrays of sensors can be treated according to their relation to space as well as time, and deductions regarding earth structure can be made that were not possible by even the most careful study of individual records. (Also see **Richter scale, seismic.**)

—F. E. Romberg, *Texas Instruments Incorporated*

Self-actuated Regulators These devices require no external power for automatically controlling pressure, temperature, level, or flow in spaces, tanks, process lines, or equipment. There are several types and styles:

Direct-actuated (DA)—where the valve is positioned directly by the measuring element.

Pilot-actuated (PA)—where the measuring element controls a pilot valve which, in turn, admits supply-line pressure to a piston or diaphragm to move the main valve.

Self-contained (SC)—where the measuring element is contained within the valve structure. The device responds only to variations (pressure, temperature, or flow) that occur in the flowing medium.

Remote-sensing (RS)—where the measuring element is separate from the valve. The device can sense and control variations in media other than that flowing through the valve.

Generally, self-actuated regulators are obtain-

TABLE S-1. *Characteristics of Self-actuated Regulators*

Range	Size and Capacity	Special Features
Pressure Regulators—used to reduce fluid pressure to value required at point of use, to maintain constant pressure or vacuum in lines or vessels, to protect equipment against unsafe pressure or vacuum (relief applications). See Fig. S-10		
30 in. Hg. (vacuum) to 150 psi (most types, large sizes) 0 to 3,000 psi (special types)	SC-DA—¼–3 in. SC-PA—¼–6 in. RS-DA—¼–4 in. RS-PA—¼–6 in.	Differential-pressure controls Combination pressure-temperature controls (PA) Various body and trim materials
Temperature Regulators—used to control heating or cooling medium to (1) hold spaces or processed material or equipment at even temperature; (2) maintain constant discharge temperature from heat exchangers, engines, compressors; and (3) control hot and cold mixing. See Fig. S-11		
−25 to +480°F	DA—¼–4 in. (2-way valve) DA—½–6 in. (3-way valve) PA—½–6 in.	Temperature indication Fail-safe thermal elements Calibrated adjustments Packless valves Manual valve positioners Combination temperature-pressure controls
Level Regulators—used to maintain level of liquid in open or pressurized tank by control of inflow or outflow (drain). The devices cannot be used with liquids that will deposit and destroy buoyancy of float or create sticky valve action. See Fig. S-12		
Limited set-point adjustment; generally set at time of installation	½–6 in.	Various float materials, sizes, and forms
Flow Regulators—used to maintain steady flow rate in clean, low-viscosity liquids. All devices in this category are self-contained. See Fig. S-13		
Some types are not adjustable. Adjustable types—10 to 100% of maximum rating	Fixed—¾–2½ in. Adjustable—½–4 in. Capacity—0.02–550 gpm (water)	Provision for remote adjustment (pneumatic, electric). Various materials of construction

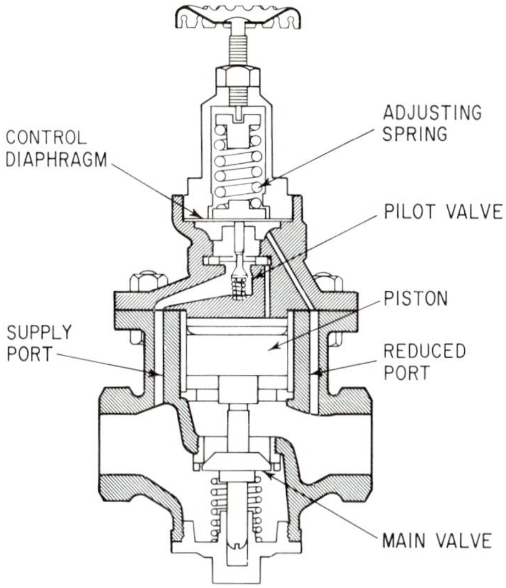

Fig. S-10. Pressure-reducing valve (internal-pilot, piston-operated type). Upper portion of device is a spring-loaded diaphragm type of self-operated pilot controller; lower portion is a piston-operated valve. The product of the pressure difference across the piston and its area is balanced by the product of the pressure difference across the main valve and its area, plus its spring load. Any change of downstream pressure causes diaphragm to reposition the pilot valve. The resulting change in pressure above the piston moves the main valve to a new position to restore the original force balance at the new conditions. Manual adjustment of outlet pressure is by handwheel. *(Leslie Co.)*

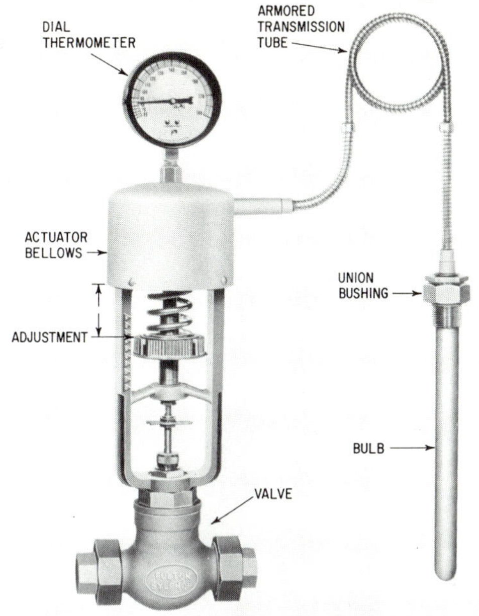

Fig. S-11. Temperature regulator (direct-actuated type, remote-sensing style). Rising temperature at the bulb produces an increase in vapor pressure of the volatile fill. Pressure is transmitted to the bellows. When the force produced equals the adjusting spring force, any further pressure increase will cause downward movement of the stem to commence closing of the valve. The reverse process occurs with a dropping temperature. *(Robertshaw.)*

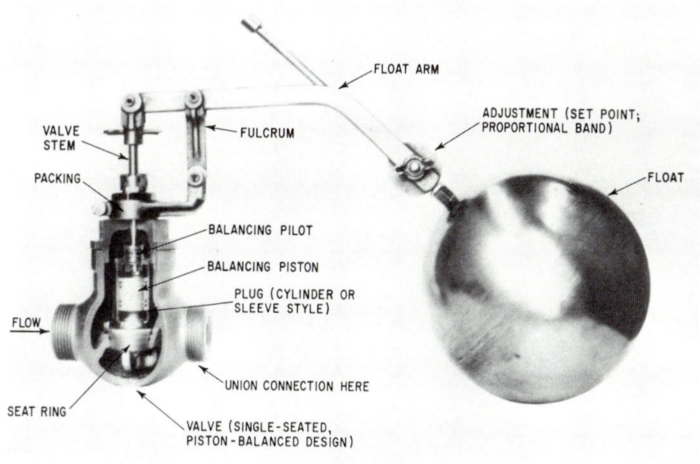

Fig. S-12. Lever regulator. Buoyant force of float caused by rise of liquid level moves valve stem down to move balancing pilot and valve plug toward seat. On falling level, weight of float and arm moves valve toward open position. *(Robertshaw.)*

able with only a proportional-control mode. The proportional band obtainable is determined by the type, size, design, and manufacturer's specifications for a particular device. Rarely are adjustments available to change the proportional band. The speed of response is fast for self-contained pressure, level, and flow regulators and medium to slow for large, remote-sensing temperature regulators.

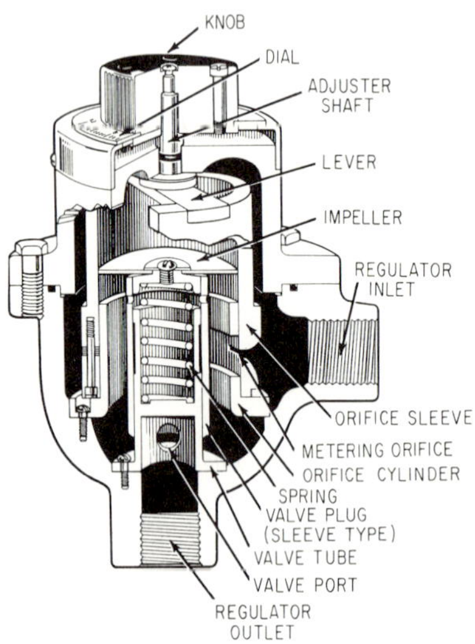

Fig. S-13. Flow-rate controller (regulator) of direct-operated type. Differential pressure across the metering orifice is always impressed across the impeller. Impeller moves in force balance with spring to position valve plug and to keep differential constant. Flow rate, corresponding to differential, thus is held constant. The set point (flow rate) is changed by rotating orifice sleeve and changing orifice length (area). Dial indicates orifice area and gpm. *(W. A. Kates Co.)*

Accuracy: Under favorable conditions, the control accuracy of self-actuated regulators may approach ±1°F for temperature, ± 0.5 psi for pressure, ± 0.5 in. for level, and ± 1 percent of rate for flow.

Advantages include (1) simplicity, (2) ruggedness, (3) ease of service, (4) low cost, (5) long useful life, and (6) no need for external power.

Limitations include (1) available accuracy, (2) available ranges, (3) size of measuring element, and (4) allowable pressure drop through valve.

The proper valve size is critical. Load or flow requirements should be very carefully estimated in advance of final selection of device.

Characteristics of self-actuated regulators are given in Table S-1. Representative devices are illustrated in Figs. S-10 through S-13.

—L. S. Dysart, *Fulton-Sylphon Division, Robertshaw Controls Company*

Self-balancing Instrument (See **Potentiometric instrument.**)

Self-organizing Control (See **Adaptive control.**)

Self-regulation, Process The inherent tendency of a process to balance its output to input. In the storage tank of Fig. S-14, a rise in level of liquid in the tank causes a larger pressure head over the outflow valve. Thus, as this pressure head increases, the flow through the valve increases. Similarly, as the pressure head decreases, the flow through the valve decreases. The changes in valve flow and head thus tend to work together to keep the level in check. Of course, there are obvious limits to this action such that should the inflow rate go beyond a limit, the increased head on the valve would not be sufficient to cause outflow equal to inflow—and thus the tank would overflow. Conversely, should the extreme change of inflow be such as to reduce the inflow to a trickle, the outflow valve—so long as it is open and even with just a small head over it—will empty the tank to the level of the valve. Consequently, the presence of self-regulation in this example must be considered a relative factor. In planning a control system, it is best to consider the phenomenon as a tendency toward self-regu-

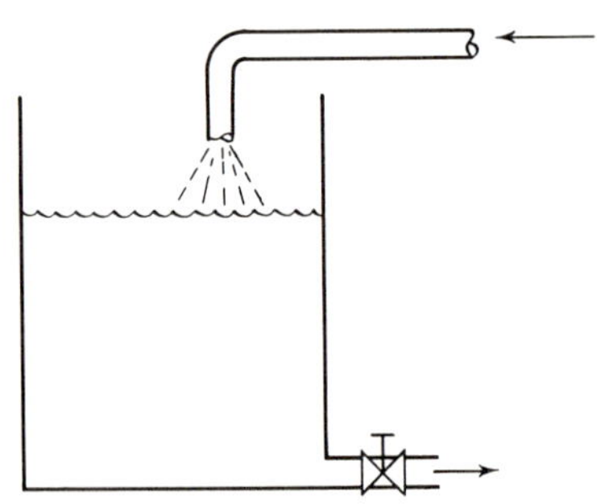

Fig. S-14. Self-regulating process.

lation and to know in advance within which limits it will be operative. Many thermal processes also have a tendency to be self-regulative. For example, a water tank that is heated by means of a steam coil normally will reach a certain temperature—even without water outflow from the tank—beyond which the water temperature will rise no further because of loss of heat by radiation or convection from the tank to the surrounding atmosphere. In a sense, any undercapacity heating system becomes self-regulating at a temperature below that which is desired because there just is not sufficient energy going into the system to make up for ambient losses and, consequently, the temperature tends to level off even with the thermostat continuing to call for more heat from a burner system that simply cannot put out any more energy. Often, such conditions are the cause of fires—where the furnace per se becomes greatly overheated and ignites flammable material near it—even though the regular heat-transfer system is inadequate to meet the needs of the total system.

Selsyn (See Synchro.)

Sensitivity Generally, sensitivity can be defined as "the ratio of a change in output to the change in input which causes it." This definition applies to the steady-state condition, and the ratio must be expressed in the units of measurement of output and input.

Usage of this term is generally limited to linear devices where the plot of output to input magnitude (transfer characteristic) is a straight line over the operating range of the device. If it is not linear, the output/input characteristic must be expressed by graphical means or by a mathematical expression (transfer function).

The word sensitivity is sometimes used to describe the maximum change in an input signal that will not initiate a response on the output. This synonym for dead band (or one-half the dead band) is considered nonstandard and seldom is used today. Amplifier sensitivity, however, frequently is meant to imply the magnitude that the input must change (from zero) to initiate an output response (for example, the sensitivity of the amplifier is 1 microvolt).

Another usage is to describe the relative degree with which a voltage-sensitive device will "load" a circuit. For example, a null-balance potentiometer would be considered sensitive as compared with a millivoltmeter utilizing a d'Arsonval movement, since its input impedance is effectively infinite at balance. It would therefore draw little or no current from the signal source, and its addition to a circuit would not change the circuit parameters (load the circuit). It is quite common, for most signal-actuated voltmeters, to publish a characteristic that is frequently referred to as "sensitivity" in a form such as "20,000 ohms per volt." The higher this value, the less current is drawn from the signal source and, therefore, the more sensitive the voltmeter.

Properly used, the definition of sensitivity applies to linear devices whose units of output are different from the units of input (an indicating millivoltmeter might express sensitivity in inches per millivolt, a pneumatic temperature transmitter in psi per °F). For devices where the output and input are expressed in the same units, such as signal-conditioning devices (amplifiers), this ratio is more properly called "gain" and the units are necessarily expressed as either a pure number or in decibels (20 × the log of the ratio). The word gain is usually preferred for cases where the input signal is sinusoidal. Where it is not sinusoidal, this ratio is determined after steady state has been reached, and the terms "static gain," "steady-state gain" and "zero-frequency gain" are commonly used.

The process-control operator is confronted with an interesting paradox with respect to this accepted definition of sensitivity. An important parameter for any process is the ratio of output change to input change expressed in their relative units (e.g., °F per Btu per minute). But this tells him relatively little about the sensitivity of his process to control action. Consider, for example, a fuel-flow control valve that will operate full-closed to full-open with a 3 to 15 (12) psi control signal, and where full valve operation will cause a change in an oven temperature of 120°F. Note that, for this process, sensitivity could be stated in terms of °F per gallon per minute, °F per Btu per minute, °F per psi, or °F per percent valve travel. But is this "high" sensitivity or "low" sensitivity? If the oven had a controlled span of 1200°F, that particular valve could exercise control over only about 10 percent of the output span (process gain = 0.1) and the operator would consider it to be "insensitive." On the other hand, if the controlled span of the oven was only 12°F, a 10 percent change in valve position would cause the process to swing over its full range (process gain = 10), and the process would be considered "highly sensitive."

The foregoing relates to the sensitivity of the process variable to changes in the controlled variable. The control-system analyst is also interested in the sensitivity of the process to changes in a particular process element. This sensitivity is the percent change in system transmittance, or transfer function, divided by the percent change in the characteristic of the element causing it. The result is a loss of system accuracy or a deviation between the process variable and the desired value (set point).

—Robert L. Wilson, *Honeywell Inc.*

Sensor A device that detects or senses the value or change of value of the variable being measured—hence the first link in the measurement-system chain. The term sensor sometimes is used instead of the terms detector, primary element, or transducer. Also see **Process variable; Transducer;** and **Variable.**

Sequence Controller (See **Timing controllers.**)

Servo A servo or servomotor may be an electric, hydraulic, or other type of motor that serves as the final controlling element in a servomechanism. The servo receives power from the amplifier element and drives the load with a linear or rotary motion. Also see **Final controlling element.**

Servomechanism A feedback control system in which one or more of the system signals represent mechanical motion. A servomechanism usually is used to control an output position mechanically in response to input-signal changes. A servomechanism also may be a regulator under certain conditions.

REFERENCES

Bower, J. L., and P. M. Schultheiss: "Introduction to the Design of Servomechanisms," John Wiley & Sons, Inc., New York, 1967.

Chestnut, H., and R. W. Mayer: "Servomechanism and Regulating System Design," John Wiley & Sons, Inc., New York, 1959.

Lauer, Lesnick, and Matson: "Servomechanism Fundamentals," 2d ed., McGraw-Hill Book Company, New York, 1960.

Zeines, Ben: "Servomechanism Fundamentals, "McGraw-Hill Book Company, New York, 1959.

Ahrendt, W. R., and C. J. Savant: "Servomechanism Practice," 2d ed., McGraw-Hill Book Company, New York, 1960.

—John Markus, "Electronics and Nucleonics Dictionary," *McGraw-Hill Book Company, New York, 1966*

Set Point In a process-control system, a fixed or constant (for relatively long time periods) command, for example, the setting of 152° F on a temperature controller for a hot-water tank, or 72° F on a room thermostat. In a program-control system, the set point can be varied continuously (or in steps) as in the case of a heat-treating process where the temperature is made to follow a gradually rising set point to a value which is held for a period of time and then made to decrease at a controlled rate to a final value.

—Richard H. Kennedy, *Bristol Division, American Chain & Cable Company, Inc.*

Setting, Controller (See **Process reaction rate;** and **Response, transient.**)

Settling Time Settling time is "the time required, following the initiation of a specified stimulus to a linear system, for the output to enter and remain within a specified narrow band centered on its steady-state value" (SAMA Standard RC 20-11-1964). Although this definition restricts the use of the term to linear systems, the term is commonly used for nonlinear systems as well. For example, the settling time for slew-rate-limited amplifiers is often quoted. In nonlinear systems, the settling-time specification should include the magnitude and other pertinent characteristics of the input stimulus.

The input stimulus is most often specified as a step or impulse function, although other input functions such as ramp, trapezoid, or sinusoid can be used. The specified band into which the output must enter and remain is often specified as ±2 percent for analog instrumentation. In high-accuracy and high-resolution digital-data-acquisition systems, the band is often much less and depends on the characteristics of the system. Typical specifications in high-resolution systems are ± (½) LSB (least significant bit) and ±0.01 percent of final value. In comparing system specifications, it is important that the band be the same for both units being compared.

The distinction between rise time and settling time should be carefully noted. As shown in Fig.

S-15, the rise time is the time for the output of a system to increase from a small percentage (typically, 5 or 10 percent, to a large percentage (typically 90 or 95 percent) of its final steady-state value. In systems which are accurately described by differential equations, a relationship between the rise time and the settling time can be derived.

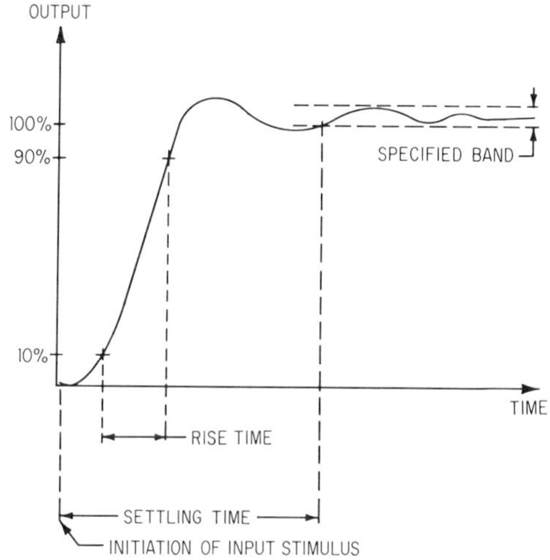

Fig. S-15. Illustration of settling time (step input).

However, few practical systems can be described by such equations over a substantial portion of their operating range with a precision which allows such a relationship to be used in the design of high-accuracy and high-resolution systems. For this reason, a settling-time specification on high-accuracy systems, networks, and components is generally preferred, since it is a more meaningful measure of performance.

—Thomas J. Harrison, *IBM Corporation*

Sextant Modern, lightweight sextants and octants are used to establish one's location quite accurately by determining the angle between the horizon and the sun or some other heavenly body. A small bubble made visible in the instrument's optical system is used as an indication of vertical if no horizon is visible. In some devices, provision is made to average the reading over a period of a minute or two—particularly useful in rough air or seas. A gyro vertical also is used as a horizontal reference (see Fig. S-16). Space sextants are described under **Space and missile-systems instrumentation.**

Shaft Position, Measurement of (See **Position-measurement and control systems; Resolver;** and **Synchro.**)

Shaker Table (See **Vibration exciter.**)

Shared-time Control Control action in which one controller divides its computation or control time among several control loops rather than acting on all loops simultaneously. See **Direct digital control; Time sharing.**

Shock, Measurement of The predominant characteristic of mechanical shock is suddenness. A shock occurs whenever any motional or loading parameter of a test item is significantly altered in a relatively short time and in a nonperiodic manner. The parameter of interest may be displace-

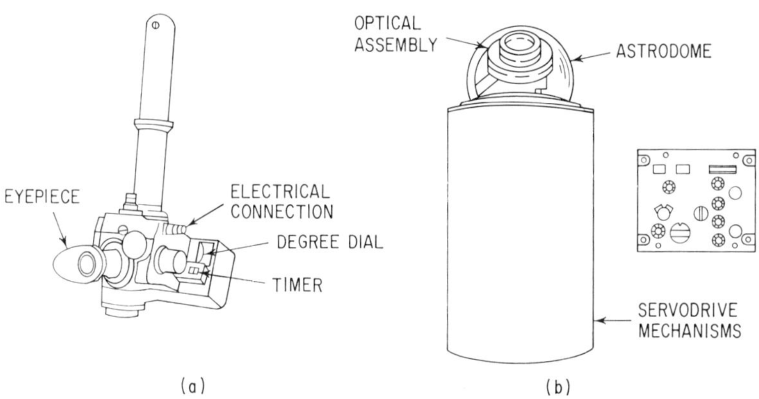

Fig. S-16. (a) Periscopic sextant; (b) photoelectric sextant tracker and control panel.

ment, velocity, acceleration, force, pressure, stress, or torque. A transducer used in shock measurement must satisfy certain general relationships. First, the natural period of the transducer must be small relative to the period of the pulse. Also, while the transducer must be damped, the damping factor should be significantly less than critical. Finally, the frequency response must be flat—as close to zero frequency as is possible; at the very least, the 3-db low-frequency cutoff point should have a frequency that is less than 0.02 of the reciprocal of the period of the pulse.

The study of shock-measurement results based on time history alone is difficult. It is often impractical to assess the uncorrelated effects which may influence or mask the results. Usually, for example, some background vibration is present. Test results, therefore, are best compared using either (1) the Fourier spectrum of the pulse or (2) the shock spectrum. The first step, therefore, is to translate the time history to the appropriate spectrum. The *Fourier* spectrum $F(f)$ is best employed when shock isolation or shock transmission is to be studied. This is obtained by taking the Fourier transform of the time history $f(t)$,

$$F(f) = \int_0^\infty f(t)\, e^{-j\omega t}\, dt$$

The Fourier spectrum is a complex quantity, having separate amplitude and phase plots.

The *shock spectrum* is used in the assessment of the damage potential of the shock pulse. The spectrum is based on the time-history response of a single-degree-of-freedom structure to the shock pulse. The pertinent quantity of interest may be displacement, acceleration, or velocity. Only the *maximum* value attained by the response is of interest. A plot of this maximum quantity against structure resonant frequency, for a fixed specific damping coefficient, constitutes the shock spectrum. Consult the alphabetical index for various transducers that are used in shock measurements.

—Alfred G. Ratz, *MB Electronics, a Textron Company*

Signal, Error (See Error signal.)

Signal Conditioning Signal conditioning is a generic term used to describe modification of an input signal in preparation for a digital-data-acquisition or instrumentation operation. The term does not have a precise meaning and is interpreted differently in different industries and even within a single industry. In one interpretation used in digital data acquisition, it refers to signal modification which takes place prior to amplification of an analog input signal. This modification might include one or more of the following operations: scaling (attenuation), filtering, conversion from current to voltage, impedance-level transformation, bridge or signal compensation, and in the case of signals derived from thermocouples, cold-junction compensation. In some industries, amplification is included as a signal-conditioning function. Furthermore, other industries include both amplification and conversion from analog to digital form as a signal-conditioning operation.

A precise definition is not possible because of this diversity of common usage. The reader is cautioned to seek a precise meaning for each encounter with this particular term; to do otherwise can result in semantic confusion.

—Thomas J. Harrison, *IBM Corporation*

Signal (Electric) Generator In the development, calibration, and testing of electronic hardware, instruments, and systems, frequently it is required to simulate certain electrical signals. A signal source provides the stimulus that creates the response to be measured. This source usually is an oscillator or a standard-signal generator of known characteristics which can be adjusted to establish a known set of conditions. These characteristics comprise (1) the frequency; (2) the output voltage and impedance; (3) the carrier-signal waveform—which typically may be sine-wave, square-wave, pulse, or random-noise; and (4) the modulation which carries the system information through variation of phase, frequency, amplitude, or timing of the carrier waveform.

Signal sources can be classified functionally as to whether the information that they yield is readily usable in *frequency-domain* or in *time-domain* analysis. Sine-wave techniques form the basis of power-generation and transmission systems and most communication systems, leading to ready frequency-domain analysis. Many newer developments in information transmission and data handling, however, such as radar systems, digital computers, telemetry, and even the more venerable wire telegraphy, are based upon pulse techniques which yield most easily to time-domain analysis.

Common to all these systems are ultimate performance limitations determined by system bandwidths and noise. Bandwidth and transient performance are closely related; they convert one into the other and can be measured as phenomena in either the frequency or the time domain at the convenience of the analyst. Noise is most easily measured by comparison with a noise source of known characteristics.

Signal-to-Noise Ratio The ratio of signal amplitude to noise amplitude. Also see **Noise.**

Silicon Controlled Rectifier (See **Electric controller**; and **Electric-motor drives, control of.**)

Silver–Silver Chloride Electrode (See **pH, measurement of**; and **Oxidation-reduction potential, measurement of.**)

Simulation, Process (See **Hybrid computer.**)

Simulator, Aerospace-flight Basically, a grounded aircraft or spacecraft in which the craft's numerous systems, such as engines, airfoils, control surfaces or thrusters, radio, radar, telemetry, and tactical-weapons systems have been replaced by computers. The validity of simulation is in direct relation to the accuracy with which the responses of each system to changes in stimuli are known (usually in the form of equations that interrelate the system variables). The computers actuate the craft's instrumentation, radar, and visual displays, radio signals, pressures on the controls, and other pilot cues exactly as would occur in actual operation of the craft with similar manipulation of its controls.

Simulator, Programming A simulator is a program which allows the simulation of a process, machine, or mathematical model. The program may be specific in nature, such as a simulator for another computer or system, or a general-purpose routine capable of taking specifications for systems or models in terms of dependent and independent functions and operators, and producing performance figures for operation of the simulated system.

Simulators are used to study system and device performance where construction of an actual device may not be feasible or economical. Often simulation studies provide guideline information for the initial prototype construction and testing procedure.

—L. E. Wise, *IBM Corporation*

Sizing, Valve (See **Control valve, sizing of.**)

Sleep, Analysis of (See **Central nervous system, instrumental analysis of.**)

Sleet, Measurement of (See **Precipitation gage.**)

Slide Valve In a slide valve, a sliding gate in continuous contact with a stationary plate is positioned by the raising or lowering of the valve stem. Ports in the plate are covered and uncovered by the sliding gate, which also has ports or orifices. When the ports in the sliding gate coincide with the plate ports, the valve is open. When the solid portion of the gate covers the ports in the plate, the valve is closed. Throttling is possible between full open or closed positions, and the ports can be shaped to obtain desired characteristics. Flow through the ports is straight through the body.

Slide valves are truly balanced; fluid velocity exerts no direct fluid thrust either up or down. Fluid thrust simply pushes the sliding gate against the plate, helping shutoff. Sliding gate and plate seats are naturally self-cleaning and self-lapping. The seats, therefore, wear in—not out. Slide valves generally are furnished in sizes through 6 in. in any castable material. Manual or automatic actuators can be used.

—Robert J. Lasher, *Fisher Controls Company*

Slidewire Potentiometer (See **Potentiometric instrument.**)

Sling Psychrometer (See **Psychrometer.**)

Slotted-line Frequency Measurements [See **Frequency (electrical), measurement of.**]

Slow Neutrons, Detection of (See **Boron thermocouple.**)

Smith III Bridge (See **Bridge circuits, electric.**)

Smoke Detection (See **Turbidimeter.**)

Snow Gage (See **Precipitation gage.**)

Software Software is a general term applied to computer programs or programming systems (see **Program, computer**). It may refer to a small component or a collection of large programming systems. References are made to software and hardware parts of a system where the hardware is the machinery itself (mechanical and electronic) supplied by the engineers and the software is any and all programming supplied by programming personnel.

In some current machines, the instructions, instead of all being wired into the control unit, are microprogrammed (see **Microprogram**) in a special control storage section of the machine using a more basic code which is actually wired into the machine. This technique allows the economic construction of various size machines which appear to have identical instruction sets. The microprogramming is not normally referred to as software.

—L. E. Wise, *IBM Corporation*

Solar Telescope Solar telescopes are usually reflectors having very long focal lengths to minimize concentration of energy. Because of the long length, coelostats or siderostats are used to feed a stationary telescope. Atmosphere must be closely controlled to produce good imagery, and spectrographs for solar work are commonly evacuated. Reflecting optics are normally used, and metal mirrors or materials with very low coefficient of expansion have been used to avoid thermal distortion.

—Joost H. Kiewiet de Jonge, *University of Pittsburgh,* and Lloyd M. Geertz, *Owens-Illinois*

Solenoid Stepping Motor (See **Stepping motor.**)

Solenoid Valve A solenoid valve consists basically of an electromagnet and a movable armature (soft-iron core) which is mechanically linked to the device to be positioned, such as a valve plug, lever, or pilot valve. Solenoids generally provide two-position operation only, but modulating solenoids are available for throttling service. When the electromagnet is energized, the armature is pulled up, and this action opens a normally closed valve and closes a normally open valve. Solenoid valves are used widely for emergency-shutoff service and as pilot valves in control systems. Many solenoid valves are of the packless type, with the actuator portion close-coupled to the valve body. Three-way and four-way valve bodies often are used with solenoid actuators to add greater versatility to the control system. Solenoids operate on standard voltages of both alternating and direct current.

—Robert J. Lasher, *Fisher Controls Company*

Sonar Acronym (sound navigation and ranging) for equipment used for sonic and ultrasonic underwater detection, ranging, sounding, and communication. Echo-ranging sonar is quite common and is an active sonar in which a sonic or ultrasonic pulse is transmitted to, reflected from an object, and received back at the transmitter location. The elapsed time interval for the pulse to travel from and back to the transmitter-receiver location is a measure of distance. The directional characteristics of the transmitting-receiving transducer provides the other necessary ingredient for range determination. For underwater communication, the pulses may be spaced according to international Morse code. Other versions are passive sonar, scanning sonar, and searchlight sonar. The British term for sonar is *asdic.*

Sound-level Meter (See **Acoustic instrumentation; Audiometer.**)

Sound Navigation (See **Sonar.**)

Space-balance System (See **Valve positioner, pneumatic.**)

Spacecraft Simulator (See **Simulator, aerospace-flight.**)

Space and Missile-systems Instrumentation Irrespective of the fundamental mission of an aerospace vehicle—as of an aircraft to transport people or materials from one city to the next or of a missile to carry destructive forces to a stationary or moving target on earth or in space or of a satellite to probe the unknowns of space or to serve an earthly reconnaissance or communications function—the vehicle obviously must travel from one point on earth or in space to another point on earth or in space.[1] The mission may terminate when the distance between the two points has been traversed, or in the case of an orbiting satellite, the journey may go on almost endlessly once the vehicle assumes a stable orbit around either the earth or some extraterrestrial body.

To guide and to control an aerospace vehicle essentially is a matter of measuring position, determining path errors, and controlling to correct

[1] Introductory information and description of orbital injection have been provided by John W. Miller, Honeywell Inc. All other descriptions have been provided by Paul F. Bechberger, Navigation and Controls Division, The Bendix Corporation.

the path. These three functions usually are called navigation, guidance, and control. The functions can be broken down further to add a fourth function, steering.

1. Navigation—the determination of position

2. Guidance—the determination of errors from the intended path

3. Steering—selecting and directing a proper series of changes to the path in order to correct the path

4. Control—changing forces on the vehicle to comply with the directed path changes

Normally, these four functions are accomplished with separate equipments. Often the equipments are located far apart, and thus a separation of their procurement has evolved. Sometimes, the equipments are procured from a single contractor and the boundaries tend to fade. In other cases, where numerous suppliers are involved, the problem of systems integration becomes paramount. However, for the purposes of editorial descriptions, the four foregoing classifications are convenient.

As in the cases of machine and process instrumentation and control systems:

1. The entire navigation-control function may be *wholly* manual, as in the situation of pilot training where the pilot, guided by his cockpit instruments and radio communications from the ground, manually adjusts the engine power and aerodynamic surfaces to achieve the desired course for the vehicle.

2. The system may be *semiautomatic* or *automatic,* as in the case of a commercial or military aircraft equipped with navigation aids and an autopilot, but it may be used with manual insertion of navigation results to a guidance computer or manual insertion of steering information to the autopilot.

3. The vehicle flight plan may be *fully programmed* prior to takeoff or lift-off, where all predictable force conditions and changes are fully anticipated and integrated into the system prior to launch. Examples are artillery shells and upper stages of ballistic missiles. Fully preprogrammed control, of course, has been well established over the years in connection with artillery firings of shells and rockets.

4. The vehicle guidance and control system may be fully *self-contained,* that is, completely functional without direction or interference from control sources outside the vehicle. For example, a missile or a satellite control system may be preprogrammed before launching to adjust for errors in a predetermined way. Inertial guidance is the best example of this.

Inertial guidance predicts a ballistic path. The system then (*a*) measures accelerations and remembers reference angles, (*b*) computes velocity and position, (*c*) predicts the destination (guidance), (*d*) determines a preferred way to correct the error (steering), and (*e*) controls the forces to change the path by way of steering instructions (controls). Alternatively, the vehicle may call upon the outside environment for information, but not for decisions. Often the navigation and guidance functions use outside data, such as stars and radio-navigation aids.

5. The vehicle may be partially or fully dependent upon *remote guidance and control.* A satellite, for example, may be fully controlled from a ground station. All force settings and controls for lift-off will be made from a combination of measurements taken from within and without the vehicle. After launching, the vehicle will be carefully observed (tracked) and further adjustments made as the result of these readings. Simultaneously, the measurement systems within the vehicle will transmit data back to ground, and based upon these measurements, further adjustments will be telemetered from ground. In other words, the total vehicle instrumentation system will be made up of some measurements and controls that are fully self-contained within the vehicle, some from data measured by the vehicle but with the closed loop closed via ground, and some

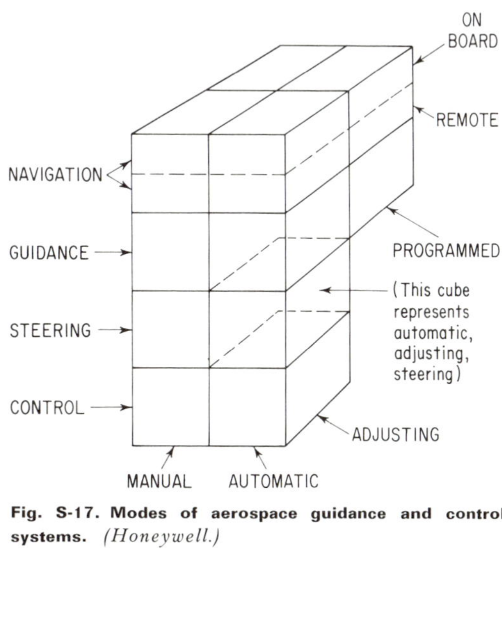

Fig. S-17. Modes of aerospace guidance and control systems. *(Honeywell.)*

from data measured by ground (such as tracking data) and adjustments telemetered to the vehicle.

Thus, from an overall control-systems and instrumentation standpoint, aerospace-vehicle systems use many degrees of automation. Almost all combinations of automatic to manual, remote to self-contained, preprogrammed, and correcting are used. The methods are varied for each function of navigation, guidance, steering, and control. Some concept of the numerous combinations may be gleaned from the matrix of Fig. S-17. A single vehicle often contains many alternative modes, using nearly all possible combinations of the matrix. Steering and control are *adjusting* by their very nature—thus, the matrix has a hole as shown. Examples of combinations might include:

Commercial airliner:

Navigation typically is	Manual, remote, adjusting
Guidance	Automatic, remote, adjusting
Steering	Manual, on-board, adjusting
But often it is	Automatic, on-board, adjusting
And it may be	Manual, on-board, adjusting

Air to air missile:

Navigation and guidance usually are combined	Automatic, on-board, adjusting
Steering and control often are combined	Automatic, on-board, adjusting

Navigation: Fundamental systems for navigation, many of which apply fully or partially to space and missile systems, are described elsewhere. See **Navigation systems.** Included here are a selected number (not exhaustive) of descriptions of important navigational and guidance methods found principally in space and missile systems.

Sextant: A space sextant measures the angle between the spacecraft and two or more known celestial bodies to determine spacecraft attitude and position. Celestial bodies which may be utilized as attitude reference sources are stars, the sun, the moon, a planet, or a pattern of stars. When a star and planet or the moon are used, the sextant essentially is a star tracker and a planet tracker—with a means of measuring the angle between the trackers combined into a single instrument. Besides being capable of tracking a planet, the sextant has capability to track landmarks on planets. The sextant and its features are shown in Fig. S-18.

As shown, the sextant essentially is a two-line-of-sight instrument providing magnification for manual visual use as well as special sensors for automatic use. One of the lines of sight identified with the landmark side of the angle is fixed to the spacecraft. The spacecraft must be turned in space to orient this line appropriately by means of commands to the attitude-control system. The

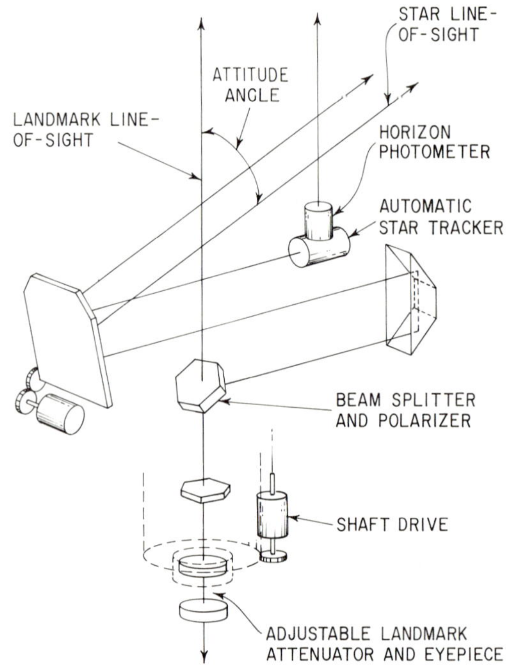

Fig. S-18. Schematic of a space sextant. *(Bendix.)*

star line can be pointed in space through the use of servomotor drives. One axis of this motion is parallel to the landmark line and changes the plane in which the angle is measured by rotating the head of the instrument as a whole. The second axis sets the angle by tilting the axis mirror. An angular encoder on this mirror provides a measure of the angle.

Light along the landmark line is polarized and then combined in the beam-splitter mirror with the starlight. This permits manual adjustment of landmark background brightness relative to the star intensity by means of an eyepiece polarizer. The sextant also uses the mirror and star-tracker sensor to provide error signals to the servodrives for automatic control. Use of the sextant on board the spacecraft for determination of vehicle orientation is shown in Fig. S-19.

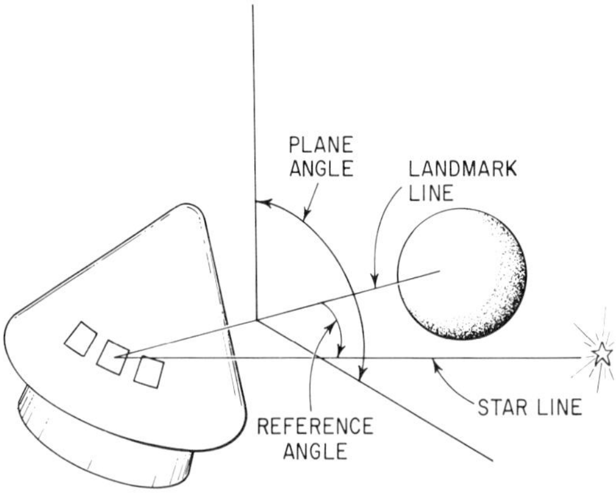

Fig. S-19. Use of sextant on board spacecraft for determination of vehicle orientation. *(Bendix.)*

Horizon Scanners: Local vertical of an orbiting vehicle can be determined by a horizon scanner. The type described here is an infrared scanner. There are basically five methods of infrared horizon scanning: (1) conical scan, (2) radiant balance, (3) horizon-sector tracking, (4) horizon-point (edge) tracking, and (5) chord scan (or long-life earth sensor, LLES).

The *conical-scan* system uses two sensing units or "heads" with an accompanying electronics package. Each head contains a sensor or telescope whose field of view (FOV) is made to describe a conical surface in space. This surface intersects the earth's surface as shown in Fig. S-20. Pitch information is derived by comparing the angles at which either FOV enters and leaves the horizon. Roll information is derived by calculating the difference in the amount of time spent by each FOV within the earth image. Roll $= 0$ if dwell time $A =$ dwell time B.

High-speed rotating parts required for the conical scan lower the reliability of this method for certain missions. In addition, if the sun enters the FOV of either telescope, roll information is no longer available, since outputs from both heads are required for roll computation.

The *radiant-balance system* uses four fixed telescopes which compare the relative infrared radiance from points on the horizon 90° apart in azimuth, as shown in Fig. S-21. Radiant balance indicates a null attitude. Because this system is static, reliability is high. However, since temperature variations in the horizon may occur—which affect the radiance level—the overall system accuracy is comparatively low. Null is indicated by a balance of radiance levels ($E_3\, dA_3 = E_2\, dA_2$, where E_3 and E_2 are the radiant energies on A_3 and A_2, respectively).

The geometry of the *horizon-sector-tracking* system is shown in Fig. S-22. This system consists of a single tracker (two may be used to prevent sun disablement) mounted on bearings so that the tracker scan plane can be rotated in azimuth while dithering in elevation. The azimuth angle reciprocates across the horizon. The time variations in the measured elevation of the horizon indicate the amount of pitch and roll present. Although very accurate, this system is unable to measure large values of pitch and roll at high altitudes and has a relatively long time constant. Of all the active sensors, this type is the simplest mechanically, but the computations involved in determining attitude are quite complex.

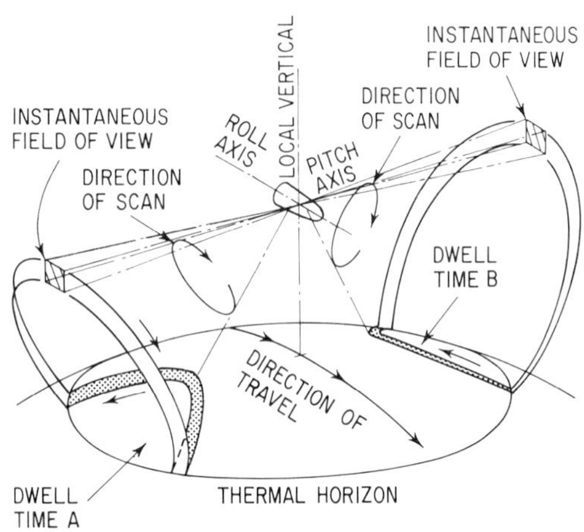

Fig. S-20. Conical-scan paths of a conical-scan horizon scanner. *(Bendix.)*

The geometry of the *horizon-point-tracking* system is shown in Fig. S-23. Four telescopes or trackers are used to derive pitch and roll information. One tracker is redundant and is used to prevent sun disablement of the system. The FOV of each tracker is oscillated or "dithered" in a vertical scan plane across the horizon by means of a small scanning mirror in front of the telescope. First, the mirror "searches" slowly down the scan plane until the proper infrared gradient is encountered. At this point, "searching" stops and dithering begins, with the dither motion centered $\pm 1\frac{1}{2}°$ from the center of the horizon gradient. Each tracker measures the horizon elevation angle in its own scan plane. These angles are summed appropriately to provide pitch and roll information. The system was designed to achieve high accuracy over a wide altitude range.

The *long-life earth sensor* consists essentially of two scanners, each containing an infrared telescope with a bolometer and a scanning mechanism to provide a positive signal when the telescope line of sight is directed to the earth. The two channels provide roll and pitch attitude error

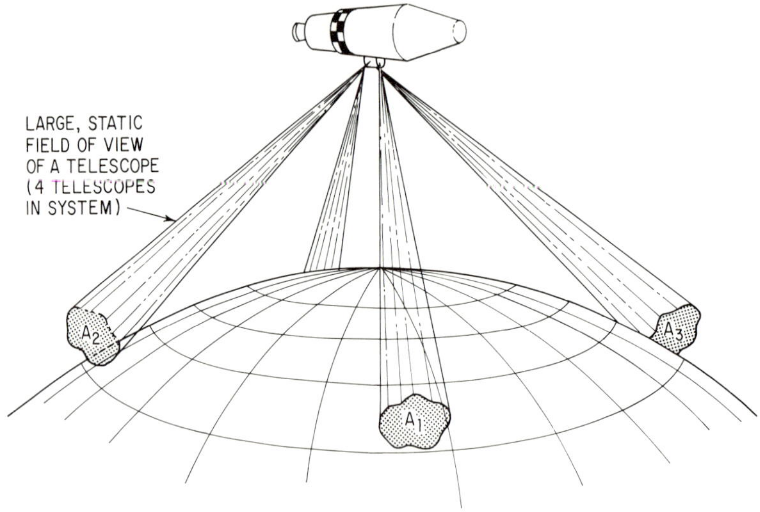

Fig. S-21. Radiant-balance system used in horizon scanner. *(Bendix.)*

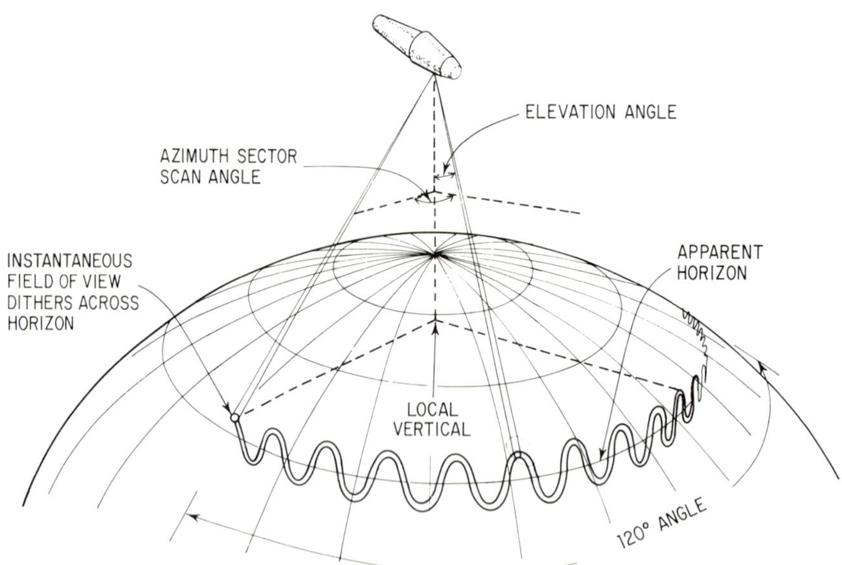

Fig. S-22. Horizon-sector-tracking system. *(Bendix.)*

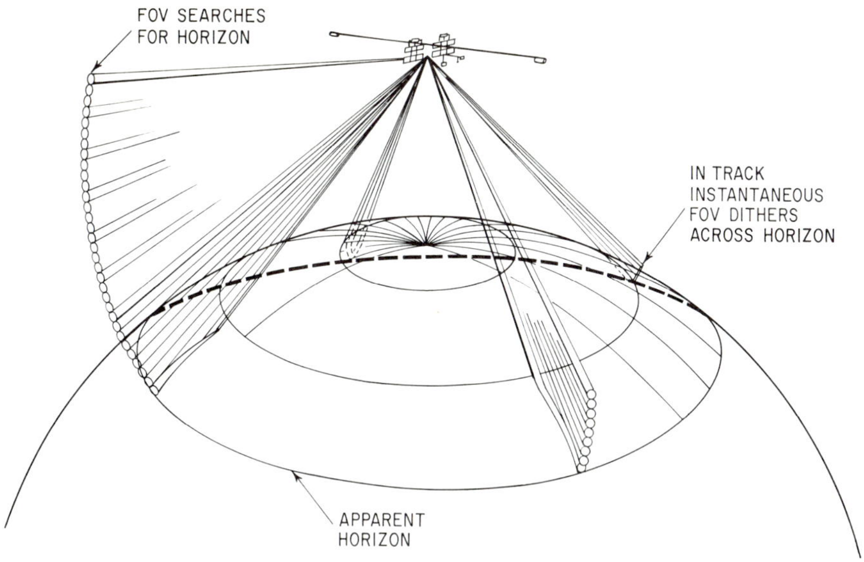

Fig. S-23. Horizon-point-tracking system. *(Bendix.)*

information to the orbiting vehicle. The sensor is aligned with respect to the vehicle so that the sensor null axis and vehicle yaw axis are collinear. The two channels scan along two orthogonal chords on the earth disk, with independent implementation of their scan motions. During scanning, the channels detect displacements, or X and Y offsets, of the vehicle yaw axis from the chord centerlines in the scan planes. These X and Y offsets are representative of roll and pitch attitude errors, respectively. Vehicle-pointing-error detection is implemented by using standard integration techniques to transform square-wave asymmetry in the detected radiance waveform into signals representative of changes in vehicle attitude. Figure S-24 depicts the FOV of the two LLES telescopes scanning across the earth target in two orthogonal scan planes. This scanning action occurs at a constant 5-Hz rate (one scan cycle is defined as two complete transversals of the earth target, that is, 136 optical degrees of scan transversal) over a constant peak scan amplitude of 68 optical degrees. System null is defined as that instant when the telescope FOV is aligned with the vehicle yaw axis. The system is mechanized so that a polarity reversal occurs in the radiance-generated waveform fed to the system integrator as a scanner passes through system null. At vehicle null, the integrator input waveforms in both scanners pass through system null. At vehicle null, the integrator input waveforms in both scanner channels are symmetrical square waves. Therefore, the integrator output is zero, indicating no X and Y pointing-error offsets.

For a pure pitch-up condition, each integrator input waveform is affected. First, the A scan is symmetrical with respect to the projection of the vehicle roll axis onto the plane of the earth disk.

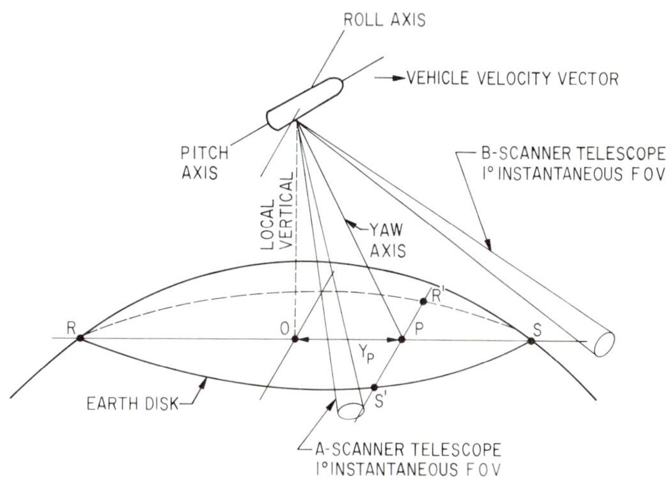

Fig. S-24. Long-life earth sensor in pure pitch-up condition. Y offset of yaw axis from center of earth disk is detected by scanner B. *($RP \neq PS$.) (Bendix.)*

Since the scan amplitude is a constant, and since the sensor scans an earth chord rather than an earth diameter, the *A* scanner is transversing a smaller earth-target dimension than at null. Consequently, the integrator input waveform covers shorter time span than it did at null. However, square-wave symmetry of this waveform has been maintained so that the integrator output remains zero, indicating zero roll.

Since the system null is aligned with the vehicle yaw axis, the *B* scanner zero cross occurs at the point in the integrator input waveform characteristic of the new pointing position of the vehicle yaw axis. The total time span of the integrator input waveform is the same as it was at null, but because of the pointing error, the *B*-scanner integrator waveform is not symmetrical. The degree of asymmetry is proportional to the offset. The direction and magnitude of the pointing error from null are contained in the sign and amplitude of the output, respectively.

Star Tracker: This device provides precise attitude information for use on board a spacecraft. The star tracker has the following characteristics: (1) adequate acquisition field, (2) small tracking field, (3) high resolution after acquisition, (4) no loss of central sensitivity, and (5) practical adequacy, including size, power, reliability, and longevity.

The tracker uses a photomultiplier tube or an image-tube detector to sense the star. The field of view of star trackers using photomultiplier tubes is oscillated. This is accomplished by oscillating a mirror or a reticle located in the field. The FOV is centered on the desired star. The optics-angle transducer determines the direction toward the star. Trackers which use image-tube detectors determine the attitude from the image position in the focal plane. The reference star is detected by means of its brightness. After being located, the tracker attempts to lock on the star. The attitude error signal, which is a function of displacement of the star from the center of the scan-arc limits, is used to drive the optics—or a field stop within the optics—such that the error signal is nulled. The vehicle attitude relative to the star then is determined by reading the angular orientation of the tracker optics relative

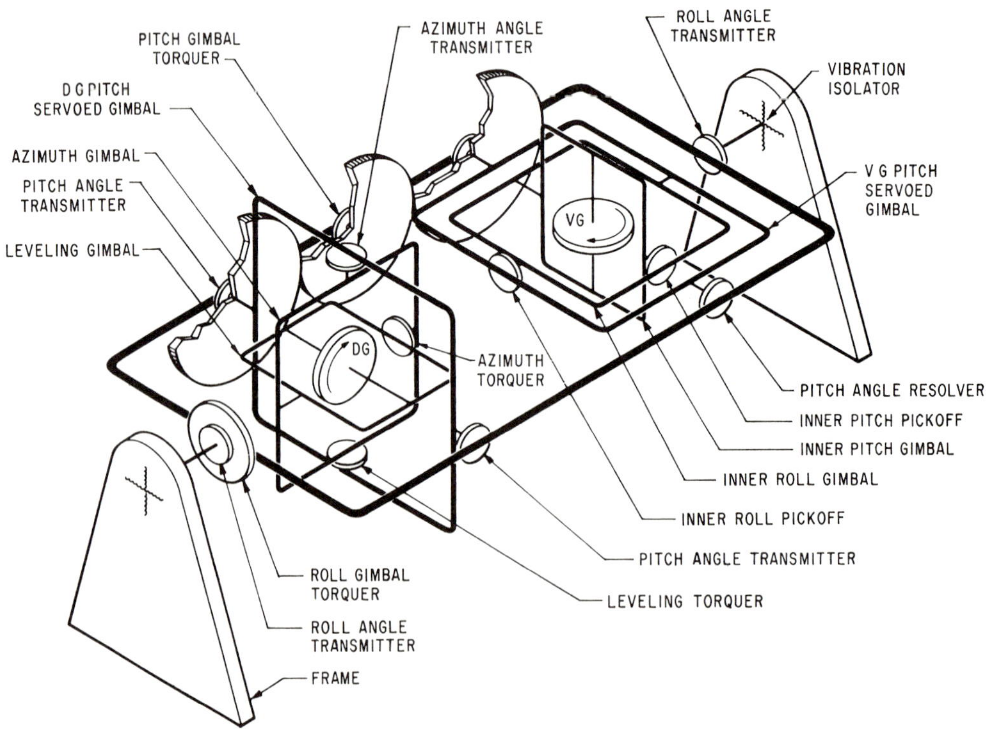

Fig. S-25. Two-gyro platform-gimbal configuration. *(Lear Siegler.)*

to the spacecraft. Reflected sunlight is a problem with this instrument. Multiple baffles for FOV restriction often are used. Proper star identification also is a problem. The combined factors of detector aging and shot noise in photomultipliers can cause star trackers to select an incorrect target star. Detector aging can be compensated by sampling the detector's sensitivity and then by increasing amplifier gain in accordance with detector fatigue. Spectral analysis may be used as a supplementary technique for star identification.

Sun Sensor: A sun sensor consists of a pair of silicon photovoltaic or cadmium sulfide photoconductive cells for each axis of operation and may be used for attitude determination. The cells are connected in a bridge configuration. The cells are physically arranged so that a pointing error in one plane produces a difference in the amount of sunlight falling on the cells in that plane. This difference produces a net output which indicates the attitude error. Coarse sun sensors have a FOV up to 360° and usually are used to bring the sun into the FOV of the fine sun sensor. Sun sensors can be used to orient a spacecraft precisely at some offset angle with respect to the sun. Precise orientation is accomplished by means of a rotatable optical slab which is placed in the FOV of the detector elements. The optical slab is rotated to a prescribed angle relative to the optical axis, thereby displacing the light rays at the detectors. The spacecraft is pointed at a corresponding offset angle relative to the sun to null the detector-bridge output. Unequal solar-cell aging, which can create an offset bias in the detector bridge, can be a problem. The application of radiation shields on the cells and the use of light baffles reduce aging. Also, the aging problem is reduced significantly by selectively choosing cells from a preirradiated lot, since continued degradation will follow the same exponential curve.

Gyroscopes: These devices are used to establish an attitude reference for spacecraft. For precision systems, the gyros are mounted on a servo-driven stabilized platform. The gyros operate as angular-error detectors over a small region near their null position. Frequently the gyros are mounted directly to the vehicle structure. The vehicle is controlled in attitude and, as such, *is* a stabilized platform. Gyroscopic devices are described in detail elsewhere. See **Flight-control systems.** See Figs. S-25 and S-26. A *control-moment gyro* also is described below.

Control Systems

Attitude-control Systems: The attitude-control system aligns the spacecraft to a desired orientation with respect to a reference-coordinate system. The various methods for sensing attitude errors have been described. In space, the disturbing moments are small but still require corrections from the control system. An angular error in vehicle attitude results in an output from

Fig. S-26. Two-gyro platform-gimbal configuration diagramed in Fig. S-25. A two-gyro platform essentially is the combination of a directional gyro and a vertical gyro, each independently gimbaled within a common outer gimbal. The vertical gyro stabilizes the directional gryo in roll and pitch to eliminate the gimbal errors in direction that occur in turns with conventional directional gyros. Provision usually is made to update the directional output continually with heading information from a magnetic or other compass. The platform shown provides directional, pitch, and roll outputs as required by other systems and displays. *(Lear Siegler.)*

the position sensor. This error signal is amplified and processed electronically to generate the appropriate command to the vehicle torquers—which may be (1) reaction jets, (2) control-moment gyros, or (3) reaction wheels. The moments on the vehicle are forcing functions which act on the vehicle dynamics and change the vehicle attitude and attitude rate, thereby closing the control loop. Block diagram of a vehicle attitude-control system is shown in Fig. S-27.

Control-moment Gyro: The CMG provides space-vehicle stabilization and control by exchanging momentum with the vehicle. Mathematically, the gyro can be represented as a spinning wheel having one or two degrees of freedom

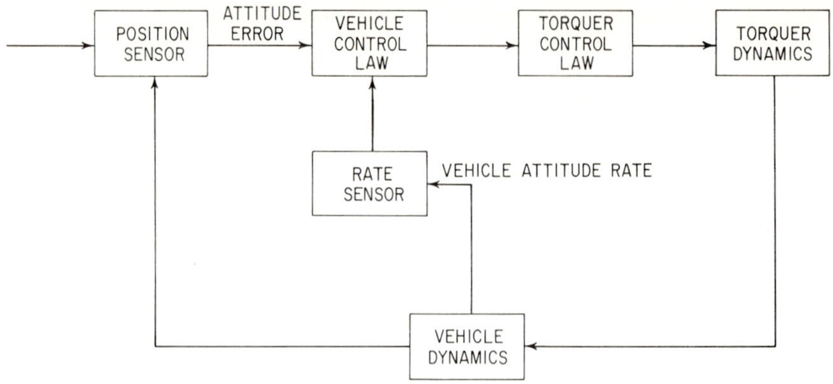

Fig. S-27. Vehicle attitude-control system. The position sensor measures the attitude error between the actual vehicle attitude and the desired vehicle attitude. The vehicle control law formulates the vehicle moment command as a function of vehicle angular position and rate. The function of the torquer control law is to convert the command moment into the proper control variable, such that the desired reaction moment will be obtained by the vehicle. Torquer dynamics generate the commanded moment to the vehicle. Vehicle dynamics are the rate and position response of the vehicle to the reaction moment. Both the rate and position commanded are such as to cancel any position error. *(Bendix.)*

relative to the vehicle in which it is mounted. Control torques on the vehicle are gyroscopic-reaction torques created by angular motion of the gyro spin vector in inertial space. The CMG achieves momentum exchange by variation of the direction of the spin vector. A two-degree-of-freedom control moment gyro is shown schematically in Fig. S-28. The device consists of a constant-speed wheel held by a housing called the inner gimbal. The inner gimbal is coupled to the outer gimbal through a pivot which is perpendicular to the wheel spin vector. The outer gimbal is held to the base through the outer pivot, which is perpendicular to the inner pivot and wheel spin vector. The inner and outer pivots are driven by geared torquers, and the dc motor is mounted on the outer frame and drives the inner frame for both gimbals. The servoamplifier drives the dc motor with a constant source and has a lag network to increase the loop static gain. A minimum of two CMGs is required to provide three-axis attitude control.

Reaction Jets: A reaction jet is a container filled with matter and energy so arranged that a portion of the matter can receive energy in kinetic form and then be ejected at a controlled rate in the desired direction. The matter is transformed into gas and escapes at a high velocity. The remainder of the system undergoes a fine rate-of-change momentum, and the container experiences a reaction force. Reaction jets are small rocket engines arrayed to provide suitable moment couples, depending upon angular-momentum transfer to the exhaust gas. The jets usually are chemical or cold-gas low-thrust engines designed for as many on-off cycles as demanded by the control loop. Control is characterized by pulsed operation of the jet and limit-cycle oscillation about the desired attitude. The fuel in space must be used sparingly since it cannot be replaced. The reaction jets often are used to stop vehicle spin after launch and to unload momentum-storage devices.

Reaction Wheel: This device stabilizes and controls a spacecraft by exchanging momentum with the vehicle. The wheel achieves momentum exchange by variation of the magnitude of the spin vector. The reaction wheel is driven by a bidirectional motor. The motor exerts a control moment on the vehicle and an equal and opposite moment on the wheel. The moment axis is fixed relative to the spacecraft. Three such wheels with mutually perpendicular spin axes provide complete attitude control. Normally, the spin axes will be parallel to the vehicle principal axes to reduce coupling. For steady pointing, the command moment is equal to the disturbance moment. For attitude changes in space, the control moment required for reasonable response time normally is much greater than the disturbance moment. The entire unit is hermetically sealed. Sensors detect deviations from the desired orientation and command the proper speed changes in the wheels to cancel disturbance moments or effect attitude changes. When a wheel is required to correct for angular-momentum changes beyond its design capabilities, the wheel will speed to its maximum value, at which point the reaction jets unload the wheel. A *reaction sphere* has the capability of providing moments about any axis. This device is a spherical shell supported by a gas bearing, a magnetic field, or an electric field. The reaction sphere may be torqued magnetically. One sphere will provide three-axis control.

Thrust-vector Control: This is a closed-loop process whereby the vehicle is prevented from tumbling under high engine forces and engine direction is changed to orient engine-caused acceleration. A generalized thrust-vector stabilization and control system is shown in Fig. S-29. The system is characterized by appropriate feedback to provide a stable control of angle or angular velocity of the thrusting vehicle. The loop also accepts input steering commands from guidance to achieve a particular desired thrust direction. Design constraints on thrust-vector stabilization and control systems vary considerably. The system is constrained by nonlinearities and by constraints on dynamic response to inputs and disturbances. The system usually is restricted by allowable limits on angular acceleration and other variables, dependent upon structural and controllability factors.

Velocity-vector Control: In a spacecraft, thrust-vector control is based upon measurements of vehicle motion, generally using self-contained instruments. Three types of measurements usually are required: (1) angular rate, (2) linear acceleration, and (3) time. The integration with time of the measured acceleration in the indicated direction—with proper consideration of the gravity-field forces—is the essential navigation portion of velocity-vector control. The processes are achieved in an on-board digital computer with the result of generating corrective steering commands to the thrust-vector control system. A system of this type can integrate only vehicle motion into changes in position, velocity, and orientation. Accurate initial conditions in these parameters are required in order to provide appropriate corrections. Initial conditions in position and velocity are obtained from the navigation system, and of orientation from the attitude-control system. A typical velocity-vector control system is shown in Fig. S-30. System errors arise from gyro

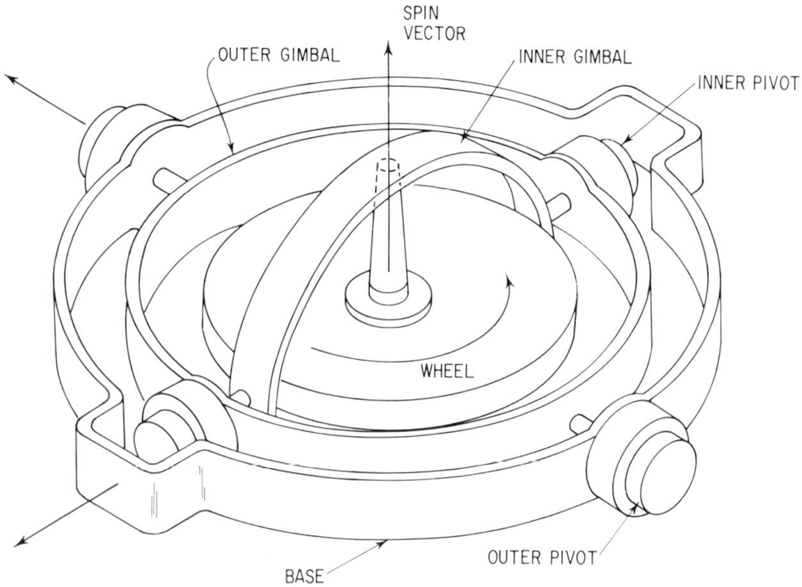

Fig. S-28. Two-degree-of-control moment gyro. *(Bendix.)*

drift. If the gyro measurements are in error, the direction of the controlled acceleration is in error. When there is an error in measured acceleration, the direction of acceleration as well as the magnitude of acceleration and the duration of motor burning are affected adversely. However, errors in the resulting velocity usually can be measured later by the navigation system and corrected.

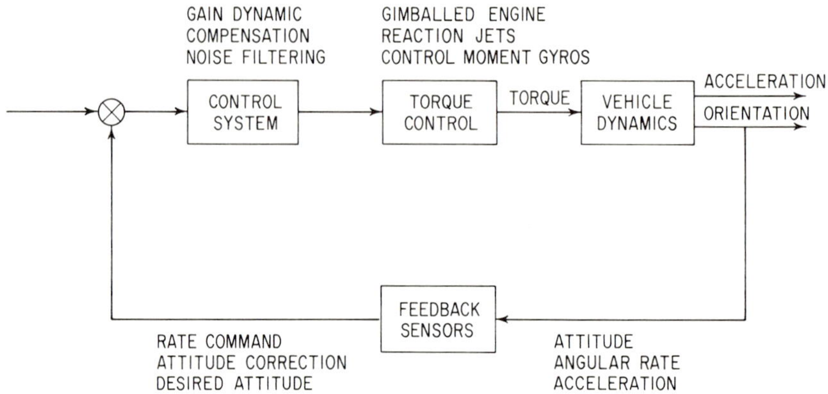

Fig. S-29. Thrust-vector control system. *(Bendix.)*

Magnetic Torquing: Forces associated with the space environment can be utilized in a passive or semipassive attitude-control system. Self-aligning mechanisms based upon utilization of the magnetic field provide weak but often adequate restoring moments to a stable orientation. The earth's magnetic field varies approximately as the inverse cube of the radius from earth's center. If the vehicle is orbiting the earth, this magnetic field can be used to produce a moment on the vehicle—except around the magnetic vector. Two methods are used: (1) a controlled current is applied to three mutually perpendicular coils; or (2) the current is applied to a gimbaled permanent magnet which turns freely except when a moment is required.

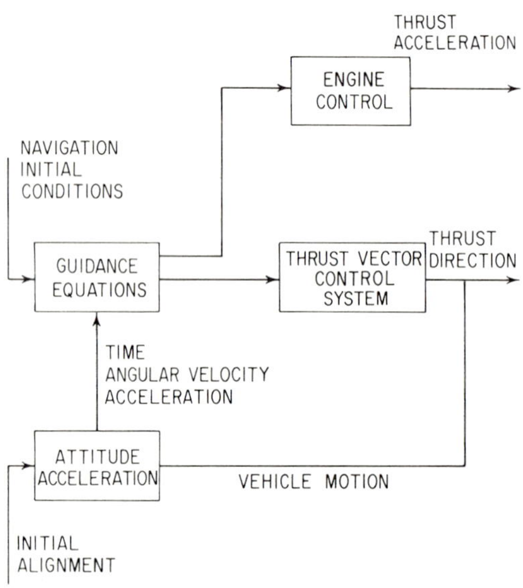

Fig. S-30. Velocity-vector control system. *(Bendix.)*

Stabilization by Gravity Gradient: A satellite assembly orbits the earth with its center of gravity at an altitude such that equilibrium exists between the gravitational force and the centrifugal force generated by orbital angular velocity. The gravitational force varies with the inverse square of the altitude. Hence, while the entire assembly may be in an equilibrium orbit, individual parts of the satellite (particularly masses located far from the center of gravity) would travel in slightly different orbits if they were free. The resulting unequalized forces acting upon the satellite produce *gravity gradient torques.*

In general, any satellite having unequal moments of inertia about its three axes will attain a preferred attitude. The minor inertia axis will align with the earth's center; the major inertia axis will be perpendicular to the orbital plane. Often, when gravity gradient torques are being used for attitude stabilization, the effect is enhanced by deploying a mass connected to the main body of the satellite by a very long, lightweight boom.

Gravity gradient torques cause oscillatory motion of the satellite—termed *libration*. Libration will continue indefinitely unless its energy can be dissipated. Generally, this is done through a damping mechanism employing one of the following energy-dissipation modes: (*a*) hysteresis loss in ferromagnetic rods interacting with the earth's magnetic field, (*b*) eddy-current loss using local permanent magnets, or (*c*) frictional loss in fluid damping. Libration energy transfer by these means is a slow process, usually requiring many orbits to achieve acceptable stabilization.

The following equations are useful in calculating gravity gradient torques:

$$M_p = \tfrac{3}{2}(I_r - I_y)\,\omega^2 \sin 2\theta_p$$
$$M_r = 2\,(I_p - I_y)\,\omega^2 \sin 2\theta_r$$
$$M_y = (I_p - I_r)\,\omega^2 \sin 2\theta_y$$

where M = torque
I = moment of inertia
ω = orbital angular velocity
$\theta_p, \theta_r, \theta_y$ = angular deflections about the pitch, roll, and yaw axes

Thrust Acceleration: This factor commonly is measured by single-degree-of-freedom accelerometers. The accelerometers are usually mounted orthogonally to each other and may be mounted directly to the spacecraft, in which case they are measuring acceleration in vehicle space. Or, the accelerometers may be mounted on an inertially stabilized gimbal, in which case they measure acceleration directly in inertial coordinates. Also see **Accelerometer.**

Orbital Injection: To inject a body into orbit, it is only necessary to control the velocity vector precisely. This is because the position cannot depart significantly from the launch point without burning too much fuel just to retain altitude while translating. Any velocity creates an orbit, but it may intersect the earth. In addition, energy is conserved by putting objects into nearly circular orbits. Such objects must have a velocity vector parallel to the earth's surface and greater than 26,000 ft/s in magnitude. Furthermore, it is convenient to direct the velocity vector due east from the launch point simply to conserve fuel. The situation further is constrained by constant-thrust engines. Once ignited, these engines will produce a certain speed profile for the rocket. Thus, it remains only to direct the added speed so that the force is in the proper direction and then to stop the engines when sufficient speed has been achieved. A typical orbital-injection guidance system thus must (1) program the vehicle attitude and (2) measure the velocity for engine cutoff.

Some very simple systems have been devised to do this. Also, some complex systems have been developed to obtain precise orbits to selectable orbital parameters. The complications arise from the earth's rotation and resulting Coriolis forces, from opposing gravity in early launch stages, from wind, from unpredicted variations in mass, and in general from the multidimensional nature of the problem. For a simple system, however, the desired velocity, attitude, and position profile can be precalculated for all times after launch. The guidance system then consists of a timer and memory to generate reference data, plus an attitude reference, velocity computations, and position measurements. The preprogrammed and measured data are compared to give an error signal in each parameter. Typically, the position errors are corrected only if they are grossly incorrect. Attitude errors are corrected continuously. Velocity errors are compensated for at the engine cutoff by modifying the attitude errors. This type of guidance system often is called "perturbation" guidance because the system acts upon perturbations from a planned trajectory.

REFERENCES

Proc. 15th Natl. ISA Aerospace Instrumentation Symposium, ISA Publ. AS691, Las Vegas, Nev., 1969.

Tutorial papers from 15th Natl. ISA Aerospace Instrumentation Symposium, ISA Publ. AS692, Las Vegas, Nev., 1969.

Instrumentation in the Aerospace Industry, developed from the 14th National ISA Aerospace Instrumentation Symposium, Boston, Mass., *ISA Publ.* AS681, 1968.

Fundamentals of Aerospace Instrumentation, developed from the 14th National ISA Aerospace Instrumentation Symposium, Boston, Mass., *ISA Publ.* AS682, 1968.

Proc. 13th Natl ISA Aerospace Instrumentation Symposium, ISA Publ. AS67, San Diego, Calif., 1967.

Aseltine, J. A.: Automatic Control in Space, *Proc. 2d IFAC Symposium, Vienna, Austria, ISA Publ.* ACS, 1967.

Aseltine, J. A.: Peaceful Uses of Automation in Outer Space, *Proc. IFAC Symposium on Automatic Control in Peaceful Uses of Space, ISA Publ.* PUAS, Oslo, Norway, 1965.

Besserer, C. W.: "Missile Engineering Handbook," D. Van Nostrand Company, Inc., Princeton, N.J., 1958.
Newman, David B.: "Space Vehicle Electronics," D. Van Nostrand Company, Inc., Princeton, N.J., 1964.
Parvin, Richard H.: "Inertial Navigation," D. Van Nostrand Company, Inc., Princeton, N.J., 1964.
Povejsil, Donald J., Robert S. Raven, and Peter Waterman: "Airborne Radar," D. Van Nostrand Company, Inc., Princeton, N.J., 1961.
Scavullo, Joseph J.: "Aerospace Ranges: Instrumentation," D. Van Nostrand Company, Inc., Princeton, N.J., 1965.
Perry, M. A. (ed.): "Flight Test Instrumentation," Pergamon Press, New York, 1965.
Locke, Arthur S.: "Guidance," D. Van Nostrand Company, Inc., Princeton, N.J., 1955.

Span (See **Range, instrument.**)

Spark Chamber [See **Radiation (nuclear), detection of.**]

Speaker-identification System (See **Voice-recognition systems.**)

Special-purpose Computer A special-purpose computer is designed to solve a specific problem or a limited set of calculations. In general, a computer of this type can be designed to perform the calculations faster and with less hardware than a general-purpose computer (see **General-purpose computer**).

An example of this type of machine might be an array processor which performs a fixed set of arithmetic operations on an array. In this case, the variables are read into the processor in a fixed format, the required calculations are performed, and the results are read out in a fixed format. Another example is a character-recognition computer which is specially built to recognize ink patterns on printed documents.

—W. W. Welz, *IBM Corporation*

Specific Gravity, Measurement of The terms *density* and *specific gravity* for fluids characterize the same physical property of the fluid. However, as defined below, the two terms cannot be used interchangeably:

Density—the mass per unit volume. This usually is expressed in grams per cubic centimeter, pounds per cubic foot, or pounds per gallon.

Specific gravity—the ratio of a fluid's density to the density of water (for liquids or solids)—and to the density of air (for gases).

Common specific-gravity scales, as can be found in various handbooks, include (1) *API* (American Petroleum Institute) for petroleum products; (2) *Balling* (used in the brewing industry); (3) *Barkometer* (used in the tanning industry); (4) *Baumé* (widely used to measure acids, light and heavy liquids, and syrups); Baumé formulas are given below; (5) *Brix* (used almost exclusively in the sugar industry); (6) *Quevenne* (used for milk testing); (7) *Richter, Sikes,* and *Tralles* (three alcoholometer scales which read directly in percent alcohol by weight in water); and (8) *Twaddle* (for industrial liquids heavier than water).

$$\text{For light liquids, degrees Baumé} = \frac{140}{\text{sp.gr.}} - 130$$

$$\text{For heavy liquids, degrees Baumé} = 145 - \frac{145}{\text{sp.gr.}}$$

Hydrometers: A *hand hydrometer* comprises a weighed float with a small-diameter stem so proportioned that a portion of its scale will be submerged during the course of measurement. The amount of submersion is an indication of the fluid's specific gravity. Hydrometers can be calibrated in terms of any of the aforementioned specific-gravity scales. Although primarily a laboratory instrument, the simple hydrometer can be located in a standpipe, equipped with an overflow at reading level, to permit visual readings of a continuously flowing liquid.

In the *inductance-bridge hydrometer,* shown in Fig. S-31, the measured liquid is held at a constant level by means of an overflow tube. The hydrometer rises and falls with specific-gravity changes. The armature of an inductance coil is attached to the lower end of a rod extending from

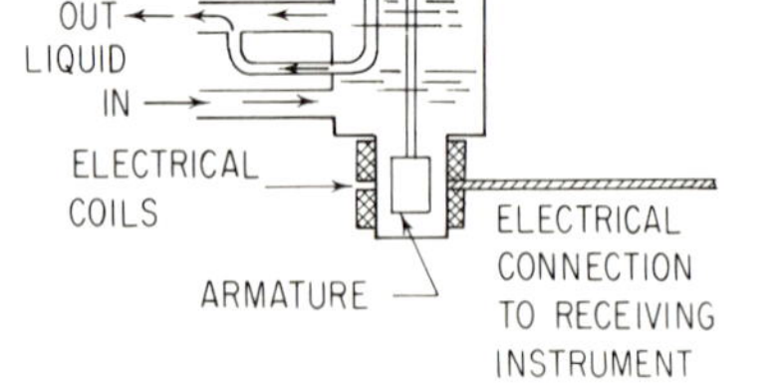

Fig. S-31. Inductance-bridge hydrometer transmitter.

the hydrometer. This moving armature and the inductance coils constitute the hydrometer transmitter. Liquid temperature usually is recorded so that temperature corrections may be made.

In the *chain-balanced float hydrometer,* shown in Fig. S-32, a self-centering hydrometer float, which operates essentially free from friction, is used. The volume of the "float" is fixed and remains entirely under the liquid surface. As the plummet moves up and down, the effective chain weight which acts upon it varies. The plummet is attached to a differential transformer to provide a related voltage output. These devices are available with automatic temperature compensation. Specific-gravity range is 0.6 to 3.5.

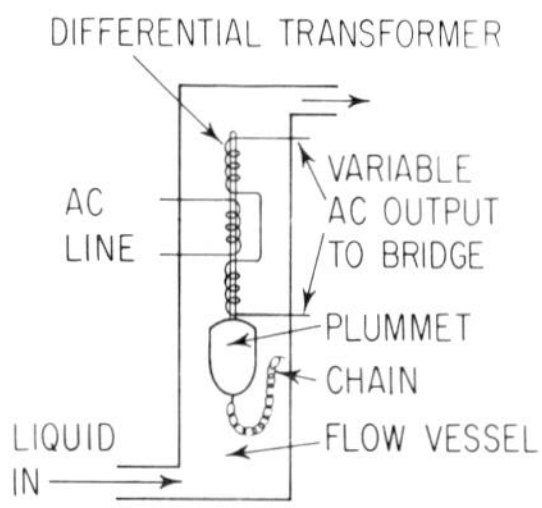

Fig. S-32. Chain-balanced float-type hydrometer.

In the *liquid-purged differential-pressure method,* two taps are installed in the side of a tank or at different elevations in a vertical pipe. These taps are led to a pressure-differential measuring device and are purged with a reference fluid, usually water. In this way, an automatically suppressed range is obtained, as well as freedom from plugged taps. In effect, the measured differential pressure is created by two equal columns, one of water and the other of the sample fluid. Purge rate is small and dilution is negligible. The system is widely used in the pulp and paper industry.

In the *air-bubbler reference-column method,* as shown in Fig. S-33, a known head of the sample liquid and of water are displaced from their respective bubbler pipes and compared with a differential-pressure measuring device. The *air-bubbler single-vessel method* is one of the simplest and most widely used methods of density measurement. As shown in Fig. S-34, two bubbler tubes are installed in the sample fluid so that the end of one tube is lower than the end of the other tube. The pressure required to bubble air into the fluid is equal to the pressure of the fluid at the ends of the bubble tubes. Since the outlet of one is lower than the outlet of the other, the difference in pressure will be the same as the weight of a constant-height column of the liquid. Thus, differential pressure is directly proportional to

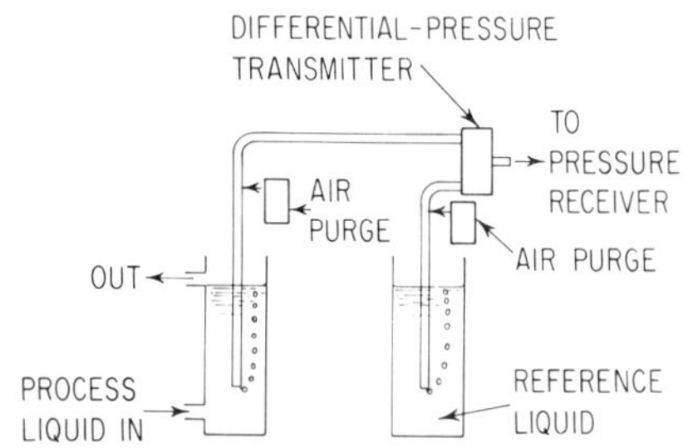

Fig. S-33. Specific-gravity-measuring system using differential bubbler and reference column.

the fluid density. The method is accurate to within 0.3 to 1 percent specific gravity when used with liquids that do not tend to crystallize in the measuring pipes.

Buoyancy Gas Balance: In this instrument, a displacer is mounted on a balance beam in the vessel. The displacer then is balanced for air

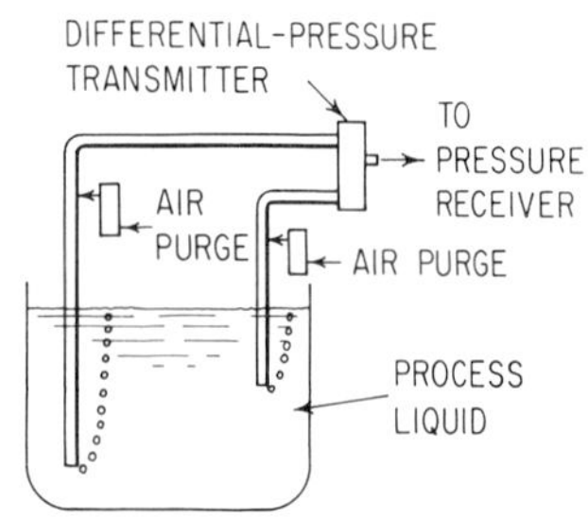

Fig. S-34. Specific-gravity-measuring system which employs differential bubbler and single vessel.

and the manometer reading (see Fig. S-35) is noted at the exact balance pressure. The air then is displaced by gas and the pressure is adjusted until balance is restored. The ratio of pressure with air to pressure with gas is the density of the gas relative to air. The device is used principally for laboratory determinations and has an accuracy to the fourth decimal place.

Gamma-ray Density Gage: See **Density meter, radiation-type.**

Boiling-point Elevation: The temperature of a boiling solution is compared with that of water boiling at the same pressure. For a particular solution, the boiling-point elevation (difference in temperature) can be calibrated in terms of density at standard temperature. Resistance thermometers generally are used for the temperature measurements.

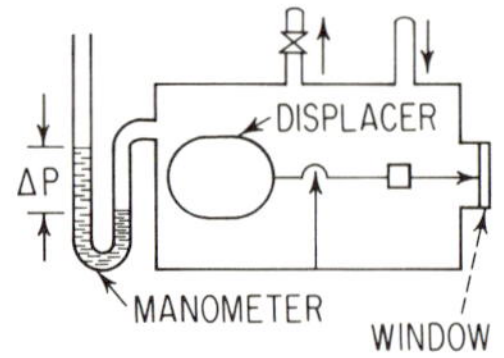

Fig. S-35. Buoyancy gas balance.

Other specific-gravity- and density-measuring systems include the viscous-drag gas-density meter, a gas densitometer which incorporates a constant-speed centrifugal blower, and a gas-density detector which compares the cooling effect of the sample gas with that of a reference gas.

Spectro Terminology The prefix *spectro* has been applied to a wide variety of instruments. For example, there are spectro*scopes,* spectro*graphs,* spectro*meters,* and spectro*photometers.* The additional compounding of names as found, for example, in spectroradiometer or spectropolarimeter tends to make the nomenclature of this broad category of instruments quite confusing. Unfortunately, the nomenclature has evolved over a long period in an unplanned and often inconsistent way. Some generalizations are pointed out here. However, it becomes clear that the makers, sellers, and users of these instruments should depend upon a detailed listing of construction and performance specifications when selling and buying rather than depend upon name categories alone.

The prefix *spectro* derives from the underlying base word *spectrum.* Initially, a spectrum referred to the rainbow series of colors, ranging from violet through indigo, blue, green, yellow, orange, and red, which are produced when white visible light is split into its component colors. A simple device for accomplishing this, such as passing sunlight through a glass prism and providing a means to note the colors visually, is a *spectroscope.* The term spectrum no longer is confined to the visible portion of the total electromagnetic radiation span but now is applied to all major portions of electromagnetic radiation, ranging from the shorter-wavelength gamma rays, x-rays, and cosmic rays—out to the longer far infrared, microwaves, and radio waves. Thus, such terms for designating instrumental techniques as infrared spectroscopy, x-ray spectroscopy, ultraviolet spectroscopy, and microwave spectroscopy have evolved. Moreover, the term spectrum no longer is confined to the designation of an electromagnetic spectrum but also is applied to almost any orderly array of phenomena. Thus, mass spectrometers sort atoms or radicals by their atomic weight, and there are spectrometers to analyze the decay "spectra" of radioactive isotopes. Some of the more important fields where the terms "spectroscopy" and "spectrometer" have been used include:

Electromagnetic spectrum	Cosmic ray, gamma ray, x-ray, ultraviolet, visible infrared, far infrared, microwave, and radiowave
Energies of particles	Beta ray (electrons), protons, neutrons, and mass
Other phenomena	Radioactive decay and Mossbauer

Generally, the use of the suffixes *scope, graph, meter,* and *photometer,* in combination with *spectro,* designates the kind of detector used in the instrument. If the spectral phenomenon produced by the instrument is predominantly readable or observable visually by the unaided eye, the instrument is a "scope"—thus *spectroscope.* If the phenomenon is recorded by means of photographic film or by a paper chart, the instrument is a *spectrograph.* If a meter is used to display data detected by a bolometer, thermocouple, thermistor, or other instrumental detector, the instrument is a *spectrometer.* If the instrument provides for measuring the intensity of various portions of spectra (not necessarily confined to the range of visible light), the designation *spectrophotometer* is used.

The designation of an instrument also may describe the basic source of radiant energy whose spectrum is measured and the manner in which that energy is processed or presented by the instrument. Some instruments display a continuous spectrum of light or other radiation. Other instruments pass the energy through filters or other absorption means to eliminate or absorb certain wavelengths. An instrument of the latter type may be termed an *absorption spectrometer* or an *absorption spectrograph,* and so on. An *emission* type of instrument generally indicates a device wherein the energy level of the specimen to be analyzed is raised by energy provided by the instrument, such as by means of a spark discharge or flame heating, thus, such terms as optical-emission spectrometer or flame photometer. In other instances, particularly absorption-type instruments, the instrument energy source will be constant as, for example, provided by an incandescent lamp, Nernst glower, source of x-rays, and so on. Instrument designations can be complicated even further by including reference to some special function. Thus, an ultraviolet-visible spectropolarimeter is an instrument used to measure the rotation of the plane of polarization, according to wavelength and light intensity. A spectrophotofluorometer is an instrument with means both for controlling the exciting wavelength and for identifying and measuring the light output of the fluorescing sample; while a fluorometer may have a filtered light input and will measure the fluorescent light from the sample, frequently by wavelength. A spectroheliograph is a spectrograph adapted for use with a telescope for obtaining photographs of the sun wherein radiation of a particular element in the sun's radiation is recorded. Also, an instrument may carry a designation that relates to a major component used, such as a prism, diffraction grating, or filter type, thus, a prism spectroscope, an x-ray diffractometer, and so on.

—Richard F. Jarrell, *Jarrell-Ash Division, Fisher Scientific Company*

Spectrograph, Doppler (See **Astronomical instrumentation.**)

Spectroheliograph An instrument for photographing the sun in monochromatic light. A solar telescope feeds light to a monochromator. The solar disk is scanned (often using the sun's motion for scanning) and film is moved past the exit slit of the monochromator at the same rate as the image motion. The result is a photograph of the sun using light of a single wavelength, usually an emission line of an element. This procedure is useful in studying solar prominences. Also see **Solar telescope**; and **Coronagraph.**

—Joost H. Kiewiet de Jonge, *University of Pittsburgh,* and Lloyd M. Geertz, *Owens-Illinois*

Spectrometer, X-ray (See **X-ray analysis.**)

Spectrophosphorimeter (See **Fluorometry.**)

Spectrophotometer (See **Spectro terminology.**)

Spectrophotometer, Flame (See **Flame photometer-spectrophotometer.**)

Spectroradiometer (See **Photometers.**)

Spectroscope An instrument to observe visually the spectrum of an emitting source, or the transmission of a sample. Frequently, a wavelength scale will be incorporated to permit the wavelength to be identified to an accuracy of about 10 millimicrons. There may be a small prism at the slit to project a reference spectrum beside that from the unknown sample so that the observer can make some judgment as to relative intensities.

Spectroscopes are simple and inexpensive instruments that can be used to introduce students to spectroscopy. Also, because of their low price and simplicity and speed of use, spectroscopes still are employed in medicine to make a quick determination of hemoglobin content of blood, or in metallurgy or chemistry for a quick identification of metallic content. For example, steels

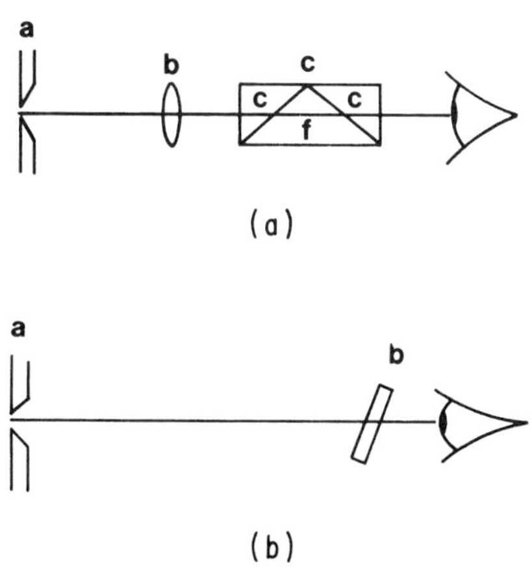

Fig. S-36. Optical diagrams of typical spectroscopes: (a) prism spectroscope: (*a*) slit, (*b*) collimating lens, (*c*) direct-vision prism of crown (*c*) and flint (*f*) glass; (b) grating spectroscope: (*a*) slit, (*b*) transmission replica grating.

are sorted routinely by a visual spectroscope when the steels become mixed or are checked for alloy type when they are received from a vendor. Optical diagrams of typical prism and grating spectroscopes used for instructional purposes are given in Fig. S-36. A comparison spectroscope

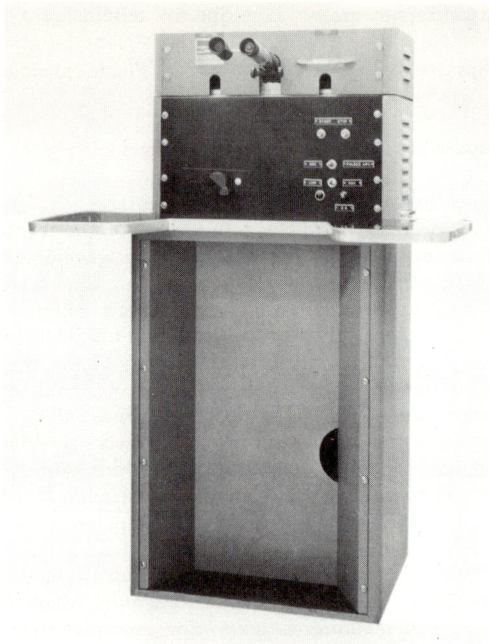

Fig. S-37. Visual spectroscope used for identifying steels. *(Spectrochem, Jarrell-Ash.)*

used to identify or sort alloys is shown in Fig. S-37.

—Richard F. Jarrell, *Jarrell-Ash Division, Fisher Scientific Company*

Spectrum Analyzer (See Wave analyzer.)

Speed, Measurement of Speed is a scalar quantity equal to the magnitude of velocity. Velocity is a vector quantity denoting both the direction and the *speed of a linear motion,* or denoting the direction of rotation and the *angular speed* in the case of rotation.[1] Although there are special cases, linear speeds generally are inferred from rotational measurements. This is the result of the manner in which machines and devices are powered, that is, by rotating electrical or mechanical equipment wherein a shaft, wheel, or gear usually is conveniently available to which a speed transducer can be coupled or attached. Special cases would include aircraft airspeed measurements, watercraft and seagoing vessels of various types, and speed measurements of various physical phenomena over long distances.

Tachometers: These are devices for measuring angular speed, usually in revolutions per minute, although they can be calibrated in many other meaningful units dependent upon application, such as feet per minute, miles per hour, or yards per minute, and even in terms of production units per unit time. Tachometers are classified as (1) mechanical or (2) electrical.

Mechanical Tachometers: These are of three principal types: (1) revolution counters, (2) centrifugal-force tachometers, and (3) resonance tachometers. A revolution counter is shown in Fig. S-38. It is used with a timing device of some form to determine the number of revolutions in a measured length of time. Thus, it measures an average rotational speed rather than an instantaneous rotational speed. A revolution counter is manually held. A worm gear on a spindle is rotated by the shaft whose rotational speed is to be measured. The worm gear meshes with a spur gear, which in turn moves a calibrated dial, indicating total revolutions. Divisions on the outer dial each represent one revolution of the spindle,

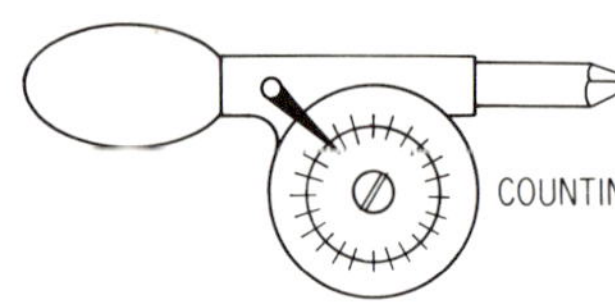

Fig. S-38. Revolution counter.

while those on the inner dial represent one revolution of the outer dial. A stopwatch, started and stopped with the counter, permits calculation of the average speed. In another form, known as a *tachoscope,* a counter is combined with a timepiece. The manual type has the two components integrally mounted and simultaneously started when the contact point is pressed against the rotating shaft. Both are stopped when the point is removed. A speed-indicator-type tachoscope has a revolution counter controlled by an automatic timer. A ratchet arrangement on a measuring wheel frees the wheel for a definite period of time, and a pointer on the wheel indicates on a dial calibrated directly in rpm. A *chronometric tachometer* is similar to the tachoscope but repeats automatically its timed period to adjust indication to the average speed for the last timed interval. The interval may be 0.5 to 30 s, with shorter pe-

[1] John Markus, "Electronics and Nucleonics Dictionary," 3d ed., McGraw-Hill Book Company, New York, 1966.

riods giving a reading nearer the instantaneous speed.

Centrifugal-force Tachometers: These also are known as governors. See **Governor.**

Resonance Tachometers: A series of consecutively tuned steel reeds (Fig. S-39) can be used to determine engine speed on the basis of predictable and repeatable vibrations caused by the machine (e.g., due to reciprocating movement of the pistons). The reed tuned to resonance with the machine-vibration frequency will respond visibly and can be arranged to indicate speed on a scale. The range of these devices is wide (600 to 100,000 rpm) and their accuracy is ±0.5 percent or better. They need contact only with a nonmoving part of the machine.

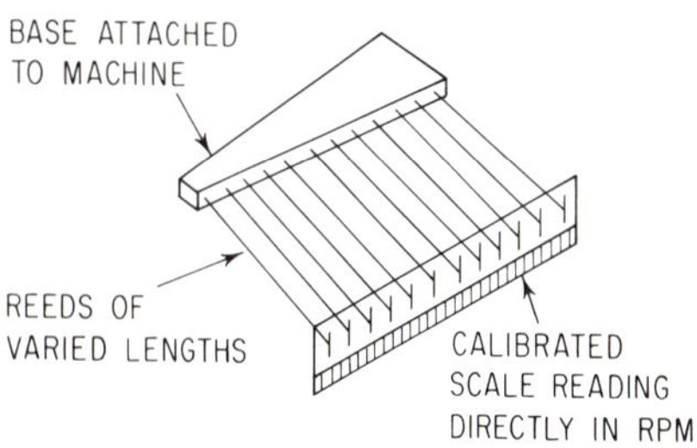

Fig. S-39. Vibrating-reed-type speed indicator.

Electric Tachometers: These consist of a transducer which converts rotational speed into an electrical signal coupled to an indicator or recorder. In proportion to speed, the transducer produces either (1) an analog signal which can be used for analog indication or (2) pulses which can be digitally counted in terms of revolutions in a unit time. Electric tachometers fall into the following major categories: (1) eddy-current or drag-type, (2) tachometer generators, (3) contactless tachometers, (4) frequency tachometers, (5) ignition-type tachometers, and (6) stroboscopic tachometers.

Eddy-current Tachometer: In this device, as shown in Fig. S-40, the transducer produces an analog signal in the form of a continuous drag due to eddy currents induced in the cup. The current is proportional to speed. An analog indicator with linear scale is used. Varieties of this tachometer have the following characteristics:

1. The magnetic field of the mechanically coupled, rotating permanent magnet sweeps over a conducting disk or cup and induces in it eddy currents, which in turn produce a torque or drag acting upon the cup against a restraining spring. Automobile *speedometers* are of this type, actually measuring angular speed of the wheels. However, the diameter of the car wheel is assumed, and thus the measurement is converted into a linear measurement in miles per hour.

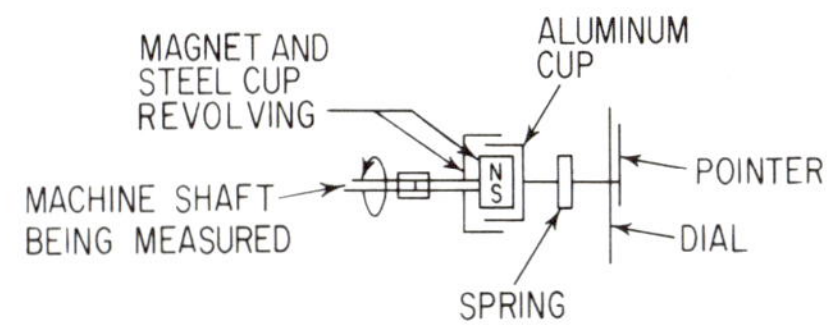

Fig. S-40. Drag-type or eddy-current tachometer.

2. Mechanical coupling of the magnet is replaced by an electric drive, consisting of a three-phase synchronous generator, driven by the machine under test, and connected to a three-phase synchronous motor, which in turn drives the magnet of the tachometer. This is a popular type of aircraft-engine tachometer.

3. The permanent magnet can be stationary, and a soft-iron rotor produces a revolving magnetic field. This type is very rugged and is used to measure locomotive speed.

4. A conducting cup or disk is located in the field of two coils, electrically 90° out of phase. Only one of these is energized by an ac source. A signal of the source frequency will appear in the other coil, in proportion to the speed of the cup. This type also is used in aircraft tachometry.

5. In an industrial version, the magnetic drag-disk torque is converted to a transmission signal by a pneumatic force-balance system. Output air pressure, 3 to 15 psi, is directly proportional to the speed of rotation. Ranges are available up to 12,000 rpm.

Electric-generator Tachometers: See **Tachometer generator.**

Contactless Tachometers: These produce pulses from a rotating shaft without being mechanically connected to it. Usually, these devices do not produce sufficient energy to actuate an indicator directly but require an amplifier of ample sensitivity. They are of three basic forms: (1) electromagnetic, (2) capacitive, and (3) optical. A *magnetic pickup head* is shown in Fig. S-41. Electrical pulses are generated each time a tooth of a rotating ferromagnetic gear passes in close proximity to the pickup head. There is no physical contact and clearance is not critical. It is available in ranges from 0 to 25 rpm to 0 to 72,000 rpm. Calibrated accuracy is ±½ percent full scale; reproducibility 0.2 percent full scale.

A *capacitive-type tachometer* is shown in Fig. S-42. A variable capacitor is formed by vanes attached to the revolving shaft. Rotation of the shaft alters the capacitance to ground. The capacitor forms a part of an oscillator tank so that the number of frequency changes per unit of time is a measure of shaft speed. The pulses thus produced go to amplifying, pulse-forming, and triggering circuits where they are converted into constant-amplitude signals. An analog dc output is produced to provide indicator deflection proportional to speed.

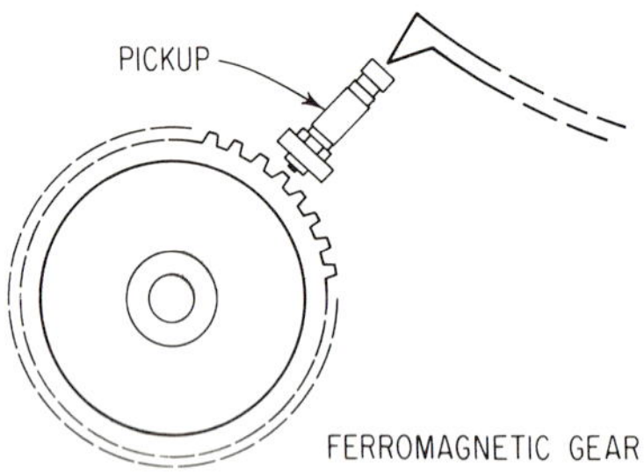

Fig. S-41. Tachometer using magnetic pickup. *(Moore Products.)*

Optical pickups of a variety of designs use the principle of shaft rotation to interrupt a beam of light falling on a photoelectric or photoconductive cell. The pulses thus obtained are amplified and either (1) counted by an electric counter or (2) shaped to an analog signal before connection to the indicator. In one design, a beam of light originating inside the tachometer case hits a white spot on the rotating shaft (or spoke of a wheel). Reflected light falls on a photoconductive cell inside the case, producing pulses in a transistorized amplifier, which in turn causes the indicator to deflect. These devices can measure speeds up to 3 million rpm.

Frequency Tachometers: The frequency of an ac signal from a suitable pickup can be measured and calibrated in terms of speed. For high resolution and accuracy, multipole generators with a large number of poles are preferable. These tachometers can be classified by the type of frequency meter used as a receiver: (1) *electronic counter*—the most accurate and with the widest range; (2) *frequency meter*—of the type used for frequency measurements except calibrated in rpm; (3) *resonance-frequency meter*—in which a polarized bridge can be used as a null detector—particularly useful in monitoring speed on an expanded scale within a very narrow range; and (4) *vibrating reed*—wherein the output from an ac generator energizes a coil which vibrates reeds at the natural vibration frequency in sympathy with the generator output.

Ignition-type Tachometers: The primary winding of the ignition coil in an internal-combustion engine can be connected in series with a battery or magneto and interrupter contacts. The contacts are opened periodically by the revolving shaft and produce pulses. These pulses are often distorted by high-frequency trains of damped oscillation. Ignition tachometers further are classified by the way use is made of the developed pulses: (1) systems using low-frequency pulses and (2) sytems using high-frequency oscillations. Tachometers of the first type usually are connected directly to the high voltage of the ignition primary winding and to ground having a low-pass filter or other means of preventing transients from affecting tachometer indications. For instance, they can have a chopper-type switch that would respond only to low-frequency pulses. The emitted pulses are carefully shaped, by either a saturable transformer or a capacitor, before being conveyed to the indicator. This kind of tachometer bleeds little power from the ignition-coil supply and does not upset operation of the ignition system.

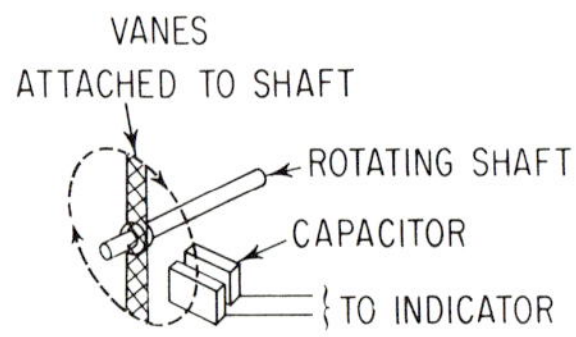

Fig. S-42. Capacitive-type contactless tachometer.

In systems using high-frequency oscillations, instead of trying to filter out the high-frequency transients, one can use them for measuring speed. Tachometers of this type are more universal because they do not have to be tailored for the particular battery, voltage, or magneto.

Stroboscopic Tachometer: See **Stroboscope.**

Speed, Standard of (See Standards, measurement.)

Speed Control (See Governor.)

Speed (Motor) Control (See **Electric-motor drives, control of.**)

Speedometer (See **Speed, measurement of.**)

Sphygmanometer (See **Cardiovascular instrumentation.**)

Spiral (See **Bourdon tubes.**)

Spirometer (See **Pulmonary system, instrumental analysis of.**)

Split-range Control A system where two manipulated variables are controlled from one controlled variable. In the system shown in Fig. S-43, it is necessary to maintain a required temperature which may alternate between a heating operation and a cooling operation. The output of the temperature controller is 3 to 15 psig or 4 to 20 mA dc for pneumatic or electronic control, respectively. The control valves are modified so that instead of modulating from fully closed to fully open for full-scale output of the controller, they are made to operate over one-half the control-signal range. The hot-water control valve operates from closed to open over a range of 9 to 3 psi or 12 to 4 mA dc; while the cooling-water valve operates from closed to open over a range of 9 to 15 psig or 12 to 20 mA dc. Therefore, when the temperature is at its set point (with 9 psig or 12 mA dc output), both valves are closed. If the temperature gets too low, the heating valve is opened; if too high, the cooling valve is opened. Other examples of split-range control include the pH control of waste streams which may alternate between being too basic or too acidic and thus require selective neutralization with acid or caustic.

—Richard H. Kennedy, *Bristol Division, American Chain & Cable Company, Inc.*

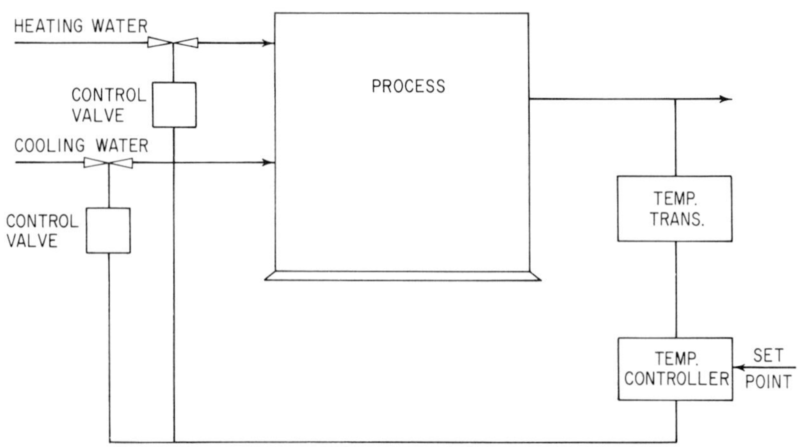

Fig. S-43. Split-range control system.

Spool Valve (See **Hydraulic relay.**)

Spot Scanner (See **Character-recognition systems.**)

Spring Scale (See **Scale, spring.**)

Stable Platform A support for inertial-guidance systems, sonar transducers, radar antennas, and other instrumentation—the attitude of which is maintained by two or more gyroscopes and their associated servosystems despite the pitch and roll of a vehicle on or in water or in space. Also see **Flight-control systems.**

Stack-type Controller (See **Pneumatic controller.**)

Stagnation Temperature [See **Air-data systems (aircraft).**]

Standards, Measurement The fundamental function of a measuring instrument is to position a pointer on a calibrated scale (or a recording pen on a chart—and so on) as a measure of an otherwise intangible quantity. Making that calibrated scale accurate, legible, and permanent in nature is a problem that is closely related to the maintenance of standards and to the achievement of instrument precision within specific tolerances. To the manufacturer of instruments, calibration is a daily task. In the course of its completion, each instrument reaches a stage where the dial or scale, hitherto blank, is marked off so that future pointer deflections can be interpreted in terms of the quantity to be measured. The process of placing a scale on the dial is called calibration. In subsequent use, the instrument may be subjected to periodic checking of the precision with which the deflections of the pointer still represent the quantities marked on the scale. This is a process of calibration checking, or of recalibration. Important to this procedure is the availability of reference standards and instrumentation for transferring measurements from the standard to the instrument in question—so that recalibration will be as accurate as the original calibration. That is why it is wise—if excellent standards capabilities and equipment are not available—to have the recalibration performed by the original manufacturer or by an organization that is approved and certified to carry out such actions. In the final essence, the government standards laboratories throughout the world, such as the National Bureau of Standards in the United States, the International Bureau of Weights and Measures at Sèvres, France, and the British Standards Institution, are responsible for maintaining the permanent reference standards and offer calibration services to instrument users and manufacturers. Like the pedigree of a noble, the calibration of an instrument should be traceable back through a series of secondary, reference, and working standards to the ultimate measurement standard. The aforementioned and other like organizations throughout the world cooperate in the establishment and maintenance of standards and hold periodic conferences in their quest constantly to improve the standards per se and the standards-maintenance system. The history of measurement standards is long and fascinating, and for further reference the reader is referred to the listing at the end of this entry.

International System of Units: This system, officially abbreviated *SI*, was established by international agreement to provide a logical and interconnected framework for all measurements in science, industry, and commerce. The system is built upon a foundation of base units and their definitions. The six base units of measurement are described here. All other *SI* units are *derived* from these base units. Multiples and submultiples are expressed in a decimal system applied to the base metric system. It is interesting to note that the use of metric weights and measures was legalized in the United States in 1866 and that the customary units of weights and measures are defined in terms of the meter and the kilogram. The only legal units for electricity and illumination in the United States are *SI* units.

In reviewing the six base units, it is interesting to note that all but one (mass) are derived from nature. Thus, aside from mass, actual physical objects (standards) no longer are the prime, ultimate standards. Of course, in establishing the conditions provided by nature to achieve the other standards, very careful and complex instrumental procedures must be followed.

Length: The meter (abbreviated m) is the base unit. The meter is defined as 1,650,763.73 wavelengths in vacuum of the orange-red line of the spectrum of krypton-86. The determination is made by means of an interferometer.

Area: The square meter (abbreviated m^2) is the base unit. Land often is measured by the hectare (10,000 square meters, or approximately 2.5 acres).

Volume: The cubic meter (abbreviated m^3) is the base unit. Fluid volume often is measured by the liter (0.001 cubic meter).

Time: The second (abbreivated s) is the base unit. The second is defined as the duration of 9,192,631,770 cycles of the radiation associated with a specified transition of the cesium atom. This condition is realized by tuning an oscillator to the resonance frequency of the cesium atoms as they pass through a system of magnets and a resonant cavity into a detector.

Frequency: The hertz (abbreviated Hz) is the base unit. One hertz equals one cycle per second.

Speed: The meter per second (abbreviated m/s) is the base unit. Dividing distance by time gives speed. A meter per second is approximately 3 feet per second.

Acceleration: The meter per second per second (abbreviated m/s^2) is the base unit. Acceleration is the rate of change in speed.

Mass: The kilogram (abbreviated kg) is the base unit. The physical standard is a cylinder of platinum-iridium alloy kept by the International Bureau of Weights and Measures at Paris. A duplicate is in the custody of the National Bureau of Standards (Washington) and serves as the mass standard for the United States. This is the only base unit still defined by an artifact.

Force: The newton (abbreviated N) is the base

unit. A force of 1 newton, when applied for 1 second, will give to a 1-kilogram mass a speed of 1 meter per second—and an acceleration of 1 meter per second per second. One newton equals approximately two-tenths of a pound of force. $1N = 1\ kg \times 1\ m/1\ s^2$.

Work and Energy: The joule (abbreviated J) is the base unit. The weight of an object is the force exerted on it by gravity. Gravity gives a mass a downward acceleration of about 9.8 m/s^2. $1\ J = 1\ N \times 1\ m$.

Power: The watt (abbreviated W) is the base unit. $1\ W = 1\ J/1\ s$.

Temperature: The kelvin (abbreviated K) is the base unit. The thermodynamic or Kelvin scale of temperature used in SI has its origin or zero point at absolute zero and has a fixed point at the triple point of water defined as 273.16 kelvins. The Celsius scale is derived from the Kelvin scale. The triple point is defined as 0.01°C on the Celsius scale, which is approximately 32.02°F on the Fahrenheit scale. The triple-point cell, an evacuated glass cylinder filled with pure water, is used to define a known fixed temperature. When the cell is cooled until a mantle of ice forms around the reentrant well, the temperature at the interface of solid, liquid, and vapor is 0.01°C. Thermometers to be calibrated are placed in the reentrant well.

Electric Current: The ampere (abbreviated A) is the base unit. The ampere is defined as the magnitude of the current that, when flowing through each of two long parallel wires separated by one meter in free space, results in a force between the two wires (due to their magnetic fields) of 2×10^{-7} newton for each meter of length.

Voltage: The volt (abbreviated V) is the base unit. $1\ V = 1\ W/1\ A$.

Resistance: The ohm (abbreviated Ω) is the base unit. $1\ \Omega = 1\ V/1\ A$.

Luminous Intensity: The candela (abbreviated cd) is the base unit. The candela is defined as the luminous intensity of 1/600,000 of a square meter of a radiating cavity at the temperature of freezing platinum (2042°K).

Light Flux: The lumen (abbreviated lm) is the base unit. A source having an intensity of 1 candela in all directions radiates a light flux of 4 π lumens. A 100-watt light bulb emits about 1,700 lumens.

Prefixes for All SI Units: The following multiples, submultiples, prefixes, and symbols have been approved for use with the international system of units:

Multiples and Submultiples	Prefixes	Symbols
$1\,000\,000\,000\,000 = 10^{12}$	tera (tĕr′ ȧ)	T
$1\,000\,000\,000 = 10^{9}$	giga (jĭ′ gȧ)	G
$1\,000\,000 = 10^{6}$	mega (mĕg′ ȧ)	M
$1000 = 10^{3}$	kilo (kĭl′ ō)	k
$100 = 10^{2}$	hecto (hĕk′ tō)	h
$10 = 10$	deka (dĕk′ ȧ)	da
$0.1 = 10^{-1}$	deci (dĕs′ ĭ)	d
$0.01 = 10^{-2}$	centi (sĕn′ tĭ)	c
$0.001 = 10^{-3}$	milli (mĭl′ ĭ)	m
$0.000\,001 = 10^{-6}$	micro (mī′ krō)	μ
$0.000\,000\,001 = 10^{-9}$	nano (năn′ ō)	n
$0.000\,000\,000\,001 = 10^{-12}$	pico (pē′ kō)	p
$0.000\,000\,000\,000\,001 = 10^{-15}$	femto (fĕm′ tō)	f
$0.000\,000\,000\,000\,000\,001 = 10^{-18}$	atto (ăt′ tō)	a

REFERENCES

Brief History and Use of the English and Metric Systems of Measurement with a Chart of the Modernized Metric System, *Natl. Bur. Std.* (*U.S.*) *Tech. News Bull.*, vol. 53, no. 1, January, 1969.

Barber C. R.: The International Practical Temperature Scale of 1968, *Metrologia*, vol. 5, no. 2, pp. 35–44, April, 1969.

McNish, A. G.: Measurement Standards, *Intern. Sci. Tech.*, November, 1965, pp. 58–64.

Actions Taken by the Twelfth General Conference on Weights and Measures, *Natl. Bur. Std.* (*U.S.*) *Tech. News Bull.*, vol. 48, no. 12, December, 1964.

McNish, A. G.: The International System of Units, *Am. Soc. Quality Control, Ann. Tech. Conf. Trans.*, 1965.

Astin, Allen V.: Standards of Measurement, *Sci. Am.*, vol. 218, no. 6, pp. 2–15, June, 1968.

Berring, Henry: The Manufacturer's Viewpoint on Instrument Calibration, *ISA J.*, vol. 1, no. 4, pp. 32–34, April, 1954.

Standard Cell The cadmium cell, developed by Dr. Edward Weston in 1893, and universally known as the Weston standard cell, has been recognized and used as the world standard of electromotive force since its acceptance by the International Committee on Electrical Units and Standards in 1908.

The Weston standard cell shown in Fig. S-44 is made in two types, the normal cell containing a saturated cadmium sulfate solution and a type used as a working standard in which the solution is less than saturated above 4°C.

The saturated cell is the basic standard for maintaining the value of the volt, and is used in this manner in all national laboratories. Its rather high temperature coefficient must always be taken into account. The legal voltage of the standard-cell group in the United States is 1.018636 absolute volts at 20°C, and this value is accepted by the international agreement.

The emf of the unsaturated cell is within 1.0188 to 1.0198 absolute volts, with the exact voltage of each cell being established by comparison with the normal or saturated cell. This cell is a useful

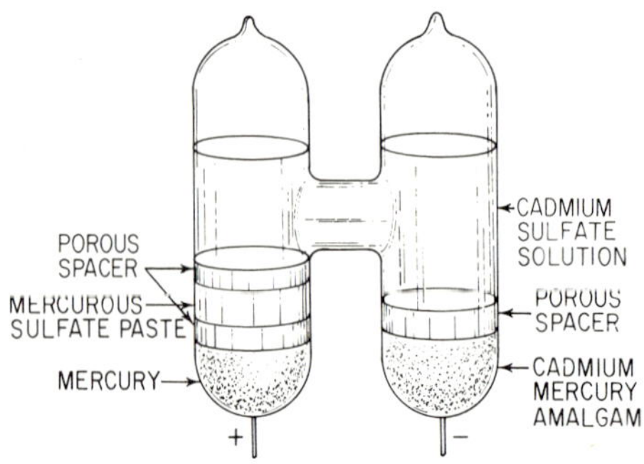

Fig. S-44. Cadmium standard cell. *(Weston.)*

working standard because of its negligible temperature coefficient.

—Adapted from "The Instrument Sketch Book," *Weston Instruments, Inc.*

Starter, Motor (See **Electric-motor drives, control of.**)

Statement, Programming The unit of information (group of words and/or symbols) supplied as the basic component of a computer program to a compiler or assembler (see **Compiler; Assembler**). The exact syntax of a statement is defined by the compiler or assembler for which it is written. A group of statements make up a program, or a segment of a program at the source level (the man definition level as opposed to the machine definition level).

A programming statement usually describes a logical unit of work for the system to do in terms of the language (see **Language, computer**). An assembler statement may specify a basic machine operation such as a register load or transfer instruction (see **Instruction**), whereas a higher-level language statement may specify a complex arithmetic operation. Both are called statements, even though the number of resulting machine-language instructions vary.

An example of a compiler-language statement is the following FORTRAN statement for the evaluation of a cubic polynomial:

$$X = A + Y*(B + Y*(C + Y*D))$$

In an assembler language the operation of loading index register 3 might be performed by the statement

```
LDXR3     LDX     I3     $TVWK
```

REFERENCE

Rosen, Saul: "Programming Systems and Languages," parts 2B, 2D, McGraw-Hill Book Company, New York, 1967.

—L. E. Wise, *IBM Corporation*

Stepping Motor An electromechanical device which rotates a discrete step angle when energized electrically. The step angle usually is fixed for a particular motor and thus provides a means for accurately positioning in a repeatable, uniform manner. Typical step angles vary from as small as 0.72° to as large as 90°. Several means for electrically energizing stepping motors include dc pulses, square-wave and fixed-logic sequence, or multiple-phase square waves. Basic design types are (1) solenoid-operated ratchet, (2) permanent magnet, and (3) variable reluctance. Variations of these basic types may be combined with gears or hydraulic amplifiers to provide increased-output-torque stepping motors. The characteristics of a limited representation of hardware are given in Table S-2.

Solenoid-operated Ratchet: A solenoid is used to operate a ratchet device or star wheel. Main advantages are low cost and high nonenergized holding torque. Limitations include short life and low stepping rates.

Permanent Magnet: Most stepping motors in use are of this type. Permanent-magnet motors offer high torque capabilities—up to 1,500 oz-in.; fast step response—2 ms possible; and high stepping rates—up to 10,000 steps per second. The accuracy of step may be within 3 percent of one step, with any error being nonaccumulative. Permanent-magnet types may be operated from pulses or square waves, dependent upon motor design. The permanent magnet, which may be in the rotor or the stator—working with electrically produced flux—causes poles and teeth to align and effect rotation. Operation of two common types is shown in Figs. S-45 and S-46.

Variable Reluctance: The variable-reluctance-type stepping motor offers essentially the same advantages as the permanent-magnet types. However, the torque capabilities have not been

TABLE S-2. *Stepping-motor Characteristics**

Normal Operating Speed at Rated Load, pps	Max Speed No Load, pps	Torque, oz-in. Operating Speed	Torque, oz-in. Detent, Zero Speed	Rotor Inertia, g-cm^2	Input Dc volts	Input Current / Power Phase	Input Current / Power Watts	Steps per Revolution
			Solenoid operated ratchet					
To 25	....	160	....	....	12, 28	1.0A	84–210	18
600	1,000	1.4–6.1	....	....	6–220	....	6	10 or 12
400	750	3.0–15.7	....	....	6–220	....	9	10 or 12
300	600	6.0–24.2	....	....	6–220	....	11	10 or 12
			Electrohydraulic					
5,000	5,000	10,000	10,000	....	2	5.4	10.8	200/500
4,000	8,000	1,440	1,440	....	28	3.0A	10	200/250
4,000	8,000	2,880	2,900	....	28	3.0A	10	200/250
4,000	4,000	8,000	8,200	....	28	3.0A	10	200/250
			Variable reluctance					
10	725	2.4	4.5	1.0	28	6.0A	15	24
18	.....	12	45	8.4	28	2.7A	75	24
....	900	...	8.0	0.74	28	1.93A	100	24
400	5,000	25	40	180	5.4	1.5A	..	200/72
250	3,000	1.0	2.7	0.004	28	355mA	20	24
300	660	2.7	6.5	3.11	28	35mA	20	24
450	6,000	0.5	0.95	0.004	28	330mA	18	24
2,000	6,000	640	1,920	...	28	20.0A	60	240
2,000	8,000	480	900	...	28	10.0A	30	240
2,000	10,000	20	30	...	28	3.0A	10	160
2,000	16,000		30	18.6	...	6.0A	36	4
			Permanent magnet					
10	130	0.1	0.22	0.15	28	90mA	2.5	4
20	80	0.2	0.2	3.7	27	150mA	4	24
40	320	0.47	0.97	2.6	24	150mA	3	48
50	...	...	4.0	0.024	26	465mA	12	8
75	190	1.5	5.5	4.02	28	43mA	13	4
100	260	1.9	4.3	4.93	28	47mA	13	8
120	145	0.6	0.3	0.77	28	280mA	7.8	3
150	250	0.45	0.13	0.45	28	93mA	5.2	4
175	200	5.0	9/7.4	7.05	28	1.25A	35	4
200	320	1.0	0.3	0.77	28	93mA	5.2	8
225	260	8.5	17/13.3	22.4	28	1.65A	46	8
250	300	2.2	3.8/3.6	1.36	28	860mA	24.1	4
300	500	0.3	0.7	0.003	26	104mA	5.4	4
360/600	...	1.3/1.8	......	34.8	20–50	.....	2.0/1.8	12/24
400	...	12	40	180	5.4	1.5mA	....	72
500	...	200	600	7,310	....	.....	...	200
600	...	100	260	1,520	4.5	3.5mA	...	200
45	...	35	90	877	3.0	4.0mA	...	500
45	...	1,200	2,500	38,010	12	3.8mA	....	200

*A limited representation of commercially available hardware. Characteristics that fall between parameters shown are probably also available.

developed to the level of the permanent-magnet types. In the variable-reluctance motor, the rotor-stator poles or teeth are aligned by the electric fields. Therefore, when the motor is deenergized, there is no residual holding torque. This is not a real disadvantage, however, in that most

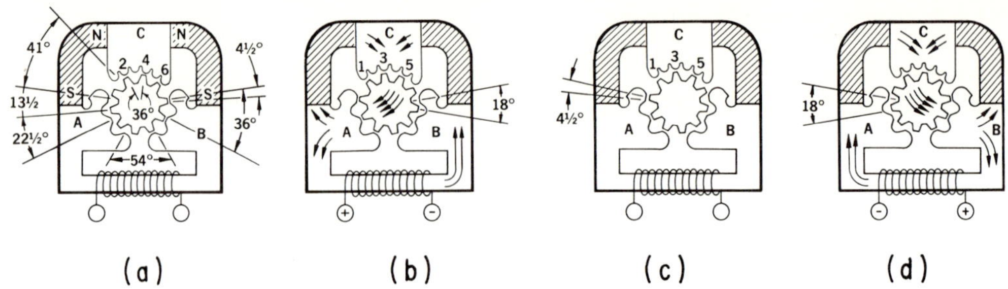

Fig. S-45. Operation of stepping motor (electromagnetic rotary incremental actuator) which converts digital pulse inputs to analog shaft-output motion. The motor diagramed starts and stops in discrete angular excursions of uniform magnitude (15°) to produce output steps always equal in number to pulse inputs. Two permanent magnets, connected in parallel for uniform flux distribution, link two portent poles (*A* and *B*) and one detent pole (*C*). Each portent pole has three sets of teeth spaced 36° apart and is positioned so that when one set of portent teeth aligns with three rotor teeth, the other set aligns with spaces between rotor teeth. Also aligned with three rotor teeth are three of six detent-pole teeth spaced 18° apart. Rotor position establishes alternate flux paths. Each pulse of alternating polarity applied to the coil advances the rotor one step by switching the magnetic flux from one path to the other. Dc pulses of plus or minus polarity overcome the permanent-magnetic detent and produce torque to advance the rotor from one detent to the next. A similar 15° motor, designed to step through 24 fixed shaft positions in one revolution, operates in a similar manner. *(Sigma Instruments.)*

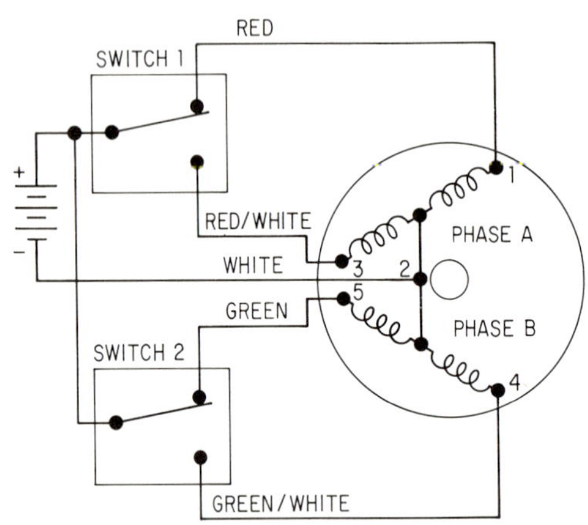

SWITCHING SEQUENCE*

STEP	SWITCH 1	SWITCH 2
1	1	5
2	1	4
3	3	4
4	3	5
1	1	5

*TO REVERSE DIRECTION, READ CHART UP FROM BOTTOM.

Fig. S-46. Connection diagram and switching sequence for a typical phase-switched stepping motor. Diagram is for a motor having a permanent-magnet stator. *(Superior Electric.)*

permanent-magnet types as well as variable-reluctance types require that power remain on the motor between step movements. The same basic drive logic as used for permanent-magnet steppers also applies to variable-reluctance steppers. A multiphase variable-reluctance stepping motor is shown in Fig. S-47.

Drives: Although most stepping motors can be driven from switches or relays, most drive circuits incorporate solid-state devices which permit high-powered, fast operation. Simple drives convert low-level pulses into power pulses or correctly phased power. More complex drives adjust the power levels to allow running at very high rates.

Step Response: At very low rates, the step movement resembles the classic damped-oscillation curve. Damping can be added to modify the curve to provide critical damping. This may be done mechanically, electrically, or with viscous fluid. A typical step function for both underdamped and approached critical damping conditions is shown in Fig. S-48.

Applications: Stepping motors provide an economical means for accurate and reliable positioning. Since steppers move in discrete steps, they avoid the problems of stability and feedback inherent in most servo devices—by allowing direct digital control in an open-loop mode of operation. Although not required, pulse feedback

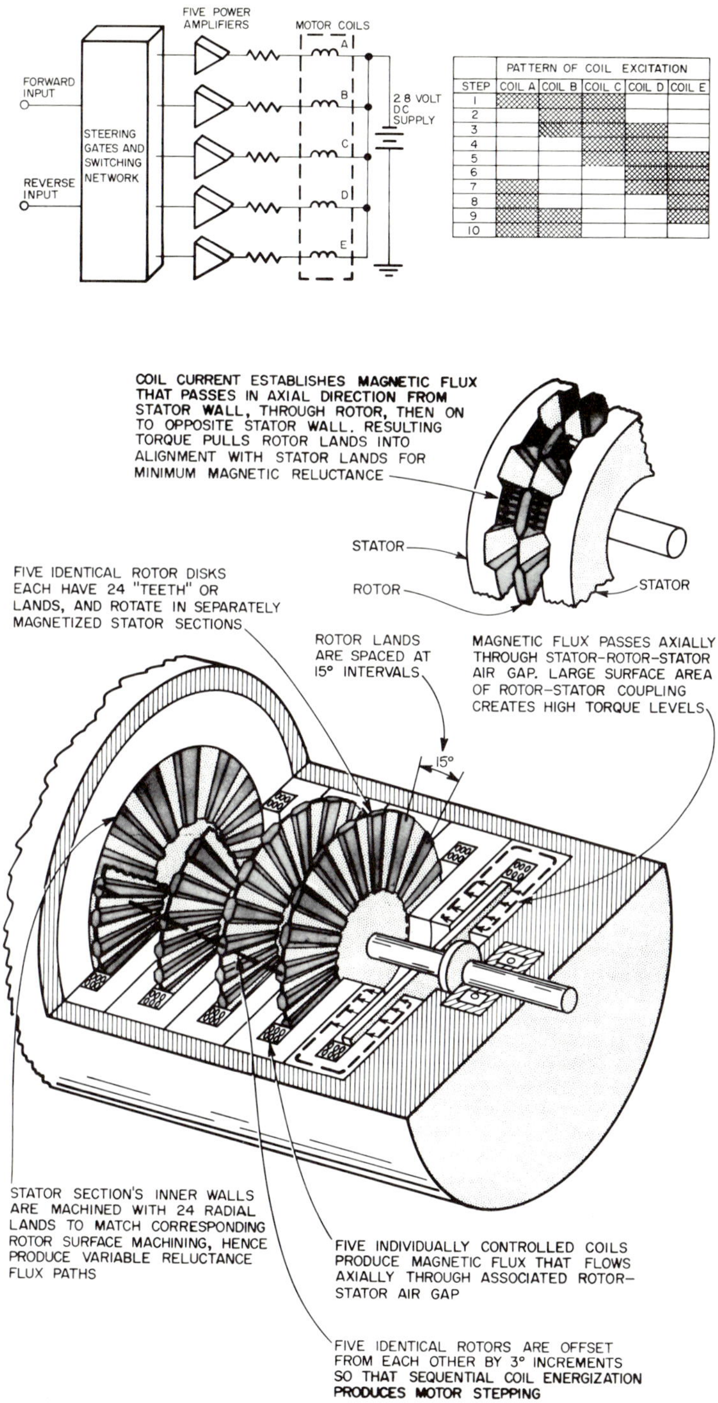

Fig. S-47. Variable-reluctance stepping motor. *(ICON.)*

readily can be coupled to a stepping motor to provide closed-loop operation. Applications range from simple indexing positioners to complex numerical contouring controls. Large inertial loads as well as small can be handled by controlling the acceleration and deceleration rates of

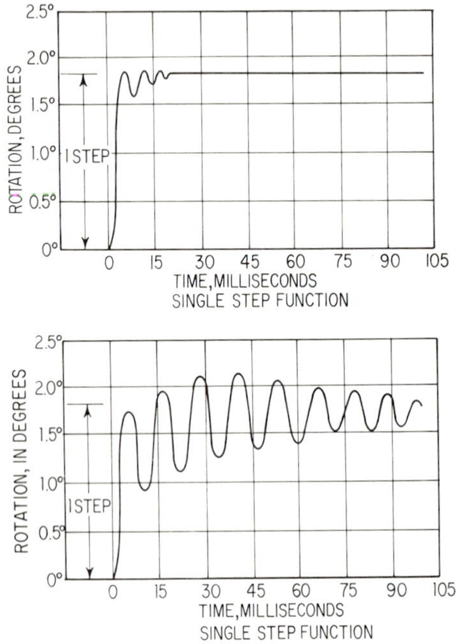

Fig. S-48. Typical step curves for a stepping motor in: (a) the undamped condition; and (b) approaching critical damping.

the input pulses. The most common applications for stepper motors include printers, tape readers, memory devices, valve actuators, positioning tables, counters, and indexers.

—Elmer W. Madsen, *The Superior Electric Company*

Stethoscope (See **Cardiography.**)

Stochastic Process A random-sampling process in which each and every member of the population (as an example) has the same opportunity of being selected.

Storage, Computer A storage device could be looked upon as any medium which is capable of storing information; using this definition, for example, a punched card could be classified as a storage medium. However, a storage unit is generally defined as a *device* on which information can be stored, read, and erased. With this definition, storage devices can be classified as immediate-access, random-access, and sequential-access storage units. The cost per bit (see **Bit**) of information is higher for immediate-access storage devices, but the access time is much faster than is achievable with random- or sequential-access storage units.

Immediate-access storage devices are storage units from which the information can be read in a few microseconds or less. Generally, this type of storage is an array of storage elements which are directly addressed, and all information in the array takes the same amount of time to be read. Some of the types of storage devices which are included in this classification are core storage (see **Core memory**), thin films (see **Thin-film memory**), and monolithic storage (see **Monolithic memory**) units. The access time (see **Access time, computer**) of the core-storage and thin-film devices is a function of the address-selection circuits and the switching time of the magnetic storage material, while the access time of monolithic storage units is primarily a function of the address-selection circuits.

Disk-storage (see **Disk storage**) units and drum-storage (see **Drum storage**) units are examples of storage devices which are generally referred to as random-access storage devices. The time required to read or write information on these devices is generally in the 10- to 200-ms range and is dependent on where the information is recorded with respect to the read/write head at the time the data are addressed and the speed of mechanical access equipment.

On a magnetic drum, the access time could be very short if data are just approaching the head, or the access time could be the rotational time of the drum if the addressed data have just passed under the head. On a drum-storage device with one set of read/write heads, the average access time is usually considered as one-half the rotational time of the drum.

The access time of disk-storage devices ranges from approximately 50 ms to several hundred milliseconds, depending upon the design of the unit. On a disk-storage device with one read/write head per disk surface, the access arm, which mechanically moves its attached read/write head across the disk surface, must first seek the disk track which contains the data, and then the logic within the device must search the track for the addressed information. If the read/write head is located on the proper track, the access time may be slightly more than one-half the rotation time of the disk. If the head is at the outermost track of the disk and the information to be read is on the innermost track, the access time is the maximum specified, since the access arm must move the head across the entire surface of the disk. The average access time on a disk-storage device is approximately the time required to seek one-half the distance across the surface of the disk plus one-half the rotational time of the disk.

A magnetic-tape storage unit is an example of a sequential-access storage device. Information stored on magnetic tape is generally blocked into records which are written sequentially on the tape. Each record has an identifier label associated with it. Using this label, the computer program (see **Program, computer**) sequentially searches the tape until the desired data are found.

Magnetic tape and disks are sometimes considered input and output devices (see **Input/output devices**), since information may be written by one computer program and then be used as input data to another computer or computer program.

—W. W. Welz, *IBM Corporation*

Storage Oscilloscope (See **Oscilloscope.**)

Storage Protect A hardware feature which allows a programming system to protect certain areas of code from alteration by other areas of code. (Fetch protect prevents any reference in a similar manner.) Several methods are used to obtain storage protection. Core may be protected by areas, each area being given a different program-settable code or key. A master key is used to allow the control program to refer to all areas.

Another system allows core to be protected on an individual-word basis, providing finer resolution but increasing the time required to protect a given area size. In both cases, instructions are provided to set the protect status by programming. Some systems provide for a disable of the feature from the console. This allows a preset protection pattern to be established, and then the program capability to change the protection status is disabled.

The protection provided is the assurance that protected locations will not be altered accidentally by store-type instructions. In single-word schemes any storage modification is prevented and in the multikey area systems-storage modification is limited to operation in areas with the same key.

The action taken in case of a storage-protect violation is the setting of an error indicator and an interrupt signal to the machine (see **Interrupt**). This signal is usually at a high priority, in the class with instruction-violation interrupts.

In a system without storage protect, it is possible to afford a measure of protection by providing code which checks any storage-modification operation for validity. The protection afforded is, of course, not perfect, and the resulting code is time-consuming. It is, however, useful in some circumstances.

—L. E. Wise, *IBM Corporation*

Storage Tube (See **Cathode-ray tubes.**)

Strain Gage A device that is used to measure the dimensional change within or on the surface of a specimen as the structure is subjected to mechanical, thermal, or a combination of both inputs. Strain gages most generally are attached to the specimen surface and may amplify the surface distortion mechanically so that the change may be measured on a simple dial indicator. Other types of strain gages measure the displacement of light rays through an optical system that is actuated by the surface strain, or convert the strain into an electrical signal. The mechanical, electromechanical, and optical strain-gage devices are considered *extensometers,* and their use generally is limited to material-properties testing or calibration tools.

Generally today the type of strain gage that immediately comes to mind when the term *strain gage* is used is the electrical type that has found applications far beyond what one normally thinks of in terms of an extensometer. The electrical-type strain gage may be based upon the measurement of a capacitance, inductance, or a resistance change that is proportional to strain. The resistance strain gage was developed in the 1930s in two basic configurations. The *unbonded* type was considered an extensometer and possessed many of the limitations of the mechanical and optical devices. The most versatile device for the experimental determination of strain for the purpose of stress analysis has been the *bonded* resistance-type strain gage. The basic principle of operation was discovered by Lord Kelvin in 1856.

The basic relationships between resistance change and strain are illustrated in Fig. S-49. When a conductor of length L and cross-sectional area A is elongated, the length increases and the area decreases by Poisson's effect to produce a resistance increase when the basic resistivity of the material remains reasonably constant. The resistance change $\Delta R/R$ then is related to the length change $\Delta L/L$ or strain ϵ by the strain sensitivity or gage factor.

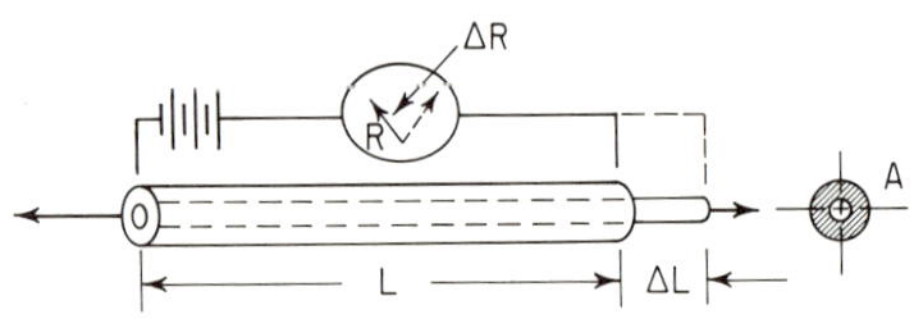

Fig. S-49. Basic relationship between resistance change and strain in resistance-type strain gage. $R = \rho(L/A) =$ resistance, ohms; $L =$ conductor length; $A =$ cross-section area; $\rho =$ resistivity constant; $S = (\Delta R/R)/(\Delta L/L) =$ strain sensitivity (gage factor when applied to a specific gage); $\Delta R/R =$ resistance change; $\Delta L/L =$ strain.

If the strain sensitivity were dependent upon dimensional change only—resulting from the usual Poisson ratio of 0.3—then all metallic conductors would have a theoretical value of 1.6 in the elastic range and 2.0 in the plastic range of the alloy used where the Poisson ratio becomes 0.5. However, the resistivity must alter with strain in order to produce the gage-factor range of 2.0 to 4.5 experienced by the alloys currently used for metallic strain gages.

The early bonded resistance strain gage developed by Dr. Arthur C. Ruge at the Massachusetts Institute of Technology was manufactured from fine strain-sensitive wire that was attached to a thin paper carrier with nitrocellulose cement for dimensional stability and to provide electrical isolation from the metal specimen. The carrier was bonded to the specimen with the same adhesive so that the specimen surface strain was faithfully transmitted into the fine-wire grid. The grid shape was conceived to provide maximum gage resistance in the smallest possible gage length and width. All bonded resistance strain gages utilize the same basic construction regardless of specific type. However, there are thousands of configurations utilizing various strain-sensing alloys, grid shapes, carrier materials, and attachment techniques designed to meet a wide variety of test and environmental conditions.

The strain gage and associated accessories, such as bonding cements, main-leadwire system between the gage and instrumentation, and protective-coating materials used to shield the gage from hostile environments, must be selected for the type of test to be performed and the temperature range expected. Static strain measurement, where all readings must be referenced to an unloaded condition at initial ambient test temperature while both load and temperature vary over extended periods of time, requires maximum electrical and dimensional stability and repeatability from the gage and accessories. Dynamic measurements, where the peak-to-peak strain magnitude under rapidly varying loads is of primary concern, reduces the demand on the gage performance except as related to fatigue life. Table S-3 shows a typical selection chart for the major gage types based upon the test application and temperature range.

In addition to the material variations, an even greater number of grid shapes are available, particularly in the foil construction to satisfy space requirements. Gage lengths vary from 6 to $\frac{1}{64}$ in. and resistance values from 1,000 to 60 Ω. Some typical configurations designed for specific measurements are shown in Fig. S-50.

Bonded resistance strain gages often are applied to structures to measure quantities other than strain. In flight testing of aircraft, gages may be installed in multiple arrangements to indicate flight loads as a function of wing-spar bending, or they may be installed on piping to measure internal pressure or on a motor drive shaft to indicate torque. There is a wide variety of commercial transducers where bonded resistance strain gages are utilized on sensing elements that are designed to respond to load, pressure, torque, and acceleration. Many of these devices are described in this encyclopedia. See the alphabetical index. Application of the strain gage essentially is limited only by the imagination of the user. Use of strain gages is increasing as the demands of automation require electrical interfacing between mechanical phenomena and computers and other elements of process-control systems.

Transducer gages usually employ the copper-nickel alloy sensing element and a reinforced phenolic or epoxy carrier. When properly com-

TABLE S-3. *Typical Bonded Resistance Strain Gages*

Grid Construction	Carrier Material and Construction	Bonding-cement Attachment Technique	Normal Temp. Range, °F*	Remarks
Copper/nickel alloys—general-purpose—static and dynamic				
Wire or foil	Paper—laminate	Nitrocellulose	−100 to +150	
Wire or foil	Phenolic—glass laminate	Epoxy or phenolic	−100 to +400	Most durable
Foil	Cast—film epoxy	Cyanoacrylate or epoxy	−100 to +250	
Foil	Cast—film polyimide	Cyanoacrylate or epoxy	−100 to +400	Widely used
Foil	Epoxy—glass laminate	Epoxy	−100 to +400	Most durable
Wire or foil	Weldable metal base	Capacitance (spot) welded	−100 to +400	Fast installation
Nickel/iron + alloys—high sensitivity—primarily dynamic				
Wire	Paper laminate	Nitrocellulose	−320 to +150	
Wire	Phenolic laminate	Epoxy	−320 to +250	
Foil	Cast—film epoxy	Epoxy	−320 to +250	
Foil	Cast—film polyimide	Epoxy	−320 to +400	
Foil	Epoxy—glass laminate	Epoxy	−320 to +500	
Nickel/chrome + alloys—extended temperature—static and dynamic				
Wire or foil	Phenolic—glass laminate	Epoxy	−452 to +600	
Wire	Removable frame	Flame-sprayed alumina	−452 to +1800	Dynamic only
Wire or foil	Weldable metal base	Capacitance (spot) weld	−452 to +600	Static
Wire	Special construction	Spot weld	−452 to +1800	Dynamic only
Foil	Epoxy—glass laminate	Epoxy	−452 to +600	
Foil	Polyimide—glass laminate	Polyamide	−452 to +700	
Foil	Removable vinyl film	Ceramic cement	−452 to +700	Static
Foil	Removable vinyl film	Ceramic cement	−452 to +1800	Dynamic
Platinum/tungsten alloys—high temperature—primarily dynamic				
Wire	Phenolic—glass laminate	Epoxy	−320 to +600	
Wire	Removable frame	Ceramic cement or flame-	−320 to +1200	Static
		sprayed alumina	−320 to +1500	Dynamic
Foil	Polyimide—glass laminate	Polyamide	−320 to +700	
p-doped silicon—high sensitivity†—primarily for transducers				
Semiconductor	Free—filament	Epoxy or ceramic	−320 to +800	
Semiconductor	Phenolic—glass laminate	Epoxy	−320 to +500	
Semiconductor	Epoxy laminate	Epoxy	−320 to +250	
n-doped silicon—high negative sensitivity†				
Semiconductor	Free—filament	Epoxy or ceramic	−320 to +800	
Semiconductor	Phenolic—glass laminate	Epoxy	−320 to +500	
Semiconductor	Epoxy laminate	Epoxy	−320 to +250	

*Conservative limits set by carrier-cement combination or alloy.

†Strain sensitivity is 20 to 60 times that of wire or foil; used in multiple arrangements to minimize temperature sensitivity.

pensated for temperature effects on the gage and transducer-element material, the unit can achieve overall accuracies of 0.10 percent of full-scale output over the normal ambient-temperature ranges from 50 to 150°F.

It is more difficult to define the accuracy that can be achieved in strain measurement for stress analysis because of the range of environments encountered and the type of test that is specified. Measurement errors can be held to ± 1 percent during short-term static and dynamic test under laboratory conditions where temperature and humidity are reasonably well controlled. Where normal earth temperatures and other environmental conditions are not controlled, measurement errors of ±5 percent are not unusual. If the test temperature extends into the cryogenic or elevated ranges beyond 400°F, the measurement accuracy will further deteriorate. Under such conditions, the data should be corrected or the

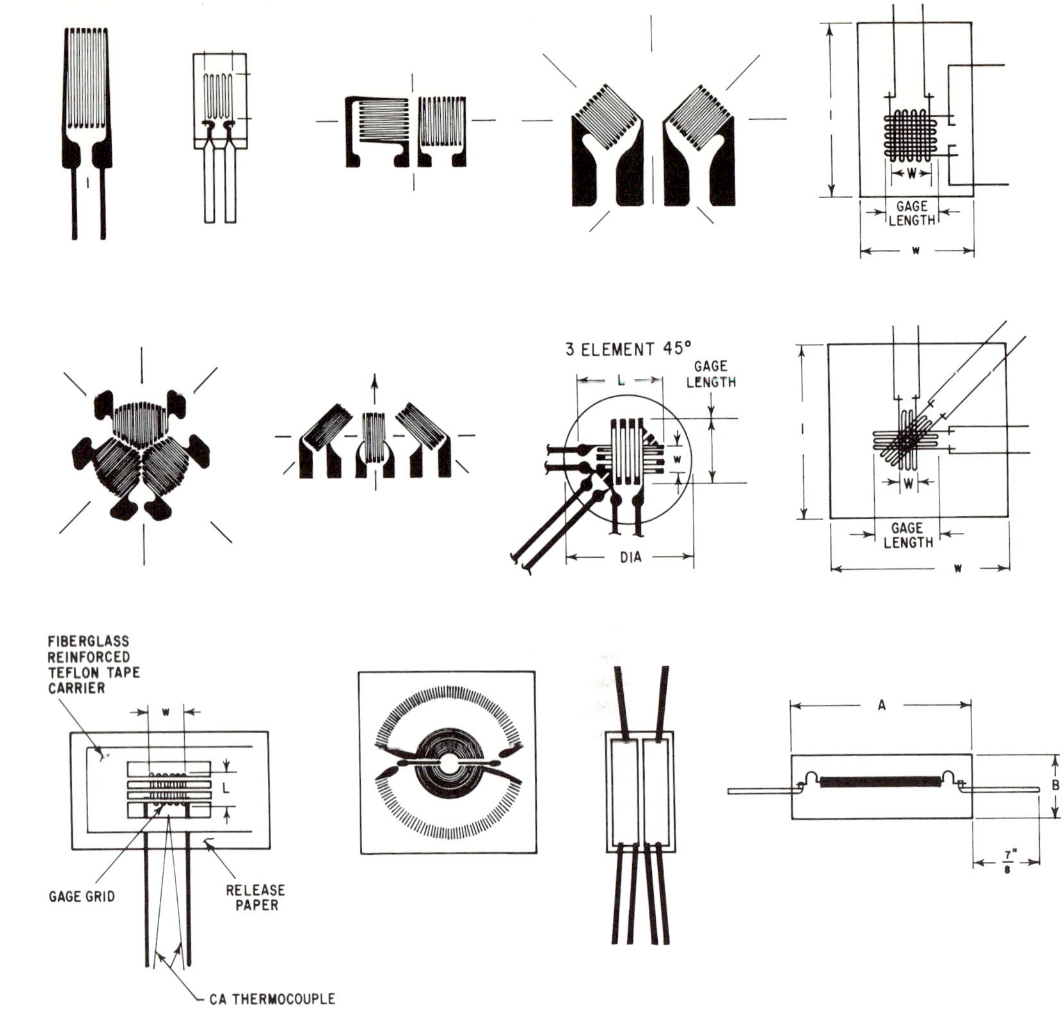

Fig. S-50. Various types of strain gages.

gage circuit compensated for the temperature effects.

Also see **Bridge circuits (electric); Load cell, strain-gage;** and **Potentiometer.**

—Leon J. Weymouth, *BLH Electronics, Inc., a subsidiary of Baldwin-Lima-Hamilton Corporation*

Strain-gage Accelerometer (See **Accelerometer.**)

Strip-chart Recorder (See **Recorder, graphic.**)

Stroboscope A device that permits intermittent observation of a cyclically moving object in such a way as to produce an optical illusion of stopped or slowed motion. Intermittency of observation can be provided by mechanical interruption of the line of sight (as with a motion-picture camera) or by intermittent illumination of the object being viewed. The modern industrial stroboscope basically is a lamp plus the electronic circuits required to turn the lamp on and off very rapidly—at rates as high as 150,000 flashes per minute. The stroboscope finds a wide variety of applications, including high-speed photography, speed measurement, and as a diagnostic tool for observing a number of physical phenomena.

An electronic stroboscope is shown in Fig. S-51.

As shown by the schematic diagram of Fig. S-52, the device includes a strobotron tube with its associated discharge capacitors, a triggering tube to fire the strobotron, an oscillator to determine flashing rate, and a power supply. The device has two basic operating modes. In one, the flashing rate is controlled by the internal oscillator and is indicated on the rpm dial. In the other mode, flashing rate is controlled by an external contactor or electrical signal. Under internal control, the device covers a flashing-rate range of from 110 to 150,000 flashes per minute in four direct-reading ranges: 110 to 690; 670 to 4,170; 4,000 to 25,000; and 24,000 to 150,000 flashes per minute. By use of harmonic techniques, speeds up to a million rpm can be measured. Accuracy is ± 1 percent of dial reading after calibration.

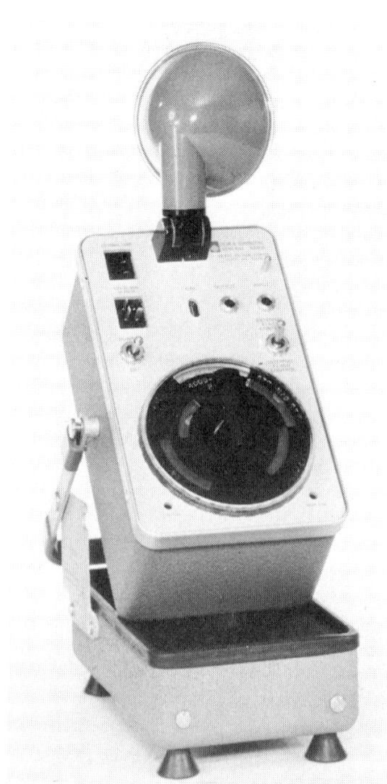

Fig. S-51. Electronic stroboscope. *(General Radio.)*

Stroboscopic Tachometer: To serve as a tachometer, a stroboscope must include its own flashing-rate control circuits and calibrated dial. If the speed to be measured is within the flashing-rate range of the stroboscope, it can be measured directly from the stroboscope dial. Otherwise, harmonic techniques must be used. In adjusting the stroboscope flashing rate for a single stationary image, one must be careful to avoid being confused by harmonics or subharmonics. Spurious images are especially confusing when the object being viewed is symmetrical. A four-bladed fan, for example, will appear stopped when the stroboscope is flashing at twice and at four times the fan speed. This particular problem is solved easily by a crayon mark placed on one of the four blades. In all speed measurements on symmetrical devices, the introduction of some asymmetrical feature is recommended in order to avoid confusion.

Photographic Applications: High-speed ob-

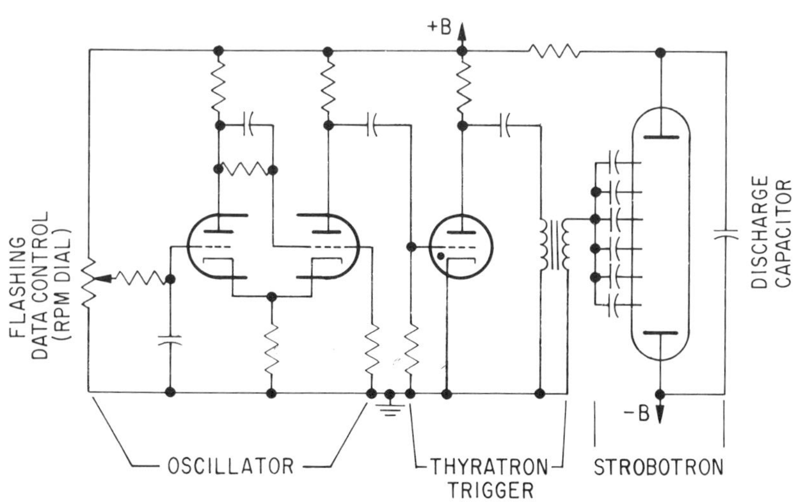

Fig. S-52. Schematic diagram of electronic stroboscope circuit.

jects, such as rifle bullets, essentially are stationary during a stroboscope flash. With typical muzzle velocities, the distance traveled by a bullet will be but a few hundredths of an inch. The high peak intensity of a strobe flash also permits photography under adverse lighting conditions and often makes possible the use of fairly inexpensive cameras and readily available film for sophisticated work.

Other Stroboscope Applications: Space here permits but a listing of numerous other uses for stroboscopic instrumentation: (1) determination of acceleration of freely falling objects, (2) loudspeaker studies, (3) studies of all types of high-speed phenomena, (4) investigation of standing waves in an elastic band, (5) determination of belt tension and horsepower, (6) inspection of lead shot in a shot tower, (7) checking registration while printing presses are running, (8) measurement of slip between two shafts, (9) eye-behavior studies, (10) inspection of many kinds of machine motions, (11) measurement of torque transmitted by shaft couplings, (12) turbine-cavitation studies, (13) investigation of air currents, (14) instrument and meter ballistics testing, (15) vibration measurements, (16) measurement of projectile velocities, (17) dynamic balancing of rotating devices, (18) calibration of camera-shutter speeds, (19) physical studies requiring a pulsed light source, as in the case of photosensitive materials investigations, (20) ripple-tank photography, (21) fuel-spray studies, and (22) stroboscopic stimulation in brain-wave studies.

REFERENCES

Kivenson, G.: "Industrial Stroboscopy," Hayden Book Company, Inc., New York, 1965.

Rutkowski, J.: "Stroboscopes for Industry and Research," Pergamon Press, New York, 1966.

Van Veen, F.: "Handbook of Stroboscopy," General Radio Company, 1966.

—Carl W. Alsen, *General Radio Company*

Stroud and Oates Bridge (see Bridge circuits, electric.)

Subprogram (See Subroutine.)

Subroutine A group of programming statements which perform a desired function and which are used as a component of a program. Subroutines are created to allow a function or group of functions to be repeatedly performed without having to recode the function each time it is needed.

Open subroutines are a collection of statements that are actually inserted into the program at the point of reference and form in-line code in the program. This insertion is usually done during the relocation linking of the program, although it may also be done at compilation time. Each reference causes an additional copy of the subroutine to be included. These routines are used where execution speed is more important than core space.

Closed subroutines are collections of statements to which control is transferred from some point in a program to perform the desired function. At the completion of the function, control is returned to the invocation point in the calling program. This technique permits a routine to be repeatedly "called" without having multiple copies in the core. Closed subroutines are often used to segment programs naturally for easier check-out and modification.

Communications with the closed subroutine may be through defined common data areas or lists (a part of the calling procedure) which specify the data to be used in the subroutine. Subroutines may manipulate the data specified in the list or common area and return control to the calling point, use the data supplied to return a value in place of the subroutine reference, do input-output, and in general, perform any other functions allowed in the language being used.

—L. E. Wise, *IBM Corporation*

Subsonic Measurements (See Acoustic instrumentation.)

Summing Point Any point at which signals are added algebraically. Or a mixing point whose output is obtained by addition, with prescribed signs, of its inputs in a feedback control system.

Supervisory Control Supervisory control implies the commanding and decision-making functions of a controlled system. These functions generally are achieved by utilization of a computer. Another definition[1] is "Control action in which the control loops operate independently subject to intermittent corrective action, e.g., set point changes from an external source."

Application of supervisory control has been found in the areas of process industries, utilities, transportation, oceanography, and even space

[1] Scientific Apparatus Makers Association.

programs. Process industries include chemical, petroleum, and petrochemical plants. Utilities include electrical power generation and distribution, water treatment and distribution, waste treatment and disposal, and gas extraction and distribution. Transportation involves railway, highway, and airway traffic controls. Oceanography includes oil exploration and search vehicles.

Most of the applications mentioned, except the process industries, involve telemetering because of great geographical extent which is common in such applications. Furthermore, a digital format is usually preferable because of the inherent resistance to noise and the security features which are not available in an analog format and finally, because of the convenience of interfacing with a computer which almost always is digital.

Therefore, a typical supervisory-control system would consist of a data transmitter (an analog-to-digital converter is required if the data happen to be in analog forms), an encoder, a communication channel (a pair of copper wires, or a pair of telephone lines, or a radio-frequency carrier), a decoder, a receiver, and finally, a computer.

However, the situation is much simplified in the process industries. A chemical plant is usually concentrated in a relatively smaller area and has no difficulties in transmitting information within reasonable accuracy without applying telemetry and security-coding schemes which appear to be essential for all the other applications.

The role of a computer in a supervisory-control system, as defined earlier, is to decide according to the information received and certain programmed calculations or logic procedures and then execute these decisions and/or manual-override commands from an operator. In the case of water-distribution industries, this might be the operation of a certain water pump or a control valve at a remote pumping station, according to certain time schedules or demands of a certain district, or some other sequential procedure. In the case of process industries, the computer would be involved in the calculations of feedforward, noninteracting, nonlinear, optimizing, and adaptive-control systems. The calculated signals are to be used for direct digital control, for setting an analog control loop, or for making some other necessary adjustments in order to improve the processes. In addition, the start-up and shutdown of a chemical plant would involve the logic controls of a computer, which is another area of supervisory application in the process industries.

—Anthony Y. Chiang, *Bristol Division, American Chain & Cable Company, Inc.*

Suppressed-zero Range (See **Range, instrument.**)

Switch, Analog Analog switch is a generic term used to refer to switches which are suited to switching analog signals, generally in high-accuracy and high-resolution applications (see **Accuracy; Resolution**). The characteristics of the switch which are important in these applications are speed, voltage and current errors, "on" and "off" resistance, and noise. The ideal switch is infinitely fast, generates no noise, introduces no voltage or current errors, has zero "on" resistance and infinite "off" resistance. Practical switches used in analog multiplexing applications approximate these ideal characteristics to varying degrees. No real switch, however, exhibits all these ideal characteristics simultaneously.

Electromechanical analog switches such as dry-reed and mercury-wetted contact relays generally have a lower "on" resistance and higher "off" resistance than solid-state switches (see **Relay**). In addition, they have no inherent voltage offset and leakage currents are very small because of the excellent isolation between the drive and signal circuits. They are, however, much slower than solid-state switches, and the switching action produces noise which persists for an appreciable length of time.

Solid-state switches are characterized by their high speed (see **Switch, FET; Switch, Bright**). "On" resistance is generally higher than in the case of electromechanical switches, and "off" resistance is sometimes lower. Switches having a *pn* junction as part of the signal path, such as bipolar transistors, exhibit an inherent voltage offset. Leakage currents between the drive and signal paths are also greater in the bipolar transistor switch because of less isolation between drive and signal. The leakages in field-effect transistors are comparable with those in electromechanical switches. Switching action produces noise primarily because of coupling of the drive signal into the signal path via interelectrode capacitance. Although the noise magnitude can be significant, it generally decays to a negligible value faster than in the case of electromechanical switches.

Both electromechanical and solid-state switches are used extensively in analog-switching applications. Relays and other electromechanical devices are used in applications involving low-level signals (less than 1 V full-scale) and relatively low multiplexing speeds, generally less than 250 samples per second. Field-effect transistors are used in low-level applications at higher speeds, but they are limited in common-mode voltage-handling capability because of lower breakdown voltages (see **Common-mode voltage**). Bipolar transistors and diodes are generally used in the analog switching of high-level signals, although with suitable circuits for the compensation of offset voltages, they can be used for low-level signals.

—Thomas J. Harrison, *IBM Corporation*

Switch, Bright The Bright switch is a solid-state switch consisting of two bipolar transistors connected in the inverted connection as shown in Fig. S-53. The switch configuration more closely approximates the ideal voltage switch than a single transistor, since offset voltage is reduced by the use of the inverted connection and the back-to-back transistor connection.

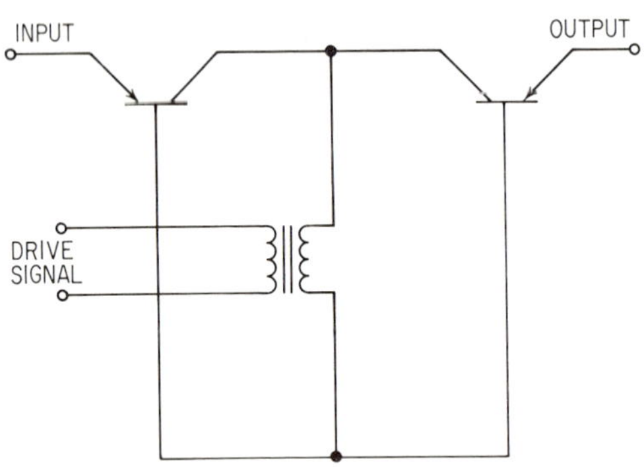

Fig. S-53. The Bright-switch configuration.

The basis for the switch operation is as follows: If a bipolar transistor is driven into saturation by a signal applied between the base and emitter, an open-circuit (i.e., collector current is zero) voltage appears between the emitter and collector. This voltage, called the offset voltage, represents an error when the switch is used in analog-switching applications, since it appears in series with the signal source being switched. If, however, the transistor is driven into saturation by a drive signal applied between the base and collector, the magnitude of the offset voltage is greatly reduced. The difference between the two offset voltages is primarily a function of the difference between the normal β (common-emitter current gain with the drive connected between the base and emitter) and the inverted β (common-emitter current gain with the collector functioning as the emitter).

R. L. Bright, in inventing the switch configuration of Fig. S-53, utilized this phenomenon and the cancellation effect of connecting the transistors in a back-to-back configuration to produce a low-offset switch. In the configuration, the offset of one transistor is partially canceled by the offset voltage of the other transistor. Net offsets of less than 50 μV can easily be obtained in this manner compared with an offset voltage of several millivolts in the typical silicon transistor.

REFERENCE

Bright, R. L.: Junction Transistors Used as Switches, *AIEE Trans. Commun. Electron.*, March, 1955, pp. 111–121.

—Thomas J. Harrison, *IBM Corporation*

Switch, FET An FET switch is generally a switch designed for use in controlling analog signals which is composed of one or more field-effect transistors. Both types of FETs, the junction FET (JFET) and the metal-oxide-semiconductor FET (MOS FET) or insulated-gate FET (IGFET), are used for analog-switching and multiplexing applications. The two outstanding characteristics of the FET which make it attractive for analog multiplexing are the absence of an offset voltage, which makes it suitable for switching low-level signals, and its high input impedance, which provides excellent isolation between the signal path and the drive voltage.

Although not as fast as bipolar switches, the FET can be used for switching at rates in excess of 100,000 samples per second. Its primary disadvantage is a relatively high "on" resistance. Typical devices have an "on" resistance ranging from 50 to 200 Ω, although a few devices are available with less than 5 Ω "on" resistance.

The absence of an inherent offset voltage is due to the fact that there is no *pn* junction in the signal path as in the case of the bipolar transistor. (compare **Switch, Bright.**) The FET is a majority-carrier device in which the conductance of a conducting channel between two electrodes, the source and the drain, is controlled by a signal applied to a third electrode, the gate. In the MOS FET, the width of the conducting channel and, therefore, its resistance are controlled by a field established between the channel and the gate which is insulated from the channel by a thin layer of material, usually silicon dioxide. In the junction FET, the width of the conduction channel is controlled by the depletion layer surrounding the gate region. In both cases, the signal path from source to drain does not contain the *pn* junctions responsible for the offset voltage in bipolar transistors.

The FET is a voltage-controlled device and, as such, has a very high input impedance. In the JFET, the input impedance is determined by the back-biased gate-to-channel *pn* junction. In the MOS FET, the input impedance is the capacitor formed by the gate and channel, separated by the insulating layer. As a result of these gate constructions, the input impedance is very high, typically in excess of $10^9\ \Omega$. Leakage currents are also very low. In the MOS FET they are determined by the dielectric properties of the insulating layer, and in the JFET they are determined by the back-biased *pn* junction. This high input impedance and low leakage provide excellent isolation between the drive signal and the signal being switched.

—Thomas J. Harrison, *IBM Corporation*

Switch, Light-coupled A light-, or photon-, coupled switch is a switch in which the switching signal is transmitted to the device by means of light energy. The switching element consists of a light-sensitive device such as a photoresistive material, a photodiode, phototransistor, or photo-FET. Light energy falling on the input of such a device effectively alters the transmission characteristics of the device to allow conduction between two terminals. That is, the resistance of the device is high when no light is present and it drops to a low value when excited by the photon energy.

Light sources for light-coupled switches are varied. Ordinary incandescent sources may be used, but more commonly, gas-discharge sources such as neon lamps and solid-state devices such as gallium arsenide light-emitting diodes are used. These sources produce less heat and have higher reliability and higher speed.

Light-coupled devices have not been widely used for analog multiplexing. The primary limitations have been switching performance, cost and reliability. Light-coupled modulators have been used in laboratory instrumentation, however, for a number of years.

The primary advantage of a light-coupled switch is the isolation which can be achieved between the signal and the drive source. This allows the control of signals without introduction of errors due to the drive source. In addition, the isolation is beneficial in maintaining a high common-mode rejection ratio in analog signal-handling equipment [see **Common-mode rejection ratio (CMRR)**].

—Thomas J. Harrison, *IBM Corporation*

Switch, Voltage (See **Switch, analog.**)

Switchboard Meter (See **Meter, electrical.**)

Synchro A general term applied to a class of variable-coupling transformers in which the variable coupling is obtained by changing the orientation of the primary to the secondary through rotation of the movable element. Specifically, the synchro transduces a rotor-position angle into a voltage or a set of voltages unique to that angle. The most common usages of synchros are in the transmission of these angular data in either torque or control (servo) systems.

Types of Synchros: The majority of synchros are divided into five classes:

Synchro Transmitter: A unit whose rotor is mechanically driven and whose stator produces a set of three output voltages corresponding to the rotor-position angle.

Synchro Receiver: The rotor is free to turn to that angular position which corresponds to the electrical input received by the stator.

Synchro Differential: The rotor is mechanically driven and the stator receives an electrical input from another synchro. The rotor output voltage corresponds to an angle which is either the sum or the difference of the electrical input angle and the mechanical rotor-position angle.

Synchro Differential Receiver: A differential unit whose rotor is free to turn to that position which is either the sum or the difference of the angles indicated by the electrical inputs to the rotor and the stator.

Synchro Control Transformer: The rotor output voltage is proportional to the sine of the angular difference between its rotor-position angle and the magnetic field of its stator. Normally, the rotor of this unit is driven back to null by a servoloop.

Construction of Synchros: Synchro rotors have windings, either one or three in number, wound on a laminated-iron structure. Transmitters and receivers have a two-pole rotor, also known as a salient pole or dumbbell rotor. The bobbin-type windings merely encircle the poles. Other synchro types employ a distributed winding in a slotted, cylindrical rotor. The rotor assembly is suspended inside the stator by means of precision ball bearings. Connections are normally made to the winding using slip rings and brushes, although flexible connectors have been used in applications where limited rotation is the rule.

The three stator windings, are distributed in a cylindrical, slotted, laminated magnetic structure. Even though these windings are sometimes referred to as being three-phase windings, it must be understood that this terminology indicates spatial displacement only. The voltages induced in these windings are all of the same time phase. The lamination assembly of either rotor or stator is skewed for one slot to minimize space harmonics caused by the slots with their concomitant errors.

Synchro Systems: The most basic and, most probably, the original synchro system was a transmitter-receiver pair. The stator leads of the transmitter are connected back to back with those of the receiver; the rotors of the two units are excited in parallel. Whenever the rotors of the synchros are not at the same angular position, currents flow in the stator windings. The magnetic interaction between the stator current and the rotor current places a torque on both rotors. Since the transmitter rotor is restrained from turning, the resultant torque acts on the receiver rotor, forcing it to assume the same angular position as the transmitter. The information concerning the transmitter rotor-position angle is thus displayed by the receiver rotor position. This information may be displayed at more than one location by paralleling additional receivers across the transmitter, or the information may be

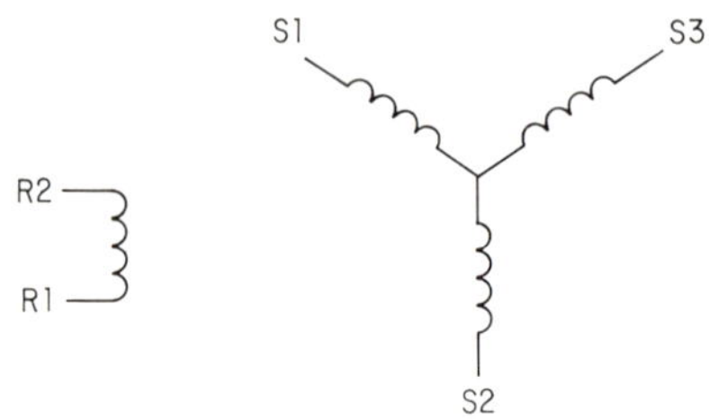

R = ROTOR LEAD S = STATOR LEAD

Fig. S-54. Schematic of synchro transmitter-receiver.

modified by interposing a differential between the transmitter and receiver; both these variations, however, are done at the expense of developed torque and hence accuracy.

A more precise method of conveying angular information is the control system, the most basic of which is a transmitter and a control transformer. The stators are tied back to back; the transmitter rotor is excited from the appropriate source. The voltage vector erected in the transmitter stator is reproduced in the control transformer stator. A voltage, proportional to the sine of the angular difference between the stator vector and the axis of the rotor winding, appears on the rotor of the control transformer. The rotor winding is connected to a servoamplifier which, in turn, is connected to the control phase of a properly excited servomotor; the motor shaft is connected through gearing to the shaft of the control transformer. As long as voltage appears on the rotor winding, the motor turns, and the motion is in such a direction as to oppose the change which caused it.

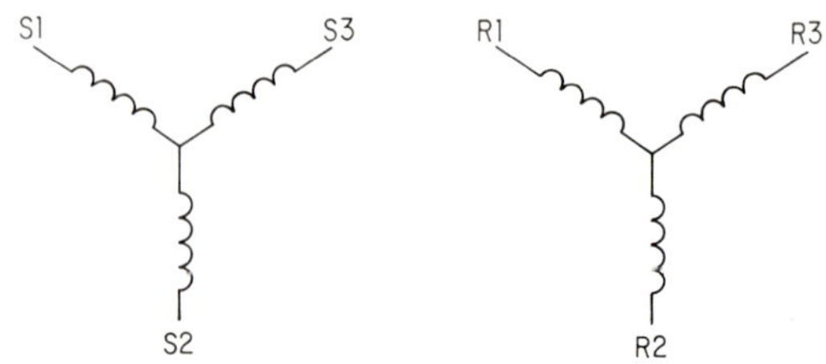

R = ROTOR LEAD S = STATOR LEAD

Fig. S-55. Schematic of synchro differential–differential receiver.

The control-transformer shaft will be turned until it is in alignment with the transmitter rotor position, so that no output voltage appears. This is the least complex of the null-seeking servosystems. A myriad of modifications are possible be-

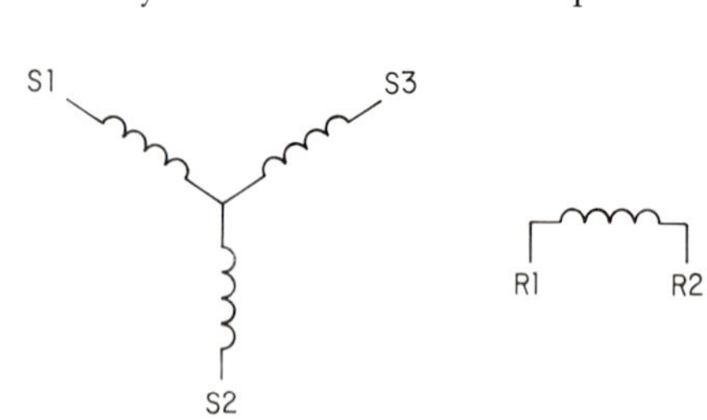

R = ROTOR LEAD S = STATOR LEAD

Fig. S-56. Schematic of synchro control transformer.

yond this simple transmitter–control transformer pair.

Synchro Characteristics: Synchros can be designed to perform over a wide range of input conditions. Input voltages range from 115 V to millivolts, input currents from approximately 1 A to fractions of a milliampere, operating frequencies from 30 Hz to 2 MHz. Accuracy of angular data transmission is usually specified as ±7 minutes of arc maximum, although ±3 minutes of arc accuracy is available. Larger-size units can have angular errors amounting to only 20 seconds of arc maximum. Schematic circuits are shown in Figs. S-54 to S-56.

Also see **Linear transformer; Resolver;** and **Transolver.**

—Joseph P. Spetzer, Jr., *Clifton, Division of Litton Industries*

Synchronism, Measurement of When two synchronous electrical machines or systems are connected in parallel, the two voltages must be equal and the speed (or synchronism) of the machines must be adjusted such that corresponding instantaneous values on the two waves are reached at the same instant, that is, when the machines will be in exact phase. With polyphase machines, the direction of phase rotation must be the same. The latter usually is assured at the time of machine installation, the phases being connected to the switches so that the phase rotation will always be right.

Lamp-synchronizing Method: This is a simple, very low cost, but somewhat limited means for determining synchronism. As shown in Fig. S-57, a and a_1 are the sources being connected in parallel and t and t_1 are transformers, the secondaries of which are connected in opposition through incandescent lamps L *and* L'. When the two sources are in synchronism, the secondary emf's neutralize each other and the lamps will remain dark. As the phase difference increases, the current through the lamps will increase, reaching a maximum at 180° of phase difference. If the machines are running at different speeds, the lamps will flicker. It is easier, of course, to determine total darkness as contrasted with maximum brightness. A voltmeter may be substituted for the lamps by connecting it so that synchronism is indicated when the reading is a maximum. With this method, it is not possible to determine which frequency of the two machines is the higher.

Synchronism Indicator: This instrument overcomes the aforementioned objection and also provides a more accurate indication of synchronism. In the *synchroscope* shown in Fig. S-58, a rotating field is produced by coils M and N connected to the busses through the reactance P and the resistance Q, respectively. An iron vane A, free to rotate, is mounted in this rotating field

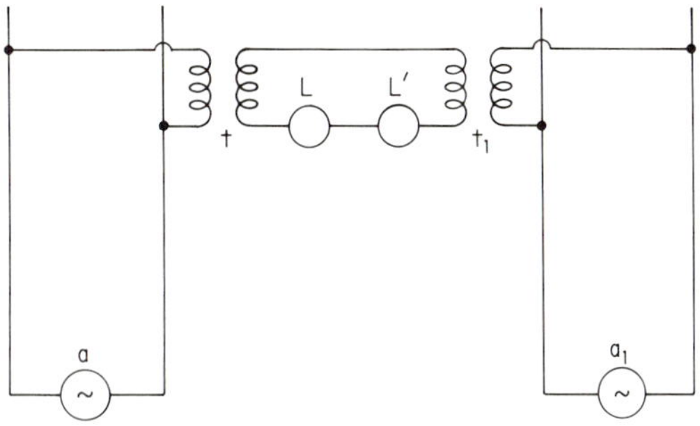

Fig. S-57. Use of lamps for synchronizing electrical equipment.

and magnetized by coil C, which in turn is connected across the incoming machine. As the vane is attracted or repelled by the rotating field from M and N, it will take up a position where this field is zero at the same instant the field from C is zero. Thus, the position at any instant indicates the phase difference. When the two frequencies are different, this position is constantly changing. The pointer will rotate in the fast or slow direction, coming to rest again at the same zero-field position when the frequencies are equal.

In another synchroscope, there is no iron in the instrument and the moving element is not allowed to rotate. The elements are practically the same as in an electrodynamometer wattmeter. The fixed coils are connected in series with a resistance and to the busses. The moving coil is

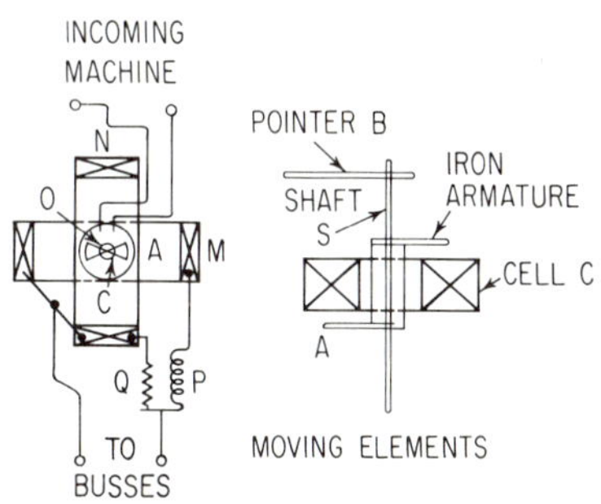

Fig. S-58. Circuit of synchroscope.

connected in series with a capacitor and the incoming machine. The two circuits are adjusted to exactly 90° difference in phase. At synchronism, there is no torque and the moving coil is held at the zero position by a control spring. If the frequencies are the same, but there is a phase difference, a torque will be exerted and the coil will move to a position of balance at the right or left (fast or slow). If the frequencies are different, the torque will vary continually and the pointer will oscillate over the dial. A synchronizing lamp illuminates the scale simultaneously.

Synchronous An operation which is performed in a fixed time interval under control of a clock is said to be synchronous (compare **Asynchronous**). For example, if a set of contacts is sampled at a fixed time interval, the operation is synchronous. This is in contrast to the contacts being sampled randomly under the control of an external signal. The read operation of a core-storage unit (see **Core memory**), for example, is generally synchronous; that is, the turning on of the X and Y selection drivers and the sampling of the core output on the sense line are controlled by a fixed-storage clock.

—W. W. Welz, *IBM Corporation*

Synchrotron (See **Accelerator.**)

Synthesizer, Frequency [See **Signal (electric) generator.**]

System, Automatic (See **Control system.**)

System Generation The procedure by which a general programming system is tailored into a specific system to support a given hardware configuration and perform a specific set of functions. The exact steps to be taken in this type of procedure are outlined in detail by the supplier of the system and generally include such things as:

Hardware description—core size, input/output units, special features, etc.

Software-option selection—accounting procedures, error exits, service routines, etc.

The software system is usually supplied in a form that will run on a minimum-configuration machine. The system-generation procedure is a job or series of jobs which performs assemblies of system modules as necessary, editing of libraries of routines supplied with the system, and the creation of a new executive supervisor (see **Executive program**) tailored for the installation. During this procedure, the auxiliary storage device(s) on which the system is to reside is configured to describe files, libraries, and work areas to be a part of the tailored system. The time required to generate a system varies from less than an hour to several hours, depending on system complexity and machine speed.

—L. E. Wise, *IBM Corporation*

Systematic Error (See **Error, measurement.**)

T

Tachometer Generator Relative perpendicular motion between a magnetic field and a conductor results in induction of a voltage in the conductor. The magnitude of the voltage is a direct function of the strength of the magnetic field and the speed with which the conductor moves perpendicular to it. Current will flow if the ends are connected to a load, such as an instrument. The polarity of the voltage and, therefore, the direction of current flow will depend upon polarity of the field and direction of conductor motion. As shown in Fig. T-1, the same effect can be obtained by rotating the magnet and holding the conductor still.

Only that portion of the conductor sweeping through the magnetic-field flux is effective in generating voltage. By forming the conductor into a loop, a greater effective length is possible and both sides of the loop will develop voltage. The polarity of the voltage changes as the loop passes through the vertical axis of the field. Rectification or commutation—change of an alternating polarity to a fixed polarity—is accomplished by the use of segments and brushes.

The magnitude of the voltage generated can be increased by the use of numerous loops properly arranged and connected. For direct current, the loops are made to rotate and are connected to segments called a commutator. For alternating current, the loops are stationary and connection to the external circuit is direct. The combination of

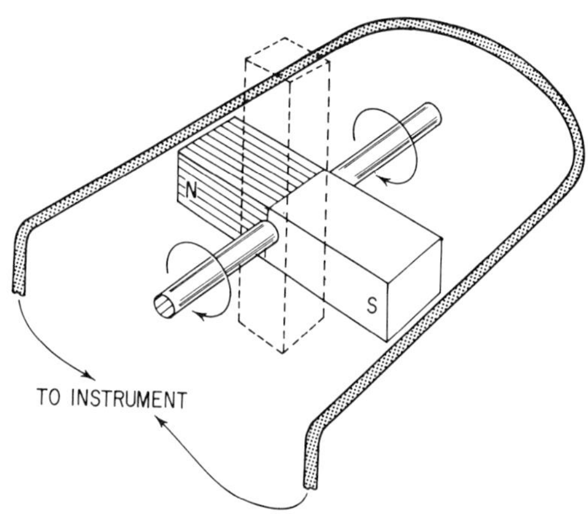

Fig. T-1. Voltage generated by rotating magnet and holding conductor still.

a tachometer generator and a suitable indicator—direct current for the commutator type as shown in Fig. T-2 or alternating current for the rotating-magnet types as shown in Fig. T-3—when suitably coupled to a shaft will indicate speed of rotation.

A variation of the alternating-current tachometer measures speed in terms of the frequency instead of the voltage generated. This system uses a transformer between generator and instrument. In any transformer, the voltage induced in the secondary winding is proportional to the rate of change of the core magnetization. By using a type of transformer iron whose magnetization reaches saturation very rapidly (during a fraction of each cycle), sharp, brief voltage excursions (pips) are obtained in the transformer output.

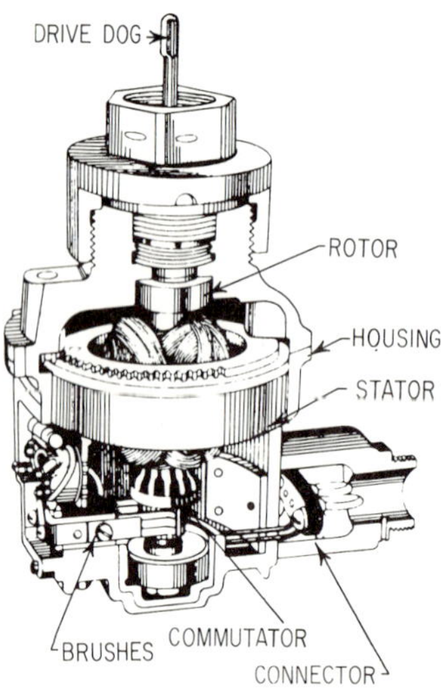

Fig. T-2. Commutator-type tachometer generator. *(Weston.)*

After rectification, these pulses are conducted into the moving coil of a permanent-magnet-type instrument where they produced pointer deflections strictly as a function of pulse frequency—hence of generator speed.

Calibration of Tachometers: The indicators or recorders used with tachometer generators can be calibrated in any unit that can be converted into a function of rotational speed, such as rpm, feet per minute, yards per minute, miles per hour, or hertz. Industrial tachometer-generator systems are capable of giving continuous direct reading of speeds up to 100,000 rpm.

Tachometer Selection: Factors which deter-

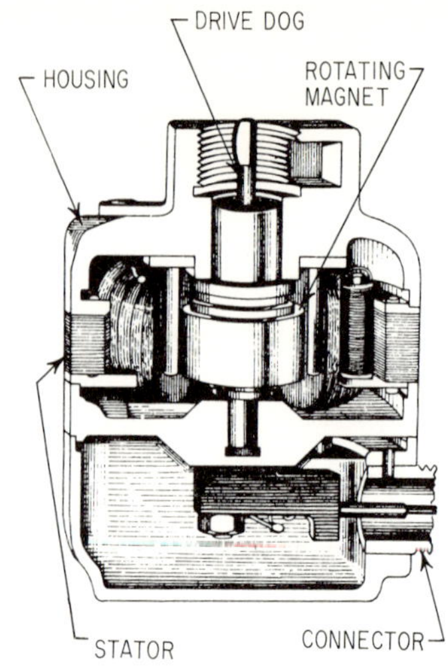

Fig. T-3. Rotating-magnet-type tachometer generator. *(Weston.)*

mine whether (1) a dc, (2) an ac voltage-responsive, or (3) an ac frequency-responsive system should be used include:

1. Top speed at which tachometer generator will operate
2. Required accuracy
3. Mechanical mounting problems

The chart of Fig. T-4 shows the top speed at which each form of generator may be used. These lim-

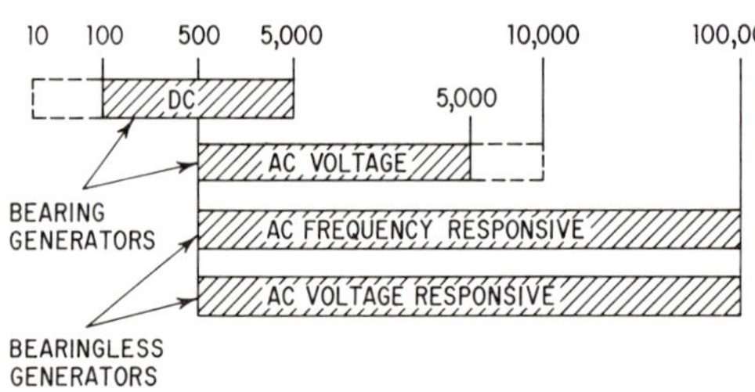

Fig. T-4. Rpm ranges of various types of tachometer generators.

itations can be modified by the proper selection of the point of measurement or by means of suitable speed-changing devices so that the generator speed can be kept within recommended limits.

Relationship of Two Speeds: This may be expressed in a variety of ways—the most common are *speed ratio* and *speed differential* (see Fig. T-5). As plotted, shafts *A* and *B* are holding a constant-speed-ratio relationship and shafts C and D are holding a constant-speed-differential relationship. The methods of measuring these two relationships are vastly different.

Measurement of speed ratio is shown in Fig.

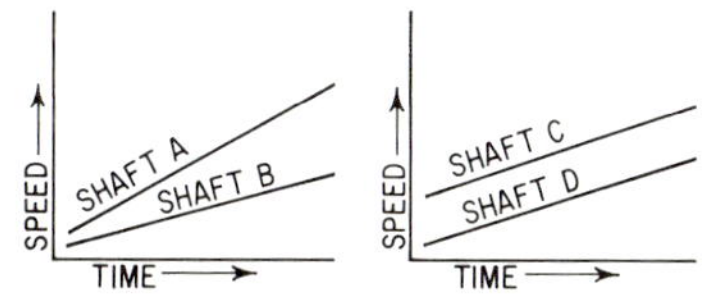

Fig. T-5. Speed ratio vs. speed differential of tachometers.

Bearingless Generators: These are ac generators of the most basic form consisting only of a permanent-magnet rotor and a stator. They have no bearings or brushes and are impervious to oil, grease, and relatively high temperatures. They have very low torque burdens of less than 1 oz-in. and are capable of speeds up to 100,000 rpm. Generally, the frequency-responsive approach is used with a bearingless generator. Since the system is solely dependent upon the frequency output of the generator, voltage variations caused by reductions in the magnetic strength of the rotor (which may result from poor handling, poor alignment of stator and rotor, or axial travel of

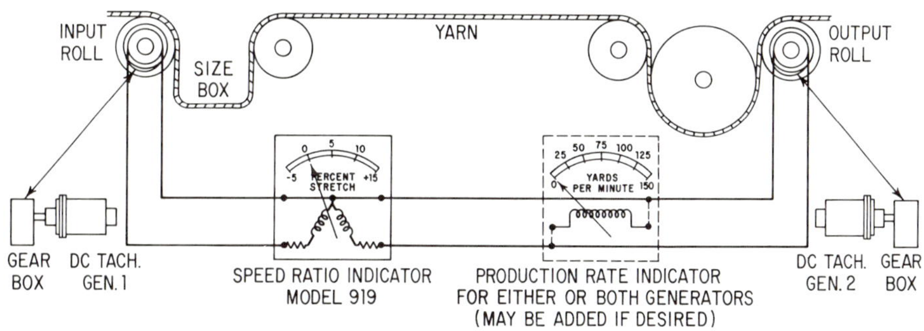

Fig. T-6. Measurement of speed ratio on textile slasher. *(Weston.)*

T-6. This system has wide application in the textile, paper, and steel industries. In the textile application shown, it is desired to measure "percent stretch" of various materials during the slashing operation.

There are several accepted methods of measuring speed differential. The first and most basic is to arrange two dc generators so that their output voltages are in opposition. By connecting a dc instrument between the two generators, a measurement of their voltage-output difference may be read in terms of rpm differential or whatever unit of measurement is required. Instruments may be connected in parallel to the generators to measure the individual speeds. In another method of measuring speed differential, two ac generators and frequency-responsive networks are arranged so that their outputs feed into a differential indicator or recorder. Typical of the operations to which this system can be applied is that of indicating the rpm difference between two engines driving a common shaft through a fluid drive.

the rotor with respect to the stator) do not affect overall accuracy.

—Adapted from "The Instrument Sketch Book," *Weston Instruments, Inc.*

Tape (See **Perforated tape.**)

Tapping Error (See **Hysteresis.**)

Target-type Flowmeter (See **Flowmeter, target-type.**)

Task A logical unit of work to be accomplished in a software control system. A task is the smallest unique unit that can contend for the computer's control and storage resources (compare **Job**).

In a multiprogramming system (see **Multiprogramming**), tasks are the elements in the system to which control is switched. Tasks are created as jobs are initialized into the system and require the control facilities. They are the active elements in asynchronous processing. Task management or task control is that portion of the control program which regulates the control switching among the contending tasks.

—L. E. Wise, *IBM Corporation*

Taste- and Odor-measurement Instrumentation Physical instruments are used in the analysis and in the controlled presentation of odor and taste substances. These instruments generally are inadequate for measuring flavor or odor quality and are insensitive to some compounds in concentrations that can be detected by the human nose. Biological detectors (mainly human), on the other hand, generally are more rapid, have a wider response range, and can be effective in chemosensory evaluation. Thus, despite the variability of chemosensory data, biological sensors will continue to play a significant role in the relevant areas of food, cosmetics, odor control, and related industries—either alone or in combination with physical instruments. Although a wide variety of biophysical and behavioral-response indicators are available, in practice the main criterion used is the *verbal response* of the human subject.

Physiochemical Instrumentation: Odor and taste substances generally can be detected and measured by the standard techniques of qualitative and quantitative chemical analysis, and several of the analytical-instrumentation techniques described elsewhere in this encyclopedia can be used. Such standard methodologies, aside from research use, however, do not suffice for flavor and odor detection and measurement in industry because of the difficult and often protracted procedures required to isolate the odor and taste substances. Further, the manufacturer of a product is interested in the behavior response of the human subject rather than a summary of chemical analyses. There are exceptions, and some of these will be described here.

Analysis comprises separation and isolation, detection, identification, and characterization. Separation commonly is achieved by distillation, crystallization, or gas chromatography. Gas chromatography can be effective with nanogram or even picogram amounts of some substances. The selectivity and sensitivity depend upon the detector used. Where applicable, gas chromatography is rapid and excellent resolution is possible. However, for identification, the most sensitive technique is mass spectrometry. The combination of a fast-scan mass spectrometer continuously sampling the output of a gas chromatograph is rapidly gaining acceptance as the method for analyzing small quantities of complex volatile mixtures. For routine use, the data should be stored and analyzed by computer.

The effectiveness of gas chromatography and other batch chemical methods tends to be reduced when the odorant to be detected is present in low concentrations in the ambient air where turbulence and diffusion occur. Under these conditions, the combination of a plasma chromatograph with a mass spectrometer has certain advantages. The manufacturer of this instrument claims that the device can detect trace organic compounds, such as those containing NO_2 and NO_3—in ambient, untreated air at concentrations down to 1 part in 10^{12} within fractions of a second, with certainty of these measurements as low as 1 part in 10^{14} under proper conditions. The speed of response is limited only by the ability to get the sample through the test cell and can be considered virtually instantaneous.

This device uses selected ion-molecule reactions to obtain the molecular trace ion which is formed at atmospheric pressure. The resulting molecular ions then are separated in accordance with their characteristic mobilities in a drift cell at atmospheric pressure. The ions thus formed then are injected directly into a quadrupole mass analyzer for mass analysis. Only those ions formed in the reaction region are seen by the mass analyzer—as the internal ionizer is used only for calibration.

The unusual sensitivity of the device[1] arises from the large cross section for certain ion-molecule reactions. Thus, those trace compounds of interest can be selectively ionized, separated, and then introduced into the mass spectrometer for mass analysis.

In addition to mass spectrometry, infrared, Raman, and nuclear magnetic-resonance spectrometry, and chemiluminescence analysis also can be used to establish chemical structure. The comparatively low sensitivity of nuclear magnetic resonance, the large sample size requirements of Raman spectrometry, and the selectivity of chemiluminescence tend to limit the range of usefulness of these techniques. In practice, two or more techniques often are used in combination.

Specialized trace-odor detectors also are available to replace or supplement the human nose. One example is the Electro-odo[2] cell developed in

[1] Franklin GNO Corp., West Palm Beach, Fla.

[2] N. N. Tanyolac, Robert College Center, Bebek, Istanbul, Turkey.

Turkey. This instrument detects and measures changes in potential or current on a dielectric surface, such as mica, when one surface of the dielectric is contaminated with odorous molecules. Sensitivity for some compounds is in the range of 10^{-8} to 10^{-9}. The device has been used to detect spoilage in meat and fish samples. With meat, changes of up to 600 μV within 3 min of initiating a test were reported. With fish, the increase in voltage was about 43 percent for 24 hr of aging.

Measurement of Biological Response: In measuring biological responses to taste substances, the stimulus is delivered at body temperature, at constant flow rate, and over a representative range of concentration. Between applications of the stimulus, the tongue is rinsed with distilled water. Flow chambers (to restrict the stimulus on the tongue) occasionally are used but are not considered essential. These specifications require no elaborate instrumentation.

In contrast, the controlled presentation of odorants presents serious difficulties—particularly if the stimulus is in the vapor phase. There are numerous designs for both static and dynamic air-dilution systems. In static systems, failure to attain equilibrium conditions within the delivery or storage units or both is a common source of error. Dynamic systems are preferable but are more complex and require a source of regulated compressed air. In one system of this type, water vapor is removed and odorant contents attenuated (if not entirely eliminated) by passing the air line through a coil immersed in liquid nitrogen. A small, metered fraction of the flow is saturated with odorant by bubbling the sample through a sparger in a controlled-temperature bath (20°C). A further fraction is used to dilute the odorant flow. A range of known concentrations can be achieved by varying the flow ratios of the streams while maintaining a constant total output. Since human odor thresholds for different substances span at least seven orders of concentration magnitude, a second and even third dilution step is necessary if the instrument is to be used to measure absolute sensitivity to odors. A small fraction of the final output can be monitored continuously by bleeding into a gas chromatograph in which the column is bypassed. A third channel of purified air can be used to flush out the outlet port of the olfactomer between tests (see Ref. 3).

While it is possible to analyze a spectral hue and specify it in internationally accepted units, no such course is available in assessing odor quality. Nor is there any assurance that the traditional, subjectively derived classification of taste qualities into *salt, sweet, sour,* and *bitter* is the most meaningful one for understanding the properties of taste receptors. Thus, the evaluation of taste and odor quality still is restricted largely to the measurement of behavioral and, to a lesser extent, biophysical indicators of response.

The biophysical approach most widely used involves recording the electrical activity from sites in the chemosensory systems. In taste experiments, the chorda typani nerve—innervating the anterior third of the tongue—is dissected out and placed on metal electrodes. Stimulation of the tongue with appropriate taste solutions initiates a burst of nerve impulses which can be amplified for display and photography on a cathode-ray oscilloscope or stored for later computer analysis.

Comparable experiments in olfaction commonly involve recording the slow voltage shifts from the surface of the olfactory organ (epithelium) using half-cell calomel electrodes, or neural activity from the olfactory receptor cells (axons) or from the olfactory bulb using platinum-iridium wire electrodes or micropipettes. Galvanic gustometry, originally developed by Krarup in 1958, is said to have value in clinical assessment of the taste function. The method is based upon the complex sensation that is elicited when a galvanic current is applied to the tongue.

Chemosensory Evaluation and Detection: The remarkable olfactory powers of some animals long have been exploited in odor detection. For example, bloodhounds have been used to detect leaks in gas lines in a plant making ethyl mercaptan and are reported to be more efficient than a team of men using conventional leak detectors. Dogs also have been used extensively in the detection of buried, concealed, and distant objects, such as land mines, booby traps, and personnel buried under snow or hidden in tunnels. The ability of dogs to follow tracks laid down by man or other animals seems to depend mainly on odor cues. In some parts of France, pigs are used to locate truffles in preference to truffle hounds, since pigs are thought to have keener olfactory sensitivity.

Laboratory techniques for assessing chemosensitivity in such animals as rats, dogs, rabbits, opossums, and blowflies have been developed.

Most chemosensory evaluation is by human subjects who form a testing panel—more costly than comparable instrumental methods in most cases. The four basic methods for obtaining estimates of taste or odor intensity or quality are (1) magnitude estimation, (2) ratio estimation, (3) rating by standards, and (4) sorting into successive intervals. In magnitude estimating, the degree of subjective response is related to a standard stimulus. The subject then is asked to assign a number to each successively presented stimulus—proportional to his subjective response to the first stimulus.

In ratio estimating, a pair of stimuli always is presented, and the subject reports the ratio of his subjective responses. In rating by standards, a series of standards is set up by the subject. These standards are scattered over the range of stimuli being measured. The standards can be assigned numbers and thus serve as fixed points. The subject is given an item to rate. He does this by finding the pair of standards which bracket his response. In sorting items into successive intervals, the subject also is presented with a set of standards. The successive pairs of items form the successive intervals along the scale of the characteristic. However, no numbers are assigned to the standard items. The subject's task is to place the item he is rating in the interval that separates those intervals to which his subjective reactions are greater—from those intervals to which his subjective reactions are weaker.

In addition to the foregoing basic methods which involve presenting the subject one item at a time, a number of methods have been devised in which the subject has to react to two net items at a time—methods of paired comparisons.

Correlation between Instrumental and Biological Measurement: Physical instrumentation does not allow an adequate evaluation of flavor quality. Conversely, the psychophysical approach is expensive and the data are highly variable. An ideal combination would be that which combined the objectivity, reliability, and low running cost of an instrument with the selective responsiveness to those aspects of quality of interest to man. A step toward achieving this goal has been made by Prof. E. von Sydow of the Swedish Institute for Food Preservation Research (SIK), Goteborg, Sweden, and his colleagues. Fractions of a complex odorant mixture eluted from a gas chromatograph are delivered to the olfactory epithelium of a frog. The slow voltage shift which this elicits in the epithelium is recorded electrophysiologically. Better correlations are found between the psychophysical and electrophysiological data than between the psychophysical and gas-chromatographic data. Thus, the frog epithelium may prove to be a useful biological detector in gas chromatography of odorous volatiles.

REFERENCES

1. American Society for Testing and Materials: "Basic Principles of Sensory Evaluation," 1968; "Manual on Sensory Testing Methods," 1968; "Correlation of Subjective-Objective Methods in the Study of Odors and Taste," Philadelphia, 1968.
2. Amerine, M. A., R. M. Pangborn, and E. B. Roessler: "Principles of Sensory Evaluation of Food," Academic Press, Inc., New York, 1965.
3. Johnston, J. W.: Quantification of Olfactory Stimuli, in C. C. Brown, (ed.), "Methods in Psychophysiology," pp. 192–220, The Williams & Wilkins Company, Baltimore, 1967.
4. Kuehner, K. L. (ed.): Recent Advances in Odor: Theory Measurement, and Control, *Ann. N.Y. Acad. Sci.,* pp. 116, 357–746, 1964.
5. Summer, W.: "Methods of Air Deodorization," Elsevier Publishing Company, Amsterdam, 1963.
6. Tanyolac, N. N. (ed.): "Theories of Odor and Odor Measurement," Technivision, London, 1968.
7. Teranishi, R.: Chemical and Instrumental Methods of Flavor Analysis, *J. Diary Sci.,* vol. 52, pp. 816–822, 1969.

—David G. Moulton,
Monell Chemical Senses Center,
University of Pennsylvania

Telemetry The transmission of instrument and control readings from one location to the next or to multiple locations. The transmission may be accomplished by means of electric solid conductors, radio and microwave, or optically or pneumatically. See **Data communications; Transmission, pneumatic;** and the **Classified Index.**

REFERENCES

Gruenberg, Elliot L. (ed.): "Handbook of Telemetry and Remote Control," McGraw-Hill Book Company, New York, 1967.
Proc. 1967 Natl. Telemetering Conf., ISA Publ. NTC67, San Francisco, Calif.
Proc. 1966 Natl. Telemetering Conf., ISA Publ. NTC66, Boston, Mass.
Proc. 1965 Natl. Telemetering Conf., ISA Publ. NTC66, Houston, Tex.
Proc. 1964 Natl. Telemetering Conf., ISA Publ. NTC64, Los Angeles, Calif.

Telephone Switch (See **Crossbar switches.**)

Telescope Generally, an optical instrument designed to assist the human eye or camera in viewing or photographing objects that are at a distance from the observer. Also see **Radio telescope.**

Optical telescopes are of two general types: (1) *refracting telescopes,* where light is collected by means of an object glass or lens and via other lenses in a total lens system forming an image of the observed object; and (2) *reflecting telescopes,* where the light is collected by means of a mirror. See **Astronomical instrumentation; Solar telescope; Telescope, reflecting;** and **Telescope, refracting.**

Optical Performance of Telescopes: Generally speaking, the optical quality of a telescope is expressed in terms of the maximum allowable wavefront error at the detector and an upper limit to the wavefront-error slope over a characteristic scale length. In the case of a single optical component, the former value defines the error of figure while the latter value defines the smoothness of figure. Diffraction-limited instruments have overall wavefront errors not exceeding one-quarter of a wavelength of light (Rayleigh limit) and small-scale wavefront error slopes not exceeding the theoretical resolving power of the instrument. The theoretical resolving power of a telescope expressed in seconds of arc is given by $4.5/A$, where A is the aperture in inches. The image diameter of a point source of light produced by a diffraction-limited objective is very nearly that predicted for a perfect objective. However, the light distribution within the image sustains, at the Rayleigh limit, a reduction in contrast in the sense that some additional 20 percent of the energy in the Airy disk is transferred to the surrounding diffraction rings. Thus, while a diffraction-limited instrument will achieve the theoretical resolving power in the case of equally bright pairs of stars, this is no longer true when the test pair consists of a bright and faint component.

Effects of Atmosphere: Under normal conditions of use, a large and perfect telescope seldom would achieve theoretical resolution because of the degrading effects of the earth's atmosphere. Atmospheric turbulence brings about image motion of stars as seen through small apertures and as greatly enlarged image diameters when stars are observed with large apertures. A typical image diameter observed in average atmospheric turbulence is about 1 second of arc. On very rare occasions, image diameters of one or two-tenths of a second of arc are observed near the zenith at well-chosen astronomical sites. Even under such conditions, direct photography further enlarges the images because of the diffusion of light in the photographic emulsion. Thus, photographic images of stars have a minimum diameter of about 30 microns no matter how small the optical image. Therefore, the resolving power of large diffraction-limited visual telescopes is limited by the atmosphere and the emulsion. Thus, at a given aperture, considerably more can be detected visually in the observation of planetary and lunar surfaces than can be photographed. For similar reasons, very close double stars are observed visually with the help of a filar micrometer.

The optical degradation due to the atmosphere becomes increasingly serious with aperture size since optical resolution increases with aperture for diffraction-limited instruments. In the case of large apertures, the optical specifications therefore are often expressed in terms of percentage of star energy collected within a disk of specified angular diameter, generally a large fraction of the stellar disk produced in the presence of atmospheric turbulence—the so-called seeing-disk diameter. However, the masking effect of atmospheric turbulence should not be construed as a justification for the relaxation of optical tolerances in large-aperture systems. Rather the very great difficulties in the manufacture of diffraction-limited telescopes of large aperture and the impossibility of achieving Rayleigh-limited reflectors of very large aperture are the principal reasons why such instruments cannot attain the resolution implicit in their apertures. It may be shown that in spite of the degrading effects of the atmosphere, high-quality instruments consistently outperform instruments of lower quality, particularly from the point of view of contrast rendition.

Astronomers therefore insist that large-aperture reflectors be as optically perfect as the state of the polishing art allows. The degree of surface perfection achieved in the optical shop is ultimately dependent on the thermal stability of the piece being worked, and the introduction of zero-expansion glass will therefore make possible more perfect mirrors through the virtual elimination of thermal ghosts. The space-borne telescope is in principle capable of the theoretical resolution given by Rayleigh's rule (it remains, however, detector-limited), and this is an important factor in attempts to improve the state of the polishing art. Optical systems for space application with resulting wavefront errors well below the diffraction limit are now called for in order to increase contrast at the theoretical resolution. This requirement implies ever more stable mechanical mounts, since the focal depth of an optical system diminishes to zero as the optical quality approaches perfection beyond the diffraction limit.

Telescope Mounts: The telescope mount must be rigid to eliminate jitter and provide an accurate and smooth drive for tracking celestial objects.

The essential feature of astronomical telescope mounts is the use of a polar axis (parallel to the earth's axis) to permit tracking of stars by driving a single axis at the sidereal rate. Additional correction of the drive is required to correct for refraction (which changes the apparent elevation of a star) and for inaccuracies in the drive, bearings, and alignment of the telescope.

There are many forms of mounts including the fork mount, where the telescope tube is suspended between two bearings at the ends of a fork; the German cross-axis mount, where the tube is offset from the polar axis; and the English mount, with an offset tube mounted between two bearings supporting the polar-axis shaft.

The offset-tube designs require counterbalancing about the polar axis but are more convenient for the Coudé focus. For very large telescopes, the yoke mount has been used. An altitude-azimuth type of mount could be used. Correct motion would be obtained by computer control of drives on both axes, plus rotation of the focal plane.

The telescope is driven about the polar axis at the sidereal rate (15 arc seconds per sidereal second time), usually by a precision worm and wheel. The rate was determined by a flyball governor driven by a falling weight on early telescopes but is more often set by a 60-Hz supply or a crystal oscillator on present-day telescopes. The oscillator provides a convenient means for altering the rate to compensate for atmospheric refraction or for lunar and planetary tracking. A fast slewing motion is normally supplied as well as setting and guiding motions that can be superimposed on the sidereal motion.

The declination axis is normally equipped with slewing, setting, and guiding motions. Guiding of most telescopes is done by manual control from visual observation, preferably through the main optical system but often through a separate guide telescope. Some telescopes have been equipped with automatic star trackers.

REFERENCES

Kuiper, Gerard P. (ed.): "Telescopes," The University of Chicago Press, Chicago, 1960.

Mickzaika, G. R., and W. M. Sinton: "Tools of the Astronomer," Harvard University Press, Cambridge, Mass., 1961.

Hiltner, W. A. (ed.): "Astronomical Techniques," vol. II of "Stars and Stellar Systems," The University of Chicago Press, Chicago, 1962.

—Joost H. Kiewiet De Jonge, *University of Pittsburgh,* and Lloyd M. Geertz, *Owens-Illinois*

Telescope, Orbiting and High-altitude The atmosphere has always provided a severe limitation to the astronomer. It degrades the image by turbulence and completely absorbs some wavelength bands. While earth-mounted telescopes must accept these limitations, observations have been made from above the atmosphere. Considerable data, particularly solar spectroscopy, have been obtained in the short-wavelength regions by instrumentation mounted on rockets. For more prolonged observations from above much of the atmosphere, balloons have been used. Planetary and solar spectra have been obtained by this method. High-altitude aircraft have been equipped for observation also. The ultimate platform for atmosphere-free observation is an orbiting satellite. Several orbiting telescopes have obtained considerable data, particularly in the ultraviolet regions that are not observable from earth.

The problems of making observations by any of these methods became complex. Generally, they include acquisition of the target, tracking of the target, thermal control, data recovery, and a myriad of detailed design problems.

The complexity of an orbiting experiment can be realized by considering the essential steps in making the first observations. The telescope and sensor must be thoroughly tested prior to being placed in orbit, including verification of the thermal design, the functioning of the sensors, etc. The experiment is then prepared for orbit. It must first survive the launch phase, which may require special packaging. Once in orbit, it is usually necessary for some time to pass to allow thermal stabilization, although with the package passing in and out of the earth's shadow, the stable condition involves transient inputs. The change in configuration from the launch phase to the operational condition must be performed remotely.

When it is time for the first viewing, the telescope must be pointed roughly in the right direction, usually by reference to the mounting spacecraft and its attitude-control system. When the sunshield is opened, the target must be acquired and tracked by automatic star trackers. The final fine guiding is preferably done through the main telescope. Any calibration of the sensor that may be required is performed, and the actual observation may then take place. Data are recorded electronically or photographically for later transmission to the ground.

All these steps must be preplanned and controlled remotely from the ground as must the sequential observations that form the entire observational program. Viewing by this method is necessarily extremely expensive, and the experiments performed by satellite telescopes are those which cannot be performed from earth (see Fig. T-7).

The role of the astronaut in astronomy has been one of direct visual observation and the use of small telescopes from spacecraft designed pri-

Fig. T-7. Orbiting astronomical observatory. *(Grumman Aircraft.)*

marily to carry men. The use of astronauts as direct observers with larger telescopes does not appear practical, although their use does seem practical for data retrieval and servicing a large, orbiting instrument.

—Joost H. Kiewiet de Jonge, *University of Pittsburgh,* and Lloyd M. Geertz, *Owens-Illinois*

Telescope, Reflecting The basic forms of reflecting telescopes are shown in Fig. T-8, all of which make use of a paraboloidal primary mirror. There are many variations, particularly of the Cassegrain form, that use slightly different surfaces, such as the Ritchey-Chretien, which has a larger coma-free field.

Reflectors provide no basic spectral limitation and provide the greatest aperture for a given investment, especially when the dome costs are considered. Reflectors have a central obscuration of a secondary mirror which reduces image contrast.

By virtue of its potentially large aperture and compact form, the reflector has become the most widely used basic observatory instrument. The largest reflectors are Newtonian or Cassegrain systems—measuring up to 200 in. in aperture. These instruments have parabolic primaries and, therefore, severely limited fields at prime focus. Direct photography of deep-sky objects is practical only at the relatively small aperture ratio available at prime focus. To improve the field at prime focus, large reflectors are equipped with Ross-type correctors placed in the converging beam. These correctors suppress coma over a widened field, but astigmatism ultimately limits the acceptable field.

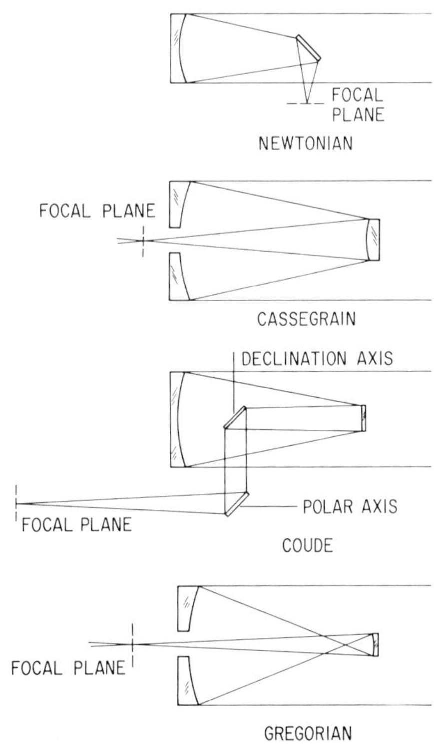

Fig. T-8. Common forms of reflecting telescopes.

The capacity of reflectors to operate in several modes (at several effective aperture ratios) has made them amenable to many specialized accessories. Thus, reflectors basically are multipurpose instruments. The aplanatic Ritchey-Chretien compound reflector has gained in stature as the ideal multipurpose reflector. The instrument is better adapted to prime-focus use than the classic Cassegrain system because in the Ritchey-Chretien instrument it is possible to suppress both coma and astigmatism over a prime-focus field by means of a suitably designed correcting system. When operated in the compound mode, the Ritchey-Chretien telescope delivers a relatively wide coma-free field which, in conjunction with a simple field flatterner, also is free of astigmatic images, although the image diameter remains a function of field angle. In this respect, the Ritchey-Chretien telescope approaches the astrograph as an instrument suitable for astrometric observation. The remaining on-axis modes at the secondary focus and Coudé focus are provided by the introduction of appropriately figured auxiliary secondary mirrors.

Giant Reflectors: As compared with refractors, reflectors are relatively unstable from a mechanical as well as thermal point of view. Developments in both the theory and application of mirror support technology and the introduction of near-zero-expansion-coefficient glass have greatly improved the stability of reflectors. Improved tube design to minimize the effects of tube flexure on optical performance also has contributed to the increasing primacy of reflectors among astronomical telescopes. Since the thermal-expansion coefficient of glass becomes increasingly critical with aperture, the availability of sufficiently large zero-expansion glass blanks, together with the aforementioned developments, will justify the consideration of new reflectors with apertures well in excess of existing telescopes. Large reflecting telescopes are installed at Mt. Palomar, California (200-in.); Mt. Hamilton, California (120-in.); Mt. Wilson, California (100-in.); Mt. Locke, Texas (82-in.); Kitt Peak, Arizona (84-in.); and Observatoire de Haute Provence, France (77-in.). Construction of a number of very large reflecting telescopes was begun in the period of 1965 to 1968, including a 240-in. instrument on an altitude-azimuth-type mounting to be installed in the U.S.S.R.; 150-in. telescopes for Australia, Chile, and Arizona; and 140-in. instruments for France and Chile.

Also see **Astronomical instrumentation; Schmidt camera; Solar telescope; Telescope;** and **Telescope, refracting.**

—Joost H. Kiewiet de Jonge, *University of Pittsburgh,* and Lloyd M. Geertz, *Owens-Illinois*

Telescope, Refracting The refracting telescope uses refracting lenses to bring the light to a focus, as seen in Fig. T-9. They are typically enclosed and are usually physically longer for the same effective focal length than are reflecting telescopes.

The most common refracting telescopes are doublets, although the larger fields for astrographs can be covered using triplets. Refractors have been preferred for astrometric measurements and for experiments where scattered light must be held to a minimum. They are limited in spectral range, and the secondary spectrum is often the limiting aberration. A refracting telescope will be color-corrected for a specific spectral region. Thus, a photographic refractor will not be corrected for visual use and vice versa. The largest visual refractor is a 40-in. aperture, and there are few refractors over 20-in. aperture, the limitation being both the availability of optical glass and the gravity sag of the lens.

Large refracting telescopes are located at Williams Bay, Wisconsin (40-in. aperture, visual); Mt. Hamilton, California (36-in., visual); Meudon, France (32.7-in, visual); Potsdam, Germany

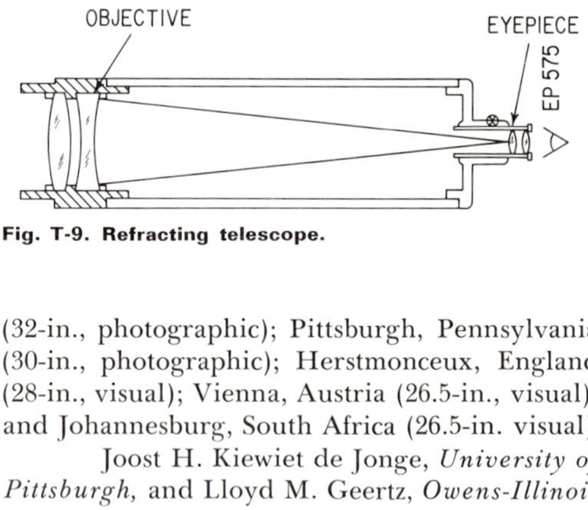

Fig. T-9. Refracting telescope.

(32-in., photographic); Pittsburgh, Pennsylvania (30-in., photographic); Herstmonceux, England (28-in., visual); Vienna, Austria (26.5-in., visual); and Johannesburg, South Africa (26.5-in. visual).

Joost H. Kiewiet de Jonge, *University of Pittsburgh,* and Lloyd M. Geertz, *Owens-Illinois*

Television A system for converting a succession of visual images into corresponding electrial signals and transmitting these signals by radio or over wires to distant receivers at which the signals can be used to reproduce the original images.

Closed-circuit television is any application of television that does not involve broadcasting for public viewing. Theater television and industrial television are examples. The presentations can be seen only on specified receivers connected to the television camera by circuits that may or may not include microwave relays and coaxial cables. Closed-circuit television is used selectively in connection with industrial and scientific instrumentation and control systems. Remote reading and monitoring is a common application—as, for example, the remote manual control of certain kinds of materials-handling operations—loading and unloading with large cranes where the operator can get a better view of what is occurring—where hazardous manipulations may be required; and in scientific teaching situations as, for example, where a group of medical students may watch an operation very close up that would be impossible for large numbers to view in situ because of space limitations. Also see **Cathode-ray tubes; Display tubes;** and **Display-command operator/machine interface.**

Television Oscilloscope (See **Oscilloscope.**)

Temperature, Measurement of There are numerous instruments for the measurement of temperature, including (1) thermometers, liquid-in-glass and filled-system; (2) pyrometers, optical, total radiation, and thermocouple; (3) electrical resistance thermometers; (4) infrared detectors, thermistor; (5) pyrometric cones; and (6) temperature-sensitive materials. Consult these key words elsewhere in this volume. Principal temperature scales include Celsius, Fahrenheit, International, Kelvin, Rankine, and Réaumur scales. Consult these key words and also the alphabetical index.

REFERENCES

Brickwedde, F. G.: Basic Concepts, Standards, and Methods of Temperature Measurement, 4th Symposium on Temperature, Its Measurement and Control in Science and Industry, *ISA Publ.* TMP1, 1961.

Dahl, A. I.: Applied Methods and Instruments for Temperature Measurement, 4th Symposium on Temperature, Its Measurement and Control in Science and Industry, *ISA Publ.* TMP2, 1961.

Hardy, J. D.: Temperature Measurements in Biology and Medicine, 4th Symposium on Temperature, Its Measurement and Control in Science and Industry, *ISA Publ.* TMP3, 1961.

Temperature Coefficient Temperature coefficient is a measure of the change of a parameter as the result of a change in ambient temperature. It is generally expressed as a percentage change in the parameter per degree of change (percent/°F or percent/°C). Some of the parameters to which the term is applied include resistance, capacitance, inductance, voltage and current offset, mean accuracy, and delay time.

There are three common methods of interpreting temperature-coefficient specifications. These are illustrated in Fig. T-10. In the "box" method, the temperature-dependent performance is bounded by a box or series of boxes whose dimensions are determined as the temperature-coefficient specification times the maximum temperature range. In this interpretation, the limit of error is the same over the specified temperature range. In the "hourglass" interpretation, the

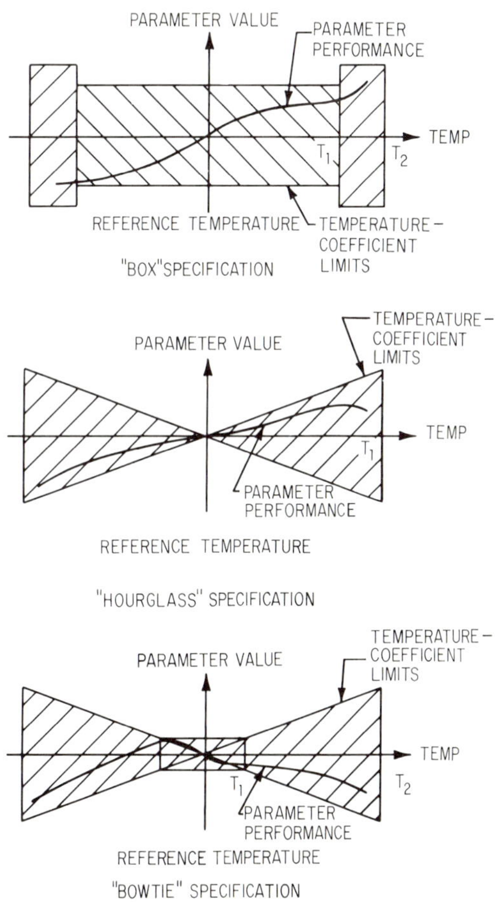

Fig. T-10. Temperature-coefficient specifications.

limits are proportional to the temperature. The "bowtie" specification is a combination of the box and hourglass interpretations. It specifies a nominal performance at reference conditions close to a reference temperature and a temperature coefficient for performance beyond these limits. Of the three interpretations, the most stringent is the hourglass method, followed by the bowtie and the box methods. In terms of utility, the bowtie is probably the most practical, since it is easier to measure and conforms more closely to environmental conditions encountered in the use of most equipment. It should be noted that the actual temperature coefficient over a small range may be considerably greater than the coefficient specified by any of these three methods.

—Thomas J. Harrison, *IBM Corporation*

Temperature Scales (See **Classified Index** under Temperature.)

Temperature-sensitive Materials (Indicators) Two broad categories of temperature indicators fall under this heading: (1) one relies on *fusion* at a precisely predetermined and reproducible temperature; (2) the other relies on a *color change* which takes place in the neighborhood of a stated temperature.

Fusible Temperature Indicators: These are widely used in research, development, and process temperature-monitoring applications and are characterized by (1) an accuracy of ±1 percent of the stated temperature rating, (2) simplicity of use, and (3) ease of interpretation. Indications of attained temperature are virtually independent of the duration of heating. Fusion-type indicators are available in the form of (1) crayons, (2) lacquer coatings, and (3) pellets. Each type is offered in a number of precalibrated, systematically spaced temperature ratings.

Crayons and liquid coatings are available in more than one hundred temperature ratings, covering the range from 100 to 2500°F. Over 90 percent of fusion-type indicators are of these two forms. A stroke or light touch with the appropriate crayon is all that is usually necessary for a determination. A dry (chalklike) mark is evidence that the labeled temperature rating of the crayon has not yet been reached. A wet (melted) mark indicates that the stated temperature rating has been attained. Accessibility of the surface (of an object whose temperature is to be determined) during measurement generally is a requirement for use of crayons, especially where temperatures above 500°F are encountered. When a workpiece is not accessible to the operator during the heating process or when it is desired to leave a mark on a very smooth or a soft surface, a liquid coating is most appropriate. The coating dries in a few seconds and leaves a dry mat mark on all surfaces, including highly polished metals, ceramics, glass, plastics, and cloth. The mark liquefies as soon as it is raised to its temperature rating. On subsequent cooling, the mark will resolidify, but it will not revert to its original dull-mat appearance but will remain glossy or crystalline as evidence of once having melted upon reaching the scheduled temperature.

The pellet form of fusible indicator is manufactured in two sizes, $\frac{7}{16}$ by $\frac{1}{8}$ in. and $\frac{1}{8}$ by $\frac{1}{8}$ in., and is used for applications which involve extended heating periods. The greater mass of the pellet favors unmistakable indication, particularly when observations must be made from a distance. Pellets are available in 114 temperature ratings—from 100 to 3200°F

Color-change Indicators: These are available in paint form and provide a scattered spectrum of indicators in the range of 175 to 1475° F. While a dramatic color change would appear to offer many advantages in a visual temperature indicator, many of the available color changes are subtle and some of them may be difficult to interpret by observers with poor color perception. The transition temperature of these indicators is markedly dependent upon the duration of heating and may occur over a range of as much as 150°F, dependent upon the "time at temperature." Color-change indicators find their greatest use in connection with temperature problems that involve relatively long heating periods under equilibrium conditions.

Temperature-indicating Adhesive Label: Temperature monitors in the form of self-adhesive labels, which exhibit a white to black change and offer an accuracy of ±1 percent, are finding increasing application. These monitors are available in a wide variety of single- and multiple-temperature indications within the range of 100 to 500°F. They consist essentially of one or more small, heat-sensitive disks, sealed under transparent, heat-resistant windows. The centers of the indicator disks or circles, as shown in Fig. T-11, turn irreversibly black at the temperature ratings shown on the label. In use, the labels are affixed to a surface and serve to register the temperature history of the workpiece. The exposed label can be made a part of a permanent

(a)

(b)

(c)

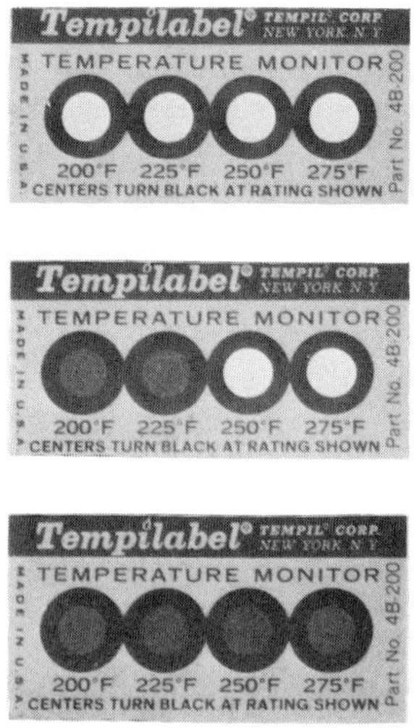

Fig. T-11. Self-adhesive, temperature-sensitive labels, each comprised of four disks with rated temperature identified under each disk. Centers of disks turn black upon reaching specified temperature. Labels shown cover range from 200 to 275°F in three 25° increments: (*a*) label has not reached lower temperature of 200°F; (*b*) label, in order, has reached 200 and 225°F but has not reached next value of 250°F; and (*c*) label has reached a temperature in excess of the top temperature rating of 275°F. *(Tempilabel.)*

record by removing it and affixing it to a production or service report.

—Elias Kimmel, *Tempil° Division, Big Three Industries, Inc.*

Temperature, Standard (See **Standards, measurement.**)

Tensile Testing [See **Testing (materials), destructive.**]

Tension, Measurement and Control of Filamentous materials such as wires, yarns and cable, magnetic tape, and sheets, and composite materials such as glass-epoxy filaments are subject to tension during processing and use. The tension that is put on papermakers' felts and the tension that is built up in rolls of paper or plastic during winding, printing, and other sheet-processing operations is an important manufacturing variable. The strength of materials under tension is determined by the structure of the materials. Thus a consideration of tension measurement and control reaches into the concept of strength of materials and the tools used to investigate that strength. This includes the measurement, recording, and control of dynamic and peak tensions that endanger control cables in naval vessels and aircraft, guy wires of antennas, and the many thousands of yarns that compose a fabric where a single yarn break can stop a loom or other high-production machine.

The measurement of tensile strength of yarns, wires, sheets, tapes, and similar materials can be determined by the use of quasi-stationary strength testers. See **Testing (materials), destructive.** These are heavy laboratory instruments. A material stretched under such conditions first is elongated and finally broken. In this manner, a stress-strain diagram, plotting tension vs. length, can be obtained. In such a graph, there is a first stage where the crimp is taken out of the material. Then there follows a region defined as that in which Hooke's law is valid. Within this range, stress is practically proportional to strain, hysteresis is a minimum, and upon removal of the strain, the material returns to its original length—undamaged. All processing, such as weaving or coil winding, must be within this range where Hooke's law is valid. Only within this region is elastic recovery and safety of manufacture that will neither deform nor change other physical aspects of the material in process possible. Once this range is transgressed, the yield point is reached. From this point on, the material is permanently deformed. A wire is drawn to a finer size. A yarn is stretch-spun, which not only increases its length and reduces its diameter but also changes its dye absorption, thus causing "shiners." Beyond this point, the tension may get to the point where separation of the micellar structure occurs, causing breakage.

The foregoing principles constitute an idealized condition. Complete elasticity and the strictly linear portion of a stress-strain curve are nonexistent. Mechanical hysteresis does take place in actuality. Our editorial concern here is primarily with tension of materials in process under the practical conditions of manufacture and anticipates certain possibilities which only a study in depth by the reader will resolve more fully.

Mechanical Shunt: In its simplest form, tension can be measured by fastening a yarn or wire to an upper support and hanging a weight on it. For a certain material of given cross section as well as chemical and geometrical state, a certain tension will be developed as a function of the weight applied. A mechanical shunt, shown in Fig. T-12, minimizes the stress added to the material. There are two reference rollers A and B and a sensing roller C located between. Pressure is applied to the sensing roller. The downward force F exerted upon the roller is $F = \frac{1}{2}t \cos \alpha$. Thus, deflection is a function of the tension t, which can be measured mechanically, electrically, or in other ways in terms of F. By adjusting the

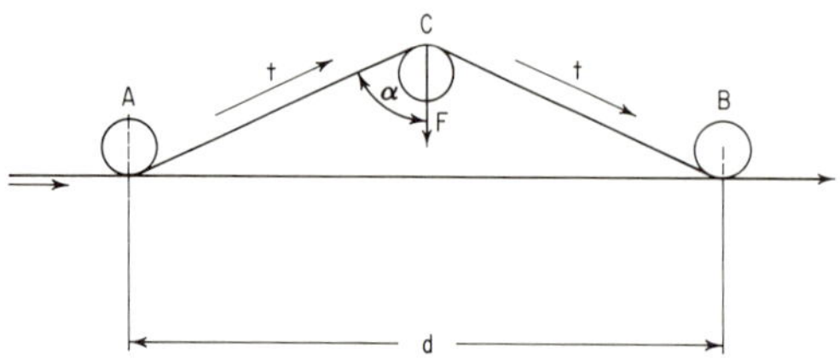

Fig. T-12. Operating principle of the mechanical shunt used for tension measurement.

distance d, the mechanical advantage of the instrument, and thus its sensitivity, can be varied. The force can be measured by having a calibrated spring yield and transfer the deflection of the center roller C onto a gear train and show tension on a dial. This becomes a trigger tension meter, as shown in Fig. T-13.

The advantage of such a mechanical tension meter is that it is small and comparatively inexpensive. The instrument can be clipped quickly to various parts of a yarn- or wire-processing operation. There never is, of course, a situation of

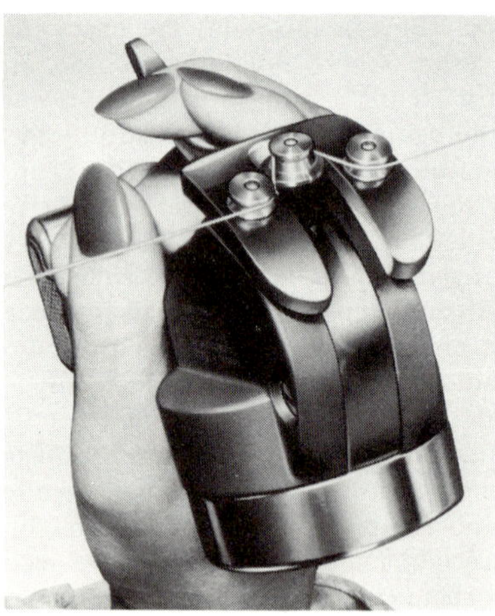

Fig. T-13. Trigger tension meter with quick trigger action. Useful in small spaces and for yarns and fine wires. *(Tensitron.)*

constant tension. But, from a practical viewpoint, the device determines "average tension" or rms by incorporating various damping means into tension meters. Damping means range from air damping to the highly efficient magnetic damping that is used in most instruments today.

The *trigger tension meter* is limited in its range by the size, which for practical reasons is determined by the outer reference rollers, which must be kept comparatively close together. Therefore, from the formula previously given, the range of a trigger-operated small tension meter is limited to the order of magnitude of a few thousand grams. To measure greater forces, there must be a higher mechanical advantage. This is ac-

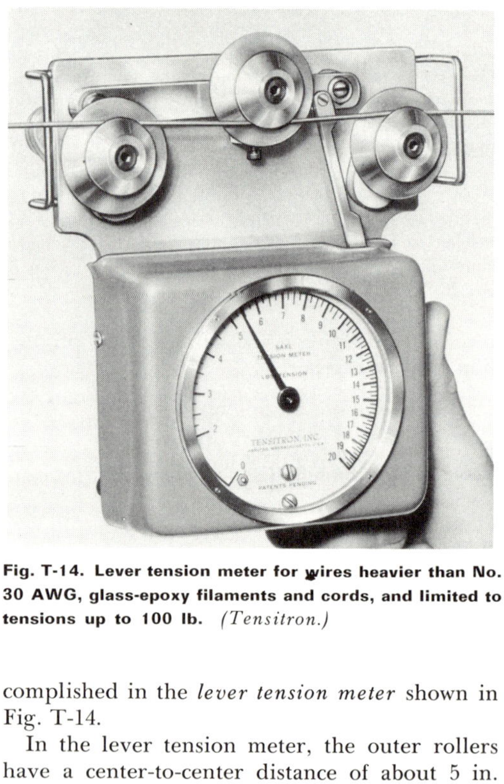

Fig. T-14. Lever tension meter for wires heavier than No. 30 AWG, glass-epoxy filaments and cords, and limited to tensions up to 100 lb. *(Tensitron.)*

complished in the *lever tension meter* shown in Fig. T-14.

In the lever tension meter, the outer rollers have a center-to-center distance of about 5 in. The rollers are not moved by a trigger but have an over-the-center movement whereby a lever puts the reference rollers in final test position beyond the 90° deflection, which in turn makes the wire or cord push them into place. Thus, regardless of counteracting forces, these meters operate under tensions several orders of magnitude higher than trigger tension meters. Lever ten-

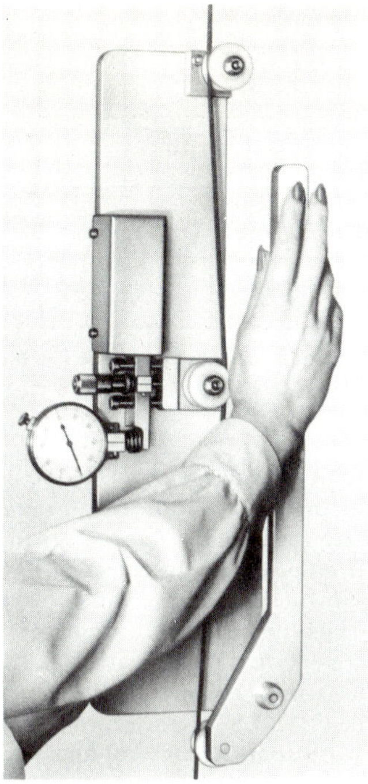

Fig. T-15. Cable tension meter for checking tension on guy wires and steering cables on aircraft and naval vessels. *(Tensitron.)*

sion meters are built to measure tensions up to 100 lb.

Tensions substantially beyond 100 lb require a different engineering approach. In the *cable tension meter,* the outer reference rollers are even farther apart. Also, a larger lever with higher mechanical advantage is used as a means of insertion. The cable exerts pressure upon the center roller, which is depressed according to the tension of the cable. The deflection of the center position is shown on a dial indicator (see Fig. T-15). A micrometer adjusts the relative position of the central sensor. This makes the same instrument adaptable to different cable diameters.

The foregoing mechanical instruments do not show critical tension changes—fluctuations which are faster than those which a mechanical gear train can follow. To measure quick tension peaks, the mass and internal friction of mechanical devices must be avoided. One solution is to place strain gages at the point of maximum stress concentration of a tension transducer.

A *three-element system* is shown in Fig. T-16. The transducer picks up the tension and converts it into an electronic signal. The signal is amplified and fed into a high-speed galvanometric recorder. With this arrangement, tension events can be studied as the high-speed phenomena that they truly are. Figure T-17, for instance, shows typical changes in tension as they occur during the start-up of a magnetic-tape recorder. Tension as recorded on a high-speed resolving device is shown to be in no way constant or uniform but to vary as a function of the material, its method of deposition, the structure of the capstan, reel, or bobbin, and many other parameters, such as mass and acceleration.

With electronic tension-measurement systems, advantage can be taken to implement more effective processing controls. Motors can be run

Fig. T-16. Electronic tension-measurement system. Transducer at left converts tension signal into electrical output. Amplifier is shown in middle. High-speed recorder is shown at right. *(Tensitron.)*

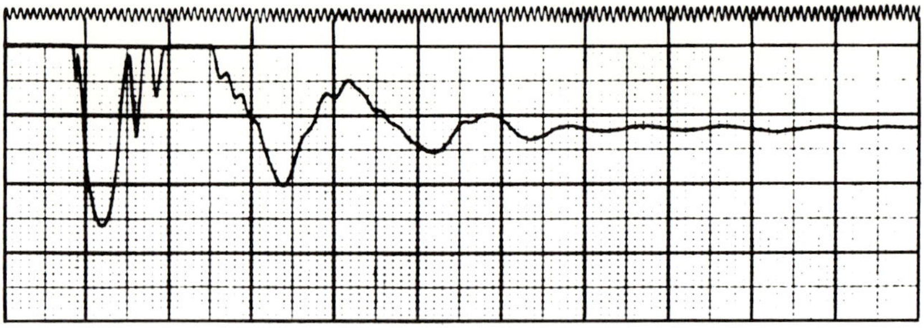

Fig. T-17. Record of tension of magnetic tape during start of a tape recorder, such as used in computers and for dictation. High-speed record clearly shows "hunting" of tension.

more quickly or slowly; dereeling brakes can be actuated so as to produce a more uniform tension of materials. Such systems are not limited to yarns and wires of modest cross section. Sensors can be located so that a cylinder can rotate between them, or a low-weight guiding tube can be attached between them. Thus, tension across the entire width of a sheet can be measured. A system in which web tension is measured and controlled is shown in Fig. T-18.

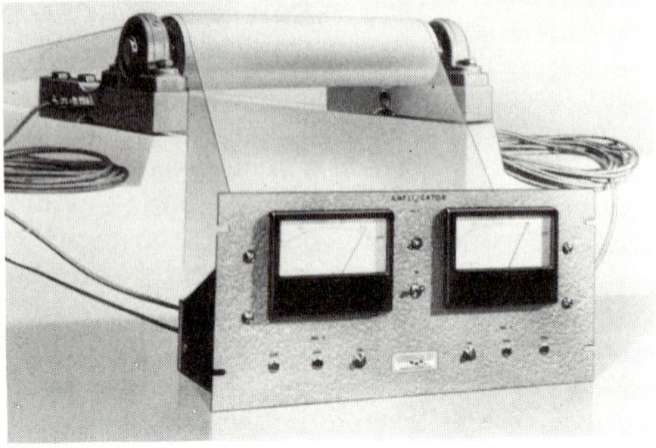

Fig. T-18. Sensing, indicating, and controlling system for sheet tension. *(Tensitron.)*

—Erwin J. Saxl, *Tensitron, Inc.*

Terminal, Data (See **Data communications.**)

Tester, Dead-weight (See **Dead-weight tester.**)

Testing (Materials), Destructive Materials-testing machines and systems commonly are divided into two classes: (1) nondestructive and (2) destructive. Also see **Nondestructive inspection.** Destructive testing usually stresses a sample of material to its breaking point to determine its mechanical properties—*tensile strength, elongation, impact resistance, hardness,* and others. Also see **Hardness, measurement of.** These tests are useful in quality control and the search for new materials for new applications.

Destructive testing systems are identified by the kinds of tests they are designed to perform and fall into the following classifications: (1) universal (Fig. T-19), (2) compression, (3) tensile, (4) creep, (5) fatigue, and (6) impact. A testing system is (1) an assembly designed to transmit force, motion, or energy from one mode or element to another in a functionally predetermined manner; and (2) a weighing or force-measuring unit. Thus, the primary elements of a testing machine are a load frame, a power unit, a force-measuring

Fig. T-19. Universal testing machine. *(Satec Systems.)*

device, appropriate controls, and the display of test results (including, in some cases, computers). The modular concept of testing-machine design permits the testing-system designer and user to regard each element in the system as a *variable* and provides more flexibility in machine capability. The concept of feedback, coupled with computer techniques, permits the automation of the testing system.

Load Frame: This must be strong and rigid. Less than adequate rigidity in the frame permits strain in the frame to appear in, and to distort, test results. More sensitive weighing systems and strain- or deflection-measuring devices have made more precise measurements possible—and have increased the requirements for rigidity in the load frame. The load frame of a universal testing machine is clearly shown in Fig. T-20.

Power Unit: Two requirements are mandatory. The power unit must be (1) sufficiently powerful to carry out the test and (2) compatible with the requirements of control. The power source usually is electrical—transmission to the load frame may be mechanical or hydraulic.

Force-measurement System: This must be independent of the power unit or any other source of outside signal that might distort test results. The signal must be capable of amplification without loss of accuracy and transmitted at command.

Controls: The modular concept requires that all elements of the control system be gathered in a single module, usually a control console, that permits controls to be added and operate the various elements in the system to perform a variety of tests. Electronic memories and logic circuitry make closed-loop automatic testing possible. In destructive testing, the significant output parameters usually are load (stress), strain, position, and time—or, in some cases, both position and time.

Universal Testing Machine: This name stems from the fact that many of the basic tests on materials can be performed within the load frame of the machine. Tests in tension or compression can be carried out without auxiliary equipment, and adapters will modify application of the load to perform other tests. The loading frame is of welded construction with a single test space (see Fig. T-20). The moving crosshead, which applies the load, is supported on two screw columns and guided by four fixed columns. This design virtually eliminates lateral movement and provides great stability, thus maintaining axial alignment over the entire power stroke.

Power units for universal testing machines may be hydraulic or screw-driven. Screw power usually is used where the stroke is greater than 8 to 12 in. and where the loads are less than 400,000 lb.

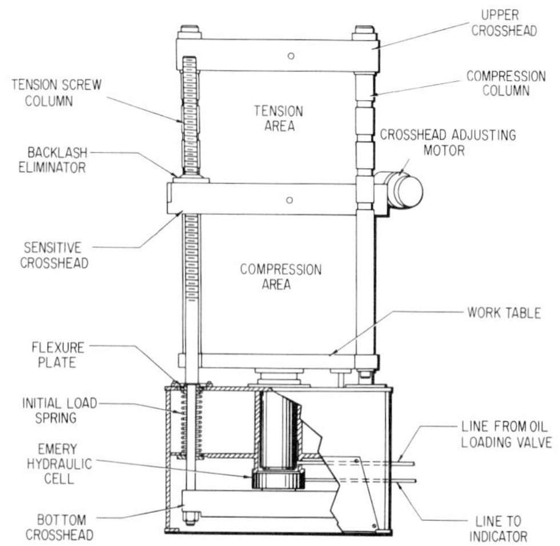

Fig. T-20. Load frame of a universal testing system. A pair of tension screw columns mounted at the ends of the bottom crosshead carry the sensitive crosshead. All motion is upward, pushing compression specimens against the sensitive crosshead or pulling tension specimens away from it. *(Satec Systems.)*

A dc motor provides power for the screws through gear reducers at the bottom of the columns, timing belts, and a clutch that are controlled electrically from the console. In combination with speed variations in the motor, the clutch provides the required range of testing speed. On larger machines, the choice of hydraulic and screw power will depend upon economics. Hydraulic systems, because of improvements in servovalves and the availability of variable-volume pumps, have gained acceptance. In a hydraulic system, power is delivered to the loading frame through a variable-volume pump (or a constant-volume pump and servovalve) to a packless ram in the loading frame. A feedback or error signal, operating through logic circuitry, compares the command signal with the actual ram speed and automatically corrects operation of the system.

Weighing or force-measurement systems fall into two classes: (1) electronic load cells and (2) hydraulic capsules. Both types can be isolated from the loading system. Quality electronic load cells offer spans (range of force measurement) of 100:1, immediate response, and high accuracy. Load cells have been used on machines of 1,000,000-lb capacity where loading is accomplished either hydraulically or with screw power.

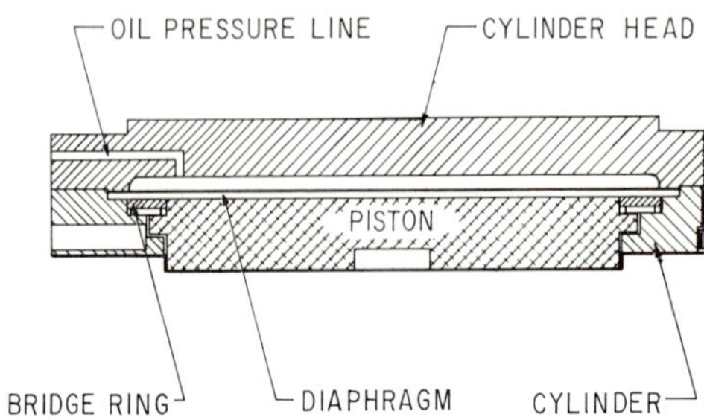

Fig. T-21. The Emery hydraulic capsule.

The hydraulic capsule shown in Fig. T-21 measures test loads with piston deflection of approximately 0.003 in. at full load. This minute deflection provides a quick response and maintains a high degree of accuracy. Even though very precise, the hydraulic capsule will stand severe abuse. Capsules, built in different sizes, provide load measurement for machines ranging from 5,000- to 12,000,000-lb capacity.

Programmed Universal Testing System: As contrasted with an open-loop conventional machine which requires two men and 26 steps to carry out a test, the programmed machine shown in Fig. T-22 permits one man to perform a standard tensile test in only nine steps. This system controls the rate of loading, analyzes test results, and prints results in tabular form. The operator simply installs the specimen, attaches the extensometer, and depresses the "start test" button. This action causes the machine to begin loading the specimen automatically at preyield-strength speed. At yield, the operator depresses the "yield trigger," which causes the load parameter in the digital display to be transferred to computer memory. The operator then removes the extensometer. After fracture (at ultimate tensile strength), the operator removes the specimen, measures the final diameter and the final length, and inserts these numbers into the computer. The operator then depresses the "return" button, which causes the machine to return to its initial

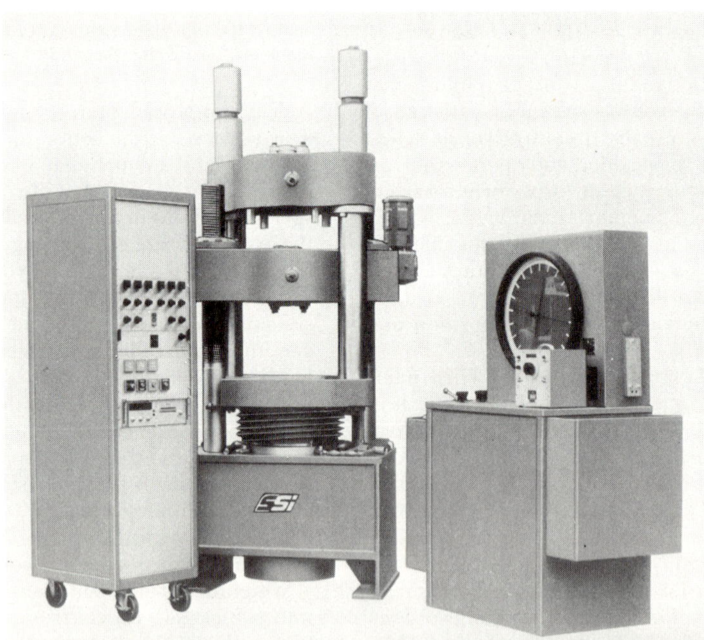

Fig. T-22. Universal testing system with associated computer, controls, and displays. The system controls the rate of loading, analyzes the test results, and prints results in tabular form. *(Satec Systems.)*

Fig. T-23. Hot-hardness tester. Computer and amplifier circuits provide a continuous readout of load and depth of indentation while controlling rate of load, amount of load, and time at load. Tests are conducted in a vacuum furnace. *(Satec Systems.)*

position and to calcuate and print yield load, yield stress, ultimate load, ultimate stress, percent elongation, and percent reduction in area. A system of this type reduces technician time and reduces computational errors.

Compression-testing Machine: This is similar to the universal testing system but simpler because the loading frame and weighing system are designed for one-way movement. Use is limited to laboratories testing materials, such as concrete, in which compressive strength is the significant property.

The *hot-hardness tester* shown in Fig. T-23 is an example of a compression machine and of the special machines that often are developed to meet new technological demands. In this system, no extraordinary demands are made on the load frame, since loads are limited to 1,000 g. The specimen is enclosed in a vacuum furnace that prevents oxidation of samples tested at temperatures up to 3200° F. The load is concentrated in an indenter that tests the hardness of the surface of the metal. The weighing system uses a proving ring that, by deflection, transforms force into voltage. See **Proving ring.** The system's computer and amplifier circuits provide a continuous readout of load and depth of indentation, while also controlling the rate of load, the amount of load, and time at load.

The computer system (see Fig. T-24) requires only three inputs: (1) *depth of indentation*—from three linear differential transformers that measure ram displacement; (2) *applied force*—sensed by a linear differential transformer mounted across the proving ring which stands between the indenter and the frame; and (3) *time*—which comes from a timer built into the system to time the test and to return the sample to its starting position. Feedback signals generated by ram movement and applied load provide the input for automatic computer control of the parameters being measured.

Tensile-testing Machine: This is a one-way version of a universal testing machine. A creep/stress-rupture machine provides an example of a specialized system for running tensile-type tests. *Creep* tests are designed to determine the ability of materials to function under load for long periods of time, often at high temperatures. Creep tests may last but a few minutes, or several years. Since creep-machine operation usually calls for long-term, constant loads, the loading system uses dead weights and multiplies the applied load through a lever arm and knife-edges. The machine used in most laboratories has a class I lever arm with vernier adjustment for precise ratios. Most machines are equipped with a leveling switch that holds the arm in a horizontal position during a test. A creep-testing machine is shown in Fig. T-25.

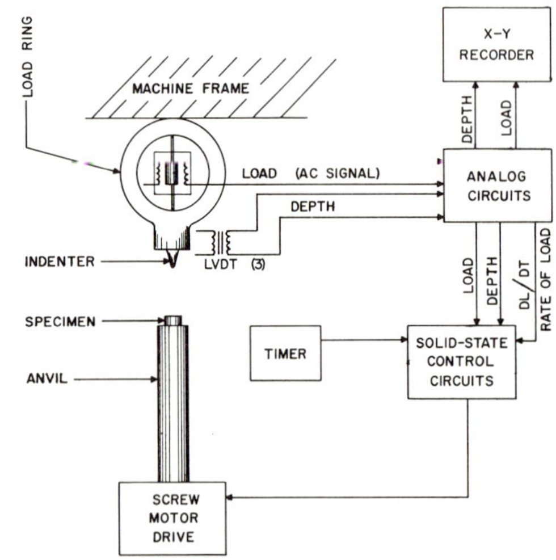

Fig. T-24. Major elements of hot-hardness tester. *(Satec Systems.)*

Where the test calls for *stress relaxation* or cyclic loading, a timer and an electronic load transducer can be added to the system. Since most creep tests search out the properties of metals at high temperatures, most standard machines carry a furnace that is capable of reaching temperatures up to 2200°F and is temperature-controlled.

While most testing systems measure loads in thousands of pounds per square inch, a *fiber-testing system* measures tensile strength in terms of grams per denier. The system includes (1) a vibrascope to determine denier and (2) a low-capacity tensile tester that records elongation and load. The vibrascope vibrates the fiber to its resonant frequency, a function of weight or denier. The fiber tensile tester draws a curve showing elongation at loads expressed in grams per denier on an *X-Y* recorder. The ability to change gain in the amplifier permits the machine to give readings of 0 to 5 or 0 to 10 g/denier full-scale. A unique circuit makes it unnecessary to zero the tester to fix the starting points of successive plots at the same level. The system automatically preloads the fiber to 0.1 g/denier to determine the fiber length before plotting. Since the loads involved are small, these measurements are taken directly from crosshead movement. The recorder pen draws the required curve and records the denier on the chart.

In contrast to most other systems, *fatigue*

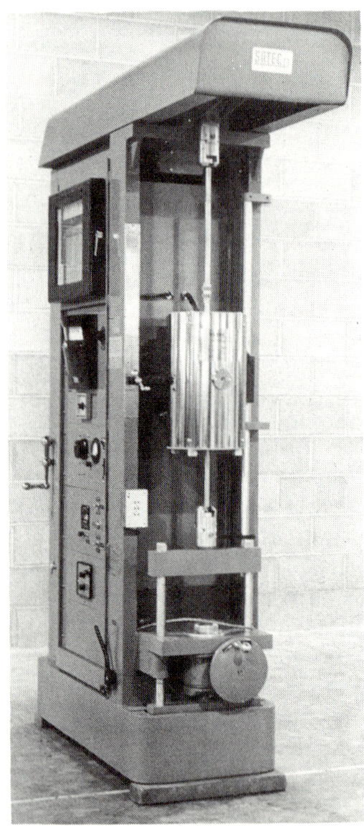

Fig. T-25. Creep-testing machine. Loading system uses dead weights and multiplies the applied load through a lever system. Sample is suspended in a high-temperature furnance in a specimen train that includes axial-alignment fixtures, studs, and adapters. *(Satec Systems.)*

testers are designed to test not only samples or specimens but fabricated parts under repeated or cyclic loads. Failures in machine parts often occur under repeated application of stresses well below those which the part could handle as a static load. It has been estimated that more than 80 percent of failures in machine parts are the result of fatigue.

Thus, a fatigue-testing system must be capable of producing completely or partially reversed loads or unidirectional reversed loads, and—to simulate actual operating conditions—it should be capable of superimposing any alternating load over a desired preload.

For many years, a constant-force inertia-compensated fatigue machine has been the standard in industrial fatigue testing, and this system still plays a major role in many laboratories. This machine operates at 1,800 Hz, and loading is accomplished by a preload spring. Obviously, the test specimen must be isolated from the machine so that loads are applied *only* to the specimen and *not* to the frame.

Preloading permits the operator to cycle the test from 0 up to machine capacity in compression, from 0 to capacity in tension, or cycles of equal amplitude in compression and tension up to half the capacity of the machine. Oscillator circuits control this fatigue system.

Probably the simplest of the destructive tests and the simplest of testing machines are those designed to determine the *impact* resistance of materials—plastics as well as metals. The most generally used impact machines are of the type in which a weighted pendulum furnishes the impact energy (see Fig. T-26).

In this test, available energy depends upon the effective mass of the pendulum and its velocity at impact. The impact point is at the bottom of the swing where kinetic energy equals potential energy at the top of the swing. Striking a specimen, the pendulum loses the amount of energy required to break the specimen and will *not* continue its swing to the full height from which it was released. The distance by which it falls short of full height provides an easy way to measure the energy absorbed in breaking the specimen. This test requires that the energy absorbed by the machine, through friction or impact, be a *known quantity* so that it can be eliminated from test results, expressed in foot-pounds.

The drop-weight tester, also used to determine nil-ductility transition temperatures, presents similar problems to the machine designer. In this system, a known weight drops from a known height to determine the number of foot-pounds required to fracture the specimen. Precision in test results requires a friction-free drop in a line normal to the specimen plane.

—Robert J. Burnup, *Satec Systems, Inc.*

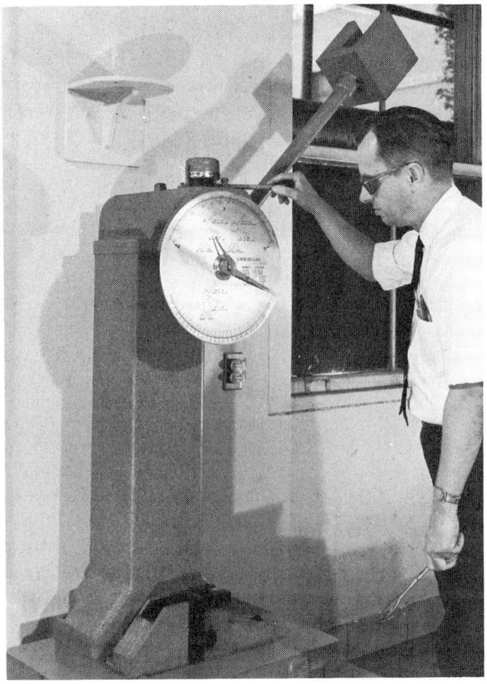

Fig. T-26. Impact tester makes use of effective mass of a pendulum and its velocity at impact. Impact point is at bottom of swing where kinetic energy equals potential energy at the top of the swing. Striking a specimen, the pendulum loses the amount of energy required to break the specimen. *(Satec Systems.)*

Theodolite A surveying instrument that has been adopted for use in metrology. A theodolite consists of a telescope in which rotation about the vertical axis is measurable. The telescope also can rotate about a horizontal axis in a true vertical plane. This rotation is capable of measurement. Theodolites normally are calibrated for (1) collimation value, that is, the centering of the crosshair to the telescope; (2) trunnion error, that is, the truth of rotation of the telescope in a true vertical plane; and (3) the centering of the axis of rotation. Accuracy of the circle, position of the plate bubble, and accuracy of the micrometers also are measurable and correctable features. Theodolites are used for alignment of large jigs and fixtures. Through triangulation, the instruments also can be used for length measurements.

Thermal-conductivity Gas Analyzer A common means of gas analysis involves measurement of the thermal conductivity of a sample as compared with a suitable reference. The detectors consist of temperature-sensitive devices such as bare or covered heated platinum, nickel, or other resistive spirals or thermistors mounted in appropriate cell cavities. Sample variations cause the measuring detector to beome hotter or colder than the reference detector. This temperature change results in resistance unbalance of the Wheatstone bridge which contains the detectors. This is measurable as an electrical voltage or current proportional to the net thermal-conductivity change of the sample.

The thermal-conductivity method is nonspecific; so the sample must be essentially binary, that is, composed of only two gases. The difference between conductivities of the two components must be sufficient to produce a significant change in detector temperature as the sample composition changes.

Some multicomponent streams may act as binary streams. Flue gas generally contains carbon dioxide, oxygen, water vapor, and nitrogen. If the sample is dried or saturated, the water vapor becomes constant. The nitrogen is constant; therefore, carbon dioxide and oxygen must interchange when compositions change, and thus the binary requirement is essentially met.

Another measurable multicomponent stream would be a ratioed mixture of natural gas in air. This stream may be treated as a binary stream because of the ratioing. A binary equivalent may also be prepared by absorbing or reacting out the measured component from a portion of sample for use as reference gas. This allows small changes in background composition to be compensated without serious loss of accuracy. The use of ultraviolet light to convert hydrogen to hydrogen chloride in a mixture of hydrogen in chlorine with trace impurities is an example.

Hydrogen and helium have very high thermal conductivities compared with air. Most other gases and vapors have conductivities below that of air. The difference between the thermal conductivity of the measured gas and its binary counterpart determines sensitivity. The reference determines electrical zero, allowing range suppression such as 90 to 100 percent, and zero-based ranges.

Thermal-conductivity detectors are used as detectors for gas chromatographs where a gas such as helium is used as the carrier gas. The output consists of time-separated series of binary mixtures of individual components of the sample in helium as separated in the chromatographic column.

—Lyle K. Barnes, *The Hays Corporation, a Milton Roy Company*

Thermal-conductivity Instruments These instruments generally fall into two categories: (1) absolute and (2) comparative. The names imply the method employed in each instrument type. In the absolute type, all quantities are measured directly; in the comparative type, the thermal conductivity of a specimen is determined by comparison with a known material in a series heat-flow arrangement. In the absolute instrument, a heat flux is produced, usually electrically, and caused to flow in a well-defined manner through a specimen. Guard heaters are employed to confine the heat flow to the selected path. Thermocouples or other suitable sensors, placed at predetermined locations within the sample, permit the temperature gradient to be established. With a knowledge of the heat-flow geometry, temperature gradient, and electrical dissipation, an absolute determination of thermal conductivity can be made. A transient-type instrument also is described briefly here.

Absolute-type Instruments: Three basic configurations are described here, namely: (1) the guarded hot plate, (2) the absolute rod, and (3) a radial-flow unit.

Guarded Hot Plate: As shown in Fig. T-27. this device consists of a flat two-zone heater in which the central-zone electrical dissipation is

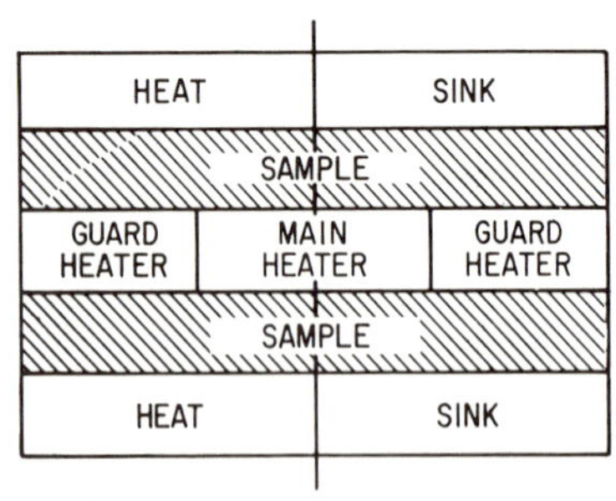

Fig. T-27. Schematic diagram of conventional guarded-hot-plate thermal-conductivity instrument.

measured and the outer zone is controlled to negate radial losses. The heater is placed between two flat specimens. Fluid-cooled heat sinks are placed against the outer-specimen surface. A unidirectional heat flux is produced equally in the two specimens, and temperature sensors located on the specimen surfaces measure the sample gradient. A guarded-hot-plate instrument constructed in accordance with ASTM Specification C 177-63 is shown in Fig. T-28.

Absolute Rod: A guard heater on one end and a fluid-cooled heat sink on the other will produce a flux along a rod specimen. Radial cylindrical guards may be used to negate radial losses. Temperature sensors spaced along the specimen measure the gradient.

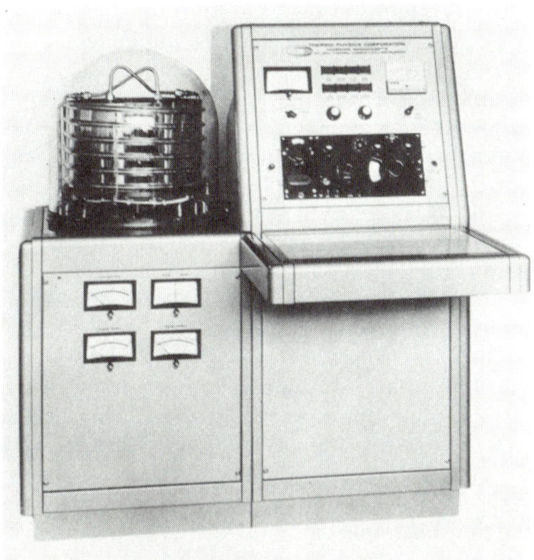

Fig. T-28. A guarded-hot-plate-type thermal-conductivity instrument. *(Thermo-Physics Corp.)*

Radial Flow: Heat generated electrically in a central element flows radially through a cylindrical specimen to a fluid-cooled external heat sink. The assembly either is sufficiently long or else utilizes end guarding to ensure true radial flow.

Comparative Instruments: An unknown specimen is located in a series heat-flow position with a known or standard material. Often, a standard is placed on each side. Radially guarding often is employed to ensure true unidirectional heat flow. The thermal conductivity is determined by comparing the temperature gradients through the unknown and known materials.

Transient Instruments: A specimen is subjected to a thermal pulse, and the resulting time-distance history permits the thermal diffusivity of the specimen to be determined. From the thermal diffusivity, density, and specific heat, the thermal conductivity can be calculated.

—James K. Sparrell, *Sparrell Engineering Research Corporation*

Thermal Well (See **Protecting tube, sensor.**)

Thermistor Detector (See **Infrared detector, thermistor;** and **Resistance thermometry.**)

Thermocouple A temperature-measuring device based on Seebeck's discovery that an electric current flows in a continuous circuit formed by wires of dissimilar metals if the junctions between the two wires are at different temperatures. A thermocouple is a commonly used industrial temperature-measuring element because it is simple and inexpensive and because combinations of thermocouple materials have been developed for use over wide temperature ranges. A simple thermocouple circuit is shown in Fig. T-29. The thermoelectric current I flows in the circuit when the junctions are at temperatures T_1 and T_2, respectively. If T_1 is the colder junction and current flows in the direction shown, A is considered to be thermoelectrically positive to B.

The Seebeck effect is a combination of two effects, namely, the Peltier and the Thompson effects. The Peltier effect, which is the larger, was discovered by Peltier in 1834 when he showed that heat is liberated or absorbed when a current flows across a junction between dissimilar metals. This is a reversible effect, measured in watts per ampere, and depends upon junction temperature. The Thompson effect, also reversible, relates current flow through a homogeneous material in which there is a temperature gradient to heat ab-

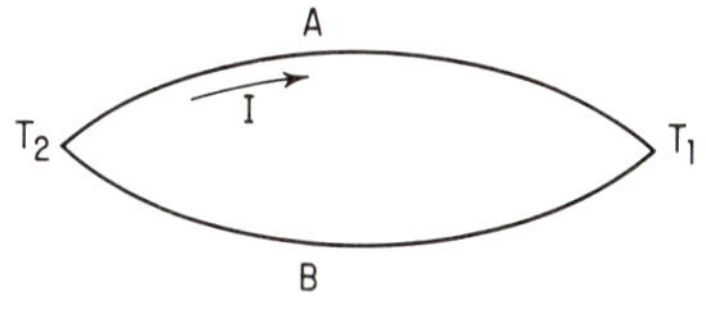

Fig. T-29. Simple thermocouple circuit. When temperature T_2 is higher than T_1, the material A is thermoelectrically positive with respect to B and the current flows as indicated by the arrow.

sorption or liberation. Although the Thompson voltage is small, this is the major reason why thermocouple temperature-emf curves are nonlinear.

Thermoelectric Laws: In the design and use of thermocouples, three laws are important:

1. *Law of the homogeneous circuit*—An electric current cannot be sustained in a circuit of a single homogeneous metal by the application of heat alone. The practical significance of this law is that in a thermocouple, constructed from two homogeneous materials, the emf is independent of the distribution of the temperature gradient but depends only on the temperature of the junctions.

2. *Law of intermediate metals*—If the temperature is uniform between a point M and a point N, in any thermoelectric circuit, the thermoelectric emf in the entire circuit is independent of any intermediate material between points M and N.

3. *Law of intermediate temperatures*—The thermal emf developed by any thermocouple of homogeneous metals with its junctions at temperatures T_1 and T_3 is the algebraic sum of the emf of the thermocouple with junctions at T_1 and T_2 and its emf when the junctions are at T_2 and T_3.

The law of intermediate metals allows connection of the measuring instrument with the thermocouple with leads of copper, as shown in Fig. T-30. So long as the temperatures at the junctions of the copper wires and the thermocouple wires are the same, the measuring device

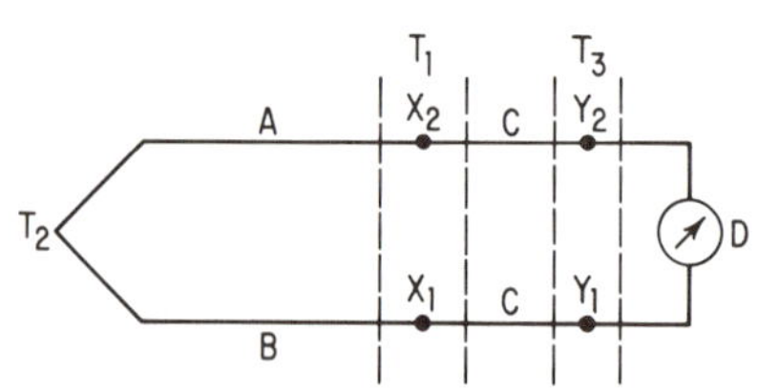

Fig. T-30. Diagram portrays how the law of intermediate metals makes possible use of nonthermocouple wires to connect the thermocouple to the measuring instrument. Thermocouple materials A and B can be connected to the instrument by use of connecting materials C and D. If temperatures at X_1 and X_2 are both at T_1 and if temperatures at Y_1 and Y_2 are both at T_3, the emf of the circuit will be independent of materials at C and D.

will have no effect on the output of the thermocouple. The law of intermediate temperatures allows one to determine the thermocouple emf when one junction is at 0°C and the other junction is at various temperatures—and then interpret in later application the emf of the thermocouple in terms of temperature when the cold junction is at some other (not 0°C) temperature.

Thermocouple Configurations: The simplest thermocouple system is a pair of insulated thermocouple wires joined to form a hot junction. For temperatures below 500° F, plastic insulations are available; in the range of 500 to 2000°F, materials such as asbestos, glass fiber, and quartz fiber are used; or ceramic insulators are slipped over bare wires to insulate them over the entire temperature range to 4000°F. The temperature limits of ceramic insulators are approximately the same as for ceramic protecting tubes (see Table T-1).

In most applications, a protecting tube or metallic well isolates the thermocouple from corrosive fluids and gases, maintains pressure in a process vessel, and protects the thermocouple from mechanical damage. A protecting tube also often serves the very practical function of supporting the thermocouple. Attached to the protecting tube or well is a connection head that contains terminals for making connection between the thermocouple and the extension wire to the measuring instrument. A typical assembly is shown in Fig. T-31.

Design Considerations: The selection of a thermocouple is a compromise among many factors, including (1) cost, (2) availability, (3) resistance to corrosion, (4) emf output of the wire at the operating-temperature range, and (5) speed of response. It also must be assumed, of course, that the thermocouple emf must increase continuously with increasing temperature over the range of use—and the melting points of the metals must be above the temperature of application.

Selection of thermocouple wire size always is a compromise of speed of response, mechanical strength, and resistance to corrosion or contamination. Small-diameter wires give high speed but break easily, corrode, kink, and readily develop inhomogeneities from mechanical working or contamination from materials that they contact.

The temperature-emf curves of common types of thermocouple wires for industrial use are shown in Fig. T-32.

Fabrication of Thermocouple Hot Junction: The hot junction may be joined by any means that will ensure good electrical continuity in use. Commonly, the two wires are twisted together and either welded or silver-soldered. The wires also may be butt-welded. Simple clamping of the wires together provides adequate connec-

TABLE T-1. *Common Thermocouple-protecting-tube Materials*

Material	Max. Operating Temp, °F	Remarks
Carbon steel	1000	For noncorrosive fluids
Wrought iron	1200	For noncorrosive fluids
Stainless steel	1800	Good resistance to corrosion
Nickel	1800	Excellent corrosion resistance for some materials
35% Ni, 15% Cr, 50% Fe	2000	Sulfurous or reducing atmospheres
80% Ni, 14% Cr	2200	Corrosion-resistant (except sulfur)
Fused silica	2300	Excellent thermal-shock resistance
Silica-alumina	3000	
Silicon carbide	3000	For continuous use in molten aluminum
Alumina, 99+%	3400	Fair thermal-shock resistance
Beryllia	4000	Excellent thermal-shock resistance. Fumes and powder toxic

tion for short-term use in clean atmospheres at low temperature.

Protecting Tubes: Selection of a thermocouple protecting tube is a compromise among cost, availability, speed of response, and mechanical and chemical characteristics. Several protecting-tube materials and their approximate upper-temperature limits are listed in Table T-1. Most thermocouples, except the platinum types, can be used successfully in metallic protecting tubes. Platinum is particularly susceptible to contamination by other metals. At high temperatures, even silica-bearing ceramic protecting tubes will poison platinum thermocouples. Silicon, reduced from the protecting tube, is absorbed by the platinum thermocouple wires. At temperatures of 1500°C and higher, it is essential to use very pure alumina insulators and protecting tubes for platinum-bearing thermocouples.

Swaged and Drawn Thermocouples: Many applications now use a swaged or drawn thermocouple construction in which the insulated thermocouple element and protecting sheath are a monolithic assembly. Thermocouple wires strung through alumina or magnesia insulators are inserted in a metallic protecting sheath. Swaging or drawing compacts the insulation and reduces the sheath to final size. Common hot-junction designs used in these thermocouples are illustrated in Fig. T-33.

Installation of Thermocouples: A thermocouple should be located as close as possible to the desired point of temperature control. To prevent conduction along the wires or to keep the protecting tube from affecting the measured temperature, the thermocouple should be immersed at least 10 diameters in the measured fluid. For measurement of the temperature of solids, the couple should be peened or otherwise firmly attached to the material. Running an insulated length of the thermocouple wire near the material will minimize conduction error and reduce the temperature gradients near the junction. Measuring the temperature of thin sections with wires of comparable diameter is extremely difficult.

A thermocouple that is measuring gas temperature in a duct whose walls are at a substantially higher or lower temperature receives or emits

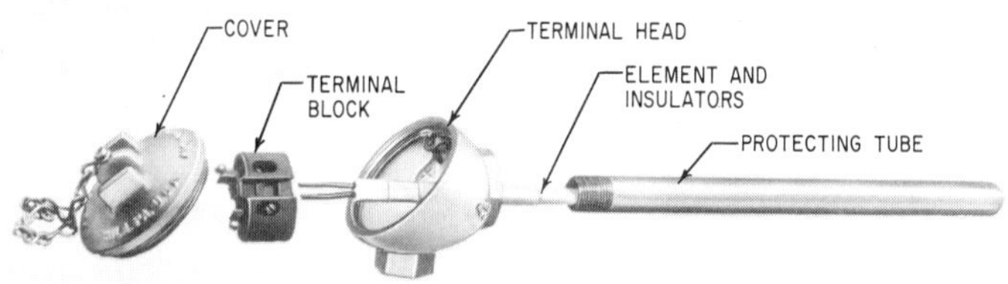

Fig. T-31. Typical industrial-thermocouple assembly.

radiation to the walls and thus substantially changes the apparent measured temperature. A radiation shield, in its simplest form a cylinder in the gas stream and completely surrounding the thermocouple junction, will attain a temperature close to that of the gas.

Reference Junctions: Either the reference junction must be maintained at a known constant temperature or automatic compensation of the thermocouple emf for changes in reference-junction temperature must be used. In the laboratory, an ice bath is the usual reference standard. For industrial applications, automatic compensation usually is provided for offsetting the effects of fluctuations in the reference-junction temperature. In millivoltmeters, a bimetallic spiral mechanically moves the pointer in response to ambient-temperature changes; or an emf is injected into the thermocouple circuit by a temperature-responsive element, such as a nickel resistor, a thermistor, or a diode. In potentiometric measurements, electrical compensation is used exclusively. In some systems, particularly in the

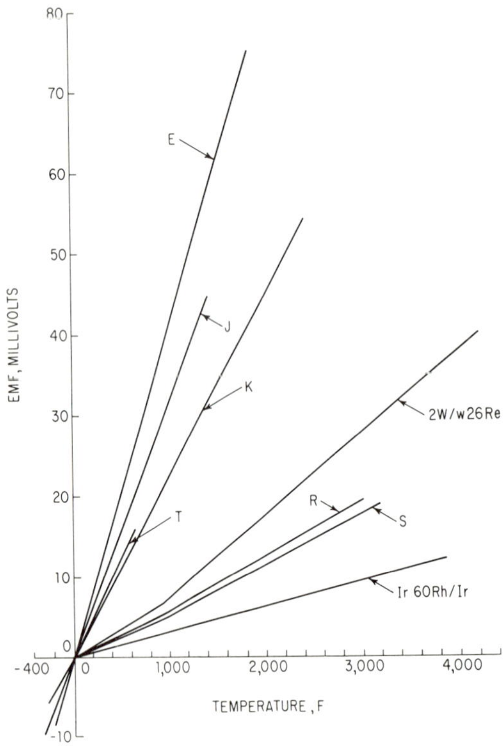

ISA type	Positive wire*	Negative wire*
E	Chromel	Constantan
K	Chromel	Alumel
J	Iron	Constantan
T	Copper	Constantan
R	Pt 13% rhodium	Platinum
S	Pt 10% rhodium	Platinum

*Other combinations of wire may match ISA calibration curves.

Fig. T-32. Temperature-emf curves of common types of thermocouple wires for industrial use.

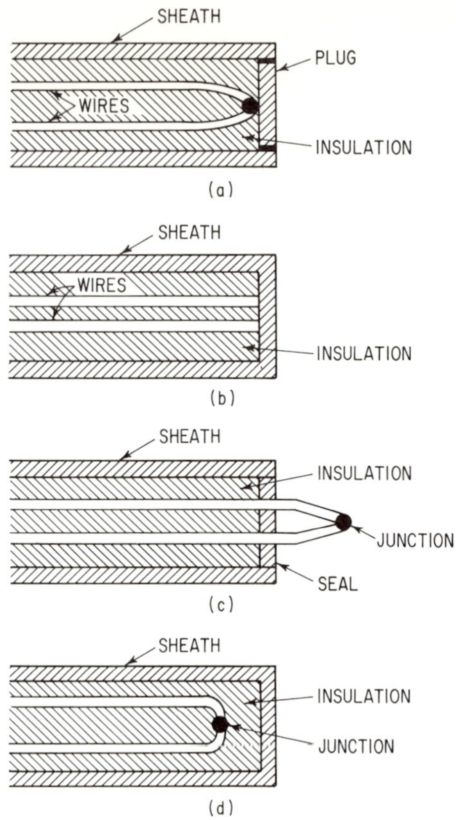

Fig. T-33. Types of junctions in swaged or drawn thermocouple elements. (*a*) Plug weld. Wires welded to make junction. Plug welded in tube end. Junction may or may not be grounded to sheath. (*b*) Integral or grounded junction. Junction, sheath, and closure formed in one operation. (*c*) Exposed. Sheath and insulation stripped. Junction is left exposed. Cement seals end of sheath. (*d*) Remote or insulated junction. Junction welded. Insulating material added before sheath closure is welded. Junction insulated from sheath.

laboratory, a controlled oven will be used to maintain a constant reference-junction temperature, See **Reference-junction compensation.**

Thermocouple Extension Wire: In most installations, the thermocouple and the instrument used to measure its emf are located some distance apart. It is uneconomical and unnecessary to run thermocouple wire from the process to the instrument measuring terminals. Typically, the difference in temperature between the instrument and the thermocouple connection head is less than 400° F. The connecting wires between the head and the instrument need be accurate only over this temperature range. Therefore, for most common thermocouples, there is extension wire which matches the standard thermocouple temperature-emf curves over a limited temperature range. These wires are supplied in several wire sizes so that a reasonable choice can be made among strength, cost, and wire resistance. With potentiometric measuring instruments the extension wire resistance is unimportant. Millivoltmeters, however, are calibrated for a specific total circuit resistance. Extension-wire resistance must be kept to a small percentage of the total circuit resistance so that changes in extension-wire resistance do not affect the indicated temperature significantly. Extension wires are available with many kinds of insulation. Common extension-wire insulations and applicable temperature ranges are given in Table T-2.

—Ernest C. Magison, *Honeywell Inc.*

TABLE T-2. *Common Thermocouple Extension-wire Insulations*

Material	Temp. Range, °F	Remarks
Polyvinyl over polyvinyl	0 to 221	Excellent general-purpose insulation
Lead sheath over polyvinyl. . . .	0 to 221	Highly corrosive locations
Buna S rubber	—65 to 140	For subzero applications
Fiberglass	To 400	Indoor use only
Asbestos over enamel	To 400	Indoor use only
Nylon	300	
Teflon	500	Excellent resistance to chemical attack

Thermocouple Vacuum Gage (See **Vacuum gaging.**)

Thermography Photography that uses radiation in the long-wavelength far-infrared region, emitted by objects at temperatures ranging from —170 to over 300°F. Also called thermal photography. An important technique used in various industrial and defense instrumentation systems.

Thermometer An instrument for determining temperature and consisting of a temperature-sensing means that is operatively connected to a measuring means to transduce the output of the sensor into an indication or record of the temperature. The range and characteristics of major types of thermometers are summarized in Table T-3 (page 680).

—Robert D. Thompson and George R. Leavitt, *Taylor Instrument Process Control Division, Sybron Corporation*

Thermometer, Filled-system The filled system is closed and consists of (1) a fluid; (2) a temperature-sensitive bulb; (3) an elastic element, such as a bourdon or bellows; and (4) a capillary connecting tube (see Fig. T-34). The elastic element moves or generates a force that is proportional to the temperature at the bulb end. All filled systems derive their useful signal from either (1) a pressure change that reflects the thermal conditions at the bulb or (2) a volume change that results from liquid expansion in the bulb.

Advantages of filled-system thermometers include: (1) no external power is required; (2) the system develops sufficient torque or force to operate a pointer, a recording pen, a controller mechanism, and in some instances a valve for process control; (3) simplicity of design and wide variety of construction forms reflect optimization of strength, corrosion resistance, response time, accuracy, and cost; (4) the system has long life and very low maintenance; (5) remote location is possible—up to 400 ft from the point of measurement; and (6) the system is safe in hazardous locations because fill materials used can be noncombustible and relatively inert chemically—and because of the absence of any electrical potential.

Liquid-expansion-type Thermal Systems: The temperature-sensitive signal from the expansion of liquid in the bulb is hydraulically transmitted to the elastic element through the capil-

TABLE T-3. *Ranges and Characteristics of Thermometers*

Instrument	Operating Principle	Range
Liquid-in-glass thermometer . . .	See **Thermometer, filled-system**	−200 to + 600°C −330 to +1110°F
Filled-system thermometer	See **Thermometer, filled-system**	−269 to + 649°C −452 to +1200°F
Thermocouple thermometer . . .	See **Thermocouple; Thermopile**	−190 to +2095°C −300 to +3800°F
Resistance thermometer.	See **Resistance thermometry**	−271 to + 660°C −456 to +1220°F
Thermistor thermometer	See **Infrared detector, thermistor**	−100 to + 400°C −150 to + 750°F
Radiation thermometer	See **Radiation pyrometer**	+ 38 to +4150°C +100 to +7500°F
Optical pyrometer	See **Optical pyrometer**	Above + 700°C Above +1300°F
Bimetallic thermometer	See **Thermostat**	−135 to + 425°F −200 to + 800°F
Thermosensitive materials.	See **Temperature-sensitive materials (Indicators)**	0 to +1370°C + 32 to +2500°F
Ultraviolet thermograph	Sensor detects change in luminescence under ultraviolet excitation. Observed visually or photographically	0 to + 200°C + 32 to + 400°F
Infrared thermograph	Sensor detects infrared radiation from a body. Photograph of body is compared with a standard of intensity	0 to + 100°C + 32 to + 212°F

lary tubing. The system is filled completely with a liquid, such as mercury—usually at an elevated pressure throughout the operating range. Liquid-expansion-filled systems are sensitive to temperature changes along the capillary and at the elastic element. A number of devices are used to negate errors which might result from this temperature effect. There are two major categories of compensation: (1) full compensation, designated as IA and VA systems by the Scientific Apparatus Makers Association; and (2) case compensation only, designated as IB and VB systems. Full compensation provides a means of reducing errors both in the case and along the capillary. Case compensation only is used when the volume of liquid in the capillary does not produce a significant error.

The error at the elastic element due to temperature change usually is reduced or eliminated with a compensating element that produces an equal and opposite force or motion. This element may be a matched bourdon spring or a bimetallic lever—with or without adjustments to minimize the temperature error. Errors resulting from temperature changes along the capillary may be reduced by running a parallel capillary to a second, opposing elastic element. Well suited to mercury actuation is the simple use of a low-expansion wire within part of the capillary bore. The wire is sized to produce a liquid-volume change that exactly matches the capillary metal-volume change when both are subjected to a temperature change. This procedure eliminates the need for a second elastic element.

Bulb and range characteristics for liquid-expansion systems are given in Table T-4. For a single elastic element, an infinite number of temperature ranges can be obtained by varying the bulb volume and the filling pressure. Bulb size varies inversely with span.

Step response of liquid-expansion systems varies from 2 to 20 s in well-stirred water and has a direct relationship to bulb size. These systems will respond to pressure changes on the bulb, but a typical error will be only 3 percent for a 1,000 psi change. The maximum difference in elevation between the bulb and the instrument case should not exceed design value. Barometric changes have no significant effect on performance. Process pressure changes usually are not reflected at the bulb because of the use of thermal wells.

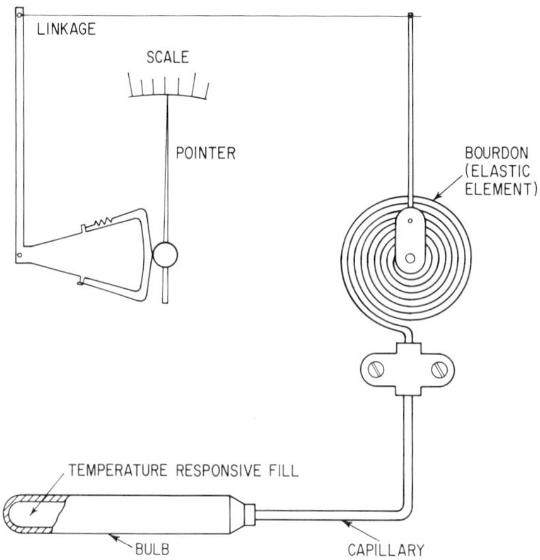

Fig. T-34. Essential elements of a filled-system thermometer.

Overrange capability usually is achieved by designing the elastic element to withstand at least one span in excess of the thermometer range—necessary because a liquid-expansion system has no cutoff point and will force liquid into the elastic element so long as the bulb temperature is increased.

Pressure-type Thermal Systems: These instruments are of two general classes: (1) gas-filled and (2) vapor-filled. Gas-filled systems operate on the principle that the absolute pressure is proportional to the absolute temperature. The vapor systems are partially filled with a pure volatile liquid so that one interface between the liquid and gas always occurs in the bulb. The pressure existing in the system is converted to a temperature indication from the relation of the vapor-pressure curve of the filling medium. Since the bulb size of gas-filled systems essentially is independent of range, the common practice for these systems is to provide a large volume ratio between the bulb and the balance of the system. This arrangement reduces capillary temperature errors and requires only that case compensation be provided.

Vapor-actuated systems respond only to the vapor pressure at the interface which must always exist in the bulb. Therefore, these systems are not subject to temperature effects when the filling medium is exposed to a temperature change in the capillary or elastic element. Liquid density changes when the temperature changes and requires careful control of the amount of liquid in the system—or of the ambient temperatures under which it will operate.

Forms of vapor thermal systems are classified by SAMA as follows:

SAMA IIA—designed to operate only when the temperature to be measured is higher than the ambient temperature of the capillary and elastic element

SAMA IIB—for operation only when the temperature to be measured is lower than the ambient temperature of the capillary and elastic element

SAMA IIC—for operation with the bulb temperature above or below the ambient temperature of the rest of the system

To accomplish SAMA IIC specifications, the bulb must be capable of containing all the liquid when below ambient, and all the vapor space when above ambient.

SAMA IID—for measurement of temperature above, below, and at the ambient

This design is accomplished by filling the capillary and the elastic element with a nonvolatile transmitting liquid to confine all the volatile actuating liquid in the bulb.

Bulb and range characteristics for gas-filled and vapor-actuated systems are given in Table T-4. Gas-filled systems are available over the widest range limits of any filled system. Usually only one bulb size is needed and different spans are obtained by filling the system to different absolute pressures. The gas-filled system is uniquely suited to the use of bulbs designed to average the temperature in an enclosure. This advantage results from the fact that bulb volume is not related to range span and there is no maximum volume limit on bulb size.

Vapor systems are the most limited in ranges and also are the only filled systems with a significant nonlinearity in their signal. These characteristics are due to the shape of the vapor-pressure curve, which is fixed for a given liquid. The

TABLE T-4. *Characteristics of Filled-system Thermometers**

Design Factor	Liquid Expansion			Pressure	
SAMA† class and type	V	1B	1A	II	III
Actuating medium	Mercury	Organic liquid (hydrocarbon)		Vapor	Gas
Low-temp limit	−38° F	−125° F	−125° F	−40° F	−400° F
High-temp limit	+1200° F compensated‡ +599° F uncompensated	+500° F	+500° F	+600° F	+1200° F
Longest span	1200° F	Linear to 280° F; nonlinear 280 to 378° F		280° F	1200° F
Shortest span	25° F	40° F§	40° F§	80° F	120° F
Typical bulb size for 200° F span	½ by 4 in.	⅜ by 2 in.	⅜ by 2 in.	11⁄16 by 3 in.	11⁄16 by 4 in.
Range top for shortest span	+550° F	+500° F	+500° F	+550° F	+100° F
Span determined by	Bulb size	Bulb size	Bulb size	Actuating medium and bourdon	Pressure of actuating medium
Chart divisions	Linear	Linear	Linear	Nonlinear	Linear
Max capillary length	25 ft uncompensated 400 ft compensated	15 ft	200 ft	200 ft	400 ft
Capillary temp compensation	Invar filler wire	None	Dual bourdons	Not required	None
Case temp compensation	Bimetallic strip	Bimetallic strip	Dual bourdons	Not usually required	Bimetallic strip
Bulb elev above/ below case	¶	No effect	No effect	Liquid head corrected by calibration	No effect
Overrange capacity % of span above range top	100%	100%	Varies as capillary length to 120%	Usually small	Depends on range span and range top to 300%

* Data for this table from Taylor Instrument Process Control Division, Sybron Corporation, Rochester, N.Y.
† Scientific Apparatus Makers Association.
‡ +1200°F range top on high-temperature design, +1000°F range top for standard-design mercury system.
§ Range top plus span must exceed 119°F to prevent excessive overrange on shipment.
¶ Must be specified if more than 30 ft above or 60 ft below.

range is independent of bulb size, and unless it is necessary to provide for the liquid expansion of large capillary volumes, very small bulbs may be used.

Vapor-actuated systems (SAMA IIA, IIC, and IID) are quite sensitive to the hydraulic-pressure head existing when the bulb and elastic element are not installed at the same level. A zero correction for bulb elevation always is recommended. Adjustment for this effect with IIB vapor and IIIA and B gas systems usually is not required. All pressure-sensitive filled systems are affected by changes in barometric pressure. Usually, elastic elements with large pressure changes across the operating range are chosen to eliminate the need for barometric compensation.

Overrange protection generally is small in vapor-actuated systems because of the nearly exponential change in vapor pressure with temperature change. A unique feature of the gas thermal system is that the maximum overrange temperature is fixed by the filling pressure and not by the range level at which it is used. Therefore, short spans will have less overrange protection.

—George R. Leavitt, *Taylor Instrument Process Control Division, Sybron Corporation*

Thermometer, Liquid-in-Glass A temperature-measurement instrument that consists of a glass envelope, a responsive liquid, and an indicating scale. The envelope is in two parts fused together: (1) a bulb completely filled with the liquid and (2) a capillary scale section that contains the liquid in excess of that required to fill the bulb. The position of the end of the liquid capillary column or index serves, by prior calibration, to indicate the temperature of the bulb. The scale may be marked directly on the capillary tube, as in the laboratory and clinical versions, or markings may be on a separate member that is mounted alongside the capillary tube, as in most domestic and industrial forms.

Filling liquids are mercury, mercury alloys, and organic liquids. The useful range is −200 to 600°C (−300 to −1110°F). The lower limit is set by the liquid used. The upper limit is determined by the type of glass used. Exposure limits for glasses used are given in Table T-5.

TABLE T-5. *Temperature Exposure Limits for Various Thermometric Glasses*

Glass	Strain Point, °C	Continuous Exposure		Intermittent Exposure	
		°C	°F	°C	°F
Corning Normal 7560.	500	370	700	430*	805*
Kimble R6.	490	360	680	420	790
Jena 16 III.	495	365	690	425	795
Corning Borosilicate 8800.	529	400	750	460	860
Jena Borosilicate 2954.	548	420	790	480	900
Corning 1720.	668	540	1005	600	1110
Jena Supremax 2955.	665	535	995	595	1100

*405°C or 760°F if Corning Standard Thermometer 0041 glass is used for stem.

For maximum accuracy, corrections should be applied for zero shift and temporary icepoint depression, both of which can be measured experimentally. Use will be according to one of three conditions: (1) *total immersion,* where all of the liquid, but none of the gas above the liquid is exposed to the medium being measured; (2) *partial immersion,* where only the liquid in the bulb and a limited, but defined, part of the capillary are exposed; and (3) *complete immersion,* where the entire thermometer is exposed.

—Robert D. Thompson, *Taylor Instrument Process Control Division, Sybron Corporation*

Thermometer, Resistance (See Resistance thermometry.)

Thermometer, Wet- and Dry-bulb (See Psychrometer.)

Thermopile Two or more thermocouples connected in series so that the emf outputs of the couples are additive. In domestic gas-burner safety systems, a thermopile actuates a relay when the pilot flame heats the hot junctions. In measuring and control systems, such series connections of thermocouples are used to measure small temperature differences or to measure average temperature. In thermoelectric coolers, series connection increases the amount of cooling available from a given current source. Thermopiles also are used in electric generators of dc power. The hot junctions may be heated by a gas or oil flame or by focused radiation from the sun.

In gas-burner safety systems and in applications where the thermopile measures average temperature, a series array of thermocouples, as shown in Fig. T-35, is customarily used. The individual two-wire thermocouples are aligned mechanically so that the hot junctions are at T_1 and the cold junctions at T_2. In safety applications, the thermocouples are in a common sheath. In averaging applications, the individual thermocouples often are mounted separately, but are electrically connected. For example, when measuring average temperature in a duct, several thermocouples whose hot junctions are located at various points in the duct are connected in series so that the summed output represents an average temperature of the duct.

A common use of the thermopile is in thermal-radiation pyrometers where it is desired to obtain a larger emf than would be possible from any single thermocouple. In a radiation thermopile, as shown in Fig. T-36, the thermocouples commonly are arranged radially so that an optical system of lenses or mirrors can focus energy from the target onto the hot junctions at the center of the radial array. The cold junctions are maintained in close thermal contact with the body of the pyrometer. Individual thermocouples in such radial thermopiles typically are flat wire that is only a few thousandths of an inch thick. Hot junctions often are shaped to provide more area for reception of radiant energy, or a nonconducting target may be cemented over the hot junctions to act as a receptor. A coating of carbon black, paint, or other material with constant absorbance over a wide range of wavelengths usually is applied to the hot junctions. Chromel-constantan thermocouple wires frequently are used. Radiation thermopiles also can be constructed by etching processes similar to those used in fabricating printed-circuit boards.

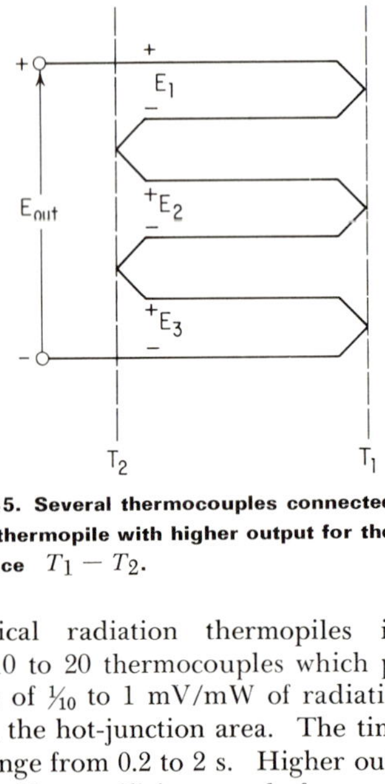

Fig. T-35. Several thermocouples connected in series to form a thermopile with higher output for the temperature difference $T_1 - T_2$.

Typical radiation thermopiles incorporate from 10 to 20 thermocouples which produce an output of ⅒ to 1 mV/mW of radiation impinging on the hot-junction area. The time constant will range from 0.2 to 2 s. Higher output can be obtained by sacrificing speed of response. Thermopiles with millisecond response have been made for heat-flux measurements where the power focused on the thermopile is sufficiently high so that very low sensitivity is not a detriment.

In thermopiles for cooling or heating (termed thermoelectric or Peltier coolers), current passing across a junction between dissimilar metals causes the junction to liberate or absorb heat. Because efficient thermal design leads to low-voltage—high-current devices, several junctions are connected in series for greater efficiency—especially when transistor modulation of current is required. It is more efficient to control a 6-V 1-A 10-junction series cooler than to control 10 junctions in parallel at 0.6 V and 10 A. Couples are formed from short blocks of *p* and *n* semiconductors, such as bismuth-telluride, lead-telluride, or ternary alloys of bismuth, tellurium, germanium, selenium, and similar metals.

—Ernest C. Magison, *Honeywell Inc.*

Fig. T-36. In radiation pyrometers, fine thermocouples, insulated by mica washers, form a thermopile. All hot junctions are in the center. Reference-junction temperatures are at the connection tabs.

Thermostat A nonindicating, nonrecording, low-cost temperature regulator mass-produced for (1) heating and air-conditioning systems for buildings, residences, and vehicles of all types—to provide comfort for occupants and to protect contents from thermal damage; (2) heating and cooling systems for machinery, engines, and motors, predominantly as protection against overheating or freezing; and (3) coarse temperature regulation for laboratory and testing apparatus. In terms of units manufactured, the thermostat is one of the most common forms of automatic controllers. The majority of thermostats operate on the expansion or contraction (with temperature changes) of a metallic element (commonly bimetal) which causes electrical or pneumatic actuation of regulatory heating/cooling equipment. Other thermostats employ liquid- or gas-filled systems much like a thermometer. Nearly all physical principles that form the basis for temperature-measuring instruments (see **Temperature, measurement of**) have or can be employed in thermostats. Cost is the predominant motivating design factor. Also see **Self-actuated regulators.**

Thickness Gaging, Nuclear-radiation Gages of this type are used to measure and control thickness of sheet materials in process. All use noncontact nuclear-gaging techniques employing radioisotopes. Several types are available: (1) single measuring assemblies for sheet-thickness control, (2) double-measuring assemblies for control of coatings, (3) a single-gage system (Prefcoat) for controlling coatings on paper and plastics, and (4) a single-gage system (Florescoat) for coating control on sheet steel.

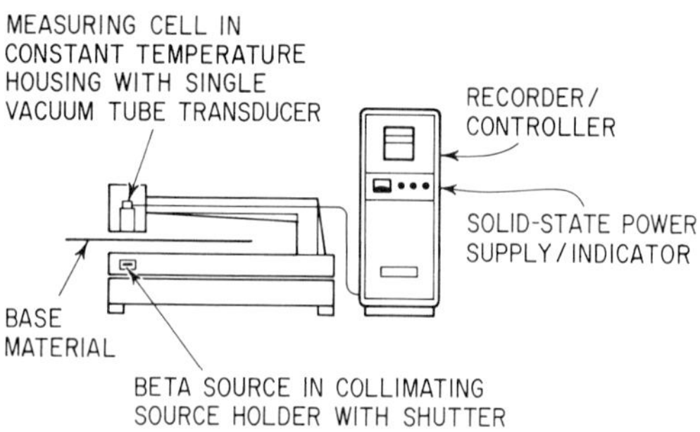

Fig. T-37. Basic single-head beta-gaging system for sheet-thickness control. *(Ohmart.)*

The basic single-head gage, shown in Fig. T-37, holds a beta radioisotope source beneath the sheet to be measured and a detector cell above. The gage can measure a single point or scan the sheet automatically. Transmitted radiation is converted into an electric current and amplified for readout or automatic control. Measurement is basis weight or weight per unit area, such as ounces per square foot. Gage configurations vary widely for specific applications. Precision is ± 1 percent of range; strontium-90 is the basic beta source; promethium-147 or krypton-85 are used for very thin materials.

The double-gage system shown in Fig. T-38 is applied to measure and control coatings on paper, textiles, and plastics. Measurement is made of the substrate before coating, and a second measurement is made after coating. The difference between the two weights determines the net coated weight. The system normally is used where the coating is 20 percent or more of the substrate.

Gages are matched on dwell cycles so that the same areas are read before and after coating. Signals are transmitted to a difference computer to derive the net coating weight, to a strip chart for readout, and to a controller to adjust the coating knife, dependent upon variations from the target value. Uncontrolled short-term

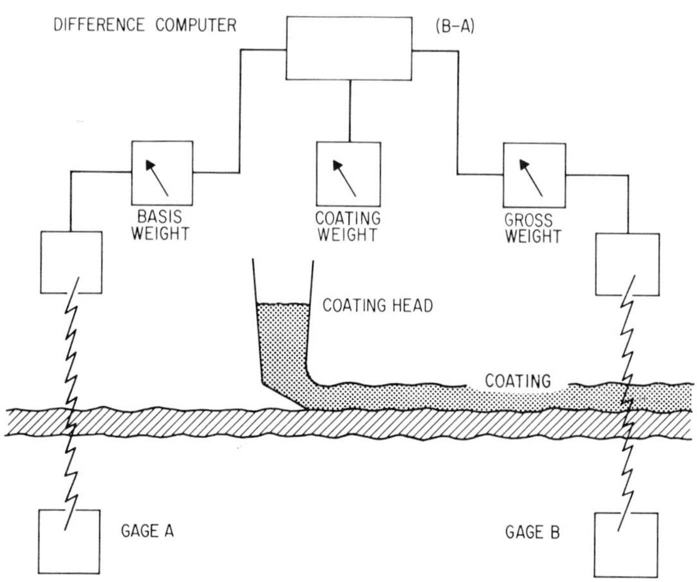

Fig. T-38. Double-gage beta-radiation system for control of coatings on paper, textiles, and plastics. *(Ohmart.)*

variations are not properly subtracted from the second signal unless a memory-delay unit is installed.

The Prefcoat system, designed to overcome the disadvantages of a two-gage system, measures and controls coatings on paper and plastics with a single gage. A photoelectric effect involving photoabsorption of energy is used. A low-energy gamma-emitting source is used. The system can be used when the atomic number of the coating material is at least twice that of the atomic number of the base material. Source and detector are placed as in single-gage system. Conversion of radiation, amplification, and control principles are the same as in beta systems.

The preferential-absorption principle detects only radiation absorbed by the ingredient with the highest atomic number—which must be the coating. Low energy (less than 0.5 million electron volts) from the radioisotope source restricts the system to limited total mass—base material and coating weight combined. The total weight per unit area and mass coefficient of the materials combination must be calculated for acceptable absorption limits.

The Florescoat system employs a phenomenon known as nuclear fluorescence to measure and control coatings on sheet steel or aluminum. These systems measure only coating weight regardless of base-material hardness or thickness. Control and readout are in absolute (not relative) terms. The system may be applied to single- or double-sided coating.

Nuclear fluorescence is produced when gamma radiation excites electrons in the metal of the coated strip. As shown in Fig. T-39, excitation is continuous, but each occurrence is only temporary. Electrons returning to the original unexcited state produce a characteristic low-energy radiation called fluorescence. The coating material also produces fluorescence in some applications. The total radiation spectrum produced is returned to the detection system. Each fluorescence has its own energy level. This is differentiated by two filters and two ion-detection chambers at the measuring point. Detectors are of opposite polarity. One filter permits all backscattered radiation to reach an ion chamber; the other filter permits only radiation from the metal strip to reach the other chamber. Hence, the base-metal-strip signals cancel one another, leaving only the coating signal as an output from the detectors. Precision is ±0.02 oz/ft². Area measured at any one point is 3 by 5 in. but can be reduced to ¾ by 5 in. in some cases.

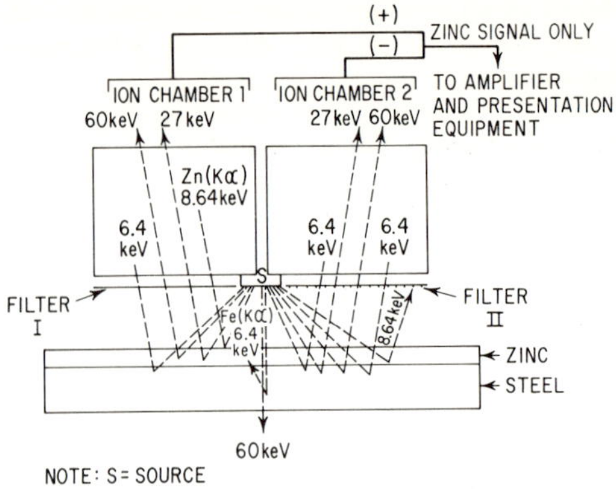

Fig. T-39. Nuclear-fluorescence gage for controlling coatings, such as zinc, on sheet metal. *(Ohmart.)*

The thickness-measurement and control systems described may be used with averaging, digital, or analog computers. All except the Florescoat system may be equipped with automatic scanning and automatic standardization. All require at least one amplifier to boost the electrical signal from the detectors.

—Gordon W. Hickenlooper, *The Ohmart Corp.*

Thickness Gaging, Pneumatic (See Gaging, pneumatic)

Thickness Gaging, Ultrasonic Ultrasonic gages find extensive application in the measurement of wall thickness. They are useful in measuring most structural materials which will transmit sound. Almost all metals, plastics, and ceramics, and various composite materials can be measured with ease. Measurements are *instantaneous,* and can be made from *one side* of the part. Suitable transducer configurations also permit readings at elevated temperatures for determining corrosion losses in chemical-process equipment, even while the equipment is "on-stream."

These gages utilize ultrasonic vibrations, most commonly in the portion of the frequency spectrum between 1 and 15 MHz. Sound waves in these frequencies have certain characteristics which make them suitable for making measurements and finding defects. The sound beam can readily pass through most structural materials. It will reflect from acoustic boundaries, either internal flaws or geometric boundaries such as the back surface of the part. Sound travels at a characteristic velocity in various materials; this velocity is dependent on the elastic properties and the density of the material, and this permits timing of the wave propagation through the material.

The basic elements of an ultrasonic-gaging system consist of a *generator* to produce the high-frequency electronic vibrations which, in turn, activate the crystal *transducer*. The transducer converts the electrical signals to mechanical vibrations, or ultrasound, and these are introduced into the part. Echoes from various boundaries in the part are picked up by the same transducer and are reversibly converted from mechanical vibrations to an electrical signal. This is further amplified and processed, and the information is displayed on one of several types of data presentation.

There are two general categories of ultrasonic gages, *resonance* and *pulse/echo*. Each has distinct electronic characteristics and readout methods. Both are dependent on transmission of the acoustic waves, reflections, and velocity in the material.

The *resonance* ultrasonic gages produce a frequency-modulated continuous-wave signal. This provides a corresponding swept frequency of sound waves which are introduced into the part. When the thickness of the part equals one-half wavelength, or multiples of one-half wavelength, standing-wave conditions or mechanical resonances occur. The frequency of the fundamental resonance, or the difference frequency between two harmonic resonances, is determined in the instrumentation. The thickness is determined by the formula

$$Th = \frac{Vel}{2F}$$

where Th = thickness of part under transducer, in.
VEL = speed of sound in the material, in./s
F = frequency, Hz.

The most common readout for ultrasonic-resonance gages is a large cathode-ray tube. Frequency indications on the tube are compared with an overlaid scale calibrated in inches, and direct thickness readings can be obtained. The readings are instantaneous, and thickness variations can be monitored as the transducer is scanned over the part.

The resonance-frequency gages generally produce accuracies of 1 percent of nominal thickness.

The *pulse/echo* technique operates somewhat like a refined sonar system. Very short electrical pulses, usually at discrete frequencies, are generated in the electronic instrument. These produce short acoustic pulses from the transducer which, in turn, pass through the material and reflect from the boundaries, just as in the resonance test. The transit *time* of the pulses through the material is measured by the instrumentation.

Thickness is determined by the formula

$$Th = \frac{Vel \times T}{2}$$

where Th = actual thickness of part under transducer or within sound beams, in.
Vel = speed of sound in material, in./s
T = transit time of sound pulse through one round trip in the material, s

The simplest and most common readout for this type of gage is a direct-reading panel meter. This is incorporated into small, portable, battery-operated instrument packages. Accuracies of between one-half of 1 percent and 1 percent of full scale are usually obtainable.

Numerical readout instruments are also available which incorporate specialized digital voltmeter readings with the ultrasonic electronic circuitry. On a standard three-digit readout, accuracies of one-tenth of 1 percent of full scale are available. The digital readouts are very easy to read and also lend themselves to incorporation in more sophisticated systems.

One limitation of ultrasonic gaging is the requirement for continuous coupling of the sound beam between the transducer and the part. Sound will not pass across an air-solid or air-liquid boundary. Liquid coupling, either a continuous thin film or some other type, is required. In some instances, complete immersion of the transducer and the material is feasible. In others, and commonly used on production lines, a bubbler or partially contained water column provides the continuous coupling path.

The total accuracy of the test system depends on both the reading accuracies noted above and the test conditions. Surface roughness, curvatures, and material-property variations may slightly reduce the total accuracy.

Both testing methods can be applied to automatic gaging systems. Electronic accessories such as high/low limit switches and strip-chart recorders to provide thickness profiles can be included.

—John E. Bobbin, *Branson Instruments Co.*

Thickness Gaging, X-ray Noncontact x-ray thickness gages measure the thickness or density of hot or cold materials while in motion or when stationary. Steel, aluminum, brass, copper, glass, paper, and rubber—in continuous strip or sheets—as well as plastic films, foils, and material coatings are typical of materials that can be gaged by this method. Different styles of x-ray gages will span a thickness range from 0.00025 in. in plastic through 2 in. in steel. The capacities of various gages are shown in Table T-6.

TABLE T-6. *Typical Thickness Capacities (in Inches) of X-ray Gages*

Gage Power, kV	Materials Measured: Steel	Hot Steel	Brass and Copper	Aluminum	Plastic	Glass
25	....	....	....	0.0004–0.030	0.00025–0.100	
40	0.0001–0.040	....	0.0001–0.030	0.001–1.0	0.100–0.500	0.050–1.0
65	0.005–0.1500	....	0.005–0.100	0.060 and up	0.500 and up	0.060 and up
100	0.005–1.000	0.010–1.000	0.005–0.500			
200	0.005–2.000	0.010–2.000	0.010–1.000			

All materials absorb x-rays to varying degrees, dependent upon thickness and density. Thickness is determined by measuring the amount of x-ray energy that a material absorbs as the material passes between an x-ray emitter and a receiver (see Fig. T-40). The gage is set to the desired thickness standard, and deviation in material thickness from nominal is indicated in percent or thousandths or millionths of an inch on a meter or digital display. A sample piece of the material to be measured or a reference standard is used to calibrate the gage. The gaging signals can be used to operate a variety of accessory instrumentation, including automatic controllers, recorders, totalizers, classifiers, markers, and sorters.

An x-ray gaging system comprises three basic units: (1) a scanning unit that contains the x-ray

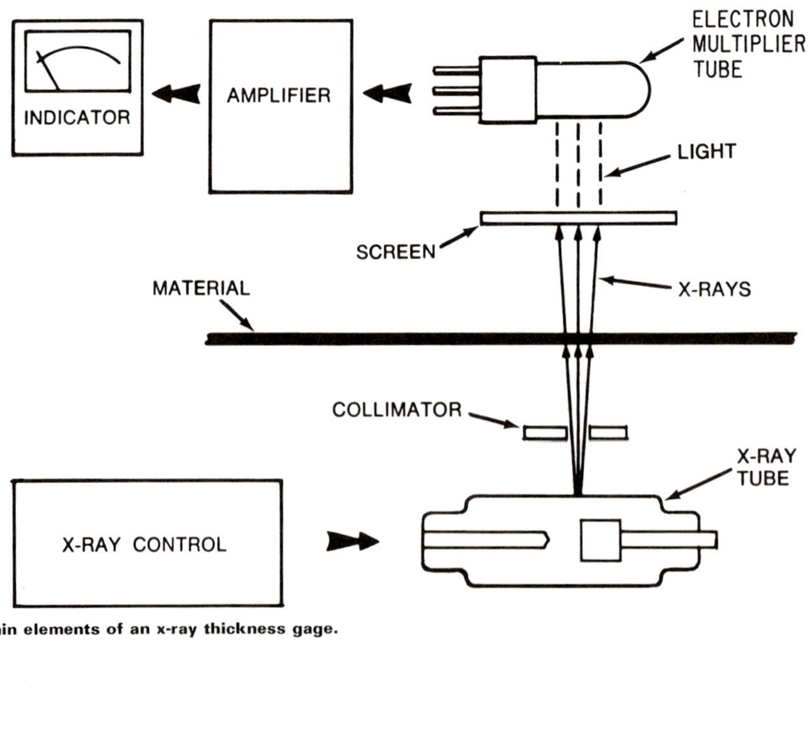

Fig. T-40. Main elements of an x-ray thickness gage.

generator and a detecting unit, (2) an operator control station, and (3) a power unit. The scanning unit generally is a C or O frame that is mounted in a stationary position or on a traversing track—as on a rolling mill, process, or inspection line.

X-ray Gage Characteristics: Systems are accurate to ±0.25 percent of the thickness being measured—or better. Repeatability of readings is 0.01 percent. The speed or response time in which an x-ray gage will react to a change in material thickness is important in keeping fast-moving strip and sheet stock within tolerance. For example, in a mill running 6,000 ft/min, a gage with 50-ms response time reacts to a change in material thickness within 5 ft of material passing—in contrast with a 250-ms response time that permits 25 ft of material to pass before indicating a thickness change.

Continuous-process x-ray gaging is an integral part of a computer-operated rolling mill. The gaging system includes an interface that receives commands for thickness settings from the computer and, in return, verifies that such commands have been accomplished. Such systems also may include digital display of gage setting, compensation for varying temperatures and alloys of steel, fully automatic calibration to maintain accuracy, and electric-motor-operated positioning systems to locate the scanning unit over the steel. A system also may record the difference in thickness between the edge and the center or "crown" of the steel. Anticipatory gaging systems will anticipate an out-of-tolerance trend or change in material thickness and use the signals to actuate screwdown, speed, and tension controls to keep material on gage. Automatic control permits rolling to close tolerances, maximum on-gage footage per ton, and less scrap or out-of-gage material at both ends of a coil or run.

—Henry L. Boppel, *Automation and Measurement Division, The Bendix Corporation*

Thin-film Memory The thin-film memory is a storage device (see **Storage, computer**) which utilizes the properties of a ferromagnetic material deposited on a substrate under controlled conditions. If a magnetic field is applied parallel to the surface of the substrate during deposition, the thin film becomes easier to magnetize in the direction of the field than it is at right angles to that direction.

With reference to Fig. T-41, current in the word line rotates the magnetization in the hard direction, causing a voltage to be generated in the bit (see **Bit**) line. This voltage is interpreted by discriminating logic in the sense amplifier to be a 0 or 1 depending on its polarity. The bit-line current, depending on its polarity, then rotates the

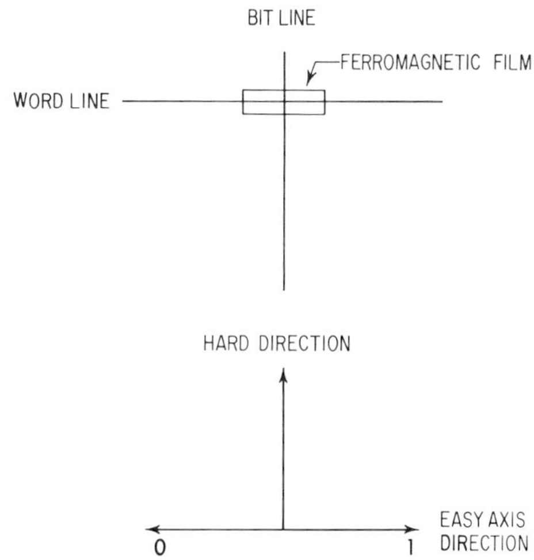

Fig. T-41. Thin-film storage.

magnetization in either the 1 or 0 easy-axis direction (see **Binary state**).

—W. W. Welz, *IBM Corporation*

Thin-layer Chromatography (See **Chromatography.**)

Thompson Effect (See **Thermocouple.**)

Three-way Valve An adaptation of a conventional two-way globe body to obtain a third flow opening. Three-way valves are known best either as diverting valves or as mixing valves. When used as a diverter, the three-way valve causes one inlet stream to split between two outlets, and in the extremes of valve travel, the incoming flow is entirely diverted to one or the other of the two outlet ports. The diverting three-way valve body is a modification of a conventional double-port globe body.

Tide and Wave Recorder For years, oceanographers have relied on visual techniques for measuring ocean swells, waves, and tide levels. Wave and tide measurements are required in basic coastal studies for developing harbors, marinas, wharfs, piers, and in general, to determine the extent of the influence of the ocean environment upon man-made structures. Oceanographers recognize the profound influence that the sea and air have on each other. Therefore, measurements of the atmosphere, such as wind speed, direction, and temperature, are related to the state of the sea. Thus, it is necessary to distinguish between wind-generated waves and swells that are created by forces far removed from the areas where the swells were generated. Present techniques utilize pressure transducers which detect pressure changes associated with swells, waves, and tides.

Wave and tide measurements normally are made in shallow water by placing a special pressure sensor at a nominal depth of 50 ft. The pressure changes are converted from a mechanical movement of a diaphragm to an electrical signal derived from a resistive or inductive element. The signal then is transmitted to a strip-chart recorder or further processed by electronic means to analyze such parameters as average wave periods, average wave heights, and tides. Instruments are designed as *remote-reading* using a cable between the sensor and readout module or as *in situ* instruments with self-contained data-recording systems capable of recording unattended for long periods of time. See Fig. T-42.

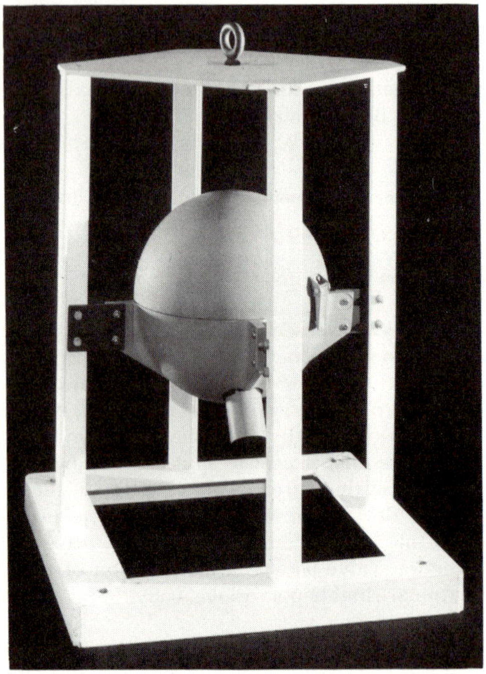

Fig. T-42. An *in situ* wave- and tide-measuring system wherein all components, including recorder, are contained within a 10-in. spherical pressure housing. The top is removable for quick change of batteries and recorder paper. *(Hydro Products..)*

The overall accuracy of wave- and tide-measuring systems is within ±3 percent of full scale. Wave-amplitude measurements normally are ±10 ft. Wave-period measurements are from 1 to 30 s, averaged over a long time constant. Tide measurements normally are ±10 ft. For locations and conditions where these values may be greater, the equipment can be modified. Trends in the design of tide- and wave-measuring equipment include the use of an acoustic link between the sensor and the surface station to eliminate costly cabling and to further the reliability of the system. New developments also include different techniques for obtaining wave and tide data, such as accoustic reflection or special buoys with displacement-sensitive transducers that relay information via microwave transmission to shore stations for data processing.

—Leon J. Corcoran, *Hydro Products, a division of Dillingham Corporation*

Time and Time Measurement Time-reference systems (or clocks) include the *frequency rates* or *cycles* of (1) the earth, (2) the pendulum, (3) the balance wheel of a watch, (4) controlled alternating electric current, (5) the tuning fork, (6) radioactive decay, and (7) the oscillations of the atom of cesium-133. *Time is what a clock reads.* There is no absolute time standard.

Apparent Solar Time: This is measured by the diurnal motion of the sun as observed from a point on the earth's surface. Apparent solar days vary in length because of the earth's rotational and orbital variables.

Mean Solar Time: This is apparent solar time averaged out so that the mean solar day is the average of all apparent solar days in the orbital year. The mean solar day is divided by 86,400 (60 s × 60 min × 24 hr) to determine the *mean solar second.* The *equation of time,* with a numerical value ranging from 0 to about 16 min, describes the difference between the apparent and the mean solar time. The value is at a maximum in February and in November.

Universal Time (UT): This also is called Greenwich Mean Time (GMT) because it is referenced to the prime (or zero) meridian which passes through Greenwich, England. UT is based on the earth's rotation, and if uncorrected, its units are equal to the mean solar second. As a time scale, the uncorrected units are designated as UT_0. The UT_1 scale is UT_0 corrected to allow for polar motion or nutation of the earth. UT_2 is UT_1 corrected for seasonal variations in the earth's rotation. UT is announced in International Morse Code each 5 min by radio stations WWV (Fort Collins, Colo.) and WWVH (Puunene, Maui, Hawaii) of the National Bureau of Standards. Other time signals broadcast by the stations are kept in close agreement with UT_2.

Civil Time: UT also is expressed in terms of the 24 *standard time zones* that serially cover the earth's surface at coincident intervals of 15 degrees longitude and 60 minutes UT, as agreed at the Washington Meridian Conference of 1884. Local *civil time* zones, with irregular statutory boundaries, were chosen so that noon occurs when the sun appears approximately overhead at the zone's central meridian. *Standard time* zones were established in the United States in 1883.

Daylight Saving Time: Since 1967, federal legislation requires that states observing DST must do so uniformly by moving the clock forward an hour 1 h after midnight on the last Sunday in April and by moving the clock back an hour 1 h after midnight on the last Sunday in October. Cities and other communities are not bound by this legislation.

International Date Line: The 1-hr shift in standard time for each 15 degrees longitude requires a 1-day jump forward in date when moving from America to Asia and vice versa. The date boundary runs irregularly by treaty from the poles through the Pacific.

Apparent Sidereal Time: This time scale, generally used by astronomers, is measured by the apparent diurnal motion of a distant star (not the sun) as observed from a point on the earth's surface. *Mean sidereal time* varies only slightly from apparent sidereal time because distant stars, unlike the sun, appear in relatively fixed positions as the earth rotates.

Ephemeris Time (ET): This is based on astronomical measurements of the orbital motions of the earth, moon, and planets in the solar system. In practice, ET is computed from observations of the moon's motion about the earth by a dual-rate moon-position camera at the U.S. Naval Observatory.

Atomic Time: This is based on the atomic second, officially defined in 1967 by the International Bureau of Weights and Measures as 9,192,631,770 oscillations of the atom of cesium-133. This value expresses the ET second as closely as possible in terms of an atomic standard and was derived by a joint experiment of the U.S. Naval Observatory and Great Britain's National Physical Laboratory, using a dual-rate moon-position camera and a cesium-beam clock. See **Clock, atomic.** Corrections to the frequencies transmitted by NBS radio stations are determined with regard to the NBS atomic standard and are published monthly (*Proc. IEEE*). The stations' carrier and modulation frequencies are offset intentionally from the atomic-frequency standard by a precisely known amount so that, on average, the time signals broadcast remain as close as possible to UT_2. In 1966, 1967, and 1968, the offset was minus 300 parts in 10^{10}.

Primary Clock: The cesium-beam clock of NBS is referred to as a primary clock because independently of any other reference it provides highly precise and accurate time to all users.

Secondary Clock: Unlike a primary clock, the secondary clock must be referenced to more accurate clocks for calibration. Today, quartz-crystal clocks are the most frequently used secondary time standards.

Master Clock: These clocks are used to control subsidiary clocks directly. For example, the 60-Hz ac plug-in electric clock employs a synchronous motor controlled by the alternating current acting as a master clock system. A wrist timepiece, in contrast, has its own frequency standard—either a 2.5- to 5-Hz balance wheel, or a 360-Hz tuning fork.

The International Astronomical Union and the General Conference of Weights and Measures officially define international time standards; while the Bureau Internationale de l'Heure, at the Paris Observatory, provides and receives data to assist in the uniform international observation of time standards.

Also see **Chronograph; Chronometer; Clock, atomic; Crystal oscillator;** and **Signal (electric) generator.**

REFERENCES

"Explanatory Supplement to the Astronomical Ephemeris and the Nautical Almanac," H. M. Stationery Office, London, 1961.

"NBS Standard Frequency and Time Services," National Bureau of Standards, Washington, D.C., 1967.

"Supplement to the Astronomical Ephemeris 1968," U.S. Naval Observatory, Washington, D.C., 1966.

Beehler, R. E.: A Historical Review of Atomic Frequency Standards, *Proc. IEEE,* vol. 55, no. 6, June, 1967.

Dostal, Frank: "Increasing Applications of Tuning Forks

in Control Systems," Electronics Division, Bulova Watch Co., Inc., Woodside, N.Y., 1966.

Markowitz, William: The Atomic Time Scale, *IEEE Trans. Instr.*, vol. I–11, nos. 3, 4, December, 1962.

Rawlings, A. L.: "The Science of Clocks and Watches," 2d ed., Pitman Publishing Corporation, New York, 1948.

—William O. Bennett, John J. Carpenter, Frank Dostal, and E. Van Haaften, *Bulova Watch Company, Inc.*

Time Constant The value T in an exponential response term $A \exp(-t/T)$ or in one of the transform factors $1 + sT$, $1 + j\omega T$, $1/(1 + sT)$, $1/(1 + j\omega T)$, where s = complex variable, t = time, s, T = time constant, $j = \sqrt{-1}$, and ω = frequency, rad/s.

For the output of a first-order (lag or lead) system forced by a step or an impulse, T is the time required to complete 63.2 percent of the total rise or decay; at any instant during the process, T is the quotient of the instantaneous rate of change divided into the change to be completed. In higher-order systems, there is a time constant for each of the first-order components of the process. In a Bode diagram, break points occur at $\omega = 1/T$.

Time Lag (See **Dead time, process; Delay time, transducer; Lag, controlled-system; Process reaction rate; Response, steady-state; Response, transient;** and **Transfer function, theoretical.**)

Time Optimal Control (See **Adaptive control.**)

Time-proportioning Control (See **Proportional-input control.**)

Time Response (See **Response time, transducer.**)

Time-schedule Controller (See **Timing controllers.**)

Time Sharing The process of using core memory for the concurrent execution of more than one job by temporarily storing all jobs but the one in control on auxiliary storage. This technique is used to allow a given area of prime computer memory to be utilized by several independent programs. This method of operation is most often associated with a computer-controlled terminal system used in an interactive mode.

The objective in using an interactive time-sharing system is to allow more efficient use of the computer system in proceeding from initial specification of a program to a final checked-out operating revision and the more efficient use of people in the solution of problems by providing answers and services in a short time on demand.

The traditional batch-processing type of operation requires the coding and preparation of a program as a separate first step. Submission of the job to a computer control center for inclusion in a batch run usually results in at least several hours' delay in a busy shop. The problem of writing and checking out a large complex program in this manner is not efficient use of the programmers' time because of loss of continuity in the problem. The time-sharing systems are designed to allow a user to operate the computer in an interactive manner, through a variety of console devices, getting immediate responses to requests for answers, and receiving immediate indications of errors in programming (see **Error procedures, computer**). The user efficiency is increased, since continuity of thought can be maintained.

Since the computer speed of performing operations requested by the user is much faster than even the time required for transmitting information down the comunications lines, and since the user may spend some time in thought between commands to the system, it is necessary to utilize the central facilities for other users or more terminal-oriented work to maintain economic utilization of the system.

Modern systems have the capability of handling many terminal users concurrently and still use available time to process batch jobs in the normal manner. The magnitude of the capability for this type of operation depends on computer speed, core size, auxiliary-core size and speed, the operating system used to manage and control these facilities, and the type and amount of terminal service to be offered.

A common method of allocating time for each user in the terminal system is time slicing. The time interval is chosen short enough to give good response to each terminal but long enough to give a usable work interval (see **Time slicing**).

In nonterminal systems, time sharing is often used to share a portion of the system memory between jobs of different priority. In some control and data-acquisition systems, a portion of the core (a partition) is allocated to run nonprocess-related jobs (background jobs) in the conventional batch-processing manner. When the system is notified that a process-related program requires the partition, the background job is saved on auxiliary storage until the higher-priority program has completed execution. The background job is then restored and resumed if no other higher-priority jobs are waiting for service. A system using this type of time sharing provides for utilization of time which would otherwise be unused in a real-time or data-acquisition system.

REFERENCE

Rosen, Saul, "Programming Systems and Languages," part 5I, McGraw-Hill Book Company, New York, 1967.

—L. E. Wise, *IBM Corporation*

Time Slicing A technique used by a software control system to control the allocation of facilities of a computer to tasks requesting service (see **Task**). The allocation basis is a fixed time interval.

This control method is used in time-sharing and multiprogramming systems to allow several jobs to be run in the same span of time (see **Job; Time sharing; Multiprogramming**). Each job is executed for the time period used in the time slice. The job is then temporarily stored on an auxiliary storage device (time sharing) or suspended (multiprogramming) while another job is being run. Each job is therefore run in an incremental fashion until complete.

This technique allows several users to utilize a computer facility as though each had complete control of the machine. Because of the relative speed between computer operation and human response, several users can be serviced one at a time, unaware of each other. The faster the computer and the storage units used for temporary storage of the programs, the more users can be handled by a single system.

—L. E. Wise, *IBM Corporation*

Timing Controllers The relation of events with time is a very important aspect of industrial and scientific instrumentation and control. Probably the most common time-event relationship is the simple function of numerous instruments, such as recording, charting, plotting, logging, and printing devices that capture measured data of many kinds and give that information a time reference. Perhaps this is visualized most easily in terms of a circular- or strip-chart recorder in which the chart is driven by an accurate timing mechanism and where the chart is calibrated along one coordinate in terms of time and along the other coordinate in terms of the one or more other variables measured. These time-coordinate instruments are not considered, however, as timers or timing controllers.

A timing controller causes one or more events to take place in accordance with some form of time schedule or program and can be as simple as an alarm clock or as complex as controlling a long sequence of operations where scores of final controlling elements may be involved and where numerous instrument set points may be made to change in accordance with a predetermined schedule. Nomenclature for timing devices has not been fully standardized. There is a vast variety of designs.

Basic Elements of Timing Controllers: These are shown in block form in Fig. T-43 (page 694). The major elements are defined as follows:

1. *Time-measuring or driving means*—the way in which timing is achieved. This element determines the limits of the range and the accuracy of the timing device.
2. *Starting means*—the impulse, manual or automatic, whereby timing action is initiated.
3. *Setting and control means*—the parts of the device which actuate the load contacts and which are adjustable to obtain the desired time intervals.
4. *Load contacts*—the means for controlling external circuits, usually comprising some form of electric or pneumatic switch. This is the link between the timer and the process, machine, or operation that it is controlling.
5. *Indicating means*—for convenience of supervision and adjustment, most timing devices incorporate a display of the present time in relation to immediate past or future events controlled.

Functional Characteristics of Timing Controllers: When the driving or control means of a timer is operating so as to produce the desired time interval, the device is said to be timing. At the conclusion of the interval, the device is said to be timed out. At this point, the timer may or may not have returned to its starting condition. The complete course of operation is termed a cycle.

Repeat Cycle: A repeat-cycle timer performs its function over and over until stopped. Most repeat-cycle timers are motor-driven; the control means, which may be a disk with trippers or a shaft with cams, operates continuously, although the motion may be intermittent. The same operations, at the same time intervals, are repeated indefinitely.

Stop Cycle: A stop-cycle timer performs its function once and then stops. The device may be returned manually or automatically to its starting position.

Reset: A reset stop-cycle timer returns to its starting position automatically upon the comple-

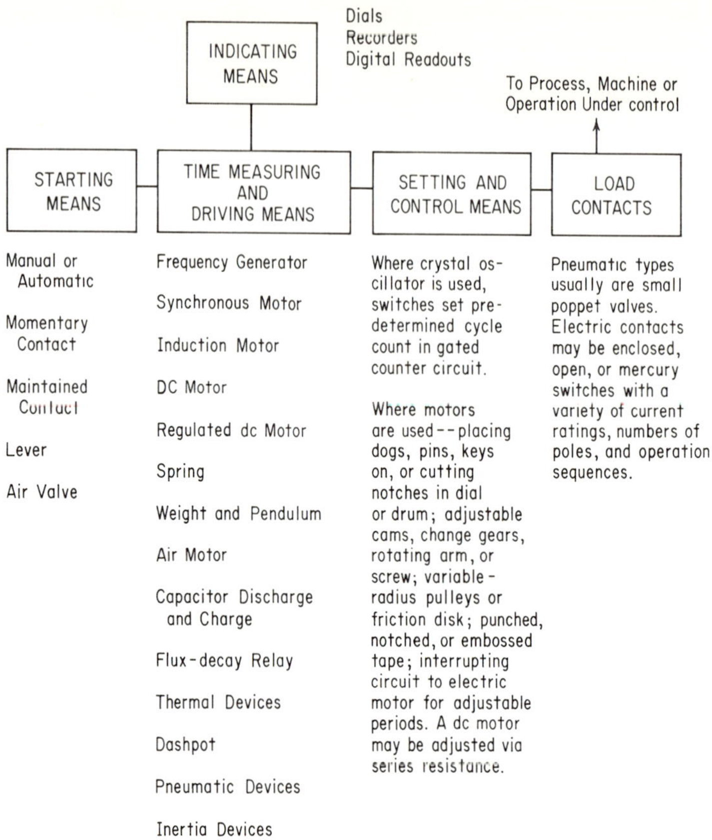

Fig. T-43. Basic elements of timing controllers.

tion of timing or upon the making or breaking of an external control circuit.

Specific Designs: A *time switch* generally is used to open and close electric circuits according to the time of day on a continuously repeating schedule. These devices frequently are used in connection with lighting, heating, and ventilating equipment and for turning processing equipment on a specified period before commencement of the workday.

A *time-delay relay* may be electric, mechanical, or electromechanical and is a contact-making device in which there is a time lag between the energizing or deenergizing of the control circuit and the subsequent opening or closing of the load circuit or circuits. A major function is to delay the start-up, continuation, recycling, or shutdown of an operation until optimum conditions have been produced.

An *interval timer* will hold circuits open or closed for a predetermined interval and then return them to their original positions. When the timer is energized, it will control the duration of an operation or process. These devices are used widely in controlling the operating time of plastic-molding presses, die-casting machines, conveyors, centrifuges, pumps, bake ovens, resistance-welding equipment, induction heaters, chemical feeders, x-ray equipment, and tire presses. An interval timer can be quite complex—built up either as a series of individual timers or as a master timer with many actuators. A complex interval timer sometimes is termed a time-cycle controller.

Application of a *time-cycle controller* is shown in Fig. T-44. Diaphragm motor valves for heating and cooling the platens of a hydraulic press, applying high and low pressure, and draining are operated in the proper sequence for the desired interval. Pneumatic pilot valves or solenoid valves actuated by electric load contacts can be used. The program for a six-circuit time-cycle controller is shown in Fig. T-45. The flexibility of a time-cycle controller is readily apparent—a multiplicity of separate circuits, each operating a number of times and in any sequence during any required interval, can be controlled.

A *time-schedule* controller links time to the control of a process variable. The value of the quantity or condition involved, such as temperature, is regulated in accordance with a predetermined time schedule. Devices for accomplishing this action have three basic functions: (1) automatic control of the variable, (2) timing, and (3) changing the set point of the controller. In essence, a time-schedule controller is an automatic controller whose set point is automatically positioned for a preselected time interval. A time-schedule controller may control a variable at a number of rates of change, as well as at different fixed values, for a series of time intervals. Cam-following techniques are used wherein a roller rides on a cam whose contour reflects the rate of rise, hold, and rate of fall (and more complexities thereof) and is linked mechanically to adjust the controller's set point. Or controlled rates of rise, hold, and fall may be programmed by using a series of interlocked interval timers which open and close circuits to achieve the desired heating, holding, and cooling rates. There are several design variations. The term *program controller* often is used in connection with the interval-timer approach, although it also can be applied to the cam or other continuous approach.

Instead of cutting a contoured cam for time-cycle control as described earlier, a system can be used wherein the desired program can be marked boldly on a circular chart—with this record being followed automatically by means of a photocell pickup. Or a punched card of the type shown in Fig. T-46 can be used as the input to a time-cycle controller. The pattern of punched holes in the card is a logical display of the program intelligence. The card illustrated is a 25-step card ca-

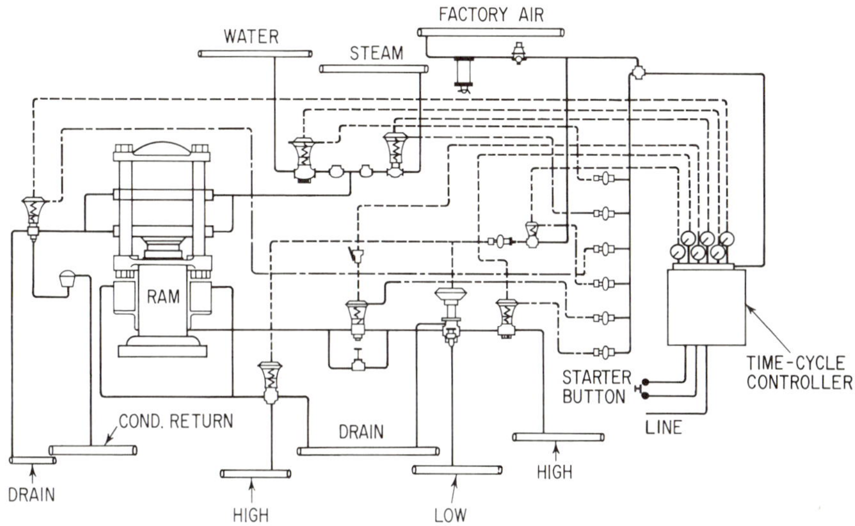

Fig. T-44. Automatic timing control of hydraulic press.

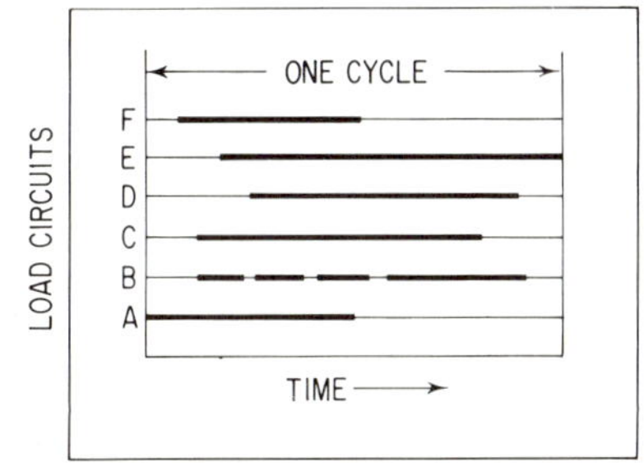

Fig. T-45. Timing diagram for a time-cycle controller.

pable of programming up to 24 outputs over a sequence of 25 steps. The horizontal dots represent the sequence of events that take place during the whole program. A vertical row of dots indicates the outputs and time intervals that occur during a single step. A single vertical row of dots, from

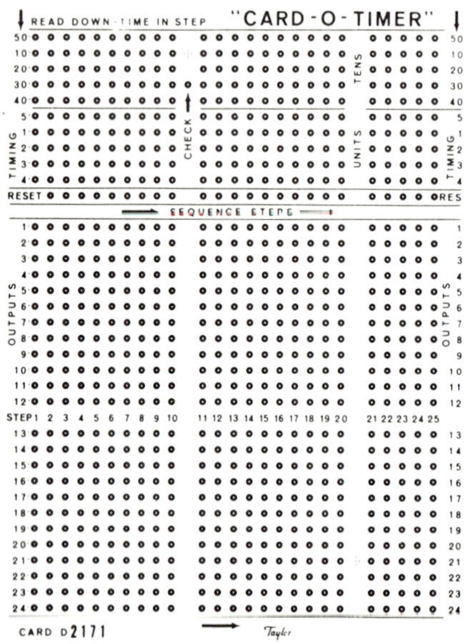

Fig. T-46. Operation card for a card-type time-cycle controller. *(Taylor.)*

top to bottom, represents the following functions: (1) the first five dots are the "tens" intervals of time that may be required in a step, (2) the next five dots are the "units" intervals of time, and (3) the remaining series of 24 dots represents the number of outputs that may be programmed during the time established for this particular step. Also see **Electric controller.**

Titrator, Automatic Although an automatic titrator can take several forms, essentially the instrument is an automated version of the familiar manual laboratory methods. End points are measured instrumentally, and the reagents used are metered and recorded. Automatic titrators have been adapted to on-line use in some cases. In one instrument, as the sample stream enters the instrument, the sample is fed into the analysis system at a constant rate by a metering pump. At the same time, one, two, or three reagents, depending on the analysis, also are metered into the system at a constant rate. As the reagents are mixed with the sample, a chemical change takes place. This change is manifested in the form of either a change in color or a change in electrical conductivity. If the change is one of color, the change is detected by a photometer; if a change of conductivity is involved, a conductivity cell is used for detection. Either form of detector produces an electrical output that is proportional to the concentration of the component being monitored. This output is used to drive a recorder which provides a direct readout of the component's concentration. In some cases, the output may be used to activate controls for corrective measures. A multipoint-analysis system is available which permits sample flows to enter the analyzer from as many as six different locations within a plant or laboratory. Average accuracy expected ranges from ± 2 to 5 percent of full scale; response speed is 90 percent in 5 to 30 min depending on particular analysis. Sample flow rates vary from 5 to 20 ml/min for liquid analysis, from 50 to 500 ml/min for gas analysis.

These instruments are used in commercial, municipal, and industrial power plants for monitoring boiler feedwater to detect silica from 0 to 250 ppb, dissolved iron from 0 to 2 ppm, phosphates from 0 to 20 ppm, and hardness from 0 to 3 ppm. Residual chlorine in water and sewage-treating plants can be determined from 0 to 1 ppm. In metal-plating plants, the instrument can be used to monitor highly toxic hexavalent chromium in waste water at the 0- to 250-ppb level. In oil refineries and chemical plants, hydrogen sulfide and mercaptans from 0 to 5 ppm in various gas streams can be monitored. Also see **Coulometry.**

Tool Changer, Automatic (See **Numerical control.**)

Torque, Measurement of (See **Dynamometer.**)

Torque-tube Level Gage (See **Liquid level, measurement of.**)

Tracers, Medical Use of (See **Radioisotope-tracer instrumentation.**)

Tracing Control Systems In certain production processes, it is required to reproduce a two-dimensional figure or pattern, or a three-dimensional model. The characteristics of the subject to be reproduced or to cause guidance of a machine can be reduced to numbers—as is done in numerical control. Tracing methods are another approach wherein master patterns or curves are followed optically (line following) as is done in some flame-cutting and fabric-cutting machines for garment production. A similar technique is used in some automated drafting machines. Dimensionally, the relationship between the pattern instruction and the action of the machine, such as cutting, may be in a one-to-one ratio; or it may be multiplied where a scaled-down pattern instruction is used. Mechanical styli also may be used to sense the master pattern, template, or model. Where the dimensions are not scaled down or where simple mechanical levers (pantograph) are used to connect the sensing side of the instrument system to the cutting or reproducing side, the system sometimes is referred to as a "monkey see/monkey do" system. Also see **Drafting system, automatic;** and **Numerical control.**

Transducer Generally, a device that converts energy from one form to another, such as acoustic energy to mechanical or electrical energy. Various measuring devices, loudspeakers, microphones, phonograph pickups, and other energy-converting devices may be termed transducers. Particularly, from the standpoints of measurement and instrumentation, a transducer is an energy-converting device that (1) receives stimulation from a physical situation or condition that is the object of measurement (the measurand) and (2) converts that stimulation into a definitely associated signal that is more appropriate or convenient to use as the input to a measurement system.

Originally, the term transducer was used to describe a device that converted mechanical stimuli, such as force, displacement, and acceleration, into a useful electrical output, such as a change in voltage or current. In modern usage, the term has broadened to include the sensing of all forms of stimuli, such as heat, radiation, and sound, and the generating of pneumatic, hydraulic, and other forms of output signals in addition to electrical outputs. Thus, with the exception of minor shades of differences in meaning that persist because of their traditional usage, transducer, sensor, detector, and primary element essentially are synonymous terms.

Occasionally, the terms *active* and *passive* are used to describe transducers. The active transducer directly generates an electric current or voltage in response to stimulation by the measurand. For example, a thermocouple converts heat energy into a related millivoltage. The passive transducer produces a change in a passive electrical parameter, such as capacitance, resistance, or inductance, as the result of stimulation by the measurand. For example, movement of the armature of a differential transformer that may result from a change in the measurand (displacement) alters the coupling of the transformer such that the ac output signal of the device (transducer) is definitely related to the displacement. Passive transducers almost always require a supply of additional external electrical energy so as to convert a change in the measurand to a change in voltage, current, or frequency. All transducers that produce pneumatic or hydraulic pressure outputs, of course, require external energy sources.

There are hundreds of different kinds of transducers for sensing scores of different physical and chemical phenomena. All the more important transducers are described in this encyclopedia; they generally can be located by looking up the measurand involved. Liberal use of the comprehensive index at the end of this volume also should be made—where the common names of transducers are listed as well as the various transduction principles, such as piezoelectric, photoelectric, thermoelectric, electrochemical, radioactive, electromechanical, electrohydraulic, electrostatic, vibrating-wire, and force-balance.

REFERENCES

Harvey, G. F. (ed): Transducer Compendium, 2d ed., *ISA Publ.* ISATC2, 1963.

Guide for Specifications and Tests for Piezoelectric Acceleration Transducers for Aerospace Testing, *ISA Publ.* RP37.2, 1964.

Guide for Specifications and Tests for Strain Gage Pressure Transducers for Aerospace Testing, *ISA Publ.* RP37.3, 1964.

Specifications and Tests of Potentiometric Pressure Transducers for Aerospace Testing, *ISA Publ.* S37.6, 1967.

Terminology and Specifications for Turbine-type Flow Transducers, *ISA Publ.* RP31.1, 1961.

Nomenclature and Specification Terminology for Aerospace Test Transducers with Electrical Output, *ISA Publ.* RP37.1, 1963.

Transfer Characteristic (See **Transfer function, theoretical.**)

Transfer Function, Theoretical A mathematical, graphical, or tabular statement of the influence which a system or element has on a signal or action compared at the input and at the output terminals. Or, the theoretical relationship between measurand and output values as determined by inherent principles of operation. It is expressed as a common function, e.g., linear, parabolic, logarithmic, sinusoidal, and so on.

Similarly, the term *transfer characteristic* (which is less frequently used) may be defined as—a description of the character of a device, transducer, sensor, instrumentality, or system as it pertains to the manipulation of an input signal, condition, or state by the device to produce an output or response. The description may be verbal, tabular, graphic, or mathematical and may or may not include time-dependent characteristics.

The transfer characteristic generally is a broad description of the character of a sensor, transducer, or system component. Thus, by way of example, the transfer characteristic of an amplifier is its gain, and the amplifier is presumed to be linear.

The transfer function tends to be a more detailed description with time-dependent characteristics being accounted for. Thus, the transfer function of another amplifier may be described by a gain term and a single-time-constant delay, i.e., $e_o/e_i = g/(1 + sT)$, where e_o is the output response, e_i is the input condition, g is the static gain, and $1/(1 + sT)$ describes the single-time-constant delay brought about by the amplifier.

The theoretical transfer function is derived through the application of analytical powers from a knowledge of the principles of operation of a sensor or transducer. The factor represents a performance expectation and not necessarily a performance achievement. In deriving the theoretical transfer function, simplifying assumptions are made. Typically, linear behavior is presumed for constituent assemblies. Hysteresis is presumed to be absent and the performance is thought to be independent of signal or measurand amplitude. These assumptions and others that might be made, while useful, may be only partially true. The theoretical transfer function thus derived may have limited applicability and lead to performance expectations which are not met in actual practice.

Depending on the true nature of a device, transducer, or sensor, the transfer function may or may not be dimensionless. In the case of the previously cited amplifier, no dimensions are attached to the transfer function. In the case of a pressure transducer or sensor, the dimensions might be volts per psi. In the development of theoretical transfer functions, it often is convenient to separate the description into two parts: (1) a static constant which has some dimension and (2) a dynamic description which is dimensionless. Thus, the transfer function of a pressure sensor or transducer might be

$$\frac{e_o}{P_i} = K_1 \frac{1 + sT_1}{(1 + sT_2)(1 + sT_3)} \qquad \frac{\text{volts}}{\text{psi}}$$

where K_1 is the static constant; s is the Laplace operator; and T_1, T_2, and T_3 are time constants of the device being described. e_o is the output response and P_i is the input condition.

—Otto Muller-Girard, *Consulting Engineer, Rochester, N.Y.*

Transformer, Linear (See **Linear transformer.**)

Transformer, Linear Variable-differential [See **Linear variable-differential transformer (LVDT).**]

Transformer Bridge (See **Bridge circuits, electric.**)

Translator A software program (see **Software**) which takes input in a specified syntax and converts it into another form according to a predetermined algorithm (see **Algorithm**). A major use of translators is as compilers for higher-level languages such as FORTRAN, ALGOL, COBOL, MAD, and PL/I (see **Language, computer; Compiler**). These programs allow the definition and solution of a problem without becoming involved in the details of machine operation and machine coding.

Translators may also be used to convert program statements in one language syntax into another language syntax. This may be useful when specialized routines are available in one language but not in another.

—L. E. Wise, *IBM Corporation*

Transmission, Electric The major methods of electrically transmitting instrument information are (1) voltage, (2) current, (3) position or ratio, (4) impulse, and (5) frequency. The first three methods are limited to wire channels; the other two types are applicable to any kind of channel.

Voltage Telemetering: Once widely used, this method is essentially of historical interest for many applications. In voltage telemetering, the measured variable is converted to an ac or dc voltage. Many systems use primary elements which produce voltage as an intrinsic part of their measuring function. Thermocouples, tachometers, and differential transformers are examples. When widely used, applications were mainly for industrial-plant distances of about 1,000 ft which thus often were not really thought of as telemetering installations, but rather as localized transmission systems. Both self-balancing potentiometers and bridges and deflectional indicators were used as receivers. Deflectional indicators had to be calibrated for line resistance when used. Some of the problems associated with straight voltage transmission included the effects of line resistance, leakage, and noise, especially where low voltages were used. The signal-to-noise ratio had to be relatively large—on the order of two or greater. See **Signal-to-noise ratio.** To avoid these difficulties, other means of transmitting signals were employed. For example, voltage-controlled oscillators can put out a variable frequency that is proportional to voltage input and, in addition, can be multiplexed over the same transmission channel.

Current Telemetering: Many industrial measuring instruments have a current output that is varied by means of an adjustable resistance in the line and consequently are not suitable for long-distance telemetering. Other instruments, however, incorporate circuits having resistance values of several thousand ohms. Most of these instruments use a mechanism to feed back the current to balance the input quantity and some type of detector-amplifier combination to detect any unbalance and correct for it. Direct currents of a few milliamperes normally are used.

The feedback loop provides a convenient locus for applying compensation to offset the effects of ambient-temperature changes on the primary element which actuates the transmitter. In addition, the use of feedback reduces the need for voltage regulation of the power supply to the transmitter. Current systems can develop higher voltages than most voltage systems and, conse-

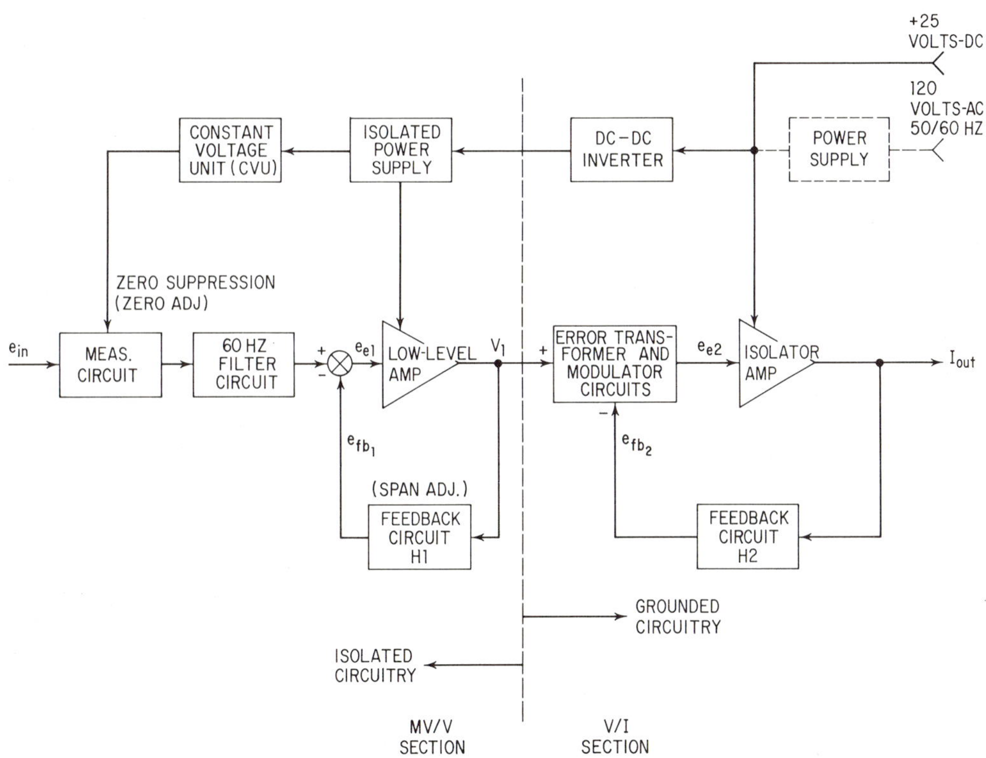

Fig. T-47. Block diagram of a millivolt-to-current transmitter. *(Honeywell VutroniK MV/I.)*

quently, can be made more immune to the effects of thermal and inductive voltages in the interconnecting leads as well as line resistance. The transmitters usually are more expensive than voltage transmitters because of the feedback amplifier required, but inexpensive milliammeters can be used as receivers without calibration for line resistance. Current systems are adaptable to adding the output of several transmitters, operating several receivers at once, and operating multipoint receivers. Speed of response is almost instantaneous with many types. Most transmitters deliver some definite minimum current when the measured variable is at its limit and thus provide a means of detecting an open circuit in the channel at the receiver. Ground return is rarely used.

The instrument[1] schematically diagramed in Fig. T-47 is a millivolt-to-current transmitter. The millivolt input signal e_{in}, whether obtained from a thermocouple or resistance bulb or some other millivolt source, is fed into the measuring circuit where the zero suppression, which includes cold-junction compensation in the case of thermocouple inputs, is added and then fed into the 60-Hz filter circuit. The filtered, zero-suppressed dc input signal then is compared with the dc feedback signal e_{fb1} at the summing junction. The difference or error between these two signals is modulated by a MOS FET[2] chopper and the resultant ac error signal e_{e1} is fed into the high-

[1] Honeywell VutroniK MV/I and R/I transmitter.
[2] Metal oxide semiconductor field-effect transistor.

Fig. T-48. Front view of MV/I transmitter (millivolt-to-current) shown at left. Device with cover removed to show terminal blocks is shown at right. *(Honeywell Inc.)*

gain ac amplifier designated as the low-level amplifier in Fig. T-47. The ac error signal is amplified and then demodulated back into a dc voltage V_1. This voltage normally is in the range of 1 to 5 V dc for 0 to 100 percent of output signal. A percentage of V_1 is fed back as e_{fb1}. The input span in millivolts is determined by the percentage of V_1 that is fed back.

The 1- to 5-V dc signal V_1 then is the input signal to the isolated *V/I* circuitry. This signal is modulated in phase with the modulated output feedback signal and fed into the isolator error transformer. The difference of these two signals is sensed by a third winding on the error transformer and is the ac error signal e_{e2}, which then is fed into the isolator amplifier. This signal then is amplified and converted into a 4- to 20-mA or 10- to 50-mA current signal by a current amplifier in the output of the isolator amplifier circuit. This current is converted into a voltage by the feedback network *H*2 to provide the feedback voltage e_{fb2}. If a 1- to 5-V dc output signal is desired, the 4- to 20-mA current signal is passed through a precision 250-Ω resistor to develop the 1- to 5-V dc signal.

The dc-dc inverter circuit consists of a saturating-core inverter circuit which provides a symmetrical square wave to drive all modulators and demodulators. This circuit also provides an isolated dc supply for the *MV/V* section of the circuitry. The inverter and isolator amplifier operate from an external + 25-V dc power supply or from an internal 120-V ac-powered dc supply. The transmitter is shown in Fig. T-48.

Position or Ratio Telemetering: This classification includes many devices which may appear quite different but which have in common the characteristic of varying the relative magnitude of two or more quantities in accordance with the measured variable. The quantities may be resistances, inductances, ac voltages, alternate half cycles of a single ac voltage, and several others. Resistance slidewire bridges, inductance bridges, and ac self-synchronous motors are examples. Also see **Linear transformer; Resolver; Transolver;** and **Synchro.** It is particularly difficult to generalize about this broad class, but most ratio systems are intended for distances below a few miles, many for distances below a few hundred feet. Most of these devices, but not all, require more than two interconnecting wires. The transmitters usually are inexpensive and require only low forces from the primary element. Many systems can be used either with low-cost deflectional receivers or with servo-actuated receivers if higher torque is needed.

Other forms of telemetry and electric transmission are described elsewhere in this encyclopedia. Particularly, see **Data communications; Frequency modulation; Pulse-code modulation; Pulse-duration telemetry; Pulse-position modulation;** and **Radio telemetry.** Also see entries listed under *Communications* in the **Classified Index.**

Transmission Densitometer (See Densitometer.)

Transmission, Pneumatic A means that uses compressed air to communicate the values of instrumental-measurement information from one location to the next. Sometimes referred to as pneumatic telemetering. Particularly in the process industries, it is impractical to extend sufficiently a basic measuring device, such as fluid-filled thermometer, over a relatively long distance from the point of measurement because of the introduction of measuring lags and errors. Pneumatic signals also can be used in the reverse direction, that is, for communicating set-point information from a central control house to on-line controlling instruments.

Pneumatic-transmission systems are comprised of a source of compressed air, a transmitter which modulates the air supply, output tubing or piping to carry the modulated air signal to the receiving point, and one or more receivers, connected in parallel. Such a system is shown in Fig. T-49. A receiver might be a simple pressure gage or

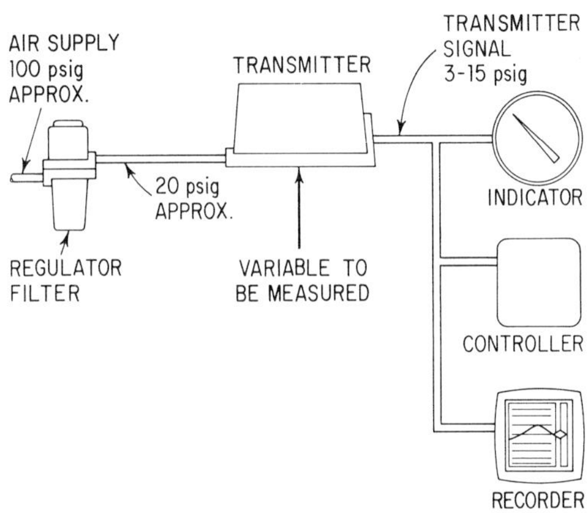

Fig. T-49. Pneumatic-transmission system.

other indicator, an automatic controller, a recorder, or some other suitable device.

The air supply for pneumatic transmitters must be clean, dry air at constant pressure. Normally, a motor or steam-driven compressor supplies cooled, dried, and filtered air at about 100 psig pressure. Pipe or tubing leads this air to field locations where transmitters are located. At the transmitters, the air is passed through a fine filter to remove any entrained dirt particles and is then reduced to 20 or 35 psig to serve as the air supply for the transmitters. Care in providing a clean, dry air supply increases the reliability of pneumatic-transmission systems.

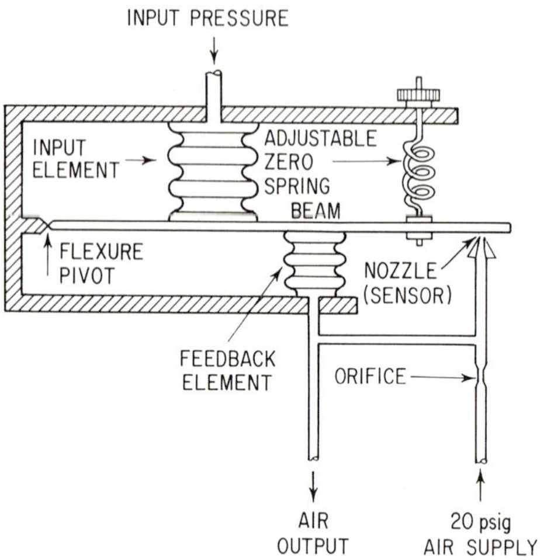

Fig. T-50. Simple pressure transmitter.

A pneumatic transmitter provides air output pressure proportional to the value of the variable being measured. An output-pressure range of 3 to 15 psi is most common, and usually requires use of a 20-psi air-supply pressure. Measurements of temperature, differential pressure, density, liquid level, and others may be transmitted by suitably designed transmitters. The transmitter is adjusted so that its output is 3 psig when the variable is at the lowest value to be measured, say 100°F. As the variable increases, the transmitter output increases in proportion, linearly, to be 15 psig when the variable has increased to its maximum value, say 200°F. Normally, a linear relationship between the value of the variable and the value of the transmitter output pressure is desired.

A simple pneumatic transmitter consisting of three main elements is shown in Fig. T-50. The first element is a sensing element, such as a bellows, bourdon, diaphragm, or other means of converting the variable into a mechanical displacement, force, or torque. The second is a sensor, frequently a direct pilot or a flapper nozzle. The third is a feedback element, usually a bellows or diaphragm. If the transmission distance is to be very large and involves a considerable amount of connecting tubing or a large volume in the receiver, then a relay similar to that used in a pneumatic controller may be employed. The relay is inserted between the nozzle and the output tubing as shown in Fig. T-51. Besides serving as a power booster, providing greater airflow into the tubing, the relay may also somewhat improve the linearity of the transmitter in the same manner as for a pneumatic controller. (See **Pneumatic controller**.)

The transmitter may include rate or derivative action to compensate for inherent lags in the measurement, such as the lag of a bulb in a filled temperature-measuring system. This dynamic compensation is achieved by an adjustable time constant in the feedback, as shown in Fig. T-51. The time constant is field-adjusted to match the measuring lag. The approximate effect of such compensation is shown in Fig. T-52. The output of the compensated transmitter more nearly agrees with the actual value of the variable during the transient than does that of the uncompensated transmitter.

The distance over which pneumatic signals may be transmitted is limited by the amount of added lag which the tubing will contribute to the system.

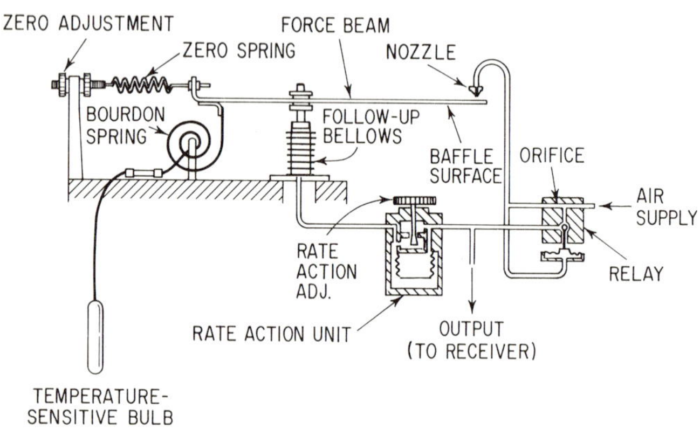

Fig. T-51. Pneumatic transmitter with relay and lag-compensating rate action. *(Taylor.)*

The output tubing includes capacitance proportional to the volume of the tube and distributed along the length of the tubing. The tubing also includes resistance, similarly distributed along the tube. These distributed parameters make precise calculation of the lag tubing or transmission lag quite tedious. It is customary to rely upon approximations made from experimental data. Usually the dynamic response of transmission tubing is represented by combination of a dead-period lag and a simple time constant which most nearly represents the data. Most often, transmission tubing is approximately $\frac{3}{16}$ in. inside diameter and made of copper, aluminum, or plastic. For such tubing, the dynamic response for a 350-ft length terminating in a 1 in.^{3} volume may be approximated by a time constant of 1 s and a dead period of $\frac{1}{3}$ s. For a length of 1,000 ft, the time constant would be 7 s and the dead period 1 s.

If the receiver volume in which the transmission system terminates is very large, the dynamic response of the tubing and its receiver load would be slow. A pneumatic booster, shown in Fig. T-53, is sometimes added at the end of the tubing and adjacent to the receiver. The booster must be of high quality and with a gain of 1 so that its output equals its input, thereby not adding an error to the measurement. The large ports of the booster allow a comparatively great airflow into or out of the connected volume in response to changes in input pressure. Dynamic response of the tubing and receiver is thus changed from a dead-period lag and a single long time constant to the same dead-period lag but with two short time constants.

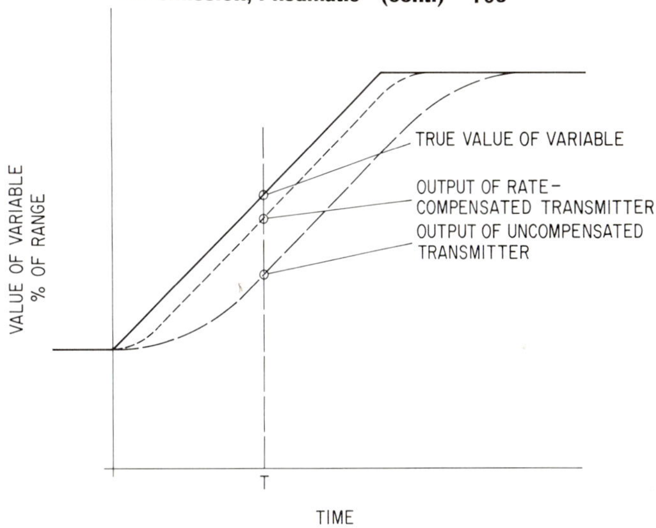

Fig. T-52. Effect of compensating a pneumatic transmitter for measuring lag.

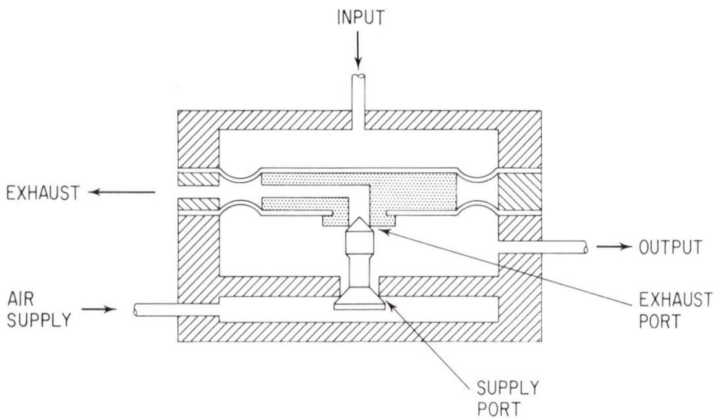

Fig. T-53. Pneumatic booster in pneumatic-transmission system. *(Taylor.)*

REFERENCES

Caldwell, Coon, and Zoss: "Frequency Response for Process Control," McGraw-Hill Book Company, New York, 1959.

Eckman, Donald P.: "Automatic Process Control," John Wiley & Sons, Inc., New York, 1958.

Harriott, Peter: "Process Control," McGraw-Hill Book Company, New York, 1964.
Nichols, N. B.: The Linear Properties of Pneumatic Transmission Lines, *ISA Trans.*, vol. 1, no. 1, p. 5, 1962.

—Howard R. Jaquith, *Taylor Instrument Process Control Division, Sybron Corporation*

Transolver A hybrid synchro—a cross between the two-phase and the three-phase units. One element of the transolver has a three-phase winding; the other element is two-phase. Some manufacturers differentiate between the two types by limiting the term *transolver* to units with a three-phase stator and the term *differential resolver* to those units with a three-phase rotor. This difference in terminology is not widespread, and the term transolver will probably survive to represent both types (see Figs. T-54 and T-55).

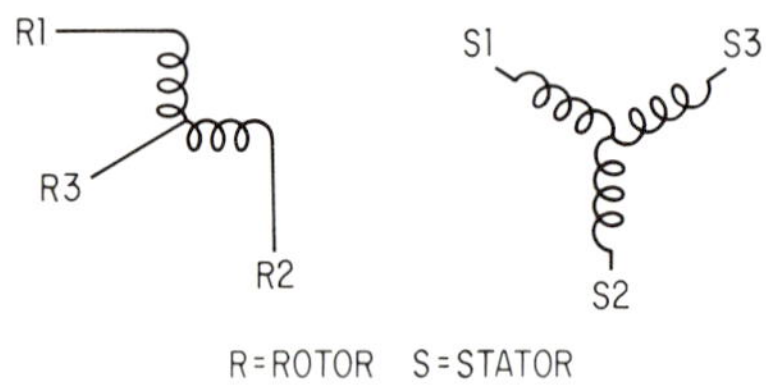

Fig. T-54. Schematic of transolver with three-phase stator.

Transolvers are used to transform three-wire information to four-wire and vice versa. A transolver, through proper switching, may act as both a transmitter and a control transformer. In the transmitter mode, one rotor phase is excited; the other is shorted to ground, thus providing electrical saliency. In the control-transformer mode, one rotor phase is fed to the servoamplifier; the other is symmetrically loaded to avoid errors due to impedance unbalance. In either mode, there is no loss of accuracy from that of the conventional synchro type that is replaced. Also see **Linear transformer; Resolver;** and **Synchro.**

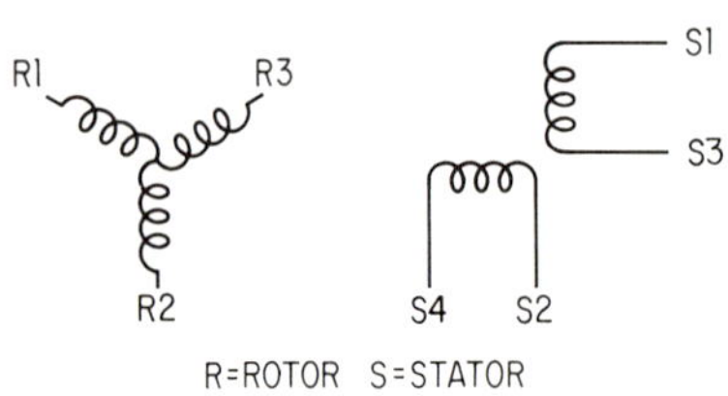

Fig. T-55. Schematic of transolver with three-phase rotor.

—Joseph P. Spetzer, Jr., *Clifton, Division of Litton Industries*

Transponder A device that is interrogated by a radio signal from a remote location and automatically transmits a response signal on either the same or a different frequency. When used in conjunction with an interrogator, a transponder may assist in the determination of bearing, range, or both. One form of transponder is the radar beacon. Also see **Air-traffic control.**

Transverse-mode Voltage (See **Differential-mode voltage.**)

Transmissometer (See **Oceanographic instrumentation.**)

True Air Speed [See **Air-data systems (aircraft).**]

True Air Temperature [See **Air-data systems (aircraft).**]

Turbidimeter An instrument designed to detect discrete inhomogeneities in a heterogeneous system, such as smoke and suspended dust particles in the atmosphere or suspended particles in a liquid. Turbid conditions can be described as cloudy, hazy, and smokey—as contrasted with completely clear and translucent. Typical applications include the control of boiler feedwater, the monitoring of process-waste effluents, measuring the haze of beer and other beverages, controlling the quality of raw water, determining the efficiency of filters, controlling filter breakthroughs—in a variety of industries. Sensitivities as low as 100 ppb with a meter accuracy of ± 0.5 percent are obtainable.

Turbidimeters are based upon (1) light-absorption measurement, (2) light-scattering measure, or (3) a combination of these two phenomena. When a beam of light is directed at particles suspended in a fluid or deposited on a transparent medium, some of the incident radiation is scattered in all directions, some of the radiation is absorbed, and the remainder is transmitted.

Purely absorption or light-scattering measurements are subject to certain limitations, including color dependency, temperature effects, light-source variations, and the refractive index and shape of the particles. Most instruments that operate on the scattering principles measure the scattering at 90°. Measurement of low-angle scattering, especially in the 0 to 45° range, has a number of advantages.

The optical system of a turbidimeter that combines absorption and low-angle scattering is shown in Fig. T-56. This device consists of a stainless-steel sampling chamber (for pressures up to 500 psi), a light source, an optical system, a special dual-photocell detector, and a display in the form of a readout meter or recorder. The light source *L* (a low-voltage flat-ribbon incandescent point source) is energized by a filament transformer. The lens system *C* collimates the light from the source and projects a beam of parallel light through the sampling chamber *SC*. Diaphragm *D* in front of the entrance window prevents stray light from entering the sample chamber.

Dual photocell *P*, which consists of sections *Pc* and *Po*, is mounted on the opposite side of the sampling chamber, directly behind the exit window. Both photocell sections are connected over special circuitry with potentiometers R_1 and R_2 to the indicating meter *M*. With no obstruction in the light pass, as would be the case when clear liquid is in the sampling chamber, all light of the parallel beam will fall upon section *Pc*, producing a positive deflection on the meter. Potentiometer R_1 then is set to read zero turbidity. By substituting the clean liquid with a liquid of known turbidity, part of the light of the parallel beam passing through the turbid liquid will be scattered. This reduces the amount of light that falls upon photocell section *Pc* and results in a decreased meter deflection. Since most of the scattered light is diverted into the angular directions from 0 to 45°, the light will fall upon section *Po* of photocell *P* to decrease further the deflection on the meter. Potentiometer R_2 now is used to set the meter reading to full scale for a given turbidity value. The meter thus is calibrated for a given type of sample.

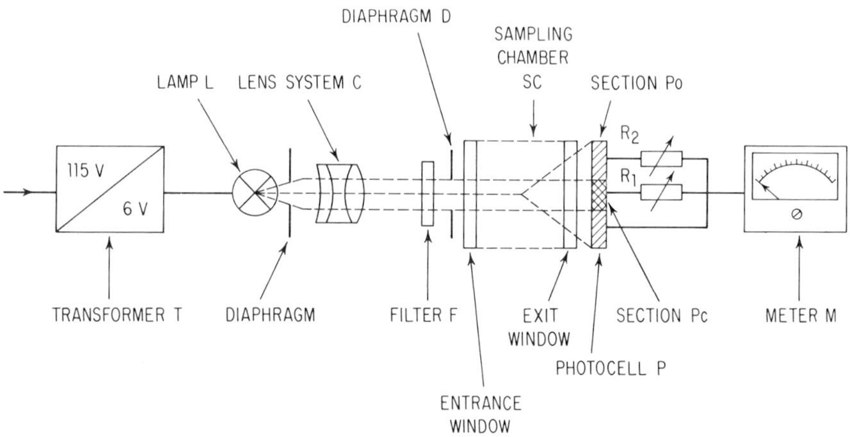

Fig. T-56. Optical system of a combination absorption and low-angle scattering turbidimeter. *(Photronic.)*

—R. Michel, *Photronic, Inc.*

Turn-and-Bank Indicator (See **Flight-control systems.**)

Two-color Pyrometer (See **Optical pyrometer.**)

U

Ultrasonics, Medical Instrumental Use of The visualization of body tissue by reflection of an ultrasonic beam is safer than radiography. Furthermore, it can show boundary layers between soft tissues which are exceedingly difficult to see by conventional radiographic techniques.

A sound-generating piezoelectric crystal is coupled through water or oil to the body surface, and sound pulses in the range of 2 to 20 MHz, with a repetition rate of 1 to 3 kHz and a typical time duration of approximately 1 μs, are projected into the body with power levels between 2 and 20 mW/cm^2. The sound is absorbed by air within the body, passes through fluid with no reflection, and produces reflections from interfaces between fluid and tissue or between interfaces of tissues, like muscle and fat, that transmit sound at different rates.

The image is displayed on an oscilloscope in the form of an "*A* scan" or a "*B* scan." In the *A*-scan presentation, the oscilloscope trace is started at the time of transmission. At some later time, the echo from an interface produces a pulse on the oscilloscope face at a distance across the trace corresponding to the time required for the echo to be received. Since this time is related to distance, a series of pulses across the oscilloscope face will correspond to a series of interfaces between dissimilar tissues within the body. This "*A* scan" is useful for measuring the dimensions of a structure known to be within the sound beam (e.g., the heart). It is also useful for following the motion of a structure (e.g., a heart valve), provided one knows what is being observed.

B-scan presentation is achieved by moving the source, the receiver unit, and the recorder in rectangular coordinates, generating a two-dimensional picture. The speed with which this can be done, however, is limited, and mechanical motion makes coupling to the body difficult and somewhat variable; so this technique cannot usually be used to study any internal structure in motion.

One system eliminates mechanical motion by using the "phased-array" principle used in radar scanning. A fixed group of transducers is arranged in an array. The transducers are then pulsed out of phase with one another in a predetermined pattern which will give directionality to the resultant beam. In the receiver mode, these same transducers are individually connected to delay lines having variable delays so that the same directionality is once again achieved.

A simpler way to avoid some of the mechanical problems is simply a rotate the transducer instead of translating it, thus producing a wedge-shaped presentation with the narrow part of the wedge at the surface.

These ultrasonic techniques are used for locating brain tumors, viewing the ventricles in the brain, and viewing the cerebral-pulse pressure wave. In studies of the heart, valve motion, pericardial effusion, and size are measured. Tumors of the breast are located by the method, and the size and shape of the eye can be measured with a fair degree of precision. The method also has been used to measure the heart rate of the fetus and to delineate fetal structures sufficiently well to diagnose twins.

REFERENCES

Dig. 7th Intern. Conf. Medical and Biological Engineering, session 21 on Ultrasonics, pp. 313–326, Stockholm, 1967.

White, D. N., J. M Clark, and M. N. White: Studies in Ultrasonic Echoencephalography—VII: General Principles of Recording Information in Ultrasonic B- and C-scanning and the Effects of Scatter, Reflection, and Refraction by Cadaver Skull on This Information, *Med. Biol. Eng.,* vol. 5, no. 1, pp. 3–14, January, 1967.

Holmes, J. H.: Medical Ultrasonic Techniques, *Proc. 2d Natl. Biomedical Sciences Symposium,* vol. 2, pp. 11–28, Plenum Press, Plenum Publishing Corporation, New York, 1964.

—J. Malvern Benjamin, Jr.,
Bionic Instruments, Inc.

Ultrasonic Sensory Aid (See **Vision, sensory aids for.**)

Ultrasonic Testing (See **Nondestructive inspection.**)

Ultrasonic Thickness Gage (See **Thickness gaging, ultrasonic.**)

Ultraviolet Spectrophotometry This analytical technique is based upon the selective absorption of radiation exhibited by molecular substances in the wavelength range from about 160 to 780 nm. Actually, this wavelength range encompasses the visible and part of the far-ultraviolet regions as well as the near-ultraviolet. Measurements below 160 nm also are feasible but require more specialized instrumentation. Ultraviolet spectrophotometry is used widely for chemical analysis and is considered indispensable in many laboratories. Partly this can be attributed to the fact that most molecular substances exhibit useful absorption in this region, and an equally important factor has been the availability of reliable instrumentation at reasonable cost.

Origin of Absorption: The interaction with ultraviolet radiation mainly involves valence electrons and therefore provides useful information on the chemical structure of the molecular system. For example, organic compounds exhibit unique absorption bands that are characteristic of the type of unsaturation present. The wavelength, or frequency, at which each band is observed provides a means for identification, while intensities can be used for making quantitative evaluations.

Further consideration of organic substances reveals that saturated compounds, such as hexane, are reasonably transparent throughout this region. Only when certain groups, called chromophores, are present does the molecule exhibit useful absorption. For example, the simple olefin hexene differs from hexane by only a single double bond but exhibits an intense absorption band near 180 nm. Other typical chromophores are listed in Table U-1.

If two or more chromophores are present in the same molecule, the observed absorption depends upon their position relative to each other. If separated by two or more carbon atoms, the resulting absorption is a simple summation of the individual chromophores. If the chromophores are adjacent to each other, i.e., conjugated, the absorption band is shifted to longer wavelengths and intensified. Finally, if the chromophores are attached to the same carbon atom, the resulting absorption is intermediate between the two extremes.

By far the majority of compounds exhibiting useful ultraviolet absorption are those containing two or more chromophores, particularly if conjugated. These are often called complex chromophores. Two types are found, those which give rise to K bands and those which give B bands. Examples of the K chromophores are polyenes having the structure $CH_3—(CH=CH)_n—CH_3$ where n is an integer greater than 1. The B bands are associated with the benzene ring.

TABLE U-1. *Typical Ultraviolet Chromophores*

Chromophores	System	Example	λ_{max}, nm	ϵ*
Carbonyl	$R(R')C{=}O$	Acetone	271	16
Carbonxyl	$R{-}C(O)OH$	Acetic acid	200	60
Ethylene	$R(R')C{=}C(R'')R''$	Ethylene	193	10,000
Diene	$R(R')C{=}CH{-}CH{=}C(R'')R'''$	Butadiene	217	21,000
Benzenoid		Benzene	198 255	8,000 230
Nitro	$-NO_2$	Nitromethane	271	19
Nitrate.	$-ONO_2$	Ethyl nitrite	270	12

*ϵ = molar absorptivity.

Finally, simple chromophores such as those listed in Table U-1 give rise to the so-called R bands. Typically the intensities of R bands are weaker than those for B bands, which in turn, are weaker than K bands.

Applications: Although the absorption characteristics of an organic compound often are calculated, this is relatively difficult and time-consuming. Thus, much work is done through empirical data with application of the various guidelines previously mentioned. This empirical approach has been of considerable help in studies on molecular structure whereby specific chemical groups can be identified and their position within the molecule determined. Ultraviolet measurements also provide useful information on steric effects and can be used to differentiate between *cis* and *trans* isomers as well as tautomeric species.

For analytical purposes, ultraviolet spectrophotometry can be used for both identification and determination. The technique is simple, precise, and for many substances, highly sensitive. It has two major limitations: (1) absorption bands are few in number and are relatively broad for most compounds, and (2) individual members within a given class of compounds exhibit nearly identical absorption characteristics. Both of these detract from the usefulness of ultraviolet spectra for identifying specific compounds. In this regard, the technique is less effective than infrared spectrophotometry, although in practice the two are complementary.

Probably of greater importance is the use of ultraviolet methods for quantitative work. This is based upon Beer's law, $A = abc$, which relates absorbance A with concentration c and path length b. The other term, absorptivity a, is a constant related to the probability of absorption at the analytical wavelength. Many substances, such as aromatic compounds, can be determined directly through their own intrinsic absorption. Others can be measured after conversion into a suitably absorbing form. For example, many steroids exhibit little or no useful absorption in their original form but give intense, characteristic spectra after treatment with sulfuric acid. Likewise, many metal ions are reasonably transparent in the ultraviolet but are readily determined as organic complexes. In this case, absorption is actually due to the organic ligand but is influenced by the metal ion.

One of the more important areas of use is the identification and determination of biologically active substances. The fact that ultraviolet methods can be applied to both aqueous and nonaqueous samples is an important advantage here. Many components in body fluids can be determined either directly or through colorimetric methods. Included are such vital components as hemoglobin, porphrins, glucose, cholesterol, and enzymes. In addition, many drugs and narcotics can be measured both in the body as well as in formulations. Vitamin assay is another related activity and actually was one of the first major applications of ultraviolet spectrophotometry.

Nearly all metals and nonmetals can be determined through their ultraviolet absorption or by colorimetric methods. Well-documented procedures have been published on countless samples including such materials as alloys, beverages, foods, minerals, plastics, biological fluids, petroleum, glass, and water.

Other common applications of ultraviolet spectrophotometry include (1) determination of common pollutants such as aldehydes, phenolics, and ozone; (2) color measurement and analysis of dyestuffs; (3) studies on polyaromatics and other carcinogenic substances; (4) determination of food additives; (5) analysis of petroleum fractions; and (6) detection and determination of pesticide residues. These, of course, are but a few of the many examples which could be cited.

Instrumentation for these measurements ranges from simple, inexpensive colorimeters to versatile, automatic spectrophotometers. The latter provide greater capabilities, but colorimeters are well suited for routine analysis and are used extensively.

The essential components of a spectrophotometer include a light source, monochromator, sample holder, detector system, and readout device. Typically, tungsten lamps are used as the light source for the visible region while hydrogen (or deuterium) discharge lamps are used in the ultraviolet. Both types provide a continuum over their respective regions of use. Common detectors include photoemissive cells and photomultiplier tubes. These give better response at some wavelengths than at others, and careful selection must be made in this regard.

The monochromator includes the dispersing element and associated optics required for isolating narrow bands of wavelengths from the source continuum. Either prisms or gratings can be used as the dispersing element, and advantages can be found for both types. Monochromator design is quite varied and accounts for many of the differences in performance found between instruments.

The sample compartment of a spectrophotometer normally accommodates liquid cells of varying path length with provision for gas cells and special accessories as well. Many instruments also include temperature regulation and, in some cases, provision for heating or cooling.

The readout system has progressed from simple null meters to potentiometric recorders and, more recently, to digital displays. There is a definite trend toward automatic digital printing and computer operations, particularly where high workloads involving routine measurements are found.

The optical path of a double-beam spectro-

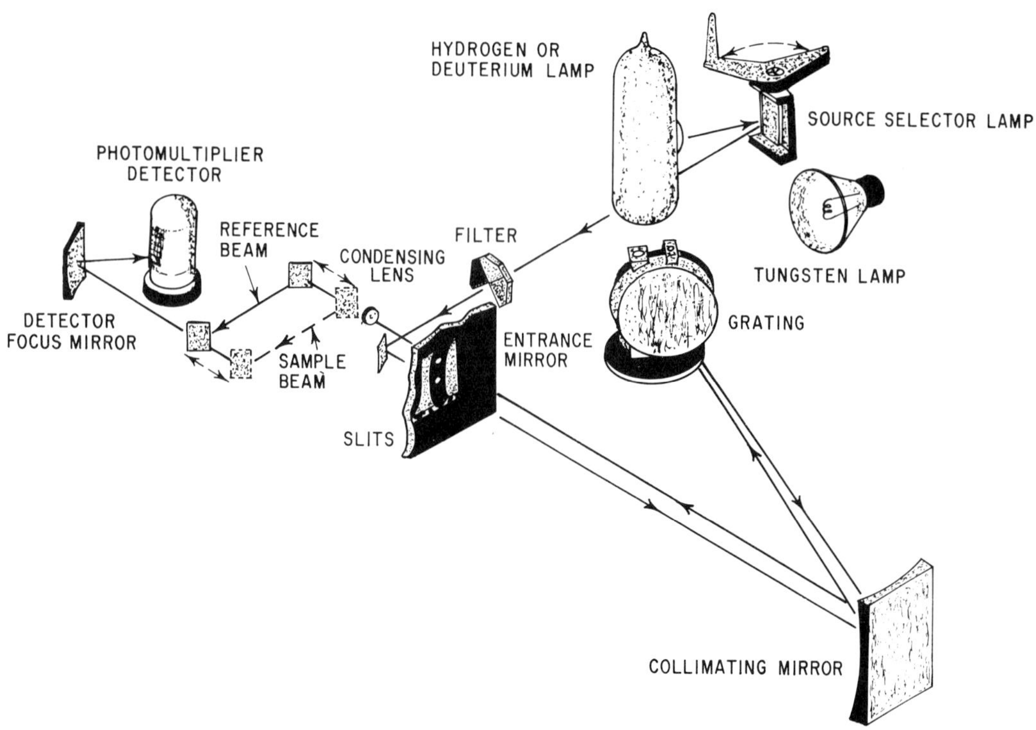

Fig. U-1. Optical diagram for double-beam grating ultraviolet spectrophotometer. *(Beckman.)*

photometer is illustrated in Fig. U-1. Each band of wavelengths isolated by the grating monochromator is split into two beams which pass alternately through the sample and reference paths. The two beams are then recombined along a single path but separated in time. Thus, the detector receives an alternating optical signal consisting of radiant power P through the sample and P_0 through the reference. This output is converted into an electrical signal that is related to the transmittance of the sample P/P_0. The transmittance is either displayed on the readout device or converted into another function, such as absorbance, which is displayed.

Automatic double-beam instruments are used widely, although manual single-beam units also continue to find widespread usage. Some of the factors involved in the selection of the proper instrument are given in Table U-2.

TABLE U-2. *Spectrophotometer Parameters*

1. Performance factors
   a. Wavelength range
   b. Wavelength repeatability
   c. Wavelength accuracy
   d. Resolution
   e. Stray light
   f. Photometric range
   g. Photometric reliability
2. Operating features
   a. Ac or dc electronics
   b. Single- or double-beam optics
   c. Manual or automatic operation
   d. Variable slit width and response controls
   e. Time drive and repetitive scanning
   f. Meter, recorder, or digital readout
   g. Scale expansion (ordinate and abscissa)

Process Analyzers: Although more limited in use than infrared analyzers, ultraviolet units also are available for continuously monitoring process streams. These devices are related more to laboratory colorimeters than to spectrophotometers, but they are more rugged and have better provision for long-term stability. Sampling is achieved with flow cells.

The simplest, and perhaps most successful, applications for ultraviolet analyzers are those involving a strongly absorbing analyte in a transparent matrix. For example, trace aromatic compounds can be monitored in water streams. If other absorbing species are present, the measurement is more difficult, but often it can be made by a differential technique. In this case, the sample absorption is measured against a reference stream which includes the interfering substances but not the analyte. This is possible if the stream is available both before and after the analyte is introduced. Another approach involves measurement of the sample before and after chemical treatment whereby the analyte is determined by the change in absorption. An example of this is the determination of phenols in water through their wavelength shift when the pH is changed.

In addition to the determination of components in liquid streams, these analyzers are also used to monitor color and turbidity. Gas samples can also be handled as required in analyzing stock gases for control of air pollutants.

Also see **Colorimeter** and **Photometers.**

—W. F. Ulrich, *Beckman Instruments, Inc.*

Unbonded Strain Gage (See **Strain gage.**)

Universal Testing Machine [See **Testing (materials), destructive.**]

USASCII Code American Standard Code for Information Interchange ANSI X3.4-1968. A binary code adopted by the American National Standards Institute (ANSI) to be used as a standard code for intercommunication between various types of devices. The code exists in a 7-bit form, although consideration is being given to a revised and expanded code of 8 bits. The latter version is referred to as USASCII-8 code. The USASCII code was formally adopted as a Federal standard by the United States in March, 1968 (see Table U-3).

—L. E. Wise, *IBM Corporation*

TABLE U-3. *USASCII Table*

				7	0	0	0	0	1	1	1	1
Bits				6	0	0	1	0	0	1	1	1
				5	0	1	0	1	0	1	0	1
4	3	2	1	Col Row	00	01	02	03	04	05	06	07
0	0	0	0	0	NUL 12-0 9-8-1	DLE 12-11 9-8-1	SP NO PCH	0 0	@ 8-4	P 11-7	8-1	p 12-11-7
0	0	0	1	1	SOH 12-9-1	DCI 11-9-1	! 12-8-7	1 1	A 12-1	Q 11-8	a 12-0-1	q 12-11-8
0	0	1	0	2	SIX 12-9-2	DC2 11-9-2	″ 8-7	2 2	B 12-12	R 11-9	b 12-0-2	r 12-11-9
0	0	1	1	3	ETX 12-9-3	DC3 11-9-3	# 8-3	3 3	C 12-3	S 0-2	c 12-0-3	s 11-0-2
0	1	0	0	4	EOT 9-7	DC4 9-8-4	$ 11-8-3	4 4	D 12-4	T 0-3	d 12-0-4	t 11-0-3
0	1	0	1	5	ENQ 0-9-8-5	NAK 9-8-5	% 0-8-4	5 5	E 12-5	U 0-4	e 12-0-5	u 11-0-4
0	1	1	0	6	ACK 0-9-8-6	SYN 9-2	& 12	6 6	F 12-6	V 0-5	f 12-0-6	v 11-0-5
0	1	1	1	7	BEL 0-9-8-7	ETB 0-9-6	′ 8-5	7 7	G 12-7	W 0-6	g 12-0-7	w 11-0-6
1	0	0	0	8	BS 11-9-6	CAN 11-9-8	(12-8-5	8 8	H 12-8	X 0-7	h 12-0-8	x 11-0-7
1	0	0	1	9	HT 12-9-5	EM 11-9-8-1	) 11-8-5	9 9	I 12-9	Y 0-8	i 12-0-9	y 11-0-8
1	0	1	0	10	LF 0-9-5	SUB 9-8-7	* 11-8-4	: 8-2	J 11-1	Z 0-9	j 12-11-1	z 11-0-9
1	0	1	1	11	VT 12-9 8-3	ESC 0-9-7	+ 12-3-6	; 11-8-6	K 11-12	[12-8-2	k 12-11-2	{ 12-0
1	1	0	0	12	FF 12-9 8-4	FS 11-9 8-4	, 0-8-3	< 12-8-4	L 11-3	\ 0-8-2	l | 12-11-3	12-11
1	1	0	1	13	CR 12-9 8-5	GS 11-9 8-5	– 11	= 8-6	M 11-4	] 11-8-2	m 12-11-4	} 11-0
1	1	1	0	14	SO 12-9 8-6	RS 11-9 8-6	. 12-8-3	> 0-8-6	N 11-5	^ 11-8-7	n 12-11-5	~ 11-0-1
1	1	1	1	15	SI 12-9 8-7	US 11-9 8-7	/ 0-1	? 0-8-7	O 11-6	— 0-8-5	o 12-11-6	DEL 12-9-7
					Controls		Numerics and special graphics		Uppercase Latin letters		Lowercase Latin letters	

TABLE U-3. *USASCII Table (Continued)*

Card hole pattern	
Zone punches	Digit punches
12-9	-4
0	-8-4
11	-9
12-0	-1
12-11-0-9	-8-1

The 7-bit USASCII has its bits numbered as follows: $b_7\,b_6\,b_5\,b_4\,b_3\,b_2\,b_1$

Control-character Representations:

NUL Null
SOH Start of heading (CC)
STX Start of text (CC)
ETX End of text (CC)
EOT End of transmission (CC)
ENQ Enquiry (CC)
ACK Acknowledge (CC)
BEL Bell
BS Backspace (FE)
HT Horizontal tabulation (FE)
LF Line feed (FE)
VT Vertical tabulation (FE)
FF Form feed (FE)
CR Carriage return (FE)
SO Shift out
SI Shift in
DLE Data link escape (CC)
DC1 Device control 1
DC2 Device control 2
DC3 Device control 3
DC4 Device control 4
NAK Negative acknowledge (CC)
SYN Synchronous idle (CC)
ETB End of transmission block (CC)
CAN Cancel
EM End of medium
SUB Substitute
ESC Escape
FS File separator (IS)
GS Group separator (IS)
RS Record separator (IS)
US Unit separator (IS)
DEL Delete

Special Graphic Characters:

SP Space
! Exclamation point
" Quotation marks
Number sign
$ Dollar sign
% Percent
& Ampersand
' Apostrophe
(Opening parenthesis
) Closing parenthesis
* Asterisk
+ Plus
, Comma
- Hyphen (minus)
. Period (decimal point)
/ Slant
: Colon
; Semicolon
< Less than
= Equals
> Greater than
? Question mark
@ Commercial at
[Opening bracket
\ Reverse slant
] Closing bracket
^ Circumflex
_ Underline
` Grave accent
{ Opening brace
| Vertical line (this graphic is stylized to distinguish it from logical OR
} Closing brace
~ Tilde

V

Vacuum Gaging Pressure-measurement systems fall into three categories: (1) positive pressures ranging from atmospheric up to 100,000 psi and higher, (2) compound positive/vacuum pressures from about +30 to −30 in. water, and (3) vacuums ranging downward from a few inches of water column to 10^{-14} torr. Low to medium vacuums like those of category 2 are measured by bellows, diaphragms, manometers, and other devices wherein the pressure of the system acts directly upon an exposed area interface. These components are identical or at least very similar to the same components used in positive-pressure gages and are not described here. See **Manometer; Pressure gage, elastic-element; Pressure transducer, vibrating-wire.**

The vacuum range further is broken into five segments: (1) *low vacuums* from atmospheric pressure to 10 torr, (2) *medium vacuums* between 10 and 10^{-3} torr, (3) *high vacuums* between 10^{-3} and 10^{-6} torr, (4) *very high vacuums* from 10^{-6} to 10^{-9} torr, and (5) *ultrahigh vacuums* from 10^{-9} to 10^{-17} torr. A *torr* is the international standard unit of atmospheric pressure or vacuum. One torr is defined as $\frac{1}{760}$th of a standard atmosphere. The torr differs from the earlier millimeter of mercury unit by only 1 part in seven million. Another unit still commonly used in connection with vacuum measurement is the *micron,* which is 10^{-3} mm Hg.

With the exception of the McLeod gage, instruments for measuring vacuum are essentially electronic or nucleonic in nature. Unfortunately, a vacuum gage of single design that is practical and reliable for detecting vacuums, commencing at atmospheric pressure and going to the ultrahigh-vacuum range, does not exist. Thus, the engineer or scientist who works in such ranges will have to switch from one gaging method to the next as he may proceed from atmospheric pressure to a pressure 10^{-1} torr with a roughing pump, on down to below 1×10^{-6} torr with the diffusion-pump heaters activated and the liquid-nitrogen cold trap filled, further down to 10^{-10} torr with the bakeout cycle and circulation of cryogenic liquid between the double walls of the system, and finally on down to 10^{-12} torr or lower by cleaning up the remaining gases with a liquid-helium-cooled absorbent. It is possible, of course, to design an instrument system wherein appropriate vacuum gages for specific ranges can be automatically switched in and out of the system as required, but such systems are costly and complex.

McLeod Gage: There are several variations of the McLeod-gage principle, including designs to make the gage semiautomatic. The gage essentially is a compression manometer in which a piston (liquid mercury) compresses a known volume of gas at an unknown pressure into a measurable smaller volume at a measurable higher pressure. The gage is an absolute gage in the sense that pressure can be computed from the physical dimensions of the gage. Thus, the McLeod gage is particularly useful as a standard for the calibration of other, more convenient instruments. The accuracy ranges in practice from the order of 1 percent in the middle- and high-pressure ranges to considerably less near the low-pressure limit. Since the gage, two versions of which are shown in Fig. V-1, operates on the basis of Boyle's law, it is necessary that the gases involved obey this law over the required range of compression. If gases become sorbed on the walls of the gage, erroneous readings will be given, always in the direction of too low a pressure indication. Since a sequence of manipulations is required to make a pressure measurement, the gage is inherently confined to intermittent rather than continuous readings.

For a single-range square-law McLeod gage, the pressure is proportional to the square of the head of mercury in the two capillary tubes. These capillaries usually are made of equal bore (for compensation of capillary depression), although this expedient fails if and when the capillary becomes dirty. In the Hastings Work/McLeod gage, there is a control valve in the mercury reservoir and an automatic zeroing system which tends to reduce errors and lower the operator-

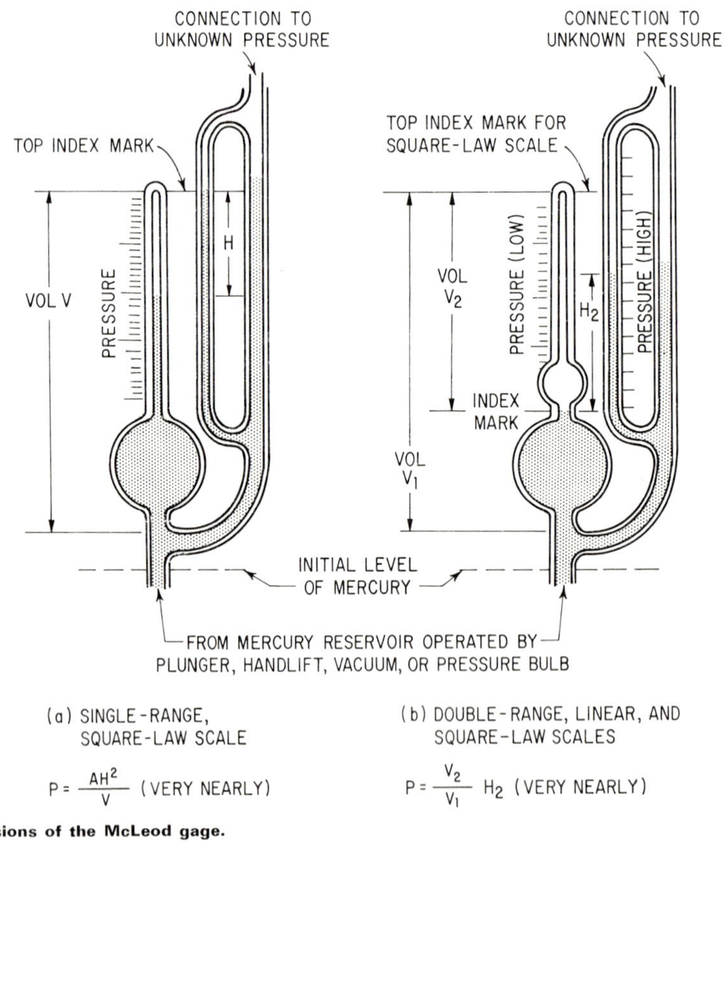

Fig. V-1. Two versions of the McLeod gage.

skill requirement. This gage is available with a dual range: 10^{-5} to 0.2 torr and 10^{-3} to 20 torr.

Thermal-type Vacuum Gages: These devices operate on the principle that the heat conductance of a gas changes with the pressure of the gas. Thus, a thermal gage is comprised of two major elements: (1) a heated surface and (2) a temperature sensor. In a *thermocouple* vacuum gage, a thermocouple is attached to a heater filament that is heated by a constant-energy dc input. To increase sensitivity, a thermopile may be used. In another version,[1] a noble-metal thermopile circuit is used. The hot junctions of the thermopile are heated directly by an alternating current while the cold junctions are kept at the ambient temperature by heavy mounting studs. Thus, a dc voltage is generated between the hot and cold junctions. As the pressure decreases, the lowering of the thermal conductivity of the gas surrounding the hot junctions tends to increase the temperature of the hot junctions, thus increasing the output of the thermopile. A third, unheated couple is connected in opposition to the heated couples and responds only to sudden ambient-temperature changes. This gage is available in three ranges: (1) 0 to 20 mm Hg; (2) 0 to 100 microns Hg; and (3) 0 to 1,000 microns Hg.

In the *resistance-type gage,* or *Pirani gage,* the same heating principle is used, but the temperature is sensed by a change in the resistance of a wire or semiconductor that normally is part of a Wheatstone bridge. Instead of using a separate heating element, a sufficient voltage is applied to the bridge to cause self-heating of the resistance element. Compensation for ambient-temperature changes is effected by using a second resistance and sensor which are enclosed in a reference vacuum. The instrument readout is the bridge-unbalance voltage or current that is needed to return the bridge to balance. By using a feedback network and servosystem, balance can be accomplished automatically.

Sometimes a thermocouple gage is combined with an ionization gage to provide a wider range of vacuum measurement by one instrument. In one instrument of this type,[2] the thermocouple-gage range is 3 to 10^{-3} torr and the ionization-gage range is 10^{-3} to 2×10^{-11} torr.

Ionization-type Vacuum Gages: These instruments measure the electric current that results from ionization of the gas whose pressure is to be measured. The differences between various ionization gages relate principally to the manner in which the ions produced. To convert a gas molecule into a positive ion, an electron must be removed by supplying energy equal to the ionization potential of the given gas molecule. The required 4 to 25 electron volts may be generated in several ways. If the energy is supplied at a constant rate, the rate of ion production will be constant. Fortunately, this is very nearly proportional to the pressure of the gas. The ion currents produced are measured by traditional electrometer circuits.

In *hot-filament gages,* sufficient energy is supplied to ionize the gas by means of electron bombardment. Electrons are derived from thermionic emission as in an ordinary vacuum triode. There are several special designs, including the Bayard-Alpert gage. In *cold-cathode gages,* a high-potential field withdraws electrons from a cold-cathode surface. The electron-emission rate is less for the cold-cathode type. Hence, the collision frequency between electrons and gas molecules must be increased. This can be done by placing a magnetic field around the gage so as to deflect the electrons. The resulting path is cycloidal, not linear. Cold-cathode gages sometimes are referred to as *discharge gages.* The Philips gage is a cold-cathode device.

Hot ionization-gage tubes measure pressures from 1×10^{-3} to 2×10^{-11} torr. A cold-cathode discharge gage may have three nonlinear overlapping ranges to cover the vacuum region between 25×10^{-3} and 1×10^{-7} torr.

Partial-pressure Gage: The instrument shown in Fig. V-2 is a type of magnetic mass spectrometer with two ion collectors arranged so that it operates as a 135° instrument for masses 1 to 10 atomic mass units and as a 90° instrument for masses 10 to 70 atomic mass units. The gage incorporates a high-sensitivity ion source with an open-ended Bayard-Alpert-type structure. An extractor electrode surrounding this structure focuses the ions into a beam and directs the beam toward the magnetic analyzer. Further, an electrode behind the ion source can be switched to collect ions produced by the ion source and thus provide measurement of total pressure in the same manner as an ionization gage. The gage can be used on all types of vacuum systems operating below 10^{-4} torr, such as thin-film vacuum evaporators, vacuum furnaces, space environmental-

[1] Hastings-Raydist.

[2] Kinney ionization-thermocouple gage.

simulation chambers, electronic-tube-processing stations, particle accelerators, and vacuum-gage-calibration systems. A partial-pressure gage can provide considerable information on the operation and effectiveness of vacuum systems and can function as a leak detector.

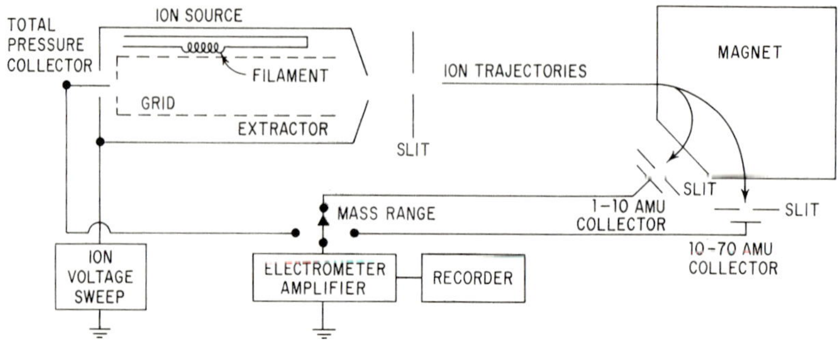

Fig. V-2. Partial-pressure gage. *(Varian.)*

Residual-gas Analyzer: One type of analyzer used in connection with vacuum systems is the quadrupole assembly shown in Fig. V-3. A quadrupole field composed of an rf and a dc component is established between four parallel rods arranged in a precise, square cross section. An electron-beam ion source is located at one end of this structure and an ion detector (electron multiplier or Faraday cup) at the other. Ions entering the quadrupole structure undergo oscillations governed by Mathieu's equations. The trajectory of an ion through the field depends, briefly, on quadrupole electric-field parameters and on the mass-to-charge ratio (mass number) of the ion. By selecting appropriate field parameters, and by varying these electronically in a predetermined manner, specific ions can be selected and a complete mass spectrum can be "scanned." The quadrupole tube is bakable to 400°C and can be mounted at almost any desired location within the vacuum chamber. The scan rate is continuously variable and may be set anywhere between 200 μs and 6 min per atomic mass unit or 100 ms to 30 min for an entire mass range.

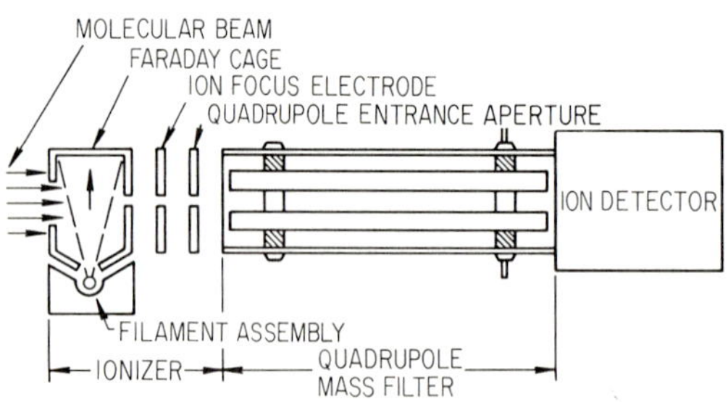

Fig. V-3. Quadrupole-type residual-gas analyzer. *(Varian.)*

Calibrated Gas Leak: These devices provide sources of specific masses of designated gases, per unit time, under stated conditions of temperature and pressure. Leaks are supplied with gas reservoirs that are filled to the pressure required to provide the calibrated flow rate. Two types of elements are available: (1) a glass-membrane permeation type and (2) an orifice (tapered glass capillary). Glass membranes are useful only with stable helium isotopes. Orifices may be used with most gases that do not corrode glass. Gas leaks of this type (1) simulate a leak in leak detection, (2) provide a calibration standard for residual-gas analyzers, (3) maintain a constant partial pressure for the calibration of a vacuum gage, (4) establish a constant flow rate of gas into a vacuum system to determine pumping speed, and (5) provide a gradual pressure rise at a predictable rate. Leak ranges are from 10^{-4} to 10^{-9} standard cubic centimeters per second. Gases used include air, argon, carbon dioxide, carbon monoxide, deuterium, Freon, helium, hydrogen, krypton, methane, neon, nitrogen, oxygen, sulfur hexafluoride, and xenon.

REFERENCES

Lewin, Gerhard: "Fundamentals of Vacuum Science and Technology," McGraw-Hill Book Company, New York, 1965.

Bancroft, G. H.: "Ninth National Symposium on Vacuum Technology Transactions," Pergamon Press, New York, 1963.
Thomas, E.: "Advances in Vacuum Science and Technology," Pergamon Press, New York, 1960.
Considine, D. M., and S. D. Ross (eds.): "Handbook of Applied Instrumentation," sec. 4-2, McGraw-Hill Book Company, New York, 1964.

Valve, Control (See **Control valve.**)

Valve, Reducing (See **Self-actuated regulators.**)

Valve, Sizing of (See **Control valve, sizing of.**)

Valve Positioner, Pneumatic A pneumatic actuator-mounted device that compares an input signal from an automatic controller with the actuator stem position—and supplies to the actuator the necessary power to make the stem position agree with the input signal. Properly applied, the valve positioner ensures exact stem positioning regardless of packing friction, valve unbalance, and actuator hysteresis. Thus, the positioner produces the desired relationship between controller signal and stem position.

Guidelines for determining whether a valve positioner may be desirable for a given application include:

1. Under many conditions, a direct-loaded actuator will perform adequately without the aid of auxiliary devices. There are three situations which indicate the need for a valve positioner or booster. A booster is nothing more than a pneumatic relay with a small input volume and large flow capacities.

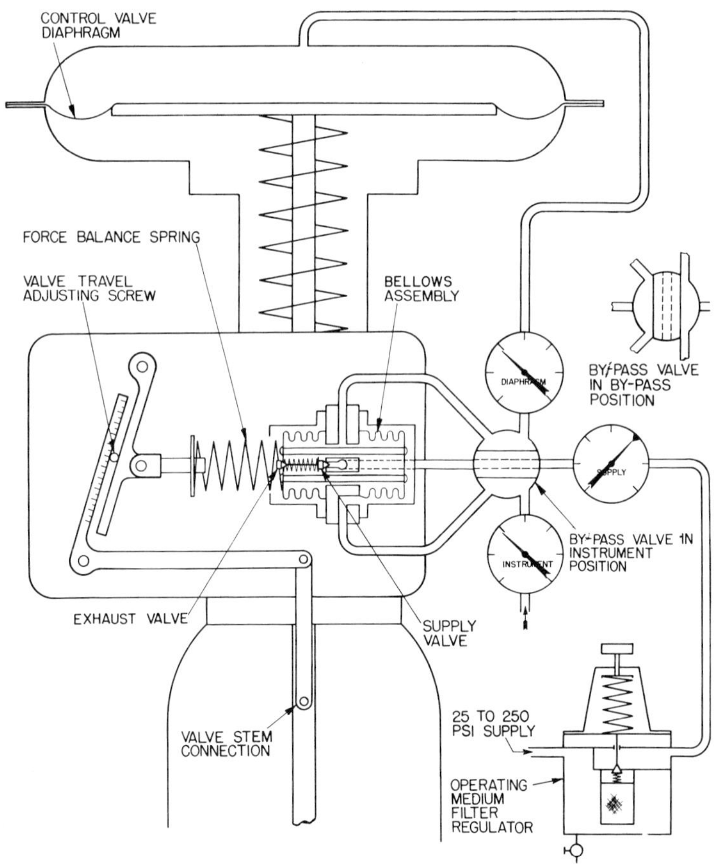

Fig. V-4. Schematic illustration of operating principle of force-balance valve positioner.

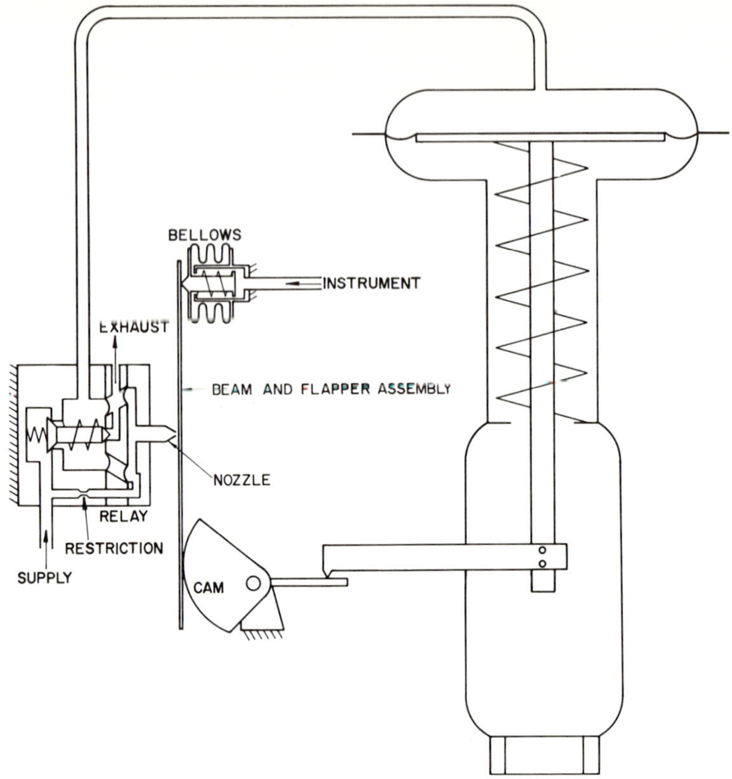

Fig. V-5. Schematic illustration of operating principle of motion-balance valve positioner.

Applications requiring the controller signal to be split-ranged between two or more control valves.

Situations where the controller signal must be amplified to obtain sufficient actuator thrust.

Systems where the best control is required with minimum overshoot from system disturbances and the fastest possible recovery. This need may result from particularly stringent process requirements or from undesirable control-equipment properties, such as high stem friction or long transmission lines.

2. The choice between a valve positioner or a booster is entirely a function of system dynamics, considering the following factors:

Positioners should be used in slow systems, those involving thermal or level processes, most mixing processes, and a few large-capacity gas processes.

Boosters should be used in fast processes, those involving liquid pressure, most gas-pressure applications, and most flow processes.

As a reference point, a slow system is defined here as one whose oscillatory frequency is at least five to ten times lower than the bandwidth of the instrument which is driving the actuator. Where this condition is not met, the system is defined as fast.

Many valve positioners are available commercially. Most differ only in design details; the principles essentially are the same. Two basic designs are in common use: (1) the force balance (Fig. V-4) and (2) the motion balance (Fig. V-5). In the force-balance positioner, the unit is in equilibrium whenever the force produced by the controller signal acting in a bellows is balanced by the force produced by a spring attached to the valve stem. For every force produced by the bellows, there is one corresponding force produced by the spring that returns the positioner to equilibrium conditions. Whenever equilibrium does not exist, a relay or pilot in the positioner is loading or unloading the actuator to stroke the valve until equilibrium returns.

In the motion-balance type, the bellows receiving the controller signal moves a beam which is fixed to the bellows at one end and to the valve stem at the other end through linkages. A pilot or relay in the positioner is activated by the beam movement either to admit loading air to or to bleed it from the actuator until the valve-stem position corresponds to the controller signal. In the motion-balance type, there is one valve-stem position for every corresponding bellows position when equilibrium conditions exist. The motion of the beam at the valve-stem end always is exactly proportional to the motion of the bellows end of the beam.

Valve positioners must be actuator-mounted and must be mechanically connected to the valve stem. A separate air supply is provided for operation. Valve positioners operate on any of the normally used instrument-pressure ranges, such as 3 to 15 psi or 6 to 30 psi—with a normal output pressure of 20 or 35 psi. A maximum supply pressure of 50 psi generally is allowed. Positioners for piston actuators accept the same input signals, but the outlet pressures are usually between 100 and 150 psi.

Models are available with a choice of feedback cams that can be used effectively to match the positioner to the control process. For example, if instability in the system exists with one cam installed in the positioner, a quick change to a special characterized cam often can be a simple solution. Models with a bypass arrangement are widely used. Bypassing the positioner allows the controller signal to operate the control valve directly and permits shop servicing without disrupting control-valve operation. In most positioners, reverse action can be obtained with a few minor adjustments and without additional parts. A direct-acting unit creates an increasing output pressure as the controller signal increases. In reverse action, the positioner output pressure decreases when the signal from the controller increases.

—Robert J. Lasher, *Fisher Controls Company*

Van de Graaff Generator (See Accelerator.)

Vapor Pressure, Measurement of (See Psychrometer.)

Variable The quantity or characteristic that is the object of measurement in an instrumentation/control system. Synonymous terms include measurement variable, instrumentation variable, and process variable. The latter term is commonly used in the manufacturing industries. See also **Process variable.**

TABLE V-1. *Classification of Variables by Physical Characteristics*

Class of Variables	Examples
Thermal variables—relate to the condition or character of a material dependent upon its thermal energy	Temperature Specific heat Thermal-energy variables, (enthalpy and entropy) Calorific value
Radiation variables—relate to the emission, propagation, and absorption of energy through space or through a material in the form of waves; and by extension, corpuscular emission, propagation, and absorption	Nuclear radiation Electromagnetic radiation, such as radiant heat, infrared, visible, and ultraviolet light; x- and cosmic rays; gamma radiation
Force variables	Total force Moment or torque Force per unit area, such as pressure, vacuum, and unit stress
Rate variables—are concerned with the rate at which a body is moving toward or away from a fixed point. Time always is a component of a rate variable	Flow Speed Velocity Acceleration
Quantity variables—relate to the total quantity of material that exists within specific boundaries	Mass Weight
Physical-property variables—are concerned with the physical properties of materials with the exception of those properties which are related to mass or chemical composition	Density and specific gravity Humidity Moisture content Viscosity Consistency Structural characteristics, such as hardness, ductility, and lattice structure
Chemical-composition variables—relate to the chemical properties and analysis of substances	A very abridged list of analysis targets would include CO_2, CO, H_2, N_2, O_2, H_2O, H_2S, NO_x, S, SO_x, C_2H_2, CH_4, H^+ (pH), and various solvents
Electrical variables	Electromotive force Electric current Resistance Conductance Inductance Capacitance Impedance

TABLE V-2. *Classification of Variables by Measurement Signals*

Variable	Mechanical Motion	Liquid Displacement	Motion of Light or Electron Beam	Total Force	Pressure	Voltage or Current	Voltage/Current Relationships	Pulse Duration	Frequency	Pulse Code
Temperature	x	x		x	x	x			x	
Specific heat[a] . . .										
Thermal-energy variables[a]										
Calorific value[a] . .										
Nuclear radiation						x	x			x
Electromagnetic radiation. .	x					x	x			
Photometric variables[b] . .										
Acoustic variables[b]										
Total force. . . .	x	x	x	o	x	x	x			
Moment or torque . .	x	x	x	x		x	x			
Pressure or vacuum. . . .	x	x	x	x	o	x	x			
Flow	x			x	x	x	x		x	
Speed or velocity	x			x	x	x			x	
Acceleration	x		x	x	x	x	x			
Mass[c] . . .										
Weight . .	x			x	x		x			
Elapsed time. . . .	x					x	x			x
Frequency	x					x	x		o	x
Position. .	x	x	x	x	x	x	x	x	x	x
Dimension[d] . . .										
Contour[d]										
Level. . . .	x	x		x	x		x		x	
Density or specific gravity	x			x	x		x			x

TABLE V-2. *Classification of Variables by Measurement Signals (Continued)*

Variable	Mechanical Motion	Liquid Displacement	Motion of Light or Electron Beam	Total Force	Pressure	Voltage or Current	Voltage/ Current Relationships	Pulse Duration	Frequency	Pulse Code
Humidity[e]										
Moisture content	x			x			x			
Viscosity	x			x	x					
Structural characteristics[f]										
Chemical composition[g] . . .										
Electrical voltage. .	x		x	x		o	x	x	x	x
Electric current	x		x			o	x	x	x	x
Electrical resistance						x	x	x	x	
Electrical inductance . . .						x	x	x	x	
Electrical capacitance . . .						x	x	x	x	
Electrical impedance. . . .						x	x	x	x	

x denotes that a standard transducer is commercially available for converting the variable to the signal.

o denotes that the variable already is in measurement-signal form.

a No simple transducers exist for the measurement of specific heat, calorific value, entropy, enthalpy, and the like. Either measurements are derived from computations based upon measurements of other variables, or specialized equipment is used to maintain a number of conditions constant so that one variable, usually temperature, varies in a predetermined relation to changes in the measured variable.

b Photometric variables and acoustical variables each include several distinct variables. The sensor for most photometric variables is some type of photocell; the sensor for most acoustic variables is some type of microphone. Both have electrical outputs.

c Mass is almost invariably measured by the gravitational effect, that is, by weight.

d Dimension and contour, defined as the relative position between points, are almost invariably measured by a position measurement, with a selected point of the dimension or contour maintained in a predetermined relationship to a reference point on the position-measuring system.

e Humidity is measured by dew-point temperature. This applies to direct measurement of temperature of pure water whose vapor pressure is equal to the vapor pressure to be measured and also to saturated lithium chloride vapor pressure in saturated lithium chloride elements.

f Structural-characteristic variables include so wide and varied a group of characteristics that there appears to be no useful common basis for discussion of their measurement.

g For most chemical-composition measurements, no simple direct transducer exists. Measuring equipment in general is elaborate. In most cases, the measurement signal is an electrical voltage or current or an electrical voltage and current relationship.

Variables generally are classified in two ways: (1) by the physical characteristics of the variable (Table V-1) and (2) by the measurement signal involved (Table V-2).

—Douglas M. Considine, *Editor-in-Chief*

Variable-area Flowmeter (See Flowmeter, variable-area.)

Variable-head Flowmeter (See Flowmeter, variable-head.)

Variable-permeance Transducer (See Electromagnetic transducer.)

Variable-reluctance Transducer (See Electromagnetic transducer.)

Variable-reluctance Linear Transducer (See Linear displacment transducer, variable-reluctance.)

Varley Bridge (See Bridge circuits, electric.)

Varmeter [See Electrical quantities, measurement of; and Power (electrical), measurement of.]

Vector Cardiography (See Cardiography.)

Velocity (Air/Gas), Measurement of (See Anemometer.)

Velocity-limiting Control Control action in which the rate of change of a specified variable will not exceed a predetermined limit.

Velocity Meter (See Accelerometer.)

Venturi Flume (See Flume.)

Venturi Tube The oldest of the recovery-cone type of primary flowmetering devices, first made commercially in 1893 under the Clemens Herschel patents. Advantages of the venturi tube include: (1) high accuracy, (2) proved long-term performance, (3) small sensitivity to flow-pattern disturbances due to poor approach piping, and (4) consequently, requirement of the shortest approach piping of all the primary metering devices for a given flow-pattern defect. The venturi tube has a very long laying length, making it the most expensive primary device on the market.

The main features of the venturi tube are (1) a true static-pressure pickup at the inlet and throat and (2) a long recovery cone with a very small cone angle reaching the full pipe diameter (see Fig. V-6). The value of the discharge coefficient, as established by ASME, is 0.984, regardless of its throat-to-inlet ratio (beta ratio). See **Flowmetering devices, primary.** The coefficient becomes constant over a Reynolds

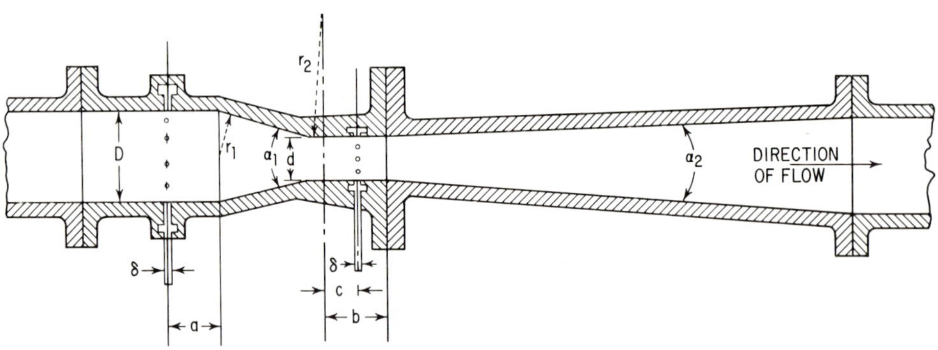

Fig. V-6. Proportions of Herschel-type venturi tubes recommended by the Committee on Flow Measurement of the International Organization for Standardization. D, pipe diameter inlet and outlet; d, throat diameter as required; a, 0.25 to 0.75 D for 4 in. $\leqq D \leqq$ 6 in; 0.25 to 0.50 D for 6 in. $<$ D $\leqq$ 32 in.; b, equal to d; c, equal to $d/2$; δ, 3/16 to ½ in. according to D, annular pressure chamber with at least four piezometer vents; r_2, 3.5 to 3.75 d; r_1, 0 to 1.375 D; a_1, 21° ± 2°; a_2, 5° to 15°.

number of 200,000. The coefficient has been well investigated, established, and defined by ASME.

The Y factor is based on calculations published in the ASME "Fluid Meters Committee Handbook." The uncalibrated tolerance on the coefficient is ± 0.75 percent, and the tolerance on the Y changes from 0 to 0.5 percent dependent upon the beta ratio of the tube, the ratio of differential pressure to inlet pressure, and the specific-heat ratio and condition of the gas to be metered (moist-dry). The venturi tube can be constructed of any suitable material so long as the hydraulic shape is preserved. Typical head losses are:

Percent of Differential at a	*Beta Ratio* of
10	0.55
11	0.75
13	0.3

The *universal venturi tube* is the most recent development and incorporates the following improvements over the classical venturi designs: (1) shorter laying length; (2) no annular chamber; (3) constant coefficient at low Reynolds number; (4) greater installed accuracy; (5) lower cost; (6) lower head loss; and (7) insensitivity to upstream disturbances, because of built-in flow conditioning.

—Edmond B. Lynch, *B I F, Unit of General Signal Corporation*

Vernier Scale As shown in Fig. V-7, a small auxiliary scale (lower) adjacent to the main scale (upper) of an instrument that provides a means to read accurately between the smallest graduations of the main scale. First conceived by Pierre Vernier in 1631, vernier scales are commonly used in hand instruments, such as micrometer calipers. In Fig. V-7, the main scale has 40 graduations per inch. Twenty-five divisions of the vernier scale are equivalent to 24 divisions of the main scale, or $24 \times 0.025 = 0.600$ in. Thus, one vernier division $= 0.600/25 = 0.024$ in. Therefore, the difference between vernier and scale divisions $= 0.025 - 0.024 = 0.001$ in. If the zero on the vernier does not coincide with a scale division, the dimension being measured is not exact to 0.025 in. It then is necessary to find which vernier division coincides with a scale division and to add that number of thousandths to the scale reading. The reading of Fig. V-7 is $2.000 + 0.300 + 0.050 + 0.018$ in. $= 2.368$ in. In recent years, a wider interpretation of the term *vernier* has developed so that it includes any device, control, or scale that is used to achieve a fine adjustment or to obtain better precision from a measurement.

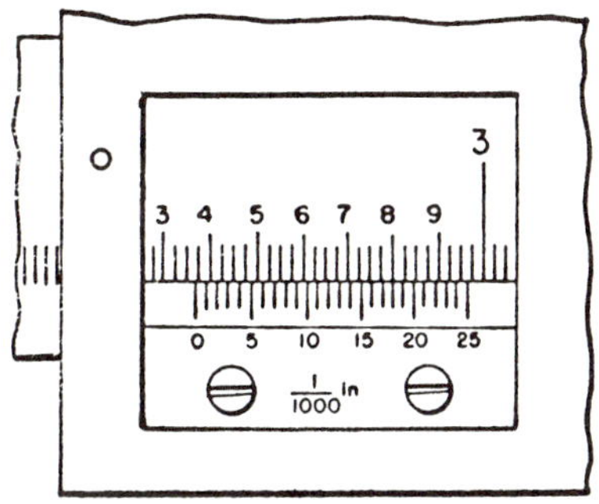

Fig. V-7. Vernier scale of the type used on hand instruments.

Vertical Gyro (See **Flight-control systems.**)

Vibrascope [See **Testing (materials), destructive.**]

Vibrating-string Accelerometer (See **Accelerometer.**)

Vibrating-wire Pressure Transducer (See **Pressure transducer, vibrating-wire.**)

Vibration, Measurement of Vibration-testing equipment is an important part of a properly equipped environmental laboratory. Such equipment is essential to the development and proof testing of aircraft and missiles as well as other vehicles of all kinds—and their components and subsystems. These testing techniques have become increasingly precise and now are extended to deal with problems in such fundamental research areas as materials development and earthquake studies, and they are giving new insight into such fields as the design of machines and the construction of bridges. See also **Vibra-**

tion pickup, piezoelectric; and **Vibration pickup, velocity.**

Use of Accelerometer: The linear piezoelectric accelerometer is the most commonly used instrument for measuring vibration (see Fig. V-8). The piezoelectric element, which is acted on by the inertial force of the mass, may be made of

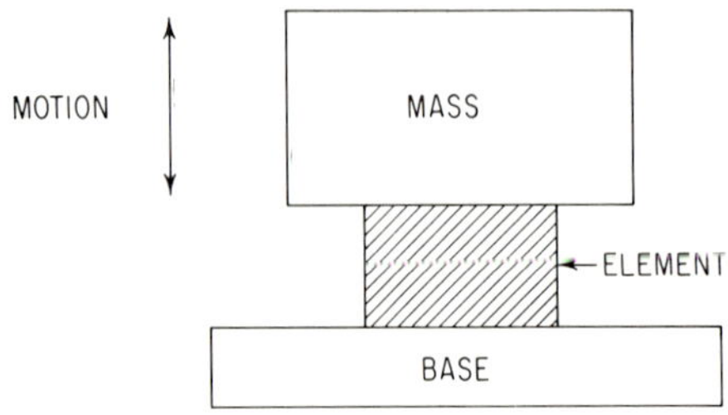

Fig. V-8. Schematic representation of accelerometer.

polycrystalline ceramics, such as barium titanite or lead zirconate. The frequency response of the accelerometer is shown in Fig. V-9. The resonance is sharply underdamped. Over the useful frequency range, the equivalent circuit of the accelerometer is shown in Fig. V-10.

Because of its high impedance, the accelerometer must be isolated from the data-conditioning equipment, and the signal must be amplified. If a voltage amplifier is used, the sensitivity of the accelerometer is expressed as

$$S_v = \frac{E}{\ddot{x}} \tag{1}$$

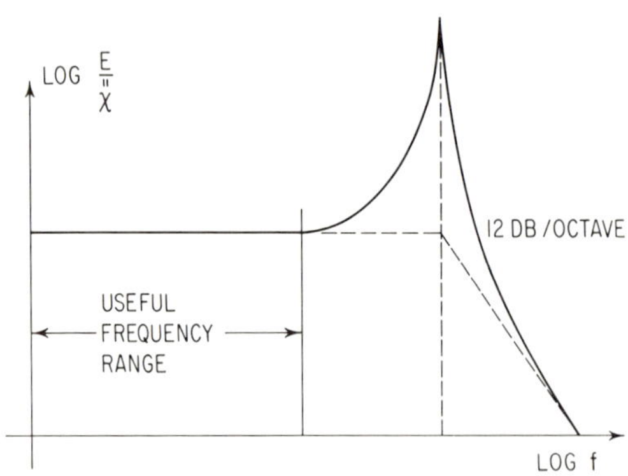

Fig. V-9. Frequency response of an accelerometer.

If a charge amplifier is used, the sensitivity is

$$S_{ch} = \frac{Q}{\ddot{x}} \tag{2}$$

where Q = charge developed across the accelerometer electrodes.
Note that $Q = CE$ and that

$$S_{ch} = (C + C_c)S_v \tag{3}$$

where C_c = capacitance of the connecting cable.

S_v is attenuated by the accelerometer cable, while S_{ch} is not. With charge amplification, however, the noise level increases linearly with cable length.

The input resistance R of either the voltage or charge amplifiers determines the low-frequency response f_L.

$$f_L(3\text{ db}) = \frac{1}{2\pi R(C + C_c)} \tag{4}$$

Hence, R usually must be very large.

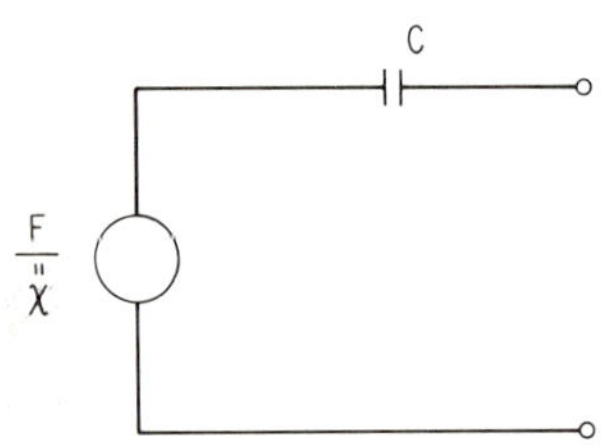

Fig. V-10. Equivalent circuit of accelerometer over useful frequency range.

A zero-drive amplifier eliminates triboelectric noise and gives an overall noise level that is independent of cable length. Triboelectric noise is reduced by $\sqrt{R_Z/R}$, where R_Z is the input resistance of the zero drive (less than 1 Ω).

Sine Testing: The manner in which an accelerometer is used to control closely sine-wave vibration testing is shown in Fig. V-11. An electrodynamic vibration exciter is used. An oscillator puts out a sinusoidal signal which is fed via a variable gain (AGC) to a power amplifier (PA). The latter, in turn, drives the vibration exciter *Ex*. The accelerometer *A* has its output amplified by the signal-conditioning circuit *H*. The signal then is detected by *D*, and a dc voltage proportional to signal level is used to control the gain of the AGC stage. Thus, independent of the usual resonant effects at the exciter or specimen, the vibration level at A can be carefully controlled.

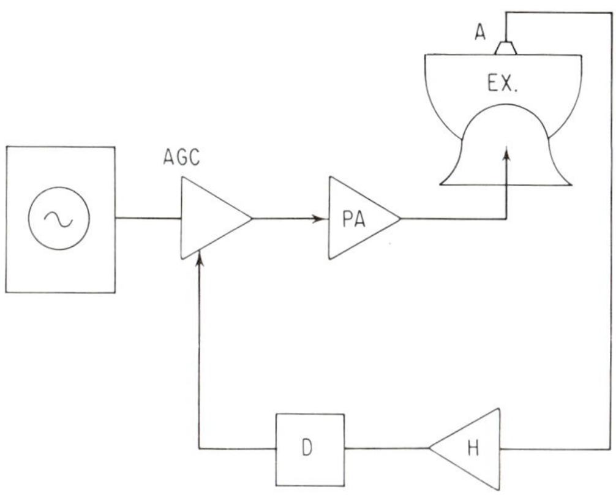

Fig. V-11. Use of an accelerometer to control closely sine-wave vibration testing.

Random Testing: Use of an accelerometer to control a random vibration is shown in Fig. V-12. An automatic random equalizer is used. A radom-noise generator supplies the input to a set of contiguous bandpass filters. In the most modern form of apparatus, these filters are crystal in nature. Each filter output is controlled by an AGC stage so as to maintain its spectral power at a desired level. The reference voltage at each servoloop summing point is used to determine the vibration level of the corresponding filter channel. A large loop gain makes it possible to calibrate the potentiometer supplying the reference voltage so as to follow any desired spectral-density plot. There is one such control for each filter channel.

Vibration testing using an accelerometer can be extended to force-impedance testing or mechanical-impedance testing, using an impedance head. The latter consists of an accelerometer plus a force gage—measuring at the same point in the same direction. The force gage consists of a piezoelectric element between two plates and pro-

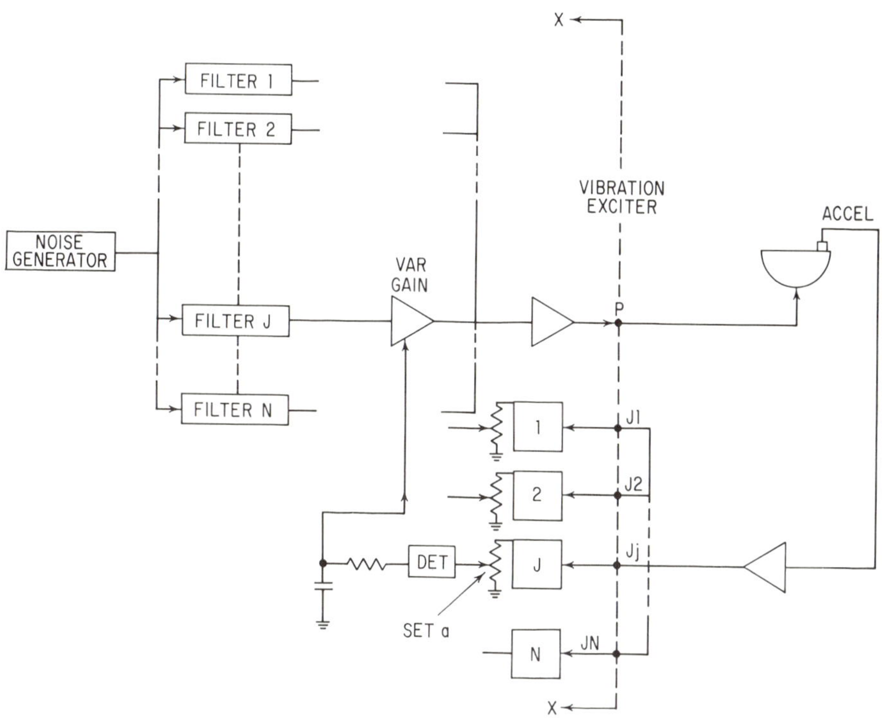

Fig. V-12. Use of an accelerometer to control a random vibration.

vides an output proportional to the pressure between the plates. The stiffness of the impedance head should be kept large. The mass of the impedance head should be kept small. But even with these precautions, mass cancellation is essential. A pseudo-force signal proportional to the acceleration is developed in a separate amplifier which then is subtracted from the force signal. The force needed to move the impedance head thus is canceled.

Impedance Testing: Use of an automatic random equalizer for impedance testing is shown in Fig. V-13. With this equalizer, not only can one set in the spectrum desired, but one also can set in on potentiometers the foundation impedance (amplitude and phase) through which the spectrum must act. The equalizer of Fig. V-13 adjusts the drive level of each channel to solve the equation

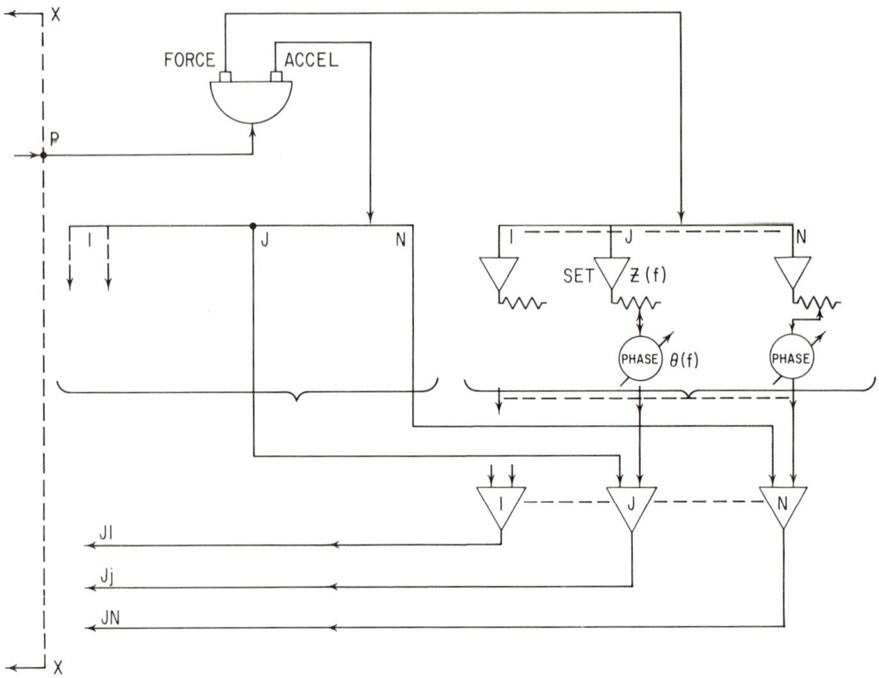

Fig. V-13. Use of an automatic random equalizer for impedance testing.

$$a_1 = \frac{a}{Z_1/Z + 1} \tag{5}$$

where a = unloaded acceleration spectrum of the foundation
Z = foundation mechanical impedance
Z_1 = mechanical impedance of the specimen

The system of Fig. V-13 adjusts the level in each channel as determined by Eq. (5). A smiliar arrangement is possible with sine testing.

Impedance Measurement: An impedance head also can be used as part of an analysis system to measure mechanical impedance. The specimen is driven sinusoidally at the point of interest with a vibration exciter. The impedance head is

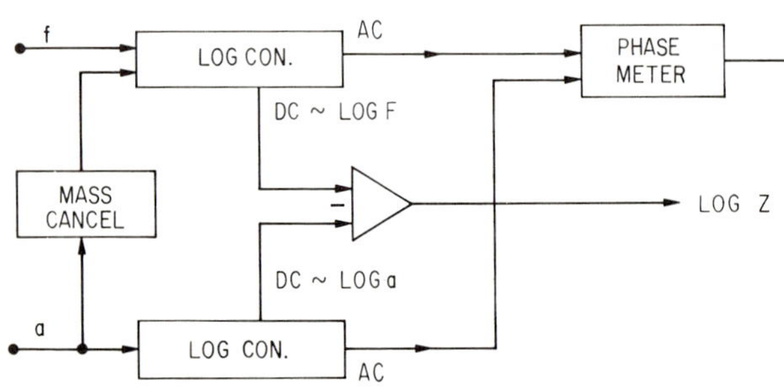

Fig. V-14. Simple impedance-computing circuit.

placed between the exciter and the specimen. A simple impedance-computing circuit is shown in Fig. V-14.

The ac force signal from the force transducer is fed to a log converter which adjusts its gain always to put out a full-scale ac signal. The gain adjustment is measured by a dc signal in a logarithmic manner. The ac force signal is phase-compared with the corresponding acceleration signal. The corresponding dc signals are subtracted.

$$\log F - \log a = \log Z(f) \qquad (6)$$

Thus, the final output yields the impedance $Z(f)$ arc θ in decibel form. Mass cancellation usually forms part of such a system.

REFERENCES

Maki, C. E.: Vibration Testing System, U.S. Patent 3157045, Nov. 17, 1964.

Ratz, A. G.: An Impedance-compensated Random Equalizer, *Proc. IES Natl. Conf.,* 1966.

Bouche, R. R.: Instruments and Methods for Measuring Mechanical Impedance, 30th Symposium on Shock and Vibration, 1961.

Mason, W. P.: "Piezoelectric Crystals and Their Applications to Ultrasonics," D. Van Nostrand Company, Inc., Princeton, N.J., 1950.

Everitt, W. L., and G. E. Anner: "Communication Engineering," 3d ed., McGraw-Hill Book Company, New York, 1956.

—Alfred G. Ratz, *MB Electronics, a Textron Company*

Vibration Exciter A transducer used in simulation of predetermined mechanical-vibration environments. Sometimes also referred to as a shaker or shaker table. A mounting or table surface usually is provided for threaded attachment of the article to be vibrated.

Vibration exciters can be classified into four main groups: (1) mechanical, (2) electromechanical, (3) electrohydraulic, and (4) pneumatic—dependent upon the source of vibration-producing energy.

Vibration Pickup, Piezoelectric A transducer containing a piezoelectric element that generates an electrical output signal proportional to acceleration. Hence, the term accelerometer also is applied to this device. The piezoelectric element may be a natural or synthetic crystal which generates an electric charge when deformed. The pickup is constructed so that the crystal is one spring of a spring-mass system. Deformation of the crystal occurs because of inertial force of the mass as it is accelerated (see Fig. V-15).

Commerical accelerometers usually are small and cylindrical in form—weighing between 0.001 and 0.2 lb. Accelerometer sensitivity generally increases with size. The resonant frequency commonly is from 25,000 to 100,000 Hz, a point two to three times the maximum operating frequency. The resonant frequency (hence, usable frequency range) increases as the accelerometer size decreases. Accelerometers may be attached to specimens by cement, removable studs, or screws. Coaxial-cable connectors usually are integral with the case at top or side.

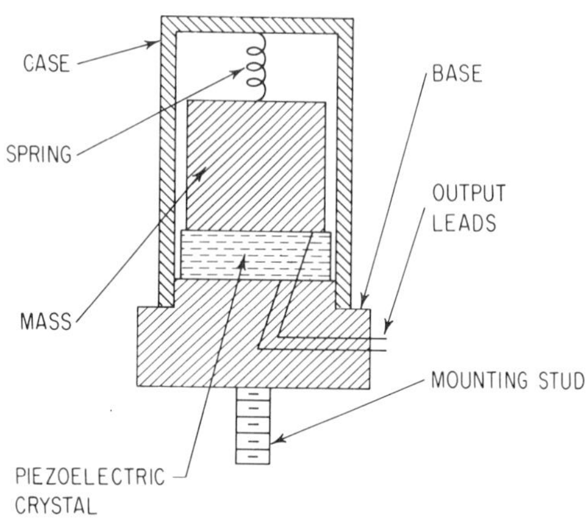

Fig. V-15. Schematic diagram of typical piezoelectric vibration pickup.

The sensitivity of an accelerometer is the electrical output per unit of applied acceleration, expressed in g (acceleration of gravity). Charge sensitivity is expressed in coulombs per g. Voltage sensitivity is expressed as millivolts per g with open circuit (infinitely high load impedance). Because of the low value of electrical output, charge, voltage, or zero-drive amplifiers often are used to raise the signal above noise level and into the range of indicating and recording instruments. Some accelerometers contain integral electronics for signal amplification and conditioning.

—O. John Zamparo, *MB Electronics, a Textron Company*

Vibration Pickup, Velocity A self-generating transducer having an output voltage proportional to velocity. The device senses the relative velocity between a coil and the magnetic field imposed on the coil by a self-contained magnetic source. In some types, a seismically suspended coil moves relative to a uniform magnetic field. A variation of this is a seismic magnet that moves relative to an anchored coil. Other types function by means of a coil that senses variations in the magnetic field. A typical pickup is shown schematically in Fig. V-16.

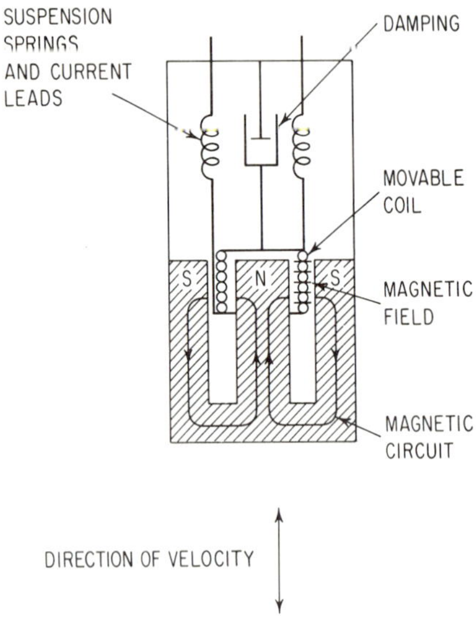

Fig. V-16. Schematic diagram of typical velocity vibration pickup.

The output signal of a velocity pickup normally is large enough so as not to require amplification. The velocity signal can be integrated and differentiated to obtain displacement and acceleration, respectively. The natural frequency of the suspended moving element usually is low—ranging from 2.5 to 10 Hz in various models. This natural frequency establishes the lower operating-frequency limit. Low velocities associated with high-frequency vibration and pickup structural resonances limit the upper frequency to approximately 2,000 Hz.

The velocity pickup is useful over a range of amplitudes determined by the sensitivity and friction at low displacements and by space limitations of moving parts at high displacements.

—O. John Zamparo, *MB Electronics, a Textron Company*

Vibrometer One type of vibrometer simply is a dial indicator which is held by hand and has associated with it quite a heavy block (mass). The actuating tip of the dial indicator is placed against the vibrating member, and the indicator is read by noting the extremities of motion of the indicating pointer. The useful range of this instrument is from approximately 10 to 100 Hz. The device will measure amplitudes from approximately 0.001 in. to slightly greater than 0.100 in. The device is limited to from 2 to 3 *g*. One commercially available instrument incorporates a mirror and light source. The prod actuates the mirror. In another instrument, the amplitude changes a capacitor and this, in turn, is amplified. Still another modification uses a moving element suspended in a seismic mass and actually scribes a plate so that a record of the vibration amplitude can be made. A further modification uses a plate that moves at a known rate, thereby indicating the frequency and waveshape of the vibration. General use for these devices lies in probing machinery for undesirable vibrations. Generally, the devices are not used for laboratory analysis of equipment that is mounted on a vibration exciter. Also see **Vibration, measurement of.**

—Harold C. Jones, *University of Maryland*

Viscosity, Measurement of A wide variety of instruments is available for the measurement and control of viscosity and related rheological properties of fluids. Descriptions of rheological measurements tend to be complex because of (1) the differences of fluids to be measured wherein sub-

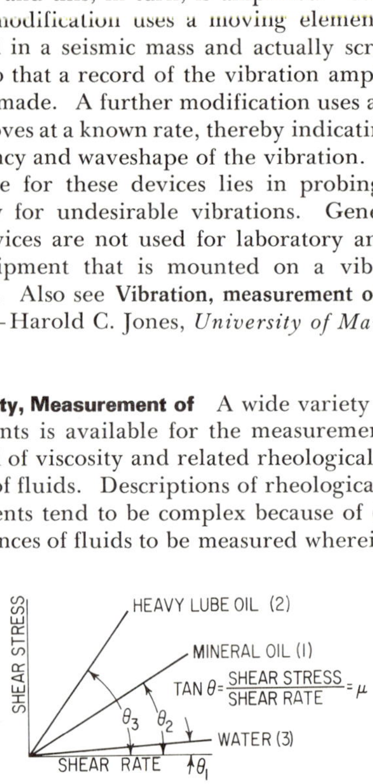

Fig. V-17. Rheograms of typical newtonian materials (constant slope, linear lines with origins at zero).

stances fall into two general classifications—newtonian and nonnewtonian, (2) the several different definitions of viscosity and the resulting different scales for calibrating viscometers, and (3) the number of different physical principles used in instruments. See Figs. V-17 and V-18 and Table V-3. Also see **Consistency, measurement of.**

Newtonian Substances: Liquids or suspensions in liquids which, when subjected to a shear stress, undergo deformation wherein the ratio of shear rate (flow) to shear stress (force) is con-

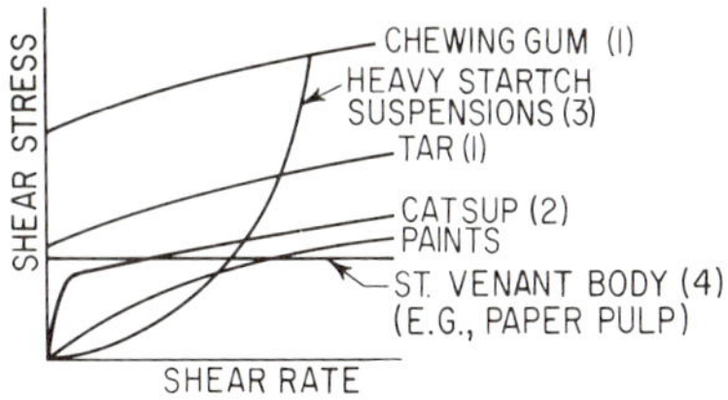

Fig. V-18. Rheograms of typical nonnewtonian materials.

stant—are said to be newtonian. The basic relationship, as given by Newton, is

$$\frac{F}{A} = \mu \frac{V}{L} \tag{1}$$

where F/A = shear stress (force per unit area)
V/L = shear rate (velocity per unit thickness of layer)
μ = proportionality constant

Thus, viscosity came to be defined as

$$\mu = \frac{\text{shear stress}}{\text{shear rate}} \tag{2}$$

Hagen-Poiseuille Law: Hagen (Germany) and Poiseuille (France) described viscosity as the ratio of shear stress vs. shear rate at the wall of a capillary tube.

$$\mu = \frac{PR/2L \text{ (shear stress)}}{4Q/\pi R^3 \text{ (shear rate)}} = \frac{\pi PR^4}{8QL} \tag{3}$$

where P = pressure differential across liquid in tube
R = inside radius of tube
L = length of tube
Q = volume rate of flow of liquid
μ = absolute (or dynamic) viscosity

This equation is limited to conditions of laminar or viscous (streamline) flow.

Dynamic (Absolute) Viscosity Units: From Eq. (3) came the absolute or dynamic unit of viscosity in cgs units, the *poise,* which is equivalent to 1 dyne-s/cm². Smaller values of viscosity commonly are expressed in *centipoises.* The viscosity of water, for example, at 68°F is 1 cp.

Kinematic-viscosity Units: Many laboratory viscometers measure ***kinematic viscosity,*** which is defined as the quotient of dynamic viscosity divided by the density of the fluid (at the same temperature). The unit for kinematic viscosity is the *stoke* (ν); for smaller values, the *centistoke.* The relationship between kinematic and absolute viscosity is

$$\text{Kinematic viscosity, stokes} = \frac{\text{absolute viscosity, poise}}{\text{density, g/cm}^3} \tag{4}$$

Viscosity Index: This term has been used with reference to petroleum products. It is an empirical number that indicates the effect of change of temperature on the viscosity of an oil. A low index number signifes an oil that changes viscosity by a large amount with a given temperature change.

Viscosity Scales: A few time-based viscosity scales are in common use: (1) Saybolt scales (in the United States), (2) Redwood scales (in Great Britain), and (3) Engler scales (in Europe). All three scales are based on the Hagen-Poiseuille law and indicate the time of efflux, under speci-

TABLE V-3. *Examples of Common Fluids Exhibiting Diverse Rheologic Characteristics*

	Nonnewtonian				
Newtonian	Pseudoplastic	Plastic	Thixotropic*	Rheopectic	Dilitant*
Water Most mineral oils Gasoline Kerosene Most aqueous salt solutions Light suspensions of dyestuffs	Catsup Printer's ink Paper pulp	Chewing gum Tar High concentrations of Asbestine in oil	Silica gel Most paints Glue Molasses Lard Fruit juice concentrate Asphalts	Bentonite soils Gypsum in H_2O	Quicksand Peanut butter Many candy compounds

*Some liquids may change from thixotropic to dilatant or vice versa as the temperature or concentration changes.

fied conditions, for a fixed volume of fluid through a specific capillary or aperture. Kinematic viscosity (and from it, absolute viscosity) can be determined from such scales by the empirical formula

$$\nu = At - \frac{B}{t} \qquad (5)$$

where A and B = constants applicable to the viscometer
t = time of efflux, s
ν = kinematic viscosity, centistokes

Commonly accepted values of A and B are given in Table V-4. Various viscosity-conversion factors are given in Table V-5.

used in Saybolt, Redwood, Engler, Scott, Ubbelohde, and Zahn viscometers and in the Parlin cup and the Ford cup. A similar type utilizes a capillary tube in place of the orifice or nozzle. The latter type includes the modified Ostwald, Bingham, and Zeitfuchs.

These are laboratory or batch-type instruments commonly used in the petroleum and allied industries. The Zahn, Parlin, and Ford types are widely used in the paint and varnish industry. With changes in orifice or capillary dimensions, these instruments are suitable for low to moderately high viscosity ranges and for viscosity approximations. Viscosity is reported in seconds and can be converted to absolute-viscosity units only if fluid is newtonian.

TABLE V-4. *Commonly Accepted Values of A and B* [Eq. (5)]

Viscometer Type	Value of Constants: A	B	Efflux Time t, s
Saybolt Universal	0.226	195	32–100
	0.220	135	Above 100
Saybolt Furol	2.24	184	25–40
	2.16	60	Above 40
Redwood No. 1 (Standard)	0.260	179	34–100
	0.247	50	Above 100
Redwood No. 2 (Admiralty)	2.46	100	32–90
	2.45	...	Above 90

Viscosity-measurement Systems: Descriptions here are confined to those physical principles which now are used in the major types of viscometers.

Time to discharge—a given volume of fluid through an orifice or nozzle. This principle is

Timed fall of ball or rise of bubble—through liquid confined in tube is proportional to absolute viscosity because, in both cases, the liquid flows in viscous flow through a restriction. This is primarily a laboratory or batch-type instrument commonly used by the petroleum industry for

TABLE V-5. *Viscosity Conversion Factors*

1 poise = 1 dyne-second/square centimeter
1 poise = 1 gram/centimeter-second
1 pound(force)-second/square foot = 1 slug/foot-second
1 pound(mass)/foot-second = 1 poundal second/square foot
ν (stokes or cm^2/sec) $\times$ 0.001076 = ν (ft^2/s)

To Convert to	Multiply the Number of: Poise by	Pound(force)-second/Square foot by	Pound(mass)/Foot-second by
Poise	1	478.8	14.88
Pound(force)-second/Square foot	0.002089	1	0.03109
Pound(mass)/Foot-second	0.06721	32.174	1

high-viscosity oils. The ball method can be timed with considerable accuracy by field coils at the start and finish point.

Drag torque on stationary element in rotating cup—in which the cup rotates at a constant speed. The stationary element, which may be a cylinder or a paddle, is restrained by a calibrated spring (see Fig. V-19). The spring deflection indicates torque which can be converted to absolute-viscosity units. This is principally a laboratory or batch-type instrument but may be equipped to record. The bowl may be equipped to control sample temperature. The instrument can be used with both newtonian and nonnewtonian fluids, and by change of rotational speed, a shear stress vs. shear rate diagram can be obtained. The instrument is suitable over a wide range of viscosities.

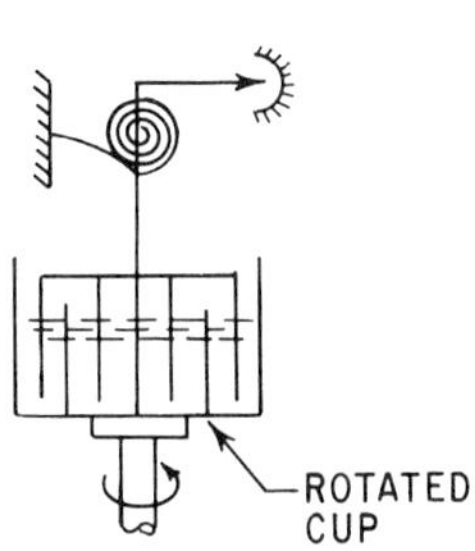

Fig. V-19. Viscometer employing principle of drag torque on a stationary element in a rotating cup. *(Brabender.)*

Timed fall of piston in cylinder—in which a submerged plunger is raised automatically, then dropped in the cylinder, forming an orifice. The piston and cylinder of a Norcross instrument are shown schematically in Fig. V-20. The time of fall is recorded automatically in terms of absolute viscosity. The instrument is available with a recorder and can be used in automatic control systems as found in the paint, oil, soaps, and plastics industries. The device is suitable for both open or pressurized service and can be designed to handle viscosity ranges from 0.1 to 1,000,000 cp with high repeatability at operating temperatures up to 800°F.

Similar types, utilizing a holed disk or holed cone and designed for laboratory use, report results in seconds, representing the time for the piston to fall a given distance through the sample.

Pressure drop through a friction tube—wherein the sample is pumped through the friction tube in viscous flow. Pressure drop across the tube is measured by a differential-pressure transmitter. The transmitter output is a direct solution to the Hagen-Poiseuille equation for absolute viscosity.

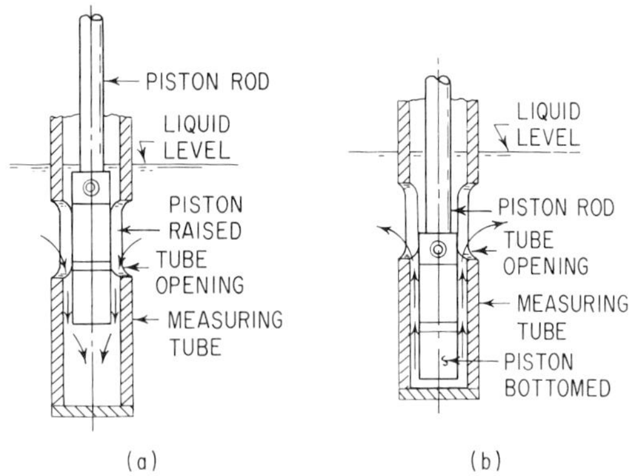

Fig. V-20. Viscometer employing timed fall of piston in a cylinder. Position of piston during filling phase is shown in *(a)*. The piston is raised, drawing liquid into the space formed under the piston. Measuring phase is shown in *(b)*. The piston falls, expelling the liquid through the same path that it entered. Time of fall is directly proportional to viscosity. *(Norcross.)*

Torque required to rotate a torque element in a liquid—As shown in Fig. V-21, the Brookfield viscometer uses a synchronous motor to drive a vertical spindle with disk, paddle, or cylinder that is submerged in the test liquid. The drive is through a calibrated spring. Angular lag of the spindle behind the motor is proportional to viscosity and is sensed in a number of ways. This

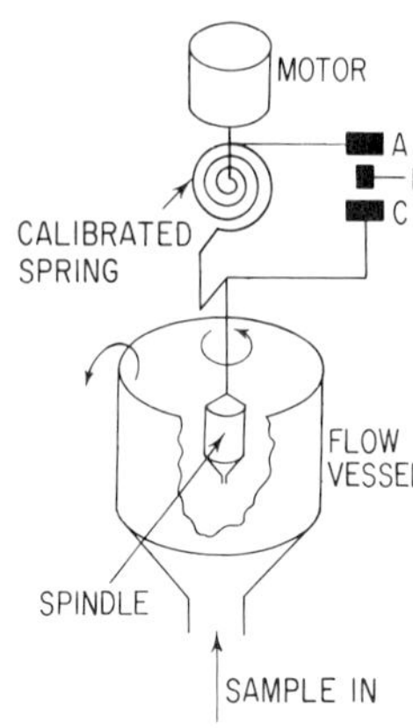

Fig. V-21. Viscometer in which torque required to rotate a torque element in a liquid is a measure of viscosity. *(Brookfield.)*

instrument can be used in open or closed vessels under pressure or vacuum and for both newtonian and nonnewtonian liquids or suspensions. By change of spindle speeds and types, the device can measure over a wide viscosity range.

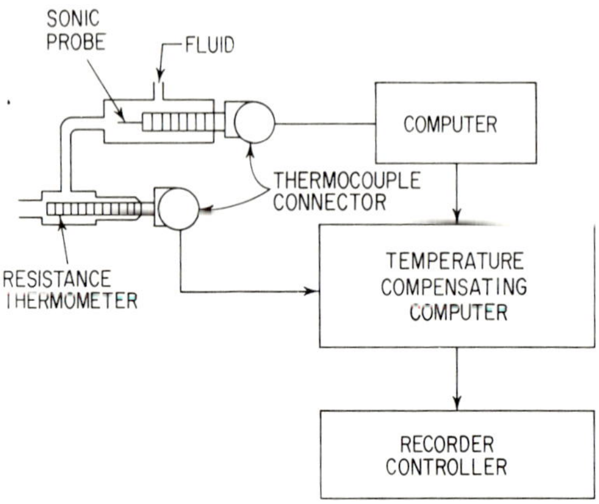

Fig. V-22. Ultrasonic-type viscometer. *(Bendix)*

Viscosity-sensitive rotameter—wherein the rotameter float is designed to be either viscosity-immune or viscosity-sensitive. If flow rate is held constant, with either a flow regulator or a constant-volume pump, the float elevation in a tapered tube is a function of viscosity. In another configuration, the unit employs a viscosity-insensitive float in conjunction with a viscosity-sensitive float to enable manual flow-rate adjustments with a valve prior to reading the instrument. These devices are suitable for recording and control and can be used in open or closed systems under pressure or vacuum. The ranges are from 3 to 30,000 cp with a 10:1 rangeability. They are useful for both newtonian and nonnewtonian liquids. The device is used to control the viscosity of fuels to burners and nozzles and for blending of fuel oils. The system is available with temperature compensation.

Ultrasonic probe—in which ultrasonic waves are applied to a thin magnetostrictive alloy-steel blade. The rate at which the blade oscillations are damped is a measure of the viscosity of the sample. A Bendix unit is shown schematically in Fig. V-22. The relationship between the damping and viscosity is solved by a relatively simple computer. The system is useful for recording and controlling and for both newtonian and nonnewtonian fluids. Ranges go up to 50,000 cp.

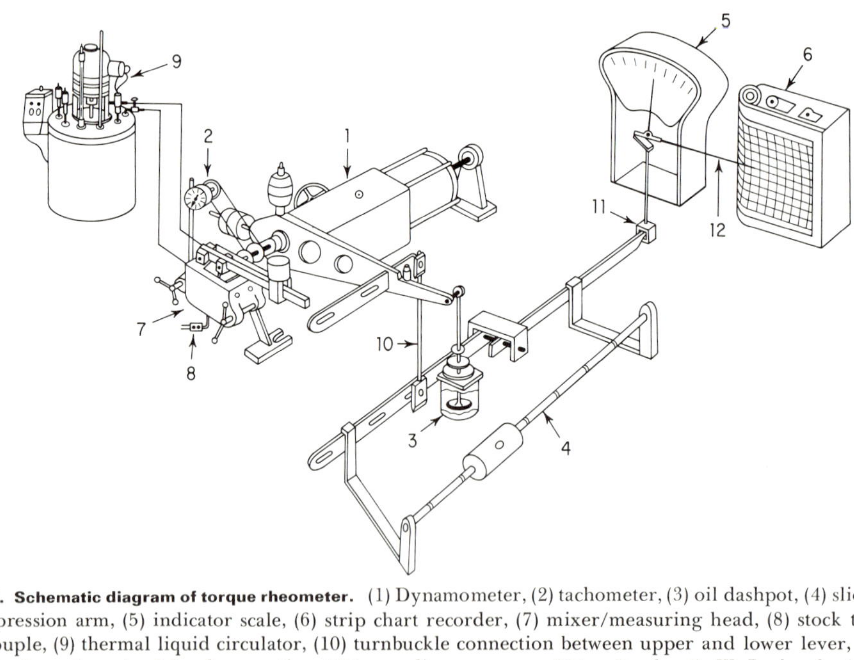

Fig. V-23. Schematic diagram of torque rheometer. (1) Dynamometer, (2) tachometer, (3) oil dashpot, (4) sliding weight zero suppression arm, (5) indicator scale, (6) strip chart recorder, (7) mixer/measuring head, (8) stock temperature thermocouple, (9) thermal liquid circulator, (10) turnbuckle connection between upper and lower lever, (11) stirrup bearing for lower lever/scalehead connection, (12) recording pen arm. *(Visco-Corder, C. W. Brabender.)*

Torque rheometer—measures viscosity on a batch basis while mixing in miniature, or on a continuous basis while extruding in miniature. The mixing or extruding drive is a recording dynamometer that operates on the principle of the Prony brake. The recorded reaction torque is a measure of viscosity at a specific shear rate. The shear rate can be varied over the range of processing interest (see Fig. V-23).

REFERENCES

Wohl, Martin H.: Instruments for Viscometry, *Chem. Eng.*, Jan. 15, Feb. 19, Mar. 25, Apr. 8, 1968.

Aronson, Milton H., and Robert C. Nelson: "Viscosity Measurement and Control," Instruments Publishing Co., Inc., Pittsburgh, 1964.

Brookfield, Donald W., and Richard A. Minard: Viscosity Measurement as Used for Automatic Process Monitoring, *Ann. N.Y. Acad. Sci.*, vol. 91, art. 4, pp. 838–849, June 2, 1961.

Minard, R. A.: Viscosity as a Process Variable, *Instr. Automation*, July, 1958.

Schowalter, W. R.: Rheological Equations of State, *Ind. Eng. Chem.*, August, 1964, p. 52.

Van Wazer, J. R.: Rheology, in "Encyclopedia of Chemical Technology," vol. 11, p. 730, Interscience Publishers, Inc., New York, 1953.

Manual Force Viscometers, *Instr. Control Systems*, vol. 36, pp. 115–124, May, 1963.

Weaver, E. Paul: The Prediction of ABS Extruded Sheet Surface Quality by Rheometer Behavior, SPE ANTEC meeting, Boston, Mass., March, 1965.

Goodrich, Judson E., and Roger S. Porter: A Rheological Interpretation of Torque Rheometer Data, ACS meeting, Phoenix, Ariz., January, 1966.

Viscous-drag Accelerometer (See **Accelerometer.**)

Vision, Sensory Aids for A series of devices exist to simplify everyday tasks for the blind, including several versions of a light probe which produces a sound—the pitch of which varies as the intensity of the light reaching the probe (Refs. 1, 2). Since the probe can be pointed in any direction, it can be used for locating windows and doors that open into the light, for telling whether a light in a room is on or off, and for generally exploring the light and shadow pattern in an environment. While some intellectually curious blind play with the instrument, it is not in extensive use. An agency called Science for the Blind (Ref. 4) designs and furnishes special instruments to enable the blind to perform special tasks required for a particular job. For example, the light probe allows a blind telephone switchboard operator to perform much more effectively. A meter has been developed that is brailled and emits a tone to allow a blind electronics technician or engineer to make circuit measurements.

Reading Aids: Work, sponsored primarily by the Veterans Administration, has been under way for many years toward the development of electronic reading aids (Ref. 5). These fall into three categories:

1. A letter-recognition device to be used for reading short sections of print to make it possible to check bank statements, proof typed letters, or recognize a particular book or magazine by reading its title. The device (Ref. 6) consists of a probe with 8 to 11 photocells arranged in a linear array, each individually connected to an oscillator the frequency of which increases in pitch from the bottom to the top of the array. When a photocell "sees" dark, it keys on its oscillator. Thus, as the probe is moved along a line of letters, each letter generates a characteristic sequence of chords, which can be learned—but with considerable difficulty. Several people using this instrument have been able to achieve reading speeds of 20 words per minute after 200 hr of practice.

2. The second category is character recognition. A device of this type takes the outputs from an array of specially placed photocells, puts them through a recognition matrix, and triggers a section of magnetic tape that speaks the appropriate letter. Under development by Mauch Laboratories (Ref. 6), the device, with its "spelled speech" output, should allow reading speeds of up to 50 words per minute. This speed still will not allow reading for pleasure but will facilitate reading for information.

3. In the third category is a device that is a complete automatic computer-controlled print-to-speech converter which does the equivalent of a human being reading printed material onto magnetic tape (Ref. 5). This computerized instrument, when fully developed, will be too expensive to be replicated often for a number of years. Thus, the logistics of a communication and distribution system will have to be worked out, probably utilizing local libraries and telephone lines.

Mobility Aids for the Blind (Ref. 7): It would be highly desirable to aid the mobile blind person by giving him distant early warning of obstacles, terrain discontinuities, and overhangs that lie a few feet in front of him. Two ultrasonic and one optical device to achieve this objective have been developed to the point of practical use. The ultrasonic aid sends out a beam of sharply focused ultrasound from a portable, hand-carried flashlight-type device (Ref. 8). Each burst of sound rapidly increases in pitch from the beginning to the end—so that the pitch of sound reflected back from a target will vary with the distance to the target. The original sound frequency range is ultrasonic, but the reflected sound is heterodyned with the instantaneously transmitted signal, the pitch of which is a function of range and the intensity of which is a function of the sound reflectivity of the target.

This presentation results in a very information-rich signal that tells a great deal about the environment. For example, if it is pointed toward the ground, a cement walk will sound "harsh," while grass will sound "soft and fuzzy." The trunk of a tree will sound "hard," while the leaves will sound "swishy." Evaluation of the device thus far indicates that only a relatively few blind people are able (or perhaps willing) to learn the meanings of the sounds of the device. Furthermore, the device will not easily detect curbs and downsteps, which many blind people feel to be a very serious drawback, since the traveler also must carry a cane.

Another sonic device (Ref. 9) resembles a small box camera that is worn around the neck. The device simply indicates the presence or absence of an obstacle with an audible sound. The device is much easier to use but less rich in its information presentation.

A third device (shown in Fig. V-24) is an optical ranging instrument that uses lasers fitted to a cane (Ref. 10). Three GaAs semiconductor lasers in the upper nacelle of the cane point to a spot at head height just in front of the cane, a variable distance straight ahead, and a point on the ground approximately 3 ft in front of the cane tip. The range of the straight-ahead beam can be adjusted from the cane tip out to 12 ft beyond the tip by moving an adjustment lever on the side of the cane. Light reflected from a target is picked up by one of three receiving systems in a lower nacelle and, after suitable amplification, triggers the appropriate output. A tactile stimulator warns of an obstacle straight ahead by poking the index finger of the hand carrying the cane, a high-pitched tone warns of an overhang, and a low-pitched tone warns of the ground's dropping away.

REFERENCES

1. Clark, L. L. (ed.): *Proc. Intern. Cong. Technology and Blindness,* 3 vols., American Foundation for the Blind, New York, 1963.
2. Zahl, Paul A.: "Blindness," Princeton University Press, Princeton, N.J., 1950.
3. *Catalog on Aids and Appliances,* American Foundation for the Blind, New York, 1963.
4. Science for the Blind, 221 Rock Hill Road, Bala Cynwyd, Pa.
5. Freiberger, H., and E. F. Murphy: Reading Machines for the Blind, *IEEE Trans. Human Factors in Electronics,* vol. HFE-2, no. 1, pp. 8–19, March, 1961.
6. Smith, G. C., and H. A. Mauch: The Development of a Reading Machine for the Blind, *Bull. Prosthetic Res.,* BPR 10-6, pp. 98–124, Fall, 1966.
7. *Bibliography of Mobility Research and Mobility Instrumentation Research,* American Foundation for the Blind, New York.
8. Kay, L; Binaural Ultrasonic Mobility Aid for the Blind, in L. L. Clark (ed.), *Proc. Rotterdam Mobil-*

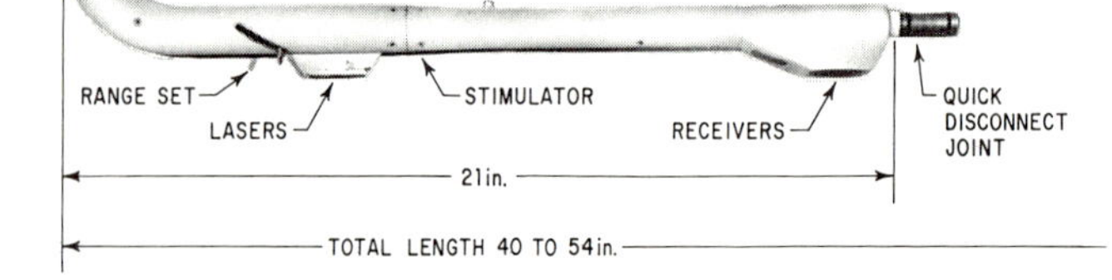

Fig. V-24. Laser cane.

ity Research Conf., American Foundation for the Blind, New York, 1964.

9. Russell, L.: Travel Path Sounder, *Proc. Rotterdam Mobility Research Conf.*, American Foundation for the Blind, New York, 1964.
10. Benjamin, J. Malvern, Jr.: A Review of the Veterans Administration Blind Guidance Device Project, *Bull. Prosthetics Res.*, BPR 10-9, pp. 62–90, Spring, 1968.

—J. Malvern Benjamin, Jr.,
Bionic Instruments, Inc.

Voice-recognition Systems Since speech is man's basic means of communication, the question often is asked, "Why not talk to machines?" Possible future applications of speech recognition are limited only by one's imagination. Although practically all applications of speech recognition still are in the future, one must not forget that the telephone, when first invented, was considered a rich man's toy.

For a limited library, voice recognition potentially may be applied (1) as an aid to paraplegics, (2) for parcel sorting, (3) for air-traffic control, (4) as an aid to jet pilots or astronauts, and (5) for remote inquiry via telephone or radio to a computer. Some possible applications of a speaker-identification system include (1) personalized door lock, (2) access to a restricted area, (3) memory protection in time-shared systems, (4) credit-card verification, and (5) verification of identity for bank transactions.

Speech has much to offer as a means of input to a system. Speech requires only a small amount of energy and leaves the hands free for other activities. Since speech is radiated outward by curvilinear waves, line of sight is not required. Because speech is independent of lighting, voice recognition can be used in the dark. Finally, speech requires no extra writing implement. A thought becomes a word or phrase with little conscious effort.

Speech Sounds: There are two classes of speech sounds: (1) *voiced* and (2) *unvoiced.* Voiced sounds use the regular "buzz" of the vocal cords as the source; while unvoiced sounds are derived from the hiss of the breath passing through various narrow constrictions formed by lips, teeth, and palate. Both sources or carriers are modulated by adjustment of resonance chambers in the throat and mouth cavities. Complex sound waves are produced that carry information encoded in frequency, intensity, and time. The human ear in conjunction with a significant portion of the brain, abstracts messages from the spoken words that fall upon it.

Speech Information: The information carried by speech may be divided into two parts: (1) "what is said" and (2) "who said it." The first category usually is referred to as *word recognition,* the second category as *speaker identification.* Opinions vary as to the best methods of extracting information from speech sounds to obtain each type of information. Generally it is agreed that some form of frequency analysis is required for both.

Word Recognition: Presently, word recognition is limited to a capability to recognize a small vocabulary of words spoken in isolation and not run together as in ordinary speech. In addition, most recognizers are designed to accept inputs from the one speaker for which the device is "tuned." Attempts to accept a number of speakers usually result in a degradation of accuracy. Work on the problem of segmentation of continuous speech and improved feature extraction, coupled with more powerful decision mechanisms, will ultimately provide a practical word-recognition system.

A typical word-recognition system is shown in Fig. V-25. This system consists of a frequency analyzer, a feature extractor, sequential feature storage, and a word categorizer. A "word code" is produced which may be used in any way desired. For example, the code may find the correctly spelled word in an auxiliary storage and cause it to be typed out.

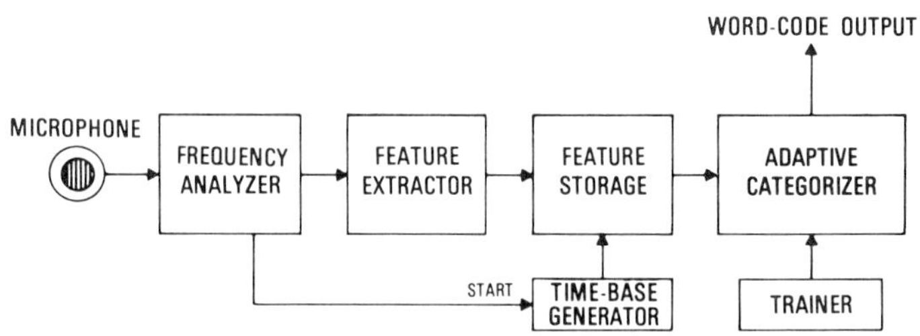

Fig. V-25. Word-recognition system.

The *frequency analyzer* subdivides complex voice waves into frequency components. One type is a bank of frequency selectors or filters. These are usually arranged so that the filter characteristics overlap somewhat to cover a band of frequencies in the desired portion of the frequency range from 100 to 10,000 Hz. A relatively small number of filters having desirable broad bandpass characteristics can be used for word recognition. A minimum practical filter bank for a frequency analyzer is shown in Fig. V-26. Individual attenuator adjustments are used to compensate for the fall-off in voice energy with increasing frequency. Filter outputs are rectified

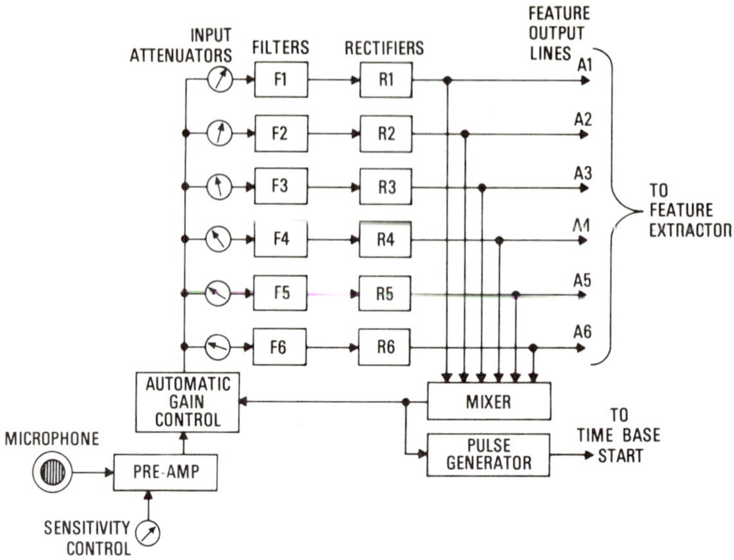

Fig. V-26. Frequency analyzer used in word-recognition system.

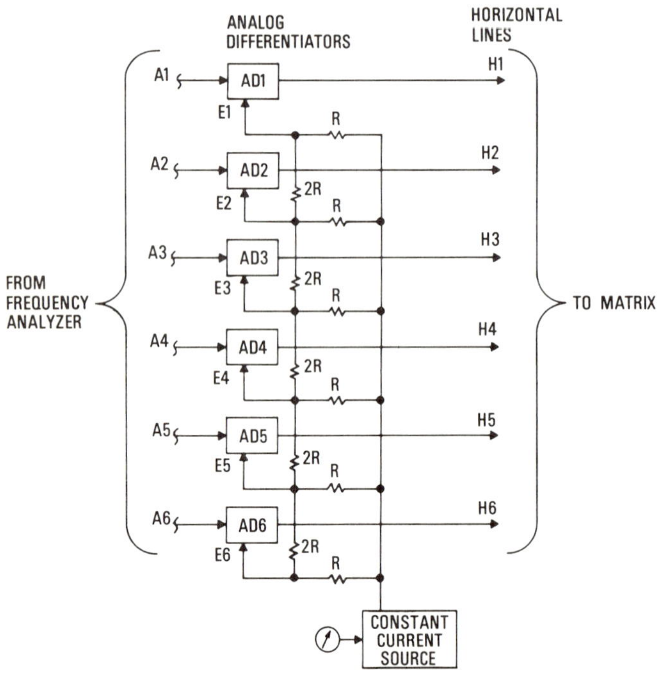

Fig. V-27. Feature extractor used in word-recognition system.

and smoothed to produce outputs $A1$ to $A6$ that vary in amplitude with time in response to the varying modulation of the voiced and unvoiced carriers of speech.

The *feature extractor* separates out identifiable elements of the frequency components. It usually includes some type of threshold, such as the floating-differential threshold system shown in Fig. V-27. The analog differentiators are integrating pulse shapers that operate when the voltage at the A input is greater than the threshold point E. The number of units that can be operated at any given time is set by the constant-current source. Detected features appear on output lines $H1$ to $H6$.

Feature storage temporarily holds serial sound elements for simultaneous observation. It is usually a matrix of storage units as shown in Fig. V-28. Each storage unit in the 6×6 matrix represents one of six features in a given time period $T1$ to $T6$. A bilateral flip-flop may be used for the storage unit.

The *time-base generator* determines the order of features being stored. It is often provided by a gated multivibrator driving an open-ended ring. The ring is reset to the "talk" position and moves through the stages $T1$ to $T6$, after a rise in the automatic-gain-controlled voltage indicates the start of the word. The rate of the multivibrator can be changed so that it becomes slower as the end of the word is approached. After recognition, the stored pattern is reset. This may be done automatically.

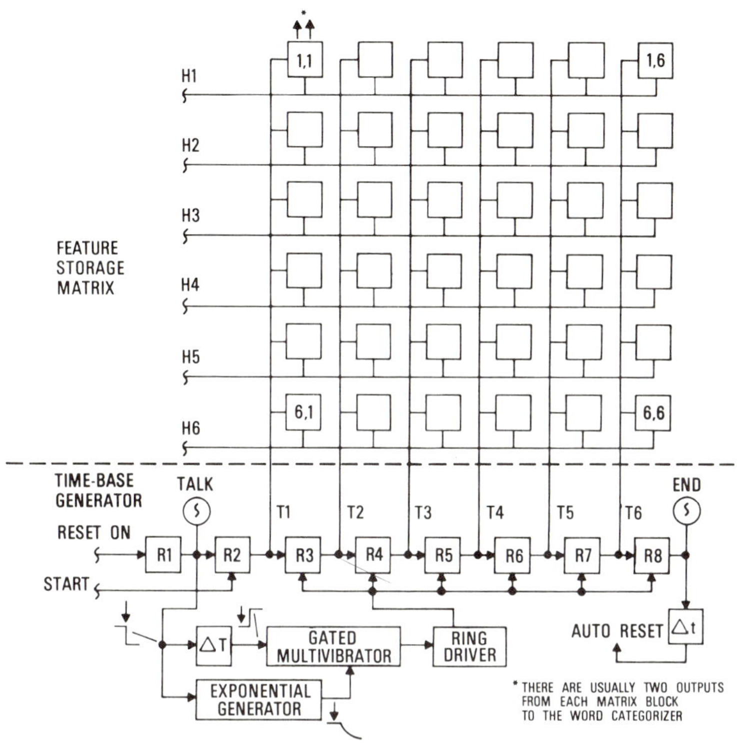

Fig. V-28. Feature storage used in word-recognition system.

Representative patterns are shown in Fig. V-29 for short words. With the wideband filters, speaker differences are minimized. Speaker *A* has a Middle Western background, while speaker *B* has a Southern accent.

The *word categorizer* identifies spoken short words or word segments by pattern matching. It can be of a trainable type, so that the decision function is adapted to the patterns of any given speaker. Both the pattern and its inverse may be used to enhance separability. A category is chosen by the "best match" technique. (An associative or content-addressable memory can also be used to categorize the patterns.) One simple, but powerful technique which has been used could be called an "electronic template." In this system, each "one" or "zero" in a given word pattern is adapted to give a unit vote to the desired decision unit. One or more samples of each word to be recognized may be used to form the word templates. When training is completed, new samples of the words in the training "library" will be recognized with accuracies on the order of 99 percent, if spoken by the individual who trained the categorizer. With no training, speakers will vary from 50 to 90 percent.

Speaker Indentification: The differences that exist between speakers may be used to identify who is speaking. A set phrase, such as "My account number is . . ." may precede the name (or number) of the individual seeking entrance to a system. A frequency analysis is made in greater detail than that which has been described for word recognition, because a finer screen is needed to separate the voices of different speakers enunciating the same phrase. This is followed by feature summation, feature ranking and encoding, and adaptive categorization of the resultant speaker pattern. When it is trained for a given speaker, he alone will be able to match the pattern

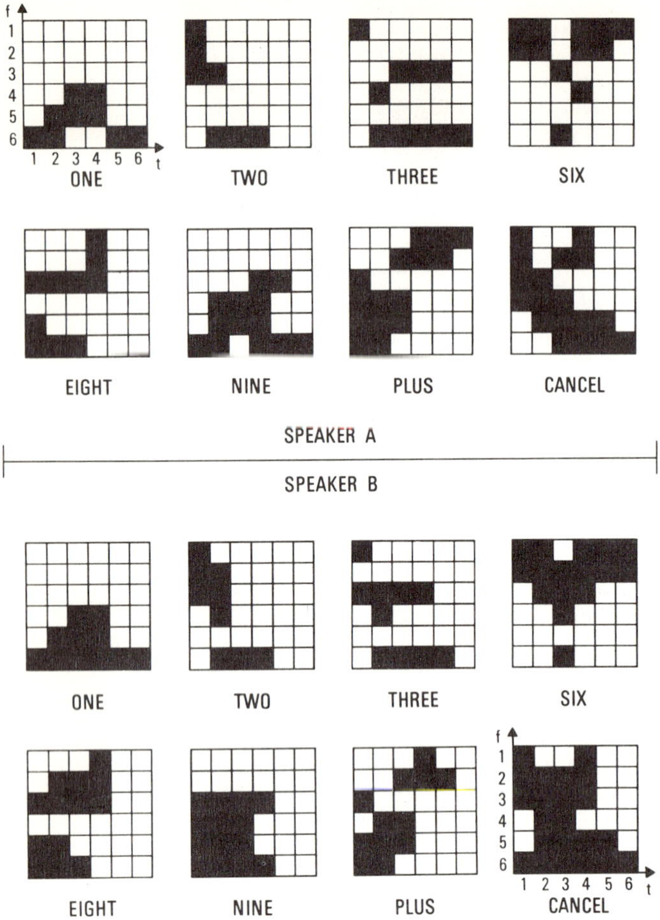

Fig. V-29. Typical digitized word patterns stored in a matrix used in a word-recognition system.

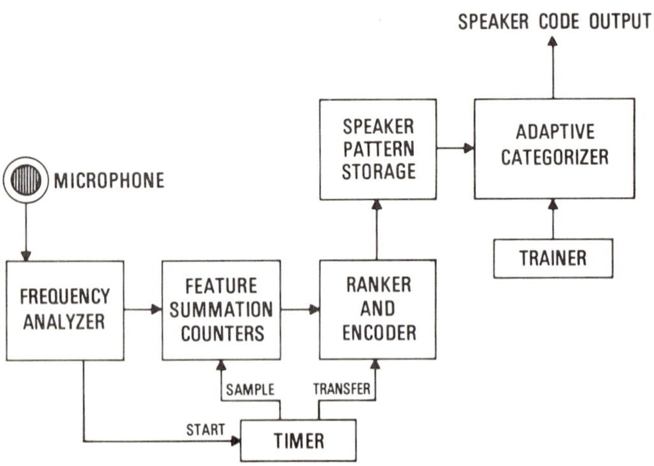

Fig. V-30. Speaker-identification system.

to the necessary degree to gain admittance. A block diagram is shown in Fig. V-30. The range of frequencies analyzed for speaker identification should include the lower frequencies of the male fundamental pitch, that is, 70 to 80 Hz and upward. Also, an absolute rather than floating threshold is used to identify the features and control summation. Another difference is that time sequence is retained and used in the pattern for word recognition; whereas integration is used to remove sequential information for speaker identification—since the identification of speech characteristics is more important than when they occurred. Many computer programs are being used in the study of fundamental problems in speech processing. These studies will be reflected in future voice-recognition systems.

REFERENCES

Clapper, G. L.: Digital Techniques for Speech Analysis, *IEEE Trans. Commun. Electron.*, no. 66, pp. 296–305, May, 1963.

Lindgren, N.: Machine Recognition of Human Language—Part I, *IEEE Spectrum,* vol. 2, no. 3, pp. 114–136, March, 1965.

King, J. H., Jr., and C. J. Tunis: Some Experiments in Spoken Word Recognition, *IBM J. Res. Develop.*, vol. 10, no. 1, pp. 65–78, January, 1966.

—Genung L. Clapper, *IBM Corporation*

Voltage, Measurement of (See **Potentiometer;** and **Voltmeter.**)

Voltage, Standard of (See **Standard cell;** and **Standards, measurement.**)

Voltage Divider (See **Bridge circuits, electric.**)

Voltammeter (See **Coulometer.**)

Voltmeter The absolute volt may be defined as the difference of potential that exists across an absolute ohm when traversed by an absolute ampere. See **Standard cell.** The term *voltage* is synonymous with the term *electromotive force* (emf). Instruments are available for measuring minute emf's of microvolt magnitude up to thousands of volts—for both alternating and direct currents.

Dependent upon the magnitude of voltage to be measured and the precision required, one of the following kinds of instruments will be used:

Potentiometers: For emf's of 1.5 V and down into the millivoltage and microvoltage magnitudes, a potentiometer may be used—particularly in the laboratory where precision is required.

However, potentiometer-type instruments of rugged construction for industrial applications in connection with numerous kinds of emf-producing transducers are available. Potentiometers are passive (resistive) networks, permitting the division of a voltage and the comparison of the partial voltage with a voltage standard. (See **Standard cell.**)

Permanent-magnet Moving-coil Instruments: Commercial and industrial panel and switchboard voltmeters are obtainable for dc voltage measurements from 0 to 1.5 V to 0 to 1 kV; and for ac voltage measurements from 0 to 1 V to 0 to 500 V. Scale lengths from 1½ to 7½ in. are obtainable in standard models—in both circular- and rectangular-front formats. Panel meters wherein the scale (PMS) is optically projected or wherein the pointer (PMP) is optically projected are obtainable. The accuracy of stock panel voltmeters ranges from ±2 to ±3 percent of full scale. Permanent-magnet moving-coil voltmeters also are obtainable in portable form for

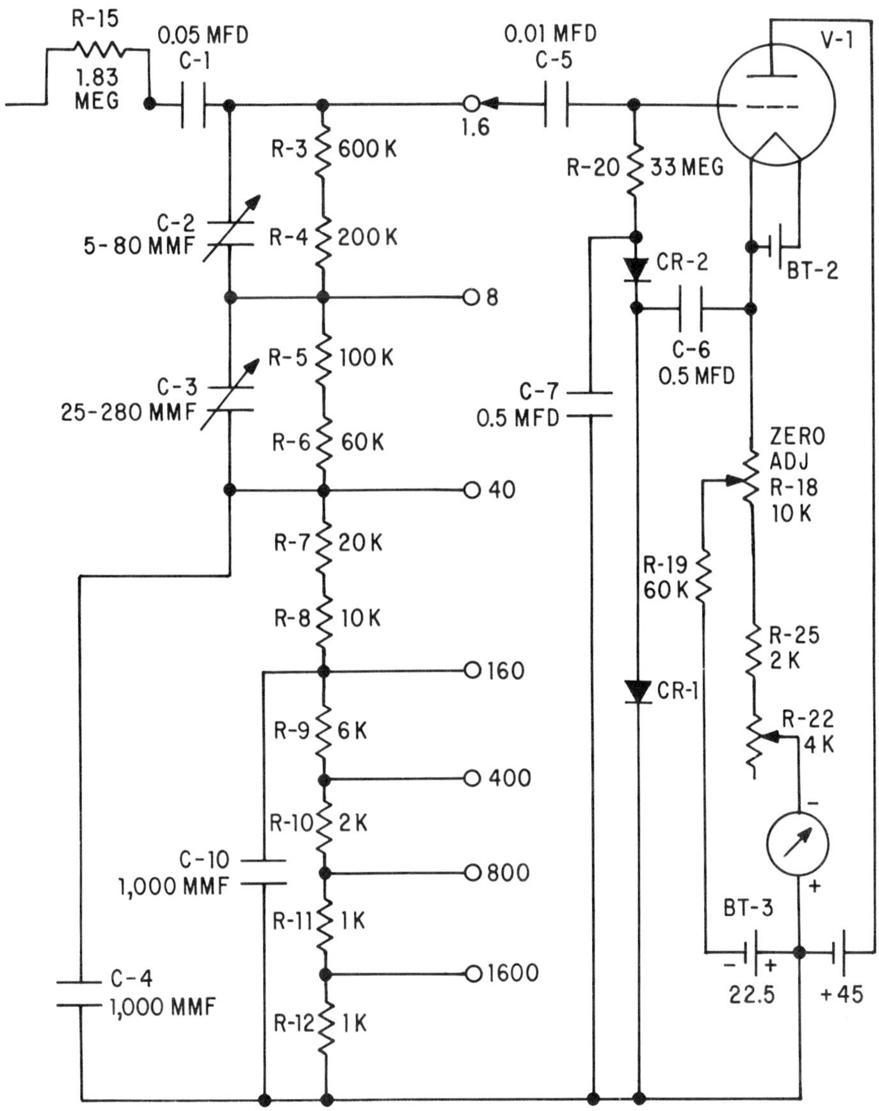

Fig. V-31. Battery-operated vacuum-tube voltmeter arranged to read ac volts peak to peak.

testing and laboratory use where the accuracy generally will range from ±1 to ±2 percent of full scale, although special higher-cost instruments are available with greater accuracy. Voltmeters also are obtainable in multimeter form where current and resistance as well as voltage may be measured with the same meter. See **Electrical quantities, measurement of;** and **Meter, electrical.**

Moving-iron Voltmeter: These meters are available in formats, ranges, and characteristics similar to the permanent-magnet moving-coil instruments just described. They are less complex in design but require more power. They are mostly used for ac voltage measurement of modest accuracy. See **Electrical quantities, measurement of;** and **Meter, electrical.**

Thermocouple Voltmeter: These meters employ a permanent-magnet moving-coil millivoltmeter to show the emf set up in the thermocouple attached to the heater wire (through which the unknown current is passing). With a suitably constructed resistance multiplier, the instrument may be used at frequencies up to 15 kHz or higher and for voltages from 10 V to several hundred volts.

Electrostatic Voltmeter: These meters measure ac voltages from 150 to 3,500 V. See **Electrical quantities, measurement of.**

Vacuum-tube Voltmeter: These meters are used for measuring ac emf's from a few microvolts up to several hundred volts. With capacitance potential dividers, the range can be increased to even higher voltages. VTVMs draw practically no current from the source. Input impedance is of the order of megohms. With careful design, these meters can give correct indications at frequencies up to 100 MHz. The circuit of a battery-operated VTVM arranged to read ac volts peak to peak is shown in Fig. V-31.

Rectifier Voltmeter: Generally, these meters employ solid-state rectifiers in which germanium and silicon units are used because of their higher working temperature and operating frequency. Full-wave ac rectification may be obtained by connecting a bridge network formed of separate half-wave rectifier units to a permanent-magnet moving-coil instrument (see Fig. V-32). Appropriate series resistances are used to provide the desired ranges. Deflections are proportional to the average value of the current flowing through the indicator, and thus a practically uniform scale is achieved. Errors result if the waveshape departs appreciably from a sine curve.

Digital Voltmeter: These are obtainable for ac and dc voltage measurements in ranges from a few millivolts up. Also see **Electrical quantities, measurement of.**

Electrometer: A device for measuring electric charge. The familiar attraction and repulsion of the suspended-gold-leaf (see **Electrical quantities, measurement of**) electroscope exemplifies the fundamental principle of the instrument. Numerous forms of electrometers have been devised.

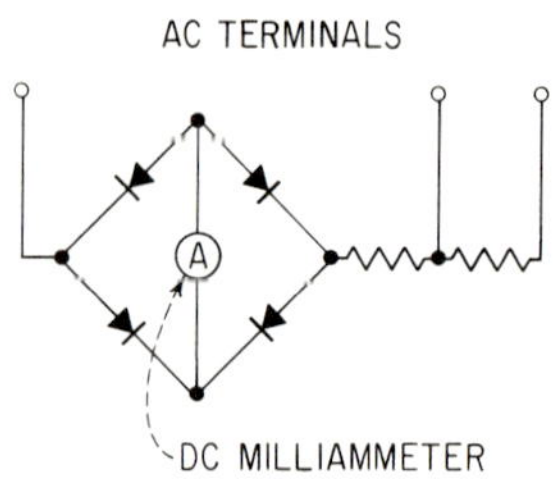

Fig. V-32. Rectifier-type voltmeter.

In the Wulf electrometer, two silvered or platinized quartz fibers are placed side by side such that they tend to bulge apart upon being charged. Similar fibers are used in the string electrometer in which the fibers are lightly stretched and deflect laterally when in an electric field. The behavior of the fibers in either design can be observed with a micrometer microscope. Such devices have been used in connection with cosmic-ray research.

Sphere-gap Voltmeter: For very high voltages, the measured distance between two electrodes and the beginning of a corona discharge provide a measure of voltage. This is now the standard method for the measurement of the crest value of alternating voltages. Numerous investigators have carried out calibrations of sphere gaps with sphere diameters ranging from a few centimeters up to 2 m.

Differential Voltmeter: See **Electrical quantities, measurement of.**

—Adapted from "The Instrument Sketch Book," *Weston Instruments, Inc.*

Volume, Standard (See **Standards, measurement.**)

Volumetric Feeder (See **Feeder, volumetric.**)

Volumetric Flowmeter (See **Flowmeter, positive-displacement.**)

Vor (See **Navigation systems.**)

W

Waste-disposal Instruments [See Analytical instrumentation (chemical).]

Water (Moisture), Measurement of (See Moisture, measurement of.)

Water Meter (See Flowmeter, positive-displacement.)

Watthour Meter A watthour is the energy consumed in 1 hr when the power or rate of power consumption is 1 W. This is the practical unit of electrical energy. A watthour meter is an instrument designed to measure the electrical energy consumed in a given time by integrating it during that time. The watthour meter generally is calibrated in watthours or kilowatthours. Essentially, watthour meters are small motors. Motor speed is proportional to the power passing through the meter. Generally, dc watthour meters operate on the principle of the electrodynamic dc motor. Ac watthour meters utilize the principle of the rotating magnetic field of the induction motor.

Dc Watthour Meters: There are two basic types: (1) the commutator-type meter shown in Fig. W-1 and (2) the mercury-motor type shown in Fig. W-2. Both types must be compensated for friction. In the commutator-type meter, a coil of about the same diameter as the field, connected in series with the armature and mounted so that the magnetic field it produces is added to that of the main field coils, is employed. Individual

Fig. W-1. Commutator-type watthour meter. Light metal disk *d*, mounted on a steel shaft which turns in a jewel bearing, rotates between the poles of two or more permanent magnets. In the diagram, a = armature, *c* = commutator, *f* = field coil, *s* = light-load compensation coil, and *r* = resistance. The armature is connected to the external circuit by means of very light silver-tipped brushes.

meter adjustment is accomplished by altering the position of the coil with respect to the field coils. The first electric-energy meter was developed by Elihu Thomson in 1889 and was of the commutator type. Although developed principally

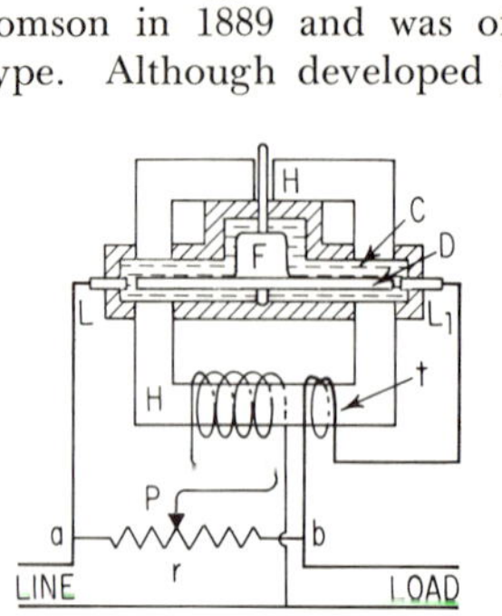

Fig. W-2. Mercury-motor-type watthour meter. *D* is a copper disk having a number of radial slots and floating in mercury, *F* is a float which supports the shaft, *H* is a laminated-iron core, and *C* is a chamber filled with mercury. Flux produced in core *H* by the shunt coil traverses the disk at two points which are diametrically opposite. The line current passes from *L* to L_1 through the mercury and diametrically through the disk because of its slots. Since the disk is cut by flux, a torque is produced which is proportional to the current and voltage.

for ac measurements, the meter has been used primarily for the metering of dc energy. In the mercury-motor-type meter—to compensate for friction due to the disk rotating in the mercury—an Ayrtron shunt (shown as *ab* in Fig. W-2) may be used so that a fraction of the voltage-circuit current can be sent through the current chamber; or a slider resistor regulates the amount of current sent through the current chamber LL_1 by a thermocouple the junction of which is heated by a coil in series with the potential circuit.

Ac Watthour Meters: The essential features of an induction-type watthour meter are shown in Fig. W-3. A compensating or lag coil is used in these meters to compensate for their tendency to run slow—resulting because the ohmic resistance

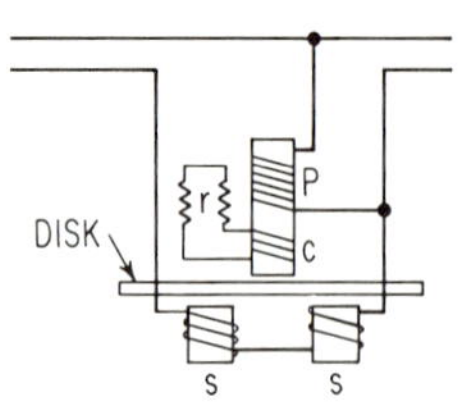

Fig. W-3. Induction-type watthour meter. In the diagram, $P =$ potential coil, $s =$ series coils, and $c =$ compensating coil.

of the potential circuit is such that the current never is exactly 90° behind the impressed voltage. Because of the absence of a commutator and armature, the friction in an induction-type meter is much less. However, the torque also is less, and thus the effect of friction at light loads must be compensated. Compensation is effected by producing a flux at the potential pole face slightly out of phase with the main flux. The resultant eddy currents produced in the disk—which will be in phase with a small component of the main flux—give rise to a slight torque which can be made ample to overcome the friction torque. When the capacity of the circuit is over 200 A, instrument-current transformers generally are used to step down the current to 5 A. If the potential is over 440 V, current transformers almost invariably are used, irrespective of the magnitude of the current, in order to insulate the meter from the line. In such cases, voltage transformers also are used to reduce the voltage to 110 V.

Var-hour Meters: Reactive volt-ampere-hour meters generally are conventional watthour meters in which the current coil is inserted in series with the load in the usual manner while the potential coil is arranged to receive a voltage at right angles to the load voltage. The total volt-ampere-hours, or "apparent energy" expended in a load, is of interest because it determines the heating of generating, transmitting, and distributing equipment. The apparent energy may be computed, if the power factor is constant, from the observed watthours *P* and the observed reactive volt-ampere-hours *Q* such that

$$\text{Volt-ampere-hours} = \sqrt{P^2 + Q^2}$$

Accuracy of Watthour Meters: Meter manufacturers assign and mark each meter with a *watthour constant,* such that the indicated watthours will be $K_h \times R$, where $K_h =$ watthour constant and $R =$ number of revolutions of meter. (Also see **Demand meter;** and **Electrical quantities, measurement of.**)

Wattmeter [See **Electrical quantities, measurement of;** and **Power (electrical), measurement of.**]

Wave Analyzer An instrument used for measuring the components of, or analyzing the spectra of, complex electrical signals, including those which are replicas of acoustic noise or mechanical vibrations. A wave analyzer of the type shown in Fig. W-4 can separate and measure individual components of periodic complex waveforms, such as harmonic or intermodulation distortion. An analyzer of this type also can be used as a tunable filter so that the individual components of a complex input signal can be used to drive other in-

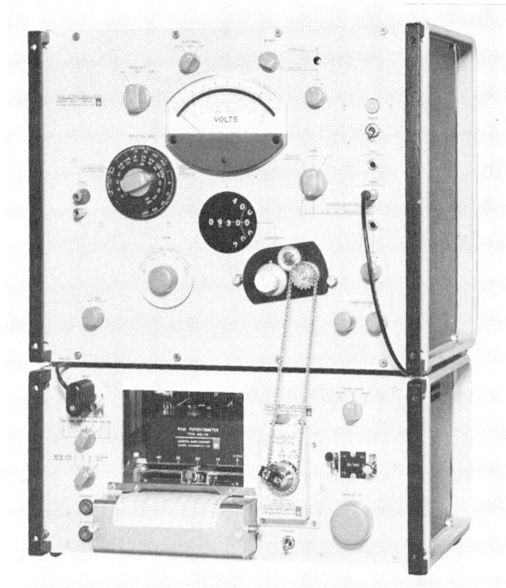

Fig. W-4. Recording wave analyzer. *(General Radio.)*

struments, such as frequency counters, when a highly accurate measure of the component frequencies is desired, or to drive earphones. When a wideband noise generator drives the analyzer, the output is a tunable narrow band of noise. Such a signal is useful in a number of psychological and architectural-acoustics tests.

In the "tracking generator" mode of operation, the output is a sine-wave signal tunable over the 54-kHz range and in tune with the analyzer. When this signal is used to drive a bridge or other network, the output can be measured by the analyzer, whose selectivity reduces the interference from extraneous noise, hum, and distortion. For automatic waveform analysis, outputs are provided for driving the graphic level recorder (at the bottom of Fig. W-4) as well as dc recorders.

The analyzer is a heterodyne type of voltmeter. The intermediate-frequency amplifier at 100 kHz includes a highly selective quartz-crystal filter whose bandwidth can be switched to 3, 10, and 50 Hz. The use of a heterodyne system makes it possible to vary the response frequency although the filter frequency is fixed. The 100-kHz output of the filter is indicated on a meter. In one mode of operation, the output also is heterodyned back to the original frequency. In another mode, the local oscillator beats with a 100-kHz quartz-crystal oscillator to function as a beat-frequency oscillator. The receiver output may be displayed on the vertical axis of an oscilloscope whose horizontal sweep is related to the receiver sweep rate, thus giving a visual indication of the frequency and relative strength of signals present.

Other analyzers are available with bandwidths that are a constant percent of the center frequency. Thus, a 1 percent bandwidth is 0.2 Hz wide when tuned to 20 Hz; it is 10 Hz wide at 1 kHz; and so on. Also in this class are octave-band and one-third-octave analyzers of especial value in the measurement of acoustic waves.

—Carl W. Alsen, *General Radio Company*

Waveform Monitor (See **Oscilloscope.**)

Wavemeter [See **Frequency (electrical), measurement of.**]

Wave Recorder (See **Ocean-wave gage;** and **Tide and wave recorder.**)

Weather Balloon A device used to carry weather instruments into the upper atmosphere. The balloon is filled with lighter-than-air gas just prior to release and is tracked by optical or radio theodolites to measure wind speed and direction. As the balloon rises, it follows the horizontal motion of the winds at its location independent of its speed of ascent. Periodically, azimuth and elevation-angle readings are made with the theodolite and converted to position. The distance and direction moved since the last reading are a measure of wind speed and direction. If the balloon is transporting a radiosonde, temperature, humidity, and pressure altitude also may be checked.

Since the surface area of the balloon is large and the inertia is small, acceleration to the speed of the wind is nearly instantaneous. The balloons are made of rubber or artificial-rubber compounds and inflated with hydrogen or helium. Helium is preferred since it is not explosive and, although twice as dense as hydrogen, has a buoyancy that is only about 7 percent less than hydrogen by volume. Buoyancy, or lift, is determined by the weight, diameter, and volume of the particular gas used.

REFERENCE

Albright, John G.: "Physical Meteorology," Case Institute of Technology, Cleveland, Ohio, 1943.

—James R. Cosby, *Environmental Science Division, The Bendix Corporation*

Weighing, Industrial (See Scale.)

Weighing, Radiation Methods for Nuclear scales weigh materials moving on belts and on vibrating, apron, drag-chain, or air-slide conveyors without contacting either materials or conveyor. Uses are found in the sand and gravel, coal and mining, cement, food, fertilizer, and chemical industries.

The operating principle, as shown in Fig. W-5, is based on the proportional absorption of low gamma radiation by materials in transit. Transmitted radiation is converted into a proportional electric current by detector cells and, when amplified and used with a tachometer, indicates the weight delivered. The same signal can be used for automatic controls to start and stop motors and for proportional control with auxiliary equipment.

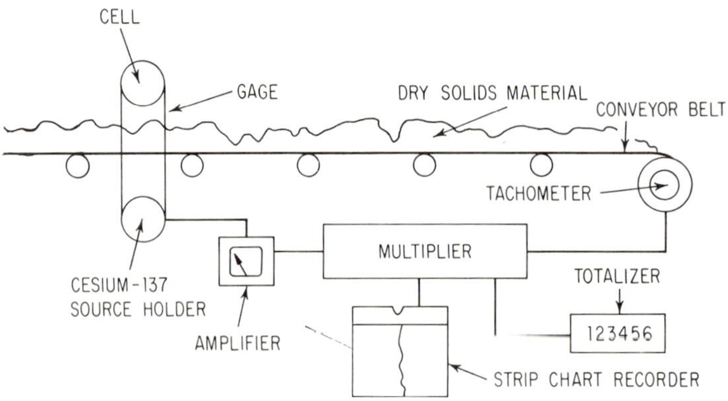

Fig. W-5. Gamma-radiation-type scale. *(Ohmart.)*

The measuring assembly consists of a C frame with one arm located beneath the conveyor and the upper arm extending across the top. The lower arm holds the gamma radioisotope and the upper arm holds the detector cells. The radiation beam is collimated to the cells. The system may be applied to conveyance widths from 8 to 72 in. The assembly may be mounted in place without stopping the conveyor. Long-term accuracy is from one-half of 1 percent.

The amplifier is an ac solid-state feedback-stabilized type with vibrating capacitor input. A unit linearizes the nuclear-gage signal (1) to represent belt loading in pounds per linear foot, (2) to multiply the belt-loading signal from linearizer and tachometer, and (3) to totalize the flow for a given time period. A recorder or five-digit printer can be added to the system.

Proportioning control includes one-, two-, or three-mode time-proportioned control for a constant-speed valve or feeder motor; one-, two-, or three-mode proportioning control of the air-operated valve or feeder; and a 200-W proportioning signal to operate a typical variable-speed drive motor on the feeder or conveyer drive.

Batching control is done by a counter with adjustable electrical contact to give a signal to indicate when a preset weight has been delivered. The counter can repeat the operation or stop the feeder.

—Gordon W. Hickenlooper, *The Ohmart Corporation*

Weight, Integration of (See **Integrator.**)

Weight, Measurement of Weight-measurement applications can be best classified by grouping them into functional categories as follows:

Process control—Weighing becomes a process-control function when material is weighed into a process at a certain rate. Where output is low, weighing is often a manual operation. High-volume production calls for automatic weighing scales; there are two kinds: draft (or increment) weighers and continuous weighers.

Formulation control—Weighing becomes a formulation-control function when two or more ingredients are fed into a hopper in preset proportions. For low-output operations and few ingredients the weighing can be manual, or semiautomatic by using cutoff switches to stop and start scale feeders. For continuous and high-output requirements, fully automated systems are required which include options for formula input and weight readout as well as control of interfacing nonweighing equipment. (See **Batching systems, automatic-weight.)**

Receiving/shipping—Weighing becomes a checking function in buying and selling when commodities are shipped from supplier to user. Bulk commodities are weighed either in the carrier (truck or railroad car) or before loading. Single items such as coils and bales are also weighed individually. Loaded trucks and cars, as well as crates and packages, are weighed to determine shipping charges.

Packaging—Weighing becomes a control function in filling packages. Consumer packaging requires small-increment (to 5 lb) high-output equipment, completely automated, and usually integral with the packaging machine. Industrial packaging is in the 10- to 100-lb range using paper, textile, or plastic bags, also drums up to 200 lb.

Check weighing—Weighing provides a means to sort out packages or machined items outside preset weight tolerances. Automatic check weighers are used in high-output consumer and industrial packaging and manual platform scales for low-output operations. In automated systems the check weigher can trigger a reject device to move out-of-tolerance packages or items off the line. Automatic check weighers are used also to detect missing inserts in cartons.

Parts counting—Weighing becomes a counting function in the mass-production industries. An operator uses a counting scale to determine the total number of pieces in a container from the weight of a sample of pieces. Automatic counting systems do all the computing for the operator and provide a printed record of the count.

Scientific applications—Weighing in the laboratory calls for equipment having a high degree of discrimination for the minutest increments of weight in their weighing range. Typical weighing accuracy of scientific balances is on the order of 0.1 g; however, for highly precise work accuracies of 0.1 to 10 mg can be obtained. Capacities of these balances range from 120 to 4,500 g.

Miscellaneous—In addition to the above basic categories there are other specialized in-plant weighing applications, commercial applications such as in retailing and post offices, domestic applications such as person weighers (baby and health scales), medical applications such as physician's and physical director's scales, and others.

Weighing Methods: Weight measurement commonly is performed by measuring the force reaction caused by a given weight. This force measurement can be done in several ways. Also see **Scale, beam; Scale, industrial; Scale, pendulum**

Mechanical measurement—using a lever and counterweight, the force reaction is counterbalanced by a calibrated weight. Advantages: simplicity, can be self-powered or manually manipulated, is unaffected by most environmental conditions. Disadvantages: requires more space than other methods, can be cumbersome in complex control systems.

Strain measurement—using a device to measure the strain resulting from stress in a member that supports the weight. For example: *Mechanical spring scale*—Advantages: compact, self-powered, low cost. Disadvantages: more susceptible than other methods to environmental conditions such as vibration and temperature, lower accuracy potential. *Strain-gage load cell (electronic)*—Advantages: compact, low energy input, provides more convenient method than mechanical measurement for very low and very high capacity weighing, well-suited for complex weight-control systems. Disadvantages: more susceptible than other methods to environmental conditions such as vibration, temperature, and electrical disturbances; not self-powered.

Pressure measurement—using a device to measure the change in pressure caused by the force reaction to the weight in the pressurized vessel which supports the weight. For example: *Hydraulic load cell*—Advantages: compact, rugged, self-powered, self-damping. Disadvantages: limited accuracy potential, developed only for specialized applications to date. *Pneumatic load cell*—Advantages: fast response, good damping characteristics, suitable for use with pressure control. Disadvantages: not self-powered, developed only for specialized applications to date.

Weighing Terminology: A complete glossary of weighing terms would number hundreds of entries. Characteristics of scales generally follow those of other instruments, and therefore, such terms as accuracy and sensitivity will not be given here. Also, definitions of scale terms found elsewhere in this volume are not given. See

Batching systems, automatic-weight; and **Scales;** and the alphabetical index of terms.

Apothecaries' System: Used for weighing and measuring drugs and compounding prescriptions; mass units: 1 ounce = 8 drams, 1 dram = 3 scruples, 1 scruple = 20 grains; volumetric units: 1 fluid ounce = 8 fluid drams, 1 fluid dram = 60 minims.

Automatic Indicating Scale: A scale that indicates automatically the weight value of any applied load in its weighing range without manipulation of the indicating elements by the operator.

Automatic Recording Scale: A scale incorporating a mechanism that imprints automatically on a ticket or tape the weight value of the applied load.

Avoirdupois System: The system of mass units for commercial weighing of all commodities except precious stones, precious metals, and drugs: 1 pound = 7,000 grains = 16 ounces, 1 ounce = 437.5 grains, 1 pound = 453.59237 grams.

Beam Indicator: A weight indicator which by manipulation of poises and/or counterpoise weights along a graduated bar balances out the applied load and indicates its value by the position of the poise; this is called mechanical indication.

Carat: A unit of mass for precious stones, especially diamonds and pearls, and equal to 200 milligrams.

Check Weigh: To reweigh by another scale (and person if manually done) a previously filled and weighed package.

Class of Scale: A classification for specifications and tolerances set forth by the National Conference on Weights and Measures. Examples include crane scales and prescription scales.

Class of Weights: A classification adopted by the National Bureau of Standards based upon design, material, and accuracy; includes state primary and secondary standards, local and industrial reference standards, test weights, industrial working standards, scientific reference standards, and laboratory working standards.

Computing Scale: A scale that indicates some function of the applied load, either with or without its weight value, such as a price-computing scale, parcel-post scale, or counting scale.

Counterbalance: Material applied at the butt end of a weighbeam or lever to balance the scale at zero load; also called back balance.

Counterpoise Weight: An adjusted, removable weight made to counterbalance a given applied load.

Dial Indicator: A weight indicator having the numerical weight information arranged in a graduated scale around a flat circular chart and using a pointer-type indicator; this is called automatic indication because no manipulation of the weight-indicating elements is necessary.

Digital Indicator: A weight indicator which presents the weight information in numerical form similar to an odometer; this is called automatic indication because no manipulation of the weight-indicating elements is necessary.

Draft: A discrete amount of bulk material which is weighed in one operation, usually on a hopper or tank scale; also termed a "weighment."

Dribble: In weighing bulk materials on a hopper or tank scale, it is that material feeding into the receptacle at the cutoff.

Dribble Compensator: A device that compensates for the lag in stopping the dribble feed at cutoff.

Equal-arm Lever: A lever whose arms on either side of the fulcrum are of equal length, that is, whose ratio is 1.

Fan Scale: An automatic indicating scale whose reading face is a fan-shaped chart, usually a price-computing counter scale.

Fractional Bar: An auxiliary beam bar whose capacity usually equals the difference between successive graduations on the main bar to which it is the auxiliary.

Full-capacity Weighbeam: A weighbeam which balances through the entire capacity of the scale on its bars without the use of loose or hanging weights.

Gross Weight: The weight of an item or quantity of material including the container, package, or the carrying vehicle; the total weight of material without deduction for dunnage, tare, or waste.

Hooke's Law: The established principle that a body acted on by an external force will deform in proportion to the stress developed, provided such stress does not exceed the proportional limit of the body; this principle is exemplified in a spring scale.

Knife-edge: The honed and hardened edge of a scale pivot fixed on a lever which makes linear contact with an opposing bearing surface.

Larry Scale: Usually a hopper scale, mounted on or in a structural framework on wheels which allows it to service several bins.

Lever: A bar provided with pivots or plate fulcrums for translating external forces.

Lever Ratio: The ratio of the lengths of the arms of a lever; also called "multiple."

Lever System: In a compound lever scale, the levers and related live parts between the load-receiving element and the steelyard rod which ties it to the weight indicator.

Long Ton: A ton of 2,240 pounds avoirdupois, equal to 20 long hundredweights, also called a gross ton.

Metric Ton: 1,000 kilograms.

Multiple: The ratio of the applied load to the counterforce required at a given knife-edge in a lever train.

National Conference of Weights and Measures: An organization of weights and measures officials of the United States which meets annually at the National Bureau of Standards, Washington, D.C.

Net Ton: A short ton.

Net Weight: The weight of an item or quantity of

material exclusive of its container, package, or carrying vehicle.

Ounce: The avoirdupois ounce = 1/16 avoirdupois pound = 437.5 grains = 28.3495 grams; the Troy or apothecaries' ounce = 1/12 troy or apothecaries' pound 480 grains = 31.1035 grams.

Packing Scale: Any scale adapted for weighing loose, bulky material, such as flour or bran, and then packing it into bags or barrels by means of compression (auger), vibration, or shaking.

Pivot: That component of the scale lever which serves as a fulcrum and knife-edge.

Platform: The horizontal, rectangular load-receiving element of a static-type scale (such as a floor scale or portable platform scale).

Poise: A weight moved along the graduated beam (or weighbeam); the point on the beam at which it counterbalances the applied load indicates the weight of the load.

Pound: The avoirdupois pound = 7,000 grains = 453.59237 grams = 16 ounces; the Troy or apothecaries' pound = 5,760 grains = 373.242 grams = 12 ounces.

Poundal: The absolute unit of force that, when acting for 1 second on a mass whose weight is 1 pound in London, will produce a velocity of 1 foot per second.

Prescription Scale: Usually an equal-arm scale of small capacity for the weighing of ingredients of medicinal and other formulas prescribed by physicians and others and entering into the ordinary trade of pharmacists and chemists.

Printer: A device for recording weight information on a ticket or tape.

Programmer: A device which selects and controls the sequence of events in an automatic weighing system.

Remote Indicator: Any type of weight indicator which is placed at a location remote from the point of weighing.

Remote Printer: A weight printer which is placed at a location remote from the dial head or the point of weighing.

Sensibility Reciprocal: The change in applied load required to change the position of equilibrium of a weighbeam or indicator a definite amount.

Standard Weight: A weight expressly designed and constructed for the testing of weights of lower classes; also used in high-precision weighing operations.

Tare: The weight of an empty container, package, or vehicle; or the deduction from gross weight made on account thereof.

Tolerance: A value fixing the limit of allowable error or departure from true performance or value.

Torsion: The stress in a part which resists twisting force.

Torsion Balance: A type of equal-arm scale in which horizontal steel bands are used instead of pivots and bearings.

Troy System: A mass-unit system where 1 ounce = 20 pennyweights, 1 pennyweight = 24 grains.

Unit-weight Scale: An automatic indicating scale whose elements comprise a reading face and indicator and one or more unit or drop weights; in practice the entire mechanism is enclosed in a cabinet and the unit weights are successively applied or removed manually or automatically to match the applied load; each weight, when applied, represents an increment of load about equal to the capacity of the reading face.

Weighbridge: In a large-capacity scale, the structural frame carried by the main bearings which supports the load-receiving element (usually the platform).

Weight: The force with which a mass is attracted toward the center of the earth by gravity; true weight of an object is its weight determined in a vacuum; apparent weight in air of an object is its weight determined in air and is less than true weight by the true weight of air displaced by the object; in scale terminology a weight is usually a metal object having a definite mass and used for weighing, testing, as a counterpoise weight, and for other purposes.

Weight Ticket: A ticket upon which the weight of an item or quantity of material is registered or written at the time of weighing.

REFERENCES

Considine, D. M.: "Industrial Weighing," Reinhold Publishing Corporation, New York, 1948.

Considine, D. M.: Process Weighing, *Chem. Eng.*, Aug. 17, Sept. 14, 1964.

—Arthur J. Burke, *Howe Richardson Scale Co.*

Weighted Network (See Decoding network, digital-to-analog coverter.)

Weights and Measures (See Weight, measurement of.)

Weir A device for measuring liquid flow in an open channel. A weir essentially is a dam over which, or through a notch in which, the liquid flows. The terms rectangular weir, triangular weir, trapezoidal weir, etc, generally refer to the shape of the notch used. Most weirs have flat upstream faces that are perpendicular to the bed and walls of the channel. Notched weirs are usually sharp-edged. As shown in Fig. W-6, the head h_0 on a weir is the liquid-level height above the crest or base of the notch. The head must be measured sufficiently far upstream to avoid the drop in level occasioned by the overfall which begins at a distance about $2h_0$ upstream from the weir. Surface-level measurements should be made a distance of $3h_0$ or more upstream, preferably using a stilling box equipped with a high-precision level gage.

With sharp-edged weirs, the sheet of discharging liquid, called the *nappe,* contracts as it leaves the opening, and free discharge occurs. Round-

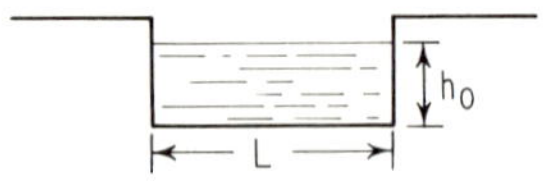

Fig. W-6. Rectangular weir.

ing the upstream edge will reduce the contraction and increase the flow rate for a given head. A clinging nappe may result if the head is very small, if the edge is well rounded, or if air cannot flow in beneath the nappe. This, in turn, results in an increase in discharge rate for a given head as compared with that for a free nappe. Flow through a rectangular weir is given by $q = 0.415 (L - 0.2h_0)h_0^{1.5} \sqrt{2g}$, where q = volumetric flow rate, ft^3/s; L = crest length, ft; h_0 = weir head, ft; g = local acceleration due to gravity (approximately 32.2 ft/s^2). This is known as the modified Francis formula for a rectangular sharp-edged weir with two end corrections; it applies where the velocity-of-approach correction is small. The Francis formula agrees with experiments within 3 percent if (1) L is greater than $2h_0$, (2) the velocity of approach is 2 ft/s or less, (3) the height of the crest above the bottom of the channel is at least $3h_0$, and (4) h_0 is not less than 0.3 ft.

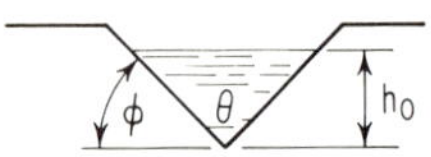

Fig. W-7. Triangular weir.

Narrow rectangular notches ($h_0 > L$) have been found to give about 93 percent of the discharge predicted by the Francis formula. Thus,

$$q = 0.386\, Lh_0^{\,1.5} \sqrt{2g}$$

In this case, no end corrections are applied, even though the formula applies only for sharp-edged weirs. The triangular-notched weir (Fig. W-7) has the advantage that a single notch can accommodate a wide range of flow rates, although this in turn reduces its accuracy. The discharge for sharp- or square-edged weirs is given by

$$q = \frac{0.31 h_0^{\,2.5} \sqrt{2g}}{\tan \phi}$$

The Cipolletti weir is a trapezoidal weir in which the sides of the notch slope one horizontal to four vertical. Its popularity is based on the belief that these sides slopes are just sufficient to correct for the side contractions of the nappe and that the flow, therefore, is proportional to the length of the weir crest.

Well, Thermal (See **Protecting tube, sensor.**)

Wet- and Dry-bulb Thermometer (See **Psychrometer.**)

Wheatstone Bridge (See **Bridge circuits, electric.**)

Wien Bridge (See **Bridge circuits, electric.**)

Word A unit of computer storage space. Some computers have a fixed number of bits per word (see **Bit**), and this word is the smallest addressable unit of storage. Word sizes on various machines range from 8 to over 36 bits. The word is the basic unit used in arithmetic and character manipulation on this type of machine. Machines of this type are usually referred to as fixed-word-length machines.

A type of machine known as a variable-word-length or character machine has as its basic addressable element a 6- or 8-bit byte (see **Byte**). Special program-settable indicators called word marks or flags are used to specify the length of words or fields. The word (or field) is the basic unit for arithmetic and transfer operations in this type of machine.

In some byte-oriented machines, the term word is applied to a fixed number of bytes, typically four. Binary-arithmetic operations are usually performed on words or multiples of words. Decimal-arithmetic and character-move operations, however, are performed on a variable-byte-length field whose length is defined in the instruction.

—L. E. Wise, *IBM Corporation*

Word-recognition System (See **Voice-recognition systems.**)

Work, Standard of (See **Standards, measurement.**)

X

X-ray Analysis X-rays occupy that portion of the electromagnetic spectrum between 0.01 and 100 angstroms in wavelength. Their range of approximate quantum energy is from 2×10^{-6} to 2×10^{-10} erg or from 10^6 to 100 electron volts. Important x-ray analytical methods are based on (1) fluorescence, (2) emission, (3) absorption, and (4) diffraction—and are used to determine qualitatively and quantitatively the element content of complex mixtures and to determine exactly the atomic arrangement and spacings of crystalline materials. Also see **X-ray analyzer, electron-microprobe**; and **X-ray fluorescence analysis.**

Source of X-rays: X-rays are emitted by atoms which are bombarded with energetic electrons. This results from two separate effects: (1) deceleration of high-speed electrons as they pass through matter and (2) ionization of individual atoms which abruptly stop the electrons. The first effect results in a continuous-type spectrum; the second effect results in characteristic line spectra.

Continuous Spectrum: The bulk of x-radiation arising from electron bombardment is the continuous spectrum. If an individual electron is abruptly decelerated, but not necessarily stopped, in passing through or near the electric field of a target atom, the electron will lose some energy ΔE, which appears as an x-ray photon of frequency $\nu = \Delta E/h$, where h is Planck's constant. An electron may experience several such decelerations before it is finally stopped, emitting x-ray photons of widely different energy and wavelength. A few electrons will be stopped in a single process, losing their entire energy and emitting an x-ray photon having the exact energy of the incident electron.

X-ray Spectral Lines: These result when the incident electrons knock orbital electrons out of an atom. If an ejected electron is from one of the inner orbits of the atom (Fig. X-1), an electron from an outer shell will fall to the inner orbit to fill the vacancy. The decrease in potential energy of this electron in approaching the nucleus results in the emission of an x-ray photon having an energy exactly equal to that lost by the electron. The wavelength λ for such photons is related to ΔE by $\lambda = ch/\Delta E$, where c is the velocity of electromagnetic energy and h is Planck's constant. Because the energy or orbital electrons is quantized, the x-ray photons can have only certain definite wavelengths which are characteristic of the atom. This situation is somewhat analogous to the more familiar ultraviolet and visible-emission spectra of materials, the difference being that the optical spectra are the result of electron transitions between energy levels of just the outermost electrons of the atoms.

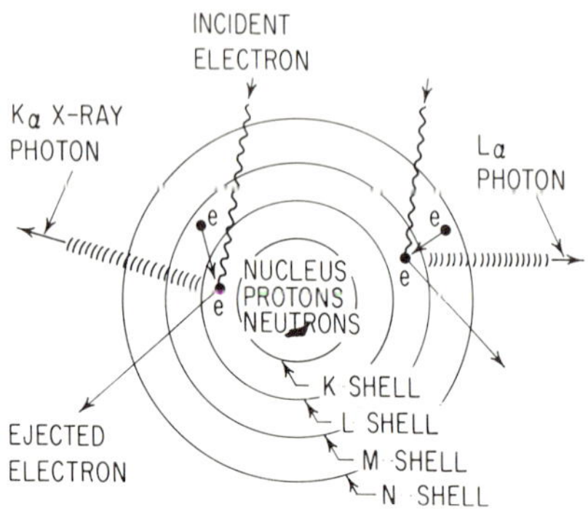

Fig. X-1. Origin of x-ray spectra due to electron bombardment.

X-rays resulting from an electron transition filling an electron vacancy in the innermost shell of an atom are known as K x-rays or K lines. X-rays formed by electron transitions to the second, or L, shell are known as L lines, and so on. Transitions to the K shell in the very light elements or to the M or N shell of the heavier elements represent a smaller energy change. Consequently, the photons which are emitted may have an energy and wavelength corresponding to the ultraviolet or visible region. To fill a vacancy in a K shell, an electron can come from either an L or M shell; the transitional energy will be slightly different depending upon the source of the electron, giving rise to K lines of slightly different energy or wavelength. These are known as K_α and K_β when they come from the L and M shells, respectively. The K_β line has a shorter wavelength than the K line, since it corresponds to a greater change in potential energy of the electron. Under high resolution, these lines can be shown to consist of several other lines (denoted $K_{\alpha 1}$, $K_{\alpha 2}$, $K_{\beta 2}$, and so on) corresponding to the exact initial and final electron position or energy level within the shells.

The energy change of a valence electron moving to fill a vacancy in any given shell increases with the atomic number of the atom. Shorter-wavelength x-rays will be emitted when an electron makes a transition to the K shell in a heavy element, such as lead, in contrast to a transition in aluminum. This is borne out by the fact that the K_α line in lead corresponds to an x-ray of 0.18 angstrom, whereas for aluminum, the K_α line is at 8.32 angstroms. X-rays, like gamma rays, sometimes are emitted when an atom disintegrates radioactively by the ejection of an alpha particle. In such disintegration, the mass of the products is slightly less than the mass of the parent material. The loss of mass m is converted into energy mc^2 (where c is the velocity of light) in the form of gamma or x-rays. Similarly, in disintegrations involving the ejection of a beta particle (an electron), emission of gamma or x-rays may result.

Generation of X-rays: An important component in an x-ray analytical device is an x-radiation generator. A high-vacuum Coolidge-type tube, wherein electrons are emitted from a heated tungsten filament and accelerated by a high voltage to an anode (target) is a common source of x-rays. The target usually is placed at a slight angle to the impinging electron beams so that the focal spot can be viewed from a side window on the tube. This window through which the x-rays are emitted commonly is made of a thin sheet of beryllium foil, mica, aluminum, or a special low-absorption glass. Only a portion of the x-rays emitted by the target actually fall on the exit window. By proper choice of target material and voltage, it is possible to generate either an essentially continuous spectrum or a continuum with superimposed line spectra (see Figs. X-2 and X-3). A wide variety of tubes is available. All high-power (high-current) commercial tubes employ a water-cooled anode, although some special high-current tubes have rotating anodes so that no one spot on the anode becomes unduly heated. Tubes of this type have been built with ratings up

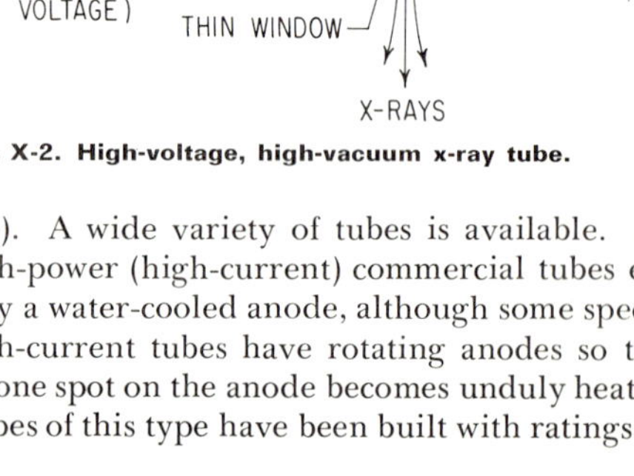

Fig. X-2. High-voltage, high-vacuum x-ray tube.

to 10 kW. Tubes are available for the generation of either very weak, long-wavelength x-rays or extremely hard x-rays, such as 2,000,000-V tubes. Tubes also are available for the generation of extremely intense beams of very short duration. For personnel safety, all tubes must be shielded. For further information on x-ray generators and tubes, see Ref. 1.

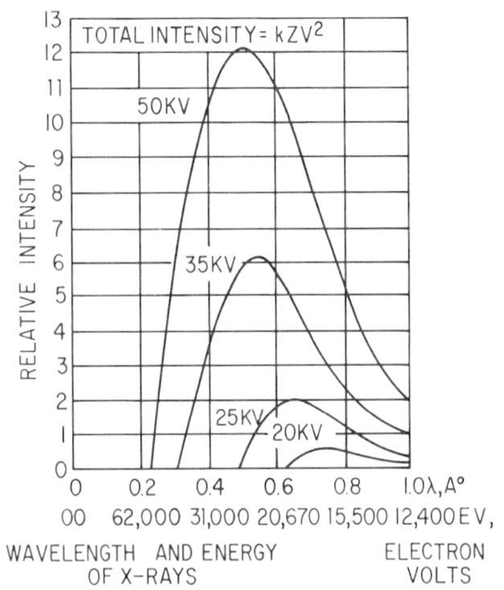

Fig. X-3. Continuous x-ray spectrum of tungsten ($Z =$ 74) at various tube voltages.

Detection of X-rays: Another important component of an x-ray analytical device is the means to detect x-rays. In early x-ray work, photographic techniques were widely used to measure intensities, and they are still used for some diffraction work. When properly used, the technique is quite sensitive. The photographic method is complicated by the fact that the x-ray absorption of the film emulsion varies with the wavelength of the x-rays—extremely large variations being encountered when the x-ray wavelengths correspond to the K or L absorption edges of silver and bromide ions. Other x-ray detectors include (1) the Geiger-Mueller tube, (2) ionization chambers, (3) scintillation counters, (4) proportional counters, and (5) electron-multiplier tubes. Photoconductive cadmium sulfide crystals have been shown to be reasonably sensitive, although noisy, x-ray detectors. See **Geiger-Mueller tube; Ionization chamber; Radiation (nuclear) detection of; Proportional counter;** and **Scintillation counter.**

X-ray Crystallography: X-rays penetrating below the surface of crystalline materials are scattered by the individual parallel layers of atoms; each atomic layer acts as a new, although weak, source of x-rays. Most of this scattered energy destructively interferes. For any portion of it to reinforce, it is necessary that the radiation from each atom layer be in phase with that from every other parallel layer. Thus, to be reinforced in a given direction at an angle θ (Fig. X-4), the spacing d between crystal planes must be rigorously related to the wavelength of the radiation. At a given angle, x-rays of one definite wavelength will be constructively reinforced; at a different angle, radiation of another wavelength can be efficiently diffracted. These variables are related by Bragg's law

$$n\lambda = 2d \sin \theta$$

where n is an integer. Note that θ is measured relative to the crystal face rather than to the perpendicular.

In addition to fulfilling this wavelength requirement, the energy will be diffracted only at an angle equal to the angle of the incident x-rays, independent of wavelength; otherwise, destructive interference is possible and the "reflected" energy is negligible. Thus, a properly oriented crystal will specularly "reflect" (diffract) monochromatic x-rays. For a given angle of incidence relative to a principal crystal plane, the "reflected" x-rays must be of a definite fixed wavelength which depends upon the spacing between lattice planes. This is the basis of operation of x-ray crystal monochromators, as shown in Fig. X-5. A suitable crystal, cleaved along a major lattice plane, is mounted so that it can be rotated on an axis along its face and perpendicular to the incident x-rays. The crystal mount is mechanically linked to an arm which moves a detector along a circle centered on the crystal face. The linkage is such that the crystal rotates at an angular rate one-half that of the detector. Thus, the detector always is at the proper angle to receive the reflected radiation, and the wavelength of the x-rays received by the detector will vary as it is rotated. If the crystal-lattice spacing is known, the wave-

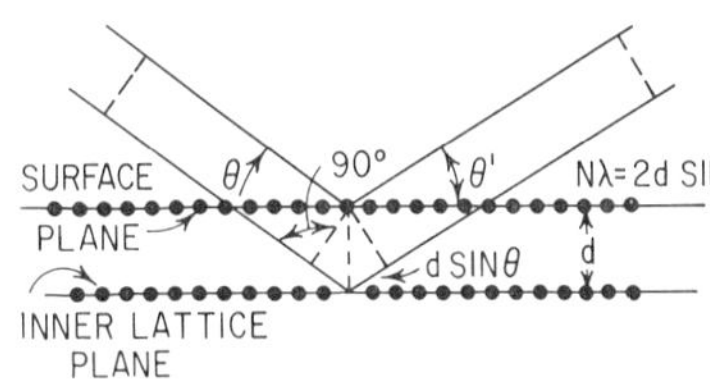

Fig. X-4. Reflection of x-rays from internal crystal plane.

length of the radiation at the detector can be calculated as a function of θ by Bragg's law.

X-ray-absorption Methods: Like the absorption of visible or ultraviolet radiation, the absorption of x-radiation at any given wavelength

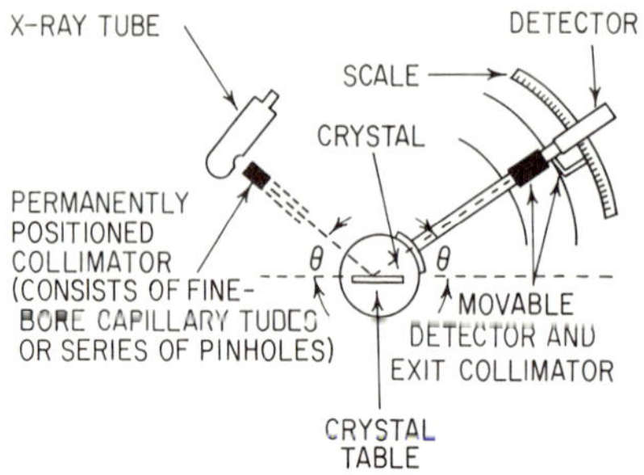

Fig. X-5. Crystal monochromator for x-rays. Crystal table is geared to detector arm so that the angular motion of crystal is one-half that of detector.

follows Lambert's law, which for the visible and ultraviolet region is usually given by

$$I = I_0 e^{-\mu_1 x}$$

where I = intensity of radiation passing through material
I_0 = intensity of incident radiation
e = natural-logarithm base
μ_1 = fraction of energy absorbed per centimeter of absorber traversed
x = thickness of absorbing material

Because of one striking difference between the absorption of x-rays and optical radiation, this equation commonly is modified for the x-ray region. Because the absorption of x-rays when passing through a fixed mass of a given material is completely independent of the physical state of that material, the absorption coefficient of Lambert's law usually employed for the x-ray region is a mass-absorption coefficient μ. This coefficient is constant for any given element or mixture of elements irrespective of whether these are present in the gas, liquid, or solid state. For example, the absorption in passing through 18 g of water is the same as if the rays pass through 18 g of ice, 18 g of steam, or a mixture of 16 g of oxygen and 2 g of gaseous hydrogen. Hence, as applied to x-rays, Lambert's law is usually written

$$I = I_0 e^{-\mu x \rho}$$

where μ is the mass-absorption coefficient μ_1/ρ, ρ being the density of the absorbing material. Lambert's law holds only if monochromatic x-rays are used. For further information on mass-absorption coefficients, see Ref. 1.

Techniques employing both the mass-absorption coefficients of the elements and their critical absorption edges have been used for chemical analysis.

Total-absorption Spectrometers: The simplest instrument is one which continuously measures the total amount of absorption from an x-ray beam as it traverses the sample. An instrument of this type is shown in Fig. X-6. The device is applicable to liquids, gases, or solids. In essence, the instrument is a photometer that compares the absorption of the sample with that of a reference sample. Also see **Photometers.** X-rays from the source pass through a mechanical chopper which permits the detector, of the scintillation type, to receive alternately the radiation transmitted by the sample material and by the reference material. An associated servounit positions a tapered wedge in the reference beam

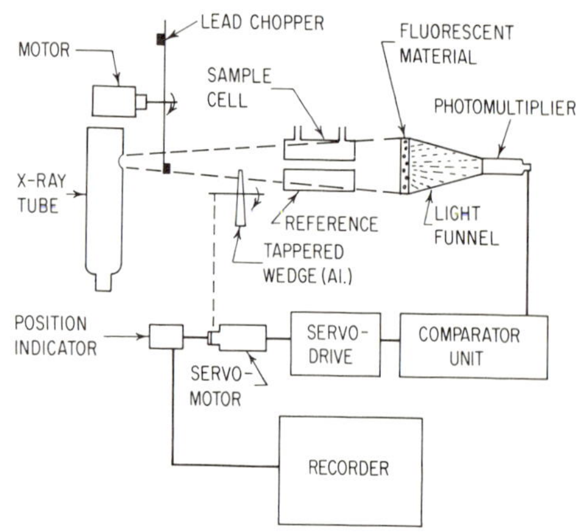

Fig. X-6. X-ray-absorption photometer for analysis of gases, liquids, and solids.

so that the amount of radiation reaching the detector from both beams is the same. The thickness of the wedge material (usually aluminum) required to balance the radiation in the two beams is directly related to the sample absorption. Similar instruments have been built for manual operation, employing differential ion chambers as detectors, omitting the servomechanism, and providing for manual adjustment of the wedge. Since this instrument measures the total sample absorption, it is suited best for the analysis of binary mixtures where one component has an element of distinctly higher atomic number than the other component. Uses include the analysis of tetraethyl lead in gasoline, chlorine content of hydrocarbons and plastics, sulfur content of oil, and the heavy-metal content of organics and glasses.

Critical-edge Techniques: A very selective x-ray-absorption technique entails making absorption measurements just above and just below the critical-edge wavelength specific to the element to be determined. The difference in absorption between a wavelength just shorter than this critical-edge wavelength and another wavelength just longer is proportional to the concentration of that element in the sample. To obtain maximum sensitivity, it is important to select the proper edge, which may be either the K edge or one of the L or M edges. Maximum sensitivity is obtained when the difference between the mass-absorption coefficients across the edge is as large as possible. Figure X-7 shows the difference in

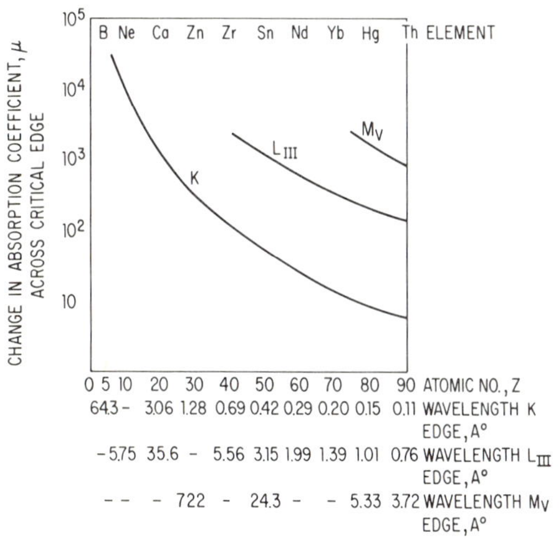

Fig. X-7. Height of absorption edge $\Delta\mu$ as a function of atomic number.

mass-absorption coefficient as a function of atomic number for the K, L_{III}, and M_V edges. In general, it is found that when the L_I edge can be used, it will be stronger than the K edge. Similarly, the L_{II} edge is to be preferred to the L_I edge whenever that line falls at a practical working wavelength. The M lines are still stronger than the L lines. The critical-edge method using an x-ray spectrometer is shown in Fig. X-8.

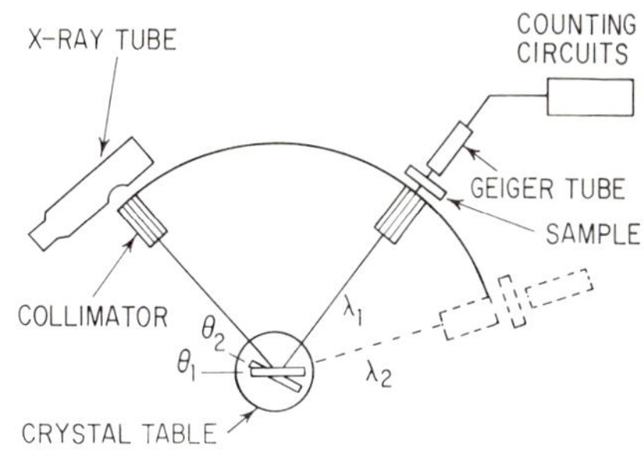

Fig. X-8. Apparatus for critical-edge-absorption measurements. Transmission of sample is measured at two wavelengths: λ_2, a wavelength just above critical edge; and λ_1, a wavelength just below critical edge.

REFERENCES

1. Friel, D. D.: Analytical Methods Employing X-rays, in Considine, D. M. (ed.): "Process Instruments and Controls Handbook," sec. 6, McGraw-Hill Book Company, New York, 1957.
2. Clark, G. L.: "Applied X-rays," 4th ed., McGraw-Hill Book Company, New York, 1955.

—W. G. Shequen,
Applied Research Laboratories

X-ray Analyzer, Electron-microprobe When a material is exposed to a highly focused beam of electrons, emission or radiant energy of various forms occurs, including the emission of x-rays. The x-radiation emitted is a measure of the chemical composition of the target material. This is the fundamental principle of the electron-probe x-ray analyzer.

Advantages of this analytical instrument include: (1) analysis can be confined to very small (microsamples) amounts of material; (2) the particular material to be analyzed need not be physically separated from its surrounding materials, as is so often required with many analytical methods; and (3) through the development of associated instrumentation, diagnostic techniques, and information displays, the method can be quite fast. Elements heavier than lithium (atomic weight = 6.939) may be studied. Limits of detection in solid solution are from approximately 0.005 to 0.05 percent, dependent upon the elements and sample matrixes involved (see Fig. X-9). Concentrations as low as 10^{-16} g may be measured.

Mainly used for metallurgical studies, nonmetallics also may be analyzed when samples are properly prepared. Biological applications include tooth and bone samples, cytochemical problems and staining techniques, physicochemical problems, and studies in pathology. Relative weight-fraction-detection limits for most elements in biological specimens are in the general range of 0.01 to 0.10 percent. Electronics-industry applications include studies of diffusion phenomena, electrical-contact surfaces, interfaces on transistors, and microcircuitry analysis.

Operation: As shown by Fig. X-10, electrons from an electron gun are directed to the sample through an electron optical system. Once the electron beam strikes the sample, a number of signal sources are activated, including (1) high-energy backscattered electrons, (2) low-energy secondary electrons, (3) cathodoluminescence, and (4) x-rays. Some heat also is generated within the sample. Volume d_3 of the specimen is that *volume from which x-rays are emitted.* This volume is determined by (1) the critical-excitation potential of the matrix elements and (2) the voltage of the electron beam. For any given sample, the total x-radiation intensity emitted for a particular element is a function of (1) the elemental concentration, (2) the electron-beam current, and to a lesser extent, (3) the excitation potential.

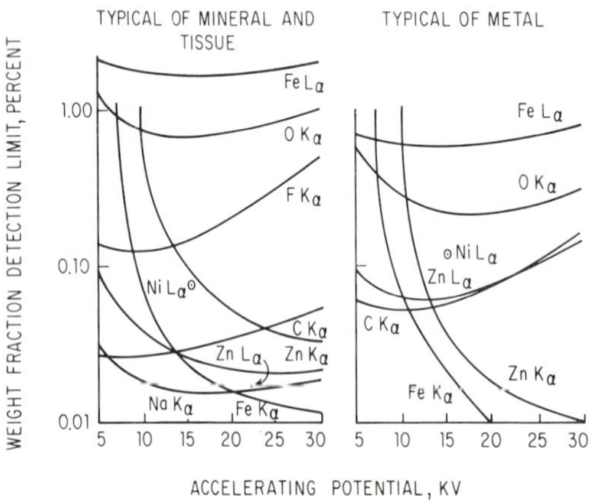

Fig. X-9. Typical weight-fraction-detecting limits of electron-microprobe x-ray analyzer.

The x-rays produced may be detected nondispersively by a proportional counter whose output may be separated as a function of energy by a pulse-height analysis system into the various wavelength components. However, better detection sensitivities can be obtained through the use of a fully focusing diffracting-crystal spectrometer in conjunction with a proportional detector and the necessary pulse-height analyzer. As shown by Fig. X-11, the necessary condition for fully focusing optics is to have the x-ray source, the crystal, and the detector slit all placed on a common circle. This geometry requires that the diffracting-crystal planes be bent to the diameter of the Rowland circle.

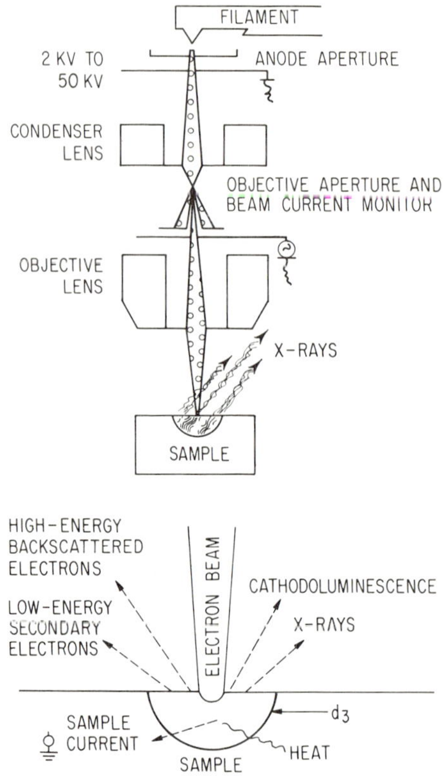

Fig. X-10. Electron-gun and probe-forming lens system of an integrated electron probe.

Spherical aberration at the detector slit is minimized by further grinding the crystal surface to fit the radius of the Rowland circle. With the resultant Johannson optics, the crystal radius is fixed and the 2θ range of the spectrometer is scanned by moving the crystal radially away from the source and, at the same time, rotating it into the detector to achieve a true focus throughout the spectrometer range. X-rays, particularly of a wavelength greater than 2 angstroms, and electrons are highly absorbed in an air atmosphere. Thus, the spectrometer system must be enclosed in a vacuum of the order of 10^{-5} torr. The present wavelength range of interest extends from approximately 1 to 100 angstroms. Diffracting crystals to cover this range must provide broad wavelength coverage, high diffraction efficiences for high peak-count intensities, good resolution, and good resulting peak-to-background ratios. Crystals that meet these objectives include LiF, ADP, KAP, PbSD, EDT, quartz, and NaCl.

Detectors: Of the three commonly used x-ray detectors—(1) Geiger counter, (2) scintillation counter, and (3) proportional counter—the latter is used most frequently for electron-probe microanalysis. In the wavelengths from 1 to 10 angstroms, sealed proportional counters may be used. For longer-wavelength analysis—in the range from 10 to 93 angstroms—the thinnest possible detector window is required to limit spectral attenuation. Nitrocellulose windows have proved successful.

An *optical microscope* is required in the system to provide the analyst with a means of reference to identify various sample areas for analysis. Sample stages may hold single or multiple samples and are provided with means for moving the sample in *x*, *y*, and *z* planes without breaking the system vacuum. After the point of interest is located on the specimen, the data may be read out in a number of ways: (1) quantitative and semiquantitative information may be obtained by processing the x-ray detector signal through a rate meter to a strip-chart or *X-Y* recorder; (2) scaler systems also provide direct readout of quantitative data integration; and (3) for operational convenience, a data translator and typewriter or teletype printout system may be connected directly to record digital-counter information as hard copy.

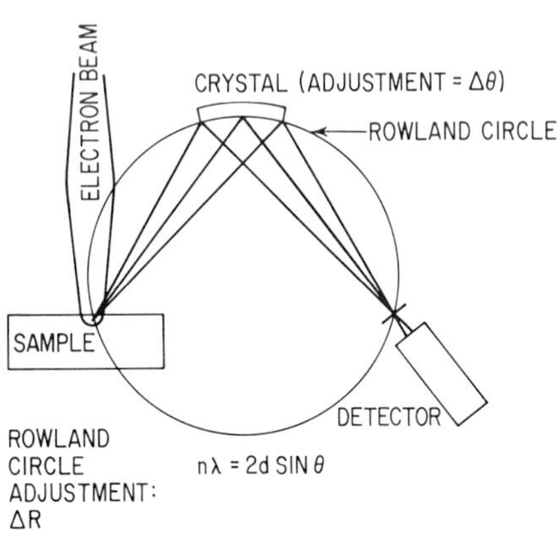

Fig. X-11. Geometry of a fully focusing diffracting-crystal spectrometer.

REFERENCES

McKinley, T. D., and K. F. J. Heinrich: "The Electron Microprobe," John Wiley & Sons, Inc., New York, 1966.

Andersen, C. A.: Electron Probe Analysis, *Chemistry*, vol. 40, pp. 16–21, April, 1967.

—W. G. Shequen,
Applied Research Laboratories

X-ray Fluorescence Analysis One of several types of spectrochemical techniques now used for laboratory analysis. The method is nondestructive, relatively high-speed in providing reliable results, compatible with computer technology and automatic control, applicable to batch or continuous processes, and has the ability to analyze a wide range of elements in the periodic table. The characteristic x-ray spectrum of each element bears a simple direct relationship to the atomic number. The relation of the wavelength λ to the atomic number Z is

$$\frac{1}{\lambda} \alpha Z^2 \qquad \text{Mosley's law}$$

Since the x-ray spectral lines come from the inner electrons of the atoms, the lines are not related to the chemical properties of the elements or to the compounds in which they may reside. Because the characteristics of the x-ray spectra are associated with energies released through transitions of electrons within the inner shells of the atom, the spectra are simple. (Also see **X-ray analysis.**) Most practical x-ray fluorescence analysis involves the detection of radiation release through electron transitions from outer shells to the K shell (K spectra), outer shells to the L shell (L spectra) and, in very few cases, from outer shells to the M shell (M spectra). Although various sources of energy, such as an electron beam, proton beam, or a gamma beam, may be used to excite the specimen, x-radiation obtained from an x-ray tube is the simplest form available for commercial instrumentation. For samples containing predominantly low-atomic-number elements, as in cement raw mix, the most efficient excitation is accomplished by using an x-ray-tube target material of relatively low atomic number, such as chromium. Elements having higher atomic numbers are most effectively excited by high-atomic-number targets, such as tungsten or platinum. An optimum target material is rhodium for the analysis of a broad range of elements.

Once x-ray fluorescence is produced from the sample by means of an x-ray tube, appropriate components of the instrument separate this radiation into its characteristic wavelengths, detect the energy emitted from each excited atom, and produce a signal that is representative of the number of atoms (concentration) of the elements in the sample. Typical excitation conditions (x-ray tube) are 50 kV, 35 mA. Bragg's law of x-ray diffraction is satisfied by the condition $N\lambda = 2d \sin \theta$, where λ is the wavelength and θ is the angle of incidence and diffraction of x-rays from a crystal whose lattice spacing is defined by d and N, the order or harmonic of the diffraction. For almost all fluorescence analysis, $N = 1$. The usable x-ray spectrum normally extends from 0.1 to about 20 angstroms. A helium or vacuum path is required for x-ray analysis of elements with atomic numbers lower than $Z \cong 24$ (wavelengths longer than 2.3 angstroms).

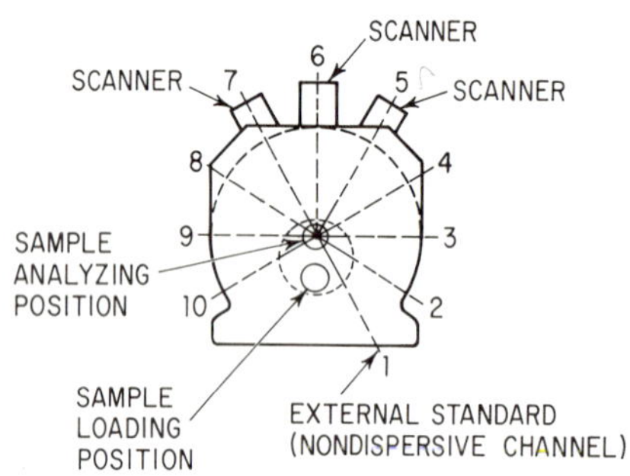

Fig. X-12. Disposition of nine Johannson x-ray monochromators in x-ray fluorescence analyzer. *(APL Quantometer.)*

Instrumentation provides for simultaneous elemental analysis using fixed preselected x-ray detection channels and scanners or goniometer to provide one or more channels that can be tuned to a wide wavelength coverage. Figure X-12 indicates a typical plan view of a multichannel array. Typical crystal materials covering the practical wavelength range of 1 to 20 angstroms are LiF, SiO_2, NaCl, EDT, ADP, and KAP. Optimum analytical data for the elements of interest are obtained by using fully focusing Johannson curved and ground crystals. Figure X-13 shows an optical diagram of a Johannson curved-crystal spectrometer. The lattice planes of a crystal can be curved to a radius equal to the diameter of the focal circle, and its front surface can be ground to the radius of the focal circle. The crystal then refocuses the width of a primary slit on the focal circle where the secondary slit is placed and diffracts the radiation passing through the primary slit in accordance with Bragg's law. Each spectrometer of an x-ray quantometer may be equipped with optimum crystal-detector combinations for specific determinations

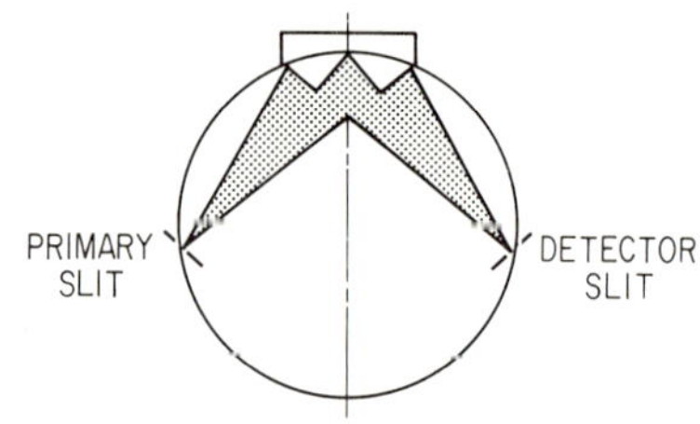

Fig. X-13. Optical diagram of Johannson curved-crystal spectrometer.

in a wide variety of matrixes, including steel, aluminum, copper-base materials, ores, cement, and slags—in both liquid and solid states.

The diffracted x-radiation is detected by Geiger, proportional, or semiproportional detectors. As in the case of crystals, detectors should be chosen on the basis of their response characteristics for the wavelengths of interest (see Fig. X-14). Geiger detectors are satisfactory for operation with Johannson x-ray spectrometers. Proportional detectors are best where pulse-height analysis is required to cope further with severe background or interfering-line effects. The proportional detector requires a solid-state amplifier in conjunction with the pulse-height analyzer.

The x-ray detector or pulse-height analyzer output is fed to an integrating device, such as a scaler or high-precision capacitor. The capacitor (integrator approach) is preferred for production and process-control instruments because of simplicity, stability, and economy. To optimize

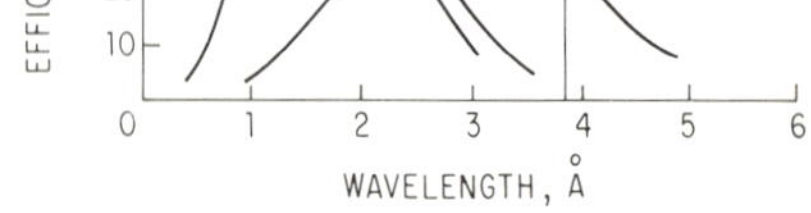

Fig. X-14. Relative efficiency of x-radiation detectors.

system stability, the effects of fluctuating x-ray-tube power on the analysis must be eliminated. One method is that of placing a standard brass disk in a position to intercept a portion of the primary x-ray beam at a point adjacent the sample. The resulting nondispersed x-rays are detected by a sealed detector—the output integrated across a capacitor simultaneously with all other detector signals being measured and subsequently being divided into intensity values obtained in elemental capacitors. This ratioing technique may be automated. At the completion of a given integration period, a high-impedance amplifier is scanned from integrator to integrator at a rate of approximately one per second. At each step, the integrator voltage is read on either a strip-chart record, a tape printer, or a typewriter. In a computer mode of operation, either digital or analog signals are made available to the computer through the same impedance amplifier at a rate of up to 50 channels per second (see Fig. X-15). Complete data transfer takes only a few tenths of a second in the analog mode.

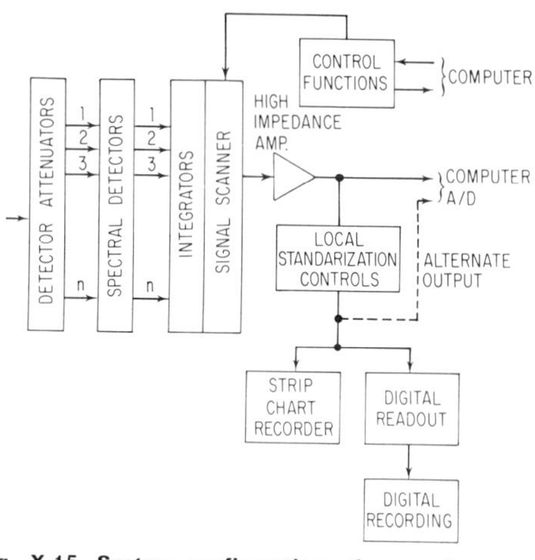

Fig. X-15. System configuration of x-ray fluorescence analyzer.

The transition from laboratory to automated instrumentation to achieve high-speed continuous analysis of dry or wet materials primarily involves the sample-handling and presentation hardware. Integration of this specialized hardware with the basic spectrometer and readout equipment results in total process-control capabilities. A typical application involves the handling of wet slurries where as many as twenty streams can be programmed to a given instrument with up to nine simultaneous analyses per stream.

Limits of detectability for the desired elemental analyses vary, depending upon the matrix, elements, methods of sample preparation, and quality of instrumentation applied. Generally, these are on the order of 1 to 100 ppm. The limit of detectability, however, is only one criterion in evaluating methods of analysis. The time of analysis is important, particularly in production and process-control laboratories, especially those equipped with computers. In multielement spectrometers, it is possible to perform as many as 22 elemental determinations in from 1 to 2 min.

REFERENCES

Newkirk, John B., Stephen R. Mallett, and Heintz G. Pheiffer: "Advances in X-ray Analysis," vols. 1–11, Plenum Press, Plenum Publishing Corporation, New York, 1958–1968.

Birks, L. S.: "X-ray Spectrochemical Analysis," Interscience Publishers, Inc., New York, 1959.

Davidson, E., W. G. Shequen, and A. Gilkerson: Rapid Analysis with X-ray Polychromators, *Spectrographer's Newsletter*, vol. 19, no. 1, 1966.

—W. G. Shequen,
Applied Research Laboratories

X-ray Thickness Gage (See Thickness gaging, x-ray.)

***X-Y* Recorder** (See Recorder, graphic.)

Y

Yaw, Aircraft (See Flight-control system.)

Yield Stress [See Testing (materials), destructive.]

Z

Zero Suppression (See Range, instrument.)

Zone Control In some equipment, particularly ovens and kilns, a large otherwise contiguous piece of processing equipment will be broken down into two or more heating or cooling zones —with separate instrumentation and control of temperature for each zone. Similarly, large spaces as found in offices and warehouses will be temperature-zone-controlled. The use of multiple thermostats in residences is another example of zone control. Obviously, each zone tends to interact with adjacent zones, and the balancing of such a system often is painstaking. Of course, the control engineer can overcome such difficulties by engineering the equipment as a total system and providing for the interfacing effects between zones.

classified index

The structure of this classified index is shown in the following table. Each entry of this encyclopedia is listed once and only once in the classified index. Cross references are used extensively throughout the book, and these references, coupled with the detailed alphabetical index at the end of the volume, provide a webbing to hold all topics together even though all topics are entered into the book alphabetically. Admittedly, some of the decisions involved in placing topics under the major categories of the table were arbitrary. This situation could have been avoided only by listing some entries under several categories. Such multiple listings would have resulted in a much longer and less convenient topical classification.

STRUCTURE OF CLASSIFIED INDEX

Applications

Aerospace

Astronomy

Design and drafting

Geophysics

Machine control

A substantial portion of this encyclopedia is devoted to this topic. Machine-control principles and hardware will be found listed in this classified index under the categories of Measurands; Measurement systems; Instrument data processing; *and* Control systems. *Particularly, see* Numerical control; *and* Position-measurement and control systems.

Medicine and biochemistry

Meteorology and hydrology

Navigation

Oceanography

Process control

A substantial portion of this encyclopedia is devoted to this topic. Process-control principles and hardware will be found listed in this classified index under the categories of Measurands; Measurement systems; Instrument data processing; *and* Control systems.

Testing and inspecting

Measurands

Chemical composition

Electrical variables

Control media

Final controlling elements

Switching elements

subject index